# HANDBOOK OF OCCUPATIONAL SAFETY AND HEALTH

# HANDBOOK OF OCCUPATIONAL SAFETY AND HEALTH

THIRD EDITION

Edited by

**S. Z. Mansdorf**
Consultant, Boca Raton, Florida

WILEY

This edition first published 2019


*Registered Office*
John Wiley & Sons, Inc., 111 River Street, Hoboken, NJ 07030, USA

*Editorial Office*
111 River Street, Hoboken, NJ 07030, USA

For details of our global editorial offices, customer services, and more information about Wiley products visit us at www.wiley.com.

Wiley also publishes its books in a variety of electronic formats and by print-on-demand. Some content that appears in standard print versions of this book may not be available in other formats.



*Library of Congress Cataloging-in-Publication data applied for*

Hardback ISBN: 9781118947265

Cover Design: Wiley
Cover Image: © GrAl/Shutterstock

Set in 10/12pt TimesTen by SPi Global, Pondicherry, India

Printed in United States of America
V10008932_032119

# CONTENTS

## CONTRIBUTORS

**Peter M. Bochnak**, Environmental Health and Safety, Harvard University, 46 Oxford St., Cambridge, MA, 02138*

**William A. Burgess,** CIH, Massachusetts Institute of Technology, Bldg. 56-235, 77 Massachusetts Ave., Cambridge, MA, 02139-4307*

**Charles Chelton,** CIH, 30 Radtke St., Randolph, NJ, 07869*

**David E. Cohen,** MD, Dermatology Department, NYU Medical Center, 560 First Ave., New York, NY, 10016

**Ralph Collipi, Jr.**, CIH, AT&T, 40 Elwood Rd., Londonderry, NH, 03053*

**Patricia Cruz,** PhD, Department of Environmental and Occupational Health, School of Community Health Sciences, University of Nevada, Las Vegas, 4505 S. Maryland Parkway, Las Vegas, NV, 89154

**Thea Dunmire,** JD, CIH, CSP, ENLAR Compliance Services, 3665 E Bay Dr. #204C, Largo, FL, 33771

**Timur S. Durrani,** MD, MPH, MBA, School of Medicine, University of California, San Francisco, 2789 25th St., San Francisco, CA, 94110

**Mitchell S. Galanek,** CHP, Massachusetts Institute of Technology, Bldg. 56-235, 77 Massachusetts Ave., Cambridge, MA, 02139-4307*

**Deborah Gold,** MPH, CIH, California Division of Occupational Safety and Health (retired), P.O. Box 501, Pacifica, CA, 94044

**Pamela Greenley,** CIH, Massachusetts Institute of Technology, Bldg. 56-235, 77 Massachusetts Ave., Cambridge, MA, 02139-4307*

**Thomas Grumbles,** CIH, FAIHA, 17806 Paintbrush Pass Ct., Cypress, TX, 77433

**Donald L. Haes, Jr.**, CHP, Massachusetts Institute of Technology, Bldg. 16-268, 77 Massachusetts Ave., Cambridge, MA, 02139-4307*

**Marilyn F. Hallock,** CIH, Massachusetts Institute of Technology, Bldg. 56-235, 77 Massachusetts Ave., Cambridge, MA, 02139-4307*

**Robert Herrick,** ScD, CIH, Department of Environmental Health, Harvard T.H. Chan School of Public Health, 665 Huntington Ave., Boston, MA, 02215

**Martin R. Horowitz,** CSP, CIH, Analog Devices, 21 Osborn St., Cambridge, MA, 02139*

**David L. Johnson,** PhD, Professor and Chair, Department of Occupational and Environmental Health, Associate Dean for Academic Affairs, College of Public Health, University of Oklahoma, Health Sciences Center, P.O. Box 26901, Oklahoma City, OK, 73126-0901

**James S. Johnson,** PhD, CIH, JSJ & Associates, Pleasanton, CA, 94588

**Jennifer L. Kyle,** PhD, MSPH, MT (ASCP), Division of Communicable Disease Control, Microbial Diseases Laboratory Program, California Department of Public Health, 850 Marina Bay Parkway, Richmond, CA, 94804

**Chris Laszcz-Davis,** MS, CIH, FAIHA, AIC Fellow, President, The Environmental Quality Organization, LLC, Affiliated, Aluminium Consulting Engineers, 74 Overhill Rd., Orinda, CA, 94563

**Charles P. Lichtenwalner,** CIH, PE, CSP, Lucent Technologies, Bell Laboratories, 600 Mountain Ave., Murray Hill, NJ, 07974

**Jerry Lynch,** CIH (deceased).

**Janet M. Macher,** ScD, MPH, Division of Environmental and Occupational Disease Control, Environmental Health Laboratory Branch, California Department of Public Health (retired), 850 Marina Bay Parkway, Richmond, CA, 94804

**S. Z. Mansdorf,** PhD, CIH, CSP, QEP, Consultant, 7184 Via Palomar, Boca Raton, FL, 33433

**Fred A. Manuele,** CSP, PE, President, Hazards Limited, 200 W Campbell #603, Arlington Heights, IL, 60005

**Andrew McIntyre,** 2150 North First Street, Suite 450, San Jose, CA, 95131

**Kevin Michael,** PhD, Michael & Associates, 246 Woodland Dr., State College, PA, 16803*

**Philip R. Morey,** PhD, CIH (deceased)

**Richard J. Nill,** CIH, CSP, Genetics Institute, 1 Burtt Rd., Andover, MA, 01810*

**Graciela M. Perez,** ScD, CPE, Department of Work Environment, College of Engineering, University of Massachusetts at Lowell, Lowell, MA, 01854

**Frank M. Renshaw,** PhD, CIH, CSP, Bayberry EHS Consulting, LLC, 6 Plymouth Dr., Cherry Hill, NJ, 08034

**Anne M. Venetta Richard,** MS, Lucent Technologies, 1600 Osgood St., 21-2L28, North Andover, MA, 01845*

**Harmony Scofield,** Manager, BSI EHS Services and Solutions, 1400 NW Compton Dr., Suite 203, Hillsboro, OR, 97005

**Richard Shaughnessy,** PhD, Department of Chemical Engineering, University of Tulsa, Tulsa, OK, 74104

**Dennis Shusterman,** MD, MPH, Division of Occupational and Environmental Medicine (Emeritus), School of Medicine, University of California, San Francisco, Campus Box 0843, San Francisco, CA, 94143

**James Stewart,** PhD, CIH, CSP, Department of Environmental Health, Harvard T.H. Chan School of Public Health, Landmark Center, 401 Park Dr., Boston, MA, 02215

**Ralph Stuart,** CIH, CCHO, Office of Environmental Safety, Keene State College, 229 Main St., Keene, NH, 03435

**Victor M. Toy,** CIH, CSP, FAIHA, Insyst OH&S, 7 West 41st Ave., #508, San Mateo, CA, 94403

**Steven Trammell,** PE, CSP, CCPSC, CHMM, BSI EHS Services and Solutions, 110 Wild Basin Rd., Suite 270, Austin, TX, 78746

*** Original affiliation from the Second Edition.**

# FOREWORD

It has been almost two decades since Louis J. DiBerardinis edited the second edition of the *Handbook of Occupational Safety and Health*. The intent of this third edition is to add to some of the existing well-written chapters from a number of pioneers in the safety and health field. For the existing chapters that have been reprinted, the format for those chapters was not changed, while the chapters that were revised do not follow the original outline style. The third edition also includes some current developments in our field such as management systems. Finally, the third edition has also been rearranged into topic sections to better categorize the flow of the chapters.

The quality of the work in the handbook is a reflection of the dedication of the authors in contributing to future generations of those practicing one of the most noble of professions protecting the health and safety of all of us.

*Boca Raton, Florida* S. Z. Mansdorf

PART I

# RECOGNITION AND CONTROL OF HAZARDS

CHAPTER 1

# RECOGNITION OF HEALTH HAZARDS IN THE WORKPLACE

MARTIN R. HOROWITZ

*Analog Devices, 21 Osborn St., Cambridge, MA, 02139*

and

MARILYN F. HALLOCK

*Massachusetts Institute of Technology, Bldg. 56–235, 77 Massachusetts Ave., Cambridge, MA, 02139-4307*

Adapted from "Potential exposures in the manufacturing industry – their recognition and control" by William A. Burgess, Chapter 18 in F. Clayton and D. Clayton (eds.), *Patty's Industrial Hygiene and Toxicology*, 4e, Vol. IA. New York: Wiley, 1991, pp. 595–674.

## 1.1 INTRODUCTION

Although employment in the United States is shifting from manufacturing to the service sector, manufacturing continues to employ in excess of 20 million workers in workplaces that present both traditional and new occupational health hazards. To understand the nature of these hazards, the occupational health professional must understand not only the toxicology of industrial materials but also the manufacturing technology that defines how contaminants are released from the process, the physical form of the contaminants, and the route of exposure. Physical stresses including noise, vibration, heat, and ionizing and nonionizing radiation must also be evaluated. Twelve specific unit operations representing both large employment and potential health hazards to the worker have been chosen for discussion in this chapter; these unit operations occur in many different industrial settings. This chapter is based on the previous reviews of the subject (Burgess 1991, 1995) as well as original scientific literature. The purpose of this chapter is to help the reader recognize potential health hazards that may exist in specific operations and industries. Other chapters in this text cover the evaluation and control of the recognized hazards.

## 1.2 ABRASIVE BLASTING

Abrasive blasting is practiced in a number of occupational settings, including bridge and building construction, shipbuilding and repair, foundries, and metal finishing in a variety of industries. The process is used in heavy industry as an initial cleaning step to remove surface coatings and scale, rust, or fused sand in preparation for finishing operations. Abrasive blasting is used in intermediate finishing operations to remove flashing, tooling marks, or burrs from cast, welded, or machined fabrications and to provide a matte finish to enhance bonding of paint or other coatings.

Various abrasives are used in blast cleaning operations. The most commonly heavy-duty abrasives for metal surfaces are silica sand, metal shot and grit, coal and metallurgical slags, and synthetic abrasives such as aluminum oxide and silicon carbide. For light-duty cleaning of plastic and metal parts where erosion of the workpiece is of concern, a number of ground organic products based on corn, oat, and fruit pits are available as well as baking soda, glass beads, plastic chips, and solid carbon dioxide pellets.

*Handbook of Occupational Safety and Health*, Third Edition. Edited by S. Z. Mansdorf.

Three major methods of blasting are used to deliver the abrasive to the workpiece. In pressure blasting, compressed air is used to either aspirate or pressurize and deliver the abrasive from a storage "pot" to a nozzle where it is directed to the workpiece at high velocity by the operator. This type of process may be used in either open-air blasting or in blasting enclosures. In hydroblasting, a high-pressure stream of water conveys the abrasive to the work surface; this process is used principally for outdoor work. Finally, in the centrifugal wheel system, a high-speed centrifugal impeller projects the abrasive at the workpiece; this method is used primarily in some types of blasting enclosures.

### 1.2.1 Application and Hazards

Blasting operations may be performed either in a variety of enclosures or in open-air operations such as bridge and ship construction. The types of enclosures used in industrial applications include blasting cabinets, automatic tumble or barrel and rotating table units where the operator controls the process from outside the operation, and exhausted rooms, where the operator is inside the enclosure. In most cases, these industrial units have integral local exhaust ventilation systems and dust collectors.

In open-air blasting in construction and shipyard applications, general area contamination occurs unless isolation of the work area can be achieved. As a result, the blasting crew, including the operator, "pot man," and cleanup personnel as well as adjacent workers, may be exposed to high dust concentrations depending on the existing wind conditions.

The most obvious health hazard of abrasive blasting operations is airborne dust contamination. Dust exposures may include the abrasive in use, the base metal being blasted, and the surface coating or contamination being removed. In the United States, the widespread use of sand containing high concentrations of crystalline quartz is still a major hazard to workers. Though elevated rates of silicosis were identified in blasters by the US Public Health Service in the 1930s, a National Institute for Occupational Safety and Health (NIOSH) alert published as recently as 1992 identified 99 cases of silicosis in abrasive blasters with 14 deaths or severe impairment (NIOSH 1992). In the United Kingdom, the use of sand was prohibited for in-plant abrasive blasting in 1949; a ban on the use of sand for both inside and outside applications has been adopted by the European Community. The use of abrasives based on metallurgical slags warrants attention owing to the presence of heavy metal contamination.

In most cases, the base metal being blasted is iron or steel, and the resulting exposure to iron dust presents a limited hazard. If blasting is carried out on metal alloys containing such materials as nickel, manganese, lead, or chromium, the hazard should be evaluated by air sampling. The surface coating or contamination on the workpiece frequently presents a major inhalation hazard. In the foundry it may be fused silica sand; in ship repair, the surface coating may be a lead-based paint or an organic mercury biocide. Abrasives used to remove lead paint from a bridge structure may contain up to 1% lead by weight. Used abrasives will often need to be disposed of as hazardous waste depending upon the level of heavy metal contamination.

Physical hazards of abrasive blasting include noise exposure and safety hazards from high-velocity nozzle discharge. The release of air by a blasting nozzle generates a wideband noise that frequently exceeds 110 dBA. This noise is greatly attenuated in a properly designed blasting enclosure, but open-air blasting requires the implementation of a hearing conservation program and frequently the use of both ear plugs and muffs (NIOSH 1975b).

### 1.2.2 Control

The ventilation requirements for abrasive blasting enclosures have evolved over several decades, and effective design criteria are now available (American Conference of Governmental Industrial Hygienists [ACGIH] 1995; American National Standards Institute [ANSI] 1978). The minimum exhaust volumes, based on seals and curtains in good condition, include 20 air changes per minute for cabinets with a minimum of 500 fpm through all baffled inlets. Abrasive blasting rooms require 60–100 cfm $ft^{-2}$ of floor for downdraft exhaust. Dust control on abrasive blasting equipment depends to a large degree on the integrity of the enclosure. All units should be inspected periodically, including baffle plates at air inlets, gaskets around doors and windows, gloves and sleeves on cabinets, gaskets at hose inlets, and the major structural seams of the enclosure.

Operators directly exposed to the blasting operation, as in blasting room or open-air operations, must be provided with NIOSH-approved type C air-supplied abrasive blasting helmets. NIOSH has assigned a protection factor of 25 for type C hooded respirators operating in a continuous flow mode and 2000 operating in the positive-pressure mode. The latter respirator is recommended for open-air blasting with crystalline silica-containing sand (NIOSH 1992). The intake for the air compressor providing the respirable air supply should be located in an area free from air contamination, and the quality of the air delivered to the respirator must be checked periodically. The respirator must be used in the context of a full respirator program (see Chapter 16).

## 1.3 ACID AND ALKALI CLEANING OF METALS

After the removal of major soils and oils by degreasing, metal parts are often treated in acid and alkaline baths to condition the parts for electroplating or other finishes. The principal hazard in this series of operations is exposure to acid and alkaline mist released by heating, air agitation, gassing from electrolytic operation, or cross-contamination between tanks.

### 1.3.1 Acid Pickling and Bright Dip

Pickling or descaling is a technique used to remove oxide scales formed from heat treatment, welding, and hot forming operations prior to surface finishing. On low- and high-carbon steels, the scale is iron oxide, whereas on stainless steels it is composed of oxides of iron, chromium, nickel, and other alloying metals (Spring 1974). The term *pickling* is derived from the early practice of cleaning metal parts by dipping them in vinegar.

Scale and rust are commonly removed from low and medium alloy steels using a nonelectrolytic immersion bath of 5–15% sulfuric acid at a temperature of 60–82 °C (140–180 °F) or a 10–25% hydrochloric acid bath at room temperature. Nitric acid is frequently used in pickling stainless steel, often in conjunction with hydrofluoric, sulfuric, and hydrochloric acids. The most common stainless steel descaling process uses nitric acid in the concentration range of 5–25% in conjunction with hydrochloric acid at 0.5–3%. For light scale removal, the concentrations are 12–15% of nitric acid and 1% of hydrofluoric acid by volume at a bath temperature of 120–140 °F; for heavy oxides, the concentration of hydrofluoric is increased to 2–3%. Pickling operations on nonferrous metals such as aluminum, magnesium, zinc, and lead each have specific recommended acid concentrations.

Acid bright dips are usually mixtures of nitric and sulfuric acids employed to provide a mirrorlike surface on cadmium, magnesium, copper, copper alloys, silver, and, in some cases, stainless steel.

The air contaminants released from pickling and bright dips include not only the mists of the acids used in the process but also nitrogen oxides, if nitric acid is employed, and hydrogen chloride gas from processes using hydrochloric acid. Extensive information is available on the effects of exposure to inorganic acids. Accidental contact with the skin and eyes produces burns, ulcers, and necrosis. For most acids, the acute effects of contact exposure are rapid and detected immediately by the affected individual. However, hydrogen fluoride penetrates the skin, and the onset of symptoms may be delayed for hours, permitting deep tissue burns and severe pain. Airborne acid mists produce upper and lower respiratory irritation. Chronic exposure to nitrogen oxides can produce pulmonary edema. One series of epidemiological studies found excess laryngeal cancer in steel workers who conducted pickling operations using sulfuric and other acids (International Agency for Research on Cancer [IARC] 1992).

An extensive listing of bath components and potential air contaminants for all common pickling and bright dip baths is provided in ACGIH Ventilation Manual (ACGIH 1995). The extent of exposure will depend on bath temperature, surface area of work, current density (if bath is electrolytic), and whether the bath contains inhibitors that produce a foam blanket on the bath or that lower the surface tension of the bath and therefore reduce misting. It is a general rule that local exhaust ventilation is required for pickling and acid dip tanks operating at elevated temperature and for electrolytic processes. Extensive guidelines for local exhaust ventilation of acid dip tanks have been developed (ACGIH 1995; ANSI 1977; Burgess et al. 1989).

Minimum safe practices for pickling and bright dip operators have been proposed by Spring (1974): (i) Hands and faces should be washed before eating, smoking, or leaving plant. Eating and smoking should not be permitted at the work location. (ii) Only authorized employees should be permitted to make additions of chemicals to baths. (iii) Face shields, chemical handlers' goggles, rubber gloves, rubber aprons, and rubber platers' boots should be worn when adding chemicals to baths and when cleaning or repairing tanks. (iv) Chemicals contacting the body should be washed off immediately, and medical assistance obtained. (v) Supervisor should be notified of any change in procedures or unusual occurrences. Because the symptoms of hydrofluoric acid exposure can be delayed and the consequences so severe, any suspected exposure should be reported to a responsible medical authority immediately.

### 1.3.2 Alkaline Treatment

***1.3.2.1 Alkaline Immersion Cleaning*** Acid and alkaline cleaning techniques are complementary in terms of the cleaning tasks that can be accomplished. Alkaline soak, spray, and electrolytic cleaning systems are superior to acid cleaning for removal of oil, gases, buffing compounds, certain soils, and paint. A range of alkaline cleansers including sodium hydroxide, potassium hydroxide, sodium carbonate, sodium meta- or orthosilicate, trisodium phosphate, borax, and tetrasodium pyrophosphate are used for both soak and electrolytic alkaline cleaning solutions.

The composition of the alkaline bath may be complex, with a number of additives to handle specific tasks (Spring 1974). In nonelectrolytic cleaning of rust from steels, the bath may contain 50–80% caustic soda in addition to chelating and sequestering agents. The parts are immersed for 10–15 minutes and rinsed with a spray. The usual temperature range for these baths is 160–210 °F, and alkaline mist and steam are potential air contaminants. Guidelines for local exhaust ventilation of these baths have been provided (ACGIH 1995).

Electrolytic alkaline cleaning is an aggressive cleaning method. The bath is an electrolytic cell powered by direct current with the workpiece, conventionally the cathode, and an inert electrode as the anode. The water dissociates; oxygen is released at the anode, and hydrogen at the cathode. The hydrogen gas generated at the workpiece causes agitation of the surface soils with excellent soil removal. The gases released at the electrodes by the dissociation of water may result in the release of caustic mist and steam at the surface of the bath.

Surfactants and additives that provide a foam blanket are important to the proper operation of the bath. Ideally, the foam blanket should be 5–8 cm thick to trap the released gas bubbles and therefore minimize misting (NIOSH 1985b). If the foam blanket is too thin, the gas may escape, causing a significant alkaline mist to become airborne; if too thick, the blanket may trap hydrogen and oxygen with resulting minor explosions ignited by sparking electrodes.

***1.3.2.2 Salt Baths*** A bath of molten caustic at 370–540° (700–1000 °F) can be used for initial cleaning and descaling of cast iron, copper, aluminum, and nickel with subsequent quenching and acid pickling. The advantages claimed for this type of cleaning are that precleaning is not required and the process provides a good bond surface for a subsequent finish.

Molten sodium hydroxide is used at 430–540 °C (800–1000 °F) for general-purpose descaling and removal of sand on castings. A reducing process utilizes sodium hydride in the bath at 370 °C (700 °F) to reduce oxides to their metallic state. The bath utilizes fused liquid anhydrous sodium hydroxide with up to 2% sodium hydride, which is generated in accessory equipment by reacting metallic sodium with hydrogen. All molten baths require subsequent quenching, pickling, and rinsing. The quenching operation dislodges the scale through steam generation and thermal shock.

These baths require well-defined operating procedures owing to the hazard from molten caustic as well as safety hazards from the reaction of metallic sodium with hydrogen. Local exhaust ventilation is necessary, and the tank must be equipped with a complete enclosure to protect the operator from violent splashing as the part is immersed in the bath. Quenching tanks and pickling tanks must also be provided with local exhaust ventilation. Ventilation standards have been proposed for these operations (ACGIH 1995; ANSI 1977).

## 1.4 DEGREASING

For many decades the principal application of degreasing technology has been in the metalworking industry for the removal of machining oils, grease, drawing oils, chips, and other soils from metal parts. The technology has expanded greatly prompted by advances in both degreasing equipment and solvents. At this time it is probably the most common industrial process extending across all industries, including jewelry, electronics, special optics, electrical, machining, instrumentation, and even rubber and plastic goods. The significant occupational health problems associated with cold and vapor-phase degreasing processes are described below.

### 1.4.1 Cold Degreasing

The term *cold degreasing* identifies the use of a solvent at room temperature in which parts are dipped, sprayed, brushed, wiped, or agitated for removal of oil and grease. It is the simplest of all degreasing processes, requiring only a simple container with the solvent of choice. It is widely used in small production shops, maintenance and repair shops, and automotive garages. The solvents used have traditionally included low-volatility, high-flash petroleum distillates such as mineral spirits and Stoddard solvent to solvents of high volatility including aromatic hydrocarbons, chlorinated hydrocarbons, and ketones. The choice of degreasing agent has changed significantly since the 1987 Montreal Protocol. The ozone depletion potential (ODP) of a given degreaser has become an important variable in choice.

Skin contact with cold degreasing materials should be avoided by work practices and the use of protective clothing. Glasses and a face shield should be used to protect the eyes and face from accidental splashing during dipping and spraying. Solvent tanks should be provided with a cover, and, if volatile solvents are used, the dip tank and drain station should be provided with ventilation control.

Spraying with high-flash petroleum distillates such as Stoddard solvent, mineral spirits, or kerosene is a widely used method of cleaning oils and grease from metals. This operation should be provided with suitable local exhaust ventilation (ACGIH 1995). The hood may be a conventional spray booth type and may be fitted with a fire door and automatic extinguishers. The fire hazard in spraying a high-flash petroleum solvent is comparable with spraying many lacquers and paints.

### 1.4.2 Vapor Degreasing

A vapor degreaser is a tank containing a quantity of solvent heated to its boiling point. The solvent vapor rises and fills the tank to an elevation determined by the location of a condenser. The vapor condenses and returns to the liquid sump. The tank has a freeboard that extends above the condenser to minimize air currents inside the tank (Figure 1.1). As the parts are lowered into the hot vapor, the vapor condenses on the cold part and dissolves the surface oils and greases. This oily condensate drops back into the liquid solvent at the base of the tank. The solvent is continuously evaporated to form the vapor blanket. Because the oils are not vaporized, they remain to form a sludge in the bottom of the tank. The scrubbing action of the condensing vapor continues until the temperature of the part reaches the temperature of the vapor, whereupon condensation stops, the part appears dry, and it is removed from the degreaser. The time required to reach this point depends on the particular solvent, the temperature of the vapor, the weight of the part, and its specific heat. The vapor-phase degreaser does an excellent job of drying parts after aqueous cleaning and prior to plating; it is frequently used for this purpose in the jewelry industry.

***1.4.2.1 Types of Vapor-Phase Degreasers and Solvents*** The simplest form of vapor-phase degreaser, shown in Figure 1.1, utilizes only the vapor for cleaning. The straight vapor-cycle degreaser is not effective on small, light work because the part reaches the temperature of the vapor before the condensing action has cleaned the part. Also, the straight vapor cycle does not remove insoluble surface soils. For such applications, the vapor-spray-cycle degreaser is frequently used. The part to be cleaned is first placed in the vapor zone as in the straight vapor-cycle degreaser. A portion of the vapor is condensed by a cooling coil and fills a liquid solvent reservoir. This warm liquid solvent is pumped to a spray lance, which can be used to direct the solvent on the part, washing off surface oils and cooling the part, thereby permitting final cleaning by vapor condensation (Figure 1.2).

A third degreaser design has two compartments, one with warm liquid solvent and a second compartment with a vapor zone. The work sequence is vapor, liquid, and vapor. This degreaser is used for heavily soiled parts with involved geometry or to clean a basket of small parts that nest together. Finally, a three-compartment degreaser has vapor, boiling and warm liquid compartments with a vapor, boiling liquid, warm liquid, and vapor work sequence. Other specialty degreasers encountered in industry include enclosed conveyorized units for continuous production cleaning.

Ultrasonic cleaning modules installed in vapor degreasers have found broad application for critical cleaning jobs. In an ultrasonic degreaser, a transducer operating in the range of 20–40 kHz is mounted at the base of a liquid immersion solvent tank. The transducer alternately compresses and expands the solvent forming small bubbles that cavitate or collapse at the surface of the workpiece. The cavitation phenomenon disrupts the adhering soils and cleans the part. Ultrasonic degreasers use chlorinated solvents at 32–49 °C (90–120 °F) and aqueous solutions at 43–71 °C (110–160 °F). These degreasers commonly employ refrigerated or water-chilled coils for control of solvent vapors; the manufacturers claim that local exhaust ventilation is not needed in this configuration.

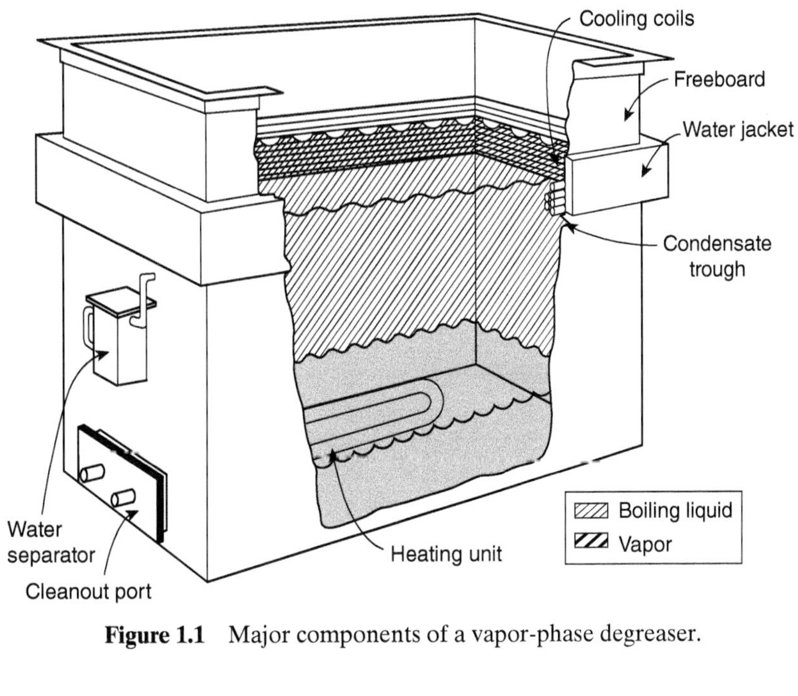


**Figure 1.1** Major components of a vapor-phase degreaser.

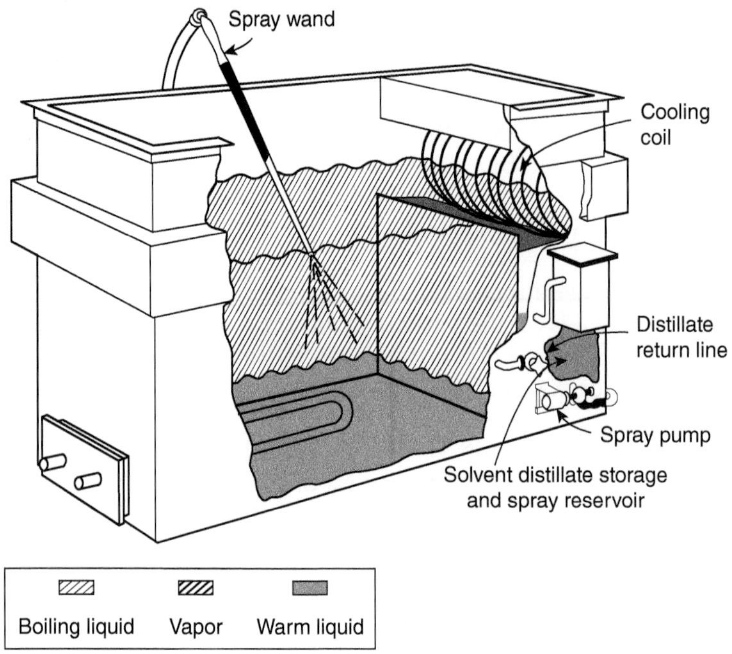


**Figure 1.2** A vapor-phase degreaser with a spray wand.

The solvents commonly used with vapor-phase degreasers have traditionally included trichloroethylene, perchloroethylene, 1,1,1-trichloroethane, methylene chloride, and a series of Freon® solvents (Table 1.1) (Dow Chemical Co. 1978; DuPont 1987). The degreaser must be designed for and used with a specific solvent. Most chlorinated degreasing solvents sold under trade names contain a stabilizer present in a concentration of less than 5%. The purpose of the stabilizer is to neutralize any free acid that might result from oxidation of the degreasing liquid in the presence of air, hydrolysis in the presence of water, or pyrolysis under the influence of high temperatures. The stabilizer is not a critical issue in establishing health risk to the worker owing to its low concentration; the solvent itself is usually the predictor of risk.

The emphasis on using non-ozone-depleting chemicals in degreasing operations has greatly influenced the choice of degreasers. The use of 1,1,1-trichloroethane and many of the Freons® has diminished, while trichloroethylene is being used more frequently. The flammability and toxicity of the new generation of materials requires careful review before use. New materials, such as D-limonene, have been found to have their own set of issues including dermatitis and low odor thresholds. Semiaqueous processes using terpenes, dibasic esters, *n*-methyl pyrrolidone, and other materials have been used in conjunction with surfactants to provide effective systems. In many cases, nonchemical cleaning substitutes have been chosen, such as bead blasting with plastic pellets or frozen carbon dioxide pellets.

The loss of degreaser solvent to the workplace obviously depends on a number of operating conditions, including the type and properties of contaminants removed, cleaning cycle, volume of material processed, design of parts being cleaned, and, most importantly, work practices. The effort to dry parts quickly with an air hose, for example, will lead to higher exposures. Maintenance and cleaning of degreasers can yield high acute exposures and may require the use of respiratory protection.

***1.4.2.2 Control*** The vapor level in the degreaser is controlled by a nest of water-fed condensing coils located on the inside perimeter of the tank (Figure 1.1). In addition, a water jacket positioned on the outside of the tank keeps the freeboard cool. The vertical distance between the lowest point at which vapors can escape from the degreaser machine and the highest normal vapor level is called the *freeboard*. The freeboard should be at least 15 in. and not less than one-half to three-fifths the width of the machine. The effluent water from the coils and water jacket should be regulated to 32–49 °C (90–120 °F); a temperature indicator or control is desirable.

Properly designed vapor degreasers have a thermostat located a few inches above the normal vapor level to shut off the source of heat if the vapor rises above the condensing surface. A thermostat is also immersed in the boiling liquid; if overheating occurs, the heat source is turned off.

There is a difference of opinion on the need for local exhaust ventilation on vapor-phase degreasers. Authorities frequently cite the room volume as a guide in determining if ventilation is needed: local exhaust ventilation is needed if there is less than 200 ft$^3$ in the room for each square foot of solvent surface or if the room is smaller than 25 000 ft$^3$. In fact, ventilation control requirements depend on the degreaser design, location, maintenance, and operating practices. Local exhaust increases solvent loss, and the installation may require solvent recovery before discharging to outdoors.

**TABLE 1.1 Properties of Vapor Degreasing Solvents**

| | Trichloroethylene | Perchloroethylene | Methylene chloride | Trichlorotrifluoroethane[a] | Methyl chloroform (1,1,1-trichloroethane) |
|---|---|---|---|---|---|
| Boiling point | 87 | 121 | 40 | 48 | 74 |
| °C | 188 | 250 | 104 | 118 | 165 |
| °F | | | | | |
| Flammability | | Nonflammable under vapor degreasing conditions | | | |
| Latent heat of vaporization (b.p.). Btu $lb^{-1}$ | 103 | 90 | 142 | 63 | 105 |
| Specific gravity | | | | | |
| Vapor (air = 1.00) | 4.53 | 5.72 | 2.93 | 6.75 | 4.60 |
| Liquid (water = 1.00) | 1.464 | 1.623 | 1.326 | 1.514 | 1.327 |

[a] Binary azeotropes are also available with ethyl alcohol, isopropyl alcohol, acetone, and methylene chloride.

To ensure effective use of local exhaust, the units should be installed away from drafts from open windows, spray booths, space heaters, supply air grilles, and fans. Equally important are the parts loading and unloading station. When baskets of small parts are degreased, it is not possible to eliminate drag out completely, and the unloading station usually requires local exhaust.

Later in this chapter under the discussion of welding, reference is made to the decomposition of chlorinated solvent under thermal and UV stress with the formation of chlorine, hydrogen chloride, and phosgene. Because degreasers using such solvents are frequently located near welding operations, this problem warrants attention. In a laboratory study of the decomposition potential of methyl chloride, methylene chloride, carbon tetrachloride, ethylene dichloride, 1,1,1-trichloroethane, *o*-dichlorobenzene, trichloroethylene, and perchloroethylene, only the latter two solvents decomposed in the welding environment to form dangerous levels of phosgene, chlorine, and hydrogen chloride (Dow Chemical Co. 1977). All chlorinated materials thermally degrade if introduced to direct-fired combustion units commonly used in industry. If a highly corroded heater is noted in the degreaser area, it may indicate that toxic and corrosive air contaminants are being generated.

Installation instructions and operating precautions for the use of conventional vapor-phase degreasers have been proposed by various authorities (American Society for Testing and Materials 1989). The following minimum instructions should be observed at all installations:

1. If the unit is equipped with a water condenser, the water should be turned on before the solvent is heated.
2. Water temperature should be maintained between 27 and 43 °C (81 and 110 °F).
3. Work should not be placed in and removed from the vapor faster than 11 fpm (0.055 m $s^{-1}$). If a hoist is not available, a support should be positioned to hold the work in the vapor. This minimizes the time the operator must spend in the high exposure zone.
4. The part must be kept in the vapor until it reaches vapor temperature and is visually dry.
5. Parts should be loaded to minimize pullout. For example, cup-shaped parts should be inverted.
6. Overloading should be avoided because it will cause displacement of vapor into the workroom.
7. The work should be sprayed with the lance below the vapor level.
8. Proper heat input must be available to ensure vapor level recovery when large loads are placed in the degreaser.
9. A thermostat should be installed in the boiling solvent to prevent overheating of the solvent.
10. A thermostat vapor level control must be installed above the vapor level inside the degreaser and set for the particular solvent in use.
11. The degreaser tank should be covered when not in use.
12. Hot solvent should not be removed from the degreaser for other degreasing applications, nor should garments be cleaned in the degreaser.
13. An emergency eyewash station would be located near the degreaser for prompt irrigation of the eye in case of an accidental splash.

To ensure efficient and safe operation, vapor-phase degreasers should be cleaned when the contamination level reaches 25%. The solvent should be distilled off until the heating surface or element is 1½ in. below the solvent level or until the solvent vapors fail to rise to the collecting trough. After cooling, the oil and solvent should be drained off and the sludge removed. It is important that the solvent be cooled prior to draining. In addition to placing the operators at risk, removing hot solvent causes serious air contamination and frequently requires the evacuation of plant personnel from the building. A fire hazard may exist during the cleaning of machines heated by gas or electricity because the flash point of the residual oil may be reached and because trichloroethylene itself is flammable at elevated temperatures. After sludge and solvent removal, the degreaser must be mechanically ventilated before any maintenance work is undertaken. A person should not be permitted to enter a degreaser or place his or her head in one until all controls for entry into a confined space have been put in place. Anyone entering a degreaser should wear a respirator suitable for conditions immediately hazardous to life, as well as a lifeline held by an attendant. Anesthetic concentrations of vapor may be encountered, and oxygen concentrations may be insufficient. Such an atmosphere may cause unconsciousness with little or no warning. Deaths occur each year because of failure to observe these precautions.

The substitution of one degreaser solvent for another as a control technique must be done with caution. Such a decision should not be based solely on the relative exposure standards but must consider the type of toxic effect, the photochemical properties, the physical properties of the solvents including vapor pressure, and other parameters describing occupational and environmental risk.

## 1.5 ELECTROPLATING

Metal, plastic, and rubber parts are plated to prevent rusting and corrosion, for appearance, to reduce electrical contact resistance, to provide electrical insulation as a base for soldering operations, and to improve wearability. The common plating metals include cadmium, chromium, copper, gold, nickel, silver, and zinc. Prior to electroplating, the parts must be cleaned, and the surfaces treated as described in Sections 1.2 and 1.3.

There are approximately 160 000 electroplating workers in the United States; independent job shops average 10 workers, and the captive electroplating shops have twice that number. A number of epidemiologic studies have identified a series of health effects in electroplaters ranging from dermatitis to elevated mortality for a series of cancers.

### 1.5.1 Electroplating Techniques

The basic electroplating system is shown in Figure 1.3. The plating tank contains an electrolyte consisting of a metal salt of the metal to be applied dissolved in water. Two electrodes powered by a low-voltage DC power supply are immersed in the electrolyte. The cathode is the workpiece to be plated, and the anode is either an inert electrode or, most frequently, a slab or a basket of spheres of the metal to be deposited. When power is applied, the metal ions deposit out of the bath on the cathode or workpiece. Water is dissociated, releasing hydrogen at the cathode and oxygen at the anode. The anode may be designed to replenish the metallic ion concentration in the bath. Current density expressed in amperes per unit area of workpiece surface varies depending on the operation. In addition to the dissolved salt containing the metallic ion, the plating bath may contain additives to adjust the electrical conductivity of the bath, define the type of plating deposit, and buffer the pH of the bath.

Anodizing, a common surface treatment for decoration, corrosion resistance, and electrical insulation on such metals as magnesium, aluminum, and titanium, operates in a different fashion. The workpiece is the anode, and the cathode is a lead bar. The oxygen formed at the workpiece causes a controlled surface oxidation. The process is conducted in a sulfuric or chromic acid bath with high current density, and because its efficiency is quite low, the amount of misting is high.

In conventional plating operations, individual parts on a hanger or a rack of small parts are manually hung from the cathode bar. If many small pieces are to be plated, the parts may be placed in a perforated plastic barrel in electrical contact with the cathode bar, and the barrel is immersed in the bath. The parts are tumbled to achieve a uniform plating.

In a small job shop operation, the parts are transferred manually from tank to tank as dictated by the type of plating operation. The series of steps necessary in one plating operation (Figure 1.4) illustrates the complexity of the operation. After surface cleaning and preparation, the electroplate steps are completed with a water rinse tank, isolating each tank from contamination. In high production shops, an automatic transfer unit is programmed to cycle the parts from tank to tank, and the worker is required only to load and unload the racks or baskets. Automatic plating operations may permit exhaust hood enclosures on the tanks and therefore more effective control of air contaminants. Exposure is also limited because the worker is stationed at one loading position and is not directly exposed to air contaminants released at the tanks.

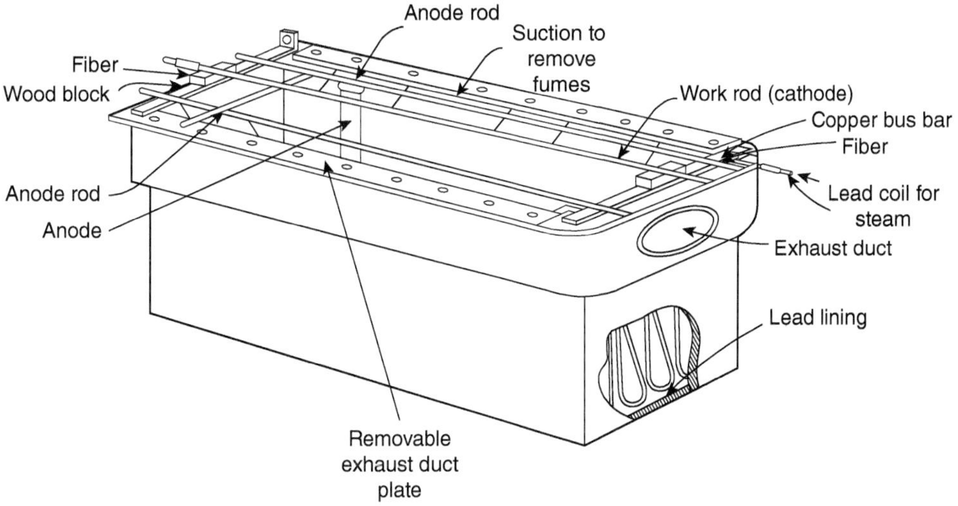


**Figure 1.3** An electroplating tank.

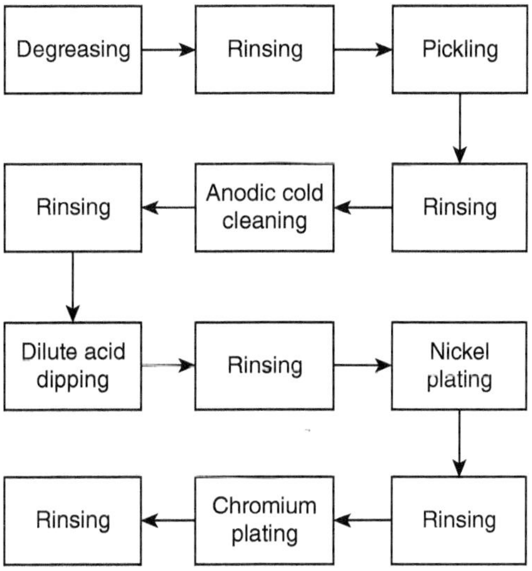


**Figure 1.4** The steps involved in chromium plating of steel.

### 1.5.2 Air Contaminants

The principal source of air contamination in electroplating operations is the release of the bath electrolyte to the air by the gassing of the bath. As mentioned above, the path operates as an electrolytic cell; so water is dissociated, and hydrogen is released at the cathode and oxygen at the anode. The gases released at the electrodes rise to the surface of the bath and burst, generating a respirable mist that becomes airborne. The mist generation rate depends on the bath efficiency. In copper plating, the efficiency of the plating bath is nearly 100%; that is, essentially all the energy goes into the plating operation and little into the electrolytic dissociation of water (Blair 1972). Nickel plating baths operate at 95% efficiency; so only 5% of the energy is directed to dissociation of water, and misting is minimal. However, chromium plating operations are quite inefficient, and up to 90% of the total energy may be devoted to dissociation of the bath with resulting severe gassing and resulting potential exposure of the operator to chromic acid mist. Although the contamination generation rate of the bath is governed principally by the efficiency of the bath, it also varies with the metallic ion concentration in the bath, the current density, the nature of the bath additives, and bath temperature. Air or mechanical agitation of the bath used to improve plating quality may also release the bath as droplets.

The health significance of the mist generated by electroplating processes depends, of course, on the contents of the bath. The electrolyte mist released from the bath is alkaline or acidic, depending on the specific electroplating process. A majority of the alkaline baths based on cyanide salt solutions are used for cadmium, copper, silver, brass, and bronze plating. Acidic solutions are used for chromium, copper, nickel, and tin. The exact composition of the baths can be obtained from an electroplater's handbook or, if proprietary, from the supplier. An inventory of the nature of the chemicals in the common electroplating baths, the form in which they are released to the air, and the rate of gassing has evolved over the past several decades, drawing heavily on the experience of the state industrial hygiene programs in New York and Michigan (Burgess et al. 1989). These data are useful in defining the nature of the contaminant and the air sampling procedure necessary to define the worker exposure. As noted below, these data are also valuable in defining the ventilation requirements for various plating operations.

### 1.5.3 Control

Proprietary bath additives are available to reduce the surface tension of the electrolyte and therefore reduce misting. Another additive provides a thick foam that traps the mist released from the bath. This agent is best used for tanks that operate continuously. A layer of plastic chips, beads, or balls on the surface of the bath also traps the mist and permits it to drain back into the bath. Where possible, tanks should be provided with covers to reduce bath loss.

Although the use of the above mist suppressants is helpful, they will not alone control airborne contaminants from plating tanks at an acceptable level. Local exhaust ventilation in the form of lateral slot or upward plenum slotted hoods is the principal control measure.

The design approach described in the Industrial Ventilation Manual (ACGIH 1995) provides a firm basis for the control of electroplating air contaminants. This procedure permits one to determine a minimum capture velocity based on the hazard potential of the bath and the rate of contaminant generation. The exhaust volume is based on the capture velocity and the tank measurements and geometry. Owing to the severe corrosion of duct work, periodic checks of the exhaust systems in plating shops are necessary. Qualitative assessment of the ventilation is possible using smoke tubes or other tracers. In many cases, the use of partitions to minimize the disruptive effects of drafts may greatly improve the installed ventilation.

In addition to the proper design and installation of good local exhaust ventilation, one must provide adequate replacement air, backflow dampers on any combustion devices to prevent carbon monoxide contamination of the workplace, and suitable air cleaning.

Because a low-voltage DC power supply is used, an electrical hazard does not exist at the plating tanks. A fire and explosion risk may result from solvent degreasing and spray painting conducted in areas contiguous with the plating area. The major chemical safety hazards are due to handling concentrated acids and alkalis and the accidental mixing of acids with cyanides and sulfides during plating, bath preparation, and waste disposal with the formation of hydrogen cyanide and hydrogen sulfide.

The educated use of protective equipment by electroplaters is extremely important in preventing contact with the various sensitizers and corrosive materials encountered in the plating shop. The minimum protective clothing should include rubber gloves, aprons, boots, and chemical handler's goggles. Aprons should come below the top of the boots. All personnel should have a change of clothing available at the workplace. If solutions are splashed on the work clothing, they should be removed, the skin should be washed, and the worker should change to clean garments. A shower and eyewash station serviced with tempered water should be available at the workplace. The wide range of chemicals presents a major dermatitis hazard to the plater, and skin contact must be avoided. Nickel is a skin sensitizer and may cause nickel itch, developing into a rash with skin ulcerations.

A summary of the health hazards encountered in electroplating shops and the available controls is presented in Table 1.2. More information on controls can be found in the references (American Electroplaters and Surface Finishers Society [AESF] 1989; NIOSH 1985a).

## 1.6 GRINDING, POLISHING, AND BUFFING

These operations are grouped together for discussion because they all involve controlled use of bonded abrasives for metal finishing operations; in many cases, the operations are conducted in the sequence noted. This discussion covers the nonprecision applications of these techniques.

### 1.6.1 Processes and Materials

Nondimensional application of grinding techniques includes cutoff operations in foundries, rough grinding of forgings and castings, and grinding out major surface imperfections in metal fabrications. Grinding is frequently done with wheels and disks of various geometries made up of selected abrasives in different bonding structural matrices. The common abrasives are aluminum oxide and silicon carbide; less common are diamond and cubic boron nitride. A variety of bonding materials are available to provide mechanical strength and yet release the spent abrasive granules to renew the cutting surface. Vitrified glass is the most common bonding agent. The grinding wheel is made by mixing

**TABLE 1.2 Summary of Major Electroplating Health Hazards**

| Exposure | How contamination occurs |
|---|---|
| | *Inhalation* |
| Mist, gases, and vapors | |
| Hydrogen cyanide | Accidental mixing of cyanide solutions and acids |
| Chromic acid | Released as a mist during chrome plating and anodizing |
| Hydrogen sulfide | Accidental mixing of sulfide solutions and acids |
| Nitrogen oxides | Released from pickling baths containing nitric acid |
| Dust | Released during weighing and transferring of solid bath materials, including cyanides and cadmium salts |
| Fumes | Generated during on-site repair of lead-lined tanks using torch-burning techniques |
| | *Ingestion* |
| Workplace particles | Accidental ingestion during smoking and eating at workplace |
| | *Skin contact* |
| Cyanide compounds | Absorption through the skin |
| Solvents | Defatting by solvents |
| Irritants | Primary irritants contacting the skin |
| Contact allergens | Sensitization |
| | *Control technology* |
| Local exhaust ventilation | |
| Mist reduction | |
| Reduce surface tension | |
| Coat surface | |
| Tank covers | |
| Isolation of stored chemicals | |

clay and feldspar with the abrasive, pressing it in shape, and firing it at high temperature to form a glass coating to bond for the abrasive grains. Resinoid wheel bonds, based on thermosetting resins such as phenol–formaldehyde, are used for diamond and boron nitride wheels and are reinforced with metal or fiber glass for heavy-duty applications including cutoff wheels. Other bonding agents are sodium silicate and rubber-based agents.

The abrasive industry utilizes a standard labeling nomenclature to identify the grinding wheel design (ANSI 1978) that includes the identification of the abrasive and bonding agent as well as grain size and structure and other useful information. Table 1.3 lists grinding wheel specification nomenclature.

Exact information on the generation rate of grinding wheel debris for various applications is not available. However, the wheel components normally make up a small fraction of the total airborne particles released during grinding; the bulk of airborne particles are released from the workpiece. After use, the grinding wheel may load or plug, and the wheel must be "dressed" with a diamond tool or "crushed dressed" with a steel roller. During this brief period, a significant amount of the wheel is removed, and a small quantity may become airborne.

Polishing techniques are used to remove workpiece surface imperfections such as tool marks. This technique may remove as much as 0.1 mm of stock from the workpiece. The abrasive, again usually aluminum oxide or silicon dioxide, is bonded to the surface of a belt, disk, or wheel structure in a closely governed geometry, and the workpiece is commonly applied to the moving abrasive carrier by hand.

The buffing process differs from grinding and polishing in that little metal is removed from the workpiece. The process merely provides a high luster surface by smearing any surface roughness with a lightweight abrasive. Red rouge (ferric oxide) and green rouge (chromium oxide) are used for soft metals and aluminum oxide is used for harder metals. The abrasive is blended in a grease or wax carrier that is packaged in a bar or tube form. The buffing wheel is made from cotton or wool disks sewn together to form a wheel or "buff." The abrasive is applied to the perimeter of the wheel, and the workpiece is then pressed against the rotating wheel mounted on a buffing lathe.

### 1.6.2 Exposures and Control

The hazard potential from grinding, polishing, and buffing operations depends on the specific operation, the workpiece metal and its surface coating, and the type of abrasive system in use. A NIOSH-sponsored study of the ventilation requirements for grinding, polishing, and buffing operations showed that the major source of airborne particles in grinding and polishing is the workpiece, whereas the abrasive and the wheel textile material represent the principal sources of contamination in buffing (NIOSH 1975a).

**TABLE 1.3 Grinding Wheel Specification Nomenclature**

| | |
|---|---|
| *Abrasives (first letter in specification)* | |
| A | Aluminum oxide |
| C | Silicon carbide |
| D or ND | Natural diamonds |
| SD | Synthetic diamonds |
| CB or CBN | Cubic boron nitride |
| *Bond (last letter in specification)* | |
| V | Vitrified |
| B | Resinoid |
| R | Rubber |
| S | Silicate |

The health status of grinders, polishers, and buffers has not been extensively evaluated. One study did not find elevated cancer mortality in metal polishers (Blair 1980), while two other studies did (Jarvolm et al. 1982; Sparks and Wegman 1980). However, each of the polishing operations studied was unique; so it is difficult to generalize to the industry as a whole. A listing of the metal and alloys worked and information on the nature of the materials released from the abrasive system are needed in order to evaluate the exposure of a particular operator. In many cases, the exposure to total dust can be evaluated by means of personal air sampling with gravimetric analysis. If dusts of toxic metals are released, then specific analysis for these contaminants is necessary.

The need for local exhaust ventilation on grinding operations has been addressed by British authorities (United Kingdom Department of Employment 1974), who state that control is required if one is grinding toxic metals and alloys, ferrous and nonferrous castings produced by sand molding, and metal surfaces coated with toxic material.

The same guidelines can be applied to polishing operations. The ventilation requirements for buffing, however, are based on the large amount of debris released from the wheel, which may be a housekeeping problem and potential fire risk.

The conventional ventilation control techniques for fixed location grinding, polishing, and buffing are well described in the ACGIH (1995) and ANSI (1966) publications. It is more difficult to provide effective ventilation controls for portable grinders used on large castings and forgings. Flexible exterior hoods positioned by the worker may be effective. High-velocity, low-volume exhaust systems with the exhaust integral to the grinder also are suitable for some applications (Fletcher 1988). The performance of the conventional hoods on grinding, polishing, and buffing operations should be checked periodically.

The hazard from bursting wheels operated at high speed and the fire hazard from handling certain metals such as aluminum and magnesium are not covered here, but these problems affect the design of hoods and the design of wet dust collection systems.

The hazard from vibration from hand tools leading to VWF can result from portable grinders and also pedestal mounted wheels. A strategy for the protection of workers from VWF has been developed by NIOSH (1989a) and includes tool redesign, using protective equipment, and monitoring exposure and health.

## 1.7 HEAT TREATING

A range of heat treating methods for metal alloys is available to improve the strength, impact resistance, hardness, durability, and corrosion resistance of the workpiece (American Society for Metals 1982). In the most common procedures, metals are hardened by heating the workpiece to a high temperature with subsequent rapid cooling. Softening processes normally involve only heating or heating with low cooling.

### 1.7.1 Surface Hardening

Case hardening, the production of a hard surface or case to the workpiece, is normally accomplished by diffusing carbon or nitrogen into the metal surface to a given depth to achieve the hardening of the alloy. This process may be accomplished in air, in atmospheric furnaces, or in immersion baths by one of the following methods.

***1.7.1.1 Carburizing*** In this process, the workpiece is heated in a gaseous or liquid environment containing high concentrations of a carbon-bearing material, that is, the source of the diffused carbon. In gas carburizing, the parts are heated in a furnace containing hydrocarbon gases or carbon monoxide; in pack carburizing, the part is covered with carbonaceous material that burns to produce the carbon-bearing gas blanket. In liquid carburizing, the workpiece is immersed in a molten bath, that is, the source of carbon.

In gas carburizing, the furnace atmosphere is supplied by an atmosphere generator. In its simplest form, the generator burns a fuel such as natural gas under controlled conditions to produce the correct concentration of carbon monoxide, which is supplied to the furnace. Because carbon monoxide concentrations up to 40% may be used, small leaks may result in significant workroom exposure. To control emission from gas carburizing operations, the combustion processes should be closely controlled, furnaces maintained in tight condition, dilution ventilation installed to remove fugitive leaks, furnaces provided with flame curtains at doors to control escaping gases, and self-contained breathing apparatus available for escape and repair operations. In liquid carburizing, a molten bath of sodium cyanide and sodium carbonate provides a limited amount of nitrogen and the necessary carbon for surface hardening.

***1.7.1.2 Cyaniding*** The conventional method of liquid carbonitriding is immersion in a cyanide bath with a subsequent quench. The part is commonly held in a sodium cyanide bath at temperatures above 870 °C (1600 °F) for 30–60 minutes. The air contaminant released from this process is sodium carbonate; cyanide compounds are not released, although there are no citations in the open literature demonstrating this. Local or dilution ventilation is frequently applied to this process. The handling of cyanide salts requires strict precautions including secure and dry storage, isolation from acids, and planned disposal of waste. Care must be taken in handling quench liquids because the cyanide salt residue on the part will in time contaminate the quench liquid.

***1.7.1.3 Gas Nitriding*** Gas nitriding is a common means of achieving hardening by the diffusion of nitrogen into the metal. This process utilizes a furnace atmosphere of ammonia operating at 510–570 °C (950–1050 °F). The handling of ammonia in this operation is hazardous in terms of fire, explosion, and toxicity.

### 1.7.2 Annealing

Annealing is a general term used to describe many heating or cooling cycles that change the metallurgical properties of the workpiece. The process varies depending on the alloy and the use of the part, but in all cases it involves heating at a given temperature for a specific time and then cooling at a desired rate, frequently at slow rates. Different types of salt baths may be used, including a blend of potassium nitrate, sodium nitrate, and nitrite for low temperatures or a blend of sodium and potassium chloride and barium nitrite for high-temperature baths. Careful handling and storage procedures must be used for nitrate salts, because of their reactivity and explosivity.

### 1.7.3 Quenching

The quench baths may be water, oil, molten salt, liquid air, or brine. The potential problems range from a nuisance problem due to release of steam from a water bath to acrolein or other thermal degradation products from oil. Local exhaust ventilation may be necessary on oil quench tanks.

### 1.7.4 Hazard Potential

The principal problems in heat treating operations are due to the special furnace environments, especially carbon monoxide, and the special hazards from handling bath materials. Although the hazard potential is significant from these operations, few data are available.

The fire and safety hazards of these operations are considerable and have been extensively reviewed (National Fire Protection Association [NFPA] 1991). Salt bath temperature controls must be reliable, and the baths must be equipped with automatic shutdowns. Venting rods need to be inserted in baths before shutdown and reheating to release gases when the bath is again brought up to temperature. If this is not done and gas is occluded in the bath, explosions or blowouts may occur. Parts must be clean and dry before immersion in baths for residual grease, paint, and oil may cause explosions. Where sprinkler systems are used, canopies should be erected above all oil, salt, and metal baths to prevent water from cascading into them. Dilution ventilation for fugitive emissions as well as local exhaust ventilation for baths is often provided, but specific standards have not been developed.

The physical hazards for employees include heat stress from furnace processes and noise from mechanical equipment and combustion air. Personal protective equipment may include goggles, face shields, heavy gloves and gauntlets, reflective clothing, and protective screens. The Wolfson Heat Treatment Centre (WHTC 1981, 1983, 1984) reports provide detailed information on hazards and safe practices.

## 1.8 METAL MACHINING

The fabrication of metal parts is done with a variety of machine tools, the most common of which are the lathe, drill press, miller, shaper, planer, and surface grinder. The occupational health hazards from these operations are similar; so they are grouped together under conventional machining. Two rapidly expanding techniques, electrochemical and electrical discharge machining (EDM), are also discussed in this section.

### 1.8.1 Conventional Metal Machining

The major machining operations of turning, milling, and drilling utilize cutting tools that shear metal from the workpiece as either the workpiece or the tool rotates. A thin running coil is formed that normally breaks at the tool to form small chips. Extremes of temperature and pressure occur at the interface between the cutting tool and the work. To cool this point, provide an interface lubricant, and help flush away the chips, a coolant or cutting oil is directed on the cutting tool in a solid stream (flood) or a mist.

The airborne particles generated by these machining operations depend on the type of base metal and cutting tool, the dust-forming characteristics of the metal, the machining technique, and the coolant and the manner in which it is applied. Each of these concerns is addressed briefly in this section.

The type of metal being machined is of course of concern. The metals range from mild steel with no potential health hazard as a result of conventional machining to various high-temperature and stainless alloys incorporating known toxic metals including lead, chromium, nickel, and cobalt, which may present low airborne exposures to toxic metals depending on the machining technique. Finally, highly toxic metals, such as beryllium, do present significant exposures that require vigorous control in any machining operation. Under normal machining operations, excluding dimensional grinding, the airborne dust concentration from conventional metals and alloys is minimal.

A range of specialized alloys has been developed for use in the manufacture of cutting tools. These materials include (i) high-carbon steels with alloying elements of vanadium, chromium, and manganese; (ii) high-speed steels containing manganese and tungsten; (iii) special cobalt steels; (iv) cast alloys of tungsten, chromium, and cobalt; and (v) tungsten carbide. The loss of material from the cutting tool is insignificant during conventional machining, and therefore airborne dust concentrations from the tool do not represent a potential hazard. However, preparing the cutting tools may involve a significant exposure to toxic metal dusts during grinding and sharpening, and such operations should be provided with local exhaust ventilation.

Coolants and cutting fluids are designed to cool and lubricate the point of the cutting tool and flush away chips. These fluids are currently available in the form of (i) soluble (emulsified) cutting oils based on mineral oil emulsified in water with soaps or sulfonates; (ii) straight cutting oils based on a complex mixture of paraffinic, naphthenic, and aromatic mineral oils with the addition of fatty acids; (iii) synthetic oils of varying composition; and a mixture of (i) and (iii). A description of the composition of the three types of cutting fluids including special additives is shown in Table 1.4 (NIOSH 1978).

***1.8.1.1 Health Effects*** Cutting fluids present two potential health problems: extensive skin contact with the cutting fluids and the inhalation of respirable oil mist. It has been estimated that over 400 000 cases of dermatitis occur in the United States each year from contact with coolants and cutting fluids. Soluble oils frequently cause eczematous dermatitis, whereas the straight oils (insoluble) cause folliculitis. One also notes occasional sensitization to coolants.

There continues to be a difference of opinion on the role of bacterial contamination of fluids in dermatitis; however, there is agreement that maintenance of coolants is of hygienic significance. A coolant sampling procedure is now available that permits evaluation of aerobic bacteria, yeasts, and fungi concentration in coolants using simple dip slides.

The application of the cutting fluid to hot, rotating parts releases an oil mist that causes the characteristic smell in the machine shop. The health effects from extended exposure to airborne mists of mineral oil and synthetic coolants are not clear. The association between mineral oil-based fluids and squamous cell carcinoma was observed in the United Kingdom in the 1800s. Unrefined or partially refined mineral oil has been classified as a carcinogen by IARC (1987); one suspected causative agent is the polycyclic aromatic hydrocarbons (PAHs) that are not removed by mild refining techniques. A study conducted in the automotive parts fabrication industry found a twofold increase of larynx cancer for straight oil exposure (Eisen et al. 1994). Other studies have found increased digestive tract and other malignancies (Decoufle 1978; Jarvolm et al. 1982; Vena et al. 1985). The exact agent(s) responsible for the increased cancer rates have not been determined. OSHA has recommended the use of severely refined mineral oils since the mid-1980s in order to minimize PAH exposure. Also the use of synthetic fluids containing both ethanolamines and nitrites has been eliminated. These mixtures have been shown to contain nitrosamines, known animal carcinogens.

**TABLE 1.4 Composition of Cutting Fluids**

I. Mineral oil
  1. Base 60–100%, paraffinic or naphthenic
  2. Polar additives
     a. Animal and vegetable oils, fats, and waxes to wet and penetrate the chip/tool interface
     b. Synthetic boundary lubricants: esters, fatty oils and acids, poly or complex alcohols
  3. Extreme pressure lubricants
     a. Sulfur-free or combined as sulfurized mineral oil or sulfurized fat
     b. Chlorine, as long-chain chlorinated wax or chlorinated ester
     c. Combination: sulfo-chlorinated mineral oil or sulfo-chlorinated fatty oil
     d. Phosphorus, as organic phosphate or metallic phosphate
  4. Germicides

II. Emulsified oil (soluble oil) – opaque, milky appearance
  1. Base: mineral oil, comprising 50–90% of the concentrate; in use the concentrate is diluted with water in ratios of 1:5 to 1:50
  2. Emulsifiers: petroleum sulfonates, amine soaps, rosin soaps, naphthenic acids
  3. Polar additives: sperm oil, lard oil, and esters
  4. Extreme pressure lubricants
  5. Corrosion inhibitors: polar organics, e.g. hydroxylamines
  6. Germicides
  7. Dyes

III. Synthetics (transparent)
  1. Base: water, comprising 50–80% of the concentrate; in use the concentrate is diluted with water in ratios of 1:10 to 1:200. True synthetics contain no oil. Semisynthetics are available that contain mineral oil present in amounts of 5–25% of the concentrate
  2. Corrosion inhibitors
     a. Inorganics: borates, nitrites, nitrates, phosphates
     b. Organics: amines, nitrites (amines and nitrites are typical and cheap)
  3. Surfactants
  4. Lubricants: esters
  5. Dyes
  6. Germicides

*Source:* NIOSH 1978.

Bronchitis and asthma may also be associated with machining fluid exposure; cross-shift changes in pulmonary function were demonstrated by Kennedy et al. (1989).

***1.8.1.2 Control*** An excellent pamphlet on lubricating and coolant oils prepared by Esso outlines procedures to minimize exposure to cutting fluids (Esso Petroleum Company 1979). Selected work practice recommendations are the following: (i) Avoid all unnecessary contact with mineral or synthetic oils. Minimize contact by using splash guards, protective gloves, protective aprons, etc. (ii) Encourage workers to wear clean work clothes, since oil-soaked clothing may hold the oil in contact with the skin longer than would otherwise occur. (iii) Remove oil from the skin as soon as possible if contact does occur; this means the installation of easily accessible wash basins and the provision of mild soap and clean towels in adequate supply. (iv) Do not allow solvents to be used for cleansing the skin. Use only warm water, mild soap, and a soft brush or in combination with a mild proprietary skin cleanser.

In many cases, the machining operations on such metals as magnesium and titanium may generate explosive concentrations of dust. Frequently, these operations must be segregated, and the operations must be conducted with suitable ventilation control and air cleaning. High-density layout of machine tools frequently results in workplace noise exposures above 85 dBA, the level that triggers the OSHA hearing conservation program.

Since the 1950s, exposures to machining fluid particulates in large production facilities has decreased due to the installation of enclosures, local exhaust ventilation, general air cleaning, and probably the increased use of soluble and synthetic fluids (Hallock et al. 1994). Today most high-volume machine tools are designed with complete enclosures and local exhaust ventilation. As of this writing, the PEL for mineral oil remains 5 mg m$^{-3}$ total particulate, but OSHA has announced its plan to reexamine this value.

### 1.8.2 Electrochemical Machining

The electrochemical machining (ECM) process utilizes a DC electrolytic bath operating at low-voltage and high current density. The workpiece is the anode; and the cutting tool, or cathode, is machined to reflect the geometry of the hole to be cut in the workpiece. As electrolyte is pumped through the space between the tool and the workpiece, metal ions are removed from the workpiece and are swept away by the electrolyte. The tool is fed into the workpiece to complete the cut. The electrolyte varies with the operation; one manufacturer states that the electrolyte is 10% sulfuric acid.

Because the ECM method is fast, produces an excellent surface finish, does not produce burrs, and produces little tool wear, it is widely used for cutting irregular-shaped holes in hard, tough metals.

In the operation, the electrolyte is dissociated and hydrogen is released at the cathode. A dense mist or smoke is released from the electrolyte bath. Local exhaust ventilation must be provided to remove this mist and ensure hydrogen concentrations do not approach the lower flammability limit.

### 1.8.3 Electrical Discharge Machining

A spark-gap technique is the basis for the EDM procedure, which is a popular machining technique for large precise work such as die sinking or the drilling of small holes in complex parts. In one system, a graphite tool is machined to the precise size and shape of the hole to be cut. The workpiece (anode) and the tool (cathode) are immersed in a dielectric oil bath and powered by a low-voltage DC power supply. The voltage across the gap increases until breakdown occurs, and there is a spark discharge across the gap, which produces a high temperature at the discharge point. This spark erodes a small quantity of the metal from the workpiece. The cycle is repeated at a frequency of 200–500 Hz with rather slow, accurate cutting of the workpiece. In more advanced systems, the cutting element is a wire that is programmed to track the cutting profile. The wire systems can also be used to drill small holes.

The hazards from this process are minimal and are principally associated with the oil. In light cutting jobs, a petroleum distillate such as Stoddard solvent is commonly used, whereas in large work a mineral oil is the dielectric. When heavy oil mists are encountered, local exhaust ventilation is needed. The oil gradually becomes contaminated with small hollow spheres of metal eroded from the part. As in the case of conventional machining, these metals may dissolve in the oil and present a dermatitis problem. An ultrahigh-efficiency filter should be placed in the oil recirculating line to remove metal particles.

## 1.9 NONDESTRUCTIVE TESTING

With the increase in manufacturing technology in the last decade of the twentieth century, a need has developed for in-plant inspection techniques. The most common procedures now in use in the metalworking industries are discussed in this section.

### 1.9.1 Industrial Radiography

Radiography is used principally in industry for the examination of metal fabrications such as weldments, castings, and forgings in a variety of settings. Specially designed shielded cabinets may be located in manufacturing areas for in-process examination of parts. Large components may be transported to shielded rooms for examination. Radiography may be performed in open shop areas, on construction sites, on board ships, and along pipelines.

The process of radiography consists of exposing the object to be examined to X-rays or gamma rays from one side and measuring the amount of radiation that emerges from the opposite side. This measurement is usually made with film or a fluoroscopic screen to provide a visual, two-dimensional display of the radiation distribution and any subsurface porosities.

The principal potential hazard in industrial radiography is exposure to ionizing radiation. This section deals with the minimum safety precautions designed to minimize worker exposure to radiation (X-rays and gamma rays) sources.

***1.9.1.1 X-Ray Sources*** X-rays used in industrial radiography are produced electrically and therefore fall into the category of "electronic product radiation." For this reason the design and manufacture of industrial X-ray generators are regulated by the Food and Drug Administration, Center for Devices and Radiological Health. ANSI has developed a standard for the design and manufacture of these devices (ANSI 1976). These standards specify maximum allowable radiation intensities outside the useful beam. They require warning lights on both the control panel and the tube head to indicate when X-rays are being generated.

The use of industrial X-ray generators is regulated by OSHA. Additionally, many states regulate the use of industrial X-ray generators within their own jurisdictions. Areas in which radiography is performed must be posted with signs that bear the radiation caution symbol and a warning statement. Access to these areas must be secured against unauthorized entry. When radiography is being performed in open manufacturing areas, it is essential to instruct other workers in the identification and meaning of these warning signs in order to minimize unnecessary exposures.

Radiographic operators are required to wear personal monitoring devices to measure the magnitude of their exposure to radiation. Typical devices are film badges, thermoluminescent dosimeters, and direct-reading pocket dosimeters. It is also advisable for radiographic operators to use audible alarm dosimeters or "chirpers." These devices emit an audible signal or "chirp" when exposed to radiation. The frequency of the signal is proportional to the radiation intensity. They are useful in warning operators who unknowingly enter a field of radiation (*Isotope Radiography Safety Handbook* 1981).

The operators must be trained in the use of radiation survey instruments to monitor radiation levels to which they are exposed and to assure that the X-ray source is turned off at the conclusion of the operation. A wide variety of radiation survey instruments is available for use in industrial radiography. When using industrial X-ray generators, especially relatively low-energy generators, it is imperative that the instrument has an appropriate energy range for measuring the energy of radiation used.

***1.9.1.2 Gamma Ray Sources*** Gamma rays used in industrial radiography are produced as a result of the decay of radioactive nuclei. The principal radioisotopes used in industrial radiography are iridium-192 and cobalt-60. Other radioisotopes, such as ytterbium-169 and thulium-170, are much less common but have some limited applicability. Radioisotope sources produce gamma rays with discrete energies, as opposed to the continuous spectrum of energies produced by X-ray generators. Owing to radioactive decay, the activity of radioisotope sources decreases exponentially with time.

Unlike X-ray generators, radioisotope sources require no external source of energy, which makes their use attractive in performing radiography in remote locations such as pipelines. However, because they are not energized by an external power supply, these sources cannot be "turned off," and they continuously emit gamma rays. For this reason, certain additional safety precautions must be exercised.

In industrial radiographic sources, the radioisotope is sealed inside a source capsule that is usually fabricated from stainless steel. The radioisotope source capsule is stored inside a shielded container or "pig" when not in use to reduce the radiation intensities in the surrounding areas. In practice, the film is positioned, and the radioactive source is then moved from the pig to the desired exposure position through a flexible tube by means of a mechanical actuator. At the end of the exposure, the operator retracts the source to the storage position in the pig.

The design, manufacture, and use of radioisotope sources and exposure devices for industrial radiography are regulated by the U.S. Nuclear Regulatory Commission (NRC). However, the NRC has entered into an agreement with "agreement states" for the latter to regulate radioisotope radiography within their jurisdictions. Organizations that wish to perform radioisotope radiography must obtain a license from either the NRC or the agreement state. In order to obtain this license, they must describe their safety procedures and equipment to the licensing authority. Check with your state agency or the NRC to see if your state is an agreement state.

Radiographic operators must receive training as required by the regulatory bodies including instruction in their own organization's safety procedures, a formal radiation safety training course, and a period of on-the-job training when the trainees work under the direct personal supervision of a qualified radiographer. At the conclusion of this training, operators must demonstrate their knowledge and competence to the licensee's management.

Areas in which radiography is being performed must be posted with radiation warning signs, and access to these areas must be secured as described earlier. An operator performing radioisotope radiography must wear both a direct-reading pocket dosimeter and either a film badge or a thermoluminescent dosimeter. Additionally, the operator must use a calibrated radiation survey instrument during all radiographic operations. In order to reduce the radiation intensity in the area after an exposure, the operator must retract the source to a shielded position within the exposure device. The only method available to the operator for assuring that the source is shielded properly is a radiation survey of the area. The operator should survey the entire perimeter of the exposure device and the entire length of guide tube and source stop after each radiographic operation to assure that the source has been fully and properly shielded. Some current types of radiographic exposure devices incorporate source position indicators that provide a visual signal if the source is not stored properly. The use of these devices may also reduce the frequency of radiation exposure incidents.

Occasional radiation exposure incidents have occurred in industrial radioisotope radiography. These incidents generally result because the operator fails to return the source properly to a shielded position in the exposure device and then approaches the exposure device or source stop without making a proper radiation survey. The importance of making a proper radiation survey at the conclusion of each radiographic exposure cannot be overemphasized. The estimated mean annual dose equivalent for radiography workers is 290 mRem $yr^{-1}$.

### 1.9.2 Magnetic Particle Inspection

This procedure is suitable for detecting surface discontinuities, especially cracks, in magnetic materials. The procedure is simple and of relatively low cost; as a result, it is widely applied in metal fabrication plants.

The parts to be inspected first undergo vigorous cleaning and then are magnetized. Magnetic particles are applied as a powder or a suspension of particles in a carrier liquid. The powders are available in color for daylight viewing or as fluorescent particles for more rigorous inspection with a UV lamp. Surface imperfections, such as cracks, result in leakage of the magnetic field with resulting adherence of the particles. The defined geometry of the imperfection stands out in a properly illuminated environment or with a handheld UV lamp.

Worker exposures using this technique are minor, consisting of skin contact and minimal air contamination from the suspension field, usually an aliphatic hydrocarbon, exposure to the magnetic field during the inspection/process, and exposure to the lamp output. General exhaust ventilation usually is adequate for vapors from the liquid carrier, and gloves, apron, and eye protection should be used to minimize skin and eye contact. A proper filter on the UV lamp minimizes exposure to UV B radiation. The hazard from magnetic fields is now under investigation.

### 1.9.3 Liquid Penetrant

This procedure complements magnetic particle inspection for surface cracks and weldment failures because it can be used on nonmagnetic materials. A colored or fluorescent liquid used as the penetrant is flowed or sprayed on the surface, or the part is dipped. The excess penetrant is removed in some systems by adding an emulsifier and removing with water or by wiping off the excess. A developer is applied to intensify the definition of the crack, and the inspection is carried out by daylight or UV lamp depending on the penetrant system of choice. The penetrant is then removed from the workpiece by either water or solvent such as mineral spirits.

The inspectors' exposures are to the solvent or carrier used with the penetrant, emulsifier, developer, and the final cleaning agent. The solvent is water, mineral spirits, or, in some cases, halogenated compounds.

### 1.9.4 Ultrasound

Pulse echo and transmission-type ultrasonic inspection have a wide range of application for both surface and subsurface flaws. This procedure resolves voids much smaller than all other methods, including radiographic procedures. The procedure commonly involves immersing the part in water to improve coupling between the ultrasound transmitter/receiver and the workpiece. The reported response on other applications of higher-energy ultrasound systems include local heating and subjective effects. Health effects have not been reported on its application in nondestructive testing of metals.

## 1.10 METAL THERMAL SPRAYING

A practical technique for spraying molten metal was invented in the 1920s and has found application for applying metals, ceramics, and plastic powder to workpieces for corrosion protection, to build up worn or corroded parts, to improve wear resistance, to reduce production costs, and as a decorative surface. The health hazards from thermal spraying of metals were first observed in 1922 in the United Kingdom when operators suffered from lead poisoning (Ballard 1972).

The application head for thermal spraying takes many forms and dictates the range of hazards one may encounter from the operation. The four principal techniques currently in use in industry are described below.

### 1.10.1 Spraying Methods

In the combustion spraying or flame method, the coating material in wire form is fed to a gun operated with air/oxygen and a combustible gas such as acetylene, propane, or natural gas. The wire is melted in the oxygen-fuel flame and propelled from the torch at velocities up to 240 m s$^{-1}$ (48 000 fpm) with compressed air. The material bonds to the workpiece by a combination of mechanical interlocking of the molten, platelet-form particles and a cementation of partially oxidized material.

In the thermal arc spraying or wire method, two consumable metal wire electrodes are made of the metal to be sprayed. As the wires feed into the gun, they establish an arc as in a conventional arc welding unit; the molten metal is disintegrated by compressed air, and the molten particles are projected to the workpiece at high velocity.

In the plasma method, an electric arc is established in the controlled atmosphere of a special nozzle. Argon is passed through the arc, where it ionizes to form a plasma that continues through the nozzle and recombines to create temperatures as high as 16 700 °C (30 000 °F). Metal alloy, ceramic, and carbide powders are melted in the stream and are released from the gun at a velocity of 300–600 m s$^{-1}$ (60 000–120 000 fpm).

In the detonation method, the gases are fed to a combustion chamber, where they are ignited by a spark plug. Metal powder is fed to the chamber, and the explosions drive the melted powder to the workpiece at velocities of approximately 760 m $s^{-1}$ (150 000 fpm). Tungsten carbide, chromium carbide, and aluminum oxide are applied with this technique, which is used on a limited basis.

### 1.10.2 Hazards and Controls

A major hazard common to all techniques is the potential exposure to toxic metal fumes. If the vapor pressure of the metal at the application temperature is high, this will be reflected in high air concentrations of metal fume. The deposition efficiency, that is, the percent of the metal sprayed deposited on the workpiece, varies between application techniques and the coating metal and has a major impact on air concentrations in the workplace. Because the airborne level varies with the type of metal sprayed, the application technique, the fuel in use, and the tool-workpiece geometry, it is impossible to estimate the level of contamination that may occur from a given operation. For this reason, if a toxic metal is sprayed, air samples should be taken to define the worker exposure.

The principal control of air contaminants generated during flame and arc spraying is local exhaust ventilation such as an open or enclosing hood (Hagopian and Bastress 1976). The operator should also wear a positive-pressure air-supplied respirator while metallizing with toxic metals. One must also consider the exposure of other workers in the area, because systems with 50% overspray obviously represent a significant source of general contamination.

All thermal spraying procedures result in significant noise exposures and require effective hearing conservation programs. Appreciable noise reduction can be achieved by changing the operating characteristics of the spray equipment, which may or may not affect the quality of the surface coating. If the resulting changes are unacceptable, the normal approach to noise control, including isolation, the use of hearing protectors, and various work practices, must be used.

The thermal arc and plasma procedures present additional hazards reflecting their welding ancestry, including UV exposure and the generation of ozone and nitrogen dioxide. If ventilation is effective in controlling metal fumes, it will probably control ozone and nitrogen dioxide, although this fact should be established by air sampling. NIOSH (1989b) study demonstrated excess visible, IR, and UV radiation from thermal arc spraying.

In addition to air-supplied respirators, operators handling toxic metals such as cadmium and lead should wear protective gloves and coveralls and be required to strip, bathe, and change to clean clothes prior to leaving the plant.

Special fire protection problems are encountered in routine metallizing operations. Unless ventilation control is effective, particles will deposit on various plant structures and may become a potential fire hazard. If one uses a scrubber to collect the particles, hydrogen may form in the sludge, resulting in a potential fire and explosion hazard.

## 1.11 PAINTING

Paint products are used widely in industry to provide a surface coating for protection against corrosion, for appearance, as electrical insulation, for fire retardation, and for other special purposes. The widespread application of this technology from small job shops to highly automated painting of automobiles includes more than half a million workers. Although the health status of painters has not been well defined, the studies that have been completed suggest acute and chronic central nervous system effects, hematologic disorders, and excess mortality from cancer. In addition, respiratory sensitization may occur from two-part urethane systems, and the amine catalyst in spray paints may cause skin sensitization. The hazards associated with the industrial application of paint products are discussed here.

### 1.11.1 Types of Paints

The term *paint* is commonly used to identify a range of organic coatings including paints, varnishes, enamels, and lacquers. Conventional paint is an inorganic pigment dispersed in a vehicle consisting of a binder and a solvent with selected fillers and additives. The conventional varnish is a nonpigmented product based on oil and natural resin in a solvent that dries first by the evaporation of the solvent and then by the oxidation of the resin binder. Varnishes are also available based on synthetic resins. A pigmented varnish is called an enamel. Lacquers are coatings in which the solvent evaporates, leaving a film that can be redissolved in the original solvent.

In the past decade, paint formulations have been influenced by environmental regulations limiting the release of volatile organic solvents. Although conventional solvent-based paints will continue to see wide application, the major thrust is to convert to formulations with low-solvent content. Such systems include solvent-based paints with high-solids content (>70% by weight), nonaqueous dispersions, powder, two-part catalyzed systems, and waterborne paints. The use of UV-curable paints may aid efforts to use powder coatings at lower temperatures for a greater variety of substrates.

### 1.11.2 Composition of Paint

The conventional solvent-based paints consist of the vehicle, filler, and additives. The vehicle represents the total liquid content of the paint and includes the binder and solvent. The binder, which is the film-forming ingredient, may be a naturally occurring oil or resin including linseed oil and oleoresinous materials or synthetic material such as alkyd resins.

The fillers include pigments and extenders, which historically have presented a major hazard in painting. The common white pigments include bentonite and kaolin clay, talc, titanium dioxide, and zinc oxide. Mineral dust used as extenders to control viscosity, texture, and gloss include talc, clay, calcium carbonate, barite, and both crystalline and amorphous silica. The pigments and extenders do not represent an exposure during brush or roller application, but when sanded or removed preparatory to repainting, these materials are released to the air and may present significant exposures. A group of pigments including lead carbonate, cadmium red, and chrome green and yellow do represent a critical exposure, both during spray application and surface preparation.

Additives to hasten drying, reduce skin formation in the can, and control fungus are added to paints. The fungicides warrant special attention because the ingredients, including copper and zinc naphthenate, copper oxide, and tributyltin oxide, are biologically active.

The solvent systems are varied and complex. The most common *organic solvents* include aliphatic and aromatic hydrocarbons, ketones, alcohols, glycols, and glycol ether/esters. These solvents have high vapor pressure and represent the critical worker exposure component in most painting techniques. High-solids paint systems (low solvent) represent a major contribution to reduced solvent exposure.

The ultimate high-solids, low-solvent formulation is a dry powder paint. This coating technique provides a high-quality job while eliminating solvent exposure for the worker. The application technique, to be described later, utilizes a dry powder formulated to contain a resin, pigment, and additives. Thermosetting resins, used for decorative and protective coatings, are based on epoxy, polyester, and acrylic resins. Thermoplastic resins are applied in thick coatings for critical applications.

Waterborne paints now represent 15–20% of construction and industry paints, and expanded use is anticipated. The present systems use binders based on copolymers of several monomers including acrylic acid, butyl acrylate, ethyl hexyl acrylate, and styrene. The other major ingredients and their maximum concentrations include pigments (54%), coalescing solvents (15%), surfactants (5%), biocides (1.1%), plasticizers (2.2%), and driers in lesser amounts. The pigments include the conventional materials mentioned under solvent-based paints. Coalescing solvents, including hydrocarbons, alcohols, esters, glycol, and glycol ether/esters, although present in low concentrations, may present an inhalation hazard. The proprietary biocides in use frequently result in exposure to significant airborne formaldehyde concentrations. The surfactants in use include known skin irritants and sensitizers (Hansen et al. 1987).

A single study of waterborne paints and worker exposure confirms that these systems represent an improvement over solvent-based systems. The authors of this study recommend that paint formulations be designed to eliminate formaldehyde, minimize the unreacted monomer content in the various polymers, eliminate ethylene glycol ethers because of their reproductive hazard, and minimize the ammonia content (Hansen et al. 1987).

Special attention must be given to two-component epoxy and urethane paint systems. The urethanes consist of a polyurethane prepolymer containing a reactive isocyanate; the second component is the polyester. Mixing of the two materials initiates a reaction with the formation of a chemically resistant coating. In the early formulations, the unreacted toluene diisocyanate (TDI) resulted in significant airborne exposures to TDI and resulting respiratory sensitization. Formulations have been changed to include TDI adducts and derivatives that are stated to have minimized the respiratory hazard from these systems. For ease of application, single package systems have been developed using a blocked isocyanate. Unblocking takes place when the finish is baked and the isocyanate is released to react with the polyester.

Epoxy paint systems offer excellent adhesive properties, resistance to abrasion and chemicals, and stability at high temperatures. The conventional epoxy system is a two-component system consisting of a resin based on the reaction products of bisphenol A and epichlorohydrin. The resin may be modified by reactive diluents such as glycidyl ethers. The second component, the hardener or curing agent, was initially based on low-molecular-weight, highly reactive amines.

Epoxy resins can be divided into three grades. The solid grades are felt to be innocuous; however, skin irritation may occur from solvents used to take up the resin. The liquid grades are mild to moderate skin irritants. The low-viscosity glycidyl ether modifiers are skin irritants and sensitizers and have systemic toxicity. The use of low-molecular-weight aliphatic amines such as diethylenetetriamine and triethylenetetramine, both strong skin irritants and sensitizers, presented major health hazards during the early use of epoxy systems. This has been overcome to a degree by the use of low-volatility amine adducts and high-molecular-weight amine curing agents.

### 1.11.3 Operations and Exposures

In the industrial setting, paints can be applied to parts by a myriad of processes including brush, roller, dip, flow, curtain, tumbling, conventional air spray, airless spray, heated systems, disk spraying, and powder coating. Conventional air spraying is the most common method encountered in industry and presents the principal hazards owing to overspray. The use of airless and hot spray techniques minimizes mist and solvent exposure to the operator, as does electrostatic spraying. The electrostatic technique, now commonly used with many installations, places a charge on the paint mist particle so it is attracted to the part to be painted, thereby reducing rebound and overspray.

In powder coating, a powder, which represents the paint formulation including the resin, pigment, and additives, is conveyed from a powder reservoir to the spray gun. In the gun, a charge is imparted to the individual paint particles. This dry powder is sprayed on the electrically grounded workpiece, and the parts are baked to fuse the powder film into a continuous coating.

The operator is exposed to the solvent or thinner in processes in which the paint is flowed on, as in brushing and dipping, and during drying of the parts. However, during atomization techniques, the exposure is to both the solvent and the paint mist. The level of exposure reflects the overspray and rebound that occur during spraying.

A common exposure to organic vapors occurs when spray operators place the freshly sprayed parts on a rack directly behind them. The air movement to the spray booth sweeps over the drying parts and past the breathing zone of the operator, resulting in an exposure to solvent vapors. Drying stations and baking ovens must therefore be exhausted. The choice of exhaust control on tumbling and roll applications depends on the surface area of the parts and the nature of the solvent.

### 1.11.4 Controls

Most industrial flow and spray painting operations utilizing solvent-based paints require exhaust ventilation for control of solvent vapors at the point of application and during drying and baking operations (NIOSH 1984; O'Brien and Hurley 1981). Water-based paints may require ventilation only when spray application is utilized. Flow application of solvent-based paints requires local exhaust ventilation depending on the application technique. The usual ventilation control for spray application of solvent-based paints is a spray booth, room, or tunnel provided with some type of paint spray mist arrestor before the effluent is exhausted outdoors. The degree of control of paint mist and solvent varies with the application method, that is, whether it is air atomization, airless, or electrostatic painting. The latter two techniques call for somewhat lower exhaust volumes. The advent of robotic spray painting permits the design of exhausted enclosures and minimizes worker exposure.

The design of the conventional paint spray booth is shown in Figure 1.5. The booths are commonly equipped with a water curtain or a throwaway dry filter to provide paint mist removal. The efficiency of these devices against paint mist has not been evaluated. Neither of these systems, of course, removes solvent vapors from the air stream.

In industrial spray painting of parts, the simple instructions contained in several state codes on spray painting should be observed:

1. Do not spray toward a person.
2. Automate spray booth operations where possible to reduce exposure.
3. Maintain 2-ft clearance between the sides of the booth and large flat surfaces to be sprayed.
4. Keep the distance between the nozzle and the part to be sprayed to less than 12 in.
5. Do not position work so that the operator is between the exhaust and the spray gun or disk.
6. Locate the drying room so that air does not pass over drying objects to exhaust food past the breathing zone of the operator.

Controls in the application of two-component urethane and epoxy paint systems must include excellent housekeeping, effective ventilation control, and protective clothing; moreover, in applications not effectively controlled by ventilation, the operators should wear air-supplied respirators. Adequate washing facilities should be available, and eating, drinking, and smoking should be prohibited in the work area.

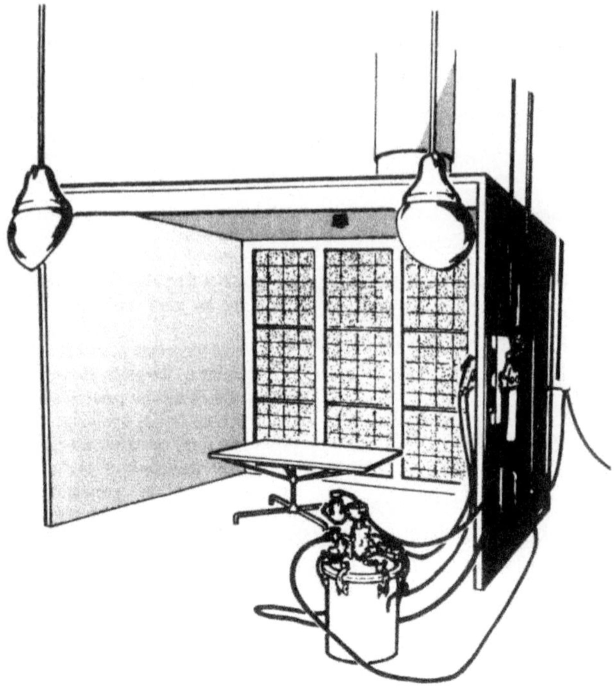

**Figure 1.5** Paint spray booth.

Dermatitis due to primary irritation and defatting from solvents or thinners as well as sensitization from epoxy systems is not uncommon. Skin contact must be minimized, rigorous personal cleanliness encouraged, and suitable protective equipment used by the operator.

## 1.12 SOLDERING AND BRAZING

Soft soldering is the joining of metal by surface adhesion without melting the base metal. This technique uses a filler metal (solder) with a melting point less than 316 °C (600 °F); hard solder is used in the range of 316–427 °C (600–800 °F). These temperature ranges differentiate soldering from brazing, which utilizes a filler metal with a melting point greater than 427 °C (800 °F).

### 1.12.1 Soldering

To understand the potential health hazards from soldering operations, one must be familiar with the composition of the solder and fluxes in use and the applicable production techniques.

***1.12.1.1 Flux*** All metals including the noble metals have a film of tarnish that must be removed in order to wet the metal with solder effectively and accomplish a good mechanical bond. The tarnish takes the form of oxides, sulfides, carbonates, and other corrosion products (Manko 1979). The flux, which may be a solid, liquid, or gas, is designed to remove any adsorbed gases and tarnish from the surface of the base metal and keep it clean until the solder is applied. The molten solder displaces the residual flux and wets the base metal to accomplish the bond.

A range of organic and inorganic materials is used in the design of soldering fluxes as shown in Table 1.5. The flux is usually a corrosive cleaner frequently used with a volatile solvent or vehicle. Rosin, which is a common base for organic fluxes, contains abietic acid as the active material.

The three major flux bases are inorganic, organic nonrosin, and organic rosin, as shown in Table 1.5. A proprietary flux may contain several of these materials. Organic nonrosin is less corrosive and therefore slower acting, and in general, these fluxes do not present as severe a handling hazard as the inorganic acids. This is not true of the organic halides whose degradation products are very corrosive and warrant careful handling and ventilation control. The amines and amides in this second class of flux materials are common ingredients whose degradation products are very corrosive.

**TABLE 1.5 Common Flux Materials**

| Type | Typical fluxes | Vehicle |
|---|---|---|
| | *Inorganic* | |
| Acids | Hydrochloric, hydrofluoric, orthophosphoric | Water, petrolatum paste |
| Salts | Zinc chloride, ammonium chloride, tin chloride | Water, petrolatum paste, polyethylene glycol |
| Gases | Hydrogen-forming gas: dry HCI | None |
| | *Organic, nonrosin base* | |
| Acids | Lactic, oleic, stearic, glutamic, phthalic | Water, organic solvents, petrolatum paste, polyethylene glycol |
| Halogens | Aniline hydrochloride, glutamic acid hydrochloride, bromide derivatives of palmitic acid, hydrazine hydrochloride or hydrobromide | Water, organic solvents, polyethylene glycol |
| Amines and amides | Urea, ethylenediamine, mono- and triethanolamine | Water, organic solvents, petrolatum paste, polyethylene glycol |
| | *Organic, rosin base* | |
| Superactivated | Rosin or resin with strong activators | Alcohols, organic solvents, glycols |
| Activated (RA) | Rosin or resin with activator | Alcohols, organic solvents, glycols |
| Mildly activated (RMA) | Rosin with activator | Alcohols, organic solvents, glycols |
| Nonactivated (water-white rosin) (R) | Rosin only | Alcohols, organic solvents, glycols |

*Source: Isotope Radiography Safety Handbook* (1981).

The rosin base fluxes are inactive at room temperature but at soldering temperature are activated to remove tarnish. This material continues to be a popular flux compound because the residual material is chemically and electrically benign and need not be removed as vigorously as the residual products of the corrosive fluxes.

The cleaning of the soldered parts may range from a simple hot water or detergent rinse to a degreasing technique using a range of organic solvents. These cleaning techniques may require local exhaust ventilation.

***1.12.1.2 Solder*** The most common solder contains 65% tin and 35% lead. Traces of other metals including cadmium, bismuth, copper, indium, iron, aluminum, nickel, zinc, and arsenic are present. A number of special solders contain antimony in concentrations up to 5%. The melting point of these solders is quite low; and at these temperatures, the vapor pressure of lead and antimony usually do not result in significant air concentrations of metal fume. The composition of the common solders encountered in industry can be obtained from the manufacturers.

***1.12.1.3 Application Techniques*** Soldering is a fastening technique used in a wide range of products from simple mechanical assemblies to complex electronic systems. The soldering process includes cleaning the base metal and other components for soldering, fluxing, the actual soldering, and postsoldering cleaning.

***1.12.1.4 Initial Cleaning of Base Metals*** Prior to fluxing and soldering, the base metal must be cleaned to remove oil, grease, wax, and other surface debris. Unless this is done, the flux will not be able to attack and remove the metal surface tarnish. The procedures include cold solvent degreasing, vapor degreasing, and ultrasonic degreasing. If the base metal has been heat treated, the resulting surface scale must be removed.

In many cases, mild abrasive blasting techniques are utilized to remove heavy tarnish before fluxing. In electrical soldering, the insulation on the wire must be stripped back to permit soldering. Stripping is accomplished by mechanical techniques such as cutters or wire brushes, chemical strippers, and thermal techniques. Mechanical stripping of asbestos-based insulation obviously presents a potential health hazard, and this operation must be controlled by local exhaust ventilation. The hazard from chemical stripping depends on the chemical used to strip the insulation; however, at a minimum, the stripper will be a very corrosive agent. Thermal stripping of wire at high production rates may present a problem owing to the thermal degradation products of the insulation. Hot wire stripping of fluorocarbon insulation such as Teflon may cause polymer fume fever if operations are not controlled by ventilation. Other plastic insulation such as polyvinyl chloride may produce irritating and toxic thermal degradation products.

**1.12.1.5 Fluxing Operations** Proprietary fluxes are available in solid, paste, and liquid form for various applications and may be cut with volatile vehicles such as alcohols to vary their viscosity. The flux may be applied by one of 10 techniques shown in Table 1.6 depending on the workpiece and the production rate. Because flux is corrosive, skin contact must be minimized by specific work practices and good housekeeping. The two techniques that utilize spray application of the flux require local exhaust ventilation to prevent air contamination from the flux mist.

**1.12.1.6 Soldering and Cleaning** In two of the fluxing techniques listed in Table 1.6, the flux and solder are applied together. One need merely apply the necessary heat to bring the system first to the melting point of the flux and then to the melting point of the solder. In most applications, however, the flux and solder are applied separately, although the two operations may be closely integrated and frequently are in a continuous operation.

The soldering techniques used for manual soldering operations on parts that have been fluxed are the soldering iron and solder pot. A number of variations of the solder pot have been introduced to handle high production soldering of printed circuit boards (PCB) in the electronic industry. In the drag solder technique, the PCB is positioned horizontally and pulled along the surface of a shallow molten solder bath behind a skimmer plate that removes the dross. This process is usually integrated with a cleaning and fluxing station in a single automated unit. Another system designed for automation is wave soldering. In this technique, a standing wave is formed by pumping the solder through a spout. Again, the conveyorized PCBs are pulled through the flowing solder.

The potential health hazards to the soldering operators are minimal. A thermostatically controlled lead–tin solder pot operates at temperatures too low to generate significant concentrations of lead fume. The use of activated rosin fluxing agents may result in the release of thermal degradation products that may require control by local exhaust ventilation. The handling of solder dross during cleanup and maintenance may result in exposure to lead dust.

After soldering operations are carried out, some flux residue and its degradation products remain on the base metal. Both water-soluble and solvent-soluble materials normally exist so that it is necessary to clean with both systems using detergents and saponifiers in one case and common chlorinated and fluorinated hydrocarbons in the other. CFCs are being replaced with cleaners such as limonene, which are less threatening to the environment. The processing equipment may include ultrasonic cleaners and vapor degreasers.

**1.12.1.7 Controls** The fluxes may represent the most significant hazard from soldering. The conventional pure rosin fluxes are not difficult to handle, but highly activated fluxes warrant special handling instruction. A range of alcohols, including methanol, ethanol, and isopropyl alcohol, are used as volatile vehicles for fluxes. The special fluxes

**TABLE 1.6 Soldering Flux Application Techniques**

| Method | Application technique | Use |
|---|---|---|
| Brushing | Applied manually by paint, acid, or rotary brushes | Copper pipe, job shop printed circuit board, large structural parts |
| Rolling | Paint roller application | Precision soldering, printed circuit board, suitable for automation |
| Spraying | Spray painting equipment | Automatic soldering operations, not effective for selective application |
| Rotary screen | Liquid flux picked up by screen and air directs it to part | Printed circuit board application |
| Foaming | Work passes over air agitated foam at surface of flux tank | Selective fluxing, automatic printed circuit board lines |
| Dipping | Simple dip tank | Wide application for manual and automatic operations on all parts |
| Wave fluxing | Liquid flux pumped through trough forming wave through which work is dipped | High-speed automated operations |
| Floating | Solid flux on surface melts providing liquid layer | Tinning of wire and strips of material |
| Cored solder | Flux inside solder wire melts, flows to surface, and fluxes before solder melts | Wide range of manual operations |
| Solder paste | Solder blended with flux, applied manually | Component and hybrid microelectric soldering |

*Source: Isotope Radiography Safety Handbook* (1981).

should be evaluated under conditions of use to determine worker exposure. Rosin is a common cause of allergic contact dermatitis, and the fume causes an allergic asthma.

### 1.12.2 Brazing

Brazing techniques are widely used in the manufacture of refrigerators, electronics, jewelry, and aerospace components to join both similar and dissimilar metals. Although the final joint looks similar to a soft solder bond, it is much stronger, and the joint requires little finish. As mentioned earlier in this section, brazing is defined as a technique for joining metals that are heated above 430 °C (800 °F), whereas soldering is conducted below 430 °C. The temperature of the operation is of major importance because it determines the vapor pressure of the metals that are heated and therefore the concentration of metal fumes to which the operator is exposed (Handy and Harman 1985).

***1.12.2.1 Flux and Filler Metals*** Flux is frequently used; however, certain metals may be joined without flux. The flux is chosen to prevent oxidation of the base metal and not to prepare the surface as is the case with soldering. The common fluxes are based on fluorine, chlorine, and phosphorus compounds and present the same health hazards as fluxes used in soldering; that is, they are corrosive to the skin and may cause respiratory irritation. A range of filler metals used in brazing rods or wire may include phosphorus, silver, zinc, copper, cadmium, nickel, chromium, beryllium, magnesium, and lithium. The selection of the proper filler metal is the key to quality brazing.

***1.12.2.2 Application Techniques*** Brazing of small job lots that do not require close temperature control is routinely done with a torch. More critical, high production operations are accomplished by dip techniques in a molten bath, by brazing furnaces using either an ammonia or hydrogen atmosphere, or by induction heating.

The brazing temperatures define the relative hazard from the various operations. As an example, the melting point of cadmium is approximately 1400 °C. The vapor pressure of cadmium and the resulting airborne fume concentrations increase dramatically with temperature. The filler metals with the higher brazing temperatures will therefore present the most severe exposure to cadmium.

The exposure to fresh cadmium fume during brazing of low-alloy steels, stainless steels, and nickel alloys has resulted in documented cases of occupational disease and represents the major hazard from these operations. This is especially true of torch brazing, where temperature extremes may occur. On the other hand, the temperatures of furnace and induction heating operations may be controlled to ±5 °C.

One study of brazing in a pipe shop and on board ship included air sampling for cadmium while brazing with a filler rod containing from 10 to 24% cadmium. The mean concentrations during shipboard operation were 0.45 mg $m^{-3}$ with a maximum of 1.40 mg $m^{-3}$.

***1.12.2.3 Controls*** Controls on brazing operations must obviously be based on the identification of the composition of the filler rod. Local ventilation control is necessary in operations where toxic metal fumes may be generated from the brazing components or from parts plated with cadmium or other toxic metals. In the use of common fluxes, one should minimize skin contact owing to their corrosiveness and provide exhaust ventilation to control airborne thermal degradation products released during the brazing operation.

## 1.13 WELDING

Welding is a process for joining metals in which coalescence is produced by heating the metals to a suitable temperature. Over a dozen welding procedures are commonly encountered in industry. Four welding techniques have been chosen for discussion because they represent 80–90% of all manufacturing and maintenance welding. These procedures are used by more than half a million welders and helpers in the United States.

In the nonpressure welding techniques to be discussed in this section, metal is vaporized and then condenses to form initially a fume in the 0.01–0.1 μm particle size range that rapidly agglomerates. The source of this respirable metal fume is the base metal, the metal coating on the workpiece, the electrode, and the fluxing agents associated with the particular welding system. A range of gases and vapors including carbon monoxide, ozone, and nitrogen dioxide may be generated depending on the welding process. The wavelength and intensity of the electromagnetic radiation emitted from the arc depends on the welding procedure, inerting gas, and base metal.

The common welding techniques are reviewed assuming that the welding is done on unalloyed steel or so-called mild steel. The nature of the base metal is important in evaluating the metal fume exposure, and occasionally it has impact in other areas. When such impact is important, it is discussed.

### 1.13.1 Shielded Metal Arc Welding

Shielded metal arc (SMA) welding (Figure 1.6) is commonly called stick or electrode welding. An electric arc is drawn between a welding rod and the workpiece, melting the metal analog a seam or a surface. The molten metal from the workpiece and the electrode form a common puddle and cool to form the bead and its slag cover. Either DC or AC power is used in straight (electrode negative, work positive) or reverse polarity. The most common technique involves DC voltages of 10–50 and a wide range of current up to 2000 A. Although operating voltages are low, under certain conditions an electrical hazard may exist.

The welding rod or electrode may have significant occupational health implications. Initially a bare electrode was used to establish the arc and act as filler metal. Now the electrode covering may contain 20–30 organic and inorganic compounds and perform several functions. The principal function of the electrode coating is to release a shielding gas such as carbon dioxide to ensure that air does not enter the arc puddle and thereby cause failure of the weld. In addition, the covering stabilizes the arc, provides a flux and slag producer to remove oxygen from the weldment, adds alloying metal, and controls the viscosity of the metal. A complete occupational health survey of SMA welding requires the identification of the rod and its covering. The composition of the electrode can be obtained from the American Welding Society (AWS) classification number stamped on the electrode.

The electrode mass that appears as airborne fume may include iron oxides, manganese oxide, fluorides, silicon dioxide, and compounds of titanium, nickel, chromium, molybdenum, vanadium, tungsten, copper, cobalt, lead, and zinc. Silicon dioxide is routinely reported as present in welding fume in an amorphous form or as silicates, not in the highly toxic crystalline form.

***1.13.1.1 Metal Fume Exposure*** The potential health hazards from exposure to metal fume during SMA welding obviously depend on the metal being welded and the composition of the welding electrode. The principal component of the fume generated from mild steel is iron oxide. The hazard from exposure to iron oxide fume appears to be limited. The deposition of iron oxide particles in the lung does cause a benign pneumoconiosis known as siderosis. There is no functional impairment of the lung, nor is there fibrous tissue proliferation. In a comprehensive review of conflicting data, Stokinger (1984) has shown that iron oxide is not carcinogenic to man.

The concentration of metal fume to which the welder is exposed depends not only on the alloy composition but also on the welding conditions, including the current density (amperes per unit area of electrode); wire feed rate; the arc time, which may vary from 10 to 30%; the power configuration, that is, DC or AC supply; and straight or reverse polarity. The work environment also defines the level of exposure to welding fume and includes the type and quality of exhaust ventilation and whether the welding is done in an open, enclosed, or confined space.

Many of the data on metal fume exposures have been generated from shipyard studies. The air concentration of welding fume in several studies ranged from less than 5 mg m$^{-3}$ to over 100 mg m$^{-3}$ depending on the welding process, ventilation, and degree of enclosure. In these early studies, the air samples were normally taken outside the welding helmet. Recent studies have shown the concentration at the breathing zone inside the helmet ranges from one-half to one-tenth the outside concentrations. If SMA welding is conducted on stainless steel, the chromium concentrations may exceed the TLV (0.5 mg m$^{-3}$); in the case of alloys with greater than 50% nickel, the concentration of nickel fume may exceed 1 mg m$^{-3}$ (VanDerWal 1990).

***1.13.1.2 Gases and Vapors*** SMA welding has the potential to produce nitrogen oxides; however, this is not normally a problem in open shop welding. In over 100 samples of SMA welding in a shipyard, Burgess et al. (1989) did not identify an exposure to nitrogen dioxide in excess of 0.5 ppm under a wide range of operating conditions. Ozone is also

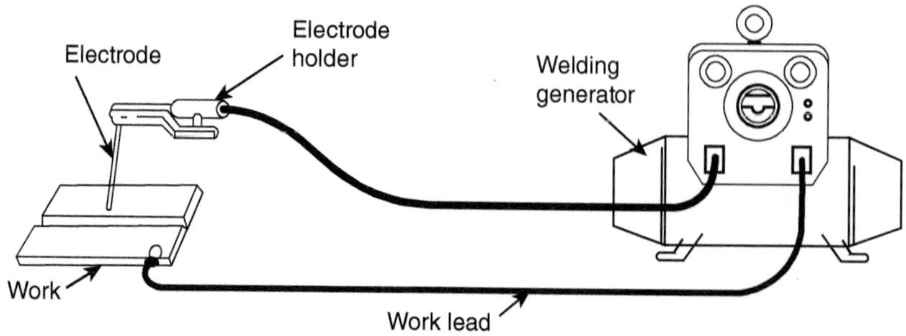


**Figure 1.6** Shielded metal arc welding (SMAW).

fixed by the arc, but again this is not a significant contaminant in SMA welding operations. Carbon monoxide and carbon dioxide are produced from the electrode cover, but air concentrations are usually minimal.

Low hydrogen electrodes are used with conventional arc welding systems to maintain a hydrogen-free arc environment for critical welding tasks on certain steels. The electrode coating is a calcium carbonate–calcium fluoride system with various deoxidizers and alloying elements such as carbon, manganese, silicon, chromium, nickel, molybdenum, and vanadium. A large part of this coating and of all electrode coatings becomes airborne during welding (Pantucek 1971). In addition to hydrogen fluoride, sodium, potassium, and calcium fluorides are present as particles in the fume cone.

Exposure to fumes from low hydrogen welding under conditions of poor ventilation may prompt complaints of nose and throat irritation and chronic nosebleeds. There has been no evidence of systemic fluorosis from this exposure. Monitoring can be accomplished by both air sampling and urinary fluoride measurements.

***1.13.1.3 Radiation*** The radiation generated by SMA welding covers the spectrum from the infrared C range of wavelengths to the UV C range. The acute condition known to the welder as "arc eye," "sand in the eye," or "flash burn" is due to exposures in the UV B range (see Figure 1.7). The radiation in this range is completely absorbed in the corneal epithelium of the eye and causes a severe photokeratitis. Severe pain occurs 5–6 hours after exposure to the arc, and the condition clears within 24 hours. Welders experience this condition usually only once and then protect themselves against a recurrence by the use of a welding helmet with a proper filter. Skin erythema or reddening may also be induced by exposure to UV C and UV B as shown in Figure 1.7 (Sliney and Wolbarsht 1980).

### 1.13.2 Gas Tungsten Arc Welding

Although SMA welding using coated electrodes is an effective way to weld many ferrous metals, it is not practical for welding aluminum, magnesium, and other reactive metals. The introduction of inert gas in the 1930s to blanket the arc environment and prevent the intrusion of oxygen and hydrogen into the weld provided a solution to this problem. In gas tungsten arc (GTA) welding (Figure 1.8) also known as tungsten inert gas and Heliarc welding, the arc is established between a nonconsumable tungsten electrode and the workpiece producing the heat to melt the abutting edges of the metal to be joined. Argon or helium is fed to the annular space around the electrode to maintain the inert environment. A manually fed filler rod is commonly used. The GTA technique is routinely used on low-hazard materials such as aluminum and magnesium in addition to a number of alloys including stainless steel, nickel alloys, copper-nickel, brasses, silver, bronze, and a variety of low-alloy steels that may have industrial hygiene significance.

The welding fume concentrations in GTA welding are lower than in manual stick welding and gas metal arc (GMA) welding. High-energy GTA arc produces nitrogen dioxide concentrations at the welder's position. Argon results in higher concentrations of nitrogen dioxide than helium (Ferry and Ginter 1953).

The inert gas technique introduced a new dimension in the welder's exposure to electromagnetic radiation from the arc with energies an order of magnitude greater than SMA welding. The energy in the UV B range, especially in the region of 290 nm, is the most biologically effective radiation and will produce skin erythema and photokeratitis.

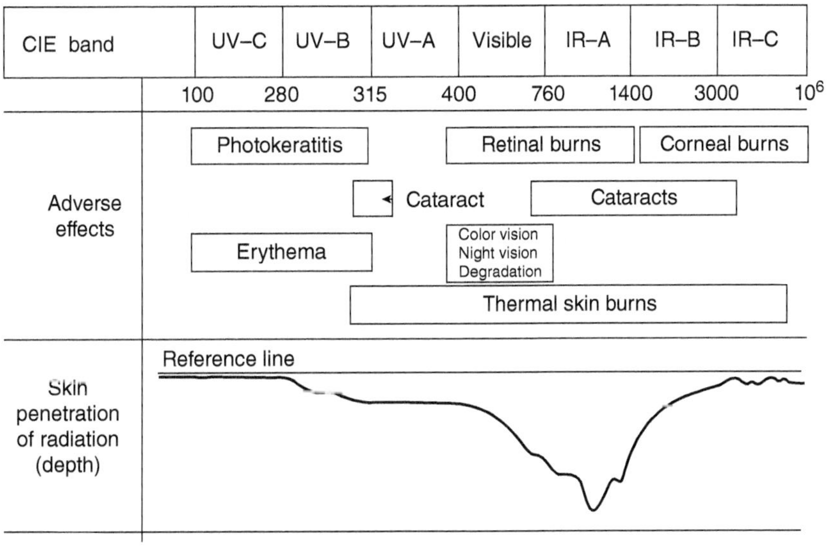


**Figure 1.7** Effects of electromagnetic radiation on the eye and skin.

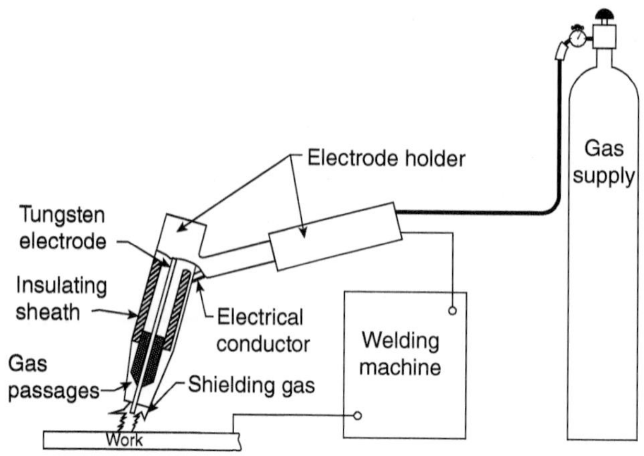


**Figure 1.8** Gas tungsten arc welding (GTAW).

The energy concentrated in the wavelengths below 200 nm (UV C) is most important in fixing oxygen as ozone. The GTA procedure produces a rich, broad spectral distribution with important energies in these wavelengths. The ozone concentration is higher when welding on aluminum than on steel, and argon produces higher concentrations of ozone than helium owing to its stronger spectral emission. The spectral energy depends on current density. Ozone concentrations may exceed the TLV under conditions of poor ventilation.

### 1.13.3 Gas Metal Arc Welding

In the 1940s, a consumable wire electrode was developed to replace the nonconsumable tungsten electrode used in the GTA system. Originally developed to weld thick, thermally conductive plate, the GMA process (also known as manual inert gas welding) now has widespread application for aluminum, copper, magnesium, nickel alloys, and titanium, as well as steel alloys.

In this system (Figure 1.9), the welding torch has a center consumable wire that maintains the arc as it melts into the weld puddle. Around this electrode is an annular passage for the flow of helium, argon, carbon dioxide, nitrogen, or a blend of these gases. The wire usually has a composition the same as or similar to the base metal with a flash coating of copper to ensure electrical contact in the gun and to prevent rusting.

An improvement in GMA welding is the use of a flux-cored consumable electrode. The electrode is a hollow wire with the core filled with various deoxifiers, fluxing agents, and metal powders. The arc may be shielded with carbon dioxide, or the inert gas may be generated by the flux core.

Metal fume concentrations from GMA welding frequently exceed the TLV on mild steel. If stainless steel is welded, the chromium concentrations may exceed the TLV, and a significant percent may be in the form of hexavalent chromium. As in the case of SMA, welding on nickel alloys produces significant concentrations of nickel fume. Nitrogen dioxide concentrations are on the same order of magnitude as SMA welding; however, ozone concentrations are much higher with the GMA technique. The ozone generation rate increases with an increase in current density but plateaus rapidly. The arc length and the inert gas flow rate do not have a significant impact on the ozone generation rate.

Carbon dioxide is widely used in GMA welding because of its attractive price; argon and helium cost approximately 15 times as much as carbon dioxide. The carbon dioxide process is similar to other inert gas arc welding shielding techniques, and one encounters the usual problem of metal fume, ozone, oxides of nitrogen, decomposition of chlorinated hydrocarbons solvents, and UV radiation. In addition, the carbon dioxide gas is reduced to form carbon monoxide. The generation rate of carbon monoxide depends on current density, gas flow rate, and the base metal being welded. Although the concentrations of carbon monoxide may exceed 100 ppm in the fume cone, the concentration drops off rapidly with distance, and with reasonable ventilation, hazardous concentrations should not exist at the breathing zone (note that the 1996 TLV for carbon monoxide is 25 ppm).

The intensity of the radiation emitted from the arc is, as in the case of GTA welding, an order of magnitude greater than that noted with SMA welding. The impact of such a rich radiation source in the UV B and UV C wavelengths has been covered in the GTA discussion. With both GTA and GMA procedures, trichloroethylene and other chlorinated hydrocarbon vapors are decomposed by UV from the arc-forming chlorine, hydrogen chloride, phosgene, and other compounds. The solvent is degraded in the UV field and not directly in the arc. Studies by Dahlberg showed that hazardous concentrations of phosgene could occur with trichloroethylene and 1,1,1-trichloroethane even though the solvent vapor concentrations were below the appropriate TLV for the solvent (Dahlberg 1971).

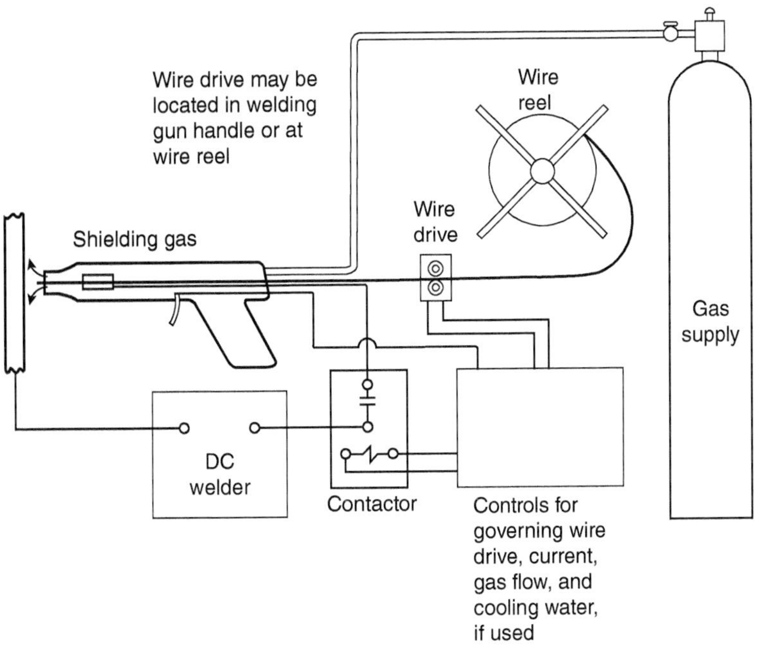


**Figure 1.9** Gas metal arc welding (GMAW).

The GMA technique produces higher concentrations of phosgene than GTA welding under comparable operating conditions. Dahlberg also identified dichloroacetyl chloride as a principal product of decomposition that could act as a warning agent because of its lacrimatory action. In a follow-up study, perchloroethylene was shown to be less stable than other chlorinated hydrocarbons. When this solvent was degraded, phosgene was formed rapidly, and dichloroacetyl chloride was also formed in the UV field.

In summation, perchloroethylene, trichloroethylene, and 1,1,1-trichloroethane vapors in the UV field of high-energy arcs such as those produced by GTA and GMA welding may generate hazardous concentrations of toxic air contaminants. Other chlorinated solvents may present a problem depending on the operating conditions; therefore, diagnostic air sampling should be performed. The principal effort should be the control of the vapors to ensure that they do not appear in the UV field. Delivery of parts directly from degreasing to the welding area has, in the authors' experience, presented major problems owing to pullout and trapping of solvent in the geometry of the part.

### 1.13.4 Gas Welding

In the gas welding process (Figure 1.10), the heat of fusion is obtained from the combustion of oxygen and one of several gases including acetylene, methylacetylene-propadiene (MAPP), propane, butane, and hydrogen. The flame melts the workpiece, and a filter rod is manually fed into the joint. Gas welding is used widely for light sheet metal and repair work. The hazards from gas welding are minimal compared with those from arc welding techniques.

The uncoated filler rod is usually of the same composition as the metal being welded except on iron where a bronze rod is used. Paste flux is applied by first dipping the rod into the flux. Fluxes are used on cast iron, some steel alloys, and nonferrous work to remove oxides or assist in fusion. Borax-based fluxes are used widely on nonferrous work, whereas chlorine, fluorine, and bromine compounds of lithium, potassium, sodium, and magnesium are used on gas welding of aluminum and magnesium. The metal fume originates from the base metal, the filler metal, and the flux. The fume concentration encountered in field welding operations depends principally on the degree of enclosure in the work area and the quality of ventilation.

The principal hazard in gas welding in confined spaces is due to the formation of nitrogen dioxide. Higher concentrations occurred when the torch was burning without active welding. Strizerskiy (1961) found concentrations of nitrogen dioxide of 150 ppm in a space without ventilation and of 26 ppm in a space with ventilation (note the 1996 TLV is 3 ppm). These investigators caution that phosphine may be present as a contaminant in acetylene and that carbon monoxide may be generated during heating of cold metal with gas burner.

The radiation from gas welding is quite different from arc welding. The principal emissions are in the visible and IR A, B, C wavelengths and require the use of light-tinted goggles for work. Ultraviolet radiation from gas welding is negligible.

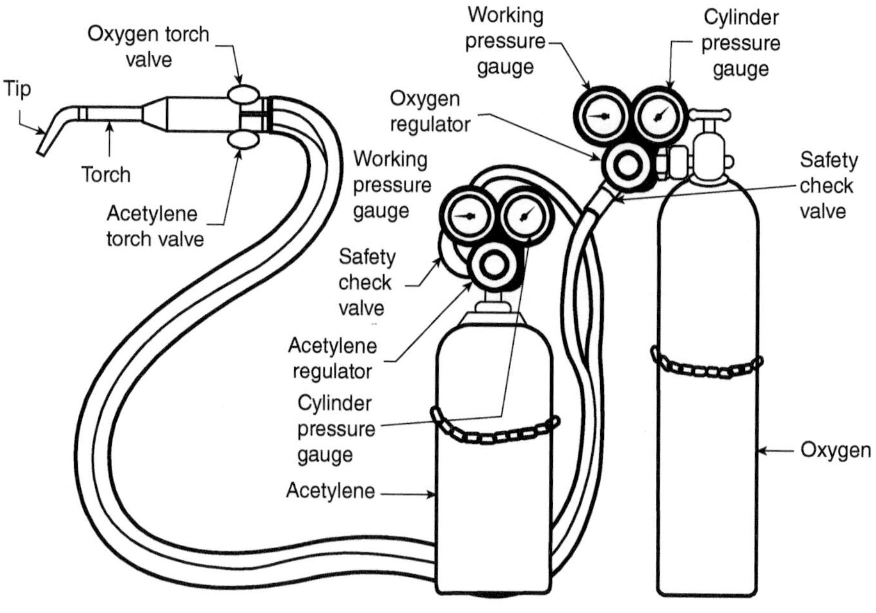


**Figure 1.10** Gas welding.

### 1.13.5 Control of Exposure

In addition to the fume exposures that occur while welding steel alloys, nonferrous metal, and copper alloys, metal coatings and the welding electrodes also contribute to the metal fume exposure. Table 1.7 shows the wide range of metal fumes encountered in welding. Recent interest in the possible carcinogenicity of chromium and nickel fume has prompted special attention to these exposures during welding. Stern et al. (1986) have suggested that welders who work on stainless and alloy steels containing chromium and nickel may be at elevated risk from respiratory cancer. Lead has been used as an alloy in steels to improve its machinability, and welding on such material requires rigorous control. This is also the case with manganese used in steel alloys to improve metallurgical properties. Beryllium, probably the most toxic alloying metal, is alloyed with copper and warrants close control during welding and brazing operations.

Welding or cutting on workpieces that have metallic coatings may be especially hazardous. Lead-based paints have been used commonly to paint marine and structural members. Welding on these surfaces during repair and shipbreaking generates high concentrations of lead fume. Cutting and welding on structural steel covered with lead paint can result in concentrations exceeding $1.0\,mg\,m^{-3}$ in well-ventilated conditions out of doors.

Steel is galvanized by dipping in molten zinc. Air concentrations of zinc during the welding of galvanized steel and steel painted with zinc silicate range from 3 to 12 times the zinc TLV under conditions of poor ventilation. Concentrations are lower with oxygen–acetylene torch work than with arc cutting methods. When local exhaust ventilation is established, the TLV for zinc is seldom exceeded.

The hazard from burning and welding of pipe coated with a zinc-rich silicate can be minimized in two ways. Because pipes are joined end to end, the principal recommendation is to mask the pipe ends with tape before painting (Venable 1979). If it is necessary to weld pipe after it is painted, the first step before welding is to remove the paint by hand filing, power brush or grinding, scratching, or abrasive blasting. If the material cannot be removed, suitable respiratory protection is required; in many cases, air-supplied respirators may be required.

A variety of techniques to reduce ozone from GMA welding, including the use of magnesium wire, local exhaust ventilation, and addition of nitric oxide, have been attempted without success. A stainless steel mesh shroud positioned on the welding head controlled concentrations to show below the TLV ceiling level of 0.1 ppm (Faggetter et al. 1983; Tinkler 1980).

Hazardous concentrations of nitrogen dioxide may be generated in enclosed spaces in short periods of time and therefore require effective exhaust ventilation.

***1.13.5.1 Radiant Energy*** Eye protection from exposure to UV B and UV C wavelengths is obtained with filter glasses in the welding helmet of the correct shade, as recommended by the AWS. The shade choice may be as low as 8–10 for light manual electrode welding or as high as 14 for plasma welding. Eye protection must also be afforded other workers in the area. To minimize the hazard to nonwelders in the area, flash screens or barriers should be installed. Semitransparent welding curtains provide effective protection against the hazards of UV and infrared from

**TABLE 1.7 Contaminants from Welding Operations**

| Contaminant | Source |
|---|---|
| Metal fumes | |
| Iron | Parent iron or steel metal, electrode |
| Chromium | Stainless steel, electrode, plating, chrome-primed metal |
| Nickel | Stainless steel, nickel-clad steel |
| Zinc | Galvanized or zinc-primed steel |
| Copper | Coating on filler wire, sheaths on air-carbon arc gouging electrodes, nonferrous alloys |
| Vanadium, manganese, and molybdenum | Welding rod, alloys in steel |
| Tin | Tin-coated steel |
| Cadmium | Plating |
| Lead | Lead paint, electrode coating |
| Fluorides | Flux on electrodes |
| Gases and vapors | |
| Carbon monoxide | $CO_2$ shielded, GMA, carbon arc gouging, oxy-gas |
| Ozone | GTA, GMA, carbon arc gouging |
| Nitrogen dioxide | GMA, all flame processes |

*Source:* ACGIH (1995).

welding operations. Sliney et al. (1981) have provided a format for choice of the most effective curtain for protection while permitting adequate visibility. Plano goggles or lightly tinted safety glasses may be adequate if one is some distance from the operation. If one is 30–40 ft away, eye protection is probably not needed for conventional welding; however, high current density GMA welding may require that the individual be 100 ft away before direct viewing is possible without eye injury. On gas welding and cutting operations, infrared wavelengths must be attenuated by proper eye protection for both worker health and comfort.

The attenuation afforded by tinted lens is available in National Bureau of Standards reports, and the quality of welding lens has been evaluated by NIOSH. Glass safety goggles provide some protection from UV; however, plastic safety glasses may not.

The UV radiation from inert gas–shielded operations causes skin erythema or reddening; therefore, welders must adequately protect their faces, necks, and arms. Heavy chrome leather vest armlets and gloves must be used with such high-energy arc operations.

***1.13.5.2 Decomposition of Chlorinated Hydrocarbon Solvents*** The decomposition of chlorinated hydrocarbon solvents occurs in the UV field around the arc and not in the arc itself. The most effective control is to prevent the solvent vapors from entering the welding area in detectable concentrations. Merely maintaining the concentration of solvent below the TLV is not satisfactory in itself. If vapors cannot be excluded from the workplace, the UV field should be reduced to a minimum by shielding the arc. Pyrex glass is an effective shield that permits the welder to view his or her work. Rigorous shielding of the arc is frequently possible at fixed station work locations, but it may not be feasible in field welding operations.

***1.13.5.3 System Design*** The advent of robotics has permitted remote operation of welding equipment, thereby minimizing direct operator exposure to metal fumes, toxic gases, and radiation.

## BIBLIOGRAPHY

American Conference of Governmental Industrial Hygienists (ACGIH) (1995). *Industrial Ventilation: A Manual of Recommended Practice*, 22e. Lansing, MI: ACGIH.

American Electroplaters and Surface Finishers Society (AESF) (1989). *Safety Manual for Electroplating and Finishing Shops*. Orlando, FL: AESF.

American Foundrymen's Society, Inc. (AFS) (1985). *Industrial Noise Control – An Engineering Guide*. Des Plaines, IL: AFS.

American National Standards Institute (ANSI) (1966). *Ventilation Control of Grinding, Polishing, Buffing Operations*, Z 43.1-1966. New York: ANSI.

American National Standards Institute (ANSI) (1976). *Radiological Safety Standard for the Design of Radiographic and Fluoroscopic Industrial X-Ray Equipment*, ANSI/NBS 123-1976. New York: ANSI.

American National Standards Institute (ANSI) (1977). *Practices for Ventilation and Operation of Open-Surface Tanks*, ANSI Z9.1-1977. New York: ANSI.

American National Standards Institute (ANSI) (1978). *Safety Requirements for the Use, Care, and Protection of Abrasive Wheels*, B7.1-1978. New York: ANSI.

American Society for Metals (1982). *Heat Treaters Guide*. New York: ASM International.

American Society for Testing and Materials (1989). *Manual on Vapor Degreasing*, 3e. Philadelphia: ASTM International.

Ballard, W.E. (1972). *Ann. Occup. Hyg.* **15**: 101.

Bastress, E.K. et al. (1973). *Ventilation Requirements for Grinding, Polishing, and Buffing Operations*, Report No. 0213 (June 1973). Burlington, MA: IKOR Inc.

Bingham, E., Trosset, P., and Warshawsky, D. (1979). *J. Environ. Pathol. Toxicol.* **3**: 483.

Blair, D.R. (1972). *Principles of Metal Surface Treatment and Protection*. Oxford: Pergamon Press.

Blair, A. (1980). *J. Occup. Med.* **22**: 158–162.

Burge, P.S., Harries, M.G., O'Brien, I.M. et al. (1978). *Clin. Allergy* **8**: 1.

Burgess, W.A. (1991). Potential exposures in the manufacturing industry – their recognition and control. In: *Patty's Industrial Hygiene and Toxicology*, vol. **I**, Part A (ed. G.D. Clayton and F.E. Clayton). New York: Wiley.

Burgess, W. (1995). *Recognition of Health Hazards in Industry: A Review of Materials and Processes*. New York: Wiley.

Burgess, W.A., Ellenbecker, M.J., and Treitman, R.D. (1989). *Ventilation for Control of the Work Environment*. New York: Wiley.

Dahlberg, J. (1971). *Ann. Occup. Hyg.* **14**: 259.

Danielson, J. (ed.) (1973). *Air Pollution Engineering Manual*, 2e, Publication No. AP-40. Washington, DC: U.S. Government Printing Office.

Decoufle, P. (1978). *J. Natl. Cancer Inst.* **61**: 1025–1030.

Dow Chemical Co. (1977). *Modern Vapor Degreasing and Dow Chlorinated Solvents*, Dow Bulletin, Form No. 100-5185-77. Midland, MI: Dow Chemical Co.

Dow Chemical Co. (1978). *How to Select a Vapor Degreasing Solvent*, Bulletin Form 100-5321-78. Midland, MI: Dow Chemical Co.

Dreger, D.R. (1982). *Machine Design* **47** (25 November).

DuPont (1987). *Freon Solvent Data*, Bulletin No. FST-1. Wilmington, DE: DuPont.

Eisen, E., Tolbert, P.E., Hallock, M.F. et al. (1994). Mortality studies of machining fluid exposures in the automotive industry. II. Risks associated with specific fluid types. *Am. J. Ind. Med.* **24**: 123.

Elkins, H.B. (1959). *The Chemistry of Industrial Toxicology*. New York: Wiley.

Esso Petroleum Company (1979). *This Is About Health, Esso Lubricating Oils and Cutting Fluids*. Southampton, UK: Esso Petroleum Company, Ltd.

Faggetter, A.G., Freeman, V.E., and Hosein, H.R. (1983). *Am. Ind. Hyg. Assoc. J.* **44**: 316.

Ferry, J. and Ginter, G. (1953). *Weld. J.* **32**: 396.

Fletcher, B. (1988). *Low-Volume-High-Velocity Extraction Systems*, Report No. 16. Bootle, UK: Health and Safety Executive, Technology Division.

Forging Industry Association (FIA) (1985). *Forging Handbook*. Cleveland, OH: FIA.

Goldsmith, A.H., Vorpahl, K.W., French, K.A. et al. (1976). *Am. Ind. Hyg. Assoc. J.* **37**: 217–226.

Gutow, B.S. (1972). *Environ. Sci. Technol.* **6**: 790.

Hagopian, J.H. and Bastress, E.K. (1976). *Recommended Industrial Ventilation Guidelines*, Publication No. (NIOSH) 76-162. Cincinnati, OH: National Institute for Occupational Safety and Health.

Hallock, M.F., Smith, T.J., Woskie, S.R., and Hammond, S.K. (1994). Estimation of historical exposures to machining fluids in the automotive industry. *Am. J. Ind. Med.* **26**: 621–634.

Handy and Harman (1985). *The Brazing Book*. New York: Handy and Harman.

Hansen, M.K., Larsen, M., and Cohr, K.-H. (1987). *Scand. J. Work Environ. Health* **13**: 473.

Health and Safety Executive (1982). *Portable Grinding Machines: Control of Dust*, Health and Safety Series Booklet HS(g), vol. **18**. London, England: H. M. Stationary Office.

HM Factory Inspectorate (1974). *Dust Control: The Low Volume–High Velocity System*, Technical Data Note 1 (2nd rev.). London, England: Department of Employment.

Hogapian, J.H. and Bastress, E.K. (1976). *Recommended Industrial Ventilation Guidelines*, NIOSH Publication No. 76-162. Cincinnati, OH: Department of Health, Education and Welfare.

International Agency for Research on Cancer (IARC) (1987). Overall evaluations of carcinogenicity: an updating of IARC Monographs volumes 1 to 42. *IARC Monogr Eval Carcinog Risks Hum Suppl* **7**: 1–440.

International Agency for Research on Cancer (IARC) (1992). *Occupational Exposure to Mists and Vapours from Strong Inorganic Acids and Other Industrial Chemicals, IARC Monographs on the Evaluation of Carcinogenic Risks to Humans*, vol. **54**, 44–130. Lyon, France: IARC.

International Molders and Allied Workers Union (AFL-CIO-CLC) (n.d.). *No-Bakes and Others: Common Chemical Binders and Their Hazards*. AFL-CIO-CLC.

*Isotope Radiography Safety Handbook* (1981). Burlington, MA: Technical Operations, Inc.

Jarvolm, B., Thuringer, G., and Axelson, O. (1982). *Br. J. Occup. Med.* **39**: 197–197.

Kennedy, S.M., Greaves, I.A., Kriebel, D. et al. (1989). *Am. J. Ind. Med.* **15**: 627–641.

Knecht, U., Elliehausen, H.J., and Woitowitz, H.J. (1986). *Br. J. Ind. Med.* **43**: 834.

Manko, H.H. (1979). *Solders and Soldering*. New York: McGraw-Hill.

National Fire Protection Association (NFPA) (1991). *NFPA86C, Standard for Industrial Furnaces Using a Special Processing Atmosphere*. Quincy, MA: NFPA.

National Institute for Occupational Safety and Health (NIOSH) (1975a). *Ventilation Requirements for Grinding, Buffing, and Polishing Operations*, NIOSH Publication No. 75–105. Cincinnati, OH: NIOSH.

National Institute for Occupational Safety and Health (NIOSH) (1975b). *Industrial Health and Safety Criteria for Abrasive Blasting Cleaning Operations*, DHHS (NIOSH) Publ. No. 75–122. Cincinnati, OH: NIOSH.

National Institute for Occupational Safety and Health (NIOSH) (1978). *Guidelines for the Control of Exposures to Metalworking Fluids*, NIOSH Publ. No. 78–165. Cincinnati, OH: NIOSH.

National Institute for Occupational Safety and Health (NIOSH) (1984). *Recommendations for Control of Occupational Safety and Health Hazards Manufacture of Paint and Allied Coating Products*, DHHS Publication No. 84–115. Cincinnati, OH: CDC, NIOSH.

National Institute for Occupational Safety and Health (NIOSH) (1985a). *Recommendations for Control of Occupational Safety and Health Hazards Foundries*, DHHS Publication No. 85–116. Cincinnati, OH: NIOSH.

National Institute for Occupational Safety and Health (NIOSH) (1985b). *Control Technology Assessment: Metal Plating and Cleaning Operations*, DHHS Publication No. 85–102. Cincinnati, OH: CDC, NIOSH.

National Institute for Occupational Safety and Health (NIOSH) (1989a). *Criteria for a Recommended Standard: Occupational Exposure to Hand-Arm Vibration*, NIOSH Publication No. 89–106. Cincinnati, OH: NIOSH.

National Institute for Occupational Safety and Health (NIOSH) (1989b). *Health Hazard Evaluation Report 88-136-1945, Miller Thermal Technologies, Inc., Appleton, WI*. Cincinnati, OH: NIOSH.

National Institute for Occupational Safety and Health (NIOSH) (1992). *NIOSH Alert, Request for Assistance in Preventing Silicosis and Deaths from Sandblasting*, DHHS (NIOSH) Publication No. 92–102. Cincinnati, OH: NIOSH.

O'Brien, D. and Frede, J.C. (1978). *Guidelines for the Control of Exposure to Metalworking Fluids*, Department of Health, Education and Welfare, NIOSH Publication No. 78–165. Cincinnati, OH: NIOSH.

O'Brien, D.M. and Hurley, D.E. (1981). *An Evaluation of Engineering Control Technology for Spray Painting*, DHHS (NIOSH) Publication No. 81–121. Cincinnati, OH: USDHHS, CDS.

Oudiz, J., Brown, A., Ayer, H.A., and Samuels, S. (1983). *Am. Ind. Hyg. Assoc. J.* **44**: 374.

Pantucek, M. (1971). *Am. Ind. Hyg. Assoc. J.* **32**: 687.

Pelmear, P.L. and Kitchener, R. (1977). The effects and measurement of vibration. In *Proceedings of the Working Environment in Iron Foundries* (22–24 March 1977). Birmingham, England: British Cast Iron Research Association.

Sliney, D. and Wolbarsht, M. (1980). *Safety with Lasers and Other Optical Sources*. New York: Plenum.

Sliney, D.H., Moss, C.E., Miller, C.G., and Stephens, J.B. (1981). *Appl. Optics* **20**: 2352.

Sparks, P.J. and Wegman, D.H. (1980). *J. Occup. Med.* **22**: 733–736.

Spring, S. (1974). *Industrial Cleaning*. Melbourne: Prism Press.

Starek, J., Farkkila, M., Aatola, S. et al. (1983). *Br. J. Ind. Med.* **40**: 426.

Stern, R.M., Berlin, A., Fletcher, A.C., and Jarvisalo, J. (1986). *Health Hazards and Biological Effects of Welding Fumes and Gases*, Excerpta Medica ICS, vol. **676**. Amsterdam: Elsevier.

Stokinger, H. (1984). *Am. Ind. Hyg. Assoc. J.* **45**: 127.

Strizerskiy, I. (1961). *Weld. Prod.* **7**: 40.

Tinkler, M.J. (1980). Measurement and control of ozone evolution during aluminum GMA welding. In: *Colloquium on Welding and Health*, Brazil (July 1980).

Toeniskoetter, R.H. and Schafer, R.J. (1977). Industrial hygiene aspects of the use of sand binders and additives. In: *Proceedings of the Working Environment in Iron Foundries* (22–24 March 1977). Birmingham, England: British Cast Iron Research Association.

United Kingdom Department of Employment (1974). *Control of Dust from Portable Power Operated Grinding Machines (Code of Practice)*. London: United Kingdom Department of Employment.

VanDerWal, J.F. (1990). *Ann. Occup. Hyg.* **34**: 45.

Vena, J.E., Sultz, H.A., Fiedler, R.C., and Barnes, R.E. (1985). *Br. J. Ind. Med.* **42**: 85–93.

Venable, F. (1979). *Esso Med. Bull.* **39**: 129.

Verma, D.C., Muir, F., Cunliffe, S. et al. (1982). *Ann. Occup. Hyg.* **25**: 17.

WHTC (1981). *Guidelines for Safety in Heat Treatment, Part 1, Use of Molten Salt Baths*. Gosta Green, Birmingham, England: Wolfson Heat Treatment Centre, University of Aston.

WHTC (1983). *Guidelines for Safety in Heat Treatment, Part 2, Health and Personal Protection*. Gosta Green, Birmingham, England: Wolfson Heat Treatment Centre, University of Aston.

WHTC (1984). *Guidelines for Safety in Heat Treatment, Part 3, Quenching, Degreasing, and Fire Safety*. Gosta Green, Birmingham, England: Wolfson Heat Treatment Centre, University of Aston.

CHAPTER 2

# INFORMATION RESOURCES FOR OCCUPATIONAL SAFETY AND HEALTH PROFESSIONALS

RALPH STUART

*Office of Environmental Safety, Keene State College, 229 Main St., Keene, NH, 03435*

JAMES STEWART

*Department of Environmental Health, Harvard T.H. Chan School of Public Health, Landmark Center, 401 Park Dr., Boston, MA, 02215*

and

ROBERT HERRICK

*Department of Environmental Health, Harvard T.H. Chan School of Public Health, 665 Huntington Ave., Boston, MA, 02215*

## 2.1 INTRODUCTION

Since 2000, the accelerating development of and access to electronic information resources has made information management an increasingly important part of all professions. The evolution of the Internet has increased the importance of both paper and electronic information in the daily life of professionals and organizations alike. This is certainly true for occupational safety and health (OSH) professionals, for whom gathering, organizing, prioritizing, and sharing data and information are core functions. It is difficult to imagine managing the variety and amount of information that a twenty-first-century OSH professional works uses without strong computer skills.

At the same time, information available on the Internet is of variable quality and not necessarily peer-reviewed. For this reason, it is important to understand the strengths and weaknesses of both online and more traditional sources (books and journals) in information used by the OSH professional. In this chapter, we will provide an overview of professional information as well as a list of key resources in these areas.

This chapter describes three uses of information by OSH professionals:

1. *As a reference tool.* Convenient access to a wide variety of information resources is of high value to the OSH professional. This section describes the use of these resources at the conceptual level, as well as listing specific resources that are likely to be of daily value to the OSH professional.
2. *As a data collection tool.* A key aspect of a successful safety program is collection and organizing information that describes and assesses workplace conditions effectively. This involves not only collecting workplace observations and developing reports based on these observations but also delivering this information in a form that is useful to the audience of interest.
3. *As a safety program management tool.* In addition to collecting information about their workplaces, OSH professionals need to share this information with a variety of audiences, including workers, management, and regulators. This can mean reformatting the information in a variety of ways, depending on the specific question being asked. For example, with regard to a specific chemical exposure, workers may ask: "Is this safe for me?" Government inspectors may ask: "Is this situation compliant with our regulations?" Upper management may ask: "Is this the best we can do?" Answers to these questions are likely to use the same data in different ways.

*Handbook of Occupational Safety and Health*, Third Edition. Edited by S. Z. Mansdorf.

## 2.2 THE INTERNET AND OTHER REFERENCE TOOLS

One of the biggest challenges facing safety and health professionals is finding information when they need it. In the twenty-first century, changing regulations, creation of new data sources, and newly emerging technologies make researching occupational safety and health issues an ongoing challenge. Fortunately, the Internet provides a readily available, complementary source of information that can supplement more traditional books, manuals, and professional journals and magazines. Some traditional sources are also available on the Internet in electronic format, although many are not. This section describes the tools that you can use to become familiar with the various types of information available.

### 2.2.1 Traditional Texts and Journals

Traditional paper-based information faces obvious disadvantages when compared with the Internet. It can be expensive to acquire, is difficult to update and hard to reuse, and may be awkward to find and acquire in a prompt way. However, there are certain key advantages to these media that should be kept in mind:

- ***Quality:*** while the development of peer-reviewed information can be a time-consuming process, this process includes steps for the contents of the document to be improved and edited. These documents often include peer review before publication that provides a level of quality control that may not be included in many Internet resources. Many Internet resources on health and safety topics do not identify the credentials of the person writing the content or how much research they conducted in developing the information provided.
- ***Provenance:*** data provenance provides a clear path to the source of the information so that the strength of the information can be evaluated. This consideration is a critical part of the research process for the OSH professional, as information on a safety data sheet or in a government document may or may not be transferable to the situation at hand. For this reason, the original circumstances for which this information was developed are an important concern. The publishing process for traditional sources of information tends to provide better organized "metadata" than electronic resources. An example of this type is the title and copyright pages found in paper books and journals.
- ***Organization:*** many electronic search engines have hidden criteria (for example, advertising considerations or limited scopes of search) that can confuse interpretation of the results of a search query. The organizational tools of paper resources (i.e. table of contents and subject indices) tend to be clearer to the reader, and the scope of the information is presented more clearly. On the other hand, Internet information is generally assumed to be open to anyone, and thus customization of the information to a specific audience is often not clear.
- ***Coverage:*** while the Internet includes information on a wide variety of topics, there is not systematic effort to cover all of the topics discussed in the paper literature. Reviewing the references in the appendix, it is noticeable that certain rapidly emerging fields, such as biosafety and chemical safety, are better suited for development of electronic information, while others, such as industrial hygiene, rely on older information resources. This difference in part reflects the complexity of the field; Internet resources are not well suited for describing complex processes that involve significant amounts of professional judgment. This difference is also affected by the time period when the information was developed; for example, biosafety as a field has evolved rapidly since 2000, and the predominance of electronic information in this field reflects this.

An effective research strategy to develop the best possible answer will require a mix of paper and electronic information. For this reason, in the appendix to this chapter, we have included both paper and electronic resources in our list of resources.

### 2.2.2 Exploring Health and Safety Web Resources

Given the easy availability of search engines and information consolidation services like Wikipedia, it is tempting to rely on an Internet search engine to collect whatever information is available on a particular topic as the need arises. However, this approach may result in finding information that does not meet the quality or reliability objectives for the OSH professional. It is a worthwhile investment of time to explore occupational safety and health-specific web directories and specialized search engines to identify reliable sources of information before a specific question arises.

Fortunately, there are websites with subject guide that list a number of Internet sites related to occupational health and safety subjects. A good example of one such website for occupational safety and health professionals is the NIOSH website at http://www.cdc.gov/niosh. This website organizes a large amount of technical information that is directly applicable to issues that OSH professionals face on a daily basis. Another helpful website to explore is that of the Canadian Centre for Occupational Health and Safety (CCOHS). This site includes a page that organizes a large number of occupational safety and health links (http://www.ccohs.ca/oshlinks). By using these sites as a starting point, you can get a sense of the types of resources on the Web. These are two examples of the many available sites that can serve as similar starting points for the information gathering process.

### 2.2.3 Develop an A Priori Internet Search Strategy

To conduct research on the Internet effectively, it is important to have a search strategy in mind while you are looking. A description of such a strategy is provided here.

***2.2.3.1 Refine the Question and Decide What Information Is Needed*** The first step in using the Internet successfully is to clearly state the question you are trying to answer. It is important to know where you are going before you start. For example, looking for a specific piece of data (e.g. *the flashpoint of acetone*) is different from looking for a technical interpretation of that data (e.g. what is "adequate ventilation" when using acetone in a laboratory?), both of which are different than a third type of information – "informal knowledge" (e.g. *use acetone in a fume hood if using more than 500 ml*). These different types of information will be found in different sources and locations on the Internet.

*2.2.3.1.1 Formal Databases* For specific pieces of data, formal databases are often the best places to look. These databases are carefully organized into specific well-defined fields that enable efficient searching of the data they contain. Public formal databases are often maintained by government entities, such as PubChem and TOXNET from the US National Library of Medicine. It is important to keep in mind that some of these databases provide access to commercial information, for example, data from safety data sheet collections, without assessing the quality of the data of the information they contain; an example of such a database is the Household Products Database (householdproducts.nlm.nih.gov). Other databases such as the Hazardous Substances Data Bank (http://toxnet.nlm.nih.gov/cgi-bin/sis/htmlgen?HSDB) are technically reviewed before being published and provide references for the source of the data.

For these reasons, it is important to understand the database being accessed and to confirm the reliability of the data before taking action on it. This can be done by comparing information from multiple sources and determining which is the most recent and cited by other sources. For example, assessing whether a chemical should be considered a carcinogen should include information from the various bodies that comment on this topic, including the American Conference of Governmental Industrial Hygienists (ACGIH), the International Agency for Research on Cancer (IARC), and the US National Toxicology Program (NTP) information on this topic. Unfortunately, these organizations have different classification systems for this question, and their determinations were made at different times, based on the evidence available at that time. Careful comparison and consideration of these differences is an important step in interpreting the information on their websites.

*2.2.3.1.2 Professional Interpretations* Professional interpretations can be useful because different views of the same technical issue can be helpful in developing solutions for your unique set of circumstances. Two well-recognized government-sponsored sources in the United States for information of this type are the "OSHA letters of interpretation" (www.osha.gov) and the "Consultations" section of the United Kingdom's Health and Safety Laboratory website (www.hsl.gov.uk/home). Other similar sites exist around the globe, e.g. the World Health Organization (http://www.who.int/topics/occupational_health/en), WorkSafe Australia (www.safeworkaustralia.gov.au/sites/SWA), and Institut de recherche Robert-Sauvé en santé et en sécurité du travail (IRSST) in Quebec (http://www.irsst.qc.ca/en). In addition, nongovernmental and private websites for environmental health and safety departments at educational research institutions or companies often provide useful information and models for safety program implementation and measuring safety performance. Lastly, there are commercial knowledge databases such as the DuPont SafeSPEC website. When using these websites, it is important to assess the assumptions used in developing the information and the impact those assumptions may have on the usefulness of the information for your specific situation.

*2.2.3.1.3 Informal Knowledge* The Internet has many informal information collections available, for example, the archives of electronic mailing lists or blog posts on OSH topics. Even if you do not find the information you are after there, you may find a reference to another Internet resource that has the information you are looking for. The challenge associated with using these websites is that identifying the best collection of keywords associated with the information you are seeking will be source dependent. For example, a biosafety information resource will discuss laboratory safety issues in a different way than chemical safety resource. It is also important to return to your original question and check the information you have found and how it applies to your question before making a decision based on it.

**2.2.3.2 *Select Keywords to Use for the Search*** Refining your research plan involves phrasing your question in a way that clearly identifies the information you are seeking. This process can sometimes be relatively straightforward ("Does OSHA have any regulations that cover the use of this chemical?"), but other times this can be more difficult (e.g. "Is this worksite a confined space?"). Finding OSHA's definition of a confined space is easy, but interpreting it in a specific instance requires collecting a significant amount of on-site information to answer this question.

A four- or five-word phrase is a good place to start your search. For example, in the confined space example, a technical magazine article on the general attributes of a confined space could help you answer your question more efficiently than a direct reading of the OSHA regulation.

When you are selecting keywords to search, it is important to consider other possible meanings for the words you select. For example, "*safety*" may refer to chemical exposures in your mind, while it refers to law enforcement issues in many other people's minds. An ambiguous word such as this is usually a poor choice to include as a search term.

To address this issue, it is best to test your initial set of keywords by using it in occupational safety and health websites and assessing the quality of the results. If the results are not what you are looking for or are too broad or narrow, refine the keyword list, and retry until you obtain acceptable results. By searching through these websites, you can see if other OSH professionals use the terms you chose to describe the situation you are researching. Once the keywords are refined in this search, they are likely to be more effective when using general Internet search engines.

The result of refining your question should be a set of keywords that you want to use for your search. You can use these keywords in performing searches at various websites that are likely to contain appropriate information. For example, if you are simply after the flash point of acetone, "flash point" and "acetone" are appropriate keywords. On the other hand, if you are concerned about ventilation requirements for using acetone, "flash point" is not likely to be helpful, and "flammable liquid" may be a reasonable substitute for "acetone." Using a combination of terms such as "acetone adequate ventilation cfm" would refine the search further.

**2.2.3.3 *Distribute a Question to a Discussion Group of Professionals*** If your search fails to produce the information you are after or you are looking for more informal information than is available at websites, it is often helpful to post a request for information to an appropriate e-mail list. To increase your chances of success when you ask a question of a group, be sure to follow Internet etiquette guidelines appropriate to that group. An example of such a resource is the technical discussion list of the American Chemical Society's Division of Chemical Health and Safety. While subscription to this list is a membership benefit of the Division, more than 10 years of the archives of the discussions are publically available at http://www.ilpi.com/dchas. These messages are keyword searchable and include both etiquette guidelines and technical discussions about issues germane to chemical safety issues.

If possible, review the archives of the group's discussions to see if it is the right group of which to ask the question and try to determine if the question has been asked and answered previously. When framing the question, be as specific as possible in asking the question, so those who read it can determine what type of answers is appropriate (i.e. general pointers to the professional literature versus specific interpretations of your information). It is most helpful to monitor the e-mail exchanges on the list for a period of time before asking your question so that you can see what sorts of questions are appropriate for the list.

**2.2.3.4 *Confirm the Quality and Reliability of Information Retrieved*** After each information search and before taking action, be sure to assess the information you obtained from the Internet searches before you base a decision or take action on it. Remember that the information available on the Internet was written based on someone else's assumptions and without knowledge of the details of your particular situation. There may be specific, critical differences between the situations that you face and that of the person writing the information. The effort involved in confirming Internet information may range from asking yourself "Does this make sense?" to checking an authoritative source or confirming with multiple Internet sources.

## 2.3 COLLECTING AND ORGANIZING DATA

Another key aspect of computer use by OSH professionals is the collection and analysis of safety program data. These data may include exposure sampling records, training program administration information, real-time sampling data acquisition, or electronic collection of workplace inspection results. Computer and Internet solutions must be carefully evaluated not only for their current functionality but also for their scalability, data transfer capability, and long-term technical and financial sustainability.

Advances in electronic technologies have enabled data from specific environmental sampling instruments to be collected and organized more effectively and at lower costs than a few short years ago. Air quality monitors, digital cameras and recorders, geographic positioning systems, bar coding, and other related techniques enable OSH professionals to provide a more accurate and meaningful assessment of workplace hazards than could be previously considered. A variety of safety issues ranging from chemical exposure determinations, documentation of accidents, oversight of workplace inspections, and emergency response protocols have all benefited from this trend. However, as with other forms of electronic information, using these capabilities must be carefully planned and frequently reviewed for the full potential to be realized.

Addressing this challenge requires development of a clear information architecture for the data being collected; development of these systems will benefit from partnerships with information professionals who have experience in addressing these concerns. Three aspects of an information collection system to consider carefully include the following:

1. The ability to collect information remotely through the Internet. Many types of sensing/sampling devices can be directly connected to the Internet and can provide real-time data about the environment as it changes in time and space. At the same time, this capability raises information security concerns and data storage and management challenges. Careful planning of what data should be captured and how it should be organized and reported is a key process of starting the data collection process.
2. Support for remote collection data through robotic and/or wireless communication systems. Many times, monitoring of conditions of concern to the OSH professional requires access to hazardous areas. Robots and wireless communication systems are becoming increasingly important in meeting this need. Achieving this goal successfully requires careful coordination of a variety of technical experts, ranging from OSH professionals to information technology engineers to data management specialists.
3. The ability to incorporate the data reports from the collection system into the safety program management system. As mentioned above, safety and health performance data are becoming increasingly available, but managing the data is becoming increasingly problematic. Careful documentation of where and how the data were collected needs to be incorporated into the data management system. It is important to remember that future OSH professionals may be asked to reconstruct environmental conditions based on the data you collect today when questions about long-term health impacts arise.

## 2.4 MANAGEMENT SYSTEM DEVELOPMENT

As electronic information proliferates, it is easy to be swamped by data collected and stored. A useful tool in selecting which data is appropriate to be collected and how it should best be stored is a safety management system (SMS). SMSs define the indicators of success for the safety program and identify the stakeholders involved in defining and achieving the goals of the program. SMSs can be based on a variety of models, such as the ANSI Z10 Standard for Occupational Health and Safety Management Systems, OSHA's Voluntary Protection Program, or the International Organization for Standardization's ISO 45001. Whatever standard is selected to use to develop an SMS, the system should define the architecture of the data necessary to manage and evaluate the safety program as it evolves over time.

Part of this SMS is likely to be the organization's website, whether the website is focused on the internal audience (i.e. an intranet) or publically available on the Internet. This website can serve as a reference source for both operational (procedures and specific facility information) and strategic (policies and plans) safety information. In developing such a library of information, the usability of the information posted is the key concern. This means that the information should be adapted to the applicable electronic medium. For example, web pages are not well suited to delivering multiple screens of detailed information on the same page. Research has consistently shown that people using web pages browse text on the page quickly, scanning for specific details they need rather than reading carefully for comprehension. Therefore, web pages must be carefully planned and linked together in ways that allow the logic of the system to help rather than challenge the reader's use of the system.

The development of an SMS will greatly benefit from being connected to other management systems within an organization, such as facility information, financial data, and human resources systems. Implementing these connections is more challenging than first anticipated because the administrative software involved is designed for other purposes. For this reason, the organization's information technology (IT) professionals should be consulted on how best to succeed in selecting a safety management support system. Oftentimes, much of this other information has significant security considerations involved, so using the data fully is not always possible. For example, there can be restrictions on who has access to the data and how they can work with it. However, as a safety information management system is piloted and as it grows, attention should be given to opportunities for synergy between electronic safety information and other organizational data.

## 2.5 CONCLUSION

In addition to more traditional technical skills such as industrial hygiene, hazard assessment, and risk management, the twenty-first-century OSH professional needs to understand the strengths and weaknesses of the information systems they use to conduct their work. Happily, high-quality information resources continue to become broader in scope and easier to access. However, making the best use of the opportunities they present requires careful planning and practice.

# APPENDIX A

## KEY SAFETY AND HEALTH INFORMATION RESOURCES

In the lists below, 58 key information resources are organized alphabetically into 10 primary areas of interest. For paper resources, ISBNs are provided; for websites, URLs and a short description are provided. The topic areas included are:

1. Biosafety
2. Chemical safety
3. Environmental health
4. Emergency response
5. Ergonomics
6. Government resources
7. Industrial hygiene
8. Occupational and environmental medicine
9. Physical hazards
10. Safety management

This list is not intended to be exhaustive; many other useful resources can be found by consulting these resources first and following references in them.

## BIOSAFETY

1. **American Biological Safety Association (ABSA);**
   www.absa.org
   ABSA was founded in 1984 to promote biosafety as a scientific discipline and serve the growing needs of biosafety professionals throughout the world. This web page provides links to information on a variety of biosafety issues, including its journal and a variety of training resources (accessed 8 November 2016).
2. **Centers for Disease Control and Prevention (CDC);**
   www.cdc.gov
   This US government site contains a large variety of health and disease information. There are numerous documents and data sets, most of which are related to public health, biosafety, occupational health and safety, and infectious diseases (accessed 8 November 2016).
3. Bailey, H.S. (2005). *Fungal Contamination – A Manual for Investigation, Remediation and Control.* Cincinnati, OH: ACGIH.
4. West, K.H. (2001). *Infectious Disease Handbook for Emergency Care Personnel*, 3e. Cincinnati, OH: ACGIH.
5. **International Centre for Genetic Engineering and Biotechnology (ICGEB), Biosafety Web Pages;**
   www.icgeb.org
   This site, based in Italy, provides access to the ICGEB bibliographic database on biosafety studies, documents on biosafety produced by international agencies, and scientific findings, articles, proceedings, and workshops on international biosafety regulations (accessed 8 November 2016).
6. **Pathogen Safety Data Sheets and Risk Assessment; Public Health Agency of Canada;**
   http://www.phac-aspc.gc.ca/msds-ftss
   These MSDS are produced for people working in the life sciences as quick safety reference materials relating to infectious microorganisms (accessed 8 November 2016).

## CHEMICAL SAFETY

7. **Agency for Toxic Substances and Disease Registry (ATSDR);**
   www.atsdr.cdc.gov
   The ATSDR website has a variety of information about hazardous chemicals in the environment. This information includes ToxFAQs, fact sheets on the hazards associated with a variety of chemicals; a list of the top 20 hazardous substances, based on Superfund experience; and a Science Corner (accessed 8 November 2016).
8. **American Chemical Society Division of Chemical Health and Safety;**
   www.dchas.org
   This website provides information related to chemical safety issues in general, with a focus on laboratory safety. It also includes links to many other ACS chemical safety resources (accessed 8 November 2016).
9. Urban, P. (2006). *Bretherick's Handbook of Reactive Chemical Hazards*, 7e. Oxford: Academic Press.
10. Lewis, R.J. (2007). *Hawley's Condensed Chemical Dictionary*. Hoboken, NJ: Wiley.
11. Lewis, R.J. (2008). *Hazardous Chemicals Desk Reference*. Hoboken, NJ: Wiley.
12. **Haz-Map: Information on Hazardous Chemicals and Occupational Diseases from the National Institutes of Health;**
   hazmap.nlm.nih.gov
   Haz-Map is an occupational health database designed for health and OSH professionals and for consumers seeking information about the health effects of exposure to chemicals and biologicals at work. Haz-Map links jobs and hazardous tasks with occupational diseases and their symptoms (accessed 8 November 2016).
13. **Journal of Chemical Health and Safety;**
   http://www.journals.elsevier.com/journal-of-chemical-health-and-safety
   The website for the peer-reviewed journal of the Division of Chemical Health and Safety of the American Chemical Society (accessed 8 November 2016).
14. Hathaway, G.J. and Proctor, N.H. (2014). *Proctor and Hughes' Chemical Hazards of the Workplace, 5e*. Hoboken, NJ: Wiley.
15. **Where to Find Material Safety Data Sheets on the Internet;**
   http://www.ilpi.com/msds/index.html
   This commercial site provides a free list of sites useful in finding MSDS and related software (accessed 8 November 2016).

## ENVIRONMENTAL HEALTH

16. **Environmental Protection Agency (EPA);**
   www.epa.gov
   The US EPA's website contains a wide variety of information related to the environment and public health. Some of the many categories of available information include technical documents, research funding, and more, assistance for small businesses and entire industries, projects and programs, news and events, laws and regulations, databases and software, and publications (accessed 8 November 2016).
17. **National Institute of Environmental Health Sciences (NIEHS);**
   www.niehs.nih.gov
   The NIEHS conducts basic research on environment-related diseases. Its web pages outline the Institute's history and research highlights; it also provides complete contact and visiting information (accessed 8 November 2016).

## EMERGENCY RESPONSE

18. **CAMEO Chemicals;**
   cameochemicals.noaa.gov
   CAMEO Chemicals is a tool designed for people who are involved in hazardous material incident response and planning. This tool is part of the CAMEO software suite, and it is available as a website and as a downloadable desktop application that you can run on your own computer (accessed 8 November 2016).

19. **North American Emergency Response Guidebook (ERG);**
    http://www.phmsa.dot.gov/hazmat/library/erg
    The Emergency Response Guidebook was developed jointly by the US Department of Transportation, Transport Canada, and the Secretariat of Communications and Transportation of Mexico for use by firefighters, police, and other emergency services personnel who may be the first to arrive at the scene of a transportation incident involving a hazardous material (accessed 8 November 2016).

## ERGONOMICS

20. Karwowski, W. and Marras, W. (2006). *The Occupational Ergonomics Handbook*, 2e. Boca Raton, FL: CRC Press.
21. **OSHA Health and Safety Topics on Ergonomics;**
    https://www.osha.gov/SLTC/ergonomics/controlhazards.html
    Many industries have successfully implemented ergonomic solutions in their facilities. These interventions include modifying existing equipment, making changes in work practices, and purchasing new tools. Simple, low-cost solutions are often available to solve problems. Use the information on this page to see what has worked for others in your industry or in other industries (accessed 8 November 2016).

## GOVERNMENT RESOURCES

22. **Canadian Centre for Occupational Health and Safety (CCOHS);**
    www.ccohs.ca
    The CCOHS is a Canadian federal government agency based in Hamilton, Ontario, which serves to support the vision of eliminating all Canadian work-related illnesses and injuries. Its website includes a variety of information – some free, some available on a subscription basis (accessed 8 November 2016).
23. **Consumer Product Safety Commission (CPSC);**
    www.cpsc.gov
    CPSC is an independent US federal regulatory agency. CPSC works in a number of different ways to reduce the risk of injuries and deaths by consumer products: it develops voluntary standards with industry, issues and enforces mandatory standards or banning of consumer products, obtains the recall of products, conducts research on potential product hazards, and informs and educates consumers (accessed 8 November 2016).
24. **European Agency for Safety and Health at Work;**
    osha.europa.eu
    This site links to more than 30 national websites maintained by the Agency's focal points (usually the lead occupational safety and health [OSH] organization in the EU member states, candidate countries, and other international partners). This is a single entry point to an overview of information that the network has to offer, from current campaigns to popular links. It is a database-driven multilingual portal providing access to OSH information in your preferred language, personalize the site and access the European and international network (accessed 8 November 2016).
25. **National Ag Safety Database (NASD);**
    nasdonline.org
    The information contained in NASD was contributed by OSH professionals and organizations from across the nation. Specifically, the objectives of the NASD project are (i) to provide a national resource for the dissemination of information; (ii) to educate workers and managers about occupational hazards associated with agriculture-related injuries, deaths, and illnesses; (iii) to provide prevention information; (iv) to promote the consideration of safety and health issues in agricultural operations; and (v) to provide a convenient way for members of the agricultural safety and health community to share educational and research materials with their colleagues (accessed 8 November 2016).
26. **International Agency for Research on Cancer (IARC);**
    www.iarc.fr
    The IARC is part of the World Health Organization. IARC's mission is to coordinate and conduct research on the causes of human cancer and to develop scientific strategies for cancer control. It is involved in both

epidemiological and laboratory research and disseminates scientific information through meetings, publications, courses, and fellowships (accessed 8 November 2016).

27. **Institut de recherche Robert-Sauvé en santé et en sécurité du travail (IRSST) of Quebec;**
    http://www.irsst.qc.ca/en
    One of the leading OSH research centers in Canada, the Institut de recherche Robert-Sauvé en santé et en sécurité du travail (IRSST), conducts and funds research activities aimed at eliminating risks to worker health and safety and at promoting worker rehabilitation (accessed 8 November 2016).
28. **National Institute for Occupational Safety and Health (NIOSH);**
    http://www.cdc.gov/niosh
    This website provides access to information resources, programs, and news from NIOSH, the federal agency responsible for conducting research and making recommendations for the prevention of work-related injury and illness (accessed 8 November 2016).
29. **Occupational Safety and Health Administration (OSHA);**
    www.osha.gov
    The website for the primary workplace regulatory body in the United States includes access to its regulations, information about safe work practices for a variety of industries, and access to statistics about its enforcement programs and actions (accessed 8 November 2016).
30. **PubChem;**
    https://pubchem.ncbi.nlm.nih.gov
    PubChem is maintained by the National Center for Biotechnology Information (NCBI), a component of the National Library of Medicine, which is part of the US National Institutes of Health. It includes a collection of over 5000-laboratory chemical safety summary for chemicals for which GHS information is publically available (accessed 8 November 2016).
31. **ToxNet;**
    toxnet.nlm.nih.gov
    ToxNet, provided by the US National Library of Medicine, is a collection of databases in the areas of toxicology, hazardous chemicals, environmental health, and toxic releases (accessed 8 November 2016).
32. **United Kingdom Health and Safety Executive;**
    www.hse.gov.uk
    The Health and Safety Executive ensures that risks to people's health and safety from work activities are properly controlled. As the website says, "the law says employers have to look after the health and safety of their employees; employees and the self-employed have to look after their own health and safety; and all have to take care of the health and safety of others, for example, members of the public who may be affected by their work activity. Our job is to see that everyone does this." The site includes a variety of useful tools for the safety professional (accessed 8 November 2016).
33. **WorkSafe Australia;**
    www.safeworkaustralia.gov.au
    Safe Work Australia leads the development of national policy to improve work health and safety and workers' compensation arrangements across Australia (accessed 8 November 2016).
34. **Young Worker Awareness, Workplace Health and Safety Agency, Ontario, Canada;**
    https://www.whsc.on.ca/Resources/For-Young-Workers/Young-Worker-Awareness-Program
    This site contains health and safety information for young workers, their parents, teachers, principals, employers, and others. Although the information is specific to the province of Ontario, Canada (the Young Worker Awareness schools program is only available to Ontario high schools), many OSH professionals likely find the information here useful (accessed 8 November 2016).

## INDUSTRIAL HYGIENE

35. **American Conference of Government Industrial Hygienists (ACGIH);**
    www.acgih.org
    The ACGIH provides access to technical information through this website. Of specific interest, this web page provides access to information about the ACGIH threshold limit values (accessed 8 November 2016).

36. **American Industrial Hygiene Association;**
www.aiha.org
Founded in 1939, the American Industrial Hygiene Association (AIHA) is an association of occupational and environmental health and safety professionals. Members represent a cross section of industry, private business, labor, government, and academia (accessed 8 November 2016).
37. Anna, D. (2014). *Protective Clothing*, 2e. Falls Church, VA: AIHA.
38. ACGIH (2016). *Industrial Ventilation: A Manual of Recommended Practice for Design*, 29e. Cincinnati, OH: ACGIH
39. **Journal of Occupational and Environmental Hygiene;**
http://www.tandfonline.com/loi/uaih20
Formerly known as the *American Industrial Hygiene Association Journal*, this website provides access to journal articles since 1940 (accessed 8 November 2016).
40. Hansen, S. (2011). *Managing Indoor Air Quality*, 5e. Lilburn, GA: The Fairmont Press.
41. Gudgin Dickson, E. (2012). *Personal Protective Equipment for Chemical Biological and Radiological Hazards Design, Evaluation and Selection*. Hoboken, NJ: Wiley.
42. Forsberg, K., Van den Borre, A., Henry III, N., and Zeigler, J.P. (2014). *Quick Selection Guide to Chemical Protective Clothing*, 6e. Hoboken, NJ: Wiley.

## OCCUPATIONAL AND ENVIRONMENTAL MEDICINE

43. **Journal of Occupational and Environmental Medicine;**
http://journals.lww.com/joem/pages/default.aspx (accessed 8 November 2016).
44. Tarlo, S. (2010). *Occupational and Environmental Lung Diseases: Diseases from Work, Home, Outdoor and Other Exposures*. Hoboken, NJ: Wiley.
45. Bingham, E. (2012). *Patty's Toxicology, Volumes 2–9*. Hoboken, NJ: Wiley.

## ORGANIZATIONAL WEBSITES

46. **Inchem;**
inchem.org
Produced through cooperation between the International Programme on Chemical Safety (IPCS) and the Canadian Centre for Occupational Health and Safety (CCOHS), IPCS INCHEM directly responds to one of the Intergovernmental Forum on Chemical Safety (IFCS) priority actions to consolidate current, internationally peer-reviewed chemical safety-related publications and database records from international bodies for public access (accessed 8 November 2016).
47. **Society for Chemical Hazard Communication (SCHC);**
www.schc.org
SCHC is a nonprofit organization with a mission to promote the improvement of the business of hazard communication for chemicals (accessed 8 November 2016).
48. **University of Minnesota Environmental Health and Safety;**
www.dehs.umn.edu
Many colleges and universities have health and safety department websites with valuable information available on it. Larger universities have nearly every general safety or health hazard associated with some part of their operation. Thus, policies and procedures for many different situations can be found on their sites. The University of Minnesota site is typical of a well-maintained university website (accessed 8 November 2016).
49. **World Health Organization;**
http://www.who.int/topics/occupational_health/en
Occupational health deals with all aspects of health and safety in the workplace and has a strong focus on primary prevention of hazards. The health of the workers has several determinants, including risk factors at the workplace, leading to cancers, accidents, musculoskeletal diseases, respiratory diseases, hearing loss, circulatory diseases, stress-related disorders and communicable diseases, and others (accessed 8 November 2016).

## PHYSICAL HAZARDS

50. Parsons, K. (2014). *Human Thermal Environments – the Effects of Hot, Moderate and Cold Environments on Human Health Comfort and Performance*, 3e. Boca Raton, FL: CRC Press.
51. Barat, K. (2014). *Laser Safety – Tools and Training*, 2e. Boca Raton, FL: CRC Press.
52. Berger, E. (2003). *The Noise Manual*. Falls Church, VA: AIHA.

## SAFETY MANAGEMENT

53. Manuele, F.A. (2014). *Advanced Safety Management: Focusing on Z10 and Serious Injury Prevention*, 2e. Hoboken, NJ: Wiley.
54. Center for Chemical Process Safety (2008). *Guidelines for Hazard Evaluation Procedures*, 3e. Hoboken, NJ: Wiley.
55. Center for Chemical Process Safety (2008). *Inherently Safer Chemical Processes: A Life Cycle Approach*, 2e. Hoboken, NJ: Wiley.
56. Perrow, C. (1984). *Normal Accidents: Living with High Risk Technologies*. New York: Basic Books.
57. National Research Council (2014). *Safe Science: Promoting a Culture of Safety in Academic Chemical Research*. Washington, DC: The National Academy Press.
58. *Understanding Risk Communication Theory: A Guide for Emergency Managers and Communicators* (2012). U.S. Department of Homeland Security. http://www.start.umd.edu/sites/default/files/files/publications/UnderstandingRiskCommunicationTheory.pdf (accessed 8 November 2016).

CHAPTER 3

# ERGONOMICS: ACHIEVING SYSTEM BALANCE THROUGH ERGONOMIC ANALYSIS AND CONTROL

GRACIELA M. PEREZ

*Department of Work Environment, College of Engineering, University of Massachusetts at Lowell, Lowell, MA, 01854*

## 3.1 INTRODUCTION

Healthy work systems require a balance between the task, the technology, the organization, the environment, and the individual (Figure 3.1). When any one of these basic connections is not functioning optimally, the work system is impaired (Hosey 1973). For example, such a system would recognize where humans excel and where machines excel, and the work system would then be balanced accordingly (Table 3.1). When humans work in environments that are not compatible with their strengths, the workers may become injured or ill. Thus a healthy work system is designed to optimize the human–machine interface and keep workers safe and healthy.

In essence, this is the science called *ergonomics*. The word ergonomics derives its meaning from the Greek roots of *ergon*, meaning work, and *nomos*, meaning law, or the *work law*. In plain English, ergonomics has been defined as applying the laws of work to design the work and work environment to fit the capabilities of the people who perform the work. The benefits of a work system, which is ergonomically designed, include a healthy, productive, and efficient work environment. Conversely, the results of system imbalance may result in poor production, poor product quality, and additional hidden costs due to preventable musculoskeletal and cardiovascular disorders (MSDs). Perhaps that is why international quality standards, such as ISO 9000, require similar system balances.

When people are involved in the work system, and musculoskeletal disorders occur, or other indicators of system imbalance are present (i.e. low quality, decreases in productivity, absenteeism, etc.), an analytical procedure needs to be implemented to identify the cause of the system imbalance. Only after analysis can procedures be put into place to bring the system back into balance. If an analysis is not performed, the repair of the system is reduced to guesswork, at best. "Guesswork" will not be able to justify the capital investment in workstation and work redesign to a skeptical CEO or thrifty purchasing manager.

When dealing with musculoskeletal disorders, the job improvement should be a process rather than a "Band-Aid." Job analysis is part of the job improvement cycle.

Recently, the National Institute for Occupational Safety and Health (NIOSH) has suggested the following seven elements of an effective program that comprise a "pathway" for addressing system balance (NIOSH 1997):

Step 1. Look for signs of potential MSDs in the system, such as frequent worker reports of aches and pains, or job elements that require repetitive, forceful exertions.

Step 2. Show management commitment in addressing possible problems, and encourage worker involvement in achieving system balance.

Step 3. Offer training to expand management and worker ability to evaluate potential system imbalances associated with MSDs.

*Handbook of Occupational Safety and Health*, Third Edition. Edited by S. Z. Mansdorf.

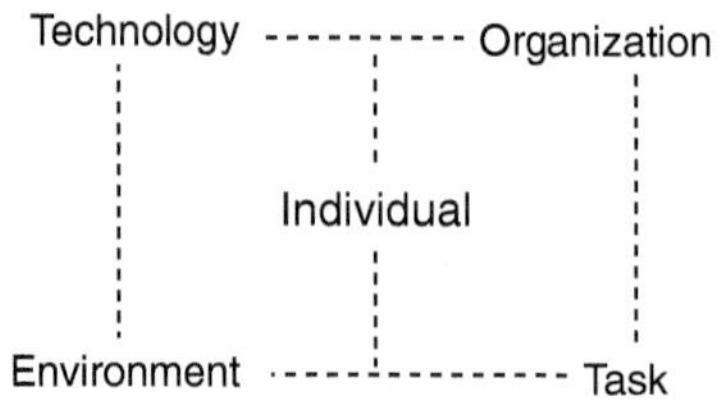


**Figure 3.1** Balance of health work systems.

**TABLE 3.1 Basic Strengths of Humans and Machines**

| |
|---|
| Human strengths over machines |
| • Sensitive to a wide variety of stimuli |
| • Ability to react to unexpected, low-probability events |
| • Ability to exercise judgment where events cannot be completely defined |
| • Perception of patterns and ability to make generalizations about them |
| Machine strength over humans |
| • Perform routine, repetitive, or very precise operations |
| • Exert great force, smoothly and with precision |
| • Operate in environments that are toxic or hazardous to humans or beyond human tolerance |
| • Can be designed to be insensitive to extraneous factors |
| • Can perform many different functions at the same time |

Step 4. Gather data to identify jobs or work conditions that are most problematic. Sources such as injury and illness logs, symptom surveys, medical records, first reports of injuries, worker compensation data, and job analysis may be used.

Step 5. Identify effective controls for tasks that pose a risk of MSDs. Evaluate these approaches once they have been implemented to see if they have reduced or eliminated the problem.

Step 6. Establish a solid healthcare management process that integrates early detection and conservative treatment to prevent impairment and disability. This should include return-to-work programs and job accommodation.

Step 7. Minimize risk factors for MSDs during the design phase of new work processes, products, tools, and operations. Thus the design team should be part of the ergonomic process to ensure system balance.

The purpose of this chapter is to provide the reader with some basic tools and references so that the above steps can be carried out in a systematic fashion. In particular, methods of ergonomic job analysis will be presented so that the professional can develop a basic understanding of how to work toward system balance.

The etiology of musculoskeletal disorders is beyond the scope of this chapter. However, it should be noted that medical management, including conservative return to work and early detection of MSDs, is paramount in an ergonomic program. In some cases, solutions to system imbalances will be intuitive, while others may require the services of a qualified ergonomist who may assist with the identification of risk factors, controls, and the evaluation of the controls. The references at the end of this chapter are offered so that the safety and health professional may build a library of resources for training, education, job analysis, and solutions.

## 3.2 JOB ANALYSIS

Ergonomic job analysis has been described as the systematic investigation of jobs to describe individual and combined work factors and work activities for the purpose of designing task demands to match the human capabilities of those who perform the task. Job analysis has also been used to identify reasons (root causes) for increased risk of MSDs. It should be stated up front that this type of analysis is not possible without resource and time commitment from top management. Resource allocation, in the form of a budget, should be secured prior to the commencement of analysis. Too often, analyses are performed only to have the resulting recommendations collect dust. This type of scenario only breeds mistrust among employees and compromises the reality of a balanced system. In the same breath, it is imperative that workers be involved in the job analysis process. After all, they are the true experts of their jobs. Workers are also a vital component for the ultimate success of the implementation of controls.

Important areas for consideration when performing job analyses include:

- Identify the population who does the job.
- Determine why the job is done (consider its relation to the whole work process and organizational design).
- Break the job into tasks, and then break the tasks into elements; determine the physical demands and mental information demands associated with each element of the tasks.
- Identify equipment, tools, and information that are needed to accomplish each task (technology).
- Locate where the tasks are done in relation to the person who performs the tasks.
- Identify the environmental conditions under which the tasks are performed (consider seasonal and product changes).

Under all circumstances, the frequency, duration, and level of exposure to work factors can be determined. Where available, these data can be compared with normative data (quantitative or qualitative) to assess the risk of MSDs. This in turn can contribute to finding integrated solutions that work toward a balanced system. Job analysis procedures are then used to evaluate the effectiveness of implemented control measures.

The following sections will present work factors to consider such as when to conduct a job analysis, as well as the steps and methods involved in such a job analysis. In addition, the selection and evaluation of solutions will be explained.

## 3.3 WORK FACTORS TO CONSIDER WHEN PERFORMING JOB ANALYSES

There are many work factors that should be considered during the ergonomic job analysis process. Some of these are more obvious and easier to identify than others.

### 3.3.1 Forces Required to Perform the Task

Force requirements have a direct impact on the muscular effort that must be expended by the worker to perform an action.

The amount of force exerted by the worker depends upon:

- Work postures (awkward postures decrease the ability to exert force).
- Speed of movement.
- The weight handled.
- The friction characteristics (ability to grip) of the objects handled.

### 3.3.2 Contact Stresses

Contact stress occurs when soft tissue of the human body comes in contact with a hard object. High contact forces (due to hard or sharp objects and work surfaces) may create pressure over one area of the body and inhibit nerve function and blood flow.

### 3.3.3 Postures Assumed During the Task

The position of the body has a direct effect on the moment (torque) about a joint. The moment is the rotation produced when a force is applied. The smaller the moment, the less force required by the muscles that support movement about the joint. Neutral postures tend to optimize the ability of the body to exert maximum force:

- Greater force is required when awkward postures are used because muscles lose their "mechanical advantage" and cannot perform efficiently.
- Prolonged postures (e.g. holding the arm out straight to hold a fixture) produce a moment (torque) around a joint. This is referred to as "static posture." Static postures may decrease the time to muscle fatigue, especially when force is required by the working muscles. This fatigue occurs as muscles continue to work with impaired blood flow caused by the static posture.
- The farther away from the body a load is carried, the greater the compressive forces on the spine.

- Shear and compressive forces on the spine increase when lifting, lowering, or handling objects with the back bent or twisted and when objects are bulky and/or slippery.
- Uneven work surfaces or surfaces that are slippery or difficult to walk on may also affect forces required to perform the job.

### 3.3.4 Frequency of Muscle Contraction

Frequent repetition of the same work activities can exacerbate the effects of awkward work postures and forceful exertions. Tendons and muscles can often recover from the effects of stretching or forceful exertion if sufficient time is allotted between exertions. However, if movements involving the same muscles are repeated frequently, without rest, fatigue and strain can accumulate, producing tissue damage. It is important to note that different tasks may require repetitive use of the same muscle groups even though the tasks are not similar. This should be taken into consideration when workers are rotated among different tasks.

### 3.3.5 Duration of Muscle Contraction

The amount of time a muscle is contracted can have a substantial effect on both localized and general fatigue. In general, the longer the period of continuous work (muscle contraction), the longer the recovery or rest time will be required. An example of prolonged muscle contraction is standing for long periods without a foot rest, chair, or sit/stand stool. Another example would include constantly gripping a tool.

### 3.3.6 Vibration

Hand–arm (segmental) vibration from handheld power tools increases force requirements and has been associated with the development of hand–arm vibration syndrome (vibration white finger or Raynaud's disease) and carpal tunnel syndrome. Guidelines for measuring continuous, intermittent, impulsive, and impact segmental vibration have been provided in the ISO standard ISO5349 (1986), entitled "Guide for the Measurement and the Assessment of Human Exposure to Hand Transmitted Vibration," and the ANSI standard ANSI S3.34-1986 entitled "Guide for the Measurement and Evaluation of Human Exposure to Vibration Transmitted to the Hand."

Whole-body vibration exposures have been observed in truck drivers, large metal stamping operations, and with the use of large vibrating tools, such as jackhammers. These exposures have been associated with back and neck disorders, including microfractures of vertebral endplates, and urinary and digestive discomfort. Currently, whole-body vibration standards are under consideration by the ISO and ANSI.

### 3.3.7 Cold Temperatures

Cold temperatures can reduce the dexterity and sensory perception of the hand due to vasoconstriction (a shunting of the blood away from the extremities to the organs to maintain a core temperature of over 96.8 °F). This may cause workers to apply more grip force to tool handles and objects than would be necessary under warmer conditions and may increase risk for accidents due to the loss of hand dexterity. There is also evidence that cold tends to exacerbate the effects of segmental vibration. The following guidelines have been provided by the American Conference of Governmental Industrial Hygienists:

- If fine work is performed with the bare hands for more than 10 minutes in an environment below 60.8 °F (16 °C), special provisions should be established for keeping the workers' hands warm.
- Metal handles on hand tools and control bars should be covered by thermal insulating materials at temperatures below 30.2 °F (−1 °C).
- If the air temperature falls below 60.8 °F (16 °C) for sedentary work, 39.2 °F (4 °C) for light work, 19.4 °F (−7 °C) for moderate work, and fine manual dexterity is not required, then gloves should be used by the workers. If gloves are not appropriate (see below), then warm air jets, radiant heaters (fuel burner or electric radiator), or contact warm plates may be utilized.
- Avoid cold exhaust exposure to hands when using hand tools.
- Avoid cold exposures to the cooling effect associated with handling evaporative liquids such as gasoline, alcohol, or cleaning fluids.

At temperatures below 39.2 °F (4 °C), an evaporative cooling effect may occur and exacerbate other musculoskeletal risk factors listed in the previous sections.

### 3.3.8 Poorly Fitted Gloves

Poorly fitted gloves can reduce sensory feedback from the fingertips and result in increased grip force. Force is also increased by working against tight-fitting gloves or trying to get a tight grip with big bulky gloves. In addition, loose gloves may present a safety hazard around moving parts and tools, such as chain drives and drills.

### 3.3.9 Obstructions

Obstructions such as rails, jigs, and fixtures, which hamper smooth free lifting and reaching motions, should be evaluated to analyze their effect on awkward postures and the forces required to accomplish the job. An example of obstructions is seen in warehouses, where the size of the first slot may not allow the worker to use safe lifting practices while they select cases. Workers may not be able to stand up in the slot when the height of the second slot is at 50 in. Thus they have to lift and carry in a stooped or bent position due to the obstruction created by the second slot.

### 3.3.10 Standing Surfaces

Prolonged standing on hard surfaces increases back and leg fatigue. In addition, standing on inclines or different levels may put stress on the spine and legs. The use of a foot pedal while standing may also increase the stress on the back, hip, and ankle. Standing on slippery surfaces may increase the risk of sudden force exertion while trying to prevent slips and falls.

To reduce the risk factors associated with prolonged standing, employers can provide foot rests or rest bars, as long as they do not pose a safety hazard. Chairs may also reduce the stress associated with standing only if the chairs are well suited for the job and the workers understand how to easily adjust them. Sit/stand stools have also been used in environments where seated postures increased other risk factors such as excessive reaching. It is imperative to analyze the effect of a chair on other risks factors before purchasing the chair. Workers are usually able to determine how a chair will affect their work environment, and they should always be included in the purchase decision process.

### 3.3.11 Prolonged High Visual Demands

Prolonged visual demands (vigilance), processing information quickly over time, making inspection decisions rapidly, and visually scanning complicated displays for slight evidence of malfunction may increase fatigue, blood pressure, and muscle tension. Inadequate lighting or poorly designed displays may result in a decreased time to fatigue. In addition to increased musculoskeletal load, these health effects are accompanied by increases in errors and decreases in quality.

### 3.3.12 Physical Energy Demands

Fatigue is believed to increase a worker's risk for musculoskeletal injury and illness. Oxygen consumption has been accepted as one of the best measures of energy demands of the job. Once the energy demands are known, the appropriate work–rest cycles can be established. Heart rate and ratings of perceived exertion have also been used with success. It is recommended that a professional with experience in the measurement of energy demand and the design of work cycles be employed to address fatigue in the work environment.

## 3.4 WHEN TO CONDUCT JOB ANALYSES

Job analyses can be conducted on jobs in a reactive or proactive manner. In the reactive approach, the health and safety professional responds to illnesses or injuries that are associated with occupational hazards. Conversely, the proactive approach commands that risk factors are actively sought through job analyses, even in the absence of injuries and illnesses. Job analyses can be as simple as checklists or as complex as electromyography to measure muscle force. Since this chapter is written for the safety and health professional who may have little or no experience with ergonomic design, the simpler methods will be presented with suggestions of how to obtain resources for more complex analyses.

### 3.4.1 The Reactive Approach

In the reactive approach, records review, single incidents of MSDs, first aid cases, reports of near misses, reports of pain, and workers compensation cases can prompt a response from the health and safety professional in the form of a job analysis. The analyst may investigate the root cause of a single case or review OSHA 200 logs for trends and incidents of many cases that are occurring in a facility. If a facility is very large, records review may provide a better picture of where the problems exist and will allow for the identification of high-, medium-, and low-risk areas.

For example, the statistics of incidence rate and severity rates may be used. The *incidence rate* is usually calculated for the period of one calendar year and represents the number of new cases per 100 workers that occurs for a specific industry (SIC code), a specific company, a specific department, a specific job, and/or a specific type of injury. Incidence rates are calculated as follows:

$$\frac{\text{Number of new cases} \times 200000}{\text{Total number of hours worked by the exposed population (SIC, company, dept., etc.)}} \times 100$$

The 200000 represents 2000 full-time hours of 40 hours per week for 50 weeks in a year, multiplied by 100 workers. This number normalizes the data so that companies, departments, and jobs may be compared despite the number of employees they have. For comparisons with fewer than 20 workers, it may be better to use the absolute number of injuries and illnesses instead of the incidence rate. Thus the incidence rate for 1995 for the nursing home industry could be described as "16.8 cases per 100 workers in 1995." This number could be compared with other industries to determine if specific nursing homes are high risk, despite their size, because the data have been normalized per 100 workers.

Another statistic that is frequently used is the *severity rate*. The severity rate is usually calculated for a period of one year and represents a severity measure (the amount of medical costs, the number of lost workdays, the number of restricted workdays, the dollars of workers compensation, etc.) per 100 workers for a given calendar year. The severity is calculated as follows:

$$\frac{\text{Severity measure} \times 200000 \text{ (lost work days, medical costs, etc.)}}{\text{Total number of hours worked by the exposed workers}} \times 100$$

Since there is usually an initial increase in the reporting of MSDs when a company increases awareness among its employees, severity can be used as the real indicator of a program's success. Since workers are encouraged to report earlier, the severity of the injuries should be decreased, even though reporting (i.e. the number of cases) may increase. Workers will report injuries when they first notice symptoms instead of waiting until they are impaired (when a more costly treatment may then be needed). Thus, as the incidence rate rises, initially, the severity rate should decrease accordingly. The combination of these statistics allows a safety and health professional to prioritize ergonomic intervention in the work environments when there are too many workers to address every job in a short time frame. In addition, these statistics may be used to explain the long-term benefits associated with system balance.

## 3.5 JOB ANALYSIS STEPS

Before the job analysis is initiated, it is important to secure management commitment. This commitment includes a budget, a commitment to fund the cost of controls, a commitment to a continuous improvement process that includes a follow-up, and the inclusion of responsibility for the ergonomic program in the job performance reviews of managers. Employees should also be included as a vital part of the job analysis team. This inclusion will necessitate that they be trained and educated in ergonomic principles related to their work and work environment.

### 3.5.1 Who Does the Analysis?

The job analysis can be done by an individual or team who has knowledge of ergonomics, how the body works, and who is knowledgeable about potential control measures for the identified exposures. It is advisable that the analyst also be familiar with the type of job that is analyzed and the types of controls that are available to abate the hazards. If a consultant is used, it is preferable that they have had previous experience in the industry in which they are working. For example, if a consultant is hired to evaluate an office environment, he/she should provide references from other office environments.

Management, the affected employees, healthcare professionals, engineering, and maintenance are all vital contributors to the process. In talking with employees from many companies, their biggest impression was formed early in the hiring process from plant and office managers who were actively involved in the job analysis and abatement procedures. At the American Saw Company, in East Longmeadow, Massachusetts, the plant manager is the person who walks each new employee through the facility to introduce the importance of various safety and health aspects in the work environment. In interviewing employees of that company, the employees stated that it is the dedication from top management that formed their safety and health "culture" from their first day on the job. Subsequently, the company has consistently maintained injury rates below their industry average.

In addition to management commitment, informal discussions with workers and supervisors can provide useful information for the job analysis. Workers who perform a job on a daily basis are often the best source of information about the specific elements that may pose an increased risk for MSDs. The use of symptoms surveys or postural discomfort surveys (Figure 3.2) may also be used to identify specific tasks that are causing discomfort or pain and can be quite useful in the analysis. These surveys can also be used for before/after comparisons after measures are implemented.

Supervisors can provide a macro view of their operations, as well as insight into the feasibility of proposed changes. Interviews also provide workers with the opportunity to participate and to provide input to the job analysis process. Soliciting worker input early in the process may also lead to the identification of more feasible and accepted control strategies. Kennebec Nursing Home in Maine found that when their employees were given the opportunities to choose controls to reduce injuries associated with resident handling, injuries were reduced. The workers were receptive to the interventions they had chosen to reduce risk factors, and unlike other nursing homes who purchased lift devices without employee input, the nursing aides actively followed the risk reduction strategies.

### 3.5.2 Basic Equipment Needed for Conducting Most Job Analyses

Equipment for recording occupational exposures can be quite sophisticated, such as 3D static biomechanical models, dynamic measures of torsion, compressive and shear force models, and the kinematic analysis of a person in motion using light sensors. These tools include electrogoniometers, lumbar motion monitors, and other electrical equipment (pocket protectors and slide rules are optional). However, in simple non-research-oriented cases, the following basic instrumentation is sufficient:

1. Cameras and film for recording workers' postures and motions during job activities. This equipment is also very useful for recording "before and after" improvements and for developing training materials (see notes on filming below).
2. Tape measures and rulers for measuring workstation dimensions, tool dimensions, and reach distances. In addition, a goniometer may be used to measure angles of the body, materials handled, viewing angles, and workstation attributes. This information may be used as inputs to computer programs.
3. Force gauges or spring scales (fish scales) for measuring the force of exertions (e.g. the force needed to push or pull a hand cart) and the weight of tools or objects in manual handling tasks.
4. Timers (stopwatches) for measuring the duration of work activities, breaks, etc. (care should be taken to assure employees that the stopwatch is not being used to establish production standards). A real-time simulation program works quite well in the job analysis process (Keyserling 1986). Real-time analysis can measure the percentage and absolute time spent in various postures. ROPEM is a technique that allows the analyst to record postures verbally into a microcassette for transfer to a computer at another time and works well in environments where the use of a computer would be difficult (Perez-Balke 1993).
5. A checklist that is designed for the task at hand and provides feedback as to what needs to be done in response to the answers. For example, a checklist for the analysis of a computer workstation would be different from one used to evaluate jobs on a construction site and might stipulate that all negative responses of "no" signify the need for further analysis. An example of a checklist from Lifshitz and Armstrong is included in Figure 3.3, and examples of checklists that have been proposed by the Occupational Safety and Health Administration (OSHA) are found in Appendix B.
6. Borg scales (Figure 3.4) provide the analysis with a tool to subjectively measure the force requirements of the job as well as the effect of controls on the effort needed to perform the job. These scales will be discussed in further detail below.

Symptoms Survey: Ergonomics Program

Date ____/____/____/

____ Plant ____ Dept no. Job name ________________________

____ Shift ________________ Hours worked/week ____ years ____ months Time on THIS job

*Other jobs you have done in the last year (for more than 2 weeks)*

____ Plant ____ Dept no. ________________ Job name ____ months ____ weeks Time on THIS job

____ Plant ____ Dept no. ________________ Job name ____ months ____ weeks Time on THIS job

(if more than two jobs, include those you worked on the most)

Have you had any pain or discomfort during the last year?

☐ Yes ☐ No (if No, stop here)

If YES, carefully shade in area of the drawing which bothers you the MOST.

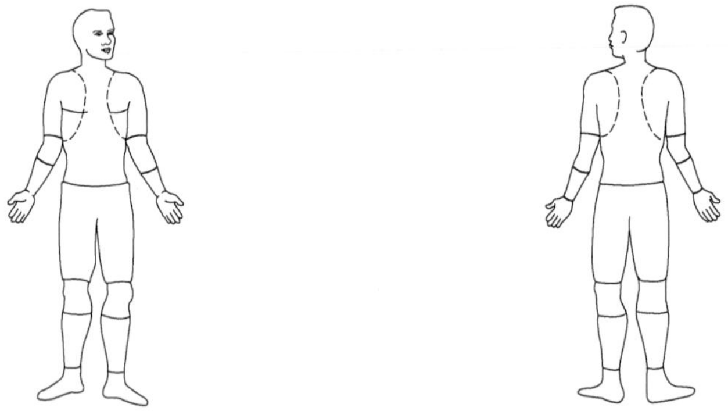

Front Back

(continued)

**Figure 3.2** Symptoms survey form.

### 3.5.3 Guidelines for Videotaping for Ergonomic Analysis

Guidelines for recording work activities on videotape are available from the National Institute for Occupational Safety and Health (USA) at 1-800-35NIOSH and are summarized as:

- If the video camera has the ability to record the time and date on the videotape, use these features to document when each job was observed and filmed. Recording the time on videotape can be especially helpful if a detailed motion study will be performed at a later date (time should be recorded in seconds). Make sure the time and date are set properly before videotaping begins.
- If the video camera cannot record time directly on the film, it may be useful to position a clock or a stopwatch in the field of view.
- At the beginning of each recording session, announce the name and location of the job being filmed so that it is recorded on the film's audio track. Restrict subsequent commentary to facts about the job or workstation.
- For best accuracy, try to remain unobtrusive; that is, disturb the work process as little as possible while filming. Workers should not alter their work methods because of the videotaping process.

Symptoms Survey Form (continued)

*(Complete a separate page for each area that bothers you)*

Check area: ☐ Neck ☐ Shoulder ☐ Elbow/forearm ☐ Hand/wrist ☐ Fingers
☐ Upper back ☐ Low back ☐ Thigh/knee ☐ Low leg ☐ Ankle/foot

1. Please put a check by the word(s) that best describe your problem

| | | |
|---|---|---|
| ☐ Aching | ☐ Numbness (asleep) | ☐ Tingling |
| ☐ Burning | ☐ Pain | ☐ Weakness |
| ☐ Cramping | ☐ Swelling | ☐ Other |
| ☐ Loss of color | ☐ Stiffness | |

2. When did you first notice the problem? ________(month) ________ (year)
3. How long does each episode last? (Mark an X along the line)

   ________/________/________/________/________/
   1 hour 1 day 1 week 1 month 6 months
4. How many separate episodes have you had in the last year? ____________
5. What do you think caused the problem? ____________

   ____________
6. Have you had this problem in the last 7 days? ☐ Yes ☐ No
7. How would you rate this problem? (mark an X on the line)

   *NOW*

   ____________

   None Unbearable

   *When it is the WORST*

   ____________

   None Unbearable
8. Have you had medical treatment for this problem? ☐ Yes ☐ No

   8a. if NO, why not? ____________

   8b. if YES, where did you receive treatment?

   ☐ 1. Company medical Times in past year ____________

   ☐ 2. Personal doctor Times in past year ____________

   ☐ 3. Other Times in past year ____________

   Did treatment help? ☐ Yes ☐ No ____________
9. How much time have you lost in the last year because of this problem? ____days
10. How many days in the last year were you on restricted or light duty because of this problem?_______ days
11. Please comment on what you think would improve your symptoms

    ____________

    ____________

    ____________

    ____________

**Figure 3.2** (*Continued*)

- If the job is repetitive or cyclic in nature, film at least 10–15 cycles of the primary job task. If several workers perform the same job, film at least two to three different workers performing the job to capture differences in work method.
- If necessary, film the worker from several angles or positions to capture all relevant postures and the activity of both hands. Initially, the worker's whole-body posture should be recorded (as well as the work surface or chair on which the worker is standing or sitting). Later, close-up shots of the hands should also be recorded if the work is manually intensive or extremely repetitive.
- If possible, film jobs in the order in which they appear in the process. For example, if several jobs on an assembly line are being evaluated, begin by recording the first job on the line, followed by the second, third, etc.
- Avoid making jerky or fast movements with the camera while recording. Mounting the camera on a tripod may be useful for filming work activities at a fixed workstation where the worker does not move around much.

**Risk Factors**

1. Physical stress
   a. Can the job be done without hand/wrist contact on sharp edges?
   b. Is the tool operating without vibration?
   c. Are the worker's hands exposed to >21 °C?
   d. Can the job be done without using gloves?
2. Force
   a. Does the job require less than 4.5 kg of force?
   b. Can the job be done without using a finger pinch grip?
3. Posture
   a. Can the job be done without wrist flexion or extension?
   b. Can the tool be used without wrist flexion or extension?
   c. Can the job be done without deviating the wrist from side to side?
   d. Can the tool be used without deviating the wrist from side to side?
   e. Can the worker be seated while performing the job?
   f. Can the job be done without a "clothes wringing" motion?
4. Workstation hardware
   a. Can the orientation of the work surface be adjusted?
   b. Can the height of the work surface be adjusted?
   c. Can the location of the tool be adjusted?
5. Repetitiveness
   a. Is the cycle time longer than 30 seconds
6. Tool design
   a. Are the thumb and finger slightly overlapped in a closed grip?
   b. Is the span of the tool's handle between 5 and 7 cm?
   c. Is the hand of the tool made from material other than metal?
   d. Is the weight of the handle below 4 kg?
   e. Is the tool suspended?

Note: "No" responses are indicative of conditions associated with the risk of cumulative trauma disorders.

**Figure 3.3** Checklist for work-related musculoskeletal disorders of the upper extremities (Lifshitz and Armstrong 1986).

Videotapes can then be reviewed using slow-motion or real-time playback more accurately to measure task durations or detect subtle or rapid movements. As mentioned, more advanced techniques such as "lumbar motion monitors," "peak performance models," and "joint motion analysis," to name a few methods, are available. These types of analysis are often used where the cost of a less than perfect abatement plan could cost many more times the amount of money spent on ergonomic controls. It is sometimes worthwhile to quantify the root cause so that design criteria can be implemented early in the setup stage of manufacturing processes.

The videotape analysis can also be invaluable as a tool for documenting visible risk factors for MSDs in the workplace. It is imperative to gain worker consent before videotaping or analyzing a work task. If the worker is not involved, the analyst may not get an accurate "picture" of the task. Perhaps an equal mistake is the use of surrogate workers for ergonomic analyses. Since people are different sizes and use different techniques to perform the same job, workers who perform the task on a regular basis, and who will be affected by the worksite changes, should be an integral part of the analytical process.

## 3.6 JOB ANALYSIS METHODS

The primary objective of job analyses is to collect sufficient information to allow the analyst to completely describe tasks as they are currently being performed (i.e. what the worker is doing and how it is being performed over time) as well as the system attributes that interact with the worker. This information allows for the assessment of risk for MSDs

**Ratings of Perceived Scales**

| Rating of pereceived exertion: whole body effort (Borg 1962) | Category scale for rating of perceived exertion: large-muscle-group activity (Borg 1980) |
|---|---|
| 20 | *Maximal |
| 19 Very, very hard | 10 Very, very strong (almost max) |
| 18 | 9 |
| 17 Very hard | 8 |
| 16 | 7 Very strong |
| 15 Hard | 6 |
| 14 | 5 Strong (heavy) |
| 13 Somewhat hard | 4 Somewhat strong |
| 12 | 3 Moderate |
| 11 Fairly | 2 Weak (light) |
| 10 | 1 Very weak |
| 9<br>Very light<br>8 | 0.5 Very, very weak (just noticeable) |
| 7 Very, very light | 0 Nothing at all |
| 6 | |

*Source*: [Borg, G.A.V. (1962). *Physical Performance of Perceived Exertion*. Lund, Sweden: Gleerups. Borg, G.A.V. (1980). A category scale with ratio properties for intermodal and interindividual comparisons. Paper presented at the International Congress of Psychology, Leibig, West Germany(1980)

**Figure 3.4** Borg scales.

and describes the system balance. Generally, this requires the analyst to observe the job during a "typical" work period under "normal" operating conditions. It may be that there is only one task ("cut thigh from chicken" or "enter tax form data into computer"). However, in general, a job consists of a group of tasks that can be defined by related task elements (Niebel 1989). Figure 3.7 depicts examples of work elements.

A job may consist of several regular and several irregular tasks (tasks not done with any regularity). It is important to identify the frequency and duration of both regular and irregular tasks as well as the frequency and duration of any "recovery" periods. The more complex the job tasks (including rotation between tasks or jobs), the more complex the analysis strategy. With the advent of cell manufacturing, there are new challenges presented for job analysis, especially when rotation is "informal" (Armstrong et al. 1986; Rodgers 1992).

Workers who perform the same job may use different methods due to differences in training, stature, or strength. Therefore, several workers who perform the same job should be observed. The analyst should also attempt to observe the same task at different times during the work shift to determine if fatigue affects workers' performance or if the workload changes during the workday. However, sometimes production records may provide this information. The common steps used in ergonomic job analysis are listed below.

### 3.6.1 Document the Job

1. Describe the job briefly, the goal of the job, and what is to be accomplished.
2. Describe the workers: how many workers are employed in each job? What are the characteristics of the workforce (e.g. gender, age, education level, seniority, union)?
3. Define the work schedule. What is the work–rest cycle? How frequently are their breaks? How many hours do employees work per week? Is work organized into shifts? How much overtime is worked per week?

4. Determine production information (work pace). Is there an established work rate? If so, how is the work rate determined (e.g. machine paced, piece rate, time standards, etc.)?
5. Take measurements (see discussion above on tools). A sketch or drawing of the workstation or worksite is useful for identifying the location of fixtures, equipment items, etc. Sketches should be labeled with dimensions indicative of work surface heights, reach distances, clearance, walking distances, etc. (Keyserling et al. 1991; Putz-Anderson 1988). If workers handle tools or objects, the location, size, shape, and weight of these items should be recorded. Force measurements (e.g. using spring scales or force gauges) should be recorded for push, pull, and lift activities.
6. Measurements may also be taken to apply to the NIOSH lifting equation (Figure 3.A.1) to evaluate exposures associated with manual material handling.
7. Describe how the work is organized. How much control do workers have over the way they perform their job tasks, and what is the level of job demand? For example, is there an opportunity for workers to communicate, receive feedback quickly, and provide input regarding problems in production quality or health and safety? Do workers perform the same tasks over and over throughout the work shift, or do workers perform a large number of different tasks? Is there an opportunity for workers to rotate to other jobs? What type of pay system is utilized (e.g. hourly wage, piece rate)?

### 3.6.2 Identify and Evaluate Exposure to Work Factors

Record work factors, such as those mentioned above, associated with specific tasks or subtasks. Describe the reason that the work factor(s) are present (root cause). Perhaps the biggest error seen in job analysis is the failure to perform root cause analysis. For example, a safety and health professional might list the cause of a hazard of bent wrists with a pinch grip as "poor worker training." The resulting recommendation would then be "train the worker," which would not address the hazard.

A root cause analysis would question the underlying exposures. For example, is it the orientation of the part that causes the worker to need to bend the wrist? If so, then a jig with a clamp might be recommended. A common error seen in warehouse and distribution centers is to retrain workers in "safe lifting procedures" or mandate that their workers "wear a back belt" when they injure their backs. Safe lifting procedures would tell workers to bend their knees even though they cannot possibly lift safely given the constraints of their warehouse design. For example, the last boxes on a pallet in the first tier, or slot, of a warehouse may be located 48 in. inside an area only 44 in. high. There would be no conceivable way for the worker to select the box "using safe lifting techniques" unless crawling was acceptable (which it is not), even if a back belt was used. Instead, a root cause analysis would show that the physical constraints of the environment result in poor work postures and lifting methods. This finding might result in the recommendation that pallets be rotated by forklift operators when the front of the pallet has been picked or that slots be reconfigured to provide a safe lifting environment.

### 3.6.3 Summarize Findings

For each task (subtask) list each body part and work factor present, including the frequency and duration of the exposure as quantitatively as possible. List the root cause for the work factor present. Figure 3.6 presents an example of a form that may be used for a summary of findings.

### 3.6.4 Compare Findings to Normative Values Where Available

In some cases, it will be obvious that task demands exceed the capabilities of the workers performing the job. For example, most analysts will recognize a job as being too repetitive if the worker has difficulty keeping up with the required pace or if the body is in constant motion. However, in other instances it may not be as clear that force or postural demands are appropriate for the population of workers performing a task. This is especially true if a new job is in the design stages. However, it is possible to estimate the percentage of workers for whom a job may be difficult. Information that can be used for this purpose is available from the following sources. In addition, the reader is encouraged to peruse the list of resources at the end of this chapter for more information:

1. For lifting analysis, use the NIOSH Work Practices Guide (NIOSH 1981; Waters et al. 1993).
2. For push/pull/carry problems, use Snook tables (Snook 1978) or computerized biomechanical models (Chaffin 1992). ANSI standard B-11-1994 also has this information as well as recommendations on human machine interface.
3. For workstation and tool design, use anthropometry tables such as those in Figure 3.5.

**Standing Body Dimensions**

| | Fifth percentile | | | | Ninety-fifth percentile | | | |
|---|---|---|---|---|---|---|---|---|
| | Ground troops | | Females | | Ground troops | | Females | |
| Standing body dimensions | cm | in. | cm | in. | cm | in. | cm | in. |
| 1. Stature | 162.8 | 64.1 | 152.4 | 60.0 | 185.6 | 73.1 | 174.1 | 68.5 |
| 2. Eye height (standing) | 151.1 | 59.5 | 140.9 | 55.5 | 173.3 | 68.2 | 162.2 | 63.9 |
| 3. Shoulder (acromale) height | 133.8 | 52.6 | 123.0 | 48.4 | 154.2 | 60.7 | 143.7 | 56.6 |
| 4. Chest (nipple) height[a] | 117.9 | 46.4 | 109.3 | 43.0 | 136.5 | 53.7 | 127.8 | 50.3 |
| 5. Elbow (radial) height | 101.0 | 39.8 | 94.9 | 37.4 | 117.8 | 46.4 | 110.7 | 43.6 |
| 6. Waist height | 96.6 | 38.0 | 93.1 | 36.6 | 115.2 | 45.3 | 110.3 | 43.4 |
| 7. Crotch height | 76.3 | 30.0 | 68.1 | 26.8 | 91.8 | 36.1 | 83.9 | 33.0 |
| 8. Gluteal furrow height | 73.3 | 28.8 | 66.4 | 26.2 | 87.7 | 34.5 | 81.0 | 31.9 |
| 9. Kneecap height | 47.5 | 18.7 | 43.8 | 17.2 | 58.6 | 23.1 | 52.5 | 20.7 |
| 10. Calf height | 31.1 | 12.2 | 29.0 | 11.4 | 40.6 | 16.0 | 36.8 | 14.4 |
| 11. Functional reach | 72.6 | 28.6 | 64.0 | 25.2 | 90.9 | 35.8 | 80.4 | 31.7 |
| 12. Functional reach extended | 84.2 | 33.2 | 73.5 | 28.9 | 101.2 | 39.8 | 92.7 | 36.5 |
| 13. Vertical arm reach sitting | 128.6 | 50.6 | 117.4 | 46.2 | 147.8 | 58.2 | 139.4 | 54.9 |
| 14. Sitting height erect | 83.5 | 32.9 | 79.0 | 31.1 | 96.9 | 38.2 | 90.9 | 35.8 |
| 15. Sitting height relaxed | 81.5 | 32.1 | 77.5 | 30.5 | 94.8 | 37.3 | 89.7 | 35.3 |
| 16. Eye height sitting erect | 72.0 | 28.3 | 67.7 | 26.6 | 84.6 | 33.3 | 79.1 | 31.2 |
| 17. Eye height sitting relaxed | 70.0 | 27.6 | 96.2 | 26.1 | 82.5 | 32.5 | 77.9 | 30.7 |
| 18. Midshoulder height | 56.6 | 22.3 | 53.7 | 21.2 | 67.7 | 26.7 | 62.5 | 24.6 |
| 19. Shoulder height sitting | 54.2 | 21.3 | 48.9 | 19.6 | 65.4 | 25.7 | 60.3 | 23.7 |
| 20. Shoulder–elbow length | 33.3 | 13.1 | 30.6 | 12.1 | 40.2 | 15.8 | 36.6 | 14.4 |
| 21. Elbow–grip length | 31.7 | 12.5 | 29.6 | 11.6 | 36.3 | 15.1 | 35.4 | 14.0 |
| 22. Elbow–fingertip length | 43.8 | 17.3 | 40.0 | 15.7 | 52.0 | 20.5 | 47.5 | 18.7 |
| 23. Elbow rest height | 17.5 | 6.9 | 16.1 | 6.4 | 26.0 | 11.0 | 26.9 | 10.6 |
| 24. Thigh clearance height | — | — | 10.4 | 4.1 | — | — | 17.5 | 6.9 |
| 25. Knee height sitting | 49.7 | 19.6 | 46.9 | 18.5 | 60.2 | 23.7 | 55.5 | 21.8 |
| 26. Popliteal height | 39.7 | 15.6 | 36.0 | 15.0 | 50.0 | 19.7 | 45.7 | 18.0 |
| 27. Buttock–knee length | 54.9 | 21.6 | 53.1 | 20.9 | 65.8 | 25.9 | 63.2 | 24.9 |
| 28. Buttock–popliteal length | 45.5 | 17.9 | 43.4 | 17.1 | 54.5 | 21.5 | 52.6 | 20.7 |
| 29. Functional leg length | 110.6 | 43.5 | 96.6 | 38.2 | 127.7 | 50.3 | 118.6 | 46.7 |

*Source*: Adapted from DOD (1981).
[a]Bustpoint height for women.

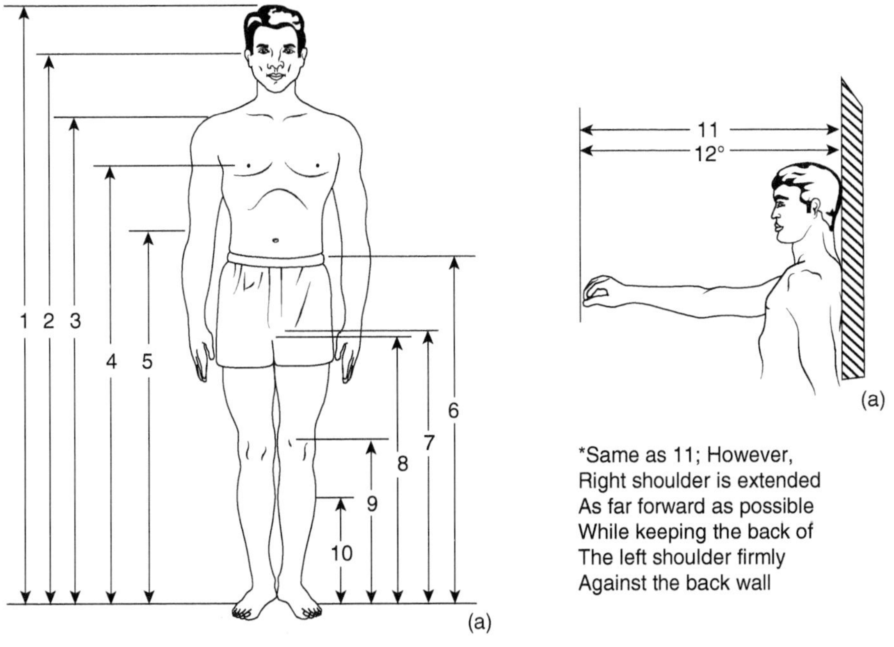
1 2 3
4 5
6
7
8
9
10
(a)
11
12°
(a)
*Same as 11; However,
Right shoulder is extended
As far forward as possible
While keeping the back of
The left shoulder firmly
Against the back wall

**Figure 3.5** Body dimensions and anthropometric data.

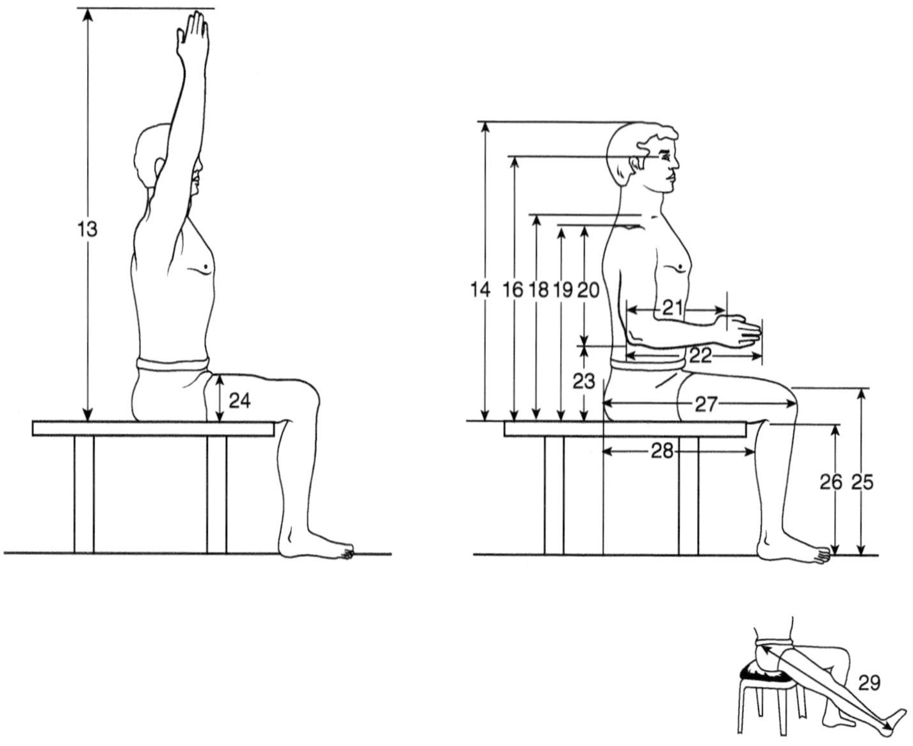


**Anthropmetric data for working positions[a]**

| Position | Fifth percentile Men cm | in. | Fifth percentile Women cm | in. | Ninety-fifth percentile Men cm | in. | Ninety-fifth percentile Women cm | in. |
|---|---|---|---|---|---|---|---|---|
| 1. Bent torso breadth | 40.9 | 16.1 | 36.8 | 14.5 | 48.3 | 19.0 | 43.5 | 17.1 |
| 2. Bent torso height | 125.6 | 49.4 | 112.7 | 44.4 | 149.8 | 59.0 | 138.6 | 54.6 |
| 3. Kneeling leg length | 63.9 | 25.2 | 50.2 | 23.3 | 75.5 | 29.7 | 70.5 | 27.8 |
| 4. Kneeling leg height | 121.9 | 48.0 | 114.5 | 45.1 | 136.9 | 53.9 | 130.3 | 51.3 |
| 5. Horizontal length, knees bent | 150.8 | 59.4 | 140.3 | 55.2 | 173.0 | 66.1 | 163.8 | 64.5 |
| 6. Bent knee height, supine | 44.7 | 17.6 | 41.3 | 16.3 | 53.5 | 21.1 | 49.6 | 19.5 |

[a]See Figure 3.4 for illustration.

**Figure 3.5** (*Continued*)

4. For frequency/duration analysis, use work–rest cycle data based on energy expenditure models that use oxygen consumption or heart rate data. Tables for these data can be found in textbooks (Astrand and Rodahl 1986; Eastman Kodak Co. 1996).
5. For visual demand/information processing and display analysis, use signal detection theory and guidelines such as those published by the American National Standards Institute (ANSI-HFES 100).
6. For vibration exposures, use NIOSH, ACGIH, or ISO guidelines on segmental and whole-body vibration. Guidelines for measuring continuous, intermittent, impulsive, and impact segmental vibration have been provided in the ISO standard ISO5349 (1986), entitled "Guide for the Measurement and the Assessment of Human Exposure to Hand Transmitted Vibration" and the ANSI standard ANSI S3.34-1986, entitled "Guide for the Measurement and Evaluation of Human Exposure to Vibration Transmitted to the Hand."

## 3.7 ASSESSING MANUAL HANDLING TASKS

Factors that need to be considered in evaluating the risk of manual handling tasks include the relationship between the task, the environment, the object handled, and the person (people) performing the task (Imada 1993; Waters et al. 1993).

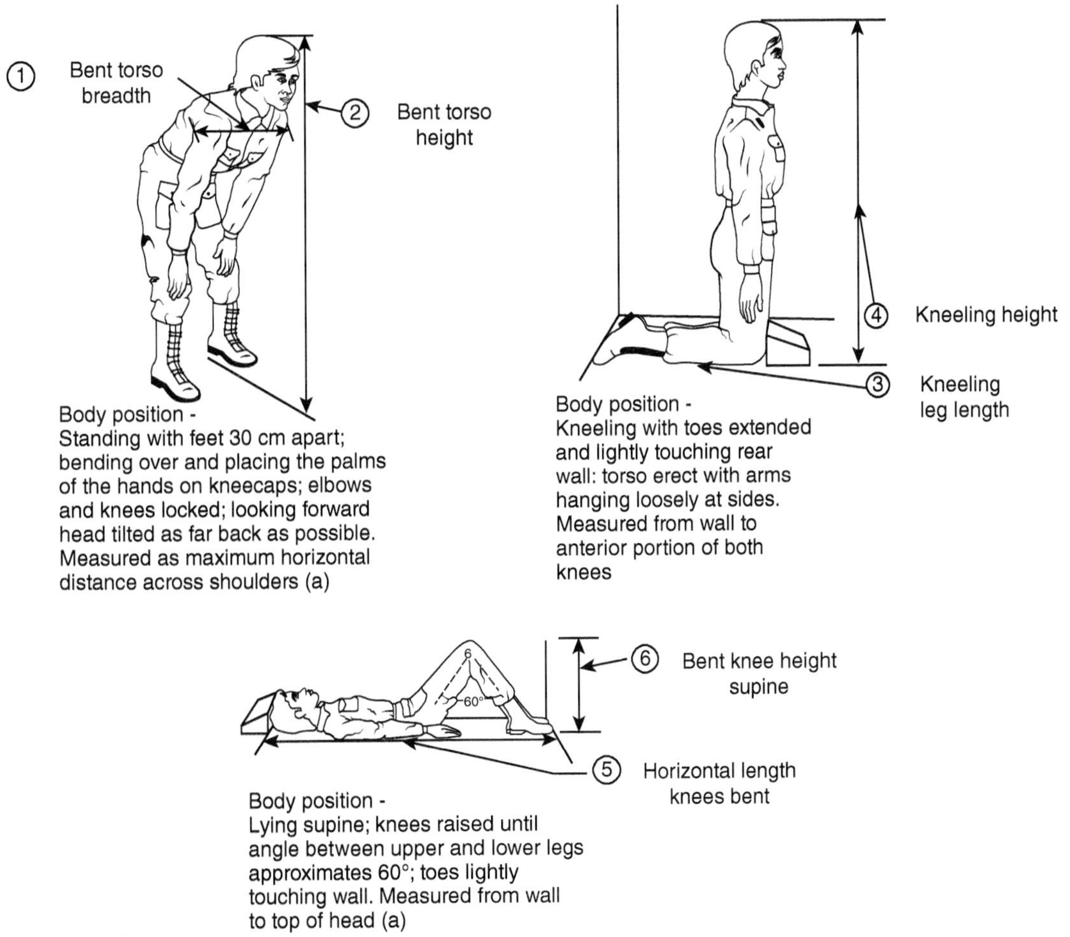


**Figure 3.5** *(Continued)*

| Considerations | Factors to consider |
|---|---|
| The tasks | Location/position of the object; frequency and duration of handling; precision, type of handling; velocity and acceleration; time constraints, pacing, incentives; work–rest cycle, shift work, job rotation; availability of assistance |
| Environment | Size and layout of workplace (obstructions); terrain; lighting/visibility/humidity/temperature; motion (vibration/transport) |
| Object handled | Nature (animate/inanimate); resistance to movement; size and shape; center of gravity; physical/chemical hazards; mechanical status; handling interface; information/instructions |
| Worker | Gender, age, strength, health status; physical status; motivation; skill/knowledge, perception; size and shape; handedness; protective equipment |

Evaluation techniques that integrate biomechanical, physiological, and psychophysical considerations to assess the appropriateness of job tasks are available. Probably the best known method for evaluating the demands of material handling tasks is the NIOSH lifting equation. The NIOSH lifting equation was first published in 1981 to assist safety and health practitioners evaluating sagittal plane lifting tasks (Habes and Putz-Anderson 1985; NIOSH 1981). The equation has recently been revised to reflect new research findings and provide methods for evaluating asymmetrical lifting tasks, lifts of objects with less than optimal hand–container couplings, and jobs with a larger range of work durations and lifting frequencies (Waters et al. 1993).

Using criteria from the fields of biomechanics, psychophysics, and work physiology, the equation defines a recommended weight limit (RWL) based on specific task parameters (e.g. the location of the load relative to the body and the floor, the distance the load is moved, the frequency of the lift). The RWL represents a load that nearly all healthy workers can lift over a substantial period of time without placing an excessive load on the back, causing excess fatigue or otherwise increasing the risk of low back pain. The actual weight of lift can be compared with the RWL for a given task to derive an estimate of the risk presented by the task and to determine if measures to reduce the risk of injury to workers are needed.

NIOSH (Waters et al. 1993) has recommended no more than 51 pounds be lifted in a "perfect" lift. The perfect lift includes using two hands, holding the load close to the body, the load having good handles, the load is located between knuckle and shoulder height, and no awkward postures or obstructions. This load limit is reduced:

- The further the object is from the body.
- The distance the object has to be lifted (distance).
- The lower or higher the lift is from knuckle to shoulder.
- The more frequent the lift.
- The more difficult it is to handle (coupling).
- The more twisted the torso is during the lift.

An example of the NIOSH lifting equation and its application is presented in Appendix A.

### 3.7.1 Strength Data

Strength data for different populations and muscle groups have been published in a number of sources (Kamon and Goldfuss 1978; Mathiowetz et al. 1985; Ulin and Armstrong 1992). If the force requirements of a task are known, it may be possible to compare these requirements against existing strength data to estimate the percentage of the worker population for which the job may be difficult. Successfully applying strength data requires knowledge of the subject population upon which the data are based, the posture in which the measurements were made, whether the measurement was static or dynamic, and how long the effort was sustained (Keyserling et al. 1991).

Computerized biomechanical models have been developed to predict the percentage of males and females capable of exerting static forces in certain postures. The advantage of these models is that most recognize that a worker's capacity for force exertion is rarely dependent on the strength of a single muscle group. Rather, the capacity for exerting force is dependent on the moment created at each joint by the external load and the muscle strength at that joint. The models compute the moment created at each joint by an exertion and compare the moments against static strength data to estimate the percent of the population capable of performing a specific exertion for each joint and muscle function (Ulin et al. 1990). Currently, both two- and three-dimensional models are available for these analyses (Chaffin 1992).

### 3.7.2 Design Criteria: Anthropometry

Anthropometric data on body size and range of joint motion can be used to assess the appropriateness of workplace, equipment, and product designs relative to the capacities for reach, grasp, and clearance of the workforce. Compilations of anthropometric data for different populations are available from numerous sources (Eastman Kodak Co. 1983; NASA 1978). An example of these data can be seen in Figure 3.5.

### 3.7.3 Physiological Data

In activities such as repetitive lifting and load carrying, large muscle groups perform submaximal, dynamic contractions. During these activities, a worker's endurance is primarily limited by the capacity of the oxygen transporting and utilization systems (maximum aerobic power) (Astrand and Rodahl 1986). Data on the maximum aerobic capacities of working populations and the energy demands of common industrial tasks have been compiled by a number of sources (Astrand and Rodahl 1986; Eastman Kodak Co. 1986; NIOSH 1981). In general, maximum aerobic capacity generally declines with age, increases with physical fitness level, and is 13–30% lower for women compared with men (Durnin and Passmore 1967). Several researchers suggest that the maximum energy expenditure rate for an 8-hour workday should not exceed 33% of maximum aerobic power (Eastman Kodak Co. 1986). Limits of 5.2 kcal $min^{-1}$ for average healthy young males, or 3.5 kcal $min^{-1}$ for populations containing women and older workers, have been proposed based on this recommendation (Bink 1962; Bonjer 1962; Waters et al. 1993).

Because table values provide only a rough approximation of the metabolic costs of a given job, models to predict metabolic energy expenditure for simple tasks based on a combination of personal and task variables have also been developed (Givoni and Goldman 1971; VanderWalt and Wyndham 1973). It has been demonstrated that the energy expenditure rate of complex jobs can be predicted if the energy expenditure rates of the simple tasks that comprise the job and the time duration of the job are known. Comparison of measured and model-predicted rates for 48 tasks indicated that models can account for up to 90.8% of the variation in measured metabolic rates (Garg et al. 1978).

### 3.7.4 Psychophysical Data

It is difficult to apply strength data to dynamic tasks involving more than one muscle group. In addition, motivational factors play an important role in determining an individual's capacity for physical work. This has resulted in a "Gestalt" method of analysis of human capability, called "psychophysics," to develop guidelines for the evaluation and modification of repetitive work tasks (Gamberale 1990; Putz-Anderson and Galinsky 1993; Snook and Irvine 1969). Psychophysical limits are generally based on data derived from laboratory simulations of a specific task in which the participants are allowed to adjust their workload to a level subjectively defined as the maximum acceptable. Limiting workload in this manner should allow workers to perform work tasks without overexertion or excessive fatigue (Snook 1978).

Although there is little data to indicate how psychophysically derived limits relate to the risk of injury during work, many researchers believe that use of these limits may be the most accurate method of determining if a given task is acceptable (Chaffin and Andersson 1991). Psychophysical limits for various lifting, pushing, and pulling tasks and other manual operations have been developed and are widely available for application (Snook and Ciriello 1991; Snook et al. 1996a, b).

An example of two psychophysical scales that have been used to rate the effort required to do a job is presented in Figure 3.4. This is the Borg scale. The scale on the left has been used to estimate whole-body exertion, since it has been validated against heart rate. The scale has been used to represent a rating of "7" as the baseline heart rate of approximately 70 beats per minute, and a rating of "20" has been proposed as the maximum heart rate of 200 beats per minute. The Borg scale on the right has been used to estimate maximum voluntary contraction (MVC) of large muscle groups. A rating of 1 corresponds to approximately 10% of MVC. In general, subjects tend to be more accurate in their ability to approximate their heart rate and MVC at the tail ends of the scale. However, these scales are excellent tools for looking at before and after results of the implementation of controls.

## 3.8 IDENTIFY POSSIBLE SOLUTIONS

Ultimately, the results of a job analysis should provide facts that allow the ergonomics team to recommend controls to eliminate or reduce identified work factors. This may be achieved through the modification of equipment and tools, workstations, or work methods that contribute to excessive work demands. In all cases, the best ergonomic solutions are those in which safe work is a natural result of the job design and are independent of specific worker capabilities or work techniques. However, in situations where design changes are infeasible, it may be possible to limit exposure to work factors by the identification of administrative controls, such as rest breaks and worker rotation. In some situations, reducing work factors may require a combination of engineering or administrative controls.

## 3.9 SELECTION AND EVALUATION OF SOLUTIONS

When possible, the proposed solution should originate from a team that includes the workers who will be affected by the solution and by management who will be responsible for providing funds to purchase new equipment (if needed) or to approve of changes to work practices. All solutions should be tested among a small group of workers, to allow adjustments to be made before widespread changes are implemented. A common flaw seen in industry is that large capital investments dare often made, usually due to artifacts such as fiscal year budget deadlines and purchase order minimums, before the solutions have been tested. Once changes are implemented, follow-up job analyses should be performed to ensure that the solutions have effectively reduced the work factors without imposing new demands on the worker.

Beyond the specific controls, a multifaceted abatement program is often necessary to properly control hazards leading to musculoskeletal disorders and to proactively protect employees. Some elements of an ergonomic solution process to bring the work system into balance are summarized below and were introduced early in this chapter as NIOSH recommendations.

### 3.9.1 Analysis

If in-house expertise does not exist, a consultant should assist an in-house team to perform a systematic evaluation with regard to existing and new work practices and workstation design. The consultant should work with the ergonomic team to recommend engineering and administrative controls to reduce or eliminate ergonomic stressors. The consultant can also assist in the implementation of the recommendations, the evaluation of the effectiveness of the controls implemented, and can make new recommendations if necessary.

### 3.9.2 Controls

Hazards associated with the development of MSDs can be controlled through proper engineering design of the job, workstation, and equipment, so the work can be performed independent of specific worker characteristics and techniques. *This requires the job to be designed to fit the worker and not* vice versa. Engineering controls attempt to reduce extreme postures, excessive forces, and repetitive motions. To be effective, employee input is necessary, since improperly designed workstations and controls will not be used if employees believe they interfere with their work. Also, after installation, the effectiveness of the controls must be evaluated and modified if necessary to ensure their effectiveness. Appendix C contains a list of examples and references from NIOSH that demonstrate the effectiveness of engineering controls for reducing exposures to ergonomic risk factors.

Administrative controls can include but are not limited to training of new employees in safe work techniques including lifting, working with minimum strain on the body, and minimizing the application of forces with the fingers, job rotation and job enhancement, adequate mandatory rest breaks, and implementation of an exercise program.

### 3.9.3 Medical Management Program

A medical management program is necessary to monitor employees and prevent early symptoms from progressing to injuries and illnesses. This program should include:

1. Determining the extent of injuries and illnesses; determining if injuries and illnesses are caused or aggravated by work.
2. Educating all employees and supervisors on early signs of injuries and disorders and encouraging early reporting.
3. Instituting a formal documented tracking and surveillance program to monitor injury trends in the plant.
4. Providing adequate treatment of ergonomic-related cases (including not reassigning employees to a job until it has been modified to minimize the hazards that resulted in the injury).
5. Allowing adequate time off for recovery after surgery or other aggressive intervention.
6. Preventive measures, including early physical evaluation of employees with musculoskeletal symptoms.
7. Allowing adequate time off after a cumulative trauma disorder is diagnosed.
8. Providing access to trained medical personnel for development and implementation of conservative treatment measures upon detection of cumulative trauma disorder symptoms.

### 3.9.4 Training Programs

A training program is necessary to alert employees on the hazards of cumulative trauma disorders and controls and work practices that can be used to minimize the hazards. This includes designing and implementing a written training program for managers, supervisors, engineers, union representatives, health professionals, and employees on the nature, range, and causes and means of prevention of ergonomic-related disorders. The training program for new and reassigned workers should allow the following:

1. Demonstrations of safe and effective methods of performing their job.
2. Familiarization of employees with applicable safety procedures and equipment.
3. The new or reassigned employee to work with a skilled employee and/or provide on-the-job training for specific jobs.
4. New or reassigned employees to condition their muscle/tendon groups prior to working at full capacity rate, which has been determined to be safe and will not cause adverse effects.

Workers should be instructed in the basics of body biomechanics and work practices to minimize the ergonomic hazards associated with their jobs. This should include, but not be limited to:

Avoid postures where:

- The elbow is above midtorso.
- The hand is above the shoulder.
- The arms must reach behind the torso.

Avoid wrist postures where there is:

- Inward or outward rotation with bent wrist.
- Excessive palmar flexion or extension.
- Ulnar or radial deviation.
- Pinching or high finger forces with above postures.

Avoid mechanical stress concentrations on the elbows, base of palm, and backs of fingers.
General lifting guidelines include:

- Keep the load close to the body.
- Use the most comfortable posture.
- Lift slowly and evenly (do not jerk).
- Do not twist the back.
- Securely grip the load
- Use a lifting aid or get help.

## 3.10 AN EXAMPLE OF JOB ANALYSIS FORMS

The forms in Figures 3.6 and 3.7 provide an outline for how the methods and information discussed above can be used to identify work factors present in jobs in the work environment so that system balance can be restored. The implementation plan is especially useful for planning which solutions will be implemented and when. Appendix C provides a list of selected studies from NIOSH on the various control strategies for reducing MSDs and discomfort.

| Implementation plan | | | Date initiated |
|---|---|---|---|
| Job name | Department | | Tracking no. |
| Process | | | |
| Number of employees | 1st shift | 2nd shift | 3rd shift |
| Musculosskeletal disorders in last year<br>Neck ___ Wrist ___ Back ___ Ankle ___ Shoulder ___ Hand ___<br>Hip ___ Elbow ___ Fingers ___ Knee ___ Other ___ | | | |
| Action | Responsibility | Date assigned | Date completed |
| Checklist score<br>Upper Lower | | | |
| Discomfort survey | | | |
| Employee input<br>No Yes (attach) | | | |
| Job analysis | | | |
| Videotape | | | |
| Employee input | | | |

| Risk factor | Cause | Solution (controls) | Status | Responsible | Finish date |
|---|---|---|---|---|---|
| A | | | | | |
| B | | | | | |
| C | | | | | |
| D | | | | | |
| E | | | | | |
| F | | | | | |
| G | | | | | |
| H | | | | | |
| I | | | | | |

**Figure 3.6** Sample forms for job analysis summary.

<table>
<tr><td colspan="4">Implementation plan</td><td colspan="2">Date initiated</td></tr>
<tr><td colspan="6">Job name      Department      Tracking no.</td></tr>
<tr><td colspan="6">Process</td></tr>
<tr><td colspan="6">Number of employees    1st shift    2nd shift    3rd shift</td></tr>
<tr><td colspan="6">Musculosskeletal disorders in last year<br>Neck ___ Wrist ___ Back ___ Ankle ___ Shoulder ___ Hand ___<br>Hip ___ Elbow ___ Fingers ___ Knee ___ Other ___</td></tr>
<tr><td>Action</td><td>Responsibility</td><td>Date assigned</td><td colspan="3">Date completed</td></tr>
<tr><td>Checklist score<br>Upper Lower</td><td></td><td></td><td colspan="3"></td></tr>
<tr><td>Discomfort survey</td><td></td><td></td><td colspan="3"></td></tr>
<tr><td>Employee input<br>No Yes (attach)</td><td></td><td></td><td colspan="3"></td></tr>
<tr><td>Job analysis</td><td></td><td></td><td colspan="3"></td></tr>
<tr><td>Videotape</td><td></td><td></td><td colspan="3"></td></tr>
<tr><td>Employee input</td><td></td><td></td><td colspan="3"></td></tr>
</table>

| Risk factor | Cause | Solution (controls) | Status | Responsible | Finish date |
|---|---|---|---|---|---|
| A | | | | | |
| B | | | | | |
| C | | | | | |
| D | | | | | |
| E | | | | | |
| F | | | | | |
| G | | | | | |
| H | | | | | |
| I | | | | | |

**Figure 3.6** (*Continued*)

Job name: Sew uppers
Task: Sew elastic and liner together for shoe upper
Cycle time: 20 seconds
Task elements: The task consists of the following six elements.

1. Get unit from the stack (left hand, wrist flexed).
2. Place unit over sewing mount (both hands, pinch grip 1 in 10 requires more than 2 lb).
3. Push through sewing machine (both hands, finger press, low force).
4. Turn knob to lift needle (left hand, pinch grip less than 2 lb, wrist extended, forearm rotation).
5. Cut thread with scissors (right hand, wrist flexed, fingers contact hand surface).
6. Discard unit into finished bin (left hand, shoulder extended).

---

Job name: Order picker
Task: Fill order
Cycle Time: 15 seconds
Task elements: The task consists of the following six elements.

1. Remove backing strip from label (right hand).
2. Retrieve stock unit from shelf (either hand).
3. Place stock unit on cart shelf (either hand).
4. Place label on stock unit (right hand).
5. Push stock unit into cart (left hand).
6. Push cart to next pick (both hands).

---

**Figure 3.7** Examples of job, tasks, work cycle, and elements.

## 3.11 SUMMARY

The information in this chapter has provided a basic framework, with examples, for ergonomic analysis and the control of poorly designed work environments. Since ergonomics is a dynamic science, the reader is encouraged to consult the resource list below to maintain a working library of information specific to the exposures of interest. In addition, the references provided by NIOSH in Appendix C provide an excellent source of networking opportunities and examples of companies that have successfully worked toward healthy work systems. Perhaps the most important "take-home message" that should be gained from this chapter is that management and employees need to work together in a concerted effort to control workplace exposures through a continuous process. It is hoped that the information provided in this chapter will facilitate that process.

# APPENDIX A

## NIOSH LIFT GUIDELINES

***Example:*** Loading Punch Press Stock

- Requires control at the destination of lift.
- Duration <1 hour.
- Weight of the supply reel = 44 lb.

| Origin | Destination |
|---|---|
| $H = 23''$ | $23''$ |
| $V = 15''$ | $64''$ |
| $D = 49''$ | $49''$ |
| $A = 0$ | 0 |
| $C$ = Fair | Fair |
| $F$ = <0.2 | <0.2 |

where:

$H$ = horizontal distance of hands from midpoint between the angles. Measure at the origin and the destination of the lift (cm or in.).
$V$ = vertical distance of the hands from the floor. Measure at the origin and destination of the lift (cm or in.).
$D$ = vertical travel distance between the origin and the destination of the lift (cm or in.).
$A$ = angle of asymmetry – angular displacement of the load from the sagittal plane. Measure at the origin and destination of the lift (degrees).
$F$ = average frequency rate of lifting measured in lifts per minute. Duration is defined to be ≤1 hour, ≤2 hours, or ≤8 hours assuming appropriate recovery allowances.

### A.1 Origin

$$
\begin{aligned}
\text{HM} &= (10/23) = 0.43 \\
\text{VM} &= (1 - 0.0075|15 - 30|) = 0.89 \\
\text{DM} &= (0.82 + 1.8/49) = 0.86 \\
\text{AM} &= (1 - 0.0032 \times 0) = 1 \\
\text{CM} &= 0.95 \\
\text{FM} &= 1 \\
\text{RWL} &= 51 \times \text{HM} \times \text{FM} \times \text{DM} \times \text{AM} \times \text{CM} \times \text{FM} \\
&= 51 \times 0.43 \times 0.89 \times 0.86 \times 1 \times 0.95 \times 1 \\
&= 15.9\,\text{lb}
\end{aligned}
$$

## A.2 Destination

$$
\begin{aligned}
\mathrm{HM} &= (10/23) = 0.43 \\
\mathrm{VM} &= (1 - 0.0075|64 - 30|) = 0.75 \\
\mathrm{DM} &= (0.82 + 1.8/49) = 0.86 \\
\mathrm{AM} &= (1 - 0.0032 \times 0) = 1 \\
\mathrm{CM} &= 1 \\
\mathrm{FM} &= 1 \\
\mathrm{RWL} &= 51 \times \mathrm{HM} \times \mathrm{VM} \times \mathrm{DM} \times \mathrm{AM} \times \mathrm{CM} \times \mathrm{FM} \\
&= 51 \times 0.43 \times 0.75 \times 0.86 \times 1 \times 1 \times 1 \\
&= 14.1 \text{ lb}
\end{aligned}
$$

$$
\begin{aligned}
\mathrm{RWL} &= \text{lower of } 15.9 \text{ and } 14.1 \text{ lb} \\
&= 14.1 \text{ lb} \\
\mathrm{LI} &= \frac{44}{14.1} \\
&= 3.12
\end{aligned}
$$

AL = 15.8 lb based on $H = 21''$

## A.3 Recommendations

- Engineering controls: use a mechanical device to load supply reel (hoist, small crane, etc.).
- Administrative controls: use two workers and load the reel from the side.
- Education and training of workers.
- Enforcement.

## A.4 Calculation for Recommended Weight Limit

$$\mathrm{RWL} = \mathrm{LC} \times \mathrm{HM} \times \mathrm{VM} \times \mathrm{DM} \times \mathrm{AM} \times \mathrm{FM} \times \mathrm{CM}$$

Recommended weight limit

| Component | | Metric | US customary |
|---|---|---|---|
| LC = load constant | = | 23 kg | 51 lb |
| HM = horizontal multiplier | = | $(25/H)$ | $(10/H)$ |
| VM = vertical multiplier | = | $(1-(0.003\lvert V-75\rvert))$ | $(1-(0.0075\lvert V-30\rvert))$ |
| DM = distance multiplier | = | $(0.82+(4.5/D))$ | $(0.82+(1.8/D))$ |
| AM = asymmetric multiplier | = | $(1-(0.0032A))$ | $(1-(0.0032A))$ |
| FM = frequency multiplier | | | |
| CM = coupling multiplier (from Table 6) | | | |

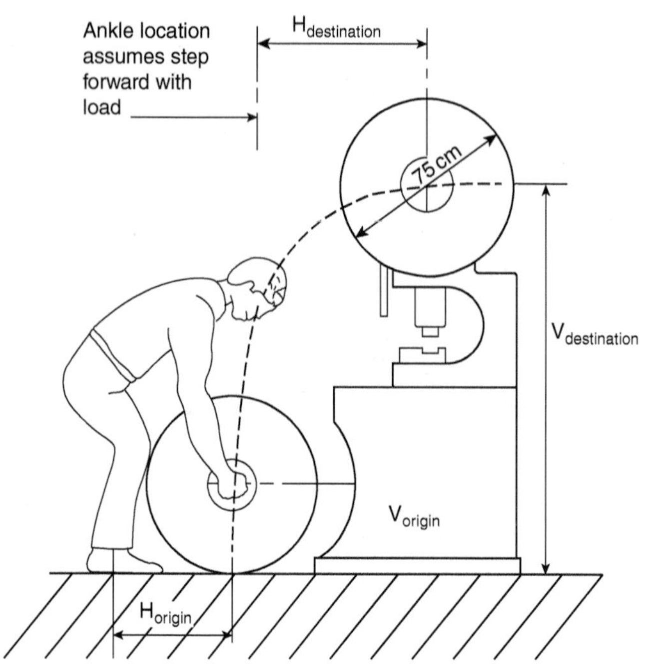


**Revised NIOSH Guide for Manual Lifting analysis of Entire Job**

Analyst: Arun Garg

Job: Loader in Receiving Area

Frequency Weighted RWL (FWRWL) 42.5 lbs.

FW Average Weight (FWW) 13.9 lbs.

Composite Recommended Weight Limit (CRWL) 7.6 lbs.

Composite Lifting Index (CLI) 1.83

Recommended Rest Allowances 0 Minutes

Estimated percent capable population excluding compressive force.

Male = 86% Capable

Female = 28% Capable

Help

Print Screen

Print

Edit Job

| Task | | Frequency Independent (Origin) | Using Task Frequency (Origin) | Frequency Independent (Dest) | Using Task Frequency (Dest) |
|---|---|---|---|---|---|
| Product A | RWL | 22.3 | 16.7 | | |
| | LI | 1.48 | 0.96 | | |
| Product B | RWL | 31.4 | 20.4 | | |
| | LI | 1.40 | 0.98 | | |
| Product C | RWL | 51.0 | 17.9 | | |
| | LI | 0.43 | 0.61 | | |

**System Requirements**

- IBM 80286 or Better Compatible Computer
- 5.25" or 3.5" Floppy Disk Drive
- 1 MB of Ram
- VGA Color Monitor
- High Quality Dot-Matrix or Laser Printer
- Microsoft Windows 3.x
- Mouse

**Four Different Windows**

- Analyst, Job, Duration, and Task List Windows
- Individual Task Data Window
- Simple Task Analysis Window
- Multiple Task (Entire Job) Analysis Window

**Figure 3.A.1** Example of a computer model to analyze lifting exposures using NIOSH lifting formula. *Source:* Courtesy of Dr. Arun Garg.

# APPENDIX B

## ASSESSMENT OF AND SOLUTIONS TO WORKSITE RISK FACTORS

### B.1 Worksite Assessment Checklists

The following are sample checklists that you may wish to use as a guide in developing your own worksite assessment checklist. These checklists cover:

- Manual handling.
- Task/work methods.
- Video display unit and keyboard issues.
- Workstation layout.
- Hand tool use.

Note: You should consider modifying these checklists to include the specific concerns of your site. Other areas to assess may include:

- Information displays and controls (such as in control rooms).
- Environmental concerns (such as lighting, noise, or heat).
- Facility concerns (such as floor condition, ventilation, or vibration).

#### B.1.1 Manual Handling Checklist

| Yes | No | |
|---|---|---|
| ☐ | ○ | 1. Is material moved over a minimum distance? |
| ☐ | ○ | 2. Is the horizontal distance between the middle knuckle and the body less than 4 in. during manual material handling? |
| ☐ | ○ | 3. Are obstacles removed to minimize reaches? |
| ☐ | ○ | 4. Are walking surfaces level, slip resistant, and well lit? |
| ☐ | ○ | 5. Are objects: |
| ☐ | ○ | • Able to be grasped by good handholds? |
| ☐ | ○ | • Stable? |
| ☐ | ○ | • Able to be held without slipping? |
| ☐ | ○ | 6. When required, do gloves improve the grasp without bunching up or resisting movement of the hands? |
| ☐ | ○ | 7. Is there enough room to access and move objects? |
| ☐ | ○ | 8. Are mechanical aids easily available and used whenever possible? |
| ☐ | ○ | 9. Are working surfaces adjustable to the best handling heights? |
| ☐ | ○ | 10. Does material handling avoid: |
| ☐ | ○ | • Movements below knuckle height and above shoulder height? |
| ☐ | ○ | • Static awkward postures? |
| ☐ | ○ | • Sudden movements during handling? |
| ☐ | ○ | • Twisting of the trunk? |
| ☐ | ○ | • Excessive reaching while holding/moving the load? |
| ☐ | ○ | 11. Is help available for heavy or awkward lifts? |
| ☐ | ○ | 12. Are high rates of repetition avoided by: |
| ☐ | ○ | • Job rotation? |
| ☐ | ○ | • Self-pacing? |
| ☐ | ○ | • Sufficient rest pauses? |
| ☐ | ○ | 13. Are pushing/pulling forces reduced/eliminated by: |
| ☐ | ○ | • Casters that are sized correctly and roll freely? |
| ☐ | ○ | • Handles for pushing/pulling? |
| ☐ | ○ | • Availability of mechanical assists? |
| ☐ | ○ | 14. Are objects rarely carried more than 10 ft? |
| ☐ | ○ | 15. Is there a prevention maintenance program for manual handling equipment? |

### B.1.2 Task/Work Methods Checklist

| Yes | No | |
|---|---|---|
| ☐ | ○ | 1. Does the design of the task reduce or eliminate: |
| ☐ | ○ | • Bending or twisting of the trunk? |
| ☐ | ○ | • Squatting or kneeling? |
| ☐ | ○ | • Elbows above midtorso? |
| ☐ | ○ | • Extending the arms? |
| ☐ | ○ | • Bending the wrist? |
| ☐ | ○ | • Static muscle loading? |
| ☐ | ○ | • Forceful pinch grips? |
| ☐ | ○ | 2. Are mechanical devices used to lift or move objects that are heavy or require repetitive lifting? |
| ☐ | ○ | 3. Can the task be performed with either hand? |
| ☐ | ○ | 4. Are the materials: |
| ☐ | ○ | • Able to be held without a forceful grip? |
| ☐ | ○ | • Easy to grasp? |
| ☐ | ○ | • Free from sharp edges? |
| ☐ | ○ | 5. Do containers have good handholds? |
| ☐ | ○ | 6. Are fixtures used to reduce or eliminate hard grasping forces (e.g. part held by the fixture, not by the hands)? |
| ☐ | ○ | 7. If gloves are needed, do they fit properly? |
| ☐ | ○ | 8. Does the task avoid contact with sharp edges or corners? |
| ☐ | ○ | 9. Is exposure to repetitive motions reduced by: |
| ☐ | ○ | • Job rotation? |
| ☐ | ○ | • Self-pacing? |
| ☐ | ○ | • Sufficient rest pauses? |

### B.1.3 Video Display Unit (VDU) and Keyboard Issues Checklist

| Yes | No | |
|---|---|---|
| ☐ | ○ | 1. Can the workstation be adjusted to ensure proper posture by: |
| ☐ | ○ | • Adjusting knee and hip angles to achieve comfort and variability? |
| ☐ | ○ | • Supporting heels and toes on the floor or a footrest? |
| ☐ | ○ | • Placing arms comfortably at the side and hands parallel to the floor (plus or minus 2 in.)? |
| ☐ | ○ | • Holding wrists nearly straight and resting them on a padded surface? (Note: wrists should not rest on the padded surface while keying) |
| ☐ | ○ | 2. Does the chair: |
| ☐ | ○ | • Adjust easily from the seated position? |
| ☐ | ○ | • Have a padded seat pan (soft but compresses about 1 in.)? |
| ☐ | ○ | • Have a seat that accommodates the worker? |
| ☐ | ○ | • Have a back rest that provides lumbar support and can be used while working? |
| ☐ | ○ | • Have a stable base with casters that are suited to the type of flooring? |
| ☐ | ○ | 3. Does the chair manufacturer offer different seat pan lengths that have a waterfall design? |
| ☐ | ○ | 4. Does the seat pan adjust for both height and angle? |
| ☐ | ○ | 5. Is there adequate clearance for the feet, knees, and legs relative to the edge of the work surface? |
| ☐ | ○ | 6. Is there sufficient space for the thighs between the work surface and the seat? |
| ☐ | ○ | 7. Are the keyboard height from the floor and the slope of the keyboard surface adjustable? |
| ☐ | ○ | 8. Is the keyboard prevented from slipping when in use? |
| ☐ | ○ | 9. Is the keyboard detachable? |
| ☐ | ○ | 10. Does the keyboard meet ANSI/HFS 100-1988 (or ISO 9241) standards? |
| ☐ | ○ | 11. Is the mouse, pointing device, or calculator at the same level as the keyboard? |
| ☐ | ○ | 12. Are the head and neck held in a neutral posture? |
| ☐ | ○ | 13. Are arm rests provided for intensive or long-duration keying jobs? |
| ☐ | ○ | 14. Is the screen clean and free from flickering? |
| ☐ | ○ | 15. Is the top of the screen slightly below eye level? (For non-bifocal users) |
| ☐ | ○ | 16. Can the screen swivel horizontally and tilt or elevate vertically? |

| Yes | No | |
|---|---|---|
| ☐ | ○ | 17. Does the monitor have brightness and contrast controls? |
| ☐ | ○ | 18. Is the monitor between 18 and 30 in. from the worker? (Some workers may need glasses specifically for computer use.) |
| ☐ | ○ | 19. Is there sufficient lighting without glare on the screen from lights, windows, and surfaces? |
| ☐ | ○ | 20. Are headsets used when frequent telephone work is combined with hand tasks such as typing, use of a calculator, or writing? |
| ☐ | ○ | 21. Is the job organized so that workers can change postures frequently? |
| ☐ | ○ | 22. Does the worker leave the workstation for at least 10 min after every hour of intensive keying and for a least 15 min after every 2 h of intermittent keying? |
| ☐ | ○ | 23. Is intensive keying avoided by: |
| ☐ | ○ | • Job rotation? |
| ☐ | ○ | • Self-pacing? |
| ☐ | ○ | • Job enlargement? |
| ☐ | ○ | • Adequate recovery breaks? |
| ☐ | ○ | 24. Is there the possibility of alternating tasks during the shift (e.g. intensive keying or mouse work, filing, copying telephone calls, intermittent keying)? |
| ☐ | ○ | 25. Are employees trained in: |
| ☐ | ○ | • Healthy work postures? |
| ☐ | ○ | • Safe and healthy work methods? |
| ☐ | ○ | • How to make adjustments to the workstation? |
| ☐ | ○ | • Awareness of risk factors for musculoskeletal disorders? |
| ☐ | ○ | • How to seek assistance with concerns? |
| ☐ | ○ | 26. Are workers able to set their own pace, without electronic monitoring or incentive pay? |

### B.1.4 Workstation Layout Checklist

| Yes | No | |
|---|---|---|
| ☐ | ○ | 1. Is the workstation designed to reduce or eliminate: |
| ☐ | ○ | • Bending or twisting of the trunk? |
| ☐ | ○ | • Squatting or kneeling? |
| ☐ | ○ | • Elbows above midtorso? |
| ☐ | ○ | • Extending the arms? |
| ☐ | ○ | • Bending the wrist? |
| ☐ | ○ | • Hands behind the body? |
| ☐ | ○ | 2. Are mechanical aids and equipment available? |
| ☐ | ○ | 3. Can the work be performed without repetitive (or static) bending or reaching? |
| ☐ | ○ | 4. Are awkward postures reduced by: |
| ☐ | ○ | • Providing adjustable work surfaces and supports (such as chairs or fixtures)? |
| ☐ | ○ | • Tilting the surface? |
| ☐ | ○ | 5. Are all job requirements visible without awkward postures? |
| ☐ | ○ | 6. Is an arm rest provided for precision work? |
| ☐ | ○ | 7. Is a foot rest provided for those who need it? |
| ☐ | ○ | 8. Are cushioned floor mats and foot rests provided for employees who are required to stand for long periods? |
| ☐ | ○ | 9. Have jobs been reviewed to determine if they are best suited for sit/stand, seated, or standing work? |
| ☐ | ○ | 10. Are seated workers provided chairs that: |
| ☐ | ○ | • Adjust easily from the seated position? |
| ☐ | ○ | • Have a padded seat pan (soft but that compresses no more than 1 in.)? |
| ☐ | ○ | • Have a seat that is wide enough to accommodate the worker? |
| ☐ | ○ | • Have a back rest that provides lumbar support and can be used while working? |
| ☐ | ○ | • Have a stable base with casters that are suited to the type of flooring? |
| ☐ | ○ | 11. Do chair manufacturers offer different seat pan lengths that have a waterfall design? |
| ☐ | ○ | 12. Does the seat pan adjust for both height and angle? |
| ☐ | ○ | 13. Are body parts (e.g. hands, arms, legs) free from contact stress from sharp edges on work surfaces? |
| ☐ | ○ | 14. Is there a preventive maintenance program for mechanical aids, tools, and other equipment? |

### B.1.5 Hand Tool Use Checklist

| Yes | No | |
|---|---|---|
| ☐ | ○ | 1. Are tools selected to: |
| ☐ | ○ | • Minimize exposure to localized vibration? |
| ☐ | ○ | • Reduce hand force? |
| ☐ | ○ | • Reduce/eliminate bending or awkward postures of the wrist? |
| ☐ | ○ | • Avoid forceful pinch grips? |
| ☐ | ○ | • Avoid the use of the hand as a hammer? |
| ☐ | ○ | 2. Are tools powered where necessary (e.g. to reduce forces, repetitive motions)? |
| ☐ | ○ | 3. Are tools evenly balanced? |
| ☐ | ○ | 4. Are heavy tools suspended or counterbalanced? |
| ☐ | ○ | 5. Does the tool allow adequate view of the work? |
| ☐ | ○ | 6. Does the tool grip/handle prevent slipping during use? |
| ☐ | ○ | 7. Are tools equipped with handles that: |
| ☐ | ○ | • Do not press into the palm area? |
| ☐ | ○ | • Are made of textured, nonconductive material? |
| ☐ | ○ | • Have a grip diameter suitable for most workers, or are different size handles available? (Women tend to have smaller hands than men) |
| ☐ | ○ | 8. Can the tool be used safely with gloves? |
| ☐ | ○ | 9. Can the tool be used with either hand? |
| ☐ | ○ | 10. Is there a preventive maintenance program to keep tools operating as designed? |
| ☐ | ○ | 11. Can triggers be operated by more than one finger to avoid static contractions? |
| ☐ | ○ | 12. Does the tool design or workstation minimize the twist or shock to the hand (in particular, observe the reaction of power tools after the torque limit is reached)? |

Revised NIOSH Guide for Manual Lifting
Analysis of Individual Task

| | | | | |
|---|---|---|---|---|
| Analyst: | | Object Weight: | Average: | 20 lbs. |
| JOB: | Loader in Receiving Area | | Maximum: | 44 lbs. |
| Task: | Product B | | | |

**ORIGIN** DESTINATION

| | |
|---|---|
| Horizontal Multiplier. HM | 0.83 |
| Vertical Multiplier. VM | 0.78 |
| Distance Multiplier. DM | 1.00 |
| Asymmetric Multiplier. AM | 1.00 |
| Frequency Multiplier. FM | 0.65 |
| Coupling Multiplier. CM | 0.95 |

Help | Print Screen | Print | Edit Task | Edit Job | Calculate Job

| | |
|---|---|
| RWL | 20.4 lbs. |
| Lifting Index.LI | 0.98 |
| Rest Allowance | 0 Minutes |

Estimated percent capable population excluding compressive force.
Male = More than 99% Capable
Female = 76% Capable

NOTE: RWL and LI are based on the Average Weight, Maximum Weight was ignored.

Individual Task Data

| | |
|---|---|
| Analyst: | Arun Garg |
| JOB: | Loader in Receiving Area |
| Task: | Product B |
| Duration: | 8 Hour (s) 0 Minute(s) |

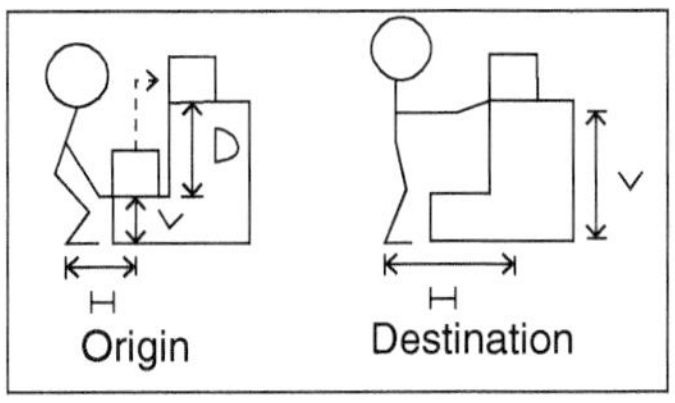


| | | |
|---|---|---|
| Average Weight | 20 lbs. | |
| Maximum Weight | 44 lbs. | |
| | ORIGIN | ☐ DESTINATION |
| Horizontal Location. H | 12 in. | in. |
| Vertical Location. V | 0 in. | in. |
| Travel Distance. D | 6 in. | |
| Asymmetric Angle, A(deg.) | 0 | |
| Frequency. F(lifts/min) | 2 | |
| Coupling | ○ Poor ◉ Fair ○ Good | ○ Poor ○ Fair ○ Good |

Help | Print Screen | Print | Calculate Task | Edit Job

# APPENDIX C

## TABLES OF ERGONOMIC STUDIES

### C.1 Select Studies Demonstrating Effectiveness of Engineering Controls for Reducing Exposure to Ergonomic Risk Factors

| Study | Target population | Problem/risk factor | Control measure | Effect |
|---|---|---|---|---|
| Miller et al. (1971) | Surgeons (use of bayonet forceps) | Muscle fatigue during forceps use, frequent errors in passing instruments | Redesigned forceps (increased surface area of handle) | Reduced muscle tension (determined by EMG) and number of passing errors |
| Armstrong et al. (1982) | Poultry cutters (knives) | Excessive muscle force during poultry cutting tasks | Redesigned knife (reoriented blade, enlarged handle, provided strap for hand) | Reduced force grip during use, reduced forearm muscle fatigue |
| Knowlton and Gilbert (1983) | Carpenters (hammers) | Muscle fatigue, wrist deviation during hammering | Bent handle of hammer and its diameter | Smaller decrement in strength, reduced ulnar wrist deviation |
| Habes (1984) | Auto workers | Back fatigue during embossing tasks | Designed cutout in die to reduce reach distance | Reduced back muscle fatigue as determined by EMG |
| Goel and Rim (1987) | Miners (pneumatic chippers) | Hand–arm vibration | Provided padded gloves | Reduced vibration by 23.5–45.5% |
| Wick (1987) | Machine operators in a sandal plant | Pinch grips, wrist deviation, high repetition rates, static loading of legs and back | Provided adjustable chairs and bench-mounted armrests; angled press, furnished part bins | Reduced wrist deviation, compressive force on lumbar–sacral disks from 85 to 13 lb |
| Little (1987) | Film notchers | Wrist deviation, high repetition rates, pressure in the palm of the hand imposed by notching tool | Redesigned notching tool (extended, widened, and bent handles, reduced squeezing force) | Reduced squeezing force from 15 to 10 lb, eliminated wrist deviation, productivity increased by 15% |
| Johnson (1988) | Power hand tool users | Muscle fatigue, excessive grip force | Added vinyl sleeve and brace to handle | Reduced grip force as determined by EMG |
| Fellows and Freivalds (1989) | Gardeners (rakes) | Blisters, muscle fatigue | Provided foam cover for handle | Reduced muscle tension and fatigue buildup as determined by EMG |
| Andersson (1990) | Power hand tool users | Hand–arm vibration | Provided vibration damping handle | Reduced hand-transmitted vibration by 61–85% |

## C.2 Reducing Exposure to Ergonomic Risk Factors

| Study | Target population | Problem/risk factor | Control measure | Effect |
|---|---|---|---|---|
| Radwin and Oh (1991) | Trigger-operated power hand tool users | Excessive hand exertion and muscle fatigue | Extended trigger | Reduced finger and palmar force during tool operation by 7% |
| Freudenthal et al. (1991) | Office workers | Static loading of back and shoulders during seated tasks | Provided desk with 10° incline, adjustable chair and tables | Reduced moment of force on lower spinal column by 29%; by 21% on upper part |
| Powers et al. (1992) | Office workers | Wrist deviation during typing tasks | Provided forearm supports and a negative slope keyboard support system | Reduced wrist extension |
| Erisman and Wick (1992) | Assembly workers | Pinch grips, wrist deviation | Provided new assembly fixtures | Eliminated pinch grips, reduced wrist deviations by 65%, reduced cycle time by 50% |
| Luttmann and Jäger (1992) | Weavers | Forearm muscle fatigue | Redesigned workstation (numerous changes) | Reduced fatigue as measured by EMG, improved quality of product |
| Fogleman et al. (1993) | Poultry workers (knives) | Excessive hand force, wrist deviation | Altered blade angle and handle diameter | Wrist deviation reduced with altered blade angle |
| Lindberg et al. (1993) | Seaming operators | Awkward, fixed (static) neck and shoulder postures, monotonous work movements, high work pace | Automated seaming task | Freer head postures during automated seaming; loads on neck/shoulder muscles reduced as indicated by EMG; perceived exertion was reduced |
| Nevala-Puranen et al. (1993) | Dairy farmers | Whole-body fatigue, bent and twisted back postures, static arm postures | Installed rail system for carrying milking equipment | Heart rate decreased; bent and twisted back/trunk postures decreased by 64%; above shoulder arm postures cut in half; mean milking time per cow decreased by 24% |

## C.3 Select Studies of Various Control Strategies for Reducing Musculoskeletal Injuries and Discomfort

| Study | Industry | Study group | Intervention method | Summary of results | Additional comments |
|---|---|---|---|---|---|
| Itani et al. (1979) | Film manufacturing | 124 film rollers in two groups | Reduced work time, increased number of rest breaks | Reduction in neck and shoulder disorders and low back complaints, improved worker health | Productivity after the intervention was found to be 86% of the pre-intervention level |
| Luopajarvi et al. (1982) | Food production | 200 packers | Redesigned packing machine | Decreased neck, elbow, and wrist pain | Not all recommended job changes implemented; workers still complained |
| Drury and Wick (1984) | Shoe manufacturing | Workers at six factory sites | Workstation redesign | Reduced postural stress, increased productivity | Trunk and upper limbs most affected by changes |
| Westgaard and Aaras (1984, 1985) | Cable forms production | 100 workers | Introduced adjustable workstations and fixtures, counterbalanced tools | Turnover decreased, musculoskeletal sick leave reduced by 67% over an 8-year period, productivity increased | Reductions in shoulder, upper back muscle load verified by EMB |
| McKenzie et al. (1985) | Telecommunication equipment manufacturing | 660 employees | Redesigned handles on power screwdrivers and wire wrapping guns, instituted plant-wide ergonomics program | Incidence rate of repetitive trauma disorders decreased from 2.2 to 0.53 cases per 200 000 work hours; lost days reduced from 1001 to 129 in 3 years | |
| Echard et al. (1987) | Automobile manufacturing | | Redesigned tools, fixtures, and work organization in assembly operations | Reduced long-term upper extremity and back disabilities; CTS surgeries reduced by 50% | |
| LaBar (1992) | Household products manufacturing | 800 workers | Introduced adjustable workstations, improved the grips on hand tools, improved parts organization | Reduced injuries (particularly back) by 50% | Company also had a labor management safety committee to investigate ergonomics |
| Orgel et al. (1992) | Grocery store | 23 employees | Redesigned checkout counter to reduce distances, installed a height-adjustable keyboard, and trained workers to adopt preferred work practices | Decreased self-reported neck, upper back, and shoulder discomfort; no change in arm, forearm, and wrist discomfort | Study lacked a reference group not subject to the same interventions for making suitable comparisons |
| Rigdon (1992) | Bakery | 630 employees | Formed union management committee to study cumulative trauma problems, which led to workstation changes, tool modifications; improved work practices | Cumulative trauma cases dropped from 34 to 13 in 4 years; lost days reduced from 731 to 8 over the same period | Union advocated more equipment to reduce manual material handling |
| Garg and Owen (1992) | Nursing home | 57 nursing assistants | Implemented patient transferring devices | IR of back injuries decreased from 83 to 43 per 200 000 work hours following the intervention; no lost or restricted workdays during the 4 months following the intervention | |

| | | | | | |
|---|---|---|---|---|---|
| Halpern and Davis (1993) | Office | 90 office workers | Adjusted workstations according to the workers' anthropometric dimensions | Body part discomfort decreased; perceived efficiency and usability of the equipment increased | |
| Lutz and Hansford (1987) | Medical products manufacturing | More than 1000 workers | Introduced adjustable workstations and fixtures, mechanical aids to reduce repetitive motions, job rotation | Medical visits reduced from 76 to 28 per month | Employees also expressed enthusiasm for exercise program introduced with other interventions |
| Jonsson (1988) | Telephone assembly, printed circuit card manufacturing, glass blowing, mining | 25 workers | Job rotation | Job rotation in light-duty tasks not as effective as in dynamic heavy-duty tasks | Measured static load on shoulder upper back muscles with EMG |
| Geras et al. (1988) | Rubber and plastic parts manufacturing | 87 plants within one company | Ergonomics training and intervention program introduced, added material handling equipment, workstation modifications to eliminate postural stresses | Lost-time prevalence rates at two plants reduced from 4.9 and 9.7 per 200000 hours to 0.9 and 2.6, respectively, within 1 year and maintained over a 4-year period | Success attributed to increased training, awareness of hazards, and improved communication between management and workers |
| Tadano (1990) | Office | 500 VDT operators | Provided training, redesigned workstations, and incorporated additional breaks and exercises into the work schedule | Cumulative trauma disorder cases reduced from 49 in the 6 months preceding the intervention to 24 in the 6 months following the intervention | |
| Hopsu and Louhevaara (1991) | Office | Eight female cleaners | Provide training and greater flexibility in their work and eliminated strictly proportioned work areas and time schedules | Average sick leave decreased from 20 days per year before the intervention to 10 days per year 2 years after intervention | Mean maximum $VO_2$ rate increased, mean heart rate decreased after intervention |
| Aaras (1994) | Telephone exchange manufacturing, office | 96 workers (divided into four groups) | Provided adjustable workstations and additional work space | Significant reduction in intensity and duration of neck pain reported after intervention | Reductions in static loading on the neck and shoulder muscles after intervention |
| Moore (1994) | Automotive engine and transmission manufacturing | Five workers | Eliminated manual flywheel truing operation by implementing a mechanical press | 29% decrease in musculoskeletal disorders, 78% decrease in upper extremity CTDs, 82% reduction in restricted or lost work time | Used participatory (team) approach to select intervention method |
| NIOSH (1994) | Red meatpacking | Three beef/pork processing companies | Implemented participatory (labor management) ergonomics program | Results varied: only two teams able to introduce changes to address identified problems; some evidence that incidence and severity of injury was reduced following introduction of an ergonomics program | Additional follow-up needed to evaluate intervention effectiveness |

(*Continued*)

**(Continued)**

| Study | Industry | Study group | Intervention method | Summary of results | Additional comments |
|---|---|---|---|---|---|
| Narayan and Rudolph (1993) | Medical device assembly plant | 316 employees | Redesigned workstation to reduce reach distances, provided adjustable chairs and footrests, provided fixtures and pneumatic gripper to eliminate pinch grips | Plant-wide CTD incidence rate reduced from 13.7 to 11.3 per 200000 worker hours after intervention; plant-wide severity rate reduced from 154.9 lost-time days to 67.8 lost-time days per 200000 worker hours | Not all jobs in plant affected by changes |
| Parenmark et al. (1993) | Chainsaw assembly plant | 279 workers | Increased number of workers and tasks, provided training, reduced work pace, adopted new wage system and flexible working hours | Sick leave dropped from 17 to 13.7 days per worker per year, labor turnover dropped from 35 to 10%; assembly errors cut by 3 to 6%; total production cost reduced by 10%; productivity not affected | Difficult to pinpoint which factor had biggest impact |
| Shi (1993) | County government (various occupations represented) | 205 workers | Education, back safety training, physical fitness activities, equipment/facility improvements (e.g. additional material handling equipment) | Back pain prevalence declined modestly; significant improvement in satisfaction and a reduction in risky lifting behaviors were reported; a savings of $161 108 was realized, giving a 179% return in the investment | |
| Reynolds et al. (1994) | Apparel manufacturing | 18 operators | Introduced height- and tilt-adjustable work stands, additional jigs, anti-fatigue mats, and automatic thread cutters | Body part discomfort reduced in shoulders, arms, hands, and wrists; no injury costs incurred in 5 months following intervention | Used worker participation approach; productivity significantly increased after intervention |

## APPENDIX BIBLIOGRAPHY


Aaras, A. (1994). Relationship between trapezius load and the incidence of musculoskeletal illness in the shoulder. *Int. J. Ind. Ergon.* **14** (4): 341–348.

Andersson, E.R. (1990). Design and testing of a vibration attenuating handle. *Int. J. Ind. Ergon.* **6** (2): 119–126.

Degani, A., Asfour, S.S., Waly, S.M., and Koshy, J.G. (1993). A comparative study of two shovel designs. *Appl. Ergon.* **24** (5): 306–312.

Drury, C.G. and Wick, J. (1984). Ergonomic applications in the shoe industry. *Proceedings of the International Conference on Occupational Ergonomics*, pp. 489–493.

Echard, M., Smolenski, S., and Zamiska, M. (1987). Ergonomic considerations: engineering controls at Volkswagen of America. In: *Ergonomic Interventions to Prevent Musculoskeletal Injuries in Industry*, Industrial Hygiene Science Series, 117–131. Cincinnati, OH: American Conference of Governmental Industrial Hygienists.

Erisman, J. and Wick, J. (1992). Ergonomic and productivity improvements in an assembly clamping fixture. In: *Advances in Industrial Ergonomics and Safety IV* (ed. S. Kumar), 463–468. Philadelphia, PA: Taylor & Francis.

Fellows, G.L. and Freivalds, A. (1989). The use of force sensing resistors in ergonomic tool design. *Proceedings of the Human Factors Society, 33rd Annual Meeting*, pp. 713–717.

Fogleman, M.T., Freivalds, A., and Goldberg, J.H. (1993). An ergonomic evaluation of knives for two poultry cutting tasks. *Int. J. Ind. Ergon.* **11** (3): 257–265.

Freudenthal, A., van Riel, M.P.J.M., Molenbroek, J.F.M., and Snijders, C.J. (1991). The effect on sitting posture of a desk with a ten-degree inclination using an adjustable chair and table. *Appl. Ergon.* **22** (5): 329–336.

Gallimore, J.J. and Brown, M.E. (1993). Effectiveness of the C-sharp: reducing ergonomics problems at VDTs. *Appl. Ergon.* **24** (5): 327–336.

Garg, A. and Owen, B. (1992). Reducing back stress to nursing personnel: an ergonomic intervention in a nursing home. *Ergonomics* **35** (11): 1353–1375.

Geras, D.T., Pepper, C.D., and Rodgers, S.H. (1988). An integrated ergonomics program at the Goodyear Tire and Rubber Company. In: *Advances in Industrial Ergonomics and Safety* (ed. A. Mital), 21–28. Bristol, PA: Taylor & Francis.

Goel, V.K. and Rim, K. (1987). Role of gloves in reducing vibration: an analysis for pneumatic chipping hammer. *Am. Ind. Hyg. Assoc. J.* **48** (1): 9–14.

Habes, D.J. (1984). Use of EMG in a kinesiological study in industry. *Appl. Ergon.* **15** (4): 297–301.

Halpern, C.A. and Davis, P.J. (1993). An evaluation of workstation adjustment and musculoskeletal discomfort. *Proceedings of the 37th Annual Meeting of the Human Factors Society*, Seattle, WA, pp. 817–821.

Hopsu, L. and Louhevaara, V. (1991). The influence of educational training and ergonomic job redesign intervention on the cleaner's work: a follow-up study. *Designing for Everyone: Proceedings of the Eleventh Congress of the International Ergonomics Association*, Paris, Vol. **1**, pp. 534–536.

Itani, T., Onishi, K., Sakai, K., and Shindo, H. (1979). Occupational hazard of female film rolling workers and effects of improved working conditions. *Arh. Hig. Rada Toksikol.* **30** (Suppl.): 1243–1251.

Johnson, S.L. (1988). Evaluation of powered screwdriver design characteristics. *Hum. Factors* **30** (1): 61–69.

Jonsson, B. (1988). Electromyographic studies of job rotation. *Scand. J. Work Environ. Health* **14** (1): 108–109.

Knowlton, R.G. and Gilbert, J.C. (1983). Ulnar deviation and short-term strength reductions as affected by a curve handled ripping hammer and a conventional claw hammer. *Ergonomics* **26** (2): 173–179.

LaBar, G. (1992). A battle plan for back injury prevention. *Occup. Hazards* **54**: 29–33.

Lindberg, M., Frisk-Kempe, K., and Linderhed, J. (1993). Musculoskeletal disorders, posture and EMG temporal pattern in fabric seaming tasks. *Int. J. Ind. Ergon.* **11** (3): 267–276.

Little, R.M. (1987). Redesign of a hand tool: a case study. *Semin. Occup. Med.* **2** (1): 71–72.

Luopajarvi, T., Kuorinka, I., and Kukkonen, R. (1982). The effects of ergonomic measures on the health of the neck and upper extremities of assembly-line packers – a four year follow-up study. *Proceedings of the 8th Congress of the International Ergonomics Association*, Tokyo, Japan (ed. K. Noro), pp. 160–161.

Luttmann, A. and Jager, M. (1992). Reduction in muscular strain by work design: electromyographical field studies in a weaving mill. In: *Advances in Industrial Ergonomics and Safety IV* (ed. S. Kumar), 553–560. Philadelphia, PA: Taylor & Francis.

Lutz, G. and Hansford, T. (1987). Cumulative trauma disorder controls: the ergonomics program at Ethicon, Inc. *J. Hand Surg.* **12A** (5, Part 2): 863–866.

McKenzie, F., Storment, J., Van Hook, P., and Armstrong, T.J. (1985). A program for control of repetitive trauma disorders associated with hand tool operations in a telecommunications manufacturing facility. *Am. Ind. Hyg. Assoc. J.* **46** (11): 674–678.

Miller, M., Ransohoff, J., and Tichauer, E.R. (1971). Ergonomic evaluation of a redesigned surgical instrument. *Appl. Ergon.* **2** (4): 194–197.

Moore, J.S. (1994). Flywheel tuning: a case study of an ergonomic intervention. *Am. Ind. Hyg. J.* **55** (8): 236–244.

Narayan, M. and Rudolph, R. (1993). Ergonomic improvements in a medical device assembly plant: a field study. *Proceedings of the 37th Annual Meeting of the Human Factors Society*, Seattle, WA, pp. 812–816.

Nevala-Puranen, N., Taattola, K., and Venalainen, J.M. (1993). Rail system decreased physical strain in milking. *Int. J. Ind. Ergon.* **12** (4): 311–316.

NIOSH (1994). *Participatory Ergonomics Interventions in Meatpacking Plants*. DHHS [NIOSH] Pub. No. 94-124. Cincinnati, OH: U.S. Department of Health and Human Services, Public Health Service, Centers for Disease Control and Prevention, National Institute for Occupational Safety and Health.

Orgel, D.L., Milliron, M.J., and Frederick, L.J. (1992). Musculoskeletal discomfort in grocery express checkstand workers: an ergonomic intervention study. *J. Occup. Med.* **34** (8): 815–818.

Parenmark, G., Malmvisk, A.K., and Ortengren, R. (1993). Ergonomic moves in an engineering industry: effects on sick leave frequency, labor turnover and productivity. *Int. J. Ind. Ergon.* **11** (4): 291–300.

Peng, S.L. (1994). Characteristics and ergonomic design modifications for percussive rivet tools. *Int. J. Ind. Ergon.* **913** (3): 171–187.

Powers, J.R., Hedge, A., and Martin, M.G. (1992). Effects of full motion forearm supports and a negative slope keyboard system on hand–wrist posture while keyboarding. *Proceedings of the Human Factors Society 36th Annual Meeting*, Atlanta, GA, pp. 796–800.

Radwin, R.G. and Oh, S. (1991). Handle and trigger size effects on power tool operation. *Proceedings of the Human Factors Society 35th Annual Meeting*, pp. 843–847.

Reynolds, J.L., Dury, C.G., and Broderick, R.L. (1994). A field methodology for the control of musculoskeletal injuries. *Appl. Ergon.* **25** (1): 3–16.

Rigdon, J.E. (1992). The wrist watch: how a plant handles occupational hazard with common sense. *The Wall Street Journal* (28 September).

Snook, S.H., Campanelli, R.A., and Hart, J.W. (1978). A study of three preventive approaches to low back injury. *J. Occup. Med.* **20**: 478–481.

Tadano, P. (1990). A safety/prevention program for VDT operators: one company's approach. *J. Hand Ther.* **3** (2): 64–71.

Westgaard, R.H. and Aaras, A. (1984). Postural muscle strain as a causal factor in the development of musculoskeletal illnesses. *Appl. Ergon.* **15** (3): 162–174.

Westgaard, R.H. and Aaras, A. (1985). The effect of improved workplace design on the development of work-related musculoskeletal illnesses. *Appl. Ergon.* **16** (2): 91–97.

Wick, J.L. (1987). Workplace design changes to reduce repetitive motion injuries in an assembly task: a case study. *Semin. Occup. Med.* **2** (1): 75–78.

Wick, J.L. and Deweese, R. (1993). Validation of ergonomic improvements to a shipping workstation. *Proceedings of the 37th Annual Meeting of the Human Factors Society*, Seattle, WA, pp. 808–811.

## TEXT BIBLIOGRAPHY

ANSI (1986). *Guide for the Measurement and Evaluation of Human Exposure to Vibration Transmitted to the Hand*. ANSI S3.34. New York: ANSI.

ANSI (1993). *Ergonomic Guidelines for the Design, Installation, and Use of Machine Tools in the Work Environment*. ANSI B11 TR1. New York: ANSI.

ANSI (1995). *Control of Work-Related Cumulative Trauma Disorders* (working draft). ANSI Z 365 (17 April). New York: ANSI.

Armstrong, T.J., Radwin, R.G., Hansen, D.J., and Kennedy, K.W. (1986). Repetitive trauma disorder: job evaluation and design. *Hum. Factors* **28** (3): 325–336.

Astrand, P.O. and Rodahl, K. (1986). *Textbook of Work Physiology. Physiological Bases of Exercise*, 3e. New York: McGraw-Hill.

Bink, B. (1962). The physical working capacity in relation to working time and age. *Ergonomics* **5** (1): 25–28.

Bonjer, F.H. (1962). Actual energy expenditure in relation to the physical working capacity. *Ergonomics* **5** (1): 29–31.

Chaffin, D.B. (1992). Biomechanical modeling for simulation of 3D static human exertions. In: *Computer Applications in Ergonomics, Occupational Safety and Health* (ed. M. Mattila and W. Karwowski), 1–11. Amsterdam, the Netherlands: Elsevier Science Publishers.

Chaffin, D.B. and Andersson, G.B.J. (1991). *Occupational Biomechanics*. New York: Wiley.

Durnin, J.V.G.A. and Passmore, R. (1967). *Energy, Work and Leisure*. London: Heineman Educational Books.

Eastman Kodak Co. (1983). *Ergonomic Design for People at Work*, vol. **1**. New York: Van Nostrand Reinhold.

Eastman Kodak Co. (1996). *Ergonomic Design for People at Work*, vol. **2**. New York: Van Nostrand Reinhold.

Gamberale, F. (1990). Perception of effort in manual materials handling. *Scand. J. Work Environ. Health* **16** (Suppl. 1): 59–66.

Garg, A., Chaffin, D.B., and Herrin, G.D. (1978). Prediction of metabolic rates for manual materials handling jobs. *Am. Ind. Hyg. Assoc. J.* **39**: 661–674.

Givoni, B. and Goldman, R.F. (1971). Predicting metabolic energy cost. *J. Appl. Physiol.* **30**: 429–433.

Habes, D.J. and Putz-Anderson, V. (1985). The NIOSH program for evaluating biomechanical hazards in the workplace. *J. Safety Res.* **16**: 49–60.

Hosey, A.D. (1973). General principles in evaluating the occupational environment. In: *The Industrial Environment – Its Evaluation and Control* (ed. G.D. Clayton), 95. Cincinnati, OH: National Institute for Occupational Safety and Health.

Imada, A.S. (1993). Macroergonomic approaches for improving safety and health in flexible, self-organizing systems. In: *The Ergonomics of Manual Work* (ed. W.S. Marras, W. Karwowski, J.L. Smith and L. Pacholski), 477–480. Philadelphia, PA: Taylor & Francis.

Kamon, E. and Goldfuss, A. (1978). In-plant evaluation of muscle strength of workers. *Am. Ind. Hyg. Assoc. J.* **39**: 801–807.

Keyserling, W.M. (1986). Postural analysis of the trunk and shoulders in simulated real time. *Ergonomics* **29** (4): 569–583.

Keyserling, W.M., Armstrong, T.J., and Punnett, L. (1991). Ergonomic job analysis: a structured approach for identifying risk factors associated with overexertion injuries and disorders. *Appl. Occup. Environ. Hyg.* **6**: 353–363.

Lifshitz, Y. and Armstrong, T.J. (1986). A design checklist for control and prediction of cumulative trauma disorders in hand intensive manual jobs. *Proceedings of the 30th Annual Meeting of the Human Factors Society*, pp. 837–841.

Mathiowetz, V., Kashman, N., Volland, G. et al. (1985). Grip and pinch strength: normative data for adults. *Arch. Phys. Med. Rehabil.* **66**: 69–72.

NASA (1978). *Anthropometric Source Book, Volumes I, II, and III*. Reference Publication 1024. Yellow Springs, OH: NASA Scientific and Technical Information Office.

Niebel, B.W. (1989). *Motion and Time Study*, 8e. Homewood, IL: Irwin.

NIOSH (1981). *Work Practices Guide for Manual Lifting*. NIOSH Technical Report No. 81-122. Cincinnati, OH: U.S. Department of Health and Human Services, National Institute for Occupational Safety and Health.

OSHA (1990). *Ergonomics Program Management Guidelines for Meatpacking Plants*. OSHA 3123. Washington, DC: U.S. Department of Labor, Occupational Safety and Health Administration.

Perez-Balke, G.M. (1993). The recording of observed physical exposures (ROPEM): a simple and low cost method of analyzing upper extremity postures (Abstract). *Proceeding of the American Industrial Hygiene Association Conference*, New Orleans, LA.

Putz-Anderson, V. (1988). *Cumulative Trauma Disorders: A Manual for Musculoskeletal Diseases of the Upper Limbs*. London: Taylor & Francis.

Putz-Anderson, V. and Galinsky, T.L. (1993). Psychophysically determined work durations for limiting shoulder girdle fatigue from elevated manual work. *Int. J. Ind. Ergon.* **11** (1): 19–28.

Rodgers, S.H. (1992). Functional job analysis technique. In: *Occupational Medicine: State of the Art Reviews* (ed. J.S. Moore and A. Garg), 680. Philadelphia, PA: Hanley & Belfus.

Snook, S.H. (1978). The design of manual handling tasks. *Ergonomics* **21** (12): 963–985.

Snook, S.H. and Ciriello, V.M. (1991). The design of manual handling tasks: revised tables of maximum acceptable weights and forces. *Ergonomics* **21**: 1197–1213.

Snook, S.H. and Irvine, C.H. (1969). Psychophysical studies of physiological fatigue criteria. *Hum. Factors* **11** (3): 291–300.

Snook, S.H., Vaillancourt, D.R., Ciriello, V.M., and Webster, B.S. (1996a). Psychophysical studies of repetitive wrist motion: part I: two day per week exposure. *Ergonomics*.

Snook, S.H., Vaillancourt, D.R., Ciriello, V.M., and Webster, B.S. (1996b). Psychophysical studies of repetitive wrist motion: part II: five day per week exposure. *Ergonomics*.

Ulin, S.S. and Armstrong, T.J. (1992). A strategy for evaluating occupational risk factors of musculoskeletal disorders. *J. Occup. Rehabil.* **2** (1): 35–50.

Ulin, S.S., Armstrong, T.J., and Radwin, R.G. (1990). Use of computer aided drafting for analysis and control of posture in manual work. *Appl. Ergon.* **21** (2): 143–151.

VanderWalt, W.H. and Wyndham, C.H. (1973). An equation for prediction of energy expenditure of walking and running. *J. Appl. Physiol.* **34**: 559–563.

Waters, T.R., Putz-Anderson, V., Garg, A., and Fine, L. (1993). Revised NIOSH equation for the design and evaluation of manual lifting tasks. *Ergonomics* **36** (7): 749–776.

## RESOURCES (PUBLICATIONS AVAILABLE FROM AREA OSHA OFFICES, EXCEPT AS NOTED)

*Ergonomics Program Management Guidelines for Meatpacking Plants* (1991) (reprinted). OSHA 3123 (free).

*Working Safely with Video Display Terminals* (1997) (reprinted). OSHA 3092*, Order number 029-016-00127-1.

*Ergonomics: The Study of Work* (1991). OSHA 3125*, Order number 029-016-00124-7.

"Back Injuries – Nation's Number One Workplace Safety Problem." This is one of a series of fact sheets highlighting USDOL programs. Fact Sheet Number OSHA 89-09.

*ErgoFacts*. OSHA's 1-page fact sheet on ergonomics hazards in the workplace, their solutions and benefits. Available from the OSHA publications office.

*OSHA Handbook for Small Businesses* (1996) (revised). OSHA 2209*.

*These publications cost one to six dollars. They are available from your local OSHA area office or US government bookstore or from the Office of Information and Consumer Affairs at:

OSHA Publications Office

200 Constitution Avenue N.W.

Room N-3647

Washington, DC 20210

*Job Safety & Health Quarterly*. The agency's quarterly magazine contains various articles on health and safety, including ergonomic issues. Available for $5.50/year from the GPO.

The *Federal Register* published the Advanced Notice of Proposed Rulemaking on 3 August 1992 and will publish the proposed standard. Order forms for the *Federal Register* can be obtained from the GPO.

Superintendent of Documents

Government Printing Office

Washington, DC 20402-9325

202-275-0019 (to fax orders and inquiries)

202-783-3238 8 a.m. to 4 p.m. Monday–Friday EST for charge orders

OSHA FAX on demand provides a variety of information available, much of which is less expensive to access via the Internet.

900-555-3400

$1.50 per minute

Internet addresses to get you started:

www.ergoweb.com

www.osha.gov

https://www.osha.gov/SLTC/ergonomics/index.html

For direct dial to the Labor News Bulletin Board, which has a gateway to the Salt Lake City Technical Lab and OSHA Computerized Information System (OCIS), dial 202-219-4784 (this is a toll call outside Washington, DC).

## ADDITIONAL RESOURCES

- National Institute for Occupational Safety and Health (NIOSH)
  Public Dissemination, DSDTT
  4676 Columbia Parkway
  Cincinnati, OH 45226
  USA
  513-533-8573 (Fax)
  1-800-35-NIOSH

  NIOSH has a free help line (8 a.m. to 4:00 p.m. EST) to answer technical questions and disseminate publications on ergonomics and other occupational safety and health questions such as blood-borne pathogens and indoor air quality. NIOSH and DHHS publications noted in the reference section of this chapter may be available through the help line.
- Human Factors and Ergonomics Society
  PO Box 1369
  Santa Monica, CA 90406-1369
  USA
  310-394-2410 (Fax)
  310-394-1811

  The Human Factors and Ergonomics Society (HFES) is the primary society that represents human factor engineers, engineering psychologists, and ergonomists in the United States. HFES publishes an annual list of consultants and a list of ergonomic and human factors programs and maintains an active employment service for ergonomists and employers who seek ergonomists. The society also publishes a newsletter and a journal entitled *Human Factors*.

- The Ergonomics Society
  Department of Human Sciences
  University of Technology
  Loughborough, Leicestershire LE11 3TU
  United Kingdom
  Like HFES, the Ergonomics Society is a professional society that assists in the publication of an international journal. The journal is entitled *Applied Ergonomics*. The journal reports on applied studies of people's relationships with equipment, environments, and work systems.
- Massachusetts Coalition on New Office Technology
  (CNOT)
  1 Summer Street
  Somerville, MA 02142
  617-776-2777
  CNOT provides information on risk factors in the office environment and a checklist for work analysis for VDU workstations. They have ongoing support groups for workers who suffer from WMSDs and provide speakers for related topics.
- Center for Workplace Health Information
  *CTDNEWS*
  410 Lancaster Avenue, Suite 15
  PO Box 239
  Haverford, PA 19041
  610-896-2762 (Fax)
  610-896-2770 or 1-800-554-4283
  *CTDNEWS* is a monthly newsletter that provides up-to-date information on legal cases, policy, journal articles, case studies of ergonomic intervention, dates of ergonomic conferences and seminars, and ergonomic resources. *CTDNEWS* will send a sample of their newsletter upon request. Their website can be found at www.ctdnews.com on the Internet.
- Ergoweb www.ergoweb.com
  Ergoweb is a place to meet and discuss ergonomics with ergonomists and those with ergonomic issues. Other useful resources on Ergoweb include a full copy of the OSHA draft proposed ergonomic protection standard (in the reference section of Ergoweb), a copy of the RULA method of analysis for the upper extremities and neck (as described in the methods section of this chapter), six different programs to analyze hazards associated with lifting, and the UTAH Process Diagnostician (an "expert" system for evaluating and designing systems to minimize ergonomic risk factors).
- Board of Certification in Professional Ergonomics (BCPE)
  PO Box 2811
  Bellingham, WA 98227-2811
  360-671-7681 (Fax)
  360-671-7601
  BCPE is a nonprofit organization responsible for the certification of human factor and ergonomic professionals. Certification requires a master's degree in ergonomics or human factors, or, equivalent, four years of full-time practice as an ergonomist practitioner with emphasis on design involvement, documentation of education, employment history, ergonomic project involvement, a passing score on an eight-hour written examination, and payment of fees for the certification process. BCPE publishes a newsletter and a directory of board-certified ergonomists and human factor practitioners.

CHAPTER 4

# EVALUATION OF EXPOSURE TO CHEMICAL AGENTS

JERRY LYNCH† and CHARLES CHELTON

*30 Radtke St., Randolph, NJ, 07869*

Adapted from "Measurement of worker exposure" by Jeremiah R. Lynch, Chapter 2 in R. Harris, L.J. Cralley, and L.V. Cralley (eds.), *Patty's Industrial Hygiene and Toxicology*, 3e, Vol. IIIA. New York: Wiley, 1994, pp. 27–80.

† Deceased.

## 4.1 SAMPLING STRATEGY

This chapter explains why workplace measurements of air contaminants are made, discusses the options available in terms of number, time, and location, and relates these options to the criteria that govern their selection and the consequences of various choices.

A person at work may be exposed to many potentially harmful agents for as long as a working lifetime, upward of 40 years in some cases. These agents occur singly and in mixtures, and their concentration varies with time. Exposure may occur continuously, at regular intervals, or in altogether irregular spurts. The worker may inhale the agent or be exposed by skin contact or ingestion. As a result of exposure to these agents, they come in contact with or enter the body of the worker, and depending on the magnitude of the dose, some harmful effects may occur. All measurements in industrial hygiene ultimately relate to the dose received by the worker and the harm it might do.

Changes in working conditions, in technology, and in society have changed old methods of measurement:

- With few exceptions, workplace exposure to toxic chemicals is much below what is commonly accepted as a safe level.
- As a consequence of the reduction of exposure, frank occupational disease is rarely seen. Much of the disease now present results from multiple factors, of which occupation is only one.
- Workers have the right to know how much toxic chemical exposure they receive, and this often results in a need to document the absence of exposure.
- Technology provides enormously improved sampling equipment that is rugged and flexible. This equipment, used with analytical instruments of great specificity and sensitivity, has largely replaced the old "wet" chemical methods.

As a consequence of these changes in the workplace and advances in technology, it is now both necessary and possible to examine in far more detail the way in which workers are exposed to harmful chemicals. Personal sampling pumps permit collection of contaminants in the breathing zone of a mobile worker. Pump–collector combinations are available for long and short sampling periods. Passive dosimeters, which do not require pumps, are available for a wide range of gases and vapors. Systems that do not require the continual attention of the sample taker permit the simultaneous collection of multiple samples. Data loggers can continuously record instrument readings in a form easily transferable to a computer. Automated sampling and analytical systems can collect data continuously. Sorbent-gas chromatography techniques permit the simultaneous sampling and analysis of mixtures and, when coupled with mass

*Handbook of Occupational Safety and Health*, Third Edition. Edited by S. Z. Mansdorf.

spectrometers, the identification of obscure unknowns. Analytical sensitivities have improved to the degree that tens and hundreds of ubiquitous trace materials begin to be noticeable.

As a starting point for a complete assessment of the risk to health posed by an occupational environment, it is necessary to know the substances to which workers are exposed. Systematic recognition of all possible hazards requires a review of inventories of the materials brought into the workplace, descriptions of production processes, and identification of any new substances, by-products, or wastes. However, these sources of information may not be enough to identify all substances, particularly those present as trace contaminants or substances generated by production process, either inadvertently or as unknown by-products. To complete the identification of all substances present, before going to the next step of evaluating exposure and risk, it may be necessary to make some substance recognition measurements. Since these measurements, which are typically made by such techniques as gas chromatography–mass spectrometry (GC-MS), are not intended to evaluate exposure, they may be area rather than personal samples and may be large volume samples for maximum sensitivity.

The term *sampling strategy* as used here means the analytical reasoning used to decide how to make a set of measurements to represent exposure for a particular purpose. The measurements should yield data that are logically and statistically adequate to provide information to describe the environment under assessment (see Table 4.1). An optimum strategy is the selection of options under the control of the exposure assessor, which efficiently achieves the objective given the physical circumstances and environmental variability (Health and Safety Executive [HSE] 1989). The occupational exposure assessment charts in Figures 4.1 and 4.2 provide tools that can be used to accomplish this end.

### 4.1.1 Purpose of Measurement

The development of a sampling strategy requires a clear understanding of the purpose of the measurement. Rarely are data collected purely for their own sake. Even when data are collected because of a demand by others, such as the government or employees, the use of the data should be considered:

> What questions will be answered by the data? What decisions could depend on those answers? For example, do we want to know whether a group of workers are overexposed? And if so, will a decision to take action to control the exposure follow? Is the control likely to be a minor change in a work practice or an expensive engineering modification? Or are the numbers to be assembled to answer the question: What level of control is currently being achieved in this industry? And may this answer lead to new decisions regarding what control is feasible? Last, will workers use the result to find out whether their health is at risk and, as a consequence, decide to change jobs or seek changes in the conditions of work?

Often there are several questions that need answers, and thus data are collected for multiple purposes. Table 4.1 summarizes the most common objectives of sample gathering. As the discussion that follows indicates, the purpose of the data determines the design of the measurement scheme. All too often data intended for multiple purposes turn out not to be suitable for any purpose. Thus it is usually necessary to focus on the prime need to be sure the strategy will meet this requirement. If possible, minor adjustment or additions can be made to meet other needs. The optimum sampling strategy is that which combines the choice of method and sampling scheme with respect to sampling location, time, frequency, and sample number so that we are confident that the data are adequate for the decisions that follow. Most employers attempt to meet the most common purpose of exposure measurement through a simple routine monitoring program. The US National Institute for Occupational Safety and Health (NIOSH) has developed a scheme for a logical stepwise analysis of data to arrive at decisions (Leidel et al. 1977).

**TABLE 4.1 Objectives of Exposure Assessment**

| | |
|---|---|
| Hazard recognition | Identification of the presence of trace by-product or waste stream hazards that are not identifiable through review of materials inventories and process descriptions |
| Exposure evaluation | Comparing measured exposure to reference exposure levels to assess risks to employee health |
| Assessment of control method effectiveness | Evaluating changes in exposure levels resulting from changes to process operations work products or modifications to engineering controls |
| Model validation | Comparing field measurements with predicted values to evaluate the powers and limitations of a model to predict responses |
| Method research | Validation of new and alternative techniques of sample collection and analysis |
| Source evaluation | Point source identification and determination of magnitude of emissions |
| Operational evaluation | Identification of sources and tasks that contribute to employee exposure |
| Epidemiology | Provide employee exposure data for epidemiological investigation into temporal and population trends in occupational illness |

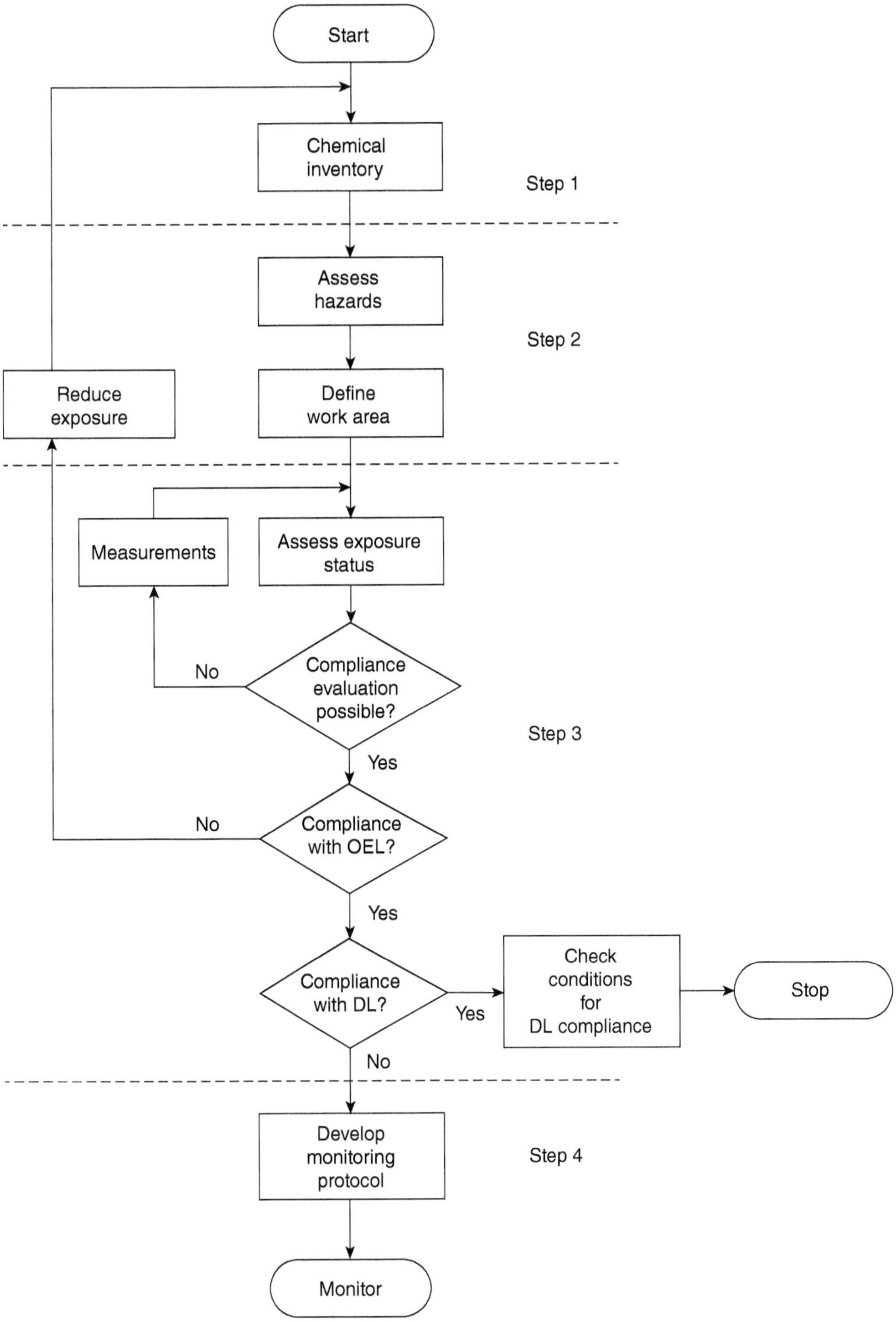


**Figure 4.1** CEFIC occupational exposure analysis chart. The occupational exposure level (OEL) and decision level (DL) used depend on the substance being evaluated.

### 4.1.2 Environmental Variability

An important factor in the design of any measurement scheme, whether it be a process quality control program, the measurement of an analyte in a lab, or an industrial hygiene exposure assessment, is to recognize the inherent patterns of the data due to natural phenomena and to match the data to an appropriate mathematical model. We are familiar with linear relationships in our daily lives where one variable increases or decreases in direct proportion to another. In our school careers, we were introduced to the use of a bell-shaped curve to model the distribution of grades in a

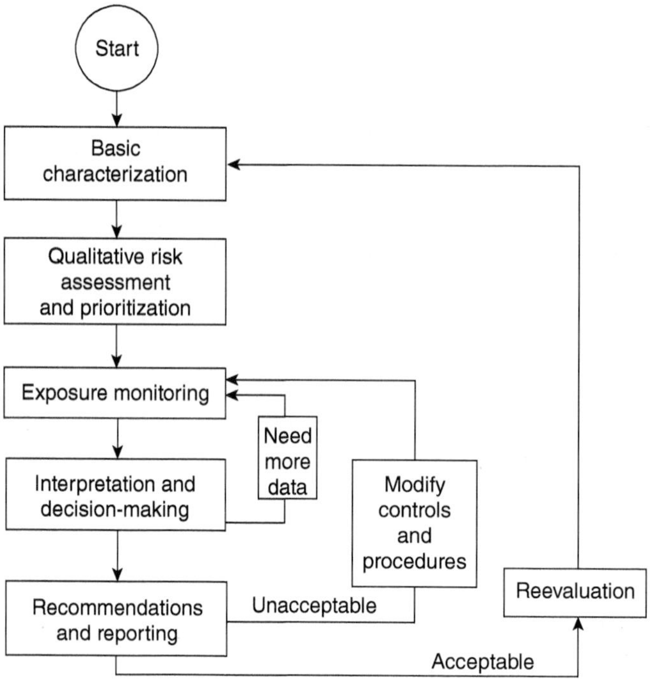


**Figure 4.2** Overall flow diagram of an exposure assessment strategy.

classroom setting. In this circumstance, student performance clusters around a central value with small proportions of poor and exceptional performances, creating a tail at each end of the curve. Industrial hygiene monitoring data and environmental measurements, in general, also tend to be described by a bell curve. The data cluster toward a central value; however, high excursions from the average, which occur infrequently, create a long tail. At the same time, the extreme variability in measurements causes a very broad-shaped curve. This type of data is best characterized by a lognormal distribution. Statistical mathematics are applied, not to the data points themselves, but to the logarithmic values that represent them. Exposure data are usually described by the mean value of a group of readings collected from a similar population of workers and by the statistic that characterizes the variability in the data, the standard deviation. The range of values that are the points of the lognormal curve from two standard deviations below the mean to two standard deviations above the mean is a data set, called the 95% confidence interval. This range is reported because it is highly probable (95% probability) that the true answer for the exposure being evaluated lies somewhere within this range (Figure 4.3).

An important factor in the design of any measurement scheme in advance of data gathering is to recognize the degree of variability likely to exist in the data set. This variability has a primary effect on the number of samples to be taken and the accuracy of the results that can be expected. Generally speaking, the fewer the number of samples, the less the accuracy. The greater the variability, the larger the number of samples required for assurance of a minimum accuracy. There is substantial documentation that the variability in employee exposure measurements tends to be quite high. During the course of a day, there are minute-to-minute variations. Daily averages vary widely from day to day.

One can speculate on the probable causes for this variability in worker exposure. The volume of space that a worker moves through in performing a task can be viewed as having an exposure zone centered on one or more point sources. Fugitive emissions from these points occur randomly like frequent small accidents rather than being the main consequence of the production process. Production rates change within a day or between days, affecting emission rates. Overlapping multiple operations within the exposure zone shift irregularly. The distribution pattern of contaminants within the zone by bulk flow, random air turbulence, and eddy diffusion is uneven in both time and space. Through all this, the target system, the worker, moves in a manner that is not altogether predictable. These and other uncertainties are the probable causes of the variability typically observed. While a goal of a monitoring effort is often to identify the most likely causative sources of exposure, the situation is far too complex to pinpoint each source of variation and to calculate its consequences. Variance, the statistical measure of variability, can be divided into within-worker variance and between-worker variance. Within-worker variance is the result of changes in the exposure of a worker from day to day when the tasks on each day are nominally similar; between-worker variance is the result of differences in procedures practiced by individuals who are essentially doing the same job. Between-worker variation is at least

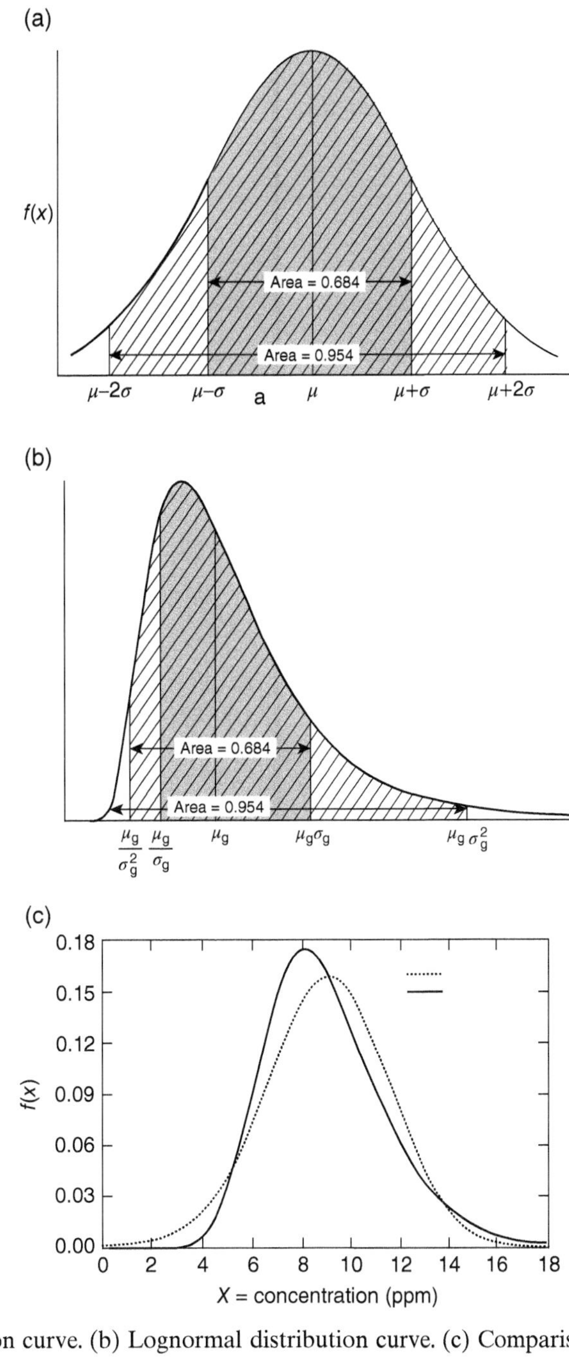


**Figure 4.3** (a) Normal distribution curve. (b) Lognormal distribution curve. (c) Comparison of normal (dotted) and lognormal (solid) distribution curves.

theoretically controllable by the way workers are grouped for sampling purposes. Tasks, observations, or review of job description can allow the health and safety professional to draw some conclusion on defining nominally homogeneous groups for sampling purposes. Since it is not usually practical to measure the exposure of all the workers all the time, it is customary to sample a subset of the homogeneous group and to apply the mean exposure value to all employees who are members of the same grouping. However, when the between-worker variance is seen to be quite large with respect to within-worker variance, it may be necessary to use information from historical sampling data or to conduct some preliminary sampling to group workers.

Sampling schemes may be designed that deal with the variability from all sources as a single pool and derive whatever accuracy is required by increasing the number of samples. Alternatively, one can postulate that a large part of the variability is due to an observable factor or factors. No hard and fast rules can be made regarding the choice of sampling schemes except that it seems logical to expect the factors that have been identified to account for the statistically significant fraction of the variance and to design a plan to collect and analyze the data in such a way that defensible description of the exposure results. Typical factors may include rate of production, shift, season of the year, and wind velocity and can be singled out as affecting worker exposure. Even after constructing sampling plans that create nominally homogeneous exposure groupings, it is likely that the residual or error variance will still be quite large. This source of variance will normally outweigh the error contributed to the analysis from other sources such as analytical or instrumental inaccuracies.

### 4.1.3 Location

The most common purpose of measurement of exposure of workers is to estimate the dose so as to prevent or predict adverse health effects. These health effects result from a substance entering the body by some route. Industrial hygienists most often estimate inhaled dose by measurement of the concentration of a substance inhaled air. Although air samples are sometimes collected from inside respirator face pieces, it is generally not possible to sample the air being inhaled directly. Therefore, the location of the sample collected inlet in relation to the subject's nose and mouth is important. We categorize sampling methods in terms of their closeness to the subject and the point of inhalation as personal samples, breathing zone or vicinity samples, and area or general air samples. A personal measurement is collected from an individual's immediate environment using a device that is worn and travels with the individual. Personal measurements are most often collected form the envelope or "breathing zone" around the worker's head, which is thought, based on observation and the nature of the operation, to have approximately the same concentration of the contaminant being measured as the air breathed by the worker. Area or general air samples are the most remote and are collected in fixed locations in the workplace. Area samples are useful for such purposes as evaluating controls or to characterize emissions, but they are at best a crude estimate of exposure.

Obviously, personal samples are the preferred method of estimating dose by the inhalation route since they have most closely measure inhaled air. OSHA enforcement operations "reflect a long-standing belief that personal sampling generally provides the most accurate measure of an employee's exposure..." (OSHA 1982).

If personal sampling cannot be used, some other means of estimating exposure must be accepted. Breathing zone measurements, made by the person who follows the worker while collecting the sample, can come close to measuring exposure. However, this intrusive measurement method may influence worker behavior, and the inconvenience of the measurement will limit the number of measurements and therefore reduce accuracy, as discussed below.

When fixed station samples are used, knowledge of the quality of the relation between their measurements and the exposure of the workers is necessary if worker exposure is to be estimated. The important question in the use of general air measurements is: What confidence can be placed in the estimate of work exposure?

In studying the relation between area and personal data with respect to asbestos, the British Occupational Hygiene Society concluded (Roach et al. 1983).

The relationship between static and personal sampling results varies according to the characteristics of the dust emission sources and the general and individual work practices adopted in a particular work area:

1. When identical sampling instruments are deployed simultaneously at personal and static sampling points and the distances between them are reasonably small, at least two-thirds of the personal sampling results obtained in a given working location are higher than those obtained from static sampling.
2. The differences found between the two types result tend to be particularly great where the static sampling points are relatively remote from dust emission points, as, for example, when "background" static testing is adopted.
3. In certain cases, results from personal sampling may be lower than those from static sampling, owing to factors such as the positioning of the sampling point with respect to air extraction systems.
4. The correlation coefficient between the personal and static measurements is statically significant, but, even so, no consistent relationship of great practical utility could be found in the limited data available.
5. Although worker exposure measurements are most often used in relation to health hazards, not all measurements made for the protection of health need to be measurements of exposure. When it has been established that an industrial operation does not produce unsafe conditions when it is operating within specified control limits, fixed station measurements that can detect loss of control may be the most appropriate monitoring system for worker's protection. Local increases in contaminant concentration caused by leaks, loss of cooling in a degreaser, or fan failure in a local exhaust system can be detected before important worker exposure occurs. Continuous air monitoring equipment that detects leaks or monitors area concentrations is often used in this way. All such systems should be validated for their intended purpose, and a performance maintenance program established with their deployment.

### 4.1.4 Sampling Period

Free of all other constraints, the most biologically relevant time period over which to measure or average worker exposure should be derived from the time constants of the uptake, action, and elimination of the toxic substance in the body (Roach 1966, 1977; Droz and Yu 1990). These periods range from minutes in the case of fast-acting poisons such as chlorine or hydrogen sulfide to days or months for slow systemic poisons such as lead or quartz. In the adoption of guides and standards, such as PELs or TLVs, this broad range has been narrowed, and the periods have not always been selected based on speed of effect. Measurement of the long-term average (multishift) exposure is much more efficient than measuring single-shift "peaks." However, standards have been developed or have been interpreted as single-shift limits. For most substances, a time-weighted average over the usual work shift of eight hours has been accepted since it is long enough to average out extremes and short enough to be measured in one workday. Several systems have been proposed for adjusting limits to novel work shifts (Brief and Scala 1975; Mason and Dershin 1976; Anderson et al. 1987).

Once the time period over which exposure is to be averaged has been decided for either biological or other reasons (Calabrease 1977; Hickey and Reist 1977), there are several alternate sampling schemes to yield an estimate of the exposure over the averaging time. A single sample could be taken for the full period over which exposure is to be averaged (Figure 4.4). If such a long sample is not practical, several shorter samples can be strung together to make up a set of full-period consecutive samples. In both cases, since the full period is being measured, the only error in the estimate of the exposure for that period is the error of sampling and analytical method itself. However, when these full-period measurements are used to estimate exposure over other periods not measured, the interperiod variance will contribute to the total error. When a measurement is made of a quantity that is varying, the uncertainty in the measurement caused by the variability becomes part of the error.

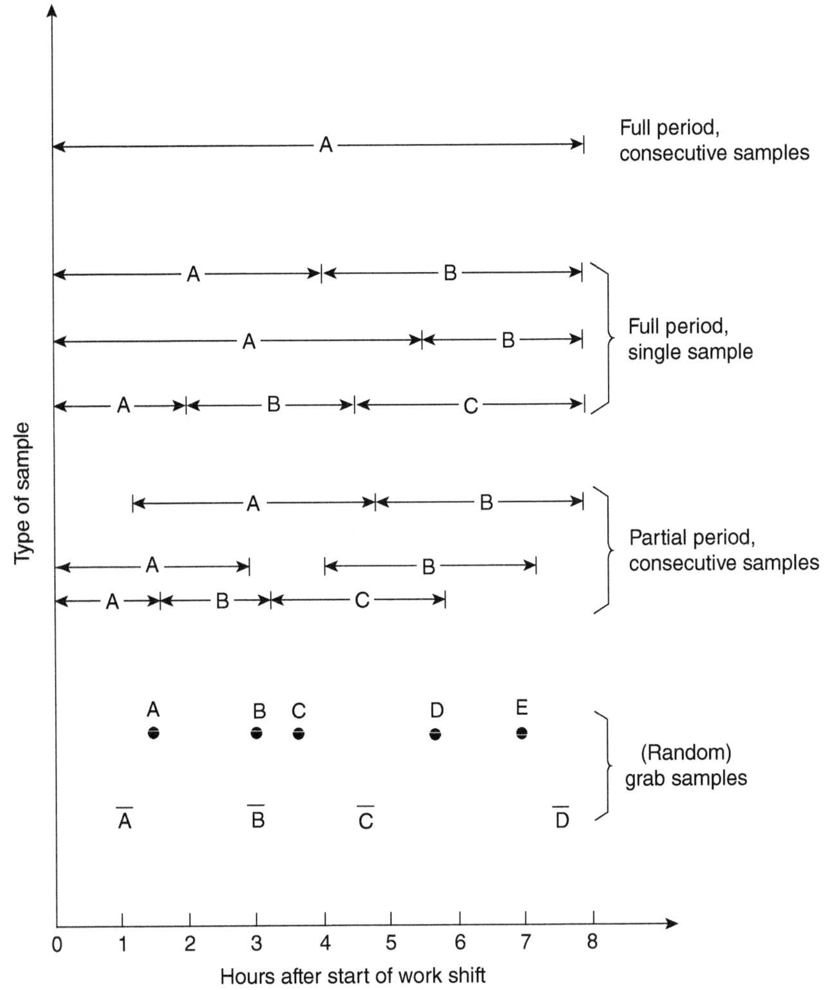


**Figure 4.4** Types of exposure measurements that could be taken for an eight-hour average exposure standard.

It is often difficult to begin sample collection at the beginning of a work shift, or an interruption may be necessary during the period to change sample collection device. Several assumptions may be made with respect to the unsampled period. Observation of employee activities may allow the assumption that exposure was zero during this period. Alternatively it could be assumed that the exposure during the unmeasured period was the same as the average over the measured period if employee activities remained unchanged. This is the most likely assumption in the absence of information that the unsampled period was different. However, it is difficult to calculate confidence limits on the overall exposure estimate since the validity of the assumption is a source of error, there is no internal estimate of environmental variance, and the statistical situation is complex.

When only very short-period or grab samples can be collected, a set of such samples can be used to calculate an exposure estimate for the full period. Such samples are usually collected at random; thus each interval in the period has the same change of being included as any other and the samples are independent. This sampling scheme of discrete measurements within a day is analogous to a set of full-period samples used to draw inferences about what is happening over a large number of days. In both cases the environmental variance, which is usually large, has a major influence on the accuracy of the results.

Short-period sampling schemes can be useful with dual standards. For example, if a toxic substance has both a short-term, say, 15-minute, limit and an 8-hour limit, 15-minute samples taken during the 8-hour period could be used to evaluate exposure against both standards. This involves some compromise, however, since samples taken to evaluate short-period exposure are likely to be taken when exposure is likely to be at a maximum rather than at random.

The traditional method of estimating full-period exposure is by the calculation of the "time-weighted average." In this method the workday is divided into phases based on observable changes in the process or worker location. It is assumed that concentration patterns are varying with these changes and are homogeneous with each phase. A measurement or measurements, usually shorter than the length of the phase, are made in each phase, and the exposure estimate $E$ is calculated:

$$E = \frac{C_1T_1 + C_2T_2 + \cdots + C_nT_n}{8}$$

where $C_n$ is the concentration measured in phase $n$ and $T_n$ the duration (hours) of phase $n$ ($\Sigma T = 8$).

When an averaging time longer than a full shift is needed, that long-term average (LTA) is usually calculated from some number of full-shift samples. For example, the coal mine dust standard is based on the average of five full-shift measurements. Single samples of multiple shifts, excluding nonshift periods, are not widely used. When workplace measurements are made for purposes other than the estimation of worker exposure, different considerations apply. While a single eight-hour sample may be an accurate measure of a worker's average exposure during that period, the exposure was probably not uniform, and the single sample gives no information on the time history of contaminant concentration. To find out when and where peaks occur, with the aim of knowing what to control, short-period samples or even continuous recordings are useful. Similarly, when a control system is evaluated or sampling methods are compared, measurements need be only long enough to average out system fluctuations and provide an adequate sample for accurate analysis. As in the case of the decision on location, the purpose of the measurement is a primary consideration in the selection of a time period of a measurement.

### 4.1.5 Frequency

By increasing the number of measurements made over a period of time or in a sampling session, the width of the error envelope as described by the confidence limits narrows, and the mean result is better defined. Also the maximum number of periods in which a limit could be exceeded (the exceedance) is reduced. With narrower (tighter) confidence limits, it becomes easier to arrive at a decision with a given degree of certainty or to be more confident that a decision is correct. The choice of the number of samples to be collected rests on three factors: the magnitude of the error variance associated with the measurement, the size of a difference between the result and the standard or guide that would be considered important, and the consequence of the decision based on the result.

The error variance associated with the measurement depends in most cases on the environmental variance. An exception is the rather limited instance of evaluating the exposure of a worker over a single day by means of a full-period measurement. In that case the error variance is determined by only the sampling and analytical error, and confidence limits tend to be quite narrow. Usually, however, our concern is with the totality of a worker's exposure, and we wish to use the data collected to make inferences about other times not sampled. There is little choice, unless the universe of all exposure occasions is measured, we must "sample," that is, make statements about the whole based on measurement of some parts.

As discussed earlier, the universe has a large variance quite apart from the error of the sampling and analytical method. In terms of our decision-making ability, the error of the sampling and analytical method may have very little impact.

The American Industrial Hygiene Association (AIHA) has addressed the issue of appropriate sample size (Hawkins et al. 1991) and recommends in the range of 6–10 random samples per homogeneous exposure group. Fewer than 6 leave a lot of uncertainty, and more than 10 result in only marginal improvement in accuracy. Also, it is usually possible to make a reasonable approximation of the exposure distribution with 10 samples. The difference between the mean and the standard necessary to achieve confidence in the conclusion decreases sharply as the number of samples used increases from 3 to 11. An important conclusion is that for a fixed sampling cost and level of effort, many samples by an easy but less accurate method may yield a more accurate overall result than a few samples by a difficult but more accurate method due to the effect of increasing sample numbers on the error of the mean in highly variable environments.

In selecting a sample size, it must be kept in mind that it is possible to make a difference statistically significant by increasing the number of samples even though the difference may be of small importance. Thus given enough samples, it may be possible to show that a mean of 1.02 ppm is significantly different from a TLV of 1.0 ppm, even though the difference has no importance in terms of biological consequence. Such a statistically significant difference is not useful. Therefore, in planning the sampling strategies, first decide how small a difference is important in terms of the data and then select a frequency of sampling that could prove this difference significant, if it existed.

The consequences of the decision made on the basis of the data collected should be the deciding factor in selecting the level of confidence at which the results will be tested. Although the common 95% (1 in 20) confidence level is convenient because its bounds are two standard deviations from the mean, it is arbitrary, and other levels of confidence may be more appropriate in some situations. When measurements are made in a screening study to decide on the design of a larger study, it may be appropriate to be only 50% confident that an exposure is over some low trigger level. On the other hand, when a threat to life or a large amount of money may hang on the decision, confidence levels even beyond the three standard deviations common in quality control may be appropriate. To choose a confidence limit, first consider the consequence of being wrong and then decide on an acceptable level of risk.

Since sampling and analysis can be expensive, some thought should be given to ways of improving efficiency. Sequential sampling schemes in which the collection of a second or later group of samples is dependent on the results of some earlier set are a possibility. This common quality control approach results in infrequent sampling when far from decision points but increases as a critical region is neared. Another means of economizing is to use a nonspecific direct reading screening method, such as a total hydrocarbon meter, to obtain information on limiting maximum concentrations that will help to reduce the field of concern of exposure to a specific agent.

Few firm rules can be provided to aid in the selection of a sampling strategy because data can be put to such a wide variety of uses. However, the steps that can be followed to arrive at a strategy can be listed:

1. Decide on the purpose of the measurements in terms of what decisions are to be made. When there are multiple purposes, select the most important for design.
2. Consider the ways in which the nature of the environmental exposure and of the agent relates to measurement options.
3. Identify the methods available to measure the toxic substance as it occurs in the workplace.
4. Select an interrelated combination of sampling method, location, time, and frequency that will allow a confident decision in the event of an important difference with a minimum of effort.

### 4.1.6 Method Selection

The selection of a measurement method depends on the sampling strategy considerations discussed above. Should a full-shift, partial-shift, task-specific, or grab sample be taken? Should the sample be representative of individual personal exposure or of the area adjacent to the task being assessed? Should the source be assessed independently of the tasks that may place the employees at risk, or should equipment emissions or a confined space be sampled to verify the absence of risk?

Is the contaminant of concern a gas, liquid, solid, or in some cases, in more than one phase? Does the environment contain only the contaminant of concern, or are other substances present as either additional potential hazards or potential interferences? The final consideration in developing a sampling plan is often the immediate availability of sampling equipment. This last decision is constrained by the original efforts to outfit an industrial hygiene field office with sampling equipment that anticipates the answers to the questions raised above. Appendices A–C contain information to assist in setting up this function. Also various trade magazines who publish annual buyers' guides can be a source of vendors (Chiltons 1996; *Ind. Hyg. News* 1996).

The discussion of sampling devices that follows will reflect the most common or typical application in assessing exposures. However, each type of device has been developed and improved over time so that they can be applied interchangeably. The notable exceptions are the direct reading instruments, which are too large in most cases to be worn for a personal exposure assessment. Very often the selection of equipment begins with a choice between collecting a sample over an extended period of time (integrated sample) and performing an abbreviated instantaneous test of the environment (grab sample). The choice of sample time is often driven by the desire to compare employee exposures or source emissions to a regulatory (i.e. OSHA) or advisory (i.e. American Conference of Governmental Industrial Hygienists [ACGIH]) exposure limit. When the limits are based on exposures over 8-hour 30- or 15-minute averaging times, it is necessary to sample the workplace in question for a similar period of time for comparison.

*Grab or short-term sampling* is driven most often by the need to perform quick and convenient measurements of contamination as a first response to an employee complaint; as a screening tool to identify target populations or activities for more intensive investigation; to assess engineering or administrative controls; to localize sources of contamination; and to confirm the appropriate selection of respiratory protection or to access acute hazards to the general population. Many of the long-term sampling techniques discussed earlier can be modified for application as short-term sampling techniques; however, generally three techniques are most often used because of the immediacy of the data. They are (i) length-of-stain detector tubes, (ii) portable direct reading instruments, and (iii) bag or container sampling.

*Integrated samples* are collected in a variety of media. These are selected based on considerations of chemical interactions that capture and stabilize the contaminant. These samples most often are sent to a laboratory for analysis. For this reason, the chemistry of the contaminant, its collection method, and analysis technique are closely associated. Table 4.2 provides a general list of techniques for sampling and analysis of such contaminants. These general methods have been evaluated and developed into specific field sampling and lab analysis protocols by the NIOSH (1984) and are updated and republished periodically. Part of planning a successful sampling program would be to select the best media to collect the contaminants of greatest concern based on the operations and activities covered by the industrial hygienists' scope of responsibilities and to arrange support from a laboratory qualified to perform the required analysis. Laboratory qualification and certification are discussed in Section 4.4.2. An added issue in selecting a support lab would be to determine its capabilities in providing technical advice and sampling supplies for circumstances when the industrial hygienists must evaluate unique or unusual exposures.

## 4.2 METHODS FOR GASES AND VAPORS

As the terms are used in this field, gases are substances such as carbon monoxide whose vapor pressure is greater than atmospheric under normal conditions, while vapors are the gaseous phase of a substance such as benzene whose vapor pressure is less than atmospheric under normal conditions and which can therefore be present in both liquid and gaseous phase. Both gases and vapors are capable of mixing completely with air, although they are generally not uniformly mixed. Sampling methods for gases and vapors must have some means of capturing the contaminant for subsequent analytical evaluation. A typical sampling system consists of a portable battery-operated vacuum pump connected to a collection medium by soft plastic tubing (usually 1/8 or 1/4 in. ID). The pump is designed to be clipped to the worker's belt and to be light enough to be worn a full shift with a minimum of discomfort. A glass tube containing solid sorbent (Figure 4.5) is attached to the plastic tube and pinned to the worker's lapel.

### 4.2.1 Absorbent Tubes

Although Table 4.2 lists a variety of choices in methods for sample collection, the most frequently used method is the concentration of the contaminant on a solid sorbent contained in a glass tube (Figure 4.6). This technique employs a sampler that is relatively small and easily worn by an employee and is dry and also convenient to handle, store, and ship. The glass tube contains a main section of solid sorbent followed by a porous plug and then a second backup section of sorbent. The backup section, which is commonly half the mass of the front section, is analyzed to determine if the analyte has broken through the front section and compromised the sample results.

The main drawback to this sampling method is that the contaminated air must be drawn across the sorbent with a battery-driven portable pump. The pump and a connecting tube are usually clipped to the employee's belt, and the sampling tube is attached to a shirt collar. The pumps are designed to allow adjustments to the rate of air volume sampled, the overall volume collected, and the total amount of contaminant collected. These parameters are critical in

**TABLE 4.2 Sampling and Analysis Method Attributes**

| Method[a] | Sampling period[b] | Ability to concentrate contaminant[c] | Ability to measure mixtures[d] | Time to result[e] | Intrusiveness[f] | Proximity to nose and mouth[g] |
|---|---|---|---|---|---|---|
| Personal sampler/solid sorbent | | | | | | |
| Sorption only | Medium to long | Yes | Yes – gases | After analysis | Medium | Very close |
| Sorption plus reaction | Medium to long | Yes | No | After analysis | Medium | Very close |
| Personal sampler/filter | | | | | | |
| Gross gravimetric | Medium to long | Yes | Yes – particulate | After weighing or analysis | Medium | Very close |
| Respirable gravimetric | Long | Yes | Yes – particulate | After weighing or analysis | Medium | Very close |
| Count | Medium to long | Yes | Yes – particulate | After counting | Medium | Very close |
| Combination filter and sorbent | Medium to long limited | Yes | Yes | After analysis | Medium | Very close |
| Passive dosimeter | Long | Yes | Yes – gases | After analysis | Low | Very close |
| Breathing zone impinger/bubbler | | | | | | |
| Analysis | Medium–limited | Yes | Yes | After analysis | High | Close |
| Count | Medium–limited | Yes | Yes – particulate | After counting | High | Close |
| Detector tubes | | | | | | |
| Grab | Short | NA | No | Immediate | High | Close |
| Long period | Medium to long | NA | No | Immediate | Medium | Very close |
| Gas vessels | | | | | | |
| Rigid vessel | Short to long | No | Yes – gases | After analysis | High | Medium |
| Gas bag | Short to long | No | Yes – gases | After analysis | High | Close |
| Evacuated/critical orifice | Medium to long | No | Yes – gases | After analysis | Medium | Distant |
| Direct reading portable meters | | | | | | |
| Nonspecific (flame ion, combination gases) | Instantaneous or recorder | NA | Yes | Immediate | High | Slightly distant |
| Specific (carbon monoxide, hydrogen sulfide, ozone, sulfur dioxide, etc.) | Instantaneous or recorder | NA | No | Immediate | Medium | Slightly distant |
| Multiple compounds (infrared, gas chromatography, etc.) | Instantaneous or recorder | Some | Yes | Almost immediate | High | Slightly distant |
| Mass monitor ($\beta$-absorber, piezoelectric) | Short | Yes | No | Almost immediate | High | Slightly distant |
| Particle counters (optical, charge) | Short | No | No | Almost immediate | High | Slightly distant |
| Sensor with data logger | Short or long | No | No | Hours | Medium | Close |
| Fixed station | | | | | | |
| High volume | Medium to long | Yes | Yes – particulate | After analysis | Low | Remote |
| Horizontal or vertical elutriator | Long to short | Yes | Yes – particulate | After analysis | Low | Remote |
| Installed monitor | Short to long | Some | No | Almost immediate | Low | Remote |
| Freeze trap | Medium | Yes | Yes – vapors | After analysis | Low | Remote |
| FTIR | Instantaneous | No | Yes – gases | Immediate | Low | Remote |

(*Continued*)

**TABLE 4.2 (Continued)**

| Method[a] | Specificity[h] | Convenience rating[i] | Sample transportability[j] | Recheck of analysis possible[k] | Accuracy[l] |
|---|---|---|---|---|---|
| Personal sampler/solid sorbent | | | | | |
| Sorption only | High by analysis | High | Good | Elution – yes; thermal des. – no | Good |
| Sorption plus reaction | High by analysis | High | Good | Yes | Good |
| Personal sampler/filter | | | | | |
| Gross gravimetric | None for weight only – high by analysis | High | Fair | Yes | Good |
| Respirable gravimetric | High by analysis | Medium | Fair | Yes | Fair |
| Count | Fair – depends on particle identification | High | Good | Yes | Poor |
| Combination filter/sorbent | High by analysis | Medium | Good | Yes | Fair |
| Passive dosimeter | High by analysis | Very high | Good | Yes | Fair |
| Breathing zone impinger/bubbler | | | | | |
| Analysis | High by analysis | Low | Poor | Yes | Fair |
| Count | Fair – depends on particular analysis | Low | Poor | Yes | Poor |
| Detector tubes | | | | | |
| Grab | Medium – some interference | High | No sample | No | Fair |
| Long period | Medium – some interference | High | No sample | No | Fair |
| Gas vessels | | | | | |
| Rigid | High by analysis | Low | Fair | Yes | Good |
| Gas bag | High by analysis | Low | Fair | Yes | Good |
| Evacuated/critical orifice | High by analysis | Low | Good | Yes | Good |
| Direct reading portable meters | | | | | |
| Nonspecific (flame ion, combination gases) | None – total of measured class | High | No sample | No | Good |
| Specific (carbon monoxide, hydrogen sulfide, ozone, sulfur dioxide, etc.) | Medium – some interference | High | No sample | No | Good |
| Multiple compounds (infrared, gas chromatography, etc.) | Medium – frequency overlap | Medium | No sample | No | Fair |
| Mass monitor ($\beta$ absorber, piezoelectric) | Mass only | High | No sample | No | Fair |
| Particle counters (optical, charge) | Count/size only | High | No sample | No | Fair |

| | | | | | |
|---|---|---|---|---|---|
| Sensor with data logger | Medium – some interference | High | No sample | No | Good |
| Fixed station | | | | | |
| High volume | High by analysis | Low | Fair | Yes | Good |
| Horizontal or vertical elutriator | High by analysis | Low | Fair | Yes | Good |
| Installed monitor | Medium – may be interferences | High | No sample | No | Good |
| Freeze trap | High by analysis | Very low | Poor | Yes | Fair |
| FTIR | Medium – may be interferences | High | No sample | No | Fair |

This table shows that not all sampling strategies are possible, since for some strategies the sampling and analytical method with the necessary combination of attributes may not exist. The technology gaps thus revealed are fruitful areas for future research and development.

[a] The methods listed include both sampling and direct reading methods. For the sampling methods the ratings of attributes that follow assume the usual range of analytical methods that can be applied to the size and type of sample collected.

[b] By *short* is meant essentially instantaneous or grab samples, while long means eight hours or longer in a single sample.

[c] Sampling methods that extract a contaminant from the air and collect it in a reduced area or volume are potentially able to improve analytical sensitivity by several orders of magnitude. However, the concentrating mechanism (filtration, sorption) may introduce errors.

[d] Most sampling methods provide a sample that can be analyzed for more than one gas or vapor, but usually not for both gases and vapors or particulates.

[e] Certain decisions (vessel entry) must be made immediately, while others can wait until after the sample is transferred to a laboratory and analyzed.

[f] When the method requires the presence of a person to collect the sample or the wearing of a heavy or awkward sampling apparatus, this intrusion of the sampling system into the work situation may affect worker behavior and exposure.

[g] As discussed earlier, locating a sampler inlet even a small distance from a worker's mouth may bias the exposure measurement. Samplers remote from the worker may not be measuring the air inhaled at all.

[h] Some methods give only nonspecific information like total weight of all dust particles or concentration of all combustible gases, while others measure a specific substance directly or provide a sample that can be analyzed for any species or element.

[i] These are estimates of the amount of work or difficulty involved in collecting samples.

[j] If the sample must be transported to a distant laboratory for analysis, the ability to withstand shock, vibration, storage, and temperature and pressure changes without being altered or destroyed is important.

[k] Some samples may only be analyzed once, while others are in a form such that rechecks, reanalyses at different conditions, or analysis for other substances is possible.

[l] Given all the possibilities for error form sampler calibration, sample collection, transport, and analysis, an overall coefficient of variation (CV) of 10% is considered good. Some count methods are subject to such counter variability that poor accuracy is usual. Method inaccuracy should not be judged alone but should be seen in combination with the inaccuracy caused by environmental variability, which is usually larger, in making decisions whether a method is sufficiently accurate for a purpose.

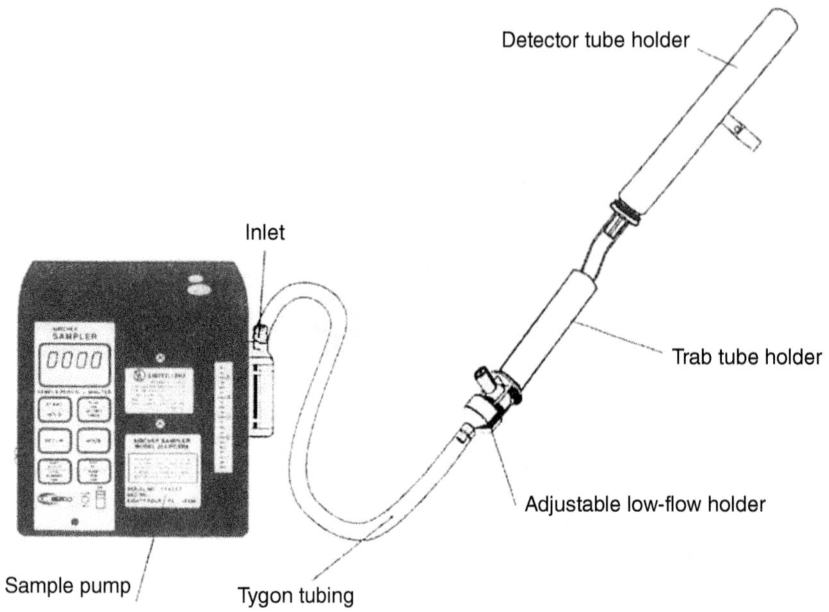


**Figure 4.5** Sampling train. *Source:* Courtesy SKC Inc.

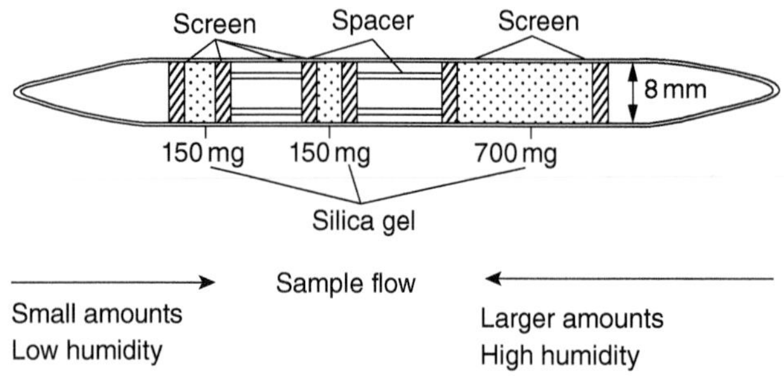


**Figure 4.6** Sampling tube for aromatic amines.

efficiently collecting an accurate representation of the environment being assessed. Limitations of absorbent tubes include the fact that not all gases and vapors are efficiently collected (e.g. ethylene) or desorbed (e.g. PNAs). Also the volume of air sampled must be carefully selected so as to yield adequate material on the sorbent for analysis without exceeding the collection capacity of the tube and causing breakthrough. The sampling and analytical method will determine these factors.

### 4.2.2 Passive Badges

As the sensitivity of analytical tools in the laboratory has improved, smaller and smaller amounts of material have been required for quantification. This has made possible the development and use of lightweight passive sampling badges (Figure 4.7). In these devices, the contaminant migrates across a narrow turbulent free space to the adsorbent material in the same way that fragrant aromas originating in a kitchen spread throughout a home. The rate of diffusion of the material represents its sampling rate and controls the total material deposited on the sorbent, which is available for analysis. Thus passive diffusion replaces the sampling pump as the means of bringing the contaminant and the sorbent into contact. Diffusion rates are determined for each material under laboratory conditions and may require adjustments if field conditions of temperature, humidity, and pressure vary greatly from the typical lab setting. Either the vendor or the analytical laboratory is the best source of information on the performance of these devices. While passive samplers offer greater convenience over pumped samplers, their effectiveness is poor in situations where significant air turbulence or the source causes rapid fluctuations in contaminant concentrations. Also the sampling time must be long enough to yield an adequate sample size. If these limitations apply, pumps and sampling tubes should be utilized.

Activated charcoal has found the widest application as an adsorbent material for collecting organic contaminants and is available packed in glass tubes and in passive badges. Other sorbents such as silica gel, alumina, and molecular

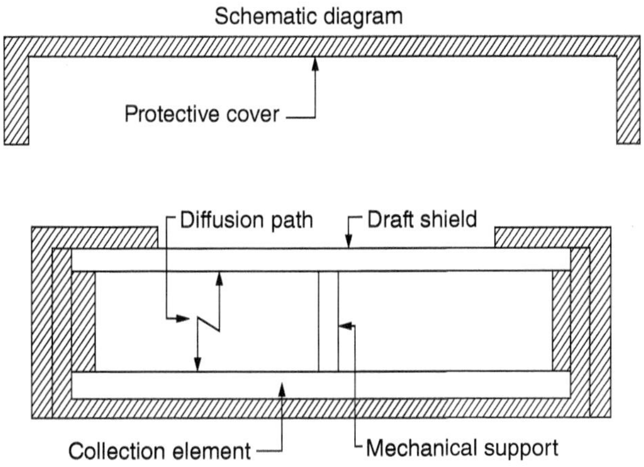


**Figure 4.7** Passive sampling dosimeter.

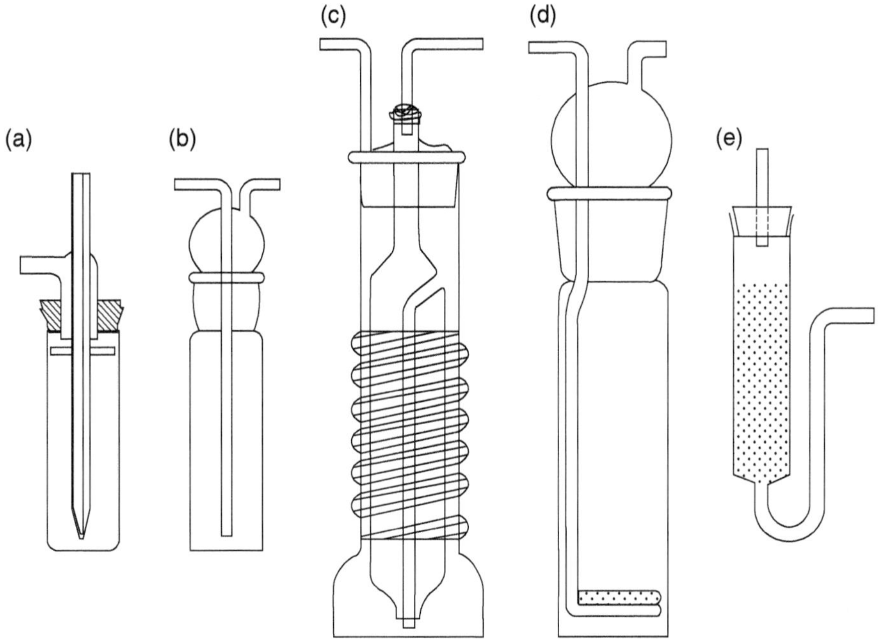


**Figure 4.8** (a)–(c) Three types of wet impingers. Smith-Greenberg impinger samples air at 28.3 lpm, $\delta P$ = 3-in Hg, and terminal velocity is 100 m s$^{-1}$. Water in impinger may be approximately 75 cc. For midget impinger, sampling rate is 2.83 lpm, $\delta P$ = 12-in $H_2O$, terminal velocity 70 m s$^{-1}$, and typical liquid = 10 cc $H_2O$. (d) Bubbler. (e) Packed absorber.

sieves have been developed for collecting polar compounds, and in some cases materials that have been developed for use in chromatography, such as XAD, have also been applied to sampling contaminants that are not more easily collected. The recommended absorbent sampling rate, total volume, and recommended techniques for sample preservation are summarized in NIOSH's established methods (NIOSH 1984). Sample calculations of these variables will be discussed in Section 4.4.1.

### 4.2.3 Impingers and Liquid Traps

In some cases where the contaminant is unstable, is reactive, or cannot be collected on solid media, solutions contained in small bubblers are employed (Figure 4.8). Air is drawn by a sampling pump through a hollow glass tube drawn to a fine tip into an impinger filled with a liquid to produce small bubbles. The bubbles provide a large gas to liquid surface, allowing the contaminants to migrate into the liquid phase. In the liquid they are trapped and in some cases stabilized by reacting with a reagent selected for this purpose. Some impingers have been designed to minimize spillage while they are attached to an employee's lapel. An additional concern is loss of the trapping solution during the sampling period due to evaporation.

### 4.2.4 Length-of-Stain Tubes

Length-of-stain detector tubes are a popular method of grab sampling because of the wide variety of tubes that have been developed for specific chemical contaminants, their ease of use, and their low cost and storage stability. These tubes provide an inexpensive method of anticipating the broadest possible range of demands for a quick initial exposure assessment. Detector tubes are sold as sealed glass tubes with a solid granular matrix such as silica gel, alumina, or pumice that has been impregnated with a dye that will change color as air containing a specific contaminant is drawn through the tube. These tubes are most commonly used for short-duration grab samples using hand-operated piston or bellows pumps. The tubes are calibrated by the manufacturer such that when a fixed volume of air is drawn across the tube, the length of the coloration corresponds to the concentration in air.

Standard detector tubes are of limited value in assessing exposure relative to a full-shift standard due to their short sampling time. For certain reactive gases and vapors, a length-of-stain indicating tube has been developed to measure full-shift exposures. In some select cases low-flow sampling pumps are used for long-period samples. Passive diffusion may be used when it allows enough contaminant to migrate onto the tube that a stain can be developed. Some applications are listed in Table 4.3.

All detector tubes are subject to limitations of specificity and sensitivity. Specificity is the ability of a method to detect and measure a desired contaminant in the presence of other chemicals. Chemicals that react similarly in a procedure are said to "interfere." Sensitivity is the smallest detectable change in analyte concentration that is measurable by the method. These issues are discussed in detail in the manufacturers' literature, and the restrictions must be strictly observed for the measurement to be considered valid.

### 4.2.5 Evacuated Containers and Bags

A technique that is often used for grab sampling is slow-filling containers that collect both contaminant and air without any method of separating or concentrating the contaminant at the time of sampling. This technique can only be considered when the analytical procedure has sufficient sensitivity to detect the contaminant at ambient concentrations (usually parts per million). Gas chromatography and infrared are best suited to this purpose. Samples are most often collected using a gas bag that is slowly filled by a low-flow pump or by using an evacuated glass or metal container. The container can be opened and filled instantly, or the mouth can be equipped with a critical orifice that will allow for slower filling at a steady rate. Syringe needles obtained from a chromatography lab can be used as critical orifices. Needles of different diameter openings can be used to vary the sampling rate. The evacuated container is

**TABLE 4.3 Long-Term Colorimetric Detector Tubes**

| |
|---|
| Acetic acid |
| Acetone |
| Ammonia |
| Benzene |
| Butadiene |
| Carbon dioxide |
| Carbon disulfide |
| Carbon monoxide |
| Chlorine |
| Ethanol |
| Ethyl acetate |
| Hydrochloric acid |
| Hydrocyanic acid |
| Hydrofluoric acid |
| Hydrogen sulfide |
| Methylene chloride |
| Nitrogen dioxide |
| Nitrous fumes |
| Perchloroethylene |
| Sulfur dioxide |
| Toluene |
| Trichloroethylene |
| Vinyl chloride |
| Water vapor |

sealed with a one-hole stopper with a hollow glass tube passing through it. The tube and needle are connected by a short run of plastic tubing that is tightly clamped. In some instances where a transportable field gas chromatograph is nearby, the gastight syringe normally used to inject samples into the instrument can become the grab sampling device itself.

With these techniques the loss of contaminant to the surface of the container is a common concern. Bags made of inert materials such as Mylar, Tedlar, and PTFE are available to minimize absorption or permeation through the bag walls. Prior to application it is important to monitor the rate of decay under controlled conditions as part of the preuse selection and calibration. Cross-contamination and memory from surface residue are also of concern. Even liquid washing may only result in the wash liquid itself leaving a residue. Gentle heating under a fresh air or inert gas flow followed by testing of the container provides the best means of avoiding this problem.

### 4.2.6 Direct Reading Instruments

Direct reading instruments provide a number of advantages over sampling and analytical methods and detector tubes, but these devices tend to be more expensive than most other equipment. They are most often purchased with specific applications in mind. A disadvantage is that they usually cannot be worn by the worker; so they are not suitable for personal sampling. These instruments are capable of "instantaneous" measurements, and they can also make integrated measurements when used in connection with a data logger (see Section 4.4.8). These instruments have evolved over the past 25 years from larger bench-scale models originally developed for the analytical chemistry lab to small hand-held devices. The drive toward miniaturization and portability in the electronics industry has resulted in weight and size reductions in instrumentation that have been remarkable. Instruments often provide a quicker response, greater accuracy, and greater specificity than detector tubes, although they are not as accurate or specific as analytical methods. With the proper technical support, direct reading instruments are amenable to fairly rapid development of field methods for analyzing unusual materials where published protocols do not exist. Table 4.4 summarizes the most common sensing technologies employed in today's instruments. Manufacturers combine the detectors with electronics and other hardware to provide for many applications:

1. Point and measure
2. Continuous area or process monitors
3. Personal dosimeter
4. Complex environment analyzers

Point-and-measure devices are used to identify point sources of contaminant release (leak meter), to assess general background conditions, or to perform a preliminary identification and assessment of an exposed population. They also serve a critical purpose in identifying environments that contain acute hazards such as carbon monoxide, hydrogen sulfide, or explosive gases. Permitted entry into confined space that may contain such hazards relies heavily on measurements by these types of instruments.

Stationary area monitoring instruments that identify process leaks are often equipped with alarms or are linked to emergency shutdown systems to control the release of acutely hazardous materials such as flammable vapors, hydride gases, and acutely toxic gases (i.e. hydrogen fluoride). They can be positioned in areas adjacent to the processes or within process enclosures to monitor for upsets.

Personal dosimeters consist of a sensor that produces a continuous signal whose output fluctuates in real time with concentration. These are equipped with either alarms that activate at hazardous concentrations or dosimeters that record and store a time history profile of the exposure.

Many instruments are available that employ chromatographic techniques to separate complex environments into individual components and quantify a single toxic substance within a group of similarly toxic materials.

The wide range of selection and breadth of applications of direct reading instruments make it unlikely that one would be able to afford or anticipate all the possible applications of these devices. However, in the past few years, an equipment rental business has developed that will supply an instrument calibrated and ready to use. Several companies provide just-in-time delivery of industrial hygiene equipment either to supplement existing equipment such as pumps or calibrators or to serve as a source of more expensive specialty equipment. This resource is helpful when a specific project or a high-visibility exposure assessment justifies a onetime rental fee but not the purchase of the equipment. Vendors should be selected who demonstrate sufficient understanding of both the sampling problem and instruments to be capable of configuring the device to the application (i.e. selecting proper columns and conditions). The supplier should also be capable of calibrating the instrument in the working range anticipated for the project since the

**TABLE 4.4 Direct Reading Instrument Sensors and Their Most Common Applications**

| Detector type | Measurement principle | Most common applications |
|---|---|---|
| Solid-state pellistor | Detects changes of heat of reaction on the surface of a catalytic solid (pellistor), which alters the pellistor's sensitivity | Explosive atmosphere; high concentration of total hydrocarbon |
| Semiconductor (1) | Detects a change in conductivity of the semiconductor material produced by trapping or release of charge carriers at the surface. The shift in conductivity is proportional to concentration | Explosive atmospheres; high concentration of total hydrocarbon |
| Semiconductor (2) | Semiconductor materials made of *n*-type metal oxides are doped or mixed with other metal oxides to react selectively with contaminants that donate electrons or remove adsorbed oxygen from the solid matrix. An electric current is produced that is proportional to concentration | Inorganic reactive gases, such as oxygen, carbon monoxide, hydrogen sulfide, sulfur oxides and nitrogen oxides, hydride gases |
| Infrared analyzers | Molecules that vibrate at the atomic level absorb a characteristic infrared wavelength in proportion to the total concentration of material in the beam path | Wide variety of gases, principally organic compounds |
| Ultraviolet analyzer | Measurement principle is the same as in infrared; however, only mercury vapor has a specific absorption wavelength in the UV range | Mercury |
| Electrochemical | Gases diffuse into an electrochemical cell consisting of solid, liquid, or gel electrolyte, an anode, and a cathode. The contaminant undergoes an electrochemical reaction, which produces a current proportional to concentration | Inorganic reactive gases as above and oxygen |
| Impregnated paper tape sulfide, isocyanates | Contaminant is drawn through a paper tape impregnated with a specific color developing reagent. The intensity of color is proportional to concentration | Hydride gases, TDI, hydrogen phosgene, chlorine organic ammonia |
| Photoionization (PID) | Ultraviolet light impinging on contaminant molecules causes the species to ionize. The ions migrate across an applied electric field to a collecting electrode when a current is produced that is proportional to concentration | Many hydrocarbons |
| Flame ionization (FID) | Organic compounds are ionized in a hydrogen/air flame, which reduces the electrical resistance of the flame in proportion to the concentration of hydrocarbon | Many hydrocarbons |

capability to generate challenge concentrations can be quite difficult to achieve in a field office for specialty assessments. Finally, the supplier should be prepared to supply a replacement instrument in the case of equipment failure during a project.

## 4.3 METHODS FOR AEROSOLS

Aerosols are solid particles or liquid droplets suspended in air. Sampling methods based on filtration, impaction, and impingement were some of the earliest techniques developed to aid in evaluating exposures in the dusty trades and over the past 75 years have evolved to be applicable as both area samplers and personal samplers. To understand their use, it is necessary to consider the properties of aerosols that affect sampling.

Sampling for aerosols differs in three significant ways from gas/vapor sampling:

1. Unlike gas/vapor environments that quickly become homogeneous after leaving the source, materials suspended in air tend to vary in concentration because of large particle settling, small particle agglomeration, or turbulence in the air itself. Droplets may evaporate while in suspension or condense as the vapor cools. Particulates may become resuspended as a result of worker activities or air turbulence.
2. Particle size is a factor in determining health hazard. Particles above 10 μm are all trapped in the nasal passages and have little probability of penetration to the lung and would not be of interest if the lung were the only target organ. However, particles such as toxic metal that are swallowed after being trapped in the throat, larynx, and upper bronchia can be dissolved in the stomach and migrate to internal target organs or, as is the case with acid mist and some allergens, cause harm at the point of deposition.

3. Finally, the dynamic interrelationship between the particles and the air that keeps them in suspension requires sampling methods that take into account the physical behavior of both. For example, it is critical in sampling to attempt to maintain nonturbulent airflow patterns at the point of interface between the sampler and the ambient environment to avoid collecting a sample that is not representative of the exposure. Therefore, sampling rate must be more specifically defined for particulates than for gases.

It is well recognized that some particulate material exerts its effects at the locus of deposition in the lung. Therefore, regulatory standards set by OSHA for substances such as quartz, cristobalite, tridymite, coal dust, and cotton dust specify exposure limits expressed as the mass concentration or total particle count of the sampled fraction of dust that is respirable. The ACGIH's Chemical Substances TLVs are currently examining particulate substances to better define the size fraction most closely associated with the health effect of concern. Future TLVs for these chemicals will be based on the mass concentration of the specified fraction. The "Particle Size Selection" TLVs are expressed in three forms (ACGIH 1996):

1. Inhalable particulate mass (IPM-TLVs) will be applied to those materials that are hazardous when deposited anywhere in the respiratory tract.
2. Thoracic particulate mass (TPM-TLVs) will be applied to those materials that are hazardous when deposited anywhere within the lung airways and the gas exchange region.
3. Respirable particulate mass (RPM-TLVs) will be applied to those materials that are hazardous when deposited in the gas exchange region.

Airborne fibrous particles such as asbestos are collected on open-faced filters also to provide a homogeneous sample for microscopic analysis. Health standards are reported in terms of fiber concentration.

### 4.3.1 Method Selection

The equipment that has been developed to measure aerosol concentrations falls into three categories:

1. Devices that are small enough that they can be worn as personal samplers either for full-period integrated samples or as grab samples.
2. Devices that can be used as either full-period or grab samples but whose size or bulk limits their application to fixed-area monitoring.
3. Handheld portable and transportable instruments that can be used for grab or full-period sampling as either area or personal samplers.

The most common sampling methods and least expensive in categories 1 and 2 depend either on filtration, impaction, or impingement of the particles as the means of collection.

### 4.3.2 Filter Sampling Methods

Personal sampling is most often accomplished by passing the contaminated environment through filtering media that capture the solids. Table 4.5 lists the most commonly used filters and typical applications. The sampling assembly will consist of an air moving pump, filter cassette, and connecting tubing. Where a respirable sample is to be collected, the cassette is attached to a cyclone size selector.

The *filter cassette assembly* (Figure 4.9) consists of three pieces: a base upon which a cellulose pad (sometimes a metal screen) is placed to provide structural support for the filter, a center section that holds the filter against the support pad for "open-faced" sampling (sampling in which the entire top of the cassette is removed), and a top cover that protects the filter and seals the cassette for storage and shipment. The base and top cover have sampling ports in the center with plastic plugs. In most sampling operations, the center section is not used. The top cover fits against the filter, and the sample is taken by drawing air through the sampling ports. The joints where the base, center section, and top cover of the cassette join should be sealed when sampling (plastic electrical tape works fine), and the center section of top cover should be pressed firmly against the filter to ensure that air passes through the filter and not around the edges.

Many occupational health limits for particulates are based on measurements of total dust. Therefore, the filter cassette is most frequently used with either the top lid in place (closed faced) or removed (open faced) for more uniform

**TABLE 4.5 Common Applications of Filters**

| Filter matrix | Common applications | Most common pore size |
|---|---|---|
| Cellulose ester | Asbestos counting, particle sizing, metallic fumes, acid mists | 0.8 μm |
| Fibrous glass | Total particulate, all mists, coal tar pitch volatiles | — |
| Paper | Total particulate, metals, pesticides | — |
| Polycarbonate | Total particulate, crystalline silica | — |
| Polyvinyl chloride | Total particulate, crystalline silica, all mists, chromates | 5.0 μm |
| Silver | Total particulate, coal tar pitch volatiles, crystalline silica, PNAs | 0.8 μm |
| Teflon | Special applications (high temp) | — |

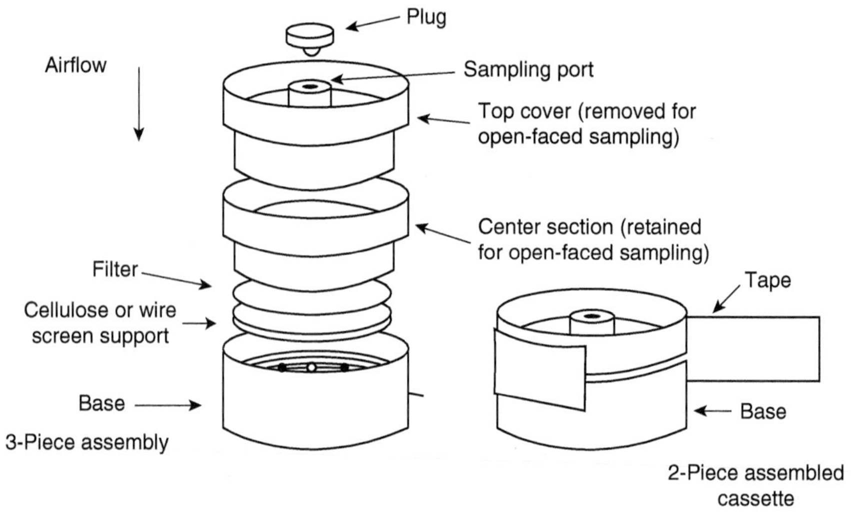


**Figure 4.9** Filter cassette assembly.

distribution of dust particles. Open-faced sampling is advisable when the method of analysis is either qualitative identification of the materials by microscopy or when the concentration is being determined by particle counting. Asbestos fibers are collected using open-faced cassettes. These cassettes are specifically designed to eliminate undersampling due to fiber adhesion to the walls of the cassette caused by static charge.

The cyclone separator is used when only the respirable fraction (the particles <10 μm in aerodynamic diameter) is to be collected. Since this size range represents the particles likely to be deposited deep in the lungs, this respirable dust sample is required to assess the hazard to employees exposed to particulates that are capable of producing lung damage using the ACGIH or OSHA limits for respirable dust (ACGIH 1996). The cyclone works by directing the stream of sampled air into a vortex. The larger particles are unable to make the turn with the curved airstream and hit the walls of the cyclone. The result is that the particles fall out of the stream before reaching the filter. The smaller particles remain in the airstream and are collected on the filter. The cyclone (Figure 4.10) consists of a tube for collecting the particles that have fallen out of the stream (grit pot), a section that fits into the top of that tube and converts the incoming airstream to a vortex, and a supporting framework for holding the cassette in place.

Because of the construction of the cyclone samplers, the calibration (see Section 4.4.3) setup is not quite the same as either cassettes or charcoal tubes because there is no easy way to attach plastic tubing to the cyclone. Figure 4.11 shows the cyclone and cassette assembly configured to allow calibration.

Another technique, elutriation, has been applied to fixed-area sampling for dust. In an elutriator the air speed and kinetic energy are controlled in such a way that nonturbulent laminar airflow is produced. In a horizontal elutriator, which is used for coal dust sampling in the United Kingdom but not in the United States, the larger particles are drawn by gravity downward and settle out at varying distances from the entrance as a function of their mass. Large particles drop out quickly, and small particles continue to be carried along with the airstream and are collected on the filter at the end. In a vertical elutriator, used for dust measurement in cotton textile mills, air enters the base of the sampler, and small particles are carried upward in the airstream to a filter, while heavier debris settles to the bottom of the sampler.

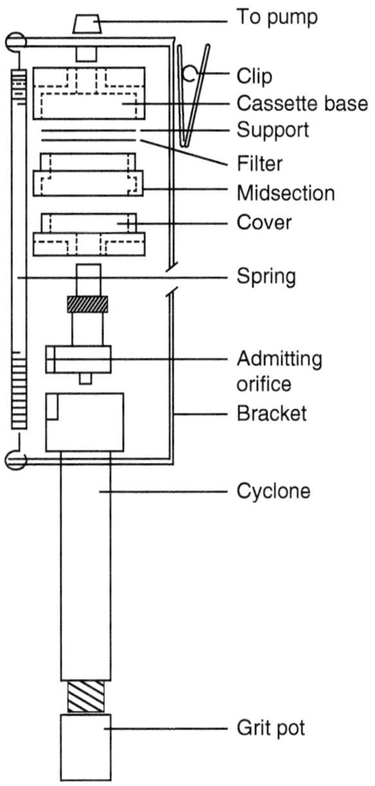


**Figure 4.10** Cyclone separator.

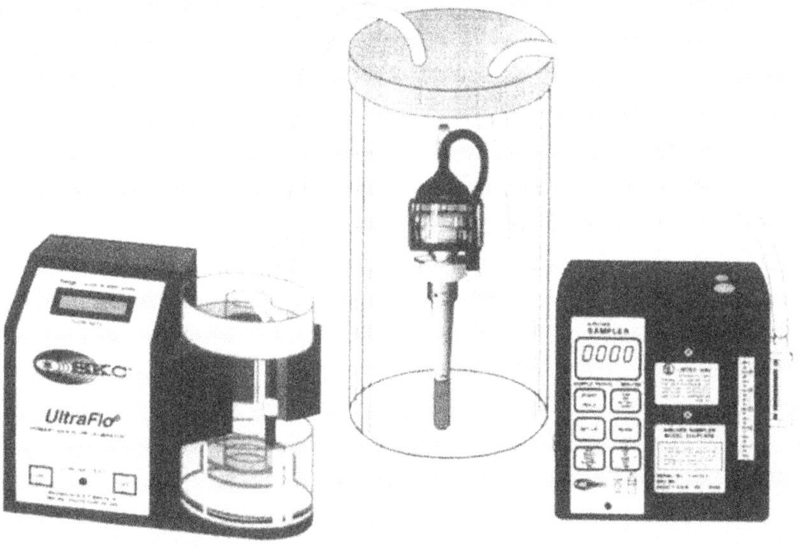

**Figure 4.11** The cyclone is calibrated by placing it in a one-liter vessel attached to an electronic bubble meter. *Source:* Courtesy SKC Inc.

### 4.3.3 Impactors

Impactors were initially designed to collect and characterize the size distribution of dusty environments. They are designed to create sudden change in direction in airflow and the momentum of dust particles so that the particles impact against a flat plate. Particles are retained on the plate's surface by an adhesive coating that minimizes bounce. Small particles traveling in the airstream contain relatively less momentum than larger particles; therefore,

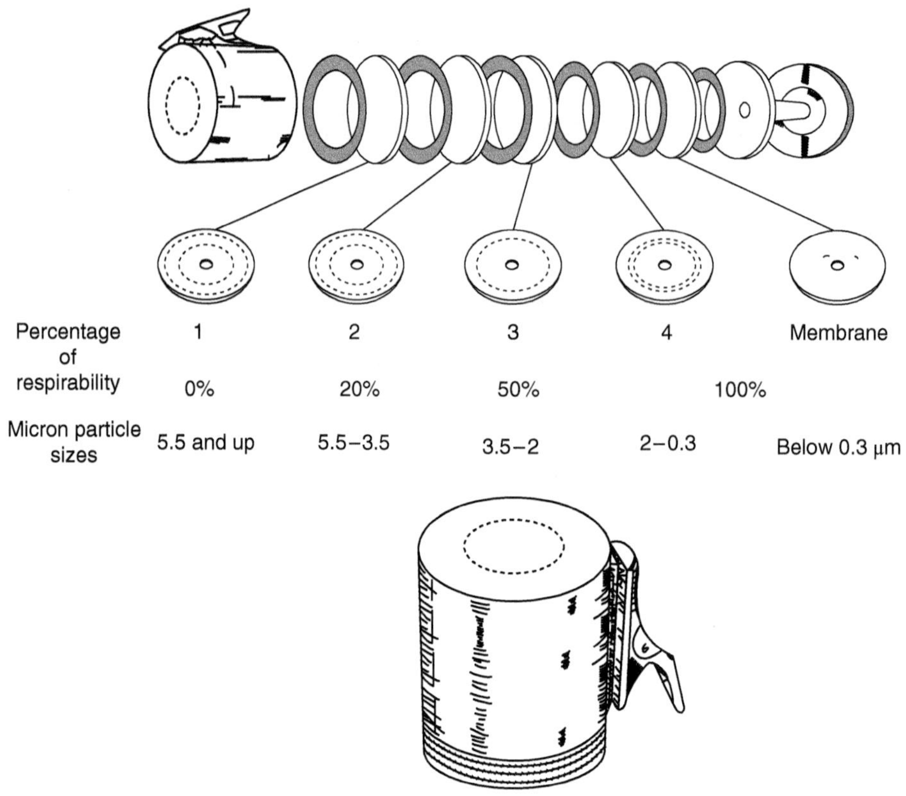


**Figure 4.12** Anderson mini-sampler.

as the airstream is redirected at a right angle to its flow, the larger particles cross the airstream and strike the flat plate, where they are captured. In a multistage impactor, a series of plates are configured in a stack, and air speed is reduced in stages. In this way, particles of progressively smaller size ranges are captured on separate plates. Devices of this design can be fairly large and were originally deployed as area samplers. However, devices such as the Anderson mini-sampler (Figure 4.12) have been developed that are small enough to be worn on an employee's lapel to permit personal sampling to characterize inhalable particle size.

### 4.3.4 Impingers

Impingers are apparatus that accelerate the dust-laden airstream to a high velocity through a small orifice. They were used in the early part of this century to capture dust particles. The particles would strike a plate immersed in a liquid and be trapped. Particles were then counted by microscope. The most familiar impingers are the Greenberg-Smith impinger and the midget impinger (Figure 4.8). These methods, however, have largely been replaced by the use of filter and impactor methods today.

### 4.3.5 Direct Reading Instruments

Instruments that can provide real-time data on particulate concentrations have continued to become available in both handheld and transportable versions and have improved in parallel with the evolution in illumination sources and microelectronics. Many bench-scale instruments, which have been developed to support air pollution monitoring and aerosol research efforts, will not be discussed here. An in-depth summary of these devices can be found in the most current edition of *Air Sampling Instruments for Evaluation of Atmospheric Contaminants*, published by the ACGIH (1978).

Instruments that are both lightweight and battery operated are available for workplace monitoring either as handheld instruments or as transportable area monitors (Table 4.6). These instruments rely on indirect methods of sensing aerosols, and frequently the calibration of signal response is developed using an artificial dust. As a result field measurements with these instruments should be viewed as approximations of exposure because the materials in the environment being monitored may differ in signal response characteristics and size distribution from the calibration dust.

**TABLE 4.6 Direct Reading Aerosol Instruments**

| Detector | Flow rate ($l\,min^{-1}$) | Size range (μm) | Lower concentration limit |
|---|---|---|---|
| | *1. Number concentration measurement* | | |
| Condensation nuclei (particle) counter (CNC or CPC) | 0.3–1.0 | 0.01–2.0 | <0.001 particle $cm^{-3}$ |
| Condensation nuclei counter ultrafine (UCNC or UCFC) | 0.03–0.30 | 0.003–3.0 | <0.001 particle $cm^{-3}$ |
| Optical particle counter, white light and laser | 0.01–28 | <1.0–2.0 | 0.001 particle $cm^{-3}$ |
| Cloud condensation nuclei counter | 1–5 | 0.08–1 | 1 particle $cm^{-3}$ |
| Ice nuclei counter | | 10 | 0.001 particle $cm^{-3}$ |
| | *2. Mass concentration measurement* | | |
| Quartz crystal microbalance | 1–5 | 0.01–20 (electrostatic) 0.3–20 (impactor) | 0.01 $mg\,m^{-3}$ |
| Vibrating sensor | 1–5 | 0.01–20 | <1 $mg\,m^{-3}$ |
| $\beta$-Attenuation sensor | 1–12 | 0.01–20 (filtration) 0.3–20 (impactor) | 0.01 $mg\,m^{-3}$ |
| Photometer, nephelometer | 1–100 | 0.1–2 | 0.01 $mg\,m^{-3}$ |
| Electrical aerosol detector | 0.5–20 | 0.1–2 | 5 $cm^{-3}$ at 1 μm |

Devices are available that measure total particle count and/or total mass concentration. Some of the same devices came with fractionating cyclones that allow for determining respirable and total particulate concentration. Other instruments incorporate size separation with particulate measurement to allow detailed real-time measurement of size distribution. The devices most often used measure a parameter of the aerosol over its entire size range. A commonly used instrument measures the concentration of the total number of particles by counting condensation nuclei. In other instruments total mass concentration is measured by the alteration in frequency of a vibrating quartz crystal caused by mass loading of particulates or by attenuating a $\beta$-radiation source. Forward scattering photometers have been developed that measure both mass and number concentration. In one application of this method, an instrument has been miniaturized, and data logging electronics have been incorporated to introduce a personal real-time monitor that also records a time history exposure profile.

In another instrument particles are deposited by electrostatic precipitation onto an oscillating quartz crystal, thereby increasing its mass and decreasing its frequency. The difference in frequency between this crystal and a reference crystal is sensed and converted into mass concentration, which is automatically displayed. The device is sold with a variety of size-selecting inlets that can be chosen to measure particulates below a certain cutoff size, including the respirable fraction. The measurement range is from 10 $\mu g\,m^{-3}$ to 10 $mg\,m^{-3}$. The vibrating crystal is somewhat sensitive to temperature and humidity fluctuations and should be allowed to equilibrate to the surrounding environment before actual measurement begins. The instrument's accuracy is believed to improve as the target dust becomes stickier. A β-attenuating analyzer operates by drawing a known quantity of air through a critical orifice and depositing its particulate contents on a surface. This surface can be a rotatable glass disk. The particulate deposits are small pinpoints on the outer periphery of the glass disks. After a sampling cycle is completed, the instrument activates a radiation emitter beneath the glass disk. Radioactive particles then travel through the pinpoint particulate deposits. A sensor on the other side of the glass disk determines how much radiation passes through the deposit and relates the degree of radiation penetration to mass of particulate collected. By placing a cyclone over the instrument inlet, it is possible to obtain a readout of mass respirable particulates.

Instruments that employ incandescent light or lasers as sources and dark-field microscopy optics have been developed. In these instruments a narrow beam of light is focused on an aerosol cloud, and light falling on photoreceptor in the near forward field is measured. These instruments can measure aerosol concentration as either total mass or particle concentration. Measurements using these devices are less sensitive to variations in the refractive index and size range of the target aerosol relative to the calibration dust than photometers. These types of instruments are available as a handheld lightweight device and also as a personal time history monitor.

Fibrous aerosols can be measured using an instrument that continuously draws air into a horizontal cylindrical cell, where it is illuminated by a helium–neon laser source. The fibers are subjected to a rotating electrical field that allows their variable scattering to be detected. Fibers that are 2–3 μm long and 0.2 μm in diameter can be measured in concentrations of 10–25 $f\,cm^{-3}$.

To determine particle size distribution, size-selecting devices are attached upstream of the sensing chambers as modifications to the instruments discussed above. Instruments that can provide real-time particle size analyses are generally large lab models requiring greater power sources, which place them in the category of transportable fixed-area monitors that are not frequently used in routine industrial hygiene evaluations. These types of instruments are well described and cataloged in the ACGIH (1978) publication.

## 4.4 GENERAL CONSIDERATIONS

### 4.4.1 Planning the Collection of a Sample

To successfully collect a representative sample, three criteria must be considered:

1. An adequate amount of contaminant must be collected to provide the lab with a large enough quantity to be analyzed. Normally a total quantity that allows for analysis in the midpoint of the sensitivity range of the analytical method is ideal.
2. The maximum capacity of either air volume or contaminant mass that the sampling matrix can retain must not be exceeded. Exceedance of this factor is usually recognizable by a lab report indicating breakthrough to the backup section of the sampler. (See Section 4.2.1 for a discussion of sampling tube design.)
3. Sampling rate must be chosen such that (1) and (2) are satisfied while sample collection takes place over the desired period of monitoring. This period closely approximates the 8-hour and 15-minute exposure limits or the duration of the task if it is less.

Information on criteria (1) and (2) will be provided in the sampling and analytical method (NIOSH 1984). Minimum sampling times and volumes can be calculated as follows:

$$\text{Minimum volume of air sampled} = \frac{10 \times \text{Analytical detection limit}}{\text{Hygiene standard}}$$

$$\text{Minimum duration of sample} = \frac{\text{Minimum volume}}{\text{Flow rate}}$$

The analytical sensitivity is normally discussed in units of mass such as milligrams or micrograms; therefore, parts per million should not be used to express the hygiene standard but rather milligrams per cubic meter. Normally one begins by calculating the two minimum values and then adjusting the parameters to the particulars of the situation.

***Example***

Collect an eight-hour sample for benzene to detect at 50% of the hygiene standard of 1 ppm (3 mg m$^{-3}$) using a laboratory that can detect 0.005 mg and a sampling tube with breakthrough capacity of 100 l.

$$\begin{aligned}
\text{Minimum volume of air sample} &= \frac{10 \times 0.005\,\text{mg}}{0.5 \times 3\,\text{mg}\,\text{m}^{-3}} \\
&= 0.033\,\text{m}^3 \text{ of air} \\
\text{Converting m}^3 \text{ to liters} &= 1000 \times 0.033 \\
&= 33\,\text{l of air} \\
\text{Minimum duration of sample} &= 8\,\text{h} = 480\,\text{in} = \frac{33\,\text{l}}{\text{flow rate}} \\
\text{Flow rate} &= 3.3\,\text{l} / 480\,\text{min} \\
\text{Converts units to common cc}\,\text{min}^{-1} &= \frac{33\,\text{l} \times 1000\,\text{cm}^3\,\text{l}^{-1}}{480\,\text{min}} \\
&= 70\,\text{cm}^3\,\text{min}^{-1}
\end{aligned}$$

The calculation indicates approximately 70 cm$^3$ min$^{-1}$ as required for the example conditions. Typically a sample such as benzene will be collected between 100 and 200 cm$^3$ min$^{-1}$ without adversely affecting the outcome of the results. The total volume required to cause breakthrough is roughly three times the volume to be sampled.

The risk of breakthrough only exists if either of two circumstances exists alone or in combination: first, if sampling at the highest rate ($200\,cm^3\,min^{-1}$) exceeds a full shift or if the concentration of benzene in the environment is substantially above the collection capacity of the charcoal. Normally total air volume to breakthrough at a given concentration is determined as part of the evolution of a sampling method, while mass loading is not. As a rule of thumb, a flow rate closest to the minimum specified range should be selected when sampling environments are suspected of high concentrations. Alternatively, the sampling period can be divided into segments, and tubes can be replaced at the end of each period (e.g. four tubes sampled consecutively for two hours each during an eight-hour operation), and the results averaged to give a full-period exposure. The maximum sampling time can be calculated as shown below:

$$\text{Maximum sampling time} = \frac{1}{70\,\text{cm}^3\,\text{min}^{-1} \times \frac{1}{1000\,\text{cm}^3} \times \frac{1}{1001}}$$
$$= 1430\,\text{min or } 24\,\text{h}$$

### 4.4.2 Analytical Lab Services and Chain of Custody

Industrial hygienists most often rely on support from laboratories who are capable of analyzing samples collected in the field. These labs can be small independent service providers, part of larger environmental testing labs, or captive, internal company labs, or, in some cases where unique tests are required, university or research labs. In selecting a lab, the health and safety professional should be convinced that the laboratory has the technical expertise to support the problems most likely encountered in the field and can demonstrate competence. Labs may have available a sampling protocols manual, which assures that the techniques used to collect material are compatible with their established analytical procedures. They may also provide collecting media. The selected laboratory should demonstrate general competence by participating in appropriate external quality assurance and certification programs. The most familiar program is the American Industrial Hygiene Association Proficiency Analytical Testing program. Laboratories that demonstrate analytical and quality assurance competency are accredited by AIHA. A list of accredited labs is published biannually in the *AIHA Journal* (usually April and September issues). The selected laboratory should be capable of producing a set of standard operating procedures, chain of custody practices, sample handling and control procedures, instrument calibration and maintenance records, internal and external testing programs, and quality control of individual methods. Frequently, these laboratories will follow protocols published by the NIOSH (1984); however, on some occasions, novel sampling and analytical procedures will be developed on request. In these cases, the client health and safety professional should review the methodology used to verify the accuracy, precision, sensitivity, temperature, humidity, and storage effects of the procedure.

Both the health and safety professional and the laboratory management should be concerned with maintaining the integrity of the sample through a chain of custody procedure. This practice assures that the analyzed samples are not confused with each other and that the sample was not damaged or in some way altered as it passes through various individuals' or organization's areas of responsibility. In the field, samples should be dated, labeled, and, if necessary, stored to remain fresh as soon as possible after collection. Sampling data sheets should be updated continuously with pre- and post-sampling flow rates, date and duration of sampling, type of sample, individual collecting the sample, and critical information such as the individual location or operation being investigated as well as pertinent field notes. Sample and data sheets should be linked by a common identification number. The laboratory and field personnel should maintain sample logs. Examples of forms that can be used for this purpose can be found in Appendices D–F.

Frequently, bulk samples are requested by the service lab. These often need to be shipped separately to avoid leakage and contamination of the field samples. The health and safety professional should also submit blank and spiked samples to determine if unanticipated circumstances have compromised the sample. Spiked samples are most easily prepared by injecting a minute volume of the pure contaminant onto the sampling media. This can be in the form of a saturated vapor obtained with a gastight syringe from the space above the liquid in a shaker bottle. In some cases, the liquid material itself can be injected. The health and safety professional should calculate the mass loading to be approximately in the midrange of the anticipated exposure measurements.

There are two types of blank samples. In one case, a sealed tube or filter is analyzed to identify any positive interferences due to contamination of lab agents or zero drift in the lab procedure. In a second case, a blank tube is broken or filter cassette uncovered in the field but not employed to actively collect a sample. At the completion of sampling this "field blank" is treated as all the other samples. This field blank identifies potential sampling errors due to loading from the environmental conditions and/or shipping and handling procedures. For example, solvents from magic markers used to identify individual passive badges have been reported to be evaporating during shipment and collecting on the charcoal of blanks and samples to produce false-positive data.

### 4.4.3 Instrument Calibration, Verification, and Maintenance

There are two degrees of calibration. A laboratory-grade calibration is designed to determine within specified limits the true values associated with scale readings on an instrument under critical challenge conditions that reflect field applications. A field-grade calibration tests the instrument's ability to indicate a correct reading. The laboratory calibration is most often performed by the manufacturer or a third-party lab and is discussed in Section 4.4.4. Field calibration is performed prior to a monitoring activity to verify instrument performance. The simplest field calibration is to challenge an instrument with a prepared gas mixture purchased from a vendor. The challenge contaminant is certified by the supplier to be at a given concentration, and the span setting of the instrument is adjusted to conform to this value. A second source of calibration mixture can be produced by evaporating calculated quantities of challenge liquid into clean static containers such as a bag, bottle, or drum in such a way that a uniform environment is produced. In some cases, the headspace above a liquid stored in its bottle serves as a most convenient source of this type of a challenge vapor. A third technique bubbles clean, dry air through a heated liquid in order to create a saturated vapor stream. This stream is then diluted in a mixing chamber with a second source of clean or humidified air to produce a challenge concentration. Each of these techniques has advantages and disadvantages and is usually chosen based on convenience, space limitations, and access to materials. References discussing calculations and setup of calibration systems can be found in the bibliography (Chelton 1993).

Field verification should also include a test to confirm that sampling tubes are not leaking. Sample collecting pumps should be calibrated for proper flow using bubble meters or flow sensors. This is critical for sampling pumps connected to tubes, impingers, and filters, and pumps internal to direct reading devices should be checked periodically. In addition, the performance of batteries, filters, flashback arrestors, sensor life, and electronics performance should also be confirmed before beginning a sampling exercise. All these evaluations should be recorded and maintained for field quality control purposes. Example forms can be found in Appendices D–F.

The above evaluation discussed can be performed either prior to use or as part of an ongoing scheduled maintenance program. The most common equipment failures are loss of battery rechargeability, leaking sample tubes, clogged prefilters, failed or clogged flashback arrestors (this should be tested and serviced by the vendor), sensor aging, electronics drift, or oxidation of contacts. Scheduled servicing or field parts replacement should be discussed with each equipment supplier on an individual basis.

### 4.4.4 Sampling Equipment and Instrument Certification

It is not uncommon for the health and safety professional to be expected to assess environmental conditions in a setting where the possibility of an explosive or flammable atmosphere of a vapor, gas, or dust exists. For this reason a majority of the equipment discussed in this chapter is designed and certified by their companies as intrinsically safe. The National Fire Protection Association (NFPA) in Article 500 of the National Electric Code describes intrinsically safe equipment as being incapable of releasing sufficient electrical or thermal energy under normal and abnormal conditions to cause ignition of specific flammable or combustible atmospheric mixtures in their most easily ignitable condition. NFPA has developed a classification scheme of hazardous environments consisting of seven groupings of gases, vapor, and dusts according to their explosive/flammable potential (NFPA 1987). Manufacturers may seek intrinsic safety certification by having their product tested in all or selected challenge atmospheres that represent these groupings. Independent testing laboratories that are listed by OSHA as Nationally Recognized Testing Laboratories are qualified to perform challenge tests in the United States applying either the ANSI/UL 913-1988 (American National Standards Institute [ANSI] 1988) standard or similar protocol. These labs also test and certify instrument performance applying the procedures found in ANSI/ISA-S12-13.1-1986 (NESVIG 1986). In this procedure instruments are evaluated for accuracy, temperature effects, response to concentration change, humidity effects, and ambient air velocity effects. Instruments that have successfully passed an evaluation are permitted to carry the certifying lab's label. Greater details on label information can be obtained by contacting the certifying lab directly. Samples of certification labels are shown in Figure 4.13.

### 4.4.5 Radio-Frequency Effects

Radio-frequency interference from two-way radios and other sources has been reported to affect the readings obtained from electronic instruments (Cook and Huggins 1984). The greatest concern is interference to direct reading devices from two-way radios during field monitoring. Devices with simple circuity and no amplification are less likely to be affected. Some vendors shield critical circuits to prevent this problem.

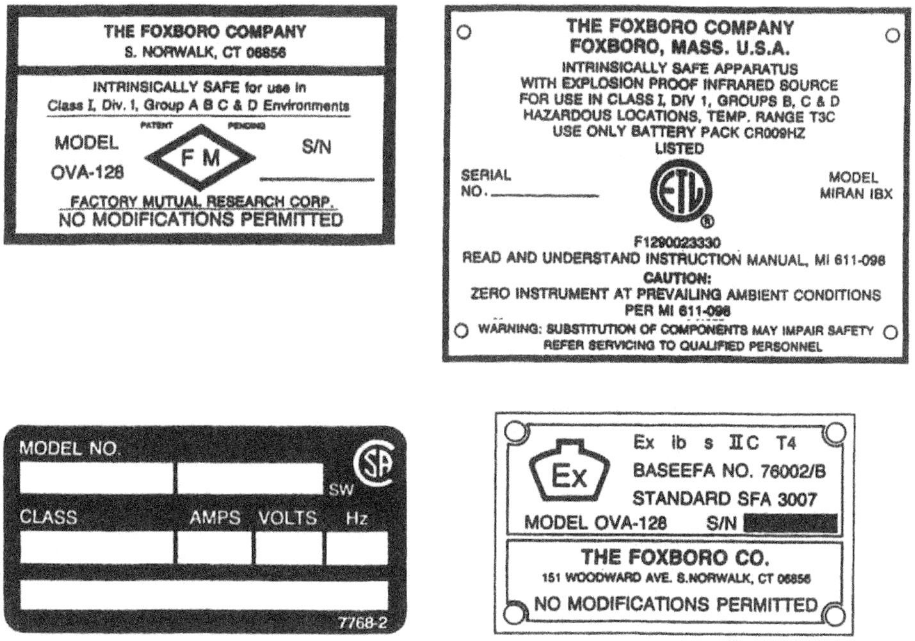


**Figure 4.13** Equipment certification labels.

### 4.4.6 Shipping

After collection, the samples are packaged and shipped to the lab for analysis. Shipment of liquid samples will require proper Department of Transportation labeling. Most samples will be classified as flammable, poisonous, and/or corrosive. Information on proper labeling and packaging can be found by contacting the lab chosen for the analysis. Another resource may be the supplier of reagents to a company's research, quality control, or process lab since the collecting solutions are common lab reagents. A review of the material safety data sheet that accompanies the reagent bottle will also prove to be a useful source of shipping information.

### 4.4.7 Records

The creation and maintenance of a good record is a critical element of any industrial hygiene program. The records on employee exposure can be the basis for epidemiological studies; the source of predicting proper respiratory selection when sampling is not feasible; a response to employee or union concerns in support of medical surveillance programs; and/or the basis to determine the need or effectiveness of engineering controls. They can also be used to advise management on the impact of proposed regulations and in litigation. Finally, a good recordkeeping system exhibits a commitment to quality and professional performance. Discussions of this topic can be found in many references (Wrench 1990; Confer 1994). Ideally, records should include a standard operating procedures manual that includes procedures for calibrating, testing, and maintaining equipment, routine sampling protocols, example calculations of flow determination, and data analysis. Equipment maintenance, repair, and calibration logs should also be included. Chain of custody records also verify the validity of exposure data. Finally, data on exposure assessments should be maintained that define the nature of the characteristics that make a population homogeneous (Tait 1997): the number of sampling sessions, the details of sampling protocol (i.e. specified flow rate, time, and volume), analytical lab results, and the results of calculations. Included in this database should be a narrative of individual or group performance during the sampling session, the types of personal protection or administrative practices in practice, and influencing environmental conditions. Many databases have been developed and discussed over time, and individuals or organizations have created their own customized systems over time. A joint effort of the ACGIH and AIHA is underway to publish guidelines discussing occupational databases in greater detail (Joint ACGIH/AIHA Task Group on Occupational Exposure Databases 1997).

### 4.4.8 Data Logging

Data loggers are electronic devices that record the output of an instrument for later analysis. The output is linked to a data logging record, which both averages the signal over brief periods of time and records the average as data points over the duration of the sample. This produces a time history of the exposure that identifies peak periods of exposure as well as the integrated overall average exposure. Data logging capabilities are often included in the overall instrument package; however, stand-alone data logging devices are also available on the market. In many cases, data loggers have added another level of convenience to the field application of direct reading instruments by replacing the larger, more inconvenient strip chart recorder.

## 4.5 SURFACE SAMPLING

### 4.5.1 General

The terms *wipe sampling, swipe sampling*, and *smear sampling* are synonyms that describe the techniques used to assess surface contamination. Wipe sampling is most often used to determine the presence of materials such as asbestos, lead and other metals, aromatic amines, and PCBs. Wipe sampling techniques are applied to:

1. Evaluate potential contact with surface contaminants by wiping surfaces that workers can touch.
2. Surfaces that may come into contact with food or other materials that are ingested or placed in the mouth (e.g. chewing tobacco, gum, cigarettes) may be wipe sampled (including hands and fingers) to detect contamination.
3. Effectiveness of personal protective gear (e.g. gloves, aprons, respirators) may sometimes be evaluated by wipe sampling the inner surfaces of the protective gear (and protected skin).
4. Effectiveness of decontamination of surfaces and protective gear (e.g. respirators) can sometimes be evaluated by wipe sampling.

When accompanied by close observation of the operation in question, wipe sampling can help identify sources of contamination and poor work practices.

Generally, two types of filters are recommended for taking wipe samples. Glass fiber filters (GFF) (37 mm) are usually used for materials that are analyzed by high-performance liquid chromatography (HPLC) and often for substances analyzed by gas chromatography (GC). Paper filters are generally used for metals. For convenience, the Whatman smear tab (or its equivalent) or polyvinyl chloride filters for substances that are unstable on paper-type filters are commonly used. Normally a 100 cm$^2$ area of the surface is wiped while wearing gloves. The filter can be dry or wetted with distilled water; however, organic solvents are not normally used (OSHA 1995). Collected filters are normally stored in glass vials for shipment.

### 4.5.2 Dermal Exposure Assessment

Many substances, especially fat-soluble hydrocarbons and other solvents, can enter the body and cause systemic damage directly through the skin when the skin has become wet with the substance by splashing, immersion of hands or limbs, or exposure to a mist or liquid aerosol. Some substances, such as amines and nitriles, pass through the skin so rapidly that the rate at which they enter the body is like that of substances inhaled or ingested. The prevention of skin contact to phenol is as important as preventing inhalation of airborne concentrations. A few drops of dimethylformamide on the skin can contribute to a body burden similar to inhaling air at the TLV for eight hours. The "skin" notation in the TLV list identifies substances for which skin absorption is potentially a significant contribution to overall exposure. However, the TLV list offers no precise definition of a "significant" contribution.

Note that the skin notation relates to absorption or toxicity via route and not to the potential to cause skin damage and dermatitis. These later effects are important, however, and should be indicated on such hazard data sources as material safety data sheets (MSDSs).

Dermal exposure assessment is a complex matter that has not received much attention in the practice of industrial hygiene. Estimation of dermal exposure involves exposure scenarios and pathways, contact duration and frequency, body surface area in contact, and substance adherence. The amount of contaminant that crosses the skin barrier and enters the body is influenced by the properties of the substance and the properties and condition of the skin at the exposed site. Models of dermal absorption have been developed (Berner and Cooper 1987; Flynn 1990). These models start with the partition coefficients of the substance, its mole weight and solubility, and the diffusivity of the compound in the lipid and protein phases of the skin. Complex theoretical models have been simplified and combined with empirical models to yield an estimate of the chemical-specific dermal permeability constant. The dermally absorbed dose (mg per kg day) can then be calculated from this constant and the skin contact area, exposure time, frequency and duration, and body weight. Since these quantities are rarely known, they must be assumed. When conservative assumptions are used, the conservatisms may compound to yield a very conservative estimate of the dermal dose.

## BIBLIOGRAPHY

American Conference of Government Industrial Hygienists (ACGIH) (1978). *Air Sampling Instruments for Evaluation of Atmospheric Contaminants*, 5e. Cincinnati, OH: ACGIH.

American Conference of Government Industrial Hygienists (ACGIH) (1996). *Threshold Limit Values for Chemical Substances and Physical Agents in the Work Environment with Intended Changes for 1996–1997*. Cincinnati, OH: ACGIH.

American National Standards Institute (ANSI) (1988). *ANSI/UL 913-1988, Intrinsically Safe Apparatus and Associated Apparatus for Use in Class I, Class II and III, Division I, Hazardous (Classified) Locations*. New York, NY: ANSI.

Anderson, M.E., MacNaughton, M.G., Clewell, H.J., and Paustenbach, D.P. (1987). Adjusting exposure limits for long and short exposure periods using a physiological pharmacokinetic model. *Am. Ind. Hyg. Assoc. J.* **48**: 335–343.

Brief, R.S. and Scala, R.A. (1975). Occupational exposure limits for novel work schedules. *Am. Ind. Hyg. Assoc. J.* **36**: 467–469.

Calabrease, E.J. (1977). *Am. Ind. Hyg. Assoc. J.* **38**: 443–446.

Chelton, C.F. (1993). *Manual of Recommended Practice for Combustible Gas Indicators and Portable Direct-Reading Hydrocarbon Detectors*. Fairfax, VA: American Industrial Hygiene Association.

Confer, R.G. (1994). *Workplace Exposure Protection: Industrial Hygiene Program Guide*. Boca Raton, FL: Lewis Publishers, CRC Press.

Cook, C.F. and Huggins, P.A. (1984). Effect of radio frequency interference on common industrial hygiene monitoring instruments. *Am. Ind. Hyg. Assoc. J.* **45**: 740–744.

Droz, P. and Yu, M. (1990). Biological Monitoring Strategies. In: *Exposure Assessment for Occupational Epidemiology and Hazard Controls* (ed. S.M. Rapport and T.J. Smith). Chelsea, MA: Lewis Publishing Co.

Hawkins, N.C., Norward, S.K., and Rock, J.C. (eds.) (1991). *A Strategy for Occupational Exposure Assessment*. Akron, OH: American Industrial Hygiene Association.

Health and Safety Executive (HSE) (1989). *Monitoring Strategies for Toxic Substances*, Guidance Note, EH42. Bottle, Merseyside, UK: HSE.

Hickey, J.L.S. and Reist, P.C. (1977). Application of occupational exposure limits to unusual work schedules. *Am. Ind. Hyg. Assoc. J.* **38** (11): 613–621.

*Ind. Hyg. News* (1996). Buyers Guide, *Industrial Hygiene News* **19**(3), Rimbach Publishing Co.

Joint ACGIH/AIHA Task Group on Occupational Exposure Databases; Lippman, M. et al. (1997). Data elements for occupational exposure databases.

Leidel, N.A., Busch, K.A., and Lynch, J.R. (1977). *Occupational Exposure Sampling Strategy Manual*, Publication (NIOSH) 77-173. Cincinnati, OH: Department of Health, Education and Welfare.

Mason, J.W. and Dershin, H. (1976). Limits to occupational exposure in chemical environments under novel work schedules. *J. Occup. Med.* **18**: 603–606.

National Institute of Occupational Safety and Health (NIOSH) (1984). *NIOSH Manual of Analytical Methods*, 4e. Washington, DC: U.S. Department of Health and Human Services. https://www.cdc.gov/niosh/docs/2003-154/default.html (accessed December 2018).

NESVIG (1986). *Performance Requirements, Combustible Gas Detectors (ISA-S12.13.1)*. Pittsburgh, PA: Instrument Society of America.

NFPA70 (1985). *National Electrical Code*. Quincy, MA: National Fire Protection Association.

Occupational Safety and Health (1996). Waco, TX. Medical Publications, Inc.

OSHA (1982). Use of personal sampling devices during inspection. *Fed. Regist.* **47**: 55478.

OSHA (1991). *Analytical Methods Manual*, Pub. 1985 with updates through 1991. Cincinnati, OH: ACGIH.

OSHA (1995). Chapter 2. Sampling for surface contaminants. In: *Technical Manual*. Washington, DC: Occupational Safety and Health Administration.

Roach, S.A. (1966). A more rational basis for air sampling programs. *Am. Ind. Hyg. Assoc. J.* **27**: 1–12.

Roach, S.A. (1977). A most rational basis for air sampling programmes. *Ann. Occup. Hyg.* **20**: 65–84.

Roach, S.A. et al. (1983). *Am. Occup. Hyg.* **27**: 1–13.

Tait, K. (1993). The workplace exposure assessment workbook (workbook). *Appl. Occup. Environ. Hyg.* **8** (1): 55–68.

Wrench, C. (1989). *Data Management for Occupational Health and Safety: A User's Guide to Integrating Software*. New York, NY: Van Nostrand Reinhold.

## APPENDIX A

### CONSIDERATIONS IN ESTABLISHING AN INDUSTRIAL HYGIENE FIELD OFFICE

- Review the operations and activities within the hygienist's scope of responsibilities to determine the most likely hazardous materials exposures to be evaluated.
- Assess the emerging concerns in occupational risk that may impact the employee population in the near term.
- Determine if integrated and/or grab sampling will be necessary.
- Select the most probable sampling procedures to be used.
- Select a support industrial hygiene laboratory service based on analytical capabilities, certification of performance, and backup technical support.
- Develop an inventory of equipment to be maintained on-site (see Appendix B).
- Obtain or develop sampling protocols and gather them into a centralized file or binder.
- Collect and centralize equipment manuals.
- Develop field calibration capabilities for sampling pumps, direct reading instruments, and other equipment as needed or arrange for external support.
- Establish an equipment calibration and repair log.
- Develop exposure sampling data.
- Develop an exposure data management system.
- Allocate space for a minimum of simple field laboratory functions (evaluate the need for lab hood and flammable materials storage cabinet).
- Collect a file of relevant MSDS.

# APPENDIX B

## INDUSTRIAL HYGIENE FIELD OFFICE INVENTORY CHECKLIST

Sampling pumps

- Low flow
- High flow
- Combination high/low flow

Adsorbent tubes

- Charcoal
- Silica gel
- Alumina
- Chromatographic packings

Passive badges
Colorimetric detector tubes
35/27-mm filter cassettes

- Cellulose acetate
- PVC
- Gold film
- PTFE

Microscope
Asbestos ID kit
Sound level meter
Noise dosimeters
Radiation survey meter
Microwave survey meter
Flow calibrator/bubble meter

## APPENDIX C

### DIRECT READING INSTRUMENT INVENTORY

| Instrument/device | Application | Common brands |
|---|---|---|
| Hand pumps | Detector tube sampling | SKC, MSA, Gastech, National Draeger |
| Pumps low flow | Charcoal tube sampling | Gilian, SKC, Du Pont, MSA |
| Pumps medium flow | Filter sampling | Same |
| Flow calibrator | Pump calibration | Gilian |
| Fibrous aerosol monitors | Asbestos/fibers | MIE, MDA, PPM, TSI |
| Double range meters | Oxygen, combustible gas | Gastech, MSA |
| Triple range meters | Carbon monoxide, oxygen combustible gas, hydrogen sulfide | Many brands |
| Carbon monoxide dosimeter | Indoor air quality | Interscan, Scott, Draeger |
| Carbon dioxide meter | Indoor air quality | Gastech |
| Infrared analyzers | Organics, carbon monoxide carbon dioxide | Foxboro |
| Hydrogen cyanide monitors | Hydrogen cyanide | MDA, Monitox |
| Hydrogen sulfide monitors | Hydrogen sulfide | Draeger, Interscan |
| Mercury vapor | Mercury | Jerome |

## APPENDIX D

### PRELIMINARY INDUSTRIAL HYGIENE EVALUATION REPORT

DATE: _________ TIME CONTACTED: _________ UNIT/AREA: ______________________________

WHO REQUESTED INVESTIGATION/SURVEY: ____________________________________________

______________________________________________________________________________

______________________________________________________________________________

______________________________________________________________________________

DATE IH RESPONDED: __________________ TIME: ____________________________________

RESPONDING IH NAME(S): __________________________________________________________

OTHER PEOPLE INVOLVED IN INVESTIGATION/SURVEY: ___________________________________

FINDINGS AT TIME OF FIELD INVESTIGATION/SURVEY: ___________________________________

______________________________________________________________________________

______________________________________________________________________________

______________________________________________________________________________

______________________________________________________________________________

FOLLOW-UP NEEDED: _____ YES _____ NO IF YES, WHAT AND WHEN TO BE DONE:

______________________________________________________________________________

______________________________________________________________________________

WAS FOLLOW-UP COMPLETED: _____ YES _____ NO IF NOT, WHY: _______________________

______________________________________________________________________________

______________________________________________________________________________

RESULTS OF INVESTIGATION/SURVEY: ________________

______________________________________________________________________________

______________________________________________________________________________

______________________________________________________________________________

______________________________________________________________________________

REPORT BY: ____________________

## APPENDIX E

### FIELD SAMPLING FORM

**1. Employee and company information**

Name ______________________ SS# ______________________ Company ID# ______________________

Company name ______________________ Billing department ______________________
Address ______________________ Supervisor ______________________
______________________ Report sent ______________________
______________________ To ______________________

**2. Task description**

Identifying task name ______________________ Shift ______________________

Job title ______________________

Equipment or location identified ______________________

Operations (circle one) routine, maintenance, upset conditions, other

Task length (circle one) continuous, intermittent, extension, partial shift, other.

Task duration ________

Narrative description (engineering controls, task performance, and characteristics)

Signature ________ Date ________

**3. Personal protective equipment**

Respiratory protection (circle one): None, supplied air, SCBA: _____ Air purifying: Cartridge _____

Body protection: Nomex coveralls, disposable coveralls, slicker suit, nitrile suit, acid suit, encapsulated suit

| | | | |
|---|---|---|---|
| | Cotton: | Long sleeve | Short sleeve |
| Gloves Y or N | Type ________ | Boots Y or N | Type ________ |
| Eyewear Y or N | Type ________ | Hearing Y or N | Type ________ |

**4. Sampling strategy**

| | |
|---|---|
| Sample type (circle one) | typical; worst case; random |
| Sample period (circle one) | full shift; partial period; short-term grab |
| Representation (circle one) | breathing zone; employee area; point source, area |

Relative humidity _____ Barometric pressure _____ Other relevant weather conditions _____

**5. Measurement**

| Sample ID | Start time | Stop time | Elapsed time | Sample rate | Total value | Sampling media |
|---|---|---|---|---|---|---|

## APPENDIX F

### INDUSTRIAL HYGIENE MONITORING RESULTS

Report to supervisor: ____________________
Department name and #: ____________________
Date sent: ____________________
IH log #: ____________________
Date sampled: ____________________
Process/equipment: ____________________
____________________
____________________
____________________

Employee name: ____________________
Employee #: ____________________
Exposure: ____________________
Permissible exposure limit (PEL): ____________________
Threshold limit value (TLV): ____________________
Sample results: ____________________
Comments: ____________________
____________________
____________________

Inform employee and return to Safety Manager

by ____________________

____________________
Safety Manager

____________________

____________________ was informed of the above information

(Employee name)

__________
(Date)

Employee/supervisor's comments: ____________________
____________________
____________________
____________________

____________________
Supervisor's Signature

CHAPTER 5

# STATISTICAL METHODS FOR OCCUPATIONAL EXPOSURE ASSESSMENT

DAVID L. JOHNSON

*Department of Occupational and Environmental Health, College of Public Health, University of Oklahoma, Health Sciences Center, P.O. Box 26901, Oklahoma City, OK, 73126-0901*

## 5.1 INTRODUCTION

Exposure assessment is a primary function of the occupational safety and health professional. Airborne gas and vapor concentrations, aerosol concentrations and particle size distributions, sound pressure levels, radiation levels, and other potential hazards are measured to assess exposures as to their ranges, maxima, averages, and frequency distribution. Exposure guidelines and standards may then be consulted to make decisions about the adequacy of workplace exposure controls. The likelihood of overexposure and regulatory noncompliance can be explored by modeling the exposure frequency distribution to determine how often unacceptably high exposure might occur.

Measurements might be made to address specific questions such as:

- At what levels are workers being exposed?
- Are exposures different for different worker groups?
- Are exposures in compliance with occupational exposure limits?
- Has an intervention changed worker exposures?
- How likely is it that workers will be overexposed?
- How much uncertainty is there in the measures and the values calculated from them?

In this chapter, statistical techniques useful in answering such questions will be reviewed. Both parametric and non-parametric techniques are discussed, with emphasis on using the Microsoft Excel® statistical functions and Analysis ToolPak® utilities wherever possible. Other spreadsheet applications will have similar tools, and of course more advanced software applications specifically designed for statistical data analysis are also available but require some training to use.

## 5.2 VARIABILITY IN EXPOSURES AND EXPOSURE MEASUREMENT

### 5.2.1 Exposures Are Variable

The question "At what levels are workers being exposed?" is not as straightforward to answer as many might think. Occupational exposures to potentially harmful chemical and physical agents will typically vary over a wide range from a lower limit of zero to some upper limit determined by the chemistry and physics of the substance, the conditions of

*Handbook of Occupational Safety and Health*, Third Edition. Edited by S. Z. Mansdorf.

its production or use in the work, the work environment, and the interaction with the worker with the work. For example, exposures during operations involving volatile organic solvents will be influenced by:

- The temperature of the process and the vapor pressure of the solvent at that temperature.
- The potential for aerosolization of liquid solvent droplets and the size distribution of those droplets.
- The volume of the work space and the effective fresh air dilution ventilation rate in the space.
- Provisions for vapor or droplet capture by local exhaust ventilation at the point of generation.
- The proximity of the worker to the work and how long they remain there during the workday.

The maximum possible exposure in this case would be limited by the solvent's vapor pressure, and it would be expected that exposures would vary over the course of the workday as breathing zone contaminant concentrations went up and down with process variations and with worker movements.

Exposure variability is influenced by temporal factors such as work cycles, time of day, day of week, and season of year, as well as by spatial factors such as the location of the process or processes in the space, variations in contaminant concentration in the work space, and the location of the worker in the space. Repeated measurements for a given worker will indicate a range of exposures but with some exposure levels occurring more frequently than others. This is termed within-worker variability. In addition, different workers may experience different exposures in the same type of work or even the same work space due to these temporal and spatial variations as well as to individual differences in gender, stature, activity level, the specific work they are doing, and the way they go about it. This is between-worker variability.

Given that exposures vary due to temporal, spatial, and individual influences, there will be not just one exposure level, but rather a *distribution* of exposures spanning some range of values. How often different exposure levels occur within the range, i.e. their probability of occurrence, determines the exposure frequency distribution. It should be clear then that the answer to the question "At what levels are workers being exposed?" must reflect this distribution. Variants of the question such as "What is the worker's average exposure?," "What are the highest expected exposures?," and "What is the most likely exposure?" require different responses, but each is based on examination of the exposure distribution. By applying probability models it is often possible to describe the distribution mathematically, so that statistical tests based on the model can then be used to answer a variety of questions.

### 5.2.2 Measurements Have Variability

Workplace exposures to chemical and physical hazards vary over time and space and between individuals as previously discussed, resulting in an exposure distribution. However, even when repeatedly measuring an unchanging quantity, there is a potential for random measurement variation due to the combined effects of perhaps many small random influences. For example, when measuring a time-weighted-average (TWA) chemical vapor exposure (assume for the moment that it is unchanging) using an air sampling pump and sorbent tube, there may be slight and random variations in the pump flow rate, sample timing, solvent or thermal extraction of collected contaminant from the sorbent media, and chemical analysis. These combine to result in a net random measurement variability termed random error, though there is no suggestion of a mistake being made. Here, "error" only refers to the difference between the measurement value and the true value being measured.

The variation of results around the true value is expected to be randomly distributed around the value, with some measures less than and some greater than the true value. Positive and negative deviations are equally likely to occur; however, the randomly positive and negative influences that combine to give the net error tend to cancel each other out to varying degrees. As a result, it is more likely that there will be small deviations from the true value than that there will be large deviations. This can be visualized as shown in Figure 5.1, in which the deviations and how often they will occur are represented by a bell-shaped curve.

The amount of random measurement variation reflects the precision of the measurement technique. Precise measurements are those that have little variation, whereas imprecise measurements have large variation. It will be seen that measurement precision has a major influence on how many measurements are needed for a statistical analysis. Measurement precision as presented here is often incorrectly called measurement "accuracy," but accuracy has a different meaning. Whereas precision refers to the similarity of repeated measures to each other, accuracy refers to how close *on average* the measurements are to the true value. A measurement technique is inaccurate if the average of repeated measurements is systematically different, i.e. off in one direction, from the true value. This might occur, for example, if an improperly calibrated instrument is used – repeated measurements may be similar but will be wrong. In contrast to the random errors that result in imprecision, systematic errors result in bias. Fortunately, bias can usually be avoided through careful selection of the measurement technique, proper instrument maintenance and calibration, training and experience in conducting measurements, and other quality assurance steps. Random error cannot be completely avoided, but can be minimized by using high precision instruments and analytical procedures. In this chapter it is assumed that all data result from accurate measurements.

### 5.2.3 Statistical Samples

In occupational health and safety work, the term "sample" typically refers to a physical quantity collected from the environment, such as an air, dust, or bulk chemical sample. In statistics, however, "sample" refers to one or (usually) more measurements taken as a representative subset of all possible measurements that might have been made. Thus, 10 Tenax tube air samples, each of five-minute duration, collected at randomly selected times over the course of a workday would be a statistical sample of the 96 possible five-minute intervals that might have been measured. In this chapter "sample" should be interpreted in the statistical sense.

In statistical sampling the measurement data are used to infer something about the thing being measured. A key requirement, then, is that the measurements are representative of the thing being measured. The number and timing of measurements should ensure that any inferences drawn from the results are based on adequate information. If, for example, several short-term measurements were being made to estimate average worker exposure over the course of a workday and it is known that exposures vary over the course of the day, it would be important to make measurements at different times during the day so that the whole day is represented. In contrast, if exposure only during a specific operation is of interest, then measurements should only be made during the operation. Determining the *who*, *what*, *when*, *where*, *how*, and *how many* of measurements is driven by the *why* of the measurements, i.e. the question one is trying to answer. Sampling strategies must be designed to ensure representativeness, or else any inferences drawn from the data are likely to be incorrect.

## 5.3 THE NORMAL AND LOGNORMAL DISTRIBUTIONS

### 5.3.1 The Normal Distribution

The deviations shown in Figure 5.1 are said to be "normally distributed" about zero, in that the shape of the curve can be mathematically expressed using the normal probability distribution model (also called the Gaussian distribution). Probability is the mathematical expression of how likely an event is to occur and ranges from 0 (certain not to occur) to 1.0 (certain to occur). For the random deviations discussed, if experience tells us that a deviation of, say, −2.0 occurs about 18 times in every 1000 measurements, then its observed probability of occurrence would be $p = 18/1000 = 0.018$ or about 1.8% of the time. For normally distributed errors a deviation of +2.0 would have the same probability because the distribution is symmetric. The normal distribution of probabilities is graphed in a manner similar to Figure 5.1, except that the $y$-axis is probability density rather than frequency of occurrence. The height of the curve, $y$, for any point $x$ on the horizontal axis, i.e. the probability density at $x$, is given by

$$y = \frac{1}{\sigma\sqrt{2\pi}} e^{-(x-\mu)^2/2\sigma^2} \tag{5.1}$$

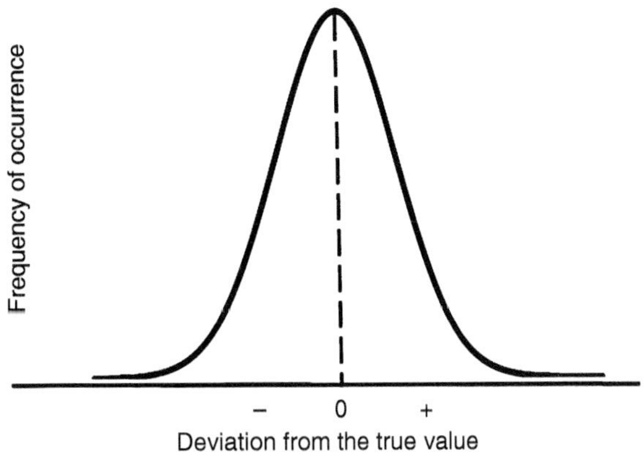

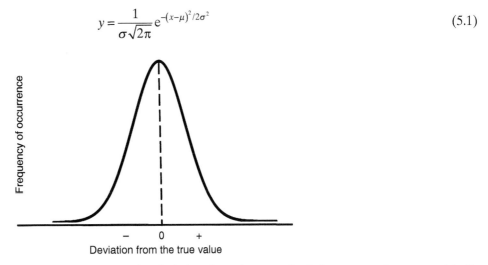

**Figure 5.1** Distribution of random errors. Positive and negative deviations from the true value being measured are symmetrically distributed about zero (no error), with smaller deviations being more likely to occur than larger deviations.

where $\mu$ is the distribution's mean (arithmetic average) value and $\sigma^2$ is the variance of other values about the mean. The square root of the variance is the distribution's standard deviation, $\sigma$. A normal distribution with $\mu = 0$ and $\sigma = 1$ is called the standard normal distribution, abbreviated $N(0, 1)$. For the $N(0, 1)$ distribution the $x$-values are called $z$-values, or $z$-scores. A normal distribution of values $x_i$ with mean $\mu$ and standard deviation $\sigma$, i.e. $N(\mu, \sigma)$, can be converted to a normal distribution $N(0, 1)$ of $z$-values using the transformation

$$z_i = \frac{x_i - \mu}{\sigma} \tag{5.2}$$

Equation (5.1) is termed the probability density function (pdf) of the normal distribution. The function is unbounded, i.e. the $x$-values range from $-\infty$ to $+\infty$, and the area under the curve represents probability. The total area under the curve sums to 1.0, and for any value of $x$, the area to the left of $x$ is the cumulative probability $p$ that a value less than or equal to $x$ will be observed. The probability that a value greater than $x$ will be observed must then be $1-p$ since these are the only two possibilities and the probability must total 1.0. The area under the curve between two values $x_1$ and $x_2$ is the probability that a value between these limits will be observed. A partial table of cumulative probabilities of the standard normal distribution from $-\infty$ to various $z$-values is shown in Table 5.1. A similar table for probabilities between $\pm z$ is shown in Table 5.2. Note that in Table 5.1, the probabilities start with 0.5000, i.e. all of the area to the left of center (half of the total) is already accounted for; that is, the table is for positive values of $z$. Complete tables can be found in any basic statistics book or online or can be conveniently obtained from one of Excel's statistical functions. The cumulative probability associated with a given $z$-value can be obtained in Excel using the NORM.S.DIST($z$, 1) function, e.g. for $z = 1.645$, NORM.S.DIST(1.645,1) = 0.95 is the cumulative probability to the left of $z = 1.645$. The 1 in the function statement lets Excel know that the cumulative probability is being requested; a 0 value would return the probability density at $z = 1.645$, i.e. the height of the curve at that point. The probability between $\pm z$ is obtained from NORM.S.DIST($z$, 1)–NORMS.S.DIST($-z$, 1), e.g. the area under the curve between $z = \pm 1.96$ is obtained from NORM.S.DIST(1.96,1)–NORMS.S.DIST(−1.96,1) = 0.95.

Data distributions such as the normal distribution can be described by their mean $\mu$ and standard deviation $\sigma$ as well as other indicators such as the data range, median, mode, and percentiles. The true mean and standard deviation

**TABLE 5.1 Table of Cumulative Probability ($-\infty$ to $+z$) of the Standard Normal Distribution**

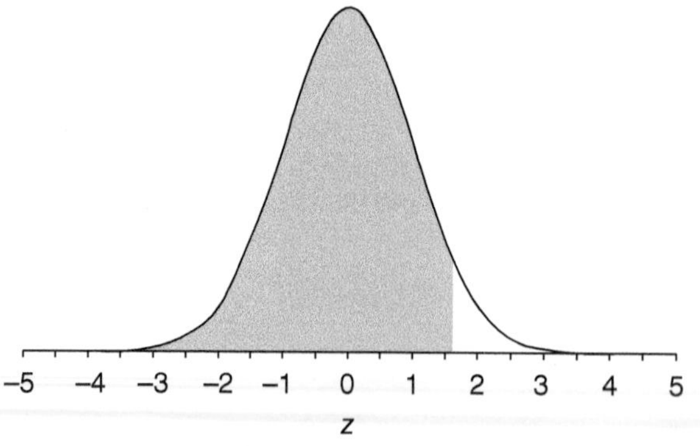


| z | 0.00 | 0.01 | 0.02 | 0.03 | 0.04 | 0.05 | 0.06 | 0.07 | 0.08 | 0.09 |
|---|---|---|---|---|---|---|---|---|---|---|
| 0.00 | 0.5000 | 0.5040 | 0.5080 | 0.5120 | 0.5160 | 0.5199 | 0.5239 | 0.5279 | 0.5319 | 0.5359 |
| 0.10 | 0.5398 | 0.5438 | 0.5478 | 0.5517 | 0.5557 | 0.5596 | 0.5636 | 0.5675 | 0.5714 | 0.5753 |
| 0.20 | 0.5793 | 0.5832 | 0.5871 | 0.5910 | 0.5948 | 0.5987 | 0.6026 | 0.6064 | 0.6103 | 0.6141 |
| 0.30 | 0.6179 | 0.6217 | 0.6255 | 0.6293 | 0.6331 | 0.6368 | 0.6406 | 0.6443 | 0.6480 | 0.6517 |
| — | — | — | — | — | — | — | — | — | — | — |
| — | — | — | — | — | — | — | — | — | — | — |
| 3.10 | 0.9990 | 0.9991 | 0.9991 | 0.9991 | 0.9992 | 0.9992 | 0.9992 | 0.9992 | 0.9993 | 0.9993 |
| 3.20 | 0.9993 | 0.9993 | 0.9994 | 0.9994 | 0.9994 | 0.9994 | 0.9994 | 0.9995 | 0.9995 | 0.9995 |
| 3.30 | 0.9995 | 0.9995 | 0.9995 | 0.9996 | 0.9996 | 0.9996 | 0.9996 | 0.9996 | 0.9996 | 0.9997 |
| 3.40 | 0.9997 | 0.9997 | 0.9997 | 0.9997 | 0.9997 | 0.9997 | 0.9997 | 0.9997 | 0.9997 | 0.9998 |

Values in this table were obtained from the NORM.S.DIST function in Excel, e.g. NORM.S.DIST(1.645,1) = 0.95.

**TABLE 5.2 Table of Probability Between ±z for the Standard Normal Distribution**

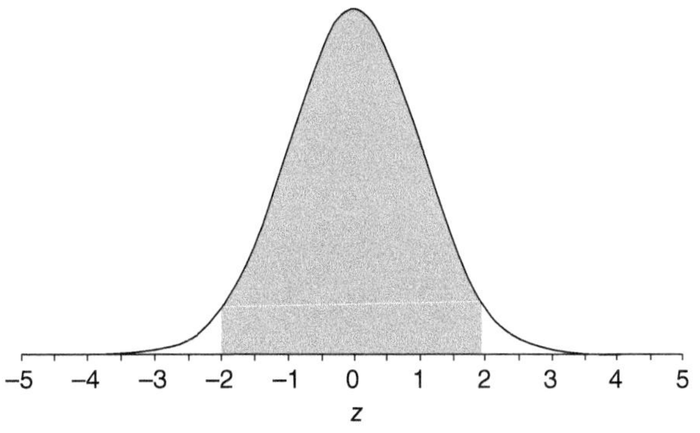


| z | 0.00 | 0.01 | 0.02 | 0.03 | 0.04 | 0.05 | 0.06 | 0.07 | 0.08 | 0.09 |
|---|---|---|---|---|---|---|---|---|---|---|
| 0.00 | 0.0000 | 0.0080 | 0.0160 | 0.0239 | 0.0319 | 0.0399 | 0.0478 | 0.0558 | 0.0638 | 0.0717 |
| 0.10 | 0.0797 | 0.0876 | 0.0955 | 0.1034 | 0.1113 | 0.1192 | 0.1271 | 0.1350 | 0.1428 | 0.1507 |
| 0.20 | 0.1585 | 0.1663 | 0.1741 | 0.1819 | 0.1897 | 0.1974 | 0.2051 | 0.2128 | 0.2205 | 0.2282 |
| 0.30 | 0.2358 | 0.2434 | 0.2510 | 0.2586 | 0.2661 | 0.2737 | 0.2812 | 0.2886 | 0.2961 | 0.3035 |
| — | — | — | — | — | — | — | — | — | — | — |
| — | — | — | — | — | — | — | — | — | — | — |
| 3.10 | 0.9981 | 0.9981 | 0.9982 | 0.9983 | 0.9983 | 0.9984 | 0.9984 | 0.9985 | 0.9985 | 0.9986 |
| 3.20 | 0.9986 | 0.9987 | 0.9987 | 0.9988 | 0.9988 | 0.9988 | 0.9989 | 0.9989 | 0.9990 | 0.9990 |
| 3.30 | 0.9990 | 0.9991 | 0.9991 | 0.9991 | 0.9992 | 0.9992 | 0.9992 | 0.9992 | 0.9993 | 0.9993 |
| 3.40 | 0.9993 | 0.9994 | 0.9994 | 0.9994 | 0.9994 | 0.9994 | 0.9995 | 0.9995 | 0.9995 | 0.9995 |

Values in this table were obtained from NORM.S.DIST($z$, 1)−NORM.S.DIST($-z$, 1).

of a distribution is generally not known and must be estimated from sample data. The sample mean $\bar{x}$ estimates the "population" mean $\mu$ and is the arithmetic average of the individual measurement values:

$$\bar{x} = \frac{1}{n}\Sigma x_i = \frac{\Sigma(f_i x_i)}{\Sigma f_i} \tag{5.3}$$

where $x_i$ is any one of the $n$ measured values and $f_i$ is the number of times that value occurs, i.e. its frequency. The sample standard deviation $s$ estimates the population standard deviation $\sigma$:

$$s = \sqrt{\frac{1}{n-1}\Sigma f_i (x_i - \bar{x})^2} \tag{5.4}$$

The variance is the square of the standard deviation.

The median is the middle value when the measures are ordered smallest to largest (or largest to smallest), and the mode is the most frequently occurring value, i.e. the peak value in a unimodal (single peak) distribution such as the normal distribution. In the normal distribution the mean, median, and mode are all the same value because it is unimodal and symmetric. The range is just the difference between the largest and smallest values.

### 5.3.2 The Lognormal Distribution

Exposure quantities such as gas and vapor concentrations, dust concentrations and particle sizes, and radiation levels are bounded on the low end by zero but are limited on the upper end only by the physics of the situation. For example, an upper limit on solvent vapor concentration – its saturation concentration – is determined by its vapor pressure. Exposure levels can get quite large, but cannot be less than zero. As a result, an exposure frequency distribution such as shown in panel (a) of Figure 5.2 is commonly seen, in which the distribution is skewed to the right. However, often the *logarithms* of the data values are normally distributed. Such data are said to be lognormally distributed.

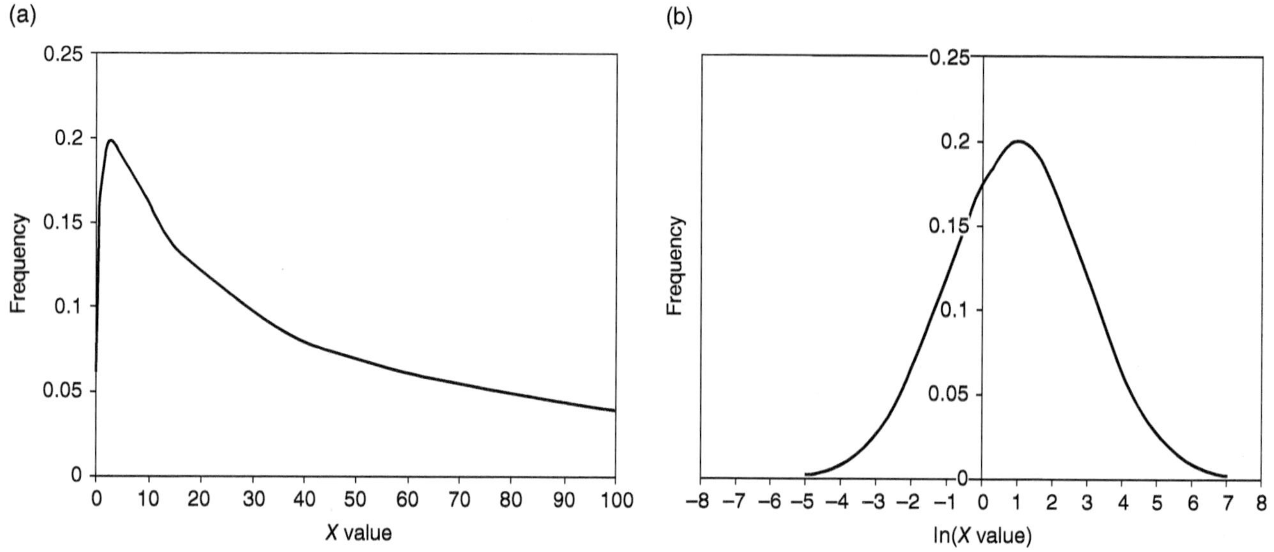


**Figure 5.2** Lognormally distributed values: (a) the distribution of measured values is skewed to the right; (b) the logarithms of the measured values are normally distributed.

**5.3.2.1** ***Descriptive Measures for the Lognormal Distribution*** The lognormal distribution is described by its geometric mean and geometric standard deviation. The geometric mean $\bar{x}_g$ can be calculated using either the natural (base-$e$) or common (base-10) logarithms:

$$\bar{x}_g = e^{\left[\frac{1}{n}\Sigma \ln(x_i)\right]} = 10^{\left[\frac{1}{n}\Sigma \log(x_i)\right]} \tag{5.5}$$

The same value for $\bar{x}_g$ is obtained using either type of logarithm. The geometric standard deviation $s_g$ can also be calculated using either natural or common logarithms. For natural logarithms it is obtained from

$$s_g = \exp(s_y) \tag{5.6}$$

where $s_y$ is the standard deviation of the natural logarithms of the data values:

$$s_y = \sqrt{\frac{1}{n-1}\Sigma\left(\ln x_i - \overline{\ln x}\right)^2} \tag{5.7}$$

$\overline{\ln x}$ is the average of the natural logarithms. Using common 10 logarithms,

$$s_g = 10^{s_y} \tag{5.8}$$

where $s_y$ is the standard deviation of the common logarithms of the data values:

$$s_y = \sqrt{\frac{1}{n-1}\Sigma\left(\log x_i - \overline{\log x}\right)^2} \tag{5.9}$$

$\overline{\log x}$ is the average of the common logarithms. The same value for $s_g$ is obtained using either type of logarithm. Geometric mean is in the units of measurement, but geometric standard deviation has no units.

Converting the data to the logarithmic form is an example of data transformation. Transformations, and especially the "log transformation," are often useful in statistical analyses. If non-normal data can be transformed to achieve a normal distribution, powerful statistical techniques can be applied to the data analysis. Due to the nature of exposure data, the log transformation is commonly used.

Exposures are typically lognormally distributed (Esmen and Hammad 1977), with a geometric standard deviation less than 3 or 4 (Leidel et al. 1977, appendix B). Lognormal data distributions with geometric standard deviations less than about 1.4 are approximately normally distributed (Leidel et al. 1977, appendix M). With a small geometric standard deviation, it is difficult to distinguish the lognormal distribution from normal distributions unless large numbers of measurements are available (Day et al. 1999).

**5.3.2.2** ***Estimating the Arithmetic Mean of Lognormal Exposures*** As discussed by Rappaport and Selvin (1987), the arithmetic mean of a worker's TWA exposures may be most clearly related to long-term chronic health hazards and disease risk. The arithmetic mean $\hat{\mu}$ of a lognormal measurement data distribution may be estimated approximately using the equation

$$\hat{\mu} = \exp\left(\bar{y} + \frac{s_y^2}{2}\right) \tag{5.10}$$

where $\bar{y}$ and $s_y$ are the mean and standard deviation, respectively, of the logarithms of the measures (Gilbert 1987, p. 167). A more precise estimate $\hat{\mu}_1$ can be obtained using the minimum variance unbiased (MVU) estimator approach described by Gilbert (1987, pp. 165–167) as

$$\hat{\mu}_1 = \Psi_n\left(\frac{s_y^2}{2}\right)\exp(\bar{y}) \tag{5.11}$$

where the value of $\Psi_n$ is taken from a table (see, e.g. Gilbert 1987) or calculated from the infinite series:

$$\Psi_n(t) = 1 + \frac{(n-1)t}{n} + \frac{(n-1)^3 t^2}{2!n^2(n+1)} + \frac{(n-1)^5 t^3}{3!n^3(n+1)(n+3)} + \frac{(n-1)^7 t^4}{4!n^4(n+1)(n+3)(n+5)} + \cdots \tag{5.12}$$

where $t = \left(s_y^2/2\right).\Psi_n(t)$ accurate to three decimal places are obtained with 10 terms in $t$ for sample sizes up to $n = 500$ and $t = s_y^2/2$ up to 2.00.

## 5.4 CONFIDENCE INTERVALS

Due to random variation there is always some uncertainty in measurements, so the question of "How much uncertainty is there in the measures and the values calculated from them?" should be asked. It is recognized that when exposures are sampled and a mean is calculated, the resulting value (calculated from the original measures) should not be taken as exact but rather should be considered as one of a distribution of possible means that might have resulted. As will be seen, the spread of the distribution reflects the uncertainty in the sample estimate of the true exposure and is determined by the number of measurements and their precision. The major portion of the total range of possible estimates, usually taken to represent 95% of possible outcomes, can be characterized using confidence intervals. Confidence intervals can then be used to test hypotheses about the exposures.

### 5.4.1 Two-Sided Confidence Intervals

**5.4.1.1** ***Interval on the Mean of the Normal Distribution*** It is often of interest to compare a measured quantity with some known or reference quantity. For example, if one knew from long experience that the average airborne contaminant concentration near a process was 50 parts per million (ppm) but a sample of 10 measurements after a process change indicated a mean concentration of 75 ppm, one would wonder whether the airborne contaminant concentration has changed. The question "Is the measured mean value different from the reference value?" might be asked. A concentration of 75 ppm certainly looks different from 50 ppm, but if a second sample of 10 measurements were taken, it would almost certainly yield a mean similar to, but not exactly the same as, the first sample mean due to random variation. In fact, if a very large number of samples were to be taken, it would be seen that the resulting sample means are normally distributed about some value that is the best estimate of the true value being measured.

What must be determined is whether the 75 ppm value really is different from the 50 ppm value or is just one of the many means that might have been obtained when measuring a true concentration of 50 ppm. This can be tested using a confidence interval on the mean. A confidence interval is a range of values defined by upper and lower confidence limits and represents the range within which the reference value would be highly likely to fall if one were actually measuring that value. If the reference value falls outside the confidence interval, it would be concluded that it is unlikely that the measured mean actually represents the reference value. That is, it would be concluded that what was being measured was "significantly different from" the reference value.

The interval is restricted to some fraction of the possible values, and a 95% interval is usually used. A 95% confidence interval is constructed, and if the reference or "hypothesized" value does not fall in the interval, one concludes "with 95% confidence" that the measured quantity is different from the hypothesized value. In this example a 95% confidence interval centered on 75 ppm would be constructed, and if the interval did not include 50 ppm, it would be concluded that the measured quantity (represented by the 75 ppm sample mean) was significantly different from the hypothesized quantity (50 ppm). This is one form of hypothesis testing. The confidence level is $(1-\alpha)$ where $\alpha$ is the acceptable type I error rate or the acceptable probability that one might incorrectly conclude that there is a difference when in fact there is no difference. If the sample mean was indeed one of the possible but extreme means that could have been measured, a conclusion of difference would be incorrect or a "false positive." A type I error rate of $\alpha = 0.05$ is nearly always used in hypothesis testing, resulting in a 95% confidence level, i.e. there is a 5% probability of making a type I (false positive) error for any given test.

A confidence interval on a normally distributed variable such as a sample mean is constructed using a value from the $t$ probability distribution that is selected for the specified $\alpha$ (usually 0.05) and the number of degrees of freedom ($df$) of the measurement. For an interval on a mean, the degrees of freedom is just $(n-1)$ where $n$ is the number of measurements making up the sample. The upper and lower confidence limits on the interval are

$$\mathrm{UCL} = \bar{x} + t_{1-\alpha/2,df}\frac{s}{\sqrt{n}}$$

$$\mathrm{LCL} = \bar{x} - t_{1-\alpha/2,df}\frac{s}{\sqrt{n}} \tag{5.13}$$

Tables of the $t$-distribution are available in any basic statistics text or online, but $t$-values may be conveniently obtained in Excel using the T.INV function, specifying T.INV($1-\alpha/2, n-1$ $df$). For a sample size of 10 measurements and $\alpha = 0.05$, $t$ = T. INV(0.975,9) = 2.262. $t$ gets larger as $n$ gets smaller, so means for small samples will have wider confidence intervals than means for large samples even if the two sample standard deviations are the same. For a sample size of only 3, $t$ = T. INV(0.975,2) = 4.303, and the confidence interval would be nearly three and a half times as wide as for $n$ = 10 (i.e. $\pm 4.303\,s/\sqrt{3} = 2.48$ vs. $\pm 2.262\,s/\sqrt{10} = 0.72$) (assuming the same sample standard deviation for each).

Wider confidence intervals make it more likely that the reference value will be included in the interval even when a true difference exists, so that the test will fail to demonstrate the true difference. This is termed a type II error or "false negative" – failing to conclude that there is a difference when in fact there is one. The acceptable type II error rate is represented by the symbol $\beta$ and should normally not exceed 0.2. From a sampling strategy perspective, then, "more is better" definitely applies – more data means smaller confidence intervals and a greater ability to correctly conclude that there is a difference when there really is a difference. This is termed the statistical power of the test. The desirable level of power for a hypothesis test is specified to be $(1-\beta)$, typically 0.80 (80%) or better. The relationship of sample size, the sample variance, and the amount of true difference to the power of a test are discussed later in the chapter.

The confidence interval discussed was a two-sided confidence interval because it extended to both sides of the mean. It is useful when asking questions such as "Is the sample mean *different from* the reference value?," where the direction of difference is not relevant.

**5.4.1.2 *Interval on the Arithmetic Mean of a Lognormal Distribution*** Confidence limits defining the confidence interval around the arithmetic mean of lognormally distributed exposures may be calculated as described by Gilbert (1987, pp. 169–171). Such calculations are unlikely to be needed in practice and also require reference to tables of values used in the calculations. For details on these calculations, the reader is referred to Gilbert (1987) and Land (1971, 1975).

### 5.4.2 One-Sided Confidence Intervals

If the direction of difference *is* relevant, as in "Is the sample mean *greater than* the reference value?," then a one-sided confidence interval is appropriate. A one-sided interval is structured using one of the two equations in Equation (5.13) except that the $t$-value is for $(1-\alpha)$ and $(n-1)$ $df$, not $(1-\alpha/2)$ and $(n-1)$ $df$. If the question for the sampling example had been "Is the 75 ppm sample mean significantly *greater than* 50 ppm?," the lower confidence limit would be calculated as $LCL = \bar{x} - t_{1-\alpha,df}\, s/\sqrt{n}$ where $t$ = T. INV(0.95,9) = 1.833. If 50 ppm was less than this limit, then it would be concluded that the sample mean was significantly greater than 50 ppm, indicating an adverse effect due to the process change.

Note that the one-sided interval is smaller than the corresponding side of the two-sided confidence interval because the $t$-value is smaller. This makes it more likely that a true difference will be demonstrated, i.e. the one-tailed test is more powerful than the two-tailed test. However, a one-tailed test should only be used if it can be justified in advance of the actual analysis. One-sided confidence intervals have direct application in several areas of occupational safety and health work.

#### 5.4.2.1 *One-Sided Confidence Intervals in Compliance Decisions*

Consider a sample of five 8-hour TWA measurements that indicates the average worker hazardous dust X exposure to be 35 micrograms per cubic meter ($\mu g\, m^{-3}$). If the occupational exposure limit (OEL) for this dust is 25 $\mu g\, m^{-3}$, one might want to perform a test using a one-sided lower confidence interval to determine whether the measured exposure was significantly greater than the OEL. If the LCL is above the OEL, one concludes with 95% confidence that the exposure is out of compliance with the OEL. Alternatively, if the sample mean had been smaller than the OEL, one might construct an *upper* confidence limit on the mean to determine whether the measured exposure is significantly *lower than* the OEL. If the UCL is less than the OEL, one concludes with 95% confidence that the exposure is in compliance with the OEL. In each case the confidence limit would be calculated using the sample standard deviation and sample size along with the appropriate $t$-value.

OSHA employs a version of the one-sided confidence interval testing approach in which the sample standard deviation, sample size, and a $t$-value need not be used. Rather, the measured value – there might be only one rather than a mean of several – is divided by the applicable permissible exposure limit (PEL) and from this is subtracted a sampling and analytical error (SAE) value that represents the combined variability of the sampling and analysis techniques used. For an 8-hour TWA exposure and PEL, if the resulting test statistic TWA/PEL – SAE is >1, then OSHA concludes with 95% confidence that the exposure was out of compliance. The SAE value is provided by the OSHA laboratory performing the chemical or other analysis. Fortunately, OSHA does not usually base compliance decisions on a single sample. For more detail on the SAE approach, see the OSHA Technical Manual, 1999, section II, chapter 1, appendix N (available online at the OSHA website, www.osha.gov) (OSHA 1999).

#### 5.4.2.2 *One-Sided Confidence Intervals for Measurement LOB, LOD, and LOQ*

Many contaminant exposure measurements have a laboratory analysis component. Filters must be weighed, sorbent tubes and diffusion badges must be extracted and chemically analyzed, and so on. Laboratory analytical methods are subject to lower limits of detection determined by the sensitivity of the instrument and any electronic "noise" in the instrument response. These are characterized using method limit of blank (LOB), limit of detection (LOD), and limit of quantitation (LOQ) descriptors, each of which is determined using a one-sided confidence interval.

When sending samples to a lab, most require sampling media that has not been used. This is called a "blank" sample. A "field blank" is sampling media, such as an air sampling filter cassette or sorption tube that is handled in exactly the same way as the sample media except that no air (in this case) is drawn through it. Blanks account for contaminant that might have been on the media as received from the manufacturer as well as any contaminant that might be introduced through media handling, shipment, or storage. This allows these potential sources of bias to be detected and accounted for in the laboratory analysis.

Even a blank containing none of the analyte of interest may cause an analytical instrument to give a weak response that seems to suggest that something is there (Armbruster and Pry 2008). The LOB may be described as the highest measurement result that is likely to be observed for a blank sample, with some specified level of probability. Replicate measurements of samples known to contain none of the analyte of interest are made, and a one-sided 95% upper confidence limit on the mean response is calculated as $LOB = \bar{x}_{blanks} + 1.645 s_{blanks}$. The 1.645 is the $z$-value for a one-sided 95% probability from a table such as Table 5.1 or from NORM.S.INV(0.95) in Excel.

The LOD is the lowest amount of analyte in a sample that can be detected with a stated level of probability, although perhaps not quantified as an exact value (Shrivastava and Gupta 2011). The instrument can distinguish a difference between the measurement and the LOB, but not well enough to establish a reliable number. The LOD is determined from the LOB and replicate measurements of a sample with a known low concentration of analyte, again employing a one-sided 95% upper confidence limit, as $LOD = LOB + 1.645\, s_{\text{low conc. samples}}$. Measured values below the LOD are said to be left censored or just censored. These are reported by laboratories as "below the limit of detection," "BDL," or similar and have no definite value. If they make up only a small fraction of the data set, perhaps less than 10%, values are sometimes assigned to them, typically either LOD/2 or $LOD/\sqrt{2}$. For lognormally distributed data having a geometric standard deviation less than 3, $LOD/\sqrt{2}$ is recommended (Hornung and Reed 1990; Hewett 2006).

The LOQ is the lowest amount of analyte in a sample that can be determined with a stated precision *and accuracy* under stated experimental conditions and is also a one-sided 95% upper confidence limit. Wilson and Poole (2009, p. 617) discuss a rule-of-thumb approach in which the LOD and LOQ are determined from signal-to-noise ratios (*S*/*N*), with *S*/*N* ~ 2:1 or 3:1 used for the LOD and *S*/*N* ~ 10:1 for the LOQ. The ASTM method for characterizing errors in filter weighing includes this guideline (ASTM International 2011). Doing in-house pre- and post-sampling filter weighing, including a number of field blanks during each sampling event, allows calculation of the weighing LOD and LOQ as 3 and 10 times the standard deviation of blank pre- and post-weight differences, respectively. Measurement weight differences less than the LOD are censored, and those between the LOD and LOQ should be viewed with caution. Results greater than the LOQ are the goal.

## 5.5 PARAMETRIC HYPOTHESIS TESTS

The concepts underlying the simple hypothesis testing approach using one-sided and two-sided confidence intervals are applied in several types of parametric hypothesis tests based on the normal probability distribution. These include Student's *t*-test, the two-sample *t*-test, and the paired-sample *t*-test.

### 5.5.1 Student's *t*-Test for Comparing a Mean to a Known Value

In the previous example a sample mean of 75 ppm was compared with a hypothesized value (50 ppm) using one- and two-sided confidence intervals. A confidence interval was constructed using a *t*-value for the chosen $\alpha$ and the sample *df*, and a conclusion of difference was reached if the hypothesized value did not fall in the interval. Student's *t*-test takes a different approach to this hypothesis test. A test statistic $t^*$ is calculated from the sample mean (75 ppm in the example), the sample standard deviation and sample size, and the hypothesized value $\mu_0$ (50 ppm) as

$$t^* = \frac{|\bar{x} - \mu_0|}{s/\sqrt{n}} \tag{5.14}$$

$t^*$ is compared with the critical value $t_{\text{crit}}$ of the *t*-distribution corresponding to the confidence interval limit. This is the same *t*-value used to calculate the confidence interval (i.e. $t_{1-\alpha/2, n-1\,df}$ for a two-sided test or $t_{1-\alpha, n-1\,df}$ for a one-sided test). $t^* > t_{\text{crit}}$ indicates a significant difference at the $(1-\alpha)$ confidence level (usually 95%).

No Excel cell function or Analysis ToolPak utility is available for performing Student's *t*-test.

### 5.5.2 Two-Sample *t*-Test for Comparing Two Independent Sample Means

**5.5.2.1 *The Two-Sample Test*** Answering questions such as "Are exposures different for different worker groups?" and "Has an intervention changed worker exposures?" involves comparing two sample means for difference. Each mean has a normal distribution of variability, but placing a confidence interval on each is not a good way to compare the means for difference. A better approach is to place a confidence interval on the *difference* of the two means $(\bar{x}_1 - \bar{x}_2)$ and see if it contains zero. If it does, then it cannot be concluded that there is a difference in the means. The two-sided confidence interval would be calculated as

$$\text{CL} = |\bar{x}_1 - \bar{x}_2| \pm t_{(1-\alpha/2, n_1+n_2-2)} s_{\text{p}} \sqrt{\frac{1}{n_1} + \frac{1}{n_2}} \tag{5.15}$$

where the | brackets indicate absolute value of the difference, $n_1$ and $n_2$ are the two sample sizes, and $s_{\text{p}}$ is a pooled estimate of the standard deviation assumed to be shared by the two samples:

$$s_{\text{p}} = \sqrt{\frac{(n_1-1)s_1^2 + (n_2-1)s_2^2}{n_1+n_2-2}} \tag{5.16}$$

Note that in this case the degrees of freedom for the difference in means is the sum of the *df* for the two means, i.e. $df = (n_1 - 1) + (n_2 - 1) = (n_1 + n_2 - 2)$.

Similar to the Student's $t$-test, another approach is to calculate a test statistic:

$$t^* = \frac{|\bar{x}_1 - \bar{x}_2|}{\sqrt{\left(\frac{1}{n_1} + \frac{1}{n_2}\right)\left(\frac{(n_1-1)s_1^2 + (n_2-1)s_2^2}{n_1+n_2-2}\right)}} \quad (5.17)$$

$t^*$ is compared with the critical value $t_{\text{crit}}$ of the $t$-distribution corresponding to the confidence interval limits for the difference in means. This is the same $t$-value used to calculate the confidence interval (i.e. $t_{1-\alpha/2,n_1+n_2-2df}$ for a two-sided test or $t_{1-\alpha,n_1+n_2-2df}$ for a one-sided test). $t^* > t_{\text{crit}}$ indicates a significant difference in means at the $(1-\alpha)$ confidence level (usually 95%).

Both one-sided and two-sided two-sample $t$-tests can be performed in Excel using either the T.TEST cell function or the $t$-test: Two-Sample Assuming Equal Variances utility in Analysis ToolPak. The T.TEST cell function provides only the probability $p$ of obtaining a result at least as large as that observed. This is termed the $p$-value for the test. A significant difference is shown if $p < \alpha$. Both one-sided and two-sided tests can be performed with this cell function. The two-sample test utility provides much more information, as shown in Figure 5.3. Note that while both the two- and one-tailed test results are shown, the one-tailed result should be used only if justified in advance.

**5.5.2.2 *Sample Size Required for Comparing Two Means*** The ability of a two-sample $t$-test to demonstrate a significant difference when one truly exists is the power of the test $(1-\beta)$ where $\beta$ is the allowable type II error rate previously discussed. Power increases with higher precision, larger sample size, and larger true difference. The smaller the true difference, the fewer the measurements, or the worse the measurement precision, the more difficult it is to statistically demonstrate the difference. In planning for sampling events, it is wise to ask the question "How many measures are needed to answer the question with the necessary statistical power?," whatever the question might be. The number required will be determined by the minimum difference considered important and the expected variability of the measurements.

Estimating the sample size for detecting a minimum specified difference $\delta$ between sample means is an iterative process. For the case of equal sample sizes, $n$, the iteration is performed using (Zar 2010, pp. 147–151)

$$n \geq \frac{2s_p^2}{\delta^2}\left(t_{1-\alpha/2,df} + t_{1-\beta,df}\right)^2 \quad (5.18)$$

where $s_p^2$ is the pooled estimate of variance for the two samples (see Eq. (5.16)) and the $t$-values are selected from a table of the $t$-distribution or are obtained from the T.INV function in Excel. If the expected sample variance $(s_p^2)$ is not known, pilot measurements may be needed to estimate it. The degrees of freedom for the $t$-values is $(2n-2)$. An initial "guess" for $n$ is made, the $t$-values are obtained, and a revised sample size estimate $n'$ is calculated. This is fed back into the equation as the new $n$, the appropriate $t$-values are selected, a second revised $n'$ is calculated, and so on iteratively until the calculated value converges to the input value.

| Variable 1 | Variable 2 | | t-Test: Two-Sample Assuming Equal Variances | | |
|---|---|---|---|---|---|
| 7.70 | 6.80 | | | | |
| 11.70 | 9.00 | | | *Variable 1* | *Variable 2* |
| 12.20 | 9.50 | | Mean | 12.4333333 | 11.5 |
| 12.20 | 9.90 | | Variance | 4.33 | 9.26444444 |
| 12.20 | 11.30 | | Observations | 9 | 10 |
| 13.10 | 12.20 | | Pooled Variance | 6.94235294 | |
| 13.50 | 12.20 | | Hypothesized Mean Difference | 0 | |
| 14.40 | 12.60 | | df | 17 | |
| 14.90 | 13.50 | | t Stat | 0.77095297 | |
| | 18.00 | | P(T<=t) one-tail | 0.22565889 | |
| | | | t Critical one-tail | 1.73960673 | |
| | | | P(T<=t) two-tail | 0.45131777 | |
| | | | t Critical two-tail | 2.10981558 | |

**Figure 5.3** Example Excel output for a two-sample $t$-test. The input data are shown on the left, and the test output on the right. Results for both two-sided and one-sided tests are provided.

For two samples of equal size $n$, the equation can be rearranged to allow post hoc estimation of the minimum detectable difference $\delta$ for a given set of sample data and specified type I and type II error rates $\alpha$ and $\beta$:

$$\delta \geq \sqrt{\frac{2s_p^2}{n}}\left(t_{1-\alpha/2,df} + t_{1-\beta,df}\right) \tag{5.19}$$

Similarly, the equation can be rearranged to allow post hoc estimation of the actual power to detect a specified $\delta$ for a given set of sample data and specified type I error rate (Zar 2010, p. 150):

$$t_{1-\beta,df} \leq \frac{\delta}{\sqrt{2s_p^2/n}} - t_{1-\alpha/2,df} = \frac{\delta}{s_p\sqrt{2/n}} - t_{1-\alpha/2,df} \tag{5.20}$$

where $\beta$ is determined as 1 minus the one-tailed probability associated with $t_{1-\beta,df}$. This probability can be obtained from the T.DIST.RT function in Excel or estimated from a table of the $t$-distribution. The quantity $\delta/s_p = |\bar{x}_1 - \bar{x}_2|/s_p$ is often referred to as the effect size.

### 5.5.3 Paired-Sample *t*-Test for Dependent Measures

The two-sample $t$-test assumes the two sets of measurements to be independent of one another. Measures that are paired in some way, as with exposure measurements on a particular worker before and after a process change, are not independent. Such "paired" samples change, and actually make more powerful, the analysis approach. Consider a situation in which exposures of each of six workers are measured, then the exposures of these same six workers are measured again after a process change is instituted (e.g. a new piece of equipment is installed). While the means of the six samples before and after the change could be compared with a two-sample $t$-test, a more powerful approach is to take the *difference* between the two measures for each of the six workers individually and use those as the exposure metric. If there really is no difference in exposures before and after the process change, then the individual differences should be due simply to random variation, with some being positive and some negative and *averaging close to zero*. The average difference can be compared to zero with the paired-sample $t$-test. This is an alternate approach to answering the question: "Has an intervention changed worker exposures?"

The paired-sample $t$-test is essentially a Student's $t$-test using the pair differences $d_i$ as the measures and setting the hypothesized difference $\mu_0 = 0$. The differences $d_i$ of the $n$ data pairs, the mean difference $\bar{d}$, and the standard deviation of the differences $s_d$ are used to calculate the test statistic $t^*$ as

$$t^* = \frac{|\bar{d}|}{s_d/\sqrt{n}} \tag{5.21}$$

$t^*$ is compared with a critical $t$-value taken from a $t$-table or from the Excel T.INV function for $(n-1)$ $df$ and the appropriate probability. In this example $t$ = T. INV(0.975,5) for a two-sided test or $t$ = T. INV(0.95,5) for a one-sided test when $\alpha = 0.05$. $t^* > t_{crit}$ indicates a significant difference in the paired measures at the $(1-\alpha)$ confidence level (usually 95%). Differencing paired measures has the effect of eliminating the within-worker variability, thereby providing greater statistical power compared with the usual two-sample $t$-test in which within-worker variability plays a role. For this reason, paired measures are usually helpful.

The paired $t$-test can be performed in Excel using either the T.TEST cell function or the $t$-test: Paired Two Sample for Means utility in Analysis ToolPak. The utility provides much more information. An example of the Excel output for a paired-sample test is shown in Figure 5.4.

### 5.5.4 Verifying the Underlying Assumptions of *t*-Tests

The $t$-tests are based on the normal probability distribution, i.e. they are parametric techniques, so they can only be used to analyze normally distributed measures. Practitioners often neglect to verify normality of the data before using $t$-tests, and this can lead to incorrect inferences if the data are non-normal. Exposure data in particular are often

| Before | After |
|---|---|
| 56 | 53 |
| 60 | 59 |
| 48 | 50 |
| 44 | 44 |
| 52 | 55 |
| 46 | 47 |
| 58 | 56 |
| 38 | 38 |
| 56 | 55 |
| 51 | 53 |
| 32 | 30 |
| 21 | 21 |

| t-Test: Paired Two Sample for Means | | |
|---|---|---|
| | *Before* | *After* |
| Mean | 46.83333 | 46.75 |
| Variance | 135.0606 | 135.2955 |
| Observations | 12 | 12 |
| Pearson Correlation | 0.987587 | |
| Hypothesized Mean Difference | 0 | |
| df | 11 | |
| t Stat | 0.157578 | |
| P(T<=t) one-tail | 0.438822 | |
| t Critical one-tail | 1.795885 | |
| P(T<=t) two-tail | 0.877645 | |
| t Critical two-tail | 2.200985 | |

**Figure 5.4** Example Excel output for a paired-sample *t*-test. The input data are shown on the left, and the test output on the right.

lognormally distributed, so applying a *t*-test to the measurement values would be incorrect. For lognormally distributed data the test should be applied to the *logarithms* of the data rather than the original values.

Techniques such as the Shapiro–Wilk *W* test (for sample sizes $n \leq 50$) and D'Agostino's test (for $n > 50$) are available to formally test the "fit" of the data to the normal distribution (Shapiro and Wilk 1965; D'Agostino 1971); the Lilliefors extension of the Kolmogorov–Smirnov test (Lilliefors 1967) is also available but is more difficult to perform. Fortunately, qualitative graphical methods usually suffice. Simple graphical approaches include the normal equivalent deviations (NED) plot and the quantile–quantile (Q-Q) plot, both of which are easily produced in Excel (though regrettably not available as standard chart options). An NED or Q-Q data plot with a linear pattern indicates normally distributed data. Normality should be verified for the one data set used in Student's *t*-test and separately for each of the two data sets in two-sample *t*-tests. In a paired-sample *t*-test, the distribution of the pair differences should be checked for normality.

An additional assumption underlying the two-sample *t*-test is that the two samples have similar variance, which should be the case if the two samples are actually just two measures of the same thing. Similarity of variance, or homoscedasticity, is checked with the *F*-test, one of the utilities available in the Excel Analysis ToolPak. A nonsignificant *F*-test ($p > 0.05$) indicates similar variances. If it should turn out that the two variances are significantly different, Welch's test, a variant of the two-sample *t*-test, can still be performed in Excel using the *t*-test: Two-Sample Assuming Unequal Variances utility.

## 5.6 NONPARAMETRIC EQUIVALENTS TO THE *t*-TESTS

Measurement data are not always normally or lognormally distributed. When they are not and cannot be made so through some sort of transformation, *or* if they are heavily censored or are ordinal measures such as Likert scale values, parametric techniques cannot be used. Instead, nonparametric or "distribution-free" techniques are used.

The nonparametric equivalent of the two-sample *t*-test is the Mann–Whitney *U* test (Mann and Whitney 1947), which operates on the ranks of the data rather than the original values. The data are ordered smallest to largest and assigned ranks, and these become the input to the analysis. A useful ranking tool in Excel is the RANK.AVG function (*not* the RANK function). The nonparametric equivalent of the paired-sample *t*-test is the Wilcoxon matched pairs test, which operates on the ranks of the pair differences. These tests are discussed in detail in texts such as Siegel (1956), Conover (1999), and Johnson (2017).

Parametric techniques are generally a bit more powerful than their nonparametric counterparts because they take advantage of more information about the data distribution. This is one reason why values are often assigned to censored data – so that a parametric analysis technique can then be used. Parametric techniques are also more familiar to most practitioners than are their nonparametric equivalents and are readily available in common software applications such as Excel.

## 5.7 COMPARING MORE THAN TWO MEANS WITH ONE-WAY ANOVA

### 5.7.1 Parametric One-Way ANOVA

**5.7.1.1 *The ANOVA*** Two means of normally distributed measures were compared for difference using the two-sample *t*-test. Comparing three or more means for difference could be done using a series of these tests, e.g. by comparing means $A$, $B$, and $C$ using *t*-tests for $A$ vs. $B$, $A$ vs. $C$, and $B$ vs. $C$. However, with each test there is an $\alpha$ probability of making a type I (false positive) error, and the overall type I error probability for three tests together would be $p = 1-(1-\alpha)^k$ where $k$ is the number of tests. For three tests, $p = 1-(1-\alpha)^k = 1-(1-\alpha)^3 = 0.143$ or about 14% for $\alpha = 0.05$ (this is 1 minus the probability of *not* making a type I error on any of the tests). As the number of means goes up, the overall type I error probability goes up rapidly. Comparing four means requires $k = n!/2!(n-2)! = 4!/2!(4-2)! = 6$ tests, and comparing five means requires $k = 5!/2!(5-2)! = 10$ tests (recall that "!" indicates "factorial"). The corresponding overall type I error rates would be 25.5 and 40.1%, respectively, so this is not an optimal approach.

Three or more means can be simultaneously compared using the analysis of variance (ANOVA) technique. For example, ANOVA could be used to compare the mean TWA exposures for a task during the three 8-hour shifts of a 24-hour operation. This is termed a one-way ANOVA because there is only one factor (task) that has three treatment levels (shift), and the means for the three shifts are being compared.

ANOVA compares the variability of measurements at each treatment level (the within-sample variance) to the variability of the treatment level means relative to the overall or grand mean of all the measurements (the between-sample variance). If the three means are all really just estimates of the same thing, i.e. they are truly not different, the between-sample variance and the within-sample variance should be similar. If the means actually represent measures of *different* things, the between-sample variance will be inflated. The two variances are compared with an *F*-test, as was performed to verify homoscedasticity in the two-sample *t*-test, and a resulting $p<\alpha$ indicates that at least one of the means is significantly different from at least one other mean.

One-way ANOVA is performed in Excel using the ANOVA: Single Factor utility in Analysis ToolPak. An example output is shown in Figure 5.5. The variance estimates being compared with the *F*-test are the mean square (MS) values shown in the table. MS values are calculated from sums of squares and their degrees of freedom. The between-sample sum of squares is

$$SS_{\text{between}} = \sum_{i=1}^{k} n_i \left(\bar{x}_i - \bar{x}\right)^2 \tag{5.22}$$

where the $n_i$ are the sample sizes, the $\bar{x}_i$ are the sample means, $\bar{x}$ is the mean of all the $N$ measurements together (the grand mean), and $k$ is the number of samples (means). The total sum of squares is the sum of the squares of the $N$ individual measurement deviations from the grand mean:

$$SS_{\text{total}} = \sum_{i=1}^{k}\sum_{j=1}^{n_i} \left(x_{ij} - \bar{x}\right)^2 \tag{5.23}$$

| A | B | C | | | | | | | |
|---|---|---|---|---|---|---|---|---|---|
| | | | Anova: Single Factor | | | | | | |
| 64 | 40 | 54 | | | | | | | |
| 60 | 32 | 57 | SUMMARY | | | | | | |
| 58 | 43 | 47 | *Groups* | *Count* | *Sum* | *Average* | *Variance* | | |
| 53 | 38 | 45 | A | 5 | 285 | 57 | 31 | | |
| 50 | 45 | 57 | B | 5 | 198 | 39.6 | 25.3 | | |
| | | | C | 5 | 260 | 52 | 32 | | |
| | | | | | | | | | |
| | | | ANOVA | | | | | | |
| | | | *Source of Variation* | *SS* | *df* | *MS* | *F* | *P-value* | *F crit* |
| | | | Between Groups | 802.5333 | 2 | 401.2667 | 13.63307 | 0.000815 | 3.885294 |
| | | | Within Groups | 353.2 | 12 | 29.43333 | | | |
| | | | | | | | | | |
| | | | Total | 1155.733 | 14 | | | | |

**Figure 5.5** Excel output for a one-way ANOVA to compare three sample means. The input data are shown on the left, and the output on the right. The *p*-value is much less than 0.05, indicating a significant difference in these means.

The within-sample sum of squares is just the difference between the total and between-sample sum of squares:

$$SS_{\text{within}} = SS_{\text{total}} - SS_{\text{between}} \tag{5.24}$$

These values, scaled by the degrees of freedom associated with each, provide the mean square variance estimates that are compared with the $F$-test. The $df$ for $SS_{\text{between}}$ is $(k-1)$, the $df$ for $SS_{\text{total}}$ is $(N-1)$, and the $df$ for $SS_{\text{within}}$ is $(N-1)-(k-1)=(N-k)$. These are shown in the table along with the resulting mean squares and the $F$-test. Note that only one $F$-test is conducted to compare the between-group variance with the within-group variance. A $p$-value less than $\alpha$ indicates a significant difference in the means.

**5.7.1.2 *Multiple Simultaneous Pairwise Comparisons*** Comparing means with one-way ANOVA is a two-step process if the ANOVA is significant, because it remains to be determined which means are significantly different from which others. A multiple simultaneous pairwise comparison approach such as Tukey's minimum significant difference (MSD) test (Tukey 1949) for equal sample sizes or the Tukey–Kramer MSD test (Kramer 1956) for unequal sample sizes should be used. These tests make all possible pairwise comparisons with an overall type I error rate $<\alpha$. Details on these tests can be found in, e.g. Johnson (2017).

**5.7.1.3 *Verifying the Assumptions of Parametric One-Way ANOVA*** Like the two-sample $t$-test, ANOVA assumes both normality of the data and similar variances (homoscedasticity), and these must be verified. An NED plot of each of the sample data sets is usually sufficient to explore normality, but if there is any doubt, a Shapiro–Wilk $W$ test or D'Agostino $D$ test for normality can be performed.

A relatively simple test for homoscedasticity is Levene's test (Levene 1960). This is an ANOVA on the *absolute values* of the residuals of the data, i.e. the absolute values of the differences between the values and their respective sample means. The absolute values of the differences between the data values and the sample mean for each sample are calculated and are then compared with an ANOVA. A nonsignificant result ($p>\alpha$) indicates similar variances. The results of a Levene's test on the data from Figure 5.5 are shown in Figure 5.6.

### 5.7.2 Nonparametric One-Way ANOVA

**5.7.2.1 *The ANOVA*** Just as the nonparametric Mann–Whitney $U$ test was required in place of the parametric two-sample $t$-test when the $t$-test assumptions could not be met, a nonparametric ANOVA technique must be used when the parametric ANOVA assumptions cannot be met. This will be the Kruskal–Wallis one-way ANOVA (Kruskal and Wallis 1952). Like the Mann–Whitney $U$ test, Kruskal–Wallis operates on the ranks of the data. All of the data are combined into one ordered list, keeping track of which measure goes with which sample, and ranks are assigned. The ranks are then separated and summed separately for each sample. These sums and the sample sizes are used to calculate a test statistic $H$ that is compared to a critical value taken from the chi-square distribution. The critical chi-square value $X^2_{\text{crit}}$ is taken from a table of the chi-square distribution or from Excel using the CHISQ.INV function

| | Data: | | |
|---|---|---|---|
| | A | B | C |
| | 64 | 40 | 54 |
| | 60 | 32 | 57 |
| | 58 | 43 | 47 |
| | 53 | 38 | 45 |
| | 50 | 45 | 57 |
| means: | 57.00 | 39.60 | 52.00 |

| Absolute values of the residuals: | | |
|---|---|---|
| A | B | C |
| 7.00 | 0.40 | 2.00 |
| 3.00 | 7.60 | 5.00 |
| 1.00 | 3.40 | 5.00 |
| 4.00 | 1.60 | 7.00 |
| 7.00 | 5.40 | 5.00 |

Anova: Single Factor

SUMMARY

| *Groups* | *Count* | *Sum* | *Average* | *Variance* |
|---|---|---|---|---|
| A | 5 | 22 | 4.4 | 6.8 |
| B | 5 | 18.4 | 3.68 | 8.372 |
| C | 5 | 24 | 4.8 | 3.2 |

ANOVA

| *Source of Variation* | *SS* | *df* | *MS* | *F* | *P-value* | *F crit* |
|---|---|---|---|---|---|---|
| Between Groups | 3.221333 | 2 | 1.610667 | 0.263009 | 0.773053 | 3.885294 |
| Within Groups | 73.488 | 12 | 6.124 | | | |
| | | | | | | |
| Total | 76.70933 | 14 | | | | |

**Figure 5.6** Levene's test for similarity of variance for three samples: A, B, and C. The one-way ANOVA is performed on the absolute values of the residuals for each sample. A nonsignificant $p$-value indicates similar variances.

for the chosen $\alpha$ and $(k-1)$ *df*. Here *k* is the number of groups being compared, i.e. $X^2_{crit} = \text{CHISQ.INV}(1-\alpha, k-1df)$. A significant difference is shown if $H > X^2_{crit}$. There is no utility in Excel for performing Kruskal–Wallis ANOVA, but third-party software add-ons can be purchased that will have this capability, or it can be done manually.

**5.7.2.2 *Multiple Simultaneous Pairwise Comparisons*** When the Kruskal–Wallis ANOVA is significant, multiple simultaneous pairwise comparisons of the average sample ranks can be performed using Nemenyi's MSD test (Nemenyi 1961) for equal sample sizes or the Bonferroni–Dunn test (Dunn 1961) for unequal sample sizes. As with the parametric Tukey and Tukey–Kramer tests, these tests control the overall type I error rate to $<\alpha$ (see, e.g. Johnson 2017, chapter 6).

## 5.8 TWO-WAY ANOVA

### 5.8.1 Parametric Two-Way ANOVA

**5.8.1.1 *ANOVA with Replication*** Often more than one factor can influence exposure. For example, grinder speed (high, medium, low), grinding wheel type (coarse, medium, fine), and local exhaust ventilation status (on, off) might well influence the amount of respirable silica dust in a concrete worker's breathing zone. Which factors have an influence can be explored using multi-way ANOVA. The present discussion is limited to two-factor or two-way ANOVA. Three-way and higher ANOVA must be performed using advanced statistical software or hierarchical multiple regressions and will not be discussed here.

Two-way ANOVA can be readily performed in Excel using the ANOVA: Two-Factor with Replication utility. Most analyses will be to compare sample means, i.e. the data are replicated within each (factor 1, factor 2) combination. Unlike one-way ANOVA in which sample sizes can be different, for two-way ANOVA, each condition must have the same number of measures. The ANOVA will determine which of the factors exerts a significant effect and whether there is any interaction between of the factors, i.e. whether they are not acting completely independently of one another. These are termed the main effects and the interaction effect. Like one-way ANOVA, two-way ANOVA compares variances using *F*-tests to determine whether any of the means are different from any of the others. If a significant result is obtained and there is no interaction, pairwise comparisons are conducted using a variation on the Tukey test.

An example Excel two-way ANOVA (with replication) output is shown in Figure 5.7 for a comparison of dust levels ($mg\,m^{-3}$) in three stone-working shops while using or not using wet methods for dust suppression. The two main effects of shop and wetting condition are both significant ($p \ll 0.05$), but their interaction is not ($p = 0.96$).

**5.8.1.2 *ANOVA Without Replication*** Two-way ANOVA can also be performed with only one measure at each condition, as can occur with repeated measures-type studies, but in that case the potential interaction effect cannot be assessed. For single measures the Excel ANOVA: Two-Factor without Replication utility is used. An example of a repeated measures study might be to assess worker thermal comfort for each of four optional protective equipment ensembles, with each subject worker being assessed while wearing each ensemble. If each worker is assessed only once with each ensemble, with the order of ensemble wear randomized individually for each worker, this would be a "randomized block repeated measures" study without replication. The measures are said to be "blocked" on subject (the individual worker). When interpreting the results, the "sample" *p*-value corresponding to the blocking variable (worker) is ignored. By convention, the blocking factor is always the row variable, and the treatment level is the column variable.

### 5.8.2 Nonparametric Two-Factor Analysis

**5.8.2.1 *Rank Tests*** Two-factor nonparametric analysis may be performed using a rank-based test such as the rank transform test (Conover and Iman 1981). This test is just a two-way ANOVA on the data ranks. The data are combined (keeping track of which value goes with which sample), then ordered smallest to largest, and ranked, with average ranks given to tie values. A parametric two-way ANOVA is then applied to the data ranks. The significance of main effects and interaction are assessed from the *p*-values.

**5.8.2.2 *Repeated Measures: Friedman's Test*** When the data are dependent measures and do not meet the requirements for parametric repeated measures two-way ANOVA, nonparametric repeated measures two-way ANOVA can be performed on data ranks using Friedman's test (Friedman 1937). ANOVA with or without replication

| | Shop A | Shop B | Shop C |
|---|---|---|---|
| Dry | 2.48 | 3.33 | 2.30 |
| | 2.89 | 3.93 | 2.71 |
| | 3.22 | 4.08 | 2.89 |
| | 3.56 | 4.09 | 3.00 |
| | 3.91 | 4.23 | 3.22 |
| | 4.01 | 4.30 | 3.40 |
| Wet | 1.95 | 2.20 | 1.79 |
| | 2.30 | 3.26 | 2.08 |
| | 2.56 | 3.43 | 2.30 |
| | 2.83 | 3.53 | 2.30 |
| | 3.09 | 3.56 | 2.56 |
| | 3.18 | 3.58 | 2.71 |

Anova: Two-Factor With Replication

| SUMMARY | Shop A | Shop B | Shop C | Total |
|---|---|---|---|---|
| *Dry* | | | | |
| Count | 6 | 6 | 6 | 18 |
| Sum | 20.06886 | 23.97408 | 17.51681 | 61.55975 |
| Average | 3.34481 | 3.995681 | 2.919469 | 3.419986 |
| Variance | 0.35336 | 0.122532 | 0.150723 | 0.391684 |
| *Wet* | | | | |
| Count | 6 | 6 | 6 | 18 |
| Sum | 15.91575 | 19.55454 | 13.74937 | 49.21966 |
| Average | 2.652626 | 3.259089 | 2.291562 | 2.734426 |
| Variance | 0.226112 | 0.284521 | 0.108649 | 0.35088 |
| *Total* | | | | |
| Count | 12 | 12 | 12 | |
| Sum | 35.98461 | 43.52862 | 31.26618 | |
| Average | 2.998718 | 3.627385 | 2.605515 | |
| Variance | 0.394065 | 0.332997 | 0.225423 | |

ANOVA

| *Source of Variation* | *SS* | *df* | *MS* | *F* | *P-value* | *F crit* |
|---|---|---|---|---|---|---|
| Sample | 4.229942 | 1 | 4.229942 | 20.3706 | 9.17E-05 | 4.170877 |
| Columns | 6.376193 | 2 | 3.188097 | 15.35327 | 2.56E-05 | 3.31583 |
| Interaction | 0.017916 | 2 | 0.008958 | 0.04314 | 0.957837 | 3.31583 |
| Within | 6.229482 | 30 | 0.207649 | | | |
| | | | | | | |
| Total | 16.85353 | 35 | | | | |

**Figure 5.7** Excel two-way ANOVA output. The two main effects (wetting condition and shop) are significant, but their interaction is not.

can be performed. Because the data are repeated measures, they will be blocked on one of the variables, e.g. on individual if repeated measures were made on individuals.

The test operates on the data ranks, but in this case the data are not combined for ranking but instead are ranked *within each block* separately. The ranks for the columns (treatments) are then summed, a test statistic is calculated, and comparison is made to a critical value from the *F*-distribution. When the measures are replicated within each (block, treatment) combination (e.g. for each individual at each treatment level), ranks are again assigned within each block separately. If the result is significant, multiple comparisons can be made to determine which treatments are significantly different from which others (see, for example, Conover 1999, p. 371).

## 5.9 EXPLORING THE UPPER TAIL OF THE EXPOSURE DISTRIBUTION

Occupational exposures are often lognormally distributed, i.e. they are skewed to the right. This upper tail of the distribution can represent a small fraction of the potential exposures, but they are the worst ones. For both acute toxicity and regulatory compliance reasons, these upper-end exposures should be of considerable concern. One might ask the questions "How frequently do these high exposures occur?," "What fraction of exposures are likely to exceed allowable levels?," and "What is the exposure concentration below which the major fraction of exposures (perhaps 90, 95, or 99%) is expected to fall?" These questions can be answered by first estimating the shape of the lognormal exposure distribution using sample data and then examining its upper tail. Alternatively, data combined with professional judgment can be used to estimate how adequately the highest exposures will be controlled. The first approach using data only involves upper tolerance limits and exceedance fractions, while the second approach employing both data and judgment involves Bayesian decision analysis (BDA).

### 5.9.1 Upper Tolerance Limits

The shape of a lognormal exposure distribution can be estimated from sampling data, i.e. the shape can be verified as lognormal using an NED plot or a formal test (e.g. Shapiro–Wilk), and the geometric mean and geometric standard deviation can be calculated. Using these values any percentile of the distribution can then be determined using probability tables. However, the shape of the distribution is only estimated, so the chosen percentile is also only an estimate of the true exposure percentile. This uncertainty is explicitly recognized by placing an upper confidence limit on the percentile estimate, usually an upper 95% confidence limit. This is termed an upper tolerance limit (UTL). For lognormally distributed data it is calculated as

$$\text{UTL} = \exp\left[\ln(\bar{x}_g) + K\ln(s_g)\right] \quad (5.25)$$

where $\bar{x}_g$ is the geometric mean of the data values and $s_g$ is the geometric standard deviation or by

$$\text{UTL} = \exp\left[\bar{y} + Ks_y\right] \quad (5.26)$$

where $\bar{y}$ is the average logarithm of the data values and $s_y$ is the standard deviation of the logarithms. $K$ is a factor taken from a table of values for this test and is determined by the chosen percentile of the distribution (usually 90, 95, or 95%), the chosen confidence level (usually 95%), and the number of measurements used to estimate the distribution. Tables of upper tolerance limit $K$ values can be found online or in statistics texts (see, e.g. Natrella 1963; Howe 1969; Johnson 2017).

By convention the 95% confidence limit on the 90th percentile would be expressed as $\text{UTL}_{95,90}$. Confidence intervals get narrower as sample sizes get larger, and the effect of this on the $\text{UTL}_{95,90}$ is shown in Figure 5.8 for a lognormal distribution with $\bar{x}_g = 5\ \text{mg m}^{-3}$ and $s_g = 2.5$ for sample sizes of 12 and 25. There is much more uncertainty when the smaller sample size is used. This illustrates the need for a solid estimate of the exposure distribution, using as many measurements as possible, before quantitatively examining the upper tail (Selvin et al. 1987).

It should be noted that it has been argued that the one-sided upper tolerance limit approach for assessing exposures in relation to an OEL such as an OSHA PEL is not optimal because tolerance limits are not correlated clearly with long-term health effects (Rappaport and Selvin 1987; Selvin et al. 1987). Rappaport and Selvin (1987) note that the mean of a worker's TWA exposures is related most clearly to long-term chronic health hazards and disease risk and have proposed a hypothesis testing approach using the arithmetic mean of TWA exposures instead.

### 5.9.2 Exceedance Fractions

A question such as "How likely is it that a worker will be exposed above an EG on any given day?" might be asked, where the EG might be an action level, PEL, ACGIH TLV, etc. If the guideline is a regulatory standard such as a PEL, this would be the probability of noncompliance with the PEL. This percentile of the exposure distribution is termed the exceedance fraction of the distribution.

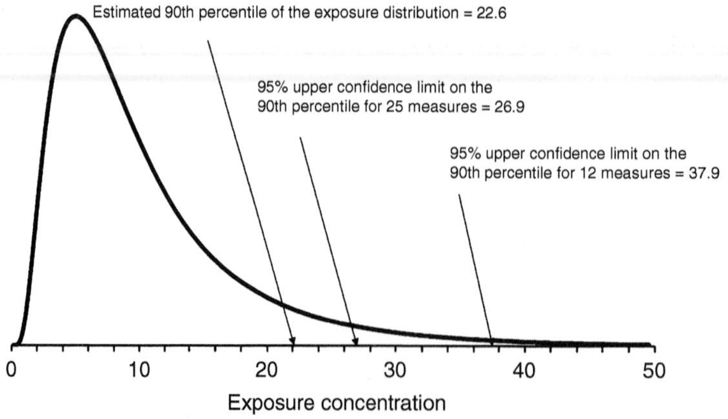


**Figure 5.8** Upper tolerance limits on the estimated 90th percentile of a lognormal exposure distribution having geometric mean = 5 mg m$^{-3}$ and geometric standard deviation = 2.5, when estimated from only 12 measures compared with from 25 measures.

Like the UTL, the exceedance fraction estimate is based on geometric mean and geometric standard deviation estimates from measurement data, so again a one-sided confidence limit must be applied. It will be a lower confidence limit in this case. For lognormally distributed exposures, an EG $z$-score $Z_{EG}$ is first calculated using the mean logarithm of the measurements $\bar{y}$, the standard deviation of the logarithms $s_y$, and the logarithm of the EG value (Hewett and Ganser 1997)

$$Z_{EG} = \frac{\ln(EG) - \bar{y}}{s_y} \tag{5.27}$$

$Z_{EG}$ is just a $z$-score, and the cumulative probability associated with this $z$-value is the "most likely" estimate of the probability percentile associated with the EG's position in the estimated exposure distribution. The best estimate of the exceedance fraction is 1 minus the probability associated with this $z$-score, which can be taken from a table of cumulative probabilities for the standard normal distribution (such as Table 5.1) or from the NORM.S.DIST function in Excel. However, an upper confidence limit must then be applied to this value. Hewett and Ganser (1997) published a chart from which the upper confidence limit can be obtained.

The influence of sample size on the exceedance fraction estimate is similar to that for the UTL. For measures of a lognormal exposure having a sample mean logarithm $\bar{y} = \overline{\ln(x)} = 4.71$ and logarithm standard deviation $s_y = 0.306$, the "most likely" estimate of the exceedance fraction is 2.7%. For a sample size of 10 measurements, the upper confidence limit on the exceedance fraction is 16%, but for a sample size of 20 measurements, it would only be 7%. Like the UTL, the exceedance fraction estimate should be based on a well-characterized exposure distribution.

### 5.9.3 Bayes' Theorem and Exposure Control Banding

Tolerance limits and exceedance fractions utilize an estimate of the true shape of the lognormal exposure distribution to model the upper tail of the distribution. These techniques are strictly data based and require substantial numbers of measures to provide reliable estimates as previously discussed. An alternative approach to assessing exposures in the upper tail of the distribution is to approach the problem from a Bayesian perspective.

Bayesian methods are based on Bayes' theorem, which describes the probability of a future event given some prior information related to the event. BDA in occupational exposure control decision-making employs the Bayesian approach to categorize the upper tail (perhaps the top 5%) of exposures for a similar exposure group (SEG) into exposure control categories. This is sometimes termed exposure control banding. The American Industrial Hygiene Association (AIHA) category scheme is widely used and establishes five categories based on comparison of the occupational exposure limit, which could be a PEL, TLV, REL, or other guideline, to the estimated 95th percentile exposure (Mulhausen and Damiano 1998, p. 64). The categories are shown in Table 5.3.

With enough data to get a precise estimate of the exposure distribution, and in particular the 95th percentile, one could use tolerance limits or exceedance fractions to characterize the level of control. Often if not usually this will not be the case, so a BDA technique approach might be taken. BDA involves combining actual exposure data (which may be somewhat limited) with subjective information based on past experience. The latter takes the form of professional judgment about the likelihood of the highest exposures, occurring under a given set of control conditions. A prior probability distribution, which is a best estimate of how often the highest exposures might occur in each of the control categories in Table 5.3, is established based on this judgment. The prior distribution is then updated with measurement data to obtain a posterior distribution. The updating is a mathematical process that requires specialty software. Examples of output from such software are often seen in the literature (see, e.g. Hewett et al. 2006; Banerjee et al. 2014; Torres et al. 2014; Xu and Stewart 2016).

**TABLE 5.3 AIHA Exposure Control Rating Scheme**

| Category | Control level | Control interval |
|---|---|---|
| 0 | De minimis | 95th percentile <0.01 OEL |
| 1 | Highly controlled | 0.01 × OEL<95th percentile <0.1 × OEL |
| 2 | Well controlled | 0.1 × OEL < 95th percentile <0.5 × OEL |
| 3 | Controlled | 0.5 × OEL < 95th percentile <OEL |
| 4 | Uncontrolled | 95th percentile >OEL |

## 5.10 CONCLUSION

Exposure measurement is at the heart of occupational safety and health practice, and the ability to properly interpret measurement results is critical to worker protection efforts. Statistical techniques provide the tools needed to account for measurement variability in order to draw valid inferences about exposures and make defensible decisions in exposure control planning and program management.

## REFERENCES

Armbruster, D.A. and Pry, T. (2008). Limit of blank, limit of detection and limit of quantitation. *Clinical Biochemist Reviews* **29** (Suppl 1): S49–S52.

ASTM International (2011). *ASTM D6552-06: Standard Practice for Controlling and Characterizing Errors in Weighing Collected Aerosols*. Conshohocken, PA: ASTM International.

Banerjee, S., Ramachandran, G., Vadali, M., and Sahmel, J. (2014). Bayesian hierarchical framework for occupational hygiene decision making. *Annals of Occupational Hygiene* **58** (9): 1079–1093.

Conover, W.J. (1999). *Practical Nonparametric Statistics*, 3e. New York, NY: Wiley.

Conover, W.J. and Iman, R.L. (1981). Rank transformations as a bridge between parametric and nonparametric statistics. *The American Statistician* **35** (3): 124–129.

D'Agostino, R.B. (1971). An omnibus test of normality for moderate and large size samples. *Biometrika* **58** (2): 341–348.

Day, G.A., Esmen, N.A., and Hall, T.A. (1999). Sample size-based indication of normality in lognormally distributed populations. *Applied Occupational and Environmental Hygiene* **14** (6): 376–383.

Dunn, O.J. (1961). Multiple comparisons among means. *Journal of the American Statistical Association* **56** (293): 52–64.

Esmen, N.A. and Hammad, Y. (1977). Lognormality of environmental sampling data. *Journal of Environmental Science and Health, Part A* **12** (1–2): 29–41.

Friedman, M. (1937). The use of ranks to avoid the assumption of normality implicit in the analysis of variance. *Journal of the American Statistical Association* **32** (200): 675–701.

Gilbert, R.O. (1987). *Statistical Methods for Environmental Pollution Monitoring*. New York, NY: Wiley.

Hewett, P. (2006). Analysis of censored data. In: *A Strategy for Assessing and Managing Occupational Exposures*, 3e (ed. J.S. Ignacio, S. Ignacio, and W.H. Bullock), 415–421. Fairfax, VA: AIHA Press.

Hewett, P. and Ganser, G.H. (1997). Simple procedures for calculating confidence intervals around the sample mean and exceedance fraction derived from lognormally distributed data. *Applied Occupational and Environmental Hygiene* **12** (2): 132–139.

Hewett, P., Logan, P., Mulhausen, J. et al. (2006). Rating exposure control using Bayesian decision analysis. *Journal of Occupational and Environmental Hygiene* **3**: 568–581.

Hornung, R.W. and Reed, L.D. (1990). Estimation of average concentration in the presence of nondetectable values. *Applied Occupational and Environmental Hygiene* **5**: 46–51.

Howe, W.G. (1969). Two-sided tolerance limits for normal populations – some improvements. *Journal of the American Statistical Association* **64**: 610–620.

Johnson, D.L. (2017). *Statistical Tools for the Comprehensive Practice of Industrial Hygiene and Environmental Health Sciences*. Hoboken, NJ: Wiley.

Kramer, C.Y. (1956). Extensions of multiple range tests to group means with unequal numbers of replications. *Biometrics* **12** (3): 307–310.

Kruskal, W.H. and Wallis, W.A. (1952). Use of ranks in one-criterion variance analysis. *Journal of the American Statistical Association* **47**: 583–621. Addendum (1953) 48:907–911.

Land, C.E. (1971). Confidence intervals for linear functions of the normal mean and variance. *Annals of Mathematical Statistics* **42**: 1187–1205.

Land, C.E. (1975). Tables of confidence limits for linear functions of the normal mean and variance. In: *Selected Tables in Mathematical Statistics*, vol. **III**, 385–419. Providence, RI: American Mathematical Society.

Leidel, N.A., Busch, K.A., and Lynch, J.R. (1977). *Occupational Exposure Sampling Strategy Manual*. NIOSH Publication No. 77-173. Cincinnati, OH: National Institute for Occupational Safety and Health (NIOSH).

Levene, H. (1960). Robust tests for the equality of variance. In: *Contributions to Probability and Statistics* (ed. I. Olkin), 278–292. Palo Alto, CA: Stanford University Press.

Lilliefors, H.W. (1967). On the Kolmogorov-Smirnov test for normality with mean and variance unknown. *Journal of the American Statistical Association* **62** (318): 399–402.

Mann, H.B. and Whitney, D.R. (1947). On a test of whether one of two random variables is stochastically larger than the other. *The Annals of Mathematical Statistics* **18** (1): 50–60.

Mulhausen, J. and Damiano, J. (1998). *A Strategy for Assessing and Managing Occupational Exposures*, 2e. Fairfax, VA: AIHA Press.

Natrella, M.G. (1963). *Experimental Statistics*. Washington, DC: US Government Printing Office.

Nemenyi, P. (1961). Some distribution-free multiple comparison procedures in the asymptotic case (abstract). *The Annals of Mathematical Statistics* **32** (3): 921–922.

Occupational Safety and Health Administration (OSHA) (1999). *OSHA Technical Manual*, 5e, TED 01-00-015 (20 January 1999). Washington, DC: US Government Printing Office. https://www.osha.gov/dts/osta/otm/otm_toc.html (updated 11 February 2014, accessed 26 February 2015).

Rappaport, S.M. and Selvin, S. (1987). A method for evaluating the mean exposure from a lognormal distribution. *American Industrial Hygiene Association Journal* **48** (4): 174–379.

Selvin, S., Rappaport, S.M., Spear, R.C. et al. (1987). A note on the assessment of exposure using on-sided tolerance limits. *American Industrial Hygiene Association Journal* **48** (2): 89–93.

Shapiro, S.S. and Wilk, M.B. (1965). An analysis of variance test for normality (complete samples). *Biometrika* **52**: 591–611.

Shrivastava, A. and Gupta, V.P. (2011). Methods for the determination of limit of detection and limit of quantitation of the analytical methods. *Chronicles of Young Scientists* **2** (1): 21–25.

Siegel, S. (1956). *Nonparametric Statistics for the Behavioral Sciences*. New York: McGraw-Hill.

Torres, C., Jones, R., Boelter, F. et al. (2014). A model to systematically employ professional judgment in the Bayesian decision analysis for a semiconductor industry exposure assessment. *Journal of Occupational and Environmental Hygiene* **11**: 343–353.

Tukey, J. (1949). Comparing individual means in the analysis of variance. *Biometrics* **5** (2): 99–114.

Wilson, I.D. and Poole, C. (2009). *Handbook of Methods and Instrumentation in Separation Science*, vol. **1**. London: Academic Press.

Xu, W. and Stewart, E.J. (2016). A comparison of engineering controls for formaldehyde exposure during grossing activities in health care anatomic pathology laboratories. *Journal of Occupational and Environmental Hygiene* **13** (7): 529–537.

Zar, J.H. (2010). *Biostatistical Analysis*, 5e. Upper Saddle River, NJ: Prentice Hall.

CHAPTER 6

# EVALUATION AND MANAGEMENT OF EXPOSURE TO INFECTIOUS AGENTS

JANET M. MACHER

*Division of Environmental and Occupational Disease Control, Environmental Health Laboratory Branch, California Department of Public Health (retired), 850 Marina Bay Parkway, Richmond, CA, 94804*

DEBORAH GOLD

*California Division of Occupational Safety and Health (retired), P.O. Box 501, Pacifica, CA, 94044*

PATRICIA CRUZ

*Department of Environmental and Occupational Health, School of Community Health Sciences, University of Nevada, Las Vegas, 4505 S. Maryland Parkway, Las Vegas, NV, 89154*

JENNIFER L. KYLE

*Division of Communicable Disease Control, Microbial Diseases Laboratory Program, California Department of Public Health, 850 Marina Bay Parkway, Richmond, CA, 94804*

TIMUR S. DURRANI

*School of Medicine, University of California, San Francisco, 2789 25th St., San Francisco, CA, 94110*

and

DENNIS SHUSTERMAN

*Division of Occupational and Environmental Medicine (Emeritus), School of Medicine, University of California, San Francisco, Campus Box 0843, San Francisco, CA, 94143*

## 6.1 INTRODUCTION

The evaluation and management of workplace exposures to infectious agents have received increasing attention in recent years. This interest is due in part to (i) the emergence of new organisms, such as the novel coronaviruses that cause severe acute respiratory syndrome (SARS) and Middle East respiratory syndrome (MERS); (ii) changes in climate, animal/human habitats, and travel that expose people to pathogens typically hosted by animals or that are present only in limited geographic areas in the human population, such as the Ebola and Zika viruses; (iii) the resurgence of vaccine-preventable diseases (e.g. measles, mumps, and pertussis [whooping cough]); and (iv) better recognition of existing hazards (e.g. hepatitis C virus [HCV] infection). Biological hazards in the workplace may cause acute or chronic infectious diseases as well as toxic and hypersensitivity reactions. Worldwide, an estimated 320000 workers die annually from work-related infectious disease, and morbidity from work-related infections is even higher although difficult to estimate (Haagsma et al. 2012).

This chapter addresses infectious diseases seen in the United States as a result of work-related exposures (Dieckhaus 2007; Stürchler 2006; Wald and Stave 2001). In some cases, sport and recreational activities overlap with work-related ones, because many of the former involve full-time or part-time workers as well as volunteers. The chapter includes three case

*Handbook of Occupational Safety and Health*, Third Edition. Edited by S. Z. Mansdorf.

studies that illustrate important aspects of recognizing and investigating work-related infections. Biological hazards not covered here include hypersensitivity reactions (e.g. hypersensitivity pneumonitis and allergic dermatitis, rhinitis, sinusitis, and asthma) and organic dust toxic syndrome (ODTS) for which the reader is referred elsewhere (Ladou and Harrison 2014; Levy et al. 2006; Rosenstock et al. 2005; Wald and Stave 2001).

Health and safety professionals are playing an expanding role in anticipating, recognizing, evaluating, and preventing occupational infections. This is so even though the role these professionals play in infectious disease control often differs, for substantive and historical reasons, from the role they play in preventing other illnesses. Many occupational diseases related to chemical or physical agents are chronic progressive processes often involving exposures over long periods of time, whereas infectious diseases often follow a single exposure incident, whether the disease that results is acute (e.g. measles) or chronic (e.g. hepatitis C). Understandably, few health and safety professionals have extensive training in microbiology, infection control, or epidemiology.

Traditionally, infection control professionals have handled worker surveillance for person-to-person transmission in healthcare facilities, services, and operations. Veterinarians and agricultural healthcare workers have managed zoonoses (i.e. those animal-borne infections that affect humans), and biosafety specialists have monitored laboratory-acquired infections. However, over the past three decades, particularly since the passage of the Occupational Safety and Health Administration's (OSHA) Bloodborne Pathogens Standard (29 CFR 1910.1030 1991a), occupational health and safety professionals increasingly have had a role in the prevention of the occupational transmission of disease. Health and safety professionals are also involved in specifying, evaluating, and implementing programs for personal and respiratory protection as well as specialized ventilation systems to protect employees from infectious diseases.

Health and safety professionals often rely on quantitative workplace monitoring to evaluate exposure risks for chemical hazards, but that approach is not used for infectious pathogens. To begin with, there are no permissible exposure limits for infectious agents. For most biological agents, there is little information about infectious dose and pathogen viability in the environment, and information on routes of transmission also may be limited. Testing for pathogenic microorganisms in the work environment is seldom done, even in outbreak investigations, except when necessary to verify the effectiveness of environmental disinfection or to establish the presence of a specific organism.

Occupational infectious diseases range from the familiar and relatively benign (e.g. the common cold) to rare and usually or frequently fatal diseases (e.g. rabies). Besides the damage they cause, these diseases may require treatment with toxic drugs, e.g. antibiotics for multidrug-resistant tuberculosis. Mass vaccination programs have almost eliminated previously common infections, such as measles, rubella, polio, and diphtheria, in some countries. For example, reported US measles cases decreased from over 700 000 in 1954 (Koo et al. 1994) to 188 in 2015, of which 26 were indigenous and 162 were imported (Adams et al. 2017). However, in recent years there have been outbreaks of a number of vaccine-preventable diseases due to decreases in vaccination rates, waning immunity, changes in vaccines, and changes in pathogens (Lynfield and Daum 2014). Affected workers can also become sources of infection for their co-workers, clients, families, and the public, and job restrictions may be necessary to prevent this. Job restrictions may also be prudent for workers with underlying risk factors (e.g. immunodeficiency or pregnancy) that may make them unusually susceptible to infectious diseases or make the consequences of infection more serious.

Infectious agents in the workplace include a wide range of microorganisms that affect a diverse variety of occupations. The most common means of preventing infectious diseases in the workplace is simple sanitation – provision of clean drinking water, elimination of common-use cups or dippers, and provision of toilets, handwashing facilities, and clean eating areas. In some cases, such as in animal handling, showers and clothes changing areas are necessary to contain pathogens. Sanitary facilities are common in offices, laboratories, and industrial environments; however, even in industrialized agriculture, people who work on farms and in food processing operations (such as egg washing) eat in areas where animal wastes may be present, or they may transfer animal feces between barns and lunchrooms on contaminated footwear (see Section 6.1.3).

There are many ways to categorize infectious disease hazards. Pathogens may be classified by the source of human exposure to the microorganism, e.g. humans (seasonal influenza), animals (rabies), vectors (Lyme disease), food or water (salmonellosis, leptospirosis), and the environment (coccidioidomycosis or Valley fever, Legionnaires' disease). Infectious diseases also can be classified by the type of agent (virus, protozoa or helminth, fungus, or bacterium) or by the mode of transmission (e.g. contact or aerosol) (see Section 6.2.2 for a discussion of modes of transmission).

Certain work environments are considered to be at elevated risk of infectious diseases. For example, all employees who have significant public contact, including bus drivers, school teachers, and retail clerks, may be at increased risk of exposure during the flu season or an influenza pandemic. This risk may be thought of as an elevated public health risk, not specific to the occupation, but increased by the frequency of opportunity to contact an infected person.

This elevated public risk is different from the risk to employees in healthcare. Sick people come to healthcare facilities for diagnostic procedures or for care and treatment. Some diagnostic procedures, such as bronchoscopy, can significantly increase exposure to respiratory pathogens. In addition to symptomatic infections, some patients may harbor disease agents such as HIV or hepatitis B or C viruses (HBV, HCV) to which employees are exposed during procedures such as injections or surgery. Immunocompromised patients also may be a source of unusual infections. For these reasons, it has long been recognized that healthcare environments must have enhanced infection control measures to protect employees, as well as the public (Siegel et al. 2017). The infectious disease risks to healthcare workers were highlighted by the outbreak of SARS in Toronto in 2003, in which 40% of cases were among healthcare workers (Grace et al. 2005; Ofner-Agostini et al. 2006), by the 2014 experience of Ebola in Africa (Elshabrawy et al. 2015), and by the 2015 experience of MERS in Korea (Lee and Wong 2015).

Biomedical and microbiological laboratories are another type of workplace in which there is a recognized occupational hazard from infectious agents. Clinical, academic, and research laboratories may intentionally amplify and concentrate infectious agents, and laboratory procedures often create potential routes of transmission that do not exist outside the laboratory (see Section 6.1.4 for further discussion of laboratory hazards and controls).

Exposures to animals similarly create the potential for infections that are not commonly found in people. For example, studies in the Netherlands (Bosman et al. 2004) and Canada (Tweed et al. 2004) found that farmers and people working around poultry had antibodies indicating that they had been exposed to avian influenza viruses that are not typically spread from person to person.

The sections 6.1.1–6.1.7 give examples of infectious disease risks in many types of work environments. However, health and safety professionals should consider whether there are potential infectious disease hazards whenever they evaluate a process or work environment, and they should develop tools to investigate reports of infectious diseases that may be associated with the workplace (29 CFR 1904.5 2001).

### 6.1.1 Healthcare

Historically, hospitals and the healthcare professions have been recognized as posing a risk of infection to both workers and patients (Haagsma et al. 2012; Sepkowitz and Eisenberg 2005). Healthcare workers are employed in a wide variety of settings, including hospitals and long-term care facilities, physicians' and dentists' offices, blood banks, outpatient clinics, emergency medical operations, and correctional facilities, or they may provide care in the patient's home.

In 1847, Semmelweis discovered the transmission mode (and means of prevention through handwashing) of puerperal fever, an often fatal streptococcal infection following childbirth. This discovery came only after a professor died of the disease from a scalpel wound received during an autopsy. Following the pioneering efforts of Florence Nightingale and Joseph Lister along with the advent of the antibiotic era, the public perception of hospitals as hazardous places decreased, as did the frequency of nosocomial (hospital-acquired) infections, now referred to as healthcare-associated infections (HAIs). However, even today, a survey found that 4% of US patients became infected while hospitalized (Magill et al. 2014), and healthcare workers attending these patients may be at risk for a variety of pathogens present in their workplaces.

The AIDS epidemic resulted in an increased focus on risks to healthcare workers. The Centers for Disease Control and Prevention (CDC) reported that from 1985 to 2013 there have been 58 confirmed and 150 possible cases of occupationally acquired HIV infection among healthcare workers, and few confirmed cases have been reported since the late 1990s (Joyce et al. 2015). These numbers are likely to be underestimates because the definition of a confirmed case includes a temporally associated seroconversion following a reported exposure incident, and healthcare workers do not report all exposures (Kessler et al. 2011). During the late 1980s, CDC recommended "universal precautions," a system utilizing barriers and other precautions to minimize or prevent exposure to blood as well as certain body fluids and tissues (Hughes 1989). In January 1996 the Healthcare Infection Control Practices Advisory Committee (HICPAC) issued new guidelines for infection control in hospitals (Garner 1996). These recommendations incorporated universal precautions as well as aspects of "body fluid isolation" into *standard precautions for all patients, to be supplemented for specific suspected or confirmed infections with transmission-based precautions* (Joyce et al. 2015). Reasons for these precautions include the following: (i) HBV, HCV, and HIV are present in the blood of a significant percentage of patients; (ii) carriage of these blood-borne viruses is often silent; (iii) screening for risk factors of infection (e.g. injection drug use, high-risk sexual behavior, and hemophilia) will identify only a fraction of infected patients in advance of providing them care; (iv) although the viruses are transmitted primarily through blood and certain other bodily fluids, some blood-borne pathogens may not be detectable in blood but be sequestered in other body compartments; and (v) additional blood-borne agents may suddenly emerge (Ho et al. 2007; Lewis et al. 2015). There are many challenges in implementing these precautions in settings such as emergency care and rescue work, and specific guidelines have been developed for these operations (FEMA 2002; Reissman and Piancentino 2010; Woodside 2013). Home healthcare is among the fastest-growing healthcare sectors. However, infection control in a residential setting can be limited by space and resource issues, and care may be provided by family members or close friends not formally trained in infection control (Shang et al. 2014).

The CDC's 1987 guidelines (Mason 1987) were incorporated into the *OSHA Bloodborne Pathogens Standard* (29 CFR 1910.1030 1991a), the first OSHA regulation pertaining to specific infectious agents (see Section 6.3.1). This standard applied to any workplace where workers were "reasonably anticipated" to be occupationally exposed (with certain exceptions such as construction). Also, employers were required to make available, free of charge, hepatitis B vaccine to covered employees. In the event of an exposure incident, employers were required to make immediately available a post-exposure evaluation and to provide post-exposure prophylaxis (PEP) if medically indicated (see Sections 6.3.2.2 and 6.4.7). Because application of the standard was not limited to hospitals or even traditional healthcare facilities, it resulted in attention to occupational exposure to blood-borne and other pathogens in nursing homes and other long-term care facilities, outpatient facilities, patients' homes, emergency medical services, public safety work, schools and childcare facilities, correctional facilities, and laboratories.

The emergence of the AIDS epidemic was followed shortly by an increase in tuberculosis cases and multidrug-resistant *Mycobacterium tuberculosis*. As a group, healthcare workers (especially nurses and laboratory workers) are at higher than average risk for tuberculosis (Baussano et al. 2011; Uden et al. 2017). CDC and other groups have published comprehensive guidelines for preventing tuberculosis in healthcare facilities (Behrman et al. 2013; Cleveland et al. 2009; Jensen et al. 2005).

In 1997, OSHA released a proposed rule and notice of public hearing on occupational exposure to tuberculosis, but this rulemaking was terminated on 31 December 2003 (OSHA 2003). Soon thereafter, California began a rulemaking project on tuberculosis and other diseases spread through infectious aerosols (particles suspended in air), i.e. aerosol transmissible diseases (ATDs). These efforts resulted in adoption of ATD standards addressing healthcare and other high-risk environments, such as correctional institutions and biological laboratories, as well as prevention of zoonotic disease transmission due to exposure to animals (California Department of Industrial Relations 2009; Gold 2015). The 2003 SARS epidemic disproportionately affected healthcare workers and drew attention to the role of infectious aerosols in healthcare settings and the need to provide airborne infection isolation and respiratory protection for novel pathogens (Shaw 2006). The 2014–2015 Ebola epidemic, in which over 10000 Africans died, is, at the time of this writing, leading to reevaluation of personal protective equipment (PPE) and engineering controls in healthcare settings in both industrialized and non-industrialized countries (Elshabrawy et al. 2015; OSHA 2015).

In 1998, CDC published a guideline for infection control in healthcare personnel (Bolyard et al. 1998). General recommendations for immunization of healthcare workers have also been published (Shefer 2011; Siegel et al. 2017). The CDC recommends that all healthcare workers with patient contact be vaccinated annually for influenza to prevent transmission to persons at high risk of influenza-related complications (Shefer 2011). CDC also recommends that healthcare workers be vaccinated against measles, mumps, and rubella (MMR); tetanus, diphtheria, and acellular pertussis (Tdap); and varicella (chickenpox). Susceptible workers exposed to patients with chickenpox, measles, mumps, or rubella often must be restricted from the workplace for two or more weeks because the agents can be transmitted before an infected worker shows signs of disease. Appropriate use of vaccines can often eliminate these restricted periods and thus may be cost-effective in addition to protecting worker health (see Section 6.4.1).

Current infection control guidelines recommend standard precautions (an expansion of universal precautions) and three types of enhanced transmission-based precautions – airborne, droplet, and contact. It is important for healthcare workers and the occupational health professionals responsible for them to appreciate the many transmission modes and the wide range of pathogenicity of the infectious agents that may be found in healthcare settings (see Sections 6.2.2 and 6.2.3). Many of the agents that patients carry (including antibiotic-resistant bacteria and fungi) are pathogenic only for persons with compromised immune function or underlying medical conditions, but employers are not always aware of individual employee risk factors.

Infection control committees at hospitals and healthcare centers oversee the development and implementation of mandated infection control programs. Institutional health and safety professionals may contribute to these activities in the areas of exposure prevention and investigation; facility operation and maintenance, e.g. of heating, ventilating, and air-conditioning systems; water supply; PPE; and waste disposal. Additional areas where health and safety professionals may play a role in infection control are in implementing engineering controls to prevent the spread of infectious agents, conducting environmental inspections and testing, evaluating exposures, and ensuring general safety and security (Erickson 2014; Siegel et al. 2017). Section 6.4.4.3 discusses the use of ventilation to reduce exposures to ATD agents.

### 6.1.2 Childcare

Childcare settings tend to promote infectious disease transmission because such operations closely group together young children who lack effective hygienic behavior and who have not yet developed immunity to common infectious agents (Nesti and Goldbaum 2007). In the United States and many other industrialized countries, the demand for and utilization of out-of-home childcare has expanded and fewer extended families are available to provide childcare. In 2007 in the United States, over half of children under five years were in childcare for 10 or more hours per week (U.S. Department of Health and Human Services 2011). Compared with children cared for at home, youngsters in childcare settings are more likely to be associated with illness outbreaks and to experience increased infectious disease rates and severity (Nesti and Goldbaum 2007).

Many gastrointestinal and respiratory tract diseases affect both childcare staff and children. Some infections affect primarily children but not adults (e.g. otitis media or inflammation of the middle ear), some of which are vaccine preventable (Zhou et al. 2014). If contracted in childhood, many agents confer lifetime immunity; therefore, few adults contract these infections. A few infectious agents, notably the hepatitis A virus (HAV), cause disease in adults but primarily inapparent infection in children. Other agents, principally cytomegalovirus (CMV) and to a lesser extent parvovirus B19, cause mild or inapparent infections in children but may have serious consequences for a fetus. Pertussis (whooping cough) can be transmitted between children and between children and adults, and infection causes serious illness in infants and young children as well as a prolonged cough often misdiagnosed as bronchitis in adults. Pertussis immunity wanes over time. It is therefore important that childcare workers be immunized according to current recommendations (Hayney and Bartell 2005). Besides infections transmitted from person to person, some diseases have been contracted from handling pets and other animals, which may be kept at childcare centers and in primary school classrooms, e.g. salmonellosis from handling infected chicks, ducks, and reptiles.

Commonsense measures can minimize infectious agent transmission and reduce the burden of illness in childcare facilities. A national health and safety performance standard has been published with specific recommendations for personal hygiene (handwashing), disinfection, exclusion policies, management of ill children, requirements for immunization, exposure to blood, and reporting of infectious disease outbreaks to a health agency (AAP/APHA/NRCHSCCEE 2013). Following these recommendations (particularly in regard to personal hygiene and disinfection) and the recommended childhood immunization schedule has been shown to be effective at preventing and controlling infectious diseases in childcare, including among childcare workers (Hayney and Bartell 2005; Lennell et al. 2008).

### 6.1.3 Agriculture, Food Processing, and Animal-Associated Occupations

Agricultural employment (including farming), food processing (e.g. slaughterhouse or abattoir work and butchering), and veterinary care are among the most hazardous occupations. Although less frequent, less serious, and less studied than injuries, infectious diseases are common in these professions. Animals are the most common sources of exposure to infectious agents for such workers (Bosch et al. 2013), but soil, water, and plants are also important.

Fieldworkers are often at increased risk from enteric pathogens due to lack of sanitation facilities and clean drinking water. Social conditions also place migrant farm workers at increased risk of interpersonally transmitted diseases, e.g. tuberculosis and sexually transmitted diseases (Baron et al. 2010). Workers with direct soil contact, particularly barefoot farmers in less developed countries, are at risk for parasitic infections including hookworm, ascariasis, cutaneous larva migrans, and visceral larva migrans.

Operations that involve the moving of soil place workers at risk for fungal infections, which often have specific geographic distributions, e.g. histoplasmosis in the Ohio and Mississippi river valleys and coccidioidomycosis in the Southwest (Lower Sonoran Life Zone) (Das et al. 2012; de Perio et al. 2015; Wilken et al. 2014). See the case study on a coccidioidomycosis outbreak among the cast and crew at an outdoor television event. This study illustrates the value of a Doctor's First Report of Occupational Injury or Illness (DFR) because it was through the DFR that workers not directly involved in soil disturbance were found to have contracted coccidioidomycosis (Wilken et al. 2014). Disturbing compost and moldy grain or other plant material may also place workers at risk for fungal (e.g. *Aspergillus* spp.) or bacterial (e.g. actinomycete) infections as well as for hypersensitivity diseases and ODTS.

Farm and ranch workers along with field researchers spend much of their time outdoors, placing them at risk for vector-borne diseases, e.g. Lyme disease, ehrlichiosis, plague, and Rocky Mountain spotted fever (in the United States) as well as malaria, viral encephalitis, and other diseases (elsewhere). Farmers (and their guests) often acquire campylobacteriosis, salmonellosis, and listeriosis from drinking unpasteurized milk and also are at risk of acquiring bovine tuberculosis. Food handlers (and cooks) may taste uncooked items such as seasoned meat or sausage, placing themselves at risk for bacterial and parasitic infections.

Since its emergence in 1996 in China, avian influenza (or bird flu caused by avian [bird] influenza type A viruses) has spread across three continents (Fournié et al. 2012a). The World Health Organization (WHO) considers the avian influenza A/H5N1 virus a public health risk with pandemic potential (Thomas and Noppenberger 2007). Sporadic, large-scale spread appears to originate from a small number of areas of virus persistence, particularly in Asia. The trade of live birds and poultry products is likely to be a major pathway for virus dissemination within and beyond endemic areas, and live-bird markets are likely to contribute to virus persistence (Fournié et al. 2012a; Refaey et al. 2015).

Among poultry workers, butchering and exposure to sick poultry have been risk factors for viral antibodies (Katz et al. 2009). As of December 2017, there have been only a few reports of human-to-human transmission of avian influenza A viruses, and that transmission has been limited and non-sustained (CDC 2017c). Serologic studies have not demonstrated evidence of seropositivity in household contacts or healthcare workers, but future strains may be more transmissible between people (Thomas and Noppenberger 2007).

Some avian influenza strains have shown greater ability to cause both asymptomatic infection and clinical disease in humans. In 2003, a veterinarian who sampled birds for an H7N7 strain in the Netherlands contracted the disease and died (Fouchier et al. 2004). A subsequent public health investigation found that 453 of 4500 persons exposed to infected birds reported health complaints, predominantly conjunctivitis (349, 77%), influenza-like illness (90, 20%), or other symptoms (Bosman et al. 2004; Koopmans et al. 2004). Eighty-nine infections were confirmed, and H7N7 was found in the tear fluid of 78 persons with conjunctivitis only. Antibodies against H7N7 were found in 51% of workers exposed to infected poultry and in 63% of farmers whose farms had infected poultry. An investigation of an H7N3 avian flu epidemic in British Columbia found 57 reported suspected cases of illness, of which two were confirmed as H7N3 (Tweed et al. 2004). WHO continues to monitor H5, H7, and other avian influenza viruses to detect human disease risk (WHO 2017b).

Recommendations for protecting poultry workers include the use of good hygiene and work practices, personal protective clothing and equipment, vaccination for seasonal influenza viruses, antiviral medication, and medical surveillance (MacMahon et al. 2008). An employer also can consider the following: reducing the number of persons exposed to the hazard, decreasing the time that workers are exposed to the hazard, instituting a maintenance program for protective equipment or clothing, and introducing better work practices and systems (Halpin 2005). Additional protection is recommended for workers who will be exposed to infected poultry, including respiratory protection, disposable clothing, formal decontamination procedures, and enhanced medical surveillance.

Underdiagnosis of illness is common in agricultural and food and animal processing workers due to their acceptance of illness as part of the job, limited medical care access (particularly for farm workers), and the need for serologic (blood) tests to diagnose many zoonoses. Infectious disease outbreaks due to organisms not previously identified as human pathogens (e.g. avian influenza strains such as H5N1 in persons exposed to birds) point out the evolving nature of work and workplaces and the continuing need to look for emerging pathogens (Ho et al. 2007). Vaccination against rabies is recommended primarily for veterinarians and animal control personnel. Often, clinically inapparent infections in young, healthy workers (e.g. Q fever) or frequent low-dose exposures protect workers against the development of more serious disease later in life. Many zoonotic exposures can be reduced through the use of protective clothing and shoes; providing handwashing, bootwashing, and other sanitary facilities; and, to the extent that agricultural practices allow, keeping animals away from living and eating areas. Some previously common zoonoses (e.g. brucellosis, anthrax, leptospirosis, and psittacosis) have been controlled to a large extent in countries with more resources through vaccination, chemoprophylaxis, and identification and elimination of infected animals, thus reducing potential exposures.

### 6.1.4 Laboratories

The type of work conducted in microbiology laboratories inherently includes potential exposures to infectious agents because laboratory workers handle pathogens and infected animals. Some of the same concerns apply to workers in biotechnology laboratories. These operations involve the large-scale production of microorganisms for many uses, e.g. in the manufacture of vaccines, antibiotics, and microbial products (such as enzymes and hormones), as animal feed, as pesticides, for environmental management of oil spills, and as biodegradation agents at hazardous waste sites (81 CFR 15315 2016; Coelho and Garcia Diez 2015; Roomes 2010). Biosafety practices aim (i) to ensure the health and safety of laboratory workers and the surrounding community and (ii) to protect the environment (USDHHS 2009; WHO 2004). Laboratory workers handling infectious agents are clearly at increased risk of laboratory-associated infections. However, other laboratory staff (e.g. glassware washers, janitors, maintenance personnel, and clerical workers) as well as non-laboratory occupants of the same building may also be at greater risk than the public.

Accurate data on laboratory-associated infections are difficult to obtain, but clearly such infections do occur and are a concern especially for women who comprise over three-quarters of laboratory technicians (Singh 2011). A survey of clinical laboratory directors conducted by the American Society for Microbiology in 2002–2004 found that 33% of laboratories reported at least one laboratory-acquired infection (Baron and Miller 2008). The survey revealed the highest incidence of infections was with *Brucella* spp. and *Neisseria meningitidis* (641 and 25 cases per 100 000 laboratorians, respectively, compared with 0.08 and 0.62 cases per 100 000 in the general population, respectively). Adding to the difficulty of there being no systematic reporting system for laboratory-acquired infections (Singh 2011), such infections are often subclinical and may have atypical exposure routes as well as incubation periods (Garcia 2010;

Pedrosa and Cardoso 2011; Singh 2009, 2011). The most frequently identified laboratory-associated infections worldwide are those caused by *M. tuberculosis*, *Coxiella burnetii*, hantavirus, arboviruses, HBV, *Brucella* spp., *Shigella* spp., *Salmonella* spp., HCV, *Cryptosporidium* spp., and *N. meningitidis* (see the case study on laboratory-acquired *N. meningitidis* infection) (Harding and Byers 2006; USDHHS 2009). Many laboratory-associated infections result from obvious accidents, e.g. spills, splashes, needlesticks, cuts, and animal bites. However, in many cases the incident that led to exposure of a laboratory worker is unclear. Often the affected person is only known to have handled an agent or to have worked near someone who was working with the agent. In these cases of unidentified exposure, transmission is assumed to be by aerosol, i.e. droplet or droplet nuclei exposure (see Section 6.2.2).

Practices have been developed over the years that allow the safe laboratory handling of even the most hazardous agents as well as the large-scale production of microorganisms and cells (81 CFR 15315 2016; USDHHS 2009; WHO 2004). Basic precautions include posted biohazard warning signs; limited laboratory access; careful manipulation of infectious fluids; biological safety cabinets to contain accidentally generated aerosols; hand-pipetting devices for transferring infectious materials; proper disposal of needles, syringes, culture material, and animal carcasses; handwashing following laboratory activities; decontamination of work surfaces before and after use as well as immediately after spills; and prohibition of eating, drinking, smoking, and food storage in laboratories. In addition, researchers working with recombinant DNA or synthetic DNA molecules are required to adhere to guidelines established by the National Institutes of Health (NIH) (81 CFR 15315 2016).

Other texts discuss regulations and guidelines related to laboratory biosafety practices as well as the safe use of pathogenic and oncogenic microorganisms (Gaudioso et al. 2012; Ned-Sykes et al. 2015; WHO 2004). These references also provide lists of standard setting or credentialing groups as well as professional associations in these fields. The National Sanitation Foundation (NSF) (Ann Arbor, MI) Standard 49 describes requirements for the design, construction, and performance of Class II biological safety cabinets, which are widely used in microbiology laboratories to contain aerosols (NSF/ANSI 2016). The CDC and the NIH have published biosafety guidelines geared toward protecting workers and preventing laboratory exposure (Ned-Sykes et al. 2015). Work with high-risk pathogens may require the use of Class III biosafety cabinets (glove boxes), respiratory and PPE, and laboratories that are constructed so that any release will be contained to the laboratory (secondary containment) and not enter the community.

Increased attention has recently been placed on the safe transport of biohazardous agents and verification that a receiving laboratory is qualified to handle them properly (CDC 2014). This concern has resulted in more stringent shipment regulations in the United States (42 CFR 72 2007) (see also Section 6.3.1.6). The mission of the US Department of Transportation (DOT) is to protect people and the environment by advancing the safe transportation of energy and other hazardous materials (PHMSA 2017). Additionally, all known biological agents, as well as human clinical samples being tested for such agents, are subject to guidelines published by the International Air Transport Association (IATA) when shipped by air, either domestically or globally (IATA 2015). Detailed guidelines for the safe transport of infectious substances are also available from the WHO (2017a).

Infectious agents may receive the most attention in microbiology laboratories, but typically they are not the only biological hazards present. Workers in microbiology and biotechnology laboratories may also need engineering controls and training in proper work practices to avoid exposure to endotoxin from gram-negative bacteria, mycotoxins from fungi, biologically active pharmaceutical products, and proteins and other allergenic materials from whole cells, cell fragments, and cell products that may cause hypersensitivity diseases, e.g. hypersensitivity pneumonitis and asthma. Laboratory work with oncogenic (tumor-producing) viruses and some recombinant DNA molecules may carry a risk of cancer induction. Appropriate training and supervision in chemical, electrical, and radiation safety may also be necessary.

Immunizations for some infectious agents (e.g. *N. meningitidis*, the rabies virus, and the Q fever agent) may be recommended for laboratory workers at high risk of exposure even though these immunizations are not recommended for adults in general (Folaranmi et al. 2015; Weigler et al. 2012). Certain poxviruses (e.g. variola major and minor, vaccinia, and monkeypox) are infectious for humans, and laboratory workers handling these viruses should be considered for smallpox vaccination (Heptonstall 2010). Engineering controls can reduce laboratory-associated infections, but the individual behavior of laboratory workers plays a large role in their safety and that of their co-workers. Prevention of laboratory-associated infections depends on how each worker perceives the risks and consequences of exposure to pathogenic agents as well as on workers' awareness of, belief in, and adherence to good laboratory practices (Kimman et al. 2008).

### 6.1.5 Occupational Travel and Exotic Diseases Closer to Home

***6.1.5.1 Occupational Travel*** Global travel and trade have evolved dramatically during the past two centuries, with ever-escalating speed, distance, and volume (Chen and Wilson 2008). Because the geographic distribution of diseases is dynamic, travel allows humans to interact with microbes and introduce pathogens into new locations and populations (Barnett and Walker 2008; Chen and Wilson 2008). Military personnel are often required to travel worldwide and may spend considerable time abroad. With the globalization of industry, job-required travel is increasing in other occupations and may involve extended residence, often in areas outside usual tourist routes.

Every two years a team of more than 200 experts updates the *CDC Health Information for International Travel*, commonly known as the *Yellow Book* because of its yellow cover (CDC 2017a). The *Yellow Book* provides the latest official CDC recommendations to keep international travelers safe and healthy. It includes a complete catalog of travel-related diseases and up-to-date vaccine and booster recommendations. The 2018 version (released in 2017) includes updated sections on emerging infectious diseases such as Zika, Ebola, and MERS. The *Yellow Book* is written primarily for health professionals but is a useful resource for anyone interested in healthy international travel. A mobile app is available for travelers and clinicians who want to take the *Yellow Book* with them. The content is also available at CDC's *Travelers' Health* website (CDC 2017a). In addition to providing the *Yellow Book*, the *Travelers' Health* website lets travelers search by destination and find information about basic travel health preparations and what to do if sick or injured while traveling. It is updated as travel health threats emerge and new information becomes available.

Some workers in travel-related and import industries have close contact with passengers and cargo, as do emergency personnel (House and Ehlers 2008). WHO prepared the *Guide for Public Health Emergency Contingency Planning at Designated Points of Entry* to bridge the gap between the legal requirements of the International Health Regulations (Gostin et al. 2015) and the pragmatic readiness and response capacity for public health emergencies at designated points of entry (WHO 2012, 2014). However, workers with passenger and cargo contact generally do not appear to be at greater risk for infectious diseases than others. Tuberculosis transmission has occurred on aircraft in which people may share air for many hours at close quarters (Dowdall et al. 2010). However, the risk does not appear to be greater on aircraft than in other enclosed spaces. Practices are under development to deal with infectious diseases on aircraft (Evans and Thibeault 2009; Lippmann et al. 2002; Mangili et al. 2015). Border patrol and coast guard workers occasionally have close, possibly physical, contact with documented and undocumented travelers who may have communicable diseases. The infectious agents of concern are those related to food, water, and sanitation; respiratory pathogens; blood-borne pathogens; and vector-borne agents (U.S. Department of Homeland Security, U.S.C.G. 2006).

Precautions that may be indicated for travelers include (i) vaccine or immune serum globulin (ISG) for hepatitis A; (ii) vaccines or chemoprophylaxis for mosquito-borne diseases such as yellow fever, malaria, and Japanese encephalitis; and (iii) insect repellents and general protective measures to avoid dengue and other mosquito-borne diseases for which prophylaxis is not available. As needed, travelers should request information on (i) reliable drinking water sources, proper food selection, and other measures to prevent and treat traveler's diarrhea; (ii) the need for wading and swimming precautions (to avoid schistosomiasis and leptospirosis); and (iii) the possible need for rabies and other prophylaxis. Although not generally work related, travelers should be reminded of the risk for acquiring sexually transmitted diseases, AIDS, and hepatitis from unprotected sex, particularly in areas where seroprevalence is high (Matteelli et al. 2013).

Immunizations recommended for travelers may depend on whether a person will make only a short visit or reside in an area. The most common vaccine-preventable diseases recorded in travelers are enteric (typhoid or paratyphoid B) fever, acute viral hepatitis, influenza, varicella, measles, pertussis, and bacterial meningitis (Gautret et al. 2012). Travelers should obtain those vaccinations recommended for all adults as well as those recommended for their particular occupations (see Section 6.4.1). Readministration of MMR vaccines has been advised for young adults who travel. Travelers should ensure protection against diphtheria and tetanus by receiving a tetanus–diphtheria toxoid (Td) booster whenever ≥10 years have elapsed since previous administration. Rabies vaccine is recommended for travelers staying ≥30 days in countries with ineffective control of animal rabies. The uniformed services have recommendations for active service members, which typically include immunization against diphtheria, tetanus, and polio with documented protection against MMR (Immunization Healthcare Branch 2017; U.S. Department of Homeland Security, U.S.C.G. 2006). In addition, immunization against influenza, hepatitis A, and hepatitis B may be recommended or required. Some nations require immunizations for entry into the country and other precautions are recommended for travel to certain regions. The CDC maintains a list of requirements by country (CDC 2017e).

***6.1.5.2 Exotic Diseases Closer to Home*** Many of the infectious diseases discussed above will be encountered only rarely in the United States. However, recent outbreaks of pertussis, measles, and hepatitis A have underlined the reliance on adequate vaccination to keep infectious diseases in check. Migrant workers and travelers may be sources of reintroduction of infectious diseases otherwise under control. Also of increasing concern is the potential of invasive insect vectors to extend the range of so-called "tropical" diseases, such as mosquitoes carrying dengue and chikungunya viruses and humans carrying Zika virus. Whether global warming or increased cross-border trading patterns are responsible, formerly exotic diseases may become a concern not just for occupational travelers but also for residents of border states (Fredericks and Fernandez-Sesma 2014).

### 6.1.6 Personal Service Workers

Personal service work includes hairdressing and barbering, tattooing, body piercing, nail services, electrolysis, and various other aesthetic services (Ontario 2015). All of these activities involve hands-on contact with clients; therefore, transmission of bacterial, fungal, and viral skin infections may occur, as may parasitic infestations. The risk of infection is increased when a worker's skin is damaged by chemicals (such as those used in nail salons) or frequent immersion in water.

Some procedures, such as piercing and tattooing, involve intentional puncturing of the skin, and accidental punctures can occur during shaving, haircutting, and nail and cuticle grooming. Some studies have found an increased prevalence of HBV or HCV among barbers, particularly where those professions are not regulated (Adoba et al. 2015; Candan et al. 2002; Sharifi-Mood and Metanat 2006) and among people who have tattoos (Carney et al. 2013). State or local authorities in the United States typically regulate these occupations through licensing requirements (see Section 6.3.2.1), but enforcement of such regulations may vary by jurisdiction. The National Institute for Occupational Safety and Health (NIOSH) maintains a web page on body art health and safety (NIOSH 2013), and OSHA has a web page with resources for nail salons (OSHA 2017).

### 6.1.7 Wastewater and Sewage Treatment

Proximity to wastewater and sewage might be assumed to carry a risk for occupational infections because human feces are a major component of domestic wastewater in the United States. Bacteria (e.g. *Escherichia coli*) that are pathogenic to body sites outside the gastrointestinal tract are a normal constituent of human feces, which may also contain viral, protozoan, fungal, and helminthic pathogens (Lee et al. 2007). Disease outcomes associated with waterborne infections include mild to life-threatening gastroenteritis, hepatitis, skin infections, wound infections, conjunctivitis, and respiratory infections. Exposure to infectious agents may occur through aerosol inhalation, contact, or hand-to-mouth contamination (Korzeniewska 2011). Some diseases (e.g. cholera, typhoid fever, hepatitis A, and amebiasis) are of such low frequency in the United States, in contrast to many developing countries, that they pose at most a low risk of infection from wastewater exposure. Workers exposed to sewage may have an increased risk of infection by *Helicobacter pylori* and hepatitis E virus (HEV). However, the prevalence of peptic ulcer and hepatitis E was not clearly increased in a study of sewage workers and municipal manual workers, suggesting that no specific measures seemed necessary to protect workers exposed to sewage from HEV or *H. pylori* (Tschopp et al. 2009). However, the study did find an increased prevalence of diarrheal disease. Higher odds ratios of respiratory, eye, and skin irritation, neurology, and gastrointestinal symptoms have been found in wastewater treatment plant workers compared with water treatment plant workers (Lee et al. 2007). In that study, tasks related to sludge handling and plant inspection showed significant associations with memory/concentration difficulties, throat irritation, and stomach pain, suggesting that exposure control programs for specific job tasks and work locations were needed. Noroviruses (NoVs) were detected at wastewater treatment plants at concentrations above the infectious dose and could contribute to the increased frequency of gastrointestinal illness among the workers (Uhrbrand et al. 2011). Solid waste workers also may be at increased risk for some infectious diseases for which they can be protected by immunization, e.g. hepatitis A, hepatitis B, and tetanus (Tooher et al. 2005).

## 6.2 INFECTIOUS DISEASES: BASIC PRINCIPLES

The following section describes fundamental terms and concepts related to infectious diseases, their transmission, and prevention. *Infection* is the entry and development or multiplication of an infectious agent in the body of a person or animal (APHA 2015) so that a response, at least an immune response, occurs. In contrast, *colonization* is the presence of a microorganism in or on a host (e.g. a person or animal) with multiplication or growth but without evidence of an effect or an immune response. The following four elements are required for an illness to be considered an *infectious disease*: an illness due to a specific *infectious microorganism* or its *toxin* that arises through *transmission* of the agent from a *source* (i.e. the person, animal, object, or substance) to a *host* (APHA 2015). However, not all infectious diseases are communicable from person to person, e.g. tetanus and Legionnaires' disease affect only the infected person, who is not a risk to others.

Only a small fraction of all microorganisms on Earth have been identified, but they are found wherever environmental conditions allow them to survive. The majority of microorganisms inhabit soil or water where they often have important ecological functions, e.g. nutrient breakdown and recycling. Only a small portion of known microorganisms are *pathogenic* (disease causing) for humans, animals, or plants. Certain microorganisms are associated with plants, animals, or humans as normal flora, e.g. on plant leaves and roots or on animal skin and mucosal membranes as well as in digestive tracts.

Infestations may also be communicated from person to person or between animals and humans. An *infestation* of a person or animal involves the lodgment, development, and reproduction of arthropods (e.g. body lice or scabies mites) on a body surface or in clothing. Infested articles or premises are those that harbor or give shelter to animals, especially arthropods and rodents, e.g. a building may be described as infested with houseflies or mice (APHA 2015).

Some microorganisms are obligate pathogens (*parasites*) and require a living host, but many are *facultative* or *opportunistic* pathogens; i.e. they can grow either as parasites in a living host or as *saprophytes* on decaying organic matter. Saprophytic microorganisms that generally do not harm healthy persons may cause disease in those who are immunocompromised. Many bacteria and fungi can be isolated from clinical and environmental samples and be grown in laboratory culture. Some obligate pathogens can only be grown in laboratory animals or in cell or tissue cultures.

Identifying an infectious disease's causative *agent* may be critical for disease control and prevention. In the late 1800s, Koch and Henle developed a series of criteria that a microorganism must meet to be confirmed as the agent of an infectious disease. Known as *Koch's postulates*, these criteria require (i) that a specific microorganism always be found in association with a given disease, (ii) that the microorganism be isolated from an infected animal or person and grown in pure culture (i.e. separate from all other microorganisms), (iii) that the pure culture reproduce the disease when inoculated into a susceptible animal, and (iv) that the microorganism be re-isolated from the experimentally infected animal. These criteria provided a foundation for medical microbiology even though the authors themselves recognized limitations soon after formulation (e.g. Koch could not explain the recurrence of cholera epidemics in the absence of an asymptomatic carrier state or environmental reservoir) (Inglis 2007), and other approaches subsequently have been proposed (Fredericks and Relman 1996; Inglis 2007; Spinler et al. 2015).

### 6.2.1 Infectious Microorganisms and Their Sources

***6.2.1.1 Microorganisms*** Microorganisms are so small that the individual units cannot be seen with the naked eye. Table 6.1 lists physical characteristics for a variety of microorganisms including their shapes and sizes. Microorganisms can be divided into two groups for classification or taxonomic purposes: (i) prokaryotes, which lack nuclei to contain their chromosomes (e.g. bacteria and archaea), and (ii) eukaryotes, true cells with membrane-bound nuclei (e.g. fungi, protozoa, and helminths). Most prokaryotes and eukaryotes contain all the enzymes required for their own replication and have the necessary equipment to obtain or produce energy. Exceptions are certain bacteria (e.g. chlamydia and rickettsia) that are obligate intracellular parasites.

*Viruses* fall outside the above classification system because they are neither truly alive nor true cells; i.e. they consist only of nucleic acid (RNA or DNA) in a protein coat, with or without a lipid envelope. They can multiply only within host cells and generally do not retain their infectious potential very long outside a host (Raloff 2011). Host–virus interactions are often highly specific (e.g. viral pathogens of animals may also be human pathogens), but plant viruses are usually harmless for humans. Viruses are responsible for the majority of acute respiratory infections in humans (e.g. influenza and the common cold), which show seasonal incidence patterns related in part to increased indoor transmission in winter.

*Prions* are proteinaceous particles smaller even than viruses, but with some similar properties. These rare and unusual infectious agents are believed to be responsible for degenerative neurological diseases in humans and animals that are known collectively as transmissible spongiform encephalopathies, e.g. Creutzfeldt–Jakob disease (a human neurodegenerative disorder), Kuru (a human disease seen in Papua New Guinea), scrapie (a nervous system disease of sheep), and bovine spongiform encephalopathy or variant Creutzfeldt–Jakob ("mad cow") disease. If prions are

**TABLE 6.1 Physical Characteristics of Microorganisms**

| | Dimensions (μm) | | | |
|---|---|---|---|---|
| Organism | Width | Length | Shape | Reproductive or resistant form |
| Viruses | 0.020–0.25 | | Geometric, complex, helical, ovoid, spherical | — |
| Bacteria | 0.3–1 | 1–10 | Coccoid, spiral, rod shaped: branched, unbranched | Endospores |
| Fungi | 1–10 | 1–100 | Spores/conidia: spherical, ellipsoid, fusiform<br>Hyphae: branched, unbranched | Spores/conidia |
| Protozoa | 5–20 | 5–75 | Spherical, elongated, changing | Cysts |
| Helminths | | | | |
| Eggs or larvae | 15–75 | 25–150 | Spherical, ellipsoid | Eggs |
| Adult worms | <1–>5 mm | 5–>10 mm | Elongated and cylindrical, flattened, elongated and segmented | |

responsible for transmissible diseases, they are the only agents without nucleic acid that can do so. Consequently, the prion theory is not universally accepted.

*Bacteria* are grouped according to their gram stain reaction (positive or negative) and their cell shape (e.g. cocci or rods). They multiply by binary fission with generation times as short as 20 minutes. Chlamydia and rickettsia are bacteria that, like viruses, are intracellular parasites. Bacteria are found in a wide range of ecological sites in the environment (usually fairly damp ones) and also in association with animals, humans, and plants both as normal flora and, less often, as pathogens. Some bacteria (e.g. *Bacillus* and *Clostridium* spp.) produce endospores (internal dormant bodies) that enable them to survive adverse environmental conditions.

*Fungi* come in many forms, from unicellular yeasts to multicellular molds and mushrooms. They can reproduce by budding or forming (often in large numbers) sexual or asexual spores; i.e. reproductive bodies formed, respectively, in saclike structures or externally. Many fungi are found in soil and on leaf surfaces where they help decompose organic matter and recycle nutrients. Many humans suffer superficial fungal infections of the skin, nails, hair, and mucous membranes. Some airborne fungal infections are fairly common in certain parts of the country (e.g. histoplasmosis and coccidioidomycosis; see the case study on an outbreak of coccidioidomycosis in a television cast and crew), other fungal infections are somewhat rare (e.g. blastomycosis and cryptococcosis), and those caused by opportunistic fungal pathogens (e.g. aspergillosis and mucormycosis) usually only cause invasive disease in immunocompromised individuals or when a person receives an overwhelming exposure. Dimorphic fungi (e.g. the agents of blastomycosis and histoplasmosis) grow as yeasts (unicellular fungi) in the body and when incubated at 37 °C in laboratory culture, but grow as molds (filamentous fungi) in the environment or when incubated at 25–30 °C.

*Protists* are single-celled organisms that do not fall neatly into any single taxonomic group. Protozoa are composed of single cells or groups of more or less identical cells and live chiefly in water, although some are parasitic in humans and other animals. The majority of pathogenic protozoa cause hematogenous diseases transmitted by insects (e.g. malaria and babesiosis) or gastrointestinal diseases transmitted via ingestion (e.g. amebiasis, giardiasis, cryptosporidiosis, and cyclosporidiosis). Toxoplasmosis is also acquired by accidental ingestion of cysts shed in cat feces or by eating raw contaminated meats. This is of particular concern for immunocompromised persons and the offspring of women infected during pregnancy. People have become infected with free-living amebae (e.g. *Naegleria fowleri* and *Acanthamoeba* spp.) by swimming in contaminated waters, being sprayed or splashed in the eyes or on the mucous membranes of the nose or mouth with contaminated water, or wearing improperly decontaminated contact lenses.

*Helminths* are parasitic worms including cestodes (tapeworms), nematodes (roundworms), and trematodes (flatworms). Some helminths have complex life cycles, with one or more host(s), and cause intestinal disease or live in host tissues. Many human helminth infections arise from contact with animal or human feces or with feces-contaminated soil or other material, resulting in intestinal or tissue diseases.

**6.2.1.2 Sources of Microorganisms** A *source* is the person, animal, object, or substance from which an infectious agent was passed to a host; sources may also be reservoirs. A *reservoir* of an infectious agent is any person, animal, arthropod (insect, e.g. mosquito or flea, or arachnid, e.g. tick or mite), plant, soil, or substance (or combination of these) (i) in which an infectious agent normally lives and multiplies, (ii) on which it primarily depends for survival, and (iii) where it reproduces itself in such manner that it can be transmitted to a susceptible host (APHA 2015). For example, the placental tissue of a ewe that recently kidded (gave birth) or dust from dried placental tissue may be the source for the agent that caused Q fever in a laboratory researcher; however, the reservoir for the Q fever agent most likely is the flock from which the ewe was purchased (Roest et al. 2013). Identifying infectious agent reservoirs (not just their immediate sources) may be important to achieve long-term control of workplace transmission.

Sources and reservoirs of infectious microorganisms in the workplace include *humans*, most often patients in healthcare facilities or children in childcare centers and schools, but also the public, clients, and co-workers. Infectious persons may have apparent disease or be asymptomatic (without symptoms) but still able to transmit infection, e.g. they may be in the incubation period of a disease, be colonized by an infectious agent without being themselves affected, or be chronic carriers of an agent (see Section 6.2.3). *Animals* (fishes, reptiles, birds, mammals, and arthropods) are also often asymptomatic carriers of infectious agents and serve as exposure sources for agricultural, food production, and veterinary care workers as well as animal handlers/trappers (see the case study on plague in a wildlife biologist). Agents with human or animal reservoirs may not cause immediate serious disease because the microorganisms evolved as parasites and may depend on their hosts for survival. Finally, the *environment* (water, soil, and plants) can be a source or reservoir of infectious microorganisms. Inanimate objects (medical devices, toys, tools, and work materials) may become contaminated and also serve as sources of infectious agents or means of agent transmission.

**6.2.1.3 *Nomenclature*** The taxonomic system that classifies similar organisms into increasingly restrictive groups uses Latin or Greek names, of which the *genus* and *species* designations are the most familiar. A name may describe an organism, commemorate a distinguished scientist, or identify the geographic source of the first isolate. For example, the name *Francisella tularensis* recognizes Francis' work and Tulare, the California county where the bacterium was first found. Naming diseases for organisms is still recommended, but inclusion of person and location names is discouraged to avoid stigmatization (WHO 2015). A microorganism may be reclassified and renamed as more is learned about it. For example, *Pneumocystis jiroveci* (previously known as *Pneumocystis carinii*) is a major cause of illness in persons with impaired immune systems, but originally it was misidentified as a form of *Trypanosoma cruzi* and thought to be a protozoan until it was recognized to be a fungus and renamed.

The first letter of a genus name is capitalized and followed by the species name in lowercase, e.g. *Bacillus subtilis*, a rod-shaped, slender bacterium, and *Bacillus anthracis*, a related rod-shaped bacterium that produces a dark carbuncle. Only the first initial of a genus name is written when used repeatedly or when several species from the same genus are named, e.g. *B. subtilis* and *B. anthracis*. Italicized genus names are followed by the word "species" (abbreviated "sp." for singular and "spp." for plural) when no species is named, e.g. a *Bacillus* sp. or several *Bacillus* spp. Identification to the species level is often important when evaluating a microorganism's significance, e.g. *B. subtilis* is an essentially harmless soil bacterium frequently used in experimental work, whereas *B. anthracis* causes anthrax, a potentially fatal zoonotic infection. The sexual and asexual forms of a fungus may be known by different genus names (e.g. *Histoplasma* and *Ajellomyces*), which, in this case, are also the names of the mold and yeast forms of this dimorphic fungus. A genus name used as a noun or adjective is neither capitalized nor italicized, e.g. the streptococci or a streptococcal infection.

Isolates within a species may be further separated into subspecies based on minor differences. Serologic distinctions among closely related microorganisms are based on detecting different antigenic determinants on their surfaces (see Section 6.3.3). Serologic tests are used primarily in epidemiologic investigations to compare multiple isolates of the same species, e.g. from several persons or environmental sources. This comparison can help investigators determine if the isolates represent a single strain (possibly identifying a person or environment as a common source) or multiple strains (possibly indicating that the infections arose from different sources). Serologic techniques allow division of species into serogroups and serotypes (often referred to by number or letter, e.g. serogroup 1 or serogroup A) and within these into serovars (often referred to by the place of origin). However, the naming of serologically identified subspecies is not as consistent as that for genera and species.

Viruses are grouped into 122 families according to (i) the type and form of the nucleic acid present (i.e. RNA or DNA, single or double stranded) and (ii) the size, shape, structure, and mode of viral replication (International Committee on Taxonomy of Viruses 2017). Within these families, viruses are placed into genera and species. Family and genus names for viruses are italicized, but the common species names are not, e.g. the mumps and measles viruses (the common species names) are both in the family *Paramyxoviridae* but belong, respectively, to the genera *Paramyxovirus* and *Morbillivirus*.

*6.2.1.3.1 Disease Names* Names for microorganisms are chosen in many ways, but they must comply with international codes of nomenclature or taxonomy. However, there is no official international nomenclature for diseases, although the WHO is a good resource. A disease may be named for the microorganism that causes it (e.g. brucellosis from *Brucella* spp.), or an infectious agent may be known for the disease it produces (e.g. the rubella virus). Such naming usually depends on whether the disease or the agent was recognized first. Diseases may also be known by more than one name, e.g. viral hepatitis B or serum hepatitis.

Some diseases may be caused by more than one agent (e.g. viral or bacterial conjunctivitis and viral, bacterial, or fungal pneumonia) or by more than one species of a genus (e.g. shigellosis and salmonellosis). In addition, some infectious agents can cause more than one disease, e.g. the varicella-zoster virus (VZV) causes chickenpox/varicella and herpes zoster/shingles. Disease names that use a microorganisms' genus or species name end in -osis (singular) or -oses (plural), e.g. psittacosis from *Chlamydia psittaci* or the shigelloses caused by *Shigella* spp. Use of a disease name (e.g. salmonellosis or tuberculosis) implies that specific symptoms were present (e.g. diarrhea or cough and weight loss, respectively). In the absence of symptoms, a person with a stool culture positive for a *Salmonella* sp. would be reported as having a salmonella infection. Likewise, an asymptomatic person with a positive tuberculosis skin test is said to have latent tuberculosis infection rather than tuberculosis disease.

### 6.2.2 Transmission of Infectious Agents

*Transmission* describes the mechanism by which an infectious agent spreads from a source or reservoir to a person. In the source/pathway/receptor model that is often used to describe occupational and environmental hazards, an infected person or animal or a material containing a pathogen is the source, and the exposed person is the receptor (see also Section 6.2). The mode of transmission then describes how an infectious agent leaves the source, the pathway it follows, and how it arrives at a site where it can establish infection in an exposed person. Understanding these aspects of infectious disease transmission can identify opportunities to prevent it (Ciplak 2013).

Pathogens may be transferred by contact with infectious materials either directly, between the host and a receptor, or indirectly, by contact with an intermediary contaminated object, which includes medical devices. Pathogens also can be transferred through droplets that a source emits, which are transmitted to a host through the air. This type of transmission may be further divided into "droplet" and "airborne," although both may be considered to be forms of "aerosol" transmission (see the discussion in Section 6.2.3). In addition, pathogens may be transmitted by a "common vehicle," such as a shared beverage or food, or by vectors, typically insects such as mosquitoes.

The *portal of entry* for an infectious agent into the human body may be the respiratory tract (via inhalation); the mouth (via ingestion); the mucous membranes, conjunctivae, or previously broken skin (via contact); and skin broken at the time of exposure (via parenteral exposure, such as injection, an incision, or a sharps injury). These portals are not mutually exclusive; for example, the mouth and the respiratory and digestive tracts have mucous membranes, and they also provide portals through which pathogens may contact target cells. Intact skin is generally believed to be an effective barrier to microorganism entry because it is composed of tightly associated epithelial cells covered by a highly cross-linked keratin layer (Dieffenback and Tramont 2015).

Infectious agents are transmitted to portals by a variety of routes, and some microorganisms may be transmitted by more than one means, sometimes leading to distinct diseases (Muller and McGeer 2006). For example, a veterinarian in a plague-endemic area may acquire infection through a flea bite (resulting in bubonic plague) or by inhaling droplets generated while caring for an animal with plague pneumonia (resulting in pneumonic plague; see the case study on plague in a wildlife biologist). Likewise, the agent of tularemia (a zoonotic infection) can cause a skin ulcer on a worker's hand where contact occurred, systemic infection if inhaled, or pharyngitis (an inflamed throat) if ingested. In addition, a common, generally non-occupational means of acquiring an infection (e.g. ingesting *Brucella* spp. in unpasteurized milk or cheese) may differ from a less common, occupational transmission mode (e.g. inhalation of airborne bacteria from brucella cultures in a microbiology laboratory), but the same clinical illness, brucellosis, would result in either case. Viruses tend to cause the same manifestations of infection regardless of their transmission mode. For example, the pathogenicity of the Ebola virus (i.e. how it acts on the body to cause disease) and symptoms typically are the same regardless of the initial route of infection (OHSA 2017). The increased recognition of occupational blood-borne infections (such as hepatitis B and HIV infection) in addition to changes in our understanding of the behavior of infectious aerosols and concerns about skin and soft tissue infections has altered traditional classification systems for disease transmission as described below.

*Contact transmission* includes direct contact with an infected source (e.g. by touching or being bitten) and indirect contact (e.g. via contaminated tools or utensils, clothing or bedding, or surgical instruments or dressings). *Direct contact* transmission requires personal contact between a worker and a source, e.g. a patient in a healthcare setting, a fellow worker, a child in a childcare or school setting, or an animal in an agricultural or food production setting. Direct contact involves touching of a part of a susceptible worker's body to another infected or colonized body surface, with physical transfer of an infectious agent between the two. The herpes, rabies, and human papillomaviruses as well as *Staphylococcus aureus* (including methicillin-resistant *S. aureus* or MRSA) are transmitted in this way. Most sexually transmitted diseases are transmitted by direct contact, by the contact of certain bodily fluids with the mucous membranes of another person (e.g. syphilis, HIV infection, or hepatitis B), or by contact between the mucous membranes of different individuals (e.g. herpes). *Indirect contact* transmission involves contact of a susceptible worker with a contaminated intermediate object, usually inanimate, e.g. an instrument, glove, or wound dressing; a laboratory culture; or animal bedding or waste. Examples of diseases transmitted by indirect contact are bacterial infections (e.g. streptococcal skin diseases, salmonellosis, shigellosis, and Q fever), viral and bacterial conjunctivitis, and dermatophytoses (e.g. athlete's foot).

Hepatitis B infection in healthcare workers, due in large part to sharps injuries, was recognized as a significant hazard in the early 1970s. This understanding, combined with the emergence of the HIV epidemic in the 1980s, led in 1987 to a CDC recommendation for "Universal Blood and Body Fluid Precautions" (CDC 1988) (see Section 6.1.1). These universal precautions, which became a legal mandate under OSHA's Bloodborne Pathogens Standard (29 CFR 1910.1030 1991b), are specific control measures to prevent transmission of agents found in blood, certain bodily fluids, or blood-contaminated items (see also Section 6.1.1).

In addressing transmission of blood-borne pathogens, parenteral transmission is distinguished from other forms of contact transmission and involves the intentional or unintentional breaking of the skin barrier by medical sharps (e.g. needles and scalpels) or other injuries, including bites. The risk of contracting HIV, hepatitis B or C, or other blood-borne infection is considerably greater through parenteral routes. Increasingly, the term "contact transmission" is reserved for pathogens that cause infection through skin or mucous membrane contact, such as *S. aureus* or *Neisseria gonorrhoeae* (the cause of gonorrhea). This distinction is important for appropriate hazard determination and risk reduction for diseases such as HIV, HBV, and HCV infection, which are not transmitted through intact skin contact with, or kissing of, an infected individual. In healthcare and related settings, such as law enforcement, preventing parenteral transmission requires specific control measures, such as the proper use of safety devices and sharps disposal boxes.

*Droplet transmission* is sometimes considered a form of contact transmission, but the mechanism of pathogen transfer is distinct. Droplets (usually of respiratory secretions) are generated when a person coughs, sneezes, or speaks or during medical procedures such as suctioning and bronchoscopy (visually examining the tracheobronchial tree using a flexible tube). They may also be generated when liquids (e.g. process waters such as metalworking fluid, body fluids such as blood or sputum, or laboratory cultures) are spilled, splashed, or sprayed. Under current infection control guidelines (Siegel et al. 2017), "droplet transmission" is considered to occur when droplets that are larger than 5 μm and contain microorganisms are propelled a short distance through the air (often given as ≤1 or 2 m) and deposit on a worker's conjunctivae, nasal mucosa, or mouth.

Violent expiratory events can be modeled as plumes containing pathogen-laden droplets within a vaporous cloud (Halloran et al. 2012). Some of these droplets may be propelled to directly strike the eyes, skin, mouth, or other mucous membranes of another person. The droplets that do not strike a surface will remain as an aerosol, until gravity causes settling (Jones and Brosseau 2015). Larger droplets settle more quickly, while horizontal or upward air movement will keep the remaining droplets entrained in the air longer and they may travel further. The size of respiratory droplets decreases with time due to evaporation of water because they are generally warmer and moister than the surrounding air. The settling of larger droplets out of a plume may also increase its buoyancy, so that the remaining droplets may travel greater distances.

*Airborne or aerosol transmission* may be distinguished from droplet transmission in some healthcare infection control guidelines. Researchers, including W.F. Wells, H.L. Ratcliffe, and R.L. Riley, engaged in a series of experiments in the post-World War II era that found that tuberculosis was transmitted through small particle aerosols (Nardell 2004). In one experiment, Richard Riley placed guinea pigs in a penthouse on the roof of a hospital in Baltimore, and the animals contracted tuberculosis infection from air piped through the penthouse from a tuberculosis ward in the hospital (Riley 2001). This transmission was considered to occur through inhalation of "droplet nuclei," i.e. respiratory droplets that tuberculosis patients had expelled and that had evaporated to a smaller size. The nuclei containing bacteria were more likely than the original droplets to remain suspended in air and to travel through the air ducts. Infection via inhalation of airborne dusts containing viral or bacterial pathogens or fungal spores is also considered a form of airborne transmission. In the current infection control paradigm, a small number of diseases, including measles and varicella, are identified as being transmitted through inhalation, while most diseases are considered to be transmitted by "droplet contact" (Roy and Milton 2004). The term "aerosol transmission" is increasingly used to describe transmission that occurs when droplets containing pathogens enter the respiratory tract of an employee or other exposed person (Jones and Brosseau 2015). Many diseases traditionally classified as spread by "droplets" may also be spread through aerosol inhalation.

*Common vehicle transmission* is similar to indirect contact transmission, but this term applies to microorganisms transmitted by contaminated items to many persons rather than to just one. Examples of contaminated items involved in common vehicle transmission are food, water, medications, medical devices, and equipment or tools. Outbreaks of foodborne or waterborne disease may occur in many workplaces and may be unrecognized, particularly in agricultural workers and field crews.

*Vector-borne transmission* occurs when an animal serves as an intermediary, carrier, or vector and transmits an infectious agent from one host (or reservoir) to another. Examples of vectors are mosquitoes, fleas, ticks, flies, rats, and other vermin. Some vectors serve simply as a mechanical means of transmitting a microorganism, e.g. a housefly carrying bacteria from one place to another. Other vectors play a biological role and serve as a host in which a microorganism multiplies or develops, e.g. mosquitoes and *Plasmodium* spp., protozoa that cause malaria. A vector may also be an intermediate host for an agent, e.g. a tick carrying *Borrelia burgdorferi*, the bacterial agent of Lyme disease. An intermediate host transports an infectious agent between a definitive host (e.g. a rodent, deer, or other mammal) and a human who may serve as an additional reservoir. Outdoor workers (including vector and animal controllers) are at risk of vector-borne diseases as are indoor workers if they are exposed to animals that serve as reservoirs of infectious agents transmitted by vectors.

### 6.2.3 The Infectious Disease Process

An *infection* occurs when a microorganism invades a body, multiplies, and causes the body to react. To produce infectious disease, a microorganism must be (i) pathogenic (i.e. disease causing), (ii) viable (alive and able to multiply, e.g. bacteria, fungi, protozoa, and helminths) or active (able to initiate replication by a host cell, e.g. viruses and perhaps prions),

(iii) present in sufficient numbers, and (iv) successfully transmitted to a susceptible host at a suitable entry site. Both living and dead microorganisms and microbial fragments can elicit hypersensitivity (allergic) reactions in sensitive persons, but only viable or active microorganisms can cause infection and then only in susceptible individuals. The term *virulence* describes the ability of an infectious agent to invade and damage host tissues. Virulence reflects the degree of pathogenicity of an infectious agent, often indicated by case fatality rates (APHA 2015). This term is often confused with infectivity. *Infectivity* describes the tendency of an organism to be transmitted and cause infection. For example, rhinoviruses (which cause the common cold) have high infectivity but low virulence; i.e. they attack many persons, but resulting illness is never itself fatal. In contrast, influenza viruses may be both highly infectious and virulent.

Some of the effects of an infection are due to toxins that an invading microorganism produces, e.g. the tetanus toxin of *Clostridium tetani*. Some foodborne diseases are due to preformed toxins present as a result of microbial multiplication in a food before consumption, e.g. the botulinum toxin of *Clostridium botulinum*. Such foodborne diseases are referred to as intoxications rather than infections. Gram-negative bacteria produce endotoxins and some fungi produce mycotoxins (Macher et al. 2012; Wald and Stave 2001). However, the health problems these toxins may cause (e.g. asthma and inhalation fevers) are not considered communicable diseases and are not addressed here.

A person must receive a threshold number of organisms or amount of toxin to contract a disease, i.e. an *infectious* or *effective dose*. Reliable information on infectious doses for humans is available for only a limited number of agents (PHAC 2017; Singh 2009; Yezli and Otter 2011). Infectious doses vary widely among agents and may depend on the exposure route. For example, one *M. tuberculosis* cell deposited in the gas exchange region of the lungs may initiate infection. However, the tissues of the upper airways are not susceptible to this bacterium, and even large numbers of bacteria deposited there appear to be harmless. Infectious doses are often reported as $ID_{50}$, the number of organisms required to infect 50% of a test population. The infectious dose for an individual worker depends on the person's general health and may be higher or lower than the $ID_{50}$. In general, larger infectious doses are required by the oral route than by inhalation or mucous membrane contact. Doses that greatly exceed the minimum may result in more severe infection or more rapid onset.

*Resistance* describes the degree to which a person can eliminate an organism without developing disease or is able to limit the severity of illness following infection. It varies greatly from person to person. Immunization or past infection may contribute to resistance. Host factors also may render some workers more susceptible to infection. Examples of such host factors are age, race, gender, underlying disease, antimicrobial treatment, administration of corticosteroids or other immunosuppressive agents, irradiation, and breaks in the first line of defense, such as damage to the skin or respiratory tract. In addition, such risk factors may predispose individuals to more severe illness should they become infected. Other than immunizations and PEP, few medications, treatments, or dietary supplements have been demonstrated to effectively increase a person's resistance to infection.

Exposure of a susceptible person to an infectious dose of an agent may result in asymptomatic infection, subclinical infection, or clinical disease. *Asymptomatic infections* occur when an organism enters the body and multiplies, but by virtue of a healthy immune response, the person has no clinical symptoms or outward signs of illness. Asymptomatic infection is the most common form of infection for some agents, e.g. HBV. Such infections may only be recognized through specific serologic testing (see Sections 6.3.2.1 and 6.3.3). *Subclinical infections* are generally mild (e.g. a slight fever, minor aches, or transient fatigue) and are usually of short duration.

*Clinical disease* is associated with signs or symptoms of infection that may aid diagnosis, e.g. the fever, chills, night sweats, fatigue, weight loss, and hemoptysis (coughing blood) of tuberculosis. However, tuberculosis is an example of an infection that seldom leads to clinical disease in healthy infected adults and is one in which infected individuals only transmit the agent (i.e. become infectious) when they have active disease. In the United States, only approximately 1–5% of persons infected with *M. tuberculosis* develop disease in the first year following infection identified by a skin test conversion. Fewer than 10% of latently infected persons will develop active pulmonary tuberculosis disease in their lifetimes (APHA 2015). In contrast, for individuals coinfected with HIV, the risk of developing active pulmonary tuberculosis rises from an estimated 10% in a lifetime to up to 15% per year in the absence of antiretroviral treatment (APHA 2015). *Carriers* harbor infectious agents, temporarily or chronically, and may transmit them to others. Carriers may be colonized, infected but asymptomatic, or have chronic or relapsing disease.

Infectious agents typically have characteristic *incubation periods*, i.e. the time interval between initial contact and the first appearance of symptoms or immunologic evidence of infection. Incubation periods are usually a few days to a few weeks. However, incubation periods may be only a few hours (e.g. salmonellosis), extend to months or years (e.g. warts and rabies, respectively), or are unknown (e.g. dermatophytoses). During this period, an infectious agent is multiplying, and the body typically begins to mount an immune response. For many agents, the larger the dose received, the shorter the incubation period (Toth et al. 2013).

The *communicable period* is the time during which an infected person or animal can transmit an infectious agent. This period may begin before signs or symptoms of infection are evident. For many infectious agents (particularly viruses), the time of greatest communicability is during the incubation period, which is when a microorganism is multiplying at the greatest rate. Consequently, the period of greatest risk of transmitting infection is often past by the time disease is recognized.

*Antibodies* belong to a group of molecules called *immunoglobulins* (Ig), of which there are several classes represented by letters, e.g. A or IgA, G or IgG, and M or IgM. Temporally, antibody responses to an infection occur in two phases. The first phase consists of IgM antibodies, which appear as early as four days following infection, peak at approximately seven to ten days thereafter, and then decline. The second phase usually involves IgG antibodies, which appear within two to three weeks of exposure and begin declining four to six weeks after infection or may persist for many years. An antibody name may identify an infectious agent (e.g. "anti-"hepatitis A virus [anti-HAV]) or a specific antigen of the agent (e.g. "anti-"hepatitis B core antigen [anti-HBc]). Laboratory serologic reports often omit the term "anti-." For example, a laboratory result may report detection of legionella antibodies rather than detection of anti-legionella antibodies.

### 6.2.4 Modeling Infectious Disease Transmission

Modeling disease transmission can help investigators identify significant contributing factors and select and evaluate control measures (Fournié et al. 2012b; Wood et al. 2014). Health and safety professionals, familiar with exposure assessment for other workplace hazards, may recognize similarities for infectious disease prevention if they can separate and individually consider the factors involved in the process. Factors that influence aerosol transmission of infectious diseases include generation rate and plume characteristics, dispersion and distribution of pathogens, ventilation strategy, pathogen survival, aerosol size, air turbulence, respiratory deposition, heterogeneity of pathogen infectivity, pathogen–host interaction, environmental control measures, and personal protection (Sze To and Chao 2010).

The Wells–Riley model has been used to examine person-to-person airborne infectious disease transmission since the 1970s and remains a simple means to understand the roles of various parameters (Riley et al. 1978). The model requires estimations of the following: (i) the number of infectious persons or animals present; (ii) the rate at which these sources release microorganisms, which depends on factors including the concentration of infectious agents at a source and the rate at which the source generates infectious particles; (iii) the breathing rates of susceptible persons; and (iv) the exposure times for potential new cases. Transmission depends inversely on the rate at which dilution ventilation or other particle removal processes reduce the concentrations of contaminants. The model has been linked with ventilation flows to aid in the design and risk assessment of healthcare buildings (Aliabadi et al. 2011; Noakes and Sleigh 2009; Villafruela et al. 2013). However, limitations of existing models of aerosol transmission to predict infectious disease risks include assumption of steady-state ventilation conditions and uniformly mixed airspace as well as necessary simplifying assumptions about sources, pathogens, and receptors (Issarow et al. 2015).

Epidemiologists have used models to estimate rebreathed air volumes to determine the social and environmental factors associated with aerosol infection and to identify locations with high transmission potential (Wood et al. 2014). The plague case study demonstrates how information from actual cases can be used to estimate the time from exposure to an agent to symptom onset and the time from exposure to death. Theoretical models can be evaluated using experimentally produced aerosols in realistic settings as well as controlled laboratory conditions (Jones and Nicas 2014). For example, fluorescent particles were used to estimate emission rates from the environment and the occupants of a classroom and to compare emission variability between persons and activities (Bhangar et al. 2016).

Modeling also has been used to evaluate human and animal exposure to ambient pathogens characterized by wind dispersion, e.g. avian influenza virus, foot-and-mouth disease virus, *B. anthracis*, *C. burnetii*, and *Legionella pneumophila* (van Leuken et al. 2015). Van Leuken et al. (2015) found that by adding meteorological information to an atmospheric dispersion model of Q fever incidence, they could improve the correlation between modeled concentration and observed incidence. Sprigg et al. (2014) used meteorological models to determine whether prediction of haboob (dust storm) events could be used to reduce the risk of valley fever (coccidioidomycosis) transmission.

In addition to aerosol transmission, dose–response models can consider other exposure routes such as mucous membrane contact as well as ingestion of food or water (Nicas and Jones 2009; Sze To and Chao 2010). Gautreau et al. (2008) studied a metapopulation model for the spread of epidemics on a large scale in which subpopulations (cities) were linked by fluxes of airline passengers. They concluded that such models could be particularly useful for the analysis of epidemics that propagate worldwide and that such predictive tools could play an important role in preparing containment measures and policies.

## 6.3 EVALUATING EXPOSURE TO BIOLOGICAL AGENTS

Health and safety programs for control of infectious agents in the workplace should (a) identify potential work-related infections; (b) identify applicable regulations or guidelines; (c) develop an appropriate medical surveillance program, including reporting and investigation of accidents and exposure incidents; (d) develop control and prevention strategies, including emergency procedures; (e) educate workers and train them to recognize and avoid exposures and to respond correctly to emergencies; and (f) inspect the workplace for compliance with health and safety rules. The first item has already been discussed; comments on the other components follow.

### 6.3.1 Regulations, Recommendations, and Guidelines

The control of communicable diseases is one of the oldest fields of public health. Not surprisingly, there are many laws designed to prevent infectious disease transmission, e.g. federal, state, and local programs that protect the public from infectious agents by regulating food, drinking and recreational waters, solid waste handling, and sewage treatment. Vector control programs, such as mosquito abatement, are intended to reduce transmission of diseases ranging from malaria to West Nile virus infection. State and local health departments may offer mass screenings or immunizations during disease outbreaks, and health officers have the authority to quarantine or isolate people if they determine that it is necessary. However, health officers prefer to work cooperatively with employers, employees, employee representatives, and state and federal occupational health and safety agencies (see Section 6.3.4 for more information on infectious disease surveillance). There are no workplace or ambient exposure limits for infectious agents, and it is unlikely there will soon be any (ACGIH 2017). Following are summaries of some work-related regulations and guidelines that may apply to occupational infectious disease control.

***6.3.1.1 General Federal, State, and Local Regulations*** The handling of agricultural items that will become food products shipped through interstate commerce is regulated by the US Department of Agriculture (USDA) and the Food and Drug Administration (FDA). Some state agencies regulate agricultural products or operations that are exempted from federal inspections. Food preparation facilities and food handlers are regulated by state departments of public health, typically through food and drug programs, as well as by local jurisdictions. State and local air quality, water, and waste management agencies govern aspects of pollution control that impact infectious disease transmission.

Some professional licenses and credentials require evidence of immunization as well as training and examination in infectious disease control. However, these regulations and those covering the handling of food products as well as liquid and solid wastes are intended primarily to protect consumers, clients, and the public, not necessarily the workers in these industries.

Many state laws specify that a local health officer must clear an infected worker in certain occupations before that person is allowed to return to work. However, these laws generally leave to the health officer's discretion the conditions that must be met before return to work. Section 6.3.2.4 addresses return-to-work requirements for healthcare workers and food handlers. Local, state, and federal public health authorities and professional associations are the best sources of information on the regulations and recommendations that apply to particular occupations. Further descriptions of government regulations of general safety and health matters are available elsewhere (Dougherty et al. 2012).

***6.3.1.2 CDC, WHO, and Professional Associations*** CDC, WHO, public health associations, and professional groups have authored many guidelines and recommendations referenced throughout this chapter. These recommendations do not have the impact of regulations or requirements but are considered accepted current standards of practice. Activities less protective than these recommendations are generally considered unacceptable.

***6.3.1.3 The Occupational Safety and Health Administration*** Section 18 of the Occupational Safety and Health Act of 1970 encourages states to develop and operate their own job safety and health programs (state plans) and precludes state enforcement of OSHA standards unless the state has an OSHA-approved plan. State plans must be as effective as federal OSHA requirements. Twenty-two states or territories have plans that cover both private employees and state and local governmental employees; five additional states and one territory have plans that cover only state and local government employees. In states without a state plan, federal OSHA exercises jurisdiction over private sector employers, and in all states, OSHA has jurisdiction over federal agencies and over certain transportation operations.

OSHA commonly addresses workplace hazards associated with infectious agents through enforcement actions using the General Duty Clause of the Occupational Safety and Health Act of 1970 section 5(a)(1) (OSHA 1970).

This section requires employers to provide a workplace free of recognized hazards that cause or are likely to cause death or serious physical harm. OSHA often refers to CDC guidelines to establish that an infectious disease constitutes a "recognized hazard."

OSHA has also promulgated regulations that control general sanitation and specific infectious disease hazards. These standards are found in Title 29 of the Code of Federal Regulations. Standard 29 CFR 1910.22(a) (Housekeeping) requires that workplaces be kept clean, orderly, and in a sanitary condition. Standard 29 CFR 1910.141 (Sanitation) requires that employers in general industry provide potable (drinking quality) water; disposal of putrescible (decomposable) wastes; control of vermin (e.g. rodents and insects); change rooms for certain employees required to wear protective or work clothing; lavatories exclusively for washing hands, arms, faces, and heads; and toilet facilities. Similar requirements for other industries (construction, maritime, and agriculture) are contained in other sections of Title 29. Standard 29 CFR 1910.142 (Temporary labor camps) requires vector control and sanitation, openable windows, and avoidance of overcrowding in sleeping quarters as well as reporting of communicable diseases to local health officers.

Standard 29 CFR 1910.132 (General requirements for PPE) requires that employers assess the work environment and determine if there are hazards that require PPE (including PPE for eyes, face, head, and extremities; protective clothing; respiratory devices; and protective shields and barriers). If hazards are found, the employer is required to determine effective PPE and to provide and maintain the PPE in a sanitary and reliable condition. There are some exceptions in 29 CFR 1910.132 to the general requirement that employers provide PPE at no cost to the employee. Sections 1910.133–1910.138 address specific types of PPE. The use of respirators is addressed in 29 CFR 1910.134, including the use of respirators to protect against biological hazards such as tuberculosis.

Standard 29 CFR 1910.145 (Specifications for accident prevention signs and tags) establishes requirements for signs posted by employers to warn of hazards. Biohazard postings are used in healthcare, laboratories, and similar environments to identify the actual or potential presence of a biological hazard and to identify equipment, containers, rooms, materials, experimental animals, or combinations of these that contain or are contaminated with hazardous biological agents.

In 1991, OSHA promulgated a standard that addressed occupational exposure to blood-borne agents (codified under 29 CFR 1910.1030, Occupational Exposure to Bloodborne Pathogens), which included, but was not limited to, HBV and HIV. Other blood-borne infections mentioned in the rulemaking documents included arboviral infections; babesiosis; brucellosis; Creutzfeldt–Jakob disease; CMV infection and disease; hepatitis A, C, D, and E; human T-lymphotropic virus type I (HTLV I); leptospirosis; malaria; relapsing fever; syphilis; and viral hemorrhagic fever. In 2000, the Needlestick Safety and Prevention Act (Pub. L. 106-430) was signed into law, and in 2001, OSHA amended the Bloodborne Pathogens Regulation to require "sharps with engineered sharps injury protection" (SESIP), e.g. safety needles.

In 1997, OSHA issued a proposed rule on occupational tuberculosis, which included several components from CDC guidelines, i.e. written exposure control plans, procedures for early identification of individuals with suspected or confirmed infectious tuberculosis, procedures for initiating isolation of individuals with suspected or confirmed infectious tuberculosis or for referring those individuals to facilities with appropriate isolation capabilities, procedures for investigating employee skin test conversions, and education and training for employees. However, this rulemaking was terminated on 31 December 2003 (OSHA 2003). In the absence of standards, OSHA may reference CDC and other guidelines to substantiate a citation using the General Duty Clause of the OSHA Act section 5(a)(1).

Standard 29 CFR 1904 requires employers to keep records of all occupational injuries and illnesses (including those involving infectious agents) that result in death, lost workdays, restricted work, transfer to another job or termination of employment; require medical treatment other than first aid; or involve loss of consciousness. Most employers are required to maintain a log and summary of occupational injuries and illnesses. Rulemaking in 2001 established criteria for and the means of recording "privacy cases" in which the employee's name is not entered (29 CFR 1904.29), which include injuries with contaminated sharps (29 CFR 1904.8). Section 29 CFR 1904.11 addresses recording of work-related tuberculosis cases. In 2014, OSHA expanded requirements for the incidents employers are required to report to OSHA. Under this rule, 29 CFR 1904.39, employers must report work-related fatalities to the OSHA area office nearest to the incident site within 8 hours and must report an inpatient hospitalization, amputation, or loss of an eye within 24 hours. As of December 2017, some state plans had not completed adoption of equivalent reporting requirements.

**6.3.1.4 *Medical Waste Handling*** Various groups offer guidelines on the handling and disposal of waste material that may contain infectious agents, e.g. the WHO's *Blue Book* (Chartier et al. 2014). Almost all medical waste is incinerated. Treated medical waste can be disposed of in solid waste landfills (Akinsanya and Bodenberg 2014). Medical waste disposal is primarily regulated at the state level. The US Environmental Protection Agency (USEPA or EPA) maintains a website to determine what laws apply in each state (USEPA 2017a). From the information gathering required in implementation of the Medical Waste Tracking Act of 1989 (PL 100-582; 40 CFR 259), the USEPA concluded that the disease-causing potential of medical waste is greatest at the point of generation and naturally tapers off after that point. Thus, the risk to the general public of disease caused by exposure to medical waste is likely to be much lower than the risk for an occupationally exposed individual, which includes persons who handle medical waste that has been improperly discarded into the general waste stream. The act defined medical waste as any solid waste generated in the diagnosis, treatment (e.g. provision of medical services), or immunization of humans or animals, in research pertaining thereto, or in the production or testing of biological reagents. Of the total amount of waste generated by healthcare activities, WHO estimates that approximately 85% is general waste comparable with domestic waste, 10% is infectious hazardous waste, and the remaining 5% is chemical or radioactive waste (Chartier et al. 2014). Hazardous healthcare wastes include waste contaminated with blood and its by-products, cultures and stocks of infectious agents, waste from patients in isolation wards, discarded diagnostic samples containing blood and body fluids, infected animal carcasses and wastes from laboratories, and contaminated materials (e.g. swabs, bandages) and equipment (e.g. disposable medical devices) (WHO 2017c). Local and state agencies have specific requirements for handling of mixed wastes, e.g. biohazard and radiological. Mixing of waste streams is to be avoided.

***6.3.1.5 Disinfectants and Medical Devices*** The EPA registers disinfectants and other antimicrobial products that are used on surfaces. The FDA sets standards for disinfectants as well as medical devices and equipment. The FDA regulates only those products sold for the purpose of sterilizing critical devices such as surgical instruments that penetrate the blood barrier and semi-critical devices that contact mucous membranes (see also Section 6.4.4.1). Physical agents such as ultraviolet (UV) radiation may also be used to sterilize instruments. Many antimicrobial chemicals and physical agents are corrosive or irritating to the skin or eyes, and PPE must be chosen to address not only the biological hazard but chemical or other hazards as well (Chartier et al. 2014) (see Section 6.4.4).

In healthcare settings, the EPA regulates disinfectants that are used on environmental surfaces (housekeeping and clinical contact surfaces), and the FDA regulates liquid chemical sterilants/high-level disinfectants (e.g. glutaraldehyde, hydrogen peroxide, and peracetic acid) used on critical and semi-critical patient care devices (Rutala et al. 2008). Disinfectants intended for use on clinical contact surfaces (e.g. light handles, radiographic-ray heads, or drawer knobs) or housekeeping surfaces (e.g. floors, walls, or sinks) are regulated in interstate commerce by the Antimicrobials Division, Office of Pesticide Programs, EPA, under the authority of the Federal Insecticide, Fungicide, and Rodenticide Act (FIFRA) of 1947, as amended (Rutala et al. 2008).

***6.3.1.6 Transfer of Hazardous Agents*** The USDA (Veterinary Services [VS] of the Animal and Plant Health Inspection Service [APHIS], Riverdale, MD) regulates the importation of animals and animal-related materials to ensure that animal and poultry diseases are not introduced into the United States. Although some animal diseases also affect humans, these regulations are designed primarily to protect domestic and wild animals. The US Public Health Service (CDC, Atlanta, GA) has jurisdiction over importation of human and nonhuman primate materials.

The threat of the illegitimate use of infectious agents (e.g. for bioterrorism) has led to provisions to regulate the packaging, labeling, and transport of select agents shipped in interstate commerce. The CDC issued a regulation that places additional shipping and handling requirements on facilities that transfer or receive select agents that are capable of causing substantial harm to human health (42 CFR 72 2007), including designation of a responsible official. The rule was designed to (i) establish a system of safeguards to be followed when specific agents are transported; (ii) collect and provide information concerning the location where certain potentially hazardous agents are transferred; (iii) track the acquisition and transfer of these specific agents; and (iv) establish a process for alerting the appropriate authorities if an unauthorized attempt is made to acquire these agents. The rule has the following components: (i) a comprehensive list of select agents (including viruses, bacteria and rickettsiae, fungi, toxins, and recombinant organisms and molecules); (ii) registration of facilities transferring these agents; (iii) transfer requirements; (iv) verification procedures including audit, quality control, and accountability mechanisms; (v) agent disposal requirements; and (vi) research and clinical exemptions.

### 6.3.2 Medical and Biological Evaluation of Exposure

Medical evaluation of worker exposure to biological agents can occur in a variety of contexts, e.g. before job placement, periodically after placement, following exposure, and for diagnosis of possible illness. Medical surveillance for chemical exposures has some relevance to infectious disease surveillance in terms of general approaches.

***6.3.2.1 Preplacement Examinations*** The goal of a *preplacement examination* is to identify any necessary reasonable accommodations that will enable an employee to do a specific job. A preplacement examination for a worker potentially exposed to infectious agents should include, within legal limits and employee consent, (i) collection of baseline data for evaluation of any abnormalities that may later be noted, (ii) medical and occupational histories, (iii) a history of immunizations, (iv) a history of past infections, and (v) a physical examination, if indicated. Laboratory tests should be based on the specific agents to which an employee may be exposed. Intradermal skin tests can be used to establish probable immunity as a result of previous infection for diseases such as coccidioidomycosis and Q fever.

When occupational exposure to human pathogens is a risk, employers should consider collecting *blood samples* and storing them as sera (*sing*., serum, the fluid portion of blood remaining after the removal of cellular components and fibrinogen by clotting) prior to the initiation of work. The US Department of Defense began such a practice in 1989 and stores specimens for all active and reserve components of the Army, Navy, and Marines (Department of Defense 2017; Perdue et al. 2015). Such samples are collected at the beginning of work (and perhaps periodically thereafter) to establish baseline seroreactivity should additional blood samples be needed for serological testing subsequent to a recognized or suspected exposure (USDHHS 2009). A portion of a stored serum may be tested concurrently with a recent sample to detect a change in antibody concentration (see Section 6.3.3). Recent and stored sera are tested at the same time to reduce analytic variability due to the method or reagents used. This "serum banking" is particularly useful for occupations in which new pathogens may be uncovered in the future and for infectious diseases that may be associated with assignments in foreign countries. Asymptomatic or unrecognized past infections may be detected or ruled out through comparison of current and prior antibody levels.

Antibody levels may be determined at the time of worker placement if placement depends on knowing an employee's susceptibility/immunity status. The possible future need to perform a post-exposure evaluation (described below) presents a basis for conducting a preplacement test or for banking a serum sample. Without the information such tests can provide, it may not be possible to distinguish a recent occupationally acquired infection from one that predated employment. Preplacement tests could also be useful for workers' compensation purposes (e.g. to establish that a blood-borne viral infection, such as HIV, HBV, or HCV infection, was temporally associated with an occupational exposure), although a combination of acute and delayed post-exposure serologic evaluation is more commonly employed, at least among healthcare workers. Of note, preplacement examinations and tests should be recommended only for bona fide occupational needs, and the Americans with Disabilities Act as well as local laws may limit the collection of these data.

***6.3.2.2 Periodic Examinations*** Periodic examinations may be performed for purposes of (i) screening and surveillance of asymptomatic workers, (ii) early detection and treatment of work-related disease, (iii) determining the adequacy of workplace controls by detecting worker exposure, and (iv) general medical screening of conditions unrelated to specific occupational exposures, e.g. hypertension. Examples of periodic examinations include skin tests for tuberculosis and serologic tests to detect antibodies developed against specific agents known or suspected to be present in a workplace, e.g. measles, mumps, rubella, varicella, and HBV as well as the agents of tetanus, diphtheria, and pertussis (see Sections 6.3.2.1 and 6.3.3). Serologic or skin test conversions in these cases will usually not indicate the time of infection, other than that it occurred between performance of the current and previous tests. Serologic testing will also identify immunized workers whose antibody titers require boosting to remain protective, e.g. anti-rabies titer.

***6.3.2.3 Post-exposure Examinations*** Post-exposure examinations are medical evaluations following known or suspected acute exposure incidents. These exams take place in one of two time frames: (i) while a worker is within a disease's incubation period and asymptomatic (Table 6.2a) or (ii) after a worker has developed symptoms or has passed through the incubation period without developing them (Table 6.2b). The primary distinction between these exams is the focus on (i) establishing or reestablishing baseline test results and possible exclusion from the workplace during the incubation period and (ii) diagnosing and possibly treating infection that does occur.

***6.3.2.4 Return-to-Work Examinations*** Return-to-work examinations are medical evaluations required to be given before a worker with a communicable disease can resume work. Local or state regulations may require such exams for healthcare workers or food handlers (see also Section 6.3.1). For some intestinal infections, one or more negative stool cultures (e.g. for shigella or salmonella) may be required before a worker is given clearance. For other infectious diseases, documentation of treatment or passage of the communicable period may suffice. Healthcare workers infected with certain agents should be precluded from caring for extremely vulnerable patients. Where protective coverings (such as gloves for skin and nail infections) will effectively prevent an infected employee from transmitting infection, the worker may be permitted to provide patient care in selected circumstances.

***6.3.2.5 Chronically Infected Workers*** Sections 6.3.2.3 and 6.3.2.4 discussed post-exposure examinations and return-to-work criteria for certain workers acutely infected with communicable diseases. Concern about workers with chronic infections has been confined primarily to healthcare workers. All workers (especially healthcare, childcare,

**TABLE 6.2a Post-exposure Medical Evaluation: Worker Asymptomatic or Within Incubation Period**

| Step | Action |
|---|---|
| 1 | Document route and circumstances of exposure |
| 2 | Identify source of exposure (e.g. human, animal, environment), if known, and conduct appropriate testing, if possible |
| 3 | Establish or reestablish baseline tests, e.g. serologic antibody titers |
| 4 | Update history: past infections, immunizations, medical history, family history |
| 5 | Physical examination |
| 6 | Laboratory tests: based on suspected agent |
| | Blood – draw and store serum sample for baseline antibody testing |
| | Other – as indicated |
| 7 | Consider PEP or treatment, if appropriate (see Section 6.4.5) |
| 8 | Determine whether the worker needs to be excluded from the workplace during the incubation period |
| 9 | Counseling |

**TABLE 6.2b Post-exposure Medical Evaluation: Worker Symptomatic or Beyond Incubation Period**

| Step | Action |
|---|---|
| 1 | Update history: past infections, immunizations, medical history, family history |
| 2 | Symptoms: nature and pattern |
| 3 | Current illness: medical diagnosis; treatment, if any |
| 4 | Physical examination: vital signs; other, as appropriate |
| 5 | Laboratory tests: consult a laboratory early on specimens and specimen collection appropriate for the suspected agent |
| 6 | Consider PEP or treatment, if appropriate (see Section 6.4.5) |
| 7 | Counseling |

and emergency workers as well as those with public contact) should be aware that they may transmit communicable diseases to others. Healthcare workers performing exposure-prone procedures (EPPs) should know their HIV, HBV, and HCV status. EPPs are those with a risk that injury to a worker would result in exposure of a patient's open tissues to the blood of the worker, e.g. blind suturing (using a fingertip to feel for a needlepoint during surgical stitching). EPPs should be defined by the medical/surgical/dental organizations and institutions at which the procedures are performed (Holmberg et al. 2012).

Policies for dealing with chronically infected workers seek to protect the workers and their co-workers along with patients, other contacts, and the public. Programs generally do not actively attempt to identify infected workers, except perhaps as follow-up on exposure incidents. Instead, such programs rely on voluntary reporting by infected individuals. HIV-, HBV-, or HCV-positive persons in high-risk occupations may report their status to an appropriate individual or panel, who decide if work restrictions or reassignment is required. These decisions may be difficult to reach due to a lack of quantifiable data on patient risk and a lack of consensus on acceptable risks. Additionally, there is controversy over requiring informed consent from patients and the related issue of confidentiality for infected healthcare workers. Some countries have explicit restrictive regulations for specific conditions, such as healthcare workers who are HBV carriers and are positive for hepatitis B e antigen (HBeAg). In the United States, the Americans with Disabilities Act and other federal and state laws protect employees from discrimination based on their actual or perceived disease status, and any job restriction must meet legal requirements.

### 6.3.3 Laboratory Diagnosis of Infectious Diseases

A discussion of infectious disease diagnosis is beyond the scope of this text, and other thorough references are available (Bennett et al. 2015; Mandell et al. 2009; Wald and Stave 2001). The following are brief descriptions of the most common laboratory tests used specifically in the diagnosis of infection. Other diagnostic tests may indicate disease but are not specific for an infectious etiology. For example, pneumonia can be diagnosed by chest X-ray, but this condition has many infectious and noninfectious causes.

The *analytical sensitivity* of a test is the limit of detection of the assay and reflects the ability of the test methodology to identify a particular analyte or infectious agent when present in a sample. The *diagnostic sensitivity* of a test is defined as the frequency of a positive test in persons with an infection (even with perfect analytical sensitivity, an infectious agent may not be present in every specimen taken from an infected person, or there may be interfering substances that prevent the test from being positive). *Analytical specificity* reflects the ability of a test to identify only the target analyte (and thus to exclude closely related analytes or species). *Diagnostic specificity* is the frequency of a negative test in uninfected persons. The terms sensitive and specific are also used in other ways to describe tests (see below). The *predictive value* of a test may be a more useful criterion to rule an infection in or out. The predictive values of positive and negative tests are the probabilities that they accurately indicate that a person has or does not have an infection, respectively (Theel et al. 2015). This value is related to the underlying rate of infection in the population, which may not be known, particularly for subpopulations such as healthcare workers.

Clinical microbiology laboratories use a number of methods to detect infectious agents in human specimens, e.g. throat and wound swabs as well as blood, sputum, and urine samples. Laboratories use many of the same methods to detect, identify, and quantify microorganisms in environmental samples, e.g. air, surface, liquid, or bulk material samples (see Section 6.3.5.2). The primary techniques for microorganism detection and identification are (i) direct visualization by microscopic examination, (ii) isolation by culture, and increasingly (iii) molecular methods.

*Microscopic examination* is usually done on a portion of a specimen submitted for testing. Visual examination is rapid and suitable for organisms with distinctive physical features, e.g. fungal spores, protozoa, and helminths. However, the sensitivity of this method (i.e. its ability to detect small numbers of microorganisms) may be low (Khairnar et al. 2009; Lawrence et al. 2007; Uddin et al. 2013; Wiedbrauk 2015). Microorganisms in some samples can be concentrated by centrifugation or filtration to increase sensitivity. The specificity, or accuracy, of organism identification by microscopic examination is limited for many bacteria and yeasts without the use of stains (e.g. dyes and fluorescent antibody stains) or gene probes, which can substantially enhance the resolving power of light microscopy (Schmolze et al. 2011).

Many infectious agents can be grown (cultured) in the laboratory and subsequently identified using microscopy, bioassays, or immunoassays (Day et al. 2007; Payment 2007; Tanner 2007). Agent detection by *culture isolation* can be very sensitive, and subsequent identification of a cultured microorganism can be very precise. Unfortunately, test results may not be available until several days or weeks after sample collection. Culturing is routine for many clinical specimens and environmental samples when looking for bacterial or fungal pathogens. However, laboratory growth is not as often attempted to detect viruses, protozoa, or helminths. Any body fluid likely to contain a suspected infectious agent should be cultured. If a worker has a fever, a blood culture should be obtained as soon as possible. Some workplace-acquired organisms (e.g. *Brucella* spp., *Bordetella pertussis*, and *Leptospira* spp.) can usually be isolated only during the first days of illness. Only culture isolation requires that microorganisms be able to multiply under laboratory conditions for detection. The other diagnostic tests will generally detect both live and dead microorganisms as well as active and inactive viruses.

*Antibody tests* may be for a specific antibody class (e.g. IgM or IgG) or for total antibodies, not distinguishing among them (Theel et al. 2015). A *titer* is the highest serum dilution at which a test clearly detects an antibody. Dilutions are often done in twofold steps, e.g. 1:2, 1:4, 1:8, and so forth. A test may be considered positive only if it exceeds a specific titer (e.g. ≥1:64 or 1:128), the value varying from test to test. These cutpoints are needed because nonspecific low-level reactivity may occur; i.e. reactivity seen at low serum dilutions may disappear at higher dilutions.

A person who previously tested negative on a serologic examination who later tests positive is said to have *seroconverted*. A significant rise in a person's titer, usually a fourfold or greater increase, between two blood samples is evidence of recent infection. A *baseline* specimen (one collected before exposure) or an *acute* specimen (one collected ≤7 days since exposure or illness onset) should be tested at the same time as a *convalescent* specimen (one collected three to four weeks after exposure or illness onset). A positive test for specific IgM antibody, e.g. IgM anti-hepatitis A virus (IgM anti-HAV), is also indicative of recent infection. However, IgM antibodies to some agents may persist for several months. An IgG or total antibody titer that is positive acutely, but does not rise at least fourfold in a convalescent specimen, is consistent with an infection in the distant past (greater than three or four months). A single positive convalescent titer for IgG or total antibody that is very high (e.g. >1:1024) may reflect recent infection. However, the only information that a single positive test provides is that the person has been infected, but not when this occurred.

The diagnostic sensitivity and specificity of antibody tests vary among infectious diseases and are generally quite high. Analytic tests that detect low antibody concentrations are described as sensitive. Specificity describes the degree to which an antibody binds a target antigen and not others. False-positive antibody tests may occur due to sample contamination or if an antibody cross reacts with a related antigen on another organism or tissue. Occasionally, medications or drugs can cause the formation of reactive antibodies, also leading to false-positive antibody reactions. Such cross-reactivity decreases the value of a test for diagnostic purposes.

A test's predictive value depends not only on test sensitivity and specificity but also on the likelihood (prior probability) that infection is present. A significant number of false-positive results are seen with some diagnostic tests used on extremely large numbers of persons, e.g. tests for HIV and HCV infection. *Confirmatory tests* are required for all positive screening tests with these diseases because of the nontrivial consequences of diagnosing them. Confirmatory tests are particularly important for individuals with a low risk of infection, e.g. volunteer blood donors. A positive initial HIV or HCV test in a person with no risk factors for infection with these agents is often a false positive and will not be confirmed with a more definitive test (in this situation, the predictive value of the positive result was low). However, a positive initial antibody test in a person with multiple risk factors for infection (e.g. injecting drug use, multiple sexual partners, or blood transfusions that predated screening for these viruses) is likely to be due to infection and generally will be confirmed (in this case, the predictive value was high).

There are a number of different methods to test for antibodies, including neutralization, agglutination, precipitation, complement fixation, immunofluorescence, and immunoassay (Theel et al. 2015). Although more than one method may be available for any agent, a clinical laboratory will probably use only one test and not always the most sensitive or specific one. The results of antibody tests performed with nonstandardized antigens may be difficult to interpret, e.g. tests for *Legionella* spp. other than *L. pneumophila* and *L. pneumophila* serogroups other than 1 to 6. Reference laboratories (e.g. commercial laboratories and local, state, or federal public health laboratories) often can confirm test results and may offer additional tests not ordinarily available.

*Antigen tests* can also be used to diagnose infection and are run similarly to tests for antibody. However, these procedures offer the advantage of testing directly for the presence of an infectious agent (as do microscopic examination and culture isolation) or a microbial toxin. Direct detection of antigen in clinical specimens is particularly effective for identification of respiratory viral infections and permits diagnosis within hours rather than days or weeks. Antigen tests are positive as long as an organism is present and, therefore, will be positive before most antibody tests. For example, antigen from *Legionella* spp. can be detected in an infected person's urine weeks before antibodies develop. Unfortunately, antigen tests may be relatively insensitive, the availability of immunoreagents is limited, and the tests do not distinguish viable from nonviable organisms.

There are an increasing number of *molecular methods* to detect the presence of infectious agents. By far, the most common technique is nucleic acid amplification through *polymerase chain reaction* (PCR), which mimics natural DNA replication (Nolte 2015). Double-stranded microbial DNA is repeatedly separated into single-stranded DNA templates using heat. The templates are combined with (i) primers that flank a target DNA sequence (one that is specific to the microorganism of interest), (ii) nucleotides (the building blocks needed to copy the target sequence), and (iii) DNA polymerase (an enzyme that facilitates the copying). Repeated separation and replication generate an exponential increase in the number of target sequences to amounts that are easily detectable. Similarly, reverse transcriptase polymerase chain reaction (RT-PCR) is used to convert RNA molecules into their complementary DNA (cDNA) sequences by reverse transcription enzymes, followed by the PCR amplification of the newly synthesized material. PCR and RT-PCR techniques may be able to detect fewer numbers of organisms than culture methods (i.e. they are more sensitive) and can often be performed more rapidly. In addition, PCR-based genomic sequence analysis has been used to detect pathogens that cannot currently be recovered by culture techniques. For example, RT-PCR was used in the initial identification of the viruses causing hantavirus pulmonary syndrome (HPS) and would be the mode of amplification used in the detection of other RNA viruses such as HIV and hepatitis. The risk of false-positive results (from sample contamination), false-negative results (due to interfering compounds in environmental samples), and the additional complexity and cost involved may limit the availability of PCR testing; however, recent FDA approval of multiplex PCR panels has encouraged and enabled their use in diagnostic laboratories (Fairfax and Salimnia 2013; Schreckenberger and McAdam 2015).

Past infection with some agents can be detected through skin testing. *Intradermal skin tests* detect a cell-mediated delayed hypersensitivity response to an antigen. To perform this test, a small amount of test solution is injected between layers of the skin (usually on the forearm) with a fine hypodermic needle. The injection site is marked and examined 48–72 hours later for evidence of a raised, hardened (indurated) area. The diameter of the raised area may be used to judge the strength of a person's response and to categorize the results; i.e. a person may be described as "skin test positive" or "skin test negative." Factors other than degree of cell-mediated immunity can affect a person's skin test response. *Anergy* describes the false-negative reaction of individuals unable to mount a cellular response due to immunosuppression. Such a condition can be the result of medication (corticosteroids or cancer chemotherapy), overwhelming infection, or advanced age. When suspected, anergy can be detected by demonstrating a person's inability to react to agents for which humans are universally responsive, e.g. mumps, trichophyton, or candida antigens. *Two-step testing* is a form of baseline skin testing used to identify a boosted skin test reaction from that of a new infection. The procedure involves performing a second skin test one to three weeks after an initial negative test. Boosting is seen in infected persons whose sensitivity has waned since initial exposure, but is raised and becomes detectable upon subsequent testing. Repeated skin tests are used to track the status of skin-test-negative workers and to identify recent infections. Skin testing does not lead to the development of antibodies in uninfected persons who are repeatedly tested. Skin tests are most often used for workers potentially exposed to tuberculosis but may also be useful to track exposure to the agents of coccidioidomycosis, paracoccidioidomycosis, histoplasmosis, leishmaniasis, and leprosy.

### 6.3.4 Infectious Disease Surveillance

Infectious disease surveillance involves the systematic collection of information from many different sources, its assessment, and taking prompt public health action based on the conclusions (APHA 2015; Singh 2009; Souza et al. 2010). Surveillance is used to measure the occurrence of illness as well as to identify changes in trends or distributions of cases (Dwyer et al. 2014). Infectious disease surveillance may be used to direct investigative and control (preventive) measures. The information collected may not necessarily benefit individual workers, in contrast to screening activities such as those described above. However, a group of workers may benefit from surveillance through resultant prevention of future disease or injury. The case study on a coccidioidomycosis outbreak among the cast and crew at a three-day outdoor television event illustrates the value of a DFR. Such reports are required in some states when a patient or physician suspects that an injury or illness is work related. The report identifies the employer, the occupation, and worksite, and it may contain physician notes and addenda. Such a system is simple, but not especially useful, flexible, or timely. Combining a state-wide database with medical records and partnering with local health departments can greatly improve surveillance.

Strategies for evaluating and surveying occupational infections may include (i) identifying the potential presence of infectious organisms; (ii) establishing workers' baseline (pre-exposure) status; (iii) identifying susceptible workers (i.e. those at risk because of an absence of immunity); (iv) identifying workers with conditions that may increase their risk of infection or disease; (v) diagnosing occupational infections in individual workers, characterizing epidemiologic features of infections (e.g. locations, times, persons, and activities involved), and verifying a work association; (vi) comparing current attack rates with usual baseline rates; and (vii) checking the effectiveness of existing control measures and identifying workplace operations and sites needing better controls. Surveillance for certain zoonotic infectious diseases may also be conducted in animals in workplaces. Animal surveillance programs may monitor animals that are part of a process (e.g. farm animals, animals at abattoirs, and laboratory animals) as well as those incidentally present in a workplace (e.g. wild rodents) (Morse et al. 2012; Nelson 2014).

***6.3.4.1 Infectious Disease Investigation and Reporting*** Individual workers may become infected in isolated events, or more than one worker may be involved in an episode. *Epidemic* refers to an increase, often sudden, in the number of cases of a disease above what is normally expected in that population in that area (CDC 2012). *Outbreak* carries the same definition of epidemic but is often used for a more limited geographic area. The expected case number is determined relative to (i) an endemic infection rate (i.e. the constant or usual prevalence for a given area, if applicable) or (ii) the baseline rate for a workplace or occupation. A single infection is usually not considered an outbreak.

The first recognized case in an initial outbreak investigation is termed the *index case*, i.e. the case that brought the outbreak to someone's attention. However, an investigation may uncover earlier infections; i.e. an index case may not have been the first or *primary case* in an outbreak (see the coccidioidomycosis case study as an example). There may be more than one primary case in common-source outbreaks, e.g. foodborne infections. Subsequent infections contracted by exposure to a primary case are termed *secondary* and *tertiary* cases and so forth. Subsequent infections with agents transmitted from person to person may be found among the co-workers, customers, and clients of cases along with family members and other contacts.

All US states and territories participate in a national morbidity reporting system. The CDC, together with the Council of State and Territorial Epidemiologists (CSTE), establishes a list of infectious diseases and related conditions reportable by physicians and other healthcare providers (Adams et al. 2017). The National Notifiable Diseases Surveillance System publishes data on the incidence of notifiable diseases (CDC, 2017d). Table 6.3 lists the infectious diseases notifiable in the United States in 2017.

***6.3.4.2 Notifiable Diseases and Case Definitions*** Notifiable diseases are those for which regular, frequent, and timely information regarding individual cases is considered necessary for prevention and control of the disease or condition (Adams et al. 2017). The authority to require case notification resides in respective state legislatures. State health departments receive these reports and provide the data to CDC. States vary in the following: how the authority is enumerated (e.g. by statute or regulation), conditions and diseases to be reported (states and local authorities can make additional diseases or conditions reportable), time frames for reporting, agencies receiving reports, persons required to report, and conditions under which reports are required. In many states, healthcare providers are required or encouraged to report diseases directly to the local rather than the state health department. This is because investigation and intervention are often provided at the local level. Increasingly, clinical laboratories report diagnosis of some diseases (e.g. tuberculosis) directly, often electronically, to a health department. Occupational infections may also be reportable to state or federal OSHA programs (see Section 6.3.1.3) and to facility licensing agencies, particularly in the case of healthcare facilities.

**TABLE 6.3 Nationally Notifiable Infectious Diseases in the United States – 2017[a] (Not All of the Following Diseases Are Seen as Work-Related Infections)**

| Diseases |
|---|
| Anthrax |
| Arboviral diseases, neuroinvasive and non-neuroinvasive |
| Babesiosis |
| Botulism |
| Brucellosis |
| Campylobacteriosis[6] |
| Chancroid |
| *Chlamydia trachomatis* infection[1] |
| Cholera |
| Coccidioidomycosis |
| Congenital syphilis |
| Cryptosporidiosis |
| Cyclosporiasis |
| Dengue virus infections |
| Diphtheria |
| Ehrlichiosis and anaplasmosis |
| Giardiasis |
| Gonorrhea[2] |
| *Haemophilus influenzae*, invasive disease |
| Hansen's disease |
| Hantavirus infection, non-hantavirus pulmonary syndrome |
| Hantavirus pulmonary syndrome |
| Hemolytic uremic syndrome, post-diarrheal |
| Hepatitis A, acute |
| Hepatitis B, acute |
| Hepatitis B, chronic |
| Hepatitis B, perinatal virus infection |
| Hepatitis C, acute |
| Hepatitis C, past or present[3] |
| HIV infection (AIDS has been reclassified as HIV stage III)[8] |
| Influenza-associated pediatric mortality |
| Invasive pneumococcal disease |
| Legionellosis |
| Leptospirosis |
| Listeriosis |
| Lyme disease[7] |
| Malaria |
| Measles |
| Meningococcal disease |
| Mumps |
| Novel influenza A virus infections |
| Pertussis[10] |
| Plague |
| Poliomyelitis, paralytic |
| Poliovirus infection, nonparalytic |
| Psittacosis |
| Q fever |
| Rabies, animal |
| Rabies, human |
| Rubella |
| Rubella, congenital syndrome |
| Salmonellosis[5] |
| Severe acute respiratory syndrome-associated coronavirus disease |
| Shiga toxin-producing *E. coli* |
| Shigellosis[9] |

(*Continued*)

**TABLE 6.3 (Continued)**

| Diseases |
|---|
| Smallpox |
| Spotted fever rickettsiosis |
| Streptococcal toxic shock syndrome |
| Syphilis[4] |
| Tetanus |
| Toxic shock syndrome (other than streptococcal) |
| Trichinellosis |
| Tuberculosis |
| Tularemia |
| Typhoid fever |
| Vancomycin-intermediate *S. aureus* and vancomycin-resistant *S. aureus* |
| Varicella |
| Varicella deaths |
| Vibriosis |
| Viral hemorrhagic fever |
| Yellow fever |
| Zika virus disease and Zika virus infection |

[a] The 10 most common infections noted by superscript ranking for the year 2015 (Adams et al. 2017).

*Case definitions* are sets of uniform criteria used to define a disease for public health surveillance, and they enable public health officials to classify and count cases consistently across reporting jurisdictions (CDC 2017d). A case of infectious disease is described as laboratory confirmed if one or more listed diagnostic tests are positive. Case definitions are included for some infectious conditions that are no longer considered nationally notifiable or that may become so in the future, i.e. amebiasis, aseptic meningitis, meningitis (other bacterial), genital herpes, genital warts, lymphogranuloma venereum, mucopurulent cervicitis, nongonococcal urethritis, pelvic inflammatory disease, and rheumatic fever.

A case may be described as clinically compatible if the syndrome was generally similar to a disease but could not be confirmed by a laboratory test. Case definitions for surveillance work are deliberately biased toward increased sensitivity (to identify all possible cases) perhaps at the expense of specificity, i.e. some false-positive reporting may occur. Therefore, surveillance case definitions should be used with caution to interpret an individual worker's diagnosis. However, surveillance data most often are used to monitor trends. Thus, a relatively high false-positive reporting rate is of little consequence if the bias remains stable over time.

### 6.3.5 Inspections and Environmental Testing

Infectious diseases are primarily monitored in workplaces through serological and skin testing of workers and disease surveillance rather than by environmental sampling for infectious agents. Investigators use various criteria to establish cause-and-effect relationships between workplace exposures and infections. Environmental testing may help establish some of these points and confirm or fail to confirm the work-relatedness of an infection. For example, (i) an infectious agent must be known, or strongly suspected, to be or have been present in a workplace; (ii) there must be a plausible means of worker exposure; and (iii) an infected worker must be shown, or strongly suspected, to have been exposed in the workplace.

A work association is easier to accept for a disease that workers are unlikely to encounter outside their jobs (e.g. brucellosis) than for infections that are fairly common in the community (e.g. influenza or the common cold) (29 CFR 1904 2001). On the other hand, detecting an infectious agent in a workplace by environmental sampling does not necessarily mean that workers are or were at risk. Rather, risk depends on several factors including the agent, the location and concentration found, and possible exposure routes. Rather than test for exposure to infectious agents, health and safety professionals may assess worker safety indirectly by auditing worker training, work practices, and worker supervision. Evaluation of control equipment (e.g. checks of containment devices, ventilation systems, and PPE performance) is used to evaluate possible exposures.

Healthcare professionals handle most infectious disease evaluations because they consist of medical examinations and clinical laboratory tests. However, health and safety professionals may need to understand the interpretation of medical findings to help healthcare personnel design environmental inspection and testing programs and implement appropriate controls. An example of a change in approach has developed around prevention of legionellosis.

Various public health and government agencies at the federal, state, and local levels as well as professional organizations have developed guidelines and regulations. While these guidelines are similar in recommending maintenance of building water systems, they differ in the population/institutions targeted, the extent of technical detail, and the support of monitoring water systems for levels of contamination (Parr et al. 2015). Environmental sampling for legionellae often finds the bacteria because they are frequently present in water supplies, often without causing apparent illness (Lucas 2015).

***6.3.5.1 Inspections*** In some workplaces, walkthrough inspections are conducted routinely to identify infectious agent reservoirs and dissemination routes, and some state plans and federal contracts require periodic workplace hazard inspections. However, in most workplaces, inspections take place following an exposure episode (e.g. a laboratory accident) or a report of a potentially work-related infection (e.g. a confirmed case of plague in a field researcher or utility worker; see the case study on plague in a wildlife biologist) (Wong et al. 2009). Initial walkthrough inspections in these cases are similar to inspections conducted to identify other workplace hazards or to evaluate indoor environmental quality (Chapter 8). During a walkthrough, health and safety professionals (i) identify potential reservoirs or sources of infectious agents as well as possible exposure routes and (ii) formulate plans for further investigation (including environmental testing if indicated) and for remedial actions on noted problems. Inspections may identify potentially hazardous situations that merit mitigation even if these situations have not been linked to worker infections. For example, rodent infestation, poor equipment maintenance, or improper waste handling should be noted even if not related to the issues that prompted the inspection. Management and employee representatives should be included in these inspections and subsequent deliberations.

***6.3.5.2 Environmental and Personal Testing*** Sections 6.1.1–6.1.7 and 6.2.3 have described workplace monitoring (through medical surveillance) for the *effects* of infectious agents to identify their presence and measure their impact. This section briefly describes sampling for the *presence* of infectious agents. Air, surface, liquid, and bulk material sampling to detect, quantify, and identify infectious agents in workplaces is a special case of environmental testing. Environmental sampling in hospitals and laboratories (where infectious agents are known to be present) formerly was routine. However, even in these facilities such testing has been almost completely abandoned, except for some agents, e.g. legionellae following an outbreak (Stout et al. 2007). Environmental sampling is relatively rare because there may not be commercially available validated assays and because testing is often expensive and time consuming when conducted correctly, e.g. many samples may be needed to detect infectious agents that are present only sporadically or in low concentrations. In addition, certain organisms can be expected to be found in some types of samples, e.g. *Aspergillus fumigatus* at a compost operation or *Coccidioides* species in soil in endemic regions of the Southwest (Das et al. 2012) (see also the case study on a coccidioidomycosis outbreak among the cast and crew at an outdoor television event). Therefore, air concentration measurements or other test results cannot be interpreted without a database of expected concentrations and information on infectious doses.

Investigators can often obtain the information they need by means other than environmental sampling. For example, it may be more productive to perform periodic visual inspections, measure water temperatures and biocide concentrations, and initiate an employee medical surveillance program than it is to sample for infectious agents in a workplace (McCoy and Rosenblatt 2015). Environmental testing most often is undertaken to (i) identify or test potential sources or reservoirs of pathogenic microorganisms (e.g. raw materials, manufacturing supplies, or cooling fluids), (ii) determine the ability of certain organisms to survive in a given environment, (iii) test the efficacy of a cleaning or disinfecting method, or (iv) check hypotheses generated by epidemiologic or medical surveillance.

Environmental sampling for infectious agents generally focuses on one or a few microorganisms, which simplifies the selection of suitable collection and detection techniques. Culture methods may be used following detection by other analyses to confirm that agents are viable or active (able to multiply) and, therefore, should be culturable (able to grow on laboratory media) (Galvin et al. 2012). It is also necessary to culture organisms to compare environmental and clinical (worker or patient) isolates to help determine if an infection was work related or if a cluster of infections originated from one source. Detection methods other than culture are more convenient or appropriate for some organisms, e.g. immunologic methods or methods based on detection of nucleic acids (see Section 6.3.3).

Sample collection methods should be selected in consultation with an experienced microbiologist who can anticipate how sample collection, transport, and storage may affect test outcomes. For example, *air samples* to identify airborne infectious agents are generally collected directly by impaction onto culture media (agar plates) or by impingement into a suitable liquid collection medium (Grinshpun et al. 2015; Macher et al. 2012). Filters may be used less often because infectious agent isolation may be low from filters due to drying and loss of cell viability. However, filters may be compatible with assays not based on organism culturability. Furthermore, gelatin membrane filters are available that have produced accurate and reliable results in the collection and detection of viruses. Air sampling has proven useful in hospitals to test the effectiveness of physical barriers separating construction sites from patient areas (Chang et al. 2014) and to determine when patient rooms closed for construction were sufficiently clean to return patients.

*Surfaces* may be sampled by washing a measured area with a swab wetted with sterile liquid or by using special contact plates containing agar medium. Surfaces occasionally tested for the presence of infectious agents include workbenches, instruments and tools, and human and animal skin. *Personal samples* (specimens from the hands, skin, nose, or other body sites) are occasionally collected from healthcare workers to identify the person or persons who could have transmitted an infectious agent to a patient. However, this type of testing should only be done when epidemiologic data links certain workers with infections, although routine screening may be considered for pathogens where there is reason to believe that workers are sources, e.g. MRSA (Dulon et al. 2014).

*Liquid* and *bulk material samples* from potential sources of airborne infectious agents (e.g. cooling tower water, metalworking fluid, animal droppings, or soil) are often more useful than air samples because the agents are concentrated in the source material, increasing the chances of detecting them. Such samples are also appropriate for identifying sources of infectious agents transmitted by contact (e.g. amebae in eyewash stations) and by ingestion (e.g. foodborne pathogens in consumable items).

Microbiologists experienced in handling environmental samples can advise health and safety professionals about appropriate sample collection methods and can choose suitable laboratory techniques to obtain the information required from the samples. All samples should be protected from contamination and extreme temperatures to ensure that they are still representative of the original source when they reach a laboratory. Other texts discuss environmental microbial sampling outdoors as well as in laboratories, industrial workplaces, animal houses, and problem buildings (Grinshpun et al. 2015; Macher et al. 2012; Mohr 2011; Reponen et al. 2011).

## 6.4 INFECTIOUS DISEASE PREVENTION

Prevention of infectious diseases in the workplace can decrease absenteeism and the costs associated with disability, sick leave, and health insurance even if the primary source of infection is non-occupational. Infectious disease prevention should be part of a facility's health and safety program and begins with identifying possible infectious agents along with their reservoirs, modes of transmission, and risk factors. The American Public Health Association publishes a pocket reference, *Control of Communicable Diseases Manual*, that lists infectious diseases alphabetically and serves as a comprehensive but concise general reference (APHA 2015). The manual provides information in each of the following categories: clinical features, causative agent(s), diagnosis, occurrence, reservoir(s), incubation period, transmission, risk groups, prevention, management of patient, management of contacts and the immediate environment, and special considerations.

Some of the primary tools for preventing work-related infections include immunization, standard (universal) precautions for blood and body fluids, and good hygiene, e.g. frequent handwashing, wound covering, and proper food storage and handling. Prevention can be approached through the three classic strategies shown in Table 6.4, i.e. primary, secondary, and tertiary prevention. Respectively, these aim to prevent exposure, intervene after exposure or when signs or symptoms of infection are first detected, and limit the consequences of clinical illness once it occurs. Occupational health and safety professionals can use their training in recognizing sources, pathways, and receptors and the approach outlined in Section 6.2.4 to identify where and how transmission may be prevented or interrupted. Infectious agents may move from sources (e.g. humans, animals, or the environment) to workers and back again. Biosafety practices in laboratories and healthcare settings are often designed to protect both workers and the materials they handle, e.g. laboratory cultures and patients.

### 6.4.1 Laboratories

Microorganisms are not equally hazardous, and the CDC Office of Biosafety has classified agents on the basis of hazard (USDHHS 2009). Agents in microbiology and biomedical laboratories are handled at one of four *biosafety levels* (BSLs). The necessary BSL is based on the infectivity, severity of disease, transmissibility, and the nature of the work being conducted (USDHHS 2009). At the extremes are BSL-1 agents (the least hazardous with the fewest restrictions on their handling) and BSL-4 agents (the most hazardous, which are only handled in high-containment facilities). Decisions on containment levels for laboratory work at BSL-1 through BSL-3 can be made at the institutional level. Very few facilities have BSL-4 capability, and those that do need to employ biosafety personnel with expertise in this level of control (Jahrling et al. 2009).

The BSL classification system applies primarily to work in laboratories. Transmission modes may differ in laboratory settings (compared with other workplaces), and higher organism concentrations may be encountered there.

**TABLE 6.4 Infectious Disease Prevention Strategies**

| Strategy | Definition | Example |
|---|---|---|
| Primary prevention | Prevent exposure and reduce the susceptibility of potentially exposed workers | Preplacement examinations and immunization of susceptible workers<br>Establishment of policies on safe work practices and periodic evaluation of health and safety programs with employee input<br>Worker education and training<br>Engineering controls:<br>Substitute safer materials or processes – choose uncontaminated materials or disinfect materials before handling<br>Isolate or enclose sources or processes, including early identification and placement of potentially infectious patients<br>Ventilation – ventilated enclosures, such as biosafety cabinets or airborne infection isolation rooms, and local exhaust and dilution ventilation<br>Inspections and environmental monitoring<br>Personal protective equipment – hand protection, skin and clothing protection, eye and face protection, and respiratory protection |
| Secondary prevention | Intervene post-exposure, post-injury, or when signs or symptoms of infection are detected | Periodic medical examinations and surveillance<br>Post-exposure or post-injury prophylaxis or treatment |
| Tertiary prevention | Limit consequences of clinical illness once it occurs | Medical treatment for infected workers<br>Worker restrictions or removal to prevent secondary cases |

Nevertheless, the recommended BSLs include the relative hazards of infectious agents. Health and safety professionals can use an agent's classification or rating as a starting point to assess its hazard. For each BSL, recommendations are given for facility design, safety equipment, and laboratory practices and techniques. Determining an agent's BSL can help health and safety professionals identify appropriate precautions (i) to prevent routine exposure (e.g. the need to avoid or contain aerosols, to prevent ingestion, to prevent skin or mucous membrane exposure, or to protect workers from arthropod vectors) and (ii) to handle emergency spills or releases. The critical role of following recommended laboratory practices and techniques to prevent workplace-related infections is illustrated by the case study of the fatal outcome in an occupationally acquired case of *N. meningitidis* infection.

The Centre for Biosecurity of the Public Health Agency of Canada (Ottawa, Ontario) has prepared *Pathogen Safety Data Sheets* (PSDSs) for over 175 human pathogens as educational and informational resources for laboratory personnel (PHAC 2017). Each PSDS provides basic facts about an agent and information on the following: (i) hazard identification, (ii) dissemination, (iii) stability and viability, (iv) first aid and medical, (v) laboratory hazards, (vi) exposure controls and personal protection, (vii) handling and storage, and (viii) regulatory authorities.

### 6.4.2 Healthcare

Infection control in healthcare settings involves identifying the type of contact that workers may have with potentially infectious body substances (e.g. blood and body fluids) and contaminated materials (e.g. wound dressings and linens). In 2007, hospital precautions to protect both patients and workers from infectious agent transmission were revised (Siegel et al. 2017).

Compliance with standard and transmission-based precautions as well as the OSHA Bloodborne Pathogens Standard (Section 6.3.1.3) will reduce workers' risk of exposure to blood-borne agents (see Section 6.1.1). The effective application of specific precautions to patients and clients infected with organisms transmitted by modes other than blood should also reduce workers' risks. Guidelines for specific healthcare settings are also available, e.g. dentistry (Bebermeyer et al. 2005; Boyce and Mull 2008; Cleveland et al. 2009; Kohn et al. 2003).

### 6.4.3 Immunizations

Immunization against infectious agents should be provided to workers when such protection is available, safe, and effective and the benefits clearly exceed the risks, e.g. side effects such as local or systemic reactions. Immunization generally is used in addition to engineering controls and PPE. Immunity acquired as a response to vaccination or natural infection is called *active* immunity. That acquired by inoculation with sera containing specific protective antibodies is called *passive* immunity (see Section 6.4.5). Active immunity may last from several years to a lifetime, whereas passive immunity may last for only a few days or months.

The Advisory Committee on Immunization Practices (ACIP) provides expert advice to the Secretary of the Department of Health and Human Services, the Assistant Secretary for Health, and the CDC on the most effective means to prevent vaccine-preventable diseases and to increase the safe usage of vaccines and related biological products (WHO 2017b) and includes specific recommendations for laboratory personnel. In addition to CDC's *Yellow Book* for international travelers and WHO's *Blue Book* on medical wastes, CDC publishes a *Pink Book* entitled *Epidemiology and Prevention of Vaccine-Preventable Diseases* (CDC 2015). Other agencies also provide recommendations on immunizations appropriate for healthcare workers as well as healthy and immunodeficient adults in the United States and its territories (Schillie et al. 2013; USDHHS 2009). The WHO may have additional recommendations for regions other than North America (see Section 6.1.5).

Making vaccinations easily available at the workplace can help to increase vaccination rates. One of the reasons for low compliance with recommended immunizations among workers is their fear of adverse events. In fact, vaccine safety has improved in recent years, and the occurrence of serious side effects for the most widely recommended vaccines is quite low (CDC 2015). Possible contraindications to immunization include the taking of immunosuppressive agents, coincidental severe illness, and pregnancy. Employers should maintain complete records of immunoprophylaxis that workers receive on the basis of occupational requirements or recommendations. The importance of identifying the need for appropriate immunizations for workers is highlighted in the case study of *N. meningitidis* infection acquired by a researcher on the job.

### 6.4.4 Substitution and Engineering Controls

Substitution and engineering controls are preferred over PPE and administrative controls to prevent exposure to infectious agents just as these control measures are preferred to control other workplace hazards (see also Chapter 18). For example, legionellosis prevention is often addressed by controlling bacterial multiplication in water systems and reducing aerosol exposures. Bacterial multiplication can be minimized by proper water temperature control, equipment cleaning and maintenance, and appropriate biocide use. Aerosol exposures can be reduced by minimizing mist generation and dispersal.

Some workplaces can reduce exposures to pathogens by establishing clear policies that encourage employees as well as patrons and clients to stay home when they are ill. However, there are many types of workplaces, such as healthcare and correctional settings, that by their nature expose employees to persons with communicable diseases. In such establishments, it is important to use a system of administrative and engineering controls that reduce exposures. Similarly, in veterinary practices, zoos, and some farms, employees can have contact with sick animals that may transmit zoonotic diseases. See Section 6.4.4 for information on transmission-based precautions.

***6.4.4.1 Substitution: Changing the Material*** Whenever possible, potentially infectious materials that are part of a work process should be treated before workers handle the items. An example of such substitution is the use of attenuated or inactivated strains of microorganisms for certain research purposes (Sinclair et al. 2012). On occasion, materials distributed as inactivated have contained a virulent or wild-type pathogen (Kaiser 2007; USDoD 2015). Therefore, prior to reducing biosafety precautions, it is important that laboratories have procedures in place to verify that modified pathogens are not capable of replication or causing disease. Plant and animal materials used in textile production and in animal-product manufacturing also should be selected carefully or disinfected before use (as appropriate for the infectious agents that may be present). Likewise, some medical and laboratory wastes should be sterilized before transport and disposal (see Section 6.3.1.4).

Antimicrobial pesticide manufacturers test their products using standard methods of the Association of Official Analytical Chemists (AOAC) International and the American Society for Testing and Materials (ASTM). This standardized laboratory testing requires specified contact times and temperatures, the use of designated strains of microorganisms, and preparation of the use concentration of a biocide as specified on the manufacturer's label. All sanitizers, disinfectants, sterilants, fungicides, microbiological water purifiers, tuberculocides, and virucides (viricides) must be registered with EPA before they can be marketed. However, the agency does not verify a manufacturer's claims about a product. The EPA maintains lists of registered disinfectants by category from sterilizers (List A) to avian (bird) flu disinfectants (List M) (USEPA 2017b).

*Sanitization* implies achieving a significant reduction in the number of vegetative environmental microorganisms of public health importance. *Disinfection* implies elimination of many, but not necessarily all, microbial forms (e.g. bacterial spores) (USDHHS 2009). *Sterilization* implies complete elimination or destruction of all microbial life and is the surest means of rendering material noninfectious. Sterility can be achieved by incineration or steam or ethylene oxide autoclaving. Waste materials that may warrant or require special processing or disposal because they may carry a risk of infectious disease transmission include (i) laboratory cultures and stocks, (ii) human and animal pathological wastes (e.g. tissues, organs, carcasses, and body parts; body fluids and their containers; and discarded bedding and materials saturated with body fluids other than urine), (iii) blood and blood products, and (iv) sharps such as needles and scalpel blades (USDHHS 2009), see also Section 6.3.1.4.

***6.4.4.2 Containing Infectious Agents*** Successful *containment* of infectious agents can simplify worker protection. Methods designed to prevent contact spread of infectious agents from potential sources or reservoirs include the sterile techniques used in laboratories and healthcare, proper food handling, and safe practices recommended for childcare, personal service work, and animal handling. Aerosolization of contaminated materials should also be minimized, e.g. the spraying of water and other liquids should be avoided, dust generation should be suppressed, fluids should be handled to avoid drips and splashes, and people should cover their noses and mouths when coughing and sneezing. Operations for which it is difficult to prevent aerosol generation should be performed within *primary containment* devices such as the biological safety cabinets and glove boxes used in laboratories (Kraus and Mirazimi 2014). Failure to use a primary containment device such as a biological safety cabinet was a key oversight illustrated in the meningococcal meningitis case study described in this chapter. Other examples of primary containment are negative-pressure booths (used in healthcare settings for sputum collection and in machine shops to enclose cutting equipment that requires metalworking fluids) and isolation rooms (used in hospitals to house potentially infectious patients) (Siegel et al. 2017). Primary barriers protect workers in the immediate area of aerosol generation. These containment devices are ventilated and serve as source controls. *Secondary containment* provides another level of protection, principally for persons outside the immediate area or facility. The room in which a primary containment device is located and the building surrounding a primary barrier are examples of areas that can provide secondary containment.

***6.4.4.3 Ventilation*** Ventilation designs used to control infectious aerosols are similar to, and modeled on, ventilation designs used to control particulate matter. Substitution, process change, and source isolation/or enclosure are preferred methods to control exposure, but ventilation is often necessary to supplement, or increase the effectiveness of, these measures. Source control to limit contaminant dispersal and minimize worker exposure can best be achieved through local exhaust ventilation (e.g. a hood, duct, air cleaner, and fan) along with isolation or enclosure (see Section 6.4.4.2). General ventilation is used to dilute and remove contaminated air, control airflow patterns within rooms, and control the direction of air throughout a facility (Sehulster and Chinn 2003).

Dilution ventilation can reduce the concentration of infectious aerosols that cannot be controlled at their sources, e.g. because a source cannot be readily identified, is mobile, or is too large to enclose. A study of office buildings found an association between higher carbon dioxide concentration (as a surrogate for lower outside air ventilation rate) and circulating rhinovirus (Myatt et al. 2004). Stand-alone or in-room air cleaners, which return air to a space, clean the already conditioned room air. Portable air cleaners and ceiling- and wall-mounted units have been studied as economical means of reducing aerosol concentration without additional heating or cooling costs (ASHRAE 2015; Siegel 2016). This research and that with respirators confirm that filters remove infectious aerosols as efficiently as they remove other particles with equivalent aerodynamic diameters (Mostofi et al. 2010), but their effectiveness depends on maintenance and adequate air circulation.

Another means of cleaning the air (at least in terms of infectiousness) is the use of ultraviolet germicidal irradiation (UVGI or UV "light"). UVGI has a long history of use in laboratories and hospitals for surface and air disinfection within biosafety cabinets. Specially designed UV lamps can be used to disinfect the air near the ceiling in occupied rooms (Hyttinen et al. 2011), and UVGI use to prevent tuberculosis transmission has been encouraged (Reed 2010; Sehulster and Chinn 2003). The effectiveness of upper-room UVGI is still being evaluated, but may offer an appropriate means of protection against airborne infectious diseases provided that worker eye and skin exposures do not exceed safe limits (Mphaphlele et al. 2015; Nardell et al. 2008).

Specialized ventilation systems have been developed for containing aerosols generated by people with "airborne" infectious diseases, such as tuberculosis and measles, and for aerosols generated by certain procedures, such as sputum induction and bronchoscopy (Hyttinen et al. 2011) (see also Section 6.2.2). Various state building codes incorporate guidelines published by the Facility Guidelines Institute; the American Society of Heating, Refrigerating and Air-Conditioning Engineers; the American Institute of Architects; and the CDC for airborne infection isolation rooms. The California ATD Standard also incorporates many of these recommendations (California Department of Industrial Relations 2009; Gold 2015). These systems typically require an increased general ventilation rate (12 air changes per hour [ACH] exhausted to the outside or through high-efficiency particulate air [HEPA] filtration, or for older construction, 6 ACH with an additional 6 ACH achieved through air cleaning technology) and a pressure differential that ensures directional airflow into the isolation room. Anterooms may also be provided that create a buffer zone, help maintain negative pressure, and provide a staging area for PPE donning and doffing. Other types of engineering controls include ventilated booths (such as sputum induction booths) and isolation tents, all of which are intended to enclose the patient and establish a type of local exhaust ventilation (Jensen et al. 2005; Nivin et al. 2002).

### 6.4.5 Education, Training, and Administrative Controls

Among the administrative controls used to prevent occupational infections are education of workers about infectious diseases and training in good work practices. Additional administrative controls include providing the necessary means and motivation to work safely, supervising workers to ensure implementation of proper practices, and institution of appropriate programs for medical evaluation, job placement, and infectious disease surveillance. One of the means by which employers ensure that workers are aware of biohazards is compliance with the requirement to post warnings at entrances to areas where biohazards are handled. These signs inform workers and visitors of biological hazards and outline routine and emergency precautions required in the area. Section 6.3.1.3 describes recording and reporting requirements, other sanitary and safety regulations, and requirements related to PPE and blood-borne pathogens.

Some OSHA standards mandate workplace training, e.g. the blood-borne pathogen and respiratory protection standards (see Section 6.3.1.3). Standards may specify training content, frequency, recordkeeping requirements, and trainer qualifications. Worker education about infectious diseases should include a description of the biology of the infectious agents that workers may encounter, personal characteristics that may place certain workers at increased risk, and identification of symptoms that may indicate exposure. Job activities and applicable regulations should be reviewed with workers, and potential sources of infectious materials and exposure routes should be discussed. Acceptable work practices should be described clearly, and proper use of process and control equipment should be explained and demonstrated. Regular reviews and updates should be scheduled, as appropriate or required. Written injury and illness prevention programs can help management and staff understand their responsibilities. All personnel must be properly trained to carry out their assignments, know how to implement emergency procedures in case of accidents, and be motivated to work safely.

### 6.4.6 Personal Protective Equipment

PPE use to prevent infectious diseases is a last line of defense when risks cannot be sufficiently reduced through engineering and other controls (Humphreys 2007) (see also Chapters 15 and 16; and Section 6.3.1). Table 6.5 outlines the types of PPE used to protect workers against infectious agents, i.e. hand protection, skin and clothing protection, eye and face protection, and respiratory protection. Appropriate precautions may reduce the need for personal protection and should be used to prevent animal bites and scratches, arthropod bites and stings, and injury from sharp instruments and needles (see Section 6.3.1).

Concerns regarding contact exposures to blood and other potentially infectious bodily fluids have led to the development of test methods to evaluate penetration of fluids that are more similar to blood and bodily fluids than water in physical aspects, such as surface tension. In 1997 and 1998, ASTM first published test methods for penetration of materials used in PPE by synthetic blood (F1670) and a bacteriophage virus (F1671). These methods have been updated in recent years (ASTM F1670/F1670M-17a 2017; ASTM F1671/F1671M-13 2013; ASTM F1819-07 2013), and ISO has similar methods (ISO 16603 and 16604) (ISO 2004a, b, c).

Different types of medical gloves are regulated in the United States by the FDA. Submissions for new glove models include requirements for a water leak test, which is published as a federal regulation (21 CFR 800.20 1990). Existing glove models, and models that were approved through the abbreviated 510(k) clearance, may not have to submit these test results. It is prudent to request test results from manufacturers prior to purchasing any protective equipment.

The American National Standards Institute (ANSI) and the Association for the Advancement of Medical Instrumentation (AAMI) have adopted a standard for surgical and isolation gowns (ANSI/AAMI 2012). This standard defines four levels of protection for materials, based on different test methods, which range from some resistance to water spray (level 1) to blood and viral penetration resistance (level 4). Chemical-resistant coveralls may also be tested against the requirements of ASTM F1670 and F1671 (ASTM F1671/F1671M-13 2013; ASTM F1819-07 2013). It should be noted that not all areas of a surgical gown are required to meet the fluid resistance specifications of a class.

In choosing PPE to protect the skin, the construction of the garment as well as the protective ability of the materials must be considered. A garment that gaps or falls open or that has unprotected seams or zippers may allow the penetration of infectious fluids. Where there is a serious fluid exposure hazard to a high-consequence pathogen, such as occurs with Ebola patients, high levels of penetration resistance along with complete body coverage is necessary. Choice of PPE must also include effective donning, doffing, and decontamination procedures (Ringen et al. 2015).

**TABLE 6.5 Personal Protective Equipment**

| Type | Function | Examples |
|---|---|---|
| Hand protection | To prevent contact with infectious materials | Medical gloves: laboratory work, patient care<br>Utility gloves for cleaning and disinfecting must be compatible with chemical exposures |
| | To prevent cuts and bites | Puncture-resistant gloves such as leather or synthetic: animal handling (if contact transmission a concern, consider using glove with outer covering that can be decontaminated, or disposable outer glove)<br>Steel mesh: shellfish and animal handling |
| Skin and clothing protection | To prevent skin contact and transfer of infectious materials from workplace | Laboratory coats: patient care, laboratory work, and animal handling if sufficient to prevent infectious fluids from passing to an employee's skin, clothing, or underclothing<br>Operating gowns: human and veterinary surgery<br>Disposable impermeable coveralls: intense body fluid exposures, cleaning, contaminated waste removal, animal handling, animal trapping |
| Eye and face protection | To prevent exposure to infectious droplets, sprays, and splashes and to prevent eye and facial injuries | Goggles, safety glasses, and face shields: dental care, laboratory work, surgery and autopsy procedures, handling of live and dead animals, animal trapping, veterinary care |
| Respiratory protection | To prevent inhalation of infectious aerosols | Respirators: patient care, dental care, laboratory work, surgery and autopsy procedures, handling of live and dead animals, animal trapping, veterinary care |

While engineering controls should be used to protect employees from infectious aerosols, complete containment of the source (infected human or animal, or infectious materials) is often not achievable in patient care and other environments. Therefore, respirators must often be used to protect employees in the near field such as in a patient room, while patient enclosures (such as airborne infection isolation rooms) protect people outside the room. California has standards that determine minimum requirements for respirator use against infectious aerosols (California Department of Industrial Relations 2009). In other states, health and safety professionals must follow OSHA regulations on respirator selection and use (OSHA 2015), and they should follow relevant local guidelines.

*Respirators* are considered a less desirable means of controlling inhalation exposure to infectious aerosols because respirators only protect workers if the equipment is properly selected, fit tested, worn, and replaced (Colton 2012). Respirator selection must be based on the hazard(s) to which a worker is exposed as well as factors such as work rate, mobility, and work requirements, e.g. a need to communicate with others. Infectious aerosols require the use of either particle-filtering respirators or atmosphere-supplying respirators (Benson et al. 2013).

Respirators may also need to provide protection against vapors, gases, or oxygen deficiency that may result from procedures for depopulating (mass slaughtering) flocks of infected animals (e.g. avian influenza) or from disinfection or decomposition of organic matter.

There are no threshold limit values, permissible exposure limits, or other reference values on which to base selection of respiratory protection against airborne infectious agents. In addition, inhalation and deposition of a single infectious agent may be sufficient to initiate infection, and sources of infectious agents may be mobile (e.g. humans and animals) or unrecognized as potential sources (infectious agents have no warning properties). Therefore, selection of respiratory protection must be based on the hazard and the process. For example, a patient who is undergoing an aerosol-generating procedure, such as bronchoscopy or sputum induction, will emit more infectious particles than a sedentary patient. Therefore, higher levels of respiratory protection, such as a powered air-purifying respirator, may be recommended for aerosol-generating procedures. Similarly, the heat, humidity, and odors in a barn in which sick animals are housed or destroyed may make use of a disposable respirator, such as an N95 filtering facepiece respirator, unsustainable, and an elastomeric facepiece respirator with combination cartridges, including charcoal, may be a more appropriate choice.

Respirator use has been recommended to prevent exposure to the agents of tuberculosis, legionellosis, and SARS (Gamage et al. 2005; Jensen et al. 2005; OSHA 2015). Recently the CDC has recommended respirator use for exposure to aerosol-generating procedures for certain pathogens for which it does not generally recommend their use, such as the influenza and Ebola viruses (MacIntyre et al. 2014). In healthcare settings, respirator use has met some resistance and confusion (Bessesen et al. 2013) but is becoming better accepted (Bryce et al. 2008). Respirators remain an important means of worker protection from airborne infectious agents in some situations.

CDC has issued recommendations to reduce the risk of contracting HPS (Mills et al. 2002), and guidelines specifically to protect wildlife researchers are also available (Kelt and Hafner 2010). At the time of this writing, there has only been documentation of person-to-person transmission of the Andes hantavirus, which is present in Chile and Argentina (Martinez-Valdebenito et al. 2014). The causal agent for HPS in the United States is the Sin Nombre virus (SNV), which has not been associated with person-to-person transmission. Recommendations for protection from SNVs are similar to precautions recommended for workers with possible respiratory exposure to *Histoplasma capsulatum* (the agent of histoplasmosis) (Lenhart et al. 2004). For example, respirators should be used to reduce exposure to aerosols that may contain an agent, e.g. dust containing feces or urine from a host animal. These recommendations for PPE would also apply to high-risk exposures in other workplaces and are similar to requirements for work in high-containment microbiology laboratories (USDHHS 2009).

### 6.4.7 Post-exposure Prophylaxis

Exposure to infectious agents is almost unique among workplace hazards in offering opportunities for PEP, i.e. interventions to prevent or modify the course of infection following a known exposure. The blood-borne pathogen standard defines an "exposure incident" as a specific eye, mouth, other mucous membrane, non-intact skin, or parenteral contact with blood or other potentially infectious materials that results from the performance of an employee's duties (USDOL, 29 CFR 1910.1030). The most up-to-date information on PEP for exposures to specific infectious agents (e.g. passive immunization, vaccination, administration of antimicrobial agents, and prompt and thorough wound treatment) is available from the CDC and other sources (CDC 2017b; PHAC 2017).

A PEP regimen may involve passive immunization through the provision of immune globulins (IG) (see Section 6.4.1). *Passive immunization* involves injection of a serum preparation containing antibodies (globulins) that confers temporary protection from antigenically related agents (Young and Cripps 2013). In the United States, all IG are prepared from human serum except botulism and diphtheria globulins, which are prepared from equine (horse) serum and may cause serum sickness, an immune reaction to horse protein (CDC 2015). In the cases of hepatitis A (for institutional and childcare workers) and measles (for healthcare workers), pooled human serum contains sufficient antibodies to confer protection, i.e. IG or ISG. Human sera for PEP against HBV, VZV, the rabies virus, and *C. tetani* are all hyperimmune preparations, i.e. they are collected from individuals whose immunity has been boosted to increase the concentration of antibodies against a specific agent. Antibodies administered passively disappear over a matter of weeks to months. In the case of viruses with long incubation periods (e.g. HBV and the rabies virus), vaccine (*active immunization*) may be administered at the same time as IG to ensure adequate protection for the duration of an infectious diseases' incubation period (Young and Cripps 2013).

PEP other than IG and vaccines may be available for some infectious organisms through administration, shortly after exposure, of *antimicrobial agents*, e.g. antibiotics or antiviral agents. The risk of adverse effects from antimicrobial agents (and often their cost) must be balanced against their potential benefits. Antimicrobial PEP is available for exposure to HIV, e.g. healthcare workers, following body fluid exposure, and medical waste handlers (Benn and Fisher 2008; Calfee 2006; Chalupka 2013; Chartier et al. 2014). The decision to administer antimicrobial PEP after potential exposure to HIV and the choices of antiretroviral regimens are difficult and complex, but seven years' experience with the 2005 Public Health Service guidelines has led to clearer recommendations and guidance (Kuhar et al. 2013).

Post-exposure antimicrobial prophylaxis is recommended for healthcare workers and close contacts of persons with meningococcal disease, due to the potentially life-threatening effects of meningococcal infection and the relative simplicity of prophylaxis (see the case study on fatal meningococcal infection in a laboratory worker). Theoretically, antimicrobial agents could be used prophylactically after any exposure to an infectious agent. However, for diseases that can be treated effectively at illness onset, there is little benefit in treating many exposed workers to prevent illness in some fraction thereof. In contrast, antimicrobial PEP may be indicated for diseases that often follow relapsing courses with severe consequences, e.g. brucellosis. Likewise, antimicrobial PEP may be appropriate during outbreaks in which many secondary cases may occur and extensive time would consequently be lost from work, e.g. mycoplasma pneumonia and possibly shigellosis and scabies. Even in such cases, one severe adverse reaction to an antimicrobial agent may outweigh the entire potential benefit of PEP, i.e. the intervention caused more harm than it prevented. Infectious disease consultants, occupational and environmental medicine physicians, and local and state health departments can provide information and assist with decisions on providing PEP in these situations.

All PEP must be administered soon after exposure to prevent infection. The maximum recommended time for IG delivery varies by infectious agent. For example, the treatment window is known to be 96 hours for VZV, up to one week for HBV, and two weeks for HAV (CDC 2017b; PHAC 2017). Such protection is not absolute (e.g. efficacy may be only 90–95% if PEP is administered immediately after exposure), and protection decreases incrementally with time. Some agents may still cause infection even if PEP is provided. However, the infection may be reduced to an inapparent one or cause only a mild illness. Unfortunately, ineffective PEP may prolong a disease's incubation period (Schillie et al. 2013). In such a case, an employee may have resumed work before illness began due to an atypically long incubation period. This uncertainty may complicate the decision of when to allow an exposed worker to return to work.

### 6.4.8 Workers at Increased Risk from Infectious Diseases

A number of characteristics of a worker as host, noted previously, may affect the risk of becoming infected as well as the likelihood of infection if exposed or the severity of any disease that does occur. These factors include age, alcoholism, coexisting diseases (infectious and noninfectious), abnormalities in the skin or respiratory tract that provide an entry site for pathogens, nutritional status, and immune status, e.g. immunity (from infection or immunization) or lack thereof, immunodeficiency, and pregnancy. Persons without functioning spleens are at increased risk of infection from *Babesia* spp. (protozoa transmitted by tick bite) and encapsulated bacteria (such as streptococci) that are otherwise of limited occupational significance (Semel et al. 2009). Susceptibility to tuberculosis disease is greater for immunodeficient persons (e.g. those with HIV infection or other immunosuppression), persons who are underweight or undernourished, persons with diabetes mellitus and debilitating diseases (e.g. chronic renal failure, some forms of cancer, silicosis, or gastrectomy), and substance abusers (e.g. drug or alcohol abusers) (APHA 2015).

***6.4.8.1 The Immunodeficient Worker*** Causes of immunodeficiency or immunosuppression include HIV infection, organ transplantation (due to administration of immunosuppressive drugs to prevent organ rejection), high-dose chronic corticosteroid therapy, malignant disease (e.g. cancer, due to the disease process and administration of immunosuppressive chemotherapy or radiation), and congenital (inherited) immunodeficiency diseases. Immunodeficiencies can be antibody mediated or cell mediated and be primary or secondary (Duraisingham et al. 2014; Picard et al. 2015).

*Primary immunodeficiencies* are uncommon and usually so severe as to preclude working if a victim survives to adulthood. In most cases, a deficiency mild enough to allow a person to be otherwise qualified to work will not place the person at increased risk of infection on the job. Nevertheless, health and safety professionals should make available to all workers information on infectious agents potentially present in a workplace. Immunocompromised workers can then share this information with their personal physicians and decide if the work poses an unreasonable risk of infection. In rare cases, it may be advisable for workers to avoid specific occupations, jobs, or job assignments. For example, chronic granulomatous disease is an inherited disorder of phagocytes (part of the cell-mediated immune process). This condition is characterized by an increased risk for bacterial and fungal infections, e.g. aspergillus pneumonia related to exposure while shoveling moldy wood chips (Seger 2008). Physicians caring for such patients should recognize occupations with a potential for exposure to high concentrations of opportunistic pathogens and advise their patients accordingly.

*Secondary immunodeficiencies* (acquired immunodeficiencies) can result from certain malignancies as well as intentionally and secondarily immunosuppressive therapies. Medical conditions such as cancers and their treatment generally produce significant immunodeficiency for a limited time. During this period, individuals are either too ill to work or are advised to restrict their activities to limit exposure to infectious agents. HIV infection is a principal cause of acquired deficiency in cell-mediated immunity among workers (Chinen and Shearer 2010). The risk of opportunistic infection is highest in those who are severely immunosuppressed. However, persons affected to this degree may be able to continue working in some occupations. Many opportunistic pathogens are ubiquitous, thereby limiting immunosuppressed workers' options for avoidance. Additionally, immunocompromised healthcare, childcare, and laboratory workers as well as those with animal contact face particular potential risks.

Susceptible workers in healthcare settings may be exposed to aerosols containing agents that are difficult to control completely, e.g. *M. tuberculosis* and VZV (the latter from patients with chickenpox or disseminated herpes zoster/shingles) (see Section 6.1.1). Bacterial and parasitic enteric pathogens, which are acquired by contact, may be effectively controlled through barrier precautions and handwashing. However, the potentially devastating effects of these infections in severely immunocompromised persons make reliance on such precautions questionable. Similar caution is advised for exposure to laboratory-acquired infectious agents; the various zoonotic pathogens encountered in agriculture and food production work; and CMV, parvovirus B19, enteric and respiratory pathogens, and other agents present in childcare settings (see Section 6.1.2). Opportunistic infectious diseases that can be acquired from animals include toxoplasmosis, cryptosporidiosis, giardiasis, salmonellosis, and listeriosis (see Section 6.1.3).

Physicians and their immunodeficient patients should discuss the patients' susceptibility to workplace infectious agents, the advisability of continuing to work in certain occupations, and the point at which a patient should change or restrict work because of an infection risk. An institutional occupational health department may provide advice in these situations. Workers disabled as a result of medical conditions are protected from various forms of workplace discrimination. Therefore, employers should provide information on the nature and extent of known or potential infectious hazards to all workers and to individual workers and their physicians upon request.

**TABLE 6.6 Infectious Diseases of Concern for Pregnant Workers**

| Infection | Effect in pregnancy | Occupation | Prevention |
|---|---|---|---|
| Rubella | Fetal death, congenital rubella syndrome | Healthcare | Immunize susceptible workers (male and female) with patient contact |
| Measles | Fetal death, maternal morbidity and mortality | Healthcare | Immunize workers with patient contact; immune globulin for exposed, susceptible pregnant women |
| Influenza | Increased morbidity and mortality in pregnant women | Healthcare | Immunize workers. Inactivated vaccine (but not live attenuated) is safe to administer during pregnancy |
| Chickenpox/varicella-zoster (shingles) | Complications more frequent and severe in pregnant women; congenital varicella syndrome, premature birth, neonatal varicella | Healthcare, childcare | Immunize susceptible healthcare workers; varicella-zoster immune globulin for exposed, susceptible pregnant women |
| Cytomegalovirus infection | Fetal death or severe nervous system abnormalities for 25% of infants infected during pregnancy | Childcare (younger, particularly diapered, children), not healthcare | Uncertain: screen and, if susceptible, either do not work during first half of pregnancy or care only for older children; avoid "intimate contact"; frequent handwashing, glove use; disinfection of toys and surfaces has been proposed |
| Erythema infectiosum (human parvovirus B19 infection, "fifth" disease, transient anemic crisis) | Fetal morbidity (hydrops fetalis) and death (<10% case fatality rate) | Healthcare, childcare | Healthcare: inform workers of risks; routine infection control measures to minimize risks; pregnant workers should not care for patients with transient anemic crisis |
| Childcare: greatest risk of transmission occurs before symptoms appear in children limiting options for prevention; routine exclusion policy for high-risk groups not recommended; individual should make decision for risk avoidance | | | |
| Toxoplasmosis | Nervous system abnormalities if infected during pregnancy | Meat and animal handlers | Consider serologic screening before or during pregnancy; barrier precautions and handwashing; avoid mucous membrane contact; do not eat raw or undercooked meat; consider reassignment during pregnancy; prenatal fetal diagnosis for first trimester infection if abortion considered; treatment during pregnancy; avoid cats, change cat litter daily |
| Listeriosis | Pregnant women have increased risk of infection; fetus at risk for miscarriage, stillbirth, and preterm labor | Laboratory workers | Discontinue work with *Listeria monocytogenes* during pregnancy |
| Brucellosis | Spontaneous abortion, preterm labor | Laboratory workers, meat and animal handlers | Discontinue work with *Brucella abortus* during pregnancy |
| Zika virus | Microcephaly and other birth defects | Occupational travel | Avoid travel to Zika-endemic areas for two months before and during pregnancy (both workers and their sex partners) |

***6.4.8.2 The Pregnant Worker*** Infections acquired during pregnancy may cause fetal morbidity (e.g. malformations), fetal mortality (e.g. spontaneous abortions or stillbirths), or severe maternal illness. Table 6.6 summarizes those infections most often of concern for pregnant workers, related occupations, and prevention strategies (Alex 2011; Hood 2008; Lynch and Spivak 2015). The CDC can provide advice on immunoprophylaxis for pregnant workers who may be exposed to infectious agents (CDC/NIOSH 2017; Frazier and Fromer 2010; Schillie et al. 2013; USDHHS 2009).

## 6.5 CONCLUSIONS

Infectious disease evaluation and management are growing areas of involvement for many occupational health and safety professionals. A wide variety of microorganisms may cause work-related infections, and these agents are transmitted by many exposure routes and affect a number of organ systems. Certain workplaces are associated with obvious microbiological hazards, e.g. childcare, human healthcare, animal care, and microbiology laboratories. However, transmission may occur incidentally or on rare occasions in many other workplaces. The fields of infectious disease control and biosafety are continually evolving. This evolution has been influenced by increasing and increasingly rapid international travel, changes in technology and land use, the emergence and reemergence of a number of infectious diseases, rising numbers of immunocompromised individuals in the workforce, and adaptation and change among infectious agents and their vectors. Environmental sampling, a routine approach for assessing many workplace health hazards, plays a much smaller role in the control and prevention of infectious diseases in the workplace. Rather, the tools used in infectious disease control are primary prevention to avoid worker exposure, immunization of workers likely to be exposed, and surveillance to identify and treat infections early, prevent infections in co-workers and other contacts, and understand how exposure may occur. Applying fundamental occupational health and safety principles to the design, installation, operation, and testing of engineering controls and the implementation of safe work practices to reduce exposures to infectious agents are important and expanding areas of involvement for occupational health and safety professionals.

## CASE STUDIES

**Coccidioidomycosis Among Cast and Crew Members at an Outdoor Television Filming Event – California, 2012**
(Extracted from a Report in *Morbidity and Mortality Weekly Report* [Wilken et al. 2014])
From 17 to 19 January 2012, cast and crew members filmed a television program at an outdoor set in Ventura County, California. After learning of a case of coccidioidomycosis (commonly known as valley fever) in someone present during this filming, the employer sent a letter (dated February 17) encouraging anyone with symptoms to seek medical evaluation. Occupational exposure to spores of *Coccidioides* species, the causative fungi (typically *C. immitis* in California and *C. posadasii* in Arizona, Texas, and Mexico), usually is associated with soil-disrupting activities (Das et al. 2012).

Patient 1, an actor on the set, received this letter and on 28 February sought care for a two-week history of fever and cough (time to illness onset: 28 days). A Doctor's First Report of Occupational Injury or Illness (DFR)* was submitted, which included a copy of the employer's letter. Patient 2, a camera operator – not the patient referred to in the original letter – sought care on 24 February after a two-week history of cough, joint aches, and muscle pain.

Not until March 2013, during a coccidioidomycosis surveillance of DFRs, did the California Department of Public Health (CDPH) connect these cases, both of whom worked in Ventura County but resided in Los Angeles County. Patient 2 was identified from the healthcare provider's notes as having worked as a camera operator at the same outdoor filming event. However, on the basis of their job titles, neither of these cases would have been expected to have engaged in soil-disrupting activities. The Los Angeles County Department of Public Health then conducted an outbreak investigation using social media searches in addition to CDPH-provided occupational surveillance records and traditional infectious disease surveillance.

The patient referenced in the employer's letter subsequently was found to be among five persons with laboratory-confirmed illness. With another five probable cases and an employee roster of 655 workers associated with that particular television episode, the attack rate for all identified cases was 1.5%.

CDPH provided the employer's environmental health and safety manager with a fact sheet "Preventing Work-Related Coccidioidomycosis (Valley fever)" (Hazard Evaluation System and Information Service 2017) for integration into their Injury and Illness Prevention Program (IIPP). Dry, dusty conditions during the filming event were reported, but because no reliable methods for environmental coccidioides sampling are available, identifying the source of the spores was not possible. CDPH recommended a comprehensive approach to reducing the incidence and severity of work-associated coccidioidomycosis, including:

1. Limiting workers' exposure to outdoor dust by controlling dust generation at the source, e.g. continuous soil wetting.
2. Providing employee training.
3. Consistently enforcing an IIPP, which includes providing respiratory protection with particle filters.

However, most patients in this outbreak were not involved in excavation or set construction and might not have been considered at increased risk for coccidioidomycosis in the existing IIPP. Nevertheless, working at a site immediately after soil disturbance might expose workers to spores, and a comprehensive IIPP for these employees should include:

1. Covering spoils piles and wetting disturbed areas.
2. Establishing criteria for suspending work on the basis of wind and dust conditions.
3. Prompt disease recognition and referral to occupational medicine clinics for evaluation, treatment, and follow-up (Das et al. 2012; Hazard Evaluation System and Information Service 2017).

———————————

*In California, healthcare providers who believe a patient's injury or illness might be work related are required to submit a DFR to the employer or their workers' compensation insurance carrier, who forward it to the California Department of Industrial Relations (1993). DFRs are provided to the CDPH for occupational injury and disease surveillance purposes.

———————————

*What is already known on this topic?*

Work-associated coccidioides infections and outbreaks have been linked to soil-disrupting activities, including construction, in areas where coccidioides is endemic.

*What is added by this report?*

Occupational surveillance identified an outbreak of coccidioidomycosis in an unexpected industry, i.e. film and television. Employees working outdoors in any industry, even those not actively engaged in soil disruption, might be exposed to coccidioides where it is endemic.

*What are the implications for public health practice?*

Occupational injury and illness surveillance can identify outbreaks not otherwise detected by traditional infectious disease surveillance. Education about coccidioidomycosis, including signs and symptoms, and exposure prevention measures should be implemented at outdoor worksites in areas where coccidioides is endemic, including worksites of industries and occupations not typically associated with soil-disrupting activities. Healthcare providers should consider the possibility of work-relatedness among patients with coccidioidomycosis diagnoses and should note employer, work location, industry, and occupation when reporting cases.

**Occupational Exposure to *N. meningitidis* and Fatal Outcome in a Laboratory Worker – California, 2012** (Extracted from a Report in *Morbidity and Mortality Weekly Report* [Sheets et al. 2014])

Within a couple of hours of leaving work on the evening of Friday, 27 April 2012, a 25-year-old microbiologist working in a research laboratory associated with Hospital A began to feel unwell. His symptoms included fever, headache, chills, and vomiting. Late the next morning, the researcher asked a roommate for a ride to the emergency department (ED) at Hospital A. The patient lost consciousness on the way to the hospital, went into respiratory arrest on the way into the ED, and was intubated. During the initial exam, the patient was found to have a diffuse petechial rash. Upon this finding, ED personnel donned N95 masks due to a high suspicion of meningococcal meningitis. Blood cultures were drawn, and antibiotics were begun for the patient. Resuscitation efforts continued for three hours, but were unsuccessful. The patient died within 24 hours of the first reported symptoms.

Blood cultures yielded no bacterial isolate. Lumbar puncture to obtain cerebrospinal fluid had not been performed due to the patient's unstable condition. Because meningococcal disease was highly suspected, ED personnel were treated prophylactically with ciprofloxacin. The San Francisco Department of Public Health and the California Department of Public Health (CDPH) were notified, and roommates, fellow laboratorians, and close acquaintances also were referred for antibiotic treatment. A culture-negative blood specimen was subsequently tested by polymerase chain reaction (PCR) at CDPH, where DNA for *N. meningitidis* serogroup B was detected.

Follow-up investigations conducted by multiple agencies (initiated by OSHA, the California Division of Occupational Safety and Health [Cal/OSHA], and CDPH) determined that the researcher had been working with *N. meningitidis* serogroups B and Y during the week preceding illness onset. Many breaches in laboratory safety protocols for work with *N. meningitidis* were found, including working on an open benchtop and failing to offer meningococcal vaccinations to the laboratorians (Sheets et al. 2014).

Biological laboratories in California are covered by the ATD Standard, which requires following recognized biosafety practices and providing recommended immunizations to laboratory employees (California Department of Industrial Relations 2009). Because OSHA ultimately determined the employee came under federal jurisdiction, the agency issued a notice of unsafe and unhealthful working conditions.

The bacterium *N. meningitidis* is an obligate human commensal, and sometimes invasive pathogen, that has an asymptomatic carriage rate estimated to be approximately 10%, although that number can be higher among young adults living in close quarters, such as college students and military personnel (Yazdankhah and Caugant 2004). Infants under one year of age and young adults have increased rates of invasive disease, and virulence is mediated by the presence of a polysaccharide capsule, surface proteins, and the bacterial genotype. There are 12 capsular types currently recognized, with serogroups B, C, and Y causing most disease in industrialized countries while serogroups A, W, and X are more prevalent in sub-Saharan Africa (Castillo et al. 2011; Stephens et al. 2007). Invasive meningococcal disease can present as meningitis, bacteremia, or both and is often accompanied by the presence of a purpuric (measles-like) rash seen first on the lower extremities and progressing upward. Case fatality rates can vary from 5 to 18% (Dwilow and Fanella 2015).— — — — — — — — —

*What is already known on this topic?*

Microbiologists working in clinical, public health, and research laboratories are at increased risk for invasive meningococcal disease, as compared with the general population. Vaccinations are recommended and biosafety precautions for the workplace have been defined in detail for microbiologists who work with *N. meningitidis* on a regular basis.

*What is added by this report?*

Research laboratories are not subject to the same oversight and inspection requirements that apply to clinical laboratories, and even though the laboratory in this case study was associated with a hospital, compliance with biosafety oversight was lacking. In this case, multiple jurisdictions (federal and state) may have complicated assignment of responsibility for failure to implement consensus recommendations for working with *N. meningitidis*.

*What are the implications for public health practice?*

All laboratory directors, including principal investigators at research laboratories, are ultimately responsible for providing a safe working environment for microbiologists. Clear guidelines are available from the CDC and the ACIP regarding workplace precautions and immunization recommendations, respectively.

**Plague Contracted by a Wildlife Biologist from a Mountain Lion – Arizona, 2007** (Extracted from a Report in the Journal of *Clinical Infectious Diseases* [Wong et al. 2009])

On 2 November 2007, a wildlife biologist working at the Grand Canyon National Park (GCNP) Service was found dead at home one week after conducting a necropsy on a mountain lion, and the biologist subsequently was diagnosed with primary pneumonic plague. Occupational exposure to *Yersinia pestis*, the causative agent of plague, is usually associated with first responders, healthcare workers, laboratory personnel, and veterinarians. Plague is a zoonotic disease endemic to the Western United States, primarily affecting rodents and their fleas; however, primary pneumonic plague is a rare but often fatal form of infection resulting from direct inhalation of bacteria and is potentially transmissible from person to person.

The deceased, a previously healthy 37-year-old male, showed initial symptoms of fever, chills, cough with blood-tinged sputum, nausea, and myalgia on the evening of 29 October 2007. The next morning the biologist sought care at a nearby clinic, but his exposure to wildlife was not documented in the patient history. He was diagnosed with viral syndrome and asked to follow up if his symptoms worsened. On the morning of 30 October 2007, he went to work, and his last known interaction was on the evening of the following day.

The cause of death was determined by postmortem examination, and the source of infection was determined by collecting physical evidence and conducting interviews with friends and colleagues of the deceased. An ecologic investigation was also conducted to assess the local plague risk in the South Rim of the GCNP, AZ.

Job-related duties for the biologist included trapping mountain lions, collaring them, and removing rodents from buildings. The investigation revealed that a "mortality signal" (no movement after six hours) was transmitted from a radio-collared mountain lion on 25 October and that the biologist found the carcass the next day in an uninhabited location in the GCNP. He carried the carcass approximately 1 km to his vehicle and conducted a necropsy in his garage. There was no evidence that he wore a respirator or other personal protective equipment (PPE) during the estimated 2.5-hour examination. The biologist's notes indicated that he concluded that the mountain lion died from a hemorrhage as a result of an attack by another mountain lion. No evidence was found that anyone else was present with the biologist during these activities.

Archived specimens of the necropsied animal tested positive for *Y. pestis*, and subtyping revealed that isolates from the mountain lion were identical to those from the biologist. Two hours after preliminary diagnosis was available, 49 contacts of the biologist received antibiotics out of an abundance of caution, although some contacts (e.g. emergency responders) had no true exposures to the symptomatic biologist. None of the contacts developed symptoms consistent with plague, and a search found no fleas, rodent die-offs, or other suggestive evidence of a plague epizootic.

The investigators concluded that the biologist died of primary pneumonic plague following inhalation of aerosols generated while handling an infected mountain lion, most likely during the prolonged, unprotected animal necropsy. The necropsy involved several procedures that could generate aerosols, including opening the thoracic cavity and transecting the vertebral column, which may have been done with power tools found in the garage. Wild cats are considered highly susceptible to plague, and the biologist was likely aware of the potential for plague in this endemic area. However, he attributed the mountain lion's cause of death to trauma and apparently did not feel a need to protect himself while handling the animal nor did he seriously consider plague as a possible cause of his symptoms.

As a result of this case, the National Park Service issued guidelines for safe handling of wildlife, including potential routes of infectious agent transmission, appropriate use of PPE, and the relative risk of species-specific activities. Policies regarding managerial oversight also were reviewed. In addition, a national online survey of wildlife workers and biologists assessed potential occupational exposures to zoonotic diseases and the use of PPE (Bosch et al. 2013). Results of the survey showed that facilitators of PPE use included availability of PPE, especially specific PPE kits ready to use during necropsies, and having PPE stocked for use in remote locations. This study found that development and use of interventions in the workplace are necessary for zoonotic disease awareness and exposure prevention.

— — — — — — — — —

*What is already known on this topic?*

Work-associated plague cases previously have been linked to close contact with cats in areas where *Y. pestis* is endemic.

*What is added by this report?*

A wildlife biologist died of primary pneumonic plague following inhalation of aerosols while handling an infected mountain lion, most likely while conducting a prolonged and unprotected necropsy of the animal. The delayed recognition of this infection was due in part to the lack of consideration of the risk of plague as well as the lack of consideration of a possible occupational cause by the treating clinicians.

*What are the implications for public health practice?*

Clinicians need to exercise a high degree of suspicion for zoonotic diseases, especially in areas where these are endemic. In addition, a complete patient history must include the occupational and recreational exposure of not only biologists who handle wildlife but also of hunters, trappers, and taxidermists. Where *Y. pestis* is endemic, education of clinicians and individuals at an increased risk of exposure to plague must include signs and symptoms of the disease as well as good hygiene practices and use of appropriate PPE. The estimates of the time from exposure to symptom onset and from exposure to death may help efforts to model and prevent the spread of pneumonic plague during natural outbreaks or following the intentional use of *Y. pestis* as a bioweapon.

## ACKNOWLEDGMENTS

The authors thank Dr. Barbara Materna, Division of Occupational and Environmental Health, and Dr. Shua J. Chai, Division of Communicable Disease Control, at the California Department of Public Health, for reviewing this chapter. We applied the "sequence-determines-credit" (SDC) approach for author sequence.

## ABBREVIATIONS

| | |
|---|---|
| ACH | air change per hour |
| AIDS | acquired immunodeficiency syndrome |
| anti-HAV | anti-hepatitis A virus |
| anti-HBc | anti-hepatitis B core antigen |
| ATD | aerosol transmissible disease |
| BSL | biosafety level |

| | |
|---|---|
| cDNA | complementary DNA |
| CMV | cytomegalovirus |
| DFR | Doctor's First Report of Occupational Injury or Illness |
| DNA | deoxyribonucleic acid |
| ED | emergency department |
| EPPs | exposure-prone procedures |
| FIFRA | Federal Insecticide, Fungicide, and Rodenticide Act |
| HAI | healthcare-associated infection |
| HAV | hepatitis A virus |
| HBc | hepatitis B core |
| HBeAg | hepatitis B e antigen |
| HBV | hepatitis B virus |
| HCV | hepatitis C virus |
| HEPA | high-efficiency particulate air |
| HEV | hepatitis E virus |
| HIV | human immunodeficiency virus |
| HPS | hantavirus pulmonary syndrome |
| $ID_{50}$ | infectious dose, 50% of test population |
| Ig | immunoglobulin |
| IgA | immunoglobulin A |
| IgG | immunoglobulin G |
| IgM | immunoglobulin M |
| IIPP | Injury and Illness Prevention Program |
| ISG | immune serum globulin |
| MERS | Middle East respiratory syndrome |
| MMR | measles, mumps, and rubella |
| NoV | norovirus |
| ODTS | organic dust toxic syndrome |
| PCR | polymerase chain reaction |
| PEP | post-exposure prophylaxis |
| PPE | personal protective equipment |
| PSDS | Pathogen Safety Data Sheet |
| RNA | ribonucleic acid |
| RT-PCR | reverse transcriptase polymerase chain reaction |
| SARS | severe acute respiratory syndrome |
| SESIP | sharps with engineered sharps injury protection |
| SNV | Sin Nombre virus |
| Td | tetanus–diphtheria toxoid |
| Tdap | tetanus, diphtheria, and acellular pertussis |
| UVGI | ultraviolet germicidal irradiation |
| VZV | varicella-zoster virus |

## LATIN NAMES

| | |
|---|---|
| *B. anthracis* | *Bacillus anthracis* |
| *B. subtilis* | *Bacillus subtilis* |
| *C. burnetii* | *Coxiella burnetii* |
| *C. immitis* | *Coccidioides immitis* |
| *C. posadasii* | *Coccidioides posadasii* |
| *C. tetani* | *Clostridium tetani* |
| *E. coli* | *Escherichia coli* |
| *H. pylori* | *Helicobacter pylori* |
| *L. pneumophila* | *Legionella pneumophila* |
| *M. tuberculosis* | *Mycobacterium tuberculosis* |
| *N. meningitidis* | *Neisseria meningitidis* |
| *S. aureus* | *Staphylococcus aureus* |
| *Y. pestis* | *Yersinia pestis* |

## AGENCIES AND ASSOCIATIONS

| | |
|---|---|
| AAMI | Association for the Advancement of Medical Instrumentation |
| ACIP | Advisory Committee on Immunization Practices |
| ANSI | American National Standards Institute |
| AOAC | Association of Official Analytical Chemists |
| APHIS | Animal and Plant Health Inspection Service |
| ASTM | American Society for Testing and Materials |
| Cal/OSHA | California Division of Occupational Safety and Health |
| CDC | Centers for Disease Control and Prevention |
| CDPH | California Department of Public Health |
| CSTE | Council of State and Territorial Epidemiologists |
| DOT | Department of Transportation |
| EPA | Environmental Protection Agency |
| FDA | Food and Drug Administration |
| GCNP | Grand Canyon National Park |
| HICPAC | Healthcare Infection Control Practices Advisory Committee |
| IATA | International Air Transport Association |
| NIH | National Institutes of Health |
| NIOSH | National Institute for Occupational Safety and Health |
| NSF | National Sanitation Foundation |
| OSHA | Occupational Safety and Health Administration |
| USDA | US Department of Agriculture |
| USDHHS | US Department of Health and Human Services |
| USEPA | US Environmental Protection Agency |
| VS | Veterinary Services |
| WHO | World Health Organization |

## REFERENCES

21 CFR 800.20 (1990). Patient examination gloves and surgeons' gloves; sample plans and test method for leakage defects; adulteration. Food and Drug Administration, HHS. *Federal Register* **55**: 51256.

29 CFR 1904 (2001). Occupational injury and illness recording and reporting requirements. Occupational Safety and Health Administration. *Federal Register* **66**: 5916–6135.

29 CFR 1904.5 (2001). Determination of work-relatedness; recording and reporting occupational injuries and illness. Occupational Safety and Health Standards. *Federal Register* **66**: 6124.

29 CFR 1910.1030 (1991a). Occupational exposure to bloodborne pathogens. Occupational Safety and Health Administration. Final rule. *Federal Register* **56**: 64004–64182.

29 CFR 1910.1030 (1991b). Toxic and hazardous substances, bloodborne pathogens. Occupational Safety and Health Administration. *Federal Register* **56**: 64175.

42 CFR 72 (2007). Interstate shipment of etiologic agents. Public Health Service, HHS. *Federal Register* **45**: 48627.

81 CFR 15315 (2016). NIH guidelines for research involving recombinant or synthetic nucleic acid molecules (NIH guidelines). *Federal Register* **81**: 15315–15322.

AAP/APHA/NRCHSCCEE (2013). *Caring for Our Children: National Health and Safety Performance Standards; Guidelines for Early Care and Education Programs*, 3e. Elk Grove Village, IL: American Academy of Pediatrics; Washington, DC: American Public Health Association.

ACGIH (2017). *Threshold Limit Values for Chemical Substances and Physical Agents & Biological Exposure Indices*. Cincinnati OH: American Conference of Governmental Industrial Hygienists.

Adams, D.A., Thomas, K.R., Jajosky, R.A. et al. (2017). Summary of notifiable infectious diseases and conditions – United States, 2015. *MMWR. Morbidity and Mortality Weekly Report* **64**: 1–143.

Adoba, P., Boadu, S.K., Agbodzakey, H. et al. (2015). High prevalence of hepatitis B and poor knowledge on hepatitis B and C viral infections among barbers: a cross-sectional study of the Obuasi municipality, Ghana. *BMC Public Health* **15**: 1041.

Akinsanya, T. and Bodenberg, T.M. (2014). Regulatory environment. In: *Risk Management in Healthcare Institutions: Limiting Liability and Enhancing Care* (ed. F. Kavaler and R.S. Alexander), 29–59. Burlington, MA: Jones & Bartlett Learning.

Alex, M.R. (2011). Occupational hazards for pregnant nurses. *American Journal of Nursing* **111**: 28–37. quiz 38–39.

Aliabadi, A.A., Rogak, S.N., Bartlett, K.H., and Green, S.I. (2011). Preventing airborne disease transmission: review of methods for ventilation design in health care facilities. *Advances in Preventive Medicine* **2011**: 124064.

ANSI/AAMI (2012). *Liquid Barrier Performance and Classification of Protective Apparel and Drapes in Health Care Facilities*, 26. New York, NY: American National Standards Institute/Association for the Advancement of Medical Instrumentation.

APHA (2015). *Control of Communicable Diseases Manual*, 20e. Washington, DC: APHA Press.

ASHRAE (2015). *ASHRAE Position Document on Filtration and Air Cleaning*. Atlanta, GA: American Society of Heating, Refrigerating, and Air-Conditioning Engineers.

ASTM F1670/F1670M-17a (2017). *Standard Test Method for Resistance of Materials Used in Protective Clothing to Penetration by Synthetic Blood*. West Conshohocken, PA: ASTM International.

ASTM F1671/F1671M-13 (2013). *Standard Test Method for Resistance of Materials Used in Protective Clothing to Penetration by Blood-Borne Pathogens Using Phi-X174 Bacteriophage Penetration as a Test System*. West Conshohocken, PA: ASTM International.

ASTM F1819-07 (2013). *Standard Test Method for Resistance of Materials Used in Protective Clothing to Penetration by Synthetic Blood Using a Mechanical Pressure Technique*. West Conshohocken, PA: ASTM International.

Barnett, E.D. and Walker, P.F. (2008). Role of immigrants and migrants in emerging infectious diseases. *Medical Clinics of North America* **92**: 1447–1458.

Baron, E.J. and Miller, J.M. (2008). Bacterial and fungal infections among diagnostic laboratory workers: evaluating the risks. *Diagnostic Microbiology and Infectious Disease* **60**: 241–246.

Baron, S., Steege, A., Welch, L., and Lipscomb, J. (2010). Addressing health and safety hazards in specific industries: agriculture, construction, and health care. In: *Occupational and Environmental Health: Recognizing and Preventing Disease and Injury* (ed. B. Levy, D. Wegman, S. Baron, and R. Sokas), 753–778. New York: Oxford University Press.

Baussano, I., Nunn, P., Williams, B. et al. (2011). Tuberculosis among health care workers. *Emerging Infectious Diseases* **17**: 488–494.

Bebermeyer, R.D., Dickinson, S.K., and Thomas, L.P. (2005). Guidelines for infection control in dental health care settings – a review. *Texas Dental Journal* **122**: 1022–1026.

Behrman, A., Buchta, W.G., Budnick, L.D. et al. (2013). Protecting health care workers from tuberculosis, 2013: ACOEM medical Center occupational health section task force on tuberculosis and health care workers. *Journal of Occupational and Environmental Medicine* **55**: 985–988.

Benn, P. and Fisher, M. (2008). HIV and postexposure prophylaxis. *Clinical Medicine* **8**: 319–322.

Bennett, J.E., Dolin, R., and Blaser, M.J. (2015). *Mandell, Douglas, and Bennett's Principles and Practice of Infectious Diseases*, 8e. Philadelphia, PA: Elsevier Saunders.

Benson, S.M., Novak, D.A., and Ogg, M.J. (2013). Proper use of surgical N95 respirators and surgical masks in the OR. *American Association of Occupational Health Nurses Journal* **97**: 457–470.

Bessesen, M.T., Savor-Price, C., Simberkoff, M. et al. (2013). N95 respirators or surgical masks to protect healthcare workers against respiratory infections: are we there yet? *American Journal of Respiratory and Critical Care Medicine* **187**: 904–905.

Bhangar, S., Adams, R.I., Pasut, W. et al. (2016). Chamber bioaerosol study: human emissions of size-resolved fluorescent biological aerosol particles. *Indoor Air* **26**: 193–206.

Bolyard, E.A., Tablan, O.C., Williams, W.W. et al. (1998). Guideline for infection control in healthcare personnel, 1998. Hospital infection control practices advisory committee. *Infection Control and Hospital Epidemiology* **19**: 407–463.

Bosch, S.A., Musgrave, K., and Wong, D. (2013). Zoonotic disease risk and prevention practices among biologists and other wildlife worker – results from a national survey, US National Park Service, 2009. *Journal of Wildlife Diseases* **49**: 475–585.

Bosman, A., Mulder, Y., de Leeuw, J. et al. (2004). *Executive Summary Avian Flu Epidemic 2003: Public Health Consequences*, 40. Bilthoven, the Netherlands: National Institute for Public Health and the Environment.

Boyce, R. and Mull, J. (2008). Complying with the Occupational Safety and Health Administration: guidelines for the dental office. *Dental Clinics of North America* **52**: 653–668. xi.

Bryce, E., Forrester, L., Scharf, S., and Eshghpour, M. (2008). What do healthcare workers think? A survey of facial protection equipment user preferences. *The Journal of Hospital Infection* **68**: 241–247.

Calfee, D.P. (2006). Prevention and management of occupational exposures to human immunodeficiency virus (HIV). *Mount Sinai Journal of Medicine* **73**: 852–856.

California Department of Industrial Relations (1993). *Doctor's First Report of Occupational Injury or Illness*. In Chapter 7, Division of Labor Statistics and Research; Subchapter 1, Occupational Injury or Illness Reports and Records. California Code of Regulations. www.dir.ca.gov/t8/14006.html (accessed 22 December 2017).

California Department of Industrial Relations (2009). *Aerosol Transmissible Diseases*. In Title 8, Section 5199. California Code of Regulations. www.dir.ca.gov/title8/5199.html (accessed 22 December 2017).

Candan, F., Alagozlu, H., Poyraz, O., and Sumer, H. (2002). Prevalence of hepatitis B and C virus infection in barbers in the Sivas region of Turkey. *Occupational Medicine (London)* **52**: 31–34.

Carney, K., Dhalla, S., Aytaman, A. et al. (2013). Association of tattooing and hepatitis C virus infection: a multicenter case-control study. *Hepatology* **57**: 2117–2123.

Castillo, D., Harcourt, B., Hatcher, C. et al. (2011). *Laboratory Methods for the Diagnosis of Meningitis Caused by Neisseria meningitidis, Streptococcus pneumoniae, and Haemophilus influenzae: WHO Manual*. Geneva, Switzerland: WHO, CDC.

CDC (1988). Update: universal precautions for prevention of transmission of human immunodeficiency virus, hepatitis B virus, and other bloodborne pathogens in health-care settings. *MMWR. Morbidity and Mortality Weekly Report* **37**: 377–382, 387–388.

CDC (2012). *Principles of Epidemiology in Public Health Practice, An Introduction to Applied Epidemiology and Biostatistics, Section 11: Epidemic Disease Occurrence*. http://www.cdc.gov/ophss/csels/dsepd/SS1978/Lesson1/Section11.html (accessed 22 December 2017).

CDC (2014). *CDC Lab Incident: Anthrax*. U.S. Department of Health and Human Services. http://www.cdc.gov/anthrax/news-multimedia/lab-incident/index.html (accessed 22 December 2017).

CDC (2015). *Epidemiology and Prevention of Vaccine-Preventable Diseases*. https://www.cdc.gov/vaccines/pubs/pinkbook/index.html (accessed 22 December 2017).

CDC (2017a). *CDC Yellow Book 2018: Health Information for International Travel*. New York, NY: Oxford University Press.

CDC (2017b). *Diseases & Conditions A-Z Index*. http://www.cdc.gov/DiseasesConditions/az/a.html (accessed 22 December 2017).

CDC (2017c). *Highly Pathogenic Asian Avian Influenza A (H5N1) in People*. U.S. Department of Health and Human Services. https://www.cdc.gov/flu/avianflu/h5n1-people.htm (accessed 22 December 2017).

CDC (2017d). *Surveillance Case Definitions for Current and Historical Conditions*. U.S. Department of Health and Human Services. https://wwwn.cdc.gov/nndss/conditions (accessed 22 December 2017).

CDC (2017e). *Travelers' Health: Destinations*. CDC. http://wwwnc.cdc.gov/travel/destinations/list (accessed 22 December 2017).

CDC/NIOSH (2017). *Reproductive Health and the Workplace: Infectious Agents*. U.S. Department of Health and Human Services. https://www.cdc.gov/niosh/topics/repro/infectious.html (accessed 22 December 2017).

Chalupka, S. (2013). Updated recommendations for the management of health care personnel occupationally exposed to human immunodeficiency virus. *Workplace Health and Safety* **61**: 504.

Chang, C.C., Ananda-Rajah, M., Belcastro, A. et al. (2014). Consensus guidelines for implementation of quality processes to prevent invasive fungal disease and enhanced surveillance measures during hospital building works, 2014. *Medical Clinics of North America* **44**: 1389–1397.

Chartier, Y., Emmanuel, J., Pieper, U. et al. (2014). *Safe Management of Wastes from Health-Care Activities*. Geneva, Switzerland: WHO.

Chen, L.H. and Wilson, M.E. (2008). The role of the traveler in emerging infections and magnitude of travel. *Medical Clinics of North America* **92**: 1409–1432. xi.

Chinen, J. and Shearer, W.T. (2010). Secondary immunodeficiencies, including HIV infection. *The Journal of Allergy and Clinical Immunology* **125**: S195–S203.

Ciplak, N. (2013). A system dynamics approach for the determination of adverse health impacts of healthcare waste incinerators and landfill sites on employees. *Environmental Management and Sustainable Development* **2**: 7–28.

Cleveland, J.L., Robison, V.A., and Panlilio, A.L. (2009). Tuberculosis epidemiology, diagnosis and infection control recommendations for dental settings: an update on the Centers for Disease Control and Prevention guidelines. *Journal of the American Dental Association* **140**: 1092–1099.

Coelho, A.C. and Garcia Diez, J. (2015). Biological risks and laboratory-acquired infections: a reality that cannot be ignored in health biotechnology. *Frontiers in Bioengineering and Biotechnology* **3**: 56.

Colton, C. (2012). Respiratory protection. In: *Fundamentals of Industrial Hygiene* (ed. B.A. Plog and P. Quinlan), 649–685. Itasca, IL: National Safety Council.

Das, R., McNary, J., Fitzsimmons, K. et al. (2012). Occupational coccidioidomycosis in California: outbreak investigation, respirator recommendations, and surveillance findings. *Journal of Occupational and Environmental Medicine* **54**: 564–571.

Day, J.G., Achilles-Day, U., Brown, S., and Warren, A. (2007). Cultivation of algae and protozoa. In: *Manual of Environmental Microbiology* (ed. C.J. Hurst, R.L. Crawford, J.L. Garland, et al.), 79–92. Washington, DC: American Society of Microbiology.

de Perio, M.A., Niemeier, R.T., and Burr, G.A. (2015). Coccidioides exposure and coccidioidomycosis among prison employees, California, United States. *Emerging Infectious Diseases* **21**: 1031–1033.

Department of Defense (2017). *Serum Repository*. https://health.mil/Military-Health-Topics/Health-Readiness/Armed-Forces-Health-Surveillance-Branch/Data-Management-and-Technical-Support/Department-of-Defense-Serum-Repository (accessed 22 December 2017).

Dieckhaus, K. (2007). Occupational infections. In: *Environmental and Occupational Medicine* (ed. W. Rom), 708–730. Philadelphia, PA: Lippincott Williams & Wilkins.

Dieffenback, C. and Tramont, E. (2015). Innate (general or nonspecific) host defense mechanisms. In: *Mandell, Douglas, and Bennett's Principles and Practice of Infectious Diseases* (ed. J.E. Bennett, R. Dolin, and M.J. Blaser), 26–33. Philadelphia, PA: Elseview Saunders.

Dougherty, D., Edens, A., Perry, W., and Levinson, A. (2012). Government regulations. In: *Fundamentals of Industrial Hygiene* (ed. B.A. Plog and P. Quinlan), 807–828. Itasca, IL: National Safety Council.

Dowdall, N.P., Evans, A.D., and Thibeault, C. (2010). Air travel and TB: an airline perspective. *Travel Medicine and Infectious Disease* **8**: 96–103.

Dulon, M., Peters, C., Schablon, A., and Nienhaus, A. (2014). MRSA carriage among healthcare workers in non-outbreak settings in Europe and the United States: a systematic review. *BMC Infectious Diseases* **14**: 363.

Duraisingham, S.S., Buckland, M., Dempster, J. et al. (2014). Primary vs. secondary antibody deficiency: clinical features and infection outcomes of immunoglobulin replacement. *PLoS One* **9**: e100324.

Dwilow, R. and Fanella, S. (2015). Invasive meningococcal disease in the 21st century – an update for the clinician. *Current Neurology and Neuroscience Reports* **15**: 2.

Dwyer, D.M., Groves, C., and Blythe, D. (2014). Outbreak epidemiology. In: *Infectious Disease Epidemiology: Theory and Practice* (ed. K.E. Nelson and C. Williams), 105–129. Burlington, MA: Jones & Bartlett Learning.

Elshabrawy, H.A., Erickson, T.B., and Prabhakar, B.S. (2015). Ebola virus outbreak, updates on current therapeutic strategies. *Reviews in Medical Virology* **25**: 241–253.

Erickson, D.S. (2014). New standard of care. FGI's guidelines feature significant changes. *Health Facilities Management* **27**: 37–40.

Evans, A.D. and Thibeault, C. (2009). Prevention of spread of communicable disease by air travel. *Aviation, Space, and Environmental Medicine* **80**: 601–602.

Fairfax, M.R. and Salimnia, H. (2013). Diagnostic molecular microbiology: a 2013 snapshot. *Clinics in Laboratory Medicine* **33**: 787–803.

FEMA (2002). *Guide to Managing an Emergency Service Infection Control Program*. https://www.usfa.fema.gov/downloads/pdf/publications/fa-112.pdf (accessed 22 December 2017).

Folaranmi, T., Rubin, L., Martin, S.W. et al. (2015). Use of serogroup B meningococcal vaccines in persons aged ≥10 years at increased risk for serogroup B meningococcal disease: recommendations of the advisory committee on immunization practices, 2015. *MMWR. Morbidity and Mortality Weekly Report* **64**: 608–612.

Fouchier, R.A., Schneeberger, P.M., Rozendaal, F.W. et al. (2004). Avian influenza a virus (H7N7) associated with human conjunctivitis and a fatal case of acute respiratory distress syndrome. *Proceedings of the National Academy of Sciences of the United States of America* **101**: 1356–1361.

Fournié, G., De Glanville, W., and Pfeiffer, D. (2012a). Epidemiology of highly pathogenic avian influenza virus strain type H5N1. In: *Health and Animal Agriculture in Developing* (ed. D. Countries, J. Zilberman, D.R.-H. Otte, and D. Pfeiffer), 161–182. New York, NY: Springer.

Fournié, G., Walker, P., Porphyre, T. et al. (2012b). Mathematical models of infectious diseases in livestock: concepts and application to the spread of highly pathogenic avian influenza virus strain type H5N1. In: *Health and Animal Agriculture in Developing* (ed. D. Countries, J. Zilberman, D.R.-H. Otte, and D. Pfeiffer), 183–205. New York, NY: Springer.

Frazier, L. and Fromer, D. (2010). Reproductive and developmental disorders. In: *Occupational and Environmental Health: Recognizing and Preventing Disease and Injury* (ed. B. Levy, D. Wegman, S. Baron, and R. Sokas), 446–460. New York, NY: Oxford University Press.

Fredericks, A.C. and Fernandez-Sesma, A. (2014). The burden of dengue and chikungunya worldwide: implications for the southern United States and California. *Annals of Global Health* **80**: 466–475.

Fredericks, D.N. and Relman, D.A. (1996). Sequence-based identification of microbial pathogens: a reconsideration of Koch's postulates. *Clinical Microbiology Reviews* **9**: 18–33.

Galvin, S., Dolan, A., Cahill, O. et al. (2012). Microbial monitoring of the hospital environment: why and how? *The Journal of Hospital Infection* **82**: 143–151.

Gamage, B., Moore, D., Copes, R. et al. (2005). Protecting health care workers from SARS and other respiratory pathogens: a review of the infection control literature. *American Journal of Infection Control* **33**: 114–121.

Garcia, L.S. (2010). *Clinical Microbiology Procedures Handbook*. Washington, DC: American Society for Microbiology Press.

Garner, J.S. (1996). Guideline for isolation precautions in hospitals. Part I. Evolution of isolation practices, Hospital Infection Control Practices Advisory Committee. *American Journal of Infection Control* **24**: 24–31.

Gaudioso, J., Caskey, S., Hashimoto, R. et al. (2012). Biological hazards. In: *Fundamentals of Industrial Hygiene* (ed. B.A. Plog and P. Quinlan), 411–483. Itasca, IL: National Safety Council.

Gautreau, A., Barrat, A., and Barthelemy, M. (2008). Global disease spread: statistics and estimation of arrival times. *Journal of Theoretical Biology* **251**: 509–522.

Gautret, P., Botelho-Nevers, E., Brouqui, P., and Parola, P. (2012). The spread of vaccine-preventable diseases by international travellers: a public-health concern. *Clinical Microbiology and Infection: the Official Publication of the European Society of Clinical Microbiology and Infectious Diseases* **18** (Suppl 5): 77–84.

Gold, D. (2015). Protecting health care workers from infectious disease: experience with Cal/OSHA's aerosol transmissible disease standard. Paper presented at: 143rd APHA Annual Meeting and Exposition (31 October–4 November 2015). Chicago, IL: American Public Health Association.

Gostin, L.O., DeBartolo, M.C., and Friedman, E.A. (2015). The international health regulations 10 years on: the governing framework for global health security. *Lancet* **386**: 2222–2226.

Grace, S.L., Hershenfield, K., Robertson, E., and Stewart, D.E. (2005). The occupational and psychosocial impact of SARS on academic physicians in three affected hospitals. *Psychosomatics* **46**: 385–391.

Grinshpun, S.A., Buttner, M.P., Mainelis, G., and Willeke, K. (2015). Sampling for airborne microorganisms. In: *Manual of Environmental Microbiology* (ed. M.P. Buttner). Washington, DC: American Society for Microbiology Press.

Haagsma, J.A., Tariq, L., Heederik, D.J., and Havelaar, A.H. (2012). Infectious disease risks associated with occupational exposure: a systematic review of the literature. *Occupational and Environmental Medicine* **69**: 140–146.

Halloran, S.K., Wexler, A.S., and Ristenpart, W.D. (2012). A comprehensive breath plume model for disease transmission via expiratory aerosols. *PLoS One* **7**: e37088.

Halpin, J. (2005). Avian flu from an occupational health perspective. *Archives of Environmental & Occupational Health* **60**: 62–69.

Harding, A.L. and Byers, K.B. (2006). Epidemiology of laboratory-associated infections. In: *Biological Safety: Principles and Practices* (ed. D. Fleming and D. Hunt), 53–77. Washington, DC: ASM Press.

Hayney, M.S. and Bartell, J.C. (2005). An immunization education program for childcare providers. *The Journal of School Health* **75**: 147–149.

Hazard Evaluation System and Information Service (2017). *Preventing Work-related Coccidioidomycosis (Valley Fever)*. California Department of Public Health and California Department of Industrial Relations. www.cdph.ca.gov/Programs/CCDPHP/DEODC/OHB/Pages/Cocci.aspx (accessed 22 December 2017).

Heptonstall, J. (2010). Bioterrorism. In: *Hunter's Diseases of Occupations* (ed. P.J. Baxter, T.-C. Aw, A. Cockcroft, et al.), 773–780. Boca Raton, FL: CRC Press.

Ho, P.L., Becker, M., and Chan-Yeung, M.M. (2007). Emerging occupational lung infections. *The International Journal of Tuberculosis and Lung Disease* **11**: 710–721.

Holmberg, S.D., Suryaprasad, A., and Ward, J.W. (2012). Updated CDC recommendations for the management of hepatitis B virus-infected health-care providers and students. *MMWR – Recommendations and Reports* **61**: 1–12.

Hood, J. (2008). The pregnant health care worker – an evidence-based approach to job assignment and reassignment. *American Association of Occupational Health Nurses Journal* **56**: 329–333.

House, H.R. and Ehlers, J.P. (2008). Travel-related infections. *Emergency Medicine Clinics of North America* **26**: 499–516.

Hughes, J.M. (1989). Universal precautions: CDC perspective. *Occupational Medicine (Philadelphia, PA)* **4 Suppl**: 13–20.

Humphreys, H. (2007). Control and prevention of healthcare-associated tuberculosis: the role of respiratory isolation and personal respiratory protection. *The Journal of Hospital Infection* **66**: 1–5.

Hyttinen, M., Rautio, A., Pasanen, P. et al. (2011). Airborne infection isolation rooms – a review of experimental studies. *Indoor and Built Environment* **20**: 584–594.

IATA (2015). *Dangerous Goods Regulations (DGR)*, 57e. Montreal, QC: International Air Transport Association.

Immunization Healthcare Branch (2017). *Vaccine Recommendations*. U.S. Public Health Division, Defense Health Agency. https://health.mil/Military-Health-Topics/Health-Readiness/Immunization-Healthcare/Vaccine-Recommendations/Vaccine-Recommendations-by-AOR (accessed 22 December 2017).

Inglis, T.J. (2007). Principia aetiologica: taking causality beyond Koch's postulates. *Journal of Medical Microbiology* **56**: 1419–1422.

International Committee on Taxonomy of Viruses (2017). *Virus Taxonomy: 2016 Release*. https://talk.ictvonline.org/taxonomy (accessed 22 December 2017).

ISO (2004a). *Clothing for Protection Against Contact with Blood and Body Fluids – Determination of Resistance of Protective Clothing Materials to Penetration by Blood-Borne Pathogens – Test Method Using Phi-X 174 Bacteriophage*. ISO/TC 94/SC 13. Geneva, Switzerland: International Organization for Standardization.

ISO (2004b). *Clothing for Protection Against Contact with Blood and Body Fluids – Determination of the Resistance of Protective Clothing Materials to Penetration by Blood and Body Fluids – Test Method Using Synthetic Blood*. ISO/TC 94/SC 13. Geneva, Switzerland: International Organization for Standardization.

ISO (2004c). *Clothing for Protection Against Infectious Agents – Medical Face Masks – Test Method for Resistance Against Penetration by Synthetic Blood (Fixed Volume, Horizontally Projected)*. ISO/TC 94/SC 13. Geneva, Switzerland: International Organization for Standardization.

Issarow, C.M., Mulder, N., and Wood, R. (2015). Modelling the risk of airborne infectious disease using exhaled air. *Journal of Theoretical Biology* **372**: 100–106.

Jahrling, P., Rodak, C., Bray, M., and Davey, R.T. (2009). Triage and management of accidental laboratory exposures to biosafety level-3 and -4 agents. *Biosecurity and Bioterrorism: Biodefense Strategy, Practice, and Science* **7**: 135–143.

Jensen, P.A., Lambert, L.A., Iademarco, M.F. et al. (2005). Guidelines for preventing the transmission of *Mycobacterium tuberculosis* in health-care settings, 2005. *MMWR – Recommendations and Reports* **54**: 1–141.

Jones, R.M. and Brosseau, L.M. (2015). Aerosol transmission of infectious disease. *Journal of Occupational and Environmental Medicine* **57**: 501–508.

Jones, R.M. and Nicas, M. (2014). Experimental evaluation of a Markov multizone model of particulate contaminant transport. *The Annals of Occupational Hygiene* **58**: 1032–1045.

Joyce, M.P., Kuhar, D., and Brooks, J.T. (2015). Notes from the field: occupationally acquired HIV infection among health care workers – United States, 1985–2013. *MMWR. Morbidity and Mortality Weekly Report* **63**: 1245–1246.

Kaiser, J. (2007). Biosafety breaches. Accidents spur a closer look at risks at biodefense labs. *Science* **317**: 1852–1854.

Katz, J.M., Veguilla, V., Belser, J.A. et al. (2009). The public health impact of avian influenza viruses. *Poultry Science* **88**: 872–879.

Kelt, D.A. and Hafner, M.S. (2010). Updated guidelines for protection of mammalogists and wildlife researchers from hantavirus pulmonary syndrome (HPS). *Journal of Mammalogy* **91**: 1524–1527.

Kessler, C.S., McGuinn, M., Spec, A. et al. (2011). Underreporting of blood and body fluid exposures among health care students and trainees in the acute care setting: a 2007 survey. *American Journal of Infection Control* **39**: 129–134.

Khairnar, K., Martin, D., Lau, R. et al. (2009). Multiplex real-time quantitative PCR, microscopy and rapid diagnostic immuno-chromatographic tests for the detection of *Plasmodium* spp: performance, limit of detection analysis and quality assurance. *Malaria Journal* **8**: 284.

Kimman, T.G., Smit, E., and Klein, M.R. (2008). Evidence-based biosafety: a review of the principles and effectiveness of microbiological containment measures. *Clinical Microbiology Reviews* **21**: 403–425.

Kohn, W.G., Collins, A.S., Cleveland, J.L. et al. (2003). Guidelines for infection control in dental health-care settings – 2003. *MMWR – Recommendations and Reports* **52**: 1–61.

Koo, D.T., Dean, A.G., Slade, R.W. et al. (1994). Summary of notifiable diseases, United States, 1993. *MMWR. Morbidity and Mortality Weekly Report* **42**: i–xvii. 1–73.

Koopmans, M., Wilbrink, B., Conyn, M. et al. (2004). Transmission of H7N7 avian influenza a virus to human beings during a large outbreak in commercial poultry farms in the Netherlands. *Lancet* **363**: 587–593.

Korzeniewska, E. (2011). Emission of bacteria and fungi in the air from wastewater treatment plants – a review. *Frontiers in Bioscience (Scholar Edition)* **3**: 393–407.

Kraus, A.A. and Mirazimi, A. (2014). Laboratory biosafety in containment laboratories. In: *Working in Biosafety Level 3 and 4 Laboratories: A Practical Introduction* (ed. M. Weidmann, N. Silman, P. Butaye, and M. Elscher), 5–12. Weinheim, Germany: Wiley-VCH Verlag GmbH & Co. KGaA.

Kuhar, D.T., Henderson, D.K., Struble, K.A. et al. (2013). Updated US public health service guidelines for the management of occupational exposures to human immunodeficiency virus and recommendations for postexposure prophylaxis. *Infection Control and Hospital Epidemiology* **34**: 875–892.

Ladou, J. and Harrison, R. (2014). *Environmental and Occupational Medicine*, 5e. Columbus, OH: McGraw-Hill Education.

Lawrence, J., Korber, D., and Neu, T. (2007). Analytical imaging and microscopy techniques. In: *Manual of Environmental Microbiology* (ed. C.J. Hurst, R.L. Crawford, J.L. Garland, et al.), 40–68. Washington, DC: American Society for Microbiology Press.

Lee, J.A., Thorne, P.S., Reynolds, S.J., and O'Shaughnessy, P.T. (2007). Monitoring risks in association with exposure levels among wastewater treatment plant workers. *Journal of Occupational and Environmental Medicine* **49**: 1235–1248.

Lee, S.S. and Wong, N.S. (2015). Probable transmission chains of Middle East respiratory syndrome coronavirus and the multiple generations of secondary infection in South Korea. *International Journal of Infectious Diseases* **38**: 65–67.

Lenhart, S.W., Hajjeh, R.A., Schafer, M.P., and Singal, M. (2004). *Histoplasmosis, Protecting Workers at Risk*. Atlanta, GA: National Institute for Occupational Safety and Health, National Center for Infectious Diseases (U.S.).

Lennell, A., Kuhlmann-Berenzon, S., Geli, P. et al. (2008). Alcohol-based hand-disinfection reduced children's absence from Swedish day care centers. *Acta Paediatrica* **97**: 1672–1680.

Levy, B.S., Wegman, D.H., Baron, S.L., and Sokas, R.K. (2006). *Occupational and Environmental Health: Recognizing and Preventing Disease and Injury*. Philadelphia, PA: Lippincott, Williams & Wilkins.

Lewis, J.D., Enfield, K.B., and Sifri, C.D. (2015). Hepatitis B in healthcare workers: transmission events and guidance for management. *World Journal of Hepatology* **7**: 488–497.

Lippmann, M., Burge, H., Jones, B. et al. (2002). *The Airliner Cabin Environment and the Health of Passengers and Crew*. Washington, DC: The National Academies Press.

Lucas, C.E. (2015). Legionellae and Legionnaires' disease. In: *Manual of Environmental Microbiology* (ed. M.V. Yates, C.H. Nakatsu, R.V. Miller, and S.D. Pillai). Washington, DC: American Society of Microbiology.

Lynch, L. and Spivak, E.S. (2015). The pregnant healthcare worker: fact and fiction. *Current Opinion in Infectious Diseases* **28**: 362–368.

Lynfield, R. and Daum, R. (2014). The complexity of the resurgence of childhood vaccine-preventable diseases in the United States. *Current Pediatrics Reports* **2**: 195–203.

Macher, J.M., Douwes, J., Prezant, B., and Reponen, T. (2012). Bioaerosols. In: *Aerosols Handbook: Measurement, Dosimetry, and Health Effects* (ed. L.S. Ruzer and N.H. Harley), 285–343. Boca Raton, FL: CRC Press.

MacIntyre, C.R., Chughtai, A.A., Seale, H. et al. (2014). Respiratory protection for healthcare workers treating Ebola virus disease (EVD): are facemasks sufficient to meet occupational health and safety obligations? *International Journal of Nursing Studies* **51**: 1421–1426.

MacMahon, K.L., Delaney, L.J., Kullman, G. et al. (2008). Protecting poultry workers from exposure to avian influenza viruses. *Public Health Reports* **123**: 316–322.

Magill, S.S., Edwards, J.R., Bamberg, W. et al. (2014). Multistate point-prevalence survey of health care-associated infections. *The New England Journal of Medicine* **370**: 1198–1208.

Mandell, G., Bennett, J., and Dolin, R. (2009). *Principles and Practice of Infectious Diseases*, 7e. Philadelphia, PA: Elsevier Churchill Livingstone.

Mangili, A., Vindenes, T., and Gendreau, M. (2015). Infectious risks of air travel. *Microbiology Spectrum* **3**: 1–2.

Martinez-Valdebenito, C., Calvo, M., Vial, C. et al. (2014). Person-to-person household and nosocomial transmission of Andes hantavirus, Southern Chile, 2011. *Emerging Infectious Diseases* **20**: 1629–1636.

Mason, J.O. (1987). Recommendations for prevention of HIV transmission in health-care settings. *MMWR. Morbidity and Mortality Weekly Report* **36 Suppl 2**: 1s–18s.

Matteelli, A., Schlagenhauf, P., Carvalho, A.C. et al. (2013). Travel-associated sexually transmitted infections: an observational cross-sectional study of the GeoSentinel surveillance database. *The Lancet Infectious Diseases* **13**: 205–213.

McCoy, W.F. and Rosenblatt, A.A. (2015). HACCP-based programs for preventing disease and injury from premise plumbing: a building consensus. *Pathogens* **4**: 513–528.

Mills, J.N., Corneli, A., Young, J.C. et al. (2002). Hantavirus pulmonary syndrome – United States: updated recommendations for risk reduction. *MMWR – Recommendations and Reports* **51**: 1–12.

Mohr, A.J. (2011). Aerosol (aerobiology, aerosols, bioaerosols, microbial aerosols). In: *Encyclopedia of Bioterrorism Defense* (ed. R. Katz and R.A. Zilinskas), 5–10. Hoboken, NJ: Wiley.

Morse, S.S., Mazet, J.A.K., Woolhouse, M. et al. (2012). Prediction and prevention of the next pandemic zoonosis. *Lancet* **380**: 1956–1965.

Mostofi, R., Wang, B., Haghighat, F. et al. (2010). Performance of mechanical filters and respirators for capturing nanoparticles – limitations and future direction. *Industrial Health* **48**: 296–304.

Mphaphlele, M., Dharmadhikari, A.S., Jensen, P.A. et al. (2015). Institutional tuberculosis transmission. Controlled trial of upper room ultraviolet air disinfection: a basis for new dosing guidelines. *American Journal of Respiratory and Critical Care Medicine* **192**: 477–484.

Muller, M.P. and McGeer, A. (2006). Febrile respiratory illness in the intensive care unit setting: an infection control perspective. *Current Opinion in Critical Care* **12**: 37–42.

Myatt, T.A., Johnston, S.L., Zuo, Z. et al. (2004). Detection of airborne rhinovirus and its relation to outdoor air supply in office environments. *American Journal of Respiratory and Critical Care Medicine* **169**: 1187–1190.

Nardell, E.A. (2004). Catching droplet nuclei: toward a better understanding of tuberculosis transmission. *American Journal of Respiratory and Critical Care Medicine* **169**: 553–554.

Nardell, E.A., Bucher, S.J., Brickner, P.W. et al. (2008). Safety of upper-room ultraviolet germicidal air disinfection for room occupants: results from the Tuberculosis Ultraviolet Shelter Study. *Public Health Reports* **123**: 52–60.

Ned-Sykes, R., Johnson, C., Ridderhof, J.C. et al. (2015). Competency guidelines for public health laboratory professionals: CDC and the Association of Public Health Laboratories. *Morbidity and Mortality Weekly Report Supplement* **64** (Suppl 1): 1–81.

Nelson, K.E. (2014). Epidemiology of infectious disease: general principles. In: *Infectious Disease Epidemiology: Theory and Practice* (ed. K.E. Nelson and C.M. Williams), 19–44. Burlington, MA: Jones & Bartlett Learning.

Nesti, M.M. and Goldbaum, M. (2007). Infectious diseases and daycare and preschool education. *Jornal de Pediatria* **83**: 299–312.

Nicas, M. and Jones, R.M. (2009). Relative contributions of four exposure pathways to influenza infection risk. *Risk Analysis* **29**: 1292–1303.

NIOSH (2013). *Body Art*. http://www.cdc.gov/niosh/topics/body_art (accessed 22 December 2017).

Nivin, B., O'Flaherty, T., Leibert, E. et al. (2002). Sputum induction problems identified through genetic fingerprinting. *Infection Control and Hospital Epidemiology* **23**: 580–583.

Noakes, C.J. and Sleigh, P.A. (2009). Mathematical models for assessing the role of airflow on the risk of airborne infection in hospital wards. *Journal of the Royal Society, Interface* **6**: S791–S800.

Nolte, F. (2015). Molecular microbiology. In: *Manual of Clinical Microbiology* (ed. J. Jorgensen, M. Pfaller, K. Carroll, et al.), 54–90. Washington, DC: ASM Press.

NSF/ANSI (2016). *NSF/ANSI Standard 49, Biosafety Cabinetry: Design, Construction, Performance, and Field Certification*. Ann Arbor, MI: NSF International.

Ofner-Agostini, M., Gravel, D., McDonald, L.C. et al. (2006). Cluster of cases of severe acute respiratory syndrome among Toronto healthcare workers after implementation of infection control precautions: a case series. *Infection Control and Hospital Epidemiology* **27**: 473–478.

OHSA (2017). *Ebola, Medical Information*. In Safety and Health Topics. https://www.osha.gov/SLTC/ebola/medical_info.html (accessed 22 December 2017).

Ontario (2015). *Infection Prevention and Control Best Practices for Personal Services Settings*. Ontario, Canada: Minister of Health and Long-Term Care.

OSHA (1970). *Duties of Employers and Employees* (Occupational Safety and Health Act of 1970).

OSHA (2003). Occupational exposure to tuberculosis: proposed rule: termination of rulemaking respiratory protection for M. tuberculosis: final rule (revocation). *Federal Register* **68**: 75767–75775.

OSHA (2015). Enforcement procedures and scheduling for occupational exposure to tuberculosis. In: *OSHA Instruction CPL*, CPL 02-02-078. Washington, DC: Department of Labor.

OSHA (2017). *Health Hazards in Nail Salons*. https://www.osha.gov/SLTC/nailsalons (accessed 22 December 2017).

Parr, A., Whitney, E.A., and Berkelman, R.L. (2015). Legionellosis on the rise: a review of guidelines for prevention in the United States. *Journal of Public Health Management and Practice* **21**: E17–E26.

Payment, P. (2007). Cultivation and assay of animal viruses. In: *Manual of Environmental Microbiology* (ed. C.J. Hurst, R.L. Crawford, J.L. Garland, et al.), 93–100. Washington, DC: American Society of Microbiology.

Pedrosa, P.B.S. and Cardoso, T.A.O. (2011). Viral infections in workers in hospital and research laboratory settings: a comparative review of infection modes and respective biosafety aspects. *International Journal of Infectious Diseases* **15**: e366–e376.

Perdue, C.L., Cost, A.A., Rubertone, M.V. et al. (2015). Description and utilization of the United States department of defense serum repository: a review of published studies, 1985–2012. *PLoS One* **10**: e0114857.

PHAC (2017). *Pathogen Safety Data Sheets and Risk Assessment*. Public Health Agency of Canada. https://www.canada.ca/en/public-health/services/laboratory-biosafety-biosecurity/pathogen-safety-data-sheets-risk-assessment.html (accessed 22 December 2017).

PHMSA (2017). *Pipeline and Hazardous Materials Safety Administration: Mission Statement*. http://www.phmsa.dot.gov/about/mission (accessed 22 December 2017).

Picard, C., Al-Herz, W., Bousfiha, A. et al. (2015). Primary immunodeficiency diseases: an update on the classification from the International Union of Immunological Societies Expert Committee for Primary Immunodeficiency 2015. *Journal of Clinical Immunology* **35**: 696–726.

Raloff, J. (2011). Germs' persistence: nothing to sneeze at. In: *Science News*. Washington, DC: Society for Science & the Public.

Reed, N.G. (2010). The history of ultraviolet germicidal irradiation for air disinfection. *Public Health Reports* **125**: 15–27.

Refaey, S., Azziz-Baumgartner, E., Amin, M.M. et al. (2015). Increased number of human cases of influenza virus A (H5N1) infection, Egypt, 2014–15. *Emerging Infectious Diseases* **21**: 2171–2173.

Reissman, D. and Piancentino, J. (2010). Protecting disaster rescue and recovery workers. In: *Occupational and Environmental Health: Recognizing and Preventing Disease and Injury* (ed. B. Levy, D. Wegman, S. Baron, and R. Sokas), 779–797. New York: Oxford University Press.

Reponen, T., Willeke, K., Grinshpun, S., and Nevalainen, A. (2011). Biological particle sampling. In: *Aerosol Measurement: Principles, Techniques, and Applications* (ed. P. Kulkarni, P.A. Baron, and K. Willeke), 549–570. Hoboken, NJ: Wiley.

Riley, R.L. (2001). What nobody needs to know about airborne infection. *American Journal of Respiratory and Critical Care Medicine* **163**: 7–8.

Riley, E.C., Murphy, G., and Riley, R.L. (1978). Airborne spread of measles in a suburban elementary school. *American Journal of Epidemiology* **107**: 421–432.

Ringen, K., Landrigan, P.J., Stull, J.O. et al. (2015). Occupational safety and health protections against Ebola virus disease. *American Journal of Industrial Medicine* **58**: 703–714.

Roest, H.I., Bossers, A., van Zijderveld, F.G., and Rebel, J.M. (2013). Clinical microbiology of *Coxiella burnetii* and relevant aspects for the diagnosis and control of the zoonotic disease Q fever. *The Veterinary Quarterly* **33**: 148–160.

Roomes, D. (2010). Genetic modification and biotechnology. In: *Hunter's Diseases of Occupations* (ed. P.J. Baxter, T.-C. Aw, A. Cockcroft, et al.), 782–799. Boca Raton, FL: CRC Press.

Rosenstock, L., Cullen, M., Brodkin, C., and Redlich, C. (2005). *Textbook of Clinical Occupational and Environmental Medicine*. Edinburgh, Scotland: W.B. Saunders.

Roy, C.J. and Milton, D.K. (2004). Airborne transmission of communicable infection – the elusive pathway. *The New England Journal of Medicine* **350**: 1710–1712.

Rutala, W.A., Weber, D.J., and HICPAC (2008). *Guideline for Disinfection and Sterilization in Healthcare Facilities, 2008*. Atlanta, GA: CDC.

Schillie, S., Murphy, T.V., Sawyer, M. et al. (2013). CDC guidance for evaluating health-care personnel for hepatitis B virus protection and for administering postexposure management. *MMWR – Recommendations and Reports* **62**: 1–19.

Schmolze, D.B., Standley, C., Fogarty, K.E., and Fischer, A.H. (2011). Advances in microscopy techniques. *Archives of Pathology and Laboratory Medicine* **135**: 255–263.

Schreckenberger, P.C. and McAdam, A.J. (2015). Point-counterpoint: large multiplex PCR panels should be first-line tests for detection of respiratory and intestinal pathogens. *Journal of Clinical Microbiology* **53**: 3110–3115.

Seger, R.A. (2008). Modern management of chronic granulomatous disease. *British Journal of Haematology* **140**: 255–266.

Sehulster, L. and Chinn, R.Y. (2003). Guidelines for environmental infection control in health-care facilities. Recommendations of CDC and the Healthcare Infection Control Practices Advisory Committee (HICPAC). *MMWR – Recommendations and Reports* **52**: 1–42.

Semel, M.E., Tavakkolizadeh, A., and Gates, J.D. (2009). Babesiosis in the immediate postoperative period after splenectomy for trauma. *Surgical Infections* **10**: 553–556.

Sepkowitz, K.A. and Eisenberg, L. (2005). Occupational deaths among healthcare workers. *Emerging Infectious Diseases* **11**: 1003–1008.

Shang, J., Ma, C., Poghosyan, L. et al. (2014). The prevalence of infections and patient risk factors in home health care: a systematic review. *American Journal of Infection Control* **42**: 479–484.

Sharifi-Mood, B. and Metanat, M. (2006). Comparison of prevalence of hepatitis B virus infection in non official barbers with blood donors. *Journal of Medical Sciences* **6**: 222–224.

Shaw, K. (2006). The 2003 SARS outbreak and its impact on infection control practices. *Public Health* **120**: 8–14.

Sheets, C.D., Harriman, K., Zipprich, J. et al. (2014). Fatal meningococcal disease in a laboratory worker – California, 2012. *MMWR. Morbidity and Mortality Weekly Report* **63**: 770–772.

Shefer, A. (2011). Immunization of health-care personnel: recommendations of the Advisory Committee on Immunization Practices (ACIP). *MMWR – Recommendations and Reports* **60**: 1–45.

Siegel, J.A. (2016). Primary and secondary consequences of indoor air cleaners. *Indoor Air* **26**: 88–96.

Siegel, J., Rhinehart, E., Jackson, M., Chiarello, L., and HICPAC (2017). 2007 *Guideline for Isolation Precautions: Preventing Transmission of Infectious Agents in Healthcare Settings*. https://www.cdc.gov/infectioncontrol/pdf/guidelines/isolation-guidelines.pdf (accessed 22 December 2017).

Sinclair, R.G., Rose, J.B., Hashsham, S.A. et al. (2012). Criteria for selection of surrogates used to study the fate and control of pathogens in the environment. *Applied and Environmental Microbiology* **78**: 1969–1977.

Singh, K. (2009). Laboratory-acquired infections. *Clinical Infectious Diseases* **49**: 142–147.

Singh, K. (2011). It's time for a centralized registry of laboratory-acquired infections. *Nature Medicine* **17**: 919–919.

Souza, K., Davis, L., and Shire, J. (2010). Occupational and environmental health surveillance. In: *Occupational and Environmental Health: Recognizing and Preventing Disease and Injury* (ed. B. Levy, D. Wegman, S. Baron, and R. Sokas), 55–68. New York, NY: Oxford University Press.

Spinler, J., Hemarajata, P., and Verslovic, J. (2015). Microbial genomics and pathogen discovery. In: *Manual of Clinical Microbiology* (ed. J. Jorgensen, M. Pfaller, K. Carroll, et al.), 238–251. Washington, DC: ASM Press.

Sprigg, W.A., Nickovic, S., Galgiani, J.N. et al. (2014). Regional dust storm modeling for health services: the case of valley fever. *Aeolian Research* **14**: 53–73.

Stephens, D.S., Greenwood, B., and Brandtzaeg, P. (2007). Epidemic meningitis, meningococcaemia, and *Neisseria meningitidis*. *Lancet* **369**: 2196–2210.

Stout, J.E., Muder, R.R., Mietzner, S. et al. (2007). Role of environmental surveillance in determining the risk of hospital-acquired legionellosis: a national surveillance study with clinical correlations. *Infection Control and Hospital Epidemiology* **28**: 818–824.

Stürchler, D.A. (2006). *Exposure: A Guide to Sources of Infections*, 910. Washington, DC: American Society of Microbiology.

Sze To, G.N. and Chao, C.Y. (2010). Review and comparison between the Wells–Riley and dose–response approaches to risk assessment of infectious respiratory diseases. *Indoor Air* **20**: 2–16.

Tanner, R.S. (2007). Cultivation of bacteria and fungi. In: *Manual of Environmental Microbiology* (ed. C.J. Hurst, R.L. Crawford, J.L. Garland, et al.), 69–78. Washington, DC: American Society of Microbiology.

Theel, E., Carpenter, A., and Binnicker, M. (2015). Immunoassays for the diagnosis of infectious diseases. In: *Manual of Clinical Microbiology* (ed. J. Jorgensen, M. Pfaller, K. Carroll, et al.), 91–105. Washington, DC: ASM Press.

Thomas, J.K. and Noppenberger, J. (2007). Avian influenza: a review. *American Journal of Health-System Pharmacy* **64**: 149–165.

Tooher, R., Griffin, T., Shute, E., and Maddern, G. (2005). Vaccinations for waste-handling workers. A review of the literature. *Waste Management and Research* **23**: 79–86.

Toth, D.J., Gundlapalli, A.V., Schell, W.A. et al. (2013). Quantitative models of the dose–response and time course of inhalational anthrax in humans. *PLoS Pathogens* **9**: e1003555.

Tschopp, A., Joller, H., Jeggli, S. et al. (2009). Hepatitis E, *Helicobacter pylori* and peptic ulcers in workers exposed to sewage: a prospective cohort study. *Occupational and Environmental Medicine* **66**: 45–50.

Tweed, S.A., Skowronski, D.M., David, S.T. et al. (2004). Human illness from avian influenza H7N3, British Columbia. *Emerging Infectious Diseases* **10**: 2196–2199.

Uddin, M.K., Chowdhury, M.R., Ahmed, S. et al. (2013). Comparison of direct versus concentrated smear microscopy in detection of pulmonary tuberculosis. *BMC Research Notes* **6**: 291.

Uden, L., Barber, E., Ford, N., and Cooke, G.S. (2017). Risk of tuberculosis infection and disease for health care workers: an updated meta-analysis. *Open Forum Infectious Diseases* **4**: ofx137–ofx137.

Uhrbrand, K., Schultz, A.C., and Madsen, A.M. (2011). Exposure to airborne noroviruses and other bioaerosol components at a wastewater treatment plant in Denmark. *Food and Environmental Virology* **3**: 130–137.

U.S. Department of Health and Human Services (2011). *Child Health USA 2011*. Rockville, MD: Health Resources and Services Administration, Maternal and Child Health Bureau.

U.S. Department of Homeland Security, U.S.C.G. (2006). *Public Health and Disease Concerns Related to Coast Guard Operations*, 29. Washington, DC: United States Coast Guard.

USDHHS (2009). *Biosafety in Microbiological and Biomedical Laboratories (BMBL)*, 5e, HHS Publication No. (CDC) 21-1112. Washington, DC: Centers for Disease Control and Prevention and National Institutes of Health. http://www.cdc.gov/biosafety/publications/bmbl5 (accessed 22 December 2017).

USDoD (2015). *Review Committee Report: Inadvertent Shipment of Live Bacillus Anthracis Spores by DoD*, 37. Washington, DC: U.S. Department of Defense.

USEPA (2017a). *Links to Hazardous Waste Programs and U.S. State Environmental Agencies*. https://www.epa.gov/hwgenerators/links-hazardous-waste-programs-and-us-state-environmental-agencies (accessed 22 December 2017).

USEPA (2017b). *Selected EPA-registered Disinfectants*. http://www2.epa.gov/pesticide-registration/selected-epa-registered-disinfectants (accessed 22 December 2017).

van Leuken, J.P., van de Kassteele, J., Sauter, F.J. et al. (2015). Improved correlation of human Q fever incidence to modelled *C. burnetii* concentrations by means of an atmospheric dispersion model. *International Journal of Health Geographics* **14**: 14.

Villafruela, J.M., Castro, F., San José, J.F., and Saint-Martin, J. (2013). Comparison of air change efficiency, contaminant removal effectiveness and infection risk as IAQ indices in isolation rooms. *Energy and Buildings* **57**: 210–219.

Wald, P. and Stave, G. (2001). *Physical and Biological Hazards of the Workplace*, 2e. Hoboken, NJ: Wiley.

Weigler, B.J., Cooper, D.R., and Hankenson, F.C. (2012). Risk-based immunization policies and tuberculosis screening practices for animal care and research workers in the United States: survey results and recommendations. *Journal of the American Association for Laboratory Animal Science* **51**: 561–573.

WHO (2004). *Laboratory Biosafety Manual*, 3e. Geneva, Switzerland: World Health Organization.

WHO (2012). *International Health Regulations (2005): A Guide for Public Health Emergency Contingency Planning at Designated Points of Entry*. Geneva, Switzerland: World Health Organization.

WHO (2014). *Coordinated Public Health Surveillance Between Points of Entry and National Health Surveillance Systems – Advising Principles*. Geneva, Switzerland: World Health Organization.

WHO (2015). *World Health Organization Best Practices for the Naming of New Human Infectious Diseases*. Geneva, Switzerland: World Health Organization.

WHO (2017a). *Guidance on Regulations for the Transport of Infectious Substances 2017–2018: Applicable as from 1 January 2017*. Geneva, Switzerland: World Health Organization.

WHO (2017b). *Influenza at the Human-animal Interface. Summary and Assessment, 28 September to 30 October 2017*. http://www.who.int/influenza/human_animal_interface/Influenza_Summary_IRA_HA_interface_10_30_2017.pdf?ua=1 (accessed 22 December 2017).

WHO (2017c). *Safe Management of Waste from Health-Care Activities: A Summary*. Geneva, Switzerland: World Health Organization.

Wiedbrauk, D. (2015). Microscopy. In: *Manual of Clinical Microbiology* (ed. P. Murray), 5–14. Washington, DC: ASM Press.

Wilken, J.A., Marquez, P., Terashita, D. et al. (2014). Coccidioidomycosis among cast and crew members at an outdoor television filming event – California, 2012. *MMWR. Morbidity and Mortality Weekly Report* **63**: 321–324.

Wong, D., Wild, M.A., Walburger, M.A. et al. (2009). Primary pneumonic plague contracted from a mountain lion carcass. *Clinical Infectious Diseases* **49**: e33–e38.

Wood, R., Morrow, C., Ginsberg, S. et al. (2014). Quantification of shared air: a social and environmental determinant of airborne disease transmission. *PLoS One* **9**: e106622.

Woodside, J. (2013). *Guide to Infection Prevention in Emergency Medical Services*, 86. Washington, DC: Association for Professionals in Infection Control and Epidemiology.

Yazdankhah, S.P. and Caugant, D.A. (2004). *Neisseria meningitidis*: an overview of the carriage state. *Journal of Medical Microbiology* **53**: 821–832.

Yezli, S. and Otter, J. (2011). Minimum infective dose of the major human respiratory and enteric viruses transmitted through food and the environment. *Food and Environmental Virology* **3**: 1–30.

Young, M.K. and Cripps, A.W. (2013). Passive immunization for the public health control of communicable diseases: current status in four high-income countries and where to next. *Human Vaccines & Immunotherapeutics* **9**: 1885–1893.

Zhou, F., Shefer, A., Wenger, J. et al. (2014). Economic evaluation of the routine childhood immunization program in the United States, 2009. *Pediatrics* **133**: 577–585.

CHAPTER 7

# OCCUPATIONAL DERMATOSES

DAVID E. COHEN

*Dermatology Department, NYU Medical Center, 560 First Ave., New York, NY, 10016*

Adapted from "Occupational dermatoses" by Donald J. Birmingham, Chapter 10 in G. D. Clayton and F. E. Clayton (eds.), *Patty's Industrial Hygiene and Toxicology*, 4e, Vol. IA. New York: Wiley, 1991, pp. 253–287.

## 7.1 INTRODUCTION

The skin represents the largest organ in the body, encompassing 1.5–2 $m^2$ of surface area. In its role as a primary defender against external insult, the skin is particularly vulnerable to damage by physical and chemical assaults in the workplace. Barrier function represents only a fraction of the duties performed by the entire integumentary system, which participates directly in thermal, electrolyte, hormone, and immune regulation without which life is not possible. The metabolic potential of the skin is impressive, and rather than merely repelling chemical or physical assaults, the skin may compensate by metabolizing and biotransforming agents to less harmful ones. Hence, the skin is far from a passive coat of armor but rather an interactive organ that is in constant flux with its environment.

The skin's precarious location has rendered it the most commonly injured organ from chemical agents and physical conditions of the workplace. Pathologic responses of the skin can vary from excessive dryness and mild redness to more generalized exfoliative dermatitides that are life threatening. Neoplasms of the skin may occur as the result of primary skin exposures or through systemic absorption via the skin or other route of entry. Benign or malignant, such events can have catastrophic consequences to the host. Historically, occupational skin disease has been morphologically documented and often has very descriptive nomenclature that easily identifies the purported causative agent. Among the workforce, a number of more descriptive titles associated with cause are commonly used, for example, asbestos wart, cement burn, chrome holes, fiber glass itch, hog itch, oil acne, rubber rash, and tar smarts. In view of the variety of skin lesions known to result from contactants within the workplace, the term *occupational dermatoses* is preferred because it includes any abnormality of the skin resulting directly from or aggravated by the work environment (Schwartz et al. 1957).

### 7.1.1 Historical

Just when and how occupational affections of the skin first occurred is a matter of conjecture; however, if we apply our past and present knowledge of diseases associated with work, it can be reasonably suspected that skin disorders in one or another form were expressed soon after humans began to perform various types of work. Archeology has shown that ancient inhabitants invented and used primitive tools and weapons made of stone, flint, bone, and wood (History of Technology 1978). It is quite likely that abrasions, blisters, bruises, lacerations, punctures, and probably more serious traumas were incurred as part of daily living associated with hunting for food, building shelter, making clothing, and gaining protection. Poisonous plants and biological agents, including various parasites, probably took their toll; however, such harmful effects remain suspect rather than documented in medical history. Perhaps the earliest references occurred in the writings of Celsus about 100 CE (White 1934) when he described ulcers of the skin caused by corrosive metals. During later centuries, several authors enhanced the knowledge of certain occupational diseases, but cutaneous ulcerations seem to have been the major occupational skin disease of record. An explanation may

*Handbook of Occupational Safety and Health*, Third Edition. Edited by S. Z. Mansdorf.

reside in the fact that ulcerations of the skin were easily recognized, especially among those handling metal salts in mining, smelting, tool and weapon making, creating objects of art, glass making, gold and silver coinage, casting, and similar metallics. It would be strange indeed if none of these tradesmen incurred skin problems caused by a substance or condition met with at work. Nonetheless, little was recorded about occupational skin disease until Ramazzini's historic treatise on diseases of tradesmen in 1700 (Ramazzini 1864). In this tome he described skin disorders experienced by bath attendants, bakers, gilders, midwives, millers, and miners, among other tradespeople. Seventy-five years later Sir Percival Pott published the first account of occupational skin cancer when he described scrotal cancer among chimney sweeps (Pott 1775).

The Industrial Revolution of the eighteenth century changed an agricultural and guild economy to one dominated by machines and industrial expansion. As cities and industries grew, so did the study of science and the eventual discovery and use of new materials such as chromium, mercury, and petroleum, among many others. The chemical age brought enormous numbers of materials, natural and synthetic, into industrial and household use (History of Technology 1978). As a result, physicians began to recognize occupational dermatoses and publish their observations in England, Germany, Italy, and France. Similarly, industrialization within the United States led to the recognition of old and new causes of occupational skin disease. Numerous dermatologists, industrial physicians, practitioners, and allied scientists have added to the information bank dealing with clinical investigations, clinical manifestations, causal factors, diagnostic procedures, and treatment and prevention of these disorders. A number of updated texts and related publications are available (Adams 1983; Birmingham 1978; Cronin 1980; Fisher 1995; Gellin 1972; Maibach 1987; Malten and Zielhius 1964; Rycroft 1986; Samitz and Cohen 1985; Suskind 1959). Today's technology has brought about entirely new exposure patterns that directly affect the health of the skin. The astonishing expansion of medical technology coupled with the emergence of increasingly dangerous infectious agents has resulted in tens of thousands of workers potentially exposed to agents never before encountered by such large numbers of people. Today, healthcare workers not only face peril from the infectious waste they may handle but also may develop skin disease from the personal protective equipment they use as well as the chemicals used to neutralize biologic hazards.

### 7.1.2 Incidence

The National Institute for Occupational Safety and Health (NIOSH) has considered the skin a vitally important organ with respect to occupational diseases. In fact, NIOSH has characterized skin disease as one of the most pervasive problems facing workers in the United States. Since 1982, skin disease is listed in the top 10 work-related diseases based on potential for prevention, incidence, and severity (NIOSH 1996). In the 1978 edition of Patty's *Industrial Hygiene and Toxicology* text, dermatologic diseases were shown to account for about 40% of all occupational diseases reported to the US Department of Labor (OSHA 1978). In the 1984 Bureau of Labor Statistics, dermatologic disease had decreased to 34% of all reported occupational disease. For disease rates, a Bureau of Labor Statistics survey show a declining trend for occupational skin disease from a high in 1972 of 16.2 events per 10000 full-time workers to a low in 1986 of 6.9 events per 10000 workers. Recently, however, there has been an upward trend in incidence, with a rate of 7.9/10000 workers in 1990 or about 61000 new cases per year. Skin disease resulting from exposures in the agriculture and manufacturing industries was responsible for the greatest number of cases with incidence rates of 86 and 41/10000 workers, respectively (NIOSH 1996). Since skin disease is often not life threatening, many believe that the rate at which it is reported to government agencies is underrepresented 10- to 50-fold. Although the healthcare field has a relatively low rate of disease, the large number of workers in this industry results in almost 3900 cases of illness per year. As indicated, while skin disease most often does not involve a life-threatening hazard, it is often assumed that the cost of occupational skin disease to society is diminutive. It has been estimated that up to 20–25% of persons with occupational skin disease lose an average of 11 days of work annually. This translates to an economic loss of $222 million to $1 billion annually.

Occupational skin *injuries* such as thermal and chemical burns, lacerations, and blunt skin trauma are extremely common. NIOSH estimates that there are approximately 1.07–1.65 million skin injuries per year, accounting for a rate of 1.4–2.2 cases per 100 workers. These potentially disabling injuries are probably the most preventable illnesses and should not be overlooked. The economic costs of injuries to the skin have not been calculated.

### 7.1.3 Structure and Function of the Skin

Any type of acute or chronic exposure to workplace chemical or physical agents can result in a skin disease. In the overwhelming majority of cases, the skin is able to compensate adequately for such assaults and thus disease is averted. This ability to protect through an impressive reservoir of compensatory mechanisms stems from the multifunctional capabilities of the entire skin system. As will be described later, the skin is not composed of a monomorphous group of cells, but rather is a complex dynamic system of differentiating cells that closely interact with almost every other organ system in the body.

Injury to the host from occupational stressors may result from toxicity to the cells of the skin or through interference of normal homeostatic functions that the skin performs. Only short periods of malfunction of thermoregulatory and electrolyte homeostasis of the skin are compatible with life. Hence, widespread destruction of the skin may result in immediate breakdown of these mechanisms and can result in death. Such insults include chemical and thermal burns, overexposure to heat and humidity, or prolonged occlusion of the skin.

Anatomically, the skin is composed of two main levels, epidermis and dermis. They are separated from each other by a basement membrane, which forms a wavy interface between the layers. The skin appendages, which include the hair follicle unit and sebaceous, eccrine, and apocrine sweat glands, have their respective ductal structure crossing the epidermis to the surface. The concentration of these appendageal structures varies greatly by location. For most, hair follicles abound on the scalp, sebaceous glands are concentrated on the face, and eccrine glands are on the palms, soles, and axillae. Apocrine glands, which have only an incompletely characterized function, are found in the axillae, areolae, and groin.

The outer or epidermal layer varies in thickness, being most protective on the palms and soles. Because it is contiguous with the dermis, it also acts as the outer cover of the cushion of connective and elastic tissue that guards the blood and lymph vessels, nerves, secretory glands, hair shafts, and muscles. Epidermal resiliency provides protection within limits against blunt trauma, and its flexibility accounts for the return of stretched skin to its normal location (Adams 1983; Birmingham 1978; Blank 1979; Cronin 1980; Fisher 1995; Gellin 1972; History of Technology 1978; Maibach 1987; Malten and Zielhius 1964; Pott 1775; Ramazzini 1864; Rycroft 1986; Samitz and Cohen 1985; Schwartz et al. 1957; Suskind 1959; White 1934).

The epidermis has many layers; however, for the purpose of this review, it can be functionally divided into the noncornified layer and the cornified layer or stratum corneum. The principle cell of the epidermis is called the *keratinocyte* and is arranged as a stratified squamous epithelium. Keratinocytes begin as basal cells abutting the basement membrane and progress upward through the epidermis via a terminal differentiation pathway that lasts approximately 14 days (Figure 7.1). During this period there are substantial changes in cell surface markers and a concomitant accumulation of keratin proteins. By the end of the two-week cycle, the keratinocyte has lost its nucleus and intracellular organelles and has flattened. It is at this point that the nonviable keratinocytes enter the stratum corneum and are known as *corneocytes*. The cell structure of the keratinocyte will be largely replaced by a tightly linked layer of fibrous proteins. It will take another 14 days for the original corneocyte to reach the surface of the skin, where it will be sloughed (Blank 1979; Kligman 1964). The stratum corneum is highly important in resisting the mass entrance and exit of water and electrolytes. Stresses such as friction, pressure, and natural and artificial ultraviolet (UV) light can induce compensatory thickening of the stratum corneum in the form of a callous. Besides acting as a water barrier and a physical shield, the stratum corneum provides modest protection against acids and acidic substances. In contrast, it is quite vulnerable to the action of organic and inorganic alkaline materials. Such chemical substances attack the stratum corneum by denaturing the keratin proteins, thus altering the cohesiveness and the capacity to retain water, which is essential in the maintenance of the barrier layer. In short, any physical or chemical force such as lowered temperature and humidity or repetitive action of soaps, detergents, and organic solvents generally leads to impairment of the barrier efficiency because of water loss and dryness (Blank 1979).

Located also in the basal layer are *melanocytes*, which are neuroendocrine-derived cells that are responsible for the production of melanin pigment. Melanin is packaged into pigment granules called melanosomes and imparts the natural pigmentation to the skin. Differences in skin pigmentation occur mostly from differences in melanosome structures than from gross quantities of melanin. The pigment granules that arise from complex enzymatic reactions within the melanocytes are picked up by the epidermal cells and eventually are shed by way of the keratin exfoliation (Fitzpatrick 1993; Kligman 1964). Melanin acts as the principal defense against UV light, since it acts as a broad-spectrum chromophore or light absorber. Besides the natural production of melanin, certain agents such as coal, tar, pitch, selected aromatic chlorinated hydrocarbons, petroleum products, and trauma can cause excess melanin production, leading to hyperpigmentation (Birmingham 1978; OSHA 1978). In contrast, members of the quinone family and selected phenolics can inhibit pigment formation following percutaneous absorption by direct action upon the melanin enzymatic system (Adams 1983; Gellin 1972; Gellin et al. 1970; Kahn 1979; Malten et al. 1971; Schwartz et al. 1957; Suskind 1959). This could result in depigmentation, impacting a marked lightening of the skin or ivory-white appearance.

Immediately below the epidermal region lies the *dermis*, which is much thicker than the epidermis. It is composed of connective tissue made up of collagen, elastic tissue, and ground substance that constitutes an encasement for the sweat glands and ducts, the hair follicles, sebaceous glands that secrete natural sebum (fatty substance), the blood and lymph vessels, and the nerve endings (Adams 1983; Birmingham 1978; Blank 1979; Fitzpatrick et al. 1993; Suskind 1959).

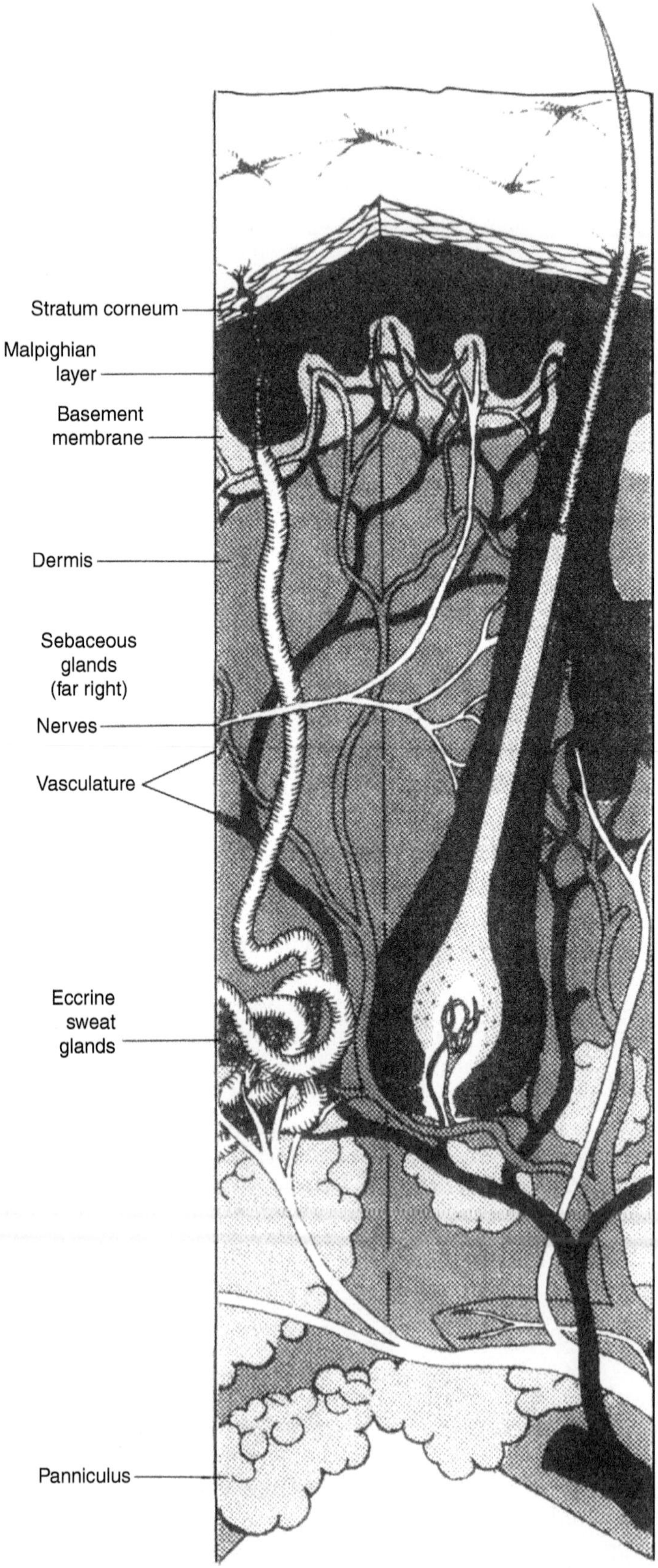


**Figure 7.1** Integration of structure, function, and pathology of human skin.

Thermoregulation is modulated by the excretion of eccrine sweat and changes in the superficial circulation of blood, all of which is controlled by the central nervous system. Thus, core body temperature and circulating blood are physiologically stabilized at a constant temperature despite climatic variations. Eccrine sweat is composed primarily of water and electrolytes and participates in temperature control through heat loss via evaporation from the surface. Simultaneously radiant heat loss is facilitated by the dilation of the cutaneous blood vessels. The opposite occurs when a decreased core and/or surface temperature causes blood vessels of the skin to constrict and shunt warm blood away from exposed surfaces thus preserving heat (Adams 1983; Birmingham 1978; Blank 1979; NIOSH 1996; Suskind 1959).

Secretory functions within the skin are relegated to sweat gland and sebaceous gland activity. Sweat gland function and sweat delivery is essential for physiological normalcy. Too much or too little sweat delivery can have deleterious effects on the whole physiological behavior. Sebaceous glands reside within the dermis as part of the hair follicle unit. Their functional product, sebum, is excreted through the hair follicle and the orifice on the skin surface. Overfunction of this gland is a frequent problem of the adolescent, but the sebaceous glands are also target sites for occupational acne resulting from working with coal tar, heavy oils, greases, and certain aromatic chlorinated hydrocarbons (Adams 1983; Birmingham 1978; Blank 1979; Samitz and Cohen 1985; Suskind 1959).

Special receptors within the skin are part of a network of nerve endings and fibers that receive and conduct various stimuli, later recognized as heat, cold, pain, and other perceptions such as wet, dry, sharp, dull, smooth, and rough (Adams 1983; Birmingham 1978; OSHA 1978; Suskind 1959).

## 7.2 PERCUTANEOUS ABSORPTION

Historically, the skin was believed to be an impervious barrier to external chemicals. More recently, however, it is understood that percutaneous absorption occurs frequently and depends on a variety of properties of the agent. The skin maintains a relatively hydrophobic character, which is imparted from the sphingolipids present in the epidermis (Elias 1992). The water content, which makes up about 20% of the weight of the epidermis, is concentrated around intracellular proteins within keratinocytes. Prolonged submersion can cause a several-fold increase in skin hydration and loss of barrier function.

Percutaneous absorption tends to follow the physics of typical membrane kinetics. Hence increasing the concentration gradient across the epidermis will result in increasing percutaneous absorption. However, two other significant factors must be considered in this complex biological membrane barrier. Molecular weight and hydrophobicity are important in determining or predicting percutaneous absorption. Low-molecular-weight molecules pass through the epidermis with greater alacrity than do larger weight compounds, and those with high octanol/water partition coefficients (more hydrophobic) will pass more easily. Hence low-molecular-weight hydrophobic compounds may pass readily through an intact epidermis and cause toxicity. Since percutaneous absorption is becoming increasingly important, pharmacokinetic models have been developed to predict percutaneous absorption of substances (Potts and Guy 1992).

Absorption may be accomplished via two routes: initially through the skin appendages such as hair follicles and sweat glands and then later through diffusion through the epidermis. Occlusion may serve to enhance penetration by hydrating the epidermis and by preventing evaporation or mechanical removal of a chemical from the skin surface. From a practical perspective, percutaneous absorption has been a particular problem for only certain chemicals such as pesticides, cyanides, aromatic hydrocarbons, mercury, lead, and others. Organophosphate pesticides like parathion are sufficiently absorbed through the skin to be fatal after moderate skin exposure. Until recently organic leads were thought to be the primary form of the heavy metal capable of percutaneous absorption. More recent work has indicated that inorganic lead in dust may result in substantial dermal levels of lead and can cause increases in total body lead burden (Stauber et al. 1994).

Body site has long been described as an important variable in percutaneous absorption. The scrotum has been identified as an area where absorption is greatest, with the face as a moderately penetrable site, and the abdomen the least (Scheuplein and Blank 1971).

### 7.2.1 Metabolism

The barrier features of the skin do not solely lie in the physical obstacle of the epidermal lipids and cross-linked proteins of the stratum corneum. The keratinocytes of the lower epidermis are capable of a legion of biotransformation reactions that may act to render toxic xenobiotics innocuous by the time they reach the superficial vascular plexes that will carry them inward. In fact, the skin possesses most of the phase 1 and phase 2 metabolic enzymes necessary for xenobiotic metabolism and biotransformation. Overall, the skin has 2% of the metabolic capacity of the liver.

Specific enzyme pathways may also be inducible and result in dramatic rises in enzyme activity. This is particularly important for biotransformation reactions when carcinogenic intermediates are formed during detoxifying metabolism of certain polycyclic hydrocarbons (Rice and Cohen 1996). Animal studies have demonstrated that when skin is exposed to polychlorinated biphenyls (PCB) or benzo[a]pyrene, the activity of the phase 1 enzyme aryl hydrocarbon hydroxylase can be increased as to account for 20% of the total body activity of the enzyme. Hence, under conditions of repeated or significant dermal exposure, the skin may be responsible for a marked amount of detoxifying and metabolism of the chemical. Exposure to these procarcinogens on the skin, however, may result in biotransformation to active carcinogens locally, without the aid of hepatic enzymes (McNulty 1985; Mukthar and Bickers 1981).

## 7.3 CAUSAL FACTORS

To detect the cause of an occupational dermatosis, it is fundamental to consider a spectrum of factors that can have indirect relationship to the disease.

### 7.3.1 Indirect Factors

In determining the cause of an occupationally acquired disease, it is important to recognize host and environmental factors that may predispose or aggravate an already existing dermatosis. These include host factors such as genetic predisposition, preexisting dermatologic disease, age, and environmental factors such as temperature, humidity, and season (Adams 1983; Birmingham 1978; Gellin 1972; OSHA 1978; Samitz and Cohen 1985; Schwartz et al. 1957; Suskind 1959; White 1934).

***7.3.1.1 Genetic Predisposition*** This particular subheading represents a wide variety of potentially confounding factors with regard to occupational skin disease. They can include genetic predisposition to skin diseases as described below or extraordinary sensitivity to routinely encountered occupational and environmental stressors. Sensitivity to chemical irritants and ability to acquire allergic contact sensitivity are both likely to be genetically determined. Extreme examples are metabolic genetic diseases that predispose individuals to disease after banal exposures. An example would be a patient with xeroderma pigmentosa (inability to repair DNA after routine sun exposure) who works outdoors and would have an enormously high risk of developing skin cancer compared with even the fairest worker without the disease.

Another genetically determined characteristic potentially impacting on the development of occupational skin disease is skin type. This nomenclature refers to the sun reactivity of the skin and can be loosely correlated to the degree of skin pigmentation. It can be particularly important for workers exposed to UV light either from sunlight or from artificial sources. The susceptibility of some skin cancers may be linked to those having lower skin types. Sunburn or UV-induced skin damage will reduce the barrier function of the skin as well as increase sensitivity to irritant chemicals. Clearly, the sunburned worker will also be less productive than a comfortable one. The use of sunscreen should be encouraged for all workers exposed to UV light on a regular basis for the prevention of skin cancer. Table 7.1 illustrates their characteristics (Fitzpatrick et al. 1993).

***7.3.1.2 Presence of Other Skin Diseases*** New and old employees with preexisting skin disease are prime candidates for a supervening occupational dermatitis or an aggravation of a preexisting disease of the skin. Hirees with adolescent acne may be at higher risk of flaring their disease if exposed to insoluble oils, tar products, greases, and certain polychlorinated aromatic hydrocarbons known to cause chloracne (Adams 1983; Birmingham 1978; Gellin 1972; Samitz and Cohen 1985; Schwartz et al. 1957; Suskind 1959; White 1934).

**TABLE 7.1 Characteristics of Skin Types**

| Skin type | Characteristics |
|---|---|
| 1 | Always burns, never tans |
| 2 | Usually burns, tans less than average (with difficulty) |
| 3 | Sometimes mild burn, tan about average |
| 4 | Rarely burns, tan more than average (easily) |
| 5 | Rarely burn, tan profusely, brown skin |
| 6 | Never burn, tan profusely, black skin |

Patients with dermatitis (inflammation of the skin) are at greater risk when exposed to any number of irritant chemical agents. The capricious skin behavior of these people does not tolerate exposure to dusty or oil-laden jobs (Adams 1983; Birmingham 1978; Gellin 1972; OSHA 1978; Schmunes 1986).

Other skin diseases known to be worsened by physical or chemical trauma, even though mild, are psoriasis, lichen planus, chronic recurrent eczema of the hands, and those skin conditions to which light exposure can be detrimental. The need of careful skin evaluation and job placement is clear (Adams 1983; Birmingham 1978; Samitz and Cohen 1985; Schmunes 1986; Suskind 1959).

***7.3.1.3 Age*** Young workers, particularly those in the adolescent group, often incur acute contact dermatitis. More often, young people are placed in service jobs, for example, fast food, janitorial, or car wash, where wet work prevails and protection is difficult. At times, disregard for safety and hygiene measures may be the reason.

Older workers usually are more careful through experience, but aging skin is often dry, and when work is largely outdoors, sunlight can cause skin cancer (Adams 1983; Birmingham 1978; OSHA 1978; Schwartz et al. 1957; Suskind 1959; White 1934).

***7.3.1.4 Environmental Factors: Temperature, Humidity, and Season*** Normal seasonal variations in temperature, humidity, wind, and incident UV light can affect the normal physiologic defense mechanisms of the body. Increased temperature coupled with high humidity can overload the normal thermoregulatory mechanisms. This may result in abnormally excessive perspiration to more extreme temperature-related diseases such as heat exhaustion and heat stroke. (See Chapter 10 for more information on heat stress.)

Excessive delivery of sweat can cause maceration of the skin in the groin, the armpits, and other sites where skin surfaces are opposed to each other. This can foster the trapping of dusts and suspended particles allowing percutaneous absorption in this hydrated and occluded environment.

Sweat can also partially solubilize nickel, cobalt, and chromium in small amounts, a situation troublesome to those individuals with cutaneous allergy to these metals. Dry chemical agents can be put into aqueous solution, which can be irritating, destructive, or harmful if absorbed. Beneficially, sweat can act as lavage to keep the skin relatively free of irritant contact (Adams 1983; Birmingham 1978; Cronin 1980; Fisher 1995; OSHA 1978; Schwartz et al. 1957; Suskind 1959; White 1934). High temperature and humidity may alone produce skin disease if pores are occluded. Miliaria or occlusive eccrine sweat gland disease may produce annoying symptoms such as itching and stinging. Widespread involvement can interfere with temperature regulation secondary to altered sweat pattern. Mild forms of the disease have been termed *prickly heat* in the vernacular. Diseases caused by temperature are elaborated on later in the chapter. Hot weather may discourage the use of protective clothing gear, thus allowing more unprotected skin for exposure to environmental contactants.

Cold weather is associated with dry skin because of lowered temperatures and humidity. Further, during cold weather, some workers simply do not like to take a shower and then go into the cold after working, allowing prolonged skin contact with potentially dangerous chemicals. Workers with inherently dry skin (xerosis and ichthyosis) usually have a worsening of their condition under the low-humidity conditions of the winter. They are at increased risk or irritation when exposed to alkaline agents, acids, detergents, and most solvents (Adams 1983; Birmingham 1978; Blank 1979; Gellin 1972; Samitz and Cohen 1985; Schmunes 1986; Schwartz et al. 1957; Suskind 1959; White 1934).

***7.3.1.5 Personal Hygiene*** Poor washing habits breed prolonged occupational contact with agents that harm the skin. Personal cleanliness is a sound preventive measure, but it depends upon the presence of readily accessible washing facilities and quality hand cleansers and the recognition by the workers of the need to use them (Adams 1983; Birmingham 1978; OSHA 1978; Samitz and Cohen 1985; Schwartz et al. 1957; Suskind 1959; White 1934).

## 7.4 DIRECT CAUSES OF OCCUPATIONAL SKIN DISEASE

Chemical agents are unquestionably the major cutaneous hazards; however, there are multiple additional agents that are categorized as mechanical, physical, and biological causalities.

### 7.4.1 Chemicals

Chemical agents have always been and will most likely continue to be a major cause of work-incurred skin disease. Organic and inorganic chemicals are used throughout modern industrial processes and increasingly on farms. They act as primary skin irritants, as allergic sensitizers, or as photosensitizers to induce acute and chronic contact eczematous

**TABLE 7.2 Factors Involved in Irritant Contact Dermatitis**

| |
|---|
| Exposure duration |
| Repeated exposures |
| Concentration |
| pH |
| Occlusion |
| Body site |

dermatitis, which accounts for the majority of cases of occupational skin disease, probably no less than 75–80% (Adams 1983; Birmingham 1978; Cronin 1980; Fisher 1995; Gellin 1972; Maibach 1987; Malten and Zielhius 1964; OSHA 1978; Rycroft 1986; Samitz and Cohen 1985; Schwartz et al. 1957; Suskind 1959; White 1934).

***7.4.1.1 Primary Irritants*** Most diseases of the skin caused by work result from contact with primary irritant chemicals. As contact dermatitis overwhelmingly represents the most common occupational skin disease, and approximately 80% of cases of contact dermatitis are of the irritant variety, irritant contact dermatitis is singularly the most common occupational disease involving the skin (CDC, NIOSH, n.d.). These materials cause an irritant dermatitis by direct action on normal skin. The eruption will occur at the site of contact if sufficient quantity, concentration, and time is allowed. In other words, any normal skin can be injured by a primary irritant. This form of dermatitis represents a nonallergic variety and is dependent on a number of factors relating to the exposure (Table 7.2). Certain irritants, such as sulfuric, nitric, or hydrofluoric acid, can be exceedingly powerful in damaging the skin within moments. Similarly, sodium hydroxide, chloride of lime, or ethylene oxide gas can produce rapid damage. These are absolute or strong irritants, and they can produce necrosis and ulceration, resulting in severe scarring. More commonly encountered are the low-grade or marginal irritants that through repetitious contact produce a slowly evolving contact dermatitis. Marginal irritation is often associated with contact with soluble metalworking fluids, soap and water, and solvents such as acetone, ketone, and alcohol. Wet work in general is associated with repetitive contact with marginal irritants (Adams 1983; Birmingham 1978; Cronin 1980; Fisher 1995; Gellin 1972; Maibach 1987; Malten and Zielhius 1964; OSHA 1978; Rycroft 1986; Samitz and Cohen 1985; Schwartz et al. 1957; Suskind 1959; White 1934).

***7.4.1.2 Primary Irritant Action on the Skin*** Clinical manifestations produced by contact with primary irritant agents are readily recognized but not well understood. General behavior of many chemicals in the laboratory or in industrial processes is fairly well known, and at times their chemical action can be applied theoretically, and in some instances actually, to chemical action on human skin. For example, we know that organic and inorganic alkalis damage keratin; that organic solvents dissolve surface lipids and remove lipid components from keratin cells; that heavy metal salts, notably arsenic and chromium, precipitate protein and cause it to denature; that salicylic acid, oxalic acid, and urea, among other substances, can chemically and physically reduce keratin; and that arsenic, tar, methylcholanthrene, and other known carcinogens stimulate abnormal growth patterns. Regardless of the specific agent eliciting an irritant contact dermatitis, multiple chemotactic cytokines are released from both the ailing keratinocytes and neighboring vascular endothelial cells. This results in infiltration of the epidermis and upper dermis with inflammatory cells and concomitant edema, resulting in the hallmark dermatitis (Adams 1983; Birmingham 1978; Cronin 1980; Fisher 1995; Gellin 1972; OSHA 1978; Rice and Cohen 1996; Rycroft 1986; Samitz and Cohen 1985; Schwartz et al. 1957; Suskind 1959; White 1934).

Irritant chemicals are commonly present in agriculture, manufacturing, and service pursuits. Hundreds of these agents classified as acids, alkalis, gases, organic materials, metal salts, solvents, resins, and soaps, including synthetic detergents, can cause absolute or marginal irritation (Table 7.3) (Birmingham 1978).

### 7.4.2 Chemical Burns

Under irritant dermatitis conditions, epidermal necrosis of scattered keratinocytes is typical (Lever and Schaumburg-Lever 1990). If, however, the damage to relatively large areas of epidermis is overwhelming to the defenses of the skin, necrosis of the epidermis may occur as a result of a chemical exposure. The mechanisms of such chemical burns are similar to those of irritant dermatitis but on a more destructive scale. For many irritants, extreme exposures can result in significant chemical burns.

Table 7.4 lists chemicals with characteristics that make them particularly corrosive to the skin. Many of their mechanisms lie in their ability to denature epidermal proteins rapidly or cause damage by their ability to cause extreme temperature changes. A more exhaustive list can be found in de Groot's text on untoward effects of chemicals on the skin.

**TABLE 7.3 Typical Primary Irritants**

| | |
|---|---|
| Acids | |
| *Inorganic* | *Organic* |
| Arsenious | Acetic |
| Chromic | Acrylic |
| Hydrobromic | Carbolic |
| Hydrochloric | Chloracetic |
| Hydrofluoric | Cresylic |
| Nitric | Formic |
| Phosphoric | Lactic |
| | Oxalic |
| | Salicylic |
| Alkalis | |
| *Inorganic* | *Organic* |
| Ammonium | Butylamines |
| Carbonate | Ethylamines |
| Hydroxide | Ethanolamines |
| Calcium | Methylamines |
| Carbonate | Propylamines |
| Cyanamide | Triethanolamine |
| Hydroxide | |
| Oxide | |
| Potassium | |
| Carbonate | |
| Hydroxide | |
| Sodium | |
| Carbonate (soda ash) | |
| Hydroxide (caustic soda) | |
| Silicate | |
| Trisodium phosphate | |
| Cement | |
| Soaps | |
| Detergents | |
| Surfactants | |
| Metal salts | |
| Antimony trioxide | |
| Arsenic trioxide | |
| Chromium and alkaline chromates | |
| Cobalt sulfate and chloride | |
| Nickel sulfate | |
| Mercuric chloride | |
| Silver nitrate | |
| Zinc chloride | |
| Solvents | |
| *Alcohols* | *Ketones* |
| Allyl | Acetone |
| Amyl | Methyl ethyl |
| Butyl | Methyl cyclohexanone |
| Ethyl | |
| Methyl | |
| Propyl | |

*(Continued)*

**TABLE 7.3 (Continued)**

| Solvents | |
|---|---|
| *Chlorinated* | *Petroleum* |
| Carbon tetrachloride | Benzene |
| Chloroform | Ether |
| Dichloroethylene | Gasoline |
| Epichlorohydrin | Kerosene |
| Ethylene chlorohydrin | Varsol |
| Perchloroethylene | White spirit |
| Trichloroethylene | |
| *Coal tar* | *Turpentine* |
| Benzene | Pure oil |
| Naphtha | Turpentine |
| Toluene | Terpineol |
| Xylene | Rosin spirit |

**TABLE 7.4 Chemicals Capable of Severe Skin Burns**

| |
|---|
| Ammonia |
| Calcium oxide |
| Chlorine |
| Ethylene oxide |
| Hydrochloric acid |
| Hydrofluoric acid |
| Hydrogen peroxide |
| Methyl bromide |
| Nitrogen oxide |
| Phosphorous |
| Phenol |
| Sodium hydroxide |
| Toluene diisocyanate |

### 7.4.3 Allergic Skin Disease

Allergic contact dermatitis represents the quintessential delayed-type hypersensitivity (type IV) allergic reaction to an external chemical. Over 2800 allergens have thus far been identified, and allergic contact dermatitis represents approximately 20% of cases of contact dermatitis (de Groot and Nater 1994). Agents capable of causing contact dermatitis are generally small-molecular-weight chemicals that act as haptens. Haptens are not intrinsically allergenic but become so when they bind to an endogenous protein, usually in the epidermis or dermis, and mark the formation of a complete allergen. As with all truly allergic reactions, initial sensitization must occur. Sensitization occurs when an allergen is incorporated in cutaneous immune cells called Langerhans cells. These cells possess surface proteins (human leukocyte antigens, HLA, class II), which allow them to directly present the digested allergen to a helper T lymphocyte. This lymphocyte will become activated and migrate to regional lymph nodes, where a clone of similarly sensitized cells will proliferate under the stimulation of cytokines like interleukins 1 and 2. These cells may now migrate back to the area of the skin that had the original contact with the allergen and cause a typical dermatitis to form. This initial sensitization may take several days to complete and last a lifetime. Now sensitized, subsequent challenges from the same allergen will result in a rapid elaboration of lymphocytes into the dermis and result in a dermatitis within 48–96 hours (Rietschel and Fowler 1995).

The ability to become sensitized appears to be strongly mediated by genetics. Hence, one is probably born with all the potential allergies that an individual will possess for a lifetime. Recent work has linked specific HLA proteins to allergy during a lifetime that will depend on whether exposure is sufficient to trigger the immune cascade necessary for sensitization.

Clinically the dermatitis caused by allergic mechanisms and irritant mechanisms may be indistinguishable. In the field, to distinguish between a primary irritant and allergic contact dermatitis, it is necessary to recognize that (i) allergic reactions usually require a longer induction period than occurs with primary irritation effects and (ii) cutaneous

**TABLE 7.5 Common Contact Allergens**

| Source | Common allergens | |
|---|---|---|
| Topical medications/hygiene products | *Antibiotics* | *Therapeutics* |
| | Bacitracin | Benzocaine |
| | Neomycin | Corticosteroids |
| | Polymyxin | α-Tocopherol (vitamin E) |
| | Aminoglycosides | |
| | Sulfonamides | |
| | *Preservatives* | *Others* |
| | Benzalkonium chloride | Cinnamic aldehyde |
| | Formaldehyde | Ethylenediamine |
| | Formaldehyde releasers | Lanolin |
| | Quaternium-15 | *p*-Phenylenediamine |
| | Imidazolidinyl | Propylene glycol |
| | Diazolidinyl urea | Benzophenones |
| | DMDM hydantoin | Fragrances |
| | Methylchloroisothiazolene | Thioglycolates |
| Plants and trees | Abietic acid | Pentadecylcatechols |
| | Balsam of Peru | Sesquiterpene lactone |
| | Rosin (colophony) | Tuliposide A |
| Antiseptics | Glutaraldehyde | Hexachlorophene |
| | Chlorhexidine | Mercurials |
| | Chloroxylenol | Thimerosal |
| | | Phenylmercuric acetate |
| Rubber products | Diphenylguanidine | Resorcinol monobenzoate |
| | Hydroquinone | Benzothiazolesulfenamides |
| | Mercaptobenzothiazole | Dithiocarbamates |
| | *p*-Phenylenediamine | Thiurams |
| | | Thioureas |
| Leather | Formaldehyde | Potassium dichromate |
| | Glutaraldehyde | |
| Paper products | Abietic acid | Rosin (colophony) |
| | Formaldehyde | Dyes |
| Glues and bonding agents | Bisphenol A | Epoxy resins |
| | Epichlorohydrin | *p*-(*t*-butyl)formaldehyde resin |
| | Formaldehyde | Toluene sulfonamide resins |
| | Acrylic monomers | Urea formaldehyde resins |
| | Cyanoacrylates | |
| Metals | Chromium | Mercury |
| | Cobalt | Nickel |
| | Gold | Palladium |

sensitizers generally do not affect large numbers of workers except when dealing with very potent sensitizers such as epoxy resin systems, phenol–formaldehyde plastics, poison ivy, and poison oak. Some other well-known sensitizers associated with occupation exposures are potassium dichromate by itself or contained in cement, nickel sulfate, hexamethylenetetramine, mercaptobenzothiazole, and tetramethylthiuram disulfide, among several other agents (Adams 1983; Birmingham 1978; Cronin 1980; Fisher 1995; Gellin 1972; Maibach 1987; Samitz and Cohen 1985; Schwartz et al. 1957; Suskind 1959). Table 7.5 lists common allergens and their likely sources of exposure (Rice and Cohen 1996).

### 7.4.4 Plants and Woods

Many plants and woods cause injury to the skin through direct irritation or allergic sensitization by their chemical nature. Additionally, irritation can result from contact with sharp edges of leaves, spines, thorns, and so on, which are appendages of the plants. Photosensitivity may also be a factor.

Although the chemical identity of many plant toxins remains undetermined, it is well known that the irritant or allergic principal can be present in the leaves, stems, roots, flowers, and bark (Barber and Husting 1977; Lampke and Fagerstrom 1968).

High-risk jobs include agricultural workers, construction workers, electric and telephone linemen, florists, gardeners, lumberjacks, pipeline installers, road builders, and others who work outdoors (Gellin et al. 1971).

Poison ivy and poison oak are major offenders. In California, several thousand cases of poison oak occupational dermatitis are reported each year. Poison ivy, oak, and sumac are members of the Anacardiaceae, which also includes a number of chemically related allergens as cashew nut shell oil, Indian marking nut oil, and mango. The chemical toxicant common to this family is a phenolic (catechol), and sensitization to one family member generally confers sensitivity or cross-reactivity to the others (Adams 1983; Cronin 1980; Fisher 1995; Gellin 1972; Gellin et al. 1971; Lampke and Fagerstrom 1968; Rycroft 1986; Schwartz et al. 1957).

Plants known to cause dermatitis are carrots, castor beans, celery, chrysanthemum, hyacinth, tulip bulbs, oleander, primrose, ragweed, and wild parsnip. Other plants including vegetables have been reported as causal in contact dermatitis (Adams 1983; Cronin 1980; Fisher 1995; Gellin 1972; Gellin et al. 1971; Lampke and Fagerstrom 1968; Rycroft 1986).

A number of woods are known to provoke skin disease. Woods do not cause as many cases as are reported from plants, but carpenters, cabinetmakers, furniture builders, lumberjacks, lumberyard workers, and model makers (patterns) can incur primary irritant, allergic dermatitis, or traumatic effects from the wood being handled. Colophony or rosin is a particularly common cause of allergic contact dermatitis and is derived from pine wood. Its color and physical properties make it a useful ingredient in many nonwood products. It may be found as a colorant in yellow soaps, and its tackiness is exploited in its use for baseball players, preparations for violin bows, and adhesives, tapes, paints, and polishes (Rietschel and Fowler 1995). Sawdust, wood spicules, and chemical impregnants in the wood may cause irritation, whereas most cases of allergic dermatitis are caused by oleoresins, the natural oil, or chemical additives. Woods best known for their dermatitis-producing potential are acacia, ash, beech, birch, cedar, mahogany, maple, pine, and spruce. Other agents capable of causing cutaneous injury are the chemicals used for wood preservation purposes, such as arsenicals, chlorophenols, creosote, and copper compounds (Adams 1983; Barber and Husting 1977; Birmingham 1978; Canadian Department of Forestry 1966).

### 7.4.5 Photosensitivity

***7.4.5.1 Phototoxicity*** Dermatitis resulting from photoreactivity is an untoward cutaneous reaction usually to the UV band of the electromagnetic spectrum. This band of light spans from 200 to 400 nm. It is further stratified into three sub-bands of light: UV-A, UV-B, and UV-C. UV-C, spanning 200–280 nm, does not penetrate the upper atmosphere and has no clinical significance from natural light sources. In artificial settings, UV-C can cause marked sunburn within hours of an exposure, faster than other bands of UV light. It has been exploited for its antimicrobial potential in sterilizing air handling systems. UV-B, 280–320 nm, has potent effects on keratinocytes and is capable of causing significant acute damage to the epidermis. It can penetrate the epidermis but cannot reach the upper portions of the dermis. It is the sub-band responsible for sunburn or UV-induced erythema. UV-A, 320–400 nm, reaches the Earth's surface in greater quantity but has less potency in causing acute epidermal damage than UV-B. It is a potent stimulator of melanin production, is capable of penetrating the skin to the upper dermis, and can cause DNA damage to keratinocytes. These combined effects result in damage to elastic tissue and dermal supporting structure, which causes wrinkling, and are likely responsible for carcinogenesis in the skin (Lim and Soter 1993).

The effect of exposure to UV light may be phototoxic, which is similar to primary irritation, or it may be allergic. While sunburn represents a phototoxic reaction, phototoxicity classically represents the interaction of the skin with a combination of a chemical and UV light. Phototoxic reactions occur when a specific chemical under the influence of UV light produces free radicals capable of inducing cell death. In that regard, the epidermal damage is similar to irritant dermatitis. Thousands of outdoor workers in construction, road building, fishing, forestry, gardening, farming, and electric and phone line erection are potentially exposed to sunlight and photosensitizing chemicals. Additionally, exposure to artificial UV light is experienced by electric furnace and foundry operators, glassblowers, photoengravers, steelworkers, welders, and printers in contact with photocure inks. Phototoxic reactions due to certain plants, a number of medications, and some fragrances have been well documented (Adams 1983; Birmingham 1968; Birmingham 1978; Cronin 1980; De Leo and Harber 1995; Emmett and Kaminski 1977; Epstein 1971; Fisher 1995; Gellin 1972; Malten and Beude 1976; Schwartz et al. 1957). In the coal and tar industry, distillation can offer exposure to anthracene, phenanthrene, and acridine, all of which are well-known phototoxic chemical agents. Related products such as creosote, pitch roof paint, road tar, and pipeline coatings have caused hyperpigmentation from the interaction of tar vapors or dusts with sunlight (see Table 7.12) (Adams 1983; Birmingham 1968; Birmingham 1978; De Leo and Harber 1995; Epstein 1971; Schwartz et al. 1957; White 1934).

Occupational photosensitivity is complicated by a number of topically applied and ingested drugs that can interact with specific wavelengths of light to produce a phototoxic or photoallergic reaction. Among such agents known to produce these effects are drugs related to sulfonamides, certain antibiotics, tranquilizers of the phenothiazine group, and a number of phototoxic oils that are used in fragrances (Birmingham 1968; Birmingham 1978; Cronin 1980; De Leo and Harber 1995; Epstein 1971; Fisher 1995; Gellin 1972).

Among the plants known to cause photosensitivity reaction are members of the *Umbellifera*. They include celery that has been infected with pink rot fungus, cow parsnip, dill, fennel, carrot, and wild parsnip (Birmingham 1968; Birmingham 1978; Emmett and Kaminski 1977; Epstein 1971). The development of classic signs of allergic contact dermatitis resulting from the combined interaction of a purported plant photoallergen and UV light is termed *phytophotodermatitis*. The photoactive chemicals in these plants are psoralens or furocoumarins and have been used for decades as therapeutic agents in the treatment of photoresponsive dermatoses like psoriasis and eczematous dermatitis.

### 7.4.6 Photoallergy

Allergic contact dermatitis may occur in the setting of UV light exposure. In photocontact allergic dermatitis, prior sensitization to the allergen is required, in contrast to phototoxic reaction, which may occur on initial introduction to a chemical. The photoallergens, in the absence of UV light, are not inherently allergenic and are incapable of causing a dermatitis in the sensitized individual. The introduction of UV light to the chemical can cause changes in the molecule that render it allergenic and hence capable of inducing a rash.

### 7.4.7 Phototesting

It is often necessary to test for sensitivity to potentially photoactive chemicals since the cause of a dermatitis may not be obvious. In such a setting photopatch testing may be performed where suspected photoallergens are placed on a patient's back in duplicate. After 24 hours one test set is exposed to UV-A light. (See patch test section for greater details.) Reactions on the irradiated side with negative reactions on the nonirradiated side confirm the diagnosis of photocontact allergic dermatitis. A majority of the chemicals listed in Table 7.6 are capable of causing photoallergic contact dermatitis and may be tested by photopatch testing methods. Those that typically cause phototoxic reaction such as tar derivative are generally not tested in clinical settings since history and physical exam are generally revealing. Table 7.6 outlines potentially photosensitizing chemicals and those used for photopatch testing at New York University Medical Center as of January 1997 (Occupational and Environmental Dermatology Unit, Department of Dermatology, New York University Medical Center). Table 7.7 lists potentially photosensitizing plants.

### 7.4.8 Mechanical

Work-incurred cutaneous injury may be mild, moderate, or severe. The injuries include cuts, lacerations, punctures, abrasions, and burns, which account for about 35% of occupational injuries for which worker's compensation claims are filed (National Electric Injury Surveillance System 1984). This translates to almost 1.5 million injuries to workers annually (American Academy of Dermatology 1994).

Contact with spicules of fiber glass, copra, hemp, and so on induces irritation and stimulates itching and scratching. Skin can react to repetitive friction by forming a blister or a callus, to pressure by changing color or becoming thickened or hyperkeratotic, and to shearing by sharp force by denudation or a puncture wound. Any break in the

**TABLE 7.6 Photosensitizing Chemicals**

| | | |
|---|---|---|
| Acridine[a] | Bithionol | Octyl methoxycinnamate |
| Anthracene[a] | Chlorhexidene | Promethazine |
| Certain chlorinated hydrocarbons[a] | Chlorpromazine | Sandalwood oil |
| Coal tar[a] | Cinoxate | Selected plant and pesticides |
| Creosote[a] | Dichlorophen | Sulfanilamide |
| Phenanthrene[a] | Diphenhydramine | Thiourea |
| Tar pitch[a] | Fentichlor | Tribromosalicylanilide |
| 1-(4-Isopropylphenyl)-3-phenyl-1,3 proandione | Hexachlorophene | Trichlorocarbanilide |
| 3-(4-Methylbenzylidene) camphor | Methyl anthranilate | Triclosan |
| 6-Methylcoumarin | Musk ambrette | |
| Benzophenone-4 | Octyl dimethyl PABA | |

[a] Not tested under normal phototesting conditions.

**TABLE 7.7 Photosensitizing Plants**

| |
|---|
| Moraceae |
| Ficus carica (fig) |
| Rutaceae |
| Citrus aurantifolia (lime) |
| Dictamnus (gas plant) |
| Ruta graveolens (rue) |
| Citrus aurantium (bitter orange) |
| Citrus limon (lemon) |
| Citrus bergamia (bergamot) |
| Umbelliferae |
| Anthriscus sylvestris (cow parsley) |
| Apium graveolens (celery, pink rot) |
| Daucus carota var. savita (carrot) |
| Pastinaca sativa (garden parsnip) |
| Foeniculum vulgare (fennel) |
| Anethum graveolens (dill) |
| Peucedanum ostruthium (masterwort) |
| Heracleum spp. (cow parsnip) |
| Compositae |
| Anthemis cotula (stinking mayweed) |
| Ranunculaceae |
| Ranunculus (buttercup) |
| Cruciferae |
| Brassica spp. (mustard) |
| Hypericaceae |
| Psoralea corylifolia (scurfy pea, bavchi) |
| Hypericum perforatum (St. John's wort) |

Fitzpatrick et al. (1993).

skin may become the site of a secondary infection (Adams 1983; Birmingham 1978; Gellin 1972; Rycroft 1986; Schwartz et al. 1957; Suskind 1959; White 1934).

Thousands of workmen use air-powered and electric tools that operate at variable frequencies. Exposure to vibration in a certain frequency range can produce painful fingers, a Raynaud-like disorder resulting from spasm of the blood vessels in the tool-holding hand. Slower-frequency tools such as jackhammers can cause bony, muscular, and tendon injury (Suvorov and Razumov 1983; Williams 1975).

### 7.4.9 Physical

Heat, cold, electricity, UV light (natural and artificial), and various radiation sources can induce cutaneous injury and sometimes systemic effects. Chemical irritants and UV light are covered in the aforementioned text.

***7.4.9.1 Heat*** Thermal burns are common among welders, lead burners, metal cutters, roofers, molten metalworkers, and glass blowers.

*Miliaria* (prickly heat) often follows overexposure to increased temperatures and humidities. Increase in sweating causes waterlogging of the keratin layer with blockading of the sweat ducts.

Excessive failure of thermoregulation under hot, humid climates may result in heat exhaustion. Symptoms include muscle cramping, nausea, vomiting, and fainting. Treatment by moving to a cooler environment and rehydrating with an electrolyte solution is necessary. Untreated heat exhaustion can progress to heat stroke. Heat stroke is characterized by elevated core temperature, neurologic symptoms, and lack of sweating. This disease has a high fatality rate if untreated. Aggressive resuscitation efforts including fluid and electrolyte replacement as well as core temperature cooling are required (De Galan and Hoekstra 1995; Schwartz et al. 1957; Suskind 1959; Samitz and Cohen 1985; Adams 1983; Meso et al. 1977; Dukes-Dobos and Badger 1977).

***7.4.9.2 Cold*** Frostbite is a common injury caused by intracellular crystallization of water. As ice crystals form, their sharp points are capable of puncturing membranes and fatally disrupting the cell's homeostatic mechanisms. Fingers, toes, ears, and nose are the usual sites of injury (policemen, firemen, postal workers, farmers, construction workers, military personnel, and frozen food storage employees are at risk). The hallmark of therapy rests in the preservation of viable tissue through rapid rewarming. Adjuvant approaches such as medication, surgery, and hemoperfusion continue to be investigated (Adams 1983; Dukes-Dobos and Badger 1977; Foray 1992; Gellin 1972; Samitz and Cohen 1985; Schwartz et al. 1957; Suskind 1959).

***7.4.9.3 Electricity*** Severe cutaneous burns of local or widespread proportions can result from electrical injury. Cutaneous signs of high-intensity electrical injury such as lightning strikes produce a pathognomonic sign consisting of ramifying fern-shaped red lesions emulating the track of the electricity. Less dramatic electrical bums can cause local necrosis of the skin, similar in nature to a chemical burn (Adams 1983; Braun-Falco et al. 1991; Gellin 1972; Schwartz et al. 1957).

***7.4.9.4 Microwaves*** Thermal burn is the major hazardous potential in contact with this radiation source (Meso et al. 1977).

***7.4.9.5 Lasers*** Laser radiation has enjoyed exponentially expanding utility in medicine but particularly in dermatology. Dozens of lasers are now available for therapeutic use in the treatment of cutaneous disease. They work by selectively destroying pathological tissue while sparing normal surrounding tissue. Lasers with light outputs in the visible spectrum are capable of destroying tissue containing complementary colors. For example, a laser producing yellow light will cause destruction of red tissue. This may be useful in eliminating disfiguring vascular birthmarks without the necessity of surgery and the resultant scar. Normal tissue is spared because the tissue without overt redness will not absorb the energy of the light. This protective phenomenon is augmented by an extremely short pulse time that allows normal surrounding tissues to cool before heat damage occurs. This is termed *thermal relaxation time*. Other lasers like the carbon dioxide laser produce nonvisible light that causes destruction to any tissue through the production of intense heat. Adverse effects of inadvertent exposure of the skin to lasers primarily rest in the thermal damage cause by absorption of the light (Wheeland 1994).

***7.4.9.6 Ionizing Radiation*** Modem industry and technology have many applications of this radiation type. It is important in the production and use of fissionable materials, radioisotopes, X-ray diffraction machines, electron beam operations, industrial X-rays for detecting metal flaws, and various uses in diagnostic and therapeutic radiology. Accidental exposures may result in severe acute cutaneous and systemic injury depending upon the level of radiation received (Gellin 1972; Meso et al. 1977; Schwartz et al. 1957). Lower-level exposures to ionizing radiation may often produce skin changes that manifest years after exposure. Skin thinning, scarring, and ulcerations occurring in sites of previous radiation exposure is termed *chronic radiation dermatitis* (Fitzpatrick et al. 1993).

***7.4.9.7 Biological*** Primary or secondary infection can happen in any occupation following exposure to bacteria, viruses, fungi, or parasites. Simple lacerations or embedment of a thorn or a wood splinter or metal slug can lead to infection. Certain occupations are associated with greater risk of bacterial infection, for example, anthrax among sheepherders, hide processors, and wool handlers; erysipeloid infection among meat, fish, and fowl dressers; and folliculitis among machinists, garage workers, candymakers, sanitation and sewage employees, and those exposed to coal tar (Adams 1983; Birmingham 1978; Samitz and Cohen 1985; Schwartz et al. 1957; Suskind 1959; Wilkinson 1982).

Fungi can produce localized cutaneous disease. Yeast infections (*Candida*) can occur among those employees engaged in wet work, for example, bartenders, cannery workers, fruit processors, or anyone who works in a wet environment. Sporotrichosis is seen among garden and landscape workers, florists, farmers, and miners. Ringworm infection of animals can be transmitted to farmers, veterinarians, laboratory personnel, and anyone in frequent contact with infected animals (Adams 1983; Birmingham 1978; Gellin 1972; Schwartz et al. 1957; Wilkinson 1982).

Certain parasitic mites inhabit cheese, grain, and other foods and will attack bakers, grain harvesters, grocers, and longshoremen. Mites that live on animals and fowl similarly are known to attack humans. In the southeastern states, animal hookworm larvae from dogs and cats are deposited in sandy soils and lead to infection among construction workers, farmers, plumbers, and, of course, anyone who works in the infected soils. Ticks, fleas, and insects can produce troublesome skin reactions and, in certain instances, systemic disease such as Rocky Mountain spotted fever, Lyme disease, yellow fever, and malaria, among other vector-borne diseases (Gellin 1972; Schwartz et al. 1957; Wilkinson 1982).

Several occupational diseases are associated with virus infections, for example, Q fever, Newcastle disease, and ornithosis. In fact, viral diseases are becoming the most important class of biological agents to cause severe illness. Well-known causes of occupational dermatoses caused by virus infection are often contracted from infected sheep, milker's nodules from infected cows, chicken pox from infected children, and herpes infections from infected patients.

Herpetic infections are particularly problematic for dentists, nurses, physicians, and others whose work occasions contact with open lesions (Birmingham 1978; OSHA 1978; Schwartz et al. 1957).

Animals such as snakes, sharks, dogs, and cats can result in aggressive necrotizing infections of the skin. Outdoor workers are particularly at risk for injury and also are placed at risk for rabies via similar circumstances. Insects such as spiders (particularly brown recluse spiders) can cause very painful and life-threatening necrotic bites in victims. Hornet, wasp, and bee stings can cause painful local reactions and are life threatening if individuals are sensitized to their venoms. Such severe allergic reactions can result in compromised breathing and blood pressure with potential shock or death (Birmingham 1978; Gellin 1972; OSHA 1978; Schwartz et al. 1957).

## 7.5 PHYSICAL FINDINGS OF OCCUPATIONAL SKIN DISEASE

The hazardous potential of the work environment is unlimited. Chemical and physical agents can produce a wide variety of clinical displays that differ in appearance and in histopathological pattern. The nature of the lesions and the sites of involvement may provide a clue as to a certain class of materials involved, but only in rare instances does clinical appearance indicate the precise cause. Except for a few strange and unusual effects, the majority of occupational dermatoses can be placed in one of the following reaction patterns. Several materials known to be causal for each clinical type are included (Adams 1983; Birmingham 1978; Gellin 1972; OSHA 1978; Samitz and Cohen 1985; Schwartz et al. 1957; Suskind 1959; White 1934).

### 7.5.1 Acute Eczematous Contact Dermatitis

Most of the occupational dermatoses can be classified as acute eczematous contact dermatitis. Heat, redness, swelling, vesiculation, and oozing are the clinical signs; itch, burning, and general discomfort are the major symptoms experienced. The backs of the hands, the inner wrists, and the forearms are the usual sites of attack, but acute contact dermatitis can occur anywhere on the skin. When the forehead, eyelids, ears, face, and neck are involved, airborne agents such as dust and vapors are suspected. More subtle clues such as upper lip, posterior ear, and mid-neck sparing may point to a photosensitivity disease since these areas are often less exposed to UV light. Generalized contact dermatitis may occur from massive exposure, wearing of contaminated clothing, autosensitization from a preexisting dermatitis, or systemic exposure.

Usually a contact dermatitis is recognizable as such, but whether the eruption has resulted from contact with a primary irritant or a cutaneous sensitizer can be ascertained only through a detailed history, a working knowledge of the materials being handled, their behavior on the skin, and a proper application and evaluation of diagnostic tests. Severe blistering or destruction of tissue generally indicates the action of an absolute or strong irritant; however, the history is what reveals the precise agent.

Acute contact eczematous dermatitis can be caused by hundreds of irritant and sensitizing chemicals, plants, and photoreactive agents. Some examples are listed in Table 7.8 (Adams 1983; Birmingham 1978; Cronin 1980; Fisher 1995; Gellin 1972; Klauder and Gross 1951; Klauder and Hardy 1946; Rycroft 1986; Samitz and Cohen 1985; Schwartz et al. 1957; Suskind 1959; White 1934). More specific agents are listed by category and type of exposure in Table 7.5.

### 7.5.2 Chronic Eczematous Contact Dermatitis

Hands, fingers, wrists, and forearms are the favored sites affected by chronic eczematous lesions. The skin is dry, thickened, and scaly with cracking and fissuring of the affected areas. Concurrent with skin thickening is the accentuation of normal skin lines and topical features. This is known as *lichenification*. Chronic nail dystrophy is a common accompaniment. Periodically, acute weeping lesions appear because of reexposure, imprudent treatment, or secondary infection. Chronic contact dermatitis occurs when exposure is perpetuated to irritant chemicals. Less often, low-level exposure to allergens can produce similar findings. In the latter case, signs and symptoms of acute contact dermatitis

**TABLE 7.8 Classes of Chemicals Causing Allergic Contact Dermatitis**

| | | |
|---|---|---|
| Acids, dilute | Herbicides | Resin systems |
| Alkalis, dilute | Insecticides | Rubber accelerators |
| Anhydrides | Liquid fuels | Rubber antioxidants |
| Detergents | Metal salts | Soluble emulsions |
| Germicides | Plants and woods | Solvents |

persist for long periods before chronic changes occur. A large number of materials (Table 7.9) have the potential to sustain the marked dryness that accompanies this chronic recurrent skin problem (Adams 1983; Birmingham 1978; Cronin 1980; Fisher 1995; Gellin 1972; Key et al. 1966; Rycroft 1986; Samitz and Cohen 1985; Schwartz et al. 1957; Suskind 1959; White 1934).

### 7.5.3 Folliculitis, Acne, and Chloracne

Hair follicles on the face, neck, forearms, backs of hands, fingers, lower abdomen, buttocks, and thighs can be affected in any kind of work, entailing heavy soilage. Comedones (blackheads/whiteheads) and follicular infection are common among garage mechanics, certain machine tool operators, oil drillers, tar workers, roofers, and tradesmen engaged in generally dusty and dirty work.

Acne caused by industrial agents usually is seen on the face, arms, upper back, and chest; however, when exposure is severe, lesions may be seen on the abdominal wall, buttocks, and thighs. Machinists, mechanics, oil field and oil refinery workers, road builders, and roofers exposed to tar are at risk. Such effects are far less prevalent than was noted in the past.

The term *folliculitis* implies inflammation of the follicles caused either by an irritant chemical or infection. *Acne* is a term reserved for a skin eruptions affecting follicles on the skin. The hallmark signs include comedones, papules (bumps <1 cm), cysts, pustules, and subsequent scars (Plewig and Kligman 1993). Often folliculitis and acne are coexistent in the same person, since chemicals known to cause one can cause the other. Site of exposure will also determine the type of follicular disease that is manifested such that the face will usually manifest with acne and the arms and legs with folliculitis. Table 7.10 lists some known causes of folliculitis and acne.

Chloracne is a disease with similarities to acne that are limited to follicular involvement and similar-sounding names. Chloracne is not merely severe acne but differs from regular acne entirely. Clinically, yellow, straw-colored cysts first appear on the sides of the forehead, around the lateral aspects of the eyelids, and behind the ears. Comedones, pustules, pigmentary disturbances, and subsequent scarring abound. Typically the head and neck, chest, back, groin, and buttocks are involved. Atypical areas may be involved if these areas are heavily or chronically exposed (Urabe and Asahi 1985).

Chloracne differs from acne in several ways. First, the histopathology of chloracne is distinct from acne. In chloracne sebaceous glands are conspicuously destroyed rather than being large and overactive (Plewig and Kligman 1993). Second, the exposure patterns of chloracne are distinct and for the most part include exposure to halogenated aromatic hydrocarbons (Table 7.11). Third, the natural history of chloracne and acne is divergent. Chemically induced acne and folliculitis will resolve without treatment within a week or two and sooner if treatment is instituted. Chloracne is classically resistant to treatment, including systemic retinoids, and may last for decades after exposure.

PCB and dioxins are epidemiologically the most frequent cause of chloracne. During the Vietnam war, chloracne occurred from exposure to a 2,3,7,8-tetrachlorodibenzo-*p*-dioxin-contaminated chlorophenoxyacetic acid herbicide

**TABLE 7.9 Chemicals Capable of Perpetuating Chronic Contact Dermatitis**

| | |
|---|---|
| Abrasive dusts (pumice, sand, fiber glass) | Chronic fungal infections |
| Alkalis | Oils |
| Cement | Resin systems |
| Cleansers (industrial) | Solvents |
| Cutting fluids (soluble) | Wet work |

**TABLE 7.10 Chemicals Capable of Inducing Acne or Folliculitis**

| |
|---|
| Asphalt |
| Creosote |
| Crude oil |
| Greases |
| Insoluble cutting oil |
| Lubricating oil |
| General-purpose petroleum oils |
| Heavy plant oils |
| Pitch |
| Tar |

**TABLE 7.11 Chloracnegens**

| |
|---|
| Hexachlorodibenzo-*p*-dioxin |
| Polybrominated dibenzofurans |
| Polybrominated biphenyls |
| Polychlorinated biphenyls |
| Polychlorinated dibenzofurans |
| Polychloronaphthalenes |
| Tetrachloroazobenzene |
| Tetrachloroazoxybenzene |
| Tetrachlorodibenzo-*p*-dioxin |

called Agent Orange. These herbicides are capable of causing a variety of neurologic and irritant dermatological findings. The chloracne was likely produced by the dioxin contaminate, a potent chloracnegen.

Domestically, chloracne most commonly occurs in the occupational setting from PCB. These dielectric compounds were commonly utilized in electrical transformers. While production of PCBs has been discontinued since the 1970s, hundreds of millions of pounds have escaped into the environment or are still present in some transformers in relatively high concentration. PCB exposure has been reported to cause illness in practically every organ system and has been described as a potential carcinogen. Despite this, rigorous epidemiologic studies in humans have failed to link PCB exposure with any specific illness other than chloracne. Hence chloracne remains the only reliable indicator of PCB exposure in humans (James et al. 1993). Further studies are necessary before conclusions can be drawn about other disease linkages.

### 7.5.4 Sweat-Induced Reactions

Miliaria is discussed above. Intertrigo occurs at sites where the skin opposes skin and allows sweat and warmth to macerate the tissue. Favored locations are the armpits, the groin, between the buttocks, and under the breasts (Adams 1983; Birmingham 1978; Gellin 1972; Rycroft 1986; Samitz and Cohen 1985; Schwartz et al. 1957; Suskind 1959).

### 7.5.5 Pigmentary Abnormalities

Color changes in skin can result from percutaneous absorption, inhalation, or a combination of both entry routes. The color change may represent chemical fixation of a dye to keratin or an increase or decrease in epidermal pigment through stimulation or destruction of melanocytes, respectively (melanin).

Hyperpigmentation from excessive melanin production may follow an inflammatory dermatosis, exposure to sunlight alone, or the combined action of sunlight plus a number of photoactive chemicals or plants. The opposite (loss of pigment or leukoderma) results from direct injury to the epidermis and melanin-producing cells by burns, chronic dermatitis, trauma, or chemical interference with the enzyme system that produces melanin. Antioxidant chemicals used in adhesives, cutting fluids, sanitizing agents, and rubber have caused complete loss of pigment (leukoderma).

Inhalation or percutaneous absorption of certain toxicants as aniline or other aromatic nitro and amino compounds causes methemoglobinemia. Jaundice can result from hepatic injury by carbon tetrachloride or trinitrotoluene, among other hepatotoxins.

Pigmentary abnormalities are caused by chemical agents listed in Table 7.12 (Adams 1983; Birmingham 1978; Gellin 1972; Gellin et al. 1970; Kahn 1979; Maibach 1987; Malten et al. 1971; Samitz and Cohen 1985; Schwartz et al. 1957; Suskind 1959).

### 7.5.6 Neoplasms

Mutagens are abundant in the workplace and are easily able to contact the skin under many circumstances. Such exposures may result in the development of benign or malignant neoplasms in the skin. These tumors, while often occurring at sites of contact with a suspected carcinogen, may also occur at distant sites or from systemic exposures. Asbestos warts and tar papillomas associated with petroleum and tar exposures are examples of benign neoplasms associated with repeated contact. Basal and squamous cell carcinomas (malignant neoplasms) are clearly associated with recurrent UV light exposure (Fitzpatrick et al. 1993; Lim and Soter 1993). Controversy exists whether malignant melanoma is associated with specific occupational exposures, but epidemiologic evidence has mounted to implicate UV light (particularly exposure during childhood) as a causal factor. Melanomas may also arise in locations classically shielded from UV light exposure (Birmingham 1978; Combs 1954; Fitzpatrick 1993; Lim and Soter 1993).

**TABLE 7.12 Chemicals Involved in Pigmentary Disturbances**

| Discoloration | Hyperpigmentation | Hypopigmentation |
|---|---|---|
| Arsenic | Chloracnegens | Antioxidants |
| Certain organic amines | Coal tar products | Hydroquinone |
| Carbon | Petroleum oils | Monobenzyl ether of hydroquinone |
| Dyes | Photoactive chemicals | Tertiary amyl phenol |
| Mercury | Photoactive plants | Tertiary butyl catechol |
| Picric acid | Radiation (sunlight) | Tertiary butyl phenol |
| Silver | Radiation (ionizing) | Bums |
| Trinitrotoluene | | Chronic dermatitis |
| | | Trauma |

**TABLE 7.13 Chemicals and Physical Agents Implicated as Cutaneous Carcinogens**

| | |
|---|---|
| Anthracene | Mineral oils containing various additives and impurities |
| Arsenic | |
| Burns | Crude oils |
| Coal tar | Radium and roentgen rays |
| Coal tar pitch | Shale oil |
| Creosote oil | Soot |
| | Sunlight |
| | Ultraviolet light |

Several chemical and physical agents are classified as industrial carcinogens, but only a few frequently cause skin cancer (Table 7.13). Admittedly, more cancers appear on the skin than at any other site; however, the number of these that are of occupational origin is not known. Sunlight is probably the major cause of occupational skin cancer, particularly among those engaged in agriculture, construction, fishing, forestry, gardening and landscaping, oil drilling, road building, roofing, and telephone and electric line installations (Berenblum and Schoental 1943; Bingham et al. 1965; Birmingham 1978; Buchanan 1962; Combs 1954; Emmett 1975; Epstein 1971; Rothman and Emmett 1988; Schwartz et al. 1957; White 1934).

In European countries, mule spinners exposed to shale oil and pressmen exposed to paraffin experienced a high frequency of carcinomatous lesions of the scrotum and lower extremities. Similar experiences with paraffin happened in the United States, but improved industrial practices and hygienic controls have all but eliminated the problem. In 1984, the International Agency for Research on Cancer determined that mineral oils containing various additives and impurities used in mule spinning, metal machining, and jute processing were carcinogenic to humans. Oils formerly in use that were responsible for cutaneous cancers including those affecting the scrotum were not as well refined as the lubricating oils being used today (Rothman and Emmett 1988).

Systemic exposures to carcinogens are clearly problematic when discussing internal malignancies such as hepatic and pulmonary cancers. For the skin there are relatively few systemic toxins capable of producing malignant neoplasms. Arsenic, however, is well known to cause a variety of benign and premalignant growths as well as basal and squamous cell carcinomas. The clinical and histopathologic appearances of these growths are often so distinct that they have been termed arsenical keratoses and arsenical carcinomas (Lever and Schaumburg-Lever 1990).

It is important to understand that skin cancers noted on workers in contact with carcinogenic agents are not necessarily of occupational origin. A certain number of people develop skin cancer irrespective of their jobs. For instance, residents in the Southwestern United States or in Australia are associated with a high frequency of skin cancer because of the exposure to sunlight. Ascertaining whether a skin cancer is truly of occupational origin in individuals residing in sun belt regions can be controversial (Berenblum and Schoental 1943; Bingham et al. 1965; Birmingham 1978; Buchanan 1962; Combs 1954; Eckhard 1959; Emmett 1975; Epstein 1971; Gellin 1972; Rothman and Emmett 1988; Schwartz et al. 1957; White 1934).

### 7.5.7 Ulcerations

Cutaneous ulcers were the earliest documented skin changes observed among miners and allied craftsmen. In 1827, Cumin (1827) reported on skin ulcers produced by chromium. Today the chrome ulcer (hole) caused by chromic acid or concentrated alkaline dichromate is a familiar lesion among chrome palters and chrome reduction plant operators. Perforation of the nasal septum also occurs among these employees, though in smaller numbers than occurred 20 years ago because many of the operations are now well enclosed (Cohen et al. 1974). Punched-out ulcers on the skin can result from contact with arsenic trioxide, calcium arsenate, calcium nitrate, and slaked lime (Birmingham et al. 1964). Nonchemical ulcerations may be associated with trauma and ulcers of the lower extremities in diabetics, pyogenic infections, vascular insufficiency, and sickle cell anemia (Birmingham 1978; OSHA 1978; Samitz and Dana 1971; Schwartz et al. 1957).

### 7.5.8 Granulomas

Cutaneous granulomas are caused by many agents of animate and inanimate nature. Such lesions are characterized by chronic, indolent inflammatory reactions that can be localized or systemic and result in severe scar formation. Granulomatous lesions can be the result of bacterial, fungal, viral, or parasitic elements such as atypical mycobacterium, sporotrichosis, milker's nodules, and tick bite, respectively. Additionally, minerals such as silica, zirconium, and beryllium and substances such as bone, chitin, coral, thorns, and grease have produced chronic granulomatous change in the skin (Birmingham 1978; OSHA 1978; Pinkus and Mehregan 1976; Schwartz et al. 1957).

### 7.5.9 Other Clinical Patterns

The clinical patterns described above represent well-known forms of occupational skin diseases. However, there are a number of other disorders affecting the skin, hair, and nails that do not fit into these patterns. Some examples follow.

***7.5.9.1 Contact Urticaria*** Contact urticaria (hives) results secondary to contact with an allergen or nonallergenic chemical urticariant. Allergic contact urticaria represents immediate-type hypersensitivity and requires sensitization to occur. Occupational contact urticaria has become an extremely important issue, particularly in the healthcare setting. In the mid-1980s during the emergence and recognition of the human immunodeficiency virus and the escalating incidence of hepatitis, standards of personal protection developed. Ultimately the term *universal precautions* arose and charged healthcare personnel to handle all blood and body fluids as if they were infected with a serious transmissible agent. As such, the use of protective gloves made of latex became widespread. This resulted in a tremendous increase in the number of gloves being used by over 5 million healthcare workers (Centers for Disease Control 1987; Turjanmaa 1987).

A recently recognized syndrome principally manifesting as contact urticaria with other symptoms such as rhinitis, conjunctivitis, asthma, and less commonly anaphylaxis and death has been associated with latex proteins found in rubber products. For healthcare workers, the main exposure is through rubber gloves and rubber material devices. The exact source of initial sensitization, however, is extremely difficult to ascertain since latex is so ubiquitous in society. The combination of extremely itchy skin and respiratory symptoms not only makes it uncomfortable for an employee to work with latex gloves but also may make it life threatening. Glove powders are known to bind latex protein allergen and become airborne when gloves are removed. Inhaled latex protein can trigger severe allergic reactions, causing compromised breathing and occasionally shock and death. Those at high risk include people who frequently use disposable latex gloves, patients with spina bifida, those with a history of multiple surgical procedures, and those with a history of hand dermatitis (Asa 1994; Charous et al. 1994; Jaeger et al. 1992; Tomazic et al. 1994). Since dermatitis of the hands is common in many occupational settings, the frequent use of disposable latex gloves should alert healthcare workers to the sentinel signals of latex allergy. Healthcare institutions are now recognizing this problem and are developing strategies to detect latex allergy in employees and cope with them. At New York University Medical Center, an algorithm has been developed to screen potentially allergic employees (Figure 7.2).

Here, employees are first screened with a latex radioallergosorbent assay test (RAST) to detect latex-specific immunoglobulin E. If negative, a "use" test utilizing a latex glove under supervised setting is performed, first with one finger and then an entire hand. If these tests are negative, eluted latex protein in solution is used for prick or scratch testing. The presence of a hive at the test site indicates latex hypersensitivity. If found to be positive, employees use alternative gloves, and the latex in the immediate worksite is minimized. To date all new latex-sensitive employees at NYU have successfully had job site or job modification protocols enacted.

Occupational contact urticaria may also occur from a variety of other allergenic sources such as foods like apple, carrot, egg, fish, beef, chicken, lamb, pork, and turkey. Other sources have arisen from animal viscera and products handled by veterinarians and food dressers and from contact with formaldehyde, rat tail, guinea pig, and streptomycin. It is likely that many of these cases go unreported and unrecognized as such (Maibach 1987; Fisher 1995, chapter 34; Odom and Maibach 1977).

***7.5.9.2 Nail Discoloration and Dystrophy*** Chemicals such as alkaline bichromate induce an ochre in nails; tetryl and trinitrotoluene induce yellow coloring; dyes of various colors may change the nail color; carpenters may have

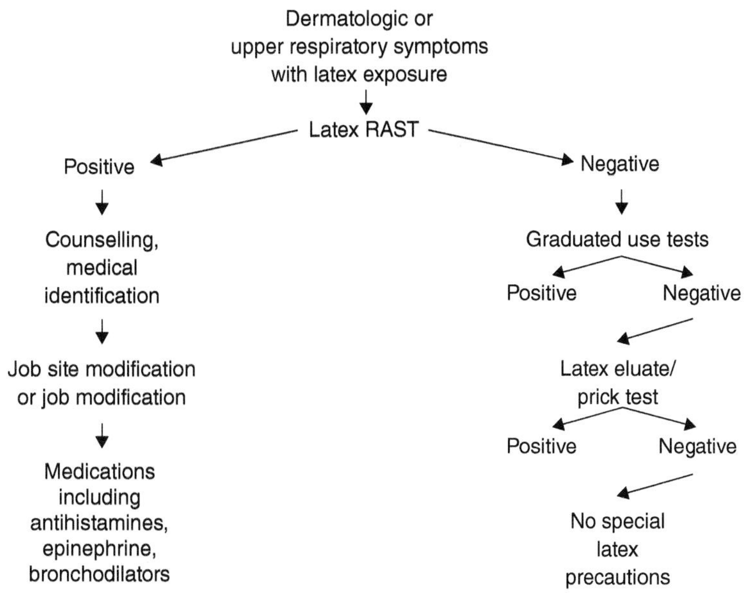


**Figure 7.2** New York University Medical Center latex energy evaluation.

wood stains on their nails. Dystrophy can follow chronic contact from acids and corrosive salts, alkaline agents, moisture exposures, sugars, trauma, and infectious agents such as bacteria and fungi. Long-standing contact dermatitis of the fingertips may disrupt the nail matrix and bed, causing dystrophic nail plates to form (Adams 1983; Fisher 1995, chapter 6, pp. 66–92; OSHA 1978; Ronchese 1948, 1965; Schwartz et al. 1957).

***7.5.9.3 Facial Flush*** This peculiar phenomenon has been reported from the combination of certain chemicals such as tetramethylthiuram disulfide, trichloroethylene, or butyraldoxime following the ingestion of alcohol. Because trichloroethylene is known to cause liver damage, the facial flush is related to the intolerance associated with alcohol ingestion (Birmingham 1978; Lewis and Schwartz 1956; Schwartz et al. 1957; Stewart et al. 1974).

***7.5.9.4 Acroosteolysis*** Several years ago, a number of workers involved in cleaning vinyl chloride polymerization reaction tanks incurred a peculiar vascular and bony abnormality involving the digits, hands, and forearms. Bone resorption of the digital tufts was accompanied by Raynaud's syndrome and scleroderma-like changes of the hands and forearms. Removal from the tank cleaning duties led to vascular bone improvement, but the skin changes did not always improve (Harris and Adams 1967; Wilson et al. 1967). Acroosteolysis has become a rare disease in the United States since the inception of strict exposure standards for vinyl chloride monomer. Polymerized vinyl chloride does not cause this illness.

## 7.6 SOME SIGNS OF SYSTEMIC INTOXICATION FOLLOWING PERCUTANEOUS ABSORPTION

A number of chemicals with or without direct toxic effect on the skin can cause systemic intoxication following percutaneous entry. Transepidermal passage may occur with or without metabolism by keratinocytes. Once into the upper dermis, chemicals have direct access to a vascular plexus and the systemic circulation, where target organs can be reached prior to excretion or metabolism. Table 7.14 lists examples of chemicals capable of significant percutaneous absorption with target organs and demonstrable effects (OSHA 1978; Proctor and Hughes 1978; Tabershaw et al. 1977). More details may be found in the Wang et al. (1993) text on percutaneous absorption.

Over a period of years, industrial and contract laboratories have developed and used tests for predicting the toxicologic behavior of a new product or one in use whose toxic action becomes questioned. Such tests on animals and humans are designed to demonstrate signs of systemic and cutaneous toxic effects, for example, primary irritation, allergic hypersensitivity, phototoxicity, photoallergenicity, interference with pigment formation, sweat and sebaceous gland activity, dermal absorption routes, metabolic markers, and cellular aberrations indicating mutation or frank carcinoma. Conducting these tests leads to prediction of toxicologic potential and diminishment thereby of the number of untoward reactions that might appear if such tests had not been done (Hood 1977; OSHA 1978).

**TABLE 7.14 Percutaneous Absorption and Target Organ Systems**

| |
|---|
| Aniline → red blood cell → methemoglobinemia |
| Benzidine → urinary bladder → carcinoma |
| Carbon disulfide → central nervous system → psychopathological symptoms and signs as peripheral neuritis + cardiac disease |
| Carbon tetrachloride → liver, kidney, and CNS → liver and kidney damage, depression |
| Chlorinated naphthalenes, biphenyls, and dioxins → skin → chloracne, hepatitis, peripheral neuritis |
| Ethylene glycol ethers → CNS → lungs, liver, and kidney damage |
| Methyl butyl ketone → CNS and peripheral nerves → polyneuritis, depression |
| Organophosphate pesticides → inhibition of cholinesterase → cardiovascular, gastrointestinal, neuromuscular, and pulmonary disturbances |
| Tetrachloroethane → CNS → depression and liver damage |
| Toluene → CNS and liver → confusion, dizziness, headache, paresthesias |

## 7.7 DIAGNOSIS

It is a common assumption among employees that any skin disease they incur has something to do with their work. At times the supposition is correct, but often there is no true relationship to the work situation. Arriving at the correct diagnosis may be quite easy, but such is not a routine occurrence. The industrial physician has a distinct advantage in being familiar with agents within the work environment and the conditions associated with contacting them. The dermatologist may find the diagnosis difficult if unfamiliar with contact agents in the work environment. The practitioner with little or no interest in dermatologic problems associated with work and also lacking dermatologic skill will find it most difficult to make a correct diagnosis. At any rate, the attending physician, specialist or otherwise, should satisfy certain basic tenets in establishing a diagnosis of occupational skin disease. The health and safety professional should provide the medical provider as much information as possible about the materials used in the workplace. An understanding of the work process is important because the offending agent may be a by-product or waste and not appear on the "list" of materials used. Likewise, workplace conditions such as temperature, humidity, and physical activities may also be important.

### 7.7.1 Patient History

Only through detailed questioning can the proper relationship between cause and effect be established. Taking a thorough history is time consuming, because it should cover the past and present health and work status of the employee. Areas requiring thorough coverage include family history, allergies, past medical illnesses, job titles, nature of work performed, materials handled, frequency of potential exposures, history and distribution of the skin findings, and any treatments used. It is important to ascertain whether the employee has had any prior dermatologic illnesses and if any systemic or topical drugs are used for any medical or surgical illness. Finally, materials used for home hobbies such as gardening, woodworking, painting, and model building may cause skin disease indistinguishable from occupationally acquired exposures (Adams 1983; Birmingham 1978; Gellin 1972; Maibach 1987; OSHA 1978; Samitz and Cohen 1985; Schwartz et al. 1957; Suskind 1959).

### 7.7.2 Appearance of the Lesions

The eruption should fit into one of the clinical types in its appearance. Although the majority of the occupational dermatoses are either acute or chronic eczematous contact dermatitis, other clinical types of disease such as follicular, acneiform, pigmentary, neoplastic, ulcerative, and granulomatous can occur. Further, one must be on the lookout for the oddities that show up in unpredictable fashion, for example, Raynaud's disease and contact urticaria (Adams 1983; Birmingham 1978; Gellin 1972; Samitz and Cohen 1985; Schwartz et al. 1957; Suskind 1959).

### 7.7.3 Sites Affected

Most cases of occupational skin disease affect the hands, digits, wrists, and forearms, for the upper extremities are truly the instruments for work. However, the forehead, face, "V" of the neck, and ears may also display active lesions, particularly when the employee is exposed to dusts, vapors, or fumes. Legs may be favored when splashing of material on the floor occurs or when suspected contactant are at ground level such as in farming. Although most of the work dermatoses are usually seen in the above sites, generalization can occur from massive exposure, from contaminated work clothing, and also from autosensitization (spread) of an already existent rash (Adams 1983; Birmingham 1978; Gellin 1972; OSHA 1978; Samitz and Cohen 1985; Schwartz et al. 1957; Suskind 1959).

### 7.7.4 Diagnostic Tests

Laboratory tests should be employed when necessary for the detection of bacteria, fungi, and parasites. Such tests include direct microscopic examination of surface specimens, culture of bacterial or fungal elements, and biopsy of one or more lesions for histopathological definition. When allergic reactions are suspect, diagnostic patch tests can be used to ascertain occupational as well as nonoccupational allergies, including photosensitization. At times, useful information can be obtained through the use of analytical chemical examination of blood, urine, or tissue (skin, hair, nails) (Adams 1983; Birmingham 1978; Cronin 1980; Fisher 1995; Gellin 1972; Maibach 1987; OSHA 1978; Rycroft 1986; Samitz and Cohen 1985; Schwartz et al. 1957; Suskind 1959).

### 7.7.5 The Patch Test

Diagnostic patch testing, properly performed and interpreted, is a highly useful procedure. The test is based on the theory that when an acute or chronic eczematous dermatitis is caused by a given sensitizing agent, application of the suspected material to an area of unaffected skin for 48 hours will cause an inflammatory reaction at the site of application. A positive test usually indicates that the individual has an allergic sensitivity to the test material. When the employee is working with primary irritants and fellow employees also are affected with dermatitis, the cause is self-evident, and patch testing is neither necessary nor indicated. Exception to this rule occurs when the employees are working with irritant agents that can also sensitize (epoxy and acrylic systems, resin hardeners of the amine group, formaldehyde, or chromates) or when an agent is highly sensitizing and many employees have become allergic. There are approximately 65 000 known irritant chemicals and 2800 described allergic chemicals, and so the patch test is invaluable in identifying a causal agent or distinguishing an irritant from allergic contact dermatitis (Adams 1983; Birmingham 1978; Cronin 1980; Fisher 1995; Rietschel and Fowler 1995; Rycroft 1986; Schwartz et al. 1957; Sulzberger and Wise 1931; Suskind 1959).

*If the test is to have relevance and reliability, it must be performed by one with a clear understanding of the difference between a primary irritant and a sensitizer.*

When patch tests are conducted with strong or even marginal irritants, a skin reaction is inevitable and usually not relevant or interpretable. However, this does not mean that a patch test cannot be performed with a diluted primary irritant. There is an abundance of published material pointing out proper patch test concentrations and appropriate vehicles considered safe for skin tests (Adams 1983; Cronin 1980; Fisher 1995; Maibach 1987). The goal of testing is to use concentrations of chemicals known to cause reactions in sensitized individuals and be negative in those not allergic. Hence, test concentrations are ideally below the threshold necessary to cause an irritant reaction but above that necessary to cause an allergic reaction. For the most part, these concentrations do not overlap for most chemicals. Further, the one performing the tests should have a working knowledge of environmental contactants, particularly those well known as potential cutaneous allergens (Adams 1983; Birmingham 1978; Cronin 1980; Fisher 1995; Maibach 1974; Rycroft 1986; Schwartz et al. 1957; Sulzberger and Wise 1931; Suskind 1959).

The technique of the test is simple. Liquids, powders, or solids are applied under occlusive conditions under stainless steel disks or in a hydrogel suspension to the back.

The test panel should include relevant chemicals based on the history and distribution of the rash. A standard allergen panel is not useful since exposure patterns across various occupations is so divergent (Cohen et al. 1997).

The North American Contact Dermatitis Group and the International Contact Dermatitis Group advocate standardized test concentrations applied in vertical rows on the back and covered by hypoallergenic tape. Contact with the test material is maintained on the skin for 48 hours, and readings are made 30 minutes or later after removal, and an additional time at the 72nd to 168th hour.

Reading the tests and interpreting them for the degree of reaction requires experience. The levels of reaction currently in use are noted in Table 7.15.

True allergic reactions tend toward increased intensity for 24–48 h after test removal, whereas irritant reactions usually subside within 24–48 h after removal (Adams 1983; Birmingham 1978; Fisher 1995; Fregert 1974; Maibach 1974; Malten et al. 1976; Sulzberger and Wise 1931).

Interpreting the significance of the test reactions is of paramount importance. A positive test can result from exposure to an irritant or a sensitizer. When specific sensitization is the case, it means that the patient is reactive to the allergens at the time of the test. When the positive test coincides with a positive history of contact, it is considered strong evidence of an allergic etiology. Conversely, the examiner must be aware that clinically irrelevant positive tests can occur if the

**TABLE 7.15 Interpretive Criteria for Patch Testing**

| | |
|---|---|
| (?) | = doubtful; faint, macular redness only |
| (+) | = weak (nonvesicular) positive reaction; redness, infiltration, possibly papules |
| (++) | = strong (vesicular) positive reaction; redness, infiltration, papules, vesicles |
| (+++) | = extreme positive reaction; bullous reaction |
| (−) | = negative reaction |
| (IR) | = irritant reaction |

patient is tested (i) during an active dermatitic phase leading to one or several nonspecific reactions, (ii) with a marginal irritant, or (iii) with a sensitizer to which the patient had developed an early sensitization, for example, nickel, but which is not relevant to the present occupational dermatitis. The patch test is incapable of testing the irritancy potential of a suspect chemical. Irritant reactions on a patch test should never be correlated with any workplace exposures. *No conclusions regarding the cause of a dermatitis can be drawn from an irritant reaction on a patch test.*

A negative test indicates the absence of an irritant or an allergic reaction. However, a negative reaction can also mean (i) testing omitted an important allergen, (ii) insufficient strength and quantity of the test allergen, (iii) poor test condition, or (iv) hyporeactivity by the patient at the time of the test.

Performing the patch test with unknown substances the employee has brought to the physician's office can be most misleading and potentially hazardous. The material could be a caustic and thus produce a strongly positive chemical burn, or perhaps no reaction will occur. In either case, the test can be misused and provide no help in diagnosing the cause. Useful information concerning unknown materials can be obtained by contacting the plant manager, physician, nurse, industrial hygienist, or safety supervisor (Adams 1983; Cronin 1980; Fisher 1995; Fregert 1974; Maibach 1974; Malten et al. 1976).

A patch test being employed more frequently has to do with suspected photoallergy. This is discussed in the photosensitivity section of the chapter. Most of the photodermatoses incurred in the work area are phototoxic and thus do not require photopatch tests for diagnosis. Some dermatologists perform photopatch tests in their offices; however, many of these patients suffering with suspected dermatoses are referred to a center where photopatch tests are performed routinely (Adams 1983; Cronin 1980; Fisher 1995; Fregert 1974; Malten et al. 1976; Williams 1975).

## 7.8 TREATMENT

Immediate treatment of an occupational dermatosis does not differ essentially from that used for a similar eruption of nonoccupational nature. In either case, treatment should be directed toward providing fairly rapid relief of symptoms. The choice of treatment agents depends upon the nature and severity of the dermatitis. Most of the cases are either an acute or a chronic eczematous dermatitis, and most of these can be managed with ambulatory care. However, hospitalization is indicated when the severity of the eruption warrants in-patient care.

Acute eczematous dermatoses caused by a contactant generally respond promptly to wet dressings and topical steroid preparations, but systemic therapy with corticosteroids should be used when deemed necessary. Corticosteroids have definitely lessened the morbidity in the acute and chronic eczematous dermatoses caused by work. Once the dermatitis is under good control, clinical management must be directed toward:

1. Ascertaining the cause.
2. Returning the patient to the job when the skin condition warrants, but not before.
3. Instructing the patient in the means necessary to minimize or prevent contact at work with the offending material.

In any contact dermatitis it is essential to establish the causal agents or situations that contributed to the induction of the disease. Follicular or acneiform skin lesions, notably chloracne, are notoriously slow in responding to treatment. Pigmentary change similarly may resist the run of therapeutic agents and remain active for months. New growths can be removed by an appropriate method and studied histopathologically. Ulcerations inevitably lead to the formation of scar tissue. Similarly, granulomatous lesions generally scar.

Almost all cases of occupational skin disease respond to appropriate therapy within two to eight weeks; however, when chloracne, pigmentary changes, or allergic contact dermatitis due to chrome or nickel are the problem, therapeutic response may take months or years. It cannot be overemphasized that contact with the causative agents must be minimized, if not eliminated; otherwise return to work is accompanied by the return of the rash (Adams 1983; Birmingham 1978; Gellin 1972; Malten and Zielhius 1964; Samitz and Cohen 1985; Schwartz et al. 1957; Suskind 1959).

### 7.8.1 Prolonged and Recurrent Dermatoses

As a rule, an occupational dermatosis can be expected to disappear or to be considerably improved within a period of two to eight weeks after initiating treatment. Yet there are cases that are recalcitrant to appropriate treatment and continue to plague the patient with chronic recurrent episodes. This situation is commonly noted when the dermatosis was caused by chromium, nickel, mercury, or certain plastics and glues. However, all cases of recurrent disease are not necessarily associated with the above materials. The following situations may be operable in prolonged and recurrent disease (Birmingham 1986; Morris 1952):

1. Incorrect clinical diagnosis.
2. Failure to establish cause.
3. Failure to eliminate the cause even when direct cause has been established.
4. Improper treatment, often self-directed.
5. Poor hygiene habits at work.
6. Supervening secondary infections.
7. Cross-reactions with related chemicals.
8. Self-perpetuation for gain.

## 7.9 PREVENTION

The key to preventing occupational skin disease is to eliminate or at least minimize skin contact with potential irritants and sensitizers present in the workplace (see Chapter 1). To do so requires:

1. Recognition of the hazardous exposure potentials.
2. Assessment of the workplace exposures.
3. Establishment of necessary controls.

Achieving these steps is more likely to occur in large industrial establishments with trained personnel responsible for the maintenance of health and safety practices. In contrast, many small plants or workplaces that employ the largest percentage of the work force have neither the money nor personnel to initiate and monitor effective preventive programs. Nonetheless, any work establishment has the responsibility of providing those preventive measures that at least minimize, if not entirely eliminate, contact with hazardous exposures.

### 7.9.1 Direct Measures

Time-tested control measures known to prevent occupational diseases are classified as primary (immediate) and indirect. The primary categories are:

1. Substitution
2. Process change
3. Isolation/enclosure
4. Ventilation
5. Good housekeeping
6. Personal protection

### 7.9.2 Indirect Measures

The indirect measures include:

1. Education and training of management, supervisory force, and employees.
2. Medical programs.
3. Environmental monitoring.

Although small plants generally lack in-house medical and industrial hygiene services, they do have access to such services through state health departments or through private consultants knowledgeable in health and safety measures.

***7.9.2.1 Substitution and Process Change*** When a particular agent or process is recognized as a trouble source, substituting a less hazardous agent or process can minimize or eliminate the problem. This has been done with a number of toxic agents, for example, substituting toluene for benzene, tetrachloroethylene, or carbon tetrachloride. Substitution has potential value in allergen replacement of known offenders such as with chromium, nickel, certain antioxidants and accelerators in rubber manufacture, and certain biocides in metalworking fluids. The addition of iron salts to cement will change hexavalent chromium, a highly allergic species to trivalent chrome, a low-grade sensitizer without affecting the quality of the cement (Turk and Rietschel 1993). When feasible, substituting a nonallergen for a hazardous agent is a recommended procedure.

***7.9.2.2 Isolation and Enclosure*** Isolation of an agent or a process can be used to minimize hours of exposure or the number of people exposed. Isolation can mean creation of a barrier, or distance, or time, as the means of isolation to lessen exposure. Enclosure of processes provides a high level of safety when hazardous agents are involved. Local enclosures against oil spray and splash from metalworking fluid lessen the amount of exposure to the machine operators.

Radiation exposures can be shielded with proper barriers and remote control systems. Bagging operations can be enclosed to lessen, if not entirely eliminate, exposure to the operators involved.

***7.9.2.3 Ventilation*** Movement of air can mean general dilution and/or local exhaust ventilation used to reduce exposures to harmful airborne agents. Local exhaust ventilation is effective in controlling vapors of degreasing tanks and in mixing, layup, curing, and tooling of epoxy, polyester, phenol–formaldehyde resin systems.

***7.9.2.4 Good Housekeeping*** A clean shop or plant is essential in controlling exposure to hazardous materials. This means keeping the workplace ceilings, windows, walls, floors, workbenches, and tools clean. Adequate storage space, properly placed warning signs, and sanitary facilities adequate in number should be available. Expeditious cleanup of spills and emergency showers for use after accidental heavy exposure to harmful chemicals are essentials of good industrial hygiene.

***7.9.2.5 Personal Protection***

*7.9.2.5.1 Clothing* It is not necessary for all workers to wear protective clothing, but for those jobs for which it is required, good quality clothing should be issued as a plant responsibility. Protective clothing against cold, heat, and biological and chemical injury to the skin is advisable. Depending upon the need, equipment such as hairnets, caps, helmets, shirts, trousers, coveralls, aprons, gloves, boots, safety glasses, and face shields should be available. Similarly, clothing and chemical screens to protect against UV light and ionizing, microwave, and laser radiation should also be readily available.

Once protective clothing has been issued, its laundering and maintenance should be the responsibility of the plant. When work clothing is laundered at home, it becomes a ready means of contaminating family members' apparel with chemicals, fiber glass, or harmful dusts.

Specific information concerned with protective equipment can be obtained from the National Institute for Occupational Safety and Health and from any of the manufacturers listed in the safety and hygiene journals. For Internet users, OSHA can be accessed via the World Wide Web at address http://www.osha.gov. NIOSH may be reached at http://www.cdc.gov/NIOSH/homepage.html.

*7.9.2.5.2 Gloves* Gloves are an important part of protective gear because the hands are most frequently exposed to chemical work. Leather gloves, though expensive, offer fairly good protection against mechanical trauma (friction, abrasion, etc.). Cotton gloves suffice for light work, but they wear out sometimes in a matter of hours. Neoprene, butadiene-nitrile, and vinyl-dipped cotton gloves are useful in protecting against mechanical trauma, chemicals, solvents, and dusts. Unlined rubber gloves and plastic gloves can cause maceration because of occlusion and contact dermatitis from the chemical accelerators or antioxidant in the material. Much time and money can be saved by reviewing the catalogs and tables provided by the manufacturers.

*7.9.2.5.3 Hand Cleansers* Of all the measures advocated for preventing occupational skin disease, personal cleanliness is paramount. Although ventilating systems and monitoring are important in controlling the workplace exposures, there remains no substitute for washing the hands, forearms, and face and keeping clean. To do this the plant must provide conveniently located wash stations with hot and cold running water, good quality cleansers, and clean towels.

Several varieties of acceptable cleansers are available on the market, and these include conventional soaps of liquid, cake, or powdered variety. Conventional soaps are used each day by millions of people and are generally considered safe. Liquid varieties including "cream" soaps are satisfactory for light soil removal. Powdered soaps are designed for light frictional removal of soil and may contain pumice, wood fiber, or corn meal.

Waterless cleansers are popular among those who contact heavy tenacious soilage such as tar, grease, and paint. They should not serve as a substitute for conventional removal of soilage. Daily use of waterless cleansers leads to dryness of the skin and, at times, eczematous dermatitis from the solvent action of the cleanser.

In choosing an industrial cleanser, the following dicta are suggested:

1. It should have good cleansing quality.
2. It should not dry out the skin through normal usage.
3. It should not harmfully abrade the skin.
4. It should not contain known sensitizers.
5. It should flow readily through dispensers.
6. It should resist insect invasion.
7. It should resist easy spoilage and rancidity.
8. It should not clog the plumbing.

*7.9.2.5.4 Protective Creams* Covering the skin with a barrier cream, lotion, or ointment is a common practice in and out of industry. Easy application and removal plus the psychological aspect of protection account for the popularity of these materials. Obviously, a thin layer of barrier cream is not the same as good environmental control or an appropriate protective sleeve or glove. There are currently no protective creams available that can provide adequate defense against irritants and allergens outside of laboratory testing conditions. Contact dermatitis may only be controlled with avoidance of offending allergens and good hygiene. The use of sunscreens offers incontrovertible protection against sunburn from UV light exposure. Their use is encouraged for all workers routinely exposed to sunlight or artificial UV light sources. DEET and citronella oil-containing lotions can offer protection for outdoor worker against arthropod assault; however, they may on occasion cause allergic contact dermatitis.

The development of protective creams has clearly been an effort to reduce the necessity for wearing gloves that may reduce manual dexterity. Current glove technology has sufficiently advanced to allow excellent dexterity for workers performing intricate duties. For decades now neurosurgeons, cardiac surgeons, and microsurgeons have successfully performed surgery requiring the highest degree of adroitness while having their hands completely enveloped in latex. Few if any occupations require higher degrees of deftness that make it impossible to wear protective gloves. However, workers exposed to strong solvent chemicals may find it difficult to locate gloves that are resistant to breakdown or provide adequate barrier protection under these harsh circumstances (Adams 1983; Birmingham 1975; Calnan 1970; Fregert 1974; Gellin 1972; NIOSH 1988; OSHA 1978; Samitz and Cohen 1985; Schwartz et al. 1957; Suskind 1959).

### 7.9.3 Control Measures

***7.9.3.1 Education*** An effective prevention and control program against occupational disease in general, including diseases of the skin, must begin with education. A joint commitment by management, supervisory personnel, workers, and worker representatives is required. The purpose is to acquaint managerial personnel and the workers with the hazards inherent in the workplace and the measures available to control the hazards. The training should be in the hands of well-qualified instructors capable of instructing the involved people with:

1. Identification of the agents involved in the plant.
2. Potential risks.
3. Symptoms and signs of unwanted effects.
4. Results of environmental and biological monitoring in the plant.
5. Management plans for hazard control.
6. Instructions for emergencies.
7. Safe job procedures.

Worker education cannot be static. It must be periodic through the medical and hygiene personnel, during job training, and periodically thereafter through health and safety meetings.

Special training courses are available at several universities, with departments specializing in occupational and environmental health and hygiene.

**7.9.3.2 *Environmental Monitoring*** Periodic sampling of the work environment detects the nature and extent of potential difficulties and also the effectiveness of the control measures being used. Monitoring for skin hazards can include wipe samples from the skin as well as the work sites and use of a black light for detecting the presence of tar product fluorescence on the skin before and after washing. Monitoring is particularly required when new compounds are introduced into plant processes (see Chapters 4 and 6).

**7.9.3.3 *Medical Controls*** Sound medical programs contribute greatly to preventing illness and injury among the plant employees. Large establishments have used in-house medical and hygiene personnel quite effectively. Daily surveillance of this type is not generally available to small plants; however, small plants do have access to well-trained occupational health and hygiene specialists through contractual agreements. At any rate, medical programs are designed to prevent occupational illness and injury, and this begins with a thorough placement physical examination, including the condition of the skin. When the preplacement examination detects the presence of or personal history of chronic eczema (atopic), psoriasis, hyperhidrosis, acne vulgaris, discoid lupus erythematosus, chronic fungal disease, dry skin, or other skin diseases, extreme care must be used in placement to avoid worsening a preexisting disease.

Plant medical personnel, full time or otherwise, should make periodic inspections of the plant operations to note the presence of skin disease, the use or misuse of protective gear, and hygiene breaches that predispose to skin injury.

When toxic agents are being handled, periodic biological monitoring of urine and blood for specific indicators or metabolites should be regularly performed.

Plant medical and industrial hygiene personnel should have constant surveillance over the introduction of new materials into the operations within the plant. Failure to do so can lead to the unwitting use of toxic agents capable of producing serious problems.

Of great importance in the medical control program is the maintenance of good medical records, indicating occupational and nonoccupational conditions affecting the skin, as well as other organ systems. Medical records are vital in compensation cases, particularly those of litigious character (Adams 1983; Birmingham 1978; Gellin 1972; NIOSH 1988; Samitz and Cohen 1985; Schwartz et al. 1957; Suskind 1959).

A well-detailed coverage of prevention of occupational skin diseases is present in the publications, *Proposed National Strategies for the Prevention of Leading Work-Related Diseases and Injuries, Part 2*, NIOSH (1988), p. 107 and the *Report of the Advisory Committee on Cutaneous Hazards to the Assistant Secretary of Labor*, OSHA (1978). Current information regarding occupational and environmental skin disease may be further studied in the texts listed in the reference section or via specialized dermatology journals such as the *American Journal of Contact Dermatitis*, the *Journal of the American Academy of Dermatology*, and *Contact Dermatitis*.

## GLOSSARY

Carter, R. L. (1992). *A Dictionary of Dermatologic Terms*. 4. Williams and Wilkins, Baltimore, MD.

**Apocrine gland** A gland that secretes a fatty substance with the rest of the discharge secretory product.

**Appendages** In dermatology, they are applied to internal structures of the skin like the nail unit, the hair unit, and sweat glands.

**Bulla** A blister of the skin.

**Bullous** Relating to many blisters.

**Chromophore** A chemical capable of absorbing various wavelengths of light.

**Comedones** Acne lesions such as whiteheads and blackheads.

**Cyst** A circumscribed nodule filled with fluid or solid matter usually located directly under the skin.

**Cytokines** Chemicals capable of specific cellular action or recruitment.

**Depigmentation** Complete loss of pigment or color.

**Eccrine gland** A term used to designate a sweat gland.

**Erythema** Relating to redness of skin.

**Exfoliative dermatitis** A red, scaling skin rash involving the majority of skin from head to toe. Causes may relate to external exposure or internal disease.

**Granuloma** An inflammatory response of specific cells intended to wall off infectious or inanimate foreign objects.

**HLA proteins (human leukocyte antigens)** Proteins on the surface of cells that identify them to the immune system as belonging to "self."

**Hyperpigmentation** Refers to excessive coloration of the skin.

**Hypopigmentation** Decreased pigmentation of the skin.

**Ichthyosis** A genetic disease of marked dryness and scaling of the skin.

**Integument** A Latin term referring to a covering, which in medical use refers to the skin.

**Keratinocyte** A principal cell making up the epidermis.

**Langerhans cells** A dendritic cell found in the skin that is capable of processing antigenic material.

**Necrosis** Death of a cell.

**Papule** A solid, raised lesion above the skin.

**Raynaud's disease/syndrome/phenomenon** A disease of blanching of the digits upon exposure to ordinary tolerable cold. It may also be accompanied by pallor, a purple color, and pain in the affected areas.

**Stratified squamous** Relating to the stacked nature of epidermal cells.

**T-lymphocyte** A specific subtype of white blood cell that regulates immune responses.

**Vesicle** A small blister.

**Xerosis** Relating to dry skin.

## BIBLIOGRAPHY


Adams, R.M. (1983). *Occupational Dermatology*. New York: Grune & Stratton.

American Academy of Dermatology (1994). *Proceedings of the National Conference on Environmental Hazards to the Skin*, 61–79. Schaumburg, IL: American Academy of Dermatology.

Asa, R. (1994). Allergens spur hospitals to offer latex-free care. *Mater. Manag. Health Care* **3** (6): 28, 30–32–34.

Barber, T. and Husting, E. (1977). Plant and wood hazards. In: *Occupational Diseases – A Guide to Their Recognition*, rev. ed. (ed. M.M. Key, A.F. Henschel, J. Butler, et al.). U.S. Department of Health, Education, and Welfare, PHS, CDC, NIOSH, DHEW-NIOSH Publications 181. Washington, DC: U.S. Government Printing Office.

Berenblum, I. and Schoental, R. (1943). Carcinogenic compounds of shale oil. *Br. J. Exp. Pathol.* **24**: 232–239.

Bingham, L., Horton, A.V., and Tye, R. (1965). The carcinogenic potential of certain oils. *Arch. Environ. Health* **10**: 449–451.

Birmingham, D.J. (1968). Photosensitizing drugs, plants, and chemicals. *Michigan* **67**: 39–43.

Birmingham, D.J. (1975). *The Prevention of Occupational Skin Disease*. New York: Soap and Detergent Association.

Birmingham, D.J. (1978). Occupational dermatoses. In: *Patty's Industrial Hygiene and Toxicology*, 3e, vol. **I** (ed. G.D. Clayton and F.E. Clayton). New York: Wiley.

Birmingham, D.J. (1986). Prolonged and recurrent occupational dermatitis. Some why and wherefores. *Occup. Med. State Art Rev.* **1** (2): 349–356.

Birmingham, D.J., Key, M.M., Holaday, D.A., and Perone, V.B. (1964). An outbreak of arsenical dermatoses in a mining community. *Arch. Dermatol.* **91**: 457–465.

Blank, I.H. (1979). The skin as an organ of protection against the external environment. In: *Dermatology in General Medicine*, 2e (ed. T.B. Fitzpatrick et al.). New York: McGraw-Hill.

Braun-Falco, O., Plewig, G., Wolff, H.H., and Winkelman, R.K. (1991). *Dermatology*, 376. New York: Springer-Verlag.

Buchanan, V.D. (1962). *Toxicity of Arsenic Compounds*. Amsterdam: Elsevier.

Calnan, C.D. (1970). Studies in contact dermatitis XXIII allergen replacement. *Trans. St. Johns Hosp. Dermatol. Soc.* **56**: 131–138.

Canadian Department of Forestry (1966). *Wood Preservation Around the Home and Farm*, Forest Products Laboratory Publication No. 1117. Ottawa: Canadian Department of Forestry.

CDC, NIOSH (n.d.). *Skin Exposure and Effects*. https://www.cdc.gov/niosh/topics/skin/default.html (accessed December 2018).

Centers for Disease Control (1987). Recommendations for prevention of HIV transmission in health-care settings. *MMWR* **36** (2): 1S–18S.

Charous, B.L., Hamilton, R.G., and Yungunger, J.W. (1994). Occupational latex exposure: characteristics of contact and systemic reactions in 47 workers. *J. Allergy Clin. Immunol.* **94**: 12–18.

Cohen, D., Davis, D., and Kozamkowski, R. (1974). Clinical manifestations of chromic acid toxicity nasal lesions in electroplate workers. *Cutis* **13**: 558–568.

Cohen, D.E., Brancaccio, R., Andersen, D., and Belsito, D.V. (1997). Utility of the standard allergen series alone in the evacuation of allergic contact dermatitis: a retrospective study of 732 patients. *J. Am. Acad. Dermatol.* **36**: 914–918.

Combs (1954). *Coal Tar and Cutaneous Carcinogenesis in Industry*. Springfield, IL: Charles C. Thomas.

Cronin, E. (1980). *Contact Dermatitis*. London: Churchill Livingstone.

Cumin, N. (1827). Remarks on the medicinal properties of Madar and on the effects of bichromate of potassium on the human body. *Edinburgh Med. Surg. J.* **28**: 295–302.

De Galan, B.E. and Hoekstra, J.B. (1995). Extremely elevated body temperature: a case report and review of classical heat stroke. *Neth. J. Med.* **47** (6): 281–287.

De Leo, V. and Harber, L.C. (1995). Chapter 23: Contact photodermatitis. In: *Contact Dermatitis* (ed. A.A. Fisher). Baltimore, MD: Williams and Wilkins.

Dukes-Dobos, F.N. and Badger, D.W. (1977). Physical hazards – atmospheric variance. In: *Occupational Disease – A Guide to Their Recognition*, rev. ed., U.S. Department HEW, NIOSH. Washington, DC: U.S. Government Printing Office.

Eckhard, L.E. (1959). *Industrial Carcinogens*. New York: Grune and Stratton.

Elias, P.M. (1992). Role of lipids in barrier function of the skin. In: *Pharmacology of the Skin* (ed. H. Muktar), 389–416. Boca Raton, FL: CRC Press.

Emmett, E. (1975). Occupational skin cancer – a review. *J. Occup. Med.* **17**: 44–49.

Emmett, E. and Kaminski, J.R. (1977). Allergic contact dermatitis from acrylates violet cured inks. *J. Occup. Med.* **19**: 113.

Emtestam, L., Zetterquist, H., and Olerup, O. (1993). HLA-DR, -DQ, and -DP alleles in nickel, chromium, and/or cobalt sensitive individuals: genomic analysis based on restriction fragment length polymorphisms. *J. Invest Dermatol.* **100**: 271–274.

Epstein, J. (1971). Adverse cutaneous reactions to the sun. In: *Year Book of Dermatology* (ed. F.D. Malkinson and R.W. Pearson). Chicago, IL: Year Book Medical Publishers.

Fisher, A.A. (1995). *Contact Dermatitis*, 4e. Baltimore, MD: Williams and Wilkins.

Fitzpatrick, T.B. (1993). Biology of the melanin pigmentary system. In: *Dermatology in General Medicine*, 4e (ed. T.B. Fitzpatrick, A.Z. Eisen, K. Wolff, et al.). New York: McGraw-Hill.

Fitzpatrick, T.B., Eisen, A.Z., Wolff, K. et al. (eds.) (1993). *Dermatology in General Medicine*, 4e. New York: McGraw-Hill.

Foray, J. (1992). Mountain frostbite. Current trends in prognosis and treatment (from results concerning 1261 cases). *Int. J. Sports Med.* **13** (1): S193–S196.

Fregert, S. (1974). *Manual of Contact Dermatitis*. Copenhagen: Munksgaard.

Gellin, G.A. (1972). *Occupational Dermatoses*. Chicago, IL: Department of Environmental, Public, and Occupational Health, American Medical Association.

Gellin, G.A., Possick, P.A., and Perone, V.B. (1970). Depigmentation from 4-tertiary butyl catechol-an experimental study. *J. Invest. Dermatol.* **55**: 190–197.

Gellin, G.A., Wolf, C.R., and Milby, T.H. (1971). Poison ivy, poison oak, and sumac – common causes of occupational dermatitis. *Arch. Environ. Health* **22**: 280–286.

de Groot, A.C. and Nater, J.P. (1994). *Unwanted Effects of Cosmetics and Drugs Used in Dermatology*, 3e. Amsterdam: Elsevier.

Harris, D.K. and Adams, W.G. (1967). Acroostereolysis occurring in men engaged in the polymerization of vinyl chloride. *Br. Med. J.* **3**: 712–714.

History of Technology (1978). *The New Encyclopedia Brittanica Macropedia*, 15e, vol. **18**.

Hood, D. (1977). Practical and theoretical considerations in evaluating dermal safety. In: *Cutaneous Toxicity* (ed. V. Drill and P. Lazar). New York: Academic Press.

Hunt, L.W., Fransway, A.F., Reed, C.E. et al. (1995). An epidemic of occupational allergy to latex involving health care workers. *J. Occup. Environ. Med.* **37**: 1204–1209.

Jaeger, D., Kleinhaus, D., Czuppon, A.B., and Baur, X. (1992). Latex-specific proteins causing immediate-type cutaneous, nasal, bronchial, and systemic reactions. *J. Allergy Clin. Immunol.* **89**: 759–768.

James, R.C., Busch, H., Tamburro, C.H. et al. (1993). Polychlorinated biphenyl exposure and human disease. *J. Occup. Med.* **35**: 136–148.

Kahn, G. (1979). Depigmentation caused by phenolic detergent germicides. *Arch. Dermatol.* **102**: 177–187.

Key, K.M., Ritter, E.J., and Arndt, K.A. (1966). Cutting and grinding fluids and their effects on skin. *Am. Ind. Hyg. Assoc. J.* **27**: 423–427.

Klauder, J.V. and Gross, B.A. (1951). Actual causes of certain occupational dermatitis – a further study with special reference to effect of alkali on the skin, effect on pH of skin, moderate cutaneous detergents. *Arch. Derm. Syphilol.*.

Klauder, J.V. and Hardy, M.K. (1946). Actual causes of certain occupational dermatitis – further study of 532 cases with special reference to dermatitis caused by petroleum solvents. *Occup. Med.* **1**: 168–181.

Kligman, A.M. (1964). The biology of the stratum corneum. In: *The Epidermis* (ed. W. Montagna and W.C. Lobitz Jr.), 387–433. New York: Academic Press.

Lampke, K.F. and Fagerstrom, R. (1968). *Plant Toxicity and Dermatitis (a Manual for Physicians)*. Baltimore, MD: Williams and Wilkins.

Lever, W.F. and Schaumburg-Lever, G. (1990). *Histopathology of the Skin*, 238–243. New York: Lippincott.

Lewis, W. and Schwartz, L. (1956). An occupational agent (*N*-butyraldoxime) causing reaction to alcohol. *Med. Ann. Dist. Columbia* **25**: 485–490.

Lim, H. and Soter, N. (eds.) (1993). *Clinical Photomedicine*. New York: Marcel Dekker.

Maibach, H.I. (1974). Patch testing – an objective tool. *Cutis* **13**: 14.

Maibach, H.I. (1987). *Occupational and Industrial Dermatology*, 2e. Chicago, IL: Year Book Medical Publishers.

Malten, K.E. and Beude, W.J.M. (1976). 2-Hydroxy-alkyl methacrylate and di and ethylene glycol D-methacrylate, contact photosensitizers in photo polymer plate procedure. *Contact Dermatitis* **5**: 214.

Malten, K.E. and Zielhius, R.L. (1964). *Industrial Toxicology and Dermatology in Production and Processing of Plastics*. New York: Elsevier.

Malten, K., Seutter, E., and Hara, I. (1971). Occupational vitiligo due to *p*-tertiary butyl phenol and homologues. *Trans. St. John Hosp. Dermatol. Soc.* **57**: 115–134.

Malten, K.E., Nater, J.P., and Von Ketel, W.G. (1976). *Patch Test Guidelines*. Nigmegan: Dekker Van de Vegt.

Mathias, C.G.T. (1986). Contact dermatitis from use and misuse of soaps, detergents, and cleansers in the workplace. *Occup. Med. – State Art Rev.* **1**: 205–218.

McNulty, W.P. (1985). Toxic and fetotoxicity of TCDD, TCDF and PCB isomers in rhesus macaques. *Environ. Health Perspect.* **60**: 77–88.

Meso, E., Murray, W., Parr, W., and Conover, J. (1977). Physical hazards: radiation. In: *Occupational Diseases – A Guide to Their Recognition*, rev. ed., U.S. Department of Health, Education and Welfare, NIOSH. Washington, DC: U.S. Government Printing Office.

Morris, G.E. (1952). Why doesn't the worker's skin clear up? An analysis of facts complicating industrial dermatoses. *Arch. Ind. Hyg. Occup. Med.* **10**: 43–49.

Mukthar, H. and Bickers, D.R. (1981). Comparative activity of the mixed function oxidases, epoxide hydratase, and glutathione-*S*-transferase in liver and skin of the neonatal rat. *Drug Metab. Dispos.* **9**: 311–314.

National Electric Injury Surveillance System (1984). U.S. Consumer Product Safety Commission, 1984; U.S. Bureau of Labor Statistics Supplementary Data Systems [SDS], 1983 data.

National Institute for Occupational Safety and Health (NIOSH) (1996). *National Occupational Research Agenda*, DHHS, Publication (NIOSH) 96-115. Cincinnati, OH: U.S. Department of Health and Human Services.

NIOSH (1988). *Prevention Planning, Implementation, Evaluation and Recommendations in Proposed National Strategy for the Prevention of Dermatological Conditions in Proposed National Strategies for the Prevention of Leading Work-Related Diseases and Injuries, Part 2, the Association of Schools of Public Health under a Cooperative Agreement with NIOSH*. Washington, DC: NIOSH.

Odom, R.B. and Maibach, H.I. (1977). Contact urticaria: a different contact dermatitis. In: *Advances in Modern Toxicology: Dermatotoxicology and Pharmacology*, vol. **4** (ed. F.N. Marzulli and H.I. Maibach). Washington, DC: Hemisphere Publishing Corp.

OSHA (1978). *Report of Advisory Committee on Cutaneous Hazards to Assistant Secretary of Labor*. Washington, DC: U.S. Department of Labor.

Pinkus, H. and Mehregan, A.H. (1976). Granulomatous inflammation and proliferation, Section IV. In: *A Guide to Dermatopathology*, 2e (ed. H. Pinkus and A.H. Mehregan). New York: Appleton-Century-Crofts.

Plewig, G. and Kligman, K.M. (1993). *Acne and Rosacea*, 2e. New York: Springer-Verlag.

Pott, P. (1775). Cancer scroti. In: *Chiruigical Works*, 734. London: 1790 ed., pp. 257–261.

Potts, R.O. and Guy, R.H. (1992). Predicting skin permeability. *Pharmacol. Res.* **9**: 663–669.

Proctor, N.H. and Hughes, J.P. (1978). *Chemical Hazards in the Workplace*. Philadelphia, PA: J. B. Lippincott.

Ramazzini, B. (1864). *Diseases of Workers* (translated from the Latin text, *De Morbis Artificum*, 1713, by W.C. Wright). New York: Hafner.

Rice, R.H. and Cohen, D.E. (1996). Toxic responses of the skin. In: *Casarett and Doull's Toxicology. The Basic Science of Poisons*, 5e. New York: Pergamon Press.

Rietschel, R.L. and Fowler, J.F. (eds.) (1995). *Contact Dermatitis*, 4e. Baltimore, MD: Williams and Wilkins.

Ronchese, F. (1948). *Occupational Marks and Other Physical Signs*. New York: Grune and Stratton.

Ronchese, F. (1965). Occupational nails. *Cutis* **5**: 164.

Rothman, A. and Emmett, E.A. (1988). The carcinogenic potential of selected petroleum derived products. *Occup. Med. – State Art Rev.* **3**: 475–494.

Rycroft, R.J.G. (1986). Chapter 16: Occupational dermatoses. In: *Textbook of Dermatology*, 4e, vol. **I** (ed. A. Rook, D.S. Wilkinson, F.J.G. Evling, and J.L. Burton). Oxford: Blackwell Scientific Publications.

Samitz, M.H. and Cohen, S.R. (1985). Occupational skin diseases. In: *Dermatology*, vol. **II** (ed. S.L. Moschella and H.J. Hurley). Philadelphia, PA: W. B. Saunders.

Samitz, M.H. and Dana, A.S. (1971). *Cutaneous Lesions of the Lower Extremities*. Philadelphia, PA: J. B. Lippincott.

Scheuplein, R.J. and Blank, I.H. (1971). Permeability of the skin. *Physiol. Rev.* **51**: 702–747.

Schmunes, E. (1986). The role of atopy in occupational skin diseases. *Occup. Med. State Art. Rev.* **1**: 219–228.

Schwartz, L., Tulipan, L., and Birmingham, D.J. (1957). *Occupational Diseases of the Skin*, 3e. Philadelphia, PA: Lea & Febiger.

Stauber, J.L., Florence, T.M., Gulson, B.L., and Dale, L.S. (1994). Percutaneous absorption of inorganic lead compounds. *Sci. Total Environ.* **145** (2): 55–70.

Stewart, R.D., Hake, C.L., and Peterson, I.E. (1974). Degreaser's flush, dermal response to trichloroethylene and ethanol. *Arch. Environ. Health* **29**: 1–5.

Sulzberger, M.B. and Wise, F. (1931). The patch test in contact dermatitis. *Arch. Derm. Syphilol.* **23**: 519.

Suskind, R.R. (1959). Occupational skin problems. I. Mechanisms of dermatologie response. II. Methods of evaluation for cutaneous hazards. III. Case study and diagnostic appraisal. *J. Occup. Med.* **1**.

Suvorov, G.A. and Razumov, I.K. (1983). Vibration. In: *Encyclopedia of Occupation and Health*, 3e, vol. **II**. Geneva: International Labor Office.

Tabershaw, I.R., Utudjian, H.M.J., and Kawahara, B.L. (1977). Chemical hazards, Section VII. In: *Occupational Diseases – A Guide to Their Recognition*, rev. ed., U.S. Department of HEW-NIOSH Publication No. 77-181. Washington, DC: U.S. Government Printing Office.

Tomazic, V.J., Shampaine, E.L., Lamanna, A. et al. (1994). Cornstarch powder on latex products is an allergen carrier. *J. Allergy Clin. Immunol.* **93**: 751–758.

Turjanmaa, K. (1987). Incidence of immediate allergy to latex gloves in hospital personnel. *Contact Dermatitis* **17**: 270–275.

Turk, K. and Rietschel, R.L. (1993). Effect of processing cement to concrete on hexavalent chromium levels. *Contact Dermatitis* **28** (4): 209–211.

Urabe, H. and Asahi, M. (1985). Past and current dermatological status of yusho patients. *Environ. Health Perspect.* **59**: 11–15.

Wang, R.G., Knaack, J.B., and Maibach, H.I. (1993). *Health Risk Assessment and Dermal and Inhalation Exposure and Absorption of Toxicants*. Boca Raton, IL: CRC Press.

Wheeland, R.G. (1994). *Cutaneous Surgery*. Philadelphia, PA: W. B. Saunders Co.

White, R.P. (1934). *The Dermatoses or Occupational Affections of the Skin*, 4e. London: H. K. Lewis & Company.

Wilkinson, D.S. (1982). Biological causes of occupational dermatoses. In: *Occupational and Industrial Dermatology* (ed. H.I. Maibach and G.A. Gellin). Chicago, IL: Year Book Medical Publishers.

Williams, N. (1975). Biological effects of segmental vibration. *J. Occup. Med.* **17**: 37–39.

Wilson, R.H., McCormick, W.I.G., Tatum, C.F., and Creech, J.L. (1967). Occupational acroosteolysis: report of 31 cases. *JAMA* **201**: 577–581.

CHAPTER 8

# INDOOR AIR QUALITY IN NONINDUSTRIAL OCCUPATIONAL ENVIRONMENTS

PHILIP R. MOREY† and RICHARD SHAUGHNESSY

*Department of Chemical Engineering, Director Indoor Air Research, University of Tulsa, Tulsa, OK, 74104*

Adapted from "Indoor air quality in nonindustrial occupational environments" by P.R. Morey, G.N. Crawford, and R.B. Rottersman, in *Patty's Industrial Hygiene*, 6e, vol. 4, edited by Vern Rose and Barbara Cohrssen. 2007. Hoboken, NJ: Wiley.

† Deceased

## 8.1 INTRODUCTION

Indoor air quality (IAQ) in nonindustrial occupational environments is not a new subject or field of study; however it carries a degree of added complexity to that of an industrial setting due to the diverse building structures that are encompassed and the broad spectrum of the population that inhabit the buildings. The approach to understanding and addressing efforts to improve IAQ in the nonindustrial setting often requires a multidisciplinary approach that involves numerous professions and expertise. The backgrounds that comprise an IAQ team investigating problems in a building may draw from medical, engineering, building science, architecture, risk communication, microbiology, psychology, indoor environmental professionals, industrial hygiene, and more. It is this principle that one must consider to effectively attend to IAQ-related problems in buildings. This chapter tracks that approach by focusing upon a logical progression of the key ingredients of an IAQ problem (sources, pathways, and people). We begin with a basic premise of the sources of pollutants in the structure, the pathways through which the pollutants move (planned and unplanned airflows), and the nature of the problem itself in terms of how it is manifested and voiced by occupants, taking into account the more vulnerable populations that are present in the nonindustrial setting. This method allows formulation of a credible hypothesis as to the etiology of the problem at hand and subsequent construction of a suitable control method to remediate the problem.

### 8.1.1 Historical

Indoor air from occupied buildings is almost always more polluted from human-sourced contaminants than the air outside the building. This is true for recently constructed buildings as well as for primitive shelters erected centuries or millennia ago. Since the biblical times it had been recognized that the building or shelter itself may be a source of contaminants (e.g. dampness causing plague, leprosy, or mold), resulting in illness (Hellea et al. 2003; Sundell 2004). It follows that a continuous source of outdoor air is required for dilution to reduce the degradation of indoor air from contaminants arising from people and their processes and activities. This was achieved historically by the inventors of roof vents for exhaust of fire smoke and by the installation of openable windows for the introduction of makeup (outdoor) air.

Historically, the concept of IAQ has included the view that outdoor ventilation air is required to both reduce adverse health effects and to provide for comfort of occupants. Thus, over two centuries ago, Benjamin Franklin wrote

*Handbook of Occupational Safety and Health*, Third Edition. Edited by S. Z. Mansdorf.

that "…I am persuaded that no common air from without is so unwholesome as the air within a closed room that has been often breathed and not changed" (Morey and Woods 1987). He further stated that outdoor air and "cool air does good to persons in the smallpox and other fevers. It is hoped, that in another century or two we may find out that it is not bad even for people in health."

By the mid-nineteenth century, it was also recognized that air consisted of at least two gases, oxygen and carbon dioxide ($CO_2$), and that staleness of indoor air in buildings was associated with the $CO_2$ level. The work of Pettenkofer indicates that the unpleasantness of stale air was not caused by warmth, humidity, $CO_2$ itself, or oxygen deficiency but rather by trace amounts of bioeffluents (malodors) associated with emissions from the skin and lungs of building occupants (Sundell 2004). Although the staleness of room air was associated with a buildup of $CO_2$, this gas itself was not recognized as an air contaminant, but rather as a surrogate indicator of human body emissions.

In the mid-nineteenth century, it was realized that tuberculosis and other airborne contagious diseases were more readily contracted in crowded places with deficient ventilation. Accordingly, ventilation requirements by the late nineteenth century recommended the provision of large amounts of outdoor air to lower the risk of disease from airborne infectious agents.

### 8.1.2 Sick Building Syndrome and Building-Related Illness

In a paper entitled "The Sick Building Syndrome" (SBS), Stolwijk (1984) described a constellation of nonspecific complaints (e.g. mucous membrane irritation, headache, fatigue, etc.) that occurred at higher prevalence rates in some building. Although these nonspecific complaints may be present at some level in all buildings, their prevalence appears highest in a subset of "sick" buildings. In the decades that followed Stolwijk's publication, most work-related occupant discomfort and annoyance complaints in nonindustrial occupational environments were referred to as SBS. Documentable clinical disease associated with building occupancy was defined as building-related illness (BRI). By definition, one of the primary differences between SBS and BRI is that with SBS, there is no single etiologic agent that can be identified within the building. SBS problems are typically multifactorial in nature that complicates defining the agent(s) responsible for the nonspecific symptoms. It is unfortunate that the term SBS focuses attention on the building itself and not the people who are the repository of the symptom complaints.

A number of factors collectively make IAQ more important now compared with 70 or 80 years ago. Construction practices today are different than those employed in the 1930s or 1940s. Windows in modern buildings generally do not open, whereas in the 1940s, natural ventilation (open windows) was common. In modern building reliance is placed on the heating ventilation and air-conditioning (HVAC) system to transport outdoor air to the breathing zone. In modern buildings, insufficient quantities of outdoor air or adequately conditioned air are often transported to the occupant breathing zone as a result of energy conservation measures that reduce HVAC system operational costs during the cooling and heating seasons. Poor maintenance and cleaning of today's complex building systems are other reasons for increased attention to IAQ issues.

Construction materials have markedly changed over the past 70 years. Stone, wood, and other "natural" construction/finishing materials have largely been replaced by synthetic building materials and moisture-sensitive pressed wood/amorphous cellulose products. The ceiling tiles, wall and floor coverings, and furniture in modern buildings can be strong sources of volatile organic compounds (VOCs), semivolatile organic compounds (SVOCs), and aldehydes. Volatile chemicals from cleaning agents, air fresheners, office machines, pesticides, and personal care products are additional sources of indoor air pollution. Additionally, alkenes from cleaning compounds and furnishings can react with ozone, resulting in by-products such as ultrafine particles and production of highly reactive compounds such as aldehydes (Shaughnessy and Weschler 2001; Sundell 2004).

SBS in nonindustrial workplaces has been associated with factors such as overcrowding, the use of fleecy finishing materials, photocopier usage, dirty HVAC system filters, and ventilation rates less than $10\,L\,s^{-1}$ or 20 cfm per occupant (Mendell 1993). Generally, there is a higher prevalence of SBS in mechanically ventilated buildings. In other words, the prevalence of SBS complaints has been found, at least in North America and European studies, to be less in naturally ventilated buildings. Most investigators consider SBS to be multifactorial in origin (Lahtinen et al. 1998). Physical factors (e.g. presence of dusts), chemical agents (e.g. SVOCs, VOCs), microbial agents (e.g. endotoxins and B-1, 3-D glucan), and psychological factors have all been considered in the etiology of SBS.

BRI occurs when occupant health problems are clearly recognizable as disease (with overt symptoms or abnormal physical signs) upon medical examination and are associated with indoor environmental exposure. Perhaps the most well-known case of a BRI was the outbreak of pneumonia (later known as Legionnaire's disease) that occurred in a Philadelphia hotel in 1976 as a result of exposure to the bacterium subsequently named *Legionella pneumonia*.

This case of BRI is still remembered by the public because of the 29 fatalities that occurred as a result of exposure in the hotel. Examples of BRIs and other air contaminants that are involved in disease etiology include:

- Cancer caused by gaseous and particulate components of ETS, some VOCs, and radon.
- Dermatitis caused by fibers from man-made insulation.
- Hypersensitivity pneumonitis (HP), humidifier fever, allergic rhinitis, adult onset asthma, and nonallergic respiratory disease associated with microorganisms, allergens, and dampness.

It is important to realize that the diagnosis of a BRI is based on medical examination of one or more occupants confirming the presence of a specific disease plus an environmental assessment showing the likelihood of a contaminant exposure that would cause the disease. Thus, the isolation of *L. pneumonia* from a pneumonia patient plus the recovery of the same serotype of *L. pneumonia* from the water system in a building where that patient worked or otherwise spent time is the basis for a diagnosis of building-related Legionnaire's disease. The diagnosis of SBS, on the other hand, is a group diagnosis based on review of the nonspecific discomfort symptoms. Since SBS symptoms remit upon exiting the building, medical examination of an individual at the physician's office is usually of little use in making SBS diagnoses. Similarly, diagnosing the building as the cause of SBS or BRI without an environmental assessment is unreliable. The same environmental agents may cause both SBS and BRI. For example, exposure to VOCs or ETS or combustion products may cause SBS symptoms. Chronic exposure to the same air containing contaminants may also elicit BRI (e.g. CO poisoning where restorative procedures such as re-equilibrating in a hyperbaric chamber may be required). The fact that BRI can be attributed to a specific building-related problem (resulting in a discrete identifiable illness which persists well beyond the actual stay in the building) makes it distinct from an SBS diagnosis, and although less commonly reported in buildings, BRI should be immediately addressed to prevent further illness spread to additional occupants in the building.

## 8.2 INDOOR AIR POLLUTANTS

Various pollutants that are important in IAQ evaluations are reviewed in this section (Table 8.1). The reader is also referred to the proceedings of international conferences on IAQ and climate (Levin 2002; Wu 2005) where numerous papers are presented on each type of pollutant covered.

### 8.2.1 Microbials

Microbial contaminants in indoor environments include viruses, bacteria, fungi, Protista, endotoxins, and mycotoxins. The surfaces of construction and finishing materials in buildings are almost always characterized by the presence of culturable and nonculturable fungi and bacteria. The indoor air in most buildings normally contains fungi similar to those found in the atmosphere outside the building. Microbial contaminants and some of the BRIs they can cause (Table 8.2) are often associated with moisture and dampness in buildings and with poor maintenance of plumbing and HVAC system components.

**TABLE 8.1 Representative Kinds of Indoor Air Pollutants and Their Sources**

| Pollutant | Sources |
|---|---|
| Glycol, ethers, and terpenoids | Cleaning products and air fresheners |
| Formaldehyde | Man-made mineral fiber products; composite products |
| Ozone | Outdoor air; intentional and unintentional generation by certain air purifiers |
| Dust/particles | Outdoor air; fleecy finishing materials including carpet; combustion-related ultrafine particles from unvented combustion devices/sources in the home |
| Phthalates | Polyvinyl chloride and other synthetics |
| Nicotine | Environmental tobacco smoke |
| Molds | Moist or damp finishing and construction materials |
| Legionella | Hot water service and cooling towers |
| Carbon monoxide/nitrogen oxide | Combustion (heating) devices; gas cooking |

**TABLE 8.2 Diseases Caused by Microbial Contaminants in Buildings**

| Disease | Symptoms | Agent and sources |
|---|---|---|
| Legionnaire's disease | Pneumonia | Legionella in hot water system; cooling towers |
| Humidifier fever | Acute influenza-like symptoms | Endotoxin in humidifier water |
| Hypersensitivity pneumonitis | Acute fever and cough, fibrosis of lung | Fungi and bacteria growing in the HVAC system |
| Asthma | Constriction of the airways | Fungi, bacteria, and house dust mites |

See Morey et al. (2008).

Most of the mold spores commonly found indoors in nonproblem buildings originate from outdoor sources. The air outdoors is usually dominated by common fungal spores such as *Cladosporium* or *Alternaria* species, which grow on the leaves of plants (phylloplane [i.e. leaf surface] fungi). When air enters a building such as through the HVAC system, through an open door, or by infiltration through cracks in the building envelope, phylloplane spores are carried indoors. Thus, the mix of mold spores in indoor air in a building without mold growth problems is normally similar to the outdoor air mix.

In buildings with highly efficient HVAC system filters, the total concentration of airborne molds indoors is generally much lower (exception, during winter snow cover) than that found outdoors. Tracking of soil indoors often leads to the accumulation of *Penicillium* and *Aspergillus* species (molds abundant in topsoil and botanical litter), especially in poorly maintained carpet. While a diversity of molds is normally present in the air in healthy buildings, the occurrence of only one of two dominating kinds of molds indoors, especially if the same kinds of molds are not present outdoors at the same time, can be an indication of a moisture problem in the building.

The growth of fungi in buildings is primarily dependent on the availability of moisture in the capillary spaces and on the surfaces of construction and finishing materials. Since fungi grow on surfaces of materials, the relative humidity in room air is less important in determining if growth will occur than the available moisture in the substrate. See the review by Fleming and Miller (2011) of moisture parameters including water activity that control microbial growth on building materials. Biodegradable materials that are damp can preferentially support the growth of xerophilic (dry-loving) fungi such as *Eurotium herbariorum* or *Wallemia sebi*. Materials that are chronically wet can preferentially support the growth of hydrophilic fungi such as *Ulocladium*, *Stachybotrys*, and *Fusarium* spp.

There are many areas in buildings where moisture problems can lead to fungal growth. These include:

- *The building envelope*. In cold climates, fungi may grow on the interior (occupied) surface of poorly insulated envelope walls when moisture in room air condenses on the cold wall surface. In depressurized air-conditioned buildings in hot, humid climates, fungi may grow on the inner surfaces of the envelope wall because moisture in infiltrating humid air from outdoors condenses on cool surfaces in contact with the interior conditioned space. This is especially true when the interior surface (adjacent to an exterior wall location) such as gypsum is lined with vinyl wall covering that further exacerbates the problem by creating an additional vapor barrier that promotes condensation due to the humid incoming air driven by the negative pressure gradient.
- *Cellulose-containing materials*. Cellulose degrading molds such as *Stachybotrys* and *Chaetomium* grow well on the amorphous cellulose (paper) of gypsum wallboard that is chronically wet. *Aspergillus versicolor* can grow on other areas of the same wallboard that is damp, but not wet. Flannigan and Miller (2011) have reviewed the literature on cellulolytic fungi including *Chaetomium globosum*, *Memnoniella echinata*, and *Stachybotrys chartarum*.
- *Water leaks and floods in buildings*. Window leaks, pipes leaks in walls, water ponding on flat roofs, and flooding can result in mold problems in buildings (Wright et al. 1999; Morey et al. 2003) and sometimes in adverse health effects among occupants. As a result of unusually heavy rains in Southern California during 1997–1998, a case of HP ("El Niño" lung) was putatively associated with the growth of the fungus *Peziza domiciliana* in a flooded basement.

The Institute of Medicine (2004) summarized the literature on moisture-impacted buildings and concluded that dampness and the presence of mold were associated with health outcomes including upper respiratory (nasal and throat) tract symptoms, cough, wheeze, and asthma symptoms in sensitized asthmatic persons. In a study reported by NIOSH (Park et al. 2004) semiquantitative indices noted during physical inspection such as the extent of water stains, visible mold, and moldy odors were found to be useful in identifying buildings with BRI. The NIOSH study is important because it shows that the physical inspection of building construction for moisture damage is the most important factor in identifying problem buildings and preventing BRI. Human activities in buildings often result in resuspension or aerosolization of settled dusts present on interior surfaces (Ferro et al. 2004; Shaughnessy and Vu 2012).

Considerable research has been carried out on the composition of fungi in nonproblem buildings. *Cladosporium* species and yeasts were found to be dominant isolates from settled dusts collected in nonproblem buildings (Horner et al. 2004). In dusts from water-damaged buildings, *Penicillium* and *Aspergillus* are frequently found in settled dust, especially near areas with visible growth on interior surfaces. Fragments of fungal spores and hyphae often of a size less than 1 μm are readily aerosolized from moldy surfaces (Gorney et al. 2002). The release of these small particles from surfaces may be controlled by different mechanisms than for the relatively larger spores and may significantly contribute to BRI because of their small size and ability to penetrate into the distal airways.

Mycotoxins are secondary metabolites produced by filamentous molds. The growth of filamentous molds on foods can result in production of mycotoxins that can result in adverse health effects such as immunosuppression and cancer when the moldy food is ingested. The ability of airborne mycotoxins to cause BRI in moldy buildings is unclear (Institute of Medicine, Board on Health Promotion and Disease Prevention, Committee on Damp Indoor Spaces and Health 2004) for a number of reasons including the large variation of many chemically distinct metabolites (Nielsen 2003) produced by fungi on moldy building substrates plus an absence of dose–response data for specific mycotoxins and specific health endpoints. Mycotoxins are found not only in spores but also in the mycelium as well as in substrates on which molds grow. Aerosolization of particulates from dry and moldy building materials provides a route of potential exposure to mycotoxins. Because mycotoxins are chemicals, the presence of mycotoxins in dusts from moldy materials is unrelated to the culturability of fungi.

Fungi growing in building materials can produce a wide range of volatile compounds known as microbial volatile organic compounds (MVOCs). MVOCs produced by fungi include 1-octen-3ol and 2 octen-1-ol. These MVOCs are almost never emitted from construction and finishing materials unaffected by biodeterioration. MVOCs are useful indicators of microbial growth. The MVOC 3-methyl furan, while a useful indicator of microbial growth in buildings, has also been shown to be a component of environmental tobacco smoke (Mehrer and Lorenz 2005).

Lukewarm water (30–40 °C) in cooling towers, evaporative condensers, and potable water service systems provides niches for the growth of *Legionella,* which are gram-negative bacteria. At present there are approximately 58 species known in the genus *Legionella*. Some *Legionella* species can be distinguished into serogroups, for example, *L. pneumophilia* serogroups 1, 4, and 6. *Legionella pneumophilia* is the species that caused the outbreak of Legionnaire's disease in Philadelphia in 1976. Legionnaire's disease is an infection (the lung of a susceptible person is the target organ) that results in pneumonia, which may be fatal. *Legionella* may also cause a milder form of illness known as Pontiac fever, which is a non-pneumonic and similar to a severe flu. Legionnaire's disease and Pontiac fever differ in the clinical reaction, incubation time, and attach rate (ASTM 2007). Legionnaire's disease is pneumonia with a 2–10-day incubation, and only about 5% of exposed subjects are affected. Pontiac fever is non-pneumonic, with symptoms that resemble influenza and start 36 hours after exposure, and affects 90% of those exposed. *L. pneumophilia* serogroup 1 is the most frequent cause of legionellosis. Identification of the environmental source of a legionellosis outbreak occurs when the species/serotype of *Legionella* from the infected person matches the species/serotype of *Legionella* in the environmental reservoir. So, it is important to always determine the species and the serotype of both clinical and environmental isolates when attempting to identify the environmental source of an infection or outbreak. Protocols for collection of *Legionella* from these sources are well known (ASTM 2007). Analysis of water samples for cultivable *Legionella* is the preferred method used in field studies. Routine sampling of building water systems for cultivable *Legionella* is considered controversial because disease may occur in the absence of positive sampling results, and positive sampling results do not imply that disease has or will occur. However, it has been suggested that *Legionella* risk is greatest when a high percentage of building water samples are positive for cultivable Legionellas, especially *L. pneumophilia* serogroup no. 1 (Stout 2007). Corrective actions to reduce *Legionella* amplification in water systems including thermal eradication, hyper-chlorination, copper-silver ionization, chlorine dioxide, etc. are reviewed elsewhere (Stout 2007; WHO 2007).

Gram-negative bacteria such as species of *Pseudomonas*, *Flavobacterium*, and *Blastobacter* grow in stagnant water (e.g. humidifier sumps) or on wet surfaces of HVAC system cooling coils or drain pans. A biofilm (slime) on a wet cooling coil surface or on a wet surface in a humidifier sump is a certain indication of microbial growth that likely includes these gram-negative bacteria. Humidifier fever, a disease characterized by influenza-like symptoms that remit after cessation of exposure, has in some cases been associated with endotoxin (from the outer membrane of gram-negative bacteria) present in water droplets emitted from some humidifiers and water spray systems with a recirculation water system or sump. These systems are prone to microbial contamination especially if maintenance is poor.

Endotoxins have been causally associated with disease, for example, byssinosis from cotton dust and organic dust toxicity syndrome in silos and barns. The endotoxin in cotton dust is derived from gram-negative bacteria that grew on the cotton plant prior to harvest and textile processing (Morey 1997). Gram-negative bacteria growing on moist sage and hay are sources of endotoxin in some agricultural settings. In general, airborne endotoxin concentrations that are two or three orders of magnitude greater than background (outdoor) levels are considered significant exposures (Milton 1999). Human-shed bacteria are found in the indoor air. These include gram-positive bacteria such as *Staphylococcus* and *Micrococcus* species present on human skin scales and *Streptococcus* species emitted as aerosols from the nasal/pharynx when a person is talking. Other than being an indicator of occurrence of people in the indoor environment, adverse health effects have not been directly ascribed to these gram-positive bacteria in nonhospital settings. Outbreaks of infective illness in the indoor environment may be caused by airborne exposure to specific human-shed influenza virus and viruses that cause the common cold. Provision of adequate amounts of outdoor air or highly filtered recirculated air to occupants in buildings is thought to reduce the risk of infection due to the common cold. The concept of provision of dilution ventilation to reduce the risk of "contagion" has been recognized in ventilation codes since the late nineteenth century (see Section 8.4).

Rhinoviruses, which are an important cause of the common cold may be spread by direct contact (e.g. nasal secretions), by indirect contact (e.g. contaminated objects), or by aerosol. Experiments by Dick et al. (1987) showed that transmission of rhinovirus infections could occur entirely by the aerosol route in experimental settings where direct and indirect contacts were excluded. Infectious agents such as in the severe acute respiratory syndrome (SARS) epidemic in 2003 are transmitted by the airborne route (Lietal 2007). The SARS outbreak in Amoy Garden (Hong Kong) was associated with transmission of the infective agent by ventilation air. A recent review (Anssen and Clark 2006) provides considerations for respiratory protection important with regard to airborne infective viruses such as SARS and avian influenza type A-H5NI viruses.

The principles of sampling for microorganism and microbial products (e.g. endotoxins) are reviewed elsewhere (AIHA 2005, 2008; Morey et al. 2007). A plan for data interpretation must be in place when sample collection occurs (Morey 2007). The objective of the sampling plan as well as appropriate controls must be clearly defined. Thus, source samples should be collected from moisture problem areas as well as from background (reference control) areas not affected by moisture or dampness. Air sampling should include outdoor controls, generally obtained from high-quality air on the roof of the building. Sampling objectives should be clearly stated. For example, sampling with a general intent to "see if there is a problem" often yields no useful information. Conversely, sampling for specific types of fungi in different areas can provide data to answer a specific question such as "Are the fungi in these areas similar enough to represent the same population?" In addition, in the last decade, scientists have further developed DNA-based sequencing analyses that are rapidly finding their way into research and, in the near future, practice. By sequencing ribosomal RNA encoding genes in bacteria and fungi and then comparing these sequences to curated databases, the different members of a fungal and bacterial community can be determined, regardless of viability or culturability. While this unbiased approach was a revolution for microbial ecology, up until the year 2007, this was a laborious and expensive process that required a high level of molecular biology expertise. The recent commercialization of new technologies that sequenced DNA by synthesis has been a game changer. DNA sequencing is independent of culturability and viability and thus circumvents the prior ecological limitations associated with culturing (Begerow et al. 2010). Thus, assessments can be made on a basis of full microbial ecology data, rather than a limited and inflexible a priori set of indicators. This vast increase and improvement in information is accompanied by a broad set of open-access tools for quantitative sequence comparisons and analyses (Caporaso et al. 2010) and opens new doors for testing a myriad of approaches, algorithms, and hypotheses for fungal and bacterial communities associated with buildings and human health (Peccia and Shaughnessy 2016).

Sampling data for culturable fungi should be interpreted as follows: (i) Building surfaces are not sterile. Visible mold should be physically removed in such a manner that spores are not dispersed into occupied or clean areas. (ii) In healthy buildings the diversity of airborne fungi indoors and outdoors should be similar. (iii) The consistent dominating presence of one to two fungal species indoors and the absence of the same species outdoors may indicate a moisture problem and degraded air quality. (iv) The consistent presence of fungi such as *S. chartarum* and *A. versicolor*, *Aspergillus fumigates*, or various *Penicillium* species over and beyond background conditions likely indicates the occurrence of a moisture problem.

### 8.2.2 Volatile Organic Compounds

VOCs are commonly associated with indoor air pollution and SBS. VOCs are characterized by boiling points ranging from about 50 to 380 °C and include alcohols, aldehydes, alkanes, aromatic hydrocarbons, halogenated hydrocarbons, terpenes, ketones, esters, and cycloalkanes. Those VOCs with boiling points less than 50–100 °C are referred to as very volatile organic compounds (VVOC). Those with boiling points above 240 °C are called SVOCs. SVOCs include pesticides, polynuclear aromatic compounds, and certain plasticizers such as phthalates (Lewis and Wallace 1988). Even in geographic areas where outdoor air pollution form sources such as heavy vehicle traffic and petroleum refineries are significant, studies in the United States in the 1980s showed that concentrations of many VOCs found

**TABLE 8.3 Median Indoor and Outdoor Concentrations of Some Individual VOCs in Japanese Homes**

| | Median concentration (μg mg$^{-3}$) | |
|---|---|---|
| VOC | Indoors | Outdoors |
| Formaldehyde | 70.6 | 7.2 |
| Acetaldehyde | 30.6 | 2.1 |
| Acetone | 23.9 | 1.0 |
| Chloroform | 4.4 | 4.2 |
| Benzene | 3.7 | 3.7 |
| Toluene | 10.4 | 5.3 |
| D-Limonene | 17.4 | <1.2 |
| *n*-Decane | 27.1 | 26.7 |
| o-Xylene | 3.8 | 2.3 |

Summertime conditions; 26 houses, 3 measurement/house adapted from Shinohara et al. (2006) and Morey et al. (2008).

**TABLE 8.4 Median Indoor and Outdoor Concentrations of Some Individual VOCs in Japanese Homes**

| | Concentration (μg mg$^{-3}$) | |
|---|---|---|
| Cleaning status | Alcohols | Alkanes |
| No cleaning[a] | 3 | 250 |
| Routine cleaning | 1030[b] | 1300[c] |

[a] Zero cleaning was performed during a six-month period.
[b] Mostly $C_9$–$C_{11}$ alkanes.
[c] Mostly isopropanol.

indoors were 2–10 times higher than the same VOCs found outdoors (Wallace et al. 1987). The variety of the VOCs found in the indoor air is almost always greater than that in outdoor air, mainly because of the very large number of possible sources in indoor environments. Table 8.3 presents results of studies in Japanese homes where median, indoor, and outdoor concentrations of VOCs were determined (Shinhara et al. 2006). The concentrations of VOCs such as formaldehyde, acetone, and k-limonene were at least an order of magnitude higher than median outdoor levels. This suggests the presence of strong indoor sources of these VOCs. By contrast, the concentrations of other VOCs both indoors and outdoors (e.g. benzene, chloroform) were similar suggesting an absence of strong indoor sources for these specific VOCs.

Major VOC sources indoors include the building and its finishing products, activities by people including cleaning and occupants themselves. Alkanes and aromatic hydrocarbons from building construction materials are commonly present in indoor air in most buildings. Alkanes such as *n*-decane and *n*-undecane may be present indoors in new buildings (and during renovation of portions older buildings) at concentrations of 100–1000 times that present outdoors. Toluene, xylene, and ethylbenzene are almost always found in indoor air because of the extensive use of these aromatic hydrocarbons in interior finishes (for example, linoleum and paints).

Cleaning activities in buildings may be associated with strong VOC emissions. Alcohols and alkanes from cleaning solvents and soaps can become predominant VOCs in buildings where original interior finishes are no longer significant emission sources. In the example in Table 8.4, isopropanol and various alkanes present in cleaning agents accounted for almost 90% of the VOCs found in indoor air in one building.

The occupants themselves may be major sources of indoor VOCs. D-Limonene (a terpene) and various siloxane compounds (e.g. decamethylcyclopentasiloxane) may be present in many personal care products including antiperspirants and deodorants (Shields 1996). Tetrachloroethylene is emitted from clothing that has been recently dry-cleaned. Hand and body lotion, moisturizing soaps, and cosmetics can be strong sources of $C_{12}$ and $C_{16}$ alkanes. Finishing materials used in newer buildings are characterized by reduced emissions of traditional solvents (e.g. toluene, xylene) and the slow emissions of higher boiling point VOCs such as monoterpenes.

Formaldehyde is often associated by the public with indoor air pollution. Although formaldehyde may be emitted from a number of sources such as from gas stoves and from smoking, its major source indoors is from construction materials such as particleboard, fiberboard, and plywood. Concentrations of formaldehyde in residential buildings are generally higher than in office buildings because of the relatively large ratio of pressed wood products to air volume in the former as compared with the latter type of building.

In the late 1980s there was a suggestion that the sum total of individual VOCs (or TVOC) in indoor air was related to the occurrence of SBS. Subsequently, the TVOC concept was discredited (Singer et al. 2006) and replaced with attention to the reaction characteristics of individual VOCs or groups of VOCs in promoting adverse health effects or irritation.

Wolkoff (1998) has reviewed the sources of VOCs in the indoor environment as follow:

- Stable, minimally reactive VOCs such as toluene, decane, etc.
- Reactive VOCs such as D-limonene, d-pinene, and alkenes that can react with ozone and form secondary aerosols.
- VOCs such as aldehydes and organic acids that are reactive by forming chemical bonds with sites in mucous membranes.
- VOCs such as pentachlorophenol and some glycol ethers including 2-butoxyethanol that are toxic.

Methods of sampling established for VOCs in industrial workplaces are not readily adaptable to nonindustrial indoor studies. Sampling methods developed for industrial workplaces are often bulky, noisy, and validated for concentrations only about an order of magnitude below the applicable industrial threshold limit value (TLV). Concentrations of VOCs found in nonindustrial indoor air are usually two or more orders of magnitude lower than industrial TLVs (Lewis and Wallace 1988). More sensitive sampling and analytical methods are therefore required for characterization of VOCs in indoor air.

Most IAQ sampling evaluations for VOCs utilize trapping of pollutants on various sorbents and subsequent laboratory analysis for specific VOCs. A variety of sorbents (often two or more sorbents per tube), including Tenax, graphitized carbon black, or carbon molecular sieves, are used to trap VOCs (Hodgson 1995). After thermal desorption in the laboratory, gas chromatography–mass spectrometry (GC/MS) are used to provide the sensitivity (generally $1\,\mu g\,m^{-3}$ for specific VOCs) required for IAQ studies. The use of passive sampling devices based on the principle of molecular diffusion into charcoal combined with GC/MS provides another method of measuring VOCs in indoor air at levels of $1\,\mu g\,m^{-3}$ or less (Shields et al. 1996).

Phthalates are SVOCs that are widely used as plasticizers in wall coverings, floor tiles, upholstery, paints, glues, hairsprays, toys, shower curtains, etc. Phthalates are released in small amounts as particulate components of settled (house) dust. Diethyl phthalate (DEP) has been found as a dominant phthalate at very low concentrations in indoor air, whereas di[2-ethylhexyl]phthalate (DEHP) was detected as the dominant phthalate in settled dusts (Zhu and Yang 2006). The approximate concentration of total phthalates in settled dusts including vacuum cleaner bag dust is in the 0.5–$1.0\,mg\,g^{-1}$ range. The mean concentrations of DEHP and DEP in settled dust reported by Zhu and Yang were 521 and $4.7\,\mu g\,g^{-1}$, respectively. There appears to be an association between the presence of phthalates in house dust and allergic symptoms in children; however this relationship is controversial (Bornehag et al. 2004).

VOCs such as alkenes including terpenoids can react with ozone to form aldehydes, which are sensory irritants (Morrison et al. 2006). In addition, the reaction of ozone with unsaturated VOCs results in the generation of a fine (0.02–0.1 μm) particulate organic aerosol. Ozone in the presence of limonene can increase fine particulate levels (particulate diameters $\leq 1\,\mu m$) by as much as 15–$20\,\mu g\,m^{-3}$ (Hubbard et al. 2005). Indoor materials that potentially are source VOCs that can react with ozone (Morrison et al. 2006) include:

- Scented cleaning compounds, essential oils, candles, etc. that contain VOCs such as D-limonene, d-pinene, and d-terpineol.
- Environmental tobacco smoke components such as nicotine that react with ozone to form aldehydes (e.g. nicotinaldehyde).
- Products such as cork and linoleum that can react with ozone to form acids and aldehydes.

### 8.2.3 Other Indoor Pollutants

The roles of pesticides, combustion products, and airborne particulate in IAQ are briefly reviewed in this section. The reader is referred to Morey and Singh (1991) and Morey et al. (2008) for a discussion on the involvement of radon and environmental tobacco smoke in IAQ.

***8.2.3.1 Pesticides*** Pesticides including insecticides, termiticides, and fungicides are often used in interior spaces to control a wide variety of organisms including wood-boring insects, moths, and fungi. Although pesticides are by definition poisons, their toxicity varies toward different types of organisms. Pesticides such as chlordane, heptachlor, aldrin, dieldrin, and chlorpyrifos have been used in the soil or under the foundations of buildings (Dingle and Tapsell 1996). The routine application of chemical pesticides as a preventative approach for pest control is being replaced by and referred to as integrated pest management (IPM). IPM programs typically include initial monitoring of an area for pests. When pests exceed a tolerance threshold, control focuses on the elimination of conditions that led to the problem. This may include sealing pathways from the outdoors and/or eliminating food and moisture sources that the pests need for indoor survival. If these actions are not successful, the use of traps, baits, or targeted pesticide application is implemented. Broadcast spray application of pesticides is considered as a last resort.

In the 1980s the EPA (1987) developed sampling and analytical procedures for over 50 specific organochlorine, organophosphate, organonitrogen, and pyrethroid pesticides using polyurethane foam sorbent with sensitivities as low as 0.01 $\mu g\,m^{-3}$. For example, among 50 residences monitored in a Florida study, 46 contained detectable levels of the pesticide chlorpyrifos (mean concentration 0.47 $\mu g\,m^{-3}$). Chlorpyrifos was detectable in outdoor air around only 9 of the 50 residences, with a mean concentration of 0.059 $\mu g\,m^{-3}$. Concentrations of chlorpyrifos as high as 37 $\mu g\,m^{-3}$ have been found in other residences (Lewis and Wallace 1988). A chlorpyrifos concentration of 0.52 $\mu g\,m^{-3}$ was found in one house four years after pest control treatment (Schenk et al. 1997). After application, pesticides may be present not only in indoor air but also in settled dusts. Thus, following application of the pesticide lindane to ethnological objects in museum store rooms, it could be found in the air (0.2–3.7 $\mu g\,m^{-3}$), in dust (up to 128 $mg\,kg^{-1}$), and on room finishes (4–32 $mg\,kg^{-1}$) (Kroos and Stolz 1996). The pyrethroid insecticides permethrin and deltamethrin used for cockroach control have been shown to be very persistent (for periods longer than 70 weeks) in settled dust (Berger-Preiss et al. 1997) on interior surfaces.

A combination of air and dust (surface) sampling is probably best for characterization of pesticide residues that may be present in indoor environments. If sampling is performed, the analytical methods used must be sufficiently sensitive (for example, 0.01 $\mu g\,m^{-3}$ for air samples) so as to detect background concentrations of specific pesticides that may be present in the outdoor air. The objective of sampling is often to determine if pesticide concentrations in areas of suspect contamination exceed background outdoor levels and average concentrations that have been previously reported in the literature (Lewis and Bond 1987).

Most pesticide-related injuries occur in the home (Kroos and Stolz 1996). Of the accidents that were not related to ingestion, inhalation, and dermal exposures were found to be equally important. Because of the potential chronic health effects from exposure to termiticides, the National Research Council (1982) has recommended that airborne concentrations in indoor air be limited as follow: chlorpyrifos (10 $\mu g\,m^{-3}$), chlordane (5 $\mu g\,m^{-3}$), heptachlor (2 $\mu g\,m^{-3}$), and dieldrin and aldrin (1 $\mu g\,m^{-3}$). It should be noted that these exposure levels are more than an order of magnitude lower than concentrations considered acceptable in industrial workplaces. However, industrial exposure limits can be exceeded following residential application of pesticides. Methyl bromide concentrations in homes following fumigation to control dry-wood termites were 5 ppm even after several aerations (Scheffrahn et al. 1992). Since the TLV for methyl bromide is 1 ppm, this study showed that better ventilation procedures following fumigation are needed.

***8.2.3.2 Combustion Products*** When combustion occurs in air, carbon dioxide ($CO_2$) and water vapor are emission products. However, carbon monoxide (CO), nitrogen dioxide ($NO_2$), sulfur dioxide ($SO_2$), and respirable particles are the by-products of combustion that can cause indoor air pollution. The type and amount of combustion by-products in indoor air of commercial buildings depend upon the type of fuel consumed (for example, diesel used by trucks contain sulfur, and $SO_2$ is an important combustion by-product) and the location of outdoor air inlets. Combustion processes that are oxygen-starved and characterized by yellow-colored flames are characterized by elevated CO emissions. Combustion under oxygen-rich conditions results in higher flame temperatures, thus emitting greater amounts of oxides of nitrogen. Considerable literature is available describing characteristics of combustion devices found in indoor environments (primarily residences) and factors affecting the emission rates of various combustion by-products (DOE 1985; Woodring et al. 1985).

Indoor air pollutants such as CO and $NO_2$ originate from multiple sources, and concentrations indoors depend on parameters such as source emission rates, the volume of air indoors, outdoor air ventilation rates, and HVAC system characteristics. Pollutants such as CO and $NO_2$ are emitted intermittently and are usually concentrated only in certain areas of the building.

CO combines with hemoglobin to form carboxyhemoglobin, thereby resulting in a decline in the oxygen-carrying capacity of the blood. Carboxyhemoglobin higher than 4–5% is known to exacerbate symptoms of individuals with preexisting cardiovascular disease. Limiting average CO exposures to 9 ppm (maximum) for eight hours or 35 ppm for one hour, as specified in the National Ambient Air Quality Standards, is intended to provide a margin of safety with regard to carboxyhemoglobin buildup in individuals with cardiovascular disease.

As a general rule the emission or transport of combustion by-products such as CO in or into a building is unacceptable. Complaints of exhaust odors coupled with occupants' symptoms such as headache, nausea, fatigue, dizziness, rapid breathing, and confusion are indicators of the presence of combustion products such as CO. The source of combustion by-products should be found and eliminated or exhausted out of the building. The measurement of relatively low CO concentrations in the range of 3–5 ppm (assume that ambient levels are lower) still indicates the occurrence of an indoor source that should be eliminated.

Establishing possible pathways for entry of combustion products into buildings is relatively simple. Thus pathways for entry of combustion products are often obvious in buildings with attached garages and loading docks or in buildings near heavily traveled roads.

HVAC outdoor air intakes near the loading dock or garage are obvious transport conduits for combustion products. Proving that combustion products can enter a building and cause occupant complaints, however, can be made difficult because of the intermittent nature of pollutant generation and HVAC and building operational variables. In one large building where a portion of an office floor had been vacated because of suspected exposure to combustion products, round-the-clock sampling for $NO_2$ for six days was necessary before it could be demonstrated that combustion products from a loading dock were being entrained in the HVAC system outdoor air inlet serving the affected office (Morey and Jenkins 1989). The concentrations of $NO_2$ at the outdoor air inlet, in the vacated office, and in the outdoor air on the roof far removed from emission sources were 2.0, 0.7, and 0.08 ppm, respectively, only at a time when a garbage truck was unloading a dumpster at the loading dock near the HVAC inlet.

In another building with a tuck-under loading dock, combustion products only entered the building interior when the lower floors were under negative pressure relative to ambient air (stack effect) during the winter. Concentrations of CO in corridors and elevator lobby areas (15–50 ppm) connecting with the dock were equivalent to those in the tuck-under dock (25–50 ppm). While tracer gas techniques can be used to demonstrate pathways of entry of combustion products into a building, smoke pencils (air current tubes), if used with care, can provide useful information on points of entry during an initial walk-through evaluation.

***8.2.3.3 Particles*** A variety of particles, most solid (for example, dust, smoke and microorganisms) and a few liquids (for example, mist form humidifiers), are present in indoor environments. Particles present in indoor air may originate in the outdoor air, in the HVAC system, or in occupied spaces. Particles in the outdoor air, for example, from heavy vehicular traffic, may enter HVAC air intakes and especially if filtration is poor and are transported to the breathing zone in ventilation supply air. Particulates from soil are tracked into buildings and can accumulate in porous finishes, especially carpet. Several studies indicate that carpet is a strong reservoir for dust accumulation with levels of 50–100 $g\,m^{-2}$ being characteristic of old carpet (Shaughnessy et al. 2005). Intense deep cleaning by HEPA vacuuming is required to remove most of the residual dust in carpet.

Major sources of particles in buildings are occupants (skin scales and fiber from clothing) and their activities (settled dust aerosolized by foot traffic). The term "personal cloud" is used to describe the fine particulate that is emitted from dusty surfaces by occupant activities such as walking (Ferro et al. 2004; Shaughnessy and Vu 2012). The occurrence of an increased amount of airborne particulate around active occupants in indoor environments suggests that personal as opposed to area air sampling is necessary to accurately determine human exposure to indoor air pollutants. Dusts from interior renovations, if not properly contained within the renovation work area, may be dispersed throughout the entire building. Maintenance activities such as moving ceiling tiles or replacing dirty filters may inadvertently lead to dispersal of particles if not carefully performed.

In general, airborne particles of a size from about 0.1 to 10 μm are of concern for human health. Particles less than 0.1 μm are exhaled, and those greater than 10 μm do not enter the lower regions of the lung. ASHRAE Standard 62.1-2013 provides guidance on the kinds of HVAC filters necessary to reduce particulate matter smaller than 10 μm (PM 10) and smaller than 2.5 μm (PM 2.5) in ventilation air. To provide more precise and usable information on selection of filters for particle removal, a new version of the American Society of Heating, Refrigeration and Air-Conditioning Engineers' (ASHRAE) Standard 52 has been published, Standard 52.2-1999. This important document establishes the minimum efficiency reporting value (MERV). Typical MERV levels range from MERV 4 to MERV 16 for commercial filters used in whole house application or as components in packaged air cleaner devices. Particulate matter filters or air cleaners with MERV of 6 or greater and 11 or greater are required to remove a substantial amount of PM 10 and PM 2.5, respectively, from the ventilation air stream (ASHRAE 62.1-2013, sections 6.2.1.1 and 6.2.1.2). On the basis of research performed on 50 commercial HVAC systems (Burroughs and Hansen 2004), MERV 6 proved inadequate to maintain system cleanliness. This three-year study revealed that MERV 11–13 was the minimum efficiency required to sustain acceptable system cleanliness. In similar work (Fisk et al. 2002), it was found that incremental improvement of contaminant control diminished above efficiencies of MERV 13. This is because of the interference of particle generation within the space and the contribution from outdoor air infiltration. Thus, optimal efficiency guidance for whole house filtration is MERV 13 for whole house HVAC systems. However, MERV 12 is the highest efficiency currently available in 1- or 2-in. pleated panel filters if the existing retention system cannot be modified. The deep-pleated, dealer-installed, in-duct housing units have a variety of MERV levels available, and MERV 12 or 13 efficiency cartridges are a convenient option.

***8.2.3.4 Residual Contamination on High Contact Surfaces*** Inadequate cleaning leads to accumulation of dust and nonvisible ecological niches, which can promote biological contaminants including viruses. CDC (2012) guidance advises that enhanced cleaning in buildings should be promoted to "encourage routine surface cleaning through education, policy, and the provision of supplies" and to "routinely clean surfaces and objects that are touched often." However, large-scale studies investigating environmental variations in this context are lacking. To facilitate an easy-to-use simplified metric for gauging the effectiveness of cleaning on a surface, adenosine triphosphate (ATP) has been of use within the last decade as a measure of biological activity. ATP serves as a marker for the detection and quantification of pollutant loads of biological origin and overall cleanliness in healthcare and food production and service industries. It can be measured through an enzymatic luciferin/luciferase reaction detected and quantified as bioluminescence. The method converts ATP into a light signal measured by an instrument providing a quantitative measurement of ATP in relative light units (RLUs). Thus, after cleaning and a drop in biomass, there is a corresponding decrease in measurable ATP. An important study, relevant to this proposed work, conducted by Shaughnessy et al. (2013) has established normal ranges for ATP readings in school environments, for both before and after cleaning, as well as percent reduction after cleaning on different types of critical or high-touch surfaces. These data also revealed a reduction of culturable bacteria coincident with ATP reduction after cleaning, as a measure of improved sanitation. It is important to note that the baseline ATP metric developed for residual surface contamination in schools from the study is the foundation for the ISSA Clean Standard for K-12 (ISSA n.d.) and has recently been extended for use beyond school facilities. Being an easy-to-use first research-based indicator of cleaning effectiveness in occupied environments, further studies are needed to discern the end effects of cleaning on occupant health.

## 8.3 PROBLEM BUILDING STUDIES

The different approaches that can be used during IAQ evaluations reflect the varied kinds of problems that can occur in buildings. Complaints may be related to inadequate ventilation or to defects in building performance. Thermal environmental parameters can interact strongly with IAQ problems in building. In some building where environmental conditions are greatly deteriorated, BRI occurs in addition to SBS. Finally, problem building can be viewed in terms of the economic losses associated with lost productivity of occupants.

### 8.3.1 Approaches to Studying SBS Complaints in Buildings

SBS is characterized by a number of nonspecific symptoms including mucous membrane irritation, eye irritation, headache, odor annoyance sinus congestion, and fatigue. Ideally, in order to understand the etiology of SBS, the prevalence of symptoms in occupants should be studied, preferably in problem and nonproblem buildings. Objectives of such case–control studies should include a determination if the percentage of dissatisfied occupants exceeds some unacceptable level and if dissatisfaction can be related to building, air contaminant, or work practice variables.

***8.3.1.1 Early NIOSH Studies*** By 1990, US National Institute for Occupational Safety and Health (NIOSH) had investigated over 500 problem buildings (Crandall and Sieber 1996). In each building NIOSH determined the single, most important factor likely relating to occupant complaints. Factors associated with complaints in descending order of frequencies were inadequate ventilation (53%), indoor pollutant sources (15%), entrainment of outdoor contaminants (10%), microbial problems (5%), building fabric contamination (4%), and unknown causation (13%).

Assessments were carried out in 104 problem buildings where complaints and building defects were both studied. Problems of the following type were identified:

(i) 63% had facility defects such as water damage, water intrusion, or poor housekeeping.
(ii) 60% had defective HVAC operation such as insufficient outdoor air ventilation, poor air distribution to the breathing zone, and flooded drain pans.
(iii) 58% had defective HVAC maintenance such as systems that were dirty and in disrepair or suffered from an absence of written operation and maintenance plans.

(iv) 51% had defective HVAC design problems such as outdoor air inlets that needed to be moved, insufficient or poorly fitting filters, or absence of minimum stops of VAV terminals.
(v) 30% had thermal environmental problems.
(vi) 24% had indoor contaminant sources such as VOCs, cooking odors, renovations dusts, and chemicals in mechanical equipment rooms.
(vii) 22% had combustion gas and restroom entrainment problems.
(viii) 12% had ergonomic (workstation design) or physical agent (lighting/glare) problems (Crandall and Sieber 1996).

While NIOSH studies do not provide information on the cause of SBS complaints, they do provide practical information on the frequency of building defects that give rise to IAQ problems.

NIOSH has also recently developed a useful tool to better characterize dampness and mold conditions in occupied environments. Currently the tool "Dampness and Mold Assessment Tool for Schools" (http://www.cdc.gov/niosh/topics/indoorenv/mold.html) is based on conditions in schools and utilizes observations in conjunction with health information collected from the schools in efforts to study the impact of existing dampness conditions on student health (and, indirectly, student performance).

***8.3.1.2 Early British Studies*** Several systematic studies on complaints related to building characteristics have been carried out in British office buildings (Burge et al. 1987; Harrison et al. 1987). In the largest study, complaint symptoms were recorded using a common protocol in 47 different groups of occupants in 42 buildings. The questionnaire used in these studies elicited information on whether the following 10 symptoms occurred during the past year and whether symptoms disappeared during periods when occupants were away from the building: dryness of eyes, itching of eyes, stuffy nose, runny nose, dry throat, lethargy, headache, fever, breathing difficulty, and chest tightness.

Ventilation systems serving the 47 different occupant groups were categorized as *natural* (open windows, no forced air), *mechanical* (forced air system without cooling or humidification), *local induction units (IUs)*, *central induction fan coil unit (FCU)*, and *variable-* or *constant-volume system*. The latter three ventilation categories were characterized by cooling of air and, in some cases, also by humidification. Questionnaire results showed that the lowest prevalence of work-related symptoms was found in the naturally or mechanically ventilated categories. Although there were considerable variations between buildings of each ventilation type, the highest symptom prevalence rates were found in ventilation systems with induction or FCUs. Somewhat intermediate symptom prevalence occurred in variable/constant air systems.

These British studies show that naturally ventilated and non-air-conditioned, mechanically ventilated buildings are the healthiest workplaces. Because the other three ventilation types examined are characterized by air conditioning and, in some cases, also by humidification, it has been hypothesized that microbiological air contaminants may be responsible for some of the higher prevalence rates of work-related symptoms found in these studies (Burge et al. 1987).

***8.3.1.3 Early European Studies*** Systematic questionnaire studies of complaints were carried out in 14 town halls and 14 affiliated buildings in the Copenhagen area (Skov et al. 1987). The buildings chosen in this study were not previously categorized as being sick or healthy (Valbjorn and Skov 1987). Environmental measurements such as the concentration of TVOCs and $CO_2$, the microbial content in floor dusts, and thermal/air moisture parameters were made in one representative office in each building. Two unique parameters, the "fleece" and "shelf" factors, were measured in each study office. The fleece factor is a measure of the surface area of all porous room furnishings (e.g. carpets, drapes, upholstery) divided by the room volume. The shelf factor is the length of open shelves in the study room divided by total volume of the room studied (Skov et al. 1987).

Work-related symptoms reported most frequently in the Danish town hall studies included eye, nose, and throat irritation, fatigue, and headache. A great variation in the prevalence of work-related symptoms was found between buildings. Workplace environmental factors such as total amount of floor dusts and the fleece and shelf factors of studied offices were related to symptom prevalence. Certain indoor environmental factors such as $CO_2$ concentrations, however, could not be related to the prevalence of work-related symptoms.

IAQ studies were performed in 56 office buildings in nine European countries during the heating season of 1993–1994 (Bluyssen et al. 1996). Measurements were made of physical factors such as temperature, relative humidity, and air movement. Chemical measurements included $CO_2$, CO, and TVOC concentration. HVAC parameters such as the rate of outdoor air ventilation were determined. Perceived air quality was determined by using trained and untrained panels. A number of important findings were reported in the 56 building European study. Outdoor air ventilation rates were on average about $25 \, l \, s^{-1}$, and yet 30% of the occupants and 50% of visitors were dissatisfied with the air quality. This finding shows that compliance with ventilation standards (see Section 8.4) does not guarantee acceptable IAQ. Source control is more important than dilution ventilation in achieving acceptable IAQ. The measurement of TVOC did not correlate with the subjective evaluation of perceived air quality. The 56 building European study suggested that occupants were a less significant source of pollutants than building materials, activities in offices, and the HVAC system itself.

***8.3.1.4 US EPA Studies*** The US EPA initiated a Building Assessment and Survey Evaluation (BASE) study in a cross section of public and private buildings in the United States. An objective of the BASE study was to obtain baseline data on chemical, physical, microbiological, ventilation, and subjective parameters often measured in IAQ evaluations in a representative population of mostly nonproblem buildings. VOCs with the highest median indoor concentrations were, in descending order, ethanol, acetone, 2-propanol, toluene, 1,1,1-trichloroethane, dichlorodifluoromethane, and *m*- and *p*-xylenes.

Forty-seven VOCs found indoors had indoor to outdoor median ratios greater than one (Hadwen et al. 1997). Questionnaire data obtained from the nonproblem buildings in the BASE study has been compared with that from the NIOSH database on complaint building. This is not surprising since NIOSH-studied buildings were known to have considerable building and HVAC defects (Crandall and Siebea 1996), whereas the BASE protocol only included buildings with no history of complaints. Some highlights of data collected in the BASE study are summarized as follows:

- *Occupant BRS*. Occupant symptoms such as upper and lower respiratory irritation, eye irritation, and headache have been related to environmental factors seen during building assessment. These factors include the height of HVAC outdoor air inlets above grade, lack of cleaning of air handling unit (AHU) drain pans and cooling coils, and low outdoor air ventilation rates (Mendell et al. 2006).
- Increased prevalence of SBS symptoms, such as headaches, appears to be related to the presence of polyester or synthetic HVAC filter composition as compared with fiberglass or cotton filters (Buchanan and Apje 2006).
- Fel d1 (cat allergen) was detectable in dusts from most BASE buildings (Macher et al. 2005). Mite allergen was detectable in approximately half of the BASE buildings.

### 8.3.2 Thermal Environmental Conditions

Human acceptance of a thermal environment is related to a number of variables such as metabolic heat production and the transfer of heat between the occupant and the environment. Heat transfer is influenced by variables including the temperature, relative humidity, and velocity of the air around the occupant. The body is at thermal equilibrium when the net heat gain or loss is zero. A number of factors in offices may adversely affect the body's thermal equilibrium and lead to thermal dissatisfaction. Examples include asymmetric thermal radiation that affects people sitting near sun-facing windows, vertical temperature differences in a room, cold drafts, and cold floors and walls.

Performance criteria for thermal environmental conditions that 80% or more of sedentary occupants in indoor environments will find acceptable have been defined in ASHRAE Standard 55-2004. Figure 5.2.1.1 in that standard provides guidance on acceptable thermal environmental conditions based on operative temperature, dew point temperature, humidity ratio (ratio of the mass of water vapor to mass of dry air in a given volume of air), and relative humidity. The upper limit for humidity ratio in figure 5.2.1.1 is 0.012, which is equivalent to a dew point temperature of 62.2 °F (16.8 °C). Consequently, according to Standard 55-2004, it is acceptable in terms of thermal comfort for the relative humidity at cool operative temperatures to exceed 70 and even 75%, conditions known to be conducive to the growth of a number of xerophilic molds.

An AIHA (2004) publication "The IAQ Investigator's Guide" identifies acceptable temperature ranges as 73–79 °F for summer months and 68–75 °F for winter. AIHA (2004) identifies an acceptable relative humidity range of 30–60%. It should be noted that many buildings would not be able to achieve the lower range of the AIHA guideline during the heating season without the introduction of supplemental humidification. Introducing moisture vapor into the air of buildings should be done with caution to avoid condensation on cool surfaces in the building envelope.

Perceived air quality is influenced by factors such as temperature, relative humidity, and air motion. In general, air at a given pollution level is perceived as more acceptable when the temperature and relative humidity are lowered (Fanger 1988). In addition, a recent study in school environments demonstrated that the combined effect of increased ventilation and reduced temperatures translates into improved student academic achievement (Haverinen-Shaughnessy and Shaughnessy 2015). Air at a temperature of 28 °C and a relative humidity of 70% is perceived as unsatisfactory by 95% of people regardless of whether the air is clean or polluted. Cool dry air is generally perceived as being freer of contaminants than warmer or humid air (Bergluno 1998). However, the potential for draft complaints increases at lower air temperatures. Lack of adequate airflow in the occupied space (often indicated by the presence of portable fans) is commonly associated with thermal environmental complaints. In these cases, airflow from diffusers may be blocked by newly added modular partitions or walls, VAV terminals may be closed, or the airflow to the zone has become unbalanced.

### 8.3.3 Building-Related Illness

Because of the long latency period and relatively small populations involved, some BRI, such a cancer, that may be caused by exposure indoors to carcinogens such as asbestos, ETS, radon, and some VOCs are estimated primarily through risk analysis (Kreiss 1989). Other BRIs attributable to specific agents such as organophosphate or CO poisoning can be successfully evaluated as a case study in a building. In the section that follows, the etiology of some BRIs of microbial origin is reviewed.

***8.3.3.1 Hypersensitivity Pneumonitis*** HP is an immunologic lung disease that occurs in some individuals after inhalation of organic dusts. HP is suspected when symptoms such as fever, cough, and chest tightness occur several hours after exposure. The diagnosis of the disease is made on the basis of review of patient symptomology plus a battery of tests (Kreiss 1989), including restrictive pulmonary function measurements, the formation of antibodies to extracts of microbial agents collected in the building, and, in rare instances, experimental inhalation of suspect antigens in a clinical setting by the patient.

Although cases of HP are rare, they are usually associated with HVAC components such as water spray systems that are heavily contaminated with microorganisms (Arnow et al. 1978) or with heavy moisture intrusion into the building. The specific microorganism(s) causing HP remains unidentified even in well-studied cases probably because of the numerous antigens present in different growth sites in the building and its HVAC system.

***8.3.3.2 Humidifier Fever*** Episodes of fever, muscle aches, and malaise with only minor pulmonary function changes have been associated with inhalation of aerosols from humidifiers or water sumps contaminated with gram-negative bacteria, bacterial endotoxins, and protozoa (MRC 1977). Symptoms of this flu-like illness generally subside within a day after exposure without any long-term adverse effects.

An example of this type of BRI occurred in an office where three of the seven occupants reported attacks of fever and chills that started late during the workday and lasted well into the night (Rylander et al. 1978). Illness was associated with the use of a humidifier on occasions when the air was considered too dry for comfort. Analysis of water from the humidifier reservoir showed that *Flavobacterium* (a gram-negative airborne *Flavobacterium*) increases from nondetectable when the humidifier was not running to about 3000 $m^{-3}$ within 15 minutes of operation.

### 8.3.4 Economic Cost of Poor IAQ

The exact number of occupants affected by SBS in nonresidential, nonindustrial buildings is unknown. However, estimates suggest that about 20–30% of the nonindustrial building stock may be classified as "problem buildings"; that is, those where SBS and impaired productivity occur because of poor IAQ (Woods 1989). Some estimates suggest that as many as one-third of problem buildings are also characterized by BRI (Woods 1989). Although this estimate may be too high, it is clear that the costs associated with each instance of BRI can be significant. Several instances where BRI has led to the evacuation of building or portion of a building and the renovation/reconstruction of the affected structure are found in the literature. The building studied by Arnow et al. (1978) originally housed about 1000 employees and was vacated for about three years during renovation. Renovations in the buildings studied by Hodgson et al. (1985, 1987) required one to three years. Costs associated with renovations of these buildings varied from $150 000 to $10 000 000. Costs of renovating two newly constructed Florida courthouses that had to be vacated because of fungal contamination have exceeded $20 000 000 in one case and $40 000 000 in the second case (Jarvis and Morey 2001).

A recent study by Harvard researchers found that doubling the ventilation rate in typical office buildings can be reached at an estimated annual energy cost of between $14 and $40 per person, resulting in as much as a $6500 equivalent in improved productivity per person per year. These findings provide strong incentive for building managers to optimize ventilation in order to maximize worker productivity, given the incremental cost of the energy as opposed to significant cost savings based on employee output (MacNaughton et al. 2015). The costs associated with improved building performance are potentially more than offset through improved productivity of building occupants. It thus makes good economic as well as environmental sense to transform a problem building into healthy building and also to prevent a healthy building from degrading into a problem building.

## 8.4 VENTILATION SYSTEMS IN NONINDUSTRIAL BUILDINGS

A HVAC system provides the occupied space with acceptable IAQ and acceptable thermal environmental conditions. Thus, the HVAC system should provide the occupied zone with an appropriate quantity of clean outdoor air and remove (dilute) air contaminants to the extent that the vast majority of people perceive the air quality as being acceptable. While dilution ventilation offers the only practical method to control odorous bioeffluents from occupants, other potential building contaminants such as VOCs, ETS, combustion products, and particulate (dusts) should also be controlled at their source.

### 8.4.1 General Description of a Typical HVAC System

The HVAC system in a large nonindustrial building usually contains one or more AHUs (also, see Morey and Shattuck 1989). In the mixed air plenum of an AHU, outdoor air is blended with a portion of the HVAC system's return air. The mixture of outdoor and return air then enters a plenum that may contain low-efficiency prefilters and more highly efficient filters located downstream of the prefilters. The prefilters trap larger particles and should be replaced frequently so as to extend the life of the more costly downstream filters that trap finer airborne dusts.

The air mixture after filtration enters the heat exchanger section, where heat can be added to or removed from the ventilation airstream as required to maintain the thermal comfort of occupants in the building. During the air-conditioning season, moisture is removed from the airstream as it passes over the dehumidification cooling coils. Moisture from the airstream passing through the heat exchanger condenses on the outside surfaces of cooling coils when the dew point temperature of the air is greater than the coil temperature. This moisture collects in drain pans beneath the heat exchanger and should exit the AHU through drain lines with deep-sealed traps. Water should not stagnate in drain pans or in other portions of the AHU.

After passing through the heat exchanger and supply fan, conditioned air is distributed to occupied spaces through a system of ducts. The main air supply duct is usually constructed of sheet metal, and it, as well as the plenum housing the fan and heat exchanger, may be internally insulated with a porous liner for both thermal and acoustic control. Internal porous liners in plenums and ducts should be undamaged and have structural integrity that prevents loose fibers from being entrained in the airstream. In some AHUs the thermal/acoustic insulation is packaged between two layers of sheet metal so that the airstream surface is readily cleanable and less likely to be subject to mold growth that occurs on the moist dust that would otherwise accumulate on or within porous insulation.

Air from the main supply ducts enters rigid branch ducts that often contain variable air volume (VAV) terminals. In a VAV system, the supply air volume delivered to the occupant zone is varied to maintain the temperature in occupied spaces, and this is accomplished by a control system including a thermostat. In a constant-air-volume system that is found in some older buildings, the interior space temperature is maintained by varying the temperature of the conditioned supply air.

Air from rigid branch ducts usually enters narrow diameter (sometimes) flexible ducts and then passes into the occupied spaced through a diffuser or supply air vent. A portion of room air mixes into supply airstream being discharged into the occupied space. If the air supply diffusers or vents and return air outlets are properly sized, selected and located outdoor air can be effectively distributed to the breathing zone (generally 3–72 in. above the floor).

In some buildings a considerable amount of the outdoor air that enters the HVAC system does not reach the breathing zone of occupants (Turner et al. 1995). The air may leak out of unsealed portions of supply ducts into wall, ceiling, attic, and other unconditioned spaces. Supply air may also short circuit into a closely positioned return air vent without benefiting persons in the occupied zone. Outdoor air may fail to reach the breathing zone because of a malfunction in the VAV terminal (Table 8.5). VAV terminals should always provide a minimum supply of outdoor air per occupant even when the thermostat setting is satisfied.

In the example in Table 8.5, a maximum of 800 cfm supply air (other measurements showed that half of the supply air was sourced outdoors) was actually being delivered through diffusers into a 1000 $ft^2$ office housing seven occupants (thermostat calling for cooling; 400 cfm outdoor air divided by seven equals 57 cfm/occupant). According to the HVAC mechanical plans for the office, 14 cfm of outdoor air per person would be provided to the office when the thermostat was not calling for cooling. Under actual minimum ventilation conditions (thermostat thermally satisfied), no outdoor or supply air was being provided to office occupants because of defective VAV terminal. It should be noted that ASHRAE Standard 62.1-2013 requires designers to provide a minimum outdoor air ventilation rate in the breathing zone of approximately 17.5 cfm per person (in the 1000 $ft^2$ area, a total of 25 cfm to dilute human bioeffluents; 60 cfm to dilute pollutants from materials in the office space itself; see Section 8.4.4).

In many buildings, the above-ceiling cavity is used for the passage of return air from the occupied space back to the AHU. The use of a ceiling plenum instead of return ducts is associated with a number of potential problems

**TABLE 8.5 Supply Airflow Rates from VAV Terminal Serving Office Zone with Seven Occupants (~1000 ft² Floor Space)[a]**

| Supply air measurement condition | Total cfm from supply diffusers | Outdoor air per occupant[b] |
|---|---|---|
| Design maximum | 750 | 54 |
| Actual maximum[c] | 800 | 57 |
| Design minimum | 200 | 14[d] |
| Actual minimum[e] | 0 | 0[d] |

[a] Morey (1999).
[b] Outdoor air comprises approximately 50% of supply air based on other measurements.
[c] Maximum flow rate measured when VAV terminal thermostat set to minimum temperature (55 °F) setting, calling for maximum cool air.
[d] Does not comply with ASHRAE Standard 62.1-2013.
[e] Minimum flow rate measured with VAV terminal thermostat set to maximum temperature (85 °F) setting, calling for minimum cool air.

such as entrainment of fibrous fireproofing particles into return air and blockage of return airflow by walls that extend to the underside of the floor above. The air in the return air plenum may then travel back to the AHU through a main return duct or riser. Some of the return air enters the AHU mixed air plenum, and some of the return air is discharged directly outdoors. Outdoor air inlets must be spatially separated from chimneys, flues from combustion appliances, cooling towers, and other contaminant sources by distances specified in table 5.5.1 of ASHRAE Standard 62.1-2013.

The amount of outdoor air brought into the HVAC system should in hot, humid climates or seasons be slightly more than the combined total of relief air and air exhausted from the building by toilet and local exhaust fans. The building as a whole is thus maintained slightly positive with respect to the atmosphere so as to retard the infiltration of unfiltered and humid unconditioned air through loose construction in the envelope and through other building openings.

In many buildings, FCUs and IUs located along exterior walls are used to condition the air in perimeter zones. FCUs contain small fans, low-efficiency filters, and small heat exchangers. These units condition and recirculate supply air (often without any outdoor air intake) in peripheral zones. IUs are generally supplied with conditioned outdoor air from a central AHU. Conditioned air from the central AHU passes through nozzles in the IU and mixes with a portion of the room air. Because a large building may contain several hundred FCUs or IUs, maintenance of this part of the HVAC system is often poor or neglected.

Pressurization can affect HVAC operation. Temperature-driven pressure differentials (stack effect) result from buoyancy of heated air (Bearg 1992). During the winter in cold climates, warm air rises to upper portions of the building causing positive pressure. Cool outdoor air then enters the negatively pressurized lower floors. When stack effect is extreme, enormous volumes of outdoor air can enter openings in the envelope (Morey 1995). Contaminants such as combustion products from loading docks and parking garages can enter the occupied space with unconditioned outdoor air. Stack effect can be reduced by designing compartmentalized HVAC systems with one AHU on each floor or by providing more return air volume for lower floors.

Negative pressurization associated with HVAC operation can also occur for other reasons. The leakage of conditioned supply air into the return air plenum or into nonconditioned spaces such as an attic can cause depressurization in the occupied space. When the occupied space becomes depressurized in air-conditioned buildings in hot, humid climates, moisture infiltrates the envelope, potentially resulting in condensation or damp conditions that may allow for fungal growth on biodegradable construction and finishing materials. Mechanical equipment rooms can also be depressurized because of leakage of air into AHUs on the suction side of the fan. VOCs from cleaning chemicals and dusts from poor housekeeping in and around the mechanical equipment room can then enter the ventilation air and eventually enter the occupied spaces. It is essential therefore that mechanical equipment rooms be fastidiously clean, not serve as work areas, and be kept free of stored materials that can contaminate ventilation air.

In several early studies $CO_2$ concentrations above 1000 ppm were associated with an increase in occupant complaints in buildings (Raijhans 1983). The normal concentration of $CO_2$ in the outdoor air typically ranges from 300 to 400 ppm. In occupied indoor environments the $CO_2$ concentration is always somewhat elevated relative to the outdoor level because a $CO_2$ concentration of about 38 000 ppm is present in the air exhaled from the lung.

Although $CO_2$ concentration remains a good surrogate for the presence of human-generated contaminants (bioeffluents) and a good predictor of over occupancy conditions, epidemiologic studies show that the prevalence of SBS symptoms is not related to $CO_2$ levels (Kreiess 1989). Studies summarized by Bluyssen (1996) indicate that air contaminants derived from nonhuman sources such as HVAC systems and interior construction and finishing materials are major sources of perceived dissatisfaction with indoor air. The concentration of $CO_2$ in indoor air would, of course, have no relation to these nonhuman emission sources.

### 8.4.2 Humidification and Dehumidification

In some types of building zones, such as hospital critical care areas, computer rooms, and animal research facilities, moisture is often added to ventilation air to maintain humidity levels usually between 40 and 50%. Moisture is preferably added to the ventilation air in commercial buildings by humidifiers that inject steam or water vapor into the AHU (often in the fan plenum) or in the main supply air ductwork. Injection nozzles should not be located near porous interior insulation that can become wet and offer a niche for the amplification of microorganisms. HVAC system humidifiers should preferably use steam as a moisture source. The supply of steam introduced into the HVAC system should be free of volatile amine corrosion inhibitors, especially morpholine (NRC 1983).

Cold water humidifiers that contain reservoirs of water can become contaminated with microorganisms that may cause BRIs such as HP, humidifier fever, and asthma. The use of biocides in these types of humidifiers may be ineffective or may in itself cause asthmatic reactions. Preventive maintenance programs for cold-water humidifiers used in HVAC systems include the following essential aspects: (i) thorough draining and cleaning of the unit so as to prevent development of biofilm (slime), (ii) physical removal of biofilm and disinfection of wet surfaces, (iii) removal of disinfectants prior to recommissioning of unit so as to avoid aerosolization of chemicals into the air stream, (iv) bleed-off of sump water at a constant rate so as to minimize the buildup of solids and impurities, and (v) installation of highly efficient upstream HVAC filters to minimize deposit of lint, fly, and particulate in the open water system.

During the air-conditioning season, as air passes around cooling coils of FCUs and AHUs, the dry bulb temperature approaches the dew point temperature. Consequently, air downstream of cooling coils has a humidity close to 100%. Organic dusts and debris is not removed during filtration can pass through the heat exchanger section and become entrained in moist, porous, internal insulation in downstream plenums and air supply ductwork. Amplification of fungi and bacteria may thus occur on the nutrients in the dusts trapped in these HVAC system locations.

During the air-conditioning season, the cooling coil's chilled water temperature may be raised in an effort to save energy costs. This results in an elevated RH in occupied spaces. As the RH in occupied spaces rises above about 65–70%, the equilibrium moisture content in organic dusts and in the capillary structure of porous material such as wall board and carpet approaches a level sufficient to support fungal spore germination and proliferation (ISIAQ 1996). As a general principle for controlling mold growth (in all buildings), it is important to prevent the surface relative humidity in building finishing materials from consistently exceeding the 65–70% level.

### 8.4.3 ASHRAE Standard 62

The current ASHRAE 62.1 standard (*Ventilation for Acceptable Indoor Air Quality* – 2016 edition) is probably the most important document in the IAQ literature. At the end of the nineteenth century, an early ASHRAE standard recommended the provision of outdoor air a minimum rate of 30 cfm ($15 \, l \, s^{-1}$) per person to lower the risk of tuberculosis in occupied spaces (Morey and Singh 1991). In the 1930s Yaglou et al. (1936) showed that from 10 to 30 cfm of outdoor air per person was required to provide acceptable dilution of ETS and bioeffluents in occupied spaces. The ASHRAE Standard for Natural and Mechanical Ventilation (1973) prescribed a "minimum" outdoor air ventilation rate of $2.5 \, l \, s^{-1}$ (5 cfm) per person to provide acceptable but not odor-free conditions and a "recommended" outdoor air ventilation rate (generally a minimum of 10 cfm ($5 \, l \, s^{-1}$) outdoor air per occupant or more) to control bioeffluent annoyance. The minimum outdoor air ventilation rate recommended in ASHRAE Standard 62-1973 (5 cfm per occupant) was intended to prevent indoor $CO_2$ levels from rising above 2500 ppm or half the recognized occupational threshold limit for this gas.

As a result of the energy crisis of the mid 1970s, ASHRAE Standard 90 (ASHARE 1975) recommended that only the "minimum" outdoor air ventilation rate of ASHRAE Standard 62-1973 (5 cfm outdoor air per occupant) be used for ventilation. Concerns about providing enough outdoor air to dilute ETS and bioeffluents to acceptable levels were overlooked.

ASHRAE Standard 62-1981 partially rectified the error of Standard 90 by recommending a minimum outdoor air ventilation rate of 20 cfm per occupant for zones where smoking was permitted. The 5 cfm outdoor air ventilation rate per occupant recommended for nonsmoking zones was intended to prevent $CO_2$ concentrations from rising above 2500 ppm and to dilute bioeffluents. ASHRAE Standard 62-1989 increased in the minimum outdoor ventilation rate compared with its predecessor standard from 5 to 15–20 cfm per person. Outdoor air requirements recommended by the ventilation rate procedure in Standard 62-1989 made no distinction between "smoking-allowed" and "smoking-prohibited" areas. A minimum of 15 cfm of outdoor air per person was recommended because it was believed that this was the minimum amount of outdoor air needed to dilute body and tobacco smoke odors are to keep $CO_2$ at acceptable levels (Janssen 1989). The outdoor air requirements specified by ASHRAE Standard 62-1989 had to be delivered to the occupant breathing zone.

The next two editions of ASHRAE Standard 62 (ASHRAE Standard 62-1999; ASHRAE Standard 62-2001) provided similar guidance on outdoor air ventilation as ASHRAE Standard 62-1989 for indoor environments.

The publication of ASHRAE Standard 62.1-2004 required fundamental changes in the procedures followed by designers in providing adequate outdoor air ventilation in nonindustrial indoor spaces. For the first time, Standard 62.1-2004 was written in code language with "shall" and "must" replacing advisory "should" and "can" verbiage. Ventilation rates were not given in Standard 62.1-2004 for smoking areas because cognizant authorities had not defined outdoor air ventilation rates, resulting in acceptable risk for ETS. In addition, a fundamental change occurred in determining the reasons why minimal outdoor air was to be provided at the breathing zone. This change involved calculation of how much outdoor air was needed to dilute people-sourced pollutants (bioeffluents) plus the amount of outdoor air required to dilute pollutants associated with occupant activities, building materials, and furnishings (Persily 2005). The required ventilation to dilute bioeffluents (generally 5–10 cfm per person) was based on the odor acceptability by persons adapted to the indoor space and not on the acceptability of indoor air to visitors. Table 6.1 of ASHRAE Standard 62.1-2004 and 62.1-2007 also contains an estimate of the minimum amount of outdoor air ventilation required to dilute sensory pollution loads from contaminants associated or related to occupant activities, building materials, and furnishings in various types of interior spaces. As an example, for office spaces, the required amount of ventilation can be calculated by using values drawn from ASHRAE 62.1-2007 which account for the requirement to address dilution for people in a space (5 cfm/person) PLUS that associated with dilution for anticipated pollutant loads (0.06 cfm/ft$^2$). For a 1000 ft$^2$ office space (typical occupancy 5 persons/1000 ft$^2$), the required ventilation would thus be (5 people × 5 cfm/person) + (0.06 cfm/ft$^2$ × 1000 ft$^2$) = 85 cfm. This translates into 17 cfm/person minimally required for ventilation. Thus, in offices more than three-fourths of the required outdoor air ventilation is needed to dilute contaminants from the building itself. Conversely, in classrooms (students ages 9 or older) with a design occupancy of 35 persons per 1000 ft$^2$, 350 of the 470 total cfm outdoor air ventilation is needed to dilute contaminants associated with people as compared with 120 cfm for the building itself (Table 8.6). In a crowded auditorium more than 90% of the required ventilation is needed to dilute occupant-sourced contaminants.

$CO_2$ concentrations have been widely used by the industrial hygiene community as a surrogate indicator of acceptable ventilation and acceptable IAQ. However, examination of table 65.112 and minimum outdoor air ventilation rate table 6.1 in ASHRAE Standards 62.1-2013 suggests that at least for some indoor environments such as offices and art classrooms, $CO_2$ concentration is an inaccurate surrogate of perceived air quality (many air contaminants are building sourced). By contrast, in crowded indoor zones such as in lecture classrooms and auditoriums where odors associated with people are the primary contaminants, $CO_2$ concentration is still a good surrogate indicator of acceptable IAQ.

ASHRAE Standard 62.1-2013 contains a number of important mandatory requirements relating to the design, operation, and maintenance of HVAC systems. For example, section 4 of this standard required that the HVAC system designer become aware of compliance of the building location with regard to local ambient (outdoor) air quality guidelines embodied in the National Primary Ambient Air Quality Standards (NAAQS). For example, ozone-cleaning devices must be installed in HVAC systems if the fourth highest maximum eight hours concentration (three years average) of this contaminant outdoors exceeds 0.107 ppm (section 6.2.1.3 of ASHRAE Standard 62.1-2013). If the $PM_{10}$ concentration in the outdoor air consistently exceeds 50 μg m$^{-3}$, the installation of a MERV filter of grade 6 or better is required to clean incoming outdoor air prior to introduction of this air into occupied spaces (section 6.2.1.1 of ASHRAE Standard 62.1-2013). Before filters installed in HVAC systems before 1999 were rated according to ASHRAE Standard 52.1-1992. This standard rated filters (air cleaners) in terms of weight arrestance and dust spot efficiency. Dust spot efficiency was used to classify filters qualitatively in terms of their ability to remove fine dusts that can visually soil interior surfaces in occupied spaces. Weight arrestance was used to refer to the ability of a filter to capture a coarse synthetic mixture of Arizona road dust, carbon black, and cotton linters. According to ASHRAE Standard 52.2-1999 filters are now rated according to 16 MERV categories describing the ability of the filter to capture particles in three size ranges, namely, 0.3–1 μm, 1–3 μm, and 3–10 μm. A prefilter, which under Standard 52.1 had an

**TABLE 8.6 Minimum Outdoor Air Ventilation Rates in the Breathing Zone Required by ASHRAE Standard 62.1-2013[a]**

| | Occupant default density | Outdoor air ventilation rate | | Total ventilation | |
|---|---|---|---|---|---|
| Type of space | Per 1000 ft$^2$ | Per person | Per ft$^2$ | Per 1000 ft$^2$ | CFM per person |
| Offices | 5 | 5 | 0.06 | 85 | 17 |
| Lecture classroom | 35 | 10 | 0.12 | 470 | 13.4 |
| Art classroom | 20 | 10 | 0.18 | 380 | 19 |
| Auditorium | 150 | 5 | 0.06 | 810 | 5.4 |

[a] These ventilation rates assume a nonsmoking condition.

arrestance of greater than 80% and less than 205 dust spot efficiency, now has a MERV rating of 6 (35–50% of 3–10 μm particles are captured). Section 5.9 ASHRAE Standard 62.1-2013 required the installation of at least a MERV 6 filter upstream of dehumidification cooling coils. A pleated filter with a 305 dust spot efficiency is equivalent to a MERV 8 rating (~70% of 3–10 μm particles are captured). Highly efficient filters capable of capturing most $PM_{10}$ and some $PM_{2.5}$ include those rated as MERV 12–13.

Section 5.9.1 of ASHRAE Standard 62.1-2013 states that "…occupied space relative humidity shall be limited to 65% or less…." This provision in ASHRAE 62.2-2013 is important because it will promote the drying of damp/wet interior surfaces where molds and mites may grow.

Section 5.13 of ASHRAE 62.1-2013 on Access for Inspection, Cleaning and Maintenance contains design requirements for installation of ventilation equipment with sufficient working space for inspection and maintenance (e.g. for filter replacement and fan belt adjustment). Thus "access doors, panels, or other means shall be provided and sized to allow convenient and unobstructed access sufficient to inspect, maintain, and calibrate all ventilation system components for which routine inspection, maintenance, or calibration is necessary." This section of Standard 62.1-2013 is important because degraded air quality in buildings is often associated with inadequate maintenance of AHUs and FCUs.

Section 7.1.4.2 of ASHRAE Standard 62.1-2013 contains requirements on control of the migration of construction-generated contaminants into occupied areas. Examples of acceptable methods to prevent migration of contaminants include plastic sheathing around the construction space and depressurization of the work area. Since construction/renovation activities in large buildings occur continuously, this portion Standard 62.1-2013 is of considerable practical importance with regard to control of construction/renovation dusts that often degrade IAQ in nearby occupied areas.

The ASHRAE Standard 62 series (Ventilation for Acceptable IAQ) through its 2001 edition (ASHRAE Standard 62-2001) applied to ventilation in both commercial and residential buildings. Beginning in 2004 a separate standard (Ventilation AND Acceptable IAQ in Low-rise Residential Buildings) was published under the number 62.2-2004. ASHRAE Standard 62.1-2004 and subsequent 62.1 standards are intended to apply to commercial, institutional, and residential (more than three stories above-grade) buildings. The scope of Standard 62.2-2004 and subsequent standards in the 62.2 series is intended to apply to single family and multifamily structures of three stories or fewer above grade. See Morey et al. (2008) for additional information on the ASHRAE 62.2 standard series.

### 8.4.4 Importance of HVAC Systems in IAQ

In mechanically ventilated buildings the quality of air delivered to occupants is often dependent upon the hygienic maintenance of the HVAC system. Studies comparing perceived air quality in naturally ventilated and air-conditioned buildings have often found higher rates of SBS symptoms in the latter (Seppanen and Fisk 2004). The perception of poor air quality associated with the provision of conditioned air is thought to be caused by pollutants originating from components of the HVAC system itself.

HVAC systems can become sources of pollutants that lead to degraded air quality for the following reasons:

- External, often malodorous air contaminants such as water droplets from cooling towers (may contain *Legionella* and water treatment chemicals) and combustion products from garages or loading docks, may enter the building through the HVAC system outdoor air intakes. Table 5.5.1 in ASHRAE Standard 62.1-2013 contains minimal separation requirements between HVAC system outdoor air inlets and external contaminant sources. For example, truck loading docks and cooling tower basins must be separated from outdoor air inlets by a minimum of 25 and 15 ft, respectively.
- Dirt on airstream surfaces in HVAC systems is a reservoir for various kinds of pollutants (e.g. VOCs, ETS, fungal spores, pollen, etc.), which can be subsequently emitted into supply air. The accumulation of dirt in HVAC systems is especially a problem where the filter efficiency is poor or where porous materials line airstream surfaces and accumulate dirt.
- Microbial contaminants can amplify in moist niches in HVAC system mechanical equipment. Fungi grow on dirty surfaces such as in moist filters and dam insulation. Standing water in drain pans and humidifiers and water droplets emitted from these sources as well as from the surfaces of dehumidifying cooling coils may contain yeasts, gram-negative bacteria, and endotoxins.
- Disinfectants and biocides used in humidifiers and drain pans may be emitted into supply air.
- VOCs from paints, solvents, and other chemicals stored in mechanical equipment rooms can enter the HVAC system through openings in negatively pressurized return air ducts and plenums as well as in AHU fan plenums.

Of particular note, the effect of HVAC systems, especially parameters such as ventilation (outdoor air) provision and temperatures, have been detailed related to studies in schools. All children spend from 7 to 10 hours per day in schools during the school year. Therefore, the quality of the air within a classroom environment has considerable effect on millions of children. Schools also provide an ideal environment for studying the effects of ventilation due to the high occupant density and the limited mobility of students in elementary school settings. Studies indicate a potential for enhanced student health and performance, which can be realized through improved classroom conditions such as adequate ventilation and thermal comfort (Mendell and Heath 2005; Daisey et al. 2003). For example, one field study from California found a statistically significant reduction in illness-related absences in response to improved ventilation rates: 1.6% reduction, per each additional liter of outdoor air per second per person (l/s-person) (Mendell et al. 2013). Another study in a large school district in the Southwestern United States demonstrated that adequate ventilation could reduce respiratory health symptoms requiring visits to school nurses (Haverinen-Shaughnessy et al. 2015). In addition, it was found that maintaining adequate ventilation and thermal comfort in classrooms could raise a student's test score from a level of "average" to "commended performance," which means high academic achievement, i.e. considerably above state passing standard (Haverinen-Shaughnessy and Shaughnessy 2015). A study based on a representative sample of Finnish elementary schools concluded that school-level indoor environmental quality (IEQ) indicators, including unsatisfactory classroom temperature during the heating season, dampness or moisture damage, and inadequate ventilation could explain a relatively large part (55–70%) of the school-level variation observed in self-reported upper respiratory symptoms and missed school days due to respiratory infections among students (Toyinbo et al. 2016). However, the majority of previous studies in this arena have been case studies or cross-sectional studies of school- and grade-level data of students' background, absenteeism, and performance. Hence, there exists limited evidence stringently juxtaposing student health and performance against classroom conditions using student-level data. Furthermore, there is a paucity of information comprehensively linking school interventions, focused on IAQ improvements, with health outcomes.

Experimental data from Denmark has also associated increased classroom ventilation rates and lower temperatures, with improved school performance (Wargocki and Wyon 2007). The authors used carbon dioxide ($CO_2$) measurements when pupils were in the classrooms to estimate the actual effective ventilation rates with a general mass balance equation (McIntyre 1980). In fact as previously noted above, $CO_2$ has for many years been recognized as a surrogate for ventilation in buildings (Persily and Dols 1990). Elevated $CO_2$ concentrations have been associated with impaired performance and increased health symptoms. However, these adverse effects have in the past been attributed to correlation of indoor $CO_2$ with concentrations of other indoor air pollutants, which are also influenced by rates of outdoor air ventilation (Seppanen et al. 1999; Wargocki et al. 2000; Shendell et al. 2004). Recent experimental research has associated low-to-moderate indoor $CO_2$ concentrations with statistically significant and meaningful reductions in decision-making performance or higher-order cognitive function among adults (Satish et al. 2012; Allen et al. 2016). The direct impacts of $CO_2$ on performance indicated by these findings may be economically important, may disadvantage some individuals, and may limit the extent to which outdoor air supply per person can be reduced in buildings to save energy.

## 8.5 IAQ ASSESSMENT PROTOCOL AND CHECKLISTS

Approaches to solving SBS problems using traditional industrial hygiene techniques are generally inadequate. The traditional industrial hygiene approach to solving occupational health problems involves recognition of the hazard (because input and output materials are known, the identity of air pollutants is generally preestablished), measuring the contaminant(s) by NIOSH or Occupational Safety and Health Administration (OSHA)-approved procedures, and interpretation of analytical data in terms of current health at its source. A common result of this approach when applied to IAQ evaluations is that compliance with industrial workplace standards is demonstrated but building-associated complaints from occupants persist (Woods et al. 1989).

For IAQ evaluation, more sensitive sampling methods and evaluation protocols are needed because comfort, occupant well-being, and general population susceptibilities are parameters that are not addressed by industrial workplace health standards. In addition, a clear understanding of how a building and its HVAC system function is required to evaluate and resolve IAQ problems. The evaluation approach described here is divided into a qualitative phase and a quantitative phase.

### 8.5.1 Qualitative Assessment

The qualitative IAQ assessment includes a telephone interview and a site visit. During this phase of the evaluation, objectives are defined, and a hypothesis as to the likely reason for occupant complaints is formulated.

At the building site, the nature of the complaints is reviewed, environmental factors that may be responsible for complaints are reviewed, environmental factors that may be responsible for complaints are visually evaluated, and an engineering assessment of the HVAC system is conducted. A primary objective of the qualitative assessment is to provide recommendations for remedial actions in a manner that avoids the necessity of costly sampling of air contaminants or measurement of HVAC system performance parameters.

A key aspect of the qualitative assessment is the analysis of a building's HVAC system. HVAC system mechanical components are visually examined for deficiencies with regard to original and current design and for operation and maintenance parameters. The control strategies that govern HVAC system operation must be understood. Items useful in the qualitative assessment of the HVAC system include a smoke pencil for visualization of airflow patterns, a complete set of mechanical plans and specifications, and the on-site availability of the facility engineer of the building being investigated.

When health complaints are reviewed, it is essential to determine if the perceived problem is one of occupant discomfort and annoyance (SBS, thermal discomfort) or if one or more occupants have a BRI. When a BRI is apparent, medical attention for the affected occupants is required along with the initiation of appropriate remedial actions.

### 8.5.2 Checklist for the Qualitative Assessment

Observations on possible contaminant sources (see checklist that follows) made during the qualitative assessment are important both in verifying the hypothesis of complaint etiology and in formulating recommendations for corrective actions. It is important during the qualitative assessment to examine not only the complaint area but also noncompliant areas and the HVAC system serving both areas.

#### *8.5.2.1 Microbial Contaminants (See Section 8.2.1; See AIHA 2008)*

______ Is there evidence of current or past water damage? How extensive is the water damage, and what are the likely sources of the water? (See EPA 2012.)

______ Is visible mold present on interior finishes and construction materials? How extensive is the surface area covered by mold? (See NYC 2008.)

______ Are moisture problems evident in the building envelope? Is there evidence of biodeterioration (condensation, mold growth) on the room (occupied) side of the envelope? Are vapor diffusion barriers correctly positioned for the buildings climatic location? (Figure 8.1)

______ Is there evidence of hidden microbial growth that may cause musty odors (MVOCs)? (Figure 8.2)

______ Do records indicate that RH consistently exceeds 70%?

______ Are porous materials such as carpet and insulation present in damp niches in the building? (Figures 8.3 and 8.4)

______ Are dead botanical materials such as bark chips used in moist locations such as in an atrium planters?

**Figure 8.1** Mold growth occurs beneath vinyl wall covering on perimeter wall because of condensation. Moldy porous materials must be removed in a manner that prevents dispersion of spores.

**Figure 8.2** Mold growth occurred on gypsum board in wall cavity because of moisture incursion. Note rusty electrical socket.

**Figure 8.3** Mold growth occurred on/in porous insulation on airstream surface in unit ventilator.

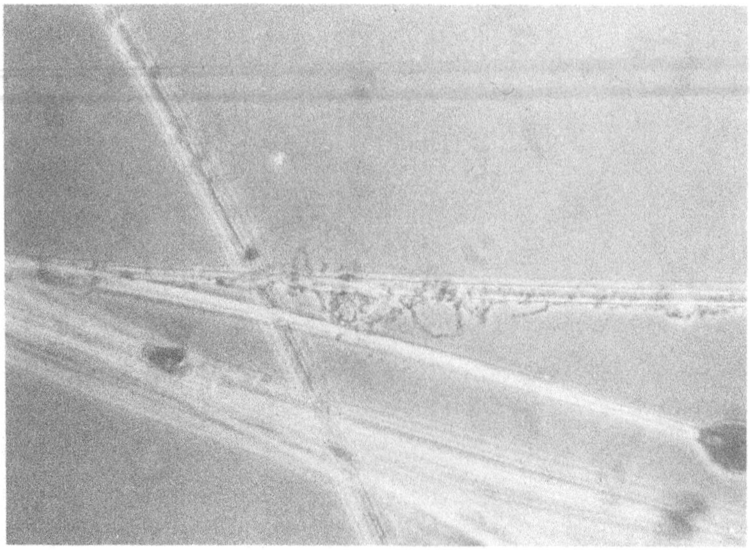

**Figure 8.4** Mold hyphae (filaments) grow around fiber glass in moldy insulation. Insulation must be discarded.

______ Are the HVAC outdoor air inlet and other building openings positioned such that bioaerosols from cooling towers are likely to be entrained? Consider horizontal and vertical separation distances as well as airflow patterns around the building and prevailing wind direction.

______ Is there standing water in the HVAC system (e.g. drain pans, humidifier sumps)? Are biofilms (slime) present on wet surfaces? (Figure 8.5)

______ Are airstream surfaces in the outdoor air intake, mixed air, filter, fan, and supply air (and supply ducts) plenums damp and dirty? Is visible mold present?

______ Are HVAC filters wet? Check both visually and by touching. Discard wet filters.

______ Does moisture (water droplets) carry over from cooling coils to downstream HVAC surfaces? (Figure 8.6)

#### *8.5.2.2 Volatile Organic Compounds and Odors (See Section 8.2.2)*

______ Are malodors present? The presence of body odors indicates inadequate ventilation. Chemical odors might be associated with cleaning compounds and restroom deodorants.

**Figure 8.5** Biofilm occurs in drain pan of AHU. Biofilm must be physically removed and maintenance program must prevent its reoccurrence.

**Figure 8.6** Rust on demister plates and fan housing forms because of water droplet carryover. Prevent carryover of water droplets from coils and spray humidifiers.

_______ Has renovation recently occurred? If yes, were VOCs and dusts exhausted directly out of the building or recirculated by the HVAC system?

_______ Are specialized processes such as printing, graphics production, and art work production occurring in occupied spaces? Are chemicals from these processes exhausted directly out of the building?

_______ Have pressed wood products that contain formaldehyde been used during renovation? If chemical odors are present, increase outdoor air ventilation. More importantly, use low VOC (including formaldehyde) emitting wood products for the next renovation.

_______ Determine who is responsible for monitoring the use of cleaning chemicals and solvents by occupants and cleaning personnel. Are protocols in place to restrict usage of toxic, irritative, and odorous chemicals in the building?

#### 8.5.2.3 *Other Indoor Pollutants (See Section 8.2.3)*

_______ Are HVAC outdoor air inlets located where they may be contaminated by combustion products from garages, loading docks, or vehicular traffic? Does the time of peak occupant complaint correlate with vehicular activity? Move the air inlet or reduce vehicular activity if combustion product entrainment occurs.

_______ Are intermittently operated sources of combustion products near HVAC outdoor air intakes (e.g. gas-fired heaters in rooftop AHUs or diesel-powered emergency generators)?

_______ Does stack effect induce infiltration of combustion gases during winter months?

_______ Are pesticides applied in the building, and what protocol is followed in their application?

_______ What kinds of practices are used by facilities and housekeeping staff to remove settled dusts from interior surfaces? When are cleaning activities performed? Are settled dusts dispersed into the air during the cleaning process? Initiate cleaning with a HEPA filter vacuum if dust control is a problem.

#### 8.5.2.4 *HVAC System Design*

_______ Are HVAC outdoor air inlets and other building openings inappropriately located near external contaminant sources?

_______ Are original HVAC plans available? Do available plans incorporate remodeling that has occurred or changes in occupant and/or thermal loads?

_______ Is porous liner located on airstream surfaces in moist niches in the HVAC system? Are liners moldy? Remove moldy liners.

_______ Was the building designed to meet the minimum outdoor air ventilation rates recommended by ASHRAE 62.1-2013?

_______ What is the design efficiency of the HVAC system filter deck? Can filtration efficiency be increased from MERV 6 to higher grades (MERV 8–10)?

_______ Do VAV terminal boxes have minimum set points that provide at least 17 cfm of outdoor air per occupant when the zone thermostat is satisfied? If not, replace or fix the equipment and/or controls.

#### 8.5.2.5 *HVAC System Operation*

_______ Are control system plans available that describe how minimum amounts of outdoor air ventilation (17 cfm/occupant for offices) are provided during the heating and cooling season and during transitional climatic periods?

_______ Do outdoor air inlet dampers open and close (check for minimum stops) according to design criteria?

_______ Is the occupied space (especially complaint area) stuffy and warm? Do VAV terminal boxes close when the thermostat is thermally satisfied? Does supply air short circuit to return air vents? Are return air vents present? Do $CO_2$ concentration measurements (generally 500–600 ppm above ambient levels) suggest inadequate outdoor air ventilation?

_______ Are thermostats properly positioned to control heating and cooling requirements in perimeter (envelope) and core (interior) areas? Are temperature recommendations of ASHRAE Standard 55-2004 being met for all occupied zones?

_______ Are exhaust systems properly removing combustion products from garage and dock areas and keeping their areas negatively pressurized relative to occupied spaces?

#### 8.5.2.6 *HVAC System Maintenance*

_______ Are written preventive maintenance protocols available for the HVAC system including FCUs and IUs? Are the protocols being followed?

_______ Is a written preventive maintenance plan available for the cooling tower including bacterial, scale, and corrosion control?

______ Are mechanical equipment rooms and HVAC plenums used for storage of solvents, paints, fertilizers, pesticides, and other indoor pollutants?
______ Are airstream surfaces in the HVAC system clean and dry? This includes dampers and pressure sensor stations, the filter plenum, plenums downstream or the heat exchanger, and air supply ductwork.

### 8.5.3 Quantitative Evaluation

In some building investigations, measurement of contaminant concentrations and HVAC system operation parameters are required to justify the cost of remedial recommendations or for litigation purposes. Many variables can affect sampling strategy. Consider first if the contaminant is from the building itself (e.g. VOC emissions from finishes), from occupant activities (e.g. VOCs from photo copied paper), or externally sourced (e.g. VOCs and combustion gases from vehicular emission). The sampling strategy used should always include contaminant collection in a worst-case location. In the case of VOCs from building finishes, the worst-case samples should be collected early in the workday before ventilation has diluted contaminant concentrations. If VOCs are likely being emitted from photocopied paper, sampling should occur in the copy room or where printed paper is stored. In the case of entrainment of vehicular emission, sampling at the outdoor air inlet and along the ventilation air stream is required. In addition to sampling in worst-case location, quantitative assessment of contaminants should occur in best-case situations (e.g. in locations where minimal pollutants are anticipated) as well as in the zone of highest-quality outdoor air (e.g. facing into the wind on the roof at a site remote from external contaminants). Data interpretation often involves a comparison of worst case with indoor and outdoor control locations.

Sampling is performed differently depending upon the objective of the quantitative assessment. If the objective is to determine occupant exposure, the samples are collected in breathing zone locations under normal building operation conditions (HVAC system is on; routine activities occur in the occupied space). However, if the objective is to demonstrate that a contaminant source is present, then the air sample, for example, for VOCs may be collected adjacent to the finishing material (or determine emission characteristics of finish materials directly in the laboratory).

Quantitative measurements are highly affected by the operation of the HVAC system. Contaminants sourced in the building will likely build up rapidly in occupied zones if a VAV terminal completely closes. Since less outdoor air is used in HBAC systems of many buildings during periods when it is very warm or very cold outdoors, the season of the year is important with regard to dilution of building sourced contaminants like VOCs. Season of the year is an important consideration when air samples of culturable fungi are obtained. Different concentrations and viability of fungi are expected in the cooling season when the building and its HVAC system are humid and moist as compared with the heating season when materials are dry.

## 8.6 CONTROL MEASURES

A major objective of both the qualitative and quantitative IAQ assessments is to provide recommendations to remediate building problems, including the causes of occupant complaints leading to SBS, BRI, or thermal discomfort. Two very different approaches can be taken when performing remedial actions to upgrade building IAQ. Remedial actions can be directed at *source control*, which involves removal of the pollutant (Morey 1995b). Alternatively, remedial actions may be directed at *exposure* control, where the concentrations of pollutants that may be present in indoor air are diluted by provision of relatively clean outdoor air or by cleaning (filtration) of indoor air. Source control is usually pollutant specific and the more effective remediation strategy because occupant exposure can be eliminated or greatly reduced.

In general, recommendations for both source and exposure control in a building are often made simultaneously. For example, the minimum outdoor air ventilation rates recommended by ASHRAE Standard 62.1-2013 are adequate for dilution of VOCs emitted from interior finishes with low source strengths. For intense sources of VOCs (e.g. liquid process photocopier), the appropriate recommendation is removal of the source (the photocopier or an artwork classroom) or effective removal of the contaminant (voc. from artwork supplies) by local exhaust ventilation.

IAQ assessments may be reactive or proactive. The reactive assessment is usually accompanied by recommendations to lessen the concentration of pollutants in occupied spaces (exposure control) rather than to eliminate exposure (source control). For example, in an existing building with a tuck-under loading dock, recommendations such as negative pressurization of the dock area, installation of double doors and air curtains, and limiting vehicle idling time are made to limit entry of combustion products into occupied spaces (exposure control). Moving the loading dock (source control) would eliminate exposure, but the cost associated with this form of remediation often inhibits action. Proactive approaches toward control of indoor air pollutants through source control are almost always easiest to implement during design of new buildings or during major retrofit of existing building.

### 8.6.1 Microbials

Sustained growth of mold in buildings is caused by moisture and dampness in building infrastructure. The primary approach toward control is achieved by eliminating sources of moisture that support growth. Since dirt dust and soiling of building materials provide nutrient for microbial growth when moisture is non-limiting, the proper cleaning of the building and its equipment is also essential for microbial control. The following actions should be taken to prevent the growth of mold in buildings:

- Eliminate sites of water accumulation and dampness.
- Keep the surface RH of interior construction and finishing material from consistently exceeding the 65–70% range.
- Dry building infrastructure as rapidly as possible following a leak or a flood so as to prevent microbial growth on susceptible materials. Porous finished such as gypsum wall board, ceiling tiles, and carpet that remain wet for periods of 24 hours or more are likely to be subject to mold colonization.
- Develop a plan of action for dealing with moisture problems such as floods and water spills that may occur in buildings; Roving moisture hidden within building infrastructure (e.g. water in wall cavities) is an important aspect of the plan.
- Incorporate a highly efficient cleaning program into facilities maintenance so that dust and dirt are physically removed from surfaces, especially those niches that may be affected by dampness.

Condensation conditions should be prevented in building envelopes in all climates. In air-conditioned buildings in hot humid climates, the following actions are intended to reduce the likelihood of moisture and mold problems:

- The building should be provided with a vapor diffusion retarder and an air barrier system, typically located toward the exterior of the envelope. Avoid the use of low-permeance materials on interior surface of external walls.
- Maintain a net positive pressurization in the building so that infiltration of humid air through the envelope is prevented.
- Prevent rain from entering the envelope.
- Avoid cooling the interior space below the average monthly outdoor dew point temperature. Keep dew point temperatures below 55 °F in indoor air (EPA 2012).

In cold climates, the following actions are intended to reduce the likelihood of moisture and mold problems in the building envelope:

- Ensure that the envelope has an adequate thermal resistance to prevent interior surfaces from becoming too cold.
- Prevent room moisture from entering the envelope. This may be accomplished by placing the vapor diffusion retarder toward the warm (interior) side.

Moisture and dirt in HVAC systems result in microbial growth. The following actions are useful for controlling moisture and microbial problems in HVAC systems:

- Locate outdoor air inlets sufficiently away from (preferably above-grade) standing water, leaves, soil, dead vegetation, and bird droppings. The outdoor air inlet should be designed so as to prevent the intake of rain, snow, and, in coastal areas, fog.
- Keep the floors, walls, and ceilings of HVAC plenums clean and dry. Air stream surfaces in HVAC plenums should be clean and dry. Air stream surfaces in HVAC systems should be smooth or easily cleanable and should be resistant to biodeterioration.
- Upgrade the efficiency of HVAC filtration from MERV 6 to MERV 8 or higher. This will reduce dust accumulation in portions of the system that are difficult to clean.
- Prevent filters from becoming wetted by rain snow, fog, or water droplets from dehumidifying cooling coils or the humidification system.
- Conditions of water stagnation (accumulation) must not occur in operating HVAC systems. Prevent the buildup of biofilm on wet surfaces of drain pans and dehumidifying cooling coils by frequent cleaning. Biocides must be physically removed after cleaning of pans and coils. Microbiocidal chemicals and disinfectants used in cleaning must not be aerosolized into occupied spaces.

- When humidifiers are used, the types that emit water vapor rather than droplets are desirable. Control of microbial growth in recirculating water humidifiers is achieved by keeping the unit clean, using high-quality water, and continuous blowdown or replacement of some of the water from the sump.
- Avoid use of unit ventilators, FCUs, and IUs unless units can be kept clean and dry through high-quality maintenance.

Controlling the growth of *Legionella* involves design, operation, and maintenance of cooling towers and potable water systems where this organism can amplify. Some general guidelines (see ASHRAE 2009) are as follows:

- Design hot water tanks and piping so that the water system can be disinfected by superheating throughout to 60 °C in the event that *Legionella* contamination occurs.
- Avoid conditions in hot water tanks and piping that are conducive to *Legionella* growth such as poor mixing, lukewarm water, and water stagnation.
- Locate HVAC outdoor air inlets and other building openings at a minimum horizontal separation distance of 25 ft (50 ft better) from cooling towers and evaporative condensers.
- Cooling tower water systems should be treated to control the growth of microbial contaminants. Microbiocidal control involves the use of biocides and/or physical methods to kill bacteria. A testing program to verify the effectiveness of microbiocidal treatment may be considered (ASTM 2007).

Consensus has been reached that the occurrence of sustained visual growth of mold on building surfaces is unacceptable (Health Canada 2004; NYC 2008). When more than about 3 $m^2$ of interior surface is covered by visible mold, the contaminated materials should be physically removed according to containment procedures similar to those used during removal of hazardous materials. General principles to be followed during removal of visible mold include the following actions:

- Fix the moisture problem that led to mold growth.
- Remove visually moldy materials under negative pressure containment in such a manner that dusts and spores are not dispersed into adjacent clean or occupied area. Biocide treatment or encapsulation of the moldy surface does not substitute for physical removal of the contaminant.
- Remove all fine dusts (particulate) from the formerly moldy area through damp wiping and/or HEPA vacuuming prior to installation of new finishes in the occupied space.

### 8.6.2 Volatile Organic Compounds

VOC emission from finishing and construction materials used in new buildings or major renovations in existing buildings is best controlled by source reduction or elimination.

VOCs may be controlled during new construction and major renovation by the following:

- Avoid use of finishes and construction materials that contain carcinogens, toxins, or teratogens.
- Avoid the use of materials containing highly odorous VOCs that are known to degrade perceived air quality at minute concentration levels (Wolkoff et al. 1996).
- Minimize the use of solvents, sealants, caulks, paints, and other products that have high emission rates for VOCs or are highly odorous.
- Provide continuous preoccupancy outdoor air ventilation at the design rates recommended by ASHRAE, preferably for 30 days, to flush out VOCs emitted from new finishes.
- Isolate areas of existing buildings undergoing major renovation by construction of critical barriers (temporary walls, plastic sheeting, or other vapor retarding layer). The construction area should be maintained at a negative pressure relative to adjacent occupied area. Air from the construction area should be locally exhausted outdoors in order to prevent VOC adsorption into finishes elsewhere in the building.

Many diverse sources of VOCs potentially occur in existing buildings not undergoing major renovation. Sources include cleaning chemicals and products, minor touchup and painting, equipment operated by occupants, and occupants' personal care products. Principles useful in controlling VOCs from diverse sources in occupied buildings include the following:

- Monitor and reduce to the lowest feasible level the usage of volatile chemicals used in cleaning agents and work processes in the building.

- Use local exhaust ventilation to remove VOCs and other air contaminants from printing machines, graphics operations, closets or rooms, where paints, solvents, and cleaning agents are stored, or from any other strong VOC sources.
- Prohibit the storage of cleaning chemicals, solvents, paints, etc. in HVAC mechanical equipment rooms or in HVAC plenums.
- Carpets, modular partitions, or other finished brought into the building during minor touchup should be previously off-gassed in a clean, well-ventilated warehouse.
- Ensure, based on emission data provided by manufacturers, that new computers, printers, copiers, etc. are not significant sources of VOCs.

### 8.6.3 Other Indoor Pollutants

The following principles should be considered for control of air pollutants including pesticide, combustion products, and particles:

- *Pesticides*. Limit pesticide usage to building areas where its application is required. Avoid environmental conditions (e.g. food in offices) that cause pest problems. Apply pesticides only during unoccupied periods. Ventilate the space (follow manufacturer's directions) prior to occupancy. Never use or store pesticides in active HVAC equipment or plenums (e.g. in FCUs or mechanical equipment rooms).
- *Combustion products*. Avoid building designs where sources of combustion products such as garages or loading docks are located beneath or within buildings or where combustion sources are located near HVAC outdoor air inlets or other building openings. An alarm should be triggered at the building central security desk when CO sensors in the garage or dock exceed high limits so that prompt action can be taken to control combustion product emissions.
- *Particulates*. Avoid activities that generate particulate contaminants. These activities may include dry mopping that disturbs settled dust and use of inefficient vacuum cleaners that allow fine particles to pass through the instrument and back into the indoor environment.
- Avoid the use of extended surface, fleecy finishes in building areas where dust control is critical or where soiling is difficult to prevent.
- Provide occupied spaces with highly filtered conditioned air at a sufficient air exchange rate so as to dilute particles that may be aerosolized by occupant activities. Air cleaning is almost always more effective when performed in the central HVAC system rather than by portable air cleaners.

### 8.6.4 EPA Guides

The IAQ Building Education and Assessment Model (I-BEAM) (EPA 2002) is a guidance tool designed for use by building professionals and other interested in IAQ in commercial buildings. I-BEAM updates and expands EPA's Building Air Quality guidance and is designed to be a comprehensive state-of-the-art guidance for managing IAQ in commercial buildings. I-BEAM consists of many individual modules, which explain different aspects of IAQ including how to manage, operate, and maintain a building for IAQ and how to ensure that energy efficiency projects are compatible with IAQ.

Guidance to school officials involved in IAQ is provided in a document entitled "Indoor Air Quality Tools for Schools" (EPA 1995). This guide is presented as an "action kit," directed primarily to kindergarten through grade 12 schools and was designed to help schools proactively address IAQ problems with minimal cost. The intent is to educate and enable existing school staff to perform qualitative IAQ assessments and to upgrade building operation and maintenance activities. Emphasis is placed on proactive actions, advising the user that expenses and efforts necessary for preventing most IAQ problems are less than those required to resolve problems after they develop. The guide includes substantial background information, primarily addressing IAQ issues pertinent to schools. Much guidance in IAQ tools for schools is presented as checklists for various categories of school staff.

### 8.6.5 Communications

Open communications between building management and occupants are necessary for the resolution of IAQ complaints, especially in schools. Successful resolution of IAQ complaints required communication and participation of occupants and/or their representatives, housekeeping and HVAC facility staff, health and safety committee representatives, and the building manager (or owner). Cooperation and early action to investigate and solve problems usually results in a successful resolution of complaints. Denial of the existence of a problem in the face of continued occupant complaints often results in a situation where problem solving is difficult and delayed.

When IAQ complaints occur in a building, the following communication actions should occur: (i) Openly discuss the complaints (or problem) with all concerned parties; solicit occupant views on reasons for the complaints; establish a policy through which all parties will be appraised of progress during complaint investigation and resolution. (ii) Develop a written procedure for recording the time, location, and nature of complaints as they occur; housekeeping and facilities staff should record building operational parameters that may coincide with the complaint (e.g. occurrence of renovation or cleaning activities); a team approach is often needed to resolve the problem; for example, the building manager may be unaware that a renovation or cleaning activity is degrading IAQ, but occupants may readily recognize the source of the problem. (iii) All parties, including occupants, should be provided with information on compliance with remedial recommendations; highly technical reports dealing with problem resolution should be accurately summarized in plain English so as to be understandable by the general public.

## REFERENCES


Allen, J.G., MacNaughton, P., Satish, U. et al. (2016). Associations of cognitive function scores with carbon dioxide, ventilation, and volatile organic compound exposures in office workers: a controlled exposure study of green and conventional office environments. *Environmental Health Perspectives* **124**: 805–812.

Begerow, D., Nilsson, H., Unterseher, M., and Maier, W. (2010). Current state and perspectives of fungal DNA barcoding and rapid identification procedures. *Applied Microbiology and Biotechnology* **87**: 99–108.

Burroughs, H.E. and Hansen, S.J. (2004). *Managing Indoor Air Quality*, 3e. Lilburn, GA: The Fairmont Press.

Caporaso, J.G., Kuczynski, J., Stombaugh, J. et al. (2010). QIIME allows analysis of high-throughput community sequencing data. *Nature Methods* **7**: 335–336.

CDC (2012). *Guidance for School Administrators to Help Reduce the Spread of Seasonal Influenza in K-12 Schools*. Last updated 17 August 2012. http://www.cdc.gov/flu/school/guidance.htm (accessed December 2017).

Daisey, J.M., Angell, W.J., and Apte, M.G. (2003). Indoor air quality, ventilation and health symptoms in schools: an analysis of existing information. *Indoor Air* **13**: 53–64.

Fisk, W.J., Faulkner, D., Palonen, J., and Seppanen, O. (2002). Performance and costs of particle air filtration technologies. *Indoor Air* **12**: 223–234.

Haverinen-Shaughnessy, U. and Shaughnessy, R.J. (2015). Effects of classroom ventilation rate and temperature on students' test scores. *PLoS One* **10** (8): e0136165.

Haverinen-Shaughnessy, U., Shaughnessy, R., Cole, E. et al. (2015). An assessment of indoor environmental quality in schools and its association with health and performance. *Building and Environment* **93**: 35–40.

Institute of Medicine, Board on Health Promotion and Disease Prevention, Committee on Damp Indoor Spaces and Health (2004). *Damp Indoor Spaces and Health*, 1e. Washington, DC: National Academies Press.

ISSA (n.d.). *Clean Standard for K-12*. The Worldwide Cleaning Industry Association, The ISSA Clean Standard: K-12 Schools. https://www.issa.com/certification-standards/issa-clean-standards/clean-standard-k-12.html (accessed December 2018).

Janssen, J. (1989). Ventilation for acceptable indoor air quality. *ASHRAE J.* (October): 40–48.

MacNaughton, P., Pegues, J., Satish, U. et al. (2015). Economic, environmental and health implications of enhanced ventilation in office buildings. *International Journal of Environmental Research and Public Health* **12**: 14709–14722.

McIntyre, D.A. (1980). *Indoor Climate*. London: Applied Science.

Mendell, M. and Heath, H. (2005). Do indoor pollutants and thermal conditions in schools influence student performance? A critical review of the literature. *Indoor Air* **15**: 27–52.

Mendell, M., Eliseeva, E., Davies, M. et al. (2013). Association of classroom ventilation with reduced illness absence: a Prospective Study in California Elementary Schools. *Indoor Air* **23**: 515–528.

Peccia, J. and Shaughnessy, R. (2016). Can the next generation of DNA tools guide the clearance of water damaged buildings. *Journal of Cleaning, Restoration and Inspection* (February 2016).

Persily, A. and Dols, W.S. (1990). The relation of $CO_2$ concentration to office building ventilation. In: *Air Change Rate and Airtightness in Buildings* (ed. M.H. Sherman), 77–84. Philadelphia, PA: ASTM International.

Satish, U., Mendell, M., Shekhar, K. et al. (2012). Is $CO_2$ an indoor pollutant? Direct effects of low-to-moderate $CO_2$ concentrations on human decision-making performance. *Environmental Health Perspectives* **120**: 1671–1677.

Seppanen, O., Fisk, W.J., and Mendell, M.J. (1999). Association of ventilation rates and $CO_2$ concentrations with health and other responses in commercial and institutional buildings. *Indoor Air* **9** (4): 226–252.

Shaughnessy, R. and Vu, H. (2012). Particle loadings and resuspension related to floor coverings in chamber and occupied school environments. *Atmospheric Environment* **55**: 515–524.

Shaughnessy, R.J. and Weschler, C. (2001). Indoor chemistry: ozone and volatile organic compounds found in tobacco smoke. *Environmental Science and Technology*.

Shaughnessy, R., Cole, E.C., and Haverinen-Shaughnessy, U. (2013). ATP as a marker for surface contamination of biological origin in schools and as a potential approach to the measurement of cleaning effectiveness. *Journal of Occupational and Environmental Hygiene* **10**: 336–346.

Shendell, D.G., Prill, R., Fisk, W.J. et al. (2004). Associations between classroom $CO_2$ concentrations and student attendance in Washington and Idaho. *Indoor Air* **14** (5): 333–341.

Toyinbo, O., Matilainen, M., Turunen, M. et al. (2016). Modeling associations between principals' reported indoor environmental quality and students' self-reported respiratory health outcomes using GLMM and ZIP models. *International Journal of Environmental Research and Public Health* **13** (4): 385.

Wargocki, P. and Wyon, D. (2007). The effects of moderately raised classroom temperatures and classroom ventilation rate on the performance of schoolwork by children. *HVAC&R Research* **13** (2): 193–220.

Wargocki, P., Wyon, D.P., Sundell, J. et al. (2000). The effects of outdoor air supply rate in an office on perceived air quality, sick building syndrome (SBS) symptoms and productivity. *Indoor Air* **10** (4): 222–236.

CHAPTER 9

# OCCUPATIONAL NOISE EXPOSURE AND HEARING CONSERVATION

CHARLES P. LICHTENWALNER

*Lucent Technologies, Bell Laboratories, 600 Mountain Ave., Murray Hill, NJ, 07974*

and

KEVIN MICHAEL

*Michael & Associates, 246 Woodland Dr., State College, PA, 16803*

Adapted from "Industrial noise exposure and conservation of hearing" by P.L. Michael, Chapter 23, in G.D. Clayton and F.E. Clayton, *Patty's Industrial Hygiene and Toxicology*, 4e, vol. IA. New York: Wiley, 1991, pp. 937–1039.

## 9.1 INTRODUCTION

Few locations or circumstances in nature produce noise loud enough or for long enough duration to damage hearing. It remained for man, himself, to produce noises capable of causing injury. The knowledge that loud noises can produce these injuries has been known for several hundred years.

The most obvious and best quantified injury from noise is deterioration of hearing ability. However, except for extremely loud sounds, noise-induced hearing loss is a slowly progressive debility that usually goes unnoticed by those affected until the loss is significant. There are no visible effects and usually no pain.

Hearing loss occurring naturally from aging is called *presbycusis*, although some have argued that much of the presbycusis seen in industrialized societies may actually be caused by ambient (non-occupational) noise exposures (Kryter 1994). Hearing losses can also be pathological from medical abnormalities, ranging from simple, easily correctable conditions such as impacted earwax and middle ear infections to severe problems such as deafness from rubella during gestation. As in many other areas of industrial hygiene, it is difficult to determine the contribution of these various elements – (i) occupational noise exposure, (ii) non-occupational noise exposure, (iii) aging, (iv) medical pathology, and (v) normal human variability – to one individual's hearing loss.

In addition to hearing loss, noise causes a number of other problems: (i) communication interference, (ii) annoyance, (iii) performance degradation, and (iv) physiological effects. High noise levels make it difficult to communicate. At moderate noise levels, telephone conversation becomes difficult. At higher levels people must raise their voices or shout to be understood. Communication problems lead to safety problems when the noise masks speech or alerting signals. Disturbing or distracting noises can lead to accidents.

In the United States, legislation has been used, first to provide compensation to those injured (Newby 1964) and then later to limit noise exposures. In 1970, the Occupational Safety and Health Administration (OSHA) used noise regulations established by the Department of Labor under the Walsh–Healey Public Contracts Act. The regulations established a permissible exposure limit (PEL) based on an eight-hour time-weighted-average (TWA) noise exposure. In 1981, OSHA amended the noise standard – adopted in final form in 1983 (Suter 1986) – establishing requirements for a hearing conservation program. From 1990 to 1995, 8106 inspections were conducted by federal OSHA and state regulators for noise, resulting in 14 138 citations and assessment of $7 703 027 in penalties.[1]

[1] Results of Freedom of Information request for inspection reports citing 29 CFR 1910.0095 and 1926.0052.

*Handbook of Occupational Safety and Health*, Third Edition. Edited by S. Z. Mansdorf.

## 9.2 PHYSICS OF SOUND

Technically, the sensation of sound results from oscillations in pressure, stress, particle displacement, and particle velocity in any elastic medium that connects the sound source with the ear. When sound is transmitted through air, it is usually described in terms of changes in pressure that alternate above and below atmospheric pressure. These pressure changes are produced when vibrating objects (sound sources) cause regions of high and low pressure that propagate from the sound source. The characteristics of a particular sound depend on the rate at which the sound source vibrates, the amplitude of the vibration, and the characteristics of the conducting medium. A sound may have a single rate of pressure/vacuum alternation or frequency ($f$), but most sounds have many frequency components. Each of these frequency components, or bands of sound, may have a different amplitude.

### 9.2.1 Frequency

Frequency is defined as the rate at which a sound is emitted from a source. Physically it is measured by the number of times per second the pressure oscillates between levels above and below atmospheric pressure (Figure 9.1). It is denoted by the symbol $f$ and is measured in Hertz (1 Hz = 1 cycle per second). It is inversely related to the period ($T$), which is the time a sound wave requires to complete one cycle. The frequency range over which normal young adults are capable of hearing sound at moderate levels is 20–20 000 Hz. Pitch or tone is the sensation associated with frequency.

A sound may consist of a single frequency (i.e. a pure tone); however, most common sounds contain many frequency components. It is generally infeasible to report the characteristics of all the frequencies emitted by noise sources; so measurements are made including the sound energy from a broad range of frequencies. As will be discussed in Section 9.5, weighting networks have been standardized for single-number assessments of sounds having properties similar to the response of the human ear. In addition to these broadband weightings, the frequency range in acoustics is frequently divided into smaller ranges. The most common range of frequencies is the octave band, where the upper edge of the band is twice the frequency of the lower band edge. One-third octave (three bands are used to measure one octave) and narrower bands are also used.

### 9.2.2 Wavelength

The distance a sound wave travels during one sound pressure cycle is called the wavelength ($\lambda$). The wavelength is related to the speed of sound by

$$\lambda = \frac{c}{f} \tag{9.1}$$

where $c$ is the speed of sound in meters per second ($\mathrm{m\,s^{-1}}$), $f$ is the frequency in Hz, and $\lambda$ is the wavelength in meters (m).

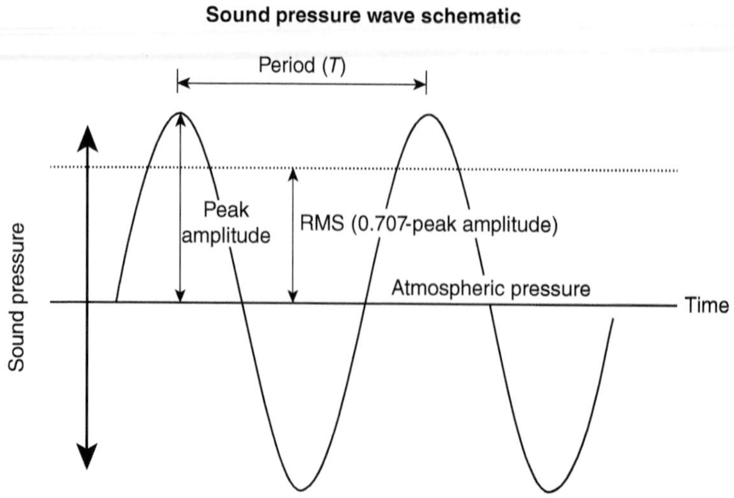


**Figure 9.1** Amplitude and frequency of a pure tone sound wave.

For most work, it is sufficient to note that sound travels in air at approximately $344\,\mathrm{m\,s^{-1}}$ at normal atmospheric pressures and room temperatures.

***Example 9.1:*** Calculate the wavelength of 20-, 1000-, and 20000-Hz waves.

At 20 Hz:

$$\lambda = \frac{c}{f} = \frac{344\,\mathrm{m\,s^{-1}}}{20\,\mathrm{s^{-1}}} = 17.2\,\mathrm{m}$$

At 1000 Hz:

$$\lambda = \frac{c}{f} = \frac{344\,\mathrm{m\,s^{-1}}}{1000\,\mathrm{s^{-1}}} = 0.344\,\mathrm{m}$$

At 20000 Hz:

$$\lambda = \frac{c}{f} = \frac{344\,\mathrm{m\,s^{-1}}}{20000\,\mathrm{s^{-1}}} = 0.017\,\mathrm{m}$$

When considering the behavior of sound, one important characteristic is the ratio of the length of the item affected by the noise to the wavelength of the sound. For instance, sound will not efficiently couple to vibrations of the entire eardrum when the ¼ wavelength of sound is less than the dimensions of the eardrum. It is not surprising that frequencies greater than 20000 Hz cannot be heard since ¼ wavelength is less than the 5-mm-diameter dimension of a typical eardrum. As will be shown later, the effectiveness of a barrier in shielding one side from a noise source on the other depends on the path differences between going around the barrier and the straight line distance in terms of the number of wavelengths.

### 9.2.3 Amplitude

***9.2.3.1 Sound Pressure and Sound Pressure Level*** The amplitude of the sound pressure is caused by the vibrations of the sound source. Loudness is the sensation from sound pressure. Pressure is measured in pascals (abbreviated Pa; $1\,\mathrm{Pa} = 1\,\mathrm{N\,m^{-2}} = 1\,\mathrm{kg\,[meter\ second^2]^{-1}}$). One atmosphere = $1.013 \times 10^5$ Pa at sea level. Sound pressures are very small variations around atmospheric pressure. Normal speech at 1-m distance from a talker averages about 0.1 Pa – one-millionth of an atmosphere.

The range of sound pressures commonly measured is very wide. Sound pressures well above the pain threshold, about 20 Pa, are found in some work areas, whereas sound pressures down to the threshold of hearing at about 20 μPa (micropascals) are used for hearing measurements. The range of common sound exposures exceeds $10^6$ Pa. This range cannot be scaled linearly with a practical instrument while maintaining the desired accuracy at low and high levels. To be able to see "just noticeable differences" in hearing at sound pressures in noisy environments would require a scale many miles long. To cover this wide range of sound pressure with a reasonable number of scale divisions and to provide a scale that responds more closely to the response of the human ear, the logarithmic decibel (dB) scale is used. The abbreviation dB is used with upper and lower case – "d" for deci and "B" for Bell – in honor of Alexander Graham Bell. The measurement unit was invented at Bell Laboratories to facilitate calculations involving loss of signals in long lengths of telephone lines.

By definition, the decibel is a unit without dimensions; it is the logarithm to the base 10 of the *ratio* of a *measured quantity* to a *reference quantity when the quantities are proportional to power* (ANSI S1.1 2013). The decibel is sometimes difficult to use and to understand because it is often used with different reference quantities. Acoustic intensity, acoustic power, hearing thresholds, electric voltage, electric current, electric power, and sound pressure level may all be expressed in decibels, each having a different reference. Obviously the decibel has no meaning unless a specific reference quantity is specified or understood. Any time a level is referred to in acoustics, decibel notation is implied.

Most sound measuring instruments are calibrated to provide a reading (called root mean square [rms]) of sound pressures on a logarithmic scale in decibels. The decibel reading taken from such an instrument is called the sound pressure level ($L_p$). The term *level* is used because the measured pressure is at a particular level above a given pressure reference. For sound measurements in air, 0.00002 Pa, or 20 μPa, commonly serves as the reference sound pressure. This reference is an arbitrary pressure chosen many years ago because it is approximately the normal threshold of human hearing at 1000 Hz. Since the eardrum responds to the intensity of the sound wave, and since intensity is proportional to pressure squared, sound pressure levels are calculated from the square of sound pressures. Mathematically, $L_p$ is written as follows:

$$L_p = 10\log\left(\frac{p}{p_r}\right)^2 \tag{9.2}$$

where $p$ is the measured sound pressure, $p_r$ is the reference sound pressure (generally in $\text{n m}^{-2}$ or pascals [Pa]), and the logarithm is to the base 10.

An equivalent form of the preceding equation is frequently found in acoustics textbooks:

$$L_p = 20\log\left(\frac{p}{p_r}\right) \tag{9.3}$$

Specifying sound pressure levels in this form masks the fact that levels are ratios of quantities equivalent to power. For technical purposes, $L_p$ should always be written in terms of decibels relative to the recommended reference pressure level of 20 μPa. Reference quantities for acoustic levels are specified in ANSI S1.8 (2011). The reference quantity should be stated at least once in every document.

Figure 9.2 shows the relationship between sound *pressure* (in pascals) and sound *pressure level* (in dB re 20 μPa). This figure illustrates the advantage of using the decibel scale rather than the wide range of direct pressure measurements. It is of interest to note that any doubling of a pressure is equivalent to a 6-dB change in level. For example, a range of 20–40 μPa, which might be found in hearing measurements, and a range of 1–2 Pa, which might be found in hearing conservation programs, are both ranges of 6 dB. Measuring in decibels allows reasonable accuracy for both low and high sound pressure levels.

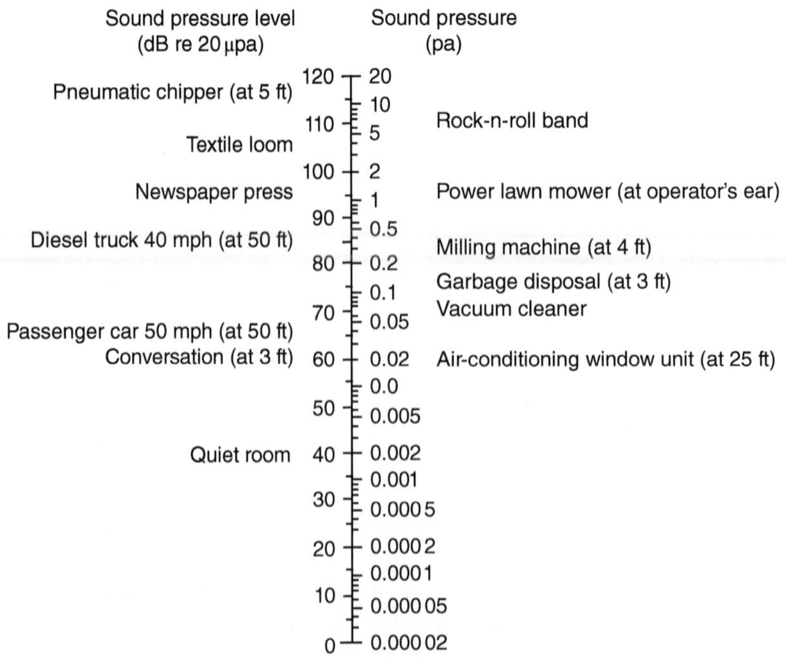


**Figure 9.2** Relation between *A*-weighted sound pressure level and sound pressure.

**9.2.3.2 *Sound Intensity and Sound Intensity Level*** Sound intensity at any specified location may be defined as the average acoustic energy per unit time passing through a unit area that is normal to the direction of propagation. For a spherical or free progressive sound wave, the intensity may be expressed by:

$$I = \frac{p^2}{\rho c} \tag{9.4}$$

where $p$ is the rms sound pressure, $\rho$ is the density of the medium, and $c$ is the speed of sound in the medium. Sound intensity units, like sound pressure units, cover a wide range, and it is often desirable to use decibel levels to compress the measuring scale. To be consistent, intensity level is defined as

$$L_{\mathrm{I}} = 10\log\left(\frac{I}{I_{\mathrm{r}}}\right) \tag{9.5}$$

where $I$ is the measured intensity at some given distance from the source and $I_{\mathrm{r}}$ is a reference intensity, $10^{-12}\,\mathrm{W\,m^{-2}}$. In air, this reference closely corresponds to the reference pressure 20 μPa used for sound pressure levels.

$$L_{\mathrm{I}} = L_{\mathrm{p}} \tag{9.6}$$

**9.2.3.3 *Sound Power and Sound Power Level*** Consider a vibrating object suspended in the air (Figure 9.3). The vibrations will create sound pressure waves that travel away from the source – decreasing by 6 dB as the distance from the source doubles. The sound *power* of this source is independent of its environment, but sound *pressure* around the source is not.

Sound power (represented by $W$) is used to describe the sound source in terms of the amount of acoustic energy that is produced per unit time (watts). Sound power is related to the average sound intensity produced in free-field conditions at a distance $r$ from a point source by

$$W = I_{\mathrm{avg}} 4\pi r^2 \tag{9.7}$$

where $I_{\mathrm{avg}}$ is the average intensity at a distance $r$ from a sound source whose acoustic power is $W$. The quantity $4\pi r^2$ is the area of a sphere surrounding the source over which the intensity is averaged. The intensity decreases with the square of the distance from the source, hence the well-known inverse square law.

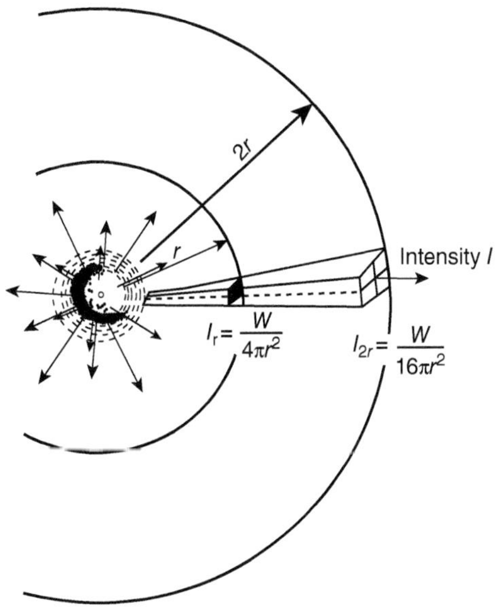


**Figure 9.3** Pulsating object suspended in air (free field). Sound intensity is inversely proportional to the square of the distance from the source.

Units for power are also usually given in terms of decibel levels because of the wide range of powers covered in practical applications. Choosing a reference surface $S_r$ of 1 m² defines power as

$$\begin{aligned}
\frac{W}{W_r} &= \frac{I_{avg} 4\pi r^2}{I_r S_r} = \frac{I_{avg}}{I_r}\frac{4\pi r^2}{S_r} \\
10\log\left(\frac{W}{W_r}\right) &= 10\log\left(\frac{I_{avg}}{I_r}\right) + 10\log\left(\frac{4\pi r^2}{S_r}\right) \\
L_p &= L_I + 10\log\left(\frac{4\pi r^2}{S_r}\right), \quad \text{and since } L_I = L_p: \\
L_{power} &= L_{pressure} + 10\log\left(\frac{4\pi r^2}{S_r}\right)
\end{aligned} \tag{9.8}$$

where $W$ is the power of the source in watts ($1\,W = 1\,N\,m\,s^{-1}$) and $W_r$ is the reference power ($10^{-12}$ W since $W_r = I_r S_r$ and $I_r = 10^{-12}\,W\,m^{-2}$ and $S_r = 1\,m^2$).

Figure 9.4 shows the relation between sound power in watts and sound power level in dB re $10^{-12}$ W. Note that the distance must be specified or inferred to determine sound pressure level from the sound power. Equation (9.8) is used to predict sound pressure levels if the sound power level of a source is known and the acoustic environment is known or can be estimated.

In a free field (no surfaces to reflect sound waves), sound waves spread out from the source, losing power as the square of the distance. Normally, however, reflecting surfaces are present or the sound source is not omnidirectional. A reflecting surface will increase the sound intensity since the volume in which the sound radiates is reduced. The *directivity factor* ($Q$) is a dimensionless quantity used to describe the ratio of volume to which a sound is emitted relative to the volume of a sphere with the same radius (Figure 9.5).

**9.2.3.4 *Sound Power Versus Sound Intensity Versus Sound Pressure*** Noise control problems require a practical knowledge of the relationship between pressure, intensity, and power. For example, consider the prediction of sound pressure levels that would be produced around a proposed machine location from the sound power level provided by the machine.

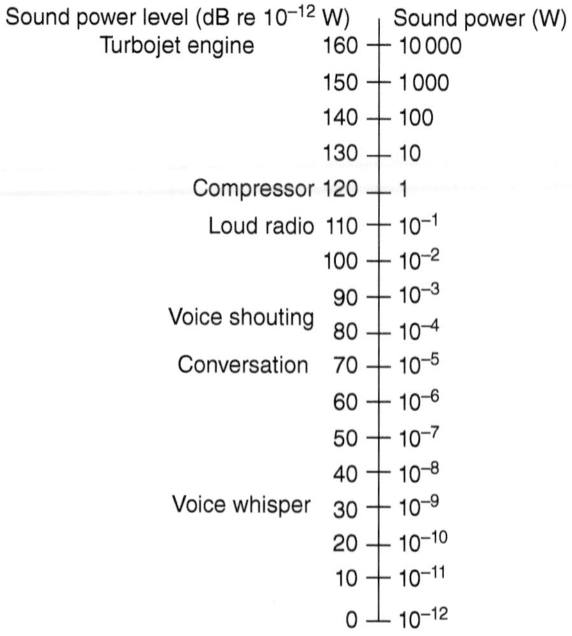


**Figure 9.4** Relation between sound power level and sound power.

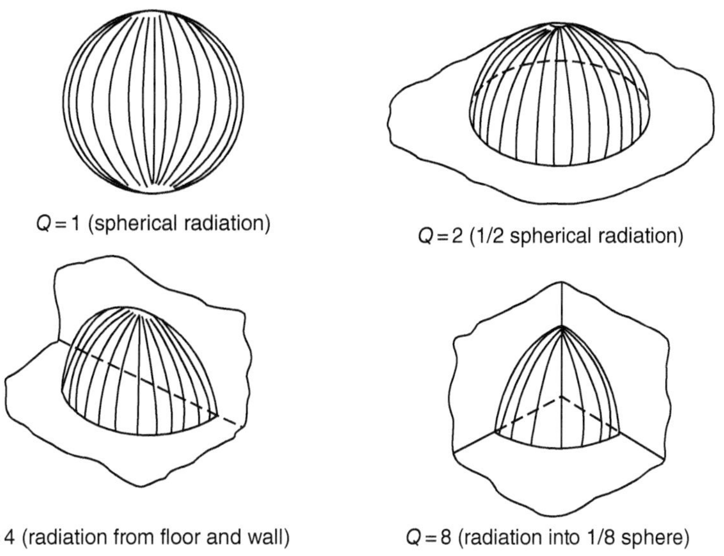


**Figure 9.5** Directivity factor ($Q$).

***Example 9.2:*** The manufacturer of a machine states that this machine has an acoustic power output of 1W. Predict the sound pressure level at a location 10m from the machine.

Answer: From Eqs. (9.4) and (9.7) in a free field and for an omnidirectional source, the sound pressure is

$$I = \frac{p^2}{\rho c} \quad \text{and} \quad W = 4\pi r^2 \quad \text{so} \quad p = (I\rho c)^{1/2} = \left(\frac{W\rho c}{4\pi r^2}\right)^{1/2}$$

$$p = \left(\frac{(1\,\mathrm{N\,m\,s^{-1}})(1.18\,\mathrm{kg\,m^{-3}})(344\,\mathrm{m\,s^{-1}})}{4\pi(10\,\mathrm{m})^2}\right)^{1/2} = 0.32\,\mathrm{Pa}$$

but since the source will be sitting on the ground, reflections will double the sound pressure level, so

$$L_{\mathrm{p}} = 10\log\left(\frac{0.64\,\mathrm{Pa}}{0.00002\,\mathrm{Pa}}\right)^2 = 90\,\mathrm{dB}(\mathrm{re}\ 20\,\mu\mathrm{Pa})$$

Admittedly, there are few truly free-field situations and few omnidirectional sources; so the above calculation can only give a rough estimation of the absolute value of the sound pressure level. However, comparison or rank ordering of different machines can be made, and at least a rough estimate of the sound pressure is available. Noise levels in locations that are reverberant, or where there are many reflecting surfaces, can be expected to be higher than that predicted because noise is reflected back to the point of measurement.

### 9.2.4 Abbreviations and Letter Symbols

Terminology abbreviations and letter symbols for reporting noise measurements are defined by consensus standards (ANSI S1.1-2013; ASME Y10.11-1984 1984) and should be used. By convention, the exponential-time-weighted, frequency-weighted sound pressure levels measured in dB re 20 μPa are known as *sound levels* (Table 9.1).

Sound levels should be reported using either the letter or symbol abbreviations. Fast, slow, and impulse exponential time weightings carry the letters *F*, *S*, and *I*, respectively, and *A*- and *C*-frequency weightings carry their respective letters. Therefore, measurements taken using the OSHA requirement for measuring with *A*-frequency weighting and slow exponential time weighting should be reported using either the letter symbol $L_{\mathrm{AS}}$ or the abbreviation SAL. Similarly, *C*-frequency-weighted measurements taken with fast exponential time weighting would be indicated with the letter symbol $L_{\mathrm{CF}}$ or abbreviated as FCL. When no frequency weighting is specified, the *A*-frequency weighting is assumed.

**TABLE 9.1 Reference Formulae and Quantities**

| Name | Definition | SI reference quantity | English reference quantity |
|---|---|---|---|
| Sound pressure level (dB) | $L_p = 20\log_{10}(p/p_0)$ | $p_0 = 20\,\mu\text{Pa}$ $= 2\times10^{-5}\,\text{N m}^{-2}$ | $p_0 = 2.9\times10^{-9}\,\text{lb in.}^{-2}$ |
| Sound power (dB) | $L_W = 10\log_{10}(W/W_0)$ | $W_0 = 1\,\text{pW}$ | $W_0 = 5\times10^{-10}\,(\text{in. lb})\,\text{s}^{-1}$ |
| Sound intensity level (dB) | $L_I = 10\log_{10}(I/I_0)$ | $I_0 = 1\,\text{pW m}^{-2}$ | $I_0 = 5.71\times10^{-15}\,\text{lb (in. s)}^{-1}$ |
| Sound exposure level (dB) | $L_E = 10\log_{10}(E/E_0)$ | $E_0 = (20\,\mu\text{Pa})^2\,\text{s}$ $= (2\times10^{-5}\,\text{Pa})^2\,\text{s}$ | $E_0 = 1\times10^{-18}\,\text{lb}^2\,(\text{in.}^4\,\text{s})^{-1}$ |

*Source:* ANSI S1.8 (ANSI S1.8 2011).

**TABLE 9.2 Letter Symbols and Abbreviations Used in Acoustics**

| Description | Letter/symbol | Meaning |
|---|---|---|
| Slow | *S* | 1.0 s exponential time weighting |
| Fast | *F* | 0.125 s exponential time weighting |
| Impulse | *I* | 0.035 s rise time, 1.5 s decay time |
| Peak | Pk | Greatest level with no exponential time weighting except as provided by frequency weighting |
| *A* | *A* | *A*-frequency weighting |
| *C* | *C* | *C*-frequency weighting |
| | EC, $L_{eq}$ | Equivalent continuous sound pressure level |
| | Mx | Greatest level for specific exponential time averaging during a specified period |
| | SEL, $L_E$ | Sound exposure level, measured in dB re $(20\,\mu\text{Pa})^2$ s |
| | $E_{A,T}$ | Sound exposure (*A*-frequency weighted) over a specified time *T*, measured in $\text{Pa}^2\cdot\text{s}$ |

*Source:* ANSI S1.8 (2011).

It is good practice, however, to specify both the frequency weighting and the duration of the measurement. For instance, a 1-hour measurement should be reported as $L_{\text{Aeq, 1h}}$. Since $L_{eq}$ is understood for time-averaged measurements, the "eq" is sometimes omitted, and a 1-hour measurement could be reported using the letter symbol $L_{\text{A, 1H}}$ or abbreviated as 1HL. Table 9.2 describes this convention.

**9.2.4.1 *Adding Sound Levels When Noise Sources Are Independent*** Most industrial noises have random frequencies and phases and can be combined as described later. In the few cases when noises have significant pure tone components, these calculations are not accurate, and phase relationships must be considered. In areas where significant tones are present, standing waves often can be recognized by the presence of rapidly varying sound pressure levels over short distances. It is not practical to try to predict levels in areas where standing waves are present.

When the sound pressure levels of two pure tone sources are added, the resultant sound pressure level $L_p$ may be greater than, equal to, or less than the level of a single source (Figure 9.6). In most cases, however, the resultant $L_p$ is greater than either single source.

At zero phase difference, the resultant of two identical pure tone sources is 6 dB greater than either single level. At a phase difference of 90°, the resultant is 3 dB greater than either level. Between 90° and 0°, the resultant is somewhere between 3 and 6 dB greater than either level. At a phase difference of 120°, the resultant is equal to the individual levels; and between 120° and 90°, the resultant is between 0 and 3 dB greater than either level. At 180° there is complete cancelation of sound. The resultant $L_p(R)$ is greater than the individual levels for all phase differences from 0° and 120°, but less than individual levels for phase differences from 120° and 180°. Also, most tonal noises are not single tones but combinations of frequencies.

Thus, at almost all points in the noise field, the pressure levels exceed the individual levels, and overall there is a 3-dB average increase in sound pressures by adding two randomly phased sources. This allows industrial hygienists to estimate resultant sound pressure levels when noise sources are added or removed from industrial environments by combining sound levels as shown below.

**9.2.4.2 *Combining Sound Levels*** It is often necessary to combine sound levels – for example, to combine frequency band levels to obtain the overall or total sound pressure level from band levels within the overall noise. Another example is the estimation of total sound pressure level resulting from adding a machine of known noise spectrum to a noise environment of known characteristics. Simple addition of individual sound pressure levels, which are logarithmic quantities, constitutes multiplication of pressure ratios; therefore, the sound pressure corresponding to each sound pressure level must be determined and added.

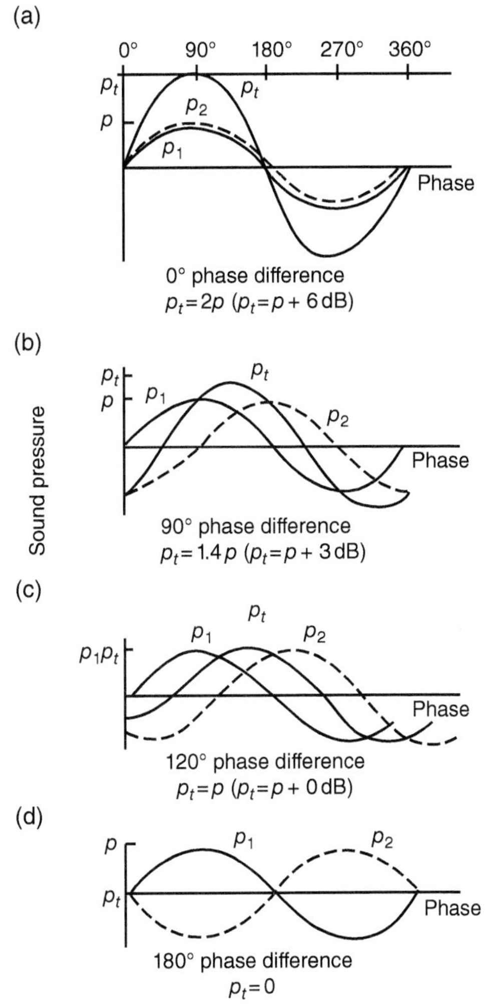


**Figure 9.6** (a–d) Combination of two pure tones of the same frequency ($p_1$ and $p_2$) with various phase differences.

For the most part, industrial noise is made up of sources having many frequencies (broad band), with nearly random phase relationships. Sound pressure levels of different random noise sources can be added by (i) converting the levels of each to pressure units, (ii) converting to intensity units that may be added arithmetically, (iii) then converting the resultant intensity to pressure, and finally (iv) converting the resultant pressure value to sound pressure levels in decibels:

$$L_{\text{total}} = 10\log\left(\sum_{i=1}^{N} 10^{L_i/10}\right) \tag{9.9}$$

***Example 9.3:*** If the octave band sound pressure levels measured for a random noise source are as shown in the table, what is the overall sound pressure level?

| Frequency (Hz) | 31.5 | 63 | 125 | 250 | 500 | 1000 | 2000 | 4000 | 8000 |
|---|---|---|---|---|---|---|---|---|---|
| $L_i$ (dB) | 85 | 88 | 91 | 94 | 95 | 100 | 97 | 90 | 88 |

$$L_{\text{total}} = 10\log\left(10^{85/10} + 10^{88/10} + 10^{91/10} + \cdots\right)$$
$$L_{\text{total}} = 10\log\left(10^{8.5} + 10^{8.8} + 10^{9.1} + \cdots\right) = 10\log\left(24.5\times 10^{10}\right) = 103.9\,\text{dB}$$

Acceptable accuracy is found in most cases, however, by using Table 9.3. Begin by adding the highest levels so that calculations may be discontinued when there is no significant difference when lower values are added to the total.

**TABLE 9.3 Simplified Table for Combining Decibel Levels of Noise with Random Frequency Characteristics**

| Numerical differences between levels | Amount to be added to the higher level |
|---|---|
| 0–1 | 3 |
| 2–4 | 2 |
| 5–9 | 1 |
| >10 | 0 |

***Example 9.4:*** Using Example 9.3 measurements, determine the overall sound pressure level using Table 9.3.

The difference between 100 and 97 dB is 3 dB; therefore, opposite the range 2–4 in the left-hand column of the table, read a value of 2 in the right-hand column, and add this value to the higher of the two levels: 100 + 2 = 102 dB. This resultant is now added to the next highest level by repeating the process: 102 − 95 = 7; from the table, read an amount to be added of 1. Thus 102 + 1 = 103 dB. The next highest level is 94, which is 9 dB lower than the total; so another 1 dB is added, giving 103 + 1 = 104 dB. All other readings are more than 10 dB lower than the total; so the procedure is halted, giving a final $L_{total}$ = 104 dB – acceptably close to the precise calculation.

## 9.3 THE EAR

The normal human ear responds to a remarkable frequency range, ranging from about 20 to 20000 Hz at common loudness levels (AIHA 1966, 1975, 1987; ANSI S3.6-1989 1989; Anticaglia 1973; Feldmen and Grimes 1985; Glorig 1958). The characteristics of any individual ear over this wide frequency range are extremely complex, and understanding the ear's capabilities is made even more difficult because of large differences among individuals. An ear's response characteristics may change as a result of physical or mental conditions, sound level, medications, environmental stresses, diseases, and other factors.

A normal healthy human ear also effectively transduces a remarkable range of sound pressures. It is sensitive to very low sound pressures that produce a displacement of the eardrum no greater than the diameter of a hydrogen molecule. At the other extreme, it can transduce sounds with sound pressures that are more than $10^6$ times greater than the ear's lower threshold value; however, exposure to high-level sounds may cause temporary or permanent damage to the ear.

The ear is divided into three sections (Figure 9.7): the outer ear, the middle ear, and the inner ear (Berger et al. 1986). Sound incident upon the ear travels through the ear canal to the eardrum, which separates the outer and middle ear sections. The combined alternating sound pressures that are incident upon the eardrum cause it to vibrate with the same relative characteristics as the sound source(s). The mechanical vibration of the eardrum is then coupled through the three bones of the middle ear to the oval window of the inner ear. The vibration of the oval window is then coupled to the fluid contained in the inner ear.

### 9.3.1 Outer Ear

The auricle, sometimes called the pinna, plays a significant role in the hearing process only at very high frequencies, where its size is large compared with that of a wavelength. The auricle helps direct these high-frequency sounds into the ear canal, and it assists the auditory system in localizing the sound source.

Ear canals have many sizes and shapes. They are seldom as straight as indicated in the figures, and the shape and size of ear canals differ significantly among individuals and can even differ between ears of the same individual. The average length of the ear canal is about 2.5 cm. When closed at one end by the eardrum, it has a quarter-wavelength resonance of about 3000 Hz. This resonance increases the response of the ear by about 10 dB at 3000 Hz – approximately the frequency at which the ear is most sensitive and also approximately the frequency at which the greatest noise-induced hearing loss occurs (see Figure 9.8).

The hairs at the outer end of the ear canal help to keep out dust and dirt; further into the canal are the wax-secreting glands. Normally, earwax flows toward the entrance of the ear canal, carrying with it the dust and dirt that accumulate in the canal. The flow of wax is normal and may be interrupted by changes in body chemistry that can cause the wax to become hard and to build up within the ear. Too much cleaning or the prolonged use of earplugs may cause increased production of wax or dryness and irritation in the ear canal. At times, the wax may build up to the point of occluding

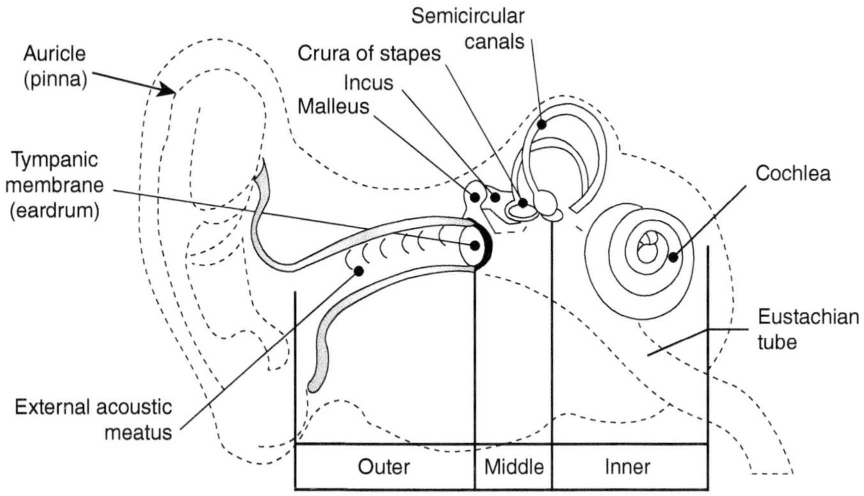


**Figure 9.7** Cross section of the ear showing the outer, middle, and inner ear configurations.

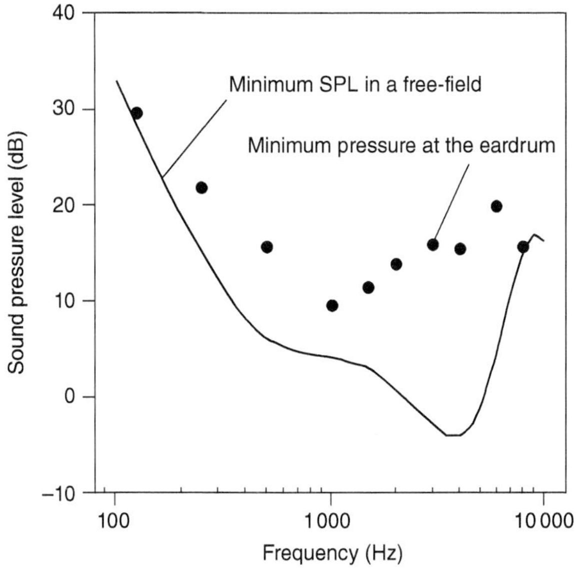


**Figure 9.8** Minimum audible sound pressure levels for young adults with normal hearing (Killion 1978). *Source:* Reproduced with permission of Acoustical Society of America.

the canal, and a conductive loss of hearing will result. Any buildup of wax deep within the ear canal should be removed very carefully, by a trained person, to prevent damage to the eardrum and middle ear structures.

During welding or grinding operations, a spark may enter the ear canal and burn the canal or a portion of the eardrum. Although very effective surgical procedures have been developed to repair or replace the eardrum, this painful and costly accident can be prevented by wearing hearing protectors.

The surface of the external ear canal is extremely delicate and easily irritated. Cleaning or scratching with matchsticks, nails, hairpins, and other objects can break the skin and cause a very painful and persistent infection. Infections can cause swelling of the canal walls and, occasionally, a loss of hearing if the canal swells shut. An infected ear should be given prompt attention by a physician.

### 9.3.2 Eardrum

The eardrum is a thin and delicate membrane that responds to the very small sound pressure changes at the lower threshold of normal hearing; yet it is seldom damaged by common continuous high-level noises. Although an eardrum may be damaged by an explosion or a rapid change in ambient pressure, the often repeated statement "the noise was so loud it almost burst my eardrums" is rarely true for common steady-state noise exposures.

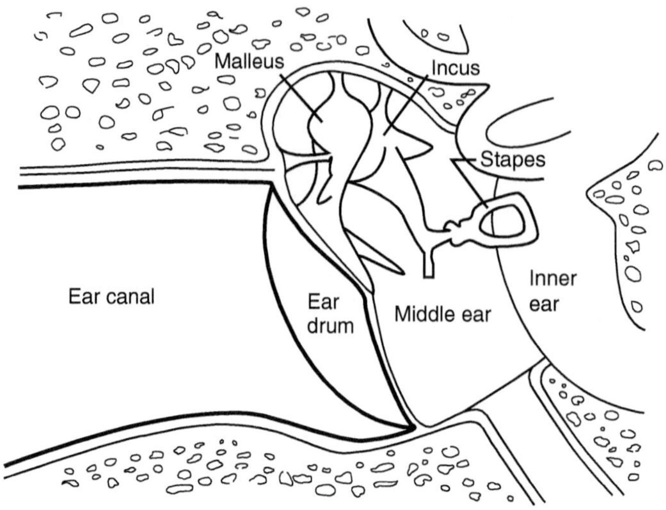


**Figure 9.9** The middle ear.

When an eardrum is ruptured, the attached middle ear bones (or ossicles) may be dislocated; thus the eardrum should be carefully examined immediately after the injury occurs to determine whether realignment of the ossicles is necessary. In a high percentage of cases, surgical procedures are successful in realigning dislocated ossicles, so little or no significant loss in hearing acuity results from this injury.

### 9.3.3 Middle Ear

The air-filled space between the eardrum and the inner ear is called the *middle ear* (Figure 9.9). The middle ear contains three small bones – the malleus (hammer), the incus (anvil), and the stapes (stirrup) – that mechanically connect the eardrum to the oval window of the inner ear.

The eardrum and middle ear act as a transformer to convert sound pressure waves in air to motion in the cochlear fluids ((Pickles 1988), p. 5). The eardrum has an area about 20 times that of the oval window, thereby providing a mechanical advantage of about 20:1. The ossicles provide an additional mechanical advantage.

The transmission through the middle ear is frequency dependent. Low-frequency waves are reduced by stiffness of the middle ear structures and compression of air in the middle ear cavity. The mass of the ossicles, less efficient modes of vibration of the eardrum, and inability of high-frequency sound waves to couple to the eardrum also reduce transmission at high frequencies. This complex system also acts as a hearing protector, since the involuntary relaxation of coupling efficiency between the ossicles can reduce the transmission of pressure from the eardrum to the oval window. The reaction time for this relaxation in the middle ear system is approximately 10 milliseconds.

The most common problem encountered in the middle ear is infection. This warm and humid air-filled space is completely enclosed except for the small eustachian tube that connects the middle ear to the back of the throat; thus it is susceptible to infection, particularly in children. If the eustachian tube is closed as a result of an infection or an allergy, the pressure inside the middle ear cannot be equalized with that of the surrounding atmosphere. In such an event, a significant change in atmospheric pressure, such as that encountered in an airplane or when driving in mountainous territory, may produce a loss of hearing sensitivity and extreme discomfort as a result of the inward displacement (or retraction) of the eardrum. Even a healthy ear may suffer a temporary loss of hearing sensitivity if the eustachian tube becomes blocked, but this loss of hearing can often be restored simply by swallowing or moving the jaw to open the eustachian tube momentarily.

### 9.3.4 Inner Ear

The inner ear is completely surrounded by bone (Figure 9.10). One end of the space inside the bony shell of the inner ear is shaped like a snail shell; it contains the cochlea. The other end of the inner ear has the shape of three semicircular loops. The fluid-filled cochlea serves to detect and analyze incoming sound signals and to translate them into nerve impulses that are transmitted to the brain. The semicircular canals contain sensors for balance and orientation.

In operation, sound energy is coupled into the inner ear by the stapes, whose base is coupled into the oval window of the cochlea. The oval window and the round window located below it are covered by thin elastic membranes to

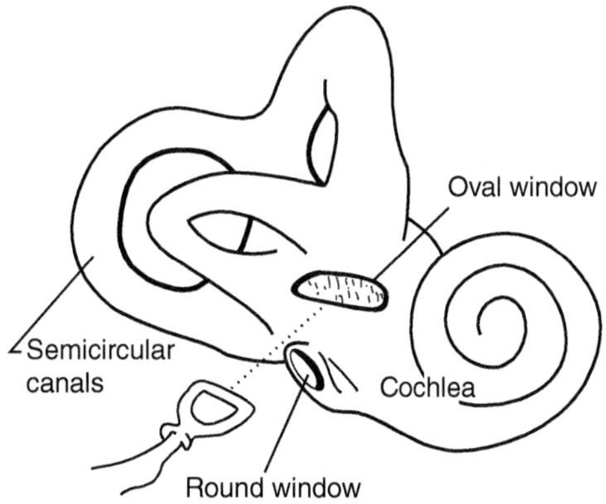


**Figure 9.10** The inner ear.

contain the fluid within the cochlea. As the stapes forces the oval window in and out with the dynamic characteristics of the incident sound, the fluid of the cochlea is moved with the same characteristic motions. Thousands of hair cells located along the two and one-half turns of the cochlea detect and analyze these motions and translate them into nerve impulses. The nerve impulses, in turn, are transmitted to the brain via the eighth nerve for analysis and interpretation.

The hair cells within the cochlea may be damaged by aging, disease, medication, blows to the head, and exposure to high levels of noise (ANSI S3.28-1986 1986). Unfortunately the characteristics of the hearing losses resulting from these various causes are often very similar, and therefore it may be difficult or impossible to determine the etiology of a particular case from an audiogram.

### 9.3.5 Conductive and Sensorineural Hearing Loss

Hearing loss can be either conductive or sensorineural. Abnormalities in the outer or middle ear create conductive losses, such as ossification of the middle ear bones, buildup of wax in the outer ear, and fluid in the middle ear. Conductive hearing loss is a result of any occurrence that changes the transfer of vibrations from the external ear to the oval window. Conductive losses frequently occur at or below 500 Hz. In many cases the loss is amenable to treatment by antibiotics or, in severe cases, by surgical intervention.

Another middle ear problem may result from an abnormal bone growth (otosclerosis) around the ossicles, restricting their normal movement. The cause of otosclerosis is not totally understood, but heredity is considered to be an important factor. The conductive type of hearing loss from otosclerosis is generally observed first at low frequencies, then extends to higher frequencies, and eventually may result in a severe loss in hearing sensitivity over a wide frequency range. Hearing aids may often restore hearing sensitivity lost as a result of otosclerosis, and effective surgical procedures have been refined to such a point that they are often recommended. An important side benefit of an effective hearing conservation program is the early detection of such hearing impairments as otosclerosis.

Hearing loss in the cochlea or auditory nerve is known as *sensorineural hearing loss*. Tumors are frequently the cause of hearing losses arising in the auditory nerve. Cochlear hearing losses can be caused by drugs or infections or may be congenital. In older age, senile changes produce a progressive incurable impairment (Pickles 1988, p. 297) known as *presbycusis*, but the major cause of sensorineural loss is from acoustic trauma or noise exposure. The sensitive hair cells of the cochlea are usually destroyed, no treatment for replacement is known, and hearing aids are frequently of limited efficacy.

### 9.3.6 How Noise Damages Hearing

Noise-induced hearing loss may be temporary or permanent depending on the level and frequency characteristics of the noise, the duration of exposures, and the susceptibility of the individual. A temporary loss of hearing sensitivity is usually restored within about 16 hours (ANSI S3.28-1986 1986); however, temporary losses may last for weeks in some cases. When temporary losses do not recover completely before other significant exposures, permanent losses may be produced. Permanent losses (noise-induced permanent threshold shifts [NIPTS]) are irreversible and cannot be corrected by conventional surgical or therapeutic procedures.

Noise-induced damage generally occurs in hair cells located within the cochlea. For common broadband noise exposures, hearing acuity is generally affected first in the frequency range from 3000 to 6000 Hz, with most affected

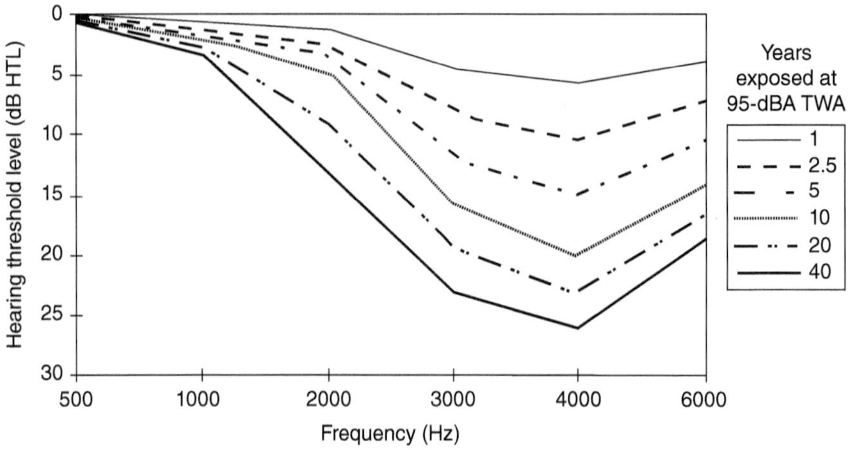


**Figure 9.11** Progression of noise-induced permanent threshold shift from ANSI S3.44-1996.

persons showing a loss or "dip" at 4000 Hz. For noise exposures having significant components concentrated in narrow frequency bands below 4000 Hz, impairments usually are found about one-half to one octave above the predominant exposure frequencies. If high-level exposures are continued, the loss of hearing generally increases around 4000 Hz and spreads to higher and lower frequencies (ANSI S3.44 2006) (see Figure 9.11).

Noise-induced hearing loss is an insidious problem because a person does not necessarily experience pain before learning that severe hearing damage has taken place. The damage may occur instantaneously, or over a long period of time, depending upon the noise characteristics. Generally, impulsive or impact noises are most likely to produce significant losses with short exposure periods, and steady-state continuous noises are responsible for impairments that develop over a long period of time.

Even after a significant amount of damage, a person with noise-induced hearing loss is able to hear common low-frequency (vowel) sounds, but the high frequencies (consonants) in speech are not heard as well. Perceived loudness levels may be nearly normal, but intelligibility will be poor in some situations. A noise-induced hearing loss becomes noticeable when speech communication is attempted in noisy, reverberant areas. Speech is masked most effectively by background noises having major frequency components in the speech frequencies, such as those found in areas where many people are talking in the background, sometimes called the "cocktail party" effect.

### 9.3.7 Occupational Noise-Induced Hearing Loss

Because of differences in the susceptibility of individuals to noise-induced hearing impairment, studies leading to the development of comprehensive damage risk criteria require complex statistical studies of large groups over long periods of time. It is therefore impossible to set a single exposure level as a dividing line between safe and unsafe conditions that would apply for all individuals. Damage risk criteria are also compromised by practical limits. Usually, exposure level limits are established as a compromise between (NIOSH 1972) the amount of hearing impairment that may result from a specified exposure dose and (U.S. Department of Labor 1969) the economic or other impact that may result from noise control expenditures.

The best estimates of the number of persons who have significant hearing impairment as a result of overexposure to noise are based on a comparison of the number of those with hearing impairments found in high-noise work areas with members of the general population, who have relatively low noise exposures (Lawther and Robinson 1987). These studies show that significant hearing impairments for industrial populations are 10–30% greater for all ages than for general nonexposed populations. At age 55, for example, 22% of a group that has had low noise exposures may show significant hearing impairment, whereas in an industrial high-noise-exposure group, the figure is 46%. In this study, significant hearing loss was defined as greater than 25-dB hearing level (HL).

Hearing impairment for an exposed population can be determined by adding the hearing threshold level associated with age (HTLA) to the NIPTS to calculate what is called the hearing threshold level associated with age and noise (HTLAN). Figure 9.11 shows the median noise-induced hearing loss expected for a group of workers exposed to $L_{Aeq} = 95$ dB for a number of years. In addition to their noise-induced threshold shifts, there will also be threshold shifts from presbycusis. Figure 9.12 shows the expected median threshold shift (and 10 and 90% ranges) for 60-year-old male workers exposed to $L_{Aeq} = 95$ dB for 40 years.

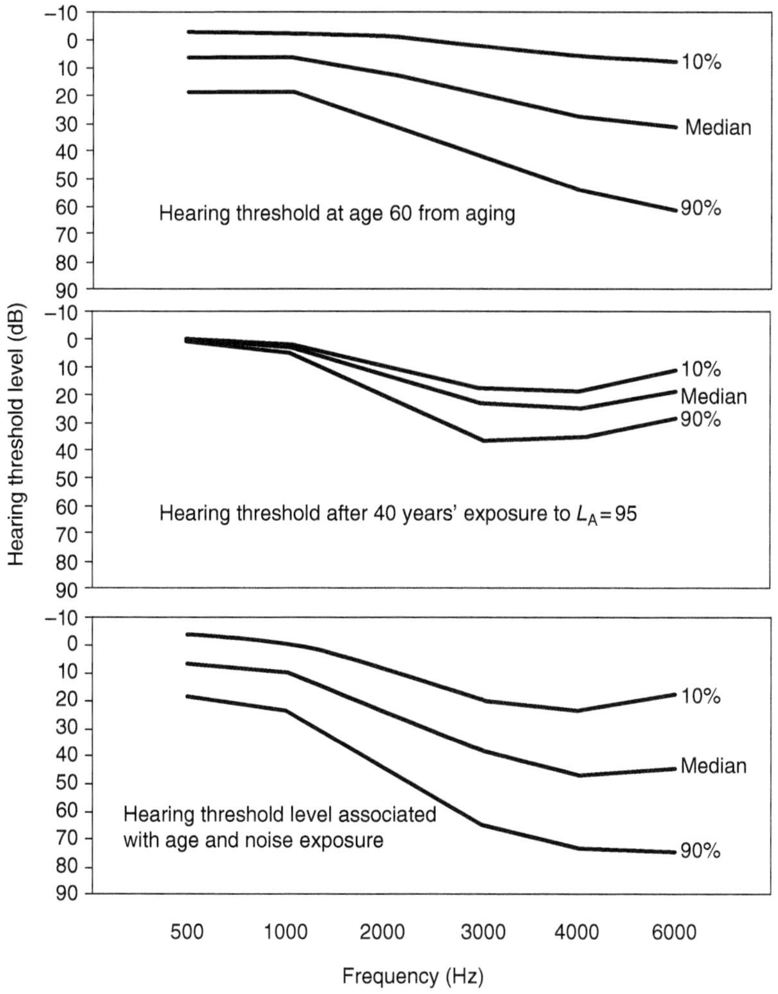


**Figure 9.12** Hearing threshold shifts (median, 10 percentile, and 90 percentile) for male workers exposed to $L_{eq} = 95$ for 40 years.

### 9.3.8 Non-occupational Hearing Loss

Most persons in our society are exposed to potentially hazardous noise away from work. These "off-the-job" exposures must be considered when studying noise-induced hearing loss. The magnitude of noise exposures varies significantly from one location to another and from one time to another. Potentially hazardous exposures away from work may result from noises from guns, chainsaws, airplanes, lawn mowers, motorboats, motorcycles, automobile or motorcycle races, farm equipment, loud music, and shop tools. Even riding in a car at legal speeds with the window down or riding in a subway may be harmful to some noise-sensitive individuals. Again, these variances must be considered in hearing loss analyses. There is a great potential for inaccuracy in studies that consider only on-the-job exposures at one work site with an assumed exposure duration.

A hearing conservation program cannot be effective unless each individual observes the rules limiting noise exposure both at work and away from work. All hearing conservation programs must emphasize the need for constant awareness of noise hazards through a continuing program of education and enforcement.

### 9.3.9 Nonauditory Physiologic Effects of Noise

Clearly, the most documented effect of noise on man is hearing loss. Other physiologic changes occur with exposure to high levels of noise in the cardiovascular, neurological, and metabolic systems. Effects have also been reported in the reproductive system and in vision and physical orientation (Miller 1974).

The vegetative nervous system is responsible for many uncontrollable reactions to both sudden and continuous high-level noise. Cardiovascular reactions include constriction of the blood vessels, an increase in diastolic blood pressure, and an increase in adrenalin level in the bloodstream. Typically, the constriction of the blood vessels diminishes over a period of about 25 minutes if the noise is discontinued or maintained at a relatively constant level. Pulse rate may increase or decrease with exposure to noise. Increased hormonal activity in humans has also been documented, with potential negative long-term effects, including tendencies toward degenerative arterial and myocardial tissues.

Neural impulses are generated in response to noise, and these impulses spread throughout the gray and white matter of the brain. It is likely that these impulses cause some of the unrelated reactions to noise exposure by affecting areas of the brain other than areas related to pure auditory function. Sudden noise causes voluntary and involuntary muscles to contract. The eyes blink, pupil size changes, facial muscles contract, and the knees bend. Familiarization with sound may or may not have an effect on the startle responses, depending on the individual.

Many of the effects associated with noise have been documented in humans from exposures to other stressors, such as bright light. Continued exposure to any stressor can lead to maladies known as diseases of adaptation. These are highly dependent on the individual and may include gastrointestinal disorders, high blood pressure, and even arthritis.

## 9.4 HEARING MEASUREMENT

The only way to monitor the overall effectiveness of a hearing conservation program is to periodically check the hearing of all persons exposed to potentially hazardous noises. Air conduction hearing thresholds must be checked at 500, 1000, 2000, 3000, 4000, and 6000 Hz. It is recommended that testing also be performed at 8000 Hz, since data at this frequency will help discriminate between noise-induced and presbycusic hearing loss.

To assure the accuracy of air conduction thresholds, audiometers must be calibrated periodically, and a quiet test environment must be maintained (ANSI S3.21 n.d.; U.S. Department of Labor 1971). A well-trained audiometric technician or hearing conservationist must be used to perform hearing threshold measurements (CAOHC 1978). Requirements for audiometer performance and for the background noise limits in test rooms have been specified in standards published by the American National Standards Institute (ANSI) (ANSI S3.1 1999; ANSI S3.6 2010; Occupational Safety and Health Administration 1983). Guidelines for training hearing conservationists have been established by the Council for Accreditation in Occupational Hearing Conservation (CAOHC 1978).

### 9.4.1 Audiometers

The audiometer is used for measuring pure tone air conduction hearing thresholds. The audiometer may be designed for manual, self-recording, or automatic operation.

The ANSI standard S3.6-1996, "Specifications for Audiometers," provides the requirements that the instruments must meet to provide accurate information. Instrumentation inaccuracies are seldom obvious, and there is a strong tendency to accept readings as being accurate. Audiometers may lose their specified accuracy very quickly if they are handled roughly. Earphones are particularly susceptible to damage from rough handling; a janitor may drop them on the floor while cleaning and not inform the person in charge of testing. Rough handling during shipment could cause unseen damage so that even a new instrument may be out of calibration when received.

Dust or dirt inside the audiometer can cause switches to produce electrical noise or to wear excessively. Poor electrical contacts may produce intermittent operation or poor accuracy. Dust covers should always be used to protect the instrument when not in use, and the exterior of the instrument should be cleaned periodically to prevent dust and dirt from getting inside the case.

High humidity, salt air, and acid fumes may corrode electrical contacts in switches within an audiometer. The increased resistance that results from corroded contacts may cause electrical noise or affect the accuracy of the instrument.

If the operating characteristics of the audiometer change significantly, the instrument must be serviced. Many changes in the instrument occur slowly, however, and may not be noticed; so instruments must be calibrated periodically. OSHA requires that biological calibrations be performed before each day's use and that acoustic calibrations be made at least annually (U.S. Department of Labor 1971). Accuracy checks should also be made any time there are any reasons to suspect a problem.

Biologic calibrations are audiograms taken on persons with known, stable hearing thresholds. These tests are typically performed at the beginning of the day before any evaluations are performed. Because hearing thresholds may increase as much as 20 dB temporarily because of allergies, colds, or other causes; it is strongly recommended that at least two normal-hearing persons be made available as biological test subjects. Relatively inexpensive artificial test heads are also available for daily biologic testing.

Audiometers that have not been calibrated for several years may not meet the ANSI standard specifications. Furthermore, differences of several decibels may be found in threshold measurement data taken with two audiometers that meet opposite extremes of the allowed calibration limits. Thus the need for dependable calibration services that will produce accurate adjustments and correction data is emphasized. The pertinent performance specifications may not be well understood by some of the laboratories offering calibration services. Therefore, a statement that the audiometer meets specifications should not be accepted without an explanation of the specific calibration procedures used and a copy of the calibration data.

### 9.4.2 Test Rooms

The sound pressure level of the background noise in rooms used for measuring hearing thresholds must be limited to prevent masking effects that cause misleading, elevated threshold values. The maximum allowable sound pressure levels for industrial audiometric test rooms specified by the OSHA (Occupational Safety and Health Administration 1983) may mask some thresholds that are better (lower hearing levels) than above 10-dB HL; so quieter test areas are recommended.

More specifically, the maximum allowable sound pressure levels in ANSI S3.1-1991 are recommended in order to measure 0-dB HL with no more than 1-dB threshold elevation for one-ear listening. These background noise levels are difficult to obtain in noisy areas and may be below the practical limit for room noise in many industrial locations. Thus it has been the accepted practice for some industries to use the 10-dB HL as the lowest hearing threshold measurement level (Occupational Safety and Health Administration 1983). If the lowest hearing level to be measured is 10-dB HL, the background noise limits shown in Table 9.4 can be adjusted upward by 10 dB.

In addition to the noise level requirements, subjective tests should be made inside the closed test booth on location to determine that no noises (e.g. talking and heel clicking) are audible. Any audible noise may distract the subject and interfere with the hearing threshold measurements. These noises must be eliminated, or the tests must be delayed until no noise can be heard. Short impulse-type noises, in particular, may be heard even though the measured sound pressure levels are below the limits in the table.

In areas where ambient noise in audiometric test rooms is of particular concern, the hearing conservationist may wish to consider continuous monitoring of background noise level. Devices are available that continuously sample the ambient noise and compare it to preset standard levels. A visual indication is provided if the noise sample exceeds the preset levels. If background noise level is measured once per day, for example, the activation of a machine or process that produces excessive noise in the booth may not be recognized. Continuous monitoring of background noise alerts the technician of this problem.

Factors to consider for audiometric test booths include the following:

1. What are the size and appearance?
2. Will opening and closing the door result in wearing of contact material (seals), necessitating frequent replacement?
3. Are the interior surfaces durable and easily cleaned?
4. Is the door easily opened from the inside, so that subjects will not feel "locked in?"
5. The observation window and seating arrangement must provide an easy view of the subject, but the subject must not be able to see the technician or audiologist.
6. If the booth is equipped with a ventilation system, the noise should be below the limits specified for the room.
7. Portability of the room is seldom an important factor, because test rooms are rarely moved.

### 9.4.3 Hearing Threshold Measurements

Normally, the purpose of hearing testing in industrial hearing conservation programs is to monitor the effectiveness of hearing conservation procedures. If this is the case, there is no need for diagnostic information that would require the use of sophisticated audiometric techniques. Pure tone air conduction hearing thresholds are usually adequate for industrial monitoring purposes (CAOHC 1978).

**TABLE 9.4 Maximum Permissible Ambient Sound Pressure Level for Audiometric Testing with Ears Covered**

| Octave band center frequency (Hz) | 125 | 250 | 500 | 1000 | 2000 | 4000 | 8000 |
|---|---|---|---|---|---|---|---|
| $L_{max}$ (dB SPL re 20 μPa) | 34 | 9.5 | 19.5 | 26.5 | 28 | 34.5 | 43.5 |

*Source:* ANSI S3.1 (1999).

Properly calibrated manual, self-recording, or automatic audiometers may be used to monitor hearing thresholds in industry. Self-recording audiometers print or store a range of hearing threshold levels between the barely perceptible sound pressure level and not perceptible sound pressure level. The range is frequently 10 dB wide, and the audiometric technician usually chooses the midpoint of the range as the subject's hearing threshold.

Microprocessor-based automatic audiometers are commonly in use in industry and in clinical environments. These units measure hearing threshold by presenting the patient with one or a series of tones and adjusting the level of the tones according to the patient responses. A common method of defining a hearing threshold with an automatic audiometer (and also via manual audiometry) is referred to as the modified Hughson–Westlake method. In this procedure, the stimulus is presented to the patient at an audible level and then decreased in 10- or 15-dB steps until it is no longer audible. After the tone is inaudible, the level is increased in 5-dB steps until the patient responds. After a positive response, the tone is lowered by 10 dB and testing resumes, with the level again raised in 5-dB steps until the patient gives a positive response. This is referred to as an ascending audiometric technique. The threshold is then defined as the level at which the patient responds to 50% of the stimuli with a minimum of two responses at a single level.

Hearing the thresholds measured by a manual or automatic audiometer are typically 2–3 dB higher than those measured with self-recording audiometers because of the differences in data analysis. For precision, and to assure consistency, it is important that the method of threshold determination remain constant throughout the hearing conservation program.

### 9.4.4 Records

The hearing conservationist's records must be complete and accurate if they are to have medicolegal significance. Records should be kept in ink, without erasures. Employee signatures should be required.

If a mistake is made in recording, a line should be drawn through the erroneous entry, and the initials of the person making the recording should be placed above the line along with the date. The entry must not be erased.

The model, serial number, and calibration dates of the instruments should be recorded on each audiogram. Records should also be kept of periodic noise level measurements in the test space.

In addition to the threshold levels, an audiogram should have space provided for the recording of pertinent medical and noise exposure information. Some typical questions are listed in Table 9.5.

### 9.4.5 Audiometric Database Analysis

The occupational noise exposure regulations require that annual audiograms of all persons included in the hearing conservation program (Section 9.7) are performed and that appropriate records are kept. A convenient measure of the effectiveness of a hearing conservation program is to compare these records with those of a non-noise-exposed population. The number of standard threshold shifts (STSs) in the industrial population is counted and compared against the number of STSs that may be recorded in the control population. An STS is defined in 29 CFR 1910.95 as a change in hearing threshold by an average of 10 dB or more at 2000, 3000, and 4000 Hz in either ear. This calculation is made after making the correction for presbycusis. Obviously, if the number of STSs in industry compares favorably with the control population, then the hearing conservation program is successful. This would indicate that either (i) the noise exposure levels are not too high, (ii) the engineering noise control activities are successful, or (iii) the use of personal hearing protection is successful, or most likely (iv) a combination of all three. ANSI S12.20, "Method for Evaluating the Effectiveness of a Hearing Conservation Program," includes additional information on this topic.

## 9.5 NOISE MEASUREMENT

Since the human ear is a pressure-sensitive device, measurements of sound pressure levels are usually sufficient to determine the hazard potential of the noise. Two broad categories of measurements are:

- Measuring noise from a specific source.
- Measuring noise to characterize a certain environment.

Personal noise exposure measurements for compliance or hearing conservation purposes would fall into the latter category. Simple sound level instruments can be used to estimate the potential risk of high noise levels. More sophisticated measurements are taken when the objective is to obtain data on which engineering changes are to be made. ANSI (ANSI S1.4 2013) and International Organization for Standardization (ISO) standards specify the accuracy of

**TABLE 9.5 Noise Exposure and Medical Information for Audiograms**

Name ______________ Date: _______ Recorded by: ___________
Employee signature __________________

| History | Yes | No | Comments |
|---|---|---|---|
| Have you had a previous hearing test? (When) | | | |
| Have you EVER had trouble hearing? | | | |
| Do you NOW have trouble hearing? | | | |
| Have you ever worked in a noisy industry? | | | |
| Do you think you can heard better in your:<br>Right ear<br>Left ear | | | |
| Have you ever heard noises in your ears? | | | |
| Have you ever had dizziness? | | | |
| Have you ever had a head injury? | | | |
| Has anyone in your family lost hearing before age 50? | | | |
| Have you ever had measles, mumps, or scarlet fever? | | | |
| Do you have any allergies? | | | |
| Are you now taking or ever regularly taken drugs, antibiotics, or medication? | | | |
| Have you ever had an earache? | | | |
| Have your ears ever run?<br>Right ear<br>Left ear | | | |
| Have you ever been in the military service?<br>Describe – especially experience with firearms. | | | |
| Have you ever been exposed to any sort of gunfire?<br>Describe. | | | |
| Do you have a second job? | | | |
| What are your hobbies? | | | |

various sound measuring equipment. Qualities range from laboratory grade (type 0), precision (type 1) to general purpose (type 2), and special purpose (type S). OSHA has specified the instrument used for industrial noise measurements to be type 2 or better.

### 9.5.1 The Sound Level Meter

The basic sound level meter consists of a microphone that converts the pressure variations into an electrical signal, an amplifier with frequency weighting, an exponential-time-averaging device, a device to determine the logarithm of the signal, and either a display or other means of storing the result in dB re 20 μPa (see Figure 9.13).

Since sound pressures are rapidly fluctuating pressure levels above and below atmospheric pressure, the average pressure is simply atmospheric pressure. Sound measuring instruments measure the pressure levels as the exponential-time-averaged square root pressure (voltage) signal multiplied by itself (voltage$^2$). Modern sound measuring instruments cover a dynamic range of more than 40–140 dB. More expensive equipment permits measurements in octave bands from about 0 dB to well above 160 dB, depending on the microphone selected.

Tolerance limits for type 2 sound level meters are less than a half-decibel change in reading for: (i) 10% change in static pressure, (ii) a 20 °C change in ambient temperature over –10 to 50 °C, (iii) a change in relative humidity between 30 and 90%, or (iv) any change within one hour of operation. Modern sound level meters are capable of repeatable, accurate, and precise measurements.

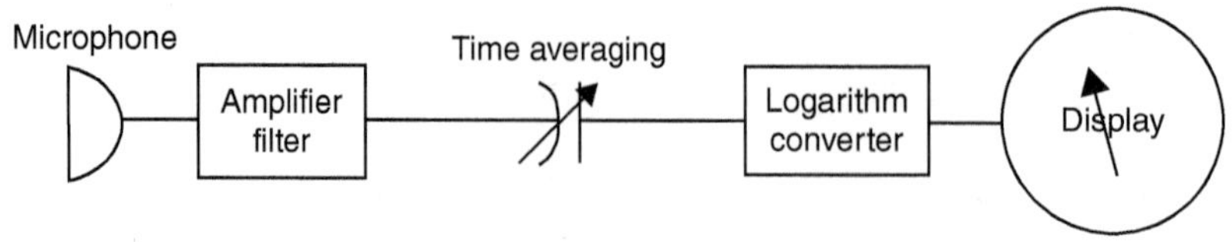


**Figure 9.13** Components of a basic sound level meter.

***9.5.1.1 Microphones and Directional Characteristics*** Most noises encountered in industry are produced from many different noise sources combined with their reflected energies. However, microphones may also be used outdoors, where conditions are closer to free field, and may be used in the near field close to the noise source. Depending on the design and purpose of a microphone, it may be calibrated for grazing, perpendicular, or random incidence sounds, or it may be calibrated for use in couplers (pressure calibration). Care must be taken to use the microphone in the manner specified by the manufacturer.

A microphone calibrated with randomly incident sound should be pointed at an angle to the major noise source that is specified by the manufacturer. An angle of about 70° from the axis of the microphone is often used to produce characteristics similar to randomly incident waves, but the angle for each microphone should be supplied by the manufacturer.

A free-field microphone is calibrated to measure sounds perpendicularly incident to the microphone diaphragm; thus it should be pointed directly at the source to be measured. A pressure-type microphone is designed for use in a coupler such as those used for calibrating audiometers; however, this microphone can be used to measure noise over most of the audible spectrum if the noise propagation is at grazing incidence (90° to the diaphragm).

Microphones used with sound measuring equipment are often said to be nearly omnidirectional, but, strictly speaking, this is generally true only for the lower frequencies. Above 1000 Hz, 1-in.-diameter microphones (above 5000 Hz for ½-in.-diameter microphones) may respond differently depending on the orientation of the microphone to the sound. When measurements are to be made of high-frequency noise produced by a directional noise source (i.e. where a high percentage of the noise energy is coming from one direction), the orientation of these microphones becomes important, or smaller-diameter microphones should be used. The tradeoff is that smaller-diameter microphones are not as sensitive as larger microphones and thus are not able to measure low sound pressure levels. For measuring levels of concern for OSHA compliance or hearing conservation, 1/2-in.-, 1/4-in.-, and even 1/8-in.-diameter microphones are capable of handling sound pressure level measurements above 70 dB.

When measuring sound pressure, it is best to position the microphone several wavelengths away from any surface that may reflect sound. In particular, this includes the body of the person conducting the measurement. For most accurate measurements, the microphone or sound level meter is mounted on a tripod with the person making the measurements standing several feet away. When measuring exposure of a machine operator, accurate measurements may be made by asking the operator to step aside and placing the microphone where the center of the operator's head is while standing in the normal operating position. However, measurements of adequate precision can usually be made by placing the microphone within 0.5–1 m of the machine operator's ear.

***9.5.1.2 Frequency-Weighting Networks*** General-purpose sound level meters are normally equipped with frequency-weighting networks that are used to characterize the frequency distribution of noise over the audible spectrum (Michael 1955). Frequency weightings, given in Table 9.6, were chosen because (i) they approximate the response characteristics of the ear at various sound levels and (ii) they can be easily produced with a few common electronic components. A linear, flat, or overall response, also included on many sound level meters, weights all frequencies equally.

The *A*-frequency weighting approximates the response characteristics of the ear for low-level sounds (below $L_A = 55$ dB). *C*-frequency weighting corresponds to the response of the ear for levels above $L_A = 85$ dB and may be used to estimate loudness when significant low-frequency sounds are present. *B*-frequency weighting, approximating the response of the ear for levels of $55 < L_A < 85$ dB, is still listed as part of ANSI S1.4, but is not commonly used today. For measuring annoyance of aircraft and other environmental noise, *D*-weighting (ANSI S1.42 2001) has been developed.

The frequency response at any frequency is specified relative to the meter's response at 1000 Hz. Positive and negative values indicate that the specific frequency contributes more or less to the overall level than a 1000-Hz tone. The values for octave and third octave frequencies are given numerically in Table 9.6 and graphically in Figure 9.14.

In use, the frequency distribution of noise energy can be approximated by comparing the levels measured with each of the frequency weightings. For example, if the noise levels measured using the *A* and *C* networks are approximately equal, it can be reasoned that most of the noise energy is above 1000 Hz, because this is the only portion of the

**TABLE 9.6 Random Incidence Relative Response as a Function of Frequency (ANSI S1.4 2013; ANSI S1.42 2001)**

| Band number | Nominal frequency (Hz) | *A*-weighting (dB) | *C*-weighting (dB) | *D*-weighting (dB) |
|---|---|---|---|---|
| 14 | 25 | −44.7 | −4.4 | −18.7 |
| 15 | 31.5 | −39.4 | −3.0 | −16.7 |
| 16 | 40 | −34.6 | −2.0 | −14.7 |
| 17 | 50 | −30.2 | −1.3 | −12.8 |
| 18 | 63 | −26.2 | −0.8 | −10.9 |
| 19 | 80 | −9.5 | −0.5 | −9.0 |
| 20 | 100 | −19.1 | −0.3 | −7.2 |
| 22 | 125 | −16.1 | −0.2 | −5.5 |
| 22 | 160 | −13.4 | −0.1 | −4.0 |
| 23 | 200 | −10.9 | 0 | −2.65 |
| 24 | 250 | −8.6 | 0 | −1.6 |
| 25 | 315 | −6.6 | 0 | −0.8 |
| 26 | 400 | −4.8 | 0 | −0.4 |
| 27 | 500 | −3.2 | 0 | −0.3 |
| 28 | 630 | −1.9 | 0 | −0.5 |
| 29 | 800 | −0.8 | 0 | −0.6 |
| 30 | 1000 | 0 | 0 | 0 |
| 31 | 1250 | +0.6 | 0 | +2.0 |
| 32 | 1600 | +1.0 | −0.1 | +4.9 |
| 33 | 2000 | +1.2 | −0.2 | +7.9 |
| 34 | 2500 | +1.3 | −0.3 | +10.4 |
| 35 | 3150 | +1.2 | −0.5 | +11.6 |
| 36 | 4000 | +1.0 | −0.8 | +11.1 |
| 37 | 5000 | +0.5 | −1.3 | +9.6 |
| 38 | 6300 | −0.1 | −2.0 | +7.6 |
| 39 | 8000 | −1.1 | −3.0 | +5.5 |
| 40 | 10000 | −2.5 | −4.4 | +3.4 |

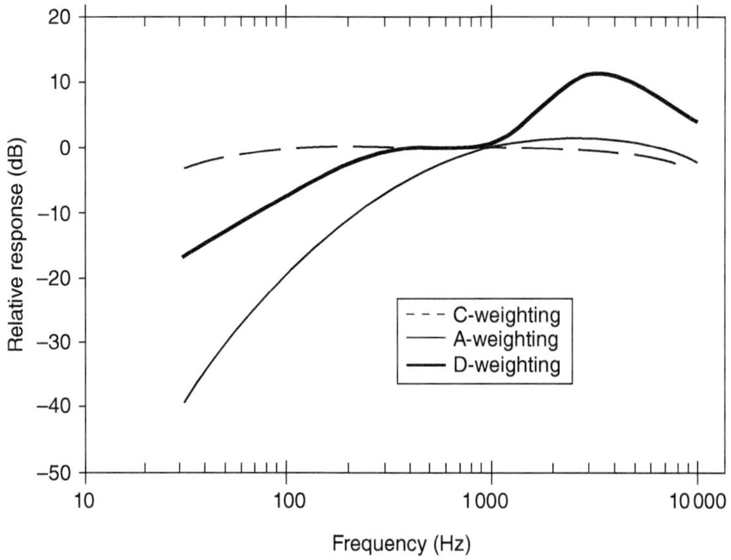


**Figure 9.14** Relative response of *A*-, *C*-, and *D*-frequency weighting filters.

spectrum in which the weightings are similar. On the other hand, a large difference between these readings indicates that most of the energy will be found below 1000 Hz.

For determining compliance with OSHA standards and to estimate the risk of noise-induced hearing loss, measurements using *A*-weightings and slow response are required. *A*-weighting has become the default for determining noise exposures for continuous, intermittent, or varying noises and is used for most work except impulsive sounds and some annoyance measurements.

**9.5.1.3 *Exponential Time Averaging*** Before the advent of digital displays, sound measuring instruments relied on galvanometers for displaying sound levels. These indicating meters had ballistic characteristics that were not constant over their entire dynamic range, often resulting in different readings, depending on the attenuator setting and the portion of the meter scale used, requiring the reader to adjust attenuators so the meter reads in a particular portion of the scale. More modem meters with electronic displays do not suffer from this limitation.

"Slow," "fast," and "impulse" time constants are specified for sound level meters. The *slow* response (one second time constant) is intended to provide an averaging effect that will make widely fluctuating sound levels easier to read; however, this setting will not provide accurate readings if the sound levels change significantly in less than 0.5 second. The *fast* response (0.125 second time constant) enables the meter to reach within 1 dB of its calibrated reading for a 0.2 second pulse of 1000 Hz; thus it can be used to measure, with reasonable accuracy, noise levels that do not change substantially in periods less than 0.2 second.

*Impulse* time averaging introduces a peak detector into the circuitry. For sound pressures that increase with time, the time constant is 0.035 second. When the sound pressure decreases, the peak detector introduces a 1.5 second decay time. Impulse exponential time averaging was introduced on earlier generations of sound level meters, so the maximum sound pressures of varying sounds could be measured. However, because of the asymmetric nature of the rise and decay times, it is not possible to calculate integrated exposures from recordings of impulse measurements. Today, many sound level meters are built with the capability of measuring the instantaneous *peak* sound level. They generally are measured with the *A*-frequency-weighting circuitry giving the equivalent of a 30-millisecond exponential time weighting.

Figure 9.15 shows fluctuations in sound pressures for a speech signal with slow, fast, and impulse time-averaged sound pressure levels. Notice how the sound pressure level changes by more than 15 dB for a fast time constant measurement. Without time averaging the fluctuations in sound pressure levels would be too fast to measure accurately. Table 9.7 shows differences between fast, impulse, and peak time-averaged levels for some commonly encountered sounds.

## 9.5.2 Sound Exposures and Long-term Average Sound Levels

Because of the large variations of typical sound levels as noted above, and because of the desire to characterize exposures with a single number, it is necessary to develop a long-term (typically eight-hour) sound level descriptor. The equivalent continuous sound pressure *level* of a time-varying sound is equal to the level of an equivalent steady

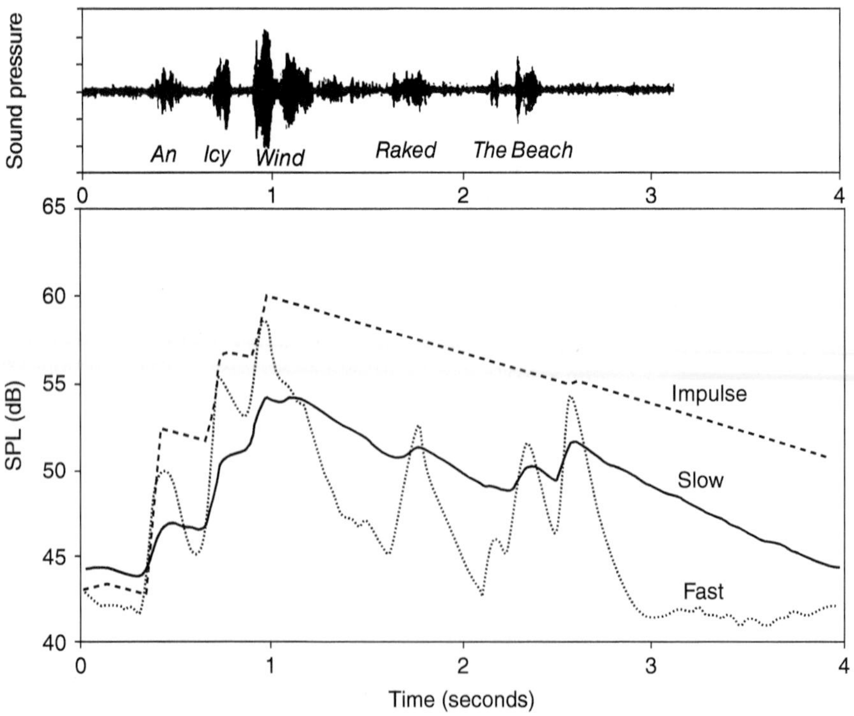


**Figure 9.15** Example of sound pressures from speech and slow, fast, and impulse time averaging.

**TABLE 9.7 Examples of Typical Sound Levels Taken with Different Exponential Time Weightings**

| Sound source | $L_{AF}$ | $L_{AI}$ | $L_{APk}$ |
|---|---|---|---|
| Sinusoidal pure tone at 1000 Hz | 94 | 94 | 97 |
| Highway traffic 15 m distance | 80 | 80 | 89 |
| Train 18 m distance | 85 | 87 | 94 |
| Noise in car 80 km h$^{-1}$ | 75 | 75 | 86 |
| Lawn mower 1 m distance | 97 | 99 | 116 |
| Diesel motor in electricity generating plant | 100 | 101 | 113 |
| Pneumatic nailing machine near operator's head | 116 | 120 | 148 |
| Air compressor room | 92 | 92 | 104 |
| Large machine shop | 81 | 82 | 98 |
| Large punch press near operator's head | 93 | 98 | 122 |
| Small automatic punch press | 100 | 103 | 118 |
| Small high-speed drill | 98 | 101 | 109 |
| Machine saw near operator's head | 102 | 102 | 113 |
| Vacuum cleaner 1.2 m distance | 85 | 88 | 105 |
| Bottling machinery in brewery | 98 | 99 | 122 |
| Pistol, 9 mm, 5 m distance from side | 111 | 114 | 146 |
| Shotgun, 5 m distance from side | 106 | 110 | 139 |

*Source:* Bruel (1977).

sound *pressure* that has the same acoustic energy as the time-varying sound actually measured. The symbol is $L_{eq}$ or $L_{Aeq}$, and the abbreviation is TEQ. Mathematically

$$L_{eq} = 10\log\left(\frac{1}{(t_2 - t_1)}\int_{t_1}^{t_2}\frac{p_A^2(t)\,dt}{p_r^2}\right) \tag{9.10}$$

where $p_A^2(t)$ is the square of the $A$-frequency-weighted sound pressure, $(t_2 - t_1)$ is the duration measured from time $t_1$ to $t_2$, and $p_r$ is the standard reference sound pressure of 20 μPa.

Note the difference between this and the response time of sound level meters, i.e. exponential-time-weighting equivalent continuous sound levels have uniform time weighting. All sounds occurring *at any time* during the measurement period are included with equal weighting. Exponential time weighting yields sound levels primarily influenced by the most recently occurring sounds.

***Example 9.5:*** What is $L_{eq}$ for an eight-hour workday consisting of four exposures as shown in the table?

| $L_{AS}$ (dB) | Duration (h) |
|---|---|
| 82 | 6.0 |
| 88 | 1.4 |
| 97 | 0.5 |
| 105 | 0.1 |

Converting levels to sound pressure

$$L = 10\log\left(\frac{p}{p_r}\right)^2$$

$$\frac{L}{10} - \log\left(\frac{p}{p_r}\right)^2$$

$$10^{L/10} = \left(\frac{p}{p_r}\right)^2$$

and hours to seconds gives the following table:

| $L_{AS}$ (dB) | $(p/p_r)^2$ | Duration (h) | Duration (s) | $(p_A/p_r)^2 dt$ |
|---|---|---|---|---|
| 82 | $1.58 \times 10^8$ | 6.0 | 21 600 | $3.42 \times 10^{12}$ |
| 88 | $6.31 \times 10^8$ | 1.4 | 5040 | $3.18 \times 10^{12}$ |
| 97 | $5.01 \times 10^9$ | 0.5 | 1800 | $9.02 \times 10^{12}$ |
| 105 | $3.16 \times 10^{10}$ | 0.1 | 360 | $1.14 \times 10^{13}$ |
| | | Total | 28 800 | $2.70 \times 10^{13}$ |

$$L_{\text{Aeq,8h}} = 10\log\left(\frac{270\times 10^{13}}{2.88\times 10^{4}}\right) = 90\,\text{dB}$$

Integrating sound level meters calculate $L_{eq}$ automatically; so the industrial hygienist is generally spared the necessity of making these calculations.

### 9.5.3 Frequency Analyzers

For noise control purposes, the rough estimate of frequency-response characteristics provided by the weighting networks of a sound level meter is not always adequate. In such cases frequency analyzers are used. The most common analyzers are octave band, one-third octave band, and narrowband analyzers using filters or fast Fourier transform (FFT) digital calculations (ANSI S1.6 2016; ANSI S1.11 n.d.).

**9.5.3.1 *Octave Band Analyzers*** The octave band analyzer is the most common type of noise filter used for many noise analyses because the octave band normally provides adequate spectral information with a minimum number of measurements. An octave is defined as any bandwidth having an upper band-edge frequency equal to twice the lower band-edge frequency.

Filters are used to distinguish the characteristics of a noise source. Octave bands "filter out" frequencies below and above a one-octave bandwidth. Therefore, the upper edge of the band of frequencies $f_u$ is twice the frequency of the lower band edge $f_l$:

$$f_u = 2f_l \tag{9.11}$$

When more detail is required, one-third octave bands are used. For one-third octaves, the upper band edge is the cube root of two times the lower band edge:

$$f_u = \sqrt[3]{2}f_l \tag{9.12}$$

Note that not all sounds outside the upper and lower frequencies are excluded. The band-edge frequencies represent the point at which the filter response is 3 dB down from the center frequency response. Specifications for octave and fractional octave filters can be found in ANSI S1.11.

When describing industrial noise, frequencies less than 20 Hz and greater than 20 000 Hz are generally not used. The bands and frequencies commonly used are shown in Table 9.6. Frequencies shown in bold are the octave band preferred frequencies – chosen to include 1000 Hz.

**9.5.3.2 *Narrowband Analyzers*** The *A*- or *C*-weighting networks are often used for hearing conservation purposes. For noise control purposes, octave band and third octave band analyzers are adequate and satisfy the requirements listed in national and international consensus standards. In rare cases, finer resolution of frequency information is needed. FFT analyzers are the most common technique used for these measurements.

FFT analyzers convert time-domain signals (amplitude versus time) to the frequency domain (amplitude versus frequency). Input signals need to be sampled at least twice the minimum frequency of interest. Thus for noise control work up to 20 000 Hz, sound pressure levels must be sampled at least 40 000 times per second. The frequency resolution of each spectral line is equal to the sampling rate divided by the number of lines chosen. Therefore, the resolution of a 1024-line FFT analysis of a signal sampled 40000 times per second would be about 40 Hz:

$$\frac{\dfrac{40000\ \text{samples}}{\text{second}}}{\dfrac{1\ \text{sample}}{1024\ \text{lines}}} = \frac{39\ \text{lines}}{\text{second}} \tag{9.13}$$

### 9.5.4 Tape Recording of Noise

It is sometimes convenient to record a noise so that it can be analyzed at a later date. This is particularly helpful when lengthy narrowband analyses are to be made or when very short transient-type noises are to be analyzed. Unfortunately, most recorders are meant for audio measurements and incorporate automatic gain control (AGC) circuitry. These circuits constantly monitor the incoming signal, increasing the gain when the levels are low, and decreasing the gain at higher sound pressure levels. Consequently, all information of the absolute level of the signal is lost. Professional or specialized recorders are needed for industrial noise measurements.

It is important to calibrate the combination of tape recorder and sound level meter at a known level, preferably throughout the frequency range. Before each series of measurements, a pressure level calibration should be made by recording the overall sound pressure level reading by stating the levels orally, along with the tape recorder dial settings. It is also good practice to state orally on each recording the type and serial numbers of the microphone and sound and surroundings and other pertinent information. This practice of noting information orally on the tape often prevents information from being lost or confused with other tapes.

### 9.5.5 Level Recording (Data Logging)

Previously, graphic level recorders were connected to the output of a sound level meter or analyzer to provide a continuous paper and ink record of the output level. Currently, most recording of levels is performed electronically by sound level meters or noise dosimeters with data logging (recording) capabilities. Data are stored inside the instruments and later printed or transferred to personal computers.

Sound pressure levels are sampled typically several times per second, and the distribution of sound pressure levels is saved in electronic memory. Depending on the mode of operation chosen by the user, either the distribution of levels is integrated over the measurement period, or, periodically, some measures are recorded from the distribution; then a new distribution is started for the next sampling period. After sampling, the user has a record of either the distribution of levels over the entire sampling period or indicators of the distribution for each period.

Distribution results are normally plotted as amount of time (or fraction of total time) sound pressure levels fall into 1-, 2-, or 5-dB bins. The data provide a discrete-valued function of sound pressure levels.

### 9.5.6 Instrument Calibration

If valid data are to be obtained, it is essential that all equipment for the measurement and analysis of sound is calibrated. When equipment is purchased from the manufacturer, it should have been calibrated to the pertinent ANSI or IEC specifications. However, it is the responsibility of the equipment user to keep the instrument in calibration by periodic checks.

Acoustic calibrators are available for checking the overall acoustical and electrical performance at one or more frequencies. These calibrations should be made according to the manufacturer's instructions at the beginning and at the end of each day's measurements. A battery check should also be made at these times. These calibration procedures cannot be considered to be of high absolute accuracy, nor will they allow the operator to detect changes in performance at frequencies other than that used for calibration. They do serve as a warning of most common instrument failures, thus avoiding many invalid measurements.

Periodically sound measuring instruments should be sent back to the manufacturer or to a competent acoustic laboratory for calibration at several frequencies throughout the instrument range. These calibrations require technical competence and the use of expensive equipment. How frequently these complete calibrations should be made depends on the purpose of the measurements and how roughly the instruments are handled. ANSI and IEC standards require a complete calibration performed at least once a year, and at any time a calibration shift greater than 1 dB is found.

### 9.5.7 Making Occupational Noise Exposure Measurements

Although noise measuring equipment may be simple to operate and read, the results will not be meaningful unless the proper measurements are taken and recorded using appropriate scales and units in a manner that can be interpreted by others.

The following guide can be used before making any acoustical noise measurements:

1. Classify the problem: Generally noise can be classified as:
   A. Generation or transmission of noise from one or more sources
      The purpose is to measure a quantity – sound pressure level or frequency spectrum – at a certain point or time relative to the source.
   B. Measurement of noise at a receiver
      Measure a quantity – integrated sound pressure level or energy – related to the effects of noise on the exposed individual.
2. Consider the type of noise:
   A. Frequency spectrum
      Continuous spectrum
         Measure overall level or octave band levels.
      Spectrum with audible tones
         Third octave band or narrow band analysis is required.
   B. Variation of level with time
      Steady noise
         For noise with small (generally <3 dB) fluctuations in level, any measurement technique may be used.
      Non-steady noise
         Measure the levels of noise and duration for each identifiable noise level. Alternatively, measure equivalent continuous sound pressure level or sound exposure.
      Impulsive
         Measure impulse sound levels and frequency of impulses or number per day, as well as background ambient noise level. Measure sound exposure with meter set to fast or peak exponential time averaging.
3. Consider the sound field (path):
   A. Free field, reverberant field, or semi-reverberant
      Document location of measurement relative to the source. Consider how sound pressure level decays with distance from source.
   B. Personal exposure monitoring
      Best accuracy for short-duration monitoring is made with microphone positioned at a location that would be the center of the employee's head while operating the machine. For noise dosimetry, place the microphone on the employees shoulder.
4. Consider the quality (ISO 2204 1979) of measurement desired:
   A. Precision – Generally not used for hearing conservation purposes. Requires a thorough description of the noise source and environment.
   B. Engineering – Useful for collecting data of existing equipment or environments for noise control purposes. Sound pressure levels are supplemented by band pressure levels. Usually need to characterize noise with sources operating and background noise.
   C. Survey – Primarily measurements of sound pressure level, OSHA noise dose, or equivalent continuous sound level. Measure the noise "as is," without modifying sources or path. Used for exposure monitoring – generally does not supply enough detail for noise control purposes.
5. Consider the purpose of the measurements:
   A. Compliance to a standard
      Make sure the equipment is capable of measuring full range of sound levels, frequency weighting, and time averaging specified by the standard. Integrating sound level meters and dosimeters must measure according to the specified exchange rate.

B. For evaluating engineering controls or source characterization
   The equipment should be capable of measuring at least octave band levels. Frequently one-third octave band or an FFT analyzer is used. Recorders – either digital or tape – permit later analysis.

A checklist of equipment usually required for acoustic evaluations would include:

Sound level meter or dosimeter(s)
Acoustical calibrator for all meters
Spare batteries for meters and calibrators
Windscreen(s) for protecting the microphones
Watch for recording measurement start and stop times
Forms for recording measurement conditions
Camera for photographic documentation
Hearing protection (and any other necessary personal protective equipment)

***9.5.7.1 Measurement Procedures*** Procedures for performing occupational noise exposure measurements can be found in ANSI S12.19 (1996), "American National Standard Measurement of Occupational Noise Exposure," or ISO 9612-1991 (1991), "Acoustics – Guidelines for the Measurement and Assessment of Exposure to Noise in the Working Environment."

Sound level meters shall meet the requirement of type 2 m per ANSI S1.4 (or IEC 651). Noise dosimeters are specified in ANSI S1.25 (IEC 942), and calibrators in ANSI S1.40 (IEC 942). All instruments should be calibrated annually by a qualified test laboratory. The battery in each meter should be checked, and the calibration of each meter checked before and after each measurement period. If there is a change of more than 1.0 dB from the initial to the final calibrations, the measurements must be rejected.

As far as possible, nothing should be done to disturb the acoustic environment at the measurement site, and measurements should be arranged so there is minimal disturbance of normal work patterns. The worker being monitored (and their supervisor) should be informed of the purpose of the measurement. At the conclusion of the measurement, input from the employee and their management should be sought to determine if there were any unusual events or work conditions that would make the measurement not representative of normal operations.

*9.5.7.1.1 Placement of Microphone of a Sound Level Meter* Measurements are taken in the employee's "hearing zone." This is a location chosen to be representative of a worker's noise exposure. When a sound level meter is used, the preferred location is in a position at the center of where the head of a worker would be while performing the task being measured. This is not always feasible; so measurements are frequently taken with the microphone placed approximately 0.1 m from the entrance to the ear canal of the ear receiving the higher sound level. For areas where there is one dominant source, care must be taken that a direct line of sight is maintained between the microphone and the source; that is, it should not be blocked by the body of the employee or the industrial hygienist taking the measurement. Usual practice is to hold the sound level meter out at arm's length – away from any reflections or absorption created by the individual taking the measurement. The sound level meter's instruction manual must be consulted to determine proper orientation of the microphone relative to the source.

*9.5.7.1.2 Dosimeter Placement* The hearing zone for a dosimeter microphone is specified as on the top of the shoulder, midway between the neck, and the outside edge of the shoulder. The shoulder facing the source should be used if the worker is consistently exposed to noise from one side. The microphone diaphragm should be oriented parallel to the plane of the shoulder. The microphone cable should be routed so it does not interfere with the worker's performance or safety. The body of the dosimeter can be placed wherever convenient. Typically, the microphone cord lies diagonally across the worker's back, leading to the electronics package, which is clipped to a belt at the employee's hip. The cable is often clipped or taped to the worker's shirt in the back to keep it from swinging loose. It may be necessary to reset the dosimeter after the initial calibration, and the dosimeter results should be recorded before performing the final calibration to avoid adding calibration signals to the worker's dose.

*9.5.7.1.3 Measurements* Measurements should be of sufficient duration to get a representative noise exposure. If the level does not vary by more than ±3 dB, it is considered steady, and an average reading from a sound level meter may be taken. Frequently the level varies by more than this; so an integrating meter should be used. If there is some periodicity to the sound level, the readings must be taken over at least one full period. If there are several distinct noise levels, each level should be measured as well as its duration and the number of times it occurs each work shift.

When dosimeters are used, all workers should be under visual observation to assure they do not perform any activities that will invalidate the measurement.

**9.5.7.2 *Recordkeeping*** Measurements, in particular those collected for determining exposure, should be recorded in a manner that assures future occupational assessments will be possible. It is important that adequate data are collected using precise and uniform definitions. The American Conference of Governmental Industrial Hygienists (ACGIH) and the American Industrial Hygiene Association have issued guidelines and recommendations (Joint ACGIH-AIHA Task Group on Occupational Exposure Databases 1996) for occupational exposure databases. Also, whenever industrial hygiene measurements are collected, some form of sampling data sheet is recommended. The sampling data sheet acts as a reminder to ensure all relevant information is collected. A sampling data sheet formed from the recommendations of the joint committee is in Table 9.8. Letter–number pairs follow from the "data groups" specified in the guidelines. Starred items are required data fields.

### 9.5.8 Making Environmental Noise Measurements

Measurements taken for community noise assessment (ISO 1996) use all the above principles. However, the position at which outdoor measurements should be taken is not always as straightforward as it is for occupational noise measurements; measurements frequently need to be taken for longer duration, and much more attention should be paid to microphone placement. A windscreen is virtually a requirement for environmental noise measurements, and a microphone cover for protection from precipitation is usually needed.

Generally the positioning of the microphone depends on the purpose of the measurements. At times the industrial hygienist will be called upon to make property line noise measurements to determine compliance with community noise criteria. The microphone should be oriented so it is most sensitive to noise from the source. It is normally placed 1.2–1.5 m above the ground and should be more than 3.5 m from any reflecting surfaces. Measurements taken near a building should be taken 1–2 m from the facade and 3 dB should be subtracted from any readings.

Measurements can be taken over a range of meteorological conditions. Choose conditions such that:

1. Wind direction is within ±45° of a line connecting the source to the microphone, blowing from the source to the microphone.
2. Wind speed is 1–5 m s$^{-1}$ at 3–11 m above the ground.
3. There is no strong temperature inversion near the ground.
4. There is no heavy precipitation.

Consideration must be paid to the duration of measurements. To cover typical human activities, reference intervals may be days and nights. It may be necessary to include intervals for evenings, weekends, and holidays. Long-term measurements are frequently performed on the order of months. In this case, seasonal variations may be important. In any case the intervals should be chosen so all significant variations are covered.

Since the noise is generally not steady, sufficient samples are needed to estimate long-term levels that characterize the noise. Frequently this is reported using percentile levels, for example, $L_1$, $L_5$, $L_{10}$, $L_{50}$, etc. – the sound pressure levels exceeded 1, 5, 10, and 50% of the time. Many dosimeters and integrating sound level meters can measure, accumulate, and report these values. Another common measurement required for community noise is brief disturbances such as aircraft flyover. Sound exposure levels (SEL) are the criteria commonly used to assess sources of this type.

Note if any tones are present. A source is considered tonal if the sound level in any one-third octave band exceeds adjacent bands by more than 5 dB. If the tonal components are clearly audible, a 5–6-dB "penalty" will added to the sound pressure level. Two to three dB is added if the tonal components are just detectable.

When it is necessary to characterize an area or a line (such as the impact of traffic noise from a highway), measurements will be taken at a number of locations on a grid. The grid spacing should be adjusted so that there is less than 5-dB difference between adjacent grid points. See Section 9.10 for more information on environmental noise.

**TABLE 9.8 Noise Monitoring Recording Form**

**Noise monitoring form**

**Facility/site information**

A-1* Company/organization:
A-2* Facility name: | A-3* Facility address
A-4* SIC code | A-5* Industrial category (descr.)
A-7* No. of employees at facility
A-6* (Contractor name, type, SIC)

**Survey tracking information**

B-1* Survey (reference) no. | B-2* Survey date
B-4 Report no.
B-3* Person performing survey
B-5 Is follow-up required? [ ] Yes, follow-up required [ ] No, follow-up is not required
B-6 Follow-up summary
B-7 Person responsible for follow-up | B-8 Date follow-up completed
B-9* Quality control reviewer name, position, SSN | B-10* Date reviewed
B-11 General survey comments

**Work area information**

C-1* Building/zone(s) | C-2* Room/area
C-3* Department
C-4* Type of work area: [ ] Open air [ ] Enclosed indoor space [ ] Confined space (descr.) [ ] Equipment cab [ ] Other
C-5 Location comments:
C-6 Climatic conditions:

**Employee information**

D-1* Employee name | D-2* Employee ID:
D-3* Administrative job title | D-4* Occupational title
D-5* Work or task description
D-6 Similar exposure groups (SEG)
D-7* Shift __:__ Start __:__ End | D-8 Union
D-9 Job safety training: [ ] Yes [ ] No
D-10 Comments:

**Process and operation information**

E-1* Process
E-2* Task
E-3* Frequency of process: [ ] Continuous [ ] Frequency: __ Times/____ (hour, day, etc.)
E-4 Comments on process:
E-5* Source(s): [ ] Single source directly associated with employee [ ] Single source associated with employee AND additional sources [ ] Single source distant from employee [ ] Multiple sources distant from employee [ ] Other (descr.)

**Exposure modifier information**

G-1* Exposure representative? [ ] Yes (identify SEG) [ ] No [ ] Unknown
G-2* Exposure representativeness comments
G-3* Exposure conditions: [ ] Typical [ ] Higher than normal [ ] Lower than normal [ ] Unknown
G-4 Basis for estimate of conditions:
G-5* Exposure pattern on day of sample [ ] Continuous exposure throughout day [ ] Continuous throughout part of day (specify) [ ] Intermittent (specify frequency and duration) [ ] Other (descr.)
G-6* Exposure frequency over extended time period: [ ] Daily [ ] Regular frequency (specify) [ ] Occasional (estimate frequency) [ ] Other (descr.)
G-9* Exposure modifier comments:

*(Continued)*

**TABLE 9.8 (Continued)**

**Noise monitoring form**

**Sample information**

H-1* Sample collected? [ ] Yes H-3* Sample no. _________ [ ] No
H-2* Reason no sample collected
H-5* Sample duration: H-4* Sample date:
H-6* Reason for sample: [ ] Baseline [ ] Scheduled [ ] Complaint [ ] Compliance [ ] Diagnostic [ ] Emergency (response) [ ] Unusual activity (descr.) [ ] Other (descr.)
H-7* Type of sample – duration: [ ] Single sample for full-shift TWA [ ] Multiple partial periods for TWA [ ] Task [ ] Peak sample [ ] Other (descr.)
H-8* Type of sample – location: [ ] Personal (outside hearing protection) [ ] Personal (inside hearing protection) [ ] Area [ ] Source (specify distance) [ ] Other (descr.)
H-9 Sample information comments:

**Sampling device information**

I-1* Sampling device type: [ ] Sound level meter [ ] Dosimeter [ ] Impact noise [ ] Other (descr.)
I-2* Sampling device identification (name, manufacturer, model no.):
I-3* Calibration documentation:
I-6 Comments:

**Administrative/engineering controls**

J-1* Administrative controls: [ ] Yes (descr.) [ ] No
J-3* Type of acoustic engineering controls: [ ] Enclosure [ ] Vibration isolation [ ] Dampening [ ] Noise absorption [ ] Noise cancelation [ ] None [ ] Other
J-4 Specific engineering controls:
J-5 Estimated effectiveness of engineering controls: [ ] Effective [ ] NOT effective (check all that apply and descr.) [ ] Improper choice [ ] Improper design [ ] Improper installation [ ] Poor condition [ ] Improperly modified [ ] Not working according to design specifications [ ] Other [ ] Not evaluated
J-6 Comments about effectiveness of engineering controls:

**Personal protective equipment information**

K-23* Hearing protection worn: [ ] Worn [ ] Not worn
K-24* Hearing protection requirements: [ ] Required [ ] Not required [ ] Determination not made
K-25* Hearing protection type: [ ] Plugs/inserts [ ] Circumaural [ ] Other (descr.)
K-26 Hearing protection specific (manufacturer, model, NRR):
K-27 Estimated effectiveness of hearing protection: [ ] Effective [ ] NOT effective (check all that apply and descr.) [ ] Improper choice [ ] Poor condition [ ] Improper use [ ] Not worn when required [ ] Other [ ] Not evaluated
K-28 Comments about effectiveness of hearing protection:

**Noise exposure results**

M-1* Noise exposure dose: _____ M-2* Exchange rate: [ ] 3 dB [ ] 5 dB
Criterion: 8 h at ____ dB = 100%
[ ] *A*-weighting [ ] *C*-weighting
[ ] Slow response [ ] Fast response
M-3* $L_{eq}$: _____ M-4 $L_{max}$ ______
M-5* Sound level measurements: $L$ = _____ dB
[ ] *A*-weighting [ ] *C*/flat weighting [ ] *D*-weighting [ ] Slow response [ ] Fast response
M-6 Impact noise measurements: _____ (dB)
M-7 Comments

## 9.6 OCCUPATIONAL NOISE EXPOSURE CRITERIA

### 9.6.1 Background: US Federal Regulations

The OSHA is tasked with enforcing the Department of Labor Occupational Noise Exposure Standard, 29 CFR 1910.95. In 1983, the Occupational Noise Exposure; Hearing Conservation Amendment; Final Rule was put into effect. The regulation stipulates aspects of the hearing conservation program such as (i) noise exposure measurement, (ii) audiometric testing programs, (iii) the identification of employees that must be included in the program, and

(iv) the recordkeeping activities that must be performed. In addition, the role of hearing protectors and training issues are addressed.

OSHA is the enforcement branch of the federal government's industrial health and safety program. The National Institute for Occupational Safety and Health (NIOSH) is the research and training division. NIOSH is responsible for performing pertinent research in the areas of industrial health and hygiene, providing training to industrial hygienists and safety inspectors, and developing recommended regulations. OSHA does not have enforcement power in the mining industry; instead the Mine Safety and Health Administration (MSHA) has this responsibility. OSHA also does not have enforcement power over the military or the railroad industry.

### 9.6.2 OSHA Noise Regulations

***9.6.2.1 Noise Exposure*** OSHA regulations state that no employee shall be exposed to greater than 90-dBA eight-hour TWA noise exposure. Any employee exposed to greater than 90-dBA TWA shall be provided protection from the effects of noise exposure through feasible engineering controls or administrative controls or with hearing protective devices.

***9.6.2.2 Noise Exposure Measurements for Hearing Conservation*** A limit of 85 dBA for eight hours was added as an action level in the 1983 amendment. The 1983 OSHA amendment states that the "employer shall administer a continuing, effective hearing conservation program, ...whenever employee noise exposures equal or exceed an eight-hour TWA of 85 dB measured on the *A*-scale (slow response), or equivalently, a dose of 50%." Monitoring must be conducted to determine if any employees exceed these limits. All those exceeding the limits must be placed in a hearing conservation program consisting of:

- Noise monitoring
- Audiometric evaluations at least annually
- Hearing protective devices
- Annual training to employees
- Maintenance of proper records

***9.6.2.3 Practical Considerations*** The limits specified in the OSHA noise exposure regulations were the most restrictive limits deemed feasible with due consideration given to other important factors, such as economic impact. These limits were not intended to provide complete protection for all persons. It is estimated that 85% of persons exposed to the OSHA limits of $L_{\text{A, 8h}} = 90\,\text{dB}$ for 8 hour/day, 5 days a week, for about 10 years, will not develop a significant hearing impairment. The other 15% of persons exposed to these limits would probably have various levels of hearing impairment depending upon their susceptibility. Exposures outside the workplace may also contribute significantly to the hearing impairment for some persons.

OSHA exposure limits are intended only as *minimum* action levels. Wherever feasible, hearing conservation measures should be extended to reduce exposure levels *below the limits specified* both at and away from the workplace. Companies may choose a more conservative approach and require the use of hearing protectors at 85 dBA.

### 9.6.3 Noise Measurements

Most occupational noise measurements are taken using meters set to *A*-weighted slow response as specified by OSHA and international standards. Usually, $L_{\text{A, S}} = 90\,\text{dB}$ is the *criterion sound level* (LC), that is, the sound pressure level that if present for a full eight-hour *criterion duration* (TC) workday would give a *dose* ($D$) of 1.0 or 100%. An *exchange rate* ($Q$) of 5 dB is used, which is the difference in sound level that would give the same dose if the exposure time were doubled or halved. The above terminology is that used in the ANSI standards.

To establish compliance to the hearing conservation amendment, *A*-weighted sound levels must be measured from 80 to 130 dB. The level $L_{\text{A, S}} = 80\,\text{dB}$ is called the *threshold level* (TL), the lowest sound pressure level used to compute the dose. If the dose exceeds 50%, the employee must be placed in a hearing conservation program. The OSHA hearing conservation noise exposure dose criteria are given in Table 9.9.

Compliance to the noise standard is established with much the same criteria except that the threshold level is set at 90 dB ($\text{TL}_{\text{AS}} = 90\,\text{dB}$). A dose exceeding 100% would be considered exceeding the noise standard. The dual nature of the standard arose from the fact that the hearing conservation program was amended to the noise regulation and subsequently interpreted by the US courts.

### 9.6.4 Exposure Calculations

When the daily noise exposure is composed of two or more periods of exposure at different levels, their combined effect is determined by adding the individual contribution as follows:

$$D_{\mathrm{OSHA}} = \sum_{i=1}^{n} \frac{C_i}{T_i} \tag{9.14}$$

This method is called the *time-weighted-average noise dose*. The values $C_1$–$C_n$ indicate the times of exposure to specified levels of noise, and the corresponding values of $T_i$ indicate the total time of exposure permitted at each of these levels (see Table 9.9). If the sum of the individual contributions exceeds 1.0, the mixed exposures are considered to exceed the overall limit value.

For $L_{A,S} > 80$ dB, the time allowed for any sound pressure level can be calculated using the formula:

$$T_{\mathrm{OSHA}} = \frac{8\mathrm{h}}{2^{(L_{A,S}-90)/5}} \tag{9.15}$$

In general the values of $T_i$ can be determined for any criterion (ANSI 12.19 1996) by

$$T_i = \frac{\mathrm{TC}}{2^{((L_i-\mathrm{LC})/Q)}} \tag{9.16}$$

where TC is the criterion duration (hour) and LC is the criterion sound pressure level (dB). Most organizations specify a criterion duration TC of eight hours, corresponding to a normal workday.

***Example 9.6:*** If a person were exposed to 90 dBA for five hours, 100 dBA for one hour, and 75 dBA for three hours during an eight-hour working day, the times of exposure are $C_1 = 5$ h, $C_2 = 1$ h, and $C_3 = 3$ h. The corresponding OSHA time limits for these exposures are $T_1 = 8$ h, $T_2 = 2$ h, and $T_3$ = infinity. Therefore, because 3 divided by infinity is zero, there is no contribution from the 75-dBA exposure:

$$D = \frac{5}{8} + \frac{1}{2} + \frac{3}{\infty} = 1.125 \text{ or } 112\%$$

Hence the time-weighted-average (TWA) noise dose for this person slightly exceeds the specified limit of 1.0 (100%).

**TABLE 9.9 Allowable Time and Eight-Hour Noise Dose per OSHA 29 CFR 1910.95, table G-16a**

| $L_{A,S}$ (dB) | Time allowed (h) | Dose (%) if exposed for 8 h | $L_{A,S}$ (dB) | Time allowed (h) | Dose (%) if exposed for 8 h |
|---|---|---|---|---|---|
| 80 | 32 | 25 | 85 | 16 | 50 |
| 90 | 8 | 100 | 95 | 4 | 200 |
| 100 | 2 | 400 | 105 | 1 | 800 |
| 110 | 0.5 | 1600 | 115 | 0.25 | 3200 |

**TABLE 9.10 Criterion Sound Level, Thresholds, and Exchange Rates**

| Organization | LC[a] | TL[b] | Q[c] |
|---|---|---|---|
| OSHA 29 CFR 1910.95 for noise exposures | 90 | 90 | 5 |
| OSHA for hearing conservation | 85 | 80 | 5 |
| ACGIH 1996 TLVs and BEIs | 85 | 80 | 3 |

*Source:* ACGIH (1996b).
[a] LC is the criterion sound pressure level (dB).
[b] TL is the threshold level (dB).
[c] Q is the exchange rate (dB).

Note that an eight-hour exposure to a continuous sound level of 90 dB would result in a dose of 100%. Similarly continuous $L_A = 95$ dB for eight hours would lead to a 200% dose, and $L_A = 85$ dB would give $D = 50\%$.

Therefore, a noise dose calculated according to the OSHA regulations can be converted to a TWA sound pressure level. Expressing dose $D$ as a fraction,

$$\mathrm{TWA} = 16.61 \log_{10}(D) + 90 \tag{9.17}$$

***Example 9.7:*** Determine the time-weighted-average sound pressure level for the 112% OSHA dose calculated above.

$$L_A = 16.61 \log(1.12) + 90 = 90.8\,\mathrm{dB}$$

### 9.6.5 Other Noise Exposure Criteria

For determining occupational noise exposure and estimating hearing impairment, the ISO (1999) and the ACGIH (1996a) have recommended use of $L_{eq,8h}$ (equivalent continuous $A$-weighted sound pressure level normalized to an eight-hour working day). ANSI S3.44-1996 recommends the same but notes that other exchange rates ($Q$) (notably 5 dB) can be used. In this case the exposure is deemed "equivalent effective level" (EEL) (ANSI S3.44 2006). ANSI 3.44 and ISO (1999) and some experts (Kryter 1994b) also use *sound exposure* $E_{A,T}$ (measured in Pa$^2$ s) to determine the risk of hearing loss from noise exposures:

$$E_{A,T} = \int_{t}^{t_2} p_A^2(t)\,dt \tag{9.18}$$

The above calculation would be done by an integrating sound level meter. When the exposure consists of only a few levels and durations, the discrete form of the above integral is used to calculate the sound exposure:

$$E_{A,T} = \sum_{i=1}^{n} \left( p_r^2\, 10^{L_A/10} T_i \right) \tag{9.19}$$

***Example 9.8:*** Using the same data for the previous example calculating $L_{eq}$, what is the sound exposure $E_{A,8h}$ for an eight-hour workday consisting of four exposures as shown in the table?

| $L_{AS}$ (dB) | Duration (h) |
|---|---|
| 82 | 6.0 |
| 88 | 1.4 |
| 97 | 0.5 |
| 105 | 0.1 |

| $L_{AS}$ (dB) | $p_A^2$ (Pa$^2$) | Duration (h) | Duration (s) | Exposure (Pa$^2$ s) |
|---|---|---|---|---|
| 82 | 0.063 | 6.0 | 21 600 | $1.37 \times 10^3$ |
| 88 | 0.252 | 1.4 | 5 040 | $1.27 \times 10^3$ |
| 97 | 2.00 | 0.5 | 1 800 | $3.61 \times 10^3$ |
| 105 | 12.64 | 0.1 | 360 | $4.55 \times 10^3$ |
| | | | Total | $1.08 \times 10^4$ |

$$E_{A,8h} = 1.08 \times 10^4\,\mathrm{Pa^2\,s}$$

Sound exposure levels, normalized to an eight-hour working day, are calculated from sound exposures using the following:

$$L_{Aeq,8h} = 10 \log\left( \frac{E_{A,T}}{E_r} \right) \tag{9.20}$$

where the reference exposure $E_r = p_r^2 T_r = (0.00002\,\mathrm{Pa})^2\, 8\,\mathrm{h} \dfrac{3600\,\mathrm{s}}{\mathrm{h}} = 1.15 \times 10^{-5}\,\mathrm{Pa^2\,s}$.

***Example 9.9:*** Show that the sound exposure $E_{A,8h} = 1.08 \times 10^4$ is equivalent to the equivalent continuous sound level $L_{Aeq} = 90$ calculated previously.

$$L_{Aeq,8h} = 10\log\left(\frac{1.08\times10^4}{1.15\times10^{-5}}\right) \cong 90\,dB$$

When a noise *dose* is given, the equivalent time-weighted-average (TWA) *level* can be determined using the generalized formula.

$$\mathrm{TWA} = \mathrm{LC} + 10\left(\frac{Q}{10\log_{10}(2)}\right)(D) \tag{9.21}$$

where LC is the criterion sound level, $Q$ is the exchange rate, and $D$ is the dose (expressed as a fraction). For the OSHA noise standard, this formula becomes that given in Eq. (9.17).

### 9.6.6 Steady-state and Impulsive Noise

The integration of continuous and impulsive noise is done automatically if exposure levels are measured using a dosimeter. The current ANSI standard S1.25-1991 requires noise dosimeters have an operating range of at least 50 dB. For measurement of impulse noises, instrumentation and suggested presentation of results is available and defined by ANSI S12.7 (1986). There is evidence (Lataye and Campo 1996) that above a critical intensity, the damage creating hearing loss involves a mechanical mechanism rather than the metabolic mechanism that is responsible for damage from steady noises of moderate intensity. A damage risk criterion that considers steady-state and impulsive sounds above a critical level and periods of quiet (Kryter 1994) may be helpful in predicting hearing loss.

## 9.7 HEARING CONSERVATION PROGRAMS

### 9.7.1 General

The objective of hearing conservation programs is to prevent noise-induced hearing loss. This obvious fact is often forgotten, or ignored, when pressures are applied for compliance with local, state, or federal rules and regulations on noise exposures. Because compliance with rules and regulations will not always prevent noise-induced hearing impairment in susceptible individuals, every effort should be made to reduce noise exposures wherever possible.

The lowest feasible noise exposure levels are obviously desirable for the health, safety, and well-being of workers. In addition, these lower limits are of significant value to employers because morale and work productivity of workers should be maximized (Cohen 1966, 1968, 1969, 1973a, b, Grether 1971; Michael and Bienvenue 1983) and the number of compensation claims for noise-induced hearing impairment should be minimized.

### 9.7.2 Requirements

An effective hearing conservation program should provide for:

1. The identification of noise hazard areas and the performance of noise exposure measurements.
2. The reduction of the noise exposure to safe levels preferably using engineering controls, or if infeasible, through administrative controls or hearing protective devices.
3. The measurement of exposed worker's hearing thresholds to monitor the effectiveness of the program.
4. The education and motivation of employees (Berger 1981) and management about the need for hearing conservation and the instruction of employees in the use and care of personal hearing protectors.
5. The maintenance of accurate and reliable records of hearing and noise exposure measurements.
6. The referral of employees who have abnormal hearing thresholds for examination and diagnosis.

**9.7.2.1 *Identification of Noise Hazard Areas*** Noise hazard areas having continuous noise characteristics can be identified with sound level meters or with dosimeters. When significant impulse-type noises are present, hazard areas may be established with noise dosimeters, integrating sound level meters, or impulse measuring devices.

Action levels for exposure must be established that are at least as low as those specified by the OSHA Rules and Regulations. In order to select the best hearing protectors for particular noise spectra, octave band sound pressure levels should be measured.

**9.7.2.2 *Reduction of Noise Exposure Levels*** As soon as a noise hazard area has been identified, hearing protective devices should be provided to reduce the exposures to safe levels. Engineering control means should then be employed where they are feasible. This will often require the use of experts in noise control to work with those at the plant, and/or others, who understand the machines and their operation. If it is not feasible to reduce exposure levels to the limits selected by engineering control means, the use of hearing protectors must be continued.

Continued monitoring of the effectiveness of hearing protector devices (hearing threshold measurements) must be maintained until engineering control procedures have reduced the noise exposures to safe levels. It is strongly advisable to continue monitoring the effectiveness of the hearing conservation program even after engineering control measures have been successfully installed if there is any reason to suspect that noise exposures at, or away from, work are significant.

**9.7.2.3 *Hearing Measurement and Recordkeeping*** Periodic hearing threshold measurements are necessary to monitor the effectiveness of a hearing conservation program. High-level noise exposures away from work may be partially responsible for any STS; hence, it is important to make a significant effort to determine the cause of any threshold shift. Affected persons should be interviewed, and their audiological historical data sheets should be updated periodically.

Hearing conservation measures away from work can be encouraged or assisted by lending or giving hearing protectors to employees for use with noisy activities. The cost of supplying hearing protectors for this use is far less than hearing loss compensation costs, and it often directs more attention to the hearing conservation program at work.

If an STS (see below) is found in an annual audiogram, a letter must be sent to the employee informing him or her of the results of the audiogram. Even though follow-up letters to employees are not required after the baseline audiogram, letters should be used to advise new employees of hearing impairments greater than would be expected for their age. These employees may be more susceptible to noise-induced hearing impairment; so this early warning should result in lower exposures at and away from work during the following year.

Records should be retained after an employee is no longer employed, as required by federal or state medical recordkeeping requirements. These records could be valuable if there is a question of legal responsibility for an employee's hearing impairment or if the employee is to return to this job. Also, *termination audiograms* are not required by OSHA, but they should be taken because of potential liability for hearing impairments after termination.

**9.7.2.4 *Baseline Audiograms*** A baseline audiogram must be established within six months of an employee's first exposure at or above an eight-hour TWA of 85 dBA or, equivalently, a dose of 50%. All subsequent audiograms are to be compared with the baseline audiogram to identify changes in hearing thresholds. If mobile audiometric testing services are used, the new baseline may be established within one year of first exposure. Hearing protectors must be worn during the second six months of this one-year period until the baseline is established.

New baselines may be established if the audiologist or physician in charge of the hearing conservation program determines that old audiograms are not valid, if an STS is established, or if the annual audiogram indicates significant improvement over the baseline audiogram.

*Note:* Not all audiologists or physicians are qualified industrial hearing conservationists. Employers should carefully investigate the professional qualifications and experience of any person employed for industrial hearing conservation. Generally, expertise only in clinical practice or certification in specialty fields other than hearing conservation is not sufficient.

**9.7.2.5 *Annual Audiograms and Standard Threshold Shifts*** Annual audiograms must be taken on all employees who are exposed to an eight-hour TWA of 85 dBA or greater. If an STS, "a change in hearing threshold relative to the baseline audiogram of an average of 10 dB or more at 2000, 3000, and 4000 Hz in either ear," is established by an annual audiogram, it must be reported to the affected employee within 21 days.

In practice, a retest is justified in most cases when an STS is found because there are many common reasons why a 10-dB STS may be temporary. All retests should be completed within 30 days. Retests performed within 21 days may show the STS was temporary and eliminate the requirement of preparing a notification letter for the employee.

Old audiograms must not be discarded when new baselines are established for an employee. The letter used to inform an employee of an STS should also be used to educate and motivate the employee to take better care of his or her hearing both at and away from work. A sample letter is presented here:

**Sample Letter to Persons Showing a Standard Threshold Shift on an Audiogram**

Dear _____:

Your hearing thresholds measured on _____ indicate a change of more than _____ dB from (worse than) your baseline audiogram. This change may be temporary, or it may be permanent. Temporary changes may be the result of colds, allergies, medications, recent exposures to high-level noises, blows to the head, etc. Depending upon the severity of these and other factors, permanent impairments may also result.

While your hearing impairment may be temporary, we recommend strongly that you have an audiological examination at the _____ Speech and Hearing Clinic (tel. xxx-xxx-xxxx). We have arranged for payment for this examination. Please continue to obey our hearing conservation rules at and away from work.

Sincerely,

### 9.7.3 Setting Up a Hearing Conservation Program

Setting up a hearing conservation program entails the development and maintenance of all the factors mentioned above. Ideally, a team made up of management, industrial hygiene, occupational medicine, safety, and production personnel will work together toward building an effective hearing conservation program. The first objective for this group should be to become aware of their problems and to consider possible ways to limit noise exposure levels as much as possible, at least to levels set by OSHA.

In practice, these committees may not work well except as advisors on general items such as company policy. Many successful hearing conservation programs have been highly dependent upon the competence and dedication of one person *who is given the time and support required to develop and maintain the program*. This person must be motivated and knowledgeable and respected by both management and workers. Extensive formal training in any specific field is not usually necessary, but he or she must have a good practical knowledge of noise-induced hearing impairment, hearing threshold measurement, and personal hearing protectors.

Outside specialty firms may be useful for specific jobs such as audiometry, or noise control projects, but it is usually to the company's benefit to have a knowledgeable employee work with the specialist. Information on individuals and detailed information on machine operation (access areas needed for operation and production flow) are often needed for the best results. Otherwise, costly mistakes can be made that may require repeating expensive projects.

Personnel and work conditions may vary widely from one plant, or work area, to another. Monitoring safety procedures is much easier where workers are concentrated in areas where they are easily visible than in situations where they are widely scattered. Significant differences are found in management/employee relationships and in motivation, education, and communication skills of hearing conservationists from one location to another. As a result, a program that works well in one place may fail in others. It is generally necessary, therefore, to customize the program for each situation.

Obviously, top management must support the program, and others in middle management should be enthusiastic and knowledgeable. Perhaps even more important are the floor supervisors and the plant nurses. These key persons often have the respect of workers, and they may be the only persons at the plant with whom some workers will communicate freely. The floor supervisor must be genuinely concerned about the health and safety of the employee and be able to answer questions about the effects of noise exposures, the use of safety equipment, and the overall importance of the program.

## 9.8 HEARING PROTECTORS

### 9.8.1 General

Personal hearing protectors can provide adequate protection against noise-induced hearing impairment in a high percentage of industrial work areas if the protectors are properly selected, fitted, and worn. In many cases, however, protectors are not worn effectively, if at all.

Even though the use of hearing protectors is one of the most important parts of a hearing conservation program, only a fraction of the total budget is spent for this purpose. Ironically, a much larger percentage of hearing conservation effort and money is often spent over long time periods on hearing threshold measurements that essentially monitor the effectiveness of poor or nonexistent hearing conservation programs.

Hearing threshold measurements are important when action is taken based on the results. In fact, the only practical means for evaluating the effectiveness of personal protectors is to monitor thresholds periodically. If no hearing losses are observed – other than those due to the aging process – the program may be considered to be successful. Noise-induced hearing impairments usually develop slowly, however; so it may take years for the results from a hearing monitoring program to become meaningful. The careful selection of protectors and a continuing hearing conservation program, including close supervision by floor supervisors, are therefore very important.

An effective hearing conservation program seldom develops automatically simply by making hearing threshold or noise measurements. Noise does not have to be painful to be potentially harmful; so many employees do not understand the need for wearing protectors. A significant effort must be made to develop and maintain an effective program.

### 9.8.2 Performance Limitations

A primary limitation of protection afforded by a hearing protector is the way it is fitted and worn. Other important limitations of a hearing protector depend upon its construction and on the physiological and anatomical characteristics of the wearer. Sound energy may reach the inner ears of persons wearing protectors by four different pathways: (i) by passing through bone and tissue around the protector; (ii) by causing vibration of the protector, which in turn generates sound into the external ear canal; (iii) by passing through leaks in the protector; and (iv) by passing through leaks around the protector. These pathways are illustrated in Figure 9.16.

Even if there are no acoustic leaks through or around a hearing protector, some noise reaches the inner ear by bone and tissue conduction or protector vibration if noise levels are sufficiently high (Berger 1984). The practical limits set by the bone and tissue conduction threshold vary significantly among individuals, and among protector types, generally from about 40 to 55 dB. Limits set by protector vibration also vary widely, generally from about 25 to 40 dB, depending upon the protector type and design and on the materials used. Contact surface area and compliance of materials are major contributing factors. The results of studies on these limitations are influenced significantly by procedures and techniques used and by the choice of subjects. These wide ranges of performance are therefore of limited value for individuals (Nixon and Knoblach 1974; Nixon and von Gierke 1969). If hearing protectors are to provide noise reduction values approaching practical limits, acoustic leaks through and around the protectors must be minimized by proper fitting and wearing.

*Perhaps the major reason why hearing protective devices fail to protect employees from noise exposure is because the protection is not always worn.* Workers fail to wear hearing protection for comfort, because they misplaced or forgot their protection, or because they want to communicate or listen to the noise without the frequency shaping caused by the device. Whatever the reason, failure to wear hearing protection devices (HPDs) for even a small fraction of a working shift will greatly reduce the effective protection. For instance, an HPD with a noise reduction rating (NRR) of 25, not worn for 15 minutes out of an 8-hour workday, will lose about 10 dB of protection (Figure 9.17).

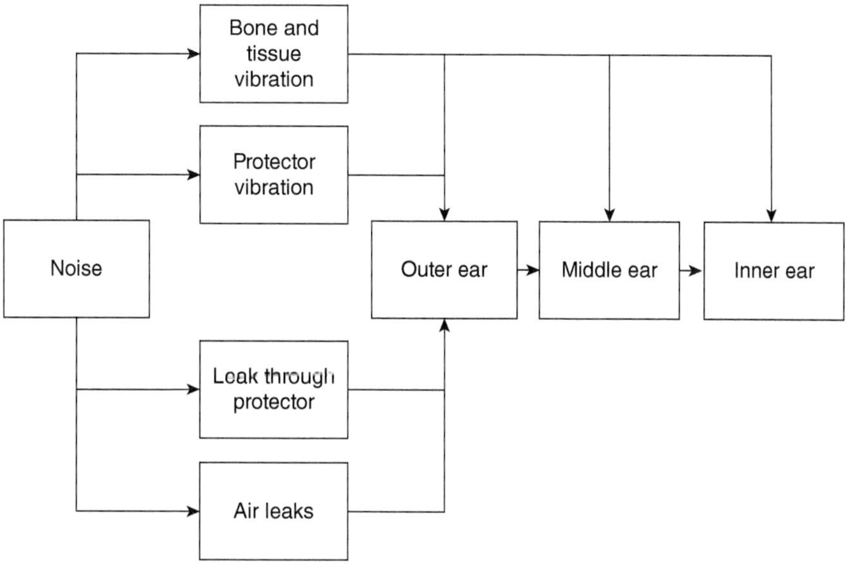


**Figure 9.16** Noise pathways to the inner ear.

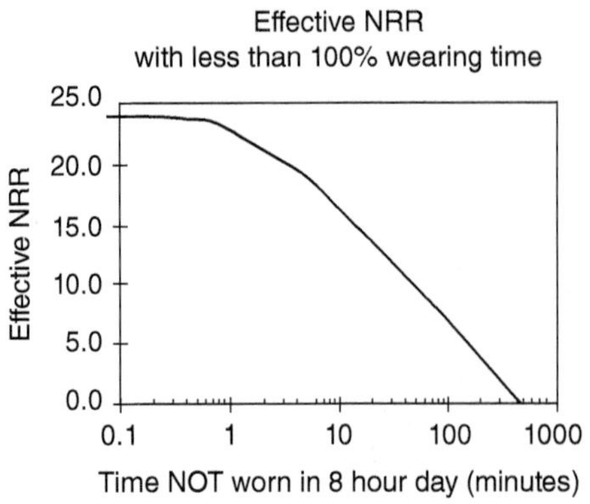


**Figure 9.17** Effective NRR vs. reported NRR when an HPD with an NRR of 25 dB is not worn full time.

### 9.8.3 Types of Hearing Protectors

Hearing protectors usually take the form of either insert types that seal against the ear canal walls or muff types that seal against the head around the ear. There are also concha-seated devices that provide an acoustic seal at the entrance of the external ear canal. There is no "best" type for all situations. However, some types are better than others for use in specific noise exposures, for some work activities, or for some environmental conditions. The 1994 publication "The NIOSH Compendium of Hearing Protection Devices" contains descriptions and laboratory attenuation data for all the HPDs currently available in the United States (Franks et al. 1994).

***9.8.3.1 Insert Types*** Ear canals differ widely in size, shape, and position among individuals and even within the same individual. Several different types of earplugs may be required to fit the wide differences in ear canal configurations. No single insert-type protector is best for all individuals, nor for all situations.

Ear canals vary in cross-sectional diameter from about 3 to 14 mm, but most are between 5 and 11 mm. Most ear canals are elliptically shaped, but some are round, and many have only a small slit-like opening that may open into a large diameter. Some canals are directed in a straight line toward the center of the head, but most are directed toward the front of the head, and they may bend in various ways.

HPDs must fit snugly and have an airtight seal to be effective. Most ear canals can be opened and straightened by pulling the pinna back, or directly away from the head, making it possible to seat many earplugs securely. For comfort and plug retention, ear canals must return to their approximate normal configuration once the protector is seated.

*9.8.3.1.1 Single- and Multiple-Sized Molded Earplugs* Molded earplugs (for instance, Figure 9.18) are usually made of a silicon-type material and are available in a variety of styles. Often, these plugs are of a "flanged" design, with one or more flanges. The single-sized molded earplugs may not provide consistently high levels of protection for a large range of ear canal sizes and shapes. Hence, multiple-sized molded plugs are available from many manufacturers.

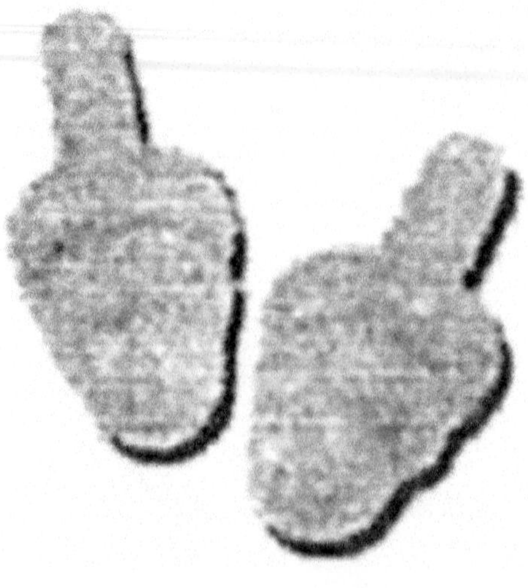

**Figure 9.18** Earplugs.

The number of sizes available usually ranges from two to five. The best molded earplugs are made of soft and flexible materials that will conform readily to the shape of the ear canal for a snug, airtight, and comfortable fit. Earplugs must be nontoxic and have smooth surfaces that are easily cleaned with soap and water. These are usually made of materials that will retain their size and flexibility over long periods of time.

The earplug size distribution for a large group of males is approximately as follows: 5% extra small, 15% small, 30% medium, 30% large, 15% extra large, and 5% larger than those supplied by many earplug manufacturers. The equal percentage of wearers for medium and large sizes indicates that many persons are fitted with earplugs that are too small.

If individuals are permitted to fit themselves, they often select a size on the basis of comfort rather than on the amount of protection provided. Some ear canals increase in size slightly with the regular use of earplugs, and earplugs may shrink in size after extended periods. Thus, if a given ear size falls between two plug sizes, it is advisable to choose the larger rather than the smaller of the two. It follows that the fit of earplugs should be checked periodically. Proper storage of molded plugs is required if long usage periods are expected. These plugs often come in zip-type plastic bags or, preferably, hard plastic cases.

*9.8.3.1.2 Malleable Earplugs* Malleable earplugs are made of materials such as silicon (swimmer's plugs), cotton, wax, glass wool, and mixtures of these materials. Typically, a small cone or cylinder of the material is hand-formed and inserted into the ear with sufficient force so that the material conforms to the shape of the canal and holds itself in position. Manufacturer's instructions should be followed regarding the depth of insertion of the plug into the canal. In any case, care should always be taken to avoid deep insertions that may cause the material to touch the eardrum.

Malleable earplugs should be formed and inserted with clean hands, because dirt or foreign objects placed in the ear canal may cause irritation or infection. They should be carefully inserted at the beginning of a work shift, and, if removed, they should not be reinserted until the hands are cleaned. For this reason, malleable plugs (and to a somewhat lesser extent, all earplugs) may be a poor choice if the work area is dirty or if the worker is subjected to intermittent high-level noises, where it may be desirable to remove and reinsert protective devices during the work period.

On the other hand, malleable plugs have the obvious advantage of fitting almost any ear canal, eliminating the need to keep a stock of various sizes, as is necessary for most molded earplugs. The cost of any disposable protector may be relatively high, depending upon how they are used.

*9.8.3.1.3 Foam-Type Plugs* Earplugs made of cylindrical or tapered foam are very popular (for instance, Figure 9.19). Generally, these plugs must be inserted about three-fourths of their length into the canal and held while they expand to fill the canal in order to provide the rated protection level. If these plugs are not inserted properly, their performance may drop significantly. It may not be possible to insert foam plugs properly in some small, rapid-bending, or slit-shaped canals.

Foam plugs are typically either composed of either vinyl or urethane-based material. The vinyl plugs are usually "stamped" out of sheets of foam material into cylinders or, less commonly, into hexagon-shaped plugs. The urethane plugs are usually tapered and manufactured in a molding process. Both types of plugs provide excellent noise attenuation *when they are fitted and worn properly*. Both types of plugs are generally comfortable to wear, although there are sometimes complaints of vinyl plugs being more "abrasive" to the ear canal. Urethane plugs are typically somewhat softer than the vinyl plugs but may be more expensive due to the more complicated manufacturing process. Both types are reusable on a limited basis, although the urethane plugs may tend to swell when wet.

**Figure 9.19** Foam-type earplugs.

**9.8.3.2 *Concha-Seated Protectors*** Protectors that provide an acoustic seal in the concha and/or at the entrance to the ear canal can be referred to as concha-seated, semi-aural, or semi-insert devices (for instance, Figure 9.20). Hearing aid-type molds used as hearing protectors are considered in this category, although some of these molds extend far into the canal. Another protector design in this class makes use of various plug shapes attached to a light-weight headband that holds the plugs in the entrance of the external ear canal. The performance and comfort of this kind of protector vary significantly among different models.

Because the materials used for hearing aid molds do not usually expand or conform after insertion, the molds must be made to fit very well in order to get the tight fit required for good attenuation. Hence significant differences in attenuation and comfort ratings may be found among these protectors made by different manufacturers.

If individually molded protectors are well made, protection levels are typically very good. Also, when made properly, these protectors can be expected to provide among the most consistent protection levels in daily use of all protector types. The initial cost of individually molded protectors is relatively high, but most should last a long period of time.

**9.8.3.3 *Earmuffs*** Manufacturers of muff-type hearing protectors (for instance, Figure 9.21) usually offer a choice of two or more models having different physical characteristics.

The performance levels of muff-type protectors depend upon many individual factors and how well these factors blend together. Generally, larger and heavier protectors provide greater protection, but the wearing comfort may not

**Figure 9.20** Concha-seated hearing protectors.

**Figure 9.21** Earmuffs.

be as good for long wearing periods. Hence workers often accept lightweight protectors more readily than the larger ones. Other important factors include the following:

1. The suspension must distribute the force uniformly around the seal, with slightly more force at the bottom to prevent leaks in the hollow behind the ear.
2. The suspension mounting should be such that the proper seal is made against the head automatically without adjustments by the wearer.
3. The earcups must be formed of a rigid, dense, imperforate material to prevent leaks and significant resonances.
4. The size of the enclosed volume within the muff shell (particularly for low frequencies) and the mass of the protectors are directly related to the attenuation provided.
5. The inside of each earcup should be partially filled with an open-cell material to absorb high-frequency resonant noises and to damp movement of the shell. The material placed inside the cup should not contact the external ear; otherwise discomfort to the wearer and soiling of the lining may result.

Ear seals should have a small circumference so that the acoustic seal takes place over the smallest possible irregularities in head contour and leaks caused by jaw and neck movements are minimized. However, a compromise between the small seal circumference and the number of persons that can be fitted properly must be made.

Earmuff cushions are generally made of a smooth plastic envelope filled with a foam or fluid material. Because skin oil and perspiration may have adverse effects on cushion materials, the soft and pliant cushions may tend to become stiff and to shrink after extended use. Fluid-filled cushions can provide superior performance, but they occasionally have a leakage problem. Foam-filled seals should have small holes to allow air to escape within the muff when mounted. Seals filled with trapped air may vibrate, causing a loss of attenuation in low frequencies. Most earmuffs are equipped with easily replaceable seals.

Earmuffs normally provide maximum protection when placed on flat and smooth surfaces; thus less protection should be expected when muffs are worn over long hair, glasses, or other uneven surfaces. Glasses with plastic temples may cause losses in attenuation of from 1 to 8 dB. In some cases, this loss of protection can be reduced substantially if small, close-fitting, wire or elastic-band temples are used. Acoustic seal covers provided to absorb perspiration may also reduce the amount of attenuation by several decibels, because noise may leak through porous materials.

Because the loss of protection is directly proportional to the size of the uneven obstructions under the seal, every effort should be made to minimize these obstructions. If long coarse hair or other significant obstructions cannot be avoided, it may be advisable to use earplugs.

The force applied by muff suspensions is often directly related to the level of noise attenuation provided. On the other hand, the wearing comfort of a muff-type protector is generally inversely related to the suspension forces; so a compromise must be made between performance and comfort. Muff suspensions should never be deliberately sprung to reduce the applied force. Not only may a loss of attenuation be expected, but the distribution of force around the seal may be changed, which may cause an additional reduction in performance. To assure the expected performance, the applied force should be measured periodically, and the muff should be visually inspected.

Muff-type protectors are sometimes chosen because their use can be monitored from greater distances than insert-type protectors. They are also easier to fit than insert types because one size fits most persons. Comfort may also be better for muff-type protectors for some persons in some work areas. However, muff-type protectors may be very uncomfortable to wear in hot work areas, particularly when the work involves a vigorous activity. Muff-type protectors are also a poor choice when work is performed in areas having limited headspace. Contact with vibrating machinery is likely to introduce high sound pressure levels inside the muff.

***9.8.3.4 Electronic Earmuffs*** Several types of electronic earmuffs are currently available. These are usually referred to as "active" muffs, and they include muffs with electronic clipping circuits and with noise-canceling circuits.

The clipping circuits typically have a microphone on the outside shell of one or both muff cups that senses the sound outside the protector and provides input to an amplifier. The amp is used with a speaker inside the cup to let the wearer hear the ambient sounds at a safe level, usually limited to about $L_A = 80$ dB. In some of these devices, equalization is performed on the signal to enhance the speech frequencies.

Noise-canceling hearing protectors have recently made a modest impact in the hearing protector market. These devices sample the noise that has penetrated the muff cup, reverse the phase of the sampled signal, and reintroduce the out-of-phase signal into the cup through a speaker. The magnitude of the attenuation provided by the noise-canceling circuits is greatest at low frequencies, and there is usually no attenuation provided at frequencies above about 1000 Hz. In fact, some of these devices unintentionally provide a small amount of amplification in a narrow frequency range around 1000 Hz. Noise canceling is an evolving technology, and it is likely the future will bring better performance and lower costs.

Electronic muffs are typically heavier and more costly than passive protectors. Electronic clipping-type protectors are particularly useful in intermittent noise exposures where communication is required. These muffs do not require removal for effective communication in this type of noise environment. Noise-canceling muffs are typically used in very-high-noise environments where traditional muff-type protectors do not provide sufficient low-frequency attenuation.

When used in passive mode (i.e. with the electronics deactivated), electronic muffs typically perform somewhat poorer than the muff would if there were no electronics present. The electronic circuit and speakers occupy space in the muff cup and therefore reduce enclosed volume of the cup. The reduced cup volume leads to poorer performance, especially in the low frequencies. Care must be taken when designing these types of protectors that all boards in the cup are solidly mounted, preventing vibration and resonance.

***9.8.3.5 Moderate or Flat Attenuation Hearing Protectors*** Employees and hearing conservationists are often attracted to protectors advertised as providing a filter that allows speech to be heard but blocks harmful noise. These filter-type devices generally provide less protection than conventional protectors, particularly at frequencies below 1000 Hz. These protectors may afford adequate protection in modest noise exposure levels, and they may provide better communication than conventional protectors. In most cases, filter-type protectors are more likely to be acceptable for use in noises having major components above 1000 Hz. The simplest of these protectors is simply vented with small holes through the protectors, permitting low frequencies (below about 1000 Hz) to pass without significant attenuation.

More recently, "flat attenuation" hearing protectors have been introduced, in both muff and insert styles. The insert types are sometimes referred to as "musician's plugs," as they are intended to provide a moderate amount of attenuation that is approximately equal at all frequencies. The "even" or flat attenuation has the benefit of not distorting the speech or music signal by attenuating some frequencies more than others. If flat attenuation protectors are not fitted properly, it is likely that the low-frequency attenuation will decrease, and the overall attenuation will not be as flat as intended.

There is a genuine need for moderate-protection hearing protectors in industry. These protectors will provide adequate protection in most work areas and allow better communication for many individuals. The hearing conservationist should not fall into the trap of thinking that "more is better" in terms of NRRs. Instead, protectors should be selected that provide adequate protection, that maximize communication, and that are worn consistently throughout the duration of the noise exposure.

***9.8.3.6 Double Protection: Protection Provided by Using Both Insert and Muff Types*** The combined attenuation from wearing both a muff-type and an insert-type protector cannot be predicted accurately because of complex coupling factors. If the attenuation values of the muff and insert protectors are about the same in a frequency band, the combined attenuation should be 3–6 dB greater than the higher of the two individual values. If one of the two protectors has an attenuation value that is significantly higher in a band, the increased attenuation by wearing both will be only slightly greater than the higher of the two.

***9.8.3.7 Summary of Advantages and Disadvantages of Protector Types*** Both insert- and muff-type hearing protectors have distinct advantages and disadvantages. Some *advantages of insert-type protectors* are as follows:

1. They are small and easily carried.
2. They can be worn conveniently and effectively without interference from glasses, headgear, earrings, or hair.
3. They are normally comfortable to wear in hot environments.
4. They do not restrict head movement in close quarters.
5. The cost of sized earplugs (except for some throwaway types and molded protectors) is significantly less than muffs.

Some *disadvantages of insert-type protectors* are as follows:

1. Almost all insert protectors require more time and effort for proper fitting than muffs.
2. The amount of protection provided by plugs is more variable between wearers than that of muff-type protectors.

3. Dirt may be introduced into the ear canal if earplugs are inserted with dirty hands.
4. The wearing of earplugs is difficult to monitor because they cannot be seen at a significant distance.
5. Earplugs should be worn only in healthy ear canals, and even then, acceptance may take time for some individuals.

Some *advantages of muff-type protectors* are as follows:

1. A good muff-type protector generally provides more consistent attenuation among wearers than good earplugs.
2. One size fits most heads.
3. Muffs are more easily seen at a distance, making program enforcement easier.
4. At the beginning of a hearing conservation program, muffs are usually accepted more readily than are earplugs.
5. Muffs can be worn despite minor ear infections.
6. Muffs are less easily misplaced or lost.

Some *disadvantages of muff-type protectors* are as follows:

1. They may be uncomfortable in hot environments.
2. They are not easily carried or stored.
3. They are not convenient to wear without interference from glasses, headgear, earrings, or hair.
4. Usage or deliberate bending of suspension band may reduce protection significantly.
5. They may restrict head movement in close quarters.
6. They are more expensive than most insert-type protectors.

***9.8.3.8 Summary*** The proper use of a good hearing protector can provide adequate protection in most work environments where it is not feasible to use engineering control measures. Special care should be taken to obtain the best protectors for a given purpose to assure adequate protection and communication. A continuing effort must also be made to ensure that protectors are used properly and to monitor hearing thresholds regularly to be sure that workers are not being exposed to harmful noises at or away from work or to ascertain that they may have other problems with their hearing.

An employer may be held responsible for any high-frequency hearing impairment regardless of the cause, because it is often impossible to apportion responsibility for a hearing impairment. If an STS is found on an annual audiogram (see Section 9.6), the cause must be investigated. If it is determined that noise exposures may be the problem, it may be necessary either for this person to use different protectors or a combination of insert- and muff-type protectors or to limit the duration of exposure by administrative means.

### 9.8.4 Selection of Hearing Protectors

Reasons for ineffective hearing protector programs often may be pinpointed by complaints from persons wearing protectors. The most common complaints are related to comfort and communication. Ideally, a hearing protector should be selected with the following objectives in mind:

1. The ear must be protected with an adequate margin of safety, but without unnecessarily reducing important communication (hearing warning signals, machines, speech, etc.). The protector's attenuation characteristics therefore should be selected to match those of the noise exposure spectra as closely as possible.
2. Environmental conditions and the kind of activity required by the job may significantly influence the comfort and general acceptability of a hearing protector. For example, insert-type protectors may be a better choice than muff types for use in high temperatures, vigorous activities, or close quarters. Muff types may be preferred to insert types in very dirty areas, for very sensitive ear canals, or for ease of monitoring.
3. Some wearers cannot be fitted properly by certain types of protectors. Performance, comfort, and/or effectiveness may be significantly enhanced by the proper choice of protector for some individuals. Wearing time is often an important consideration in making this decision.
4. It is difficult to predict or understand the acceptance of protector types at times. Important factors may include (i) the wearer's appearance or hairstyle while wearing the protector, (ii) something said about particular models, or (iii) just a desire to be different from others.

The effectiveness of a hearing conservation program, job performance, health and safety considerations, and morale may be adversely affected if attention is not given to all these factors. In particular, attention should be given to requests for changes in protector types. Obviously, there is no single protector, or protector type, that is best for all individuals or all situations. A choice of several protectors should be offered. This action also reinforces the importance of the program.

The acceptance of protectors can be helped, in some cases, by issuing selected protectors to management a few days before they are issued to workers. This procedure may cause some employees to react with demands that they be given these protectors, too, thereby creating a more positive attitude when protectors are issued. In all cases supervisory personnel and visitors must be required to obey all safety and health rules if safety equipment is to be used effectively. Exceptions cannot be tolerated.

***9.8.4.1 Hearing Protector Ratings*** Hearing protectors are often selected with consideration being given only to the magnitude of a single-number performance rating. Generally, this rating is the Environmental Protection Agency (EPA) NRR (Environmental Protection Agency 1979). Although a single-number rating is useful in initiating a program when more precise octave or one-third octave noise measurement data are not available, its use requires large safety factors because of the lack of precision.

OSHA regulations specify that a safety factor of 7 dB is to be subtracted from the NRR when $A$-weighted exposure levels are used. The calculation of the NRR includes additional safety factors of two times the standard deviation plus 3 dB for spectral uncertainty. Therefore, the total safety factor applied to laboratory data because of the poor accuracy of the NRR can easily exceed 15 dB.

Even with the large safety factors now being used with the NRR, there are situations where some individuals may not be adequately protected. For example, the NRR misleads the user of muff-type hearing protectors in typical steel industry noise exposure spectra by more than ±8 dB as compared with calculations based on the more accurate NIOSH Method 1 (American Iron and Steel Institute 1983).

**Steps for Calculating the Amount of Protection Provided by an HPD Using the NIOSH Method 1 (Lempert 1984; NIOSH 1982) ("Long Method")**

1. Measure and record the octave band exposure levels.
2. Adjust the octave band values with the $A$-frequency weightings (Table 9.6). The overall $A$-frequency-weighted sound exposure level ($L_A$) for unprotected ears is equal to the logarithmic sum (see Section 9.2.4.2).
3. Record the mean hearing protector attenuation values supplied by the protector manufacturer and two times the corresponding standard deviation value.
4. Subtract the mean attenuation values from the corresponding $A$-frequency-weighted levels, and then add the two standard deviations to the result to obtain the $A$-frequency-weighted exposure levels under the protector.
5. Combine the $A$-frequency-weighted exposure levels to obtain an estimate ($L_{A[\text{protector}]}$) of the highest sound pressure level to which a population of wearers of this protector would be exposed 97.5% of the time *if the protectors are fitted and worn properly*.

Three examples of noise exposure calculation using the NIOSH Method 1 are presented. In Example 9.10, the sound pressure levels in each octave band are equal, resulting in a "flat" exposure. This example demonstrates how the NIOSH Method 1 used with a flat noise exposure is similar to the NRR calculation. In Example 9.11, the noise exposure is "falling," with lower sound pressure levels in the higher-frequency octave bands than in the lower-frequency bands. In this case, the exposure to the ear is significantly greater than indicated by calculations using the NRR because the NRR tends to overestimate the protection afforded in primarily low-frequency exposures. Example 9.12 demonstrates the opposite situation with a "rising" noise exposure. The NRR tends to underestimate the protection afforded an HPD wearer in this type of noise exposure, and therefore the exposure to the ear is less than is expected if calculations using the NRR are used.

The NRR is calculated assuming an exposure to pink noise (equal energy in each octave band) from 125 to 8000 Hz:

$$\text{NRR} = L_{C(\text{pink noise})} - L_{A(\text{protector})} - 3 \tag{9.22}$$

where $L_{C(\text{pink noise})}$ is the $C$-frequency-weighted sound pressure level for the pink noise exposure, $L_{A(\text{protector})}$ is the $A$-frequency-weighted sound pressure level under the protector, and 3 (dB) is a safety factor to correct for spectral differences between the assumed pink noise spectrum and actual noise exposures.

***Example 9.10:*** The manufacturer of an HPD reports an NRR of 23 with the following octave band attenuation values and standard deviations.

| Frequency (Hz) | 125 | 250 | 500 | 1000 | 2000 | 4000 | 8000 |
|---|---|---|---|---|---|---|---|
| Attenuation (dB) | 18 | 20 | 30 | 39 | 34 | 32[a] | 36[b] |
| Standard deviation | 3.5 | 2.5 | 3.5 | 3 | 3 | 3[a] | 3.5[b] |

[a] Average of 3150 and 4000 Hz.
[b] Average of 6300 and 8000 Hz.

Verify the NRR.

The *C*-frequency-weighted sound pressure levels are determined from a spectrum with constant octave band levels. One hundred decibel is arbitrarily chosen.

| Frequency (Hz) | 125 | 250 | 500 | 1000 | 2000 | 4000 | 8000 |
|---|---|---|---|---|---|---|---|
| Constant octave band level | 100 | 100 | 100 | 100 | 100 | 100 | 100 |
| *C*-frequency weighting (from Table 9.6) | −0.2 | 0.0 | 0.0 | 0.0 | −0.2 | −0.8 | −3.0 |
| Calculate $L_C$ | 99.8 | 100 | 100 | 100 | 99.8 | 99.2 | 97 |

$$L_C = 10\log\left(\sum 10^{9.98} + 10^{10.0} + 10^{10.10} + \cdots\right) = 108.0$$

Using NIOSH Method 1 and the pink noise "flat" spectrum, calculate the exposure to the ear and the attenuation provided by the hearing protector.

| | Octave band center frequency (Hz) | | | | | | |
|---|---|---|---|---|---|---|---|
| Step | 125 | 250 | 500 | 1000 | 2000 | 4000 | 8000 |
| 1. Octave band exposures (from above) | 100.0 | 100.0 | 100.0 | 100.0 | 100.0 | 100.0 | 100.0 |
| 2a. *A*-frequency weighting (from Table 9.6) | −16.1 | −8.6 | 3.2 | 0.0 | +1.2 | +1.0 | −1.1 |
| 2b. $L_A$ (100 dB + step 2) | 83.9 | 91.4 | 96.8 | 100.0 | 101.2 | 101.0 | 98.9 |
| 2c. Optional (combine octave band levels) | | | | $L_A$ = 107.0 dB | | | |
| 3a. Mean attenuation (from manufacturer) | 18 | 20 | 30 | 39 | 34 | 32 | 36 |
| 3b. Standard deviation times 2 (from manufacturer) | 7 | 5 | 7 | 6 | 6 | 6 | 7 |
| 4. *A*-frequency-weighted levels in protected ear (step 2b–step 3a + step 3b) | 72.9 | 76.4 | 73.8 | 67.0 | 73.2 | 75.0 | 69.9 |
| 5. Combine octave band levels | | | | $L_{A(protector)}$ = 81.9 dB | | | |

NRR = $L_C - L_{A(protector)} - 3$ = 108.0 − 81.9 − 3 = 23.1, which is rounded to 23 dB.

***Example 9.11:*** What is the protection calculated according to NIOSH Method 1 for the same hearing protector for a noise source with the following "falling" octave band levels?

| Frequency (Hz) | 125 | 250 | 500 | 1000 | 2000 | 4000 | 8000 |
|---|---|---|---|---|---|---|---|
| SPL (dB) | 85 | 96 | 86 | 80 | 75 | 68 | 67 |

Using the procedure listed above, the following table is constructed.

| | Octave band center frequency (Hz) | | | | | | |
|---|---|---|---|---|---|---|---|
| Step | 125 | 250 | 500 | 1000 | 2000 | 4000 | 8000 |
| 1. Octave band exposures (measured) | 85 | 96 | 86 | 80 | 75 | 68 | 67 |
| 2a. *A*-frequency weighting (from Table 9.6) | −16.1 | −8.6 | −3.2 | 0.0 | +1.2 | +1.0 | −1.1 |
| 2b. $L_{protector}$ (step 1 + step 2) | 68.9 | 87.4 | 82.8 | 80 | 76.2 | 69 | 65.9 |
| 2c. Optional (combine octave band levels) | | | | $L_A$ = 89.5 dB | | | |
| 3a. Mean attenuation (from manufacturer) | 18 | 20 | 30 | 39 | 34 | 32 | 36 |
| 3b. Standard deviation times 2 (from manufacturer) | 7 | 5 | 7 | 6 | 6 | 6 | 7 |
| 4. *A*-frequency-weighted levels in protected ear (step 2b-step 3a + step 3b) | 57.9 | 72.4 | 59.8 | 47 | 48.2 | 43 | 36.9 |
| 5. Combine octave band levels | | | | $L_{A(protector)}$ = 72.8 dB | | | |

Therefore, the protection provided by the HPD in the specified "falling" noise spectrum would be 89.5 – 72.8 = 16.7 dB, which is 6.3 dB less than the NRR.

***Example 9.12:*** Again, using the same hearing protector, what would be the protection calculated according to NIOSH Method 1 for a noise source with the following "rising" octave band levels?

| Frequency (Hz) | 125 | 250 | 500 | 1000 | 2000 | 4000 | 8000 |
|---|---|---|---|---|---|---|---|
| SPL (dB) | 85 | 96 | 100 | 109 | 111 | 110 | 104 |

Using the procedure listed above, the following table is constructed.

| | Octave band center frequency (Hz) | | | | | | |
|---|---|---|---|---|---|---|---|
| Step | 125 | 250 | 500 | 1000 | 2000 | 4000 | 8000 |
| 1. Octave band exposures (measured) | 85 | 96 | 100 | 109 | 111 | 110 | 104 |
| 2a. *A*-frequency weighting (from Table 9.6) | −16.1 | −8.6 | −3.2 | 0.0 | +1.2 | +1.0 | −1.1 |
| 2b. $L_{\text{protector}}$ (step 1 + step 2) | 68.9 | 87.4 | 96.8 | 109 | 112.2 | 111 | 102.9 |
| 2c. Optional (combine octave band levels) | | | | $L_A = 116.0\,\text{dB}$ | | | |
| 3a. Mean attenuation (from manufacturer) | 18 | 20 | 30 | 39 | 34 | 32 | 36 |
| 3b. Standard deviation times 2 (from manufacturer) | 7 | 5 | 7 | 6 | 6 | 6 | 7 |
| 4. Four-frequency-weighted levels in protected ear (step 2b-step 3a + step 3b) | 57.9 | 72.4 | 73.8 | 76 | 84.2 | 85.0 | 73.9 |
| 5. Combine octave band levels | | | | $L_{A(\text{protector})} = 88.4\,\text{dB}$ | | | |

Therefore, the protection provided by the HPD in the specified "rising" noise spectrum would be 116 – 88.4 = 27.6 dB, which is 4.6 dB greater than the NRR.

The following examples demonstrate exposure calculations to be followed if octave band noise exposure data are not known and single-number estimates of exposure, such as the *A*- or *C*-weighted sound pressure level, are used.

### Steps for Estimating Worker's Noise Exposure When *C*-Frequency-Weighting Sound Pressure Level $L_C$ Is Known

Exposure is estimated by subtracting the NRR from the *C*-frequency weighting:

$$L_{A(\text{protector})} = L_C - \text{NRR} \tag{9.23}$$

where $L_{A(\text{protector})}$ is the *A-weighted* sound level exposure estimated under the protector (dB) and $L_C$ is the long-term *C-weighted* sound pressure level noise exposure *outside* the protector.

***Example 9.13:*** Calculate the estimated worker exposure for the above three frequency spectra using the *C*-frequency-weighted level.

From the calculations in Example 9.10, $L_C = 108\,\text{dB}$:

$$L_{A(\text{protector})} = 108 - 23 = 85\,\text{dB}$$

$L_C$ from Example 9.11 is:

| Frequency (Hz) | 125 | 250 | 500 | 1000 | 2000 | 4000 | 8000 |
|---|---|---|---|---|---|---|---|
| SPL (dB) | 85 | 96 | 86 | 80 | 75 | 68 | 67 |
| *C*-frequency weighting (from Table 9.6) | −0.2 | 0.0 | 0.0 | 0.0 | −0.2 | −0.8 | −3.0 |
| $L_C$ | 84.8 | 96 | 86 | 80 | 74.8 | 67.2 | 64 |

$$L_C = 10\log\left(10^{8.48} + 10^{9.6} + 10^{8.6} + \cdots\right) = 10\log\left(4.8 \times 10^9\right) = 96.8\,\text{dB}$$

$$L_{A(\text{protector})} = 96.8 - 23 \cong 74\,\text{dB}$$

$L_C$ from Example 9.12 is:

| Frequency (Hz) | 125 | 250 | 500 | 1000 | 2000 | 4000 | 8000 |
|---|---|---|---|---|---|---|---|
| SPL (dB) | 85 | 96 | 100 | 109 | 111 | 110 | 104 |
| *C*-frequency weighting (from Table 9.6) | −0.2 | 0.0 | 0.0 | 0.0 | −0.2 | −0.8 | −3.0 |
| $L_C$ | 84.8 | 96 | 100 | 109 | 110.8 | 109.2 | 101 |

$$L_C = 10\log\left(10^{8.48} + 10^{9.6} + 10^{10.0} + \cdots\right) = 10\log\left(3.10\times 10^{11}\right) = 114.9\,\text{dB}$$

$$L_{A(\text{protector})} = 114.9 - 23 \cong 92\,\text{dB}$$

**Steps for Estimating Worker's Noise Exposure When *A*-Frequency-Weighting Sound Pressure Level $L_A$ or $L_{Aeq}$ Is Known**

Accuracy is lost in calculating hearing protection when only *A*-weighted sound pressure levels are known since the NRR is based on a flat spectrum. Because of this OSHA specifies that a 7-dB safety factor must be used to compensate for the $L_C - L_A$ differences found in "typical" industrial noise. The exposure under the protector is estimated by the following:

$$L_{A(\text{protector})} = L_A - (\text{NRR} - 7) \tag{9.24}$$

where $L_{A(\text{protector})}$ is the *A*-weighted sound level exposure estimate under the protector (dB) and $L_A$ is the long-term *A*-weighted noise exposure *outside* the protector (dB).

***Example 9.14:*** Calculate the estimated worker exposure for the above three frequency spectra using the *A*-frequency-weighted levels.

In Example 9.10, $L_A$ = 107.0 dB. Therefore, the estimated exposure according to the NIOSH formula would be

$$L_{A(\text{protector})} = 107.0 - (23 - 7) = 91$$

Similarly, in Example 9.11, $L_A$ = 89.4 dB; so the estimate exposure is

$$L_{A(\text{protector})} = 89.5 - (23 - 7) \cong 74$$

From the calculations in Example 9.12, $L_A$ = 116 dB; so the estimated exposure is

$$L_{A(\text{protector})} = 116 - (23 - 7) = 100$$

| | Estimate exposure $L_{A(\text{protector})}$ | | |
|---|---|---|---|
| Spectrum | NIOSH Method 1 | Using $L_C$ | Using $L_A$ |
| Falling | 73 | 74 | 74 |
| Flat (pink noise) | 82 | 85 | 91 |
| Rising | 88 | 92 | 100 |

The table demonstrates the potential differences between the various methods of exposure calculation. The NIOSH Method 1 calculation is considered to be the most accurate of the three because it uses all available measurement data and does not rely on any single-number estimate. In these examples, the $L_C$ exposure calculation tends to be greater than the NIOSH Method 1, since it includes the 3-dB safety factor from the NRR calculation. The $L_A$ exposure calculations vary from the $L_C$ calculations because of the additional 7-dB safety factor imposed by OSHA and the significant de-emphasis of low-frequency noise in the *A*-weighting.

The large safety factors used in the single-number NRR rating may result in many workers being significantly overprotected, causing a variety of problems, ranging from injuries when warning signals are not heard to reduced work efficiency when important machine sounds are inaudible. In addition, unnecessarily inhibited communication may cause significant annoyance and stress-related effects that can in turn encourage wearers to deliberately to disable their protectors to decrease attenuation.

Another weakness of single-number ratings, such as the NRR, is that they are often based on just 1 or 2 of the 9 third octave test signals used in laboratory measurements. Generally, the controlling test signals are below 1000 Hz, and performance levels for other test signals may have little or no effect on the final NRR. For example, two protectors may have the same EPA NRR because of their limiting attenuation values at test signals centered at 250 or 500 Hz, but one of these protectors may provide more than 15 dB greater protection than the other for higher frequencies, above 1000 Hz, without this information being indicated. The NRR values in this example would be a reasonably accurate assessment of protector performance if the highest exposures were centered at 250 or 500 Hz, but if the noise exposures contain prominent high-frequency components, the NRR values may be very misleading.

The more precise NIOSH Method 1 for calculating hearing protector performance ratings can be used to obtain a more accurate estimate of protection while at the same time affording a more accurate means of maximizing communication. None of the safety factors discussed above are required when octave band sound pressure levels are used to determine exposure levels under hearing protectors.

In Europe, the most common single-number rating of hearing protector performance is the single number rating (SNR) (European Committee for Standardization 1993). The SNR has the same pitfalls as the NRR, although it is somewhat less sensitive to outlier data during the laboratory testing. The European market also utilizes a three-number rating system, called the high–middle–low (HML) system. The three numbers in the HML rating estimate the attenuation provided in the high test frequencies, middle test frequencies, and low test frequencies, respectively. The HML values are used in conjunction with measurements of *A*- and *C*-frequency-weighted sound pressure levels to obtain the HPD attenuation that can be expected in specific noise environments. The European tests are performed according to ISO 4869-1 (1990), whereas the US testing is performed according to ANSI S3.19-1974.

**9.8.4.2 *Derating the NRR*** The literature contains many references to the differences between laboratory ratings of hearing protectors and the amount of attenuation measured in the field. This discrepancy has led to derating procedures that are applied to laboratory data, which is labeled on the hearing protector packaging. Some industrial hygienists derate laboratory data by 50% before calculating exposure under the protector, and other hearing conservation programs use a straight 10-dB derating scheme. These procedures can lead to significant overprotection, which may unnecessarily inhibit communication. None of these inaccurate derating methods are necessary if the hearing conservation program utilizes (i) the laboratory ratings for guidance only, (ii) a field monitoring system as a training tool and to verify protector fit (see Section 9.8.7.1), and (iii) an HPD selection procedure that ensures that devices are selected that provide sufficient protection while avoiding overprotection. The derating schemes do not help solve the problem of poor protector usage and may exacerbate the problem of communication in noise.

### 9.8.5 Communication Without Hearing Protectors

Performance and safety aspects of a job often depend on the workers' ability to hear warning signals, machine sounds, and speech in the presence of high noise levels. The effect of noise on communication depends to a large extent on the spectrum of the noise, the hearing characteristics of the worker, and the attenuation characteristics of hearing protectors, if they are used.

For a normal-hearing person, speech communication is affected most when the noise has high-level components in the speech frequency range from about 400 to 3000 Hz. Speech interference studies (ANSI S3.14 1997) show that conversational speech begins to be difficult for a speaker and a normal-hearing listener, separated by about 2 ft when broadband noise levels approach about 88 dBA. Hearing-impaired persons have much more difficulty communicating in noise than persons having normal hearing; the degree of difficulty depends upon the amount and type of impairment.

### 9.8.6 Communication with Hearing Protectors

Few hearing protectors are selected and purchased with any thought being given to communication of any kind. Overprotection may be considered as acceptable, or even desirable, against many health and safety hazards, but overprotection against noise exposure may cause significant communication problems. Maximizing communication while wearing hearing protectors often improves safety and work efficiency conditions and can prolong the working lifetime of skilled workers having high-frequency hearing impairment. Overprotection from noise may also lead to the deliberate misuse or rejection of hearing protectors.

Hearing protectors interfere with speech communication in quiet environments for most persons, regardless of their hearing characteristics. Normal-hearing persons can often raise their voice levels to provide satisfactory communication in moderate levels of noise.

When wearing hearing protectors in noise levels between 88 and 97 dB, normal-hearing persons sometimes complain that the protectors prevent communication, although the HPDs usually do not affect the overall speech-to-noise ratio and probably do not impede communication. Above about 97-dBA background noise levels, normal-hearing persons are often able to communicate about as well with as without wearing protectors, albeit very poorly (Bienvenue and Michael 1979; Michael 1965; Michael et al. 1970). In fact, protectors may improve speech communication for some normal-hearing persons when background noise levels are higher than 97 dBA because speech-to-noise ratios are held relatively constant and distortion is reduced. Optimal communication is usually provided when the protector's attenuation characteristics are matched to those of the noise spectra. Individuals with hearing loss almost always have difficulty communicating in noisy areas.

### 9.8.7 Integration of Communication Requirements and the Use of Field Monitoring Systems

Among the problems associated with the use of hearing protective devices in industry are poor training and motivation on proper wearing techniques, differences between labeled NRRs and achievable field attenuation, and impaired communication with the use of HPDs. The use of hearing protector field monitoring systems and the careful selection of hearing protectors that maximize communication provide an opportunity for the hearing conservationist to address these issues in a methodical and well-accepted manner.

The careful selection of HPDs and verification of protector fit on the individual wearer will result in reduced incidence of noise-induced hearing loss and fewer injuries caused by a lack of communication. The comprehensive documentation of HPD selection and usage verifies effective program administration and apportions responsibility to the hearing protector wearer.

***9.8.7.1 Field Monitoring Systems*** Field monitoring systems are designed to allow the hearing conservationist to measure the attenuation provided by hearing protectors *on the individual wearer*. This capability eliminates the need for single-number ratings, such as the NRR, and misleading safety factors. The NRR is derived from laboratory test methods that are designed to estimate attenuation afforded to a *population*, not to an *individual*. Current laboratory performance ratings represent a "best-fit" situation using trained and motivated subjects under closely supervised conditions. It is likely that the attenuation achieved in the field will be significantly less than the laboratory values *unless* the wearers are properly fitted and motivated.

Field measurement systems perform several functions for the industrial hearing conservation program administrator, including (i) training of wearers in correct fitting procedures, (ii) random field sampling of protector effectiveness, (iii) documentation that training was provided and that proper protection was provided to the employee, and (iv) identification of failing or deteriorating protectors and changes in ear physiology.

Experience has shown that individual measurement systems are, in general, well received by the HPD wearers. The employees are typically interested in how the protectors function, and they appreciate the attention to their individual needs. Individual HPD attenuation measurement is particularly valuable as a training tool during the initial selection of insert-type HPDs. The hearing conservationist can assist the wearer during the initial fitting and measure the attenuation that is provided. If the attenuation is sufficient, the employee should then refit the HPD and the measurement procedure should be repeated. If the measured attenuation is sufficient after the wearer has fitted the device, documentation is provided that adequate training and protection was provided to the employee.

Portable field monitoring systems are currently available for both muffs and insert-type hearing protectors.[2]

*9.8.7.1.1 Field Monitoring of Insert-Type Protectors* The amount of noise attenuation provided to wearers of insert protectors varies widely across the general population. The attenuation provided to poorly trained plug wearers or to individuals with narrow, sharply bending, or slit-shaped ear canals may be much lower than the labeled laboratory values. Persons with large ear canals may be able to insert protectors deep into the canal without achieving a satisfactory seal, resulting in poor attenuation, especially at the lower frequencies. Visual inspection of HPD fit is not always sufficient.

For insert-type protectors, field monitoring systems are available that essentially replicate the laboratory tests as defined in ANSI S3.19-1974, except that the stimuli are presented via headphones. This test involves measuring the hearing thresholds of the HPD wearer at selected test frequencies with and without the HPD in place. The difference in hearing threshold at each test frequency is equal to the amount of noise attenuation provided by the hearing protector.

*9.8.7.1.2 Field Monitoring and Muff-Type Protectors* For muff-type protectors, existing field measurement systems include a handheld microprocessor-based unit that utilizes two microphones, one located at the entrance to the ear canal and the other located outside the cup of the muff. A digital readout displays the difference in sound pressure level between the two microphones. The unit is designed to be used at the employees' workplace; therefore, the attenuation measured will be accurate while the wearer is in *that particular noise exposure*. Use of the protector in noises with differing spectral characteristics will affect the amount of noise attenuation provided.

**9.8.7.2 *Selection of Hearing Protection Devices to Maximize Communication*** There is currently a European guidance document, EN 458, entitled "Hearing Protectors – Recommendations for Selection, Use, Care and Maintenance," that addresses the issue of selecting hearing protectors that maximize the ability to communicate. The document includes several methods of HPD selection, including the octave band method, the HML method, the HML Check, and the SNR method. All the methods are dependent on the level of the exposure at the ear (i.e. under the protectors).

A summary of the octave band method follows:

1. Using octave band noise exposure data, octave band hearing protector attenuation characteristics, and the NIOSH Method 1, calculate the $A$-weighted exposure level under the protector.
2. The allowable criteria are based on eight-hour exposures and may be selected based on the legal exposure limits (i.e. 90 dBA) or a more conservative level (i.e. 85 dBA). Compare the $A$-weighted exposure under the protector to the criteria level in Table 9.11.

For communication purposes, the careful selection of HPDs is more critical for individuals with hearing loss than for normal-hearing individuals. The range of "acceptable" or "good" HPD selections is greater for normal-hearing individuals than the range specified in EN 458. The intelligibility of noise-degraded speech does not decrease from a maximum for normal-hearing individuals until the presentation level is about 50–60-dB SPL, whereas the EN 458 definition of "overprotection" starts at 70 dBA for a criterion level of 85 dBA. As the degree of hearing loss increases, the number of "good" or "acceptable" HPD selections is reduced because of the increased possibility of overprotection.

The algorithm may be used for HPD selection for large groups of wearers. In this case, a conservative selection criteria (i.e. selection assuming the wearer has at least a moderate hearing loss) should be employed, since it is likely that at least some of the individuals in the group will be hearing impaired.

The algorithm is dependent on laboratory HPD attenuation data. The attenuation afforded to an individual HPD wearer, of course, is not necessarily equal to the lab data. The best way to implement these criteria is to use a field monitoring system to verify HPD performance and the HPD selection algorithm to determine the range of acceptable protectors. This approach documents that the employee has been fitted with protectors that provide sufficient protection while optimizing the ability to communicate.

Proper HPD selection according to these guidelines does not guarantee good communication. Obviously even the best HPD selection will not significantly improve a highly adverse communication situation. To predict the ability to communicate in noise, ANSI S3.79, "American National Standard Methods for Calculation of the Speech Intelligibility Index," should be referenced.

Properly fitted individuals are more likely to wear their protectors consistently since they will not be unnecessarily "isolated" from other workers and they will be less likely to intentionally disable protectors to decrease attenuation. Fitting HPDs using these criteria will dissuade hearing conservationists from the "more attenuation is better" attitude and create a greater demand for HPD manufacturers to develop and market comfortable HPDs with a range of

**TABLE 9.11 Recommendations for Hearing Protector Attenuation**

| Relationship between exposure under the protector (in dBA) and allowable exposure criteria (in dBA)[a] | HPD selection |
|---|---|
| Exposure > criteria | Insufficient protection |
| Criteria > exposure > (criteria −5 dB) | Acceptable |
| (Criteria −5 dB) > exposure > (criteria −10 dB) | Good |
| (Criteria −10 dB) > exposure > (criteria −15 dB) | Acceptable |
| (Criteria −15 dB) > exposure | Overprotection |

[a] Exposure (dBA) = $A$-weighted exposure level under the protector.

attenuation characteristics. The use of a field measurement system and an HPD selection procedure is educational for both the hearing conservationist and HPD wearer. The individual wearer becomes involved in the fitting process and is required to assume responsibility for the correct wearing of the protector.

***9.8.7.3 Documentation*** Continued periodic use of a field measurement system and careful selection of HPDs that provide sufficient protection without needlessly impairing communication provides the best defense possible against litigation based on hearing loss or lack of communication ability. To date, the most common type of litigation against industry in this area has been hearing loss and related physiological problems such as tinnitus. It is likely, however, that the number of cases will increase where industry is held responsible for "overprotecting" employees.

Measurement of the attenuation provided by hearing protectors, especially insert type, on the individual wearer is extremely valuable. This type of measurement should be required when the individual is initially fitted with the HPD and at least annually thereafter. Individual measurement of hearing protector performance documents that adequate and proper protection was provided to the worker and *shifts responsibility of effective HPD use to the employee.* The time and cost required for these measurements is more than justified by the reduction in liability and, more important, the protection of the individual. These measurements, used in conjunction with HPD selection criteria that optimize the ability to communicate while wearing HPDs, result in a truly comprehensive hearing protector management program.

## 9.9 NOISE CONTROL PROCEDURES

### 9.9.1 Introduction

Noise control efforts should be approached using the paradigm:

$$\text{Source} \rightarrow \text{Path} \rightarrow \text{Receiver}$$

The noise from most equipment is waste energy. For this and efficiency reasons, the best way to reduce noise is to tackle the problem at the source. Generally, reducing the noise at the source also offers the most options. Changes to the path generally involve adding barriers or enclosing the equipment, but may involve adding sound-absorbing materials. Reductions of more than a few decibels are difficult to achieve by these modifications. At the other end of the path is the receiver or affected employee. Reduction of noise exposures here are achieved by either removing the employees from the sound field, limiting time in the area, or through the use of hearing protection. While the latter is really a modification of the noise path, personal protective equipment is usually considered noise control at the receiver, and as with other industrial hygiene uses of personal protective equipment, should be the last step taken for controlling exposures – only used when other measures have proven ineffective.

### 9.9.2 Noise Control

The first step in providing quiet equipment is to make a strong effort to have purchase orders include noise limits. The desired quiet equipment will not always be available, but at least these specifications will provide an incentive for the design of quiet products. Modifications to equipment to reduce noise include closer tolerances, better assembly, balancing of rotating machinery, redesign of components, and other quality control measures. Usually these changes must be left to the manufacturer of the equipment. Users of equipment can specify the noise level that will be tolerated in new equipment purchases. Prior to the promulgation of OSHA noise regulations, few manufacturers were concerned with noise emissions. Currently, many manufacturers attempt to distinguish their products by emphasizing their lowered noise emissions. Purchasers of equipment, however, may not be aware of the cost advantages if quiet equipment is purchased – avoiding the need for hearing conservation programs and incurring less hearing loss by employees. The health and safety professional should assure that management and purchasing departments are aware of the need to purchase quiet equipment. Proper maintenance of equipment must also be stressed since virtually all mechanical equipment becomes noisier as components wear.

If the need is to reduce the noise from existing machinery, consider both generation and radiation of sound for possible noise control measures. Once generated, the noise will be from (i) the direct sound field, (ii) reverberant sound field, or (iii) a structure-borne path. Finally, the only true measure for reducing operator exposure at the receiver is to reduce the amount of time the operator spends in the sound field. Providing a noise enclosure for the operator can also be considered as a receiver noise control measure.

The following systematic approach could be used for controlling noise. Each will be discussed later in somewhat greater detail.

**Source: Generation**

1. Reduce impact noise.
2. Reduce or eliminate aerodynamically generated noise.
3. Reduce or eliminate any resonance effects.
4. Modify or replace gears or bearings.
5. Reduce unbalance in rotating systems.

**Source: Radiation**

6. Move the machinery to a new location – distant from exposed personnel.
7. Provide vibration isolation to reduce the radiation of noise from the surface on which the machinery is mounted.
8. For large heavy machinery, use an inertia block.
9. Insert flexible connectors between the machine and any ductwork, conduit, or cables.
10. Reduce or modify the surfaces that radiate noise.
11. Apply vibration isolation to machine housing.
12. Use active noise cancelation or active vibration cancelation.

**Path: Direct**

13. Provide a partial or full enclosure around the machine.
14. Use sound-absorptive materials.
15. Use an acoustical barrier to shield, deflect, or absorb energy.
16. Reduce the leakage paths permitting noise to leak from enclosures.

**Path: Reverberant**

17. Apply sound-absorptive materials to walls, ceiling, or floor.
18. Reduce reflections by moving equipment away from corners or walls.

**Path: Structure Borne**

19. Use ducts lined with absorptive materials.
20. Use wrapping or lagging on pipes to increase their sound insulation.
21. Reduce turbulence from liquid or gaseous flows.
22. Add mufflers.

**Receiver**

23. Use enclosure or control room to house operators.
24. Provide hearing protection to operators.
25. Reduce amount of time operators are allowed to work in high-noise areas.

#### 9.9.2.1 Reduce Generated Noise

*9.9.2.1.1 Reduce Impact Noise* Mechanical and materials handling devices commonly produce noise from impact. Noise can be reduced by:

1. Reducing the dropping height of goods collected in boxes or bins (Figures 9.22 and 9.23).
2. Using soft rubber or plastic to receive hard impacts.

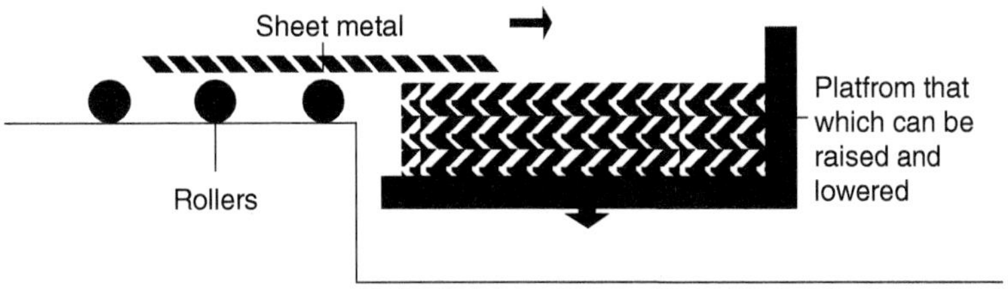


**Figure 9.22** Reduce height of drops.

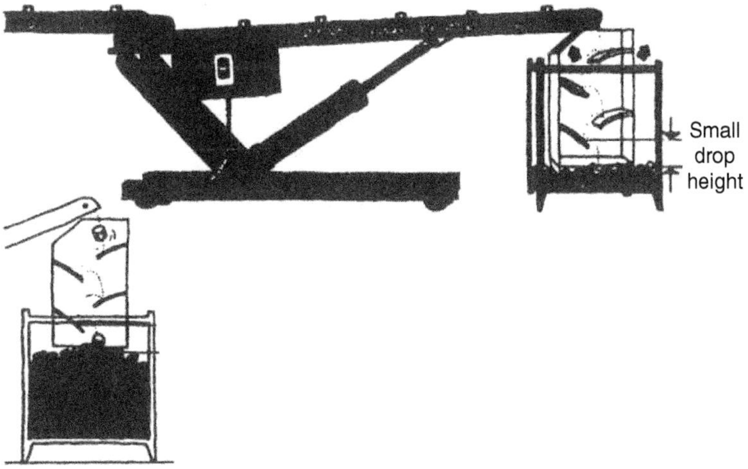


**Figure 9.23** Reduce drop height and use rubber flaps to slow fall speed.

3. Increasing the rigidity of containers receiving impact goods and add damping materials – especially to large surfaces.
4. Using belt conveyors, which are generally quieter than roller conveyors.
5. Regulating the speed or cycle time of conveyors to prevent collisions and excess noise.

*9.9.2.1.2 Reduce or Eliminate Aerodynamically Generated Noise* A wide variety of methods can be used for noise control from aerodynamically generated noise.

9.9.2.1.2.1 Change the Character of the Noise Where great distances are involved, such as outdoors, air reduces high-frequency sounds more than lower frequencies. Atmospheric attenuation for pure tones at 70% relative humidity and 10 °C is given by

$$A_{\text{atm,f}} = \frac{r}{1000\,\text{m}}\left[0.6 + 1.6\left(\frac{f}{1000\,\text{Hz}}\right) + 1.4\left(\frac{f}{1000\,\text{Hz}}\right)^2\right] \tag{9.25}$$

where $r$ is the distance between the source and receiver (m) and $f$ is the frequency (Hz).

For environmental noise control, replacing a source with one of higher frequency may reduce the sound level at typical property line distances (Figure 9.24). Note, however, that shifting to higher frequencies will likely bring the source into a frequency range where the ear is more sensitive. Therefore, careful consideration of perceived loudness must be made before applying this solution.

9.9.2.1.2.2 Reduce the Area of the Source Large surfaces when vibrating will produce high sound levels. Consider replacing solid plates, when possible, with expanded metal, wire mesh, or perforated metal (Figure 9.25).

9.9.2.1.2.3 Change the Source Dimensions So Noise Is Canceled at the Edges At the edges of large vibrating plates, the pressure and rarefaction waves tend to cancel. The same principle can be applied using long narrow surfaces instead of square surfaces (Figure 9.26).

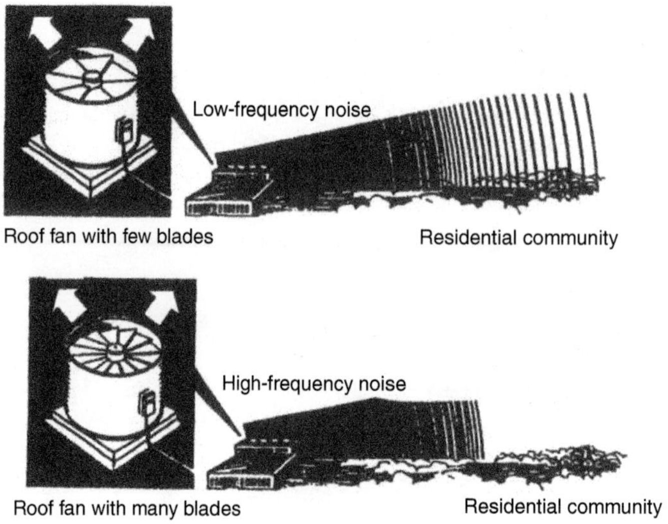


**Figure 9.24** Convert fan noise to higher frequencies, which are more attenuated by atmosphere.

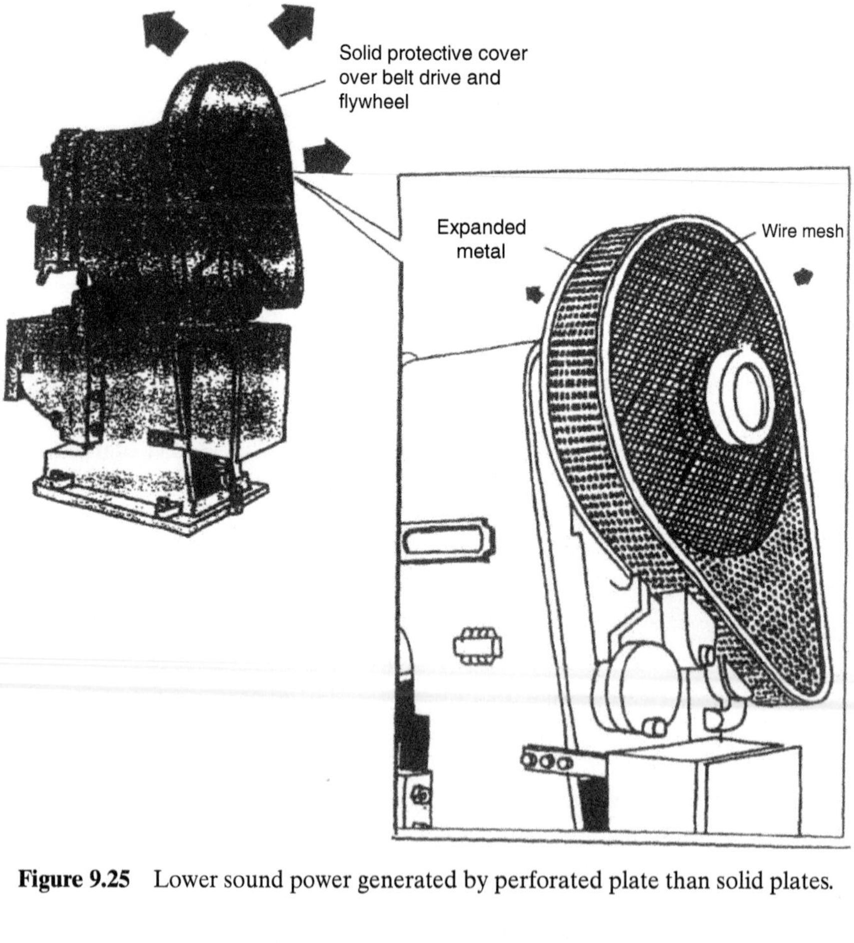


**Figure 9.25** Lower sound power generated by perforated plate than solid plates.

9.9.2.1.2.4 Reduce or Remove Interrupted Wind Noise Causing Tones Wind instruments use the principle of air blowing across an edge to produce standing-wave pressure vibrations. When tonal noise is produced by machinery, it may be possible to eliminate the wind and thus the noise (Figure 9.27).

9.9.2.1.2.5 Reduce Turbulence in Fluids Smooth laminar flow of fluids through ducts and pipes does not generate noise. The more turbulent the flow, the greater the noise. Vapor bubbles can be created by abrupt changes in the flow of fluids. Providing gradual transitional changes in cross-sectional area reduces the chance of forming these bubbles (Figure 9.28).

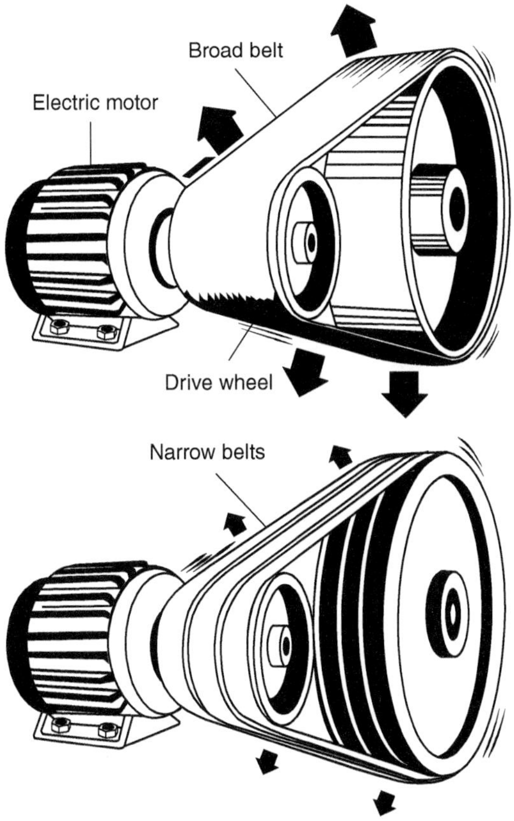


**Figure 9.26** Less sound power produced by several narrower belts than a single broad belt.

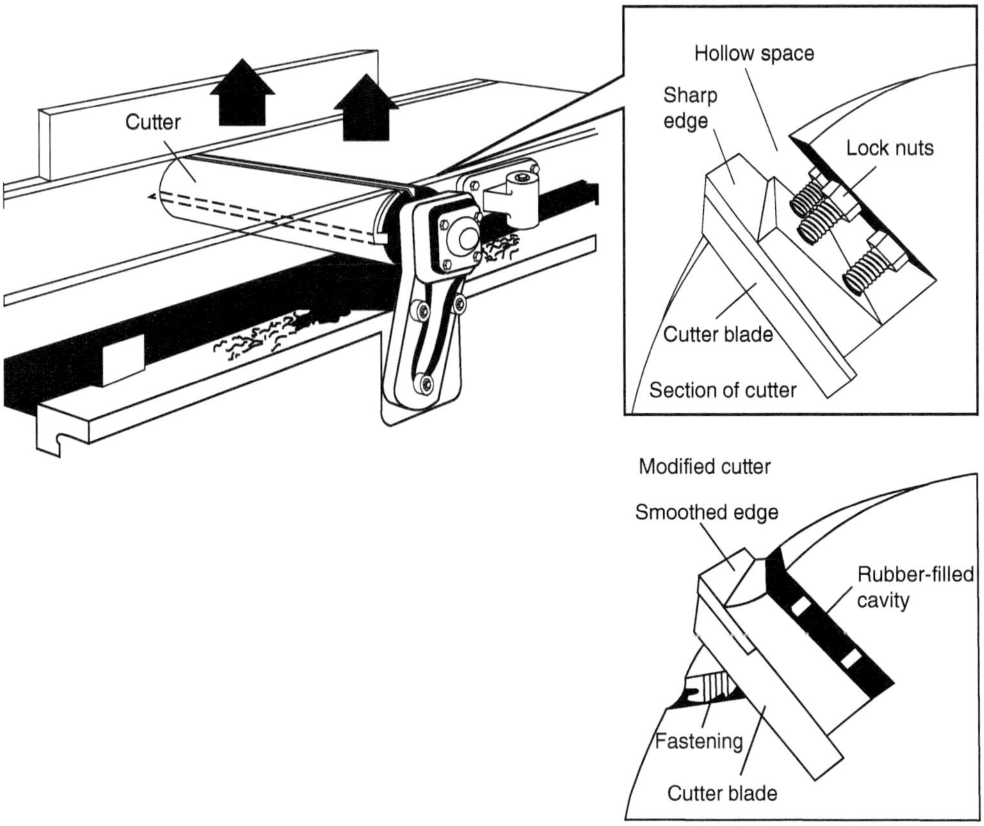


**Figure 9.27** Reduce tonal noise by filling cavities of rapidly moving parts.

Implosion of vapor bubbles or cavitation can be severely damaging to plumbing as well as causing severe noise problems. Cavitation occurs when static pressure downstream of a valve is greater than the vapor pressure, and at some point within the valve, the static pressure is less than the liquid vapor pressure. Reducing the pressure in several smaller steps prevents cavitation and reduces noise (Figure 9.29).

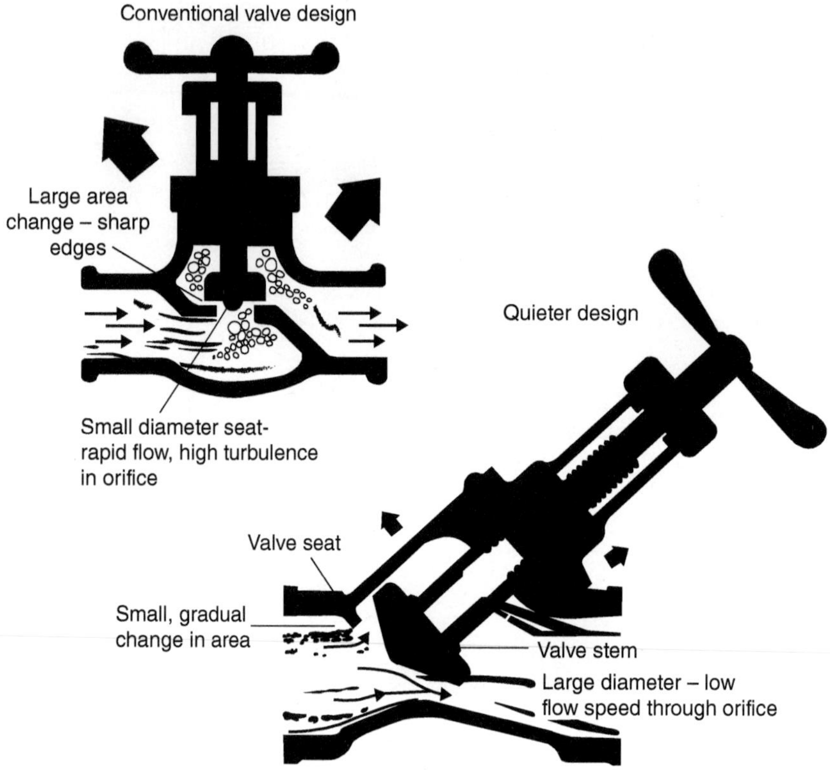


**Figure 9.28** Straighten fluid flow pathways.

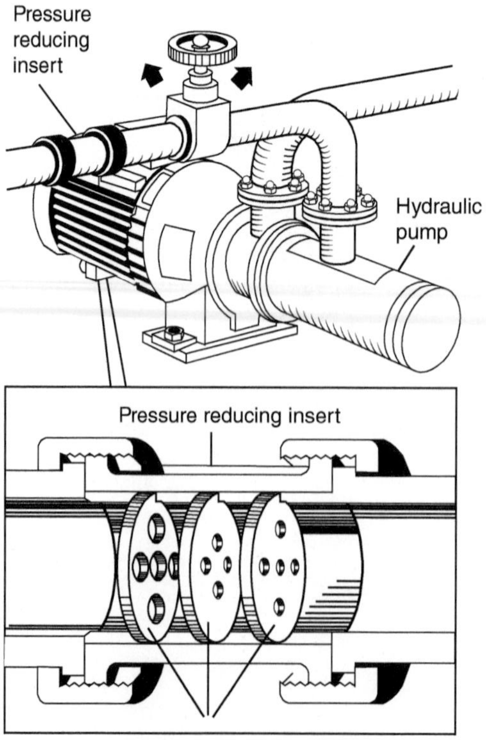


**Figure 9.29** Prevent cavitation.

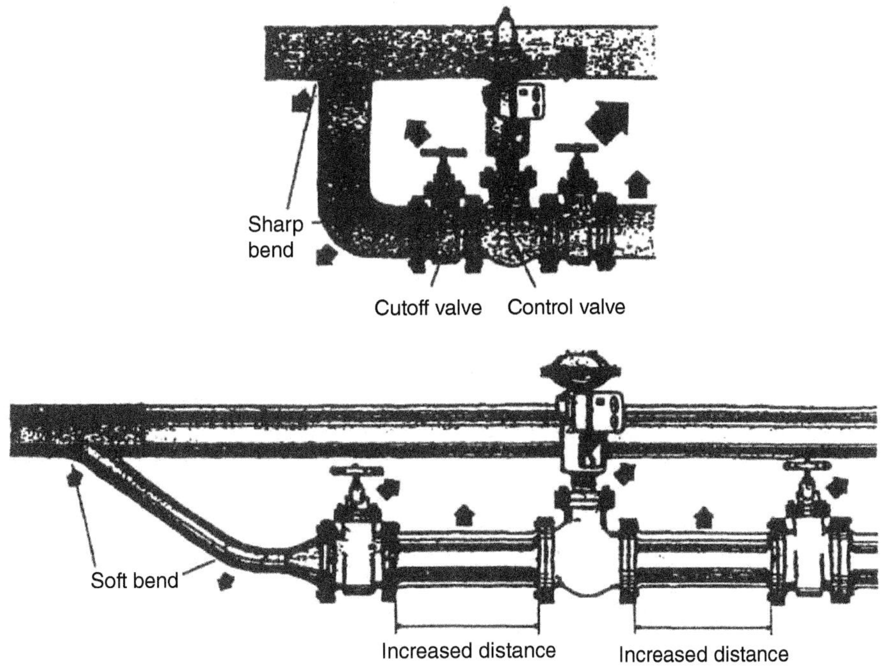


**Figure 9.30** Straighten bends and transition diameter changes in fluid flow to reduce turbulence.

Turbulence at the walls of ducts and pipes is always present. To reduce noise, interior walls should be smooth, free of protrusions at joints, and sharp bends at tees and wyes should be avoided. Turning vanes can be placed inside ductwork when construction methods utilize sharp bends. Straightening vanes can be used to smooth the flow downstream of any changes in direction, diameter, or system branches (Figure 9.30).

9.9.2.1.2.6 Methods for Reducing Fan Noise Fans, in particular, produce large amounts of turbulence, which produces noise. Since the sound power generated by a fan varies as the fifth power of rotational speed, the most cost-effective noise control is to reduce the speed when possible. When purchasing new fans, consider also quieter designs. For instance, backward-curved blades on squirrel cage fans are quieter than straight blades or forward-curved blades.

Fans mounted inside ductwork create significant noise, especially when mounted in regions where a great deal of turbulence is present. In-line duct fans should be mounted in low-turbulence regions of ductwork (Figure 9.31).

Pneumatic equipment produces noise from turbulence when the high-speed gas stream mixes with the ambient air at the tool outlet. The simplest control measure is to reduce the exit speed of the gas. Reducing the speed by a factor of two can decrease the noise level by 15 dB. Silencers can also be installed.

9.9.2.1.2.7 Silencers Silencers or mufflers can be classified into two fundamental groups – absorptive or reactive. They are made to reduce noise while permitting flow of the air or gas. Absorptive silencers contain porous or fibrous materials and use absorption to reduce noise. The basic mechanism for reactive silencers is expansion or reflection of sound waves, leading to noise cancelation.

*9.9.2.1.2.7.1 Absorptive Silencer* The simplest form of absorptive silencer is a lined duct. Generally long sections of ducts are lined with absorptive material, but lining is particularly effective along duct bends. Typically 2–5-cm acoustical-grade fibrous glass is used. Where dust is present or humid conditions could create microbiological growth, the absorptive material is covered with thin plastic or Mylar. In the past, HVAC ducts were lined with absorptive materials downstream of fans in office buildings to reduce fan noise. Several of these installations have had severe microbiological growth, causing other (nonacoustic) industrial hygiene problems (see Chapter 8).

Another form of absorptive silencer is parallel baffles. Good design includes aerodynamically streamlined entrance and exit ends with perforated spaces filled with highly absorbent acoustical materials. The first few feet of length are highly absorbent; so the attenuation is not linear with length. Thick absorbent materials with wide spaces between absorbers are effective for low frequencies, while thin absorbers and narrow spaces are effective for higher frequencies. They should be considered for cooling and exhaust air whenever sources are to be enclosed (Figure 9.32).

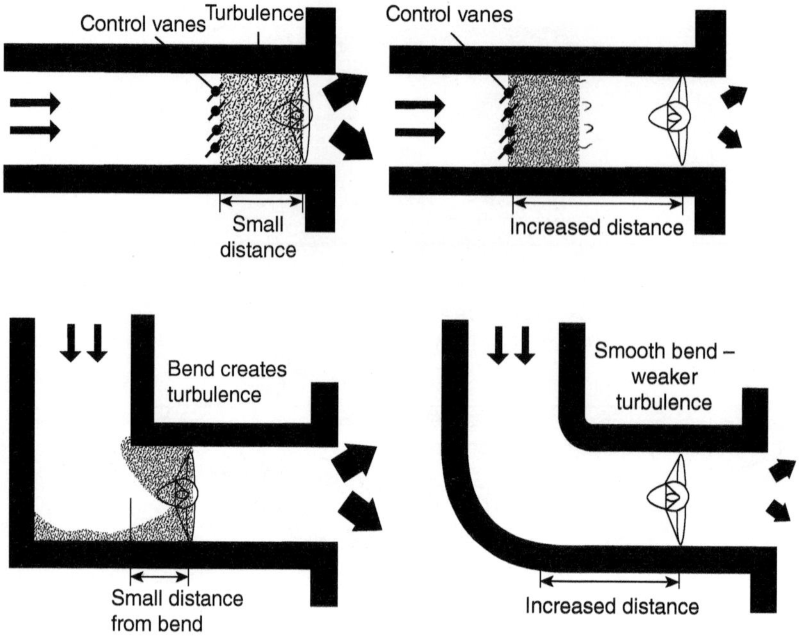


**Figure 9.31** Reduce turbulence around fans.

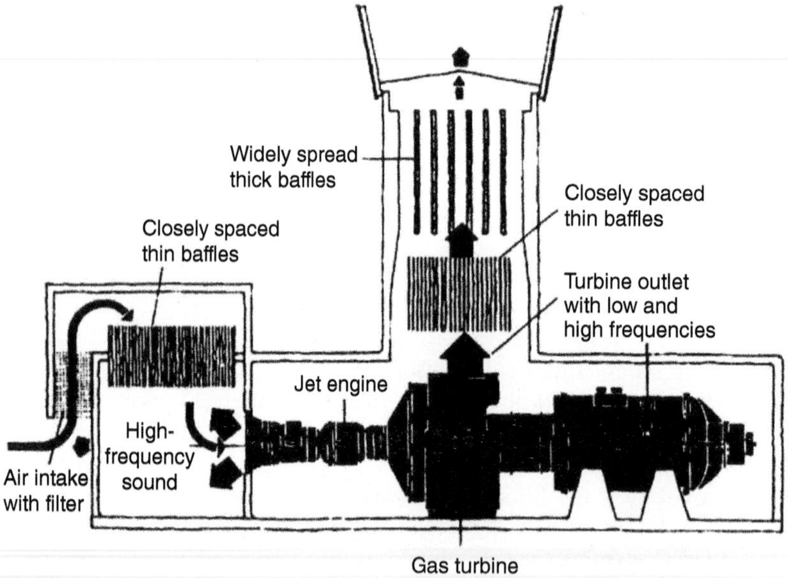


**Figure 9.32** Absorption mufflers sized for dominant frequencies.

*9.9.2.1.2.7.2 Reactive Silencer* The simplest form of a reactive silencer is a single expansion chamber. As the air enters and leaves the chamber, the expansion and contraction in pressure cause reflection of sound waves. The reflected wave added to the incoming sound wave results in destructive interference, leading to noise reduction. This reduction only occurs when

$$l = \frac{n\lambda}{4}, \quad n = 1,3,5,\ldots \tag{9.26}$$

where $l$ is the length of the muffler, $\lambda$ is the wavelength of the tone, and $n$ is an integer. This equation can be used to calculate the length needed for a reactive silencer.

*9.9.2.1.3 Reduce Vibrations* In many cases, machinery noise is created by a vibrating source coupling to a large radiating surface. When possible, large radiating surfaces should be detached from vibrating sources (Figure 9.33).

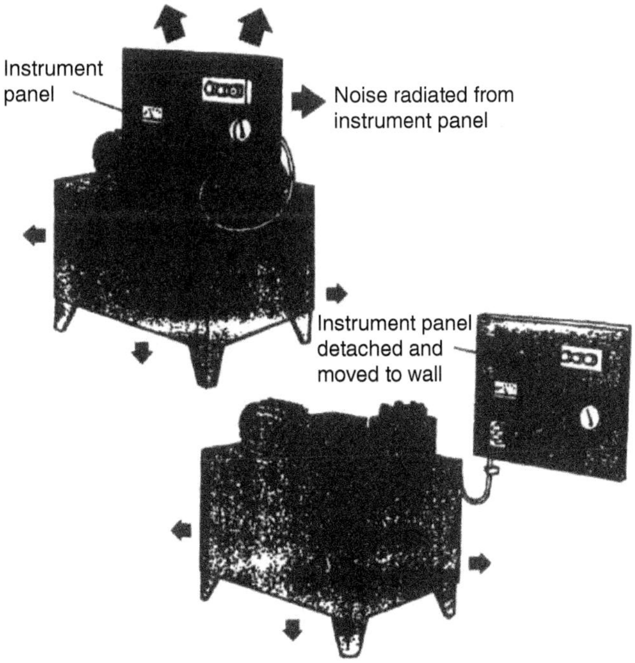


**Figure 9.33** Remove panels from vibrating equipment.

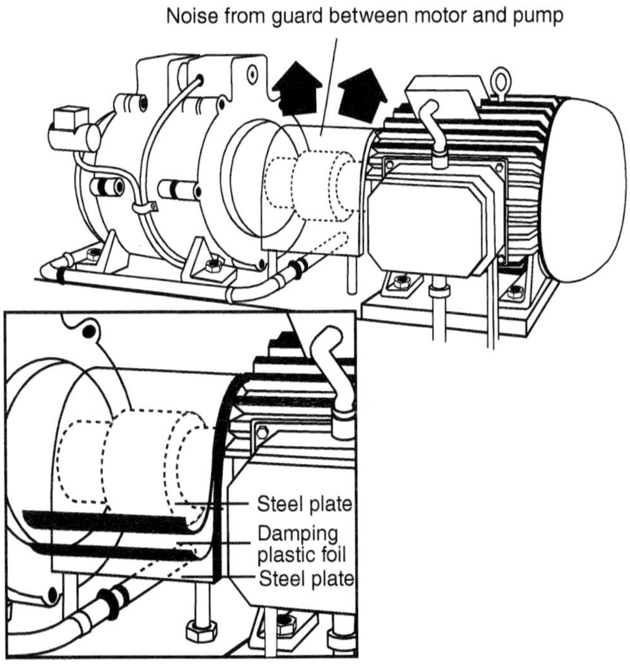


**Figure 9.34** Provide vibration damping on flexible panels.

The vibrating surface may be stiffened to limit the motion or may be covered with a damping material (Figure 9.34). Alternatively, the source might be detached from the building structure and mounted on the floor.

If possible, the equipment should be placed on a concrete base plate resting directly on the ground. More effective isolation is achieved when the base plate is not part of the building structure and is mounted directly on the ground (Figure 9.35).

When equipment is attached to the building structure, care must be taken to provide vibration isolation or the noise may be transmitted throughout the building by vibrating the building structural members. The equipment may be mounted on concrete inertia blocks or directly to steel frames. Regardless of the mounting, some forms of vibration

**Figure 9.35** Heavy vibrating equipment should be isolated from building.

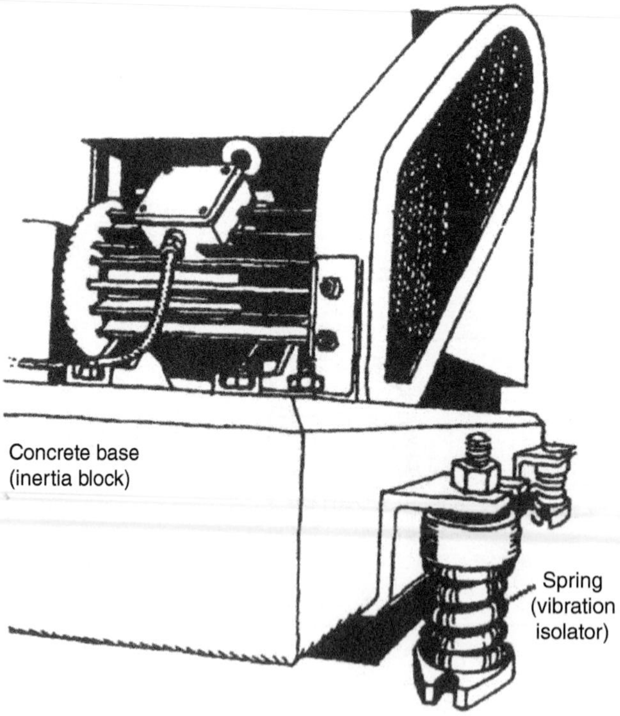


**Figure 9.36** Heavy vibrating equipment can be placed on inertial blocks with vibration isolators and dampers.

isolators are usually used (Figure 9.36). The degree of isolation achieved with vibration isolators depends on the frequency of vibrations relative to the natural frequency of the system and the amount of damping built into the isolator. Formulae for determining the isolation achievable are provided by manufacturers of vibration isolators.

Vibration isolation may not be completely effective when noise is transferred through piping or conduits from the equipment. Flexible connectors to mount the tubing to the building must also be considered (Figure 9.37).

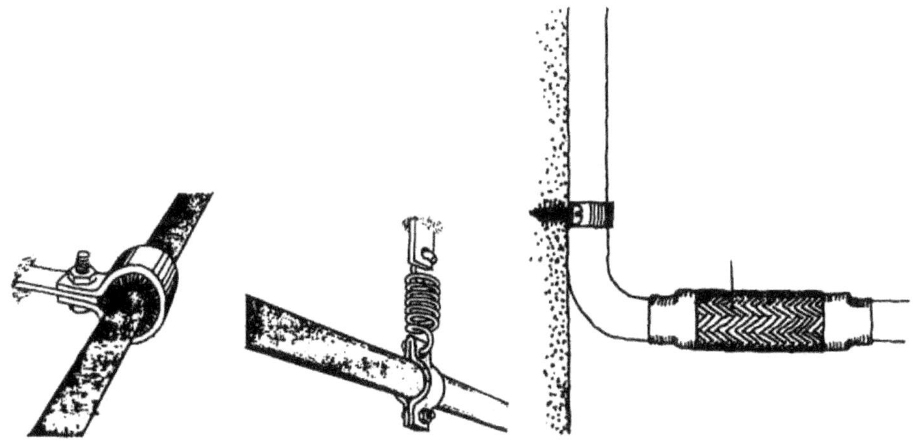

**Figure 9.37** Flexible couplings prevent transmission of vibration to building structure.

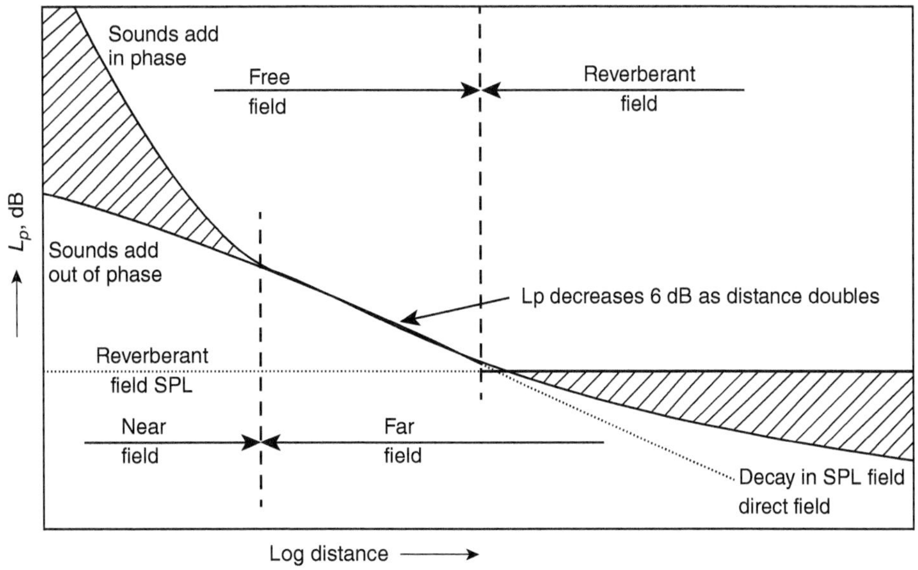


**Figure 9.38** Sound pressure levels in the near, free, and far fields.

### 9.9.3 Noise in Rooms

The sound *power* of a source is independent of its environment, but sound *pressure* around the source is not (Figure 9.38). Considering the equipment as a point source, if it could be suspended in midair away from any reflecting surfaces, including the ground, the equipment would be in a free field, and sound pressure levels would decrease by 6 dB as the distance from the source was doubled.

In practice, machinery is neither a point source nor in free-field conditions. Noise is usually emitted from all parts of equipment, and not all parts vibrate in phase with each other. Consequently, some parts will be moving in, while other parts move out, leading to a partial cancelation of sound pressure. At other parts of the equipment, parts may be moving in phase, reinforcing the sound pressure levels. Close to a machine, in the near field, sound pressure levels may be higher or lower than predicted from sound power and distance.

Outdoors, at distances about two times the longest dimension of the machine, these effects disappear, and the inverse square law or 6-dB decrease per doubling of distance becomes valid. Indoors, reverberations from walls, floors, and ceilings will result in less of a decrease in sound pressure level as the distance increases than predicted in a free-field environment.

Noise sources should be kept away from walls. The worst placement is in the corners, where the reflections are greatest (Figure 9.39).

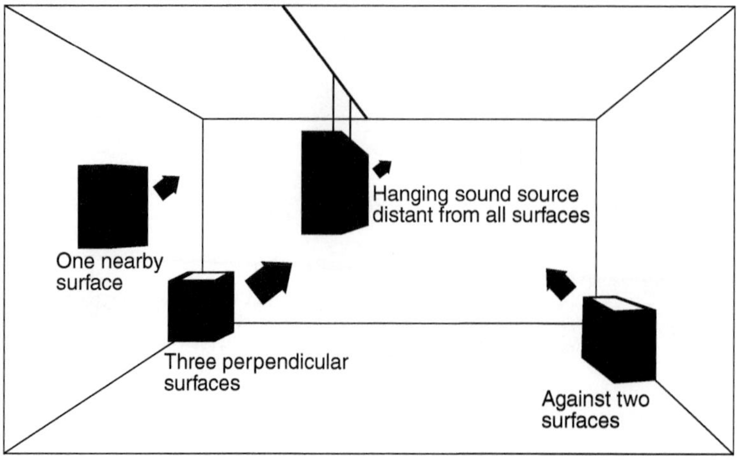


**Figure 9.39** Avoid placing equipment near walls and corners.

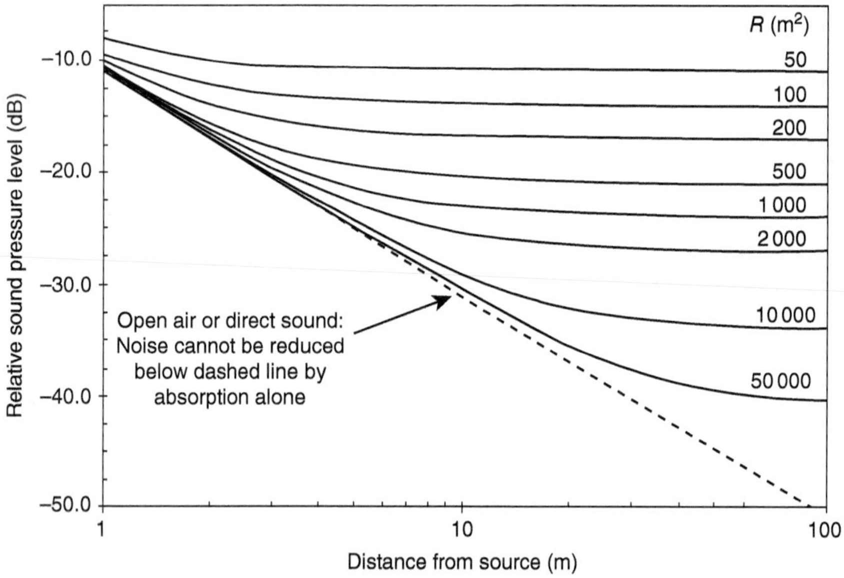


**Figure 9.40** Relative sound pressure level in a room as a function of room constant and source-receiver distance.

***9.9.3.1 Treating Rooms with Absorbent Materials*** Absorbing materials are used when it is desired to reduce the noise within a particular environment. Sound reaching the ear consists of two components: sound transmitted directly from the source (direct sound) and sound reflected from all the room's surfaces (reverberant sound). The level of direct sound depends only on distance from the source and is reduced by 6 dB for each doubling of distance from the source (inverse square law). In contrast, reverberant sound depends on the room size, shape, and absorption and does not depend on distance from the source.

The unit for sound absorption in a room is the room constant $R$, measured in $m^2$.

The relative sound pressure level in rooms can also be calculated using the Eyring theory, which for a point source is calculated (Hodgson and Warnock 1992) as

$$L_p = L_{p1} - L_{p2} = 10\log\left(\frac{Q}{4\pi r^2} + \frac{4}{R}\right) \tag{9.27}$$

where $L_p$ is the relative sound pressure level (dB), $Q$ is the directivity factor (taken to be 1), $r$ is the source–receiver distance (m), and $R$ is the room constant ($m^2$).

From Figure 9.40 we see that close to the source the sound pressure level is determined primarily by distance, whereas far from the source it is strongly influenced by the amount of absorption. Thus increasing the room acoustic absorption is not effective in reducing the sound pressure level close to the source.

**TABLE 9.12 Estimation of Room Acoustic Character from Fraction of Surface Area Covered with Absorptive Material**

| Fraction of total room surface area covered with absorption material | Room acoustic characteristics |
|---|---|
| 0 | "Live room" |
| 0.1 | "Medium-live room" |
| 0.15–0.2 | "Average room" |
| 0.3–0.35 | "Medium-dead room" |
| 0.5–0.6 | "Dead room" |

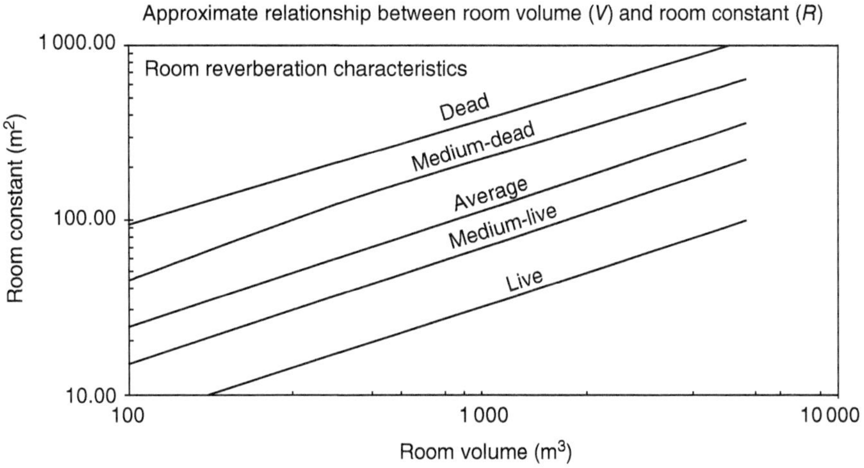


**Figure 9.41** Approximate relationship between room constant and room volume for spaces with varying acoustic characteristics.

In practice, the room constant can be difficult to calculate or requires some relatively sophisticated equipment to measure. For a rough estimation of the room constant, determine the fraction of the total room surface covered with absorption material (i.e. carpeting, drapes, acoustic ceiling tiles, absorbent wall panels, etc.). Table 9.12 can be used to estimate the acoustic characteristics of the room.

Use the room acoustic characteristic and room volume to estimate the room constant (Figure 9.41).

***Example 9.15:*** A computer room 50 ft by 50 ft with 10-ft ceiling with acoustic tile ceiling and some sound-absorptive panels on the walls has an "average" room constant ($R = 80\,m^2$). Installing carpeting was suggested to reduce the noise from a supplemental HVAC unit located about 15 ft (4.5 m) from several employee's workstations. Estimate the reduction in noise.

Heavy carpeting on the floor would increase the fraction of room surface covered by 0.36 [(floor area)/(room surface area) = $(50' \times 50')/(2 \times 50' \times 50' + 4 \times 50' \times 10') = 0.36$] – changing the room reverberation characteristic from "average" to "dead" and increasing the room constant $R$ to 300 (see Figure 9.41). Using Eq. (9.27) or Figure 9.40, the relative sound pressure level change is 5 dB.

Note the above is an extremely crude estimate and it relies on a number of simplifying assumptions. A more accurate estimation should be made measuring octave band sound pressure levels from the source and calculating the change in room constant at each octave band level from the absorption coefficients. In particular, for this example, the noise from the HVAC unit is likely to be predominantly in the low frequencies, where carpeting is not particularly absorbent. Most machinery is used inside buildings, and even outdoor equipment is seldom suspended above the ground. Consequently, free-field conditions seldom exist.

***9.9.3.2 Sound Absorption Coefficients*** Machines that contain cams, gears, reciprocating components, and metal stops are often located in large, acoustically reverberant areas that reflect and build up noise levels in the room. Frequently the noise levels in adjoining areas can be reduced significantly by using sound-absorbing materials on walls and ceilings. However, the amount of reduction close to the machines may be slight because most of the noise exposure energy is coming directly from the machines and not from the reflecting surfaces. The type, amount, configuration, and placement of absorption materials depend on the specific application; however, the choice of absorbing materials can be guided by the absorption coefficients listed in Table 9.13. Absorption coefficients of 1.0 will absorb all sound randomly impinging on the surface; coefficients close to 0 mean the material absorbs very little acoustic energy.

**TABLE 9.13 Sound Absorption Coefficients of Surface Materials**

| | Frequency (Hz) | | | | | |
|---|---|---|---|---|---|---|
| Material | 125 | 250 | 500 | 1000 | 2000 | 4000 |
| Brick: | | | | | | |
| Glazed | 0.01 | 0.01 | 0.01 | 0.01 | 0.02 | 0.02 |
| Unglazed | 0.03 | 0.03 | 0.03 | 0.04 | 0.05 | 0.06 |
| Unglazed, painted | 0.01 | 0.01 | 0.02 | 0.02 | 0.02 | 0.03 |
| Carpet: | | | | | | |
| Heavy (on concrete) | 0.02 | 0.06 | 0.14 | 0.37 | 0.60 | 0.65 |
| On 40 oz. hairfelt or foam rubber (carpet has coarse backing) | 0.08 | 0.24 | 0.57 | 0.69 | 0.71 | 0.73 |
| With impermeable latex backing on 40 oz. hairfelt or foam rubber | 0.08 | 0.27 | 0.39 | 0.34 | 0.48 | 0.63 |
| Concrete block: | | | | | | |
| Coarse | 0.36 | 0.44 | 0.31 | 0.29 | 0.39 | 0.25 |
| Painted | 0.10 | 0.05 | 0.06 | 0.07 | 0.09 | 0.08 |
| Poured | 0.01 | 0.01 | 0.02 | 0.02 | 0.02 | 0.03 |
| Fabrics: | | | | | | |
| Light velour: 10 oz. $yard^{-2}$, hung straight, in contact with wall | 0.03 | 0.04 | 0.11 | 0.17 | 0.24 | 0.35 |
| Medium velour: 14 oz. $yard^{-2}$, draped to half-area | 0.07 | 0.31 | 0.49 | 0.75 | 0.70 | 0.60 |
| Heavy velour: 18 oz. $yard^{-2}$, draped to half-area | 0.14 | 0.35 | 0.55 | 0.72 | 0.70 | 0.65 |
| Floors: | | | | | | |
| Concrete or terrazzo | 0.01 | 0.01 | 0.015 | 0.02 | 0.02 | 0.02 |
| Linoleum, asphalt, rubber, or cork tile on concrete | 0.02 | 0.03 | 0.03 | 0.03 | 0.03 | 0.02 |
| Wood | 0.15 | 0.11 | 0.10 | 0.07 | 0.06 | 0.07 |
| Wood parquet in asphalt on concrete | 0.04 | 0.04 | 0.07 | 0.06 | 0.06 | 0.07 |
| Glass: | | | | | | |
| Ordinary window glass | 0.35 | 0.25 | 0.18 | 0.12 | 0.07 | 0.04 |
| Large panes of heavy plate glass | 0.18 | 0.06 | 0.04 | 0.03 | 0.02 | 0.02 |
| Glass fiber: | | | | | | |
| Mounted with impervious backing, 3 lb $ft^{-3}$, 1 in. thick | 0.14 | 0.55 | 0.67 | 0.97 | 0.90 | 0.85 |
| Mounted with impervious backing, 3 lb $ft^{-2}$, 2 in. thick | 0.39 | 0.78 | 0.94 | 0.96 | 0.85 | 0.84 |
| Mounted with impervious backing, 3 lb $ft^{-3}$, 3 in. thick | 0.43 | 0.91 | 0.99 | 0.98 | 0.95 | 0.93 |
| Gypsum board: ½ in. thick nailed to 2 in. × 4 in. s, 16 in. on center | 0.29 | 0.1 | 0.05 | 0.04 | 0.07 | 0.09 |
| Marble | 0.01 | 0.01 | 0.01 | 0.01 | 0.02 | 0.02 |
| Plaster: | | | | | | |
| Gypsum or lime, smooth finish on tile or brick | 0.013 | 0.015 | 0.02 | 0.03 | 0.04 | 0.05 |
| Gypsum or lime, rough finish on lath | 0.14 | 0.10 | 0.06 | 0.05 | 0.04 | 0.05 |
| With smooth finish | 0.14 | 0.10 | 0.06 | 0.04 | 0.04 | 0.03 |
| Plywood paneling, ⅜ in. thick | 0.28 | 0.22 | 0.17 | 0.09 | 0.10 | 0.11 |
| Sand: | | | | | | |
| Dry, 4 in. thick | 0.15 | 0.35 | 0.40 | 0.50 | 0.55 | 0.80 |
| Dry, 12 in. thick | 0.20 | 0.30 | 0.40 | 0.50 | 0.60 | 0.75 |
| 14 lb $H_2O/ft^3$, 4 in. thick | 0.05 | 0.05 | 0.05 | 0.05 | 0.05 | 0.15 |
| Water | 0.01 | 0.01 | 0.01 | 0.01 | 0.02 | 0.02 |
| Air, per 10 $m^3$ at 50% RH (for larger spaces, air attenuates sound – particularly the higher frequencies) | | | | 0.32 | 0.81 | 2.56 |

*Source:* Federal Interagency Committee on Noise (1992).

***9.9.3.3 Noise Barriers and Enclosures*** The amount of noise reduction that can be attained with barriers depends on the characteristics of the noise source, the configuration and materials used for the barrier, and the acoustic environment on each side of the barrier. It is necessary to consider all these complex factors to determine the overall benefit of a barrier.

The noise reduction of barriers or enclosures varies significantly. Single-wall barriers with no openings may provide as little as 2–5-dB reduction in the low frequencies and a 10–15-dB reduction in the high frequencies. Higher reduction values are possible with heavier barriers with greater surface areas. Higher values may also be expected when the source and/or the persons exposed are close to the barrier. The effects of two- and three-sided barriers are difficult to predict on a general basis. However, well-designed partial enclosures may provide noise reduction values of more than twice as much as single-wall barriers. Complete enclosures from simple practical designs may provide noise reduction values in excess of 10–15 dB in the low frequencies and in excess of 30 dB in the high frequencies.

The best noise isolation is achieved with a complete enclosure constructed around the noise source. The enclosure should be lined with absorbent material, or the reverberation inside the enclosure will increase the sound pressure level, negating some of the transmission loss achieved by the enclosure walls. In addition, care should be taken to limit the openings in the enclosure. Generally openings are necessary for ventilation for cooling, piping, conduits, or material handling. However, as shown below, even small openings will reduce the transmission loss. For instance, an enclosure capable of reducing noise by 40 dB will provide less than 20 dB of attenuation if 1% of the total surface area is left open.

Partial enclosures (Figures 9.42 and 9.43) must be lined with absorptive material to obtain maximum effectiveness. If the workers are not in the direct line of the opening, a shadow effect of 3–15 dB for high-frequency sounds may be achievable (Bruce and Toothman 1986). The shadow effect is limited to high-frequency sounds where the dimension of the enclosure is several times the wavelength of the noise.

*9.9.3.3.1 Calculating Barrier Noise Reduction* If a sound source is in a room with a large amount of absorption (i.e. few reflections), blocking the direct path with a partial barrier may provide adequate noise control. (This technique is more commonly used outdoors, since even a modest amount of reverberation will destroy the effectiveness of a shield.) Based on theoretical considerations, the decrease in sound pressure level $L_p$ in the shadow

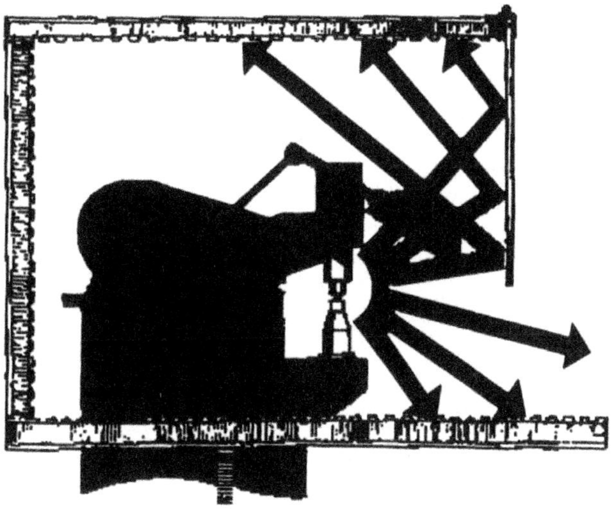

**Figure 9.42** Reducing high-frequency noise with a barrier and partial enclosure lined with absorbent material.

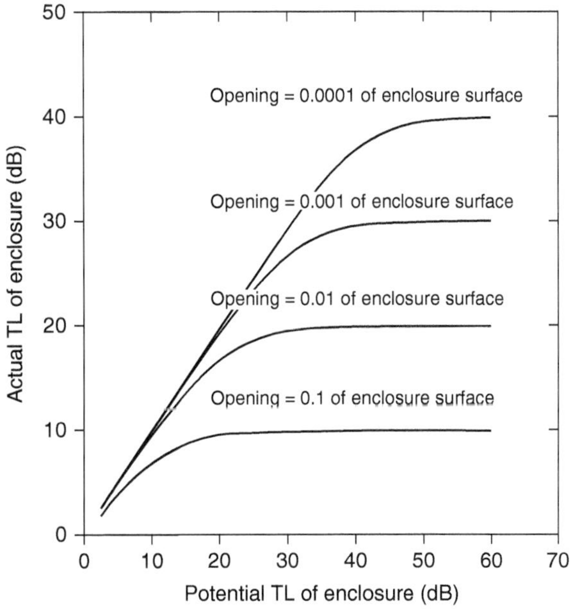


**Figure 9.43** Reduction in transmission loss if an enclosure has openings.

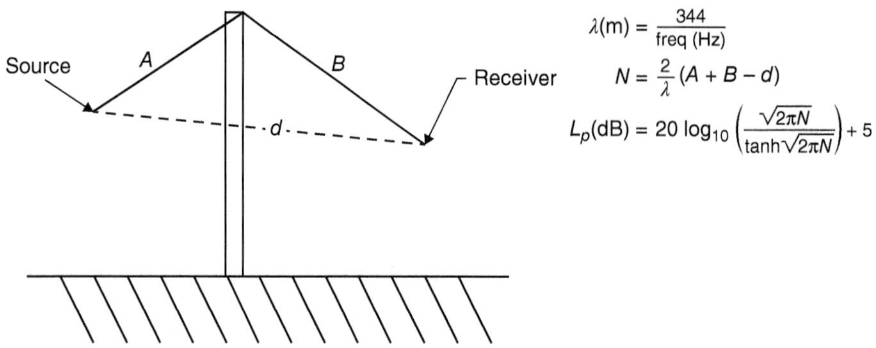


**Figure 9.44** Geometrical considerations for noise reductions $L_p$ by barriers.

of a barrier is as given in Figure 9.44, where $N$ is a geometric factor, $A + B$ is the shortest distance over the barrier between source and receiver, $d$ is the straight line distance between source and receiver, $\lambda$ is the wavelength of sound, and $L_p$ is the decrease in sound pressure level.

***Example 9.16:*** What is the expected decrease in sound pressure levels at typical speech frequencies (500, 1000, 2000 Hz) by increasing the height of a barrier between two reservation agents from 1.2 to 1.5 m.

The distance between source and receiver of seated telephone agents is taken to be 1.05 m. Assume agents are seated 1.2 m apart. Therefore, $d$ = 1.2 m, and for a 4-ft (1.22 m) barrier:

$$A = B = \sqrt{(0.6)^2 + (1.22 - 1.05)^2} = 0.624\,\text{m}$$

Similarly, for a 5-ft (1.52 m) barrier, $A = B = 0.762$ m.

Using the above formulae or graphs, the following table is constructed.

| Frequency (Hz) | $S$ | $N$ (at 4 ft) | $N$ (at 5 ft) | $L_p$ (4 ft) | $L_p$ (5 ft) | Difference (dB) |
|---|---|---|---|---|---|---|
| 500 | 0.7 | 0.1 | 1.0 | 7.0 | 12.9 | 5.9 |
| 1000 | 0.3 | 0.3 | 1.9 | 8.6 | 15.8 | 7.2 |
| 2000 | 0.2 | 0.5 | 3.8 | 10.7 | 18.8 | 8.1 |

Under ideal conditions, a decrease in sound from the adjacent reservation agent could be reduced by 6–8 dB. In practice, there will be reflections from nearby surfaces (in particular from the ceiling and floor), and end effects from the barrier will be important.

Recommendations for use of indoor barriers (Gordon and Jones 1991) are as follows:

1. The barrier should be as close to the source or receiver as possible. However, the source should not touch the barrier to prevent transferring vibrations.
2. Barrier should extend beyond the line of sight of the source plus ¼ wavelength of the lowest frequency of interest for attenuating the noise.
3. Select a solid material without any holes or openings. Barriers made of material with more than 4 lb ft$^{-2}$ are usually adequate. The TL of the barrier should be at least 10 dB greater than the required attenuation.

## 9.10 COMMUNITY NOISE

Most community noise exposures, by themselves, do not cause noise-induced hearing impairment. Noise problems from communities commonly entail complaints of communication and stress-related problems such as annoyance.

The effects of noise on communication can be measured with a reasonable degree of accuracy (Beranek 1992; von Gierke and Eldred 1993a); however, other psychological and physiological responses to noise are extremely

**TABLE 9.14 Factors Affecting Community Response to Noise and Suggested Corrections**

| Type of correction | Description | Correction added to measured $L_{dn}$ (dB) |
|---|---|---|
| Seasonal | Summer (or year-round operations) | 0 |
| | Winter only (or windows always closed) | −5 |
| Outdoor residual noise level | Quiet suburban or rural community (away from large cities, industrial activities, and trucking) | +10 |
| | Normal suburban community (away from industrial activity) | +5 |
| | Urban residential community (not near heavy traffic or industry) | 0 |
| | Noisy urban residential community (near relatively busy roads or industrial areas) | −5 |
| | Very noisy urban residential community | −10 |
| Previous exposure or community attitudes | No prior experience with intruding noise | +5 |
| | Community has had some exposure to intruding noise; little effort is being made to control noise. This correction may also be applied to a community that has not been previously exposed to noise, but the people are aware a bona fide effort is being made to control it | 0 |
| | Community has had considerable exposure to intruding noise; noise maker's relations with the community is good | −5 |
| | Community is aware that operations causing noise is necessary but will not continue indefinitely. This correction may be applied on a limited basis and under emergency conditions | −10 |
| Pure tone or impulse | No pure tone or impulsive character | 0 |
| | Pure tone or impulsive character present | +5 |

*Source:* von Gierke and Eldred (1993b).

complicated and difficult to measure in a meaningful way. Attempts to correlate noise exposure levels and annoyance, or the general well-being of humans, are complicated by many factors such as:

- The attitude of the listener toward the noise source.
- The history of individual noise exposures.
- The activities of the listeners and stresses on the listeners during the noise exposures.
- The hearing sensitivities of the listeners and other differences in individual responses.
- Whether the noise source contains tonal or impulsive components.
- Ambient noise in the environment.
- Season of the year.

Corrections may be made to measured sound levels to attempt to account for the some of these factors (Table 9.14).

### 9.10.1 Community Noise Regulations

For the most part, old rules and regulations were based on vaguely defined nuisance factors. Nuisance-type regulations often take the form "There shall be no unnecessary nor disturbing noise. ..." An obvious weakness of this form of regulation is the failure to specify the conditions of how, when, and to whom noise is unnecessary or disturbing. Innumerable arguments may result in the interpretation of these laws when an attempt is made to enforce them. On the other hand, this kind of law may be useful in some cases where it is not possible to make reasonable decisions based on exposure (sound pressure) levels.

Most recent zoning codes have specified maximum noise limits for various zones. These so-called performance zoning codes are more objective and easier to enforce than the nuisance laws. Unfortunately, many problems must be considered individually on a nuisance basis because of the many possible noise exposure situations. For example, a noise limit of $L_A = 55\,\text{dB}$ may be generally acceptable for a given neighborhood. However, it is reasonable to assume that lawn mowers should be permitted and they may produce a noise level of $L_A = 88\,\text{dB}$ at the property line. The lawn mower noise should be acceptable at certain hours when other persons' activities are not unreasonably affected. On the other hand, this same noise may not be acceptable if neighbors have guests on the adjoining lawn, if they are trying to sleep, and so on.

Obviously performance laws may be used to establish guidelines for reasonable noise exposure levels, but nuisance laws may also be needed in some instances. No law can replace the fact that all individuals must show consideration for their neighbors.

Most performance-type noise ordinances establish limiting levels that are well below the level where physiological damage may occur. Usually the limits are based on the number and level of adverse responses in the community rather than trying to measure individual annoyance reactions to noise. This decision is made in most cases because the group statistics involved in describing "community responses" to noise avoid many of the variables that are extremely difficult to account for individual annoyance reactions.

Most performance-type noise codes specify limits of sound pressure level that are based on:

1. A selected frequency weighting or noise analysis procedure.
2. The pattern of exposure times for various noise levels.
3. The ambient noise levels that would be expected in that particular kind of community without the offending noise source(s).
4. A land-use zoning of the area.

Regulations based on specific exposures – such as those from aircraft or from ground transportation – often work well at other locations having the same kinds of noise (see below). Caution must be used, however, in applying these results to other kinds of noise exposure.

*The Noise Guidebook* document (U.S. Department of Housing and Urban Development 1985) provides guidance on noise policy, attenuation of outdoor noise, and noise assessment. This document does not constitute a standard, specification, or regulation, but it does present available technical knowledge that may be used in a uniform and practical way by communities of various sizes to tailor ordinances to their specific conditions and goals.

The most common time-averaged descriptor used for community noise problems is the day–nighttime average sound level, abbreviated DNL or $L_{dn}$ (Eq. 9.28). This is the 24-hour equivalent sound level obtained after the addition of a 10-dB penalty for sound levels that occur at night between 10 p.m. and 7 a.m. The penalty is based on the fact that more people are disturbed by noise at night than any other time. Background noise is usually much lower at night, and sleep disturbance is one of the prime motivators of community reactions to noise:

$$L_{\mathrm{dn}} = 10 \log \frac{1}{24} \left( \frac{\int_{07:00}^{22:00} p_{\mathrm{A}}^2(t)\, dt}{p_{\mathrm{ref}}^2} + \frac{\int_{22:00}^{07:00} 10 p_{\mathrm{A}}^2(t)\, dt}{p_{\mathrm{ref}}^2} \right) \tag{9.28}$$

Sound level meters equipped with clocks and appropriate computing power can directly measure $L_{dn}$. It is also common practice to measure $L_{eq}$, add 10 dB for nighttime readings, and then combine the sound levels using one of the techniques shown in Section 9.2.4.2.

***9.10.1.1 Transportation Noise*** A number of social surveys of community annoyance have been analyzed used as criteria for evaluating the environmental impact of aircraft noise and recommended for predicting the effects of general transportation noise (Federal Interagency Committee on Noise 1992). An updated analysis (Finegold et al. 1994) recommended a minor change to the calculation (Figure 9.45).

Slight differences were noted in community response to various kinds of transportation noise (rail, road traffic, and aircraft). Therefore, caution should be used equating community noise created by industrial sources to transportation noise, but the above equation and modifiers such as those suggested in Table 9.14 can provide an initial estimation for planning purposes.

***9.10.1.2 A Guide for Community Noise Criteria*** Acceptable noise levels vary from one community to another depending on the history of noise exposures and other variables. In most cases, noise codes are tailored to the character and requirements of the particular community.

Annoyance, sleep interference, and speech interference are the effects used as a basis for establishing positions of acceptable community noise. Community noise exposures guidelines have been summarized (Shaw 1996) as shown in Table 9.15.

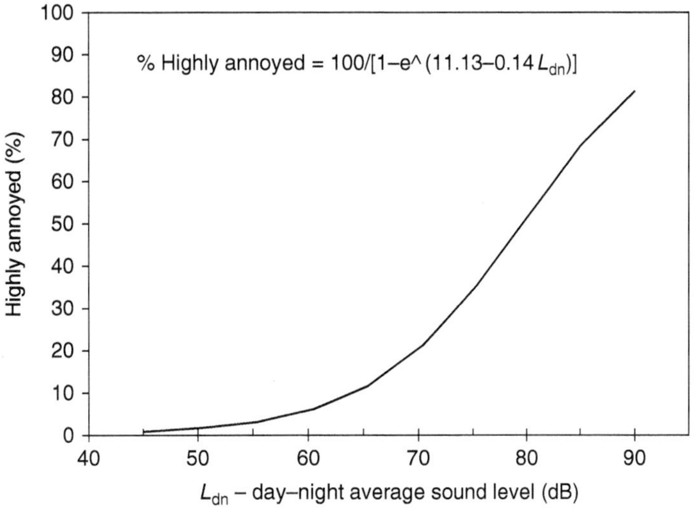


**Figure 9.45** Community annoyance in response to general transportation noise.

**TABLE 9.15 Community Noise Exposure Guidelines**

| Authority | Specified sound levels | Criterion |
|---|---|---|
| EPA levels document (U.S. Environmental Protection Agency 1974) | $L_{dn} \leq 55$ dB (outdoors) | Protection of public health and welfare with adequate margin of safety |
| | $L_{dn} \leq 45$ dB (indoors) | |
| WHO document (Berglund and Lindvall 1995) | $L_{eq} \leq 50/55$ (outdoors: day) | Recommended guideline values (Task Force Consensus) |
| | $L_{eq} \leq 45$ dB (outdoors: night) | |
| | $L_{eq} \leq 30$ dB (bedroom) | |
| | $L_{max} \leq 45$ dB (bedroom) | |
| US Interagency Committee (FICON) (Federal Interagency Committee on Urban Noise 1980) | $L_{dn} \leq 65$ dB | Considered generally compatible with residential development |
| | $65 \leq L_{dn} \leq 70$ dB (indoors) | Residential use discouraged |
| Various European road traffic regulations (Gottlob 1995) | $L_{dn} \leq 45$ dB (indoors) | Remedial measures required |

## ACKNOWLEDGMENT

Both authors need to acknowledge Paul L. Michael for the original text (Michel 1991) that formed the basis for this chapter and for his education of the industrial hygiene community in principles of hearing conservation (and, in the case of one of the authors, education in everything else). The authors are also indebted to David Byrne for his careful review of the chapter.

## GLOSSARY

**Absorption coefficient** The sound absorption coefficient of a given surface is the ratio of the sound energy absorbed by the surface to the sound energy incident upon the surface.

**Acoustic intensity** ($I$) The average rate at which sound energy is transmitted through a unit area normal to the direction of propagation. The units used for sound intensity are joules per square meter ($J\,m^{-2}\,s^{-1}$). Sound intensity is also expressed in terms of a sound intensity level ($L_I$) in decibels referred to $10^{-12}\,W\,m^{-2}$.

**Acoustic power** See **Sound power**.

**Acoustic pressure** See **Sound pressure**.

**Ambient noise** The overall composite of sounds in an environment.

**Amplitude** The quantity of sound produced at a given location away from the source or the overall ability of the source to emit sound. The amount of sound at a location away from the source is generally described by the sound pressure or sound intensity, whereas the ability of the source to produce noise is described by the sound power of the source.

**Anechoic room** A room that has essentially no boundaries to reflect sound energy generated therein. Thus any sound generated within is referred to as being in a free field (see definition).

**Atmospheric pressure** ($P$) Measured in pascals (Pa) $P = 1.013 \times 10^5$ Pa $= 1.013 \times 10^5$ N m$^{-2}$.

**Audiogram** A recording of hearing levels referenced to a statistically normal sound pressure as a function of frequency.

**Audiometer** An instrument for measuring hearing thresholds.

**Continuous spectrum** A spectrum of sound that has components distributed over a known frequency range.

**Criterion duration** (TC) Duration used a basis for noise exposure measurements – typically eight hours.

**Criterion sound level** (LC) The sound pressure level that would give 100% of the allowable noise exposure is continued for the criterion duration.

**Cycle** A cycle of a periodic function, such as a single-frequency sound, is the complete sequence of values that occur in a period.

**Cycles per second** See **Frequency**.

**Decibel** (dB) A convenient means for describing the logarithmic level of sound intensity, sound power, or sound pressure above arbitrarily chosen reference values (see text).

**Density of air** At 1 atm and 22 °C, $\rho = 1.18$ kg m$^{-3}$ (Beranek and Ver n.d.).

**Diffuse sound field** A sound field that has sound pressure levels that are essentially the same throughout, with the directional incidence of energy flux randomly distributed.

**Directivity factor** ($Q$) Measure of how much the sound is concentrated in a given direction.

**Effective sound pressure** The sound pressure at a given location derived by calculating the root mean square (rms) value of the instantaneous sound pressures measured over a period of time at that location.

**Exchange rate** ($Q$) The number of decibels change in sound level that gives an equivalent exposure when the duration is halved or doubled. Frequently called the *doubling rate*.

**Free field** A field that exists in a homogeneous isotropic medium, free from boundaries. In a free field, sound radiated from a source can be measured accurately without influence from the test space. True free-field conditions are rarely found, except in anechoic test chambers. However, approximate free-field conditions may be found in any homogeneous space where the distance from the reflecting surfaces to the measuring location is much greater than the wavelengths of the sound being measured.

**Frequency** The rate at which complete cycles of high- and low-pressure regions are produced by the sound source. The unit of frequency is hertz (Hz). 1 Hz = 1 cycle per second.

**Hertz** (Hz) Unit for measuring frequency 1 Hz = 1 cycle per second.

**Infrasonic frequency** Sounds having major frequency components well below audible range (below 20 Hz).

**Level** The level of any quantity, when described in decibels, is the logarithm of the ratio of that quantity to a reference value in the same units as the specific quantity.

**Loudness** An observer's impression of the sound's amplitude, which includes the response characteristics of the ear.

**Noise** The terms *noise* and *sound* are often used interchangeably, but generally, sound describes useful communication or pleasant sounds, such as music, whereas noise is dissonance or unwanted sound.

**Noise dose** Noise exposure expressed as a percentage.

**Noise reduction coefficient** (NRC) The arithmetic average of the sound absorption coefficients of a material at 250, 500, 1000, and 2000 Hz.

**Octave band** A frequency bandwidth that has an upper band-edge frequency equal to twice its lower band-edge frequency.

**Peak sound pressure level** The maximum instantaneous level that occurs over any specified period of time.

**Period** ($T$) The time (in seconds) required for one cycle of pressure change to take place; hence it is the reciprocal of the frequency.

**Pitch** A measure of auditory sensation that depends primarily on frequency but also on the pressure and wave form of the sound stimulus.

**Pure tone** A sound wave with only one sinusoidal change of pressure with time.

**Random incidence sound field** See **Diffuse sound field.**

**Random noise** A noise made up of many frequency components whose instantaneous amplitudes occur randomly as a function of time.

**Resonance** A system is in resonance when any change in the frequency of forced oscillation causes a decrease in the response of the system.

**Reverberation** Reverberation occurs when sound persists after direct reception of the sound has stopped. The reverberation of a space is specified by the reverberation time, which is the time required, after the source has stopped radiating sound, for the rms pressure to decrease 60 dB from its steady-state level.

**Root mean square sound pressure** The root mean square (rms) value of a changing quantity, such as sound pressure, is the square root of the mean of the squares of the instantaneous values of the quantity.

**Sound intensity** ($I$) The average rate at which sound energy is transmitted through a unit area normal to the direction of propagation. The units used for sound intensity are joules per square meter sec ($J\,m^{-2}\,s^{-1}$). Sound intensity is also expressed in terms of a sound intensity level ($L_I$) in decibels referred to $10^{-12}\,W\,m^{-2}$.

**Sound power** ($P$) The total sound energy radiated by the source per unit time. Sound power is normally expressed in terms of a sound power level ($L_p$) decibels references to $10^{-12}\,W$.

**Sound pressure** ($p$) The rms value of the pressure above and below atmospheric pressure of steady-state noise. Short-term or impulse-type noises are described by peak pressure values. The units used to describe sound pressures are pascals (Pa), newtons per square meter ($N\,m^{-2}$), dynes per square centimeter ($dyn\,cm^{-2}$), or microbars (μbar).

**Sound pressure level** ($L_p$) Sound pressure is also described in terms of sound pressure level in decibels referenced to 20 μPa.

**Standard threshold shift** (STS) Defined in 29 CFR 1910.95 as "a change in the hearing threshold relative to the baseline audiogram of an average of 10 dB or more at 2000, 3000, and 4000 Hz in either ear."

**Standing waves** Periodic waves that have a fixed distribution in the propagation medium.

**Threshold level** The lowest sound level used for measuring noise exposure.

**Transmission loss** (TL) Ten times the logarithm (to the base 10) of the ratio of the incident acoustic energy to the acoustic energy transmitted through a sound barrier.

**Ultrasonic** The frequency of ultrasonic sound is higher than that of audible sound – usually considered as frequencies above 20 kHz.

**Velocity** The speed at which the regions of sound-producing pressure changes move away from the sound source is called the velocity of propagation. Sound velocity ($c$) varies directly with the square root of the density and inversely with the compressibility of the transmitting medium; however, the velocity of sound is usually considered constant under normal conditions. For example, the velocity of sound is approximately 344 $m\,s^{-1}$ (1130 $ft\,s^{-1}$) in air, 1433 $m\,s^{-1}$ (4700 $ft\,s^{-1}$) in water, 3962 $m\,s^{-1}$ (13 000 $ft\,s^{-1}$) in wood, and 5029 $m\,s^{-1}$ (16 500 $ft\,s^{-1}$) in steel.

**Wavelength** ($\lambda$) The distance required to complete one pressure cycle. The wavelength, a very useful tool in noise control, is calculated from known values of frequency ($f$) and velocity ($c$): $\lambda = c/f$.

**White noise** White noise has an essentially random spectrum with equal energy per unit frequency bandwidth over a specified frequency band.

## BIBLIOGRAPHY

ACGIH (1996a). *1996 TLVs and BEIs Threshold Limit Values for Chemical Substances and Physical Agents Biological Exposure Indices*. Cincinnati, OH: ACGIH.

ACGIH (1996b). *1996 TLVs and BEIs Threshold Limit Values for Chemical Substances and Physical Agents Biological Exposure Indices*, 110–112. Cincinnati, OH: ACGIH.

AIHA (1966). *Industrial Noise Manual*, 2e. Akron, OH: American Industrial Hygiene Association.

AIHA (1975). *Industrial Noise Manual*, 3e. Akron, OH: American Industrial Hygiene Association.

AIHA (1987). *Industrial Noise Manual*, 4e. Akron, OH: American Industrial Hygiene Association.

American Iron and Steel Institute (1983). Steel Industry Hearing Protector Study. Vol. 1, *Applications Handbook*, Vol. 2, *References*. Washington, DC : AISI.

ANSI S1.1 (2013). *Acoustical Terminology*. New York: American National Standards Institute.

ANSI S1.4 (2013). *Specification for Sound Level Meters*. New York: American National Standards Institute.

ANSI S1.6 (2016). *Preferred Frequencies, Frequency Levels, and Band Numbers for Acoustical Measurements (ASA 53)*. New York: American National Standards Institute.

ANSI S1.8 (2011). *Reference Quantities for Acoustical Levels (ASA 84)*. New York: American National Standards Institute.

ANSI S1.11 (2004). *Specifications for Octave-Band and Fractional-Octave-Band Digital Filters (ASA 65)*. New York: American National Standards Institute.

ANSI S1.42 (2001). *Design Response of Weighting Networks (ASA 64)*. New York: American National Standards Institute.

ANSI S3.1 (1999). *Maximum Permissible Ambient Noise Levels for Audiometric Test Rooms (ASA 99)*. New York: American National Standards Institute.

ANSI S3.6 (2010). *Specification for Audiometers*. New York: American National Standards Institute.

ANSI S3.6-1989 (1989). *American National Standard Specifications for Audiometers*. New York: American National Standards Institute.

ANSI S3.14 (1997). *Rating Noise with Respect to Speech Interference (ASA 21)*. New York: American National Standards Institute.

ANSI S3.21 (1992). *Manual Pure-Tone Threshold Audiometry, Thresholds for (ASA 19) (R 1992)*. New York: American National Standards Institute.

ANSI S3.28-1986 (1986). *Methods for the Evaluation of the Potential Effect on Human Hearing of Sounds with Peak A-Weighted Sound Pressure Levels Above 120 Decibels and Peak C-Weighted Sound Pressure Levels Below 140 Decibels (Draft ASA 66)*. New York: American National Standards Institute.

ANSI S3.44 (2006). *Determination of Occupational Noise Exposure and Estimation of Noise-Induced Hearing Impairment*. New York: American National Standards Institute.

ANSI S12.7 (1986). *Method for Measurement of Impulse Noise*. New York: American National Standards Institute.

ANSI S12.19 (1996). *Measurement of Occupational Noise Exposure*. New York: American National Standards Institute.

Anticaglia, J.R. (1973). Physiology of hearing. In: *The Industrial Environment – Its Evaluation and Control*. U.S. DHEW, PHS, CDC, National Institute for Occupational Safety and Health.

ASME Y10.11-1984 (1984). *Letter Symbols and Abbreviations for Quantities Used in Acoustics* (Revision of ASA Y10.11-1953 (R1959)). New York: ASME.

Beranek, L.L. (1992). Criteria for face-to-face speech communication. In: *Noise and Vibration Control Engineering: Principles and Practice* (ed. L.L. Beranek and I.L. Vér), 618–621. New York: Wiley.

Beranek, L.L. and Ver, I.L. (1992). *Noise and Vibration Control Engineering Principles and Applications*, 32. New York: Wiley.

Berger, E.H. (1981). *EARLog 7 Motivating Employees to Wear Hearing Protective Devices*. Indianapolis, IN: Cabot Safety Corp.

Berger, E.H. (1984). *EARLog 13 Attenuation of Earplugs Worn in Combination with Earmuffs*. Indianapolis, IN: Cabot Safety Corporation.

Berger, E. H., Ward, W. D Morrill, J. C., and Royster, L.H. (1986). *Noise and Hearing Conservation Manual*, 4e, Chapter 5. Akron, OH: American Industrial Hygiene Association.

Berglund, B. and Lindvall, T. (eds.) (1995). *Community Noise, Document Prepared for the World Health Organization*. Stockholm, Sweden: Center for Sensory Research.

Bienvenue, G.R. and Michael, P.L. (1979). Digital processing techniques in speech discrimination testing. In: *Rehabilitation Strategies for Sensorineural Hearing Loss*. New York: Grune & Stratton.

Bruce, R.D. and Toothman, E.H. (1986). Engineering controls. In: *Noise and Hearing Conservation Manual*. Farfax, VA: AIHA.

Bruel, P.V. (1977). Do we measure damaging noise correctly? *Noise Control Eng. J.* **8**: 52–60.

CAOHC (1978). *Council for Accreditation in Occupational Hearing Conservation Manual*. Cherry Hill, NJ: Fischler's Printing.

Cohen, A. (1966). Effects of noise on performance. *Proceedings of the International Congress of Occupational Health*, Vienna, Austria, pp. 157–160.

Cohen, A. (1968). *Trans. N. Y. Acad. Sci.* **30**: 910–918.

Cohen, A. (1969). Effects of noise and psychological state. In: *Proceedings of National Conference on Noise as a Public Health Hazard*, 74–88. Washington, DC: American Speech and Hearing Association.

Cohen, A. (1973a). Extra-auditory effects of occupational noise. Part I: disturbances to physical and mental health. *Natl. Saf. News* **108** (2): 93–99.

Cohen, A. (1973b). *Natl. Saf. News* **108** (3): 68–76.

Environmental Protection Agency (1979). Noise labeling requirements for hearing protectors, 40 CFR Part 211, Subpart B. *Fed. Regist.* **190**: 56139–56147.

European Committee for Standardization (1993). *Hearing Protectors – Recommendations for Selection, Use, Care and Maintenance – Guidance Document*, CENprEN 458. Brussels, Belgium: European Committee for Standardization.

Federal Interagency Committee on Noise (1992). *Federal Agency Review of Selected Airport Noise Analysis Issues*. Washington, DC: Environmental Protection Agency.

Federal Interagency Committee on Urban Noise (1980). *Guidelines for Considering Noise in Land Use Planning and Control* Document 1981-338-006/8071. Washington, D.C.: U.S. Government Printing Office.

Feldmen, A. and Grimes, C. (1985). *Hearing Conservation in Industry*, 77–88. Baltimore, MD: Williams and Wilkins.

Finegold, L.S., Harris, C.S., and von Gierke, H.E. (1994). Community annoyance and sleep disturbance: updated criteria for assessing the impacts of general transportation noise on people. *Noise Control Eng. J.* **42** (1): 25–30.

Franks, J., Themann, C., and Sherris, C. (1994). *The NIOSH Compendium of Hearing Protection Devices*, NIOSH Publication 95-105. Cincinnati, OH: U.S. Department of Health and Human Services, Public Health Service, Centers for Disease Control and Prevention, National Institute of Occupational Safety and Health.

von Gierke, H.E. and Eldred, K.M. (1993a). Effects of noise on people. *Noise/News International* **1** (2): 75–76.

von Gierke, H.E. and Eldred, K.M. (1993b). Effects of noise on people. *Noise/News International* **1** (2): 79.

Glorig, A. (1958). *Noise and Your Ear*. New York: Grune & Stratton.

Gordon, C.G. and Jones, R.S. (1991). Control of machinery noise. In: *Handbook of Acoustical Measurements and Noise Control*, 3e (ed. C.M. Harris), 40.8. McGraw-Hill.

Gottlob, D. (1995). Regulations for community noise. *Noise/News International* **3** (4): 223–226.

Grether, W.F. (1971). *Noise and Human Performance*, Report AMRL-TR-70-29, AD 729 213. Wright–Patterson Air Force Base, OH: Aerospace Medical Research Laboratory.

Hodgson, M. and Warnock, A.C.C. (1992). Noise in rooms. In: *Noise and Vibration Control Engineering Principles and Applications* (ed. L.L. Beranek and I.L. Ver). New York: Wiley.

ISO (1996). *Acoustics – Description and Measurement of Environmental Noise: Part 1: Basic Quantities and Procedures. Part 2: Acquisition of Data Pertinent to Land Use. Part 3: Application to Noise Limits*. Geneva, Switzerland: International Organization for Standardization.

ISO (1999). *Acoustics – Determination of Occupational Noise Exposure and Estimation of Noise-Induced Hearing Impairment*. Geneva, Switzerland: International Organization for Standardization.

ISO 2204 (1979). *Acoustics – Guide to International Standards on the Measurement of Airborne Acoustical Noise and Evaluation of Its Effects on Human Beings*. Geneva, Switzerland: International Organization for Standardization.

ISO 4869-1 (1990). *Acoustics – Hearing Protectors – Part 1: Subjective Method for the Measurement of Sound Attenuation*. Geneva, Switzerland: International Organization for Standardization.

ISO 9612-1991 (1991). *Acoustics – Guidelines for the Measurement and Assessment of Exposure to Noise in the Working Environment*. Geneva, Switzerland: International Organization for Standardization (ISO).

Joint ACGIH-AIHA Task Group on Occupational Exposure Databases (1996). Data elements for occupational exposure databases: guidelines and recommendations for airborne hazards and noise. *Appl. Occup. Environ. Hyg.* **11** (11): 1294–1311.

Killion, M.C. (1978). Revised estimate of minimum audible pressure: where is the “missing 6 dB”? *J. Acoust. Soc. Am.* **63** (5): 1501–1508.

Kryter, K.D. (1994a). *The Handbook of Hearing and the Effects of Noise*. San Diego, CA: Academic Press.

Kryter, K.D. (1994b). *The Handbook of Hearing and the Effects of Noise*, 275–277. San Diego, CA: Academic Press.

Lataye, R. and Campo, P. (1996). Applicability of the Leq as a damage-risk criterion: an animal experiment. *J. Acoust. Soc. Am.* **99** (3): 1621–1632.

Lawther, A. and Robinson, D.W. (1987). *Further Investigation of Tests for Susceptibility to Noise-Induced Hearing Loss*. ISVR Technical Report 149. Southampton: Institute of Sound and Vibration Research, University of Southampton.

Lempert, B. (1984). *Sound Vib.* **18**: 26–39.

Michael, P.L. (1955). Noise measurements and personal protection. *Industrial Hygiene Foundation Transactions of the 20th Meeting*, Pittsburgh, PA, pp. 1–10.

Michael, P.L. (1965). Hearing protectors – their usefulness and limitations. *Arch. Environ. Health* **10**: 612–618.

Michael, P.L. and Bienvenue, G.R. (1983). Industrial noise and man. In: *Physiology and Productivity at Work – The Physical Environment*. New York: Wiley.

Michael, P.L. et al. (1970). *Intelligibility Test on a Family of Electroacoustic Devices (FEADS) – Final Report*. Contract No N00953-70-2620. U.S. Navy.

Michel, P.L. (1991). Industrial noise and conservation of hearing. In: *Patty's Industrial Hygiene and Toxicology, 4e, Vol. I, Part A, General Principles* (ed. G.D. Clayton and F.E. Clayton), 937–1039. New York: Wiley.

Miller, J.D. (1974). Effects of noise on people. *J. Acoust. Soc. Am.* **56** (3): 729.

Newby, H.A. (1964). *Audiology*, 2e. New York: Appleton-Century-Crofts.

NIOSH (1972). *Criteria for a Recommended Standard: Occupational Exposure to Noise*, NIOSH 73-11001. Cincinnati, OH: NIOSH.

NIOSH (1982). *A Report on the Performance of Personal Noise Dosimeters*. Washington, DC: NIOSH.

Nixon, C.W. and Knoblach, W.C. (1974). *Hearing Protection of Ear Muffs Worn Over Eyeglasses*. AMRL-TR-74-61. Wright–Patterson Air Force Base, OH: Aerospace Medical Research Laboratory.

Nixon, C.W. and von Gierke, H.E. (1969). Experiments in bone conduction threshold in a free sound field. *J. Acoust. Soc. Am.* **31**: 1121–1125.

Occupational Safety and Health Administration (1983). Occupational noise exposure: hearing conservation amendment, 29 CFR Part 1910. *Fed. Regist.* **48** (42).

Pickles, J.O. (1988). *An Introduction to the Physiology of Hearing*, 2e, 5. London: Academic Press.

Shaw, E.A.G. (1996). Noise environments outdoors and the effects of Community noise exposure. *Noise Control Eng. J.* **44** (3): 115.

Suter, A.H. (1986). Hearing conservation. In: *Noise and Hearing Conservation Manual* (ed. E.H. Berger, W.E. Ward, J.C. Morrill, and L.H. Royster). Fairfax, VA: American Industrial Hygiene Association.

U.S. Department of Housing and Urban Development (1985). *The Noise Guidebook*. U.S. Government Printing Office, HUD-953-CPD.

U.S. Department of Labor (1969). Safety and Health Standards for Federal Supply Contracts (Walsh–Healy Public Contracts Act). *Fed. Regist.* **34**: 7948.

U.S. Department of Labor (1971). Occupational Safety and Health Standards (Williams–Steiger Occupational Safety and Health Act of 1970). *Fed. Regist.* **36**: 10518.

U.S. Environmental Protection Agency (1974). *Information on Levels of Environmental Noise Requisite to Protect the Public Health and Welfare with an Adequate Margin of Safety*, Document EPA 550/9-74-004. Washington, DC: U.S. Environmental Protection Agency.

Witt, M. (ed.) (1980a). *Noise Control, a Guide for Workers and Employers*. U.S. Department of Labor, Occupational Safety and Health Administration, Office of Information, OSHA 3048, p. 106.

Witt, M. (ed.) (1980b). *Noise Control, a Guide for Workers and Employers*. U.S. Department of Labor, Occupational Safety and Health Administration, Office of Information, OSHA 3048, p. 37.

Witt, M. (ed.) (1980c). *Noise Control, a Guide for Workers and Employers*. U.S. Department of Labor, Occupational Safety and Health Administration, Office of Information, OSHA 3048, p. **25**.

Witt, M. (ed.) (1980d). *Noise Control, a Guide for Workers and Employers*. U.S. Department of Labor, Occupational Safety and Health Administration, Office of Information, OSHA 3048, p. 31.

Witt, M. (ed.) (1980e). *Noise Control, a Guide for Workers and Employers*. U.S. Department of Labor, Occupational Safety and Health Administration, Office of Information, OSHA 3048, p. 33.

Witt, M. (ed.) (1980f). *Noise Control, a Guide for Workers and Employers*. U.S. Department of Labor, Occupational Safety and Health Administration, Office of Information, OSHA 3048, p. 47.

Witt, M. (ed.) (1980g). *Noise Control, a Guide for Workers and Employers*. U.S. Department of Labor, Occupational Safety and Health Administration, Office of Information, OSHA 3048, p. 59.

Witt, M. (ed.) (1980h). *Noise Control, a Guide for Workers and Employers*. U.S. Department of Labor, Occupational Safety and Health Administration, Office of Information, OSHA 3048, p. 61.

Witt, M. (ed.) (1980i). *Noise Control, a Guide for Workers and Employers*. U.S. Department of Labor, Occupational Safety and Health Administration, Office of Information, OSHA 3048, p. 49.

Witt, M. (ed.) (1980j). *Noise Control, a Guide for Workers and Employers*. U.S. Department of Labor, Occupational Safety and Health Administration, Office of Information, OSHA 3048, p. 57.

Witt, M. (ed.) (1980k). *Noise Control, a Guide for Workers and Employers*. U.S. Department of Labor, Occupational Safety and Health Administration, Office of Information, OSHA 3048, p. 79.

Witt, M. (ed.) (1980l). *Noise Control, a Guide for Workers and Employers*. U.S. Department of Labor, Occupational Safety and Health Administration, Office of Information, OSHA 3048, p. 29.

Witt, M. (ed.) (1980m). *Noise Control, a Guide for Workers and Employers*. U.S. Department of Labor, Occupational Safety and Health Administration, Office of Information, OSHA 3048, p. 39.

Witt, M. (ed.) (1980n). *Noise Control, a Guide for Workers and Employers*. U.S. Department of Labor, Occupational Safety and Health Administration, Office of Information, OSHA 3048, p. 83.

Witt, M. (ed.) (1980o). *Noise Control, a Guide for Workers and Employers*. U.S. Department of Labor, Occupational Safety and Health Administration, Office of Information, OSHA 3048, p. 87.

Witt, M. (ed.) (1980p). *Noise Control, a Guide for Workers and Employers*. U.S. Department of Labor, Occupational Safety and Health Administration, Office of Information, OSHA 3048, p. 15, 92.

Witt, M. (ed.) (1980q). *Noise Control, a Guide for Workers and Employers*. U.S. Department of Labor, Occupational Safety and Health Administration, Office of Information, OSHA 3048, p. 62.

Witt, M. (ed.) (1980r). *Noise Control, A Guide for Workers and Employers*. U.S. Department of Labor, Occupational Safety and Health Administration, Office of Information, OSHA 3048, p. 21.

CHAPTER 10

# HEAT STRESS

ANNE M. VENETTA RICHARD

*Lucent Technologies, 1600 Osgood St., 21-2L28, North Andover, MA, 01845*

and

RALPH COLLIPI, JR.

*AT&T, 40 Elwood Rd., Londonderry, NH, 03053*

Adapted from "Heat Stress: Its Effects, Measurement, and Control" by John E. Mutchler, Chapter 21 in G.D. Clayton and F.E. Clayton, *Patty's Industrial Hygiene and Toxicology, 4e, Vol. IA*. New York: Wiley, 1991, pp. 763–837.

## 10.1 SIGNIFICANCE OF HEAT STRESS IN INDUSTRY

As industry has developed, through the Industrial Revolution to our present highly technological society, on-the-job potential for injury and illness from acute exposure to heat has increased far beyond that known earlier to home-centered craftsmen. Among the more dangerous original industrial vocations were those using molten materials, such as glass and metals. In these first "hot industries," the ever-present danger of bums, explosions, and spills of molten material was well known and accepted, as were potential illness and death from very hard physical work in excessively hot environments (Clayton and Clayton 1991).

The traditional hot work industries (i.e. foundries, smelting, firefighting, military, mining, construction, utilities, glass working, tire and rubber, and textile industry workers) are being augmented by industries at even greater risk. Hazardous chemical handlers, lead and asbestos abatement workers, emergency responders, nuclear and radiation containment specialists, and laboratory and hospital personnel who must wear nonpermeable personal protective equipment along with warm or hot ambient conditions are new sources that need special attention to protect them from potential heat stress (Barrett 1991). Knowledge of the process involved in the exchange of heat between the worker and the environment and the effects of heat can reduce the potential adverse effects of heat exposure.

## 10.2 PHYSIOLOGY OF HEAT STRESS

Basic to any understanding of heat stress is comprehension of the interaction that occurs between our bodies and our environment. Humans must maintain an internal body temperature within a narrow range, near 37 °C (98.6 °F), to remain healthy and efficient. If the "core temperature" of the body falls below 35 °C (95 °F), hypothermia results, and death is likely at core temperatures below 27 °C (80.6 °F) and above 42 °C (107.6 °F) (Clayton and Clayton 1991, p. 773). Internal body temperature is maintained at the regulated level when heat loss is in balance with heat production, which depends on the controlled exchange of heat between the body and the environment.

The heat that must be exchanged is a function of (i) the total heat produced by the body (*metabolic heat*) and (ii) the heat gained, if any, from the environment. The rate of heat exchange with the environment depends essentially on conditions such as air temperature and velocity, the amount of humidity or vapor pressure of water in the air, and

*Handbook of Occupational Safety and Health*, Third Edition. Edited by S. Z. Mansdorf.

radiant temperature. The body's inability to accommodate these contributions can result in heat stress. Body conditions that influence the heat equilibrium are skin temperature, amount of evaporated sweat produced, and the type, amount, and characteristics of the clothing worn. Respiratory heat loss is generally of minor consequence except during vigorous activity in very dry environments.

## 10.3 HEAT EXCHANGE AND HEAT BALANCE

There are several concepts that must be considered when performing hazard assessments for heat stress. These concepts are the basis for education and training programs. *Conduction* is the transfer of heat between two objects when they are in direct physical contact with each other. A person immersed in a cool bath will lose heat from the body via conduction as the body heat is transferred to the object of lower temperature, the water. Conduction of heat to or from solid objects usually can be ignored since workers are generally not in direct contact with surfaces hotter than normal body temperatures for any sustained length of time. However, heat loss by conduction into air occurs when the air in contact with the skin is below body temperature (negative heat load). Conversely, heat gain by conduction from the air occurs when air temperature exceeds body temperature (positive heat load).

Convective heat exchange between the skin and air immediately around the skin is a function of the difference in temperature between the skin and the air and rate of air movement over the skin. *Convection* can produce a negative or positive heat load. For example, when the air temperature is higher than the skin temperature, the contribution to heat load by convection will be positive.

Another aspect of heat exchange is *radiation*, or the transfer of heat from one object to another via electromagnetic waves through a vacuum or air. The heat is not perceived until it is absorbed by the object that the electromagnetic waves strike. A body gains heat by radiation when the temperature of the surrounding surfaces is above body surface temperature (i.e. the sun, steam pipes, and blast furnaces), and conversely a body loses heat when the surrounding objects have surface temperatures lower than the temperature of the body surface. Radiant heat can produce a negative or positive heat load and is independent of air motion.

*Evaporation* is generally the mechanism most in use by the body for the dissipation of large amounts of heat generated by working muscles. Evaporation is determined by air speed and the difference between the vapor pressure of perspiration on the skin and the partial pressure of water in the air. In industries with high levels of ambient water vapor from wet processes or escaping steam (mining, laundries), high relative humidity reduces evaporative cooling capacity. In a hot, dry environment, the maximum sweat production, maintained over eight hours, is about 1 liter per hour ($1\,h^{-1}$). Heat is also lost through respiration in negligible amounts.

### 10.3.1 Heat Balance Equation

The total heat load is a combination of the metabolic heat load and the environmental factors. This condition is often expressed by the "heat balance equation":

$$H = M + R + C + D - E$$

where $H$ is the heat load, which should be 0; $M$ is the metabolic heat gain; $R$ is the radiant heat gain or loss; $C$ is the convective heat gain or loss; $D$ is the conductive heat gain or loss; and $E$ is the evaporative loss.

If ambient conditions rise along with the metabolic load, the body has an increasingly difficult task of cooling itself and "balancing" the equation (Wildeboor and Camp 1993). The heat stress disorders resulting from the heat load in excess of "0" are explained in Section 10.5. Complicated measurements of metabolic heat production, air temperature, air water vapor pressure, wind velocity, and mean radiant temperatures (MRT) are required to compute this equation.

## 10.4 FACTORS AFFECTING HEAT TOLERANCE

### 10.4.1 Acclimatization

Acclimatization refers to a set of adaptive physiological and psychological adjustments that occur when an individual accustomed to working in a temperate environment undertakes work in a hot environment. These progressive adjustments occur over periods of increasing duration and reduce the strain experienced on initial exposure to heat. This enhanced tolerance allows a person to work effectively under conditions that they might have been unable to endure before acclimatization (NIOSH 1993). Acclimatization is marked by a lower heart rate; lower body temperature; an increase in sweat rate, which increases evaporative cooling; reduced circulatory load; and enhanced heat conduction through the skin. Not only does the sweat rate increase, but the composition of the sweat changes. Sweat has a lower osmotic pressure than water, has one-third of the electrolyte concentration of extracellular fluids, and is even more dilute in persons not acclimated to the environment (Smith and Slarinski 1987). Failure to replace the water lost in sweat will retard or even prevent the development of the physiological adaptations characteristic of acclimatization. Thus acclimated people have less strain on their bodies while working in heat. Full acclimatization is usually achieved within seven days of beginning work in a hot environment. Subjective discomfort of working in the hot environment dissipates by four to seven days, with a decrease in heart rate and body temperature and increased sweat rate. Unacclimated workers should work at 50% of required work on the first day and increase production at a rate of 10% each day thereafter until full acclimatization is achieved (NIOSH 1972a).

After heat exposure on several successive days, the individuals perform the same work with a much lower core temperature and heart rate and higher sweat rate (reduced thermoregulatory strain) and with none of the distressing symptoms that may be experienced initially. Heat acclimatization represents a dynamic state of conditioning rather than a long-term change in innate physiology. The level of acclimatization is relative to the initial level of physical fitness and the total heat stress experienced by an individual. Thus a worker who does only light work indoors in a hot climate will not achieve the level of acclimatization needed to work outdoors (with the additional heat load from the sun) or to do harder physical work in the same hot environment indoors.

### 10.4.2 Age

Older workers (40–65 years of age) are at some disadvantage when working in hot environments. As the body ages, its maximal oxygen intake decreases, and the body is more likely to reach the stage where its metabolism becomes *anaerobic* rather than *aerobic. Anaerobic* is defined as able to live and grow where there is no air or free oxygen, as certain bacteria, and *aerobic* is the ability to live, grow, or take place only where free oxygen is present. Older adults may have up to 30% less cardiovascular reserve than younger adults. Also, older workers have less efficient sweat glands, causing a delay in the onset of sweating and a lower sweat rate, and thus are less able to compensate for heat load by evaporative cooling. Underlying degenerative diseases of the heart and lungs may further compromise the older worker's circulatory capacity to move heat away from the body core and to the surface. Given these limitations, the older worker can work effectively in a hot environment if allowed to work at an independent pace. It has been suggested that with the current emphasis on fitness in today's society, physiological responses to heat stress, rather than age, should be a major consideration for working in a hot environment (Dukes–Dubos and Henschel 1980).

### 10.4.3 Gender

Men tolerate an imposed heat stress slightly better than women in hot, dry climates, but not as well as women in humid conditions, according to a study in 1980 (Shapiro et al. 1981). Actual body size is more predicative of risk. Smaller persons (<50 kg) of either sex are at a disadvantage with regard to heat tolerance due to an increased surface area to mass (SA : M) ratio and a lower aerobic capacity.

Recent retrospective epidemiological studies have associated hyperthermia during the first trimester of pregnancy with birth defects, especially in the CNS development (e.g. anencephaly). It is prudent to monitor the body temperature of a pregnant woman exposed to total heat loads above the recommended exposure limit every hour to assure that the body temperature does not exceed 39–39.5 °C (102–103 °F) during the first trimester.

Heat exposure has been associated with temporary infertility in both males and females, with the effects more pronounced in males. Workers who report problems with fertility should be evaluated to determine whether a temporary transfer is needed or whether the amount of work exposure in hot environments should be limited.

### 10.4.4 Obesity

It is well established that obesity predisposes individuals to heat disorders. The acquisition of fat means that additional weight must be carried; therefore, a greater expenditure of energy is required to perform a given task, and a greater proportion of aerobic capacity is used. In addition, obese individuals have less surface area for their weight, thus limiting the rate of heat exchange with the environment. Obesity also alters sweating and sweat gland distribution, and the increased layer of subcutaneous fat provides an insulating barrier between the skin and the deep-lying tissues. The fat layer theoretically reduces the direct transfer of heat from the muscles to the skin. The increased metabolic load due to excess weight and decreased surface area for cooling puts the obese worker at risk when performing in a hot environment.

### 10.4.5 Physical Fitness

Levels of fitness in workers are important criteria for determining their ability to work in a heat stress environment. Fitness increases heat tolerance by improving cardiovascular capacity via increasing the capillary bed and vascular tone. Also, improved cardiac output decreases heart rate during hot work with less need to accelerate the heart. Therefore, a physically conditioned person, by virtue of having a higher maximum ventilatory capacity, has a wider margin of safety in coping with the added circulatory strain of working under heat stress.

### 10.4.6 Wellness Programs

There has been a thrust to implement "wellness programs" in some business sectors. The general concept is that business needs a healthy workforce to keep pace with the daily demands of the job. It is felt that employees who are "well" have more energy, resilience, and stamina and are more likely to be part of a high-performing team and will feel respected and valued.

Healthy employees also help business to manage healthcare costs. There is strong empirical evidence that employees at lower levels of health risk use fewer health plan dollars.

Wellness programs can include such things as employee assistance programs, health and well-being survey data, smoking cessation sessions, health fitness centers or membership assistance programs, lifestyle change counseling, and lifting and back education sessions.

Employee educational and informational programs that are part of wellness programs may also increase employee awareness about potential risk factors associated with their occupational duties. These programs can address issues like ergonomics awareness associated with computer use or, in the case of heat stress, how acclimatization and proper hydration and work breaks will help to reduce the potential for illness.

### 10.4.7 Water and Electrolyte Balance

Effective work performance in heat requires a replacement of body water and electrolytes lost through sweating. If this water is not replaced by drinking, continued sweating will draw on water reserves from both tissues and body cells, leading to dehydration. Water loss, through sweat, of up to 1.4% body weight can be tolerated without serious effects (Clayton and Clayton 1991). At water loss of 3–6% of body weight, work performance is impaired; continued work under such conditions leads to heat exhaustion. Acclimated workers are better able to maintain appropriate water balance, although sweating workers should be supplied with cool drinking water and encouraged to drink small amounts every 15–20 minutes to assure adequate fluid replacement. One to 3 gallons per day per worker is typically supplied in hot environments.

The typical American diet provides enough salt for proper salt balance in most acclimatized workers. Unacclimated workers may be given salt supplements in the form of salt tablets, preferably impregnated to avoid gastric irritation, if ample water is available. A preferable practice is to use salted water (one tablespoon per gallon) or ingest a rehydration beverage that contains sodium, specifically any electrolyte replenisher. Salt supplements should be decreased or discontinued after acclimatization so as not to suppress normal hormonal mechanisms of balancing salt and water.

### 10.4.8 Alcohol and Drugs

Alcohol is a risk factor for heat strain by physiologically or behaviorally altering thermoregulatory functions. Alcohol interferes with central and peripheral nervous function and is associated with dehydration by suppressing alcohol dehydrogenase (ADH) enzyme production, leading to hypohydration. The ingestion of alcohol before or during work in the heat should not be permitted. Therapeutic drugs such as diuretics, antihypertensives, anticholinergic drugs, antihistamines, CNS inhibitors, muscle relaxants, atropine, tranquilizers, sedatives, beta-blockers, and amphetamines can interfere with thermoregulation and could potentially affect heat tolerance. Over-the-counter medications should not be overlooked, as indicated in Table 10.1. Any worker subject to heat stress taking therapeutic medications who is exposed even intermittently or occasionally to a hot environment should be under the supervision of a physician who understands the potential ramifications of medications on heat tolerance. It is difficult to separate the heat-disorder implications of drugs used therapeutically from those that are used socially. Nevertheless, there are many drugs other than alcohol that are used on social occasions. Some of these have been implicated in cases of heat disorder, sometimes leading to death (Clayton and Clayton 1991).

**TABLE 10.1 Relationship of Common Medications and Heat Stress**

1. Amphetamines – increase the metabolic needs and put additional stress on the person
2. Anticholinergic medications – inhibit perspiration
3. Antihistamines – inhibit perspiration
4. Atropine – inhibits perspiration
5. Beta-blocking agents – may enhance dehydration by promoting an inappropriate increase in sweat production
6. Central nervous system inhibitors – affect the hypothalamus' ability to regulate heat properly
7. Diuretics – inhibit necessary expansion of bodily fluid volume; lower circulating fluid volume; inhibit cutaneous vasodilation
8. Muscle relaxants – may cause postural hypotension
9. Tranquilizers and sedatives – phenothiazines, tricyclic antidepressants, monoamine oxidase inhibitors, and glutethimide (Doriden) – implicated in lower heat tolerance
10. Vasodilators – cause a relaxing action on smooth muscles; decreased peripheral resistance; reflex increase in heart rate

*Source: AAOHN J.* 39(8), 372, August 1991.

## 10.5 HEAT STRESS DISORDERS

Environmental heat and the inability to remove metabolic heat can lead to well-known reactions in humans, including increased cardiovascular activity, sweating, and increased body core temperature. A variety of heat stress disorders resulting from the body's reaction to excessive heat, in order of increasing severity, are heat rash, heat syncope, heat cramps, heat exhaustion, and heat stroke. In addition to the physiological effects of heat, overexposure to heat and accompany dehydration may precipitate psychological effects, including irritability, an increase in the frequency of errors, a higher frequency of accidents, and a reduction in efficiency in the performance of skilled physical tasks.

### 10.5.1 Heat Rash

Heat rash is a profuse, red vesicular rash accompanied by prickly sensations of areas affected by heat. It is caused by plugged sweat glands, retention of sweat, and an accompanying inflammatory reaction. This is precipitated by hot, humid conditions in which sweat cannot adequately evaporate. The most common heat rash is prickly heat (miliaria rubra), which appears as red papules, usually in areas where the clothing is restrictive, and gives rise to a prickly sensation, particularly as sweating increases. The papules may become infected unless they are treated. Another skin disorder (miliaria crystallina) appears with the onset of sweating in skin previously injured at the surface, commonly in sunburned areas, although it has been reported to occur without clear evidence of previous skin injury. Heat rashes disappear when the individuals are returned to cool environments. Heat rash may be treated by keeping the skin clean and with light applications of mild drying lotions or powders.

### 10.5.2 Heat Syncope

Heat syncope is fainting while standing immobile in a hot environment and is caused by the pooling of blood in the lower extremities. Syncope is due to poor or inadequate acclimatization, and recovery is typically prompt and complete after moving the worker to a cooler environment. Syncope can be prevented by intermittent activity to assist the return of venous blood return to the heart (Clayton and Clayton 1991, p. 786).

### 10.5.3 Heat Cramps

Heat cramps are painful spasms of the voluntary muscles (i.e. arm, leg, abdomen, or back) used during work. The onset may be during work or after several hours of heavy work or while showering. Heat cramps are caused by a reduction of the concentration of sodium chloride (salt) in blood below a certain critical level. They are caused by a continued loss of salt in the sweat and accompanying dehydration, drinking large volumes of water without appropriate replacement of salt, depletion of body electrolytes, and lack of acclimatization. The dilution of tissue fluid and resultant transfer into muscle tissue precipitate the cramps. Heat cramps can be readily alleviated by replacing the water and salt in body fluids that has been lost and resting the muscle.

### 10.5.4 Heat Exhaustion

Heat exhaustion is a milder form of heat disorder linked to depletion of body fluids and electrolytes. This is a state of collapse due to insufficient blood supply to the brain. It is precipitated by dehydration, electrolyte loss, low arterial blood pressure, widespread vasodilation, or a compromised cardiovascular system (i.e. competing demands for blood, poor physical fitness, alcohol consumption). Heavy exertion, heat, and poor acclimatization also are causative factors. The symptoms of heat exhaustion include fatigue, fainting, weak pulse, low blood pressure, headache, nausea, vertigo, weakness, giddiness, skin clammy and moist, and complexion pale, muddy, or hectic flush. Oral temperature is normal or low, but rectal temperature is usually elevated (37.5–38.5 °C, or 99.54–101.34 °F). The use of occlusive clothing limits skin evaporation, creating a rise in skin temperature. As skin temperature approaches core temperature, the blood has less capacity to transport heat from the core to the skin. Under these conditions, heat exhaustion can occur at core body temperatures less than 43 °C (109.44 °F) and with heart rates of 120–130 beats per minute.

Treatment includes rest in a cool place and liquids high in electrolytes. Administer fluids by mouth or give intravenous infusions of normal saline (0.9%) if patient is unconscious or vomiting. Keep at rest until urine volume and content indicate water and electrolyte balances have been restored.

Heat exhaustion can be prevented by proper acclimatization using a breaking-in schedule for five to seven days. Ample drinking water must be available at all times during the work shift, and saline liquids or dietary salt supplements can be taken during acclimatization only.

### 10.5.5 Heat Stroke

Heat stroke is a serious disorder that is linked to failure of the body's thermoregulatory mechanisms and can be life threatening. Sweat production, the normal body coping mechanism, is blocked, and, consequently, the opportunity for evaporative cooling is reduced. The body temperature rises and heart rate increases. The classic symptoms of heat stroke include (i) a major disruption of central nervous function (unconsciousness, convulsions, or coma); (ii) lack of sweating, causing hot, dry, flushed, or cyanotic skin; and (iii) a rectal temperature in excess of 41 °C (105.8 °F) continuing to rise if the condition goes untreated. Irreversible changes occur in the body organs beyond 40 °C (104 °F).

*Heat stroke is a medical emergency, and immediate medical action is required.* Any procedure from the onset that will cool the patient will improve the prognosis. Cooling ensures that the vasoconstriction of superficial blood vessels is avoided, as this hinders the body's cooling ability by diverting blood away from the skin. Cooling by placing the person in a shady area, removing the outer clothing, and tepid sponging and increasing air movement by fanning are all necessary and appropriate until professional methods of cooling and assessing the degree of the disorder are available. Excessive cooling with ice water should be avoided. Cooling attempts should be discontinued once the core temperature reaches 37.7 °C (99.9 °F). Treat for shock and anticipate possible cardiac arrest or cardiovascular collapse due to electrolyte imbalance. It is important that transfer to a hospital is quickly arranged so that further close monitoring can be undertaken to identify possible cardiovascular and renal problems and the extent of organ and tissue involvement. Deaths do occur, the majority within the first 12 hours but the rest within 2 weeks.

Heat stroke typically occurs in young men undertaking moderate to hard physical exercise, and it is more likely to occur in those who are unacclimatized, obese, dehydrated, and inappropriately dressed and those who have consumed alcohol. Prevention involves medical screening of workers, and placement must be based on health and physical fitness. Workers must be acclimatized for five to seven days by graded work and heat exposure and monitored during sustained work in severe heat.

## 10.6 MEASUREMENT OF THE THERMAL ENVIRONMENT

Heat stress refers to the total heat load on the body that is the contribution by both environmental and physical factors. The assessment of heat stress includes the use of environmental variables to describe the thermal environment and may also include variables to reflect how the environment exchanges heat with the worker. The degree of discomfort and the level of stress caused by environmental heat depend on air temperature, relative humidity, velocity of air movement, and the contribution of radiant heat sources.

### 10.6.1 Indexes of Heat Stress

There are several indexes that can be used for assessing heat stress. They range from simple dry-bulb and wet-bulb temperature measurements to algebraic combinations of multiple environmental variables, which may include correction factors and modifications. Several indexes of thermal stress are commonly used in the assessment of industrial heat stress. These include:

- The effective temperature (ET) index.
- The equivalent effective temperature corrected for radiation (ETCR) index, a modification of ET that helps to determine the contribution made by radiant heat.
- The predicted 4-hour sweat rate (P4SR).
- The heat stress index (HSI).
- The wet-bulb globe temperature (WBGT) index.

The development of HSI has been based on use of the thermometric scale, sweat rates, and calculations of heat loads and the evaporative capacities of the environment.

### 10.6.2 Dry-Bulb and Wet-Bulb Temperatures

The dry-bulb air temperature (TA) is the temperature of the ambient air as measured with a thermometer or equivalent instrument (see Section 10.7.1). It is the simplest climatic factor to measure.

The natural wet-bulb temperature (NWB) is the temperature measured by a thermometer covered by a wetted cotton wick and exposed only to the naturally prevailing air movement. Accurate measurement of NWB requires the use of a clean wick, distilled water, and shielding to prevent radiant heat gain.

The psychrometric wet-bulb temperature (WB) is determined by forcing air over the wetted wick that covers the sensor. The WB is generally measured with a psychrometer, which consists of two mercury-in-glass thermometers mounted alongside each other on the frame of the psychrometer. One thermometer is used to measure the dry-bulb temperature (TA), and the second is used to measure the wet-bulb temperature. The air movement can be manual, as with a sling psychrometer, or mechanical, as with a motor-driven psychrometer.

These simple temperature parameters are easy to measure; however, they do not provide sufficient information to help determine the thermal exchange between the worker and the occupational environment.

### 10.6.3 Effective Temperature

Effective temperature is an index that combines dry-bulb temperature, wet-bulb temperature, and air velocity to estimate a thermal sensation to that of a given temperature of still saturated air. When taking measurements, ensure that the dry-bulb thermometer is shielded from radiation. Determination of ET also requires an anemometer to measure air velocity in feet per minute (fpm). Refer to Section 10.7.3 for more information on taking air velocity measurements. Nomograms were developed that characterized these equivalent environments, as shown in Figure 10.1 (Powell and Hosey 1965).

To use the nomogram, connect the dry-bulb thermometer temperature to the wet-bulb temperature with a straight line. The ET can be read where this line intersects the measured corresponding air velocity. Studies have shown that the risk of fatal heatstroke begins at 28 °C ET, and this risk increased sharply above 33 °C ET. There is little risk of heat stroke at less than 26 °C ET.

### 10.6.4 Equivalent Effective Temperature Corrected for Radiation

The use of a black-globe temperature in place of a dry-bulb temperature will allow a correction for the contribution by surrounding radiation sources. This is known as the equivalent effective temperature corrected for radiation. The nomogram in Figure 10.1 is also used to determine the ETCR; however, the dry-bulb temperature reading is substituted by the black-globe temperature reading, and the ETCR is then determined in the same manner as the ET. This particular ET nomogram scale pertains to workers wearing lightweight summer clothing. There is another scale for seminude men called the "basic" scale (Powell and Hosey 1965).

The limitations of ET and ETCR as HSI are because of their basis on sedentary workers, with no consideration given toward the type of clothing worn and its inherent absorption/desorption properties (refer to Section 10.9.3).

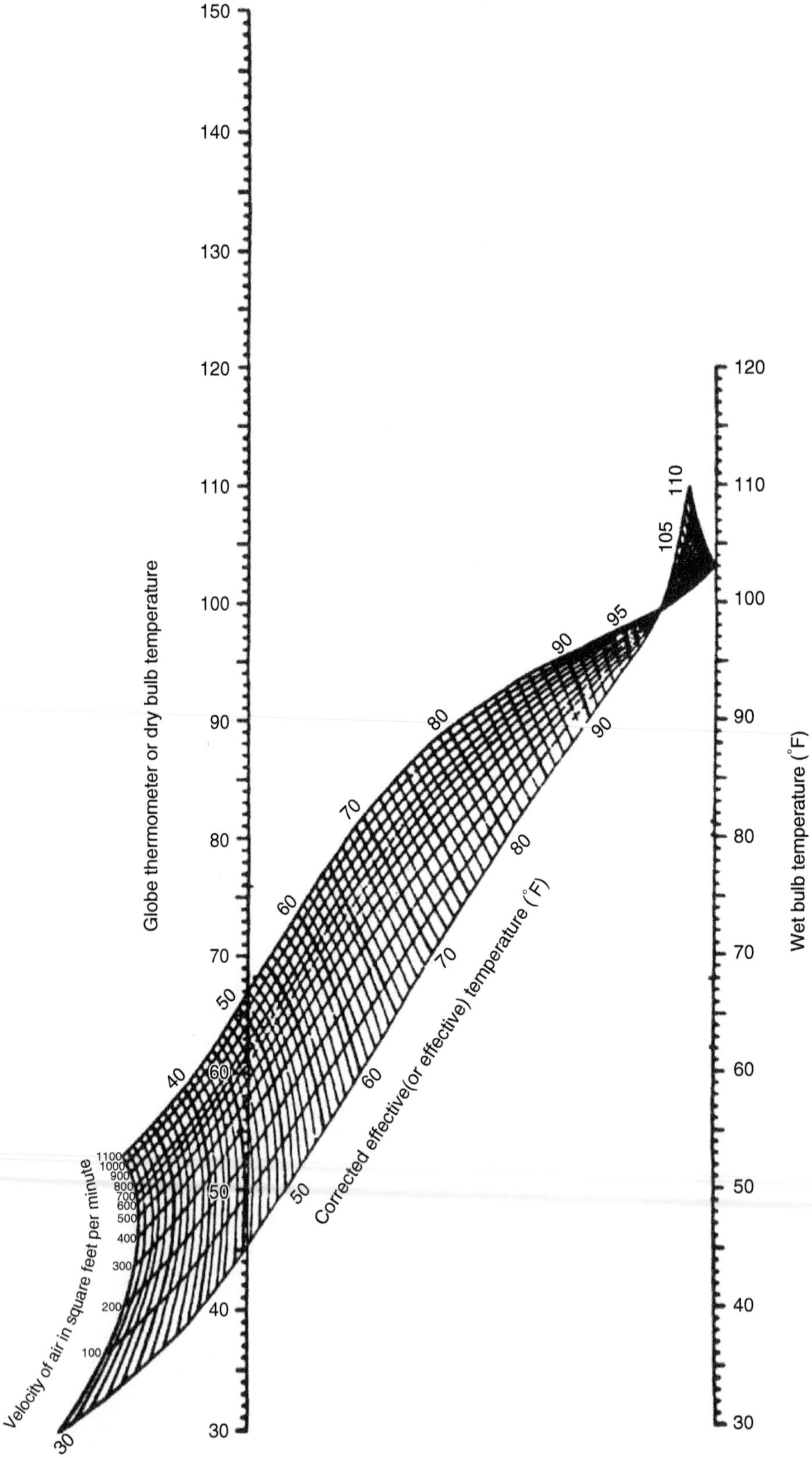


**Figure 10.1** Normal scale of corrected effective (or effective) temperature. Instructions for use: Stretch a thread or place a rule to join dry-bulb and wet-bulb temperatures. Note where this cuts appropriate air velocity line and read effective temperature at this point on the grid lines. *Source:* From Powell and Hosey (1965).

***Example 10.1:*** Given a dry-bulb temperature = 76 °F, a wet-bulb temperature = 55 °F, and air speed = 100 ft min$^{-1}$, find ET. Using the nomogram shown in Figure 10.1, use a rule to join the dry-bulb and wet-bulb temperatures. At the point where the rule intersects with the air velocity line, read the effective temperature on the grid lines. In this case the ET = 67 °F. That is, the environment in which these readings were taken would provide the same thermal sensation as one with dry-bulb and wet-bulb temperatures of 67 °F and "still" air (~25 ft min$^{-1}$). The ETCR can be determined by using the black-globe temperature reading in place of the dry-bulb temperature and following the same procedure (Powell and Hosey 1965).

### 10.6.5 Predicted 4-Hour Sweat Rate

The P4SR is an index based on observations of sweat rates under various environmental conditions. This index is expressed in liters and is a representation of the amount of sweat generated by a fit, well-acclimatized worker. It is believed that most workers will be unable to tolerate four hours of exposure as the P4SR rises above 4.5 l. It has been recommended that the safe limit for exposure of unacclimatized workers is 2.5–3 l.

The P4SR index requires globe temperature, wet-bulb temperature, air velocity, and metabolic rate of the workers. Although the index works well for temperatures that cause moderate sweating, it does not account for worker acclimatization. This index is probably of more value to physiologists than industrial hygienists:

1. *Calculation of the P4SR.* Calculation of the P4SR requires the use of a three-part nomogram, as displayed in Figure 10.2 (*Heating/Piping/Air Conditioning* 1976). The scale on the left represents the globe temperature. The five straight lines on the right constitute the psychometric wet-bulb scales, which are dependent upon air velocity. The third part of the nomogram consists of a group of curves running downward from right to left, which represent the basic 4-hour sweat rate (B4SR). The appropriate B4SR should be used depending on the air velocity of the environment:
   a. *Modified wet bulb.* The wet-bulb value requires modification in the following circumstances:
      (i) If the globe thermometer temperature differs from the dry-bulb temperature, the wet-bulb temperature is corrected by the addition of an amount equal to 0.4 (globe temperature minus dry-bulb temperature). Monitors similar to the one shown in Figure 10.3 that measure dry-bulb, wet-bulb, and globe temperature are available.
      (ii) If the energy expenditure exceeds 54 kcal m$^{-2}$ h$^{-1}$ (the metabolic rate of men sitting in chairs), an amount that is read off from the small inset chart is added to the wet-bulb temperature. The metabolic rate should be converted to an hourly rate and then divided by a factor of 1.8 to express it in kcal m$^{-2}$ h$^{-1}$.
      (iii) If the men are wearing clothing, imposing a greater stress than that imposed by shorts, an appropriate amount, dependent upon the amount of clothing worn, is added to the wet-bulb temperature. In the case of men wearing overalls over shorts, add 1.8 °F.
   b. *B4SR.* Draw a line between the globe temperature on the scale on the left and the modified wet-bulb temperature on the appropriate wet-bulb scale on the right. The B4SR is the point where this line intersects the appropriate curve for the given air speed on the B4SR scale.
   c. *P4SR*
      (i) Men sitting in shorts: P4SR = B4SR.
      (ii) Men working in shorts: P4SR = B4SR + 0.014 ($M$ − 54) where $M$ is the metabolic rate in kcal m$^{-2}$ h$^{-1}$.
      (iii) Men sitting in overalls worn over shorts: P4SR = B4SR + 0.25.
      (iv) Men working in overalls over shorts: P4SR = B4SR + 0.25 + 0.02 ($M$ − 54).
2. *Exposure limits.* As P4SR rises above 4.5 l, an increasing number of men will be unable to tolerate 4 hours of exposure. Lower P4SR values have never been very clearly related to physiological strain. It has been suggested that the safe limit for exposure of unacclimatized men may be 2.5–3 l. Use the nomogram in Figure 10.2 for the following example.

***Example 10.2:*** This sample solution demonstrates how to calculate the P4SR (Powell and Hosey 1965).

$$\text{Given: } GT = 105$$
$$WB = 80$$
$$V = 70\,\text{ft min}^{-1}$$
$$M = 100\,\text{kcal m}^{-2}\,\text{h}^{-1}$$

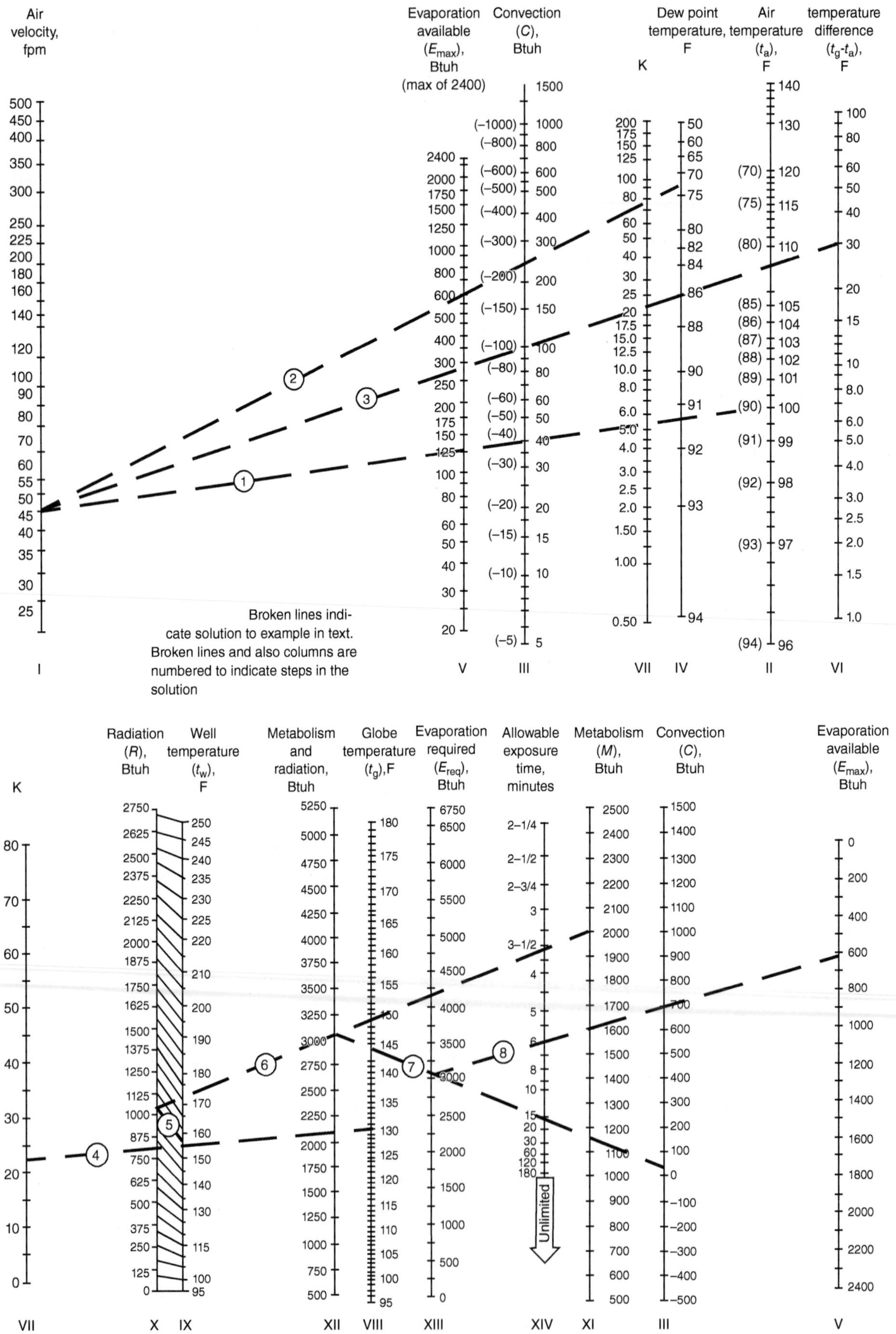


**Figure 10.2** Nomogram for the determination of heat stress index (HSI) and allowable exposure time (AET). *Source:* Reprinted from *Heating/Piping/Air Conditioning* (1976).

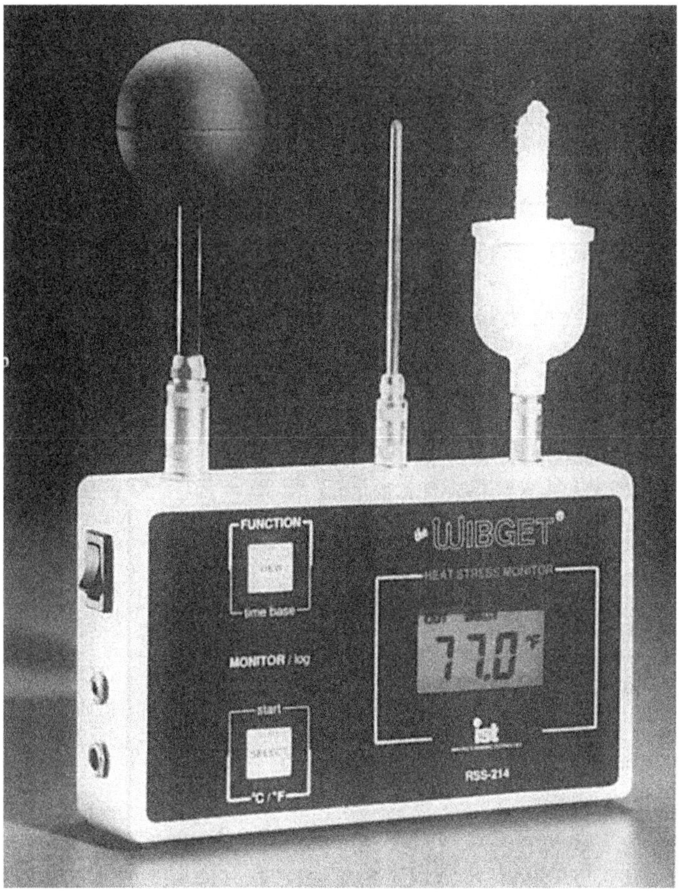


**Figure 10.3** Heat stress monitor that measures dry-bulb, wet-bulb, and globe temperature.

*Step 1*. From the small graph, find 4 °F to be added to the wet-bulb temperature to compensate for $M$ above resting.
*Step 2*. At the right side of the P4SR nomogram, enter $T_{wb}$(WB) = 84(80 + 4). Follow the 84 $T_{wb}$ line to the intersection with the 70 ft min$^{-1}$ line.
*Step 3*. Using a ruler, connect this point to $T_g$(GT) = 105.
*Step 4*. Read P4SR where this transverse line intersects with air velocity, $V = 70$. P4SR = 1.2.

This value represents conditions that pose little physiologic strain, as the upper tolerance limit for fit young men dressed in shorts is P4SR = 4.5. This index is less accurate in predicting strain as the upper level of tolerance is approached.

### 10.6.6 Heat Stress Index

Another metabolic rate method to measure heat stress is the HSI. The HSI is generally used by heating, ventilating, and air-conditioning engineers as a good diagnostic tool but does not provide an accurate assessment of heat stress in the work environment. This method was developed by Belding and Hatch at the University of Pittsburgh in the mid-1950s (Belding and Hatch 1955) and combines the heat exchange of radiation ($R$) and convection ($C$) components with metabolic heat ($M$) in terms of the required sweat evaporation ($E_{req}$). Stated algebraically,

$$E_{req} = M \pm R \pm C$$

The HSI is determined by 100 times the ratio of ($E_{req}$) to the maximum evaporative capacity of the environment ($E_{max}$):

$$\text{HSI} = \frac{E_{req}}{E_{max}} \times 100$$

**TABLE 10.2 Evaluation of Heat Stress Index**

| Index of heat stress | Physiological and hygienic implications of eight-hour exposure to various heat stresses |
|---|---|
| −20<br>−10 | Mild cold strain. This condition frequently exists in areas where individuals recover from exposure to heat |
| 0 | No thermal strain |
| +10<br>+20<br>+30 | Mild to moderate heat strain. Where a job involves higher intellectual functions, dexterity, or alertness, subtle to substantial decrements in performance may be expected. In performance of heavy physical work, little decrement is expected unless ability of individuals to perform such work under no thermal stress is marginal |
| +40<br>+50<br>+60 | Severe heat strain, involving a threat to health unless individuals are physically fit. Break-in period required for workers not previously acclimatized. Some decrement in performance of physical work is to be expected. Medical selection of personnel desirable because these conditions are unsuitable for those with cardiovascular or respiratory impairment or with chronic dermatitis. These working conditions are also unsuitable for activities requiring sustained mental effort |
| +70 | Very severe heat strain. Only a small percentage of the population may be expected to qualify for this work. Personnel should be selected (i) by medical examination and (ii) by trial on the job (after acclimation). Special measures are needed to assure adequate water and salt intake. Amelioration of working conditions by any feasible means is highly desirable and may be expected to decrease the health hazard while increasing efficiency on the job. Slight "indisposition," which in most jobs would be insufficient to affect performance, may render workers unfit for this exposure |
| +100 | The maximum strain tolerated daily by fit, acclimatized young workers |

*Source:* Adapted from Belding and Hatch (1955).

Body heating occurs when the HSI exceeds 100, and body cooling occurs when the HSI is less than 100. Table 10.2 shows the physiological and hygienic implications of eight-hour exposures to various levels of the HSI (Hertig 1973).

Values of $E_{req}$ and $E_{max}$ may be computed by means of appropriate equations or by use of the nomogram method developed by McKarns and Brief (1966) and shown in Figure 10.2.

To use the nomogram, follow these steps:

1. Determine the convective heat load by extending a line from the air velocity scale (column I) to the air temperature scale (column II). The convective load is indicated at the intersection of this line with column III. Note whether the heat load is positive or negative, as indicated by an air temperature above or below 35 °C (95 °F), respectively.
2. Obtain the maximum available evaporative cooling ($E_{max}$) from column V by extending a line from the air velocity scale (column I) to the dew point temperature scale (column IV). The dew point temperature is obtained from a psychometric chart at the intersection of the lines for wet-bulb and dry-bulb air temperatures.
3. Determine $K$ in column VII by extending a line from the air velocity scale to the temperature difference scale (column VI). Transfer the value of $K$ to column VII in the second set of alignment charts.
4. Extend the line from the $K$ scale to the globe temperature scale (column VIII), and read the radiant wall temperature in column IX.
5. Project this value to the radiation scale (column X) by extending a line parallel to the given slanting lines.
6. Estimate the appropriate metabolic rate (see Section 10.6.7) and locate this value of $M$ on the metabolism scale (column XI). Connect columns X and XI and determine the sum of metabolism and radiation in column IX.
7. Transfer the convective load determined in step 1 to column III, noting whether it is negative or positive. Connect this point with the sum in column XII and read the required rate of evaporation ($E_{req}$) in column XIII.
8. Locate the available evaporative cooling in column V. Extend a line from this point to column XIII. The approximate allowable continuous exposure time (AET) is indicated at the intersection of this line with column XIV.

The HSI provides an analysis tool for occupational heat exposure that also serves as a predictor of the "allowable exposure time" (AET) (Brief and Confer 1971). For an average man, AET is given by the equation:

$$\text{AET} = \frac{250 - 60}{E_{\text{req}} - E_{\text{max}}}$$

The HSI loses some applicability at very high heat stress conditions. It also does not identify correctly the heat stress differences resulting from hot, dry and hot, and humid conditions. The strain resulting from metabolic versus environmental heat may not be differentiated because $E_{req}/E_{max}$ is a ratio that may disguise the absolute values of the two factors (NIOSH 1986).

The HSI not only provides an excellent starting point for specifying corrective measures for heat stress but also offers information that helps to determine feasible engineering control measures.

***Example 10.3:*** This sample solution demonstrates how to use the Belding and Hatch nomogram to find $E_{req}$, HSI, and AET (Powell and Hosey 1965).

$$\text{Given: } \text{GT} = 130\,°\text{F}$$
$$\text{TA} = 100\,°\text{F}$$
$$\text{WB} = 80\,°\text{F}$$
$$V = 50\,\text{ft min}^{-1}$$
$$M = 2000\,\text{Btu h}^{-1}$$

*Step 1.* Determine $C$ (convection). Connect 50 fpm (column I) with TA = 100 °F (column II). Read $C = 40\,\text{Btu h}^{-1}$ (column III).
*Step 2.* Determine $E_{max}$ (maximum evaporative heat loss). From a psychometric chart using the dry-bulb and wet-bulb temperatures, read the dew point of 73 °F. Connect 50 fpm (column I) to the dew point = 73 °F (column IV). Read $E_{max} = 620$ (column V).
*Step 3.* Determine the constant, $K$. Connect $V = 50$ fpm (column I) with $T_g - T_a$ (GT − TA) or (130 − 100) = 30 (column VI). Read $K = 22$ (column VII).
*Step 4.* Determine $T_w$ (mean radiant temperature). Enter $K = 22$ in column VII. Connect this to $T_g = 130$, column VIII. Read $T_w = 155$ (column IX).
*Step 5.* Determine $R$ (radiant heat exchange). Follow the slanting line to column X; read $R = 1050\,\text{Btu h}^{-1}$.
*Step 6.* Determine $R + M$ (radiation and metabolism). Connect $R = 1050$ with $M = 2000$ (column XI). Read $R + M = 3050$ on column XII.
*Step 7.* Determine $E_{req}$ (requirement for evaporation of sweat). Enter $C = 30$ in column III of the lower figure. Connect with $R + M = 3050$ on column XII. Read $E_{req} = 3090\,\text{Btu h}^{-1}$ on column XIII.
*Step 8.* Determine allowable exposure time (AET). Enter $E_{max}$ in column V of the lower diagram. Connect with $E_{req} = 3090\,\text{Btu h}^{-1}$ (column XIII). Read AET = 6 minutes.

Compute the HSI:

$$\text{HSI} = \frac{3090}{620} \times 100 = 500$$

An HSI value in excess of 100 is viewed as the maximum tolerated strain tolerated daily by fit, acclimatized young men.

### 10.6.7 Job Ranking

Job ranking involves the establishment of categories for each job into light, medium, and heavy based on the type of operation. The permissible heat exposure limit for that job can be determined by use of the ACGIH table in Figure 10.4 (ACGIH 1981). Some examples of job ranking are as follows:

1. Light work (up to 200 kcal $\text{h}^{-1}$ or 800 Btu $\text{h}^{-1}$), e.g. sitting or standing to control machines, performing light hand or arm work.
2. Moderate work (200–350 kcal $\text{h}^{-1}$ or 800–1400 Btu $\text{h}^{-1}$), e.g. walking about with moderate lifting and pushing.
3. Heavy work (350–500 kcal $\text{h}^{-1}$ or 1400–2000 Btu $\text{h}^{-1}$), e.g. pick and shovel work.

Table 10.3 shows some recommended work–rest regimen strategies to help stay within the permissible heat exposure limits.

Table 10.4 also lists energy expenditures for various types of tasks.

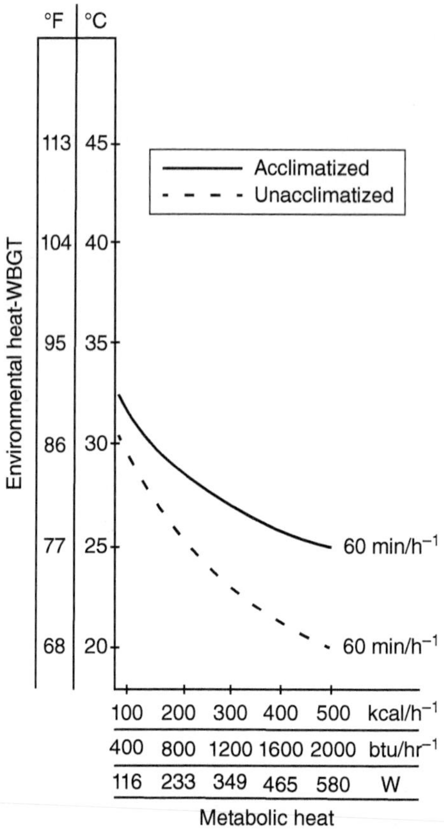


**Figure 10.4** ACGIH permissible heat exposure threshold limit values.

**TABLE 10.3 Permissible Heat Exposure Threshold Limit Values (WBGT-°F [°C])**

| Workload ($Btu\,h^{-1}$) | Continuous work | 75% work, 25% rest each hour | 50% work, 50% rest each hour | 25% work, 75% rest each hour |
|---|---|---|---|---|
| Light (800) | 86 (30.0) | 87 (30.6) | 89 (31.4) | 90 (32.2) |
| Medium (1400) | 80 (26.7) | 82 (28) | 85 (29.4) | 88 (31.1) |
| Heavy (2000) | 77 (25) | 79 (25.9) | 82 (28) | 86 (30) |

**TABLE 10.4 Energy Expenditures, *M*, for Various Activities**

| Activity | $M$ ($Btu\,h^{-1}$) | Activity | $M$ ($Btu\,h^{-1}$) |
|---|---|---|---|
| Typing, electrical[a] | 227–330 | Hoeing | 1045 |
| Typing, mechanical[a] | 300–375 | Mixing cement | 1115 |
| Lying at ease | 334–360 | Walking on the job | 1165–1610 |
| Sitting or standing at ease | 405–450 | Pushing a cart | 1190–1660 |
| Draftsman, drilling machine, light assembly line | 430 | Shoveling | 1285–2495 |
| Armature winding, printer | 525 | Digging trenches | 1425–2090 |
| Light machine work, machine wood sawing | 570 | Gardening, digging, brush clearing | 1450 |
| Medium assembly work | 650 | Sawing wood | 1500–1780 |
| Driving a car | 670 | Forging | 1520–1595 |
| Sheet metal worker | 715 | Furnace tending | 1595–3850 |
| Casual walking | 715–925 | Scrubbing, hand drilling wood | 1665 |
| Machinist | 740 | Crosscutting with bucksaw | 1780–2500 |
| Drilling rock, drilling coal | 900–2255 | Climbing stairs or ladder | 1830–3140 |
| Weeding | 905–1855 | Planing wood | 1925–2160 |
| Bricklayer | 950 | Tree felling | 1950–3020 |
| Timbering | 975–2140 | Trimming felled trees | 2070–2760 |
| Machine fitting, tractor plowing, grass cutting | 1000–1200 | Slag removal | 2520–3000 |

[a] Women.

### 10.6.8 Wet-Bulb Globe Temperature

The WBGT index has proved very successful in monitoring heat stress and minimizing heat casualties in the United States and has been widely adopted. The WBGT index provides a fast and convenient method to quickly assess conditions that pose threats of thermal strain. It has been adopted as the principal index for the TLV for heat stress established by the ACGIH (OSHA 1970). NIOSH endorses WBGT as the preferred measure of severity of occupational exposures to heat stress (NIOSH 1972b, 1986). The main criteria used by NIOSH for the selection of a suitable index were that (i) the measurements and calculations must be simple and (ii) index values must be predictive of the physiological strain of heat exposure.

The WBGT index is an algebraic approximation of the ET concept. Air velocity does not have to be measured directly to calculate the intensity of WBGT, as allowances are made for this factor by the use of the naturally convected wet-bulb sensor.

For outdoor use with solar load, the index is derived from the formula

$$\text{WBGT} = 0.7\text{NWB} + 0.2\text{GT} + 0.1\text{TA}$$

where NWB is the natural wet-bulb temperature, GT is the globe temperature, and TA is the dry-bulb (air) temperature. For indoor use, the weighted expression becomes

$$\text{WBGT} = 0.7\text{NWB} + 0.3\text{GT}$$

The WBGT has limitations as an HSI, especially at high levels of severity (Ramanathan and Belding 1973). Studies have shown clearly that environmental combinations yielding the same WBGT levels result in different physiological strains in individuals working at a moderate level (Ramanathan and Belding 1973). This problem is compounded when impermeable clothing is worn, because evaporative cooling (wet-bulb temperature) will be limited, and in this instance, the WBGT values are irrelevant. Correction factors for clothing are found in Table 10.6. Nevertheless, WBGT has become the index most commonly used and recommended throughout the world.

When taking WBGT measurements, the dry- and wet-bulb thermometers should be shielded from the sun and other radiant surfaces without restricting airflow around the bulb. The globe thermometer is a 6-in.-diameter hollow copper sphere painted on the outside with a black matte finish.

### 10.6.9 Wet-Globe Temperature

The wet-globe thermometer includes a hollow 3-in. copper sphere covered by a black cloth that is kept at 100% wetness from a water reservoir. The wet sphere exchanges heat with the environment by the same mechanism that a person with a totally wetted skin would use in the same environment. In this regard, heat exchange by convection, radiation, and evaporation is integrated into a single instrument reading (Botsford 1971).

During the past several years, the wet-globe temperature (WGT) has been used in many laboratory studies and field situations, where it has been compared with the WBGT (Ciriello and Snook 1971; Johnson and Kirk 1980; Beshir 1981; Beshir et al. 1982; Parker and Pierce 1984). In general, the correlation between the WGT and WBGT is high. Nevertheless, the relationship between the two is not constant for all combinations of environmental variables. A simple approximation of the relationship is WBGT = WGT + 2 °C for conditions of moderate radiant heat and humidity and is applicable for general monitoring in industry (NIOSH 1986).

## 10.7 INSTRUMENTATION

The basic parameters measured to help determine the level of heat stress are temperature, humidity, air velocity, and radiant heat from the sun or infrared sources. Technological advances have rendered instrumentation more compact and bundled for ease of handling and use in the field. There are many instrument manufacturers who have heat stress monitoring devices available. The type of instruments needed for assessment of heat stress depends on the HSI to be used for the assessment and the parameters needed to use the chosen index.

### 10.7.1 Thermometers

A wide variety of thermometers is available for use in measuring temperature. The liquid-in-glass thermometer is probably the most commonly used, is relatively inexpensive, and comes in many temperature ranges. Bimetallic thermometers are generally of the type used in dial thermometer configurations. Thermocouples are made of wires of two different metals and operate on the principle of electromotive force variation at the junction of the wires. A thermistor is a semiconductor that will show a significant change in resistance of the metal wire with even small temperature changes.

Air temperature measurements can easily be affected by the presence of surrounding surfaces that vary significantly from the ambient air temperature. Thermocouples and thermistors are less sensitive to this problem than liquid-in-glass thermometers because of their small size. It is possible to shield the thermometers from these sources by the use of heavy aluminum foil and/or by increasing airflow over the sensor.

### 10.7.2 Humidity

Humidity is a representation of water vapor presence in the atmosphere. For heat stress assessment, the relative humidity can be measured by a psychrometer. A psychrometer is an instrument with both a dry-bulb and a wet-bulb temperature sensor. The wet-bulb sensor is covered by a wetted wick, which is cooled by evaporation that is the result of air movement of at least 900 fpm (World Meteorological Organization 1971). This air movement can be achieved by a small fan or squeeze bulb that is part of the instrument or by the use of a sling psychrometer. The sling psychrometer is hand spun by the operator to achieve this effect. The WBGT index uses a natural wet-bulb (NWB) temperature, which is not exposed to forced air movement, making this monitoring parameter always less than or equal to the psychometric wet-bulb temperature.

Hygrometers are direct reading instruments that measure relative humidity or dew point that are generally found in humidity recorders and control instruments. As a result, they have little use in field heat stress assessments and are better suited for obtaining data in laboratory or other controlled settings.

### 10.7.3 Air Velocity

Thermal anemometers are the best choice for measuring nondirectional air movement at relatively low air velocities. This instrument passes air over a heated sensor, which responds when it is cooled by the air mass flow. Heated thermocouple anemometers operate by sensing the temperature difference between a heated and unheated thermocouple that are exposed to the air flow. Hot wire anemometers use a fine wire that is heated and measure the variation of temperature and electrical resistance of the airflow to which it is exposed. Vane anemometers are better suited for high air velocities and may be of the rotating type, which measure revolutions per elapsed time, or the swinging vane type, which measure deflection of the vane.

### 10.7.4 Radiant Heat

The black-globe thermometer is probably the best choice for measuring thermal radiation. The radiant heat is absorbed by the black surface of the sensor, causing the globe temperature (GT) to be greater than the air temperature. As the globe cools by convection, the temperature stabilizes when the heat gained by radiation equals the rate of heat lost by convection. This effect may take as long as 15 minutes to complete. The exchange of radiant heat with a worker is best represented by the MRT. This can be calculated with the globe temperature, air velocity, and air temperature with the following equation (NIOSH 1986):

$$\mathrm{MRT} = \mathrm{GT} + 1.8V \times 0.5(\mathrm{GR} - \mathrm{TA})$$

where MRT is the mean radiant temperature in °C, GT is black-globe temperature in °C, $V$ is the air velocity in m s$^{-1}$, and TA is the dry-bulb temperature in °C.

## 10.8 CONTROL MEASURES

A number of elements should be considered when attempting to control heat stress. These include bodily heat production, number and duration of exposures, heat exchange components as affected by environmental factors, thermal conditions of the rest area, and worker clothing and protective equipment (Powell and Hosey 1965). Before selecting control measures, it is crucial to identify the components of heat stress to which workers are exposed in new operations. Using the basic heat balance equation, $S = (M - W) + C + R - E$, heat stress can be reduced by modifying metabolic heat production ($M - W$) or environmental heat production. Environmental heat load ($C$, $R$, and $E$) responds to engineering controls (e.g. ventilation, air conditioning, screening, insulation, and modification of process or operation), as well as protective clothing and equipment. Metabolic heat production ($M - W$) can be modified by work practices and application of labor-reducing devices.

### 10.8.1 Control at the Source

The most fundamental approach to engineering control of heat in the workplace is eliminating heat at its point of generation. One feasible alternative is to change the operation or substitute a process component of lower temperature for one of higher temperature. One example is the use of induction heating rather than direct-fired furnaces for certain forging operations. Heat is controlled most effectively if it is regulated at the source, and the options for control of heat at the source are isolation, reduction in emissivity, insulation, radiation shielding, and local exhaust ventilation.

The most practical method for limiting heat exposure from hot processing operations that are difficult to control or for operations that are extremely hot is to isolate the heat source. Such operations might be partitioned and separated from the rest of the facility, located in a separate building, or relocated outdoors with minimal shelter (e.g. an industrial boiler, segregated from other operations in the same facility).

The rate at which heat is radiated from the surface of a hot source can often be lowered if the emissivity of the source is reduced through surface treatment. The emissivity of a hot source can be lowered if the emissivity of the source is reduced through surface treatment. The emissivity of a hot source can be lowered by painting with aluminum paint or covering the source with sheet aluminum. When an oven, boiler, or other hot surface is covered with aluminum paint or sheet aluminum, less heat is radiated to workers nearby, and heat is conserved inside the unit, representing a substantial savings in energy cost.

Insulation is not mutually exclusive of "isolation" in the context of engineering control of heat stress. Insulation also prevents the escape of sensible and radiant heat into the work environment. An example of insulation that also has implications for energy conservation is that of pipe-covering insulation on steam lines. In addition to reducing radiant heat exchange, insulation reduces the convective heat transfer from hot equipment to the work environment by minimizing local convective currents that form when air that contacts very hot surfaces is heated.

When pipe leakage occurs in urban heating systems, it is often difficult to close the circuit in order to perform the necessary repairs. In the event of an escape of liquid and steam in environments that may be confined spaces, workers may be exposed to temperatures of 80–85 °C (176–185 °F). Although these exposures are infrequent and may be short term, workers must be protected from the high heat and burn potential posed by the steam. In these instances, lockout/tagout procedures must be implemented to control hazardous energy. In the event that this is not feasible, measures should be taken to insulate or shield workers from the hazards.

Workers who operate high-temperature furnaces throughout the day may face significant radiant heat exposures. One control strategy for this type of exposure is the use of heat-reflective curtains or shields between the furnace and the worker. Workers would stand behind this barrier whenever they do not have to tend the furnace to insert or remove work. Frequent breaks away from the high heat area and electrolyte replacement are also an important part of worker protection in this type of environment.

Radiation shielding represents an extremely important control measure. Radiant heat passes through air without heating the air; it heats only the objects in its path that are capable of absorbing it. Shielding of radiant heat sources means putting a barrier between the worker and the source to protect the worker from being a receptor of the radiant energy. Radiation shielding can be classified into reflecting, absorbing, transparent, and flexible shields.

Reflective shields are constructed from sheets of aluminum, stainless steel, or other bright-surface metallic materials. The advantage of using aluminum is 85–95% reflectivity, and the successfulness of using aluminum as shielding depends on the following:

1. There must be an aluminum-to-air surface; the shield cannot be embedded in other materials.
2. The shield should not be painted or enameled.
3. The shield should be kept free of oil, grease, and dirt to maximize reflectivity.
4. The shield should be separated from the hot source by several inches when used as an enclosure.
5. Corrugated sheeting should be arranged so that the corrugations run vertically to help maintain a surface free of foreign matter.

### 10.8.2 Local Exhaust Ventilation

Canopy hoods with natural draft or mechanical exhaust ventilation are used commonly over furnaces and hot equipment. Because heat has a tendency to rise, it must be remembered that local ventilation removes only convective heat. Also, as the temperature of air increases, the volume increases as a result of expansion; therefore, the ventilation system must be designed with this in mind. Radiant energy losses, whose magnitude often overrides convective losses, are not controlled by local exhaust hooding. Radiation shielding must be used as well to control what is likely to be the larger fraction of the total heat escaping from the hot process (*Heating/Piping/Air Conditioning* 1976). More information regarding design of ventilation systems can be found in Chapter 18.

### 10.8.3 Localized Cooling at Workstations

Relief can be provided at localized areas by the introduction of cool air in sufficient quantities to surround the worker with an "independent" atmospheric environment. This local relief or "spot cooling" serves two functions, depending on the relative magnitudes of the radiant and convective components in the total heat load. If the overall thermal load is primarily convective in the form of hot air surrounding the worker, the local relief system displaces the hot air immediately around the individual with cooler air having a higher velocity. If such air is available at a suitable temperature and is introduced without mixing with the hot ambient air, no further cooling of the worker is necessary. The introduction of cooling air must not interfere with local exhaust ventilation systems used for contaminant control.

When there is a significant radiation load, the local relief system must provide actual cooling to offset the radiant energy that penetrates the mass of air surrounding the worker. The temperature of the supplied air must be low enough to make the convection component ($C$) negative in the heat exchange model to offset the radiation component ($R$).

In extreme heat, workers should be stationed inside an insulated, locally cooled observation booth or relief room to which they can return after brief periods of high heat exposure. Air-conditioned crane cabs represent one application of this concept now in common use.

### 10.8.4 General Ventilation

A common method for heat removal in the hot industries is general ventilation by utilizing wall openings for the entrance of cool outside air and roof openings for the discharge of heated air. For an ideal system of general ventilation, combined with radiation shielding, the outside air must be cooled either by an evaporative cooler (i.e. water sprays or wetted filters) or a chilled coil system before it is distributed throughout the plant. The inlet air should enter near floor level, be directed toward the workers, and flow toward the hot equipment; thus the coolest air available is received by the workers before its temperature is increased by mixing with warm building air or circulation over hot processes. This air then flows toward the hot equipment and, as its temperature increases, rises and escapes through vent openings in the roof. Provisions for proper distribution of the air supply should receive the same consideration that is given to the selection of exhaust equipment. The basic strategy when general ventilation is used to remove heated air is to position the exhaust openings, either natural draft or mechanically operated, above the sources of heat and as close to them as practical (Belding and Hatch 1955). The key point with air movement is that any air movement can promote cooling by evaporation even if the air temperature is above that of the skin.

### 10.8.5 Moisture Control

Moisture control includes both prevention of increased humidity and the use of dehumidification procedures. Effective controls such as enclosing hot water tanks, covering drains carrying hot water, and repairing leaking joints and valves in steam pipes offer direct measures to alleviate heat stress in warm-moist industries. Dehumidification can be accomplished by refrigeration, absorption, or adsorption. In the context of occupational heat exposures, refrigeration is the most widely used technique to condition the air in relief areas, operating booths, or other local or regional portions of an industrial facility.

## 10.9 MANAGEMENT OF EMPLOYEE HEAT EXPOSURE

### 10.9.1 Education and Training

The development of an education and training program is an important aspect of controlling heat stress. The program should include workers and supervisors and should be conducted prior to assignment on jobs where heat exposure is a concern and periodically as determined by the level and frequency of exposure. Workers should be educated to the dangers of work in hot conditions and the early signs and symptoms of heat fatigue, heat exhaustion, and heat stress in themselves and in their colleagues. Supervisors and selected personnel should be trained to recognize the signs and symptoms of heat disorder and emergency first aid measures. They must be trained in the safe use of any personal protective clothing with which they are supplied. Training must also explain the need to take frequent breaks and not to accumulate them with the aim of leaving work early at the end of the day. The need to report accidents, illnesses, and ill health as early as possible needs to be emphasized and that failure to report illness promptly may lead to a worsening of conditions with the potential for serious physical consequences.

The importance of acclimatization should also be stressed. The body will adapt physiologically over a few days of heat exposure so that it becomes more efficient at dealing with raised environmental temperatures. It does not, however, allow the body to tolerate a raised core temperature. Sweat output increases, and the pulse rate and deep body temperature decrease. This adaptation, however, may be lost in as little as three days away from work; so workers should be aware of the increased demands they will place on their bodies after a holiday, long weekend, or a period of illness.

During hot work, regular replacement of fluids is needed, the amount and composition of which will depend on the physical effort involved and the ambient temperatures. Physical training should be encouraged, as it aids the body's ability to cope with the increased demands that heat places on the body. Instruction on the possible combined effects of heat and alcoholic beverages, prescription and nonprescription drugs, and other physical agents should also be provided.

### 10.9.2 Medical Supervision

A health monitoring program should include a preplacement physical examination and history, with concentration on the cardiovascular, metabolic, renal, skin, and pulmonary systems. Previous intolerance of hot environments and pregnancy are reasons for special consideration. It is recommended that occupational health professionals refer to a checklist (Table 10.5) at the time of employment and review annually those clients at risk. Each facility should determine the degree of severity and the number of positive risk factors that should be considered to preclude employment in high heat environments. After employment, workers should receive periodic examinations and annual examinations after the age of 45 (Wildeboor and Camp 1993). Ongoing attention should be placed on nutritional status, weight gain, and accident and injury records. Management and healthcare professionals should anticipate and prepare for unseasonably hot weather, summer temperature changes, and heavy production requirements.

**TABLE 10.5 Heat Stress Evaluation Form (Sample) (Barrett 1991)**

| | | |
|---|---|---|
| Name ________ SS no. ____ DOB ________ | | |
| Date ________ Department/job description ________ | | |
| Yes | No | History of heat tolerance (e.g. heat stroke, heat exhaustion) |
| Yes | No | History of nonacclimatability |
| Yes | No | More than 40 years old, no history in hot environments |
| Yes | No | Obese (body fat >15%) |
| Yes | No | Weight less than 50 kg |
| Yes | No | Skin disease symptomatology over large areas |
| Yes | No | History of alcohol or substance abuse |
| Yes | No | Hypertension (>160 mm/95 mm) |
| Yes | No | Diuretics, anticholinergic, vasodilators, antihistamines, CNS inhibitors, muscle relaxants, atropine, MAOs, sedatives, beta-blockers, amphetamines |
| Yes | No | Organic disease of the heart/vascular system |
| Yes | No | COPD, active lung disease, asthma |
| Yes | No | Liver, renal, endocrine, metabolic, digestive disease |
| Yes | No | Pregnancy |
| Yes | No | Infertility problems |
| Total no. of positive risks ____ | | Recommendations ________ |
| Signature ________ Date ____ | | |

*Source:* Reproduced with permission of SAGE Publications.

### 10.9.3 Effect of Clothing on Heat Exchange

Clothing can have a profound effect on the heat exchange process. It is the insulating effects of clothing that reduce heat loss to the environment. When it is cold, of course, reduced heat loss is beneficial, but when it is hot, clothing interferes with heat loss and can be harmful.

The insulation value of most materials is a direct linear function of its thickness. The material itself (whether it is wool, cotton, or nylon) plays only a minor role. It is the amount of trapped air within the weave and fibers that provides the insulation. If the material is compacted or gets water-soaked, it loses much of its insulating properties because of the loss of trapped air.

The unit of measure for the insulating properties of clothing is the *clo*. The *clo unit* is a measure of the thermal insulation necessary to maintain in comfort a sitting, resting subject in a normally ventilated room at 70 °F (21 °C) and 50% relative humidity. Because the typical individual in the nude is comfortable at about 86 °F (30 °C), 1 clo unit has roughly the amount of insulation required to compensate for a drop of about 16 °F (9 °C). The typical value of clothing insulation is about 4 clo per inch of thickness (1.57 clo $cm^{-1}$). clo values for some typical articles of men's and women's clothing are given in Table 10.6. The formula for computing overall clo values for a clothing ensemble is

$$\text{Total clo units} = 0.8 \times (\text{sum of individual items}) + 0.8$$

Another feature of clothing that affects heat transfer is the permeability of the material to moisture. It is the permeability that permits evaporative heat transfer through the fabric. In general, the greater the clo value of the fabric, the lower is its permeability. The index of permeability (im) is a dimensionless unit that ranges from 0.0 for total impermeability to 1.0 if all moisture that could be evaporated into the air could pass through the fabric. Typical im values of most clothing materials in still air is less than 0.5. Water-repellent treatments, very tight weaves, and chemical protective impregnation can reduce the im values significantly. This is an issue when trying to protect workers wearing impermeable clothing, such as that used at hazardous waste sites or for emergency response activities. It is imperative that these workers are closely monitored and that a strict work–rest regimen is planned and enforced. In hot environments, evaporation of sweat is vital to maintain thermal equilibrium, and materials that interfere with this process can result in heat stress. In a cold environment, if evaporation of sweat is impeded, a garment can become soaked with perspiration, thus reducing its insulating capacity.

### 10.9.4 Protective Clothing

Water-cooled garments include (i) a hood that provides cooling to the head; (ii) a vest that provides cooling to the heart and torso; (iii) a short undergarment that provides cooling to the torso, arms, and legs; and (iv) a long undergarment that provides cooling to the head, torso, arms, and legs. None of these water-cooled garments provide cooling to the hands and feet. Water-cooled garments and headgear require a battery-driven circulating pump and container where the circulating fluid is cooled by the ice. The amount of ice available determines the effective time of the water-cooled garment.

Air-cooled suits and/or hoods that distribute cooling air next to the skin are available. The total heat exchange from sweat-wetted skin when cooling air is supplied to the air-cooled suit is a function of cooling air temperature and cooling airflow rate. Both the total heat exchanges and the cooling power increase with cooling airflow rate and decrease with increasing cooling air inlet temperature. Attaching a vortex tube to the worker with a constant source of

**TABLE 10.6 clo Insulation Values for Individual Items of Clothing**

| Men's clothing | clo | Women's clothing | clo |
|---|---|---|---|
| T-shirt | 0.09 | Bra and panties | 0.05 |
| Underpants | 0.05 | Half-slip | 0.13 |
| Lightweight short sleeve shirt | 0.14 | Lightweight blouse | 0.20 |
| Lightweight long sleeve shirt | 0.22 | Lightweight dress | 0.22 |
| Lightweight trousers | 0.26 | Lightweight slacks | 0.26 |
| Lightweight sweater | 0.27 | Lightweight sweater | 0.17 |
| Heavy sweater | 0.37 | Heavy sweater | 0.37 |
| Light jacket | 0.22 | Light jacket | 0.17 |
| Heavy jacket | 0.49 | Heavy jacket | 0.37 |
| Socks | 0.04 | Stockings | 0.01 |
| Shoes (oxfords) | 0.04 | Shoes (pumps) | 0.04 |

**TABLE 10.7 TLV WBGT Correction Factors in °C for Clothing**

| Clothing type | clo[a] value | WBGT correction |
|---|---|---|
| Summer work uniform | 0.6 | 0 |
| Cotton coveralls | 1.0 | −2 |
| Winter work uniform | 1.4 | −4 |
| Water barrier, permeable | 1.2 | −6 |

[a] clo: insulation value of clothing. One clo unit = 5.55 kcal $m^{-2}$ $h^{-1}$ of heat exchange by radiation and convection for each °C of temperature difference between the skin and adjusted dry-bulb temperature (the average of the ambient air dry temperature and the mean radiant temperature) (ACGIH 1981).

compressed air supplied through an air hose is a method of body cooling in many hot industrial situations. However, the vortex tube is noisy, and the hose limits the area in which the worker can operate.

Cooling with an ice packet vest will vary with time and with its contact pressure with the body surface, plus any heating effect of the clothing and hot environment. Because the ice packet vest does not provide continuous and regulated cooling over an indefinite period of time, exposure to a hot environment would require replacement of the frozen vests every three to four hours. Ice packet vests can add an additional 10–15 lb of weight, which can increase the metabolic load and negate much of the benefits. However, ice packet vests allow increased mobility of the worker unlike other control measures. The greatest potential for use of an ice packet vest is short-duration tasks and emergency repairs, and it is relatively less expensive than other cooling approaches.

Metallized reflecting fabrics can provide protection against radiant heat, but if their surfaces are damaged or become dirty, then their effectiveness is reduced. Workers need to be trained and aware of the limitations of clothing, which may also need to have other characteristics such as flame resistance. The majority of workers in hot environments will not be wearing protective clothing and will don the minimum of garments so as to reduce heat retention. Wetted cottons tend to be cooler and more comfortable to wear than man-made fabrics.

### 10.9.5 Work–Rest Regimen

Shortening the duration of exposure and increasing the frequency and length of rest periods, allowing workers to self-limit exposure, are administrative measures that may be taken to control heat stress. The length of the work/rest is dependent on the nature of the work and the environmental conditions. Workers should rest before becoming fatigued and should remain at rest until the heart rate drops below 100 beats per minute. Air-conditioned (about 24 °C), low-humidity rest areas speed the rate and degree of recovery. Workload can be modified by the use of mechanized tasks and shared workloads.

According to NIOSH, there are several ways to control the daily length of time and temperature to which a worker is exposed to heat stress conditions:

Schedule hot jobs for a cooler part of the day (early morning, late afternoon, or evening), if possible.
Schedule routine maintenance and repair work in hot areas for the cooler seasons of the year.
Alter the work–rest regimen to permit more rest time.
Provide shade or air condition for rest and recovery.
Use extra personnel to reduce the exposure time of each member of the work team.
Employ a buddy system for high-risk tasks.
Permit freedom to interrupt work when a worker feels extreme heat discomfort

## 10.10 SUMMARY

The success of any heat stress program begins with hazard assessment. Processes or operations where heat stress may be a concern should be reviewed prior to installation for engineering controls needed to minimize exposure. For operations already in place, a field assessment using one of the HSI described earlier should be performed to determine if additional engineering controls are needed. The use of personal protective equipment and work–rest regimens can be effective measures to reduce exposures until engineering controls are in place or when they are not feasible. Fitness, acclimatization, and hydration are key elements to any heat stress program; however worker education and awareness are probably the most important aspects of keeping at-risk workers healthy in heat stress situations. It is important to realize that heat stress can be an insidious occupational hazard when it is not well understood by workers and management.

## ABBREVIATIONS

| | |
|---|---|
| AET | Allowable exposure time as calculated from elements of the heat stress index (HSI) |
| $C$ | Rate of heat exchange (net) by convection between an individual and the environment |
| $C$ | Ceiling limits (WBGT) recommended by NIOSH for all workers in hot jobs |
| DS | Change in heat content of the body |
| $E$ | Rate of heat loss by evaporation of water from the skin |
| EDZ | Environment-drive zone – the range of heat load ($M+R+C$) beyond the prescriptive zone in which physiological response is affected drastically by the thermal environment |
| $E_{max}$ | Maximum evaporative heat loss by water vapor uptake in the air at prevailing meteorologic conditions |
| $E_{req}$ | Heat loss required solely by evaporation of sweat to maintain body heat balance |
| ET | Effective temperature – an index used to estimate the effect of temperature, humidity, and air movement on the subjective sensation of warmth |
| ETCR | Effective temperature corrected for radiation – an index for estimating the effect of temperature, humidity, and air movement on the subjective sensation of warmth using globe temperature rather than air temperature |
| GT | The temperature inside a blackened, hollow, thin copper globe measured by a thermometer whose sensing element is at the center of the sphere |
| HSI | An index of heat stress derived from the ratio of $E_{req}$ to $E_{max}$ |
| $M$ | Rate of transformation of chemical energy into energy used for performing work and producing heat |
| MRT | The mean radiant (surface) temperature of the material and objects totally surrounding an individual |
| NWB | The wet-bulb temperature measured under conditions of the prevailing (natural) air movement |
| PZ | Prescriptive zone – the range of environmental heat load ($M+R+C$) at which the physiologic strain (heart rate and core body temperature) is independent of the thermal environment |
| $R$ | Rate of heat exchange by radiation between two radiant surfaces of different temperatures |
| RAL | Recommended alert limits (WBGT) specified by NIOSH for unacclimated, healthy workers |
| REL | Recommended exposure limits (WBGT) specified by NIOSH for acclimated, healthy workers |
| RH | Relative humidity – the ratio of the water vapor pressure in the ambient air (VPA) to the water vapor pressure in saturated air at the same temperature |
| TA | The temperature of the air surrounding a body (also dry-bulb temperature) |
| TLV | Threshold limit values specified by the American Conference of Governmental Industrial Hygienists |
| TS | The mean of skin temperatures taken at several locations and weighted for skin area |
| ULPZ | The level of heat stress at the interface of the PZ (upper limit) and the EDZ (lower limit) |
| $V$ | Air velocity |
| VPA | The partial pressure exerted by water vapor in the air |
| VPS | Vapor pressure exerted by water on the skin |
| WBGT | Wet-bulb globe temperature – an empirical index of heat stress obtained by weighting NWB, GT, and TA (outdoors with solar load) |
| WGT | Wet-globe temperature – an empirical index of heat stress as measured within a 3-in. copper sphere covered by a black cloth kept at 100% wetness |

## BIBLIOGRAPHY


ACGIH (1981). *TLV's – Threshold Limit Values for Chemical Substances and Physical Agents in the Workroom Environment with Intended Changes for 1981*. Cincinnati, OH: ACGIH.

Barrett, M.V. (1991). Heat stress disorders, old problems new implications. *AAOHN J.* (August): 269.

Belding, H.S. and Hatch, T.F. (1955). Index for evaluating heat stress in terms of resulting physiological strains. *Heat. Piping Air Cond.* **27**: 129–135.

Beshir, M.Y. (1981). A comprehensive comparison between WBGT and Botsball. *Am. Ind. Hyg. Assoc. J.* **42**: 81–87.

Beshir, M.Y., Ramsey, J.D., and Burford, C.L. (1982). Threshold values for the Botsball: a field study of occupational heat. *Ergonomics* **25**: 247–254.

Botsford, J.H. (1971). A wet globe thermometer for environmental heat measurement. *Am. Ind. Hyg. Assoc. J.* **32**: 1–10.

Brief, R.S. and Confer, R.G. (1971). Comparison of heat stress indices. *Am. Ind. Hyg. Assoc. J.* **32**: 11–16.

Ciriello, V.M. and Snook, S.H. (1971). *Am. Ind. Hyg. Assoc. J.* **38**: 264–271.

Clayton, G.D. and Clayton, F.E. (eds.) (1991). *Patty's Industrial Hygiene and Toxicology*, 4e, vol. **1**, Part A, 763. New York: Wiley.

Dukes-Dubos, F.N. and Henschel, A. (1980). *Proceedings of a NIOSH Workshop on Recommended Heat Stress Standards*. Cincinnati, OH: USDEH, Public Health Service, CDC, NIOSH.

*Heating/Piping/Air Conditioning* (1976). Reinhold, Inc., Penton/IPC.

Hertig, B.A. (1973). Thermal standards and measurement techniques. In: *The Industrial Environment – Its Evaluation and Control*, 3e, DHEW (NIOSH) Publication No. 74-117, 413–429. Cincinnati, OH: U.S. Department of Health, Education, and Welfare, Public Health Service, Centers for Disease Control, National Institute for Occupational Safety and Health.

Johnson, A.T. and Kirk, G.D. (1980). A comprehensive comparison- the experimental differences between the WBGT and Botsball. *Am. Ind. Hyg. Assoc. J.* **41**: 361–366.

McKarns, J.S. and Brief, R.S. (1966). Nomographs give refined estimate of heat stress index. *Heat. Piping Air Cond.* **38**: 113.

NIOSH (1972a).

NIOSH (1972b). *Criteria for a Recommended Standard: Occupational Exposure to Hot Environments*, HSM-72-10269. Cincinnati, OH: U.S. Department of Health, Education and Welfare, NIOSH.

NIOSH (1986). *Criteria for a Recommended Standard: Occupational Exposure to Hot Environments, Revised Criteria, 1986* (April 1986). Cincinnati, OH: U.S. Department of Health and Human Services, NIOSH.

NIOSH (1993). *The Industrial Environment*. Chapters 30 and 31. Washington, DC: NIOSH.

OSHA (1970). *Occupational Safety and Health Act PL 91-596, 91st Congress, S.2193*. Washington, DC: U.S. Department of Labor.

Parker, R.D. and Pierce, F.D. (1984). Comparison of heat stress measuring techniques in a steel mill. *Am. Ind. Hyg. Assoc. J.* **45**: 405–415.

Powell, C.H. and Hosey, A.D. (eds.) (1965). *The Industrial Environment – Its Evaluation and Control*, 2e, Public Health Service Publication No. 614. Washington, DC: Government Printing Office.

Ramanathan, N.L. and Belding, H.S. (1973). Physiologic evaluation of the WBGT index for occupational heat stress. *Am. Ind. Hyg. Assoc. J.* **34**: 375–383.

Shapiro, Y., Hubbard, R.W., Kimbrough, C.M., and Pandolf, K.B. (1981). Physiological and hematological responses to summer and winter dry-heat acclimation. *J. Appl. Physiol. Respir. Environ. Exerc. Physiol.* **50**: 792–798.

Smith, N.J. and Slarinski, C.L. (1987). *Sports Medicine*, 161–163. Philadelphia, PA: W. B. Saunders.

Wildeboor, J. and Camp, J. (1993). Heat stress – its effect and control. *AAOHN J.* **41** (6): 268–274.

World Meteorological Organization (1971). *Guide to Meteorological Instruments and Observing Practices*. Geneva: WMO.

CHAPTER 11

# RADIATION: NONIONIZING AND IONIZING SOURCES

DONALD L. HAES, JR.,

*Massachusetts Institute of Technology, Bldg. 16-268, 77 Massachusetts Ave., Cambridge, MA, 02139-4307*

and

MITCHELL S. GALANEK

*Massachusetts Institute of Technology, Bldg. 56-235, 77 Massachusetts Ave., Cambridge, MA, 02139-4307*

## 11.1 NONIONIZING RADIATION

### 11.1.1 Introduction

While the use of nonionizing radiation (NIR) in the workplace is not new, acknowledgment for its potential hazards is still unfolding. Standards for NIR safety are numerous, but are often incomprehensible and hard to find. It is imperative that NIR sources in industry be identified and evaluated and their potential for hazards controlled. While sources of NIR can be found in the workshop, they can also be found in administrative and physical plant activities. These sources include, but are not limited to, the following:

- Static (direct current [DC]) magnets.
- Industrial alternating current (AC) line voltage.
- Radio-frequency (RF) induction ovens.
- RF heaters/sealers.
- Handheld two-way communications.
- Microwave (MW) ovens.
- Cellular telephones.
- Infrared (IR).
- Visible light (VL)
- Ultraviolet (UV) radiation.
- Lasers (visible and invisible).

Comprehending the basic physics of exposures will be central to this struggle.

### 11.1.2 Nonionizing Radiation in Industry

***11.1.2.1 Physics of Nonionizing Radiation*** *NIR* is a practical term for the frequency band of the electromagnetic spectrum that lacks the energy to break chemical and/or molecular bonds, hence being unable to cause *ionization* (ionization is the energetic removal or addition of orbital electrons from the outer shells of atoms, leading to ion production).

*Handbook of Occupational Safety and Health*, Third Edition. Edited by S. Z. Mansdorf.

There are several different classes of NIR, differentiated according to wavelength and hence by frequency (see Figure 11.1). NIR consists of electromagnetic waves composed of electric and magnetic fields, mutually orthogonal to the direction of propagation (see Figure 11.2). The waves move through a vacuum with the velocity of light. There is a mathematically inverse relationship between the frequency and wavelength:

$$\text{Frequency}(f) = (\text{velocity of light}[C] + \text{wavelength}[\lambda])(\text{in free space})$$

The frequency is usually expressed in units of hertz (Hz) with one hertz defined as one cycle per second. Figure 11.1 shows the electromagnetic spectrum from the static fields to the very high frequencies found in cosmic rays. To help simplify this chapter, the elements of the electromagnetic spectrum included in this discussion are as follows: static field (also known as DC), extremely low frequency (ELF), very low frequency (VLF), RF (including MW), IR, VL, UV, and lasers (which may use IR, VL, and/or UV). Table 11.1 gives the frequency ranges and acronyms associated with these frequency bands.

A number of devices emit a portion of the electromagnetic spectrum as either their primary product or consequences of their use. Lasers (light amplification by stimulated emission of radiation) are designed to emit light and may be dangerous because of their primary function. Video display screens (e.g. televisions and video display terminals [VDTs]) emit VL as a design function. In addition, these same products, and a legion of other man-made equipment, emit other frequencies of the electromagnetic spectrum as a consequence. The generated electromagnetic field is an unavoidable technological consequence.

All devices that employ the use of NIR can be considered to contain three basic parts: the source, a transmission pathway, and a receiver. For the confined classification, leakage may occur along the transmission pathway between the source and the receiver. For open classifications, the source itself may be broadcasting large amounts of energy through open space, the pathway. The receivers may be chambers for energy absorption or mobile devices designed to absorb the signal for processing. In either arrangement, the surveyor must be certain of all source strengths, pathways, and intended receivers.

***11.1.2.2 Static Fields*** Static fields do not vary in amplitude over time. Static *electric* fields abound (walk across a carpet in a dry room and touch something metallic), but are usually harmless. Very large sources of static electric fields are not usually found in the work environment. Static *magnetic* fields, however, can be found in the workplace in sufficient strength to present a potential for harm.

In industry, the typical sources are large magnets used for metallic coupling and research purposes. While very strong magnetic fields are present in close proximity to the magnets, the intensity of the field reduces dramatically with distance. These weaker fields are also known as *fringe fields*.

The two main sources of very strong magnetic fields found in the industrial environment are magnets used to compile metallic objects and nuclear magnetic resonance (NMR) imaging devices used for research. Near (within about 1 m) NMRs, workers can be exposed to static magnetic fields up to 0.1 T in intensity, which is about 2000 times that of the background magnetic field of the earth. For many of these devices, the magnetic field strengths can be roughly determined by maps provided by the manufacturer. These maps have contour lines that relate field strength with distance from the magnetic's center (bore). The demarcation of the 0.5 mT line (often referred to as the "5-G line") is essential for pacemaker warning. The placement of this posting should be verified with measurements by a competent health physicist or other similarly qualified individual.

Before we discuss measurements, it should be noted that varying the orientation of the measuring device within the lines of flux will greatly vary the results. This is because most measuring devices are highly directional, while the sources are too. Thus orientation becomes critical for peak field evaluation. The root mean square (RMS) value can be determined by acquiring the peak field intensities in three orthogonal directions and calculating the square root of the sum of the squares. Many measuring devices already accomplish this through design.

Electric fields can be measured by inserting a displacement sensor (a pair of flat conductive plates) into the field and measuring the electric potential between the plates. The electric field lines land on one plate and create voltage that drives a current through a meter to the other plate, where the field lines continue. The electric field strength is expressed typically in units of volts per meter ($V\,m^{-1}$).

Measuring magnetic field strength is often accomplished by loops of conducting wire. The lines of magnetic field passing through the loop induce current flow. The field can be calculated by measuring the amperes of induced current and divided by the circumference of that loop, with units of field strength expressed in amperes per meter ($A\,m^{-1}$). The relative intensity of magnetic fields can also be determined by measuring the magnetic flux density using the Hall effect. Here, an object in a magnetic field will develop a voltage in a direction perpendicular to the magnetic field, which can be measured. Hall effect sensors are less sensitive than loops and are often used for determining strong DC fields. Typical units are the tesla (T) or gauss (G), where 10 000 G = 1.0 T.

| Frequency (Hz) | | Wavelength (m) | | Function device(s) | Ionization | Spectral region |
|---|---|---|---|---|---|---|
| $10^{0}$ | 1 Hz | $3 \times 10^{8}$ | | | | |
| $10^{1}$ | | $3 \times 10^{7}$ | | Power (50/60 Hz) | | |
| $10^{2}$ | | $3 \times 10^{6}$ | 3 Mm | | | |
| $10^{3}$ | 1 kHz | $3 \times 10^{5}$ | | | | |
| $10^{4}$ | | $3 \times 10^{4}$ | | CRT/TV circuitry (17–31 kHz) | | |
| $10^{5}$ | | $3 \times 10^{3}$ | 3 km | AM radio (535–1705 kHz) | | Radio frequency (RF) |
| $10^{6}$ | 1 MHz | $3 \times 10^{2}$ | | | | |
| $10^{7}$ | | $3 \times 10^{1}$ | | VHF television (54–88 MHz)<br>FM radio (88–108 MHz) | | |
| $10^{8}$ | | $3 \times 10^{0}$ | 3 m | VHF television (174–216 MHz)<br>UHF television (470–890 MHz) | | Microwave (MW) |
| $10^{9}$ | 1 GHz | $3 \times 10^{-1}$ | | Microwave ovens (2–45 GHz)<br>AT&T telephone (3.7–4.2 GHz) | | |
| $10^{10}$ | | $3 \times 10^{-2}$ | 3 cm | Radar | ↑ Non-Ionizing | |
| $10^{11}$ | | $3 \times 10^{-3}$ | 3 mm | | | Infrared (IR) |
| $10^{12}$ | 1 THz | $3 \times 10^{-4}$ | | | | |
| $10^{13}$ | | $3 \times 10^{-5}$ | | | | Ultraviolet (UV) |
| $10^{14}$ | | $3 \times 10^{-6}$ | 3 μm | | | |
| | | | | Visible light | | |
| $10^{15}$ | 1 PHz | $3 \times 10^{-7}$ | | | | |
| $10^{16}$ | | $3 \times 10^{-8}$ | | | | |
| $10^{17}$ | | $3 \times 10^{-9}$ | 3 nm | | Ionizing ↓ | ↑ X rays ↓ |
| $10^{18}$ | 1 EHz | $3 \times 10^{-10}$ | | | | |
| $10^{19}$ | | $3 \times 10^{-11}$ | | | | ↑ Gamma rays ↓ |
| $10^{20}$ | | $3 \times 10^{-12}$ | 3 pm | | | |
| $10^{21}$ | | $3 \times 10^{-13}$ | | | | |
| $10^{22}$ | | $3 \times 10^{-14}$ | | | | |
| $10^{22}$ | | $3 \times 10^{-15}$ | 3 fm | | | ↑ Cosmic rays ↓ |
| $10^{23}$ | | $3 \times 10^{-16}$ | | | | |
| $10^{24}$ | | $3 \times 10^{-17}$ | | | | |
| $10^{25}$ | | $3 \times 10^{-18}$ | 3 am | | | |
| $10^{26}$ | | $3 \times 10^{-19}$ | | | | |
| $10^{27}$ | | $3 \times 10^{-20}$ | | | | |
| $10^{28}$ | | $3 \times 10^{-21}$ | | | | |

**Figure 11.1** Electromagnetic spectrum.

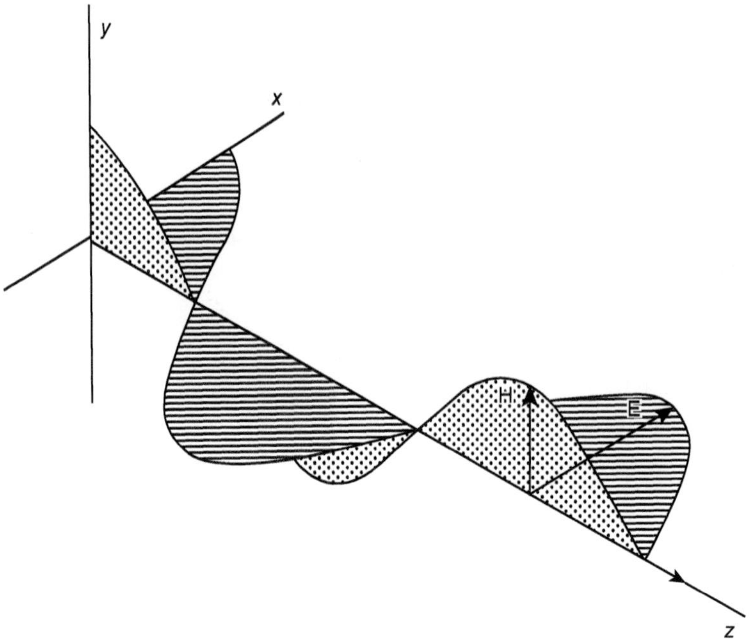


**Figure 11.2** The electromagnetic wave. The magnetic ($H$) field component is vertical, and the electric field ($E$) component is horizontal, with the direction of propagation along the $z$-axis.

**TABLE 11.1 Frequency Bands of the Electromagnetic Spectrum**

| Frequency range | Designate/acronym |
|---|---|
| 0–30 Hz | Sub-extremely low frequency/SELF[a] |
| 30–300 Hz | Extremely low frequency/ELF |
| 0.3–3 kHz | Voice frequency/VF |
| 3–30 kHz | Very low frequency/VLF |
| 30–300 kHz | Low frequency/LF |
| 0.3–3 MHz | Medium frequency/MF |
| 3–30 MHz | High frequency/HF |
| 30–300 MHz | Very high frequency/VHF |
| 0.3–3 GHz | Ultra high frequency/UHF |
| 3–30 GHz | Super high frequency/SHF |
| 30–300 GHz | Extremely high frequency/EHF |
| 0.3–3 THz | Super extremely high frequency/SEHF[a] |

[a] No "official" acronym.

Standards for exposure to static magnetic fields can be found published by the American Conference of Governmental Industrial Hygienists (ACGIH 1997). These published exposure standards are as follows:

- Whole body: 60 mT routine occupational time-weighted daily average.
- Limbs: 600 mT.
- Ceiling value: 2 T.
- Pacemakers: 0.5 mT.

Additional standards for exposure to static magnetic fields can be found published by the International Commission on Non-Ionizing Radiation Protection (ICNIRP) (1994). These published exposure standards are as follows:

- Whole body: 200 mT average over a working day.
- Limbs: 5 T.
- Ceiling value: 2 T.
- Public: 40 mT continuous exposure.

Very strong electric fields greater than 5 kV m$^{-1}$ could cause irritating sparks, while fields in excess of 15 kV m$^{-1}$ could cause painful sparks. Very strong magnetic fields have been postulated to increase blood pressure and rotate sickle cells, but this is yet to be verified scientifically. The main concern in strong magnetic fields above about one millitesla (mT, $10^{-3}$T) is the attraction of loose ferromagnetic objects (for example, iron gas cylinders, steel tools, etc.), which may be propelled toward the magnet at considerable velocity. The object will adhere to the magnet, as will anything that remained in the way. Additionally, surgical implants may be torqued, dislodged, or rotated, resulting in serious injury or death. Field strengths roughly above 0.5 mT may also interfere with unshielded electronic equipment, such as implanted cardiac pacemakers (Tenforde 1985). Even though field strengths may be "in compliance" with established pacemaker safety standards, patients must be warned that the strength of these fields may be enhanced by ferromagnetic objects held in close contact to the wearer (Hansen et al. 1990).

Control of exposure to static electric and magnetic fields can best be accomplished by the use of distance. While shielding is purported to be available for magnetic fields, the manufacturers' claims are dubious at best. As a final note, magnetic media, such as the magnetic stripes on the backs of credit cards and diskettes, can be erased by fields above 1 mT. Digital watches can also be damaged by intense magnetic fields.

***11.1.2.3 Extremely Low Frequency and Very Low Frequency*** For time-varying fields, the "wave" consists of an electric field component and a magnetic field component, each perpendicular to the direction of motion and to each other (see Figure 11.2). This is referred to as being "mutually orthogonal." The "polarity" of the field is determined by the direction of the electric field relative to the direction of propagation. In Figure 11.2, the electric field is horizontal relative to the direction of motion; thus that wave is considered to be horizontally polarized. The polarization of the incident wave plays an important part in determining the amount of energy absorbed.

ELF and VLF vary in intensity over time, as shown in Figure 11.1. While ELF fields are often referred to AC fields, it is technically incorrect to refer to them as "radiation." When the wavelength is extremely long, as in ELF, exposed objects, including people, will be enveloped within a wavelength of the source. The physical characteristics of such "confined" fields dictate that radiant energy effects will be trivial (Polk and Postow 1996). What the exposed object encounters is an "induction field," in which charges from the nearby electromagnetic source induce currents and charges on and within the exposed body. Understanding this technical difference will help avoid confusion between the very low imparted energies and the physiological changes that occur in cells exposed to induction fields.

The generation, transport, and use of common industrial and household electricity are associated with ELF fields. The ubiquitous potential source of exposure is the power line, a frequently visible (unless buried) source of ELF. Although the electric and magnetic field components are necessarily generated together, the magnetic field has been implicated to health outcomes.

As in DC electric fields, ELF electric fields can be measured by inserting a displacement sensor into the field and measuring the electric potential between the plates. The electric field strength is typically expressed in units of volts per meter (V m$^{-1}$) or even thousands of volts per meter (kV m$^{-1}$). The electric field lines are severely perturbed by the presence of anything in the field, including the survey equipment and even the surveyor. Some equipment has been developed to aid this problem, including dielectric extension arms and remote readouts coupled to the meter via fiber-optic cables.

Again as in DC magnetic fields, measuring ELF magnetic field strength is often accomplished by loops. The current can be augmented by increasing the number of turns in the loops or by putting a core of permeable material in the loop. While the appropriate units are A m$^{-1}$, field strengths are often expressed in microtesla (μT, $10^{-6}$T) or milligauss (mG). Meter readings are often converted to mG assuming the permeability of free space = $4\pi \times 10^{-7}$ H m$^{-1}$; 1 mG ≈ 80 mA m$^{-1}$. Measuring ELF magnetic fields are not as problematic as electric fields and can be accomplished accurately with little training and practice.

Standards for exposure to ELF fields can be found published by the ACGIH. These published exposure standards are as follows:

- Less than 100 Hz electric fields: 25 kV m$^{-1}$.
- 1–30 kHz magnetic fields: $60000/f\,\mu$T where $f$ is the frequency of the source in Hz. As is the case with DC fields, AC fields may interfere with the proper operation of cardiac pacemakers and defibrillators. ACGIH has set limits of exposure to 1 kV m$^{-1}$ electric fields and 0.01 mT (1 G) magnetic fields.

Despite persistent methodologic problems of dose and exposure, residence near power lines and work with or around electrical power have been associated with the development of leukemia in several epidemiologic studies of children and adults, respectively. Wertheimer and Leeper (1979) were the first to attempt to show a positive association between a higher "dose" magnetic field (albeit by a surrogate exposure assessment) and leukemia. Evidence for an association between adult occupational leukemia and ELF exposure is not compelling so far. Published data show odds ratios at or below unity as well as above. Almost all 95% confidence intervals of cohort and case–control studies included unity. Recent studies have attempted to overcome some of the deficiencies of earlier studies by measuring ELF field exposures at the workplace and taking work duration into consideration.

In apparent contrast to the epidemiology results, short-term measurements with magnetic field meters that were also used in some of the studies provided no evidence for an association between exposure to 50/60 Hz magnetic fields and the risk of any form of cancer in children. These studies prompted the National Academy of Science Committee to suggest that, confounded by some unknown risk factor for children, leukemia, which is associated with residency in the vicinity of power lines, might be the explanation (National Academy of Science (NAS)/National Research Council 1996).

Taken together, the epidemiology results from residential magnetic field exposure studies and childhood cancer is not strong enough in the absence of support from experimental research to form the basis for exposure standards. For this reason, published standards list allowable exposures several orders of magnitude above those levels specified in epidemiology studies to produce a positive effect.

Computers are associated with NIR exposures. Most computers use VDTs, which employ cathode ray tubes. Both have a source of electrons (cathode) at one end and a phosphorus coating on the inside of the viewing screen (anode) at the other. The electrons, accelerated by high voltage, are focused onto the screen. They display information when the rapidly moving electron beam strikes the screen to produce VL (the only type of NIR emitted by design!). The resultant NIR fields are in both the ELF and VLF frequency range. Extensive measurements at MIT yielded fields at 50 cm from the screen ranging from 1.9 to 87.5 mA $m^{-1}$ (Haes and Fitzgerald 1995). Larger displays allow the operator to sit farther from the screen, with correspondingly less exposure. Thus seating positions are crucial in assigning any "dose" retrospectively. In fact, it is questionable if any information from measurements at fixed locations from the VDTs can be useful assigning VDT "dose."

Epidemiologic studies of reproductive outcomes of VDT users bear the generic problem familiar to all NIR field studies. Exposures have never been quantified. Instead, reproductive outcomes have been compared retrospectively to survey reports of VDP "use" time at the keyboard. This approach introduces two important limitations. First, both the outcome and the exposure are reconstructed retrospectively, with an implied assumption that one will not bias the other. Second, any reproductive effect detected, if real, would be associated with VDT use in general and not necessarily with electromagnetic emissions specifically.

Goldhaber et al. (1988) were the first to report a statistically significant positive association of VDT use with spontaneous abortions among members attending three clinics of a large prepaid group healthcare plan. Previous, similarly designed studies had concluded that no reproductive hazard is associated with VDT use during pregnancy. Windham et al. (1990) reviewed earlier data and reported that the entire positive effect was limited to those respondents in the initial studies who were contacted by telephone after not responding to questionnaires. The latest major study was conducted by the National Institute for Occupational Safety and Health (NIOSH), with the conclusion that "The use of VDTs and exposure to the accompanying electromagnetic fields were not associated with an increased risk of spontaneous abortion in this study" (Schnorr et al. 1991). The body of accessible data does not imply a clear association between VDT use and adverse reproductive outcomes. Based on a large number of epidemiological studies, there do not appear to be adverse effects on reproductive outcomes as a result of exposure to low-frequency fields from VDTs or other sources.

Very few standards apply to VDT exposure. A number of countries and agencies have created exposure limits that include the NIR emission from a VDT. Table 11.2 shows the wide variations in frequencies covered and exposure intensities permitted.

As with static fields, the control of exposure to ELF and VLF electric and magnetic fields is best done through distance. However, shielding is also available. Electric field shielding can be sheets of grounded conductors, allowing induced charges to flow between the conductor and the earth. The material can be solid, but a mesh of at most ¼ wavelength will also do. Magnetic fields can be controlled using a permeable alloy that confines the magnetic flux lines and diverts them. This material is usually made of high nickel alloys called "mu metal," or soft iron. However, neither shielding material is easy to form into the necessary complex shapes. As another source of magnetic field control, field cancelation can be used. Here, fields of similar magnitude, but opposite orientation, can be used to cancel out of the first field, since the vector sum is near zero.

***11.1.2.4 Radio Frequency*** These bands are generally defined to include the frequencies 10 kHz to 300 GHz. Although a wide spectrum of frequencies is included, exposures are commonly grouped together under the collective term "RF exposures." RF is used in industry in varying frequency ranges and applications. The major sources of RF exposure found in industry are RF induction ovens, RF heater/sealers, two-way communication devices, MW ovens, and cellular telephones. Most all of these sources operate at a frequency assigned by the Federal Communications

**TABLE 11.2 Various Magnetic Field Exposure/Emission Standards in the Frequency Range ≤300 kHz**

| Organization or country/year | Frequency range | Magnetic field (μT)[a] | Exposure duration and/or comments |
|---|---|---|---|
| ANSI C95.1-1992 | 3–100 kHz | 205 | "Uncontrolled environment" |
| ACGIH/1996-97 | 4–30 kHz | 2.01 | "Threshold values" |
| 105 CMR[b] 122.00/1986 | 10–3000 kHz | 1.99 | Occupational |
| MPR-III/1994[c] | 5–2000 Hz | 0.250 | At 0.5 m; three planes |
| | 2–400 kHz | 0.025 | Centerline, 0.25 m above and below |

[a] Most exposure limits list magnetic field intensity ($H$) in units of ($A\,m^{-1}$). These values have been converted to magnetic flux density ($B$) in units of ($\mu T$, $1\times10^{-6}$T), $B = \mu H$, by assuming the permeability ($\mu$) of free space ($4\pi\times10^{-7}\,H\,m^{-1}$) is the same for tissue.
[b] Commonwealth of Massachusetts Department of Public Health (105 CMR 122.000: Massachusetts Department of Public Health – fixed facilities that generate electromagnetic fields in the frequency range of 300 kHz–100 GHz and microwave ovens).
[c] This standard is a "performance" standard. It is based on VDT emissions achievable with current technology and not on any "safe" level of EMF emissions.

Commission (FCC) in the industrial, scientific, and medical (ISM) band of the electromagnetic spectrum. Typical frequencies are 13.56, 27.12, 40.68, and 2450 MHz. RF exposures can be from sources, leakage along the transmission pathway, or receivers. The areas affected in the body can be whole or partial body or even from contact with a reflective object within an intense low-frequency (from 3 kHz to 100 MHz) RF field.

RF induction ovens operate as convection ovens but use RF as their initial heat source. These usually operate in the kHz range (typically 300–450 kHz) and produce intense electric and magnetic fields, which are very difficult to measure. Luckily, the thermal component of the operation of these ovens often precludes worker RF exposure above safety standards due to the impossibility to remain near the sources.

In the hospital setting, diathermy, first used in 1907, uses RF to produce a therapeutic effect. Common diathermy units use 27.12 MHz RF to stimulate circulation and promote healing. Diathermy units operating at 2450 MHz have also been used, but are not as common. A specialized form of diathermy is pulsed electromagnetic stimulation (PES). PES is a noninvasive procedure used for the palliative treatment of postoperative pain and edema in superficial soft tissue. RF ablation is a nonsurgical procedure to treat some types of rapid heart beating. In these procedures, a physician guides a catheter with an electrode at its tip to the affected area of the heart muscle. RF is transmitted to a very small area to kill the muscle cells conducting the extra impulses that cause rapid heartbeats. RF exposures in hospital settings are usually confined to the patients receiving electrosurgery or diathermy treatments. However, hospital staff may be exposed to RF from the unintentional use of the sources or the near-field region (within a few wavelengths) of portable transmitters, including portable transceivers and cellular telephones.

MW ovens are ubiquitous in industry and households. These devices use RF at 2450 MHz to heat nonmetallic materials. While MW ovens may be ordinary in the home, they have recently been used in both laboratory and industrial settings as a quick method to heat. Since MW ovens are designed to keep all the MW within the RF chamber, emission standards, vice exposure standards, apply. The present emission standard for MW ovens, 21 CFR 1030.10 from the Food and Drug Administration (FDA), allows new ovens to leak no more than 1 $mW\,cm^{-2}$ when measured no closer than 5 cm from the device and no more than 5 $mW\,cm^{-2}$ once the ovens leave the store (21 CFR 1030.10 2012).

Cellular telephones utilize wireless communication technology (operating between 800 and 900 MHz in the UHF portion of the RF band of the electromagnetic spectrum) to interlink their information to an existing wire-line telephone network. The communication network consists of many systems. The call can be initiated by any of the following sources: a portable handheld phone operating at very low power (usually <1 W), a mobile unit operating at a higher power (several watts), or a landline telephone. The call is transferred to the nearest cell. The latest entry into the wireless communications industry is personal wireless systems (PCS). PCS phones operate at much higher frequencies, typically around 1900 MHz, and send and receive digital signals. While older cellular technology used analog signals, the carriers are slowly converting to all-digital networks.

Below about 300 MHz, the field intensity cannot be suitably quantified by measurements of the electric nor magnetic fields alone; thus, in the absence of scientific evidence to the contrary, individual readings of both electric fields and magnetic fields are necessary to determine regulatory compliance. Measurements of both fields are always necessary below 30 MHz. It is beyond the scope of this chapter to attempt to train personnel to make RF measurements. It should be mentioned, however, that the probe should be held no closer than 20 cm from the source of leakage. For broadcast sources, calculations of the safe distances for both survey equipment and personnel must be performed prior to actually making measurements. For more details on the procedures to make RF measurements, refer to the National Council on Radiation Protection and Measurements (NCRP) Report number 119 (NCRP 1993) and the

Institute of Electrical and Electronics Engineers/American National Standards Institute (IEEE/ANSI) C95.3-1992 (IEEE/ANSI C95.3-1992 1992).

Electric fields at frequencies above 100kHz are measured using small dipole antennae. The electric field induces a surge of current in the dipole, which is connected to a diode. An amplified signal is applied to a meter, with results displayed in field strength units (FSUs) such as $V\,m^{-1}$, or $V^2\,m^{-2}$, or even power density (a measure of the power going through an area of space) ($S$) in watts per square meter ($W\,m^{-2}$) or milliwatts per square centimeter ($mW\,cm^{-2}$). *Note:* Power density ($S$ in $mW\,cm^{-2}$) is derived from the following relationship between the electric ($E$ in $V\,m^{-1}$) and magnetic ($H$ in $A\,m^{-1}$) field strengths:

$$\frac{E^2}{3770} = S = H^2 \times 37.7$$

Dipoles and diodes can be used at lower frequencies (below about 100kHz), but displacement sensors are also used. As before, the resultant RMS field intensity should be determined. Most field survey instruments gang three sensors together, so they are mutually orthogonal to provide a nearly isotropic response.

Magnetic fields at frequencies below 300MHz are measured with single loops, and the field orientation is critical to determine peak field strengths. Three mutually orthogonal loops have been used to allow nearly isotropic responses. As before, the magnetic field induces a current that drives a meter, with results displayed in FSUs such as $A\,m^{-1}$, or $A^2\,m^{-2}$, or even power density milliwatts per square centimeter ($mW\,cm^{-2}$). The resultant RMS field intensity should be determined. Most field survey instruments gang three sensors together, so they are mutually orthogonal to provide a nearly isotropic response.

Recently, equipment to measure the "grasping" or "contract" currents have been made commercially available. The grasping or contact current measuring device uses electrodes to measure electrically charged objects within a relatively intense RF field. Similarly, "induced current" meters have been made available. These devices measure the current induced through either foot, or both feet, to determine if current flowing through a constriction in the body, such as the ankle, exceeds the maximum permissible.

Measurements of RF fields to determine regulatory compliance should only be undertaken by a competent health physicist or other suitably trained individual. The appropriate equipment used is not only very expensive to purchase and calibrate but is often difficult to use.

RF exposure standards, including contact/grasping currents and induced current exposure limits, have been published by many organizations around the world. ANSI (IEEE/ANSI C95.1-1992 1992), NCRP (1986), and ACGIH are but a few. The RF standard most cited is that published by ANSI. Refer to Table 11.3 for RF exposure limits for "uncontrolled areas" (those areas not restricted for reasons of RF exposure, where personnel are exposed to RF as a consequence of their employment). Table 11.4 contains RF exposure limits for "controlled areas" (those areas restricted for reasons of RF exposure, where personnel are exposed to RF as a consequence of their employment).

Exposure to electric and magnetic fields that vary with time results in internal body currents and energy absorption in tissues, depending on the coupling mechanisms and the incident frequency. Table 11.5 lists the dosimetric quantities to be considered and corresponding units.

Cellular telephones operating in the frequency range of 800–900MHz were implicated in the causation and/or promotion of brain cancer through the evening syndicated "talk show" media circuit. The show reported that a lawsuit was filed due to a man's wife succumbing to brain cancer, which was detected shortly after her continual use of a cellular phone; the tumor was reported to be in the shape of an antenna. This particular show generated a flurry of controversy, continuing until the present day. Sadly, the dismissal of the lawsuit without finding was not as widely publicized. Nevertheless, numerous studies have been published attempting to determine possible carcinogenic effects of exposure to RF with frequencies in the range used by communication systems. A concise summary of laboratory research findings was recently published by the ICNIRP (1996). There is substantial evidence that RF is not mutagenic and exposure to these fields is therefore unlikely to initiate carcinogenesis. A recent study did indicate that use of cellular telephones may be detrimental; it was reported that use of cellular telephones in a moving automobile increases the risk of accidents to the level of driving drunk (Redelmeier and Tibshirani 1997).

Recent concern has been expressed about the possible interference of cellular telephone usage with the proper operation of cardiac pacemakers. The FDA has researched this concern and concluded: "Based on these preliminary findings, cellular phones do not seem to pose a significant health problem for pacemaker wearers" (Food and Drug Administration, Department of Health & Human Services 1995).

A significant difference between RF and ELF exposures is that RF exposures, especially RF exposures in the MW band, may have thermal effects. Exposure to low-frequency electric and magnetic fields in air results in a negligible amount of energy absorbed and no measurable temperature rise in the body. Exposures of greater intensity than

**TABLE 11.3 IEEE/ANSI Maximum Permissible Exposures for Uncontrolled Environments**

| Frequency range (MHz) | Electric field strength ($E$) ($V m^{-1}$) | Magnetic field strength ($H$) ($A m^{-1}$) | Power density ($S$) $E$ field; $H$ field ($mW cm^{-2}$) | Averaging time $\|E\|^2$, $S$, or $\|H\|^2$ (min) | |
|---|---|---|---|---|---|
| 0.003–0.1 | 614 | 163 | (100; 1000000)[a] | 6 | 6 |
| 0.1–1.34 | 614 | $16.3/f$ | $(100; 10000/f^2)$[a] | 6 | 6 |
| 1.34–3.0 | $823.8/f$ | $16.3/f$ | $(180/f^2; 10000/f^2)$[a] | $f^2/0.3$ | 6 |
| 3–30 | $823.8/f$ | $16.3/f$ | $(180/f^2; 10000/f^2)$[a] | 30 | 6 |
| 30–100 | 27.5 | $158/f^{1.668}$ | $(0.2; 940000/f^{3.336})$[a] | 30 | $0.0636/f^{1.33}$ |
| 100–300 | 27.5 | 0.0729 | 0.2 | 30 | 30 |
| 300–3000 | | | $f/1500$ | 30 | |
| 3000–15000 | | | $f/1500$ | $90000/f$ | |
| 15000–300000 | | | 10 | $616000/f^{1.22}$ | |

*Induced and contact RF currents*

| Frequency range (MHz) | Maximum current (mA) | | |
|---|---|---|---|
| | Through both feet | Through each foot | Contact |
| 0.003–0.1 | $900f$ | $450f$ | $450f$ |
| 0.1–100 | 90 | 45 | 45 |

[a] Values of plane wave equivalent power densities given for comparison.

**TABLE 11.4 IEEE/ANSI Maximum Permissible Exposures for Controlled Environments**

| Frequency range (MHz) | Electric field strength ($E$) ($V m^{-1}$) | Magnetic field strength ($H$) ($A m^{-1}$) | Power density ($S$) $E$ field; $H$ field ($mW cm^{-2}$) | Averaging time $\|E\|^2$, $S$, or $\|H\|^2$ (min) |
|---|---|---|---|---|
| 0.003–0.1 | 614 | 163 | (100; 1000000)[a] | 6 |
| 0.1–3.0 | 614 | $16.3/f$ | $(100; 10000/f^2)$[a] | 6 |
| 3–30 | $1842/f$ | $16.3/f$ | $(900/f^2; 10000/f^2)$[a] | 6 |
| 30–100 | 61.4 | $16.3/f$ | $(1.0; 10000/f^2)$[a] | 6 |
| 100–300 | 61.4 | 0.163 | 1.0 | 6 |
| 300–3000 | | | $f/300$ | 6 |
| 3000–15000 | | | 10 | 6 |
| 15000–300000 | | | 10 | $616000/f^{1.2}$ |

*Induced and contact RF currents*

| Frequency range (MHz) | Maximum current (mA) | | |
|---|---|---|---|
| | Through both feet | Through each foot | Contact |
| 0.003–0.1 | $2000f$ | $1000f$ | $1000f$ |
| 0.1–100 | 200 | 100 | 100 |

[a] Values of plane wave equivalent power densities given for comparison.

**TABLE 11.5 Dosimetric Quantity for RF Exposure Assessment**

| Frequency range | Dosimetric quantity | Units |
|---|---|---|
| 1 Hz–10 MHz | Current density | Amperes per meter ($A m^{-2}$) |
| 1 Hz–100 MHz | Contact currents | Amps (A) |
| 100 kHz–10 GHz | Specific absorption rate (SAR) | Watts per kilogram ($W kg^{-1}$) |
| 300 MHz–10 GHz | Specific absorption (SA) | Joules per kilogram ($J kg^{-1}$) |
| 10–300 GHz | Power density ($S$) | Watts per square meter ($W m^{-2}$) |

1 $mW\,cm^{-2}$ at frequencies above about 10 MHz can lead to significant energy absorption and temperature increases. Exposures of greater intensity than 10 $mW\,cm^{-2}$ may cause increases in whole body temperature. Heating from RF is no different in producing thermal effects than other sources of exogenous sources of heat except that the degree and distribution of heating are less predictable. The initial physiologic response in laboratory animals is cutaneous vasodilation, which reverses rapidly if the source of RF is removed. Surface evaporation, metabolism, cell growth, and cell division, as well as immune response, can be altered by increasing thermal RF exposures. Heated humans have been noted to suffer psychological symptoms, intermittent hypertension, and bums, while heated laboratory rats and monkeys suffered decreased task performance at specific absorption rates (SAR) (a measure of energy absorbed in the body) in the range of 1–3 $W\,kg^{-1}$. At levels of body temperature elevation above about 1–2°C, a large number of biological effects have been observed, including increased blood–brain barrier permeability, ocular impairment, stress-associated effects on the immune system, decreased sperm production, teratogenic effects, and alterations in neural functions.

Thermally vulnerable organs such as the eye may sustain damage from high-energy exposures. MW exposure to the eye may cause injury, including cataracts (at frequencies <2 GHz), corneal damage (at frequencies above 10 GHz), and retinal lesions. MW cataracts have been induced in experimental animals and reported, rarely, in man. The present consensus is that very-high-intensity (>1.5 $kW\,m^{-2}$) exposures are required to produce detectable eye damage. Frequent reports of an auditory sensation from powerful RF fields are due to rapid thermoelastic expansion and secondary pressure waves, which can be detected by the cochlea. This is known as a *microthermal effect*, where a high local rate of tissue absorption and expansion occurs in the absence of any measurable change in temperature.

In the frequency range between 100 kHz and 110 MHz, shocks and burns can result either from touching an ungrounded metal object that has acquired a charge in the field or by contact between a charged body and a grounded metal object. Threshold currents that result in biological effects that range from perception to pain have been measured in controlled human experiments. Generally, when the current between the point of contact of the human body and the conductor exceeds 50 mA, there is a risk of severe burns and irreversible tissue damage.

Control of RF exposures by shielding can be relatively inexpensive, but may be an arduous task for the novice. Metal-conducting wire mesh, particularly copper, is very useful, as long as the openings are less than ¼ wavelength and all sections are securely grounded. Foam can be used to gradually diminish the fields through insulators of varying impedance. Distance is perhaps the best tool for control of stray RF fields. However, it must be mentioned that the proper maintenance of equipment to prevent RF leakage is of the utmost importance. Refer to Figure 11.3 for the internationally recognized RF caution sign published by ANSI (IEEE/ANSI C95.2-1982 1996).

***11.1.2.5 Infrared, Visible Light, and Ultraviolet*** *Note:* In the part of the electromagnetic spectrum above about 1 terahertz (THz, $10^{12}$ Hz), it is traditional to refer to wavelength vice frequency. All objects with temperatures above absolute zero emit IR radiation, which extends in a band between 760 nanometers (nm, $10^{-9}$) and 1 millimeter (mm, $10^{-3}$). The human eye detects VL in the narrow band between 760 nm at the red end and 380 nm at the violet end, with peak sensitivity at 555 nm. UV light extends in a narrow band between 400 and 40 nm.

The major industrial sources of IR are heating and drying devices and lasers (to be discussed later in this chapter). Large sources of VL exist in industry, mainly for the expected application: artificial lighting. UV sources in industry are very diverse. They may be in the form of intentionally harmful UV lights, such as those used in

**Figure 11.3** The radio-frequency (RF) hazard warning symbol. Courtesy ANSI C95.2-1982.

biological safety cabinets to kill entering or exiting bacteria, or unnecessarily harmful, such as UV lasers. Welding is also a major source of harmful UV.

The units used to express IR and UV exposures are typically mW or μW $cm^{-2}$. Standards for IR, VL, and UV can be found published by the ACGIH for nonlaser sources. Standards for IR exposure vary with the coinciding presence of VL and ambient temperatures. UV standards are used to calculate "safe stay times" by using tables and/or formulae to calculate the time necessary to exceed to TLV. For UV exposures incident on the unprotected eyes or skin, between 180 and 400 nm, use the following formula to calculate the appropriate "safe" stay time, in seconds, for every eight-hour period: divide (0.003 J $cm^{-2}$) by $E_{eff}$ in W $cm^{-1}$, where $E_{eff} = \Sigma E_\lambda S_\lambda \Delta\lambda$ ($E_\lambda$ = spectral irradiance in W $cm^{-2}$ $nm^{-1}$; $S_\lambda$ = relative spectral effectiveness [unitless]; and $\Delta\lambda$ = bandwidth in nm).

The major hazard from IR exposure is the coinciding rise in the temperature of the absorbing tissue. IR is absorbed by tissue water and therefore nonpenetrating; energy transfer is directly to outer surfaces. Pain is the best warning property in normal skin for thermal injury due to IR. Conversely, the eyelid and eye do not have ample thermal warning properties. Eyelid blisters are common following heating exposures, and the eye is at risk for injury from IR exposures. IR overexposure produces corneal burns and choroid and retinal damage.

There is laboratory evidence that chronic exposure to excessively bright light may lead to premature degeneration of the cones responsible for color vision. The retina is sensitive to excess light.

The harmful effects of UV are due primarily to the range of 315–280 nm, also known as "UV-B." Acute effects upon the eye include photokeratitis and photoconjunctivitis. Human eyesight perceives UV poorly (or not at all) because of absorption by ocular media. Acute damage may occur before there is awareness of exposure. Keratitis and conjunctivitis are usually reversible within several days. Burns may occur from lateral UV-B exposure to the cornea and not merely through the pupil and lens. Exposures to UV around 297 nm induce cataract formation with a clear dose–response relationship.

Keratitis has been observed in biotechnology environments, where UV is used for sterility and for DNA sequencing operations. Irradiance levels of germicidal UV lamps vary markedly with distance from the lamp or nearby reflecting surfaces. Some UV-B exposure is necessary for vitamin $D_3$ production where nutritional supplies are inadequate. However, there is ample evidence that UV exposure causes substantial skin pathology. There has been an alarming epidemic of skin cancers in the parts of the world where increases in outdoor recreation and admiration for tanning have occurred.

Control of IR, VL, and UV sources can be accomplished through specially designed eyewear and, to some extent, protective clothing. Only qualified personnel should make a determination of the effectiveness of any particular form of personal protection.

***11.1.2.6 Lasers*** "Laser" is an acronym for light amplification by stimulated emission of radiation. An energized laser emits intense monochromatic (that is, one wavelength color) visible or invisible coherent (that is, all photons of light are in "phase" with one another) radiation continuously or in pulses.

IR lasers include the neodymium–yttrium garnet (Nd: YAG) and carbon dioxide ($CO_2$) lasers. VL lasers include dye lasers, argon, helium–neon lasers, and krypton lasers. UV lasers include "excimer" lasers, and Nd: YAG lasers tripled in frequency into the UV spectrum.

Measurements of the optical radiation from lasers is usually accomplished with either thermal or quantum detectors. Thermal detectors measure the rise in temperature in crystals, although these are best used for IR. UV, IR, and VL measurements can be obtained with quantum detectors. These detectors emit electrons in response to being struck by radiation.

The focusing ability of the eye increases the retinal hazard of any visible laser beam. Laser classification and associated hazards are described in Table 11.6. Eye protection issues are outlined in ANSI Z136.1-1993 for the safe use of lasers (ANSI Z136.1-1993 1993). Skin protection is also important for noninterlocked operations of class 4 lasers. $CO_2$–$N_2$ lasers emit invisible energy; so burns may potentially be severe before the hazard is appreciated. For the proper posting of laser warning signs (see Figure 11.4), refer to ANSI Z136.1-1993.

The most consistent hazard is not related to the radiation, but to the possibility of electric shock. Fire, cryogenic, and X-ray hazards are also associated with laser use. For these reasons, only appropriately qualified personnel should be involved with the design, installation, personnel, procedures, and any changes in conditions of laser use. Additionally, there is a significant acute hazard to the eye and skin from certain types of laser equipment. These depend upon the wavelength, intensity, and duration of exposure. Lasers may emit IR, VL, and UV light.

Personal protective equipment has serious limitations when used as the only protective measure for class 4 lasers. Eyes and skin are best kept out of the beam by interlock devices, which turn off the laser electronically or else provide physical barriers between operators and bystanders and the beam during operations. Not all research laser activities are amenable to interlock protection. Personal protective devices are then used. Eye shielding must be selected in

**TABLE 11.6 Laser Classification**

| Class of laser[a] | Potential danger |
|---|---|
| 1 | Essentially harmless |
| 2; 2a | Do not stare into the beam |
| 3; 3a; 3b | Hazardous<br>Direct viewing must not occur<br>Specular reflections are also dangerous<br>Skin exposure is harmful |
| 4 | Extremely hazardous<br>Direct viewing must not occur<br>Specular reflections are extremely dangerous<br>Diffuse reflections dangerous<br>Skin exposure extremely hazardous |

[a] When class 3 and 4 lasers are fully embedded into an enclosed system that has full safety interlocks, the system may be classified as a class 1 system (per instructions of the laser safety officer [LSO]).

**Figure 11.4** The laser warning symbol. Courtesy Laser Institute of America.

accordance with the wavelength employed and should include side shielding. Absorptive filters are generally preferred to reflective types, as absorptive filters are reliable regardless of the incident angle of the beam. A problem with optical density markings on laser protective eyewear has been uncertainty with the reliability of manufacturers' markings concerning transmittance. Independent checks are not truly achievable with available equipment; so it is important to pick the most reputable supplier. Another problem with eye protection is that complete protection renders otherwise visible beams invisible.

Laser surgery or any other tissue use of class 4 lasers will inevitably generate heat and pressure, creating smoke and fume. The smoke is malodorous. Designed ventilation is the appropriate industrial hygiene measure, but this fails to address adequately that the smoke plume may carry infectious organisms.

Calculations such as "optical density" (OD) (the protection factor necessary to reduce a given laser hazard to the level below which permanent injury is prevented) and "nominal hazard zone" (NHZ) (the distance necessary to reduce a given laser hazard to the level below which permanent injury is avoided) are necessary in determining the safe use of any class 3 or 4 laser systems. Confirmatory measurements should be made with radiometric instrumentation to ensure that appropriate eye (and skin) protection is chosen.

Medical surveillance of laser workers and support staff is highly recommended to ascertain eye injury in the event of accidental exposure. Baseline eye exam before working with the laser is essential in making this determination. Further eye examinations will be required post-incident and post-employment.

### 11.1.3 Summary

The three elements that must be considered in any effective safety program designed to control actual or potential hazards are as follows:

1. Recognition
2. Evaluation
3. Control

This chapter was designed to help the reader to *recognize* NIR hazards that may be found in the industrial setting. Upon recognition of each potential or actual hazard, a competent individual should perform a safety *evaluation*. Based on the outcome of the evaluation, *control* may or may not be necessary.

There is an immense potential for serious injury considering all the NIR sources in use by industry. It is imperative that an NIR safety program be initiated and maintained in each facility that uses NIR sources. The development of an NIR safety program, as well as the evaluation of any NIR actual or potential hazard, must be carried out by an individual properly trained and experienced in this complex field of occupational safety. Consultants are available for such work. Qualified individuals may be found by contacting the Health Physics Society (1313 Dolly Madison Blvd, Suite 402, McLean, VA, 22101).

## 11.2 IONIZING RADIATION

These sources can come in the form of sealed radioactive material used for its radiation emissions, unsealed radioactive material used in a research or manufacturing process, or machine-produced radiation in the form of an X-ray machine or the by-product of a high-voltage supply. It is important to be able to recognize potential sources of ionizing radiation and have an effective evaluation program to limit worker exposures. It is the responsibility of management to provide the necessary training, facilities, equipment, and personnel to maintain levels of radiation exposure to its employees, the general public, and the environment to as low as reasonably achievable (ALARA). The following sections outline administrative, process, and engineering controls that can be used to help achieve this goal.

### 11.2.1 Specific Licensing

Users of radioactive sources typically must obtain a specific license from an appropriate federal or state agency depending on the state where the facility will be located. States fall into two categories: agreement and nonagreement states. Agreement states are those that have entered into an agreement with the Nuclear Regulatory Commission (NRC) to license, regulate, and inspect the use of radioactive material within their boundaries. Nonagreement states are directly licensed, regulated, and inspected by NRC. See Appendix A for the current list of agreement states.

The specific license application outlines to the agency the types and quantities of radioisotopes that will be possessed and used, a description of the proposed use of the radioisotopes, and a comprehensive radiation safety program to ensure that radioisotopes will be possessed, stored, used, and disposed in a responsible and safe manner. Administrative control in the safe use of radioisotopes begins with the license application process.

### 11.2.2 General Licensing

Many radiation sources used as part of a mechanical system or process do not require a specific license to possess and use them. These sources are possessed and used under a general license agreement. Any person or company in the United States has the right to possess and use generally licensed quantities of radioactive material. Because a license is not required to buy these sources, it is often difficult to track these sources within the work environment. Furthermore, it is difficult to involve workers in a comprehensive radiation safety program if it is unknown that radioactive sources are possessed. Some examples of generally licensed radioactive sources are static eliminator bars, ionizing-type smoke detectors, gas chromatography analyzers that use a radioactive source, and density gauges. Although these sources are generally licensed, it does not exempt the employer from performing certain checks and balances on the sources during their use. These requirements are typically outlined in instructions from the device manufacturer. They typically require some form of monitoring or radiation level measurements at specified intervals.

### 11.2.3 Source Material

As with the generally licensed sources, certain quantities of naturally occurring radioactive material (source material) are exempt from licensing. Uranium- and thorium-containing materials are used widely in industry from shielding and counterweights to coating processes on high polished optical surfaces. Although these materials can be possessed and used without a license (and many times without radiation safety considerations), the potential for worker exposure, especially internal exposures from some manufacturing processes, can be significant.

### 11.2.4 Machine Radiation Sources

X-ray machines such as diagnostic machines as part of a company health clinic to analytical machines such as diffraction or fluorescence equipment used to study manufactured goods may present an exposure potential to users and surrounding environments. The emissions from these machines or the area where the machines are used must be controlled to ensure worker safety. Most state radiation control programs require a formal registration of radiation-producing machines. The registration process requires that users comply with the agency rules and regulations for the safe use of these types of machines.

### 11.2.5 Radiation Safety Officer

A qualified person should be identified as the radiation safety officer (RSO) for the facility. The responsibilities of the RSO may include:

1. Ensuring facility compliance with all applicable federal, state, and local regulations and ordinances.
2. Updating license parameters, as necessary, via amendment request to the granting agency.
3. Ensuring compliance with conditions of generally licensed sources, exempt sources, and radiation-producing machine registrations.
4. Reviewing all proposed uses of radiation, especially new procedures that require substantially increased amounts of radiation.
5. Provide radiation safety training to all radiation workers, ancillary personnel, or other employees whose duties require them to be in areas where radiation sources or machines are used and stored.
6. Approve all purchases of radiation sources or machines to ensure compliance with the license conditions or registrations.
7. Perform radiation safety audits of all radiation source and machine use.
8. Be available to assist in any emergencies, special decontamination efforts, or worker exposure evaluations.
9. Maintain all appropriate records as required by specific license conditions, general license requirements, and machine registration regulations.
10. Properly dispose of radiation sources, radioactive materials, or radiation-producing machines when they are no longer of any use to the facility.
11. Approve of all transportation of radioactive sources.

The RSO duties need not translate into a full-time position and typically will be only a small percent of a person's responsibilities. This will depend on the size of the program and the number of radiation sources and workers. Typically, certified health physics consultants are used to assist RSOs of small programs. Consultants can evaluate the adequacy of radiation shielding, perform required leak testing of radioactive sources, provide radiation worker training seminars, and assist the RSO in worker exposure evaluations. However, it is very important that ample time be scheduled to allow the RSO to audit the radiation safety program on an ongoing basis.

### 11.2.6 Radiation Safety Liaison

A person at the user level should be identified to assist the RSO in the administrative functions necessary to ensure compliance with license conditions and regulations. The liaison's responsibilities may include the following:

1. Maintaining a current inventory of radiation sources stored in the lab. Perform physical inventories of sealed sources as required.
2. Exchange of personnel monitoring devices on a timely basis.

3. Perform and maintain records of radiation surveys or source leak tests.
4. Inform the RSO when new persons join the user group to ensure the person receives proper training.
5. Be responsible for radioactive source security and use logs.
6. Inform the RSO whenever new generally licensed or exempt natural sources are purchased.
7. Act as the liaison between the user group and the RSO.

Small facilities with only a few radiation sources and a small number of radiation workers may not need a radiation safety liaison. The RSO would be directly responsible for the above-listed duties. However, if radiation sources or machines are used in the field or by various different work groups within the facility, the need for this position becomes evident.

### 11.2.7 Administrative Procedures

Licensed facilities using radiation sources or machines that produce radiation should have a set of standard operating procedures addressing the following functions:

1. Radiation worker training.
2. Ancillary personnel training.
3. Assignment of personnel monitoring devices and record of radiation exposures.
4. Radiation source ordering, receipt, and inventory.
5. Routine bioassay or *in vivo* measurements of radiation workers, as necessary.
6. Emergency medical surveillance after accidents or known exposure above allowable limits.
7. Policy for pregnant radiation workers.
8. Environmental monitoring of experiments that potentially release airborne containments.
9. Radiation survey meter calibration and records of calibration certificates.
10. Procedures for and record of safe disposal of low-level radioactive waste or final disposition of radioactive sources or machines.
11. Emergency procedures.

These standard operating procedures will be used as part of an effective radiation safety program. Administrative control will include maintenance of all records required by regulations, registrations, and license conditions. These records will be scrutinized during licensing agency inspections. Records of personnel exposure histories and radioactive waste disposal will be maintained indefinitely.

### 11.2.8 Licensed Radiation Sources or Machines

A thorough understanding of the radioisotopes used in the sealed sources or the type of radiation-producing machine to be used is essential to determine the control methods necessary for safe handling of the sources or machines. The appropriate instrumentation and calibration needed to detect the material, the potential pathways of exposure, handling tools or shielding necessary to reduce external exposures, the need for personnel monitoring devices, personal protective equipment, emergency procedures, and waste disposal requirements are safety areas that need to be addressed.

### 11.2.9 Radiation Worker Training

The most important aspect of radiation worker safety is providing an effective training program to ensure the worker understands the potential hazards involved while working with the radioactive sources or radiation-producing machines. Radiation worker training is a requirement of the NRC or state licensing agency. Radiation safety training is mandatory for all radiation workers. The following subjects should be covered in a radiation worker training seminar:

1. Units of radioactivity and radiation dose.
2. Concept of radioactive decay and half-life.

3. Radiation detection and measurement.
4. Type and amounts of radioactive material licensed.
5. Safe handling techniques and dose reduction techniques.
6. Standard operating procedures for the source or machine being used.
7. Personal protective equipment.
8. Maximum permissible exposure limits.
9. ALARA concept.
10. Biological effects and risks from occupational exposures.
11. Bioassay and *in vivo* measurement of ingested radioisotopes.
12. Radiation and contamination survey techniques.
13. Leak testing of sealed sources.
14. Emergency procedures and decontamination methods.
15. Machine calibrations and maintenance.
16. Transportation of radiation sources for fieldwork.

The radiation safety training seminar should afford the worker an opportunity to ask questions concerning the safe use of the radiation sources. Attendance at training seminars will be documented.

### 11.2.10 Safe Handling and Dose Reduction Techniques

Once information is known about the types and amounts of radiation sources to be handled, the worker can develop safe handling and dose reduction techniques. Whenever radioactive sources are handled, the worker must wear protective clothing to reduce the possibility of personnel contamination. Lab coats, gloves, and safety glasses are required during all handling. Double gloving is appropriate. The outside glove would be changed if contamination is detected. A dry run of all new procedures, especially when significant increases in activity are required, is important to preclude unforeseen handling problems. Benches should be covered with plastic-backed absorbent bench paper whenever liquid sources are handled. Continual contamination monitoring with a portable survey meter during handling will reduce the potential for widespread contamination.

Dose reduction can be accomplished by utilizing the concepts of time, distance, and shielding or a combination of the three. Radiation dose is measured as a rate, usually in millirads per hour. If one can reduce the amount of time they are exposed to a source, their cumulative radiation dose equivalent will be lowered. Of the three concepts, time may be the most difficult to change due to experimental procedures.

Distance from a source of radiation can significantly reduce the worker's exposure. The dose rate from a point source of radiation, for example, a 10-mCi sealed source of $^{137}Cs$ found in density gauges, is reduced by the square of the distance as one moves away from the source (inverse square law). Thus the dose rate at 1 cm is 100 times greater than the dose at 10 cm. Dose reduction to the hands and fingers can be significantly reduced with use of tongs to handle sealed sources.

Appropriate shielding, depending on the type and energy of the emitted radiation, can also significantly reduce worker exposures. For X-ray and gamma-ray emitters, the recommended shielding is a high-atomic-number material such as lead. The thickness of the shielding will be determined by the energy of the incident radiation.

### 11.2.11 Radiation Detection and Measurement

To work safely with sealed or unsealed sources of radioactivity or radiation-producing machines, workers must be fully trained in the use of radiation detection equipment to measure potential radiation exposure rates and contamination levels. In addition, radioactive sealed sources are required to be leak-tested at a certain intervals (usually six months). These leak-test samples must be analyzed in analytical equipment to quantify any radioactive material leakage.

### 11.2.12 Survey Instruments

Lightweight portable radiation survey instruments must be available in radiation laboratories to enable workers to monitor themselves and their work areas while handling sources. The most versatile radiation detector used is the Geiger Mueller (GM) detector. It is capable of detecting beta, X-ray, and gamma-ray radiations emitted by the most commonly used radioisotopes used in sealed sources. In addition to the GM detector, there are portable specialty detectors such as scintillation detectors or ionization chambers whose use becomes necessary depending on the source of radiation and the measurements required. However, the GM detector serves most purposes from a radiation worker exposure evaluation potential.

All survey instruments must be routinely calibrated against certified radioactive standards. In addition, small radioactive check sources should be attached to each instrument to allow the worker to check operability each time the instrument is used.

### 11.2.13 Analytical Instruments

These instruments are used to quantify amounts of radioisotopes. For most small users or users of sealed sources that require leak testing, these instruments are prohibitively expensive. Most small users contract with laboratory facilities for radioactivity analysis on such things as leak tests or contamination smears. The types of analyzer typically used in these analysis laboratories are liquid scintillation counters, gas flow proportional counters, or gamma-ray spectroscopy analyzers.

### 11.2.14 Maximum Permissible Exposure Limits

Regulations outline the maximum permissible exposure limits allowed for radiation workers, pregnant radiation workers, and the general public. The limits are reported in units of millirems. The following are the current limits of exposure.

| Area exposed | Annual limit (mrems) |
|---|---|
| 1. Total effective dose equivalent (internal and external whole body) | 5000 |
| 2. Lens of the eye | 15000 |
| 3. Skin and extremities | 50000 |
| 4. Pregnant radiation worker | 500 |
| 5. General public | 100 |

Of the commonly used radiation sources, beta emitters are classified as nonpenetrating radiation and will give the worker a skin exposure. X-ray and gamma-ray emitters, whether from radioactive sources or machine-produced, are classified as penetrating radiation and will give the worker a whole-body exposure.

### 11.2.15 ALARA Concept

Although there are limits for exposures to workers as outlined above, it is the responsibility of the RSO to operate the radiation safety program such that workers utilize safe handling and dose reduction techniques to keep radiation exposures as ALARA.

### 11.2.16 Biological Effect and Risks from Occupational Exposures

Although there is some risk associated with all exposures to radiation, workers who keep their exposures within the maximum permissible exposure limits maintain the associated risk at acceptable levels.

Exposures from uses of radiation sources or machine-produced radiation should be kept with 10% of the maximum permissible limits, and any exposures above 10% should be investigated. Typically, workers are kept within 1–5% of the maxima. The effects associated with small radiation exposures are practically unmeasurable due to the varying levels of natural background radiation received by everyone and exposures to other potentially harmful materials. Radiation workers should be encouraged to read NRC Regulatory Guide 8.29, "Risks from Occupational Radiation Exposure." This document reviews potential health effects from radiation exposures and enables the worker to compare the risk associated with occupational exposures to the risk associated with other occupations and the risk associated with normal daily life activities.

### 11.2.17 Worker Exposure Monitoring

Workers handling radioactive sources or working with radiation-producing machines that may result in potential exposure to radiation in excess of 10% of the maximum permissible exposure limits should be monitored. Typically monitoring is accomplished by use of film badge or thermoluminescent dosimeters capable of measuring exposures for X-rays, gamma rays, and beta particles. The dosimeter should be worn at all times when radiation exposures are possible. The film badge or thermoluminescent dosimeter is used to measure whole-body and skin exposures. If workers are handling large radioactive sources, extremity monitors should also be provided. The dosimeter should be exchanged on a monthly or quarterly basis. More frequent exchange of dosimeters is recommended for workers who consistently receive significant radiation exposures (>25% of the MPEs). Radiation dosimetry reports should be made available to the workers, and all results greater than 10% of the MPEs should be investigated to establish if additional dose reduction controls are needed. Radiation exposure histories must be maintained indefinitely.

Monitoring of individual workers may not be feasible or necessary for certain uses of radioactive sources or radiation-producing machines. Area monitors, either passive systems such as the film badge monitor or an active system such as a GM detector-based radiation monitor, can be used to estimate radiation doses to all workers who are present in the area.

### 11.2.18 Radiation and Contamination Surveys

Radiation workers must be trained to perform radiation and contamination surveys whenever radioactive sources or radiation-producing machines are used. Radiation surveys are simply measuring the exposure rate in millirems per hour in the areas where radiation sources are used or stored. The results of these surveys will determine the approximate amount of exposure the worker can anticipate receiving during the time spent in the area. These results can determine whether shielding may be necessary to reduce potential exposures. Radiation surveys should be made during all handling phases of the sources. The GM detector-equipped survey meter is an appropriate instrument for these measurements. An ion chamber survey instrument is a good choice for measuring potential exposures from machine-produced radiation (typically X-rays).

Contamination surveys are performed to determine if any unsealed radioactivity has been spilled or transferred to areas where it is not welcome. Such areas are bench tops, gloves, lab coats, survey instruments, hood aprons, and equipment. If detected, these areas must be decontaminated to license condition or regulatory limits for removable contamination. The GM detector-equipped survey meter is the most versatile instrument for these types of surveys.

In addition to the everyday surveys required while working with radioisotopes, routine surveys including swipe tests should be performed in all radiation laboratories. Swipe tests are performed by rubbing a piece of filter paper or cotton swab over the area to be tested. The typical area surveyed is $100\,cm^2$. The swipe tests are then analyzed in an appropriate analytical instrument such as a liquid scintillation counter or a gas flow proportional counter.

Sealed radioactive sources must be tested for radioactivity leakage at certain time intervals, typically every six months. Swipe testing of the source surface or the closest accessible surface followed by analytical analysis is appropriate. The results of the test must be less than or equal to $0.005\,\mu Ci$ of activity. If the results exceed this value, the source must be removed from service, and further testing must be done on the source.

The frequency of routine surveys can be weekly, monthly, quarterly, semiannually, or annually depending on the types and amounts of radioactivity or the radiation-producing equipment handled in the facility. The RSO or RSO liaison is usually responsible for the routine surveys.

### 11.2.19 Engineering and Environmental Controls

For the use of unsealed radioactive material, engineering control is in the form of carefully planned work space with surfaces that are easily cleaned. Benches and floors should have nonporous surfaces. Sinks used for radioisotope disposal should be stainless steel. The physical layout of the facility is very important. Workers' desks or lockers should not be in the middle of the work area. Since smoking, eating, and drinking are not permitted in radiation work areas, an area should be made available for these activities.

Work with volatile radioactive material or processes that generate airborne radioactivity such as grinding, polishing, or coating must be done in a fume hood or in a work space with localized exhaust. The exhaust from this hood/area should be filtered through an appropriate filter (HEPA/activated carbon filter) to trap any potential releases to the environment. The exhaust air from any hood/area where volatile or airborne radioactive materials are used must be sampled to prove compliance with the EPA NESHAPS regulatory limits for release of airborne pollutants. In addition, the room air should be sampled via a breathing zone sampler to alert workers to potential airborne radioactivity levels and possible internal radiation exposures.

### 11.2.20 Transportation of Radiation Sources

The transportation of radioactive sealed sources or devices containing such sources must be done in compliance with Department of Transportation rules and regulations found in Title 49 of the Code of Federal Regulations. Most devices are sold with a carrying case that is acceptable for transportation. However, there is specific documentation that must accompany each shipment. A record of all shipments should be maintained for inspections by licensing authorities. There are basic training requirements for shippers of hazardous materials that will have to be met for shipment of radioactive sources.

If radioactive sources or devices are to be used outside the jurisdiction of the licensing authority, the user may have to apply for reciprocity to use these devices in different geographical locations. Consult your state radiation control department or the radiation control department in the state where the work will be done for reciprocity details.

### 11.2.21 Low-Level Radioactive Waste Disposal and Source Disposition

The management and disposal of low-level radioactive waste and final source disposition is an integral part of any radiation safety program. There are several licensed low-level waste brokers within the United States who can assist licensees in the disposal of any wastes or sources. Current waste disposal fees are very expensive. Users of unsealed radioactive materials must incorporate a waste minimization plan as part of an effective radiation safety program. Users of sealed radioactive sources should negotiate an agreement with the source manufacturer to accept return of the source when it is no longer of use to the project. Currently, there are only three sites within the United States accepting low-level waste. These sites are located in Richmond, Washington; Clive, Utah; and Barnwell, South Carolina. Most states have appointed boards to manage the low-level wastes generated within their boundaries. These boards as well as state radiation control agencies can be a good resource for up-to-date information on waste disposal requirements.

### 11.2.22 Laboratory Surveillance and Management Audits

Routine radiation laboratory inspections should include surveys for removable contamination, radiation exposure level measurements, and observations for compliance with the licensee radiation safety program (i.e. evidence of eating or drinking in the lab). Development of a compliance checklist serves as a useful tool during radiation audits. Radiation workers should be observed handling radioactive sources to ensure good technique as well as regulatory compliance. Deficiencies should be noted and reports sent to the laboratory supervisors along with suggested corrective actions for improvement. Group retraining seminars are an effective avenue to discuss problems with the entire laboratory and to work on effective solutions. It is important to include management in periodic audits as well as all deficiency reports sent to groups. This will keep management informed of potential regulatory compliance problems.

### 11.2.23 Emergency Procedures

Whenever radioactive sources are used, the potential exists for accidents or spills that result in personnel or facility contamination. Facilities should establish emergency procedures to enable efficient and effective handling of accident situations. There are three general categories of accidents involving radioactive material: (i) radioactive contamination of personnel and facilities including personnel injury, (ii) radioactive contamination of personnel and facilities, and (iii) radioactive contamination of facilities only.

The following are general procedures to follow in the event of a radiation accident.

#### *11.2.23.1 Contamination and Personnel Injury*

1. Immediately notify appropriate emergency response personnel (police, fire, etc.).
2. Remove person from contaminated area, if possible, depending on the extent of the injury.
3. Remove contaminated clothing from person and wait for emergency medical help to arrive.
4. Isolate the contaminated area of the facility.
5. Inform emergency respondents that patient has potential radioactive contamination. Give information on radioisotope, amount, etc. if known.
6. Accompany patient to treatment area. Bring your radiation survey meter to help identify contaminated areas.
7. Notify the RSO.

Whenever a person sustains an injury in an accident involving radioactive material, first aid comes first. After appropriate medical attention is received, decontamination of the facilities can commence.

#### *11.2.23.2 Contamination of Personnel and Facilities (No Injury)*

1. Isolate the contaminated persons in a noncontaminated area.
2. Isolate the contaminated area or equipment with physical barriers.
3. Survey persons and remove any contaminated clothing articles.
4. Decontaminate personnel to minimize total radiation exposure.
5. Notify the RSO.
6. Begin decontamination of facilities.

The facilities must be decontaminated to levels acceptable to license conditions. Perform a radiation survey to determine the physical extent of the contamination. Working from a clean or noncontaminated area, begin decontamination of facility. Clean a small area and wipe test to check for remove contamination. Repeat as necessary until removable contamination levels are below license limits. Perform a thorough radiation contamination survey to ensure all areas are clean. Continue cleaning until all areas have been successfully decontaminated and the facility can be cleared for further use. Keep appropriate records in an accident or incident file. Radiation laboratories should be equipped with radioactive spill kits to aid in clean-up efforts when an accident occurs.

***11.2.23.3 Loss of Radioactive Source or Device*** An additional emergency scenario may be the loss of a radioactive sealed source or a device containing radioactivity. Such loss of control usually requires notification of competent authorities such as the state radiation control department or the NRC. It is advisable to procure the services of a competent health physics consultant to assist facility personnel in emergency situations.

## 11.3 CONCLUSIONS

There are many sources of radiation commonly found in the workplace environment. These sources span the energy spectrum from nonionizing to ionizing radiation. Each source carries its own potential for worker occupational exposure, especially if not handled or managed properly. Many facilities have a combination of sources that are used as part of an overall process but are only a small part of that process. It is sometimes difficult for a facilities or plant manager to have a handle on all the different regulatory and safety compliance considerations that must be followed to assure safe use of these sources and a safe work environment for all workers. All too often an accident or incident is the catalyst for the development of a solid radiation safety program to deal effectively with these sources. Many times facilities lack the technical resources in-house to deal with the potential problems.

Due to the wide variety and uses of these radiation sources, retaining the services of a health physics consultant certified by the American Board of Health Physics and with applicable experience in the type of radiation source being handled is a viable option for the facility.

## BIBLIOGRAPHY

21 CFR 1030.10 (2012). Food and Drug Administration (1 April).

ACGIH (1997). *Threshold Limit Values for Chemical Substances and Physical Agents and Biological Indices*, 123–125. Cincinnati, OH: American Conference of Governmental Industrial Hygienists.

ANSI Z136.1-1993 (1993). *American National Standards Institute for the Safe Use of Lasers*. Orlando, FL: Laser Institute of America.

Federal Communications Commission Rules (1996). Radiofrequency radiation; environmental effects evaluation guidelines. *Federal Register* **1** (153): 41006–41199. [47 CRF Part 1; Federal Communications Commission].

Food and Drug Administration, Department of Health & Human Services (1995). *Update on Cellular Phone Interference with Cardiac Pacemakers* (3 May 1995). Washington, DC: Food and Drug Administration, Department of Health & Human Services.

Goldhaber, M.K., Polen, M.G., and Hiatt, P.A. (1988). The risk of miscarriage and birth defects among women who use visual display terminals during pregnancy. *Am. J. Ind. Med.* **13**: 695–706.

Haes, D.L. Jr. and Fitzgerald, M.F. (1995). Video display terminal very low frequency measurements: the need for protocols in assessing VDT user "dose". *Health Phys.* **68** (4): 572–578.

Hansen, D.J., Baum, J.W., Weilandics, C. et al. (1990). Enhancement of low-level magnetic fields by ferro-magnetic objects. *Appl. Occup. Environ. Hyg.* **5** (4): 236–241.

ICNIRP (1996). Health issues related to the use of hand-held radiotelephones and base transmitters. *Health Phys.* **70** (4): 587–593.

IEEE/ANSI C95.1-1992 (1992). American National Standard). *Safety Levels with Respect to Human Exposure to Radio Frequency Electromagnetic Fields, from 3 kHz to 300 GHz*. New York, NY: The Institute of Electrical and Electronics Engineers, Inc.

IEEE/ANSI C95.2-1982 (1996). American National Standard). *RF Safety Warning Symbol* (Reaffirmed 1996). New York, NY: The Institute of Electrical and Electronics Engineers, Inc.

IEEE/ANSI C95.3-1992 (1992). American National Standard). *Recommended Practice for the Measurement of Potential Electromagnetic Fields – RF and Microwave*. New York, NY: The Institute of Electrical and Electronics Engineers, Inc.

International Commission on Non-Ionizing Radiation Protection (ICNIRP) (1994). Guidelines on limits of exposure to static magnetic fields. *Health Phys.* **66**: 100–106.

National Academy of Science (NAS)/National Research Council (1996). *Possible Health Effects of Exposure to Residential Electric and Magnetic Fields*. Washington, DC: National Academy Press.

National Council on Radiation Protection and Measurements (NCRP) (1986). *Biological Effects and Exposure Criteria for Radiofrequency Electromagnetic Fields*, NCRP Report 86. Bethesda, MD: NCRP.

National Council on Radiation Protection and Measurements (NCRP) (1993). *A Practical Guide to the Determination of Human Exposure to Radiofrequency Fields*, NCRP Report 119. Bethesda, MD: NCRP.

Polk, C. and Postow, E.P. (eds.) (1996). *Handbooks of Biological Effects of Electromagnetic Fields*. Boca Raton, FL: CRC Press.

Redelmeier, D.A. and Tibshirani, R.J. (1997). Association between cellular-telephone calls and motor vehicle collisions. *N. Engl. J. Med.* **336**: 453–458.

Schnorr, T.M., Grajewski, B.A., Hornung, R.W. et al. (1991). Video display terminals and the risk of spontaneous abortion. *N. Engl. J. Med.* **324**: 727–733.

Telecommunications Act of 1996 (1996). 47 U.S.C.; Second Session of the 104th Congress of the United States of America (3 January 1996).

Tenforde, T.S. (1985). Biological effects of stationary magnetic fields. In: *Biological Effects and Dosimetry of Static and ELF Electromagnetic Fields* (ed. M. Grandolfo, S. Michaelson, and A. Rindi), 93–128. New York: Plenum Press.

Wertheimer, N. and Leeper, E. (1979). Electrical wiring configurations and childhood cancer. *Am. J. Epidemiol.* **109**: 273–284.

Windham, G.C., Fenster, L., Swan, S., and Neutra, R.R. (1990). Use of video display terminals during pregnancy and the risk of spontaneous abortion, low birthweight, or intrauterine growth retardation. *Am. J. Ind. Med.* **18**: 675–688.

## SUGGESTED READING

The following are some suggested references that would be useful to a person responsible for ionizing radiation safety:

Cember, H. (1985). *Introduction to Health Physics*, 2e. Oxford: Pergamon Press.

National Research Council, Committee on the Biological Effects of Ionizing Radiations (1980). *The Effects on Populations of Exposure to Low Levels of Ionizing Radiation: 1980*, BEIR IV. Washington, DC: National Academy Press.

National Research Council, Committee on the Biological Effects of Ionizing Radiations (1990). *Health Effects of Exposure to Low Levels of Ionizing Radiation*, BEIR V. Washington, DC: National Academy Press.

Nuclear Lectern Associates (1984). *The Health Physics and Radiological Health Handbook*, Second Printing (June 1984). Olney, MD: Nuclear Lectern Associates.

Turner, J. (1986). *Atoms, Radiation, and Radiation Protection*. Oxford: Pergamon Press.

U.S. Codes of Federal Regulations. Title 10, Chapter 1, Nuclear Regulatory Commission Rules and Regulations. Government Printing Office.

U.S. Code of Federal Regulations. Title 40, Protection of the Environment.

U.S. Code of Federal Regulations. Title 49, Department of Transportation.

U.S. Department of Health, Education, and Welfare, Public Health Service (1970). *Radiological Health Handbook*, Revised Edition (January 1970). U.S. Department of Health, Education, and Welfare.

U.S. Nuclear Regulatory Commission (1977). Regulatory Guide 8.10, *Operating Philosophy for Maintaining Occupational Radiation Exposures As Low As Is Reasonably Achievable* (May 1977).

U.S. Nuclear Regulatory Commission (1987). Regulatory Guide 8.13, *Instruction Concerning Prenatal Radiation Exposure*, Revision 2 (December 1987).

U.S. Nuclear Regulatory Commission (1996). Regulatory Guide 8.29, *Instruction Concerning Risks from Occupational Radiation Exposure* (February 1996).

# APPENDIX A

## LIST OF AGREEMENT STATES

Currently, the following 30 states have been granted Agreement State status by the Nuclear Regulatory Commission:

| | | |
|---|---|---|
| Alabama | Kentucky | New York |
| Arizona | Louisiana | North Carolina |
| Arkansas | Maine | North Dakota |
| California | Maryland | Oregon |
| Colorado | Massachusetts | Rhode Island |
| Florida | Mississippi | South Carolina |
| Georgia | Nebraska | Tennessee |
| Illinois | Nevada | Texas |
| Iowa | New Hampshire | Utah |
| Kansas | New Mexico | Washington |

Information concerning a state's status with regard to the use of radioactive material can be obtained by calling the State Radiation Control Office.

CHAPTER 12

# ENTERPRISE RISK MANAGEMENT: AN INTEGRATED APPROACH

CHRIS LASZCZ-DAVIS

*The Environmental Quality Organization, LLC, Affiliated, Aluminium Consulting Engineers, 74 Overhill Rd., Orinda, CA, 94563*

## 12.1 INTRODUCTION

### 12.1.1 Scope of Chapter and Benefits of an Enterprise Risk Management Approach

In today's world, change and uncertainty are constants. With increased demand for greater transparency in decision-making, better educated and discerning citizens, globalization and technological advances, adapting to change and uncertainty (while striving for operating efficiency) is basic to sound decisions and the possibility of a viable future.

The primary benefit to having an overarching enterprise risk management (ERM) process is to evaluate and manage the potential risks of situations or systems with significant potential impact that require attention beyond the normal management processes of daily operations. Having this knowledge adds perspective and context, helps clarify what is important, facilitates decision-making, and helps the organization make an informed decision to reduce risk.

In its simplest form, ERM is a structured, consistent, and continuous risk management process that is applied across an entire organization that brings value by

- Proactively identifying, assessing and prioritizing material risks.
- Developing and deploying effective mitigation strategies.
- Aligning with strategic objectives and business processes.
- Embedding supporting key elements into an organization's culture for sustainability (risk ownership, governance, and oversight, reporting and communications, and leveraging technology and tools).

This chapter provides a basic introduction to an integrated ERM approach that actively identifies, analyzes, and manages risk within an organization, and the role occupational health and safety (H&S) plays within the organization's risk management efforts. Occupational H&S risks are but one set of potential risks, among others, that must be considered in an organization when discussing broader ERM. H&S risk assessment and management is not a technical stand-alone endeavor but, instead, is one that needs to be integrated into a company's overarching business to ensure that the appropriate strategic, operational, and financial decisions are made.

Discussion in this chapter will include several items: an overview of the broader enterprise elements that need to be considered when anticipating risks; the more specific traditional H&S qualitative risk assessment techniques available to the H&S professional today; and some of the quantitative risk analysis techniques that are used less frequently by the traditional H&S professional, but often in more complex situations to assess severity potentials. Severity potentials (consequence analysis) include chemical exposures, fire, thermal and radiation effects, explosion effects, chronic chemical exposure effects, and contaminant dispersion possibilities, among others.

Qualitative techniques include, among others, the safety checklist, job safety analysis (JSA), "what-if" analysis (WIA), hazard and operability (HAZOP) analysis, failure mode and effects analysis (FMEA), fault tree analysis (FTA),

*Handbook of Occupational Safety and Health*, Third Edition. Edited by S. Z. Mansdorf.

layers of protection analysis and Bowtie methodologies. Quantitative methods used in consequence analysis (severity potentials) are beyond the scope of this chapter and will not be discussed in any detail. However, the one exception in this chapter applies to chemical exposure potentials, which will be dealt with in some detail.

Environmental risks will also not be addressed within the scope of this chapter, although they must be considered by the organization in its final risk deliberations. While some occupational health-related issues will be addressed, it is suggested that one refer to the chapter on human health risk assessment in *Patty's Industrial Hygiene*, 6e, Volume 2, by Wiley, 2011, for a more elaborative discussion.

### 12.1.2 Overarching ERM Framework

The overarching ERM framework is best pictorialized in Figure 12.1.

Risk management includes several basic steps: identifying and assessing, planning and managing, and measuring, monitoring, and reporting on effectiveness. Supporting components that influence the strength of these risk management steps include governance and compliance, policies and procedures, technology and systems, and a risk-based culture.

Traditional risk silos include finance, human resources, operations, legal and information technology. H&S is often addressed within an operations, legal, or compliance silo. However, silos generate white space, which leads to miscommunication, duplicated resources, strategic misalignment, and operational misalignment, thus the need for an overarching ERM process that reduces/eliminates these white spaces.

The pictorial of traditional risk silos is provided (Figure 12.2).

ERM on the other hand spans the organization (Figure 12.3), tying it together and enabling a common leadership view of the organization's strategic and operational risks. H&S needs to be part of this broader dialogue, which spans the organization so its intersections with other functions and initiatives are considered.

An organization must use sound judgement in its identification, evaluation, assessment, and management of risk (compliance and financial reporting, operational, financial reporting, market/business environment, and strategic business). It takes all facets of an organization performing their respective risk-related functions so the overarching enterprise risk can be assessed and properly managed. While H&S staff are often best equipped to address the assessment of compliance and operational H&S risk, executive leadership must ultimately assume responsibility for all facets of the risk pyramid (ERM) and its implementation.

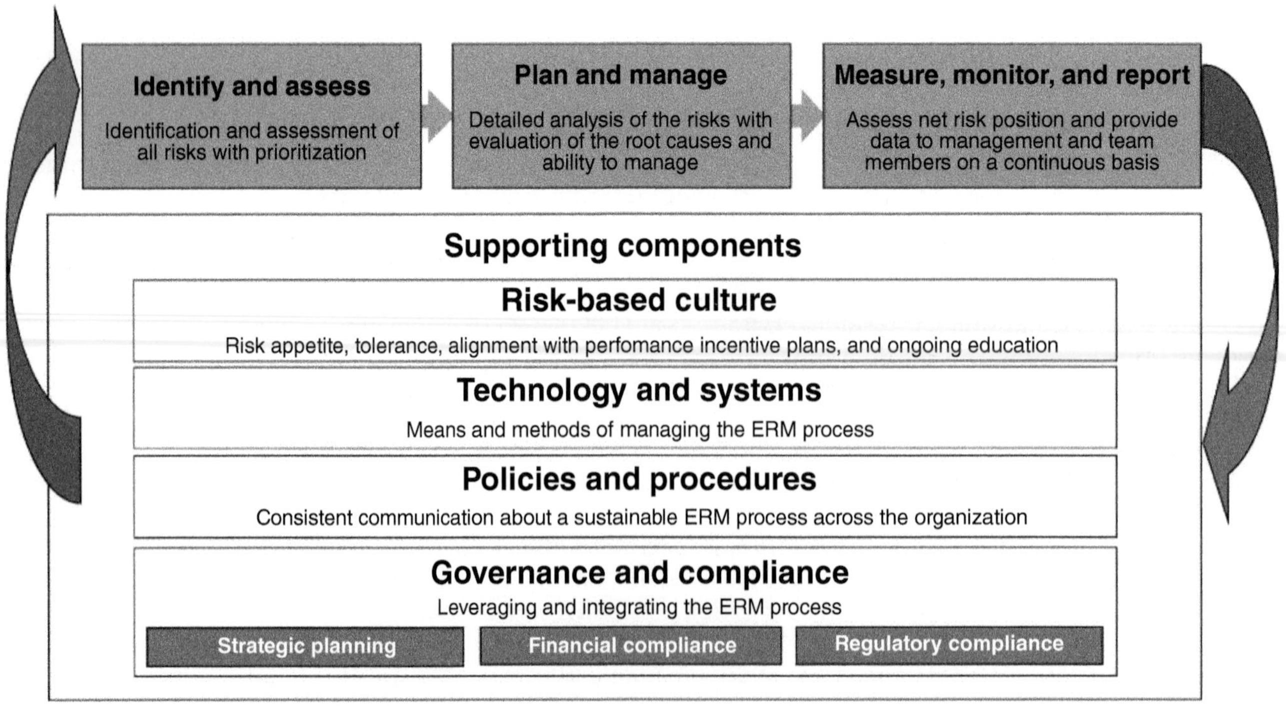




**Figure 12.1** ERM framework. *Source:* Courtesy of Lisa Kremer, Senior Vice President, Practice Leader, Marsh/Strategic Risk Consulting, San Francisco, CA.

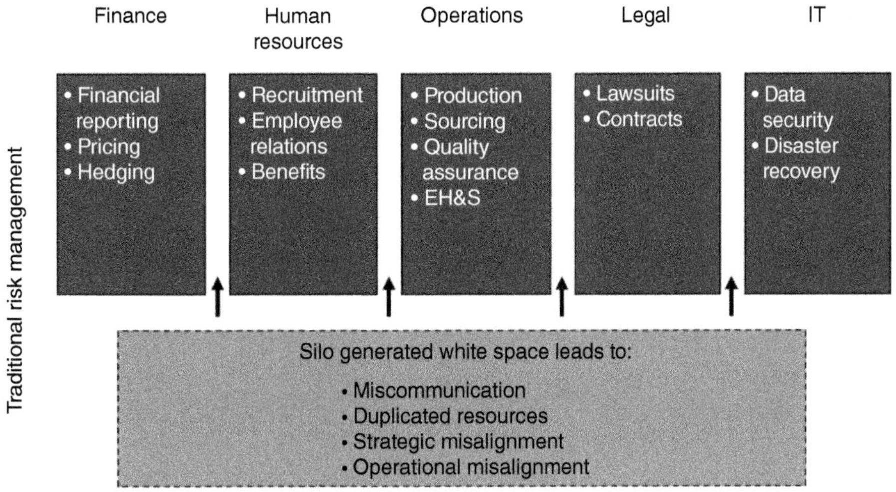




**Figure 12.2** Traditional risk silos. *Source:* Courtesy of Lisa Kremer, Senior Vice President, Practice Leader, Marsh/Strategic Risk Consulting, San Francisco, CA.

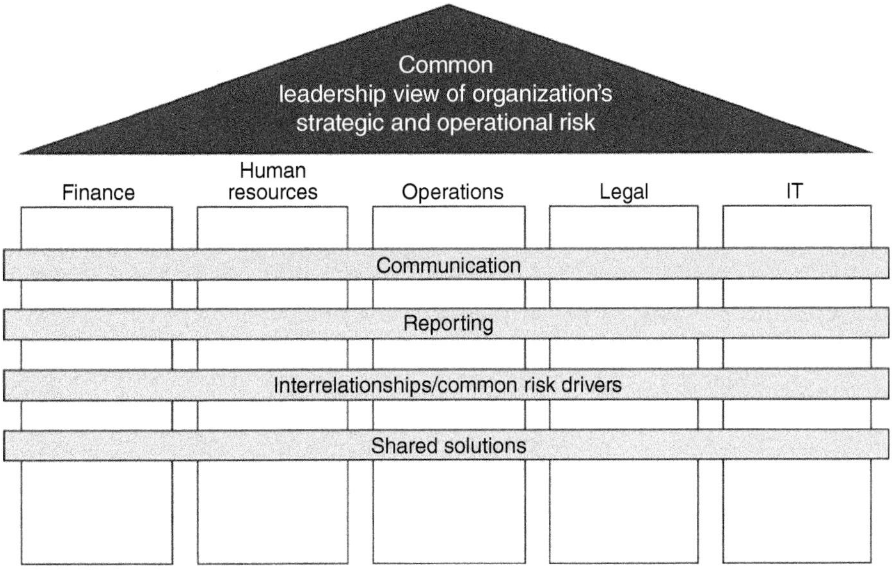




**Figure 12.3** ERM spans across the organization. *Source:* Courtesy of Lisa Kremer, Senior Vice President, Practice Leader, Marsh/Strategic Risk Consulting, San Francisco, CA.

## 12.2 ASSESSMENT AND MANAGEMENT OF H&S COMPLIANCE AND OPERATIONAL RISK

With regard to H&S risk assessment specifically, once identified and evaluated, the risks can then be prioritized in an informed way and resources applied most productively so as to provide the greatest added value. Virtually every organization that has excellent H&S performance conducts risk assessments. These organizations see reduced injury and illness rates, improved operating profits, lowered property loss insurance rates, and greater productivity improvements.

The risk assessment techniques described in this chapter are valuable tools that can be used to help build this performance. They can bridge the gaps left when traditional H&S approaches are not sufficient for understanding and controlling the hazards associated with evolving, innovative, and complex equipment and facilities.

Furthermore, a number of H&S regulations require the application of risk assessment techniques. Several examples follow:

- OSHA's Process Safety Management of Highly Hazardous Chemicals standard (29 CFR 1910.119 1992) requires certain facilities handling highly hazardous chemicals to apply formal risk assessment techniques.
- OSHA's personal protective equipment (PPE) revision (29 CFR 1910, Subpart I 1994) requires that employers to conduct and document a hazard assessment and equipment selection program.
- And finally, EPA's risk management programs (40 CFR 68 1996) requires that certain facilities handling chemicals to conduct worst-case hazard assessments.

### 12.2.1 Basic Definitions

**Risk assessment** is the process of making a determination of how safe or healthy a situation is. William Lowrance (1976) defines risk as "a measure of the probability and severity of harm to human health." So the work of risk assessment involves determining (i) what the potential risks are, (ii) how often a certain unwanted event could occur (the probability or statistical chance), (iii) how serious the consequences of that unwanted event are (the severity, impact), (iv) the acceptability of that risk, (v) what practices are in place to mitigate the risks, and (vi) where we can improve. Given the convergence of the physical and social sciences, it is often as much art as it is science.

At its simplest, **hazard** may be defined as threat of harm to a resource of value. It is important to note that, when discussing health risk assessment, **hazard** and **toxicity** are not the same. **Toxicity** is an inherent property of a chemical, whereas the **hazard** presented by a chemical includes not just its toxicity but also the ease with which humans or animals can come into contact with the chemical. The **toxicity** of a chemical can be defined as the inherent capacity of a substance to produce injury or harm. For example, a capsule of botulinum toxin, an extremely toxic chemical, can be stored away in a vault and present little **hazard**. On the other hand, less toxic slug pellets, containing mostly bait and a small percentage of metaldehyde and scattered around the garden, can present a big hazard to pets and children simply because of its accessibility. Another simple illustration relates to iron pills. Iron pills are a common source of poisoning in children. In a house with small children, 100 iron pills kept in child-proof bottles on a high shelf pose much less concern than 100 iron pills left in an open dish on the counter. The **toxicity (inherent property)** is the same, but the **hazard (inherent property plus ease of contact)** is different.

**Risk**, in this context, refers to the probability (statistical chance) that harm, injury, or loss will occur. To expand on the previous example, in households where iron pills are left out on the counter and where small children can access them, the **risk** or **probability** for poisoning is higher than in a household where pills are stored properly and out of harm's way.

**Epidemiology** is the study of patterns of disease in human populations and the investigation of the factors that affect these patterns. The discovery of an association between smoking and lung cancer is an example of a pattern discovered by epidemiology. Epidemiology is used to define the probability that certain effects are the result of specific causes. It does not prove cause and effect; rather, it can only show the statistical strength of associations between potential causes and specific effects. These associations are important in the hazard identification step of risk assessment because they can suggest to scientists that exposure to a particular substance or occupation might be hazardous to a given population. While epidemiology (in some form) has been around for at least 150 years, it gained in sophistication and complexity through the 1970s and 1980s.

**Risk management** is the process of evaluating various options to mitigate risk and aligning the organization in implementation. Risk managers must find the answers to questions like "How safe is safe?" and "How much risk is acceptable?" To answer these questions, risk managers consider political, social, economic, and engineering information, as well as scientific risk assessment information.

The notion of determining the likelihood and the severity of unwanted events or situations can be visually illustrated. The qualitative risk matrix is one example of how this risk information can be displayed and communicated (Figure 12.4).

### 12.2.2 Health and Safety Risk Assessment Methodologies

Eight risk assessment techniques (see Table 12.1) are described in this chapter and include (i) safety checklists, (ii) JSA, (iii) WIA, (iv) HAZOP analysis, (v) FMEA, (vi) FTA, (vii) layers of protection analysis (LOPA), and (viii) Bowtie methodologies. Several factors influence the decision as to which methodology to apply. These factors include the complexity of the situation, the experience of the decision-makers, the perception of the consequences, the regulatory requirements, and the capabilities of the risk assessors.

| | <--------- Severity of injury or illness consequence --------> | | | |
|---|---|---|---|---|
| Likelihood of OCCURRENCE or EXPOSURE for selected unit of time or activity | Negligible | Marginal | Critical | Catastrophic |
| Frequent | Medium | Serious | High | High |
| Probable | Medium | Serious | High | High |
| Occasional | Low | Medium | Serious | High |
| Remote | Low | Medium | Medium | Serious |
| Improbable | Low | Low | Low | Medium |

**Likelihood:**

*Frequent:* Likely to occur repeatedly
*Probable:* Likely to occur several times
*Occasional:* Likely to occur sometime
*Remote:* Not likely to occur
*Improbable:* Very unlikely – may assume exposure will not happen

**Severity/consequence:**

*Negligible:* First aid or minor medical treatment
*Marginal:* Minor injury, lost workday accident
*Critical:* disability in excess of three months
*Catastrophic:* Death or permanent total disability

**Risk level:**

*Low:* Risk acceptable, remedial action discretionary
*Medium:* Take remedial action at appropriate time
*Serious:* High priority remedial action
*High:* Operation not permissible

**Figure 12.4** Example of a risk assessment matrix. *Source:* ANSI/AIHA Z10-2005 (n.d.).

Embodied in all the eight assessment methodologies is the basic concept of determining "how likely" and "how serious." There are a wide variety of approaches to analyze risk situations systematically and formally. Understanding this simple risk matrix relationship provides the basis not only for determining the risk but also for controlling and reducing the risks, that is, how to reduce the likelihood and the severity of potential accidents.

For example, to reduce the risk of exposure to chemicals, noise, or radiation, the use of PPE will reduce the severity of the *consequences* of the incident. Using safety goggles could prevent a severe eye injury as a result of a chemical splash, but would not reduce the possibility of the splash happening in the first place. Likewise, the use of safety belts in an automobile will reduce the injury severity to occupants of a car during a crash. The safety belts, however, would not lessen the possibility or frequency of the accident.

Similarly, driving defensively and being alert could reduce the *likelihood* of an accident but would not reduce the severity of the injury. The use of machinery lockout/tagout procedures would reduce the chances of a machine inadvertently starting up with someone in the danger zone. Those practices, however, would not reduce the severity of the injury if the machine did start up. Some actions will reduce both the likelihood and the severity. By driving slower, the chances of an accident are reduced and the severity of the accident is lessened.

Once the severity and likelihood are determined, the next step in assessing risks is to make judgments regarding the "acceptability." The greater the likelihood and the more severe the accident can be, the higher the risk. For example, if it is determined that having a severe fire while using solvents is quite possible and could result in several fatalities, then that risk would be very high and unacceptable. On the other hand, if the assessment resulted in the determination that the likelihood of the fire is remote because of the existing safety precautions and would only result in moderate property loss because of the sprinkler system, then it may be judged as an acceptable risk.

The risk matrix can also be used in a quantitative manner. For example, in any one year, the chances of a situation occurring as extremely likely could be defined as one in ten, quite possible as one in a hundred, unusual as one in a thousand, unlikely as one in ten thousand, remote as one in a hundred thousand, and extremely remote as one in a million. Similarly, catastrophic consequences could be defined as a fatality or multiple fatalities, serious as permanent disabilities, and minor as temporary disabilities or less. A quantitative analysis could then be displayed numerically on the matrix.

**TABLE 12.1 Eight Risk Assessment Techniques**

| Name | Purpose | When to use | Procedure | Type of results | Nature of results | Data requirements | Limitations, comments |
|---|---|---|---|---|---|---|---|
| H&S checklist | Identification of H&S issues and concerns that need to be addressed | Early in conceptual or preliminary design phase | Check off applicable H&S items, predesigned list | Check list of items or concerns | Qualitative only | Gross knowledge of system and applicable H&S standards | Success limited to the experience of the users and breadth of list |
| Job safety analysis | Provides safety requirements for simple job tasks | For existing job procedures with annual update | Step-by-step review of job tasks | List of specific requirements to do tasks safely | Qualitative only | Written job instructions are helpful | Only good for well-defined, noncomplex job tasks |
| What-if analysis | Identification of likely things that could go wrong and possible controls | Popular approach that can be used in most situations, as system changes | Asking "What-if" questions at each step of the process | List of potential problems and recommended controls | Qualitative only | Operating instructions, flow diagrams | Depends on team members' experience with similar situations |
| Hazard and operability studies | Identification of problems that could compromise a system's ability to achieve intended productivity | Late design phase when design is nearly firm; also for an existing system when a major redesign is planned | Examine diagrams, flowchart at each critical node, identify operational deviations, causes and consequences | List of hazards and operating problems, deviations from intended functions, consequences, causes, and suggestions | Qualitative with quantitative potential | Detailed system descriptions, flowcharts, procedures, knowledge of instruments and operation | Depends heavily for its success on data completeness and accuracy of drawings |
| Failure mode and effect analysis | Identification of all the ways a piece of equipment can fail, and each failure mode's effect(s) on the system | At design, construction, or operation and reviewed every 3 to 5 years | Collect up-to-date design data on equipment and relationship to the rest of the system; list all conceivable malfunctions; describe effects | List of identified failure modes, potential effects, and needed controls | Qualitative, although can be quantified if failure probabilities for components are known | System equipment list; knowledge of equipment function; knowledge of system function | Poor at showing interactive sets of equipment failures that lead to events, not useful for errors or common-cause failures |
| Fault tree analysis | Deduction of causes of unwanted event via knowing of combinations of malfunctions | At design, operation and updated as significant changes are made in the process | Construct a diagram with logic symbols to show the logical relationships between situations | List of sets of equipment and human errors that can result in specific unwanted events | Qualitative with quantitative potential with probabilistic data on components and subsystems | Complete understanding of the system's functions | Enables ID and quantitative examination of critical factors and interrupt modes for chains of failures |

| | | | | | | | |
|---|---|---|---|---|---|---|---|
| Layers of prevention analysis | Semiquantitative risk assessment technique that uses order of magnitude categories for initiating event frequency, consequence severity, and the likelihood of failure of independent protection layers (IPLs) to approximate risk of an incident scenario | Evaluates effectiveness and independence of safety measures, especially protective systems | Team identifies independent protection layers and assigns risk reduction credits to each layer, depending on different criteria | Provides higher level of review than HAZOP for potential scenarios that can result in single or multiple fatalities to ensure adequate protection with sufficient availability in place to reduce risk | Semiquantitative | Often used in conjunction with HAZOP. Requires up to date P7ID, detailed project design criteria, equipment specifications, material specifications, and other similar engineering design information | If conducted with incomplete information or on design that is not fixed, study may take longer and result in greater number of recommendations. Results may not be accurate if inexperienced reviewers or less than optimal understanding of plant processes |
| Bowtie methods | Combination of two other techniques – fault tree analysis and event tree analysis. Bowtie diagram can be used to list preventative, controlling, and mitigating barriers that may impact incident and consequence | Particularly useful in communicating hazards and how they are managed. Typically conducted on highest ranked risks from a HAZOP or risk register | List fault tree analysis on left side, hazard in middle, and event tree on right side. The Bowtie methodology, composed of trees, offers a means to connect analysis with hazard, event, and consequences. Conducted by multi-disciplinary team | See column to the left | Structured approach using two easily understood techniques where a qualitative approach undesirable | Understanding of risks, hazards, and barriers | Analysis success depends on team experience and facilitator. Inexperienced analysts may struggle to develop logic and use data not statistically significant |

The question arises as to who is deciding what level of risk is acceptable. This can be a complicated business. The perspective of each of the involved stakeholders is quite different. Stakeholders can include the plant workers, the plant management, the corporate officers, the stockholders, the regulatory agencies, and the local neighboring community. The acceptance by a worker personally exposed to the risk of a plant fire is probably different from that of a corporate officer to that of a regulatory agency to that of the surrounding community. The nature of the risk also directly affects the perception of its acceptability. A risk assumed voluntarily such as rock climbing is perceived differently than one borne involuntarily, such as having a nuclear power plant built next to one's house. Additionally, a common hazard such as driving a car is viewed differently than a "dread" hazard such as cancer. The perceived or actual benefits, such as worker pay, taxes to the community, and social responsibility, also play an important part in risk acceptance decisions.

Kaplan and Garrick (1981) put it well: "...risk cannot be spoken of as acceptable or not in isolation, but only in combination with the cost and benefits that are attendant to that risk. Considered in isolation, no risk is acceptable! A rational person would not accept any risk at all except possibly in return for the benefits that come along with it."

The value and beauty of the risk matrix concept is to allow the various stakeholders to come to some agreement on what the risk actually is and subsequently have the argument as to the acceptability issue alone. Based on an organization's or individual's values and beliefs, there can be true differences as to what is acceptable. Use of the risk matrix, however, can at least facilitate a discussion between interested stakeholders on "how likely" and "how serious."

### 12.2.3 When Is a Health and Safety Risk Assessment Performed?

The nature and complexity of the situation will determine when to apply formal risk assessment techniques. In many cases, conforming to well-known and well-developed work practices will suffice. For example, a confined space permit guides the user through a series of steps designed to assess the likelihood of an incident occurring such as lack of adequate oxygen and the potential severity of the situation such as a fatality. Use of the permit additionally provides guidance and direction for actions necessary to control those risks. Flame permits and lockout/tagout procedures are other examples of work practice approaches designed to address specific reoccurring occupational risks. These are examples of risk assessments that have already been formalized into specific safe work practices.

Many H&S standards have evolved to the point where they are accepted as part of design standards, industry best practices, or consensus standards and indeed regulatory requirements. These standards and regulations need to be applied as an ongoing integral part of any risk management program. The formal risk assessment methodologies described in this chapter are designed to complement and enhance this ongoing effort. They are not designed to supplant or replace these well-developed H&S practices.

On the other hand, complex operations involving many hazards, having the potential for many fatalities, or perhaps impact on the community, may require the application of a very sophisticated quantitative risk assessment technique. For example, a large chemical processing facility may need to evaluate the risk of fire, explosion, or toxic exposures to their workers and the surrounding community neighbors. Also, evolving regulations for certain operations legally require the application of formal risk assessment techniques.

H&S risk assessments can and should be conducted:

- From the conceptual stage of a program or project proposal.
- Throughout the design stage, from preliminary to final design.
- To the construction stage.
- During the startup and debug stage.
- Throughout the life of the operations.
- During maintenance activities.
- Through the shutdown, dismantling, and disposal operations.

The National Safety Council (1993) suggests the following benefits of using formal risk assessments:

- It can uncover hazards that have been overlooked in the original design, mock-up, or setup of a particular process, operation, or task.
- It can locate hazards that developed after a particular process, operation, or task was instituted.
- It can determine the essential factors in and requirements for specific job processes, operations, and tasks.
- It can indicate what qualifications are prerequisites to safe and productive work performance.
- It can indicate the need for modifying processes, operations, and tasks.

- It can identify situational hazards in facilities, equipment, tools, materials, and operational events (for example, unsafe conditions).
- It can identify human factors responsible for accident situations (for example, deviations from standard procedures).
- It can identify exposure factors that contribute to injury and illness (such as contact with hazardous substances, materials, or physical agents).
- It can identify physical factors that contribute to accident situations (noise, vibration, insufficient illumination).
- It can determine appropriate monitoring methods and maintenance standards needed for safety.

### 12.2.4 Risk Management: Hierarchy of Controls

Controlling exposures to occupational hazards is the fundamental method of protecting workers. Traditionally, a hierarchy of controls has been used as a means of determining how to implement feasible and effective control solutions.

One representation of this hierarchy is shown in Figure 12.5.

The idea behind this hierarchy is that the control methods at the top of the pyramid are potentially more effective and protective than those at the bottom. Following this hierarchy normally leads to the implementation of inherently safer systems, where the risk of illness or injury has been substantially reduced.

NIOSH leads an initiative called "Prevention through Design (PtD)" (http://www.cdc.gov/niosh/topics/ptd) to prevent or reduce occupational injuries, illnesses, and fatalities through the inclusion of preventive considerations in all designs that impact workers. The hierarchy of controls is a PtD strategy.

Typical hazard control measures (risk management) include the development of operating procedures, the use of PPE, the substitution of chemicals, and the installation of engineering controls such as ventilation, guarding, and safety interlocks. Procedures can include the use of safety permits (confined spaces, hazardous work permits, and chemical line openings). Other procedures can include periodic preventative maintenance requirements on safety interlocks, sprinkler systems, or hoisting equipment. PPE requirements might include safety shoes during the handling of heavy equipment. Similarly, improvements to the facility or equipment could be needed.

### 12.2.5 Why the Increased Global Interest in ERM?

According to Lisa Kremer, Senior Vice President, Strategic Risk Consulting Practice Leader, Marsh LLC, San Francisco, the convergence of multiple forces (external developments, internal demand, and methodological advances) make ERM implementation both timely and practicable (see Figure 12.6).

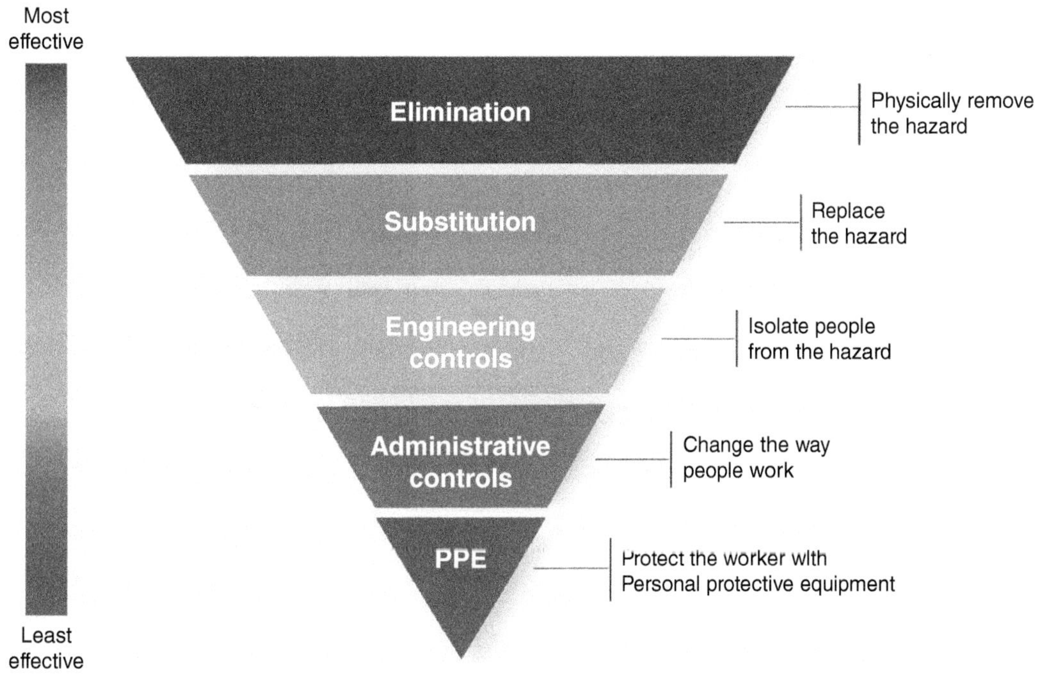


**Figure 12.5** Hierarchy of controls. *Source:* National Institute of Occupational Safety and Health (n.d.).

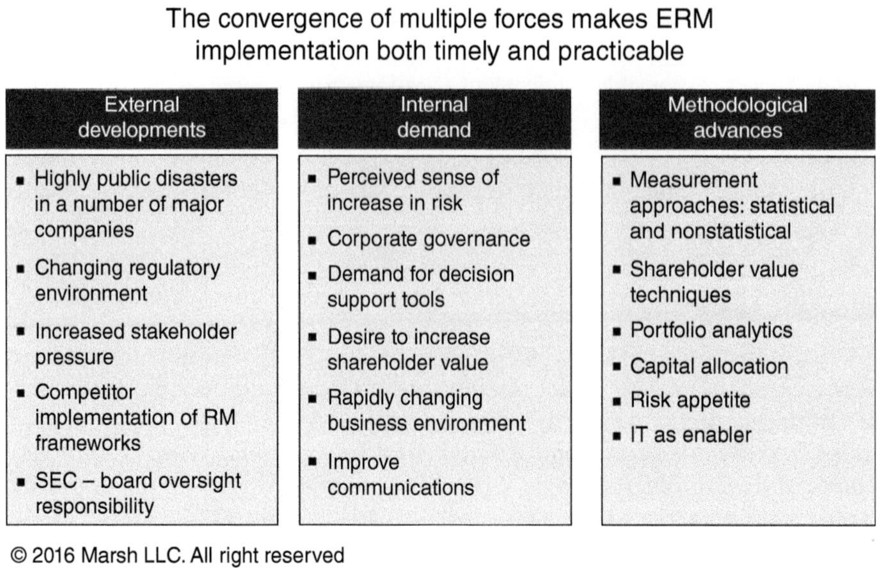


**Figure 12.6** Increased global interest in ERM. *Source:* Courtesy of Lisa Kremer, Senior Vice President, Practice Leader, Marsh/Strategic Risk Consulting, San Francisco, CA.

## 12.3 RISK-RELATED CHECKLISTS

### 12.3.1 Overview of Checklists

Checklists are generally a list of items or questions related to the situation. The main purpose is to ensure that key H&S aspects of that situation are identified so that further discussion and analysis will take place. It provides assurance that key H&S elements have not been overlooked or forgotten. Each area of concern can be checked physically, can be reviewed for compliance with regulatory requirements, can be analyzed to determine if it meets best industry practice, or can be set aside for a more rigorous hazard review. Checklists represent the most basic and simplest method for identifying hazards that *need* to be controlled.

### 12.3.2 Developing Checklists

Checklists are developed by individuals and are based upon their past experience and knowledge of engineering and design codes as well as regulatory requirements and company requirements. Traditional checklists vary widely in level of detail and purpose. Some organizations have well-developed and well-conceived checklists that must be completed before a conceptual project moves forward to initial design. Others have checklists used by developers of operating procedures to ensure that key elements of safely operating a piece of equipment are not neglected. Some organizations require a formalized sign-off on H&S checklists before new capital costs are authorized to ensure that safety costs have been considered in the request for project funding.

The level of detail within the checklist needs to be related to the potential risks associated with the situation being reviewed as well as overall complexity. The effective development and use of a checklist is directly related to the experience and skills of the preparer and user. Completing the checklist assumes the user has the knowledge of the underlying questions and answers implied in the checklist structure. Experienced individuals can play an important role in the development of checklists focused on the organization's specific needs and guide the inexperienced users on how they can be effectively applied.

### 12.3.3 When to Use H&S Checklists

A simple overview checklist can be an excellent H&S reminder at the conceptual stage of developing a new facility, designing a new product, or modifying an existing facility or piece of equipment. Before committing large sums of money to establishing a new manufacturing site or designing a new piece of equipment, a properly formatted checklist can identify major H&S-related items that need to be considered. This could include things as simple as adequate water supply for fire protection to items as complex as dealing with community acceptance of a new neighborhood risk.

While developing proposals to request funds, a H&S checklist can ensure that major H&S elements have been identified and have been included in the estimated cost to build or modify a new product, facility, or piece of equipment. For example, overlooking the cost of conducting product safety testing could result in a substantial financial overrun.

Applying a detailed H&S checklist at the design stage ensures that H&S standards, best industry practices, and regulatory requirements are identified. Even the best of designers cannot keep track of all the requirements. This is an opportunity to make sure that major design mistakes are minimized and that the cost of reengineering safety into the equipment after the fact is reduced.

During the construction phase, a H&S checklist ensures that hazards associated with a specific construction project have been identified. A well-developed checklist at this point can reduce the chance of construction errors and resultant injuries. Contractors can also be instructed on the H&S rules and expectations during this phase of the project.

Finally, an operational H&S checklist can ensure that basic H&S concerns have been identified prior to commissioning the new product or facility or equipment. The list would establish the need for items such as maintenance procedures, operational procedures, JSA, routine inspections, and overarching audits.

### 12.3.4 Pros and Cons of H&S Checklists

Safety checklists are the most basic and simplest of hazard identification techniques. They provide the impetus for further and more detailed H&S analyses. They minimize the chances of major H&S concerns being forgotten or neglected. The results of using a H&S checklist are only as good as the experience and skills of the user.

Limitations also exist. One must be aware of the limitations in using simple H&S checklists. The use of the list can become so narrow that other associated hazards are ignored or neglected. Therefore, the preparer of the checklist is challenged with making the list inclusive but not overwhelming. Additionally, the checklist could result in generalized action items rather than specific things to do. In any case, the H&S checklist is a simple, versatile, and highly resource-effective method to identify normally encountered hazards.

### 12.3.5 Examples of H&S Checklists

Checklists need to be developed for the specific situation under review. The examples that follow are intended to be illustrative in nature. Organizational experience and skills will dictate the specific character and shape of the H&S checklists that evolve for their specific needs.

Figure 12.7 illustrates a checklist that can be used to identify the hazards associated with a new piece of mechanical equipment undergoing a conceptual design review. For example, an engineer developing a conceptual design for a new assembly machine would ensure that all of the potential hazards have been identified through the use of the checklist. Then as the detailed design proceeds, appropriate accommodations for guards, interlocks, and the associated costs would have been made.

Likewise, Figure 12.8 illustrates a checklist that could be used to identify potential construction hazards and the actions needed to control those hazards. For example, an organization could use this checklist with an outside contractor to review the hazards associated with a construction project. The checklist would provide a way for documenting the agreements on how those hazards would be controlled.

Some checklists are designed to lead a H&S review of specific pieces of equipment. Figure 12.9 illustrates a checklist for pumps.

Finally, most equipment vendors have developed checklists of one form or another that provide H&S guidelines on operations, maintenance, or debugging for their specific equipment.

## 12.4 JOB SAFETY ANALYSIS (JSA)

### 12.4.1 JSA Overview

A JSA is a simple four-step hazard analysis technique used to identify the hazards associated with individual job tasks and to develop the best controls to minimize those risks. *This approach is generally used for simple, well-defined job tasks that contain injury risks.* An individual experienced in safely doing the task along with a supervisor, a team coach, or a safety representative can best conduct the analysis. A JSA is generally not appropriate for conducting design reviews or understanding the hazards of a complex process. In fact, a JSA may be the result of a recommendation from a more detailed process hazard review. As such, it is ideal for analyzing *job tasks* such as clearing jams in machinery, operating a lathe in the shop, performing appropriate lockout/tagouts, what to do in the event of specific shearing operations, what PPE to wear, moving a drum in the factory, or chemical handling, among other items.

| Mechanical Equipment Hazards Checklist | | | |
|---|---|---|---|
| Project/equipment description: | | | |
| | | | |
| | | | |
| Potential hazards | Applies | Does not apply | Actions required |
| • Rotating/moving parts | | | |
| • Nip/pinch points | | | |
| • High loads | | | |
| • Hi/low pressures | | | |
| • Hi/low temperatures | | | |
| • Noise exposures | | | |
| • Chemical exposures | | | |
| • Electrical exposures | | | |
| • Radiation exposures | | | |
| • Laser exposures | | | |
| • Fire exposures | | | |
| • Ergonomic exposures | | | |
| • Operating procedures | | | |
| Testing/debugging procedures | | | |
| • Maintenance procedures | | | |
| Applicable standards ANSI, NFPA, ASME | | | |
| • Other/miscellaneous | | | |
| Reviewed by: | | | Date: |
| | | | |
| CC: Operations, Maintenance, Safety, Equipment File | | | |

**Figure 12.7** Mechanical safety checklist example.

| Project title I scope | | Actions required |
| --- | --- | --- |
| Area/facility: | | |
| A. Contractor safety management:<br>• Competent person/construction supervisor – registered, regulatory requirement, identification, availability | | |
| • Employee training/orientation – written program, training frequency, company rules | | |
| • Subcontractor expectation/orientation – program, implementation, oversite role | | |
| • Work placement – frequency, documentation, corrective actions | | |
| • Safety rules – availability, distribution, posting of "construction area," signs with key information, training | | |
| B. General safety and health:<br>• Job–local rules, barricading and warning signs, roads, material staging, vehicle parking, trailers, utility lines, potential exposures and health hazards, radios/walkman prohibitions | | |
| • *Housekeeping* – cleanup and disposal of debris, workplace conditions, egress routes<br>• Emergencies – phone if first aid/medical, eye wash and safety showers, and injury reporting; fire alarms, evacuation, and reporting; spill response and reporting. | | |
| • Personal protective equipment – safety glasses and goggles, hard hats, safety shoes, gloves, respirators, life lines, ear protection, special clothing. | | |
| *Tools* – condition, inspection, double insulated if powered, use, guarding, training<br>*Loaning of company tools and equipment* – only on rare occasions and with a written permit<br>• Environmental – PCB, asbestos, lead paint, refrigerants, ballasts, hydraulics, cleaners, spray cans, oily rags, spills | | |
| C. Fire safety:<br>• Welding and cutting – open flames and company fire permit requirements as well as local fire department permits | | |
| • Flammable liquids – handling and storage of solvents, gasoline, other fuels<br>• Smoking – not allowed inside or on company buildings<br>• Storage of combustibles, permit for shutdown of sprinklers and fire alarm systems<br>• Emergency response – alarm locations, use of fire extinguishers, fire hose, exits, emergency marshals | | |
| | | |

**Figure 12.8** Construction safety checklist example.

| | | |
|---|---|---|
| D. Electrical safety:<br>• Lockout – de-energizing, locking and tagging out equipment prior to servicing, no hot work | | |
| • Ground fault – GFCI protection in and outside construction and potentially wet areas, custodial work | | |
| *Tools* – three wired, grounded, non-defective and double insulated powered portable hand tools<br>*Temporary wiring* – permit required for temporary power and lighting in the company - occupied buildings<br>• Licenses – electricians need for state license | | |
| E. Chemical safety:<br>• Toxic/hazardous materials – hazard communication, material safety data sheets (MSDS), labeling, use | | |
| • Poisons, explosives, pesticides – state license and local safety office approval, registration | | |
| • Process piping – line breaking permit required to work on high temperature/pressure and process piping | | |
| Confined spaces: permit required for entry into vessels, boilers, manholes and other confined spaces | | |
| G. Ladders/ staging/scaffolding:<br>• Ladders – conditions, storage, nonconducting, securing | | |
| *Staging/scaffolding* – guardrails and toe boards, scaffold-grade planking, loading, stability, competent person<br>• Lilts – powered vs. manual, stability, condition, travel | | |
| H. Excavation and trenching: dig safe, egress, slopes, soil class, barricading and shoring, competent person | | |
| 1. Floor/wall openings and roof work:<br>• Roof work – life lines, safety nets, motion stopping safety system, competent person monitoring system, melt pots, training, others | | |
| I. Floor/wall denim – size, guarding, toe boards, coverings, location | | |
| J. Rigging/hoisting/cranes:<br>• Cranes – licensed operator, inspection certificate, barricading, boom travel, load limit, overhead lines, training, load swing | | |
| • Rigging – plan, load limits, equipment condition | | |
| K.City/state permits: building, wiring, plumbing, etc. | | |
| L.Other special situations:<br>• Powder-actuated tools – prohibited in occupied company buildings | | |
| • Blasting/explosives/demolition – local safety approval<br>• *Gas cylinders* – storage, handling, use, disposal<br>• Radiation/lasers – power, precautions, training | | |
| Reviewed by/with: Date: | | |
| CC: Project File, Contractor, Construction Engineer, Safety Department | | |

**Figure 12.8** (*Continued*)

The results of a JSA should be incorporated into the formal, written procedures for that task. For example, the analysis may result in the recognition of the need to wear safety glasses with side shields while operating a certain machine. Therefore, in addition to the other requirements for quality control and production specifications, the PPE requirements for safety glasses would be included as part of the operating procedures. Similarly, defining the needed safety controls, such as lockout/tagout, would be included in written maintenance procedures.

| **Pump identification** no.: | | **Comments** |
|---|---|---|
| Process requirements: | | |
| A. Centrifugal pump:<br>1. Can casing design pressure be exceeded?<br>2. Is downstream piping/equipment adequately rated?<br>3. Is backflow prevented?<br>4. Suction piping overpressure (single pumps)<br>5. Suction piping overpressure (parallel pumps)<br>6. Is damage from low flow prevented?<br>7. Can fire be limited? | | |
| B. Positive displacement: pump<br>1. Can casing design pressure be exceeded?<br>• Pressure relief valve in discharge<br>• Set pressure = casing DP minus maximum suction pressure<br>• Pressure relief valve discharge location (viscous materials)<br>C. General requirements:<br>1. Guarding requirements – rotating parts<br>2. Environmental requirements – seal design<br>3. Noise emissions – meets standards<br>4. Electrical requirements | | |

**Figure 12.9** Pump safety checklist.

The JSA can be conducted using a form similar to the one shown in Figure 12.10. The first column shows the various steps of the job, the next column records the potential hazards of that step, and the last column identifies the required H&S controls. Notice that this technique moves beyond the H&S checklist approach of simple hazard identification. A JSA also involves the selection of actionable proper safe work practices.

The four basic steps in conducting a JSA are

1. Selecting the job or task to be reviewed.
2. Breaking the job or task into sequential or successive steps.
3. Identifying the potential hazards at each step.
4. Deciding on the required action or procedure to minimize each potential hazard.

The results of the documented analysis are then incorporated into the actual written job procedure. The procedure can then be used to assist in the training of those individuals who also perform the task. A periodic review of the written procedure should be conducted to assure that the H&S requirements are still appropriate. Additionally, occasional audits help determine if the required H&S controls are actually being used and incorporated into the day-to-day operations.

| Division: | I Machine/operation: | |
|---|---|---|
| 1. Job task description: | | |
| 2. Task steps | 3. Potential hazards | 4. Safety and health controls |
| | | |
| Date: | Review due date: | Approved: |

**Figure 12.10** Job safety analysis form.

### 12.4.2 The Benefits of a JSA

There are several major benefits that derive from conducting JSA. The JSA approach itself is easy to understand, does not require a great deal of training, and can be quickly completed by experienced individuals. The process also provides an opportunity for an individual to be recognized for his/her knowledge of the operation. The results of the review can provide a written document that can be used to train new employees, a consistent method of operation that can reduce process variables, and a common set of H&S expectations and reduced injury/illness rates. Additionally, the JSA document can be reviewed as part of planned H&S audits and provide a starting point for reviewing job procedures if an incident/accident does occur.

### 12.4.3 Step One: Selecting the Job or Task

Jobs or tasks that can be simply described are best suited for a JSA. The tasks must be defined fairly specifically to take most advantage of the power of JSAs. For example, broadly defined jobs such as constructing a building, making chemicals, or working in a laboratory are not suitable for JSAs. On the other hand, tasks such as threading a coating machine, assembling scaffolding, or charging a drum of chemicals to a reactor are suitable subjects for safety analysis. Some traditional permit-type safety practices are based on the JSA approach, such as confined space entry permits, chemical line opening permits, and temporary wiring permits.

In selecting and prioritizing the jobs to be reviewed, several factors need to be considered. Analysis of past injury/illness data may indicate that individuals did not know the safe procedure for accomplishing certain tasks. Perhaps up-to-date written procedures need to be established. The data could illustrate the existence of a low frequency of very serious injuries or a high frequency of minor injuries. This would be considered a high-risk situation on the risk matrix (Figure 12.4). Employees new to a department may need to be trained on doing the job, and, therefore, up-to-date written operating instructions may be needed. Likewise, a new process or new piece of equipment may need a JSA so that the initial operating procedures can be developed. Perhaps a more advanced risk assessment technique has uncovered the fact that the existing operating procedures need improvement.

Recording the specific task to be analyzed on the JSA form is the first step in a productive JSA. This brief written description of the task also helps further to define the boundaries of the analysis.

### 12.4.4 Step Two: Breaking the Task into Sequential Steps

After the task or job for review has been selected, the next step is to list all the discrete steps in performing the task. An experienced operator with the help of a co-worker, team leader, supervisor, or H&S representative is usually the best choice to accomplish this. The operator can walk through the task and describe what is being done at each step. The supervisor or team leader can then record the steps on the first column of the JSA form. Identifying the hazards and associated safety controls needed at each step should be held until all the steps in the job task are listed. Notice that each step starts with an action word. The intent at this point is to determine what is being done.

An overly detailed procedure that provides lots of descriptive information but little in the way of additional hazard opportunities **leads to boredom and needless paperwork. A proper balance is the challenge for selecting the steps in a JSA.** An experienced operator may find that the JSA job steps selected are too basic. On the other hand, a new operator may find that the same information developed from a JSA is exactly what is needed to do the job safely. Consider combining jobs with only few steps together, as well as breaking up jobs with too many or too complex tasks. Be sure that the work is actually observed so that key steps are not taken for granted, especially at the beginning or the end of the task.

### 12.4.5 Step Three: Identifying the Hazards Associated with Each Step

The next step is to identify all of the potential hazards associated with each step. The hazards should then be recorded on the second column of the JSA form. Identifying the proper safety controls needed at each step can be done at this point or at the next step after all the hazards are identified. If solutions are generated at this point, there is some chance of missing some hazard potentials. In any event, an experienced operator is generally able to provide excellent insights as to what is acceptable and has gone wrong at each step of the job or task.

At each step of the task, the hazard potentials would then be identified and recorded. For instance, if the particular task is pouring caustic liquid from one container to another, the potential hazards could be chemical burns to various parts of the body such as the eye, face, hands, or torso, depending on the size of the container and nature of the procedure. Additionally, another potential hazard that could be associated with this task could be the strain of the back or arm depending on the size and weight of the container.

Be sure to observe and review the actual job task. Recalling the task from memory or not including elements that begin or start the task may lead to an incomplete analysis. The "sometime quirks" of the task that do not get reviewed can result in the operator not knowing the full story and the required safe procedures. The experienced analyst can maximize the chances of ensuring a complete review by asking open-ended questions such as "Does the procedure ever change?," "Is this step ever done differently?," or "Is the task completed differently on day shift than it is on swing shift?" Having another experienced operator review the completed JSA or videotaping the procedure are other excellent approaches.

### 12.4.6 Step Four: Developing the Safety Controls for the Hazards at Each Step

Now that the hazards have been identified, the next step is to develop the necessary safety controls. Remembering the risk matrix (Figure 12.4), the idea is to move the high risks out of the unacceptable range. The risks can be reduced by lowering the possibility that something will go wrong, by reducing the severity of the consequences if it does go wrong, or by doing both. In some cases, judgments must be made about the potential risks of alternate procedures that could be proposed. Minimizing risks at each step in the job, incorporating analyses into broader risk assessments, and gauging conformance to regulatory requirements are ultimate goals.

Using the JSA form and the hazards identified at each step, develop and document the required safety controls. There are generally several controls that could do the job. The challenge is to determine the best one that is both cost-effective and that, in fact, will be implemented.

If the task appears to be highly hazardous and easy solutions to reducing the risks do not appear obvious, consider the possibilities of accomplishing the task in an entirely new way. Brainstorming ideas with operators, designers, and trades personnel can often result in cost-saving, less hazardous ways of accomplishing the task. Ensure that the proposed solutions do not result in riskier situations.

Typical hazard control measures include the development or adjustment of procedures, engineering controls such as equipment design and ventilation, and requirements to use PPE. Procedures can include the use of safety permits such as confined space entries, hazardous work, and chemical line openings. Other procedures can include periodic preventive maintenance inspections on safety interlocks, fire protection systems, or ladder conditions. Procedures could be developed to check periodically on compliance with the established H&S rules. The use of PPE requirements could be established, such as wearing safety shoes during movement of materials. Similarly, improvements to the facility or equipment can be made to reduce the chance of accidents, such as improved lighting, interlocked safety guards, and improved material storage layouts.

### 12.4.7 Summary: Making JSAs Effective

The JSA is an easy-to-use and results-oriented tool for identifying and controlling the hazards of everyday tasks. The National Safety Council (1993) suggests that it is an excellent starting point for questioning the established way of doing a job. Conducting the analysis does not require extensive training, but does require someone knowledgeable in potential hazards recognition. The results are most effective when performed by experienced individuals who are actually at risk from doing the task. The JSAs can be used for training new employees, agreeing on H&S procedures with experienced employees, and can be the basis for well-written operating procedures that integrate robust H&S features. The completed JSA document needs to be incorporated into the overall management system with appropriate approvals as well as periodic reviews and updates. When a new employee needs to be trained to do the task safely, it is also an excellent time to review the existing JSA and written job procedures. It is an opportunity for knowledgeable individuals to ensure that the JSA is actually up to date. In any event, an annual review is the best choice.

An example of a completed JSA is illustrated in Figure 12.11.

## 12.5 "WHAT-IF" ANALYSIS

### 12.5.1 "What-If" Analysis Overview

A WIA is a structured brainstorming method of determining what things can go wrong, judging the risks of those situations, and recommending corrective actions where appropriate. With little experience in methodology, a review team experienced in the process, equipment, or system under review can effectively and productively uncover the major H&S issues. Led by an energetic and focused facilitator, the review team assesses step by step what can go wrong based on their past experiences and knowledge of similar situations.

Using an operating procedure and/or a piping and instrument diagram (P&ID), the team reviews the operation or process using a form similar to the one illustrated in Figure 12.12. Team members usually include operating and maintenance personnel, design and/or operating engineers, other specific knowledgeable people as needed (chemist, structural engineer, radiation expert, etc.), and a H&S representative. At each step in the procedure or process, "what-if" questions are asked and answers generated. To minimize the chances that potential problems are not overlooked, moving to recommendations is generally held until all the potential hazards are identified.

The review team then makes judgments regarding the likelihood and severity of the "what-if" answers. If the risk indicated by those judgments is unacceptable, then a recommendation is made by the team for further action. The completed analysis is then summarized and prioritized and responsibilities assigned.

| | | |
|---|---|---|
| Division: VCR Assembly | I Machine/operation: VCR viewing mirror carrier machine | |
| 1. Job task description: In mirror carrier machine manually insert bellows and mirrors, actuate glue application and curing process, stack completed assemblies, and package and transport boxed assemblies | | |
| 2. Task steps | 3. Potential hazards | 4. Safety controls |
| (i) Insert bellows into assembly machine fixture | • Possible repetitive motion problems<br>• Danger if machine cycles while hand is in the danger zone<br>• Danger of finger/hand being cut while inserting bellows | • Request an ergonomic evaluation<br>• Safety interlocks are designed into machine operation and interlock reliability checks are integral to the machine operation<br>• Risk is minor and no additional controls are needed |
| (ii) Stack mirrors inside the viewing mirror carrier machine | • Danger of cuts to the fingers from the sharp edges | • Present mirror thickness does not present unreasonable hazard however<br>• Provide a "sweep-up" device to allow for the pick up of sharp pieces of broken mirrors |
| (iii) Place mirror onto the nest | • Danger of cuts to the fingers from the sharp edges<br>• Danger if machine cycles while hand is in the danger zone | • Same as above – provide sweep-up device<br>• Same as above – safety interlocks provided |
| (iv) Wipe glue tips | • Exposure of RTV epoxy glue to fingers, hand, and possibly to the face and eyes inadvertently | • Redesign fixture to eliminate the need to wipe the glue<br>• Provide MSDS and review irritant properties<br>• Ensure that eyes and skin are washed immediately upon exposure |
| (v) Close VCR assembly doors | • No apparent hazards | • None required |
| (vi) Activate cycle timer button | • Machine cycles while hand is in the danger zone<br>• Potential UV light exposure to eyes and skin | • Same as above – safety interlocks provided<br>• Redo UV level measurements especially at spaces around door and provide preventive maintenance checks on mounting |
| (vii) Open VCR assembly doors | • None apparent | • None required |
| (viii) Release mechanism and remove completed assembly | • Machine mis-cycles<br>• Sharp edges | • Same as above – safety interlocks provided<br>• Same as above – provide sweep-up device |
| (ix) Put completed assembly into tray | • None apparent | • None required |
| (x) Other associated hazards | • General area eye hazards<br>• General housekeeping/slipping hazards | • Require the use of safety glasses<br>• Ensure floor kept clear of parts |
| Date: 20 August 2017 | Review due date: August 2017 | Approved: Susan Supervisor |

**Figure 12.11** Job safety analysis for VCR viewing mirror carrier machine.

### 12.5.2 Getting Started: What Is Needed

The first steps in conducting an effective analysis include picking the boundaries of the review, involving the right individuals, and having the right information. The boundaries of the review may be a single piece of equipment, a collection of related equipment, or an entire facility. A narrow focus results in an analysis that is detailed and explicit in defining the hazards and specific recommended controls. As the review boundaries expand to include a large complex process or even an entire facility, the findings and recommendations can become more of an overview in nature. The analysis can include the various stages of a construction project, the procedural steps involved in the operation of the equipment or facility, the written maintenance procedures for a piece of equipment, or the major changes in staff impacting process or system. A clear understanding of the boundaries of the analysis starts the review off in an effective manner.

Assembling an experienced, knowledgeable team is probably the single most important element in conducting a successful WIA. Individuals experienced in the design, operation, and servicing of similar equipment or facilities are essential. Their knowledge of design standards, regulatory codes, past and potential operational errors, and

| Division: | Description of operation: | | By: | Date: |
|---|---|---|---|---|
| | | | | |
| What if? | Answer | Likeli-hood | Conse-quences | Recommendations |
| | | | | |

Figure 12.12 Example "what-if" hazard analysis form.

maintenance difficulties brings a sense of practical reality to the review. On the other hand, including new designers and new operators in the review team mix presents an excellent learning opportunity for subjects that are not usually taught in design school or in operating classes.

The next most important step is gathering the needed information. One important way to gather information for an existing process or piece of equipment is for each review team member to visit and walk through the operation. Videotapes of the operation or maintenance procedures or still photographs are important and often underutilized excellent sources of information. Additionally, design documents, operational procedures, or maintenance procedures are essential information for the review team. If these documents are not available, the first recommendation for the review team becomes clear: develop the supporting documentation! Effective reviews cannot be conducted without up-to-date, reliable documentation. An experienced team can provide an overview analysis, but nuances of specific issues such as interlocks, pressure reliefs, code requirements, or actual practices are not likely to be found.

### 12.5.3 Conducting the Review: How Is It Done?

Now that the team has had an opportunity to review the information package, the next step is to conduct the analysis. Generally, an experienced hazards review facilitator will lead the group through a series of "what-if" questions. A focused, energetic, and knowledgeable facilitator can keep the review moving productively and effectively. It is the job of the facilitator to keep the effort productively moving.

Step 1. **Developing the "What-If" Questions.** Using the available documents and the experience and knowledge of the review team, "what-if" questions can be formulated around human errors, process upsets, and equipment failures. These errors and failures can be considered during normal production operations, during installation, during maintenance activities, and during debug situations. The questions could address any of the following situations:

- Failure to follow procedures or procedures followed incorrectly.
- Procedures incorrect or latest procedures not used.
- Operator that is inattentive or not trained.
- Procedures modified due to operational upset.
- Processing conditions upsets.
- Equipment failure.
- Instrumentation miscalibrated.
- Debugging errors.
- Utility failures such as power, steam, and gas.
- External influences such as weather, vandalism, and fire.
- Combination of events such as multiple equipment failures.
- Change in critical resources.

Experienced personnel are knowledgeable of past failures and likely sources of information describing potential "what-if" scenarios. That experience should be used to generate credible "what-if" questions.

For example, consider a chemical manufacturing process that includes the charging of a granular-like material from a 55-gallon drum to a 1000-gallon mix vessel containing a highly caustic liquid. Some typical questions are shown in Figure 12.13 for illustration purposes only.

| Division: Chemical Ops | Description of operation: Manufacturing B mix/ drum charging operations | | | By: Review Team<br>Date: August 2017 |
|---|---|---|---|---|
| | | | | |
| What if? | Answer | Likelihood | Conse-quences | Recommendations |
| 1. Granular powder is not free flowing?<br>2. Drum is mislabeled?<br>3. Wrong powder in the drum?<br>4. Drum hoist is not used?<br>5. Two drums are added?<br>6. Drum is mis-weighed?<br>7. Drum hoist fails?<br>8. Drum is corroded?<br>9. Ventilation at mix tank is not operating?<br>10. Granular powder becomes dusty?<br>11. Powder gets on operator's skin?<br>12. Tank mix liquid level too high? | | | | |

**Figure 12.13** Example "what If" hazard analysis form – typical questions for drum charging operation.

As the "what-if" questions are being generated, the facilitator should ensure that each member of the team has an opportunity to input potential errors or failures. Determining the answer to each question as it is generated creates the danger of closing too soon on all the possible upsets. The facilitator needs to be sure that the team has really probed into all the possibilities before going to the next step of answering the questions. The analysis can be divided into smaller pieces if there is a danger of just developing questions and not having the value of them fresh in mind while responding those questions.

Step 2. **Determining the Answers.** After being assured that the review team has exhausted the most credible "what-if" scenarios, the facilitator then has the team answer the questions. What would be the result of that situation occurring? Or are there several consequences with varying impacts?

If done correctly, reviewing the potential equipment failures and human errors can point out the possibilities not only for not only H&S improvements but also for minimizing operating and quality problems. Including the operators and trades personnel in the review can bring a practical reality to the conclusions that are reached (Figure 12.14).

Step 3. **Assessing the Risk and Making Recommendations.** Having considered the answers to the "what-if" questions, the next task is to make judgments regarding the likelihood and severity of that situation. In other words, what is the risk? Remembering the risk matrix (Figure 12.4), the review team needs to make judgments regarding the level and its acceptability. The team will then continue the analysis, question by question, until the entire process or operation has been assessed. At this point, the facilitator should have the team step back and review the "big picture" and determine if they have inadvertently missed anything. Initial H&S recommendations are provided at this point. Figure 12.15 presents a matrix with suggested recommendations.

| Division: Chemical Ops | Description of operation: Manufacturing B mix/ drum charging operations | | | By: Review Team<br>Date: 8/2017 |
|---|---|---|---|---|
| | | | | |
| What if? | Answer | Likelihood | Consequ-ences | Recommendations |
| 1. Granular powder is not free flowing? | 1. Back injury potential when breaking up clumps | | | |
| 2. Drum is mislabeled? | 2. Quality issue only | | | |
| 3. Wrong powder in the drum | 3. If wet, could cause chemical exothermic reaction | | | |
| 4. Drum hoist is not used? | 4. Back injury potential | | | |
| 5. Two drums are added? | 5. Quality issue only | | | |
| 6. Drum is mis-weighed? | 6. Quality issue only | | | |
| 7. Drum hoist fails? | 7. Leg, foot, and back arm injury | | | |
| 8. Drum is corroded? | 8. Iron contamination as well as drum failure and injury | | | |
| 9. Ventilation at mix tank is not operating? | 9. Dusting and potential operator exposure | | | |
| 10. Granular powder becomes dusty? | 10. Same as above | | | |
| 11. Powder gets on operator's skin? | 11. Possible burn | | | |
| 12. Tank mix liquid level too high? | 12. Possible caustic splash as well as quality issue | | | |

**Figure 12.14** "What-if" hazard analysis, completed steps 1 and 2.

| Division: Chemical Ops | Description of operation: Manufacturing B mix/ drum charging operations | By: Review Team<br>Date: 8/2017 |
|---|---|---|

| What if? | Answer | Likelihood | Consequences | Recommendations |
|---|---|---|---|---|
| 1. Granular powder is not free flowing? | 1. Back injury potential to break up clumps | Quite Possible | Serious | Design automated de-lumping equipment |
| 2. Drum is mislabeled? | 2. Quality issue only | Remote | Serious | Improved label from vendor |
| 3. Wrong powder in the drum | 3. If wet, could cause chemical exothermic reaction | Unlikely | Minor | Include inspection in procedure |
| 4. Drum hoist is not used? | 4. Back injury potential | Possible | Serious | Train personnel and ensure use |
| 5. Two drums are added? | 5. Quality issue only | Remote | Minor | None |
| 6. Drum is mis-weighed? | 6. Quality issue only | Possible | Serious | Require second check on weight |
| 7. Drum hoist fails? | 7. Leg, foot, and back arm injury | Remote | Serious | Ensure hoist on PM program |
| 8. Drum is corroded? | 8. Iron contamination as well as drum failure and injury | Remote | Serious | None |
| 9. Ventilation at mix tank is not operating? | 9. Dusting and potential operator exposure | Unlikely | Minor | Include ventilation check in operating procedure |
| 10. Granular powder becomes dusty? | 10. Same as no.9 above | Unlikely | Minor | None beyond existing procedure |
| 11. Powder gets on operator's skin? | 11. Possible burn | *Quite* Possible | Serious | Use dust Stilt and *gloves* |
| 12. Tank mix liquid level too high? | 12. Possible caustic splash as well as quality issue | Remote | Very serious | Use goggles and apron |

**Figure 12.15** Example of completed "what-if" analysis form.

### 12.5.4 Reporting the Results: To Whom and How?

The hard work of conducting the analysis has been completed. The important work of documenting and reporting the results still remains. The makeup of the organization generally determines to whom and how the results get reported. Usually, the department or plant manager is the customer of the review. The leader of the review team will generate a cover memo that details the scope of the review and the major findings and recommendations. In some organizations, the report recommendations will also assign responsibilities and a time frame for actions. In other cases, a separate staff or function will review the recommendations and determine the actions required. A periodic report is then generated to summarize the present status of each of the recommendations. Those organizations that have a well-developed hazard review program require follow-up reassessments on some periodic basis, but often enough to accommodate the changing organization, practices, process, and operations.

### 12.5.5 "What-If" Summary: Pros and Cons

The WIA technique is simple to use and has been effectively applied to a variety of processes. It can be useful with mechanical systems such as production machines, with simple task analysis such as assembly jobs, and with reviewing complex tasks in chemical processing.

Individuals with minimal hazard analysis training can participate in a complete and meaningful way. It can be applied at any time such as during installation, acceptance testing, operations, or maintenance. The results of the analysis are immediately available and usually can be applied quickly. This is especially true if the review team members also operate or maintain the system being assessed.

On the other hand, the technique does rely heavily on operational and hands-on experience and intuition of the review team. It is somewhat more subjective than other methods such as HAZOP (see Section 12.6), which involves a more formal and systematized approach. If all the appropriate "what-if" questions are not asked, this technique can be incomplete and miss some hazard potentials. It may be appropriate to assign more complex portions of the system to a more rigorous review such as HAZOP.

## 12.6 HAZARD AND OPERABILITY (HAZOP) ANALYSIS

### 12.6.1 HAZOP Analysis Overview

A HAZOP analysis is the systematic identification of every credible deviation in a system or process, usually a chemical manufacturing process, from the design intent. Resultant adverse consequences from those deviations are then identified as well as the initiating causes. The risk of those deviations are then assessed, and if deemed unacceptable, then a set of recommended actions are determined. It requires rigorous adherence to the methodology to be sure that no potential hazards are missed. This method has its roots in the United Kingdom with Imperial Chemical Industries (ICI) in the 1960s (Knowlton 1992). The analysis requires individuals who are expert in the design requirements and the design intent of the facility, up-to-date P&IDs, a well-defined system, and a hazard review facilitator who is knowledgeable of the HAZOP technique. These HAZOPS are often performed by outsiders with internal company support.

The team reviews the plant section by section, line by line, and item by item using key guide words to initiate discussion. The guide words prompt the team members to consider deviations from the plant design intent such as more of or less of, none, reverse, and other. Those guide words are then applied to the relevant plant operating parameters under review such as flow, pressure, temperature, materials, etc. The causes and consequences of those deviations are then assessed, and the need for added risk controls is determined.

### 12.6.2 HAZOP: Getting Started

Assembling the right people to conduct the analysis is a critical step. The detailed and rigorous nature of the process probing is not something that everyone enjoys or is capable of contributing to. On the other hand, the logical, sequential, and ordered methodology of step-by-step review of a process can be enlightening to those who have been heavily involved in the design or operation of the process. It provides an opportunity to test and expand skills and knowledge. There are generally five to seven review participants whose functions are illustrated in Table 12.2.

The review team leader needs to keep the analysis focused on identifying problems and recommending solutions. To be successful, the review team needs to have the process information organized into a suitable and usable form after the scope of the hazard review is set. Typically, the information consists of line drawings, flowsheets, plant layouts,

**TABLE 12.2 HAZOP Team Members**

- Review team leader
- Process engineer/chemist
- Design engineer
- Maintenance engineer
- Supervisor/leader
- Operations/supervisor/leader
- Safety/fire/industrial hygiene
- Leader
- Scribe
- Experience in the process is helpful
- Familiarity with the process chemistry and operations
- Knowledge of the piping and instrument diagram (P&ID) as well as equipment design requirements
- Familiar with equipment deviations
- Knowledge of operating unit and deviations
- Experience with health, safety, and industrial hygiene standards and guidelines

Note taker familiar with HAZOP techniques

and P&IDs. Additionally, operating manuals, maintenance manuals, and equipment manuals are usually available. A critical step in conducting effective reviews is dividing the process into individual nodes, sections, or operating steps for analysis. The team leader will usually define the nodes prior to the meetings. The nodes can be highlighted on P&IDs where the process parameters have an identified process design intent. These can be pipe sections where pressure, temperature, and flow conditions have been established. Processing equipment components such as pumps, valves, vessels, and heat exchangers are points between nodes that could cause changes in these parameters. Key guide words associated with the process parameters are also established.

Once the team has been selected and the information gathered and distributed to the team members, the review meetings can be scheduled. Ideally, meetings should not be scheduled for more than four hours with breaks every one-and-half to two hours. The number of meetings will depend on the depth of preparation of the leader, the knowledge of the participants, and the complexity of the process. The number of hours to complete a HAZOP analysis has been estimated by several authors. The Center for Chemical Process Safety (CCPS 1992) Guidelines has suggested that simple, small systems require 8–12 hours for preparation, 1–3 days for review, and 2–6 days for documentation. Freeman et al. (1992) have suggested that the time to complete a HAZOP analysis is a function of the skill level of the team leader, the number of P&IDs, and their complexity. A suggested formula provides weighting factors. For example, a HAZOP analysis consisting of one P&ID, three nodes, and one major piece of equipment requires about 18 hours of team meeting time. Fifteen P&IDs, 81 nodes and 13 major equipment items require about 100 hours of team meeting time. Arco Chemical Company (Sweeney 1993) suggests that a "two to four week study is likely to produce between 50 and 200 findings."

For large, complex processes, it may be necessary to use several teams and team leaders to complete the review within a reasonable time frame.

### 12.6.3 Conducting the HAZOP: How Is It Done?

Now that the team has been assembled, the information gathered, and the meeting scheduled, the analysis can proceed. The analysis is systematic and includes use of the following terms:

- Nodes – The locations on P&IDs at which the process parameters are analyzed for possible deviations. In short, this constitutes the division of the process into individual sections or processes with defined boundaries.
- Deviations – Departures from the design intentions that are uncovered by systematically applying appropriate guide words to the process parameters (i.e. no flow, high pressure, low temperature).
- Intention – How the plant is expected to operate in the absence of deviations at nodes (i.e. pressures, rates, levels, conditions).
- Causes – Reasons why deviations from intentions may occur. A credible cause should be considered as meaningful and included as part of the analysis (i.e. equipment failure, human error, power failure, unanticipated situation).
- Consequences – Results of the deviations if they should occur (i.e. injury, spill, fire, explosion, release to atmosphere).
- Risk – The likelihood of the deviation occurring and the severity of the consequences.
- Guide words – Used to discover or derive the potential deviations from design intentions; common guide words and meanings are listed in Table 12.3.
- Parameters – Characteristics of the process that when deviated from the design intent could result in an injury, environmental upset, or business loss. Common chemical process parameters are

| | |
|---|---|
| Flow | Mixing |
| Temperature | Addition |
| Pressure | Substitution |
| Level | Reaction |
| Composition | pH |
| Frequency | Time |
| Viscosity | Information |
| Voltage | Speed |

Now, using the terms and definitions described, as well as their knowledge of the process, the HAZOP review team conducts the systematic and structured analysis. Although described as sequential, the actual review

**TABLE 12.3 HAZOP Guide Words and Meanings**

| Guide word | Meaning |
|---|---|
| No | Negation of the design intent – No part of the design intentions is achieved and nothing else happens |
| Less | Quantitative decrease – Refers to quantities and properties such as flow, temperature, and pressure |
| More | Quantitative increase – Same as above including quantities and properties such as flow, temperature, and pressure |
| Part of | Qualitative decrease – Only some of the design intentions are achieved; some are not |
| As well as | Qualitative increase – All the design intentions are achieved together with some additional items |
| Reverse | Logical opposite of intent – Applicable to activities such as flow |
| Other than | Complete substitution – No part of the design intention is achieved, and something different occurs |

Process reviewed: ______ Drawing no. ______ Review date:

Node reviewed: ______ Process parameter: ______ Reviewers: ______

Design intention: ______

| Guide word deviation | Causes | Consequences | Risk | Recommended action |
|---|---|---|---|---|
| | | | | |
| | | | | |

**Figure 12.16** Hazard and operability analysis sample form.

steps are closely connected. The results are documented by the assigned scribe on a form similar to Figure 12.16. The collected data include the causes, consequences, risk judgments, and recommended actions for potential deviations at each node in the process.

If the review is conducted on an existing process, the team members need to tour the facility in order to become familiar with the size and the interrelationships of the key components. Operations that are spread out over five acres at one level in a rural area provide a different sense of risk and hazards than those that are located in one five-story building in an urban location.

The steps in the HAZOP method are as follows:

1. **Describe the design intention of the line or vessel in the process.** At this point, the most knowledgeable person on the team would define the design intention. For example, this could mean a transfer line designed to transfer caustic solution from the discharge of pump A to the inlet of vessel B at 100 °F and 70 psig and at a rate of 125 gallons per minute.
2. **Select appropriate process parameter(s) associated with the line or vessel for analysis.** The team would then select process parameter guides and the associated guide words. Review the list above for potential process

parameters to investigate. In this case, the parameters of *flow rate, fluid temperature,* and *fluid pressure* are important operating parameters to investigate for possible deviations.

3. **Apply the guide word deviations to the parameter(s) of interest.** The team would select the most important and applicable deviations that would occur with that process parameter. See the list above for typical process deviations of interest. In the example concerning caustic flow rates, the team would review the deviation possibilities of no flow, more flow, reduced flow, reverse flow, and other flows.
4. **List credible consequences and causes of those deviations.** As the team moves through possible deviations from the intended design, a list of scenarios including credible causes leading to unwanted consequences will be developed. For example, no flow could result in an off-specification production batch. Several causal factors would be considered, such as closed valve on the pump discharge or an empty line feeding the supply side of the pump. Similarly, reverse flow could result in catastrophic introduction of caustic into another holding tank and could have been caused by power failure and gravity siphoning effects. In some cases, there may be gaps in the process information design intention or knowledge of the team members. This could result in delaying or deferring that portion of the analysis in order to obtain more information.
5. **Judge the risk.** Given the potential deviations and the potential consequences of those deviations, the team will then make judgments on the level of risk and the acceptability of that risk. The concepts embodied in the risk matrix (Figure 12.4) should be used in that determination. Some organizations (Greenburg et al. 1991) have developed a semiquantitative approach. That is, frequency rates and severity consequences are assigned numerical rankings and when combined add up to risks that are low enough to be either acceptable or marginal, requiring additional study, or unacceptably high, requiring aggressive interventions.
6. **Recommend actions.** If the risk was judged to be unacceptable and a solution was apparent to the team, a recommendation for action would be appropriate. If, on the other hand, a solution was not apparent, a recommendation for further study such as an engineering analysis or thermal calorimeter work may be the right choice. Some organizations will include in the report the assignment of recommendations to specific individuals with specific follow-up time frames. In some cases, an action may be recommended even though the risk may be low since the benefit of it is obvious and is obtained at a low cost. Typical risk reduction actions include (i) changing the design, (ii) changing the operating procedure, (iii) changing the processing conditions, or (iv) changing the process itself. These actions reduce the risk by reducing the severity of the resulting consequences or reducing the likelihood of the deviation or both.
7. **Repeat previous steps.** The team would then sequentially review each of the selected nodes for the entire process, analyzing each possible deviation and the resultant consequences and make appropriate recommendations. The particular section of the P&ID under review is usually marked off as that portion of the analysis has been completed. An example of a partially completed HAZOP worksheet is illustrated in Figure 12.17.

As the team works its way through the process, a series of potential deviations, consequences, risks, and recommendations are developed. To expedite the analysis, some organizations use a predetermined list (Table 12.4) of relevant deviations for process section types (CCPS 1992). That is, when reviewing pipelines, the process parameters of flow, pressure, temperature, concentration, leak, and rupture would apply. For large and complex facilities with many processes, the hazard review data can be collected on spreadsheet software or on specifically designed programs for HAZOP analysis. This helps make the job of managing the results of a large hazard analysis effort somewhat easier. Software programs are being developed that take information from a computer-generated P&ID directly into a HAZOP application program. In this case the HAZOP information generated "stays with" the process design information.

### 12.6.4 Reporting the Results

By this time the review team has generated a great deal of good information. As with other types of hazard reviews, the information needs to be communicated in an effective manner. An executive summary that highlights the important findings is directed toward the facility or department manager. Highly risky situations need to be communicated verbally as they are identified and not wait until a written report is drafted, reviewed, and issued. The person(s) who will act on the recommendations needs to know the specifics of the findings included in the report details. In some organizations, a periodic report is issued that summarizes the percent of recommendations acted upon by process, by plant, or by business unit. Well-disciplined organizations will include timetables and responsibilities. This gives top management a sense of how well the hazard review and risk management process is being managed overall. The management system should ensure that the quality of the responses are monitored or audited to ensure that the recommendations are being acted upon in a responsible and credible manner.

Process reviewed: Caustic Treatment System | P&ID#: D-2346 Rev. 3 | Review date: 20 August 2017

Node reviewed: Transfer Section "A" | Process Parameter: Flow & Pressure | Reviewers: Susan F., Frank B., John P., Mary Y.

Design intention: Transfer Caustic Solution from pump discharge A to inlet of vessel B at 100 °F, 70 psis and 124 grin

| Guide word deviation | Causes | Consequences | Risk | Recommended action |
|---|---|---|---|---|
| | | | | |
| 1A. No flow | 1. Pump not turned on<br>2. Pump impeller corroded<br>3. In-line valve closed<br>4. Pipe ruptured<br>5. Computer does not start pump | 1. No production<br>2. No production/quality<br>3. Pump motor burns out<br>4. Environmental release and possible injury<br>5. No production | 1. Moderate – acceptable<br>2. Moderate – acceptable<br>3. Low – acceptable<br>4. High – unacceptable<br>5. Moderate – acceptable | 1. Put checks in procedure<br>2. Do maintenance checks<br>3. None<br>4. Install collection berm and detection alarm<br>5. Test pump startup by input simulation |
| 1B. Reverse flow | 1. Power failure and check valve failure<br>2. Overfilling of tank B and valve A open | 1. Caustic flows into hold tank and large reaction<br>2. Same as no.1 | 1. Very high – unacceptable<br>2. Same as no.1 | 1. Provide aux. power and valve maintenance<br>2. Provide level alarm on tank B and double check on valve A in operating procedure |
| 1C. As well as | 1. City water back flow | 1. Same as 1B above | 1. Same as 1B above | 1. Install backflow preventer |
| 1D. High flow | 1. No credible cause determined | | | |
| 1E. Low flow | 1. Same as 1A above | | | |
| | | | | |
| 2A. Low pressure | 1. Same as IA above | 1. Same as 1A above | 1. Same as IA above | 1. Same as 1A above |
| 2B. High pressure | 1. No credible cause determined | | | |

**Figure 12.17** Completed hazard and operability analysis for caustic treatment system.

**TABLE 12.4 HAZOP-Predetermined Chemical Processing Deviation Types**

| | Process section type | | | | |
|---|---|---|---|---|---|
| Deviation | Distillation column | Mix vessel | Line | Heat exchanger | Pump |
| High flow | | | x | | |
| Low/no flow | | | x | | |
| High level | x | x | | | |
| Low level | x | x | | | |
| High interface | | x | | | |
| Low interface | | x | | | |
| High pressure | x | x | x | | |
| Low pressure | x | x | x | | |
| High temperature | x | x | x | | |
| Low temperature | x | x | x | | |
| High concentration | x | x | x | | |
| Low concentration | x | x | x | | |
| Reverse flow | | | x | | |
| Leak | x | x | x | x | x |
| Rupture | x | x | x | x | x |

### 12.6.5 HAZOP Summary: Pros and Cons

The HAZOP methodology is used extensively in the chemical processing industry as an effective technique to conduct hazard reviews. One of the keys to the successful application of the technique is careful preparation by the team leader in selection of nodes and ensuring that design intentions have been established. This technique provides a way to probe exhaustively into all of the potential process deviations and upsets. The technique may become laborious and tedious to an inexperienced review team, but it does ensure a complete and thorough analysis.

The HAZOP method is an advanced design review that is structured, systematic, and complete. Assuring that the design is aligned with engineering codes alone may be a technique that is simpler to understand, more manageable, and easier to apply. On the other hand, codes and best practices are based on past experience with little predictive value, are usually minimum consensus documents, and do not consider the interrelationships of process risks. Incomplete information regarding the design intent will prove to be a frustration to completing the analysis. For this reason, the HAZOP system is best applied after the process engineering is completed and the P&IDs are complete.

## 12.7 FAILURE MODE AND EFFECTS ANALYSIS

### 12.7.1 FMEA Overview

An FMEA is a type of hazard review that probes the failures of components within a process or system and the resultant effects of those failures. For example, a failure of a shutdown interlock may result in a machine continuing to run when a safety guard is moved, which could further result in an injury. This methodology is largely equipment oriented and is generally not used when human failures could be major contributors to the process risks. This technique is useful in analyzing single failure modes that result in unwanted situations but is not efficient for reviewing a large number of multiple combinations of component failures. The results of the FMEA analysis generally lead to improved equipment reliability.

Once the boundaries of the analysis have been set, the review starts with updated process drawings such as P&IDs. Reviewers who are familiar with equipment functions and common break downs list the major system components and the possible ways in which they can fail. The effects of those failures are then determined, as well as some likely known combinations of failures. The review team then assesses the risks and makes judgments regarding the need for additional safety controls. The findings are documented in a failure mode and effect tabular form, major recommendations determined, and a final report issued.

### 12.7.2 FMEA: Getting Started

The experience and skills of the review team determine to a large degree the thoroughness, the completeness, and the accuracy of the resulting analysis. Individuals skilled in equipment and component failures are especially important to this methodology. Trades personnel and plant engineering personnel who design, fix, and analyze plant problems are very helpful resource to the effort. Also, those systems that have electrical and instrument controls with associated safety interlocks can benefit from the inclusion of instrumentation engineers as part of the review team.

In addition to P&IDs, logic or ladder diagrams, instrument loop control diagrams, and wiring diagrams are helpful to the effort. If the drawings are not current, they must be marked up after a field review or walk through. The facilitator could mark the key components that the team will be reviewing prior to the first review meeting and use this listing as part of the opening overview. Scheduling of meetings follows along the lines indicated for HAZOP-type reviews, although the time requirements are slightly reduced because the analysis is focused on failures that individuals are experienced with. For instance, CCPS (1992) suggests that, for simple systems, it take two to six hours of leader preparation, one to three days for the team review, and one to three days for the write-up incorporating team comments. On the other hand, large and complex systems may take one to three days to prepare, one to three weeks to evaluate, and two to four weeks to document.

Prior to the actual review, or as part of the start of the review, the team members should physically "walk through" the facility to become familiar with the actual layout of the equipment and its relationship with other equipment. If reviewing a new design, the design engineer should mentally "walk" the team through the process describing the system and the functions of each major piece of equipment in the system.

### 12.7.3 Conducting the FMEA: How Is It Done?

To conduct the analysis, the facilitator will lead the team through a series of steps, starting with the identification of each major piece of equipment, the possible failure modes, the effects of that failure, and the associated risks and companion recommended action items. The results of the analysis are documented on a form similar to Figure 12.18.

Step 1: **Identify and Describe the Equipment.** Either prior to the meeting or as the first step in the review, the major equipment items are listed in the order in which they appear on the P&ID or process flowchart. As the review progresses, each piece of equipment can be "checked off" as each of its failure modes have been evaluated. The description usually characterizes the equipment functionality (such as air valve safety solenoid). Equipment numbers are sometime available and can provide an easy additional identifying reference, such as steam safety solenoid valve V-431. The key is to ensure that the equipment is identified in ways that the reviewers can understand and allow later usability by those who must act upon the recommendations of the FMEA.

Process/system reviewed: ______________________ Facility: Review date: __________

| Component description | Failure mode | Effects | Risk | Recommended action |
|---|---|---|---|---|
| | | | Probability/severity | |
| | | | | |

Figure 12.18 Failure mode and effects analysis.

Step 2: **Determine the Failure Modes.** For each component or piece of equipment listed, determine all of the possible failure modes that are consistent with the equipment operation. The nature of the failure should be realistic and credible. For instance, a typical failure mode for a solenoid valve could be failure to open when signaled. A solenoid valve, relief valve, and water pump in a water tempering system may fail in several different ways, as illustrated in Figure 12.19.

For relatively simple systems, all of the failure modes can be determined prior to analyzing the effects of those failures. For large, complex systems it is probably best to consider each component and its associated failures sequentially so as not to lose the meanings of the discussions, associations, and connections. Some organizations list the cause of those failures to describe more fully those situations and provide a basis for reviewing and understanding a later incident that has occurred that was not predicted by the analysts.

Step 3: **Determine the Effects of the Failures.** Now having a list of what failures can occur, the review team then determines the effects of those failures. The credible downstream effects should be determined as well as immediate effects. This reality check keeps the analysis within manageable bounds, as well as having the practical effect of keeping the review team focused on the most likely and credible scenarios.

Using the example started above, the failures of the valves and pump could credibly result in effects illustrated in Figure 12.19. The review can continue sequentially along the P&ID path of component failures and effects until all these scenarios are considered. Judgments around the associated risks can be made concurrently for complex processes or the team may decide to do this analysis after all the failure effects have been determined.

Step 4: **Judge the Risk.** At this point, the review team needs to make a judgment about the risks around each of the equipment failures and the associated effects of those failures. What is the likelihood of that failure? How severe is the resulting effect? The team will be using their experience to develop qualitative responses to each of the scenarios. Having known or estimated equipment failure rates can lead to a quantified assessment (reliability analysis) where needed. Common definitions can help the team members calibrate and align along these judgments.

Although the FMEA is not ideally suited to analyze combinations of failures, the team can selectively approach scenarios that have happened in the past or the team expects could easily happen in the future. Attempting analysis on many of these combinations may lead to an enormous number of scenarios and a task difficult to manage.

Process/system reviewed: Tempering System for Eye Wash Water Supply System — Facility: Mix Plant C — Review Date: 20 August 2017

P&ID#: -4598 Rev. B — Reviewers: John P., Claire S., Tony M., and Frances T. — Page: 4 of 7

| Component description | Failure mode | Effects | Risk | Recommended action |
|---|---|---|---|---|
| | | | Probability/severity | |
| C. V-431: Steam valve safety solenoid – normally open | 1. Closes unexpectedly while operating<br>2. Fails to open when signaled (sticks closed)<br>3. Fails to close when signaled (sticks open)<br>4. Valve leaks to surroundings | 1. Loss of tempering for incoming water. Cold water in system<br>2. Same as above<br>3. Potential of scalding eye wash water<br>4. Steam leak in small enclosed area | | |
| D. RV-36: Relief valve on heater set for 90 psig | 1. Fails to operate when needed (i.e. heater exceeds 90 psig and valve sticks closed)<br>2. Opens unexpectedly (i.e. heater at <90 psig) | 1. Potential rupture of heater<br>2. Release of hot water in small enclosed area | | |
| E. P-236: Water supply pump – normally operating | 1. Fails unexpectedly while operating<br>2. Fails to stop when signaled (keeps pumping)<br>3. Fails to start when signaled (doesn't pump) | 1. No water at all to eye wash system<br>2. Potential injury when attempting repairs<br>3. Same as 1 above | | |

**Figure 12.19** Example of a partially completed failure mode and effects analysis.

Step 5: **Recommend Actions.** Based on the level of risks determined for each of the situations, the team then makes recommendations. The discussions and related brainstorming at each step have usually generated a list of robust suggestions. Again, as with each of the hazard review methodologies, the team needs to guard against re-engineering the process or solving each of the problems. On the other hand, the experience and shared knowledge of the group can and does lead to excellent recommendations for solution.

In the FMEA method, the nature of the review generally leads to ways to improve equipment reliability. This can take the form of a better preventive maintenance program, equipment redundancy, more robust equipment selection, or additional alarms/shutdown features.

### 12.7.4 Reporting the Results

The information developed during the review needs to be communicated in a manner similar to other hazard analyses. An executive summary that highlights the important findings is directed to the facility or department manager. The person(s) who will act on the detailed recommendations need to know the specifics. The tabular nature of the worksheets with associated equipment numbers assists in clarifying who needs to do what regarding what equipment. A periodic status report is usually issued until all of the findings are considered and resolved.

### 12.7.5 FMEA Summary: Pros and Cons

The FMEA methodology is particularly well suited for equipment-oriented systems that have little or no human interface. The analysis is generally limited to single failure modes that directly result in an incident or accident. A complex mechanical or electrical system that has multiple operator interface opportunities is beyond the scope of FMEA methodology. The method is largely equipment oriented, and having good records and experience with the equipment failure rates specific to the system under review is necessary. Trades and maintenance engineering personnel can be quite helpful in this regard. The HAZOP approach provides a systematic review of all potential equipment failure modes throughout the system. The effects of each failure is considered separately and then in selected combinations of other failure modes.

## 12.8 FAULT TREE ANALYSIS (FTA)

### 12.8.1 FTA Overview

An FTA is a deductive approach to risk assessment. Whereas FMEA describes what can happen using inductive reasoning, the FTA methodology, starting with the unwanted event, deduces how it can happen. Some authors make the distinction between hazards *identification* techniques such as HAZOP and FMEA and hazard *analysis* methods such as FTA. The FTA technique was reported (Henley and Kumamoto 1981) to have been developed by H. A. Watson of the Bell Telephone Labs in the early 1960s. The fault tree itself is a graphical representation of the various combinations of events that can result in a single selected accident or incident. The FTA starts with the hypothetical unwanted event such as a fire, an amputation, or a chemical release. All the possibilities that can contribute to that event are described in the form of a tree. The branches of the tree are continued until independent initiating events are reached. The strength of the fault tree methodology lies in its visual representation of the combinations of basic equipment and human failure modes that can lead to the unwanted event. If the failure modes have known or estimated failure rates and probability data, then the data can be applied to the tree using Boolean algebra, resulting in a quantitative determination of the top event probability.

The fault tree itself is a graphical representation of Boolean algebra using logic symbols (i.e. AND gates, OR gates, INHIBIT gates) to break down the causes of the top unwanted event into basic or primary failures and errors. These basic events generally have known equipment failure rates (i.e. pump breakdowns per hour of use, switch cycle rate failures, weld failure rates) and human failure rates (i.e. errors per attempt, trained vs. untrained, high stress vs. low stress, complex vs. simple tasks). The analysis begins with the unwanted or top event. The immediate causes of that event are then identified. The relationships of these immediate causes are shown on the tree through the use of the appropriate connecting gates and events. The basic gates are "OR" and "AND." These gates denote a relationship of the state of one or more other events. Each of the immediate causes is then analyzed in a similar fashion until the basic initiating events have been determined. The relationships between these basic events and the unwanted top event are then displayed by the resulting fault tree.

The tree may then be evaluated qualitatively or quantitatively. A minimal cut set (combination of failures) is the smallest set of primary events, which must occur in order for the top unwanted event to happen. They are important because they represent the errors or faults that must be changed in order for the top event to change. Using actual or estimated failure rate data allows the analyst to quantify the minimal cut sets. Computer simulation can assist with the more complex trees.

### 12.8.2 FTA: Getting Started

The FTA methodology requires analysts skilled in the technique as well as being generally familiar with the process to be reviewed. Construction of the tree does not easily lend itself to the committee approach, but the results are reviewed by individuals who have unique knowledge of the system, process, and equipment being analyzed. In practice the analyst, with knowledge of the system, constructs the tree, the experts review the tree and associated logic, and the tree is modified and reviewed until the primary faults and associated logic is agreed upon. In complex situations, the hazard review team members are assigned to develop a particular tree or section of a tree based on their expertise. Those developed tree sections are then reviewed for correctness and logic by the entire team members.

In order to conduct an analysis, an unwanted top event is selected. This is usually accomplished by using other hazard analysis or identification techniques or by the experienced, intuitive knowledge of the process hazards. The top event analysis is usually focused on a particular piece of equipment or certain part of the process. The design parameters and operating procedures at that part of the process must be well known in order to construct a credible tree. On the other hand, a well-constructed tree can reveal or underscore the fact that there is insufficient knowledge to make a credible risk determination. More data would need to be developed to proceed further. If a quantitative analysis is desired, then failure rate data for the equipment under review need to be available or realistically estimated.

The construction of a fault tree, which focuses on one out of many possible system hazards, can be time consuming and resource dependent. Powers and Lapp (1976) suggest that a single chemical process generate over 50 hazardous events. Each event would require two to three man-days to analyze. They report that the US Atomic Energy Commission study (WASH-1400) required over 25 man-years to complete fault trees for one boiling water reactor and one pressurized water reactor. The Guidelines for Hazard Evaluation Procedures (CCPS 1992) estimates that modeling of a single top event of a simple process with an experienced team could be done in a day, while complex systems could require weeks or months. For simple/small systems, they further suggest times of one to three days of preparation, three to six days of tree generation, two to four days of qualitative evaluation, and three to five days for documentation. For complex/large systems, equivalent requirements would be four to six days, two to three weeks, one to four weeks, and three to five weeks.

### 12.8.3 Conducting the FTA: How Is It Done?

To generate a fault tree, the analyst must know and understand the symbols and conventions associated with this method of hazard analysis.

*Basic Fault Tree Symbology: Events and gates.* Table 12.5 shows some of the basic symbols and associated meanings used in the construction of fault trees.

*Basic Fault Tree Terminology*. The analyst should be familiar with the following terminology and definitions. The terms have different meanings and are used to prompt different actions during the generation and analysis of the fault tree.

*Fault* A – term indicating a human or equipment malfunction that can be "self-correcting" once the situation(s) causing the malfunction is corrected. For example, (i) the component operates at the wrong time due to an upstream command error and (ii) an electrical switch is in the wrong position because of power inadvertently applied and has sustained a fault.

*Failure* A – term indicating a human or equipment malfunction that needs to be repaired before the component can successfully operate again. The component does not operate properly when called upon to do so. For example, an electrical switch mechanically stuck in the wrong position has sustained a failure.

*Categories of Failures and Faults* – Faults and failures are grouped into three classes: primary, secondary, or command, to assist in assuring that the analysis drives to basic, primary faults and failures.

- *Primary* is a component fault in an environment for which the component was designed. It is usually attributable to a defect in the failed component, for example, failure of a compressed air receiving tank designed for 200 psig failing at 100 psig or a failure of an electrical switch, in a wet environment, designed for that situation.
- *Secondary* is a component fault or failure in an environment for which it was not designed. This is usually attributable to an external force or conditions. Similar to the discussion above, examples would be failure of the 200 psig designed air receiver at 300 psig or an electrical switch not designed for a wet environment failing during a water wash down.
- *Command is* a fault involving the proper functioning of a component although at the wrong time. Usually attributable to a fault in commanding component. An example could be a machine operator starting a piece of equipment at the wrong time.

*Fault Tree Construction.* In addition to being familiar with the symbology and terminology conventions, the analysis also includes the determination of the immediate and necessary causes for the event to occur. The logic-driven nature of the analysis requires that the immediate causes of the unwanted effects be determined in a sequential and stepwise nature. The immediate causes for the unwanted event become intermediate faults that are then sequentially analyzed until the events become basic or primary failures. Early identification of basic events could indicate that the analysis jumped over several intermediate causes or that the needed complexity of the fault tree approach was not warranted.

There are generally three major steps in the fault tree methodology. The first step is the determination of the top unwanted event and the boundaries of the analysis; the second step is the construction of the fault tree itself; and the third step is the evaluation of the tree. Managing the documentation, review, and follow-up actions are similar to other hazard evaluation techniques.

Step 1: **Top Unwanted Event and Analysis Boundaries**. Determination of the top unwanted event usually comes out of another inductive hazard evaluation technique such as HAZOP or what-if method. Those approaches will help uncover *what* can happen, whereas the fault tree approach can help specify *how* that event can occur. Experienced analysts can also intuitively determine those situations for which a more focused review is necessary. One danger in using the fault tree method exclusively lies in the possibility of not analyzing potential unwanted events because they were not identified in the first place.

The top unwanted event must be sufficiently defined so that the evaluation can be conducted efficiently and the results credible. A top event of "multiple plant fatalities" starts the analysis in a broadly scoped, poorly defined manner. On the other hand, an event defined such as "fatality due to explosion in the nitration reactor during startup operations"

**TABLE 12.5 Fault Tree Symbology**

| Symbol/name | Description |
|---|---|
| 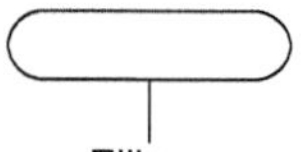 Ellipse | **Top event:** Contains description of the system-level fault or the undesired event. Input to the ellipse is from a logic gate |
| 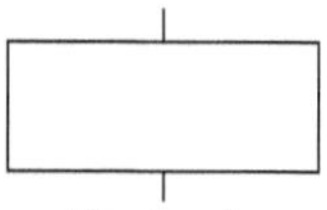 Rectangle | **Fault event:** Contains description of a lower-level fault. Fault events receive inputs from and provide outputs to a logic gate |
| 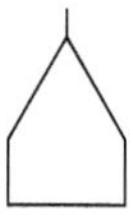 Pentagon | **Input event:** Contains a normal system operating input, which has the capability of causing a fault to occur. The input event is used as an input to the logic gate |
| 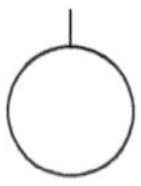 Circle | **Basic event:** Contains a failure at the lowest level of examination, which has the capability of causing a fault to occur. The basic event is used as an input to a logic gate |
| 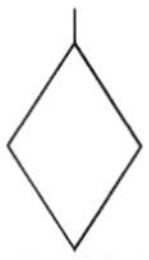 Diamond | **Undeveloped event:** Contains a failure at the lowest level of examination, which can be expanded into a separate fault tree. The undeveloped event is used as an input to a logic gate |
| Transfer In  Triangle | **Transfer function:** Signifies a connection between two or more selections of the fault tree to prevent duplicating subbranches at multiple tree locations or to signify a location on a separate sheet of the same fault tree |
| Transfer out  Triangle | |
| 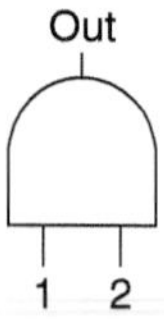  | **AND gate:** Output is to any fault event block or Transfer Out function. Inputs are from any fault event block or Transfer In function. Output occurs only if all inputs exist |
| 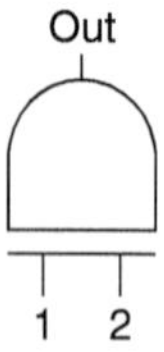  | **Ordered AND gate**: Output is to any fault event block or Transfer Out function. Inputs are from any fault event block or Transfer In function. Output occurs only if all inputs exist and the inputs occur in a specific order |
|   | **OR gate**: Output is to any fault event block or Transfer Out function. Inputs are from any fault event block or Transfer In function. Output occurs only if one or more of the input events occur |

**TABLE 12.5 (Continued)**

| Symbol/name | Description |
|---|---|
| 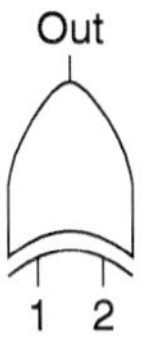 | **Exclusive OR gate**: Output is to any fault event block or Transfer Out function. Inputs are from any fault event block or Transfer In function. Output occurs when one, and only one, of the input events occur |
| 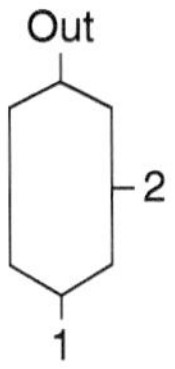 | **Inhibit gate**: Output is to any fault event block or Transfer Out function. Inputs are from any fault event block or Transfer In function. One input is a lower fault event and the other input is a conditional qualifier |

*Source:* Courtesy of David Dylis, System Reliability Center, Alion Science Reliability Center, Rome, NY, 2017.

allows the analysis to become focused quickly and the team can productively apply their knowledge and use their time efficiently.

Clearly defining the top event also helps clarify and establish the analysis boundaries. Limiting the analysis to the "fatality" described above would exclude analysis of situations like those during unattended operations, minor upsets that would not result in an explosion or shutdown/maintenance operations. Defining the physical and operational boundaries, as well as the level of detail and assumptions, also allows the team to be clear on the actual bounds and scope of the analysis. Does the analysis go beyond the physical bounds of the reactor? Beyond normal operations? Beyond simple reactor failure into modes of failure such as weld failure, corrosion effects, jacket limits, brittle failure, etc.? How about utility failures, reactor batch size changes, or status of agitator, pumps, and valves? How about mitigation possibilities such as emergency response personnel? The clearer the definition of the review, the clearer the results of the analysis. On the other hand, these clarifications may point out all the "other" situations that also may require analysis.

Step 2: **Fault Tree Construction.** Starting with the top unwanted event, the analyst then constructs in a logical, sequential fashion all the contributing fault events until the basic or primary events are uncovered. When completed, the tree should show a cause and effect relationship and trace the primary events to the top unwanted event through a series of intermediate events or faults. The construction of the tree commences in a deductive cause and effect fashion. Simplistic immediate causes of a top unwanted event such as a "fire" would be (i) fuel and (ii) oxygen and (iii) ignition source. The fire triangle (Figure 12.20)!

Figure 12.21 presents an example fault tree for electromechanical passenger elevator.

Step 3: **Analysis of Fault Tree.** With the completed fault tree, information is visually displayed showing how the individual failures and specific faults can combine and cause the incident or accident, the unwanted event. Determining all these combinations, called the *minimal cut sets,* that will result in the top unwanted event is the next step. If the minimal cut sets are readily apparent, the use of the fault three methodology is probably not warranted. The minimal cut sets are useful for listing the various ways in which the event can happen. For large FTA, minimal cut sets are best determined by computer programs that also can provide a convenient method of quantifying the tree.

The minimal cut sets then are evaluated to determine where the failures and faults can most readily result in the top unwanted event. If the basic or primary failure rates are known or can be estimated, they can then be used to quantify the resulting analysis. Typical human and equipment failure and fault probability data are available in the literature (Lees 1980) and include failure rate per hour and failure rate per cycle. Presenting the resulting quantitative analysis as an absolute number should be avoided unless presented in context and used to compare with other quantitative analyses that were generated in a similar fashion. The quantitative results have their greatest value in comparing the relative risk of different options for risk reduction measures (Greenburg et al. 1991). Additionally, the analyst needs to ensure that the fault events are independent in *nature*. In some cases, a "common cause" will increase the likelihood of the top event occurring. That is, if all events in a minimal cut set can occur due to the same event, then the top event will occur with this single cause. For example, if all the pumps in a cut set have a common power source, they are subject to the "common cause" of power failure, or similarly if all instruments in a cut set have been (mis)calibrated, they are subject to the same "common cause" failure.

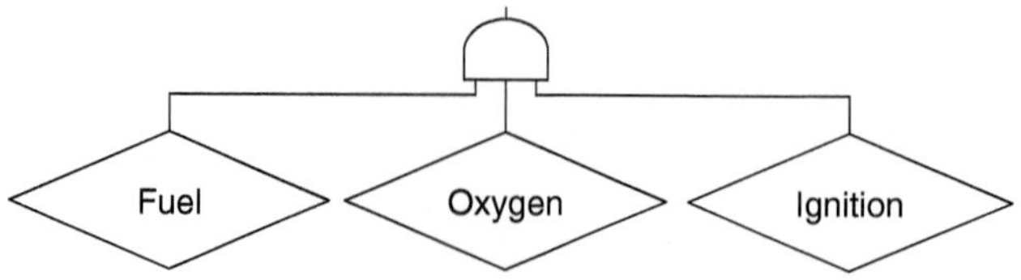


**Figure 12.20** Simplistic fire fault tree.

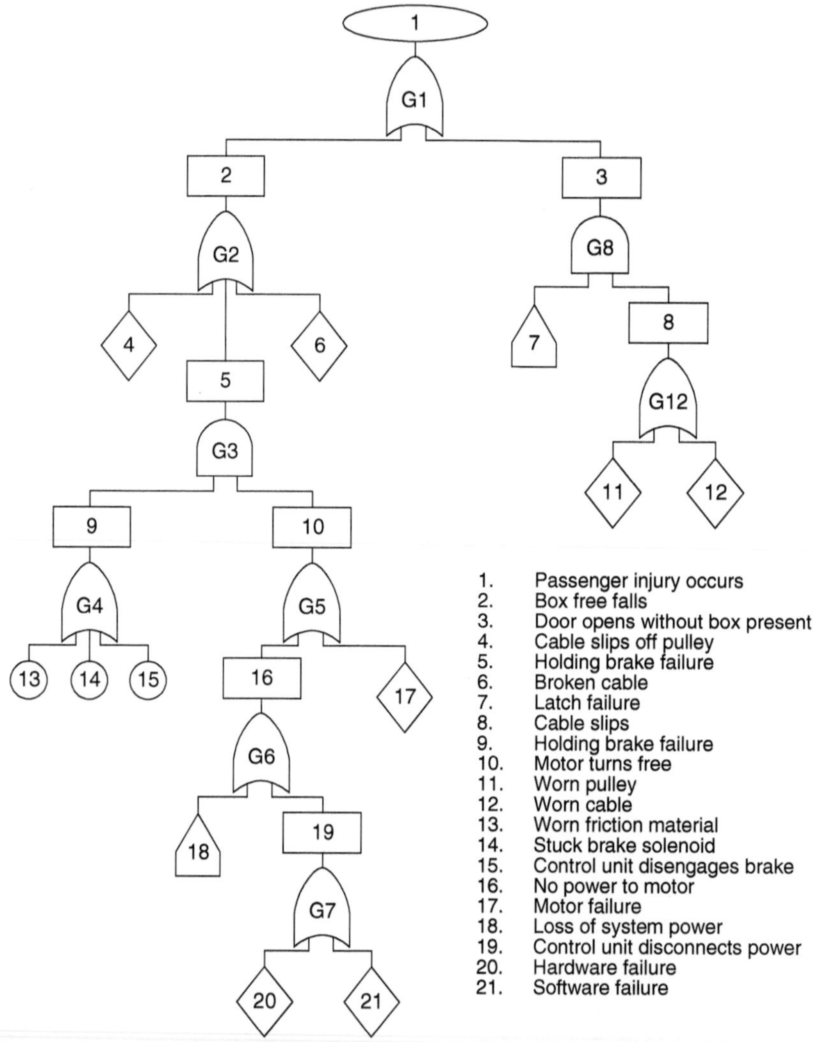




**Figure 12.21** Example fault tree for electromechanical passenger elevator. *Source:* Courtesy of David Dylis, System Reliability Center, Alion Science Reliability Center, Rome, NY, 2017.

### 12.8.4 Reporting the Results

An executive summary that highlights the important findings of the analysis is directed to the facility or department manager. Complex trees will need an explanation of the major contributors to the unwanted event including the lists of the minimal cut sets and their significance. The basic tree and associated cut sets usually does not present self-evident findings like the tabular form of WIA or HAZOP studies. The associated explanatory information therefore needs to be carefully constructed so that the results can be understood, evaluated, and critiqued. When the results are well documented, including the assumptions, boundaries, and explanatory background, the analysis can be more easily updated as the hazards and associated risks change.

Recommendations for risk reductions based on the results of the analysis can be shown in the form of a redrawn fault tree showing the proposed controls. The fault tree can then be used visually and numerically to display the effects of the recommendations on the risk.

### 12.8.5 FTA Summary: Pros and Cons

The FTA process provides a rigorous, methodical, structured way to evaluate how a specified accident or incident can occur. Through deductive reasoning, the analyst can identify the causes and the resulting effects capable of creating the undesired event. The visual nature of the analysis provides the vehicle for communicating a large amount of complex information in a fairly efficient fashion. In addition to allowing for qualitative analysis, the logically connected structure provides an ideal situation for quantification of results if failure rates and fault data are known.

The tree approach allows for the analysis to focus on a very specific, identified hazard. Virtually all the ways in which the unwanted event can occur can be identified, analyzed, quantified, and eventually controlled. It has an important side benefit of improved understanding of how a particular system works and how it fails.

On the other hand, depending on the complexity of the system and level of detail needed, this approach can be very time consuming. It also requires an experienced analyst who works alone for much of the time until the results are periodically reviewed. Keeping a team energized to construct and analyze a tree as a committee is very hard work.

Additionally, the validity of failure rate data must be established if the tree is to be quantified. Although there are many data available, the analyst must use care in applying the data directly because the equipment use, surrounding conditions, and preventive maintenance programs vary widely.

## 12.9 LAYERS OF PROTECTION ANALYSIS (LOPA)

LOPA is a semiquantitative, simplified risk assessment technique that uses order of magnitude categories for initiating event frequency, consequence severity, and the likelihood of failure of independent protection layers (IPLs) to approximate the risk of an incident scenario. The team identifies the IPLs and assigns risk reduction credits to each layer, depending on different criteria. LOPA does not suggest which safeguards to add or which design to choose, but it does assist in deciding between alternatives.

LOPA is used to evaluate the effectiveness and independence of safety measures, especially protective systems. Safety integrity levels (SILs) may be determined using the LOPA technique. It is a powerful technique that may be used to provide a higher level of review than HAZOP for potential scenarios that can result in a single or multiple fatalities to ensure that adequate protection with sufficient availability is in place to reduce the risk. The technique provides a semiquantitative review of the hazards and associated safeguards or layers of protections. It may be combined with HAZOP to evaluate the safeguards identified in a HAZOP.

Like HAZOP, LOPA is best performed by a team of five to seven members specializing in process, operations, maintenance, instrumentation, and process safety and a scribe. It is most effective if the LOPA is conducted at the same time as the HAZOP, making use of a team's knowledge. A vulnerability in the analysis exists if it is conducted with incomplete information or on a design that is not fixed. Furthermore, poor team commitment may increase the time needed to complete the study. Inexperienced or an inappropriate selection of team members may result in a lower quality effort.

Applications are many and include situations such as the following that involve risk-informed decision-making:

- Management of change.
- Evaluating facility siting risk.
- Mechanical integrity programs.
- Identification of operator roles.
- Incident investigation.
- Emergency response planning.
- Bypassing of a safety system.
- Determination of the design basis for overpressure protection.

Additional information is available in the CCPS *Guidelines for Risk Based Process Safety*, 2007 and *Layer of Protection Analysis: Simplifies Process Risk Assessment*, 2001.

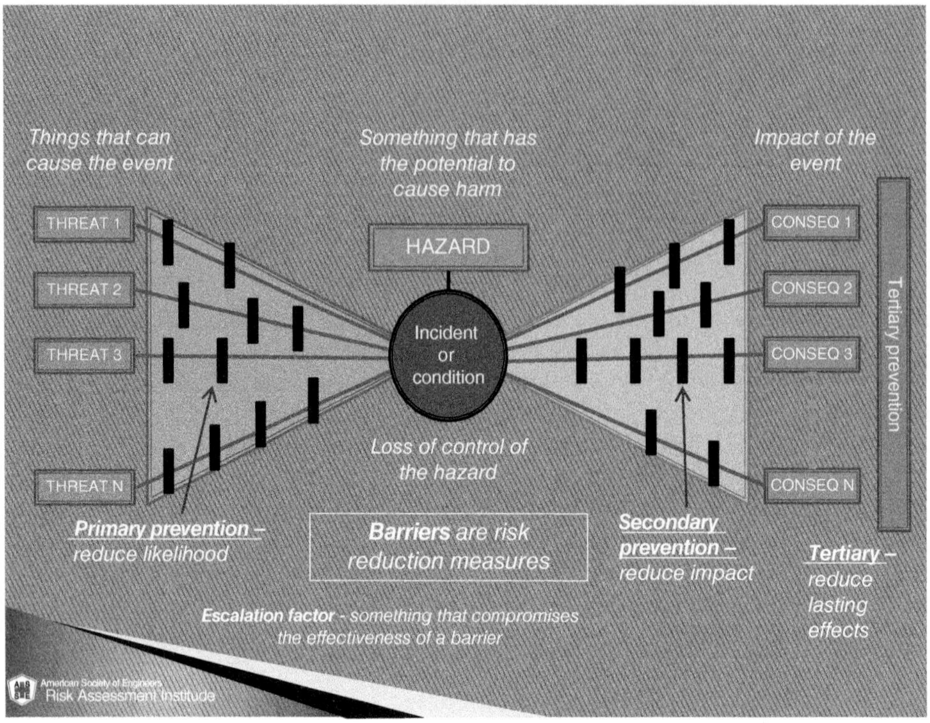


**Figure 12.22** A pictorial of an example Bow Tie diagram. *Source:* Courtesy of Ken Daigle, CSP and Vice President, BP, Safety Management, and the Risk Assessment Institute, American Association of Safety Engineers (ASSE).

## 12.10 BOWTIE ANALYSIS METHODOLOGY

Bowtie methodology is a combination of two other techniques – FTA and event tree analysis, with the fault tree on the left side, the hazard in the middle, and the event tree on the right-hand side. The bowtie diagram, composed of the trees, can be used to indicate preventive, controlling, and mitigating barriers that may impact the incident and its consequences. It is a structured method to assess risk where a qualitative approach may not be possible or desirable. The analysis is dependent on the experience of the team and the facilitator. Inexperienced analysts may struggle to develop the correct logic and may use data that is not statistically significant.

The methodology is applicable to all potential hazards and the final diagram is an effective communication tool in describing hazards and how they are to be managed. The analyses are typically conducted on the highest ranked risks from HAZOP or risk register.

Figure 12.22 presents a pictorial of an example bowtie diagram, courtesy of Ken Daigle, CSP and Vice President, Safety Management, BP, Los Angeles, CA, and the Risk Assessment Institute, ASSE, Chicago, Illinois.

## 12.11 MEASURING FOR EFFECTIVENESS (BOELTER AND LOGAN 2011)

Identifying and measuring the significance of risk and risk reduction is not easy. There are no simple answers. Since the significance of a particular risk often varies between observers, the significance of risk may best be compared to other risks which have better understood consequences or benefits.

A starting point with metrics is to ask what information is available. This information would encompass considerations such as technical and scientific knowledge, programs to ensure that controls are in place and functioning, knowledge of the law as a minimum, and expectations of the community and workforce.

Metrics and their measurement are important in any scientific and management process. One cannot manage what one cannot measure. Since the development of the scientific method starting in the early 1500s, whenever mankind has been able to measure things, it has made great progress both in understanding and controlling them. Measurement forms the basis of input and continuous improvement. If appropriate metrics are not selected, the effectiveness of the health, safety, and environmental management systems can be undermined as reliable information may be lacking to inform managers how well risks are controlled.

### 12.11.1 Measuring Performance

The primary purpose of measuring health, safety, and environmental performance is to provide information on the progress and current status of the strategies, processes, and activities used by an organization to control risks to health, safety, and the environment. Measurement information nourishes the operation and development of the risk management and control system. Measurement provides information on how the system operates in practice, identifies areas where remedial action is required, provides a basis for continuous improvement, and provides feedback and motivation. Measurement can help all levels of an organization or community determine if a system is in place across all parts of the organization and if there is a supportive culture in the face of competing demands for resources.

### 12.11.2 Different Information Needs

An effective risk management system is built on a set of linked metrics, which reflect the structure of an organization. Performance measures are necessarily derived to meet **intra**organizational needs and to efficiently measure and provide feedback to a specific risk-related issue. There will be a more limited number of metrics that can be used **inter**organizationally.

While the primary focus for performance measurement is to meet the internal needs of an organization, there is a need to demonstrate to external stakeholders (i.e. regulators, insurance companies, shareholders, suppliers, contractors, neighbors, the public, etc.) that effective controls are in place for health, safety, and environmental risks. Local community and society pressure for accountability reaches broadly through routes such as corporate social responsibility. The challenge for organizations is to communicate their performance in ways that are meaningful to their various stakeholders.

### 12.11.3 Traditional Metrics

Traditional business management performance metrics include earnings before interest, taxes, depreciation and amortization (EBITDA), return on investment, and market share. A common feature of these measures is they are generally positive (i.e. achievement) rather than negative (i.e. failure).

On the other hand, risk-related performance metrics have often been presented as trailing indicators. Some examples of trailing indicators are injuries, the recordable incident rate, workers' compensation costs, or violations and penalties. Accidents must occur to obtain a recordable and thus measurable conclusion. Such metrics look "after the fact" and are measures of failure as opposed to looking forward (or predictively).

Trailing indicators do not always reveal everything about the health of a business or risk management plan. Using traditional metrics alone may limit necessary and appropriate understanding and leadership. Nevertheless, trailing indicators can prove useful in a number of ways:

- Safety improvement opportunities
- Trends analysis
- Prioritized safety initiatives
- Verified intervention effectiveness
- Regulatory statistics

### 12.11.4 Hazard Metrics

All activities have inherent hazards and potential risks. The range of activities undertaken by an organization will necessarily create hazards, risks, and benefits, all of which will vary in nature and significance. The range, nature, distribution, and significance of the hazards will determine the risks that need to be controlled. Ideally the hazards should be completely understood and the risks should be eliminated altogether, but this is not always reasonable or practical. Information regarding hazards provides important inputs into the planning and review processes to ensure that proportionate effort, prioritization, and emphasis are allocated to the control of risks. Fundamentally, the business of the EHS professional is finding safe ways (not unacceptable risk) to work with hazards and under hazardous conditions. The management system is the process that turns uncontrolled hazards to controlled risks.

### 12.11.5 Prospective Metrics

Health, safety, and environmental risk management success is the *absence* of an outcome (injuries or ill health) rather than a *presence*. This differs from other processes that get measured. Organizations need to recognize that there is no single reliable measure of H&S performance. What is required is a "dashboard" of measures, providing information on a range of health, safety, and environmental activities. All metrics in the toolbox should meet the five criteria: (i) specific, (ii) measurable, (iii) attainable, (iv) realistic and relevant, and (v) time constrained.

Forward-looking measures of health, safety, and environmental performance can be accomplished using leading indicators such as activities, behaviors, and accomplishments. These can be measured by techniques such as

- System audits.
- Accident investigations.
- Near-miss responses.
- Exposure assessment and sampling plan progress.
- Exposure profile reduction process.
- Exposure reduction plan progress (engineering controls, substitutions, eliminations).
- Communication programs.
- Committee activities.
- JSA completions.
- Culture/climate/perception surveys.
- Behavior observations.

Leading indicators increase management involvement, active participation by employees, and engaged community participation. One can review the trailing indicators for trends, but it is more critical to focus ahead, eliminate risky exposures, and increase safe behaviors. Employers have the legal obligation to create safe and healthful workplaces and environments but cannot be successful unless employees and the surrounding community eliminate risky practices and behaviors and strengthen enabling behaviors and practices. In addition to engineering controls and inspection audits, management systems need to also measure factors or qualities that influence behavior.

Risk reductions are driven by a number of points such as

- Relevant and effective training.
- Management's visible participation in health, safety, and environmental programs.
- Engaging and empowering employees in key decision areas.
- Working relationships.
- Regular evaluation and reduction of job and environmental hazards.

When developing metrics, consider the following:

- Customize your metrics – No one approach is best.
- Start with a few small, understandable and easily measured metrics.
- Benchmark to the best metrics.
- Adapt metrics – Do not simply adopt.
- Link the metric to the risk control process.
- Examine and incorporate:
  - Business values.
  - Types and levels of exposures.
  - Drivers of performance in the organization.
- Have employees and stakeholders develop their own metrics.

Leading predictive metrics should be intent on preventing incidents or illnesses from occurring. Metric performance reports can be developed and distributed to provide a timely indication of daily performance. Reports can be circulated to provide personnel with several metrics, which represent the daily involvement of line management and staff in monitoring and improving H&S conditions. Some leading metric examples include:

- Recordable Events Ratio – measures the number of recordable injuries in relation to the number of first-aid events. Safety theory indicates many first-aid events occur before a more serious recordable injury.
- Investigation Timeliness – measures the success rate for line supervisors to promptly perform an initial investigation when safety incidents occur.
- Assigned Corrective Action Completion – measures the timeliness of completion for assigned corrective actions and preventive actions after an incident has occurred.
- Training Completion – measures the completion status of assigned EHS training.

### 12.11.6 Reactive Metrics

Failures in risk control need to be measured (reactive metrics), to provide opportunities to check performance, learn from failures and improve the risk management system.

Reactive metrics identify and report:

- Injuries and work-related ill health.
- Other losses such as damage to property.
- Incidents such as those with the potential to cause injury, ill health, or loss.
- Hazards and faults.
- Weaknesses or omissions in performance standards and systems.

Basic investigations should

- Establish what happened.
- Identify the reasons if performance was substandard.
- Identify the underlying failures in the risk management system.
- Learn from events.
- Prevent recurrences.
- Satisfy legal and reporting requirements.

Root cause analysis and preventive actions should

- Investigate incidents ranked as high risk.

Reactive monitoring should address questions such as

- Are injuries/ill health/loss/incidents occurring?
- Where are they occurring?
- How serious are they?
- What are the costs?
- What improvements in the risk management system may be needed?
- Is the trend getting better or worse?

### 12.11.7 Measuring Culture

The culture of an organization or community is an important factor in ensuring the effectiveness of risk control. The risk management system is an important influence on the culture, which, in turn, impacts the effectiveness of the risk management system. Cultural metrics, therefore, form part of the overall process of performance measuring. Many of the activities that support the development of a positive risk control culture need to be measured. These activities are control, communication, cooperation, and competence.

### 12.11.8 Management Engagement

It is critical to engage all levels of management to drive results in any organization. Line, middle, and upper management assume different roles within companies, each having a unique contribution as to how decisions are made. Developing an organizational engagement plan for risk reduction may be one of the most critical parts of any risk reduction process. Any well-designed risk assessment and control progress can fail without the proper management support.

The process of creating a plan is the first step in management engagement and should consider several elements:

- Identification of a list of important stakeholders throughout the company.
- For each stakeholder, rating of the following:
  - Current and needed level of understanding/knowledge for risk issue/mitigation.
  - Current and needed level of support for risk issue/mitigation.
- Defined specific actions that will close the gap between current and needed levels of engagement.

Executive management must help define, understand, communicate, and support the risk management metrics so that the proper expectations and behaviors are developed throughout every level of management. It is critical to work through line and middle-level management to gain understanding and support so communication efforts with executive management are successful. This process is very iterative and requires a good understanding of an organization's culture to be effective.

### 12.11.9 Dashboards, Metrics, and Performance Indicators

The use of dashboards, performance indicators, and other process feedback tools has become widely popular because of recent advancements in computers and Internet technology. It is critical that the performance metrics and indicators measure the elements that are most critical to the risk management process. Metrics should be selected based on how well they properly drive the behaviors and outcomes. Risk managers must be careful to not select measures solely because they are available.

The SMART process is a popular and effective method for assessing the quality of a specific performance metric. In short,

**S = Specific**: clear and focused to avoid misinterpretation, should include measurement assumptions and definitions, and be easily interpreted.
**M = Measurable**: can be quantified and compared to other data, should allow for meaningful statistical analysis, and avoid "yes/no" measures except in limited cases, such as start-up or systems-in-place situations.
**A = Attainable**: achievable, reasonable, and credible under expected conditions.
**R = Realistic**: fits into the organization's constraints and is cost-effective.
**T = Timely**: doable within the time frame given.

## REFERENCES

29 CFR 1910, Subpart I (1994). *Personal Protective Equipment Standards for General Industry*. Washington, DC: Occupational Safety and Health Administration.

29 CFR 1910.119 (1992). *Process Safety Management of Highly Hazardous Chemicals*. Washington, DC: Occupational Safety and Health Administration.

40 CFR 68 (1996). *Accidental Release Prevention Requirements: Risk Management Programs*. Washington, DC: Environmental Protection Administration.

ANSI/AIHA/ASSP Z10-2012 (n.d.). American National Standard for Occupational Health and Safety Management Systems.

Boelter, F.W. and Logan, P. (2011). *PATTY's Industrial Hygiene*, 6e, Volume 2: Chapter on Evaluation and Control. Wiley.

Center for Chemical Process Safety (CCPS) (1992). *Guidelines for Hazard Evaluation Procedures*, 2e, with Worked Examples. New York: CCPS of the American Institute of Chemical Engineers.

Freeman, R.A., Lee, R., and McNamara, T.P. (1992). Plan HAZOP studies with an expert system. *Chemical Engineering Progress* **88**: 28–32.

Greenburg, H.R., Cramer, J.J., and Stone and Webster Engineering Corporation (1991). *Risk Assessment and Risk Management for the Chemical Process Industry*. New York: Van Nostrand Reinhold.

Henley, E.J. and Kumamoto, H. (1981). *Reliability Engineering and Risk Assessment*. Englewood Cliffs, NJ: Prentice-Hall.

Kaplan, S. and Garrick, B.J. (1981). On the quantitative definition of risk. *Risk Analysis* **1** (1).

Knowlton, R. Ellis, (1992). *An Introduction to Hazard and Operability Studies: The Guide Word Approach*, 10th Printing. Vancouver: Chemetics International Company.

Laszcz-Davis, C., Boelter, F., Hearl, F., Jayjock, M., Logan, P., McLaughlin, C., O'Reilly, M., Radcliffe, T. Jr., and Stenzel, M. (2011). *PATTY's Industrial Hygiene*, 6e, Volume 2: Evaluation and Control, Human Health Risk Assessment. Wiley.

Lees, F.P. (1980). *Loss Prevention in the Process Industries*. London, England: Butterworths.

Lowrance, W.W. (1976). *Of Acceptable Risk*. Los Altos, CA: William Kaufman.

National Institute of Occupational Safety and Health (n.d.). *Prevention Through Design*. https://www.cdc.gov/niosh/topics/hierarchy. (Accessed on February 19, 2019).

National Safety Council (2018). *Accident Prevention Manual for Industrial Operations*, 14e.

Powers, G.J. and Lapp, S.A. (1976). Computer-aided fault tree synthesis. *Chemical Engineering Progress*. IEEE Transactions on Reliability. April. 2–13.

Sweeney, J.C. (1993). ARCO chemical's HAZOP experience. *Process Safety Progress* **12** (2): 83–91.

CHAPTER 13

# SAFETY AND HEALTH IN PRODUCT STEWARDSHIP

THOMAS GRUMBLES
*17806 Paintbrush Pass Ct., Cypress, TX, 77433*

## 13.1 INTRODUCTION

Product stewardship (PS) is of growing importance for manufacturing companies as well as others in any supply chain. Manufacturing industries face increasing expectations from multiple stakeholders including government agencies, communities, surrounding manufacturing, customers, the media, and multiple nongovernment organizations to continuously improve the health, safety, and environmental (HSE) performance of their products. PS can provide a defined process that enables business to meet or exceed these expectations and add value to the bottom line.

This chapter will discuss:

- Common definitions for PS.
- Specific program elements common to PS programs.
- Regulatory foundations for PS programs.
- Traditional HSE elements that are involved.
- Expanded HSE roles in PS.

As will be noted later in this chapter, there is no consensus definition of PS. However, one publically available definition that encompasses many of the common elements of PS programs is as follows:

*Responsibly managing the health, safety, and environmental aspects of raw materials, intermediate, and consumer products throughout their life cycle and across the value chain in order to prevent or minimize negative impacts and maximize value.* (Product Stewardship Society 2014)

## 13.2 HISTORY OF PRODUCT STEWARDSHIP

It is difficult to find the source or the time for the origin of the modern PS concept. While the first public concerns about chemicals became more widespread in the early 1960s, these concerns generally revolved around their environmental impact. The focus on products and their impacts throughout their life cycle ("cradle to grave") evolved over time as a key element in industry activities and programs. This "cradle to grave" approach was accelerated by notable catastrophic events and began to take a more organized or program-oriented shape in the mid- to late 1980s. Examples of high profile events include the Tylenol recall in 1983, the Love Canal environmental pollution issues in the mid-1970s, and the Bhopal, India, methyl isocyanate (MIC) release from a Union Carbide plant in 1984. Industry leaders quickly recognized that the management of chemical products and their potential impact on people and the environment was a continuum; it could not stop at the plant gate. Although customers and other product handlers traditionally had to take responsibility for product use and disposal, chemical companies had growing motivation that drove them to advise and assist those customers with proper handling. Motivating factors included increasing potential for liability

*Handbook of Occupational Safety and Health*, Third Edition. Edited by S. Z. Mansdorf.

cost increase in the development of risk management plans (RMPs) and activities and general commercial pressures for more product information sharing. In fact, companies might even withhold products from customers who failed to handle them appropriately, thus practicing true PS. Industry leaders also realized that the public would not understand differentiations about who had control of a product at any one time nor would the media encourage that understanding to take place.

## 13.3 DEFINITION OF PS

In general, there are no consensus definitions of PS. PS may be referred to as product safety, included in sustainability or corporate social responsibility (CSR) efforts. Depending on the industry sector or organization involved, the definition can be focused on one element of the supply chain such as the electronic industry take back programs, carpet recycling, or safe disposal practice promotions in general. The general principles involved and the specific elements of a definition vary across industry sector associations, private organizations, government agencies, and advocacy groups. However, an analysis of publicly available documents can give a clear sense of the range of common elements involved in PS programs and lead one to an understanding of the elements involved.

There are several aspects of PS we often take for granted. Examples include ingredient labels on food and other packaging, warning labels on alcoholic beverages and tobacco products, tamper-resistant packaging, recycling bins, and public service commercials. The common element in each is producer responsibility, or stewardship, by the product's manufacturer. In some instances, there are regulatory requirements driving these actions, but in some cases there are no regulations or requirements defining the specific actions needed and being taken to accomplish PS (Grumbles and Agopsowicz 2011).

An Internet search for the definition of PS will result in many hits. Some examples from company, government agency, and private advocacy groups are shown below.

### 13.3.1 Companies

***13.3.1.1 Kodak*** PS is an integrated business process for identifying, managing, and minimizing the HSE risks throughout all stages of a product's life in the best interest of society and our key stakeholders: customers, employees, and shareholders.

***13.3.1.2 Linde*** PS is the responsible and ethical management of the HSE risks of our products throughout their product life cycle.[1]

### 13.3.2 Government Agencies

Agencies at the state and federal level have varying degrees of involvement in PS activities.

***13.3.2.1 State of Connecticut*** PS is the act of minimizing HSE, and social impacts and maximizing economic benefits of a product and its packaging throughout all life cycle stages. The producer of the product has the greatest ability to minimize adverse impacts, but other stakeholders, such as suppliers, retailers, and consumers, also play a role. Stewardship can be either voluntary or required by law.[2]

***13.3.2.2 British Columbia*** In British Columbia, extended producer responsibility (EPR) (formerly referred to as industry PS) is an environmental policy approach in which the producer's responsibility for reducing environmental impact and managing the product is extended across the whole life cycle of the product, from selection of materials and design to its end of life.[3]

### 13.3.3 Private Advocacy Groups/Institutions

Private advocacy groups and institutions have also defined PS based on their activities and emphasis.

---

[1] http://www.linde-gas.com/en/sheq/product_stewardship/index.html (accessed September 2015).
[2] http://www.ct.gov/deep/cwp/view.asp?a=2708&q=447190&depNav_GID=1763 (accessed September 2015).
[3] http://www2.gov.bc.ca/gov/topic.page?id=BEBA70369C274C8FBA4FB42BE828A9EB (accessed September 2015).

***13.3.3.1 Product Stewardship Institute*** The growing PS movement in the United States seeks to ensure that those who design, manufacture, sell, and use consumer products take responsibility for reducing negative impacts to the economy, environment, public health, and worker safety. These impacts can occur throughout the life cycle of a product and its packaging and are associated with energy and materials consumption, waste generation, toxic substances, greenhouse gases, and other air and water emissions. In a PS approach, manufacturers that design products and specify packaging have the greatest ability and therefore greatest responsibility to reduce these impacts by attempting to incorporate the full life cycle costs into the cost of doing business.[4]

***13.3.3.2 Global Product Stewardship Council*** PS is an environmental management strategy guided by the principle that whoever designs, produces, sells, or uses a product takes responsibility for minimizing that product's environmental impact.[5]

### 13.3.4 Industry Associations

Many of the major chemical manufacturing, processing, and distribution companies are committed to PS performance programs through their trade associations. These include the American Chemistry Council and Canadian Chemical Producers Association Responsible Care programs, the National Paint and Coatings Responsible Coatings program, and the NACD Responsible Distribution program. All of these programs define PS elements and expectations for program implementation.

One example is shown below. These industry programs often have a management system structure with specific system elements for member company implementation.

***13.3.4.1 American Chemistry Council (ACC)*** ACC is a group that calls their efforts product safety. The Product Safety Code requires that companies include product safety and stewardship as part of their management systems. Product safety management requires an understanding of intended product uses, a science-based assessment of potential risks from products, and a consideration of the opportunities to manage product safety along the value chain. A key component of managing product safety by parties in the value chain is exchanging information regarding product hazards, intended uses, handling practices, exposures, and risks. PS is the responsibility to understand, manage, and communicate the health and environmental impacts of chemical products.[6]

## 13.4 FUNDAMENTAL PROGRAM ELEMENTS FOR PRODUCT STEWARDSHIP (GRUMBLES AND AGOPSOWICZ 2011)

In reviewing publically available definitions and industry association program contents, common elements and principles do become clear. PS is about the HSE aspects of products throughout their life cycle. PS requires that a company manage its products from their inception to disposal, commonly referred to as "cradle to grave." PS is intended to make HSE protection an integral part of the design, manufacture, distribution, use, recycle, and disposal of products. As such, PS has clear links to occupational hygiene, hazard communication, risk assessment, process safety, regulatory compliance, and overall pollution prevention efforts that HSE professionals are involved with on a routine basis.

A model PS program would include the following responsibilities in each of the typical business functions described below.

### 13.4.1 Product Design and Development

This stage includes benchtop or conceptual research as well as defining the markets or uses anticipated for the product being developed. Depending on the global region of the activity, it can include some elements of regulatory compliance. More often it requires sound professional judgment on the possible hazards of a product, developing appropriate handling procedures based on the known and suspected hazards of the products being developed and the development of appropriate hazard communication documents when full information on the product may not be available. This stage also requires consideration of possible use conditions as they relate to potential use and exposure scenarios.

---

[4] http://www.productstewardship.us/?page=Definitions (accessed September 2015).

[5] http://www.globalpsc.net/about-the-council (accessed September 2015).

[6] http://responsiblecare.americanchemistry.com/Responsible-Care-Program-Elements/Product-Safety-Code (accessed September 2015).

### 13.4.2 Purchasing

Obtaining services and products, feedstocks, and processing supplies should be considered as an important part of the life cycle of a product and included in PS practices. These practices should include assuring you are doing business with companies with good HSE practices, assuring your service providers are doing business in an environmentally sound matter, and obtaining appropriate information on the products and manufacturing supplies you buy.

### 13.4.3 Manufacturing

This is the stage most familiar to many HSE professionals. It is the full range of HSE activities involved in any manufacturing or distribution facility. This is the stage of PS with many regulatory requirements that can be considered elements of an overall PS program. It is the stage where classic HSE is practiced.

### 13.4.4 Distribution

Once a product leaves the manufacturing location, the ability to directly control the risks of the product becomes more difficult. This phase in the product life cycle includes transit, in-transit storage, packaging/repackaging, and other distribution activities. There are regulatory requirements for transportation of specific defined hazardous materials that include container specifications and some elements of hazard communication, but for good PS all products need to be considered for these elements commensurate with their risks.

### 13.4.5 Marketing

This element includes the applications and uses that are marketed and the representations of product safety made by the marketing personnel. It is important that marketing literature does not conflict with technical literature and is appropriately descriptive of the hazards and handling needs for a product without becoming a substitute for an SDS.

### 13.4.6 Use Conditions

Being aware of customer uses and commensurate with the hazard of the product, the customers' ability to handle products safely is key to accomplishing good PS. Having a formal system in place to respond to customer questions and indications of misuse are also integral to a good PS/risk management program. Using customer and supplier feedback received in the ongoing development of your hazard communication or handling documents for the customer can result in improved risk management and communication.

## 13.5 REGULATORY FOUNDATION FOR PRODUCT STEWARDSHIP

There are no regulatory requirements for chemical producers or other manufacturers to have specific "PS programs." However, if one can agree to define the elements of PS as discussed in this chapter, then there are clearly regulatory requirements for some individual elements of the concept. These are activities that most HSE professionals are familiar with, and it is important to understand how these activities impact or play a role in PS implementation.

At this time, in the US federal statutory, authority to control the overall environmental impacts of products in commerce as contrasted to manufacturing operations is limited. HSE regulations have traditionally focused on regulating manufacturing facilities. Transportation regulations regulate a portion of the products in commerce if they meet specific hazard criteria. However, on a global basis some elements of PS are becoming increasingly evident in regulations involving the life cycle aspects of products. These range in scope from individual chemicals to manufactured articles. There is a new emphasis or focus on human and environmental protection through the regulation of products. Some of this regulatory activity is outside of the North American region (e.g. REACH) but has global impact as business is increasingly global in nature.

### 13.5.1 Hazard Communication Regulations

Providing detailed hazard information to customers and/or other product handlers including the general public is an issue of growing importance.

At the heart of PS is risk management through the communication of hazards and providing recommendations to those handling chemical and other products throughout the supply chain. Hazard communication regulations exist in many parts of the world. Fundamental requirements to develop and SDSs for chemical substances and products and to supplement this information with product labeling are relatively mature and understood by many. It should be noted that even this requirement is still relatively new as evidenced by the Hazard Communication and WHIMS programs in the NAFTA region and similar regulations in Europe. These requirements as written have only existed since the mid-1980s. There are product labeling requirements administered by a range of agencies including the USEPA, consumer product safety organizations, and food and drug agencies.

The landscape for basic hazard communication is changing as regulations change to achieve alignment with the Globally Harmonized System of Classification and Labelling of Chemicals (GHS) (United Nation 2011). The GHS is a system for standardizing and harmonizing the classification and labeling of chemicals. To the extent that countries adopt the GHS into their systems, the regulatory changes would be binding for covered industries. The specific hazard criteria, classification processes, label elements, and safety data sheet requirements within an existing regulation will need to be modified to be consistent with the harmonized elements of the GHS. It is anticipated that all existing hazard communication systems will need significant changes in order to comply with the GHS. The global effort required for this ongoing transition including input of safety and health professionals will be very significant for a period of several years.

### 13.5.2 Hazard Assessment Regulations

Hazard assessment of products is a fundamental element of any PS effort. While hazard communication and transportation regulations require some specific defined hazard data to be used or generated to determine the physical and health hazards of a product, there is no specific minimum set of hazard data required in the United States for most existing chemicals. Registered pesticides are an exception to this based on requirements found in USEPA pesticide registration regulations. When developing a "new chemical" as defined by the various chemical control regulations, a producer is required to submit what data are available in the United States and specific hazard data sets in Canada and the European Union (EU), depending on the volume and other factors such as specific uses. PS takes this element much further and in essence requires companies to do what is right, "commensurate" with the anticipated hazards and risk. In other words, increased risk requires additional hazard information to be developed and made available in the appropriate manner to product users and in many cases the public.

### 13.5.3 Risk Management Regulations

In the United States section 112(r) of the Clean Air Act Amendments[7] requires USEPA to publish regulations and guidance for chemical accident prevention at facilities that use "extremely hazardous substances" as defined by the agency. These regulations and guidance are contained in the RMP rule. The information required from facilities under RMP helps local fire, police, and emergency response personnel prepare for and respond to chemical emergencies.

Making RMPs available to the public also fosters communication and awareness to improve accident prevention and emergency response practices at the local level. The RMP rule was built upon existing industry codes and standards. It requires companies that use certain flammable and toxic substance to develop a risk management program.

Each facility's program should address three areas:

1. Hazard assessment that details the potential effects of an accidental release, an accident history of the last five years, and an evaluation of worst-case and alternative accidental releases.
2. A prevention program that includes safety precautions and maintenance, monitoring, and employee training measures.
3. An emergency response program that spells out emergency healthcare, employee training measures, and procedures for informing the public and response agencies (e.g. the fire department) should an accident occur.

### 13.5.4 Chemical Data Reporting Rule

In January 2003, US Environmental Protection Agency (USEPA) issued significant amendments to the Toxic Substances Control Act (TSCA) section 8(a) Chemical Data Reporting (CDR) Rule (40 CFR Part 711 2015). These revisions significantly increased the information to be reported for all chemicals on the TSCA inventory with the addition of exposure-related information.

---

[7] http://www2.epa.gov/rmp (accessed September 2015).

USEPA requires that manufacturers report, in ranges, (i) the number of workers "reasonably likely to be exposed" to the chemical at the manufacturing site (ii), the physical form(s) in which the chemical substance is sent off-site (iii), the percentage of total reported production volume associated with each physical form, and (iv), the maximum concentration of the chemical substance at the time it leaves the submitter's manufacturing site or, if the chemical substance is site limited, the maximum concentration at the time it is reacted on-site to produce a different chemical substance. "Reasonably likely to be exposed" means an exposure to a chemical substance, which, under foreseeable conditions of manufacture, processing, distribution in commerce, or use of the chemical, is more likely to occur than to not occur. Covered exposures include exposures through any route of entry but exclude "accidental" or "theoretical" exposures. The regulation does not fully define what an accidental or theoretical exposure includes.

### 13.5.5 New Chemical Development Regulations

In the United States, new chemicals are regulated by the USEPA Toxic Substances Control Act (40 CFR Part 720 2015). When a new chemical substance, as defined by the Toxic Substances Control Act, is to be put into commerce, there are specific data and information requirements for submittal and approval to the USEPA. The USEPA must in essence approve the commercial production of the chemical based on the data provided and the known uses, exposures, and releases for the product.

For the research phase of new chemicals, there are actually a number of regulatory exceptions in terms of hazard assessment, hazard communication, and provision of information in the more limited supply chain of research and development. When a manufacturer moves to the point of filing a premanufacture notification (PMN) for a new chemical, good PS is reflected in the PMN form requirements.

In Part II of the PMN form, Human Exposure and Environmental Release, submitters have the opportunity to describe use and exposure conditions in their workplace as well as in the sites not controlled by them, specifically the customer or chemical user workplace. The more accurate and precise the information is can have a significant impact on the USEPA review of the application. Doing a good job in understanding exposure and use prior to manufacture can lower the potential for testing requirements or risk management controls such as a significant new use rule (SNUR) (40 CFR Part 721 2015).

Considering the practices of all parties in the supply chain including the customer is integral to the concept of PS. There are few regulations that require specific producer activities throughout the supply chain. Understanding chemical uses and working with users is not generally required by various country regulations. This will change as the REACH regulations are implemented in the EU, and chemical producers must determine and register general use and exposure conditions for their products, develop chemical safety assessments based on this information, and provide specific handling requirements to their customers (see REACH description below).

### 13.5.6 European Chemical Control Regulations: Registration, Evaluation, and Authorization of Chemicals (REACH)

In June 2007, a new set of regulatory requirements for the registration, evaluation, and authorization of chemical substances came into force (Regulation (EC) No. 1907/2006 2015). This is the most comprehensive set of chemical regulations to be enacted in the EU, and the impact of these regulations will be global in nature. In the future, based on a specified set of deadlines, all substances produced, sold, or imported in the EU will have to be registered. The regulations remove the distinction between new and existing substances and require the same level of comprehensive data to be submitted for all substances with the amount of information varying based on volume and inherent hazard.

For all substances produced above 10 tons annually, a chemical safety assessment will have to be submitted as part of the required information. For substances determined to be dangerous, a chemical safety report must be developed, and that chemical safety report must describe exposure scenarios. Exposure scenarios are sets of conditions that describe how substances are manufactured and used during manufacture and ultimate use by the downstream users. The chemical safety report must also have information on how the manufacturer or importer controls or recommends users to control exposures to workers and the environment. The exposure scenarios must also include recommended risk management measures for all "identified uses." These exposure scenarios and risk management measures will then need to be annexed in some way to material safety data sheets. This again highlights the growing importance of MSDS programs.

The REACH requirements further emphasize the need for exposure assessment skills and practice to be involved in complying with these chemical control regulations. Exposure evaluation, specifically for downstream uses of chemical substances will be required. Overall the need to determine exposure through the life cycle of a substance and recommended control measures is an integral part of these precedent setting regulations.

## 13.6 IMPLEMENTING PRODUCT STEWARDSHIP

### 13.6.1 Interdisciplinary/Organizational Issues

Based on the elements described in this chapter, it is clear that implementing PS takes a cross-functional approach and affects many departments or functions in a company. This requires significant effort in a company to maintain an understanding of each function or departments' role and to assure each function is performing the activities needed.

PS organizational structures and individual roles vary from one enterprise or institution to the next. In commercial enterprises, the size and geographic configuration of the company, its product portfolio, its market countries, and its PS goals will be the primary factors that drive how the PS professionals are organized. They will also affect whether that organization is one individual who, among many other duties, provides general PS support to the entire enterprise or is instead a small to very large multidisciplined team of professionals that operate under a central organization and/or that reports within a particular business function (Product Stewardship Society 2014).

Furthermore, the PS function itself can be composed of several distinct disciplines including toxicology, occupational medicine, industrial hygiene, and regulatory affairs. If specialists in these fields are not within the PS function of an organization, the different forms of expertise will need to be available, or those doing PS will need to develop a working proficiency in the mentioned areas. This is a challenge HSE professionals are familiar with. It is quite normal for an HSE professional to bring together engineering, management, hourly, and union personnel to solve a problem or implement a solution. These skills and experience translate well to implementing or working in a PS program.

### 13.6.2 Synergies with Existing Company Program Elements

Implementation of any PS program can be facilitated by finding synergies with existing company programs. Regulatory compliance programs that compose portions of a PS program have been discussed, and these programs can be reviewed for additions or revisions that further advance the goals of stewardship. For example, adding additional language to technical marketing literature to assist customers in getting answers on hazards or handling instructions or elaborating on uses that are not recommended or supported by the manufacturer can provide additional risk management measures.

In addition to compliance programs, one can look to initiatives including quality management programs. Many companies carry an ISO 9001 certification. Examples of specific synergies with the ISO 9001:2008[8] standard are shown below.

| | |
|---|---|
| **7.4.1 Purchasing process**<br>The organization shall ensure that purchased product conforms to purchase requirements. The type and extent of control applied to the supplier and the purchased product shall be dependent upon the effect of the purchased product on subsequent product realization or the final product.<br>The organization shall evaluate and select suppliers based on their ability to supply product in accordance with the organization's requirements. Criteria for selection, evaluation, and reevaluation shall be established. Records of the results of evaluations and any necessary actions arising from the evaluation shall be maintained | **Product stewardship purchasing**<br>This element of the standard has clear links to the purchasing element of product stewardship (PS). Requiring suppliers to provide appropriate HSE information and guidance on the products sold should be a basic element for any PS program.<br>Suppliers of products or services should be evaluated for HSE practices in how they do their business. Criteria for these practices should be considered as part of the procurement process |
| **7.3.2 Design and development inputs**<br>Product design and development inputs relating to product requirements shall be determined and records maintained. These inputs shall include:<br>(a) ..........<br>(b) Applicable statutory and regulatory requirements<br>(c) ..........<br>(d) Other requirements essential for design and development | **Product stewardship product development**<br>This stage includes benchtop research as well as the markets or uses anticipated for the product being developed. It can include some elements of compliance. More often, it requires sound professional judgment on the possible hazards of a product, developing appropriate handling procedures based on the known and suspected hazards of the products being developed and the development of appropriate hazard communication documents when full information on the product may not be available. This stage also requires consideration of possible use conditions as they relate to potential use and exposure scenarios |

*(Continued)*

[8] http://www.iso.org/iso/iso_9000 (accessed February 2015).

**(Continued)**

| 7.2.3 Customer communication | Product stewardship use conditions |
|---|---|
| The organization shall determine and implement effective arrangements for communicating with customers in relation to product information and customer feedback | Being aware of customer uses and commensurate with the hazard of the product, the customers' ability to handle product safely is key to accomplishing good PS. Having a formal system in place to respond to customer questions and indications of misuse are also integral to a good risk management program. Considering the feedback you have received in the ongoing development of your hazard communication or handling documents for the customer can result in improved risk management and communication |

## 13.7 PRODUCT STEWARDSHIP PROGRAM EXAMPLES

### 13.7.1 Consumer Products

International Association for Soaps, Detergents and Maintenance Products (AISE) Product Stewardship Programme for Liquid Laundry Detergent Capsules.[9]

The Product Stewardship Programme for Liquid Laundry Detergent Capsules is a voluntary initiative of AISE, launched at the end of 2012. Liquid laundry detergent capsules are a relatively new product form that have been progressively introduced in Europe. They are used daily by millions of consumers across Europe. They help consumers use just the right amount of detergent for their washing needs without waste and are an effective answer to sustainability with smaller packs and important savings in terms of transport and $CO_2$ emissions. While they are safe when used as instructed, it is important that they are handled safely and, as any other household cleaning product, kept away from children.

Due to the occurrence of several ingestion incidents with liquid laundry detergent capsules, especially involving young children, the detergent industry has introduced precautionary measures through a voluntary PS program. With a view to further reducing the number of these incidents, additional measures will be introduced, including the use of an aversive agent in the soluble film of the capsules. In addition, the use of an aversive agent in the film is also being proposed as a mandatory measure under an amendment of the CLP legislation.

### 13.7.2 Electronics Products

Electronics Product Stewardship Canada (EPSC)[10] was founded in 2003. It is a not-for-profit, industry-led organization created to design, promote, and implement sustainable solutions for the recycling of end-of-life electronics. Its membership is composed of 30 leading electronics manufacturers.

EPSC has established PS programs in British Columbia, Nova Scotia, PEI, Ontario, Saskatchewan, Manitoba, Quebec, and Newfoundland in partnership with the Retail Council of Canada (RCC). EPSC members also sit on the industry council for the Alberta Recycling Management Authority.

EPSC members want to see Canada's electronic waste properly managed. These industry leaders are aware of the pressures on municipalities for landfill management and the environmental necessity to recycle electronics products and reuse the valuable resources they contain. EPSC was created to work with both industry and government to develop a flexible, effective, and efficient Canadian solution.

Since 2004, electronics recycling programs across Canada have diverted over 500 000 tons of end-of-life electronics from landfill in British Columbia, Alberta, Saskatchewan, Manitoba, Ontario, Quebec, Nova Scotia, PEI, and Newfoundland.

[9] http://www.aise.eu/our-activities/product-stewardship-programmes/liquid-detergent-capsules/aise-product-stewardship-programme-for-liquid-laundry-detergent-capsules-122012.aspx (accessed September 2015).
[10] http://epsc.ca (accessed September 2015).

### 13.7.3 Basic Petrochemicals

***13.7.3.1 Ethylene Oxide Safety Task Group*** As a part of the chemical industry's responsible management of its products, the Ethylene Oxide Panel Ethylene Oxide Safety Task Group (STG),[11] a group composed of producers and industrial users, developed the *Ethylene Oxide Product Stewardship Manual*, 3e. The STG's mission is to generate, collect, evaluate, and share information pertinent to the safe handling of ethylene oxide during the manufacture, transportation, use, and disposal.

This manual is intended to provide general information to persons who may handle or store ethylene oxide. It includes information on the properties, health effects, environmental effects, and hazards of ethylene oxide. Also included is information on personnel protective equipment, equipment preparation for maintenance, and emergency response and selected regulations. Information on design of facilities and transportation and unloading operations are not currently available on this website. Interested persons may contact their ethylene oxide suppliers for further information or to request a copy of these specific materials.

This work is protected by copyright. Users are granted a nonexclusive royalty-free license to reproduce and distribute this manual, subject to the following limitations: (i) the work must be reproduced in its entirety, without alterations, and (ii) copies of the work may not be sold.

## REFERENCES


40 CFR Part 711. *TSCA Data Reporting Rule*. http://www.epa.gov/cdr (accessed September 2015).

40 CFR Part 720. *Premanufacture Notification*. http://www.epa.gov/oppt/newchems (accessed September 2015).

40 CFR Part 721. *Significant New Uses of Chemical Substances*. http://www.epa.gov/oppt/newchems/pubs/cnosnurs.htm (accessed September 2015).

Grumbles, T. and Agopsowicz, D.E. (2011). Chapter 46. Product stewardship: a viable practice for the industrial hygienist. In: *Patty's Industrial Hygiene and Toxicology*, 6e, vol. **4** (ed. V.E. Rose and B. Cohrssen). Hoboken, NJ: Wiley.

Product Stewardship Society (2014). *Core Competencies for the Product Stewardship Professional*. http://www.productstewards.org (accessed June 2014).

Regulation (EC) No. 1907/2006. https://europa.eu/youreurope/business/product/chemicals-reach/index_en.htm (accessed September 2015).

United Nation (2011). *Globally Harmonized System of Classification and Labelling of Chemicals*, 4e revised. New York and Geneva: United Nations.


---

[11] http://www.americanchemistry.com/ProductsTechnology/Ethylene-Oxide/EO-Product-Stewardship-Manual-3rd-edition (accessed September 2015).

PART II

# GENERAL CONTROL PRACTICES

CHAPTER 14

# PREVENTION THROUGH DESIGN

FRANK M. RENSHAW

*Bayberry EHS Consulting, LLC, 6 Plymouth Dr., Cherry Hill, NJ, 08034*

## 14.1 INTRODUCTION

Prevention through design (PtD) is a guiding principle in occupational safety and health (OSH). It is anchored in the belief that eliminating or reducing hazards and controlling residual risks at the design stage is the most effective way to prevent incidents, injuries, illnesses, and fatalities. The principle applies to the design and redesign of facilities, processes, products, and operations throughout their life cycle (National Institute for Occupational Safety and Health [NIOSH] 2010).

Hazard elimination and reduction are the preferred approach under PtD, but it is recognized that not all hazards can be eliminated or reduced. Those that cannot must be minimized and controlled at acceptable risk levels. The classical "hierarchy of controls" is an essential feature of PtD (American National Standards Institute/American Society of Safety Engineers [ANSI/ASSE] 2011). The hierarchy guides the OSH professional in choosing and applying prevention methods. Hazard elimination resides at the top of the hierarchy, followed by hazard reduction through substitution of less hazardous materials, methods, or processing conditions. Engineering controls are next, followed by warning systems, administrative controls, and personal protective equipment as the last choice in the hierarchy.

Applying prevention methods as early as possible in the design stage of an enterprise is another essential feature of PtD. The principle also calls for consideration of prevention methods at the redesign stage and throughout the life cycle of an enterprise including decommissioning (ANSI/ASSE 2011).

The most effective way to implement PtD is to incorporate the principle and its essential features into an organization's OSH policy, management systems, and work processes that cover design, redesign, and management of change (MOC). This integration theme is emphasized throughout the chapter. The technical tools and practices that bring PtD to life – hazard evaluation, risk assessment, and control strategies – are discussed with numerous reference citations to accommodate the reader's specific needs and interests. A multifaceted approach, incorporating PtD into management systems and addressing hazards and risks with state-of-the-art analytical tools and control technologies, will enable an organization to include prevention considerations in all designs that impact workers, contractors, and others impacted by an enterprise's operations (NIOSH 2010). This is the ultimate goal of PtD.

## 14.2 WHY IS PtD IMPORTANT?

The annual US burden of work-related injuries, illnesses, and fatalities is an estimated 3.55 million nonfatal injuries, 199 400 illnesses, and 4585 fatalities (Bureau of Labor Statistics [BLS] 2014, 2015). The annual direct and indirect cost of occupational injuries, illnesses and fatalities in the United States is estimated to be in the range of $128–$155 billion (Schulte 2005). While total case incidence rates have steadily declined in the United States since 2003, the number of workers who fall victim to occupational hazards and the related costs to these workers, their families, employers, and

*Handbook of Occupational Safety and Health*, Third Edition. Edited by S. Z. Mansdorf.

society remain at unacceptably high levels. Eliminating workplace hazards and controlling residual risks at the design stage is the most effective way to reduce this burden and is the impetus behind the NIOSH-led national initiative on PtD (NIOSH 2010).

There is increasing evidence that PtD is gaining importance as a preferred strategy to address the burden of injuries, illnesses, and fatalities. PtD is gaining support among business leaders "…as a cost-effective means to enhance occupational safety and health" according to NIOSH (2010, p. 5). A 2010 survey of 35 US-based companies who regularly benchmark and share OSH best practices revealed that 77% required some form of PtD in their operations and 66% of this segment required consideration of PtD by their contractors, suppliers, or both (Newell 2011). Another study in the United Kingdom gathered feedback from construction-related professionals on their beliefs about the impact of PtD on design and construction (Gambatese 2011). An overwhelming percentage (87–91%) believed PtD increased construction worker and end-user health and safety.

The emergence of national regulations, initiatives, and consensus standards that contain PtD provisions is another indication of the importance and growing support for this principle. Australia has adopted design guidance and a code of practice on safe design for new and modified structures (Australian Safety and Compensation Council [ASCC] 2006; Safe Work Australia [SWA] 2012). The United Kingdom (United Kingdom Health and Safety Executive [UKHSE] 2015) and Singapore (Singapore Workplace Safety and Health Council [SWSHC] 2011) have included PtD provisions in their regulatory requirements for new construction projects. The United States, as indicated earlier, has undertaken a multiyear national initiative on PtD (NIOSH 2010). The prevailing consensus standards on OSH management systems, ANSI/AIHA/ASSE Z10 (American National Standards Institute/American Industrial Hygiene Association/American Society of Safety Engineers [ANSI/AIHA/ASSE] 2012) and British Standards Institute OHSAS 18001 (British Standards Institute [BSI] OHSAS 18001 2007), include essential features of PtD on hazard identification, risk assessment, determination of necessary controls, design review, and MOC. A new consensus standard devoted entirely to PtD was adopted in 2011 (ANSI/ASSE 2011). The Center for Chemical Process Safety (CCPS) has published a "Concept Series" textbook on inherently safer chemical processes (CCPS 2009). The key principles of inherent safety closely parallel the essential features of PtD.

The business value of PtD is another important consideration. Biddle (2013) used a business case decision tool (Silverstein 2010) to quantify costs and benefits of PtD interventions in three case studies. The first study involved safe patient lifting and handling. Ergonomic and administrative interventions were found to be cost beneficial due to a reduction in musculoskeletal injuries and worker compensation costs. The second study involved lifting and carrying hazards associated with wine grape harvesting. Redesign of the grape picking tubs resulted in a clear reduction in ergonomic risk factors. The primary business benefit, from the managers' perspective, was the increase in worker satisfaction with the smaller tubs. The third study involved replacement of conventional dry cleaning, which uses the volatile organic solvent perchloroethylene, with a wet cleaning method using a water-based detergent. Wet cleaning proved to be the best option on the basis of operating cost alone without quantifying the potential benefits of reduced occupational illness or improved productivity. An additional benefit of wet cleaning was the reduced environmental impact due to elimination of the chlorinated hydrocarbon solvent. McLeod, in a separate study involving small business, reported favorable cost–benefit ratios for ergonomic interventions in die cast and assembly, metal parts finishing, and material handling operations (MacLeod 2011). The economic benefits of employing inherently safer designs for chemical and petrochemical processes have also been shown to be advantageous over other safe design solutions (CCPS 2009, pp. 216–218). Inherent safety strategies focus primarily on eliminating or reducing the hazard of a chemical process at its source. According to CCPS, inherently safer designs have been associated with higher initial capital costs, but lower ongoing operating costs than alternative risk management strategies.

There are many reasons PtD is important, but none is more compelling than the need and the opportunity through application of this principle to reduce the burden of incidents, injuries, illnesses, and fatalities on employees, employers, and society.

## 14.3 HISTORY OF PtD

The origin of PtD as a guiding principle is diffuse in time and geography. Key historical events that mark the emergence and development of the principle are listed in Table 14.1. The events can be grouped into three time periods: Pre-1900, 1900–1969, and 1970–Present. The periods reflect a slow start for PtD in early times, acceleration of events with the advent of the industrial revolution, and a high level of activity on many fronts coinciding with passage in 1970 of the US OSH Act. The list of historical events represents the successful application of prevention methods in design and redesign of the occupational environment. There is no attempt here to reconstruct the overall history of OSH (Blunt et al. 2011), a task much greater in scope and space than is intended in this chapter.

**TABLE 14.1 Historical Events in the Development of PtD**

| Period | Source | Event | Reference |
|---|---|---|---|
| 3000 BCE | Egypt | Early accounts of traumatic injuries and treatments | NSC (1988) |
| 2000 BCE to 200 CE | Hammurabi, Babylonia; Hippocrates, Greece; Pliny, Roman Empire | Written accounts of work related injuries, illnesses and treatment including loss of eyesight, fractured limbs, tetanus, poisoning due to lead and mercury among miners, and use of primitive respirators | Blunt et al. (2011), NSC (1988) |
| 600 CE to 1400s | Rothari, Lombard; Ellenbog, Austria | Principles emerged on compensation for specific injuries; writings on hazards of metal poisoning; warning of hazards due to coal burning in confined spaces | NSC (1988) |
| 1500s to 1700s | Agricola, Saxony; Ramazzini, Italy; Paracelsus, Switzerland | Writings on miners' diseases, need to ventilate mines and use personal protective equipment; hazards of irregular motions and unnatural postures of the body | NSC (1988) |
| 1700s to 1890s | Great Britain, France, Germany | Transition from cottage industries to machines, materials, and work methods associated with the industrial revolution; new hazards, injuries, and illnesses such as phossy jaw from inhaling the fume of white or yellow phosphorus. Enactment of Chimney Sweepers Act to address scrotal cancer hazard of this trade; enactment of British Factories Acts, which stimulated the adoption of factory safety precautions | Grimaldi and Simonds (1989), Hamilton (1943), NIOSH (1973), Patty (1958) |
| 1877 | Massachusetts | Commonwealth-passed factory acts similar to British law covering general provisions to protect against OSH hazards | Grimaldi and Simonds (1989) |
| 1908 to 1948 | US Federal and State Governments | Adoption of worker compensation laws for federal and private sector employees served as an incentive for employers to provide safe working conditions | Clayton (1973) |
| 1909 | US Match Making Industry | Substitution of phosphorus sesquisulfide in place of white and yellow phosphorus enabled the eradication of the occupational disease, phossy jaw | Hamilton (1943) |
| 1910 | US Public Health Service; US Bureau of Mines | These two federal agencies conducted exploratory OSH studies in the mining and steel industries | Clayton (1973) |
| 1911 | New York | Triangle Shirtwaist Company fire took the lives of 146 employees. Response by New York City and state legislature led to revisions of the labor code and adoption of fire safety measures that became a model for the entire country. Measures exemplified the PtD principle of reducing hazard and controlling residual risk | Cornell University, Kheel Center (2011) |
| 1911, 1938, 1939 | ACGIH, AIHA, ASSE | The American Conference of Governmental Industrial Hygienists, American Industrial Hygiene Association, and American Society of Safety Engineers formed as professional OSH organizations. Voluntary standards and guidelines with PtD provisions among member services of each organization | ACGIH (2015), AIHA (2015), ASSE (2015) |
| 1913 | US National Safety Council | Council created as a forum to share common problems and solutions in accident prevention, convened a national safety congress, formed a national safety code program that became the American Standards Association and then American National Standards Institute (ANSI) | NSC (1988) |
| 1936 | US Walsh Healey Public Contracts Act | A regulation stipulating that all contracts with the federal government for goods and services valued at \$10 000 or more must be done under safe and sanitary working conditions. Regulation also requires consideration of hazards at the contract stage before work commences and use of a hierarchy of controls in addressing exposures to atmospheric contaminants | McFarlane (2004), USDOL (1951) |

*(Continued)*

**TABLE 14.1 (Continued)**

| Period | Source | Event | Reference |
|---|---|---|---|
| 1947 | System Safety and MIL-STD_882 | Concept formally introduced by Institute of Aeronautical sciences that closely parallels PtD principle. Involves application of technical and managerial skills to systematic, forward-looking identification and control of hazards throughout life cycle of a project, program, or activity. Led to Department of Defense MIL-STD-882: System Safety Program for Systems and Associated Subsystems and Equipment | Roland and Moriarty (1990) |
| 1970 | US Occupational Safety and Health Act | Enacted in 1970 to improve consistency and performance of states in occupational safety and health | OSH (1970) |
| 1995 | US National Safety Council | Formed the Institute for Safety through Design to expand knowledge on the subject through seminars, workshops, safety conferences, and publications | Manuele (2008) |
| 1996 | Construction Industry Institute | Conducted research on the benefits of safe design for construction worksite safety and developed a computerized toolbox on this subject | CII (1996, 2010) |
| 1996 | Center for Chemical Process Safety (CCPS) | Developed an Inherently Safer Design initiative for chemical processes including a concepts text and checklist on this subject | CCPS (1996, 2009) |
| 2004 | OSHA Construction Roundtable | OSHA Alliance Program Construction Roundtable formed a Design for Safety Workgroup and website. Participants share construction-related safety information and develop compliance assistance tools and resources | OSHA (2015a, b), Toole (2015) |
| 2005 | Center for the Protection of Worker Rights (CPWR) | Sponsored a study to investigate the viability of designing for safety to improve construction worker job safety and health. Identified design opportunities and barriers to their implementation by design professionals | Gambatese et al. (2005) |
| 2007 to 2012 | ANSI/AIHA/ASSE and BSI | Consensus standards on occupational health and safety management systems. Contain essential features of PtD including: hazard identification, risk assessment, risk analysis, risk reduction following hierarchy of controls, and management of change throughout life cycle of organization | ANSI/AIHA/ASSE (2012), BSI OHSAS 18001 (2007), BSI OHSAS 18002 (2008) |
| 2007 to Present | US PtD National Initiative | NIOSH-led national initiative on PtD conducted over seven years. The initiative included goals and objectives in five focus areas: research, education, practice, policy, and small business. Outputs included technical papers, symposia, educational materials, and a consensus standard dedicated to PtD. Initiative has focused attention on the value and importance of PtD across industry sectors | ANSI/ASSE (2011), Lin (2008), NIOSH (2010, 2011, 2014), Renshaw (2013a), Schulte et al. (2008) |
| 1994 to Present | UK Health and Safety Executive | UK Construction (Design and Management) (CDM) regulations are progressive and specific with roles and duties of clients (owners), designers, contractors, and workers. General principles of prevention specified in regulations are closely aligned with concept of PtD. CDM brings together in one place responsibilities for those involved with project design and those responsible for construction | Gibb (2011), UKHSE (2015) |
| 1994 to 2012 | National Occupational Health and Safety Commission; Australian Safety and Compensation Council; Safe Work Australia | Former National Occupational Health and Safety Commission and its successors including Safe Work Australia conducted research and issued a standard, guidance, and code of practice on safe design of structures. Code provides guidance to architects, designers, builders, and others on key elements of safe design and a process for integrating design and risk management | ASCC (2005, 2006), NOHSC (1994, 1995), SWA (2012) |
| 2008 to Present | Singapore Workplace Safety and Health Council | Workplace Safety and Health Council and Ministry of Manpower launched Design for Safety (DfS) initiative to improve construction site safety. Safe design guidelines for buildings and a design review process with checklists are key features of the initiative. A DfS coordinator is also appointed for each project and maintains a risk register | SWSHC (2011) |

### 14.3.1 Pre-1900

Historical accounts of traumatic injuries and treatments such as burns and removal of foreign objects from the body date back to 3000 BCE at the time of building the Egyptian pyramids (National Safety Council [NSC] 1988). Written accounts of work-related injuries and illnesses continued through periods of Babylonian, Greek, and Roman rule and included loss of eyesight, fractured limbs, tetanus, and poisoning due to lead and mercury among miners from the first century CE to the Middle Ages. Basic principles of compensation for work-related disabilities and death were codified in ancient Lombardy in the seventh century. Writings of physicians such as Ellenbog, Agricola, Paracelsus, and Ramazzini documented the causes and nature of diseases associated with specific occupational groups and physical impairment from irregular motions and unnatural positions of the body (NSC 1988; Patty 1958, p. 2). Preventive measures to combat reported occupational injuries and illnesses were limited in nature and scope. Measures included the use of primitive respirators and ventilation in mines and warnings of hazards due to coal burning in confined spaces.

The Industrial Revolution emerged in England in the mid-eighteenth century and spread to the European continent and other countries including the United States in the nineteenth and early twentieth centuries (Industrial Revolution 2015). The Revolution transformed agricultural and small handicraft shop societies into urbanized centers with industrial factories, new materials such as iron and steel, new manufacturing methods, machinery, energy sources, and a fundamental change in the organization of work. With this transformation came new workplace hazards and the need to increase efforts in hazard recognition, evaluation, and control. The English Factory Acts were among the earliest attempts to regulate working conditions and hazards. The Mines Act of 1842 provided punitive compensation for preventable injuries caused by unguarded machinery (Grimaldi and Simonds 1989). Subsequent acts covered hazards related to ventilation, machine guarding, safety devices for steam boilers and powered lifting equipment, use of safety lamps, and handling of explosives. Extending the mine safety regulations to other sectors and establishments and broadening coverage to include special hazards such as white lead were completed by the late 1800s in England. Massachusetts in 1877 was the first state in the United States to adopt regulations similar to the English Factory Acts. The Massachusetts regulations covered machine guarding, ventilation, cleanliness, hazards connected with hoists and elevators, fire escapes, and a requirement for main doors in establishments to open in an outward direction to facilitate egress. Other states followed Massachusetts' lead by adopting and updating basic OSH standards well into the mid-twentieth century.

### 14.3.2 1900–1969

The elimination of white and yellow phosphorus from the manufacture of strike matches is one of the true PtD success stories of the early 1900s. Alice Hamilton (1943) wrote of the painful and disfiguring occupational disease, phossy jaw, which resulted from inhalation of white and yellow phosphorus fume and ingestion of these compounds by workers in the match factories. When the seriousness of the disease became public in 1909, white phosphorus was replaced with a less hazardous form, phosphorus sesquisulfide. This classic example of PtD by material substitution made it possible for the US Congress to impose a tax on white phosphorus matches. The tax provided an incentive for manufacturers to switch to the sesquisulfide, which led to the eradication of phossy jaw from this industry in the United States.

An occupational safety tragedy, the Triangle Shirtwaist Company fire (Cornell University, Kheel Center 2011) occurred close in time to the phossy jaw tragedy. The Triangle fire broke out on the 8th, 9th, and 10th floors of the textile factory building in Manhattan on 25 March 1911. The fire spread rapidly, taking the lives of 146 of the 500 employees in the building. Firemen called to the scene were unable to rescue workers who were trapped due to locked doors to the stairwells. Firemen's ladders and hoses were ineffective because they were unable to reach above the sixth floor. Many workers jumped to their death in the streets or died due to collapse of the fire escape ladder. The response to the Triangle fire was swift. The City of New York created the Bureau of Fire Prevention to end confusion and clarify fire prevention and protection requirements. The PtD principle of reducing hazards and controlling residual risks was advanced by adopting design requirements and fire protection practices in the municipal building code. The New York State legislature established a commission to investigate working conditions and ultimately adopted state-wide legislation in this area.

Two major US regulatory activities in the first half of the twentieth century advanced the use of prevention methods in workplace design. The first activity involved the adoption by Congress of worker compensation laws for federal employees in 1908 (Clayton 1973, p. 2). Passage of similar laws by state legislatures in subsequent years provided worker compensation coverage to employees in the private sector. Worker compensation laws did not prescribe safe working conditions or methods of safe design. The laws did provide an economic incentive for employers to provide a safe and healthy work environment rather than pay for post-incident compensation.

The Walsh–Healey Public Contracts Act of 1936 was the second major regulatory activity to focus attention on the duty of employers to provide safe working conditions. The act required all contracts with the federal government for goods or supplies worth at least $10000 to stipulate that "...none of the work for the contract would be done under hazardous or unsanitary working conditions" (McFarlane 2004). The act also required consideration of workplace hazards at the contract stage before work commenced, another essential feature of PtD. It included specific requirements to control hazards such as airborne contaminants, building and equipment layout, and pressure vessel design. It called for control of exposures to air contaminants by choosing from a list of engineering methods. Personal protective equipment was not permitted in lieu of engineering measures except when such measures were impracticable (USDOL 1951). This provision formalized the requirement to follow a hierarchy in choosing hazard control solutions.

Voluntary efforts to improve working conditions in America's workplaces also gained momentum in the early twentieth century. The National Safety Council was created in 1913 out of a request by the Association of Iron and Steel Electrical Engineers (O'Reilly 2009, p. 9). The Council provided a forum for identifying and sharing solutions to common problems in accident prevention. One of the Council's key initiatives in addition to its annual safety congress was the organization of a national survey of safety regulations and practices. This survey documented the need for standardization of methods and practices in industrial safety and led to the formation in 1920 of a national safety code program. This program evolved into a national program for standardization of safety and engineering practices, initially known as the American Standards Association (ASA) and today as the American National Standards Institute (ANSI). This development marked the first time a reliable, consensus-based source of safety codes and practices became available to American industry.

Other governmental and voluntary efforts advanced PtD thinking and practice in the early and mid-1900s. The US Public Health Service and US Bureau of Mines initiated OSH studies in the mining and steel industries beginning in 1910 (Clayton 1973, p. 2). Health departments in New York and Ohio formed the first industrial hygiene units beginning in 1913. The Pennsylvania State Department of Health, Division of Occupational Health as one example, published Hygienic Information Guides on airborne hazards such as asbestos and nickel carbonyl and recommended practices that followed the hierarchy of controls (Commonwealth of Pennsylvania, Department of Health, Division of Occupational Health 1964, 1965). The American Society of Safety Engineers (ASSE 2015), American Conference of Governmental Industrial Hygienists (ACGIH 2015), and American Industrial Hygiene Association (AIHA 2015) were organized during this period. All three organizations published standards and guidelines, many containing PtD provisions, as a service to their members and the public.

Standard OSH textbooks of the mid-1900s, such as the National Safety Council's *Accident Prevention Manual*, called for the inclusion of safety in the design and layout of facilities (NSC 1946). Patty (1958) emphasized the importance of presenting the basic principles of industrial hygiene to architects and designers "so that industrial hygiene will start with the blueprints of factories and machines" (p. 15). The US Public Health Service, Division of Occupational Health (USPHS 1965), and its successor, NIOSH (1973), outlined general methods for the control of occupational exposures to injurious materials or conditions in their training course manuals for industrial hygiene engineers and chemists. These methods followed a control hierarchy similar to the version supported under PtD, although strict adherence to a hierarchical approach was not prescribed in these early editions of the training manual.

The concept of system safety was formally introduced in 1947 by the Institute of Aeronautical Sciences. Systems safety is defined as:

> The application of special technical and managerial skills to the systematic, forward-looking identification and control of hazards throughout the life cycle of a project, program, or activity. (Roland and Moriarty 1990, p. 8)

This concept closely parallels the principle of PtD. The system safety movement led to promulgation by the Department of Defense (DOD) of MIL-STD-882: "System Safety" (DOD 2012). MIL-STD-882 calls for analysis of each component and operational procedure of a proposed system for its contribution to the hazard of the process. The standard requires both hazard likelihood and consequences to be taken into account and risk mitigation measures to be pursued in accordance with the classical hierarchy of controls. It applies to all Department of Defense systems and facilities and to all phases in the life cycle of the system. System safety aligns more closely with PtD than any other formalized OSH event or activity of the twentieth century.

### 14.3.3 1970–Present

The number and nature of PtD-related events increased significantly from 1970 to the present. These events involve both regulatory and voluntary efforts and have taken place in many countries around the globe.

***14.3.3.1 Occupational Safety and Health Administration (OSHA) Act*** The US Congress enacted the Occupational Safety and Health Act in 1970 (*OSH* 1970). This legislation was driven by a lack of consistency and underperformance among the states in protecting worker safety and health. PtD was not an explicit requirement in the act. However, essential features of PtD such as use of the hierarchy of controls were included in occupational health regulations issued by the OSHA in 1971 (USDOL and OSHA 1971). The adoption of consensus standards such as the National Fire Protection Association's (NFPA 2015) flammable and combustible liquids code also promoted the use of PtD methods such as engineering assessments and engineering solutions throughout industry.

***14.3.3.2 Institute for Safety Through Design*** The US National Safety Council established an Institute for Safety through Design in 1995 (Manuele 2008). The Council formed the Institute because "…in investigational reports of occupational injuries and illnesses…design causal factors were not adequately addressed" (Manuele 2008, p. 127). The key objective of the Institute was set forth in its definition of safety through design:

> …the integration of hazard analyses and risk assessment methods early in the design and engineering stages, and taking the actions necessary so that risks of injury or damage are at an acceptable level. (Manuele 2008, p. 127)

Key accomplishments of the Institute included the expansion of knowledge on safety through design by conducting seminars, workshops, and safety conferences. Christensen and Manuele (1999), two leaders of the initiative, authored a textbook on safety through design. The Institute was disbanded in 2005, but its influence including the *Safety through Design* text continues to this day.

***14.3.3.3 Inherently Safer Chemical Processes*** The Center for Chemical Process Safety (CCPS), a technology alliance of the American Institute of Chemical Engineers, undertook a program in the mid-1990s to promote the concept and benefits of inherently safer design. The Center was created in 1985 by the American chemical industry in response to the Bhopal, India, toxic vapor cloud tragedy of 1984 and other major process safety incidents. The inherent safety initiative included the publication of a concepts book: *Inherently Safer Chemical Processes – A Life Cycle Approach* (CCPS 1996, 2009). There are many similarities between the inherently safer design concept and PtD. The inherently safer approach focuses on the properties of materials and processes and eliminating or reducing the hazard at its source rather than accepting the hazard and attempting to mitigate the effects. A detailed "Inherently Safer Technology Checklist" with more than 150 inherently safer alternatives is provided in Appendix A of the CCPS concepts book (CCPS 2009, pp. 293–330).

***14.3.3.4 Construction Industry Institute: Design for Safety*** The benefit to construction job site safety of applying PtD methods in the design phase of new projects has been studied extensively by the US-based Construction Industry Institute (CII). The Institute concluded from its "Design for Safety" research that

> …incorporating safety measures into the design phase directly improves safety on the job site, and ultimately leads to lower total installed cost due to fewer dollars spent on mediating hazards and acquiring construction insurance. (CII 1996, p. v)

CII's design for safety research team in the course of conducting their research accumulated more than 400 design suggestions that focus on improving construction worker safety. These suggestions were incorporated into an interactive computer program, the "Design for Construction Safety ToolBox," to assist CII's member companies in planning and designing facilities (CII 2010).

***14.3.3.5 Designing for Safety to Improve Construction Worker Safety and Health*** The Center to Protect Workers' Rights (CPWR) sponsored research to determine the viability of designing for safety to improve construction worker safety and health (Gambatese et al. 2005). Their research included a review of the literature and OSHA construction safety standards and a pilot survey of architects and engineers engaged in construction design. The review of OSHA standards yielded a list of design modifications that, if implemented, would remove the hazard and eliminate the need for job site safety measures. The pilot survey results revealed that "…the concept of designing for safety is gaining interest in the construction community but…is not part of standard design practice" (Gambatese et al. 2005, p. 17). Key barriers to implementing PtD by designers are "limited knowledge of the concept, lack of motivation or willingness to embrace the concept, concerns about legal liability, and the perception that designing for safety can lead to increased project costs, schedule problems, and diminished design creativity" (Gambatese et al. 2005, p. 18). Counter to this latter perception is the earlier finding by Hecker et al. (2004) that designing for safety is relatively easy to implement and is effective in reducing construction worker hazards.

***14.3.3.6 OSHA Alliance Program Construction Roundtable*** OSHA formed the Alliance Program Construction Roundtable in 2004 to provide a forum for sharing construction-related safety information and develop compliance assistance tools and resources (OSHA 2015a, b). A design for safety (DFS) workgroup was convened in 2005 and operated through 2009 under the auspices of the Alliance Roundtable. Numerous tools and resources were developed and disseminated by the workgroup including a 2–4-hour training course on design for construction safety. A website was created and continues to provide valuable news, updates, web links to DFS and PtD training course offerings, presentations on the subject, and tools as aids in applying PtD principles in construction safety (Toole 2015).

***14.3.3.7 OSH Management System Standards*** The prevailing, consensus standards on OSH management systems, ANSI/AIHA/ASSE Z10 (ANSI/AIHA/ASSE 2012) and BSI OHSAS 18001 (BSI OHSAS 18001 2007), include essential features of PtD. The "Implementation" section of ANSI/AIHA/ASSE Z10 requires risk assessments to be conducted as part of the design, redesign, and minor modification of processes and for all applicable stages in the life cycle of the organization (ANSI/AIHA/ASSE 2012, pp. 17–18). Establishing a process to achieve feasible risk reduction is also a requirement of the standard (ANSI/AIHA/ASSE 2012, p. 18). A risk reduction strategy is prescribed according to the familiar hierarchy of controls: elimination, substitution, engineering controls, warnings, administrative controls, and personal protective equipment. BS OHSAS 18001 follows a similar path with requirements for hazard identification, risk assessment, determination of necessary controls, and MOC for all aspects of an organization's activities. A hierarchy is prescribed for use when determining controls to eliminate or reduce risks, and detailed guidance in applying the hierarchy is provided in the companion document to the standard (BSI OHSAS 18002 2008).

***14.3.3.8 US National Initiative on PtD*** A national initiative on PtD was launched by NIOSH in 2007 (Schulte et al. 2008). "The ultimate objective of the PtD initiative is to achieve a cultural change so that designing out occupational hazards is considered the norm" (NIOSH 2014, p. 2). The initiative includes activities that are useful in identifying, assessing, and eliminating or controlling workplace hazards throughout the life cycle of facilities, processes, work methods and operations, equipment, tools, products, materials, new technologies, and the organization of work (NIOSH 2014). The plan to implement the initiative includes goals and objectives in five focus areas: research, education, practice, policy, and small business (NIOSH 2010). In the focus area of research, a survey of 200 large US companies revealed that "…a majority of companies were aware of PtD and included PtD principles in their operations" (NIOSH 2014, p. iv). Research has also documented improvement in business value when the essential features of PtD are followed as was discussed earlier (Biddle 2013).

In the focus area of education, PtD essential features have been included in safety management and engineering textbooks and in education modules targeting students in architectural, civil, and construction engineering. Positive outcomes in the "practice" focus area have been reported in numerous peer-reviewed journal articles and two international symposia dedicated to PtD (Lin 2008; NIOSH 2011). Incorporating PtD methods into the design and redesign process has also been addressed as an output of this focus area (Renshaw 2013a).

In the policy area, PtD concepts have been added to consensus standards, which are used by engineers, safety professionals, industrial hygienists, the semi-conductor industry, and general industry. A highlight in the policy area was the issuance of a new ANSI standard on PtD with tools and guidance to assist organizations in integrating PtD into their OSH management systems (ANSI/ASSE 2011). In the focus area of small business, the NIOSH nanotechnology center has conducted exposure assessments at nanomaterial production and user facilities in support of research on control technology for nanomaterial operations (NIOSH 2014, p. 22).

The original plan for the PtD National Initiative was scheduled to be implemented over the period of 2007–2014 and was extended in 2015 for an additional five years. The initiative has had a positive impact by raising awareness and understanding of PtD and its benefits and by providing guidance and tools to implement the principle.

***14.3.3.9 UK Construction (Design and Management) Regulations*** The UK Health and Safety Executive introduced Construction (Design and Management) Regulations (CDM) in 1994 with subsequent revisions (UKHSE 1994, 2015). The regulations prescribe what those involved in construction need to do to protect themselves and others affected by the work. The regulations bring together in a single package requirements for those involved in project design and those responsible for construction. The CDM regulations define roles and duties of the principal duty holders for projects that include clients, designers, contractors, and workers. The designer's duties under CDM are of special interest as they are closely aligned with PtD thinking. The designer's duty, "When preparing or modifying designs, [is] to eliminate, reduce or control foreseeable risks that may arise during: construction, and the maintenance and use of a building once it is built" (UKHSE 2015, p. 6).

A noteworthy feature of the CDM regulations are the "General Principles of Prevention" (UKHSE 2015, p. 73), which apply to all industries, including construction. The principles of prevention are to:

1. Avoid risks.
2. Evaluate the risks which cannot be avoided.
3. Combat the risks at source.
4. Adapt the work to the individual, especially regarding the design of workplaces, the choice of work equipment and the choice of working and production methods, with a view, in particular, to alleviating monotonous work, work at a predetermined work rate and to reducing their effect on health.
5. Adapt to technical progress.
6. Replace the dangerous by the non-dangerous or the less dangerous.
7. Develop a coherent overall prevention policy which covers technology, organization of work, working conditions, social relationships and the influence of factors relating to the working environment.
8. Give collective protective measures priority over individual protective measures.
9. Give appropriate instructions to employees.

These principles apply to all industries and provide the basis to identify and control risks on construction projects. Enforcement of the CDM regulation has been recognized as a key factor in the United Kingdom's steady reduction of construction workplace fatalities since 1996 (Gibb 2011).

***14.3.3.10 Australian National Initiative*** The National Occupational Health and Safety Commission's (NOHSC 1994) "National Standard for Plant" was one of the earliest Australian initiatives on PtD. The standard described a process for systematically identifying hazards and assessing, controlling, and reviewing risks on an ongoing basis. The target audiences were designers, manufacturers, and suppliers. The standard covered design, installation, use, maintenance, and decommissioning. A follow-on publication consisted of a guide to risk management for designers, manufacturers, importers, suppliers, and installers of plants (NOHSC 1995).

Further study by the Commission focused on the contribution of design factors to occupational injuries and fatalities (ASCC 2005). Over a two year period from 2000 to 2002, the study revealed 37% of work-related fatalities involved design-related issues. Following publication of these findings the Australian Safety and Compensation Council provided guidance on eliminating workplace hazards and controlling risks at the design stage or in the modification stage for buildings, structures, equipment, vehicles, and the processes used for work (ASCC 2006).

The latest Australian publication on this subject is a code of practice on safe design of structures (SWA 2012). This code provides guidance to architects, building designers, engineers, owners, builders, and developers. The document covers key elements of safe design including (i) use of a risk management approach, (ii) consideration of the life cycle, (iii) knowledge and capability, (iv) consultation, cooperation, and coordination, and (v) information transfer. Valuable appendices to the document include roles and responsibilities, a safety in design checklist, and case studies on safety in design.

***14.3.3.11 Singapore Design for Safety Initiative*** The Singapore Workplace Safety and Health Council (SWSHC) collaborated with the Ministry of Manpower in 2008 to launch a PtD-related initiative called Design for Safety (DfS). The objective of the initiative is to improve construction site safety by reducing risk at the source. A cornerstone of DfS is the Council's *Guidelines on Design for Safety in Buildings and Structures* (SWSHC 2011). These guidelines include a design review process to assist project teams in delivering safe designs and transfer of vital safety and health information along the construction process chain. Duties of the designer, client (owner), and main contractor to facilitate risk reduction at the source are included in the guidelines as well as sample checklists for use in each of the key stages of the construction project.

A Design for Safety coordinator is required as part of the DfS process. The DfS coordinator is appointed by the client and is expected to follow the project from the concept design stage, through construction and handover to the client for maintenance. The coordinator's duties include facilitation of design reviews and maintenance of a safety and health risk register. The register provides a record of risks identified in the design stage as well as residual risks that cannot be removed through design changes. Contractors are to be apprised of these residual risks during the tendering process so they can accurately price for the project. The register is handed over to the client upon project completion.

Thus far the information in this chapter has focused on the "what", the "why," and the historical development of PtD as a guiding principle of OSH. The remainder of the chapter guides the OSH professional through the implementation process that addresses the "how" of PtD.

## 14.4 IMPLEMENTING PtD

Implementing PtD is a significant undertaking but need not be viewed as a daunting task. The best way to ensure successful implementation and sustainability is to incorporate the essential features of PtD into an organization's design and redesign process. This section guides the OSH professional through the implementation of PtD based on a recommended structure and three-step process.

### 14.4.1 Seek First to Understand PtD

OSH professionals should begin the process of implementing PtD by "seek[ing] first to understand..." (Covey 1990, p. 237) the design and redesign process of their organization. The key to such understanding lies in the answers to these three questions:

1. What are the essential features of PtD?
2. Who owns the design and redesign process and how does it work?
3. Who are the key partners that enable PtD to succeed?

***14.4.1.1 What are the Essential Features of PtD?*** The essential features of PtD are repeating themes in the historical events and initiatives cited in the preceding section. They are listed as mandates in Table 14.2 for the sake of emphasis and simplicity. These features need to be well understood and accepted in order to successfully incorporate PtD into the design and redesign process.

***14.4.1.2 Who Owns the Design and Redesign Process and How Does It Work?*** The design and redesign process normally resides at the heart of an organization's capital project delivery (CPD) process: "The mechanism used within an organization to manage the development and execution of capital projects from idea conception through start-up and process optimization" (Renshaw 2007, p. 12).

The CPD process is typically owned by the central engineering or construction function within an organization and may be co-owned by the manufacturing, operations, and commercial business functions. The CPD process is normally executed through a stages-and-gates approach. Acquiring a good understanding of CPD, the project flow, the stages and gates and maintaining a collaborative relationship with the process owners are essential for successful implementation of PtD.

***14.4.1.3 Who Are the Key Partners That Enable PtD to Succeed?*** Knowing who the key internal and external partners are and how to collaborate as an OSH professional with them is another key to PtD implementation. There can easily be twenty or more partners involved in the design and redesign process as shown by the example in Table 14.3 (adaptation of information provided by D.S. Heidel, personal communication, 18 March 2011). There are three major partners: (i) the organization's internal EHS function, (ii) the internal engineering, operations, and technical functions; and

**TABLE 14.2 Essential Features of PtD**

1. Eliminate or reduce hazards and control residual risks at the source or as early as possible in the life cycle of the enterprise
2. Include the design and redesign of all facilities, processes, products, and operations
3. Consider all phases in the life cycle of the enterprise from concept stage through design, construction, start-up, operation, maintenance, shutdown, and decommissioning
4. Apply appropriate hazard evaluation and risk assessment techniques throughout the life cycle of the enterprise
5. Manage all changes to facilities, processes, products, and operations
6. Record and track to closure all tasks identified and all follow-up actions from hazard evaluation and risk assessment studies, design reviews, and management of change reviews
7. Integrate the hazard evaluation, risk assessment, and design review process into the project flow
8. Follow the hierarchy of controls in choosing and applying hazard elimination, hazard reduction, and risk control measures
9. Ensure that prevention methods are included in all designs that impact workers, contractors, and others affected by the enterprise
10. Provide a framework for consultation, cooperation, and coordination among key stakeholders of the design and redesign process

*Note:* The term enterprise includes facilities, processes, products, substances, operations, equipment, machinery, tools, work processes, personnel, and the organization of work.

**TABLE 14.3 Key PtD Partners**

| Internal<br>Environment, health, safety | Internal<br>Engineering, operations, technical |
|---|---|
| Occupational safety | Project owner |
| Process safety | Project team leader |
| Industrial hygiene | Lead process engineer |
| Occupational health | Engineering specialties |
| Environment | Operations/site management |
| Internal fire brigade | Technical/maintenance |
| Security | Procurement |

| External |
|---|
| Contractor engineering/construction |
| Safety contractor engineering |
| EHS coordinator design |
| Safety coordinator construction |
| Notified body/official organizations |
| Public/community fire brigade |
| Insurance |
| Suppliers |

(iii) the external partners involved with design, construction, regulatory authorizations, insurance, the suppliers, and the community. Cooperation, coordination, and collaboration among the partners are essential for PtD to work. The three "C's" are explicit requirements of Singapore's national PtD initiative (SWSHC 2011). They help to ensure all partners in the CPD process work together and are "singing from the same sheet of music".

With this basic understanding of the essential features of PtD, of the design and redesign process (CPD), and of potential key partners, the OSH professional can move forward with implementation. The path forward involves three major activities:

1. Setting policy and standards.
2. Establishing work processes and procedures.
3. Applying tools and practices.

The structure and interrelationship between these activities and the overarching principle of PtD are shown in Figure 14.1.

### 14.4.2 Setting Policy and Standards

The first major activity in implementing PtD involves policy and standards setting. The importance of formally adopting PtD as a guiding principle is captured in this informative guidance from ANSI Z10 (ANSI/AIHA/ASSE 2012, p. 31):

> The OHS policy of an organization provides a starting point for everything the organization wants to achieve in occupational health and safety (ANSI/AIHA/ASSE 2012, p. 31). It also establishes an overall sense of direction, ...sets the principles of action for an organization, and demonstrates the formal commitment of an organization, particularly, ...top management, toward effective OHS management.

**Figure 14.1** Structure for implementing PtD.

**TABLE 14.4 Model Environmental, Health, and Safety Policy Adapted to Include PtD**

XYZ Company
Model Environmental, Health, and Safety Policy[a]
20XX

The XYZ Company is the world's leading manufacturer of ___ products. Our environmental, health, and safety values are of utmost importance to us. They are embodied in the guiding principles set forth in this policy. They reflect our respect and care for the environment, our employees, contractors, customers, and communities. We are committed to incorporating these values into everything we do as we seek to improve the quality of life and the environment through our products and services.
Our guiding principles are:

- We will design our businesses, processes, and products with full consideration for the needs of the present global community and the impact of our design decisions on the ability of future generations to meet their needs.
- *We will include prevention considerations in the design and redesign of all facilities, processes, products, and operations and will incorporate safe design methods into all phases of hazard elimination and risk control.*
- We will continuously review and improve our worldwide operations, processes, and products, with the goal of making them free of adverse environmental, health, and safety impacts for all of our stakeholders.
- We will meet or exceed all applicable laws, regulations, and XYZ Company standards.
- ---------
- ---------

Every employee and contractor is responsible for compliance with this policy. We will audit our performance and the Board of Directors will monitor our commitments and progress.

President/CEO/Chairman

[a] *Source:* Adapted from the former Rohm and Haas Company.

***14.4.2.1 Setting Policy*** Most major organizations have in place OSH or environmental, health, and safety (EHS) policies that document and communicate their guiding principles and commitment to responsible performance in this functional area. It is desirable to introduce PtD as one of the guiding principles in such a policy. An example of specific language that can be used for this purpose is highlighted in italics in Table 14.4 (Rohm and Haas Company 2008).

Elevating PtD to the level of a guiding principle lends the same visibility and importance to PtD as other EHS principles such as sustainability, continuous improvement, and compliance with laws and regulations. The amended EHS policy can serve as a means by which top management can communicate its adoption of PtD as a principle throughout the organization.

***14.4.2.2 Setting Standards*** Standards provide specific information on how management expects the principles and policies of an organization to be implemented. Standards may define the level of performance required and explain what should be done or how it should be done. The focus in setting a PtD standard should be on an organization's management system standard for OSH. If an organization operates within the framework of ANSI/AIHA/ASSE Z10 (ANSI/AIHA/ASSE 2012) or OHSAS 18001 (BSI OHSAS 18001 2007), PtD should be incorporated into the organization's management system standard of choice. Appropriate PtD language will need to be inserted in ANSI/AIHA/ASSE Z10 that addresses OHS policy and design review and management of change. Comparable sections of

**TABLE 14.5 OHS Management System Standard with Suggested Wording (Bold Type) to Incorporate PtD Guiding Principle and Essential Features**

| | |
|---|---|
| **3.1 Management leadership**[a]<br>**3.1.2 Occupational health and safety (OHS) policy**<br>The organization's top management shall establish a documented OHS policy as the foundation for the OHSMS. This policy shall include a commitment to:<br><br>A. Protection and continual improvement of employee health and safety<br>B. *Inclusion of prevention considerations in design and redesign of all facilities, equipment, processes, work methods, and products and incorporation of safe design methods into all phases of hazard elimination and risk mitigation*<br>C. Effective employee participation<br>D. Conformance with the organization's health and safety requirements<br>E. Compliance with applicable OH&S laws/regulations | **5.1 OHSMS operational elements**[a]<br>**5.1.3 Design review and management of change**<br>The organization shall establish a process ***that includes prevention through design methods*** to identify and take appropriate steps to prevent or otherwise control hazards at the design and redesign stages and for situations requiring management of change to reduce potential risks to acceptable levels. The process for design and redesign and management of change shall include:<br><br>A. Identification of tasks and related health and safety hazards ***as early as possible in the life cycle of facilities, equipment, processes, work methods, and products***<br>B. Recognition of hazards associated with human factors including human errors caused by design deficiencies<br>C. Review of applicable regulations, codes<br>D. **Application of control measures (hierarchy of controls – section 5.1.2)**<br>E. A determination of the appropriate scope<br>F. Employee participation |

[a] Suggested wording to incorporate PtD in ANSI/AIHA Z10-2012.

OHSAS 18001 are "OH&S Policy" and "Hazard Identification, Risk Assessment and Determining Controls". An example of appropriate PtD language is given in Table 14.5.

Organizations that have not adopted an OSH management system standard may choose to incorporate PtD language into internal technical standards on design review and MOC or address these activities through specific work processes and procedures.

### 14.4.3 Establishing Work Processes and Procedures

The second major activity in implementing PtD involves the establishment of work processes[1] and procedures,[2] or modification of existing ones, to enable ongoing consideration of prevention features in designs and redesigns. Examples include the processes and procedures associated with design reviews, hazard evaluation, risk assessment, and application of risk control measures. This is a critical step in which essential features of PtD are incorporated into two key work processes, the capital project delivery (CPD) process and the MOC process.

***14.4.3.1 Capital Project Delivery Process*** Most organizations use this work process to manage the development and execution of major projects. Capital projects normally involve major expenditures by an organization to "...acquire or upgrade fixed, physical, non-consumable assets, such as buildings, equipment, or a new business" (Whatis.techtarget.com n.d.). Such projects are ideal candidates for PtD intervention. They involve extensive planning and design work and are well suited for consideration of prevention methods early in the life cycle of the project. Smaller capital projects and non-capital projects, such as routine repairs, may also fall under the CPD process even though they may be classified as operational rather than capital expenditures (Accounting Coach n.d.). Where projects do not fall under the CPD work process, the MOC work process may be used. Application of PtD through the MOC process is discussed in Section 14.4.3.2.

*14.4.3.1.1 Stages-and-gates Approach* The CPD process typically follows a stages-and-gates approach (Dysert 2002; Lawson et al. 1999). Discreet stages cover the life cycle of a project from initiation to project closeout. Decision or gate meetings are conducted by the project team at the conclusion of each project stage. The purpose of these

[1] The term "process" refers to linked activities with the purpose of producing a product (or service) for a customer (user) within or outside the organization. The term "work processes" refers to [an organization's] most important internal value creation processes. They might include product design and delivery, customer support, supply chain management. [An organization's] key work processes frequently relate to [their] core competencies and to the factors considered important for business growth (Baldridge 2009).

[2] "Procedures" define the specific instructions necessary to perform a task or part of a process. Procedures can take the form of a work instruction, a desktop procedure, a quick reference guide, or a more detailed procedure (KCG Consultant Group, 2010).

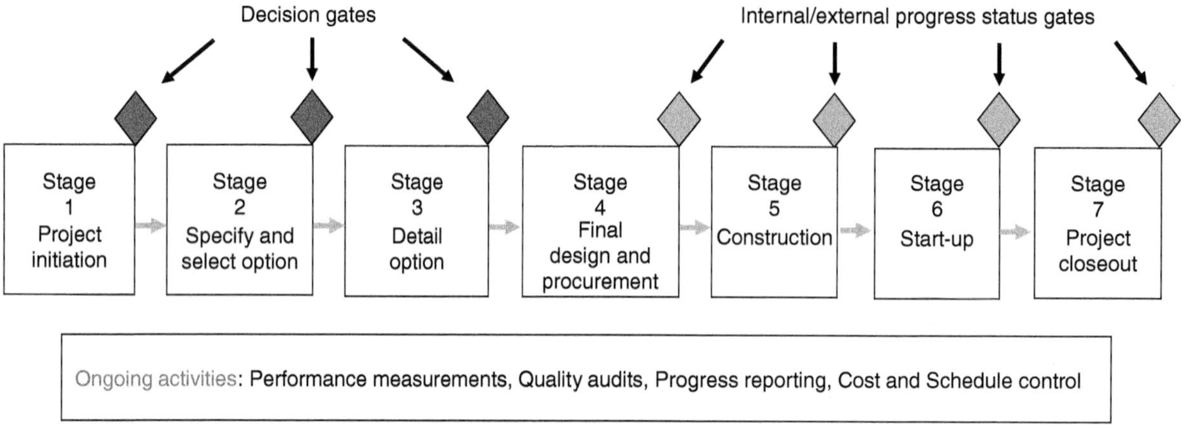


**Figure 14.2** Stages and decision gates in the capital project delivery process. *Source:* Adapted from Rohm and Haas Company.

meetings is to (i) determine the progress of the project in relation to schedule, (ii) confirm that time-sensitive milestones have been met, (iii) address unanticipated changes and obstacles to achieving the project plan, and (iv) assess whether the project should continue as planned, be altered, or terminated. A seven-stage project example is shown in Figure 14.2.

*14.4.3.1.2 Project Execution and Design Deliverables* A key feature of the stages-and-gates process is the selection of project execution and design deliverables. These are measurable outputs from major tasks carried out in each project stage. An example for a project involving a new chemical manufacturing facility is given in Table 14.6.

*14.4.3.1.3 OSH Deliverables* The CPD approach provides a unique opportunity to define OSH deliverables in parallel with execution and design deliverables for each project stage. An example of key OSH deliverables for a project involving a new chemical manufacturing facility is shown in Table 14.7. The selection and scheduling of hazard evaluation and risk assessment studies and the determination of risk control strategies can also be formalized as OSH deliverables. Further discussion of these methods and strategies is included in Section 14.4.4.

*14.4.3.1.4 Verification* The main purpose of gate meetings as noted earlier is to assess a project's progress relative to plan and address unanticipated changes and obstacles. A key aspect of the assessment is verification that an appropriate level of engineering design and documentation have been provided, that the design is valid, and that the necessary technical and OSH reviews have occurred. CPD teams perform this verification by a variety of methods. One method involves the use of an OSH Deliverables Planning Template of the type shown in Table 14.8 (Renshaw 2011). The template includes a sign-off requirement for those who are responsible for each deliverable. Additional sign-offs may be required by others in a functional approval role, such as EHS regulatory experts, and by the project manager as shown in this example.

An electronic sign-off process has been reported by a biotechnology company as part of their software expert system for conducting EHS design reviews (Bender 2011). The system assists the company's EHS professionals in identifying hazards and appropriate control measures through a series of checklist questions with links to reference standards and guidelines. The system also ensures through an operational readiness review that all gate reviews have been completed and design solutions have been incorporated into the final design and installation.

***14.4.3.2 Management of Change*** The MOC work process, like CPD, plays a key enabling role in implementing PtD. MOC has its origin in quality management where unwanted change can lead to major product quality and customer satisfaction problems. The following definition of MOC clarifies what it is and why it is of such importance to the field of OSH:

> Management of change (MOC) is a process for evaluating and controlling modifications to facility design, operation, organization, or activities – **prior to implementation** – to make certain that no new hazards are introduced and that the risk of existing hazards to employees, the public, or the environment is not unknowingly increased. (CCPS 2008a, p. 1)

**TABLE 14.6 Example of Key Project Execution and Design Deliverables for a CPD Project Involving a New Chemical Manufacturing Facility**

| **Stage 1 Project initiation** → | **Stage 2 Specify and select option** → | **Stage 3 Detail option** → | **Stage 4 Final design and procurement** → | **Stage 5 Construction** → | **Stage 6 Start-up** → | **Stage 7 Project closeout** |
|---|---|---|---|---|---|---|
| Project mission, initial goals and objectives, scope 1.1.1 | Refined goals and objectives 2.1.1 | Finalized goals and objectives 3.1.1 | Updated goals and objectives 4.1.1 | Detailed construction plan and schedule 5.1.1 | Commissioning and start-up plan and detailed schedule 6.1.1 | Closeout plan 7.1.1 |
| Generate and initial screen of options 1.1.2 | Select best option 2.1.2 | Finalized project scope document (IBL/OBL), project specifications, standards 3.1.2 | P&IDs, mechanical, electrical, civil, structural, and control system programs all complete 4.1.2 | Construction roles and communication plan 5.1.2 | Operating rates documented 6.1.2 | Final acceptance documentation 7.1.2 |
| Project execution strategy 1.1.3 | Preliminary project execution plan and schedule 2.1.3 | Project execution plan and schedule 3.1.3 | Detailed spare parts list 4.1.3 | Facility mechanically complete (approved for commissioning) 5.1.3 | Facility operating to specifications 6.1.3 | Project critique, customer feedback, closeout report 7.1.3 |
| Afford to spend model 1.1.4 | Preliminary design basis and process description, prelim. P&IDs, PFDs, control philosophy, site plan, raw material list, prelim. major/critical equipment lists 2.1.4 | Finalized design basis, process description, PFDs, P&IDs, control philosophy, site plan, list of materials, major equipment quotes/bids 3.1.4 | All equipment purchase orders issued 4.1.4 | As-built, quality assurance, spare parts, turnover documentation 5.1.4 | Revised operating and maintenance procedures and documentation for design changes (as-built) issued 6.1.4 | Appropriate learnings and design ideas communicated 7.1.4 |
| Relative cost of selected options 1.1.5 | Finalized economic analysis/conceptual cost estimate 2.1.5 | Appropriation request (AR) submitted 3.1.5 | Cost compliance report 4.1.5 | Construction cost report 5.1.5 | Revised start-up cost estimate 6.1.5 | Project and equipment files submitted to facility 7.1.5 |
| Prevention through design in scope 1.1.6 | Prevention through design addressed 2.1.6 | Prevention through design addressed 3.1.6 | Prevention through design addressed 4.1.6 | Prevention through design addressed 5.1.6 | Prevention through design addressed 6.1.6 | Prevention through design documented 7.1.6 |

*Source:* Adapted from Rohm and Haas Company.

**TABLE 14.7 Example of Key OSH Deliverables for a CPD Project Involving a New Chemical Manufacturing Facility**

| Stage 1 Project initiation | → | Stage 2 Specify and select option | → | Stage 3 Detail option | → | Stage 4 Final design and procurement | → | Stage 5 Construction | → | Stage 6 Start-up | → | Stage 7 Project closeout |
|---|---|---|---|---|---|---|---|---|---|---|---|---|
| OSH project plan 1.2.1 | | OSH design review requirements defined 2.2.1 | | OSH design review complete, recommendations issued 3.2.1 | | OSH design review recommendations resolved and actions tracked 4.2.1 | | Pre-construction safety plan 5.2.1 | | Pre-start-up OSH review completed and actions addressed 6.2.1 | | OSH project evaluation 7.2.1 |
| Regulatory and permit plan 1.2.2 | | OSH hazard analysis and risk assessment requirements defined, studies initiated as needed 2.2.2 | | Hazard analysis: risk assessment studies complete, recommendations issued 3.2.2 | | Hazard analysis: risk assessment study recommendations resolved, risk targets met, actions tracked 4.2.2 | | Construction permits 5.2.2 | | Completed site acceptance testing and checks of equipment, OSH devices, critical retrofits 6.2.2 | | Assessment of residual risk versus targets 7.2.2 |
| Early identification of significant and unique OSH hazards and risks 1.2.3 | | Regulatory and permitting requirements defined 2.2.3 | | Regulatory and permitting documents complete 3.2.3 | | Appropriate permit application and regulatory approval 4.2.3 | | Construction safety program effectiveness verification 5.2.3 | | Preliminary industrial hygiene exposure monitoring complete 6.2.3 | | Prevention through design compliance verification 7.2.3 |
| Early identification of regulatory prohibition and restrictions on hazardous substances 1.2.4 | | Insurance assessment done, fire protection security, machinery noise and guarding, ergonomic requirements finalized 2.2.4 | | Equipment and machinery noise, guarding, ergonomic specifications approved, checks and test protocols issued 3.2.4 | | Detailed drawings reviewed and approved 4.2.4 | | Regulatory permit completion and approval verification 5.2.4 | | All incidents investigated and impact of root causes on design/operation addressed 6.2.4 | | Feedback on OSH standards and modifications initiated as appropriate 7.2.4 |
| Early identification of prevention through design opportunities 1.2.5 | | Prevention through design addressed 2.2.5 | | Prevention through design addressed 3.2.5 | | Prevention through design addressed 4.2.5 | | Prevention through design addressed 5.2.5 | | Prevention through design addressed 6.2.5 | | Prevention through design documented 7.2.5 |

*Source:* Adapted from Rohm and Haas Company.

**TABLE 14.8 OSH Deliverables Planning Template**

Project:

Manager:

Date:

| OHS deliverable | Description | Date completed mm/dd/yy | Deliverable documented (Y/N) | Actions tracked (Y/N) | Project. manager reviewed (Y/N) |
|---|---|---|---|---|---|
| 1.2.1 | **OSH project plan** | | | | |
| 1.2.2 | Regulatory and permit plan | | | | |
| 1.2.3 | Early identification of significant and unique OSH hazards and risks | | | | |
| 1.2.4 | Early identification of regulatory prohibition and restrictions on hazardous substances | | | | |
| 1.2.5 | Early identification of prevention.… | | | | |
| 2.2.1 | **OSH design review requirements defined** | | | | |
| 2.2.2 | Hazard analysis and risk assessment requirements defined, studies initiated | | | | |
| 2.2.3 | Regulatory and permitting requirements defined | | | | |
| 2.2.4 | Insurance assessment done, fire protection, security, machinery noise and guarding, ergonomics requirements finalized | | | | |
| 2.2.5 | Prevention through design addressed | | | | |
| 3.2.1 | **OSH design review complete, recommendations issued** | | | | |
| 3.2.2 | Hazard analysis and risk assessment complete, recommendations issued | | | | |
| 3.2.3 | Regulatory and permitting documents complete | | | | |
| 3.2.4 | Equipment and machinery noise, guarding, ergonomic specifications approved, checks and test protocols issued | | | | |
| 3.2.5 | Prevention through design addressed | | | | |
| 4.2.1 | **OSH design review recommendations resolved and actions tracked** | | | | |
| 4.2.2 | PHA recommendations resolved, risk targets met, actions tracked | | | | |
| 4.2.3 | Appropriate permit application and regulatory approval | | | | |
| 4.2.4 | Detailed drawings reviewed and approved | | | | |
| 4.2.5 | Prevention through design addressed | | | | |
| 5.2.1 | **Pre-construction safety plan** | | | | |
| 5.2.2 | Construction permits | | | | |
| 5.2.3 | Construction safety program effectiveness verification | | | | |
| 5.2.4 | Regulatory permit completion and approval verification | | | | |
| 5.2.5 | Prevention through design addressed | | | | |
| 6.2.1 | **Pre-start-up OSH review completed and actions addressed** | | | | |
| 6.2.2 | Completed site acceptance testing and checks of equipment, OSH devices, critical retrofits | | | | |
| 6.2.3 | Preliminary industrial hygiene exposure monitoring complete | | | | |
| 6.2.4 | All incidents investigated and impact of root causes on design/operation addressed | | | | |
| 6.2.5 | Prevention through design addressed | | | | |
| 7.2.1 | **OSH project evaluation** | | | | |
| 7.2.2 | Assessment of residual risk versus targets | | | | |
| 7.2.3 | Prevention through design compliance verification | | | | |
| 7.2.4 | Feedback on OSH standards and modifications initiated as appropriate | | | | |
| 7.2.5 | Prevention through design documented | | | | |

*14.4.3.2.1 Scope* The term "modifications" is what differentiates MOC from CPD. The latter process covers primarily new designs and major facilities. MOC addresses redesigns and other modifications to existing facilities and in this role is one of the most important elements of OSH management. The MOC process can be applied to major technical changes, such as modification of a product packaging system, and basic changes such as increasing the flow rate of a pump with a higher-speed motor. The MOC process may also be applied to organizational changes such as redefining the technical qualifications or experience requirements for a control room operator in a chemical process.

*14.4.3.2.2 Importance* The MOC process is strategically important in implementing PtD. MOC enables an organization to incorporate features of PtD into changes at the design stage. MOC typically deals with existing operations and in this sense is not early in the life cycle of a process. But the MOC process provides an opportunity to influence the design of an impending change and thereby eliminate or reduce the hazard and control the residual risk of an ongoing operation.

The MOC process is recognized by the chemical and refining industries and by OSHA as an essential element of process safety management. Facilities covered by OSHA's Process Safety Management of Highly Hazardous Chemicals Rule (USDOL and OSHA 1992) and EPA's Risk Management Plan Rule (for establishments with Program 3 covered processes) (EPA 1996) are required to implement MOC procedures for all changes to facilities, process chemicals, technology, equipment, and procedures except replacement in kind (RIK) changes.

*14.4.3.2.3 How MOC Works* A well-functioning and effective MOC process includes these basic steps (CCPS 2008a, p. 3):

1. An employee first originates a change request.
2. Then qualified personnel, normally independent of the MOC originator, classify and review the request to identify any potentially adverse impacts.
3. Based on this review, and after addressing any additional requirements, a responsible party either approves or rejects the change for execution.
4. If the change is approved, it can be implemented.
5. Before start-up of the change, potentially affected personnel are either informed of the change or provided with more detailed training, if needed.
6. Affected process safety information (PSI) is modified to reflect the change.

A detailed logic diagram showing the workflow associated with the MOC process is given in the CCPS text on this subject (CCPS 2008a, pp. 125–126).

*14.4.3.2.4 MOC Critical Success Factors* Following the step-wise workflow is essential for a successful MOC process. In addition, a commitment by everyone within an organization to these basic principles are critical success factors for MOC:

- Every change must be managed.
- Changes must NOT be made without an appropriate review.
- The type of review should be consistent with the complexity of the change.
- The MOC work process should be simple to use.

The third principle is especially important for the success of MOC. Each change must be properly classified as to level (complexity), and the type of review (sophistication) needs to be matched to the complexity of the change. Figure 14.3 is a graphical representation of this relationship. There are several excellent references on MOC including the CCPS text on this subject (CCPS 2008a), "Guidelines for Managing Process Safety Risks During Organizational Change" (CCPS 2013), and the non-mandatory guidance provided in OSHA's PSM Standard (USDOL and OSHA 1992, appendix C).

### 14.4.4 Applying Tools and Practices

The third major activity in implementing PtD involves the application of tools and practices such that foreseeable hazards and risks are properly assessed and addressed at the design and redesign stage. The tools include hazard evaluation and risk assessment methods such as hazard and operability (HAZOP) studies, occupational exposure assessments,

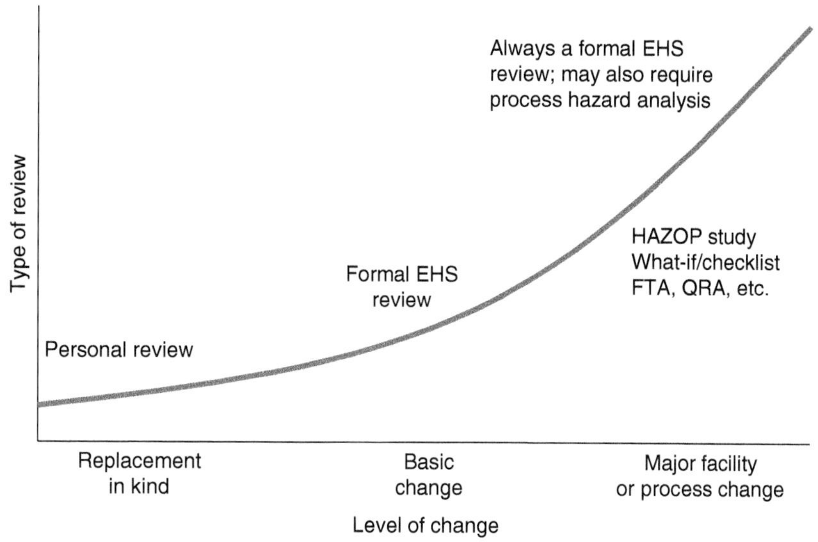


**Figure 14.3** Matching the type of OSH review to the level of change.

design reviews, and quantitative risk assessment. The practices include control strategies and solutions such as those prescribed and recommended in regulations, published standards, and industry codes of safe practice. Guidance in the selection and application of these practices is one of the most important contributions the OSH professional can make to an organization's design and redesign process. It is at this point in the life cycle of an enterprise that project teams and partners bring to life the benefits of PtD by translating safe practices into safe designs.

***14.4.4.1 Hazard Evaluation and Risk Assessment Methods*** These tools include the methods used by organizations and individuals to evaluate hazards and assess risks associated with their facilities and operations. The term "hazard evaluation" includes the identification of hazards and analysis of the consequences resulting from events or exposures to the hazards. Risk assessment incorporates the likelihood or probability of the event or exposure together with the consequence. It is this combination of consequences and likelihood that determines risk.

The number and variety of hazard evaluation and risk assessment methods have greatly increased since OSHA published its list of approved process hazard analysis (PHA) methodologies in 1992.[3]

The OSHA list addresses only process safety hazards (fires, explosions, toxic vapor cloud releases). Many additional methods are available to assess traditional occupational safety hazards such as falls, contact with objects and equipment, and traditional occupational health hazards such as workplace exposures to chemical, physical, and biological agents as well as ergonomic hazards.

A list of selected hazard evaluation and risk assessment methods is provided in Table 14.9. The primary sources in preparing the list were (i) ANSI/ASSE Z590.3-2011: *Prevention through Design: Guidelines for Addressing Occupational Hazards and Risks in Design and Redesign Processes* (ANSI/ASSE 2011); (ii) *Guidelines for Hazard Evaluation Procedures*, 3e[r] (CCPS 2008b); and (iii) standard OSH texts.

Each listing includes a brief description of the method and a reference to the source. No attempt has been made to compile a comprehensive list of all available OSH hazard evaluation and risk assessment methods. This list focuses on proven methods that are known to be relevant and applicable in addressing prevention considerations at the design and redesign stage of facilities, processes, products, and operations.

An important role of the OSH professional is to guide and assist project team members in selecting and scheduling the appropriate methods for use at the optimum stage of the design and redesign process. Figure 14.4 provides an example in which the chosen hazard evaluation and risk assessment methods have been synchronized with the different stages in the CPD process (Renshaw 2013b).

[3] The OSHA-designated methodologies for determining and evaluating the hazards of processes using highly hazardous chemicals are (i) what-if, (ii) checklist, (iii) what-if/checklist, (iv) hazard and operability (HAZOP) study, (v) failure mode and effects analysis, (vi) fault tree analysis, or (vii) an appropriate equivalent methodology (USDOL and OSHA 1992, pp. 6404, 6412).

**TABLE 14.9 Selected Hazard Evaluation and Risk Assessment Methods**

| Method | Description | Reference |
|---|---|---|
| Preliminary hazard analysis | Qualitative hazard evaluation and risk assessment procedure normally conducted in early stages of design process. Involves formulating hazard scenario, describing task, operation, system, exposures, consequence and likelihood, assigning a risk rating, and recommended risk reduction countermeasures. Origin of procedure is system safety | ANSI/ASSE (2011), p. 40, CCPS (2008b), pp. 73–78, DOD (2012) |
| Design safety review | Qualitative hazard evaluation procedure normally conducted by team engaged in design or redesign of facilities, processes, products, and operations. Reviews are normally conducted at key project stages by face-to-face meetings supplemented by completion of checklists prior to meetings. Purpose of reviews is to ensure project is in conformance with design intent, construction standards, and relevant regulatory requirements | ANSI/ASSE (2011), pp. 32–33, CCPS (2008b), pp. 79–83 |
| What-if analysis | Qualitative hazard evaluation procedure in which a group of people use a brainstorming approach to ask questions about potential, undesired events and decide if further risk reduction is warranted | ANSI/ASSE (2011), p. 40, CCPS (2008b), pp. 100–108 |
| Checklist analysis | Qualitative hazard evaluation procedure using a written list of questions to verify the status of a system with respect to standards, practices, anticipated hazards, and protective measures | ANSI/ASSE (2011), p. 41, CCPS (2008b), pp. 93–98 |
| What-if/checklist analysis | Qualitative hazard evaluation procedure combining what-if brainstorming and checklist approaches | ANSI/ASSE (2011), p. 41, CCPS (2008b), pp. 107–114 |
| Hazard and operability study (HAZOP) | Qualitative hazard evaluation procedure using a systematic, scenario-based approach and guide words to identify and evaluate the consequences and impact of deviations from desired practices | CCPS (2008b), pp. 115–134, ANSI/ASSE (2011), p. 41 |
| Failure mode and effects analysis (FMEA) | Qualitative hazard evaluation procedure to assess the reliability of equipment and response of the system to failures | ANSI/ASSE (2011), p. 42, CCPS (2008b), pp. 134–141 |
| Fault tree analysis | Quantitative hazard evaluation procedure that focuses on system failures one at a time and causes of failures. May be used in combination with consequence analyses to perform quantitative risk assessments | ANSI/ASSE (2011), pp. 42–43, CCPS (2008b), pp. 142–157 |
| Event tree analysis | Quantitative hazard evaluation procedure that identifies outcomes of protective system success or failure due to an initiating cause. Used to identify design and procedural weaknesses and countermeasures | CCPS (2008b), pp. 158–166 |
| Risk assessment matrix | Qualitative risk assessment procedure that brings together consequences and likelihood of undesired events in the form of risk scores. Results can be used for risk ranking, judging the acceptability of risks, or need for further risk reduction measures | ANSI/ASSE (2011), pp. 34–38, ANSI/AIHA/ASSE (2012), pp. 49–50, DOD (2012) |
| Layer of protection analysis | Quantitative risk assessment in simplified form to analyze risk of individual scenarios and benefit of independent protection layers for a chemical process | CCPS (2008b), pp. 223–232 |
| Quantitative risk assessment | Quantitative assessment of risk taking into account consequences and frequency of undesired events for a given process and comparing results of assessment to risk criteria | CCPS (1999) |
| Human reliability analysis | Quantitative hazard evaluation procedure to systematically assess the potential and effects of human error on the safety and reliability of a process or operation | CCPS (2008b), pp. 279–287 |
| Facility siting | Qualitative hazard evaluation procedure that assesses risks to personnel in relation to location of facility and process hazards | CCPS (2008b), pp. 288–292, API (2003, 2007) |
| Safety of machinery | Qualitative hazard evaluation and risk assessment procedures for new, modified, or rebuilt power driven machines | ANSI (2010) |
| Occupational exposure and health risk assessment | Qualitative and quantitative health risk assessment procedures for occupational exposures to chemical, physical, and biological agents | Ignacio and Bullock (2006), Mulhausen and Damiano (2011) |
| AIHA ergonomic assessment toolkit | Qualitative and quantitative hazard evaluation procedures to use in analyzing jobs for ergonomic risk factors | AIHA Ergonomics Committee (2011) |
| The AIHA® value strategy | Qualitative hazard evaluation and risk assessment procedure to develop a business case to invest in preventing occupational or environmental injury or illness | Brandt and Silverstein (2011) |

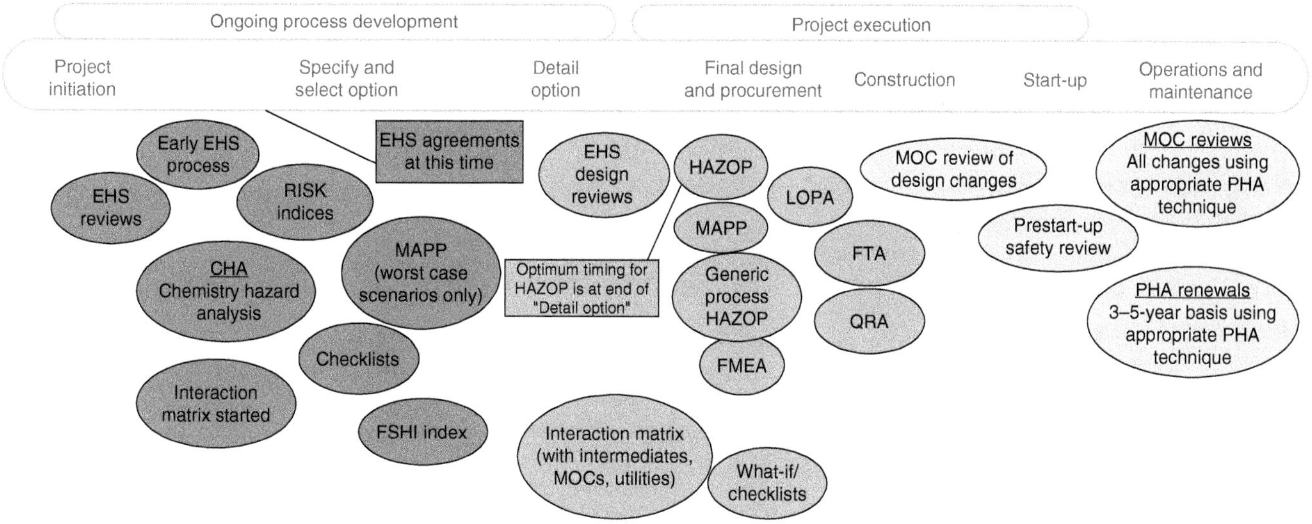


**Figure 14.4** Hazard evaluation and risk assessment methods synchronized with the CPD process. *Source:* Adapted from Rohm and Haas Company.

***14.4.4.2 Control Strategies*** The Australian, Singapore, UK, and US national PtD initiatives (ASCC 2006; NIOSH 2010; UKHSE 2015; SWSHC 2011), the OHS management system standards (ANSI/ASSE/AIHA 2012; BSI OHSAS 18001 2007), and the PtD consensus standard (ANSI/ASSE 2011) all reference the traditional hierarchy of controls as the preferred strategy to eliminate or reduce hazards and control residual risks. Understanding and applying the hierarchy and other relevant control strategies in the design stage are important competencies for OSH professionals and their organizations.

*14.4.4.2.1 Hierarchy of Controls* The NIOSH-led Plan for the National PtD Initiative supports the use of the traditional hierarchy in choosing control strategies:

> PtD utilizes the traditional hierarchy of controls by focusing on hazard elimination and substitution, followed by risk minimization through the application of engineering controls and warning systems applied during design, redesign, and retrofit activities. However, PtD also supports the application of administrative controls and personal protective equipment when they supplement or complement an overall risk minimization strategy and include the appropriate program development, implementation, employee training, and surveillance. (NIOSH 2010, p. 5)

The major categories of the hierarchy are listed with examples in Table 14.10.

The top three categories are preferred because they focus on eliminating or reducing the hazard and controlling residual risk by reliable engineering measures. The lower three categories are less preferable because they rely to a greater extent on mitigative measures and are limited in reliability due to human factors (ANSI/AIHA/ASSE 2012). An expanded discussion of the logic supporting the hierarchy of controls and guidance in its application are provided in the ANSI/ASSE Standard on PtD (ANSI/ASSE 2011, pp. 45–49).

*14.4.4.2.2 Inherently Safer Design* The CCPS inherently safer design concept was cited earlier in the chapter. "It is a philosophy for addressing safety issues in the design and operation of facilities that use or process hazardous chemicals" (Hendershot 2011, p. 48). CCPS developed the concept in 1996 to "...illustrate and emphasize the merits of integrating process research, development, and design into a comprehensive process that balances safety, capital, and environmental concerns throughout the life cycle of the process" (CCPS 2009, p. 1). The inherently safer approach is compatible with the traditional hierarchy of controls. However, inherent safety places a greater emphasis on hazard elimination and reduction than is the case for other control strategies including the traditional hierarchy of controls. The inherently safer approach is based on the following four control methods:

1. *Minimize*: Use smaller quantities of hazardous substances (also called *Intensification*).
2. *Substitute*: Replace a material with a less hazardous substance.

**TABLE 14.10 Hierarchy of Controls**

| **Most effective** | Controls | Examples |
|---|---|---|
| | 1. Elimination | • Design to eliminate hazards, such as falls, hazardous materials, noise, confined spaces, and manual material handling |
| | 2. Substitution | • Substitute for less hazardous material<br>• Reduce energy. For example, lower speed, force, amperage, pressure, temperature, and noise |
| | 3. Engineering controls | • Ventilation systems<br>• Machine guarding<br>• Sound enclosures<br>• Circuit breakers<br>• Platforms and guard railing<br>• interlocks<br>• Lift tables, conveyors, and balancers |
| | 4. Warnings | • Signs<br>• Backup alarms<br>• Beepers<br>• Horns<br>• Labels |
| | 5. Administrative controls | *Procedures*<br>• Safe job procedures<br>• Rotation of workers<br>• Safety equipment inspections<br>• Changing work schedule<br>*Training*<br>• Hazard communication training<br>• Confined space entry |
| **Least effective** | 6. Personal protective equipment | • Safety glasses<br>• Hearing protection<br>• Face shields<br>• Safety harnesses and lanyards<br>• Gloves<br>• Respirators |

*Source:* Adapted from ANSI/AIHA/ASSE (2012), p. 51.

3. *Moderate*: Use less hazardous conditions, a less hazardous form of a material, or facilities that minimize the impact of a release of hazardous material or energy (also called *Attenuation* and *Limitation of Effects*).
4. *Simplify*: Design facilities which eliminate unnecessary complexity and make operating errors less likely and which are forgiving of errors that are made (also called *Error Tolerance*) (CCPS 2009, p. 27).

The CCPS text on inherent safety provides several worked examples and case studies illustrating each of these control methods (CCPS 2009, pp. 251–286).

***14.4.4.3 Control Solutions*** The responsibility for determining a control strategy and detailing control solutions for project-specific hazards normally resides with the project team. OSH professionals may or may not be involved in this process, but they should make themselves available to project team members whenever possible. They should participate in design reviews and share their expertise and experience on safe designs, safe practices, and regulatory requirements. They also have an important role to play in verifying that committed control solutions are included in final designs and the "as-built" facility.

Table 14.11 provides a list of control solutions that have been presented at two national PtD conferences hosted by NIOSH in recent years (NIOSH 2011; Schulte et al. 2008). The variety of hazards and sectors reflects the versatility and acceptance of PtD as a guiding principle in the prevention of work-related injuries, illnesses, and fatalities.

**TABLE 14.11 Control Solutions**

| Sector | Organization | Hazard | Solution | Reference |
|---|---|---|---|---|
| AFF | University of Wisconsin, NIOSH: research on intervention for musculoskeletal hazards among berry growers | Ergonomic | Five practices to reduce ergonomic risk were publicized among growers: use of hoop houses, prone carts, portable field stools, narrow aisle platform trucks, and long-handled diamond hoes. Study results indicated an increase in awareness for four of five practices, but no increase in adoption of any | Chapman et al. (2008) |
| AFF | NIOSH: research on Alaska commercial fishing | Machinery entanglement | Emergency stop that de-energizes the capstan deck winch | Lincoln et al. (2008) |
| AFF | AgSafe | Ergonomic | Redesigned smaller wine grape harvesting tub | Wolfe (2011) |
| ASW | Washington Group International | Radioactive waste | Isolation of operator and use of remote arms and power manipulators to sort waste | Zarges and Giles (2008) |
| CON | National Asphalt Pavement Association: road building | Asphalt air emissions; silica dust | Modified process conditions, local exhaust ventilation | Acott (2011) |
| CON | OSHA: lessons learned from incident investigations | Structural collapse | Wide range of design recommendations for safety of structures including parking garage floor slabs and columns, anchorage of post tensioning beams, concrete for formwork, composite beam design, bridge girders and steel trusses, anchor points for fall protection, providing tanks or major equipment at ground level, positioning of splices for steel columns | Ayub (2011) |
| CON | East Carolina University: research on vegetative roofs as part of green building design | Safe access, falls, structural loading, relation of vegetation to other known building hazards | Safe design guidelines, roof structural loading considerations, access for maintenance considerations, parapet fall protection | Behm (2011) |
| CON | WorleyParsons: design and build | General safety | Best design practices for walkways, stairways, ladders, guards, electrical, accessibility, safety showers, occupational environment, material handling, maintenance | Borowski (2011) |
| CON | Sinclair, Knight, Merz: design and build | Falls | Telescopic canopy | Genn (2011) |
| CON | University of Colorado: research on design risk analysis | Falls | Method for identifying hazards and quantifying risks in early project phase in order to identify most efficient safety intervention | Hallowell (2011) |
| CON | University of Florida: research on industrial and power, commercial, and residential projects | Falls and other general construction-related hazards | Industrial design and build firms use a formalized PtD process including standardized designs, checklists, prefabricated work, modularization, less field work, better access, less elevated work, adequate parapet height for fall protection; only slight progress in commercial projects, little progress in residential projects | Hinze (2011) |
| CON | EPA: regulatory | Isocyanate exposure from polyurethane foam | Ventilation design | Lamba (2011) |
| CON | Innovative Technical Solutions, Inc. | Falls | Designing out ladder use in construction by replacement with moveable or stationary work platforms, properly design stairs, prefabrication building segments | Lyons (2011) |

(*Continued*)

**TABLE 14.11 (Continued)**

| Sector | Organization | Hazard | Solution | Reference |
|---|---|---|---|---|
| CON | Liberty Mutual: research on corporate and commercial real estate, buildings | Slips, trips, falls | Floor surface treatments, slip resistance standards, entrance design and mats, stairway handrail design, step design, building codes | Maynard (2011) |
| CON | University of Massachusetts Lowell: research on maintenance worker safety for green buildings | Electrical, confined space, slips, trips, falls, struck by objects, chemical exposure, lifting, awkward positions | Designing in ease of access to spaces, equipment layout, chemical substitution, equipment designs less prone to corrosion, minimized lifting hazards, improved lighting | Omar et al. (2011) |
| CON | OSHA Alliance Construction Roundtable | Falls from roof edge, floor openings, nonmoving vehicles, structural collapses during construction | Roof hatch access and hole protection, wall height specifications to allow parapets as fall protection, avoidance of building collapse through engineering design review, better construction drawings, coordination between design engineer and contractor | Quintero and Menon (2011) |
| CON | Hoffman Construction Company | General construction worksite safety and health hazards | Description of system to rate projects on the basis of importance given to construction safety and health including PtD features | Rajendran (2011) |
| CON | Bucknell University and Oregon State University: research on the owner's role in PtD | General construction worksite safety hazards | Fabrication line layout, temporary ladder, platform, and safety line, prefabricated steel stairs, concrete wall panels, and concrete segmented bridge, design safety checklist example, constructability review process, and other PtD tools outlined | Toole and Gambatese (2011) |
| CON | Bucknell University and Oregon State University | Full range of hazards in construction including falls, electrical, struck by, chemical, soil cave-ins, confined space | Increase in four design-related methods discussed: prefabrication, use of less hazardous materials and systems, application of construction engineering, and spatial investigation and consideration | Toole and Gambatese (2008) |
| CON | Oregon State University: research on PtD measures for worker safety on construction sites | Falls | Concrete fall arrest straps, increasing height of roof parapets, placing electrical wires in slabs and underground, concrete imbeds for guardrail supports, soil retention with railing | Tymvios (2011) |
| GOV | NASA: facilities | Noise | "Buy Quiet" and "Quiet by Design" initiatives, equipment design specifications, eliminate or reduce at source, engineering controls, isolation, enclosure | Cooper (2011) |
| HCS<br>AFF<br>OTH | NIOSH: research on business cases for PtD in a nursing home, wine grape harvesting, and wet garment cleaning | Ergonomic, chemical | Installed mechanical patient lifting equipment in nursing home and friction reducing sheets, substituted small tub in wine grape harvesting, and water-based cleaning agent in place of solvent in garment cleaning | Biddle (2013) |
| HCS | Kaiser Permanente | Hospital related | Patient handling equipment and digital radiology, as examples | Malcolm (2008) |
| HCS | University of Maryland, Public Employees Federation, Architectural Resources, Inc. research: workplace violence in healthcare and social services | Workplace violence | Study results focused on four areas to reduce risks: access control, safety of the premises, ability to observe patients, and activity support (layout/design, use of safe materials, proper maintenance) | McPhaul et al. (2008) |
| HCS | NIOSH: ambulance safety research | Crash-related injuries | Identified opportunities to improve compartment and seating design, worker restraints, communication access | Moore (2011) |

| | | | | |
|---|---|---|---|---|
| HCS | University of Massachusetts Lowell: research on chemical alternatives in hospitals | Chemical | Substituting less hazardous hydrocarbon solvent as fixative, reducing mercury usage, digital in place of wet chemical imaging, microfiber in place of conventional wet mopping | Quinn (2011) |
| MFG | DuPont: chemicals | Electrical | Electrical switchgear, substation, system, and appliance design | Floyd (2011) |
| MFG | Toyota: motor vehicles | Ergonomic | Work station design, manual handling, tool design | Horsford (2011) |
| MFG | Small Business: metal parts, die cast, assembly | Ergonomic | Workstation design, material handling improvements | MacLeod (2011) |
| MFG | Bayberry EHS Consulting (formerly Rohm and Haas Company): chemicals | Chemical | Elimination and control of open system chemical operations | Renshaw (2013a) |
| MIN | Rio Tinto Alcan | Mobile equipment | Decreased mobile equipment inventory, increased use of barriers for separation and isolation, eliminated process steps | Laddychuk (2008) |
| PST | Purdue University: research on nanotechnology | Pyrophoric, flammable, toxic, inert gases and oxygen, liquid chemicals in small quantities | Engineering controls for gases including building security, gas segregation, enclosure, blast protection, closed system distribution, monitoring, emergency shutoff devices, local exhaust and general ventilation, mechanical integrity of gas piping and distribution systems, gas sensing and alarm system | Weaver (2011) |
| ALL | EPA Office of Pollution Prevention and Toxics | Chemical | Developed a tool to facilitate identification and substitution of safer chemical alternatives | Anderson (2011) |
| ALL | Liberty Mutual Group | Overexertion, slips and falls, bodily reaction, falls to lower level, machine accidents | Overexertion – ergonomics; Slips and falls – tribology; Bodily reaction – architecture and work station design; falls to lower levels – anchorages, harness design, railings, stair design; machinery – design of controls and guarding systems | Braun (2008) |
| ALL | University of Massachusetts Lowell: research on chemical alternatives | Chemical | Developed a process to assess chemical and nonchemical alternatives to replace chemicals or technologies of high concern | Tickner (2011) |

Key to sector abbreviations: AFF, Agriculture, Forestry, Fishing and Hunting; ALL, All sectors; ASW, Administrative and Support and Waste Management and Remediation Services; CON, Construction; GOV, Government agency; HCS, Health Care and Social Assistance; MIN, Mining, quarrying, and oil and gas extraction; MFG, Manufacturing; OTH, Other services (except public administration); PST, Professional, Scientific, and Technical Services.

## 14.5 CONCLUSION

PtD is a principle that should guide all OSH professionals in identifying and addressing workplace hazards. It is a sensible approach and is widely recognized as the most effective way to prevent incidents, injuries, illnesses, and fatalities. The principle of PtD focuses attention on the dual responsibility of all OSH professionals to address foreseeable hazards at the design and redesign stage as well as existing hazards and risks of ongoing operations. It is far too easy to become immersed in the details of controlling an existing risk and neglect foreseeable hazards that may be introduced with a new or newly modified facility. Incorporating PtD into an organization's design and redesign process will ensure that new hazards are identified and eliminated or controlled before they enter the workplace. This is the benefit and the goal of PtD.

## BIBLIOGRAPHY

Accounting Coach (n.d.). *What Is a Capital Expenditure Versus a Revenue Expenditure?* http://www.accountingcoach.com/blog/capital-expenditure-revenue-expenditure (accessed 10 July 2018).

Acott, M. (2011). Catching lightning in a bottle. *NIOSH Conference: Prevention Through Design – A New Way of Doing Business: A Report on the National Initiative*, Washington, DC (August 2011). http://www.asse.org/assets/1/7/Michael_Acott.pdf (accessed 10 July 2018).

American Conference of Governmental Industrial Hygienists (ACGIH) (2015). *American Conference of Governmental Industrial Hygienists: About Us*. http://www.acgih.org/about-us/history (accessed 10 July 2018).

American Industrial Hygiene Association (AIHA) (2015). *American Industrial Hygiene Association: About Us*. https://www.aiha.org/about-aiha/Pages/default.aspx (accessed 10 July 2018).

American Industrial Hygiene Association (AIHA) Ergonomics Committee (2011). *Ergonomic Assessment Tool Kit*. https://www.aiha.org/get-involved/VolunteerGroups/Documents/ERGOVG-Toolkit_rev2011.pdf (accessed 10 July 2018).

American National Standards Institute (ANSI) (2000). *Risk Assessment and Risk Reduction – a Guide to Estimate, Evaluate and Reduce Risks Associated with Machine Tools* (ANSI B11.TR3-2000). McLean, VA: ANSI.

American National Standards Institute (ANSI) (2010). *Safety of Machinery: General Requirements and Risk Assessment* (ANSI B11.0-2010). McLean, VA: ANSI.

American National Standards Institute/American Industrial Hygiene Association/ American Society of Safety Engineers (ANSI/AIHA/ASSE) (2012). *Occupational Health and Safety Management Systems* (ANSI/AIHA/ASSE Z10-2012). Park Ridge, IL: ANSI/AIHA/ASSE.

American National Standards Institute/American Society of Safety Engineers (ANSI/ASSE) (2011). *Prevention Through Design: Guidelines for Addressing Occupational Hazards and Risks in Design and Redesign Processes* (ANSI/ASSE Z590.3-2011). Park Ridge, IL: ANSI/ASSE.

American Petroleum Institute (API) (2003). *API RP 752 – Management of Hazards Associated with Location of Process Plant Buildings*, 2e. Washington, DC: API.

American Petroleum Institute (API) (2007). *API RP 753 – Management of Hazards Associated with Location of Process Plant Portable Buildings*. Washington, DC: API.

American Society of Safety Engineers (ASSE) (2011). *Prevention Through Design – A New Way of Doing Business*. http://www.asse.org/professionalaffairs/ptd (accessed 10 July 2018).

American Society of Safety Engineers (ASSE) (2015). *The American Society of Safety Engineers: About Us*. http://www.asse.org/about (accessed 10 July 2018).

Anderson, C.B. (2011). Design for the environment approaches to safer chemicals. *NIOSH Conference: Prevention Through Design – A New Way of Doing Business: A Report on the National Initiative*, Washington, DC (August 2011). http://www.asse.org/assets/1/7/Cal_Baier-Anderson.pdf (accessed 10 July 2018).

Australian Safety and Compensation Council (ASCC) (2005). *Design Issues in Work-Related Serious Injuries*. http://www.safeworkaustralia.gov.au/sites/swa/about/publications/Documents/286/DesignIssues_WorkRelatedSeriousInjuries_2005_PDF.pdf (accessed 10 July 2018).

Australian Safety and Compensation Council (ASCC) (2006). *Guidance on the Principles of Safe Design for Work*. http://www.safeworkaustralia.gov.au/sites/swa/about/publications/Documents/154/GuidanceOnThePrinciplesOfSafeDesign_2006_PDF (accessed 10 July 2018).

Ayub, M. (2011). Structural collapses during construction. *NIOSH Conference: Prevention Through Design – A New Way of Doing Business: A Report on the National Initiative*, Washington, DC (August 2011). http://www.asse.org/assets/1/7/Mohammad_Ayub.pdf (accessed 10 July 2018).

Baldridge (2009). Work processes. In *Baldridge Glossary*. http://www.baldrige21.com/Baldrige_Glossary.html (accessed 10 July 2018).

Behm, M. (2011). PtD practice: making buildings safe to construct, operate, and maintain. *NIOSH Conference: Prevention Through Design – A New Way of Doing Business: A Report on the National Initiative*, Washington, DC (August 2011). http://www.asse.org/assets/1/7/Michael_Behm.pdf (accessed 10 July 2018).

Bender, D. (2011). Amgen EHS by dEHSign – project lifecycle approach. *Session Conducted at the American Industrial Hygiene Conference and Exhibition*, Portland, OR (May 2011).

Biddle, E.A. (2011). Have major US companies adopted prevention through design (PtD) practices and policies? *Presented at the National Occupational Injury Symposium, Morgantown, WV* (18–20 October 2011).

Biddle, E. (2013). Business cases: supporting PTD Solutions. *Professional Safety* **58** (3): 56–64.

Biddle, E.A., Carande-Kulis, V.G., Woodhull, D. et al. (2011). The business case for occupational safety, health, environment and beyond. In: *Occupational Health and Safety* (ed. R.J. Burke, S. Clarke, and C.L. Cooper), 47–70. Burlington, VT: Gower Publishing.

Blunt, L.A., Zey, J.N., Greife, A.L., and Rose, V.E. (2011). History and philosophy of industrial hygiene. In: *The Occupational Environment: Its Evaluation, Control, and Management*, 3e (ed. D. Anna), 3–23. Fairfax, VA: American Industrial Hygiene Association.

Borowski, J. (2011). PtD evolution – the role of safety/health professional in moving PtD forward. *NIOSH Conference: Prevention Through Design – A New Way of Doing Business: A Report on the National Initiative*, Washington, DC (August 2011). http://www.asse.org/assets/1/7/John_Borowski.pdf (accessed 10 July 2018).

Brandt, M.T. and Silverstein, B.D. (2011). The AIHA value strategy. In: *The Occupational Environment: Its Evaluation, Control and Management*, 3e (ed. D. Anna), 1575–1583. Fairfax, VA: American Industrial Hygiene Association.

Bratt, G.M., Nelson, D.I., Maier, A. et al. (2011). Occupational and environmental health risk assessment/risk management. In: *The Occupational Environment: Its Evaluation, Control and Management*, 3e (ed. D. Anna), 165–226. Fairfax, VA: American Industrial Hygiene Association.

Braun, T.W. (2008). Prevention Through Design (PtD) from the Insurance Perspective. *Journal of Safety Research* **39** (2): 137–139.

British Standards Institute (BSI) OHSAS 18001 (2007). *Occupational Health and Safety Management Systems – Requirements* (BS OHSAS 18001:2007). Herndon, VA: BSI.

British Standards Institute (BSI) OHSAS 18002 (2008). *Occupational Health and Safety Management Systems – Guidelines for the Implementation of OHSAS 18001:2007*. Herndon, VA: BSI.

Bureau of Labor Statistics (BLS) (2014). *News Release: Employer-Reported Workplace Injuries and Illnesses – 2013*. TABLE 14.5. Incidence rates and number of nonfatal occupational injuries by selected industries and ownership, 2013, TABLE 14.6b. Numbers of cases of nonfatal occupational illnesses by major industry sector, category of illness, and ownership, 2013. Bureau of Labor Statistics, U.S. Department of Labor. http://www.bls.gov/news.release/archives/osh_12042014.pdf (accessed 12 June 2015).

Bureau of Labor Statistics (BLS) (2015). *News Release: Revisions to the 2013 Census of Fatal Occupational Injuries (CFOI) Counts*. Bureau of Labor Statistics, U.S Department of Labor. http://www.bls.gov/iif/oshwc/cfoi/cfoi_revised13.pdf (accessed 12 June 2015).

Center for Chemical Process Safety (CCPS) (1996). *Inherently Safer Chemical Processes: A Life Cycle Approach*. New York, NY: American Institute of Chemical Engineers.

Center for Chemical Process Safety (CCPS) (1999). *Guidelines for Chemical Process Quantitative Risk Analysis*, 2e. New York, NY: Wiley.

Center for Chemical Process Safety (CCPS) (2007). *Guidelines for Risk Based Process Safety*. Hoboken, NJ: Wiley.

Center for Chemical Process Safety (CCPS) (2008a). *Guidelines for Management of Change for Process Safety*. Hoboken, NJ: Wiley.

Center for Chemical Process Safety (CCPS) (2008b). *Guidelines for Hazard Evaluation Procedures*, 3e. Hoboken, NJ: Wiley.

Center for Chemical Process Safety (CCPS) (2009). *Inherently Safer Chemical Processes: A Life Cycle Approach*, 2e. New York, NY: Wiley.

Center for Chemical Process Safety (CCPS) (2013). *Guidelines for Managing Process Safety Risks During Organizational Change*. Hoboken, NJ: Wiley.

Chapman, L.J., Newenhouse, A.C., Pereira, K.M. et al. (2008). Evaluation of a four year intervention to reduce musculoskeletal hazards among berry growers. *Journal of Safety Research* **39** (2): 215–224.

Christensen, W.C. (2009). Challenges of safe design, part 1: the challenge and responsibilities. *Connect Press* (21 July 2009): 3.

Christensen, W.C. and Manuele, F.A. (eds.) (1999). *Safety Through Design*. Itasca, IL: National Safety Council.

Clayton, G.D. (1973). Introduction. In: *The Industrial Environment – Its Evaluation and Control* (ed. G.D. Clayton). Washington, D.C.: DHEW, PHS, CDC, NIOSH Superintendent of Documents.

Clayton, G.D., Clayton, F.E., Cralley, L.J. et al. (1948). *Patty's Industrial Hygiene and Toxicology*, 1e. New York: Wiley.

Commonwealth of Pennsylvania, Department of Health, Division of Occupational Health (1964). *Hygienic Information Guide No. 7: Asbestos* (HH-19041 P Rev 7/64). Harrisburg, PA: Commonwealth of Pennsylvania, Department of Health, Division of Occupational Health.

Commonwealth of Pennsylvania, Department of Health, Division of Occupational Health (1965). *Hygienic Information Guide No. 39: Nickel Carbonyl* (HIH-19082 P Rev. 4/65). Harrisburg, PA: Commonwealth of Pennsylvania, Department of Health, Division of Occupational Health.

Construction Industry Institute (CII) (1996). *Design for Safety* (Research Summary 101-1). Austin, TX: CII, The University of Texas at Austin.

Construction Industry Institute (CII) (2010). *Design for Construction Safety Toolbox, Version 2.0* (IR101-2). Austin, TX: CII, The University of Texas at Austin.

Cooper, B. (2011). "Buy quiet" and "quiet by design". *NIOSH Conference: Prevention Through Design – A New Way of Doing Business: A Report on the National Initiative*, Washington, DC (August 2011). http://www.asse.org/assets/1/7/Beth_Cooper.pdf (accessed 10 July 2018).

Cornell University, Kheel Center (2011). *The 1911 Triangle Factory Fire*. http://www.ilr.cornell.edu/index.html (accessed 5 May 2015).

Covey, S.R. (1990). *The 7 Habits of Highly Effective People, Habit 5: Seek First to Understand, Then to Be Understood* (ed. Fireside). New York, NY: Simon & Schuster.

Creaser, W. (2008). Prevention through design (PtD) safe design from an Australian perspective. *Journal of Safety Research* **39**: 131–134.

Department of Defense (DOD) (2012). *Department of Defense Standard Practice – System Safety* (MIL-STD-882E). http://www.system-safety.org/Documents/MIL-STD-882E.pdf (accessed 10 July 2018).

Driscoll, T., Mitchell, R., Mandryk, J. et al. (2001). Work-related fatalities in Australia, 1989-1992: an overview. *Journal of OHS – Australia and New Zealand* **2001** (17): 45–66.

Driscoll, T.R., Harrison, J.E., Bradley, C., and Newson, R.S. (2005). *Design Issues in Work-Related Serious Injuries*. Canberra, Australia: Australian Safety and Compensation Department of Employment and Workplace Relations, Commonwealth of Australia.

Dysert, L. (2002). *A Capital Investment Review Process* (PM.09.1-PM.09.4). Morgantown, WV: AACE International Transactions.

Environmental Protection Agency (EPA) (1996). *Accidental Release Prevention Requirements: Risk Management Programs Under the Clean Air Act, Section 112(r)(7): List of Regulated Substances and Thresholds for Accidental Release Prevention*. Washington DC: EPA.

European Commission – Joint Research Centre (1997). *Guidance on the Preparation of a Safety Report to Meet the Requirements of Council Directive 96/82/EC (Sevesso II)*. Luxembourg: Office for Official Publications of the European Communities, Institute for Systems Informatics and Safety, Major Accident Hazards Bureau.

Floyd, H.L. (2011). Progress in impacting policy in workplace electrical safety. *NIOSH Conference: Prevention Through Design – A New Way of Doing Business: A Report on the National Initiative*, Washington, DC (August 2011). http://www.asse.org/assets/1/7/H_Landis_Floyd_1.pdf (accessed 10 July 2018).

Gambatese, J.A. (2011). Findings from the overall PtD in UK study and their application to the US. *Paper Presented at the NIOSH Workshop on Prevention Through Design – A New Way of Doing Business: Report on the National Initiative*, Washington, DC (August 2011). http://www.asse.org/assets/1/7/John_Gambatese.pdf (accessed 10 July 2018).

Gambatese, J.A., Hinze, J., and Behm, M. (2005). *Investigation of the Viability of Designing for Safety*. Silver Spring, MD: The Center to Protect Workers' Rights (CPWR).

Gambatese, J.A., Gibb, A.G., Bust, P., and Behm, M. (2009). Industry's perspective of design for safety regulations. *Working Together: Planning, Designing, and Building a Healthy and Safe Construction Industry, CIB W99 Conference*, Melbourne, Australia (21–23 October 2009).

Genn, K. (2011). Safe by design. *NIOSH Conference: Prevention Through Design – A New Way of Doing Business: A Report on the National Initiative*, Washington, DC (August 2011). http://www.asse.org/assets/1/7/Kelvin_Genn.pdf (accessed 10 July 2018).

Gibb, A. (2011). Construction (design and management) regulations: PtD survey of UK and UK experience. *NIOSH Conference: Prevention Through Design – A New Way of Doing Business: A Report on the National Initiative*, Washington, DC (August 2011). http://www.asse.org/assets/1/7/Alistair_Gibb.pdf (accessed 10 July 2018).

Grimaldi, J.V. and Simonds, R.H. (1989). *Safety Management*, 5e, 30–32. Homewood, IL: Irwin/American Society of Safety Engineers.

Hallowell, M. (2011). Design risk analysis: an attribute-based method. *NIOSH Conference: Prevention Through Design – A New Way of Doing Business: A Report on the National Initiative*, Washington, DC (August 2011). http://www.asse.org/assets/1/7/Matthew_Hallowell.pdf (accessed 10 July 2018).

Hamilton, A. (1943). *Exploring the Dangerous Trades: the Autobiography of Alice Hamilton, MD*. Boston, MA: Atlantic Monthly Press, Little, Brown and Company.

Hecker, S., Gambatese, J.A., and Weinstein, M. (2004). Life cycle safety: an intervention to improve construction worker safety and health through design. In: *Designing for Safety and Health in Construction: Proceedings From a Research and Practice Symposium*. Eugene, OR: University of Oregon Press.

Hendershot, D.C. (2011). Inherently safer design: an overview of key elements. *Professional Safety* **56** (2): 48–55.

Hinze, J. (2011). PtD from the project and contractor's perspective. *NIOSH Conference: Prevention Through Design – A New Way of Doing Business: A Report on the National Initiative*, Washington, DC (August 2011). http://www.asse.org/assets/1/7/Jimmie_Hinze.pdf (accessed 10 July 2018).

Horsford, W. (2011). Incorporating safety and ergonomics in the Toyota manufacturing design process. *NIOSH Conference: Prevention Through Design – A New Way of Doing Business: A Report on the National Initiative*, Washington, DC (August 2011). http://www.asse.org/assets/1/7/Bill_Horsford.pdf (accessed 10 July 2018).

Howard, J. (2008). Prevention through design – introduction. *Journal of Safety Research* **39** (2): 113.

Ignacio, J.S. and Bullock, W.H. (eds.) (2006). *A Strategy for Assessing and Managing Occupational Exposures*, 3e. Fairfax, VA: AIHA Press.

Industrial Revolution (2015). In *Encyclopedia Britannica*. http://www.britannica.com/event/Industrial-Revolution (accessed 10 July 2018).

KCG Consultant Group (2010). *Policy, Process, and Procedure – What Is the Difference*. http://www.kcggroup.com/Policies ProcessesProcedureDifferences (accessed 24 May 2015).

Laddychuk, S. (2008). Paving the way for world-class performance. *Journal of Safety Research* **39** (2): 143–149.

Lamba, A. (2011). Spray polyurethane foam (SPF): EPA considerations. *NIOSH Conference: Prevention Through Design – A New Way of Doing Business: A Report on the National Initiative*, Washington, DC (August 2011). http://www.asse.org/assets/1/7/Anjali_Lamba.pdf (accessed 10 July 2018).

Lawson, G., Wearne, S., and Iles-Smith, P. (eds.) (1999). *Project Management for the Process Industries*. Rugby, UK: Institution of Chemical Engineers.

Lin, M.-L. (ed.) (2008). Prevention through design [Special issue]. *Journal of Safety Research* **39** (2): 111.

Lincoln, J.M., Lucas, D.L., McKibbin, R.W. et al. (2008). Reducing commercial fishing deck hazards with engineering solutions for winch design. *Journal of Safety Research* **39** (2): 231–235.

Lyons, T.J. (2011). Designing out ladder use (and falls). *NIOSH Conference: Prevention Through Design – A New Way of Doing Business: A Report on the National Initiative*, Washington, DC (August 2011). http://www.asse.org/assets/1/7/Thomas_J_Lyons.pdf (accessed 10 July 2018).

MacLeod, D. (2011). Costs and benefits of ergonomic interventions for small business. *Paper Presented at the NIOSH Workshop on Prevention Through Design – A New Way of Doing Business: Report on the National Initiative*, Washington, DC (August 2011). http://www.asse.org/assets/1/7/Dan_MacLeod.pdf (accessed 10 July 2018).

Malcolm, C. (2008). Building the case for prevention through design presentation – Kaiser Permanente. *Journal of Safety Research* **39** (2): 151–152.

Manuele, F.A. (2008). Prevention through design (PtD): history and future. *Journal of Safety Research* **39** (2): 127–130.

Maynard, W. (2011). Prevention of slips, trips and falls through facility design. *NIOSH Conference: Prevention Through Design – A New Way of Doing Business: A Report on the National Initiative*, Washington, DC (August 2011). http://www.asse.org/assets/1/7/Wayne_Maynard.pdf (accessed 10 July 2018).

McFarlane, H.C. (2004). Walsh-Healey public contracts act (1936). Major acts of congress. In Encyclopedia.com. http://www.encyclopedia.com/topic/Walsh-Healey_Act.aspx (accessed 5 May 2015).

McPhaul, K.M., London, M., Murett, K. et al. (2008). Environmental evaluation for workplace violence in healthcare and social services. *Journal of Safety Research* **39** (2): 237–250.

Moore, P.H. (2011). Ambulance redesign to reduce EMS injuries: influencing design through standards development. *NIOSH Conference: Prevention Through Design – A New Way of Doing Business: A Report on the National Initiative*, Washington, DC (August 2011). http://www.asse.org/assets/1/7/Paul_Moore.pdf (accessed 10 July 2018).

Mulhausen, J.R. and Damiano, J. (2011). Comprehensive exposure assessment. In: *The Occupational Environment: Its Evaluation, Control and Management*, 3e (ed. D. Anna), 229–242. Fairfax, VA: American Industrial Hygiene Association.

National Fire Protection Association (NFPA) (2015). *NFPA 30: Flammable and Combustible Liquids Code*, 2015e. Quincy, MA: NFPA.

National Institute for Occupational Safety and Health (NIOSH) (1973). *The Industrial Environment – Its Evaluation and Control*. U.S. Department of Health, Education, and Welfare, Public Health Service, Center for Disease Control, National Institute for Occupational Safety and Health. Washington, DC: U.S. Government Printing Office.

National Institute for Occupational Safety and Health (NIOSH) (2010). *Prevention Through Design: Plan for the National Initiative* (U.S. Department of Health and Human Services, Centers for Disease Control and Prevention, National Institute for Occupational Safety and H, [NIOSH] Publication No. 2011-121). Cincinnati, OH: DHHS/CDC/NIOSH.

National Institute for Occupational Safety and Health (NIOSH) (2011). *Prevention Through Design – A New Way of Doing Business: A Report on the National Initiative* (22–24 August 2011). http://www.asse.org/professionalaffairs/ptd (accessed 10 July 2018).

National Institute for Occupational Safety and Health (NIOSH) (2014). *The State of the National Initiative on Prevention Through Design* (DHHS [NIOSH] Publication No. 2014-123). Cincinnati, OH: DHHS/CDC/NIOSH.

National Occupational Health and Safety Commission (NOHSC) (1994). *National Standard for Plant* (NOHSC 1010:1994). http://www.ascc.gov.au/NR/rdonlyres/D50590D-BB66-455F-8E80-4E5F0AD6A6AE/0/PlantStandard_NOHSC1010.pdf (accessed 10 July 2018).

National Occupational Health and Safety Commission (NOHSC) (1995). *Plant Design: Making It Safe, a Guide to Risk Management for Designers, Manufacturers, Importers, Suppliers and Installers of Plant*. http://www.safeworkaustralia.gov.au/sites/SWA/about/Publications/Documents/37/PlantDesignMakingItSafe_1995_ArchivePDF.pdf (accessed 10 July 2018).

National Safety Council (NSC) (1946). *Accident Prevention Manual*, 1e. Chicago, IL: NSC.

National Safety Council (NSC) (1988). Occupational safety: history and growth. In: *Accident Prevention Manual for Industrial Operations: Administration and Programs*, 9e, 2. Chicago, IL: NSC.

Newell, S. (2011). Benchmarking management practices related to PtD in the U.S. *Paper presented at the NIOSH Workshop on Prevention Through Design – A New Way of Doing Business: Report on the National Initiative*, Washington, DC (August 2011). http://www.asse.org/assets/1/7/Stephen_Newell.pdf (accessed 10 July 2018).

Occupational Safety and Health (OSH) Act of 1970 (1970). 84 STAT. 1590, 29 USC 651.

Occupational Safety and Health Administration (OSHA) (2015a). *Safety and Health Management Systems eTool: Hazard Prevention and Control*. Occupational Safety and Health Administration. https://www.osha.gov/SLTC/etools/safetyhealth/comp3.html (accessed 10 July 2018).

Occupational Safety and Health Administration (OSHA) (2015b). *Alliance Program Construction Roundtable: Roundtable Background, Meeting Notes* (20 October 2004). https://www.osha.gov/dcsp/alliances/roundtables/roundtables_construction.html#!1B (accessed 10 July 2018).

Omar, M., Quinn, M., Buchholz, B., and Geiser, K. (2011). Incorporating occupational safety and health of preventive maintenance workers into the design and operation of green buildings: safe is the new green. *NIOSH Conference: Prevention Through Design – A New Way of Doing Business: A Report on the National Initiative*, Washington, DC (August 2011). http://www.asse.org/assets/1/7/Mohamed_Omar.pdf (accessed 10 July 2018).

O'Reilly, J.T. (2009). Historical perspectives. In: *Accident Prevention Manual for Business and Industry, Administration and Programs*, 13e (ed. P.E. Hagan, J.P. Montgomery, and J.T. O'Reilly). Itasca, IL: National Safety Council.

Patty, F.A. (1958). Industrial hygiene – retrospect and prospect. In: *Industrial Hygiene and Toxicology: Vol. 1, General Principles* (ed. F.A. Patty). New York, NY: Interscience Publishers, Inc.

Quinn, M.M. (2011). Implementing prevention through design in hospitals: alternatives assessment. *NIOSH Conference: Prevention Through Design – A New Way of Doing Business: A Report on the National Initiative*, Washington, DC (August 2011). http://www.asse.org/assets/1/7/Margaret_Quinn.pdf (accessed 10 July 2018).

Quintero, D. and Menon, G. (2011). PtD current and future activities and challenges. *NIOSH Conference: Prevention Through Design – A New Way of Doing Business: A Report on the National Initiative*, Washington, DC (August 2011). http://www.asse.org/assets/1/7/Danezza_Quintero_and_Gopal_Menon.pdf (accessed 10 July 2018).

Rajendran, S. (2011). Sustainable construction safety and health rating system. *NIOSH Conference: Prevention Through Design – A New Way of Doing Business: A Report on the National Initiative*, Washington, DC (August 2011). http://www.asse.org/assets/1/7/Sathy_Rajendran.pdf (accessed 10 July 2018).

Renshaw, F. (2007). The role of management of change in prevention through design (PtD). *Presentation at the NIOSH Prevention Through Design Workshop*, Washington, DC (July 2007).

Renshaw, F. (2011). Incorporating prevention through design methods into the design and redesign process. *NIOSH Conference: Prevention Through Design – A New Way of Doing Business: A Report on the National Initiative*, Washington, DC (August 2011). http://www.asse.org/assets/1/7/Frank_Renshaw.pdf (accessed 10 July 2018).

Renshaw, F. (2012). Incorporating prevention through design methods into the design and redesign process. *Presentation at the American Industrial Hygiene Association 2012 Fall Conference*, Miami, FL.

Renshaw, F. (2013a). Design: methods for implementing PTD. *Professional Safety* **58** (3): 50–55.

Renshaw, F. (2013b). Incorporating prevention through design methods into the design and redesign process. *Presentation at the American Society of Safety Engineers 2013 Professional Development Conference*, Las Vegas, NV (June 2013).

Rohm and Haas Company (2008). *Environmental, Health, Safety, and Sustainable Development Principles and Policy* (1 January 2008). Philadelphia, PA: Rohm and Haas Company.

Roland, H.E. and Moriarty, B. (1990). *System Safety Engineering and Management*, 2e. New York, NY: Wiley.

Safe Work Australia (SWA) (2012). *Safe Design of Structures: Code of Practice*. http://www.safeworkaustralia.gov.au/sites/swa/about/publications/pages/safe-design-of-structures (accessed 10 July 2018).

Schulte, P.A. (2005). Characterizing the burden of occupational injury and disease. *Journal of Occupational and Environmental Medicine* **47**: 607–622.

Schulte, P.A., Rinehart, R., Okun, A. et al. (2008). National prevention through design (PtD) initiative. *Journal of Safety Research* **39**: 115–121.

Silverstein, B.D. (ed.) (2010). *Value Strategy Manual*. Fairfax, VA: AIHA.

Sinclair, R., Cunningham, T., and Schulte, P. (2013). A model for occupational safety and health intervention diffusion to small businesses. *American Journal of Industrial Medicine* **56** (12): 1442–1451.

Singapore Workplace Safety and Health Council (SWSHC) (2011). *Guidelines on Design for Safety in Buildings and Structures*. First Revision. Singapore: Workplace Safety and Health Council and Ministry of Manpower.

Spellman, F.R. and Whiting, N.E. (1999). *Safety Engineering: Principles and Practices*, 19. Rockville, MD: Government Institutes, a Division of ABS Group, Inc.

Tickner, J.A. (2011). Alternatives assessment in context. *NIOSH Conference: Prevention Through Design – A New Way of Doing Business: A Report on the National Initiative*, Washington, DC (August 2011). http://www.asse.org/assets/1/7/Joel_Tickner.pdf (accessed 10 July 2018).

Toole, T.M. (2015). *Prevention Through Design: Design for Construction Safety. Alliance – An OSHA Cooperative Program*. www.designforconstructionsafety.org (accessed 10 July 2018).

Toole, T.M. and Gambatese, J. (2008). The trajectories of prevention through design in construction. *Journal of Safety Research* **39** (2): 225–230.

Toole, T.M. and Gambatese, J. (2011). Owner's role in facilitating prevention through design. *NIOSH Conference: Prevention Through Design – A New Way of Doing Business: A Report on the National Initiative*, Washington, DC (August 2011). http://www.asse.org/assets/1/7/T_Michael_Toole.pdf (accessed 10 July 2018).

Tymvios, N. (2011). PtD construction case studies. *NIOSH Conference: Prevention Through Design – A New Way of Doing Business: A Report on the National Initiative*, Washington, DC (August 2011). http://www.asse.org/assets/1/7/Nicholas_Tymvios_-_2.pdf (accessed 10 July 2018).

United Kingdom Health and Safety Executive (UKHSE) (1994). *Construction (Design and Management) Regulations (CDM)* (SI 1994/3140 HMSO). London: Health and Safety Commission. http://www.legislation.gov.uk/uksi/1994/3140/contents/made (accessed 10 July 2018).

United Kingdom Health and Safety Executive (UKHSE) (2015). *Construction (Design and Management) Regulations 2015 (CDM 2015), Designers: Roles and Responsibilities*. http://www.hse.gov.uk/construction/cdm/2015/designers.htm (accessed 10 July 2018).

US Department of Labor (USDOL) and Occupational Safety and Health Administration (OSHA) (1971). *Occupational Safety and Health Standards; National Consensus Standards and Established Federal Standards, 29 C.F.R. §1910.93(b)*. Washington, DC: Office of the Federal Register.

US Department of Labor (USDOL) (1951). *Safety and Health Standards for Contractors Performing Federal Supply Contracts Under The Walsh-Healey Public Contracts Act, U.S.C. §50-204.50*. U.S. Wage and Hour and Public Contracts Divisions, U.S. Department of Labor, 23. Washington, DC: U.S. Government Printing Office.

US Department of Labor (USDOL) and Occupational Safety and Health Administration (OSHA) (1992). *OSHA. Process Safety Management of Highly Hazardous Chemicals, Explosives and Blasting Agents* (29 CFR 1910.119). Washington, DC: Office of the Federal Register.

US Public Health Service, Division of Occupational Health (USPHS) (1965). *The Industrial Environment – Its Evaluation and Control, Syllabus*. U.S. Department of Health, Education, and Welfare, Public Health Service, Division of Occupational Health, PHS Publication No. 614. Washington, DC: U.S. Government Printing Office.

Weaver, J. (2011). Prevention through design: capital project design and execution. *NIOSH Conference: Prevention Through Design – A New Way of Doing Business: A Report on the National Initiative*, Washington, DC (August 2011). http://www.asse.org/assets/1/7/John_Weaver.pdf (accessed 10 July 2018).

Whatis.techtarget.com (n.d.). *Capex (Capital Expenditure)*. http://whatis.techtarget.com/definition/CAPEX-capital-expenditure (accessed 10 July 2018).

Wolfe, A. (2011). Improving the work life of workers in the wine grape harvesting industry through PtD. *NIOSH Conference: Prevention Through Design – A New Way of Doing Business: A Report on the National Initiative*, Washington, DC (August 2011). http://www.asse.org/assets/1/7/Amy_Wolfe.pdf (accessed 10 July 2018).

Zarges, T. and Giles, B. (2008). Prevention through design (PtD). *Journal of Safety Research* **39** (2): 123–126.

CHAPTER 15

# HOW TO SELECT AND USE PERSONAL PROTECTIVE EQUIPMENT

RICHARD J. NILL

*Genetics Institute, 1 Burtt Rd., Andover, MA, 01810*

## 15.1 INTRODUCTION

> A school crossing guard's reflective vest. An ironworker's fall arrest system. A floor installer's kneepads. A police officer's bulletproof vest. An astronaut's spacesuit. A bridge worker's life jacket. A firefighter's breathing apparatus. A machinist's safety glasses. An NFL quarterback's helmet. A nail-machine operator's earplugs. A fish processor's cut-resistant gloves. A chemical worker's butyl rubber boots. A lead battery worker's dust respirator. A traveling salesman's seat belt. A chemist's face shield. A construction worker's hard hat. A hazmat responder's encapsulating suit...

Worn daily by millions of workers for protection against both the expected and the unexpected hazard, personal protective equipment (PPE) ranges from the simple article to the complex system. It may protect the entire body or just one part. It may be durable or disposable, low cost or expensive, fully portable or motion restricting, comfortable or annoying, or fashionable or dull. Widely varied in both form and function, all PPE has one important thing in common: it is worn by individual workers to reduce the personal risk of occupational injury and illness.

This chapter will cover many of the types of PPE in common use in general industry. It will introduce the reader to the different types available, selection and use considerations, factors affecting successful PPE use in an organizational setting, and sources of additional information.

Some PPE is covered elsewhere within this handbook, including hearing protection and respiratory protection (Chapters 9 and 16, respectively).

## 15.2 GENERAL CONSIDERATIONS

### 15.2.1 Goals of an Effective PPE Program

To deliver the desired protection, the use of PPE must be effectively managed by the organization, no matter how large or small. PPE is well managed when the following elements are in place:

- Hazardous tasks have been evaluated, and PPE has been selected where needed.
- The right PPE, properly sized, is always available to the worker.
- Users are knowledgeable in when, where, why, and how to use PPE.
- PPE is kept in good working condition.
- PPE is worn every time.

*Handbook of Occupational Safety and Health,* Third Edition. Edited by S. Z. Mansdorf.

### 15.2.2 Hazard Assessment

Shortly after assuming a safety role at the operations level, the safety and health practitioner will be asked to help determine whether PPE is required for a specific job or task in the workplace. Where do you find this information? With few exceptions, the appropriate protective equipment for a task or occupation will not be found prescribed in a textbook or in a government regulation or consensus standard. Instead, case-by-case judgments must be made using information about the workplace and job task, commonsense estimates of the potential for injury, and knowledge about the types of protective equipment available.

The need for PPE in the workplace is determined using some form of formal or informal hazard assessment. The more common types of formal hazard assessments that can uncover the need for PPE are listed below.

***15.2.2.1 Accident/Incident Investigation*** An injury should not be, but often is, the first indication that the job is not being performed as safely as it should be. Lessons from accidents are hard won and should not be wasted.

PPE is commonly prescribed after an accident as a measure to prevent reoccurrence or to reduce its severity. If you are lucky enough to experience a noninjury incident or "near miss," it can be a golden opportunity to examine work practices, including assessing the need for PPE, before harm to the worker actually occurs.

PPE requirements identified in an accident investigation should be shared with other parts of the organization having similar hazards. Knowledge about an actual injury experience can help motivate workers to wear protective equipment.

***15.2.2.2 Job Safety Analysis*** Job safety analysis, commonly referred to as JSA or task safety analysis, is a methodical, easy-to-understand evaluation of a job or task. Preparation and updating of JSAs is an excellent worker-involvement safety activity.

A three-column form can be used. Each step of the job or task is listed in the left column. The hazards associated with each task are then listed in the center column, and protective or preventive measures, often including the use of PPE, are listed in the right column. A completed JSA is a good tool for initial and ongoing worker training.

***15.2.2.3 Process Hazard Review***

What-if/checklist, failure mode and effects, and fault tree analysis are techniques used to evaluate the hazard of complex systems, especially those carrying a risk of serious injury or death. One frequent outcome of these reviews is to identify or reinforce the need for PPE. A more detailed discussion of hazard analyses is presented in Chapter 12.

In addition to the above hazard review techniques, many organizations also establish other types of formalized procedures for safety review of changes to existing, or introduction of new, processes. For example, introduction of a new chemical may require the sponsoring engineer to obtain and circulate an MSDS to a defined list of functional organizations responsible for storing, handling, and disposing the chemical, hazmat emergency response, first aid staff, and industrial hygiene, toxicology, medical, and safety advisors. The hazard review and communication that occur as part of this sign-off procedure often include specification of appropriate PPE for all those who may be at risk of exposure.

***15.2.2.4 Targeted PPE Hazard Assessment*** PPE can be targeted as the subject for a department or organization-wide hazard assessment. A PPE hazard assessment is similar to a JSA but is limited to identifying the different types of PPE as a preventive measure. Using a form like that in Figure 15.1, a member or members of the organization survey the department, looking for hazards by walking around, observing work activity, and talking to workers and management. The following hazard categories are of particular interest:

1. *Impact* that could arise from sources of motion, such as machinery or processes where any movement of tools, machine elements, or particles could exist. Impact could also occur from movement of personnel that could result in collision with stationary objects (low pipes), or from objects falling from above, or from work surface to floor.
2. *Penetration* arising from sharp objects that could pierce the hands or feet or the head if dropped from above.
3. *Compression* of feet by rolling or pinching objects.
4. *Chemical or harmful dust exposure* via inhalation or skin contact.
5. *Heat*, capable of causing contact burns or physiological heat stress.
6. *Light (optical) radiation* such as ultraviolet or laser light that could harm the eye.
7. *Noise*.

Bldg/area: ______________________ Survey by: ______________________
(Dept supervisor/team leader)

Dept: ______________________ Date: ______________________

| **Hazard type** (impact, penetration, chemical, heat, harmful dust, compression, light, radiation [e.g. welding, laser], electric shock, noise, etc. | **Location/source/task** | **Recommended PPE** |
|---|---|---|
| | | |
| | | |
| | | |
| | | |
| | | |
| | | |
| | | |
| | | |

**Figure 15.1** PPE hazard assessment and selection form.

Workers should be involved throughout the hazard assessment and PPE selection process. Experienced workers are nearly always the persons most familiar with how the job is performed, and their input is key to a thorough hazard assessment. Workers will also often have to change their behavior in order to perform the work while wearing the protective equipment; involvement in hazard assessment helps provide an understanding as to why the equipment should be worn, a key to motivating individuals to make the accommodations to use it.

### 15.2.3 PPE Can Be Hazardous

The use of PPE can itself bring new hazards to the job that are easily overlooked. PPE is often viewed as being inherently free of hazards because, after all, it is "safety equipment." When selecting or specifying PPE, watch for hazard tradeoffs.

For example, nonbreathable chemical protection garments can lead to heat stress, which can be more of a hazard than the chemical threat. Placing hearing protection on an already hearing impaired worker can make him/her unable to hear warning sounds. Loose-fitting gloves used near in-going nip points can add to the compression hazard. Ill-fitting safety shoes can injure the foot or trip the wearer. An air-supplied respirator hooked into a nitrogen pipe can be deadly. A hard hat worn during elevated work can become the most likely falling object hazard if worn without a chin strap. In all these cases, the benefits of the PPE should be weighed against the new risks, and the application of PPE will require that additional efforts, especially in worker training, be made to minimize the new risk.

### 15.2.4 PPE in the Hierarchy of Hazard Control

The use of PPE makes the most sense when the hazard cannot be fully controlled by first using other means, including the following (refer also to Chapter 18):

- Change materials or methods.
- Use engineering controls to enclose or isolate the hazard from the worker using full or partial barriers. Use exhaust ventilation.
- Develop and communicate safe work practices.

PPE may appear to be an easy way to control hazards, but it should be viewed as one of the organization's last lines of defense. Why? Because *it takes a significant effort by the organization to make sure that the correct PPE is worn by everyone, all the time.*

Each user must be properly equipped, be knowledgeable of what is required when, know how to use it, and be motivated to use and maintain it all the time. Even in a small department of four workers, that can add up to too much opportunity for noncompliance or error, especially when the equipment is uncomfortable or inhibits productivity.

There are many downsides to the use of PPE. Usually there are some workers who do not believe the equipment is necessary, and therefore will not use it, no matter how hard you try to convince them. Some individuals are never comfortable with the PPE on, while most need at least some time for familiarity and adaptation. Sizing the very large or small worker is a frequent problem. Dirty, contaminated, or damaged personal protection can be worse than using none at all. Finally, the use of inappropriate PPE can give the worker a false sense of security, possibly resulting in a more serious injury or illness.

In spite of the limitations of PPE, its own possible hazards, and the difficulty in managing its use, PPE in many cases is highly useful and essential for safe task performance.

### 15.2.5 Selecting and Acquiring Personal Protective Equipment

A hazard assessment of the task or workplace will lead into the selection and specification of protective equipment. Specific selection guidelines are provided later in this chapter. General factors to consider include the following:

- Protective capability. Can the available equipment provide the degree and type of protection needed for the particular hazard? What are the equipment's limitations? What new hazards will be introduced by using the equipment, and how can these be controlled?
- Impact on product quality. In some environments like manufacturing, research, and healthcare, some protective gloves or clothing can shed particles or elements harmful to the particular process.
- Task compatibility. Can the worker perform the task while using the equipment?
- Comfort. Is one type of equipment more comfortable than another?
- Ease of use. How difficult is it to don and operate the equipment? Is one brand more user-friendly?
- Supplier service. Is a knowledgeable supplier available to assist with selection and delivery of samples and product?
- Product identification. How are the different brands packaged? Can sizes be readily determined by labels or color coding?
- Sizes. Is a range of sizes available? Does one product provide a single size with more of a universal fit compared with another?
- Durability. Will the equipment stand up to the physical and chemical demands of the task? Is disposable equipment more practical than reusable?
- Cleaning and decontamination. If the equipment is reusable, can it be cleaned and contaminated? Who will clean it, and how often?
- Cost. Does the risk justify the more expensive PPE?
- Fashion. Will one brand be more acceptable to workers because it is more up to date?

The most important factor determining whether or not the equipment will be worn is *user comfort*. It is essential to allow workers to participate in the selection. *People like to choose what they wear* – even safety equipment. A variety of types, styles, colors, and sizes should be made available to workers, within practical limits. Note that some regulations, like the Occupational Safety and Health Administration (OSHA) asbestos standard as it relates to respiratory protection, require that workers be provided with more than one type or style (*Asbestos* 1996). Allow for a trial use of samples in the workplace before making a final selection.

PPE is a big business, and it should not be difficult to find manufacturers and distributors to help with PPE selection for most applications. *Best's Safety Directory* (1996) is one of the most complete resource guides for PPE shoppers. *Industrial Hygiene News* (1996), *Occupational Health and Safety Magazine* (1996), *Occupational Hazards* (1996), and *Chilton's Industrial Safety and Hygiene News* (1996) are periodicals that heavily advertise PPE to industry in the form of monthly and bimonthly magazines and tabloids and annual buyer's guides. Free subscriptions are generally available to the safety and health practitioner.

The annual conferences of the American Industrial Hygiene Association, American Society of Safety Engineers (now the American Society of Safety Professionals), and National Safety Council include large expositions featuring a wide variety of exhibitors demonstrating PPE products. Local and national suppliers can also be found in the yellow pages under Safety Equipment & Clothing.

### 15.2.6 User Training

The ultimate goal of a good PPE program is *to ensure that the right PPE is correctly worn every time its needed for worker protection*. Training and motivating workers expected to wear PPE is no less important, and probably more challenging, than each of the other elements of a PPE program.

User training should be provided upon initial assignment to the job, whenever the hazards or PPE type changes, or when usage indicates a lack of skill or understanding on the part of the user.

By definition, training conveys mostly "how to" as opposed to why, but a PPE training session should provide some background as to the reasons why it is needed, especially including hazard assessment information.

Training content should be based on the PPE manufacturer's recommendations and the organization's own specific requirements and should include the following:

- A brief explanation of why the particular equipment is necessary.
- Where and when it should be used.
- Where to get it, including replacement parts.
- How to properly don, doff, adjust, and wear it.
- The equipment's limitations, especially as applied in the specific workplace.
- Proper care, maintenance, useful life, storage, cleaning, and disposal of PPE.

Each worker should demonstrate an understanding of the training and the ability to use the equipment properly. A written quiz at the conclusion of the training session can be helpful. The opportunity to try the equipment on should be provided during the training session. For more complicated equipment such as self-contained breathing apparatus (SCBA), each user should be given a practical use test, such as correctly inspecting and donning the unit and negotiating a maze in the dark. New users of any PPE should be observed closely by supervisors and co-workers for proper technique. PPE use is also a common item included in periodic safety walkthrough audits.

The organization should take advantage of multiple opportunities to communicate management's expectations regarding when PPE should be used. They may include the following:

- Signs or rules posted in the workplace. PPE may be a posted requirement for an entire area and/or for specific tasks or workstations within the area.
- As part of new employee orientation.
- Periodic training classes on PPE.
- One-on-one safety contacts.
- JSAs, or other hazard reviews, kept accessible to the workplace or posted on walls.
- As a periodic topic in department safety meetings.
- Within standard operating procedures, such as batch prep sheets.
- As part of a set of general safety rules for the organization.
- As part of department or employee safety handbooks.
- As a stand-alone PPE program document.

- Floor plans with PPE requirements mapped (Figure 15.2).
- PPE matrices (Figure 15.3).
- In written job postings or advertisements.

### 15.2.7 PPE Inspection, Maintenance, and Storage

When purchased, most PPE arrives with written instructions from the manufacturer that cover inspection, maintenance, and storage. These should be covered in the user training as discussed above.

Workers have a tendency to extend the life of protective equipment beyond its capability, especially if replacements are not readily available. Worn or damaged equipment should be discarded and replaced. PPE should be stored in a clean, dry designated location and put away after use. It should not be draped around the work area and exposed to dirt or chemical contamination or degradation by sunlight or weather.

While responsibility for PPE maintenance is usually left to the individual worker, some organizations will issue PPE from a central stockroom to which the worker returns the used equipment at the end of the workday or workweek. The stockroom will then perform inspection, cleaning, and maintenance as needed before returning the equipment to service.

### 15.2.8 Standards and Regulations

The use of PPE is regulated in general industry and construction in the United States by the US Department of Labor, OSHA. For general industry, Subpart I of 29 CFR 1910 General Industry Standards holds the bulk of the PPE standards, while the subject index under Protective Clothing & Equipment makes reference to miscellaneous PPE requirements sprinkled through the various standards. The PPE standards were updated in 1994 to require employers to select PPE based on a hazard assessment and to provide user training. A brief section on hand protection was added, and some changes were made to sections covering eye and face, head, and foot protection and electrical protective equipment. A provision was added specifically prohibiting the use of defective or damaged equipment. In 1994 OSHA also updated its fall protection standards for construction, Subpart M.

OSHA enforces by reference the provisions of several PPE-related national consensus standards published by the American National Standards Institute (ANSI), which specify performance requirements for the manufacture and use of eye and face, head, respiratory, and foot protection. The American Society for Testing and Materials (ASTM) publishes testing procedures and performance requirements for electrical insulating equipment and test procedures for chemical permeation and liquid penetration of materials used to make chemical protective clothing (CPC). The National Fire Protection Association (NFPA) publishes standards applicable to firefighters' clothing, including suits worn for chemical vapor and splash protection. The US Centers for Disease Control's National Institute for Occupational Safety and Health (NIOSH), commonly referred to as OSHA's sister agency for research, must approve all occupational respiratory protection.

A list of various PPE types with applicable standards and regulations is provided in Table 15.1.

### 15.2.9 Who Must Pay for PPE

It is good business to provide protective equipment that helps keep workers on the job, ready to come to work day after day, and many employers pay for PPE required by the job. In 1994, OSHA indicated in its instructions to field compliance personnel that where PPE is required to perform the job safely in compliance with OSHA standards and the equipment is not uniquely personal in nature or usable off the job, then the employer must provide and pay for it (OSHA 1994a). This directive was challenged by the Union Tank Car Company, and in October 1997, the Occupational Safety and Health Review Commission ruled that while employers must ensure that PPE is provided, they need not pay for it because this was not explicitly stated in the OSHA PPE standard (*Secretary of Labor v. Union Tank Car Co., OSHRC. No. 96-0563, 16 October 1997*). At this writing, OSHA has stated its intention to amend the standard through a formal rulemaking requiring employers to pay.

OSHA makes an exception to the "employer must provide" rule to allow workers to provide their own equipment to accommodate work situations in which it is customary for workers in a particular trade, such as welding, to provide their own PPE. Such cases are the exception, not the norm.

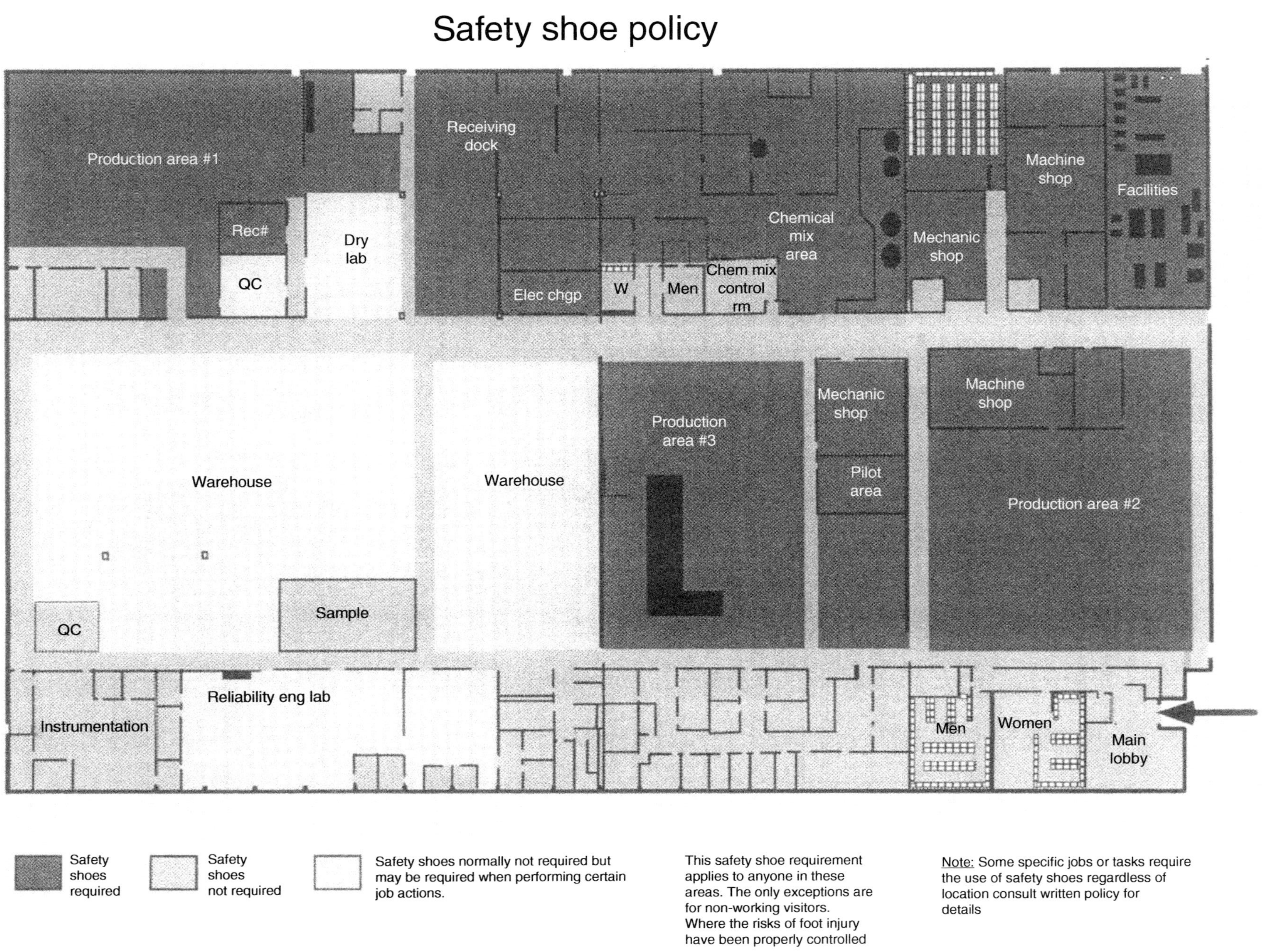


**Figure 15.2** PPE requirements mapped onto a floor plan.

Bldg/area: ______________________ Survey by: ______________________
(Dept supervisor/team leader)

Dept: ______________________ Date: ______________________

Certification of hazard assessment

Based upon an assessment of workplace hazards, the following is the minimum level of personal protection required for the locations and operations listed.

| Minimum PPE required | | Job task, area, or job function | | |
|---|---|---|---|---|
| Head | Hard hat | | | |
| | Bump cap | | | |
| Eye and face | Safety glasses w/side shields | | | |
| | Chemical goggles | | | |
| | Face shield | | | |
| | Laser glasses/goggles | | | |
| | Welder's goggles or helmet * | | | |
| | Other eye and face | | | |
| Hand | Gloves, leather | | | |
| | Gloves, chemical * | | | |
| | Gloves, thermal | | | |
| | Other hand protection * | | | |
| Body | Apron * | | | |
| | Coverall * | | | |
| | Other body protection * | | | |
| Foot | Chemical boots * | | | |
| | Safety shoes | | | |
| | Other foot protection * | | | |
| Respiratory | Respiratory protection * | | | |
| Ear | Hearing protection | | | |
| Other PPE | | | | |

*Use footnotes to fully specify, i.e. optical shade, material of construction, brand and model, or stock number.

Footnotes: ______________________

**Figure 15.3** Personal protective equipment (PPE) requirements for location or task.

## 15.3 TYPES OF PPE

### 15.3.1 Chemical Protective Clothing

Gloves and garments specifically designed for protection against chemicals are referred to as chemical protective clothing. CPC is usually intended to protect the body from direct contact with chemicals in their solid or liquid phase but is sometimes used for protection against gases or vapors as well.

***15.3.1.1 CPC Selection Considerations*** In addition to the general selection considerations identified in Section 15.2.5, the following specifics about the hazard should be considered during CPC selection:

- Which parts of the body are at risk of chemical contact?
- What is the anticipated duration of contact?
- How toxic or hazardous is the chemical? Is it corrosive to tissue? Does its threshold limit value (ACGIH 1996) have a "skin" designation, indicating that skin absorption is a potentially significant contributor to the overall dose? Is the chemical a skin sensitizer?
- Is the chemical in solid or liquid form? Will it be used at an extreme temperature?

**TABLE 15.1 Standards and Regulations Applicable to Personal Protective Equipment**

| PPE type | OSHA 29 CFR 1910 general industry standards | Consensus standards |
|---|---|---|
| General | 1910.132 General requirements | General information available through the Industrial Safety Equipment Association, (703)525-1695 |
| Chemical protective clothing (CPC) | 1910.120 Hazardous waste operations and emergency response | ASTM F739, Standard Test Method for Resistance of Protective Clothing Materials to Permeation by Liquids or Gases under Conditions of Continuous Contact<br>ASTM F903, Standard Test Method for Resistance of Materials Used in Protective Clothing to Penetration by Liquids<br>ASTM F1001, Standard Guide for Selection of Chemicals to Evaluate Protective Clothing Materials<br>ASTM F1052, Standard Practice for Pressure Testing of Gas-Tight Totally Encapsulating Chemical-Protective Suits<br>ASTM 1154, Standard Practices for Qualitatively Evaluating the Comfort, Fit, Function, and Integrity of Chemical-Protective Suit Ensembles<br>ASTM F1342, Standard Test Method for Protective Clothing Material Resistance to Puncture<br>ASTM F1359, Standard Practice for Determining the Liquid-Tight Integrity of Chemical Protective Suits or Ensembles under Static Conditions<br>NFPA-1991, Standard on Vapor-Protective Suits for Hazardous Chemical Emergencies<br>NFPA-1992, Standard on Liquid Splash-Protective Ensembles and Clothing for Hazardous Materials Emergencies<br>NFPA-1993, Standard on Support Function Protective Clothing for Hazardous Chemical Operations |
| Head | 1910.135 Head protection | ANSI Z89.1-1986, American National Standard for Personal Protection – Protective Headwear for Industrial Workers Requirements |
| Eye and face | 1910.133 Eye and face protection | ANSI Z87.1-1989, American National Standard for Occupational and Educational Eye and Face Protection<br>ANSI Z136.1-1993, American National Standard for Safe Use of Lasers |
| Hand | 1910.138 Hand protection | Some information available from the International Hand Protection Association, (301)961-8680, and the National Industrial Glove Distributors Association, (215)564-3484 |
| Foot | 1910.136 Foot protection | ANSI Z41-1991, American National Standard for Personal Protection – Protective Footwear |
| Fall protection | Subpart M of OSHA 1926 Construction Standard | ANSI A10.14-1991, Requirements for Safety Belts, Harnesses, Lanyards, Lifelines, and Drop Lines for Constructional and Industrial Use<br>ANSI Z359.1-1992, Safety Requirements for Personal Fall Arrest Systems, Subsystems, and Components |

- Does the chemical warn the body of its presence on the skin?
- Can the chemical be readily removed from skin once contaminated?

***15.3.1.2 Rating Chemical Resistance*** A key characteristic of CPC is its ability to serve as a protective barrier to specific chemical agents as determined by its resistance to permeation, degradation, and penetration.

*15.3.1.2.1 Permeation* Permeation is the resistance to chemical movement through protective clothing material on a molecular level via absorption, diffusion, and then desorption. Low-level permeation is measured using ASTM Standard Test Method F739 (ASTM F739 1991), in which a swatch of the clothing is exposed on one side to the chemical of interest while analytical equipment measures desorption on the opposite side. Permeation as measured with this method is expressed as time (minutes) to initial breakthrough (breakthrough time [BT]) and also as a steady-state permeation rate following initial breakthrough ($mg\,m^{-2}\,min^{-1}$). There are no quantitative health standards for permissible skin contact; hence breakthrough time is the parameter typically used for clothing selection.

One widely used field resource, *Quick Selection Guide to Chemical Protective Clothing* (Forsberg and Mansdorf 1993), is a lookup guide providing clothing material selection guidance based on permeation data for commonly used hazardous materials tested against 15 generic and proprietary barrier materials used to construct gloves, suits, and boots. The test data in the guide are those published by CPC manufacturers. In general, the guide considers a BT of greater than four hours to be acceptable protection, while materials with BT of one to four hours should be used with caution, and those with breakthrough at less than one hour are not recommended.

Studies have shown that BT for the "same" materials, such as natural rubber used to make gloves, can vary significantly between manufacturers. Therefore, when selecting CPC, permeation data specific to the products under consideration should be obtained from the manufacturers. Figure 15.4 shows an example of a typical table of permeation data provided by a CPC supplier. Another widely cited source of permeation data is *Guidelines for the Selection of Chemical Protective Clothing*, 3e (Schwope et al. 1987).

As a general statement, the thicker the barrier material, the longer the chemical holdback will be. For example, latex gloves typically sold as "chemical gloves" will usually be at least 11 and up to 60 mil (thousandths of an inch) thick, while surgical latex gloves may be only 6 mil thick.

*15.3.1.2.2 Degradation* Another CPC performance characteristic is degradation, which is the resistance to deterioration of the material's physical properties such as weight, thickness, elongation, tear strength, and cut and puncture resistance. Significant degradation by chemical attack will render a material unsuitable for use. While degradation tests are important, interpretation of their results is somewhat subjective.

*15.3.1.2.3 Penetration* A third important characteristic of CPC is resistance to penetration, which is the resistance to flow of a chemical on a nonmolecular, or gross, level through closures, seams, pinholes, or other material imperfections. A standard penetration test has been developed that exposes a portion of the garment to the challenge chemical, with the penetration measured as the time for a visible droplet to appear on the far side (ASTM F903 1990). Penetration is especially important when considering the way garments are stitched, glued, heat sealed, and seamed.

Penetration is an important selection factor. Excellent resistance to permeation by a material is of little use if the garment is seamed with common stitches, which in effect are holes poked throughout the material.

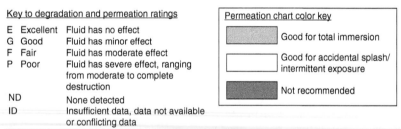

Key to degradation and permeation ratings

E Excellent — Fluid has no effect
G Good — Fluid has minor effect
F Fair — Fluid has moderate effect
P Poor — Fluid has severe effect, ranging from moderate to complete destruction

ND — None detected
ID — Insufficient data, data not available or conflicting data

Permeation chart color key

- Good for total immersion
- Good for accidental splash/intermittent exposure
- Not recommended

| | Silver shield (4 mil) | | | Viton (9 mil) | | | Butyl (17 mil) | | | Nitrile latex (11 mil) | | |
|---|---|---|---|---|---|---|---|---|---|---|---|---|
| Chemical | D | BT | PR | D | BT | PR | D | BT | PR | D | BT | PR |
| n-Hexane | E | > 6 h | ND | ID | > 11 h | ND | P | ID | ID | E | ID | ID |
| Hydrazine (70% in water) | G | > 6 h | ND | P | ID | ID | E | > 8 h | ND | G | > 8 h | ND |
| Hydrochloric acid (37%) | E | > 6 h | ND | E | ID | ID | E | ID | ID | P | ID | ID |
| Hydrofluoric acid (50%) | G | > 6 h | ND | G | ID | ID | F | ID | ID | P | ID | ID |
| Isobutyl alcohol | E | ID | ID | E | > 8 h | ND | E | > 8 h | ND | G | > 8 h | ND |
| Isobutyraldehyde | E | ID | ID | P | 4 min | 11.5 | E | > 8 h | ND | P | ID | ID |
| Methacrylic acid | ID | ID | ID | F | > 8 h | ND | G | > 8 h | ND | P | 1.7 h | 23 |
| Methacrylonitrile | E | ID | ID | F | 4 min | 462 | G | 6.8 h | 0.001 | P | 7 min | 560 |
| Methyl chloroform | ID | > 6 h | ND | E | > 15 h | ND | P | ID | ID | P | 41 min | 76.4 |
| Methyl cyanide | ID | > 8 h | ND | ID | ID | ID | E | > 8 h | ND | ID | ID | ID |
| Methyl ethyl ketone | E | > 24 h | ND | P | ID | ID | E | > 8 h | ND | P | ID | ID |
| Methyl isocyanate | ID | ID | ID | P | 4 min | 121 | P | 1.1 h | 9.0 | P | ID | ID |
| Methylamine (40% in water) | F | 1.9 h | 2.0 | E | > 16 h | ND | E | > 15 h | ND | G | > 8 h | ND |
| Methylene chloride | G | > 8 h | ND | F | 1 hr | 7.32 | P | 24 min | 133 | P | 4 min | 766 |

**Figure 15.4** Example of a glove chemical resistance guide. *Source:* Courtesy of North Safety Products, Hand Protection Division.

***15.3.1.3 Other Considerations*** In addition to permeation, degradation, and penetration, several other factors must be considered during CPC selection. These affect not only chemical resistance but also the worker's ability to perform the required task. These include the following:

- Heat transfer – Almost without exception, CPC materials are nonbreathable materials and severely inhibit the evaporation of water from the skin, which is important to shedding excess body heat. While some CPC may be thicker or have a higher insulation value than others, it is the impermeability to water vapor that most affects the level of heat stress to the worker. The choice of material is therefore not critical from a heat stress standpoint – all CPC increases heat stress.
- Durability – Does the material have sufficient strength to withstand the physical stress of the task at hand? Will it resist tears, punctures, and abrasions? Will it withstand repeated use after contamination/decontamination?
- Flexibility – Will the CPC, especially gloves, interfere with the worker's ability to perform the task?
- Vision – Can the worker see well enough in all directions with the equipment on?
- Temperature effects – Will the material maintain its protective integrity and flexibility under hot and cold extremes?
- Ease of decontamination – Can the material be decontaminated? Should disposable CPC be used?
- Duration of use – How does the estimated task duration compare with the permeation, degradation, and penetration data?

***15.3.1.4 CPC Ensembles*** CPC is worn in a wide variety of occupations – wherever hazardous chemicals are used. Frequently gloves worn alone, or gloves worn with sleeves, apron, or chemical gown (apron with attached sleeves), eye protection, and boots may be all that is needed. In some extreme cases, however, more chemical protection may be needed.

The push to clean up hazardous waste "Superfund" disposal sites begun in the early 1980s and continuing to the present day created the need to identify PPE ensembles for workers initially approaching, and then working in, environments where chemical exposure is a risk and where the hazard is often unknown and potentially high. In response to this risk, the Environmental Protection Agency (EPA) first alone, and then in conjunction with NIOSH, OSHA, and the US Coast Guard, issued recommended PPE ensembles termed "levels of protection" based on the potential hazard. These are described in Table 15.2. While the EPA levels of protection are usually applied to hazardous waste sites or hazardous materials ("hazmat") emergency response operations (OSHA 1910.120 1996), the idea of "low- to high-risk" PPE ensembles that are specified based on the potential hazard has broader applicability to most any task or occupation where chemical exposure is a risk.

The NFPA has issued standards covering manufacture and use of CPC for use by firefighters (NFPA-1991 1994; NFPA-1992 1994; NFPA-1993 1994).

Figure 15.5 depicts examples of CPC ensembles.

***15.3.1.5 Decontamination of CPC*** Much of the CPC in use today is designed to be "limited use," meaning it is intended to be disposed after a single use or after a use that results in chemical contact and/or physical wear. The reason limited-use CPC prevails over more durable garments is the difficulty associated with cleaning chemical contamination from CPC material and then assuring or measuring that its protective properties remain after being used and subjected to the decontamination process.

Some decontamination processes include aeration, that is, evaporation of volatile contaminants, water/detergent washing, water/bleach washing, and freon-based dry cleaning (Johnson and Anderson 1990).

All CPC use, including use of disposables, should be accompanied by a decontamination procedure to prevent chemical exposure to the wearer while doffing the garments, contamination spread throughout the workplace, exposure to the worker's family from taking contaminated apparel home, and contamination of workers who may handle the waste garments.

### 15.3.2 Head Protection

***15.3.2.1 Hard Hats*** The most common form of head protection is the hard hat. The hard hat is designed to protect against the hazard of solid objects falling from above, which can produce perforation or facture of the skull or brain lesions from sudden displacement (*Encyclopedia* 1983). Head protection in the form of a hard hat or other covering can also protect against electric shock, burns from splashes of hot or corrosive liquids, or molten metal.

**TABLE 15.2 Sample Protective Ensembles Based on EPA Protective Ensembles**

| Level of protection | Equipment | Protection provided | Should be used when | Limiting criteria |
|---|---|---|---|---|
| A | Recommended<br>Pressure-demand, full facepiece SCBA or pressure-demand supplied-air respirator with escape SCBA<br>Fully encapsulating, chemical-resistant suit<br>Inner chemical-resistant gloves<br>Chemical-resistant safety boots/shoes<br>Two-way radio communication<br>Optional<br>Cooling unit<br>Coveralls<br>Long cotton underwear<br>Hard hat<br>Disposable gloves and boot covers | The highest available level of respiratory, skin, and eye protection | The chemical substance has been identified and requires the highest level of protection for skin, eyes, and the respiratory system based on either measured (or potential for) high concentration of atmospheric vapor, gases, or particulates or site operations and work functions involving a high potential for splash, immersion, or exposure to unexpected vapors, gases, or particulates of materials that are harmful to skin or capable of being absorbed through intact skin; substances with a degree of hazard to the skin are known or suspected to be present, and skin contact is possible; operations must be conducted in confined, poorly ventilated areas until the absence of conditions requiring level A protection is determined | Fully encapsulating suit material must be compatible with the substances involved |
| B | Recommended<br>Pressure-demand, full facepiece SCBA or pressure-demand supplied-air respiratory with escape SCBA<br>Chemical-resistant clothing (overalls and long-sleeved jacket; hooded, one- or two-piece chemical splash suit; disposable chemical-resistant one-piece suit)<br>Inner and outer chemical-resistant gloves<br>Chemical-resistant safety boots/shoes<br>Hard hat<br>Two-way radio communications<br>Optional<br>Coveralls<br>Disposable boot covers<br>Face shield<br>Long cotton underwear | The same level of respiratory protection but less skin protection than level A; it is the minimum level recommended for initial site entries until the hazards have been further identified | The type and atmospheric concentration of substances have been identified and require a high level of respiratory protection, but less skin protection; this involves atmospheres with IDLH concentrations of specific substances that do not represent a severe skin hazard or that do not meet the criteria for use of air-purifying respirators<br>Atmosphere contains less than 19.5% oxygen<br>Presence of incompletely identified vapor or gases is indicated by direct reading instrument, but vapors and gases are not suspected of containing high levels of chemicals harmful to skin or capable of being adsorbed through the skin | Use only when the vapor or gases present are not suspected of containing high concentrations of chemicals that are harmful to skin or capable of being absorbed through the intact skin<br>Use only when it is highly unlikely that the work being done will generate either high concentrations of vapor gases or particulates or splashes of material that will affect exposed skin |

**TABLE 15.2 (Continued)**

| Level of protection | Equipment | Protection provided | Should be used when | Limiting criteria |
|---|---|---|---|---|
| C | Recommended<br>Full facepiece, air-purifying canister-equipped respirator<br>Chemical-resistant clothing (overalls and long-sleeved jacket; hooded, one- or two-piece chemical splash suit; disposable chemical-resistant one-piece suit)<br>Inner and outer chemical-resistant gloves<br>Chemical-resistant safety boots/shoes<br>Hard hat<br>Two-way radio communications<br>Optional<br>Coveralls<br>Disposable boot covers<br>Face shield<br>Escape mask<br>Long cotton underwear | The same level of skin protection as level B, but a lower level of respiratory protection | The atmospheric contaminants, liquid splashes, or other direct contact will not adversely affect any exposed skin<br>The types of air contaminants have been identified and concentrations measured, and a canister is available that can remove the contaminant<br>All criteria for the use of air-purifying respirators are met | Atmospheric concentration of chemicals must not exceed IDLH levels<br>The atmosphere must contain at least 19.5% oxygen |
| D | Recommended<br>Coveralls<br>Safety boots/shoes<br>Safety glasses or chemical splash goggles<br>Hard hat<br>Optional<br>Gloves<br>Escape mask<br>Face shield | No respiratory protection; minimal skin protection | The atmosphere contains no known hazard<br>Work functions preclude splashes, immersion, or the potential for unexpected inhalation of or contact with hazardous levels of any chemicals | This level should not be worn in the exclusion zone<br>The atmosphere must contain at least 19.5% oxygen |

*Source:* NIOSH (1985).

***15.3.2.2 When Required*** In the United States, OSHA requires that hard hats be provided and worn where there is a risk of injury from objects falling from above or where workers are exposed to electrical conductors that could contact the head. OSHA further requires that hard hats comply with the ANSI protective headwear standard (ANSI Z89.1-1986 n.d.), which specifies physical and performance requirements and test methods for helmets.

***15.3.2.3 ANSI Hard Hat Types*** ANSI type 1 helmets have a full brim, while type 2 have a baseball cap-like bill that protrudes out over the wearer's eyes. ANSI further specifies three classes of electrical insulation:

- Class A provides protection against impact and exposed low-voltage conductors (proof-tested at 2200V).
- Class B provides protection against impact and exposed high-voltage conductors (proof-tested at 20 000V).
- Class C provides protection against impact only.

The majority of helmets in use are type 2, class A or B, while class C finds use especially in the forestry industry. The wearer should be able to identify the type of helmet by looking inside the shell for the manufacturer name, ANSI standard, and ANSI class.

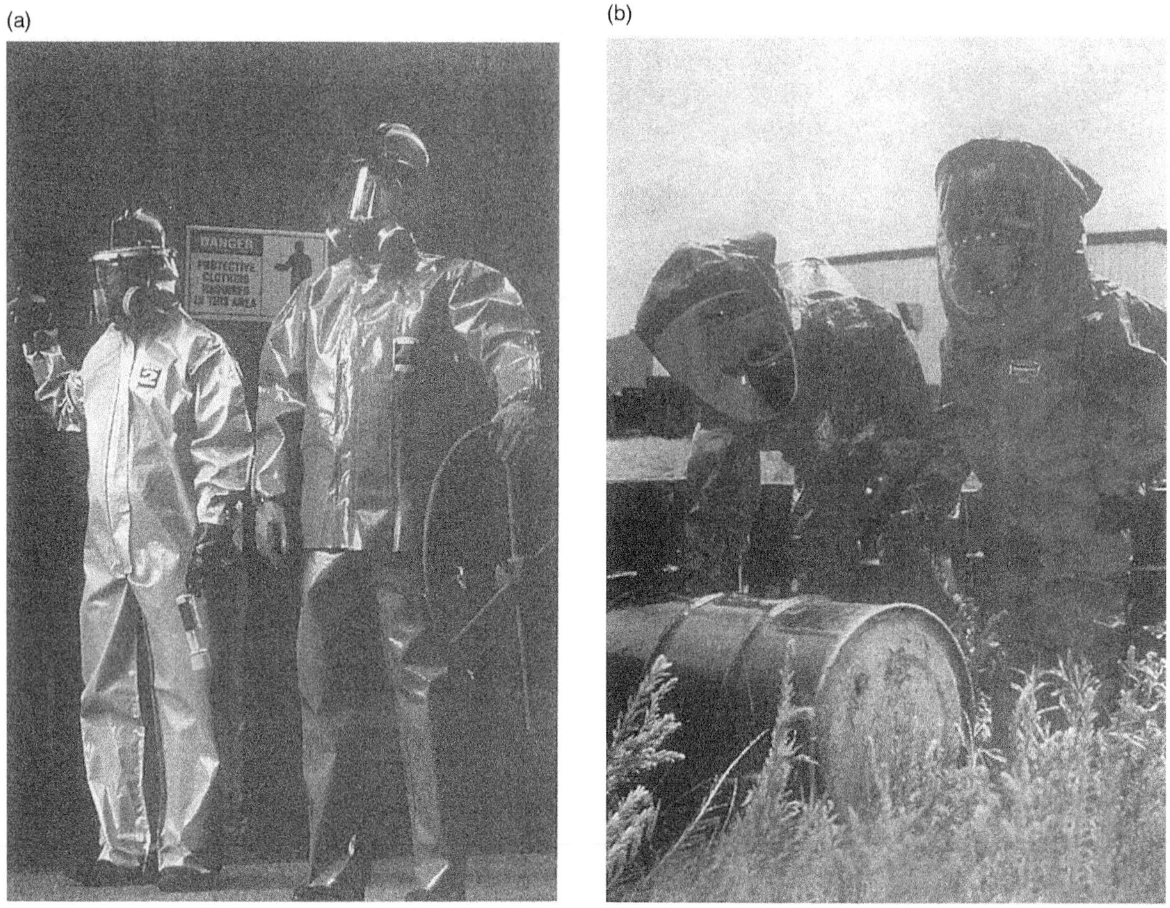

**Figure 15.5** Chemical protective clothing. (a) EPA level C ensemble. (b) Chemical encapsulating suit, available as splashproof and gastight level A or splashproof level B. (c) Chemical- and flash-resistant chemical suit, fully compliant with NFPA-1991 (gastight) or 1992 (splashproof) standards. *Source:* Courtesy of Kappler Safety Group.

***15.3.2.4 How They Work*** Impact protection is provided through the design of two key components: the hat's hard shell and the suspension system (Figure 15.6). The shell is designed to be smooth so as to deflect falling objects to the side. It is also designed to flex or deflect upon impact into the space between the skull and shell created by the suspension. The suspension itself is constructed to distribute the impact energy over a larger area and to absorb energy by stretching slightly (while still preserving the suspension-to-shell space).

High-density polyethylene is the most common hard hat shell material. Other materials in use include polycarbonate, polycarbonate glass, polyester glass, and aluminum. For high-temperature applications, polycarbonate or polyester is more appropriate than polyethylene, and the suspension should be a woven material.

***15.3.2.5 Accessories*** Hard hats are available that can be accessorized with hearing protection, visors, chin straps, winter liners, and miner's lamps. (Refer to Figure 15.7.)

#### *15.3.2.6 Use Considerations*

- Persons working overhead should wear a hard hat with chin strip to secure the hat from becoming a falling object itself.
- Plastic hard hats kept in the rear window ledge of a car are subjected to damage from heat and sunlight and can become hazardous projectiles if the car suddenly stops.
- Ventilation holes cannot be drilled in the shell because they reduce or eliminate both impact and electrical insulation protection.
- Most hard hats are not designed to provide impact protection when worn backward.
- Some adhesives in stickers applied to plastic hard hats and some chemicals used in the work environment can attack plastic shell material.
- Damaged hats should be replaced.

(c)

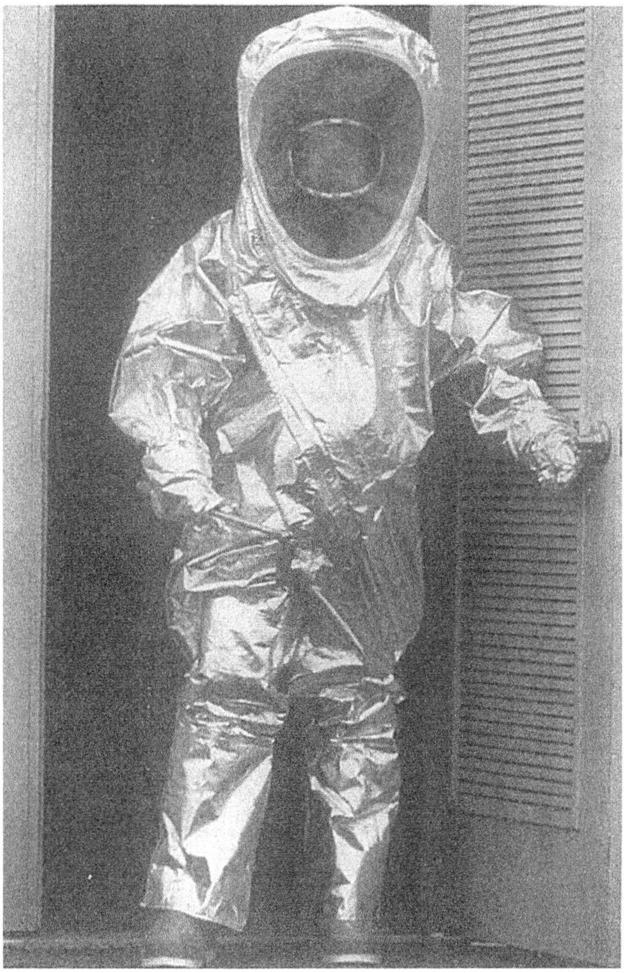

**Figure 15.5** (*Continued*)

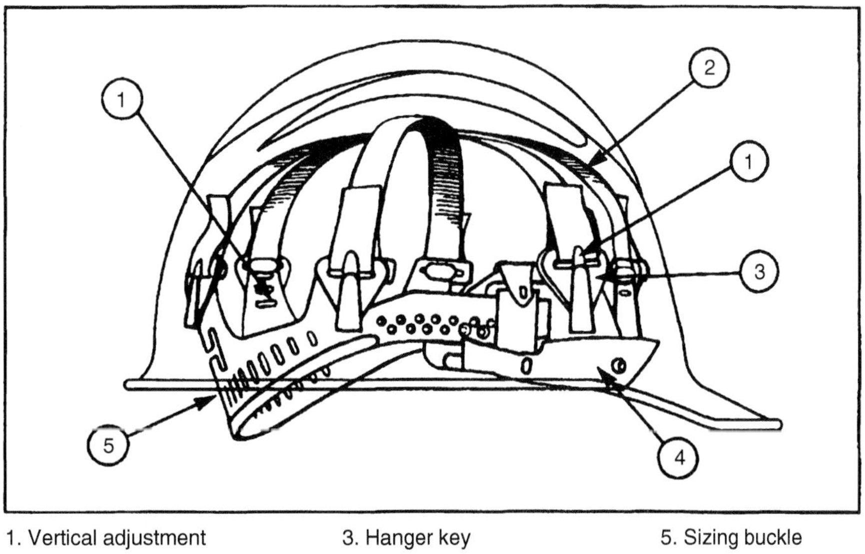


1. Vertical adjustment
2. Nylon crown strap
3. Hanger key
4. Absorbent brow pad
5. Sizing buckle

**Figure 15.6** Hard hat suspension design. *Source:* Courtesy of E. D. Bullard Company.

**Figure 15.7** Accessorized hard hat, ANSI type 2. *Source:* Courtesy of E. D. Bullard Company.

**Figure 15.8** Bump cap. *Source:* Courtesy of E. D. Bullard Company.

#### *15.3.2.7 Bump Caps*

Bump caps (Figure 15.8) are thin-shelled caps with a front bill, which are usually worn for "struck against" hazards in environments having low headroom. There are no design standards for bump caps. While they are not designed to protect against falling objects, bump caps may be sufficient protection for persons at risk of collision with low pipes or ducts while walking.

### 15.3.3 Eye and Face Protection

Eye and face protection includes safety spectacles (with or without side shields), eyecup goggles, cover goggles, and face shields. Eye and face protection is typically used for protection against the following hazards:

- Physical hazards such as flying objects or collision with stationary objects. Flying or falling objects are the most common category of hazard causing eye injury (US Department of Labor 1980).
- Chemical hazards, including dusts, mists, and splashes.
- Radiant energy, including ultraviolet light and lasers.

***15.3.3.1 When Required*** In the United States, OSHA requires that eye and/or face protection be provided as determined by a hazard analysis when workers are "exposed to eye or face hazards from flying particles, molten metal, liquid chemicals, acids or caustic liquids, chemical gases or vapors, or potentially injurious light radiation" (OSHA 1910.133 1996).

Table 15.3 provides guidance in the selection of appropriate eye and face protection. Note that spectacles and goggles are considered to be "primary" eye protection, while face shields are termed "secondary" protection. While primary protectors may be worn alone, secondary protectors are to be worn only in combination with primary protection. Figure 15.9 depicts various types of eye and face protection.

Table 15.4 provides minimum protective shade values for protection against radiant energy.

***15.3.3.2 Side Shields*** Side shields should be worn on spectacles where there is a hazard from flying objects. OSHA made this an explicit requirement with their 1994 PPE standards revisions (OSHA 1994b, 1996). Most spectacles with plano (non-corrective, nonprescription) lenses are equipped with side shields incorporated into the design and manufacture of the frame and temple bars. Wearers of prescription safety glasses, particularly those designed to look and be worn as streetwear, frequently add detachable side shields when working in an eye hazard area.

Where peripheral vision is especially important, such as in driving powered industrial vehicles, glasses without side shields may be more appropriate. The risk from decreased vision should be weighed against the risk from flying objects.

***15.3.3.3 ANSI Standard*** The ANSI occupational and educational eye and face protection standard (ANSI Z87.1-1989 n.d.) provides minimum requirements for eye and face protective devices and guidance for their selection, use, and maintenance. Spectacles, goggles, face shields, and welding helmets worn for personal protection must be manufactured in accordance with ANSI Z87, and each major component (other than lenses on spectacles) must be marked with a trademark identifying the manufacturer and "Z87."

Safety spectacles are tested as a complete assembly, with the frames being just as important as the lenses. Combinations of normal streetwear frames with safety lenses are not allowed.

***15.3.3.4 Contact Lenses*** Contact lenses are considered to offer no protection to the eye from any hazards. Safety glasses or goggles must be worn over contact lenses in environments presenting a hazard from flying objects. The original hard contacts were considered a possible hazard in dusty environments by possibly trapping abrasive particles between the lens and the cornea. Until recently, contact lenses have been considered a hazard to users in chemical use areas because of the risk of chemical absorption by the lens or "trapping" of chemicals behind the lens and against the eye tissue. While there is still agreement that the presence of a contact lens can complicate the irrigation of an eye exposed to chemical splash, the hazard of contact lenses in chemical environments is now considered to be less than once thought. Some chemical safety experts suggest allowing contact lens use in chemical environments provided that co-workers are trained to assist with lens removal in the event of a splash exposure (*Chemical Health & Safety* 1995).

In their 1998 revision to the respirator standard, OSHA deliberately made no mention of contact lenses in the text of the standard, citing in the standard's preamble that there is no evidence of an increased hazard from contact lens use when compared with the use of conventional eyeglass inserts inside full facepiece respirators (*Federal Register* 1998). In its respirator standard, ANSI allows contact lenses with respirators provided that the individual "has previously demonstrated that he or she has had successful experience wearing contact lenses" (ANSI Z88.2-1992 n.d.).

A story surfaced in the safety trade press in the late 1970s reporting a case where a contact lens user was exposed to optical radiation from welding and experienced a severe eye injury when the ultraviolet energy fused the plastic contact lens to his cornea. This widely published story was found to be untrue – an alarming rumor. This false anecdote still occasionally surfaces today, which is why it is mentioned here.

**TABLE 15.3 Eye and Face Protection Selection Chart**

| Source | Assessment of hazard | Protection |
|---|---|---|
| Impact – chipping, grinding, machining, masonry work, woodworking, sawing, drilling, chiseling, powered fastening, riveting, and sanding | Flying fragments, objects, large chips, particles, sand, dirt, etc. | Spectacles with side protection, goggles, face shields; see notes (1), (3), (5), (6), and (10); for severe exposure, use face shield |
| Heat – furnace operations, pouring, casting, hot dipping, and welding | Hot sparks | Face shields, goggles, spectacles with side protection; for severe exposure, use face shield; see notes (1), (2), and (3) |
| | Splash from molten metals | Face shields worn over goggles; see notes (1), (2), and (3) |
| | High-temperature exposure | Screen face shields, reflective face shields; see notes (1), (2), and (3) |
| Chemicals – acid and chemical handling, degreasing, plating | Splash | Goggles, eyecup, and cover types; for severe exposure, use face shield; see notes (3) and (11) |
| | Irritating mists | Special-purpose goggles |
| Dust – woodworking, buffing, general dusty conditions | Nuisance dust | Goggles, eyecup, and cover types; see note (8) |
| Light and/or radiation: Welding: electric arc | Optical radiation | Welding helmets or welding shields; typical shades, 10–14; see notes (9) and (12) |
| Welding: gas | Optical radiation | Welding goggles or welding face shield; typical shades, gas welding 4–8, cutting 3–6, brazing 3–4; see note (9) |
| Cutting, torch brazing, torch soldering | Optical radiation | Spectacles or welding face shield; typical shades, 1.5–3; see notes (3) and (9) |
| Glare | Poor vision | Spectacles with shaded or special-purpose lenses, as suitable; see notes (9) and (10) |

[1] Care should be taken to recognize the possibility of multiple and simultaneous exposure to a variety of hazards. Adequate protection against the highest level of each of the hazards should be provided. Protective devices do not provide unlimited protection.
[2] Operations involving heat may also involve light radiation. As required by the standard, protection from both hazards must be provided.
[3] Face shields should only be worn over primary eye protection (spectacles or goggles).
[4] As required by the standard, filter lenses must meet the requirements for shade designation in 1910.133(a)(5). Tinted and shaded lenses are *not* filter lenses unless they are marked or identified as such.
[5] As required by the standard, persons whose vision requires the use of prescription (Rx) lenses must wear either protective devices fitted with prescription (Rx) lenses or protective devices designed to be worn over regular prescription (Rx) eyewear.
[6] Wearers of contact lenses must also wear appropriate eye and face protection devices in a hazardous environment. It should be recognized that dusty and/or chemical environments may represent an additional hazard to contact lens wearers.
[7] Caution should be exercised in the use of metal frame protective devices in electrical hazard areas.
[8] Atmospheric conditions and the restricted ventilation of the protector can cause lenses to fog. Frequent cleansing may be necessary.
[9] Welding helmets or face shields should be used only over primary eye protection (spectacles or goggles).
[10] Non-side shield spectacles are available for frontal protection only, but are not acceptable eye protection for the sources and operations listed for "impact."
[11] Ventilation should be adequate but well protected from splash entry. Eye and face protection should be designed and used so that it provides both adequate ventilation and protects the wearer from splash entry.
[12] Protection from light radiation is directly related to filter lens density. See note (4). Select the darkest shade that allows task performance.
*Source:* OSHA 1910.133 (1996).

***15.3.3.5 Laser Eyewear*** The ANSI laser standard (ANSI Z136.1-1993 n.d.) provides guidance on selection of protective eyewear for workers potentially exposed to ANSI class 3b or 4 lasers above the maximum permissible exposure level. Sixteen selection factors are provided in the standard. Laser eyewear is clearly labeled with the optical density and wavelength for which protection is afforded.

### 15.3.4 Hand Protection

Hands are the worker's primary tools for most jobs and are at relatively high risk of injury from cuts, burns, abrasion, electrical shock, amputation, and chemical injury.

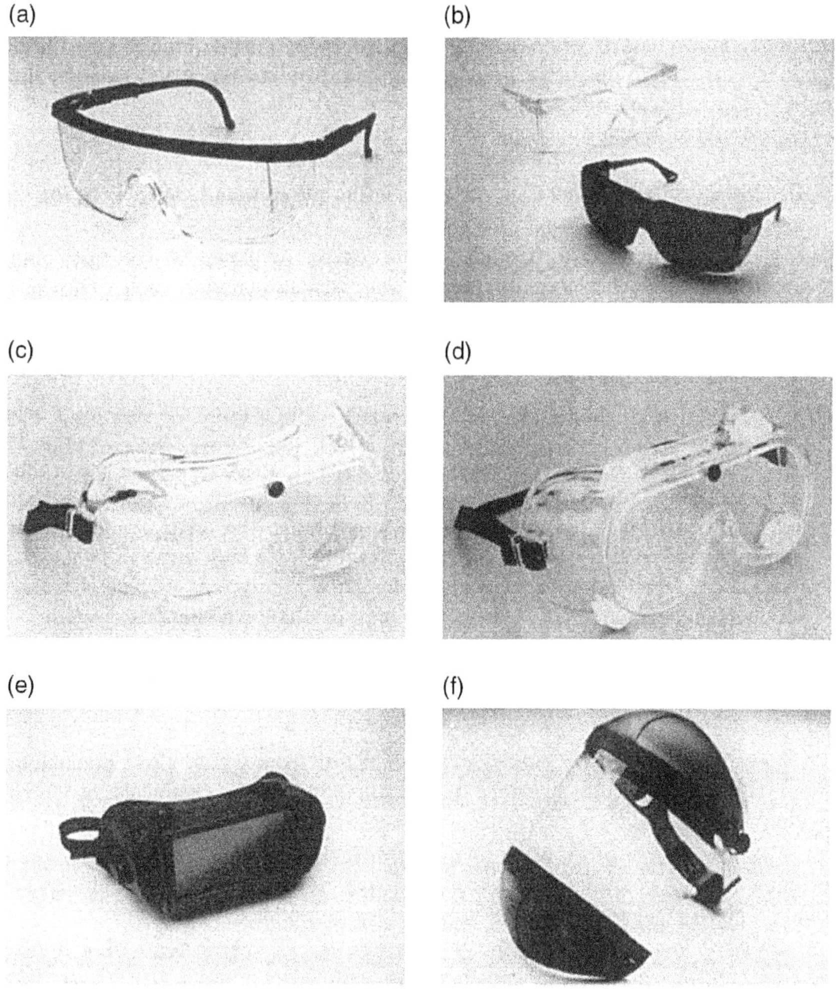

**Figure 15.9** Types of eye and face protection. (a) Basic spectacles with side shields and adjustable temple length. (b) Visitor's glasses can fit over prescription eyewear. (c) Ventilated goggle. (d) Chemical goggle with indirect vent. (e) Welder's goggle. (f) Face shield with neck and forehead protection. *Source:* Courtesy of Aearo Company.

***15.3.4.1 Types Available*** There is a wide assortment of gloves, hand pads, sleeves, and wristlets available for hand and arm protection. Each glove type has its own advantages and disadvantages relative to the protection offered and the ability of the user to perform the task while wearing the glove. Involvement of the end user in PPE selection is particularly important with gloves, and different types and manufacturers should be experimented with freely.

There are three general functional types of gloves:

1. *General-purpose leather* from cowhide is the glove traditionally worn for protection against cuts, abrasion, and heat/cold.
2. *General-purpose fabrics* include a wide variety of gloves woven from cotton and cotton/polyester blends or specialty cut-resistant materials such as Kevlar by DuPont or steel mesh. Fabric gloves are commonly coated or stippled on the palm with a polymer for better grip in specific environments or fully dipped for greater liquid or chemical resistance.
3. *Chemical-resistant* gloves are seamless types manufactured by dipping a hand-shaped solid form into a viscous solution that polymerizes upon drying. The less expensive types most commonly encountered are made of natural latex rubber, neoprene, and nitrile rubber. Less common but having specific application are butyl rubber, Viton, PVC (vinyl), and polyvinyl alcohol (PVA). A relatively new material constructed of a PE/EVAL laminate is resistant to the widest variety of chemicals. With its relatively poor grip and finger dexterity, this type is frequently worn as an underglove for applications involving high or unknown chemical hazard.

**TABLE 15.4 Filter Lenses for Protection Against Radiant Energy**

| Operations | Electric size 1/32 in. | Arc current | Minimum[a] protective shade |
|---|---|---|---|
| Shielded metal arc welding | <3 | <60 | 7 |
| | 3–5 | 60–160 | 8 |
| | 5–8 | 160–250 | 10 |
| | >8 | 250–550 | 11 |
| Gas metal arc welding and flux-cored arc welding | | <60 | 7 |
| | | 60–160 | 10 |
| | | 160–250 | 10 |
| | | 250–500 | 10 |
| Gas tungsten arc welding | | <50 | 8 |
| | | 50–150 | 8 |
| | | 150–500 | 10 |
| Air carbon | (Light) | <500 | 10 |
| Arc cutting | (Heavy) | 500–1000 | 11 |
| Plasma arc welding | | <20 | 6 |
| | | 20–100 | 8 |
| | | 100–400 | 10 |
| | | 400–800 | 11 |
| Plasma arc cutting | (Light)[b] | <300 | 8 |
| | (Medium)[b] | 300–400 | 9 |
| | (Heavy)[b] | 400–800 | 10 |
| Torch brazing | | | 3 |
| Torch soldering | | | 2 |
| Carbon arc welding | | | 14 |
| **Operations** | **Plate thickness (in.)** | **Plate thickness (mm)** | **Minimum[a] protective shade** |
| Gas welding | | | |
| Light | Under 1/8 | Under 3.2 | 4 |
| Medium | 1/8–1/2 | 3.2–12.7 | 5 |
| Heavy | Over 1/2 | Over 12.7 | 6 |
| Oxygen cutting | | | |
| Light | Under 1 | Under 25 | 3 |
| Medium | 1–6 | 25–150 | 4 |
| Heavy | Over 6 | Over 150 | 5 |

[a] As a rule of thumb, start with a shade that is too dark to see the weld zone. Then go to a lighter shade, which gives sufficient view of the weld zone without going below the minimum. In oxyfuel gas welding or cutting where the torch produces a high yellow light, it is desirable to use a filter lens that absorbs the yellow or sodium line in the visible light of the (spectrum) operation.

[b] These values apply where the actual arc is clearly seen. Experience has shown that lighter filters may be used when the arc is hidden by the workpiece.

*Source:* 29 CFR 1910.133(a)(5).

**15.3.4.2 *Selection Considerations*** The following glove selection process is suggested:

1. *Identify one or two most important hazards* from which glove protection is desired. Is it abrasion, cuts, punctures, temperatures, microbes, chemicals, or ionizing radiation?
2. *Consider the degree of exposure to the hazards.* When a glove is to be worn for protection against two different chemical operations, for example, immersion exposure should be given greater consideration than incidental splash.
3. *Identify potential impacts of the glove on the process where they will be applied.* The wrong glove can introduce contamination with serious unwanted consequences to the underlying objectives of the task. Examples include the bacteria-free, or sterile, gloves needed in an operating room and the powder-free, low-particle-shedding gloves needed in a computer chip fabrication area. Inks used to mark glove sizes may dissolve and introduce contamination.
4. *For chemical protection, use permeation and degradation information* provided by the glove manufacturer or from sources such as the *Quick Selection Guide to Chemical Protective Clothing* (Forsberg and Mansdorf 1993) to identify protective materials.

In applications where protection against chemical hazards is needed, data on permeation and degradation of the *specific* glove under consideration for the particular application should be obtained from the manufacturer if possible. Although charts of chemical resistance for generic material types such as natural rubber are useful, they do not take into account important factors such as the thickness of the product under consideration or the manufacturer-specific glove manufacturing process, which has been shown to affect chemical permeation rates.

Where test data are not available, the user can perform a simple, gross test for chemical compatibility by dipping the glove into, or filling a finger with, the challenge chemical and observing for swelling, hardening, softening, dissolving, or penetration of the material. If the chemical substance is especially hazardous, however, laboratory testing of the specific agent and glove should be either requested of the glove manufacturer or commissioned directly with a testing company.

Not only does glove thickness affect permeation rate, but it is also a factor determining resistance to tears and abrasions. Thicker is usually better. For example, vinyl (PVC) as a material in general has excellent resistance against most concentrated acids. However, thin ambidextrous vinyl gloves sold in 100 per pack dispenser boxes should generally not be worn when handling concentrated acids because they tear and abrade more easily and have quicker BT than their thicker counterparts intended for use as chemical protection. The same is true of latex and nitrile gloves, which are available in very thin (e.g. 4 mil) versions.

5. *Consider the grip and dexterity demands of the task.* Tasks requiring manipulation of fine parts or tools with the fingers cannot be performed with thicker gloves. Embossed rubber gloves will grip wet objects better than smooth rubber.
6. *Consider reusability of the glove after exposure to the task.* Gloves exposed to glue or resins that are difficult to remove, for example, will often be discarded after a single use. Conversely, chemical gloves exposed to volatile organic liquids may dry out and stay contamination-free for long periods of time.
7. *Consider price*, especially relative to reusability and numbers of individuals who will need protection.
8. *Consider cuff style and length.* Choices include band top, knit wrist, safety cuff, bell gauntlet, slip-on, and open cuff. For welding or metal fabrication jobs, choose a gauntlet cuff for greater protection. For cold environments, a knit wrist will keep cold air out. Slip-on or open cuff styles make for easier donning and doffing where gloves will be frequently removed.
9. *Consider color or marking.* The correct glove for a particular task often comes to be identified within an organization by *color* because it is the simplest way to differentiate gloves. Glove colors *are not standardized* – you cannot identify the material of an unknown glove by its color. An organization can, however, create its own glove color standard, which is very useful in communicating the correct glove for the task. Color is therefore a consideration when adding a new glove type or replacing an existing one. In addition to color, glove types and sizes can be identified by labels printed on the glove packaging or on the glove cuffs themselves.

#### 15.3.4.3 Gloves Bring Tradeoffs

All PPE, and gloves in particular, present tradeoffs or compromises to the user. Usually (but not always) protection from a physical or chemical hazard comes at the expense of dexterity. A chemistry lab or chemical plant cannot stock the most chemically resistant glove for every substance on hand; protection should be provided against the most hazardous substances and exposures. The tradeoffs need to be identified in advance and understood by the user so that accommodation becomes easier.

In some cases two pairs of gloves may need to be worn simultaneously to address the significant hazards and/or meet the demands of the task. For example, a hazardous materials incident responder may wear a thicker, more abrasion-resistant glove over one having higher chemical resistance. A meat packer may wear a waterproof glove over one made of steel mesh.

#### 15.3.4.4 Glove Size

A poor-fitting glove is awkward to wear and can lead to clumsy, costly mistakes or accidents. Gloves are often sized numerically using a measurement of the circumference around the palm of the hand at the knuckle area as follows:

| Glove size | Hand circumference (in.) |
|---|---|
| XS | 6–7 |
| S | 7–8 |
| M | 8–9 |
| L | 9–10 |
| XL | 10–11 |

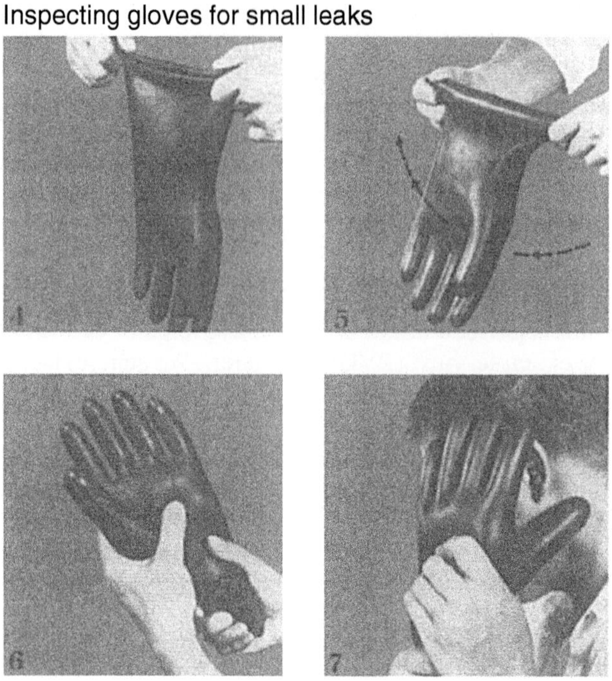


**Figure 15.10** Glove leak checking procedure. Inspecting gloves for small leaks. To air test gloves for pinholes and other damage, follow these procedures: Hold glove downward and grasp the cuff as shown in illustration 4. Twirl the glove upward toward your body to trap the air inside the glove as shown in illustration 5. Then squeeze the rolled cuff tightly into a U shape with the right hand to keep the trapped air inside. Squeeze with the other hand and look for damage exposed by inflation as shown in illustration 6. Then hold the inflated glove close to your face and ear, squeezing the glove, to feel and listen for air escaping from holes as shown in illustration 7. Important practices to follow: Inspect frequently for glove damage. Rinse gloves daily with clear water inside and out and thoroughly dry. Wear leather protectors. Replace old, worn, or damaged protectors. *Source:* Photos and corresponding procedures courtesy of North Safety Products, Hand Protection Division.

***15.3.4.5 Leak Checking*** Gloves should be visually inspected before each use. While a quick visual inspection may be all that is needed, sometimes a small or pinhole-type leak can create an unacceptable risk, such as applications involving protection against high electrical voltage or high-hazard chemicals with poor skin contact warning properties such as hydrofluoric acid. In this case a more thorough leak check should be performed, as illustrated in Figure 15.10. Some chemical glove manufacturers will use compressed air to perform leak checking of every glove sold for high-hazard applications, if specified by the customer.

***15.3.4.6 Latex Allergy*** Gloves made of natural latex have been identified in recent years as being responsible for causing or aggravating an allergic sensitivity in a significant percentage of wearers. The allergic reaction most commonly appears as a reddening or itching at the site of glove contact, often the wrists. In some cases it can progress into generalized hives, shortness of breath, or, rarely still, full anaphylactic shock and death. Certain unidentified natural proteins in latex harvested from rubber trees are believed responsible for the allergic response. In high latex use areas such as hospitals, general contamination of the area by glove powders that have themselves been contaminated by latex protein can make the environment a hazard for those already sensitized. Hypoallergenic latex or nonlatex alternatives are available.

### 15.3.5 Foot Protection

Protective footwear is available for protection against a variety of hazards. The most common hazards are impact from falling objects and compression from rolling objects.

***15.3.5.1 Types Available and When Required*** The ANSI specifies minimum performance for 6 types of protective footwear (ANSI-Z41-1991 n.d.):

- *Impact and compression resistance* is needed where heavy objects are being moved around in the workplace by hand or with material handling equipment. Most safety footwear includes the steel toe box, which provides protection against falling or rolling objects.

- *Metatarsal protection* is available for environments having a higher risk of falling object injury and features what amounts to an extended toe box protecting much of the top of the foot.
- *Conductive footwear* can reduce the hazard of static electricity buildup, which could cause ignition of flammable atmospheres. Chemical workers handling flammable liquids or gases should wear type 1 conductive footwear. Type 1 footwear is designed to be electrically conductive and should not be worn near exposed electrical circuits. Type 2 conductive footwear is intended for use by electrical linemen or others needing equalization of the electrical potential between the person and the energized equipment.
- *Electrical hazard footwear* is designed to be resistant to electrical conductance, protecting the wearer from electric shock. They should be worn by persons exposed to energized electrical equipment. They should not be worn in hazardous/flammable locations where conductive footwear is required.
- *Sole-puncture-resistant footwear* is designed to protect the wearer from penetration of the bottom of the foot by sharp objects, such as protruding nails at construction sites. This is typically accomplished by a steel insole.
- *Static-dissipative footwear* is a compromise between conductive and electrical hazard footwear, providing protection against hazardous levels of static buildup while at the same time providing resistance against electric shock. Most chemical workers exposed to flammable atmospheres also have some degree of exposure to electrical hazards, making static dissipative the footwear of choice.

While chemically resistant footwear often meets one or more of the above performance standards, there is no widely used standard for protection against chemical hazards. Only recently have footwear manufacturers begun to publish extensive charts showing chemical permeation and degradation for their chemically resistant products constructed of natural or synthetic polymers. One standard referenced by footwear manufacturers is NFPA-1991 (1994).

In many chemical environments such as chemical laboratories or semiconductor fabrication areas, impact and compression are low risks, and while the risk of chemical splash is low enough that rubber boots are not needed, something more protective than open-toed or canvas street shoes is required. In these environments some organizations will require "shoes constructed of leather or leather-like material which completely cover the foot."

One shoe of each pair designed and constructed to meet one or more of the ANSI foot protection categories must be marked on the inside with letters and numbers coded to section 1.5 of the ANSI standard. For example, the label can be used to confirm that the shoe is designed for impact and compression (PT), for electrical hazard (EH), or to be static dissipative (SD).

***15.3.5.2 Posting Foot Hazard Areas*** Many organizations post PPE requirements at the entrances to hazard areas, often with signs composed of a signal word like CAUTION or NOTICE, followed by a short command such as "Safety Glasses Required in This Area." In the case of safety glasses, all workers or visitors to an area can be accommodated with eye protection relatively easily because visitors' glasses or goggles are relatively inexpensive to keep on hand and one size fits all. The same is true for hard hats. With safety footwear, however, organizations serious about compliance with safety rules should think twice before posting the area unless the risk to all who enter is very high. While toe or metatarsal caps designed to provide temporary impact and compression protection over street shoes are available, they are rarely put into use. They can be noisy and uncomfortable and, worse still, can present a slipping or tripping hazard to the wearer. In most instances, posting the area as "safety shoes required" is an unreasonable rule frequently ignored, which ultimately works against the organization's safety credibility. A better alternative is to make safety footwear a job or task requirement for those who routinely work in the foot hazard area.

### 15.3.6 Other PPE

***15.3.6.1 Seat Belts*** All workers will travel over the road while on business at one time or another. While sales and field service personnel are most at risk of vehicular injury, all organizations, no matter what the size or type of the business, should have a policy requiring the use of seat belts while traveling on company business.

***15.3.6.2 Fall Protection*** Falls from heights have serious consequences. In the US construction industry, falls are the leading cause of worker fatalities. Each year, on average, between 150 and 200 workers are killed, and more than 100000 are injured as a result of falls at construction sites (OSHA 1995).

PPE employed as fall protection is of two general types: personal fall arrest systems and positioning device systems.

*Personal fall arrest systems* (Figure 15.11) are used to arrest a worker in a fall from a working level and consist of an anchorage, connectors, and a body belt or body harness and may include a deceleration device, lifeline,

**Figure 15.11** Personal fall arrest system with restraint lanyard. *Source:* Courtesy of Rose Manufacturing Company.

or suitable combinations. According to OSHA, if a personal fall arrest system is used for fall protection, it must do the following (OSHA 1995):

- Limit maximum arresting force on a worker to 900 lb when used with a body belt.
- Limit maximum arresting force on an employee to 1800 lb when used with a body harness.
- Be rigged so that a worker can neither free-fall more than 6 ft nor contact any lower level.
- Bring a worker to a complete stop and limit maximum deceleration distance a worker travels to 3.5 ft.
- Have sufficient strength to withstand twice the potential impact energy of a worker free-falling a distance of 6 ft or the free-fall distance permitted by the system, whichever is less.

Effective 1 January 1998, the use of a body belt for fall arrest was prohibited by OSHA. A body harness must be used.

A *positioning device system* is a body belt or harness system set up so that a worker can free-fall no farther than 2 ft. It allows a worker to be supported on an elevated vertical surface, such as a wall or tower, and work with both hands free while leaning.

For more detailed information on fall protection, consult OSHA's fall protection regulations (OSHA, Subpart M 1994), OSHA's fall protection booklet (OSHA 1995), ANSI's fall protection standards (ANSI Z359.1-1992 n.d.; ANSI A10.14-1991 n.d.), or *Introduction to Fall Protection*, by J. Nigel Ellis (1993).

***15.3.6.3 Electrical Protection*** Electrically insulating, nonconductive garments, tools, and shielding devices should be used by those who are working near exposed energized parts that might be accidentally contacted or where dangerous electrical heating or arcing might occur (OSHA 1910.335 1990). Specifications for electrical protective equipment are provided in section 1910.137 of the OSHA general industry standards and are based on applicable ASTM standards.

***15.3.6.4 Back Belts*** In spite of an absence of scientific evidence that shows they prevent injury, back belts came into widespread use during the late 1980s and early 1990s as personal protection against back injury caused by material handling. Scientific studies of Home Depot and Walmart employees who are/have been required to wear back belts are currently underway. The former study is being conducted by UCLA and California OSHA, while the latter is by NIOSH.

Until these and additional studies are completed, the safety and health professional community will continue to view back belts as capable only of helping with the healing process and preventing reinjury (as supported by scientific studies of injured workers), but not as PPE for the average worker. One concern is that the belts may actually increase risk by creating a false sense of security, which can lead to greater risk taking by the user, or by weakening of abdominal muscles from prolonged wearing, but an increase in injury among belt users has not been scientifically proven either.

## BIBLIOGRAPHY


ACGIH (1996). *1996 TLV's and BEI's, Threshold Limit Values for Chemical Substances and Physical Agents, Biological Exposure Indices*. Cincinnati, OH: American Conference of Governmental Industrial Hygienists.

ANSI A10.14-1991 (1991). *Requirements for Safety Belts, Harnesses, Lanyards, Lifelines, and Droplines for Construction and Industrial Use*. New York: American National Standards Institute.

ANSI Z41-1991 (1991). *American National Standard for Personal Protection – Protective Footwear*. New York: American National Standards Institute.

ANSI Z87.1-1989 (1989). *American National Standard Practice for Occupational and Educational Eye and Face Protection*. New York: American National Standards Institute.

ANSI Z88.2-1992 (1992). *American National Standard for Respiratory Protection, Paragraph 7.5.3.3*. New York: American National Standards Institute.

ANSI Z89.1-1986 (1986). *American National Standard for Personnel Protection – Protective Headwear for Industrial Workers – Requirements*. New York: American National Standards Institute.

ANSI Z136.1-1993 (1993). *Safe Use of Lasers*. New York: American National Standards Institute.

ANSI Z359.1-1992 (1992). *Safety Requirements for Personal Fall Arrest Systems, Subsystems and Components*. New York: American National Standards Institute.

*Asbestos* (1996). *29 CFR1910.1001, Asbestos*, Appendix C. Washington, DC: US Department of Labor, Occupational Safety and Health Administration.

ASTM F739 (1991). *Standard Test Method for Resistance of Protective Clothing Materials to Permeation by Liquids or Gases Under Conditions of Continuous Contact*. Philadelphia, PA: American Society for Testing and Materials.

ASTM F903 (1990). *Standard Test Method for Resistance of Protective Clothing Materials to Penetration by Liquids*. Philadelphia, PA: American Society for Testing and Materials.

*Best's Safety Directory* (1996). Oldwick, NJ: A. M. Best Company.

*Chemical Health & Safety* (1995). Contact lens emergencies (January/February).

*Chilton's Industrial Safety & Hygiene News* (1996).

Ellis, J.N. (1993). *Introduction to Fall Protection*, 2e. Des Plaines, IL: American Society of Safety Engineers.

Encyclopedia (1983). *Encyclopedia of Occupational Health and Safety*, 3e (Revised). Geneva, Switzerland: International Labour Office.

*Federal Register* (1998). Respiratory protection; final rule **63**(5): 1152.

Forsberg, K. and Mansdorf, S.Z. (1993). *Quick Selection Guide to Chemical Protective Clothing*, 2e. New York: Van Nostrand Reinhold.

*Industrial Hygiene News* (1996).

Johnson, J.S. and Anderson, K.J. (eds.) (1990). *Chemical Protective Clothing*, vol. **I**, 143. Fairfax, VA: American Industrial Hygiene Association.

NFPA-1991 (1994). *Standard on Vapor-Protective Suits for Hazardous Chemical Emergencies*. Quincy, MA: National Fire Protection Association.

NFPA-1992 (1994). *Standard on Liquid Splash-Protective Suits for Hazardous Chemical Emergencies*. Quincy, MA: National Fire Protection Association.

NFPA-1993 (1994). *Standard on Support Function Protective Clothing for Hazardous Chemical Operations*. Quincy, MA: National Fire Protection Association.

NIOSH (1985). *Occupational Safety and Health Guidance Manual for Hazardous Waste Site Activities*, NIOSH, OSHA, USCG, and EPA, NIOSH Publication No. 85-115. Washington, DC: NIOSH.

NIOSH (1994). *NIOSH Pocket Guide to Chemical Hazards*, US Department of Health and Human Services, National Institute for Occupational Safety and Health, Publication No. 94-116 (June 1994). Cincinnati, OH: NIOSH.

*Occupational Hazards* (1996).

*Occupational Health & Safety Magazine* (1996).

OSHA (1994a). *OSHA Clarifies Obligation of Employers to Pay for Personal Protective Equipment*. Washington, DC: Office of Information.

OSHA (1994b). Final rule revising general industry standards for personal protective equipment for eyes, face, head and feet. *Federal Register* **59** (April 6): 16334.

OSHA (1995). *Fall Protection in Construction*, Publication 3146. Washington, DC: US Department of Labor.

OSHA (1996). Technical amendment to OSHA's final rule on personal protective equipment for general industry. *Federal Register* **61** (2 May): 19547.

OSHA 1910.120 (1996). *29 CFR 1910.120 – Hazardous Waste Operations and Emergency Response*. Washington, DC: US Department of Labor.

OSHA 1910.133 (1996). *29 CFR 1910.133 – Eye and Face Protection*. Washington, DC: US Department of Labor.

OSHA 1910.335 (1990). *29 CFR 1910.335 – Safeguards for Personnel Protection*. Washington, DC: US Department of Labor.

OSHA, Subpart M (1994). *29 CFR Parts 1910 and 1926 – Safety Standards for Fall Protection in the Construction Industry*; final rule (9 August).

OSHA (1995). *Inspection Guidelines for 29 CFR 1910, Subpart I, The Revised Personal Protective Equipment Standards for General Industry*, OSHA Instruction STD 1-6.6, (16 June 1995). Washington, DC: OSHA.

Schwope, A.D., Costas, P.P., Jackson, J.O., and Weitzman, D.J. (1987). *Guidelines for the Selection of Chemical Protective Clothing*, 3e. Cincinnati, OH: *American Conference of Governmental Industrial Hygienists.*

*US Department of* Labor (1980). *Accidents Involving Eye Injuries*, Report No. 597. Washington, DC: Bureau of Labor Statistics, US Government Printing Office. Other useful personal protective equipment references:

CHAPTER 16

# RESPIRATORY PROTECTIVE DEVICES

JAMES S. JOHNSON

*JSJ & Associates, Pleasanton, CA, 94588*

## 16.1 INTRODUCTION

Respiratory protective devices (RPDs),[1] commonly called respirators, are one of many types of personal protective equipment (PPE) devices used to protect workers from a specific hazard. Some of these devices such as hard hats, steel-toed shoes, or eyeglasses are used to protect a worker from hazards that are not likely to occur on a regular basis. RPDs, however, fall into a category of devices that are being used to protect workers from an existing or recognized and real-time hazard. If the RPDs fail to provide the expected protection, it is likely that the wearer will be exposed and suffer immediate or delayed adverse consequences. RPDs are also unlike other personal safety devices such as hard hats, shoes, and safety glasses in that their proper selection for a specific hazard is more complex and requires training, education, and specific workplace monitoring and analysis.

RPDs have been used as early as Roman times, and scattered mention of them occurs in reports by physicians related to industrial processes during the Middle Ages. Until the nineteenth century, all RPDs were air-purifying devices intended to prevent the inhalation of a variety of particulates, aerosols, and gases. They varied in design, ranging from animal bladders or rags wrapped around the nose and mouth to full face masks made of glass with air inlets covered by particulate filters. During the 1800s RPDs were produced that combined aerosol filters and vapor sorbent materials. These advances were primarily made for firefighters.

Chemical warfare agents, introduced in World War I, focused attention on the need for improved and usable RPDs. The Bureau of Mines successfully carried out this work for the US Army. After the war, misuse of surplus army gas masks by civilians highlighted the need for RPD standards.[2] The Bureau of Mines developed these standards and in 1920 approved their first RPD, a self-contained breathing apparatus (SCBA) for mine rescue (Yant 1933). Out of this effort grew the federal RPD approval system that is still in effect. Many of the RPD developments made during the decade following 1920 remain in use today.

To use RPDs properly, one must implement a comprehensive RPD program, have knowledge about the federal and state regulations covering the use of RPDs, and know the types of devices available and elements of an RPD program that are critical to ensure the safe use of these devices. This chapter presents these topics in the same order, with definitions of RPD terms located at the end of this chapter.

[1] I have chosen to use the term respiratory protective device (RPD) rather than respirator throughout this chapter to draw a clear distinction between the various classes and types of RPDs available to protect workers.

[2] B. J. Held, "History of Respiratory Protective Devices in the U.S., Pre World War I." Lawrence Livermore National Laboratory, Energy Research and Development Administration, contract W-7405-Eng-48.

*Handbook of Occupational Safety and Health*, Third Edition. Edited by S. Z. Mansdorf.

## 16.2 REGULATIONS AND GUIDANCE DOCUMENTS COVERING RESPIRATORY PROTECTIVE DEVICES AND PROGRAMS

In the United States, the Occupational Safety and Health Administration (OSHA) has established regulations covering the use of RPDs in the workplace. These regulations can be found in 29 Code of Federal Regulations Part 1910.134[3] (Appendix 1) and cover the types of programs required to support the use of RPDs including training, fit testing, and medical clearance. These regulations were established in 1971 and revised by the OSHA in 1998 with the exception of assigned protection factors (APFs) that was promulgated in 2006.[4] These regulations when looked at carefully identify a partnership between regulators, RPD manufacturers, employers, and workers that form the basis for achieving proper respiratory protection (Metzler et al. 2016).

The National Institute for Occupational Safety and Health (NIOSH) has responsibility for testing and approving RPDs for use by employers. Current regulations were originally promulgated by the Bureau of Mines prior to the establishment of NIOSH by the 1971 OSH Act.[5] Responsibility was transferred to NIOSH by this act and eventually shared with the Mine Safety and Health Administration (MSHA). NIOSH has an ongoing modular approach program for reviewing and revising existing requirements covering the approval of RPDs. The first of these regulatory standard updates affecting the design of air-purifying particulate filters and RPDs was promulgated in 1995.[6] Since then NIOSH has issued RPD standards for air-purifying respirators (APRs) and air-purifying escape respirators (APERs) with Chemical, Biological, Radiological and Nuclear (CBRN) protection, powered air-purifying respirators (PAPRs) with CBRN protection, SCBA with CBRN protection, open-circuit self-contained breathing apparatus (OCSBA), and closed-circuit escape respirators (CCER). RPD standards currently under development at NIOSH are for PAPRs, closed-circuit self-contained breathing apparatus (CCSCBA) with CBRN protection, and supplied-air respirators (SAR).[7] These regulations clearly place the responsibility on the RPD manufacturer to provide products that meet all of the appropriate NIOSH requirements for initial and ongoing NIOSH RPD approval. Since OSHA requires employers to provide NIOSH-approved RPDs to their employees, nearly all RPDs in use in US workplaces are approved by NIOSH.

There are several organizations that provide important reference documents for individuals with responsibility for the development and implementation of an RPD program. The American National Standards Institute (ANSI) provides voluntary consensus standards for a variety of safety practices and topics. Within ANSI the responsibility for RPD standards has been assigned to the Z88 Committee that is managed by the American Society of Safety Engineers (ASSE). ANSI/ASSE Z88.2 "Practices for Respiratory Protection," issued originally in 1969, provided the basis for much of the original RPD program requirements found in the current OSHA standard (American National Standard Institute, ANSI Z88.2-1969 1969). The ANSI Z88.2 standard has since been revised in 1980, 1992, and 2015 (American National Standard Institute, ANSI Z88.2-1980 1980; American National Standard Institute, ANSI Z88.2-1992 1992; American National Standard Institute, American Society of Safety Engineers 2015). In fact many of the requirements from the 1969 version of ANSI Z88.2 are still included in the current MSHA RPD program requirements. Other ANSI/ASSE RPD standards available are the ANSI Z88.6 standard (American National Standard Institute, American Industrial Hygiene Association, American Society of Safety Engineers 2006) for medical evaluation of respirator wearers and the ANSI Z88.10 standard (American National Standard Institute, American Industrial Hygiene Association, American Society of Safety Engineers 2010) on respirator fit-test methods. Several new ANSI/ASSE standards are under development addressing half-mask fit-test methods (Z88.15), APFs (Z88.16), and RPD terminology (Z88.17). The reader is encouraged to obtain a copy of the most recent version of these documents when available, as they are specifically designed to provide guidance to individuals with responsibility for developing and maintaining RPD programs.

Several professional societies provide additional useful information in the area of respiratory protection. The National Fire Protection Association (NFPA) publishes a variety of documents designed to help individuals in the fire service address respiratory protection for specific fire service needs. NFPA also develops specific RPD certification standards that address unique RPD performance requirements needed to operate safely in fire and emergency response activities. These certification standard requirements are implemented in addition to the NIOSH approval requirements for SCBAs and other RPDs used by the fire service. There are currently four NFPA equipment performance standards: SCBAs (NFPA 1981)(2013); selection, care, and maintenance of SCBSs (NFPA 1852)(2013);

[3] Code of Federal Regulations, Title 29, Part 1910.134, "Respiratory Protection."

[4] Code of Federal Regulations, Title 29, Part 1910.134, "Respiratory Protection, Assigned Protection Factors; Final Rule."

[5] Code of Federal Regulations, Title 30, Part 11, "Respiratory Protective Devices; Tests for Permissibility; Fees."

[6] Code of Federal Regulations, Title 42, Part 84, "Respiratory Protective Devices."

[7] http://www.cdc.gov/niosh/npptl/RespStandards/ApprovedStandards/scba_cbrn.html.

wildland respirators (NFPA 1984)(2016); and breathing air quality (NFPA 1989) (2013). NFPA also develops standards that address specific requirements for fire service RPD programs including RPD fit testing and medical evaluation. Two additional standards under development by NFPA are a standard on respiratory protection equipment for tactical and technical operations (NFPA 1986) and a standard on combination unit respirator systems for tactical and technical operations (NFPA 1987). OSHA has published excellent references on RPD programs looking at the required program elements for small entity compliance(Occupational Safety and Health Administration 2011) and partnering with NIOSH to address hospital RPD program needs (National Institute for Occupational Safety and Health, Occupational Safety and Health Administration, 2015). Current technical articles on RPD programs and the performance and selection and use of RPDs can be found in a variety of professional journals such as the *Journal of the International Society for Respiratory Protection.*[8] The *Journal of Occupational and Environmental Hygiene*[9] and the *Annals of Occupational Hygiene.*[10]

## 16.3 TYPES OF RESPIRATORY PROTECTIVE DEVICES

### 16.3.1 Introduction

There is a large variety of RPDs available to protect workers. Some devices have very specific uses such as an abrasive blasting helmet for protection from crystalline silica exposure and dust or air-purifying mercury half or full facepiece RPDs for protection from mercury vapor. Other devices, such as air-purifying particulate RPDs or airline RPDs, can be used to protect workers from a wide variety of different contaminants. These contaminants exist in different forms (e.g. solids, liquids, gases) and different toxicities varying from highly toxic to nontoxic. Because of the variety of forms and toxicities, there is no one universal method for categorizing RPDs for their selection. One important aspect to consider is whether an RPD is being selected for emergency response or routine use. In the former case, the concentration of a contaminant or contaminants is normally unknown or difficult to estimate, while for routine use the workplace environment contaminants and concentrations may be well known and under close control. Another consideration is whether an RPD is selected to provide a safe environment using a separate source of breathing air (supplied-air RPD) or by purifying the air using a filter or absorber (air-purifying RPD). In both cases, careful selection of the device and supporting equipment is required to ensure the safety of the wearer. One scheme for classifying RPDs is shown in Figure 16.1.

### 16.3.2 Respiratory Protective Device Facepieces

Most RPDs can be considered modular units. There are facepieces containing inhalation valves, exhalation valves, headbands, and air-purifying elements or supplied-air attachments. Many manufacturers make identical facepiece molds for both air-purifying and supplied-air models. Facepieces from a manufacturer must be matched with specific air-purifying elements or supplied-air connections. NIOSH approval covers the entire RPD assembly, and any effort to arbitrarily use replacement parts, a filter, cartridge, or supplied-air connection other than what was specifically designed by the manufacturer and approved by NIOSH will void the NIOSH approval and violate OSHA requirements.

Facepieces can be classified into tight- and loose-fitting models. Tight-fitting facepieces require a good seal between the sealing surface of the device and the wearer's face; loose-fitting devices such as hoods or helmets do not. Until the 1970s RPD manufacturers usually produced tight-fitting RPD facepieces in one or two sizes only. Although each company's facepiece was different, the RPDs tended to fit males better than females. This was probably a reflection of employment trends from earlier years. Today most manufacturers supply facepieces in two or three sizes that are designed to fit the facial features of both men and women. However, some individuals who have unusual facial characteristics may have difficulty finding a model that offers both comfort and proper fit, and it is recommended that more than one manufacturer be considered when selecting RPDs.

---

[8] *Journal of the International Society for Respiratory Protection,* International Society for Respiratory Protection c/o Ziqing Zhuang, PhD., 105 Eaglebrook Court, Venetia, PA 15367, USA.

[9] *Journal of Occupational and Environmental Hygiene,* American Industrial Hygiene Association, American Conference of Governmental Industrial Hygienists, c/o T. Renee Anthony, PHD, CIH, CSP, FAIHA, College of Public Health, The University of Iowa, 1455 N. Riverside Drive, 100 CPHB, Iowa City, IA 52242, USA.

[10] *The Annals of Occupational Hygiene, British Occupational Hygiene Society*, c/o Professor Noah Seixas, School of Public Health, University of Washington, 4225 Roosevelt Way NE, Suite 100, Seattle, WA 98195.

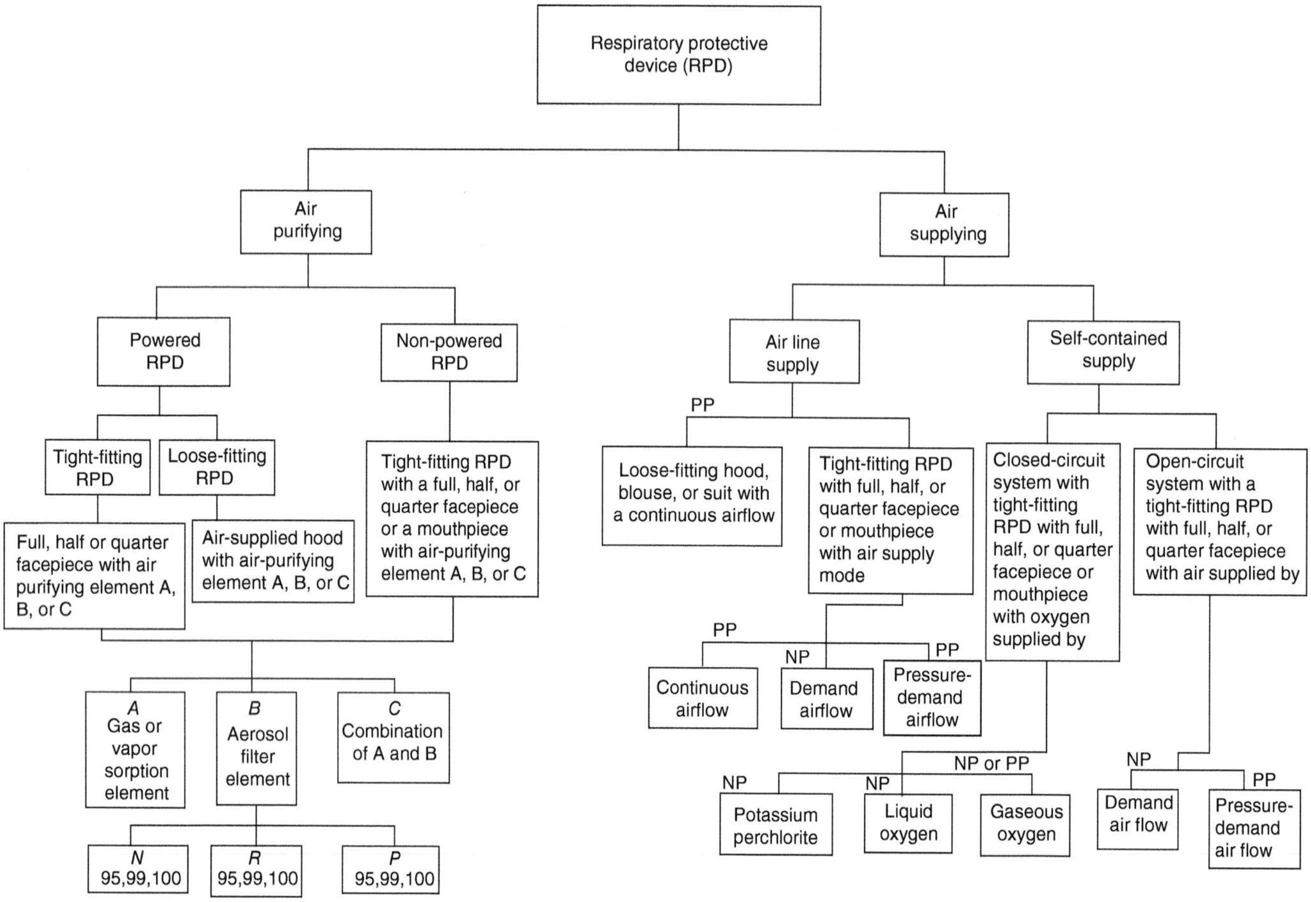


**Figure 16.1** Respirator classification scheme. PP, positive-pressure mode of operation; NP, negative-pressure mode of operation.

The tight-fitting facepiece is intended to adhere snugly and form a seal with the skin of the wearer. It can be available in three types: quarter facepiece, half facepiece, and full facepiece. The quarter facepiece covers the nose and mouth; the half facepiece covers the nose, mouth, and chin (Figure 16.2). Full facepiece APRs covers the entire face from chin to hairline and from ear to ear (Figure 16.3). The major components of the half and full facepiece RPDs are shown in Figures 16.4 and 16.5.

The facepiece is usually held in place by elastic or rubber headbands. Quarter and half facepiece RPDs may be secured by one strap attached to each side of the facepiece, that is, a two-point suspension, or two straps attached at two points on each side of the facepiece, that is, a four-point suspension. The two straps, instead of being attached to tabs on the edge of the facepiece, may be part of a yoke that is fastened to the facepiece by one or two points in the front of the facepiece. Two- or four-point suspension systems work well and are a design consideration of the manufacturer based on the facepiece configuration and weight. Full facepiece RPDs have a head harness attached to the facepiece at four, five, or six points. The large sealing surface of the full facepiece and the distribution of the headband attachment points assist in maintaining a stable facepiece with less slippage than can be experienced with quarter or half facepieces. A number of recent designs of half facepiece RPDs provide a wide sealing surface to reduce possible facepiece contaminant leakage.

An important consideration in selecting a facepiece is there can be nothing that interferes with the sealing surface of a tight-fitting RPD and the sealing surface of the face. This includes facial hair, the temple bars of eyeglasses, and piercings with various forms of jewelry. In addition, large facial scars that come into the sealing surface of an RPD can provide a place where contaminants may leak into the facepiece. Because tight-fitting RPDs are designed to carefully fit the contours of the face, fit testing is required to ensure that the RPD has been properly selected.

Loose-fitting RPDs are designed to operate without a tight sealing surface on the face. As such, there is less concern about selecting a proper size RPD or requiring that wearers be clean-shaven. There is also no requirement for fit-testing wearers of such devices. A very popular example of these devices is the supplied-air hood (Figure 16.6). The hood covers the head, neck, and upper torso and usually includes a neck cuff (seal) and some type of air diffuser

**Figure 16.2** Elastomeric half facepiece RPD in three sizes, MSA Comfo Classic. Courtesy of Mine Safety Appliance Company.

**Figure 16.3** MSA Advantage and Moldex 9000 full facepiece RPD. Courtesy of Mine Safety Appliance Company and Moldex-Metric Inc.

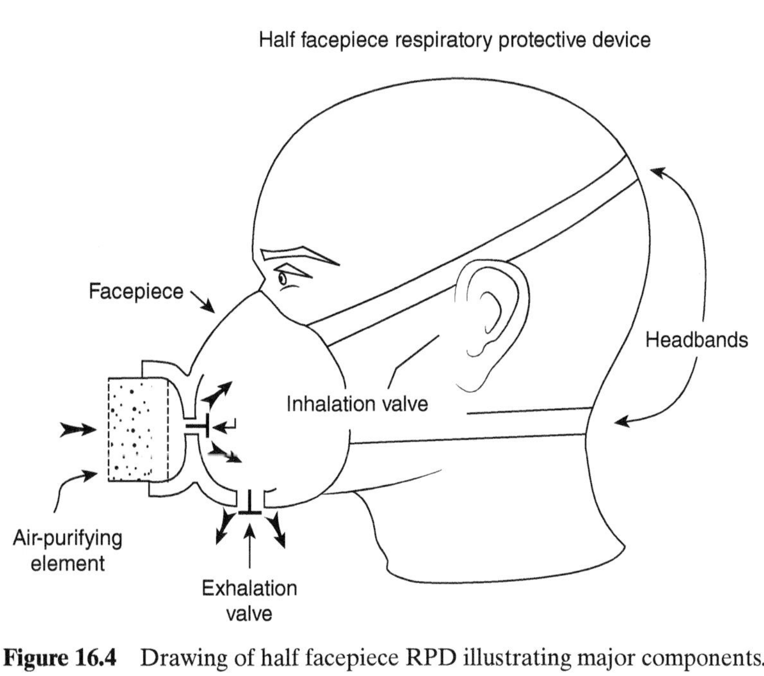


**Figure 16.4** Drawing of half facepiece RPD illustrating major components.

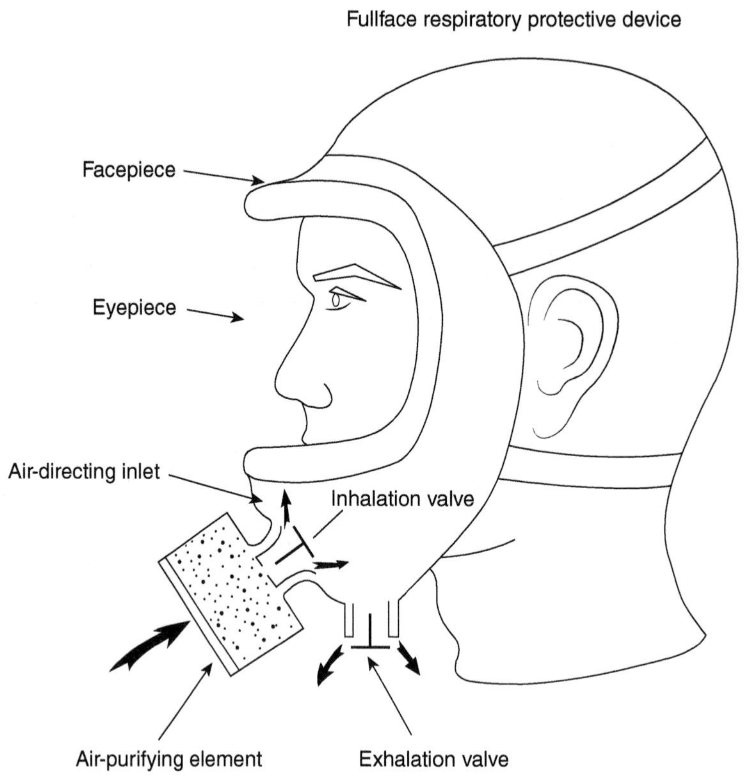


**Figure 16.5** Drawing of full facepiece RPD illustrating major components.

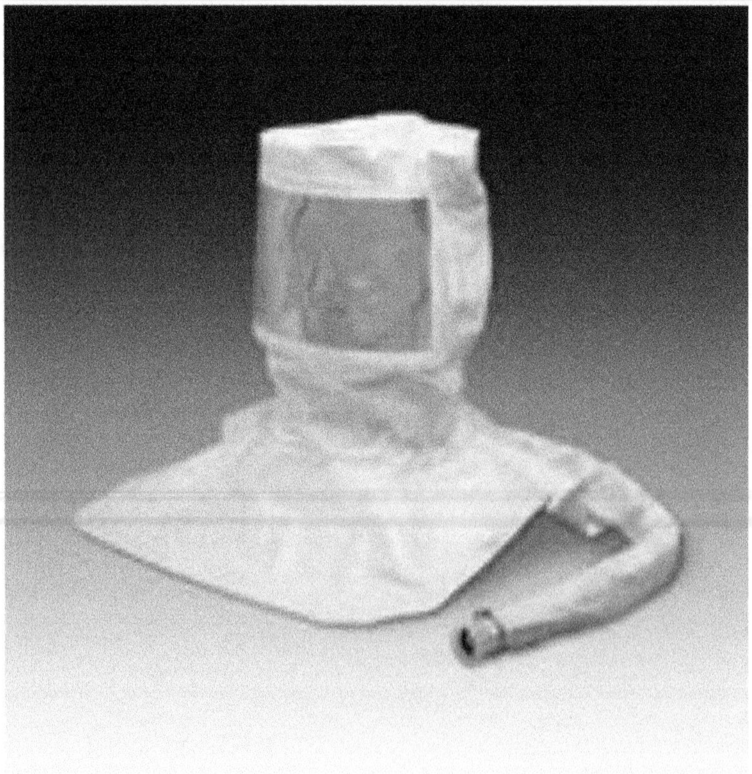

**Figure 16.6** Supplied-air hood RPD, reusable, Allegro Maintenance Free Saran™ Hood. Courtesy of Allegro Industries.

above the head. Air is provided through a hose leading into the hood. The loose-fitting facepiece hood covers at least the face and may cover the entire head and is a popular variation of the standard hood. Because the hood is not tight fitting, it is important that a sufficient quantity of air be provided to maintain an outward flow of air, e.g. positive pressure, and prevent contaminants from entering the hood.

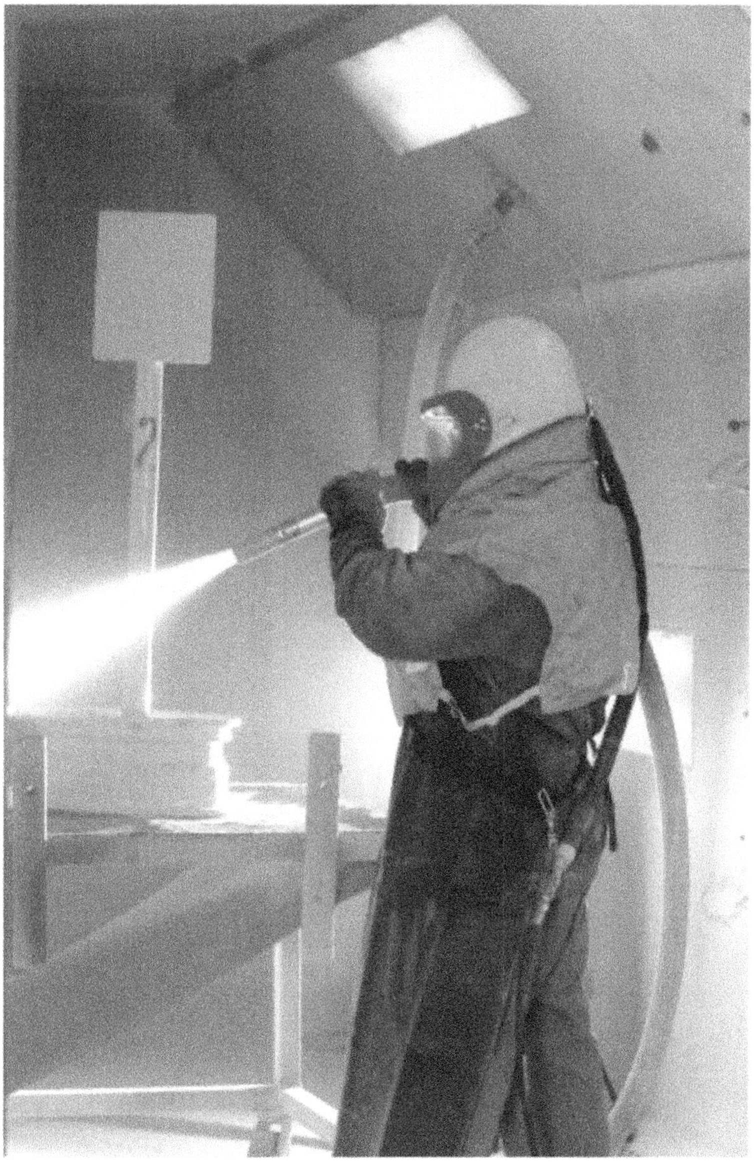

**Figure 16.7** Abrasive blaster's supplied-air hood with helmet RPD. Courtesy of E.D. Bullard.

Supplied-air RPDs can be made rugged and used as an abrasive blaster's helmet (Figure 16.7) or low profile hoods with full facepiece RPDs (Figure 16.8). NIOSH testing and approval regulations have specific requirements to ensure a durable covering to withstand the rigors of the abrasive blasting atmosphere. OSHA regulates abrasive blasting including respiratory protection in their ventilation regulations (1910.94 (a)).[11] There are lightweight plastic and Tyvek® hoods available that can be easily disposed of when used in atmospheres where decontamination may be difficult (e.g. handling pharmaceuticals) (Figure 16.9).

Some of these supplied-air devices incorporate a helmet with face shield. Air is supplied to the rear of the helmet and travels inside the top and comes down across the face shield. Improvements in this design now create an airtight seal with the helmet, a face shield, and a neck seal to provide a higher level of protection. There are also supplied-air suits that cover the entire body and provide whole-body protection. There are no NIOSH approval regulations for supplied-air suits. The American Society for Testing and Materials (ASTM) has several consensus standards for evaluating encapsulating suit performance, such as the Standard Test Method for Pressure Testing Vapor Protective Suits, ASTM F1052-14 (2014), and the Standard Test Method for Man-In-Simulant Test (MIST) for Protective Ensembles, ASTM F2588-12 (2012).

[11] Code of Federal Regulations, Title 29. Part 1910.94(a), "Abrasive Blasting."

**Figure 16.8** Abrasive blaster's supplied-air hood with full facepiece RPD, MSA Abrasi-Blast. Courtesy of Mine Safety Appliance Company.

**Figure 16.9** Disposable supplied-air hood RPD. Courtesy of Allegro Industries.

### 16.3.3 Air-Purifying Respiratory Protective Devices

RPDs can be categorized according to those that remove hazardous contaminants before entering the facepiece by filtration or absorption and those that supply the wearer with a clean source of breathable air. The most common forms of RPDs are air-purifying devices (APRs). They tend to be of a more simple construction and are less costly.

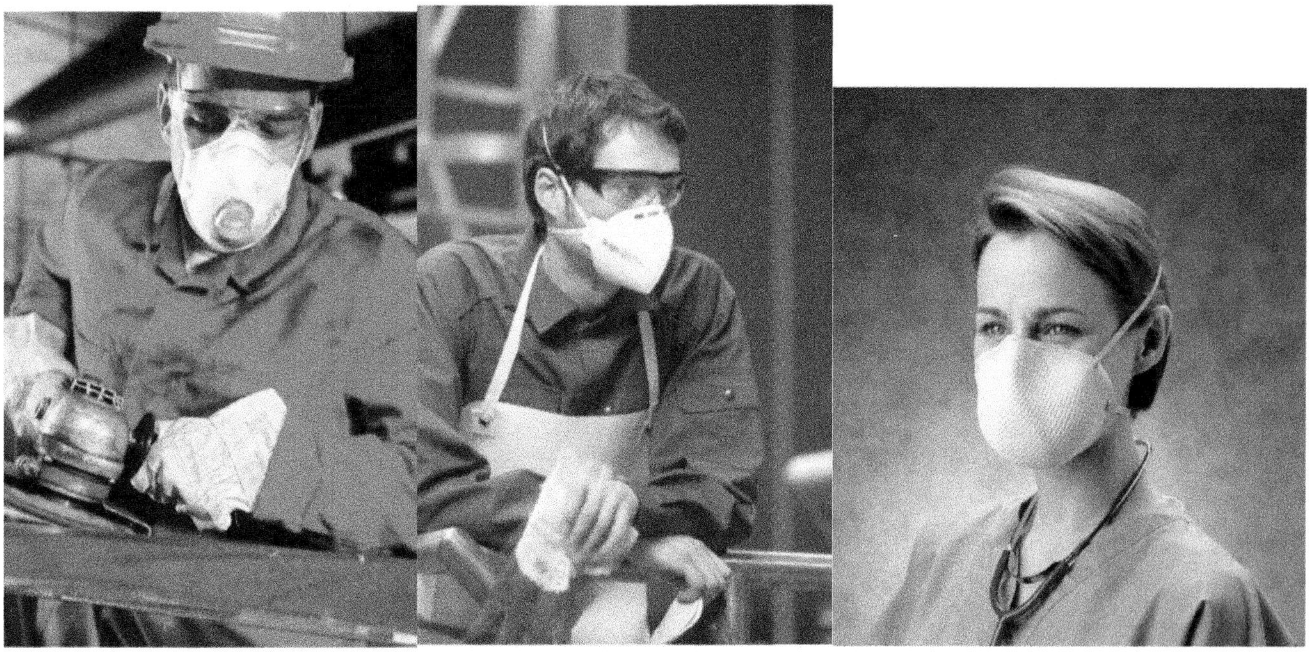

**Figure 16.10** Variety of filtering facepiece RPDs: Draeger X-plore and Draeger 1700 and Moldex N95 1500. Courtesy of Draeger Inc. and courtesy of Moldex-Metric Inc.

These devices can be found with a wide variety of tight-fitting facepieces and some loose-fitting facepieces with auxiliary blowers that use air-purifying elements. The three most important features of these RPDs are that the air-purifying element must be carefully chosen to remove the toxic contaminant(s) present, the element must be periodically replaced, and for tight-fitting RPDs the facepiece must seal to the wearer's face. Some half facepiece RPDs have the air-purifying element build directly into the RPD and are identified as filtering facepiece RPDs (Figure 16.10). They have been referred to as "disposable" RPDs since the entire device can be purchased as a "single-use" device and is discarded after each use. This is a useful feature when the RPD is difficult to clean or disinfect such as when used for protection against infectious agents or when an RPD is infrequently used and the expense and effort of having an RPD maintenance program is not cost-effective. Reusable devices on the other hand can be cleaned, serviced, and inspected cost-effectively for reuse in many RPD programs.

From their original NIOSH approval in the earlier 1970s until now the "disposable," "single-use" half facepiece RPDs have gone through a number of name changes from "maintenance-free, limited use" to now a class of half facepiece RPDs identified as "filtering facepiece RPDs." This class of RPDs now has the largest number of NIOSH approvals.

Air-purifying RPDs can be used to remove particulate contaminants, gases, vapors, and combinations of these agents. They cannot be used when there is an insufficient concentration of oxygen in the atmosphere, when the concentration of a contaminant is so high as to present an immediately dangerous to life or health (IDLH) hazard to the worker, or for contaminants that cannot be effectively removed by existing filters and sorbents.

In the 1998 revision to its RPD standard, OSHA now requires that the employer have objective data to determine when to change the cartridges or canister of an air-purifying RPD for protection against gases and vapors. The change schedule will be a function of the concentration of the contaminant(s), length of time the RPD is worn, and the efficacy of the cartridge for a specific gas, vapor, or mixture. There are currently a number of computer program tools to make the calculation simple from NIOSH and a number of major RPD manufacturers. It is recommended that the RPD manufacturer be contacted to assist in obtaining the objective data required by OSHA if the information is not available on their website. There is an ANSI standard, ANSI Z88.7 (2010), that covers the color code marking of cartridges and canisters. The current color-coding scheme used for filters and sorbents is summarized in Table 16.1. Although this allows for quick identification, each cartridge and canister will also have a written label indicating its approval and proper usage. RPD wearers should be trained to check the RPD and the color and wording of the air-purifying element prior to donning the device.

Half and full facepiece air-purifying RPDs are available with auxiliary blowers and are known as powered air-purifying respirators (Figure 16.11). These devices have several advantages over the more common non-powered air-purifying RPDs. The blower eliminates the need for the wearer to inhale through the cartridge, which eliminates a significant resistance to breathing, resulting in a decrease in the work level. This is a useful feature either for workers exerting substantial

**TABLE 16.1 COLOR ASSIGNED TO AIR-PURIFYING ELEMENTS**

| Atmospheric containment class or containment to be protected against | Color assigned |
|---|---|
| Acidic gases | White |
| Basic gases | Green |
| Organic vapors | Brown |
| Formaldehyde | Tan |
| Carbon monoxide gas | Blue |
| Mercury vapors | Orange |
| All aerosols (high-efficiency filter) | Purple |
| CBRN | Black |

**Figure 16.11** Powered air-purifying fullface RPD, MSA OptimAir PAPR. Courtesy of Mine Safety Appliances Company.

energy while wearing an RPD or for those individuals who have a subnormal pulmonary function. Tight-fitting half and full facepiece PAPRs offer significant improvements over non-powered RPDs by reducing the likelihood of contaminants' entering the facepiece from negative-pressure leaks. These devices tend to also provide some cooling for the wearer and are well suited for working in hot environments. An innovative PAPR helmet design that integrates the blower and filter/absorber into the helmet and provides head, eye, face, and respiratory protection for the wearer is shown in Figure 16.12. Loose-fitting PAPRs allow users to have facial hair or other obstacles (e.g. glasses) that would interfere with the fit of an RPD. Other brands of loose-fitting PAPRs provide full-view hoods for improved vision (Figure 16.13). There is one PAPR on the market that maintains positive pressure in the facepiece and provides a very high level of protection for the wearer (Figure 16.14).

Particulate filter RPDs protect against dusts, mists, fumes, smokes, and biological contaminants. The filter elements remove these contaminants by mechanical filtration, electrostatic attraction, or a combination of both. Mechanical filtration efficiency is affected by particle size. Aerosol is a general term used to describe all solid or liquid particles below 1 μm in aerodynamic diameter, the most difficult particle size to filter efficiently.

NIOSH traditionally classified filters and testing protocols according to dusts, mists, fumes, and high-efficiency particulate air filters (HEPA).[12] There were different tests for each type of filter, and some devices that passed multiple tests received multiple approvals (e.g. dust/mist or dust/fume/mist filters). HEPA filters can stop almost all aerosols and have a minimum filter efficiency of 99.97% against the most penetrating sized aerosol, 0.03 μm. For mechanical filtration there is a direct relationship between filter efficiency and the pressure drop through a filter. The latter translates as difficulty in breathing through a filter. Therefore, while a HEPA filter is the most protective,

[12] Code of Federal Regulations, Title 30, Part 11, "Respiratory Protective Devices; Tests for Permissibility; Fees."

**Figure 16.12** Powered air-purifying respirator RPD integrated with a helmet, visor, and electronic management system. Courtesy of Interactive Safety Products.

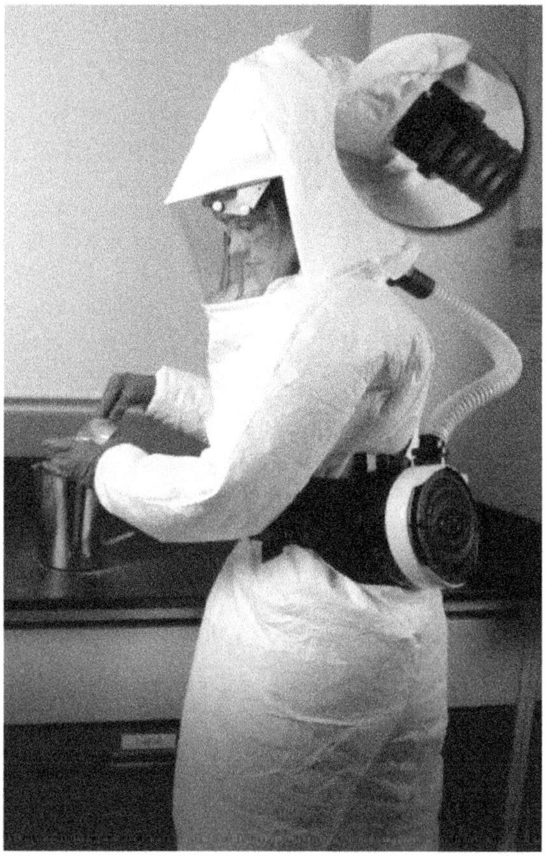

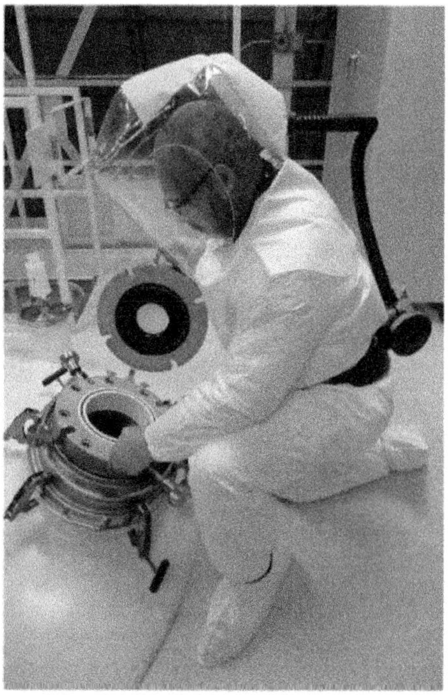

**Figure 16.13** Powered air-purifying respirator RPD, with disposable hood: T-Series Quick-Attach Respirator Hood and Dover Sentinel full-view hood. Courtesy of E.D. Bullard Company and courtesy of ILC Dover.

**Figure 16.14** Positive-pressure powered air-purifying respirator RPD, SE400 Fan-Supplied Positive-Pressure Demand Respirator. Courtesy of S.E.A. Group.

it often has a higher pressure drop than most dust filters when used for a short time period and will require replacement more frequently. Given a high enough concentration of aerosols and/or long enough use, all filters can become clogged with particulate matter, making it impossible to breathe through the media. Employees must be trained to replace the filter media before a significant increase in breathing resistance has occurred or when required by the change-out schedule.

To correctly select a filter, some knowledge about the workplace is required. Ideally information concerning the size distribution of aerosols in the workplace would be helpful. In 1995, NIOSH introduced a revision to their testing and approval requirements for particulate filters.[13] The new approval procedure identifies filters by three-letter designations and three filter efficiency designations.[14] The letters N, R, and P are used to describe the filter's ability to function in a contaminated environment, e.g. oil mist. The letter N means that no oil mist can be present, R indicates that the filter is resistant to oil mist, and P indicates that the filter is oilproof. There are three efficiencies used to describe filter performance: 95, 99, and 100% (>99.97%). All filters are tested against the most penetrating solid or liquid penetrating aerosol of approximately 0.3 μm. The traditional dust/mist filter has been replaced by the N95 filter, while P100 filters are used instead of HEPA filters.

The advantage of this revised testing and approval scheme is it will ensure that at least 95% of all aerosols are removed when an N95 filter is used in a non-oil mist environment. This revised approval process for filter performance has been in place for 20 plus years with no significant problems identified.

RPDs used to protect healthcare workers currently require additional evaluation by the manufacturer to meet Food and Drug Administration (FDA) class II medical device requirements.[15] Currently there is only one class of RPDs approved by NIOSH and FDA, the surgical N95 RPD. Discussions are currently underway to harmonize these two sets of requirements so that the NIOSH approval addresses the FDA requirements for this and other classes of RPDs routinely used to protect healthcare workers.

Gas- and vapor-removing RPDs are the second major category for air-purifying RPDs. Gases and vapors are captured by absorption, chemisorption, adsorption, or catalysis when passing through a sorbent. The most widely used material for removing gases and vapors is activated carbon. This material has a large internal surface area (typically 1000 $m^2\ g^{-1}$) and

---

[13] Ibid.

[14] Code of Federal Regulations, Title 42, Part 84, "Respiratory Protective Devices."

[15] http://www.fda.gov/downloads/medicaldevices/deviceregulationandguidance/guidancedocuments/ucm072561.pdf.

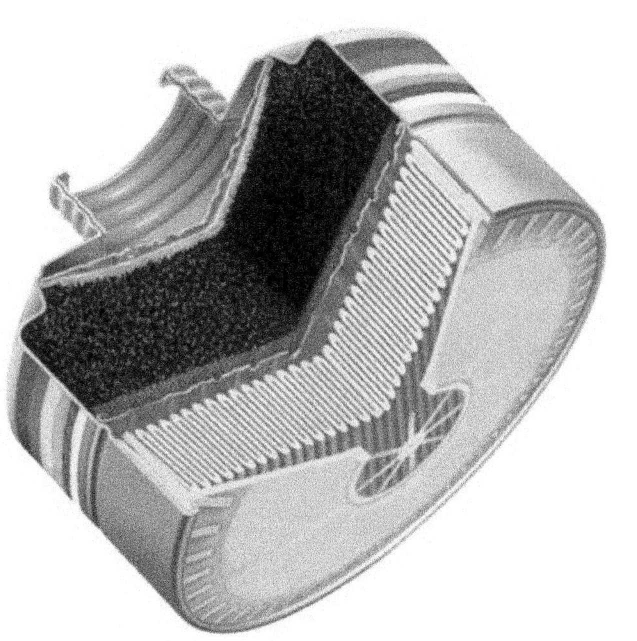

**Figure 16.15** Cutaway of a cartridge for gas, vapor, and particulate.

captures vapors and gases through adsorption and physical attraction due to molecular forces (van der Waals forces of adhesion). Alone it is an excellent adsorbent of many organic vapors. Impregnating it with specific materials increases its retention efficiency for certain gases and vapors. Sorbents are packed in either cartridges or canisters. The only difference is that canisters contain more sorbents and will last for a longer time. They are also larger, weigh more, and may have a higher resistance to airflow for the wearer. A cutaway of a typical combination organic vapor and particulate cartridge structure is shown in Figure 16.15.

There are specific cartridges and canisters manufactured for some gases, including ammonia and/or methylamine; acid gases for chlorine, hydrogen chloride, and sulfur dioxide; and carbon monoxide. The most common need for a gas and vapor air-purifying RPD is for protection from organic vapors. NIOSH has a general approval test for organic vapor cartridges and canisters using carbon tetrachloride as a test agent.[16] This test is designed to examine the quality of the activated carbon and cartridge or canister design. This test is not designed to inform the user that it can or cannot be used for protection for a specific chemical or application.

The parameters governing whether a gas or vapor will be efficiently collected include the contaminant concentration, airflow, relative humidity, and nature of the contaminant. Polar organic vapors with low molecular weights (e.g. methanol) tend to be poorly adsorbed onto activated carbon. Those with larger molecular weight and that are nonpolar tend to be efficiently collected.

There is published literature in the area of gas and vapor RPD testing that may help in determining the appropriateness of a cartridge or canister for a specific application (Woods 2015). It is recommended that the RPD manufacturer be contacted to assist in obtaining the objective data required by OSHA if the information is not available on their website. The change schedule will be a function of the concentration of the contaminant(s), length of time the RPD is worn, and the efficacy of the cartridge for a specific gas, vapor, or mixture. There are currently a number of computer program tools to make the calculation simple from NIOSH: MultiVapor™ Version 2.2.4 Application (or most current edition),[17] 3M Select & Service Life Software,[18] MSA Response Guide,[19] and other service life calculators are available on RPD manufacturers' websites. For unusual chemicals or mixtures, it is again recommended that the RPD manufacturer be contacted to assist in obtaining the objective data required by OSHA. There are also a few analytical laboratories that will test a single substance or mixture with a specific brand of RPD cartridge. This information can be used to calculate a change schedule for unusual chemicals and mixtures.

OSHA no longer allows air-purifying RPDs for protection against organic gases or vapors to be selected on the basis of sensory perception. Available end-of-service-life computer programs make the determination of an appropriate

---

[16] http://www.cdc.gov/niosh/npptl/stps/pdfs/TEB-APR-STP-0046A.pdf. http://www.cdc.gov/niosh/npptl/stps/apresp.html.
[17] http://www.cdc.gov/niosh/npptl/multivapor/multivapor.html.
[18] http://extra8.3m.com/SLSWeb/home.html.
[19] http://webapps2.msasafety.com/responseguide/Home.aspx.

**Figure 16.16** Full facepiece RPD, MSA Millennium with a CBRN canister attached. Courtesy of Mine Safety Appliances Company.

change-out schedule relatively easy to do. If a chemical has a detectable odor threshold below the threshold limit value (TLV)/action level, the wearer as an additional indicator can use this to indicate that the respirator cartridges require replacement. In addition to the computer tools and published literature previously mentioned, the RPD manufacturer is another source to provide assistance in obtaining the objective information required by OSHA. A useful reference that contains odor thresholds for a large number of compounds is *Odor Thresholds for Chemicals with Established Occupational Health Standards* (American Industrial Hygiene Association 1989)

Another but limited way to address cartridge or canister change schedule is the use of end-of-service-life indicators (ESLI). ESLIs are part of the air-purifying cartridge or canister and provide a visual color change that indicates they need replaced. There have only been a few ESLIs approved by NIOSH such as mercury (vapor), ethylene oxide, and hydrogen sulfide. Care must be exercised if these indicators are used because of employee color blindness, visual access to the ESLI while wearing the RPD, and actual performance of the colorimetric indicator. A new NIOSH-approved colorimetric ESLI system has just been released by 3M that can be used with 56 organic compounds.[20] Other major RPD manufacturers will no doubt provide ESLIs for their products to compete with this new ESLI system.

There are a number of very important uses for combination cartridges and canisters with a filter and sorbent. They include, but are not limited to, pesticide application and paint spraying. RPD manufacturers make organic vapor cartridges with either a built-in prefilter or filter that can be added to the front of a cartridge or canister. There are also combination cartridges and canisters for capturing a variety of gases and vapors. These include combination organic vapor and acid gas (sulfur dioxide, hydrogen chloride) and Type N canister that provides protection from the four major gases and vapors (organic vapors, acid gases, carbon monoxide, and ammonia) plus a HEPA filter for particulates. The CBRN canister for full facepiece RPDs and PAPRs was developed as part of the NIOSH response to provide appropriate RPD protection for emergency responders after the terrorist's attacks on 11 September 2001 (Figure 16.16). A very extensive review of the hazardous chemicals that could be used in a terroristic incident was carried out. This resulted in the identification of seven general families of toxic agents that contained the 139 toxic agents of interest. Ten test representative agents (TRA) for the toxic agent families were identified and a summary is provided in Table 16.2. The other hazardous materials considered in the CBRN designation are particulate biological agents, particulate radiological/nuclear agents, and other particulates that are effectively removed by a P100 filter. The "OSHA/NIOSH Interim Guidance (April 2005), Chemical – Biological – Radiological – Nuclear (CBRN) Personal Protective Selection Matrix for Emergency Responders" provides background and use information on CBRN RPDs.[21]

[20] http://multimedia.3m.com/mws/media/1080985O/service-life-indicator-6000i-series-niosh-tech-data-sheet.pdf.

[21] https://www.osha.gov/SLTC/emergencypreparedness/cbrnmatrix/index.html.

**TABLE 16.2 TEST REPRESENTATIVE AGENTS (TRAs) AND SUMMARY INFORMATION**

| Test representative agent (TRA) | Family (7) | Number of toxic agents in each family |
|---|---|---|
| 1. Ammonia | Basic gas | 4 |
| 2. Cyanogen chloride | Acid gas | 32[a] |
| 3. Cyclohexane | Organic | 61 |
| 4. Formaldehyde | Formaldehyde | 1 |
| 5. Hydrogen cyanide | Acid gas | [a] |
| 6. Hydrogen sulfide | Acid gas | [a] |
| 7. Nitrogen dioxide | Nitrogen oxide | 5 |
| 8. Phosphine | Hydride | 4 |
| 9. Sulfur dioxide | Acid gas | [a] |
| 10. Dioctyl phthalate (DOP) | Particulate | 32 |
| Total number of hazardous agents | | 3 |

[a] Included in the 32 toxic agents count listed in TRA number 2, cyanogen chloride.

**Figure 16.17** Self-contained breathing apparatus, Scott Air-Pak X3. Courtesy of Scott Safety, A Tyco Business.

### 16.3.4 Airline Respiratory Protective Devices

Airline RPDs provide the wearer with a fresh supply of breathable air. All devices have a facepiece, an airflow regulator, a compressed air hose connecting to the facepiece/regulator, and a source of breathing air that also has a regulator. They come with either their own supply of air directly attached to the RPD in which case they are referred to as a self-contained breathing apparatus (Figure 16.17) or require attachment to an auxiliary source of breathing air (Figure 16.18). This source may be compressed air cylinders or tanks, or it may be an air compressor located near the wearer or in some remote location. The facepieces may be a full facepiece, half facepiece, hood, helmet, or even a suit.

These devices have a number of advantages over air-purifying devices. They tend to provide the wearer with a higher degree of protection because they normally maintain a positive pressure in the facepiece and reduce the amount of contaminants that can leak into the RPD. These devices can protect against gases and vapors that are not removed by air-purifying cartridges. They can provide protection for longer periods of time since there is no need to change an air-purifying element except for the SCBA that has an air supply depending on cylinder size that can last up to an hour. They can also provide some significant cooling for workers operating in hot environments mainly from sweat

**Figure 16.18** Positive-pressure supplied-air RPD with escape bottle to supply an auxiliary source of breathing air. Courtesy of Mine Safety Appliances Company.

**Figure 16.19** Supplied-air RPD vortex cooling assemble, 3M Respirat and 3M Speedglas Fresh-Air Welding Helmet with 3M Respirat. Courtesy of 3M.

evaporation. A vortex cooling adaptor can be added to a specialized airline system for welding and other applications that separates the compressed gas into hot and cold streams that can be used to cool or heat the air by ±50°F (Figure 16.19).

The use of airline RPDs also has a number of disadvantages. They generally are more expensive to purchase as well as to operate. For fixed systems, where piping is permanently in place, careful planning must be made in designing where workers will need to work and wear RPDs. With the exception of an SCBA, all these devices require attachment to a compressed air line that limits how far a worker can be from a source of breathable air. Airline RPDs are generally not useful when employees must work on multiple levels or have far to walk typically no more than several hundred feet with a compressed air hose.

As previously mentioned, one of the advantages of airline RPDs is the ability to maintain a positive pressure inside the facepiece. Manufacturers provide either pressure-demand or continuous-flow designs to accomplish this. For pressure-demand devices, the design is such that air is bled into the RPD during inhalation to maintain a slightly positive pressure. This also requires that the wearer overcome an inch of water pressure during exhalation created by a spring-loaded exhalation valve. Continuous-flow RPDs, as the name implies, keep a steady stream of air coming into the facepiece. An advantage of continuous-flow RPDs is the absence of significant breathing resistance for the wearer. The selection of these devices is very dependent on the source of breathing air. When compressed air bottles are used,

**Figure 16.20** Positive-pressure full facepiece airline RPD with five-minute escape bottle: MSA PremAire® Supplied-Air Respirator System and Draeger Colt and Micro Airpacks. Courtesy of Mine Safety Appliances Company and courtesy of Draeger Inc.

a pressure-demand tight-fitting facepiece is often required to conserve the amount of available air. When an air compressor is available, continuous-flow loose-fitting RPDs are often the choice for comfort, and the quantity of available breathable air available is not a problem. NIOSH testing and approval requires a minimum of 4 cubic feet per minute (CFM) for tight-fitting RPDs and 6 CFM for loose-fitting RPDs.[22] These requirements are designed in general to try to maintain a positive pressure in the RPD facepiece most of the time.

Breathing air systems need to be designed, maintained, and managed to provide the quality and dependability of breathing air required for airline RPDs (Canadian Standards Association, 2013). The quality of breathing air is specified under OSHA RPD regulations as Grade D air as specified by the Compressed Gas Association (2011) Grade D air addresses contaminants such as condensed hydrocarbons, carbon monoxide, carbon dioxide, and odor. These contaminants can result from the use of a compressor and do not take into consideration additional contaminants that may be located near the intake air source of the compressor. It is important that the air intake for a compressor used for breathing air be located in a clean environment away from known sources of emissions such as vehicle exhaust or other contaminant sources.

Moisture is an important consideration when air compressors are used to provide breathing air. The presence of excess moisture in the compressed air can result in the formation of water droplets downstream of the compressor or bottle and can result in frozen airlines and regulators if low ambient temperatures are experienced. However, systems that remove nearly all the water vapor in the air also create problems if employees are expected to wear RPDs for a substantial period of the workday. The lungs require humidification from air containing water vapor, and extremely dry compressed air can cause respiratory and eye discomfort to workers.

In addition to the quality of breathing air, the dependability of the source of breathing air is an important factor. If a worker has entered a heavily contaminated environment and the air compressor fails, or if the compressed air hose leading to the RPD is compromised (e.g. accidentally crushed by a motor vehicle), how will the worker escape from this environment? This is an obvious problem for workers entering confined spaces where the environment is not breathable, but may also be a problem when working in a work environment that is IDLH. For these environments only an SCBA or airline with an auxiliary escape compressed air bottle is acceptable (Figure 16.20). The auxiliary escape system will automatically switch to the bottled air if the source of breathing air from the compressed air line is disrupted.

[22] Code of Federal Regulations, Title 42, Part 84, "Respiratory Protective Devices."

These bottles can provide from 5 to 15 minutes of air, but can only be used with tight-fitting pressure-demand facepieces. For less hazardous environments, an alternative is to have a reservoir of compressed air that allows the wearer to exit an environment should power be disrupted or the compressor fail. The use of bottled air systems solves some of the breathing air quality issues, but the availability of a sufficient quantity of air may be a concern for some operations.

The breathing hose is a very important component of this breathing air system. As was noted previously, care must be taken so that the integrity of the compressed air line is not compromised such as being pinched by a heavy vehicle or other mobile pieces of process equipment. The breathing hose can also be permeated by toxic organic liquids that can contaminate the breathing air supply. To maintain NIOSH approval of these supplied-air RPD systems, a breathing air hose supplied by the RPD manufacturer must be used.

### 16.3.5 Self-Contained Breathing Apparatus

An SCBA is an atmosphere-supplying RPD in which the wearer has a source of breathing air directly integrated into the RPD (Figure 16.21). The major advantage of these devices is that the wearer is not dependent on some remote source for breathing air.

SCBAs can be classified as either open-circuit or closed-circuit devices. All SCBAs have tight-fitting facepieces and will operate in the demand (for some closed-circuit units) or pressure-demand mode. The closed-circuit device uses either a bottle or compressed oxygen or an oxygen-generating chemical source to maintain the necessary level of oxygen in the RPD. Air is rebreathed through the unit, and the wearer can generally use such devices for much longer times. These units are designed for mine rescue, where a long service life of up to four hours is required (Figure 16.22). The open-circuit units tend to provide the wearer with breathing air for 15–60 minutes (a minimum of 30 minutes is required for use in IDLH environments). The exact service life is dependent on the size of the compressed breathing air tank, lung (vital) capacity of the wearer, and their work rate (breathing rate). Emergency response personnel, such as firefighters, and others who must spend a relatively short amount of time in a highly hazardous environment use these devices. Breathing air comes into the RPD from a compressed air tank, and the wearer exhales into the environment – hence the term *open circuit*.

Several general rules govern the use of these sophisticated devices. The wearer must be physically fit to use them since they are often 30 or more pounds in weight. These devices are used in highly hazardous work environments, and the wearer must be trained in their use to allow he or she to handle emergencies (e.g. bypass a faulty regulator on the RPD) and work in highly stressful environments while wearing an RPD. The current NFPA 1981 CBRN SCBAs have electronic heads-up monitoring systems that provide the wearer with real-time information on their critical systems. Finally, these devices themselves must be checked at the start of a shift in the fire service, and at least monthly, if stored for emergency response, to ensure that they are in operational condition and ready for immediate use. SCBAs are

**Figure 16.21** Self-contained breathing apparatus, MSA FireHawk and Draeger AirBoss. Courtesy of Mine Safety Appliances Company and Courtesy of Draeger Inc.

**Figure 16.22** Four-hour SCBA rebreather: Draeger BG 4 and Biomarine BioPak 240R. Courtesy of Draeger Inc. and Courtesy of Biomarine Rebreathers, Neutronics Inc.

often used in response to emergencies, and the wearer typically does not have the luxury of carefully inspecting the device prior to having to don it and perform their work activity. A general checklist covering items that need to be inspected is provided in reference (Occupational Safety and Health Administration 2011) along with a variety of other program element checklists.

The concept of an open-circuit CBRN SCBA used in combination with a non-powered CBRN tight-fitting RPD and/or a powered air-purifying tight-fitting RPD was evaluated by NIOSH with the opening of NIOSH Docket Number 82 and 82A in October of 2006. The Institute of Medicine (IOM) held the workshop "Developing a Performance Standard for Combination Unit Respirators" on 30 April 2015 (Gummaria et al. 2015). For this type of RPD to be developed and sold commercially in the United States, there need to be performance and approval standards, determination of the APFs, how the equipment will be switched between modes, and guidance on the use of air-purifying canisters in IDLH environments. A more detailed list of the benefits and challenges is provided in the IOM workshop report. The individual components of the combination RPD are approved and are being used separately (Figure 16.23); the challenge is joining two types of RPDs together so they can be used safely.

### 16.3.6 Escape-Only Respiratory Protective Devices

OSHA requires that some employers plan to protect their employees in the case of an emergency. For some facilities and operations, this may include the use of RPDs for escape use only. While any of the previously mentioned RPDs can be used for escape purposes, there are some devices specifically designed for this purpose. These devices are compact, are easy to use, and require no prior fit testing.

One air-purifying RPD that is approved by NIOSH for escape only is the mouthpiece RPD. A mouthpiece is held in the wearer's mouth, and a clamp is placed over the nostrils. The lips are placed tightly around the mouthpiece, and all air comes through the filtering RPD attached to the mouthpiece. These devices are small and very portable and for some industries (e.g. chlorine manufacturer) have been routinely assigned to all individuals working in a production area should an unplanned chemical release occur (Figure 16.24). A more sophisticated mouthpiece RPD is the self-contained self-rescuer that is used for mine escape. These rebreathers are available with a 30- or 60-minute duration (Figure 16.25).

Another type of escape-only device uses a small cylinder of compressed air and usually has a plastic hood that can be quickly pulled over the wearer's head. These devices have an air source that may be rated for 5 or 10 minutes (Figure 16.26). The advantage of this RPD over a mouthpiece RPD is that device performance is not dependent on the RPD cartridge being able to effectively filter or absorb whatever contaminants may be present. Therefore, it is especially well suited when unknown or highly toxic contaminants are present in high concentrations from a catastrophic release. One disadvantage is that these devices tend to be large, and while a wearer can carry them,

**Figure 16.23** Air-purifying and PAPR CBRN respirators, MSA Responder air-purifying RPD and PAOR. Courtesy of Mine Safety Appliances Company.

they are not as convenient as a mouthpiece RPD that can be clipped to one's belt. Another disadvantage of these devices is that they rely on a highly compressed air source (3000–5000 psi), and the RPDs must be regularly inspected (at least monthly) to ensure that they are full and have the required service life.

A new escape RPD that resulted from the 911 disasters is the CBRN APER (Figure 16.27). The CBRN APERs consist of a hood made of materials resistant to the permeation of chemical agents and a CBRN canister. These devices can include an oral/nasal cup, mouthpiece, or fan to further protect the respiratory system. These devices may be used to escape from IDLH environments, but not for entry into or working in an IDLH atmosphere. These APERs are designed for the general working population and provide escape protection against CBRN agents.

## 16.4 RESPIRATORY PROTECTIVE DEVICE PROGRAM

The performance of an RPD is no better than the employer's implementation of the required RPD program elements of OSHA 29 CFR 1910.134[23] (Attachment 1). This OSHA regulation was initially based on ANSI Z88.2-1969 (1969), revised twice by formal rulemaking and is now supplemented by the 2015 edition of ANSI/ASSE Z88.2 (2015). One of the first steps the employer should do in this process is the appointment of a program administrator to develop, implement, and manage the RPD program. The program administrator should then develop individual written standard operating procedures (SOPs) to address each of the following elements needed for their program:

1. Procedures based on a hazard assessment used for selecting RPDs used in the workplace.
2. Procedures for medical evaluations of employees required to use RPDs.
3. Fit-testing procedures for tight-fitting RPDs.

[23] Code of Federal Regulations, Title 29, Part 1910.134, "Respiratory Protection, Assigned Protection Factors; Final Rule."

**Figure 16.24** Mouthbit air-purifying escape RPD: North 7900 and Draeger PARAT 3200. Courtesy of North by Honeywell and courtesy of Draeger Inc.

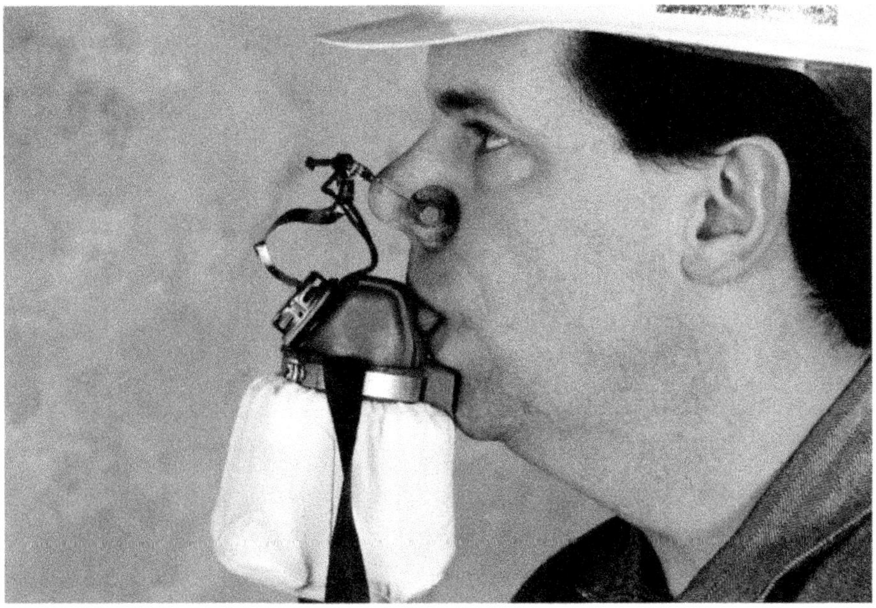

**Figure 16.25** Mouthbit mining escape rebreather RPD, MSA W65. Courtesy of Mine Safety Appliances Company.

**Figure 16.26** Supplied-air escape hoods available with 5- and 10-minute air supply bottles, MSA TransAire®. Courtesy of Mine Safety Appliances Company.

4. Procedures for proper use of RPDs in routine and reasonable foreseeable emergency situation.
5. Procedures and schedules for cleaning, disinfecting, storing, inspecting, repairing, discarding, and otherwise maintaining RPDs.
6. Procedures to assure adequate air quality, quantity, and flow of breathing air for atmosphere-supplying RPDs.
7. Procedures for training of employees in the proper use of the RPDs and the respiratory hazards to which they are potentially exposed during routine and emergency situations.
8. Procedures for the evaluation of the RPD program records.
9. Procedures for regularly evaluating the effectiveness of the RPD program.

This written RPD program shall be updated on a regular basis to reflect changes in the workplace as well as any changes in RPD equipment or other elements supporting the program.

A description of each RPD program element outlined in the OSHA requirements (29 CFR 1910.134) is discussed below. The reader is again encouraged to consult with other reference-related materials on RPD programs for additional information (Occupational Safety and Health Administration 2011; National Institute for Occupational Safety and Health, Occupational Safety and Health Administration 2015). The OSHA RPD standard focuses on RPD program elements required to ensure the safety of the wearer.[24] References Occupational Safety and Health Administration (2011) and National Institute for Occupational Safety and Health Occupational Safety and Health Administration (2015) contain examples of RPD programs to use in developing specific workplace programs.

### 16.4.1 Appointment of a Respiratory Protective Device Program Administrator

OSHA requires the employer designate an individual (or group of individuals) who is responsible to develop and manage the RPD program. Ideally this person should report to a senior manager in the organization to assure proper priority and resources are provided. The program administrator develops individual written SOPs to address each required element listed previously. The administrator may designate individuals with responsibilities for certain

[24] Ibid.

**Figure 16.27** CBRN escape hood, Scape® escape respirator, ILC Dover. Courtesy of ILC Dover.

components of the program (e.g. wearers may have responsibility for inspection and maintenance of their device). Although there is no specific training required for RPD program administrators, these individuals should be provided with formal training on this subject. A number of private consultants and professional societies offer courses on the various aspects of RPD protection.

### 16.4.2 Written Standard Operating Procedures

An important aspect of any RPD program is written evidence that it exists and that specific individuals have been identified and assigned key responsibilities. The SOPs are primarily designed to identify all of the major RPD program elements noted above, describing each of them in some detail, how they will be implemented, and who has responsibility for their operation. The SOPs will also be of value for someone outside of the organization to examine and determine whether all the required components of a successful RPD program have been identified, assigned, and implemented.

### 16.4.3 Hazard Assessment

The need for RPD protection begins with the identification and assessment of the hazard. The hazard may exist on a routine basis or infrequently due to required maintenance activities. An RPD may be selected not because a hazard exists, but because a control method may fail to work properly. For example, an RPD may be selected when breaking into a chemical line that supposedly had previously been flushed with water. The RPD would be used should the line contain some unexpected contaminant. Sometimes RPDs are selected when no health hazard is present, but employees find a material to be a respiratory irritant.

Each situation mentioned above requires a different type of hazard assessment in determining how to properly select the correct RPD. Only qualified health and safety professionals such as certified industrial hygienists, health physicists, or certified safety professionals should perform the hazard assessment.

The hazard assessment should be a written document that provides enough information so that the basis for the RPD selection is clearly documented. If workplace air sampling is required, it should be specifically identified with specific results listed that were used in the selection process. If published literature is used to estimate the expected airborne levels of contaminants, the specific article should be referenced, and where possible, a copy of articles used should be attached to the hazard assessment.

Careful attention must also be taken to consider the consequences should an RPD fail to perform its proper function due to the hazard present. Some contaminants are irritants, and employees will quickly know if they should leave a work area due to the malfunction or limitation of an RPD. Other contaminants can produce chronic organ damage, and employees may not be aware of the hazard of continuous overexposure until an irreversible and adverse health outcome has occurred. This can happen long after the worker has changed jobs or retired, e.g. asbestosis, mesothelioma, and silicosis. When needed, specific details on the hazard and the effects of human exposure should be included in the training addressing this workplace operation.

Factors that should be considered when completing the hazard assessment for toxic materials are:

- Physical and chemical properties.
- Adverse health effects.
- Occupational exposure levels.
- Results of workplace sampling.
- Work operation description.
- Time period of RPD wear.
- Work activity and stresses on the wearer.
- Warning properties.
- Interaction of other types of PPE worn with the RPD.
- Capabilities and limitations of RPD types under consideration.

Hazard assessment should also include options for other controls besides the use of RPDs (see Chapter 12). This may include permanent engineering solutions from the substitution of a material that is less toxic, reduced emissions through the use of local exhaust ventilation, or other engineering controls, along with the possibilities of reducing the contact time for an employee to a hazard through administrative or other controls. RPDs are generally uncomfortable to wear and can reduce employee productivity and ability to communicate with other employees. Even when an RPD program has been successfully implemented to control a hazard, efforts should continue to explore other control options with the goal of reducing or eliminating the use of RPD protection.

### 16.4.4 Respiratory Protective Device Selection

The advantages and disadvantages of the wide variety of RPDs available for selection are outlined in the previous section. There are three main considerations that govern the selection of an RPD. The first and by far most important is the protection afforded by the RPD. An RPD must be selected that will provide sufficient protection to ensure that a worker will not be overexposed to a contaminant. Less important are the costs associated with both the RPD and the program necessary to support its safe use and comfort of the wearer.

The protection factor assigned to an RPD is one of the more contentious issues in the field of RPD protection. The protection factor for an RPD is defined as the ambient (outside) contaminant concentration divided by the contaminant concentration inside the facepiece. A protection factor of 1 represents an RPD that allows a contaminant to readily enter the RPD and offers no protection to the wearer. A protection factor of 100 represents an RPD that allows 1% or less of the contaminants from the ambient environment to enter the inside of an RPD.

The protection afforded by an air-purifying RPD, and its resultant protection factor, is a complex function of how well the air-purifying elements work, whether the RPD and all its parts have been properly maintained, how well the RPD fits the wearer, and how the wearer uses the RPD.

In the early 1970s Edwin Hyatt and others at the Los Alamos National Laboratory (LANL) applied the Bureau of Mines quantitative DOP man testing equipment and procedures that were used to measure RPD performance for high-toxicity dusts, fumes, and mists and radionuclides to develop RPD quantitative fit-test procedures. This LANL program produced results that were used to generate a list of APFs for general RPD classes (Hyatt 1976). LANL also developed a panel of 25 individuals with specific facial dimensions that represented anthropometric specifications for male and female test subjects used for approval testing of RPDs (Hack 1978). These various facial dimensions were derived from a 1967–1968 survey of US Air Force personnel. NIOSH has proposed an upgraded panel based on extensive research to be more representative of the current ethnic composition of the US workforce. This major research endeavor has resulted in a proposed panel of 35-member bivariate panel to be used to approve half and full facepiece RPDs intended to fit the general population (Zhuang et al. 2007). Another part of this project is the development of a set of human test head forms that represent the updated facial dimensions included in the new test panel. A set of five test heads has been developed to represent the various facial measurements defined using principal component analysis (Figure 16.28). This new fit panel information when integrated into the RPD approval process will improve the fitting characteristics of future NIOSH-approved commercial models of RPDs. For now this information is available for RPD manufacturers to use if they wish.

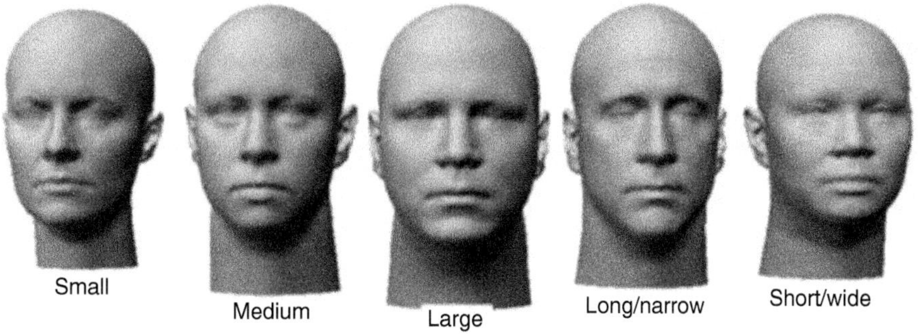


**Figure 16.28** NIOSH 5 representative head forms for the bivariate panel using principal component analysis to explain facial variations. Courtesy of NPPTL.

The selection of the correct RPD for a specific work activity involves the consideration of a number of factors. A filtering facepiece RPD is the least expensive and least protective (APF = 10) type of device to purchase, and an SCBA (APF = 10000) is the most protective and expensive RPD. A filtering facepiece RPD as compared with a reusable half facepiece RPD may make economic sense when the contaminant is difficult to remove from the RPD. A loose-fitting PAPR may be a reasonable choice when relatively few individuals need RPDs that currently have facial hair that could interfere with the fit of a tight-fitting facepiece. This approach may not be economically viable when large numbers of individuals must wear RPDs, and it is more appropriate to require RPD users to be clean-shaven. Airline RPDs may require the purchase of an expensive central compressor and piping system, or it may be possible to use an existing system. There are also small moderately priced portable compressors that are appropriate for one or two individuals (Figure 16.9).

The maximum use concentration (MUC) is defined as the maximum atmospheric concentration of a hazardous substance from which an employee can expect to be protected when wearing a specific class of RPD. It is determined by multiplying the APF of the RPD or class of RPDs times the exposure limit of the hazardous substance. The exposure limit is normally the OSHA permissible exposure limit (PEL), the American Conference of Governmental Industrial Hygienists TLV, or the NIOSH recommended exposure limit (REL). If the workplace exposure approaches the MUC of the respirator selected, the next higher class of respirators should be selected. The IDLH value for the hazardous substance should not be exceeded by the MUC, and if an IDLH concentration exists, the correct respiratory protection should be selected.

Medical clearance evaluations by a physician or licensed healthcare professional (LHCP) are required to assure workers can safely wear a specific type of RPD. This initial medical evaluation can be completed several different ways following OSHA guidance. The initial step of filling out the OSHA questionnaire can be done by the employee or with the help of a physician or LHCP. The physician or LHCP reviews the answers and orders additional tests they feel necessary for the evaluation. The employer is also required to provide the physician or LHCP with the following information:

- The type and weight of the RPD being used.
- The duration and frequency of use (including use during rescue and escape).
- The expected physical work effort.
- Additional protective clothing and equipment to be worn.
- Temperature and humidity extremes that may be encountered.
- A copy of the written RPD program.

The physician or LHCP reviews all of this information and provides the employer with a recommendation for each employee providing the following general information, no RPD usage, full RPD usage, or limited RPD usage (with specific restrictions). The ability of the worker to safely tolerate the significant physiological burden created by wearing an RPD is the purpose of the medical evaluation. The preplacement medical evaluation must be completed prior to fit testing and the use of any RPD. Additional medical evaluations are not specified on a specific schedule but shall be provided based on observations and a request by the physician or LHCP, RPD program administrator, RPD fit tester, supervisor, or the worker.

ANSI/ASSE Z88.6-2006 "American National Standard for Respiratory Protection-Physical Qualifications for Respirator Use" provides additional information related to medical evaluations for workers required to use RPDs (American National Standard Institute, American Industrial Hygiene Association, American Society of Safety Engineers 2006).

### 16.4.5 Tight-Fitting Respiratory Protective Device Fit Testing

RPD fit testing is a very important part of the RPD program because it confirms and documents that the specific tight-fitting RPD model selected for the worker fits properly. OSHA requires that a worker receive an initial fit test before beginning work using the same make, model, style, and size of RPD that will be used in the workplace. Any additional PPE worn that could impact the fit or performance of the RPD should also be worn during fit testing. Fit testing also provides a good opportunity to train the worker on how to don and wear the RPD properly. The results of improper donning can be demonstrated quickly by smell or taste or with direct readout instrumentation. There are two general types of fit-testing techniques, qualitative and quantitative. Qualitative fit testing utilizes the worker's sense of taste or smell of the test agent while carrying out a set of standardized exercises to determine if a proper RPD fit has been achieved. A typical qualitative fit-test apparatus is shown in Figure 16.29. Quantitative fit testing on the other hand uses a separate analytical instrument and the same set of standardized exercises to determine if there has been a measurable leak of the RPD (Figure 16.30).

Additional fit testing is required when a different RPD facepiece (size, style, model, or make) is used, or at least annually after the initial test. Additional fit tests shall also be provided when observations are made that identify changes in the RPD user's physical condition that could affect RPD fit. Such conditions include but are not limited to facial scarring, dental changes, cosmetic surgery, or an obvious change in body weight. Fit testing is required for all positive- or negative-pressure tight-fitting RPDs such as SCBAs, PAPRs equipped with a tight-fitting full facepiece, and APRs before a worker can use them.

ANSI/ASSE Z88.10-2010 "ANSI/ASSE Respirator Fit Testing Methods" (American National Standard Institute, American Industrial Hygiene Association, American Society of Safety Engineers 2010) provides additional useful information related to RPD fit testing of workers required to use RPDs.

Appendix A of 29 CFR Part 1910.134 (Mandatory) Attachment 1 provides specific OSHA-accepted fit-test protocols:

1. Fit-testing protocols – General requirements
2. Qualitative fit-test (QLFT) protocols
   - General
   - Isoamyl acetate protocol
   - Saccharin solution aerosol protocol
   - Bitrex™ (denatonium benzoate) solution aerosol qualitative fit-test protocol
   - Irritant smoke (stannic chloride) protocol
3. Quantitative fit-test (QNFT) protocols
   - General
   - Generated aerosol quantitative fit-test protocol
   - Ambient aerosol condensation nuclei counter (CNC) quantitative fit-testing protocol
   - PortaCount fit-test requirements
   - Controlled negative pressure (CNP) quantitative fit-testing protocol

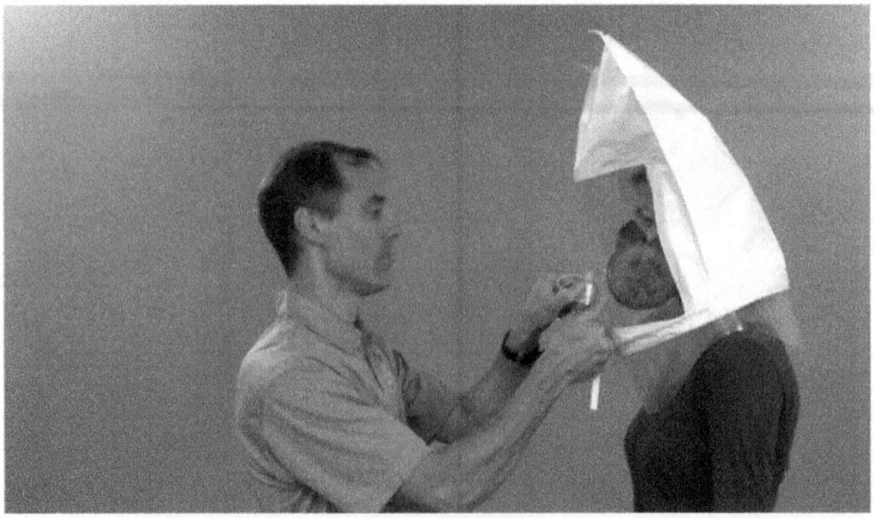

**Figure 16.29** Qualitative fit test using Bitrex. Courtesy of 3M.

**Figure 16.30** Quantitative fit test of a full facepiece RPD using room aerosol and a PortaCount respirator fit tester 8038. Courtesy of TSI Inc.

### 16.4.6 Respiratory Protective Device Use

Procedures are required to instruct workers on the proper use of RPDs in routine and reasonable foreseeable emergency situations to assure workers know how to use the assigned RPD properly to obtain the proper level of protection. A clean and unobstructed RPD sealing surface on the wearer's face is pivotal in assuring a tight-fitting RPD seals properly. Tight-fitting RPD use procedures should forbid facial hair and the use of eyeglass temples or straps that interfere with the RPD sealing surface. Controlling facial hair/beards has presented ongoing RPD program issues for decades. A personnel policy that clearly states that facial hair is not permitted for specific jobs that require the use of RPDs with tight-fitting facepieces is the best way to address this issue. Some accommodation can be made for facial hair by using a PAPR or airline equipped with a hood, but these RPDs cannot be used in an IDLH environment. When other PPE is provided, the employer shall ensure that it does not interfere with the seal of the facepiece to the face of the user. If the user requires corrective lenses, specific manufacturer spectacle insert kits with the appropriate prescription lenses shall be provided for full facepiece RPDs (Figure 16.31). The restriction on wearing contact lenses has been removed for full facepiece RPDs and hoods because of improved design and comfort of these lenses.

An adequate facepiece seal is necessary for all tight-fitting RPDs to provide the proper level of protection. OSHA requires a positive or negative user seal check test be successfully completed in the field after each donning of a tight-fitting RPD. To carry out a positive-pressure user seal check, the wearer closes off the exhalation valve or the RPD regulator opening on the SCBA RPD facepiece with the palm of the hand and exhales gently into the facepiece. For filtering facepiece RPDs, the hands are cupped over the outside surface of the RPD, and the wearer exhales. The fit is considered satisfactory if a slight positive pressure can be built up inside the facepiece without any evidence of outward leakage of air at the seal. To carry out a negative-pressure user seal check, the wearer closes off the inlet opening of the cartridges, canister, or SCBA mask-mounted regulator opening with the palm of the hand(s), inhales gently so that the facepiece collapses slightly, and holds his/her breath for 10 seconds. If the facepiece remains in its slightly collapsed condition and no inward leakage of air is detected, the tightness of the RPD is considered satisfactory. If the palm of the hand cannot effectively cover the canister or mask-mounted regulator opening, a thin latex or nitrile glove can be used to cover it. As part of this training, the specific manufacturer user instructions should be followed.

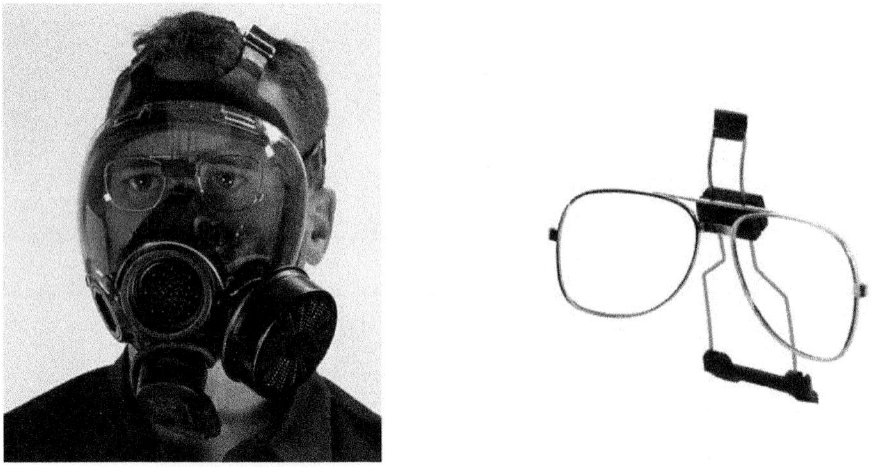

**Figure 16.31** Millennium full facepiece RPD corrective lenses insert kit and corrective lenses insert. Courtesy of Mine Safety Appliances Company and Courtesy of Draeger Inc.

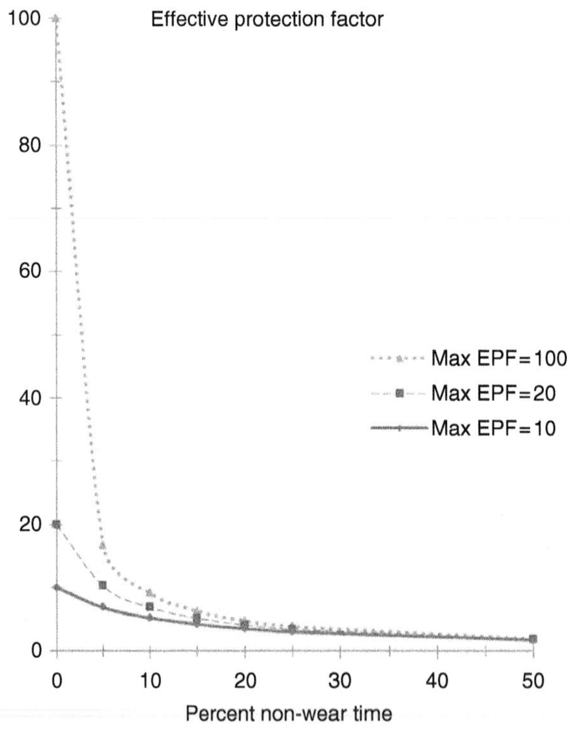


**Figure 16.32** Graph showing the effect of not wearing a respirator on protection provided the wearer as expressed by the percent non-wear time on the effective protection factor.

When carrying out a user seal check, the worker should be trained to examine the RPD for any defective parts, e.g. facepiece, head straps, valves, connecting tube, and cartridges or canisters that are not functioning properly. If any problems are identified, the RPD should be repaired or replaced.

Another required procedure should lay out an ongoing activity to determine the effectiveness of RPDs in use. This activity requires an evaluation of how the RPDs are used and maintained in the field as well as how the worker behaves when using the RPD. It is clearly recognized that not wearing an RPD for short periods of time significantly reduces the level of protection; the effective protection factor (EPF) provided the wearer. Not wearing a full facepiece RPD 5% of the time reduces the EPF from 100 to 17 and from 10 to 7 for a half facepiece RPD. Figure 16.32 illustrates the impact of percent non-wear time on the EPF.[25] When RPD performance or use problems are observed in the field, the employer shall make sure the worker leaves the RPD use area before removing the RPD to address the issue. Any issues identified should be documented and corrected. A separate procedure for entry into IDLH atmospheres must be developed that lists the specific hazards, requirements, and detailed work practices to be followed. This IDLH procedure shall require a dedicated outside standby person who has received appropriate training and equipment (SCBA or positive-pressure airline full facepiece RPD with five-minute escape bottle). This equipment for the outside standby is necessary to provide timely and effective emergency rescue and communication. The outside standby before undertaking a rescue shall inform his/her employer of the rescue action so appropriate backup actions can be immediately initiated. Checklists that summarize the requirements of these procedures make good training aids and improve the understanding, performance, and compliance.

### 16.4.7 Respiratory Protective Device Maintenance and Storage

When RPDs are used repeatedly, a procedure describing the comprehensive maintenance program should include, cleaning, disinfecting, storage, inspection, repair, and disposal. Depending on the type and quantity of RPDs used, commercial RPD maintenance services may be an option worth evaluating. RPD manufacturer's recommendations related to these specific activities should be followed where appropriate.

RPDs should be cleaned daily or after each use and disinfected when used among more than one worker. Filtering facepiece RPDs not intended to be cleaned or disinfected can be disposed after each use, daily, or after a number of days depending on the condition of the RPD. Some RPDs that are used in areas heavily contaminated or with toxic contaminants may need to be cleaned after each use, while others may require daily cleaning. Based on the contamination level, RPDs may require cleaning at the worksite before being transported to the regular cleaning location, e.g. radioactive particulates. RPDs should not be placed back into storage until they have been adequately cleaned, disinfected, and inspected when appropriate. All operations involving cleaning, rinsing, and drying of RPDs should be done in contaminant-free environments.

RPDs must be disassembled to allow for parts inspection and adequate cleaning. OSHA requires all RPDs to be inspected during cleaning. If a defective RPD part is found upon inspection, it must be replaced with the RPD manufacturer's NIOSH-approved part. RPDs that have been repaired with non-NIOSH-approved parts or other manufacturer's parts violate approval requirements and are no longer certified. NIOSH recently published a fact sheet calling attention the problems with the use of aftermarket replacement component parts.[26] There are a variety of cleaning materials available for RPDs. Some care must be taken to avoid the use of materials that may degrade the RPD components or cause skin irritation (e.g. quaternary ammonium compounds). Facilities that use a large number of reusable RPDs daily may find that a central facility with a commercial dishwasher is the most efficient means for RPD cleaning. Care must be taken to make sure that the water used for cleaning does not exceed 110 °F (43 °C) to avoid permanently distorting RPD facepieces. After an RPD is cleaned, it must be rinsed and then dried. Large volume maintenance facilities may also find it advantageous to modify a commercial dryer following the same temperatures noted above to reduce the drying time. Finally, the RPD components must be reassembled, inspected, placed in a sealed plastic bag, dated, and placed into storage.

RPD storage varies depending on whether devices are used for emergency-only use or routine wear. OSHA requires emergency devices such as SCBAs be stored in areas that are clearly identified and inspected at least monthly. A number of companies make appropriate signs and storage units that can be used to identify the location of such devices. Emergency-use RPDs should be placed in areas of a facility where they are quickly and easily accessible, yet are likely to remain in a contamination-free environment so that they can be safely donned.

Non-emergency-use RPDs are often kept either in a central supply area or issued to individual employees. In the latter case, employees must be careful to store their RPD in such a manner that does not result in the distortion or damage of the facepiece or other component degradation due to vibration, sunlight, heat, or extreme cold. Ideally, separate lockers or sections of lockers can be designated for the storage of individually assigned RPDs. Most RPDs after drying should be kept in a sealed plastic bag. This protects the device from contamination and can keep important components such as gas- and vapor-removing cartridges from being contaminated with water vapors. PAPRs that use rechargeable batteries may require separate storage areas that keep batteries charged on

---

[25] Appendix A.7, "Effective Protection factor (EPF)," American National Standard Institute, American Society of Safety Engineers, ANSI/ASSE Z88.2-2015, "Practices for Respiratory Protection," Park, IL p 50–52 (2015).

[26] http://www.cdc.gov/niosh/docs/2016-107/pdfs/2016-107.pdf.

a continuous basis until they are ready for use. Some effort is required for multiple PAPR users to distinguish batteries ready for use from those that are currently being charged.

### 16.4.8 Breathing Air Quality, Quantity, and Flow for Atmosphere-Supplying Respiratory Protective Devices

Breathing air for atmosphere-supplying RPDs is required to be of high purity, meet quality levels for content, and not exceed certain contaminant levels and moisture requirements. OSHA requires that RPD breathing air meet the requirements for Type 1-Grade D breathing air described in ANSI G-7.1-1989 (Compressed Gas Association 2011). Cylinders used to supply breathing air to RPDs shall be hydrostatically tested periodically and maintained as prescribed by DOT.[27] When cylinders are received with breathing air, a certificate of analysis is required from the supplier that confirms the breathing air meets Grade D air. The moisture content in the cylinders shall not exceed a dew point of −50 °F (−45.6 °C) at one atmosphere pressure. If a compressor is used to supply breathing air for SCBA tank filling or airline RPDs, the compressor should be located so that contaminated air does not enter into the air intakes. Moisture content should be controlled so that the dew point at one atmosphere pressure is 10 °F below ambient temperature. The compressor should also be equipped with suitable in-line air-purifying sorbent beds and filters to further ensure breathing air quality. Carbon monoxide breathing air levels shall not exceed 10 ppm for non-oil-lubricated compressors. Oil-lubricated compressors shall be equipped with a high-temperature alarm, carbon monoxide monitor, or both. If only a high-temperature alarm is used, the breathing air shall be monitored at intervals to confirm carbon monoxide levels in the breathing air are not exceeding 10 ppm. The breathing air hose coupling shall be unique to prevent the connection to non-respirable air or other in-house gas systems. Breathing air cylinders shall be marked in accordance with the NIOSH approval standard 42 CFR Part 84.[28] NFPA 1989 "Standard on Breathing Air Quality for Emergency Service Respiratory Protection" (2013) requires quarterly air sampling from breathing air systems to confirm compliance with requirements for oxygen, carbon monoxide, carbon dioxide, condensed oil and particulate content, water content, non-methane volatile organic compounds, odor, and nitrogen content. Commercial contractors are available to provide this quarterly sampling service utilizing a sampling system that is returned for analysis after the sample is collected. Sample results are available on the Web within a short period of time.

### 16.4.9 Respiratory Protective Device Training

OSHA requires that the employer provide effective training to employees who are required to wear RPDs prior to wearing the device in the workplace. The extent of the training depends on the complexity of the RPD selected and the nature of the work and hazard. For hazards that are from nuisance dusts, less training is necessary than for entering atmospheres that are IDLH. The amount of training required will vary and depend on whether wearers have to maintain their own device (cleaning, disinfecting, and storage) or whether this is done by a central service or completely avoided by the use of disposable devices. The qualitative or quantitative fit-testing requirements can also be integrated with the training program to take advantage of donning and doffing as well as actually observing the level of protection when the RPD is worn properly. In addition some supervisory personnel including those with responsibility for portions of the RPD program will also need to have RPD training. The training must be comprehensive, must be understandable, and must recur annually, and more often, if necessary. The classroom and/or computer-aided training should address the nature of the hazard, how the RPD chosen works, applicable government regulations, when the RPD must be worn, and hands-on exercises where each student has, at a minimum, a chance to handle the RPD they will wear, examine all of the working parts, remove and replace filter/absorber cartridges, don and doff the RPD several times, and carry out a user seal check. For the training to be judged successfully, the employees must be able to demonstrate knowledge and understanding in the following areas:

- The nature of the hazard resulting in the selection of RPDs
- How to recognize medical signs and symptoms that may limit or prevent the effective use of this RPD
- The applicable government regulatory requirements that the employer must follow and ensure that their employees follow
- Why the RPD is necessary and how improper fit, usage, and maintenance can compromise the protective effect of this RPD

---

[27] CFR Part 180.205, "General Requirements for Requalification of Specification Cylinders (2010).

[28] Code of Federal Regulations, Title 42, Part 84, "Respiratory Protective Devices."

- What controls are being investigated to minimize or eliminate the future need for respiratory protection
- The functions, capabilities, and limitations of the RPD(s) selected for this hazard
- The effects of overexposure to the contaminant or other signs that an RPD may be failing to provide adequate protection (e.g. odor detection)
- How to inspect, put on and remove, use, and check the seals of the RPD
- How to use this RPD effectively in emergency situations, including situations in which the RPD malfunctions
- What the procedures are for maintenance and storage of the RPD
- Where RPDs and parts (e.g. filters, cartridges) can be obtained and where employees should discard spent RPDs and used parts
- When must RPDs be worn (both for routine work activities, maintenance and emergency operations)

### 16.4.10 Respiratory Protective Device Program Audits

Regular evaluations of the workplace are required to ensure that the written RPD program is being properly implemented and workers are using the RPD properly. The RPD program administrator is responsible for these evaluations by personally doing them or assigning the task to a competent employee or consultant. Direct interaction with employees who use the RPDs is necessary to identify any concerns they have with the program. This evaluation should review, at a minimum, RPD fit, appropriate RPD selection for the hazard the employee is exposed, and proper RPD use, maintenance, and storage. A written report should be prepared summarizing the findings and corrective actions. The RPD program administrator is responsible to track the status of all outstanding findings and bring to senior management any items that are not closed out in a reasonable time frame. As a rule of thumb, these evaluations should be done at least annually or more frequently if significant issues are identified.

### 16.4.11 Respiratory Protective Device Program Records

There are a variety of records that should be kept documenting the RPD protection program. Ideally, a central location will have all records kept in in a written format or electronically to allow for easy access and program evaluation. OSHA provides the specific recordkeeping requirements for fit testing and the RPD program in 29 CFR 1910.134(m) and for exposure assessment and medical evaluations in 29 CFR 1910.1020.[29] Some records are confidential and will be in an employee's medical file or personnel exposure records. However, an RPD recordkeeping file can maintain a list of individuals who have been medically cleared to use specific RPD brands and models. It is worthwhile to cross-reference some of these records, as they are useful for more than one purpose. Many records can be stored electronically allowing access to these records in multiple files. The specific retention time for employee exposure records, medical records, and related documents is 30 years.

The following documents are considered minimum records required to be kept:

- A copy of the current SOP covering the operations of the entire RPD program including those individuals responsible for the RPD program management and implementation.
- Hazard evaluations of all areas and operations where the use of RPDs is required for either routine use or maintenance. The exposure assessment report including supporting documentation such as the safety data sheet (SDS), a description of the analytical and mathematical methods used, and a summary of the background data relevant to the interpretation of the results obtained are required to be kept for 30 years. This information should be updated whenever a reevaluation of an operation or job is completed.
- Emergency planning records indicating what emergencies have been anticipated, who will respond to them, and what RPDs will be worn.
- Records for individual wearers indicating the results of fit testing, training, the RPD (model, size and brand) specifically selected for them, and medical release information indicating that they have been cleared to wear the specific RPD they used. Note: Medical records indicating the results of medical testing must be kept confidential in employee records.
- Training records providing an outline of material used for training along with the instructor who performed the training.

[29] https://www.osha.gov/pls/oshaweb/owadisp.show_document?p_table=STANDARDS&p_id=10027.

- Inspection records for emergency-use RPDs. It is advantageous to keep these both with the device and copies in a central location. By keeping them with the device, someone can quickly determine in an emergency that the device was inspected within a month, as required by OSHA regulations.
- Documentation of any RPD problems noted and investigations and corrective action taken.
- Results of all audits (internal and external) of the RPD program and what action was taken as a result of recommendations from these audits.
- Copies of all appropriate RPD standards including OSHA 29 CFR 1910.134 and ANSI Z88.2 (current edition).
- Where appropriate historic files of each set of records should be kept to document the changes and improvements in the RPD program over time.

### 16.4.12 Respiratory Protective Device Problems

The following topics are ongoing RPD use issues that can impact RPD programs if not appropriately dealt with.

***16.4.12.1 Vision*** Many individuals need corrective lenses to perform their job adequately. It should be noted that many half facepiece RPDs do not have a good interface with eyeglasses. This means that the eyeglass frame and RPD must be carefully chosen to allow the wearer to have eye protection and proper vision while wearing an RPD. Full facepiece RPDs do not allow the use of standard glasses as the temple bars will interfere with the seal of the RPD. This can be corrected with special eyeglass kits provided by the RPD manufacturer. Most fullface RPD lenses are made from polycarbonate and meet most of the requirements in ANSI/ISEA Z87.1-2015 (2015). For specific guidance, see ISEA Eye and Face Protection Selection Tool.[30]

Most individuals can successfully wear contact lenses to correct a vision problem while wearing a fullface RPD. OSHA originally prohibited the use of contact lenses with RPDs, but changed their position following advances in contact lenses technology and the results of a study of firefighters that failed to find problems when these RPD wearers wore contact lenses (Lawrence Livermore National Laboratory 1986). NIOSH in the 2005 Current Intelligence Bulletin (2005-139) (National Institute for Occupational Safety and Health 2005) recommends that workers be permitted to wear contact lenses when handling hazardous materials in conjunction with other protective eye wear or full facepiece RPDs.

***16.4.12.2 Facial Hair*** It has been previously mentioned that the presence of facial hair, the temple bars of eyeglasses, and piercings with various forms of jewelry may interfere with the seal or function of a tight-fitting RPD. Tight-fitting pressure-demand airline and SCBA devices are designed to keep a positive pressure inside the facepiece at all times. A review by Stobbe et al. found that the pressure in devices examined remained positive despite the presence of facial hair (Stobbe et al. 1988). However, variation among workers, the variety of devices, and the physical exertion required by a task cannot guarantee that facial hair will not interfere with the proper function of these devices. OSHA, NIOSH, and ANSI have taken the position that facial hair that comes into contact with the seal of a tight-fitting facepiece shall not be allowed (American National Standard Institute, American Society of Safety Engineers 2015).[31] A number of court challenges to employers who have required employees to be clean-shaven when assigned tight-fitting RPDs have found in favor of employers based on safety considerations. In a 2012 Standards Interpretation letter, Thomas Galassi, Directorate of Enforcement Programs, OSHA, noted: "The standard (29 CFR 1910.134) states that the employer cannot permit RPDs with tight-fitting facepieces to be worn by employees who have facial hair that comes between the sealing surface of the facepiece and the face. An employee whose records show a RPD wearer passing a fit-test with facial hair in the RPD sealing surface area is not considered to be compliant with the standard. The fit achieved with a beard or facial hair is unpredictable; it may change daily depending on growth of the hair and position of the hair at the time the fit is tested."[32] When relatively few employees are required to wear RPDs and when loose-fitting PAPRs are a possible choice, employers should consider this option as opposed to requiring that all employees be clean-shaven. This is especially true when RPD usage is infrequent. To eliminate problems with facial hair/beards and wearing tight-fitting RPDs, clear and concise details in conditions of employment eliminate any confusion when an employee raises the topic.

---

[30] https://safetyequipment.org/wp-content/uploads/2015/06/Eye-and-Face-Selection-Guide-tool1.pdf.

[31] Code of Federal Regulations, Title 29, Part 1910.134, "Respiratory Protection"; http://www.cdc.gov/niosh/npptl/RespStandards/ApprovedStandards/scba_cbrn.html.

[32] https://www.osha.gov/pls/oshaweb/owadisp.show_document?p_table=INTERPRETATIONS&p_id=28997.

***16.4.12.3 Voice Communications*** It is not unusual that the use of RPDs may significantly interfere with verbal communication with other employees as well as the use of other forms of electronic communication. RPD facepieces can be equipped with either a passive speaking diagram or an electronic system that uses microphones to improve verbal communication when this is critical to the job. Specific training in this area is very important so the RPD wearer understands how he/she is heard on the receiving end of the communication.

***16.4.12.4 Oxygen Deficiency*** An ambient workplace environment should contain approximately 20.9% oxygen that equates to a partial pressure of oxygen of 159 mm Hg ($PO_2$ 159 mm Hg) at sea level. As part of the hazard assessment recommended by ANSI Z88.2 (2015) the RPD program manager should determine if there is a potential for the oxygen concentration to be below 20.9% oxygen (i.e. atmospheric $PO_2 < 159$ mm Hg). If the potential exists, the oxygen concentration should be measured, and if it is found to be below 20.9%, the cause of this deficiency must be identified and understood. Where an oxygen concentration is confirmed to be below 19.5% (at sea level), the cause of deficiency shall be determined and the environment shall be classified as oxygen-deficient IDLH. OSHA does not use the term "oxygen-deficient IDLH atmosphere" in their RPD standard. Per the preamble, OSHA augments this oxygen deficiency policy by using the ANSI Z88.2-1980 oxygen-deficient IDLH atmosphere cutoff point, which is "an oxygen partial pressure of 100 mm Hg or less in the freshly inspired air in the upper portion of the lungs, which is saturated with water vapor." The ANSI Z88.2-1980 definition of oxygen-deficient IDLH ($PO_2 \leq 100$ mmHg in the lungs) equates to a dangerously low level of oxygen in the human body (alveolar blood has only 90% oxygenated hemoglobin). Oxygen deficiency symptoms become very noticeable below this blood oxygenation level. As presented above OSHA actually has two levels of oxygen deficiency ($O_2$-deficient and $O_2$-deficient IDLH) but chose not to establish two terms in the OSHA standard to address these oxygen levels to avoid confusion. OSHA states in 1910.134(d)(2)(iii) that all oxygen-deficient atmospheres shall be considered IDLH. For entry into all IDLH atmospheres, OSHA requires either an SCBA or a combination airline RPD with auxiliary escape bottle, both of which must be full face and pressure demand. Looking closer at the OSHA $O_2$ deficiency policy, it allows for exceptions when an SCBA must be worn for entry into atmospheres below 19.5% oxygen. OSHA allows any atmosphere-supplying RPD to be worn when oxygen is strictly controlled within the $O_2$ concentration equivalent to 16.0 to <19.5% oxygen at sea level ($PO_2$ 122 to <148 mm Hg). However, according to the 1998 preamble, OSHA believes that employers will only rarely have occasion to avail themselves of the exception noted above because most atmospheres with oxygen content between 16 and <19.5% are not well controlled and a drop in oxygen content could have severe consequences. For additional information on RPD use in oxygen-deficient atmospheres, refer to References American National Standard Institute, American Society of Safety Engineers (2015) and Spelse et al. (2016a, b).

***16.4.12.5 Cold Environments*** Cold weather may cause RPD facepieces to fog, prevent valves from seating properly, or keep facepieces from being supple and forming a good face seal. Nose cups can be installed in full facepiece RPDs and special compounds can be used to minimize the facepiece lens from fogging.

It was previously mentioned that components used in airline RPDs and SCBAs could freeze if there is sufficient water vapor in the compressed breathing air being used. To prevent this, the dew point of the compressed air should be well below (10–20 °C) the lowest ambient temperatures that will be encountered.

## 16.5 CONCLUSION

If you select RPDs for worker protection, it is very important to remember the significance of this decision. Historically we can quickly identify occupational disease, shortened life spans, and premature death of workers when proper attention was not paid to the requirements of the RPD selection process. In the United States proper respiratory protection is achieved through a partnership between regulators, RPD manufacturers, employers, and workers. Each member of the partnership has specific responsibilities outlined in the following federal regulations, OSHA 29 CFR Part 1910.134 and NIOSH 42 CFR Part 84, and supplemented by national consensus standards such as ANSI and NFPA. As in any partnership, the overall effectiveness is based on partners knowing and carrying out their responsibility. This chapter has provided an overview of the OSHA and NIOSH respiratory protection regulations as well as voluntary national consensus standards such as ANSI Z88 and NFPA that provide additional guidance on various topics in these regulations. Descriptions and details of the general classes of RPDs currently available have been provided for air-purifying RPDs, airline RPDs, SCBA, and escape-only RPDs. The importance of the RPD program has been presented and the following program elements have been described: program administration; written SOPs; hazard assessment; selection; tight-fitting RPD fit testing; use; maintenance and storage; breathing air quality, quantity, and flow; training; audits; program records; problem areas; and glossary. The need for health and safety professionals to pay close attention to the proper use of RPDs is clearly illustrated by a review of OSHA's top 10 annual citations for the past 10 years, 2005–2015. Respiratory protection citations have been in the top five for all 10 years most frequently residing at the number four position. Fines can be significant, especially with the 78% increase in penalties effective 1 August 2016, but the long-term concern of worker exposure to toxic agents and related occupational disease and death is the real reason a proper RPD program should be implemented and maintained any time RPDs are used in the workplace.

## GLOSSARY OF TERMS*

**Abrasive blasting respiratory protective device:** Device designed to protect the wearer from inhalation of, impact of, and abrasion by materials used or generated in abrasive blasting.
**Aerosol:** Suspension of solid, liquid, or solid and liquid particles in a gaseous medium.
**Aerosol penetration:** Ability of particles to pass through a particle-filtering material.
**Airline respiratory protective device:** An atmosphere-supplying RPD in which the respirable gas is not designed to be carried by the wearer.
**Air-purifying respiratory protective device:** An RPD in which ambient air is passed through an element (filter, cartridge, or canister) and the contaminant(s) of concern is removed.
**Assigned protection factor (APF):** The minimum expected workplace level of respiratory protection that would be provided by a properly functioning and used class of respiratory protective devices to properly fitted and trained wearers when all elements of an effective RPD program are established and are being implemented.
**Atmosphere-supplying respiratory protective device:** A device that supplies the RPD user with breathing air from a source independent of the ambient atmosphere.
**Atmospheric dew point:** Temperature at which moisture begins to condense from a gas as the gas is cooled at standard atmosphere.
**Canister or cartridge:** A container with a filter, sorbent, or catalyst or combination of the three, which remove specific contaminants from the air passed through the container.
**Continuous-flow respiratory protective device:** An airline or other atmosphere-supplying RPD that provides for a continuous flow of breathable air into the facepiece.
**Demand respiratory protective device:** An airline or other atmosphere-supplying RPD that allows breathing air into the facepiece only during inhalation and creates a negative pressure with respect to ambient conditions.
**Doffing:** Process of removing or taking off the RPD effectively.
**Donning:** Process of putting on the RPD effectively.
**Disposable respiratory protective device**: An early term for filtering facepiece RPDs intended to be discarded and not maintained.
**Escape-only respiratory protective device:** An RPD designed for emergency use to allow the wearer to escape from a hazardous environment. It is not intended for routine use or for entering hazardous environments.
**End-of-service-life indicator:** A system that indicates the end of the overall life of an RPD.
**Filter or air-purifying element:** An RPD component used to remove aerosols.
**Filtering facepiece respiratory protective device:** A negative-pressure particulate air-purifying RPD with a filter as an integral part of the facepiece or with the entire facepiece composed of the filtering media.
**Fit factor:** A numeric expression of how well a tight-fitting RPD fits the wearer during a fit test.
**Fit test:** Use of a challenge agent and specific protocol to qualitatively or quantitatively determine an individual's ability to obtain an adequate seal with a specific make, model, and size of an RPD.
**Gas:** A fluid that is in a gaseous state at standard temperature and pressure that expands to occupy the space of the enclosure in which it is confined.
**High-efficiency particulate air filter (HEPA):** A filter that will remove at least 99.97% of monodispersed particles of 0.03 μm in diameter. The equivalent NIOSH 42 CFR Part 84 particulate filters are the N100, R100, and P100 filters.
**Hood:** A respiratory inlet covering that completely covers the head and neck and may cover portions of the shoulders.
**Immediately dangerous to life and health (IDLH):** Atmosphere that poses an immediate threat to the life, would cause irreversible adverse health effects, or would impair an individual's ability to escape from a dangerous atmosphere.
**Loose-fitting facepiece:** An RPD inlet covering that is designed to form a partial seal with the face.

* International Standards Organization (ISO) 16972 "Respiratory protective devices – Terms, definitions, graphical symbols and units of measurement" provides additional source of respirator terminology and definitions (ISO 2010)..

**Maximum use concentration (MUC):** The MUC of a hazardous substance for which an employee can be expected to be protected when wearing an RPD and is determined by the assigned protection factor of the RPD or class of RPD and the exposure limit of the hazardous substance. The MUC can be determined mathematically by multiplying the assigned protection factor specified for an APR by the required OSHA permissible exposure limit, short-term exposure limit, or ceiling limit. When no OSHA exposure limit is available for a hazardous substance, an employer must determine an MUC on the basis of relevant available information and informed professional judgment.
**Negative-pressure protective device:** An RPD that becomes negative with respect to ambient air pressure outside the RPD during inhalation.
**Partial pressure:** Pressure of a gas in a mixture equal to the pressure that it would exert if it occupied the same volume alone at the same temperature.
**Particle filter efficiency:** Degree to which a filter removes aerosols from the ambient atmosphere.
**Physician or other licensed healthcare professional (PLHCP):** An individual whose legally permitted scope of practice (i.e. license, registration, or certification) allows him or her to independently provide or be delegated the responsibility to provide some or all of the healthcare services required by the medical surveillance program element.
**Positive-pressure respiratory protective device:** An RPD in which the pressure inside the RPD inlet covering exceeds the ambient air pressure outside the RPD.
**Powered air-purifying respirator (PAPR):** An air-purifying RPD that uses a mechanical blower to force ambient air through air-purifying elements to the inlet covering.
**Pressure-demand respiratory protective device:** A positive-pressure atmosphere-supplying RPD that is specially designed to provide additional air during inhalation to maintain a positive air pressure greater than ambient atmosphere.
**Program protection factor:** An estimate of the respiratory protection provided to a worker in the context of a specific RPD program.
**Qualitative fit test:** A pass/fail method that relies on the subject's sensory response to detect a challenge agent in order to assess the adequacy of the RPD fit.
**Quantitative fit test:** A test method that uses an instrument to assess (quantify) the amount of face seal leakage into the RPD in order to assess the adequacy of its fit.
**Respiratory protective device (RPD):** A personal device designed to protect the wearer from the inhalation of hazardous atmospheres.
**Respirator inlet covering:** The portion of the RPD that protects the wearer's breathing zone. This may be a facepiece, hood, suit, or mouthpiece.
**Self-contained breathing apparatus (SCBA):** An atmosphere-supplying RPD in which the respirable air is carried by the wearer.
**Sorbent:** A material that is contained in a cartridge or canister and that removes specific gases and vapors.
**Tight-fitting facepiece:** An RPD inlet covering designed to form a complete seal with the face.
**User seal check:** An action conducted by the RPD user to determine if the RPD is properly seated to the face.
**Vapor:** The gaseous phase of a substance that normally is a liquid or solid at room temperature.
**Workplace protection factor (WPF):** A measure of the protection provided in the workplace, under the conditions of that workplace, by a properly selected, fit-tested, and functioning RPD while it is correctly worn and used.

## REFERENCES


American Industrial Hygiene Association (1989). *Odor Threshold for Chemicals with Established Occupational Health Standards*. Fairfax, VA: American Industrial Hygiene Association.

American National Standard Institute, American Industrial Hygiene Association, American Society of Safety Engineers (2006). *Respirator Use—Physical Qualifications for Personnel*, ANSI/AIHA/ASSE Z88.6-2006. Park Ridge, IL: American Industrial Hygiene Association.

American National Standard Institute, American Industrial Hygiene Association, American Society of Safety Engineers (2010). *Respirator Fit Test Methods*, ANSI/AIHA/ASSE Z88.10-2010. Park Ridge, IL: American Industrial Hygiene Association.

American National Standard Institute, American Industrial Hygiene Association, American Society of Safety Engineers (2010). *Color Coding of Air Purifying Respirator Canisters, Cartridges and Filters*, ANSI/AIHA/ASSE Z88.7-2010. Park Ridge, IL: American Industrial Hygiene Association.

American National Standard Institute, American Society of Safety Engineers (2015). *Practices for Respiratory Protection*, ANSI/ASSE Z88.2-2015. Park Ridge, IL: American Society of Safety Engineers.

American National Standard Institute, ANSI Z88.2-1969 (1969). *Practices for Respiratory Protection*. New York: ANSI.

American National Standard Institute, ANSI Z88.2-1980 (1980). *Practices for Respiratory Protection*. New York: ANSI.

American National Standard Institute, ANSI Z88.2-1992 (1992). *Practices for Respiratory Protection*. New York: ANSI.

American National Standard, International Safety Equipment Association (2015). *American National Standard for Occupational and Educational Personal Eye and Face Protection Devices*, ANSI/ISEA Z87.1-2015. Arlington, VA: International Safety Equipment Association.

American Society for Testing and Materials International (2012). *Standard Test Method for Man-In-Simulant Test (MIST) for Protective Ensembles*, ASTM F2588-12. West Conshohocken, PA: ASTM.

American Society for Testing and Materials International (2014). *Standard Test Method for Pressure Testing Vapor Protective Suits*, ASTM F1052-14. West Conshohocken, PA: ASTM.

Canadian Standards Association (2013). *Z180.1-13 - Compressed Breathing Air and Systems*. Toronto, ON: CSA Group.

Compressed Gas Association (2011). *Commodity Specification for Air*, CGA Specification G-7.1 (ANSI Z86.1-2011). Arlington, VA: Compressed Gas Association.

Gummaria, C. et al. (2015). *Developing a Performance Standard for Combination Respirators: Workshop in Brief*. Washington, DC: The National Academies Press.

Hack, A.L. (1978). Respirator protection factors: Part I – development of an anthropometric test panel. *Am. Ind. Hyg. Assoc. J.* **39** (12): 970–975.

Hyatt, E.C. (1976). Respiratory Protection Factors. Los Alamos Scientific Laboratory Report No. LA-6084-MS, January 1976.

International Organization for Standardization (2010). *Respiratory Protective Devices – Terms, Definitions, Graphical Symbols and Units of Measurement*, ISO 16972. Geneva, Switzerland: ISO.

Lawrence Livermore National Laboratory (1986). *Is it Safe to Wear Contact Lenses with a Full Facepiece Respiratory?* Livermore, CA: Lawrence Livermore National Laboratory.

Metzler, R.W., Spelce, D., Johnson, J.S., and Rehack, T.R. (2016). Effective respiratory protection – achieved through partnership, legislation, regulation and American National Consensus Respirator Standards. *J. Int. Soc. Resp. Prot.* **33** (1): 39–46.

National Fire Protection Association (2013). *Standard on Open-Circuit Self-Contained Breathing Apparatus (SCBA) for Emergency Services*, NFPA 1981-2013. Quincy, MA: National Fire Protection Association.

National Fire Protection Association (2013). *Standard on Selection, Care, and Maintenance of Open-Circuit Self-Contained Breathing Apparatus (SCBA)*, NFPA 1852-2013. Quincy, MA: National Fire Protection Association.

National Fire Protection Association (2013). *Standard on Breathing Air Quality for Emergency Services Respiratory Protection*, NFPA 1989-2013. Quincy, MA: National Fire Protection Association.

National Fire Protection Association (2016). *Standard on Respirators for Wildland Fire Fighting Operations*, NFPA 1984-2016. Quincy, MA: National Fire Protection Association.

National Institute for Occupational Safety and Health (2005). *Contact Lens Use in a Chemical Environment*, Current Intelligence Bulletin, vol. **59**, DHHS (NIOSH) 2005-139. Cincinnati, OH: Department of Health and Human Services, Centers for Disease Control, National Institute for Occupational Safety and Health.

National Institute for Occupational Safety and Health, Occupational Safety and Health Administration (2015). *Hospital Respiratory Protection Program Kit*, DHHS (NIOSH) 2015-117. Atlanta, GA: DHHS (NIOSH) Publisher.

Occupational Safety and Health Administration (2011). *Small Entity Compliance Guide for the Respiratory Protection Standard*, OSHA3348-09. Washington, DC: U.S. Department of Labor.

Spelse, D.L., Metzler, R.W., Johnson, J.S., and Rehak, T.R. (2016a). Respirator selection for oxygen-deficient atmospheres. *The Synergist* **27** (3): 28–31.

Spelse, D.L., Metzler, R.W., Johnson, J.S., and Rehak, T.R. (2016b). Respirator protection for oxygen deficient atmospheres. *J. Int. Soc. Resp. Prot.* **33** (2): 1–21.

Stobbe, T.J., da Roza, R.A., and Watkins, M.A. (1988). Facial hair and respirator fit: a review of the literature. *Am. Ind. Hyg. Assoc. J.* **49** (4): 199–203.

Woods, G.O. (2015). Correlating and extrapolating air-purifying respirator cartridge breakthrough times: a review. *J. Int. Soc. Resp. Prot.* **32** (1): 23–36.

Yant, W.P. (1933). Bureau of mines approved devices for respiratory protection. *J. Ind. Hyg.* **15**: 473–480.

Zhuang, Z., Bradmiller, B., and Shaffer, R.E. (2007). New respirator fit test panels representing the current civilian work force. *J. Occup. Environ. Hyg.* **4** (9): 647–659.

## CHAPTER 17

# HOW TO ESTABLISH INDUSTRIAL LOSS PREVENTION AND FIRE PROTECTION

PETER M. BOCHNAK

*Environmental Health and Safety, Harvard University, 46 Oxford St., Cambridge, MA, 02138*

The information provided in this chapter was current and reliable at the time of publication. Regulatory and technical requirements are subject to change; hence, the information detailed herein is only in compliance with current regulations in effect at the time of publication. Consult OSHA regulations or seek legal counsel for accurate information on current requirements. The writer disclaims any liability resulting from reliance on obsolete or other inaccurate information presented in this chapter.

### 17.1 INTRODUCTION

Today, with the real presence of the Occupational Safety and Health Administration (OSHA) and the employee's "right" to safety and health, coupled with the "age of risk management," captive insurance companies, and large self-retention of risk by major corporate entities, there is a greater need for management to familiarize itself with the general field of fire protection. In the past, there appears to have been a general feeling that the protection of plants and facilities from fire and the maintenance of their integrity could be safely left to insurance underwriter personnel or perhaps to a very small staff or an individual who might have something less than a heavy background in this area. Now, with greater retention of risk by corporations and the tremendous investments involved in many facilities, there is a need for many more people, at the very least, to have more than a superficial knowledge of the world of loss prevention.

This need has impacted not only upon plant management but also upon architects and engineers. Many architectural engineering firms have found it prudent to employ specialists in fire protection on their staffs to handle what should be considered a very important part of the overall plant design. It is important for these technical people to develop some knowledge of fire protection, since many firms of this type are not only providing fire protection concepts but in many cases are doing the actual design, although they may not possess a high degree of fire protection expertise.

Risk management, with its attempts to bring a degree of sophistication to decisions on handling and preserving corporate assets, must by necessity begin with personnel who have the knowledge of safety and fire protection necessary to evaluate the risks facing a corporation.

Where does the risk manager turn – or, for that matter, the plant manager, technician, architect, or engineer who needs to know something about fire protection? Many publications are available on this very broad subject that, for the most part, are directed at those having more than a causal interest in this field. Many of these publications are voluminous and represent complete works either on the entire scope of fire protection or on specific areas. For the professional, a library of these publications is indispensable.

However, turning back to the question as to where the busy manager should turn for some background, there appear to be few publications that give a relatively short treatment or an overview on this subject. So we now come to the objective of this chapter, which is to provide the manager, the architect, the plant engineer, the technician, and others with a background in safety and fire protection in as brief and concise a manner as possible and then point out other sources available for more in-depth information.

*Handbook of Occupational Safety and Health*, Third Edition. Edited by S. Z. Mansdorf.

This chapter is, therefore, specifically directed to management-oriented personnel with an involvement in loss prevention, to technicians who are not engineering oriented but who nevertheless must have a relatively strong background in safety, and to the architects and engineers involved with overall plant design.

### 17.1.1 Basic Process Safety

The integration of safety into any plant, especially on processing toxic or flammable liquids or gases, involves several disciplines. The initial requirement is to determine the toxicological and flammability properties of the products. The next is to determine whether the products can be processed safely. These determinations may require review by a toxicologist, industrial hygienist, process engineer, equipment design engineers, an instrument engineer, and a safety engineer, among others.

### 17.1.2 Equipment Safety

After the process has been approved as safe, then safety must be built into the plant hardware. Pressure vessels, pumps, fired heaters, and other equipment must be designed to withstand the operating conditions, which include abnormal upsets such as exothermic reactions, failure of instrument controls, and human failures. In anticipation of such failures, safety alarms and shutdown controls must be designed into the process. Wherever possible, automatic safety shutdowns should fail in the safe position. Where practical, backup manual systems should be provided.

### 17.1.3 Operator Safety

Next, operators must be trained in the safe operation of the unit, on how to detect incipient equipment problems, and finally in what action to take when an emergency arises. Procedures must be written for normal plant operation and also for any anticipated emergency. These procedures should be available to all affected personnel and periodically reviewed and updated to reflect changing conditions. Management should publicize to plant personnel that they support these procedures and take the necessary steps to assure that the procedures are being practiced.

### 17.1.4 Safety Audits and Training Programs

In addition to day-to-day safe plant operations, periodic safety audits should be made and firefighters be kept trained to respond to any emergency (see Chapter 12 for a discussion of hazard analysis and Chapter 19 for a discussion on program audits). An emergency response plan should be developed utilizing plant and municipal medical staff, plant security staff, and mutual aid provided by other local industries. The emergency response plan should be tested periodically to ensure effectiveness. Safety cannot be initially designed into a plant and then forgotten. Maintaining safety is an ongoing daily struggle. Even when it is maintained to the highest degree, there is no guarantee that accidents will not occur. However, the converse is true: *If safety is not observed, accidents are inevitable.* Plants for the most part are still run by people. Even if plants become completely automated with robots and controlled by computers, people will still be required to assure safety.

## 17.2 RISK

### 17.2.1 Definition

A risk is the possibility of an undesirable occurrence. The only justification for taking a particular risk is that the reward clearly exceeds the penalty if the associated accident takes place. In order to decide whether to take a particular risk, it must be quantified carefully.

### 17.2.2 Quantification

To quantify a risk, it is necessary first to determine what the risk is and the extent of damage in a worst-case scenario. Next it is necessary to determine the probability of the accident occurring. In evaluating the probability of an event, there are *two current popular hypotheses:* (i) A disaster that never happened is about to happen. (ii) A disaster that never happened won't happen. Both of these hypotheses are false. Three Mile Island proved that a nuclear reactor meltdown was possible, and Chernobyl made it a reality. A likely risk should not be accepted unless the penalty of the incident is also accepted. If the loss of property and production is acceptable, then the risk is acceptable. Of course, this also assumes that all reasonable precautions have been taken to eliminate or mitigate the penalty. Pure chance taking is never good business practice, and when the risk involves the possible loss of life, then if should not be accepted.

### 17.2.3 Methods for Risk Quantification

In order to quantify a risk, some of the following tools can be used:

*Past experience.* What has caused an incident, how often can it be expected to occur, and what is the expected extent of damage? To make this determination, a database should be constructed from experience within the plant or company and throughout identical and similar industries. This is a costly, time-consuming procedure, and where very few incidents have occurred, the validity of the information may be questionable. Some data on the reliability of instruments appear in Anyakora et al. (1981). Probabilities for equipment failure can be found in Browning (1969).

*Logic models.* These are concerned primarily with the "on/off" connection between variables and incidents and are useful for analyzing complex systems.

Two common techniques used in logic models are failure modes and effects analysis (FMEA) and fault tree analysis (FTA).

FMEA breaks down a system into its simplest components and asks "what-if" questions regarding all possible failure modes for each component. The probability of a failure occurring in the system is the sum of the failure probabilities for each component. The advantages of FMEA are that it highlights components with the highest probability of failure and permits improving these components. FMEA results in a systematic piece-by-piece review of each component and permits establishing inspection and test programs to monitor the performance of high-failure components.

FTA, which goes further than FMEA in that it identifies failures in subsystems, is often used in conjunction with FMEA. It determines the consequence of two or more component failures occurring simultaneously.

A more detailed discussion of these forms of hazard analysis is provided in Chapter 12.

## 17.3 HAZARD IMPACT

### 17.3.1 Identifying Potential Hazards

As stated previously, the risk-taking decision should be based not only on the probability of an incident occurring but also on the potential impact or consequences of the incident. *Consequence analysis* can be the most difficult phase of overall hazard analysis. Some of the most serious hazards involve the formation of toxic and flammable vapors. It is often extremely difficult to determine the maximum quantity of flammable vapor that may be involved in an explosion or the maximum quantity of a toxic release that could be expected. As an example, the percentage of a propane gas release that can be in the flammable range at any given time is 2.15–9.60% by volume in air. Thus the assumed amount of explosives could be incorrect by a factor of 4.5. For this example, one method of calculating peak overpressure of an explosion is to determine TNT equivalent based on the heat of combustion per pound of the chemical, related to TNT. This is a rough correlation, but is used because more is known about the pressure waves and destruction caused by the detonation of TNT than that of any other explosive.

### 17.3.2 Consequence Analysis

From the calculated extent and intensity of potential damage, the impact on both plant/personnel and the surrounding community can be determined. Using these data, plans can be formulated to cope with the results. The determination of the impact on plant and personnel should consider the following factors:

- The number of people exposed in the plant onsite areas and in administration buildings, shops, warehouses, laboratories, etc., both during the day and at night.
- The extent of property damage within the plant.
- In-plant and outside first aid response.
- The time required for cleanup and repair or replacement of damaged equipment and the lost production cost.

- The effect on employees who may lose work if a portion of the plant is shut down for an extended period.
- Backlash legislation, that is, the passing of local, state, or federal laws that would increase the cost of doing business. An example is mandating additional safety features that are not necessary, forcing the plant to relocate or shut down.

In determining the impact of an incident on the surrounding community, the following factors should be considered:

- The maximum number of people outside the plant exposed to a hazard during the day and night.
- The type of people exposed and the probability of panic, i.e. adults, children, senior citizens, and handicapped.
- Medical response: the time required to evacuate injured as casualties increase.
- Extent of property damage outside plant, including lost production and lost work time.
- Damage to the company's overall public image.
- Passage of laws that could cause considerable added cost without appreciable added safety.
- Permanent loss of some markets from boycotts and lost production.
- Personal injury and damage suits.

## 17.4 ELEMENTS OF A LOSS CONTROL PROGRAM

As new technologies are developed, hazards may be introduced, and when industry and business expand, existing hazards may increase. To remain prosperous, a corporation must minimize the possibility of casualties, interrupted production, and property losses caused by fires and explosions. Corporations are also required by law (OSHA 29 CFR 1910) to provide a reasonably safe workplace for their employees. To satisfy these responsibilities and also remain competitive, corporations must eliminate hazards where possible and minimize those that cannot be eliminated.

### 17.4.1 Policy Statement

A program of identification, evaluation, and control of hazards is known as a *loss control program*. For this program or any program involving the welfare of a corporation or its employees to be effective, the sincere and total support and the sustained interest of top management are essential. This is particularly true for all corporate loss control programs. In addition to top management support, this policy of controlling losses must be communicated to lower levels of management and to all employees. This can be accomplished during the formation of the loss control program by a written policy statement from the chief executive officer demonstrating his or her complete support of the program; outlining the procedures, objectives, responsibilities, and accountabilities; and further indicating his or her desire that all employees of the organization support those responsible for the formation and implementation of the loss control organization and function. This shows that management cares – that management is directly involved.

The objective of the fire loss control program should be fully stated, with emphasis on the protection of employees against injury and the conservation of corporate assets. Both management and labor benefit from the responsible safeguarding of profit centers and the dependable continuity of operations. This policy should emphasize to all employees that major damage to a facility by fire or explosion in most cases caused temporary or total loss of jobs.

### 17.4.2 Loss Control Organization

Some form of a loss control organization should be developed to implement any corporate loss control program. No definitive guidelines can be established for the simple reason that such an organization could vary from a very informal structure to one that is highly structured.

Under any loss control program it is important to have some control of these functions at the corporate level and to assign line responsibilities at the division and plant levels. Figures 17.1 and 17.2 indicate suggested loss control organizational functions for (i) a large, multidivision, multiplant facility with a corporate loss prevention staff and (ii) a similar organization not employing fire protection personnel, but depending upon outside sources, principally the technical services of insurance brokerage firms.

The loss control manager could be known by a variety of other names, depending on the industry – for example, fire loss control manager, fire loss prevention and control manager, or property conservation director. In some industries, the loss

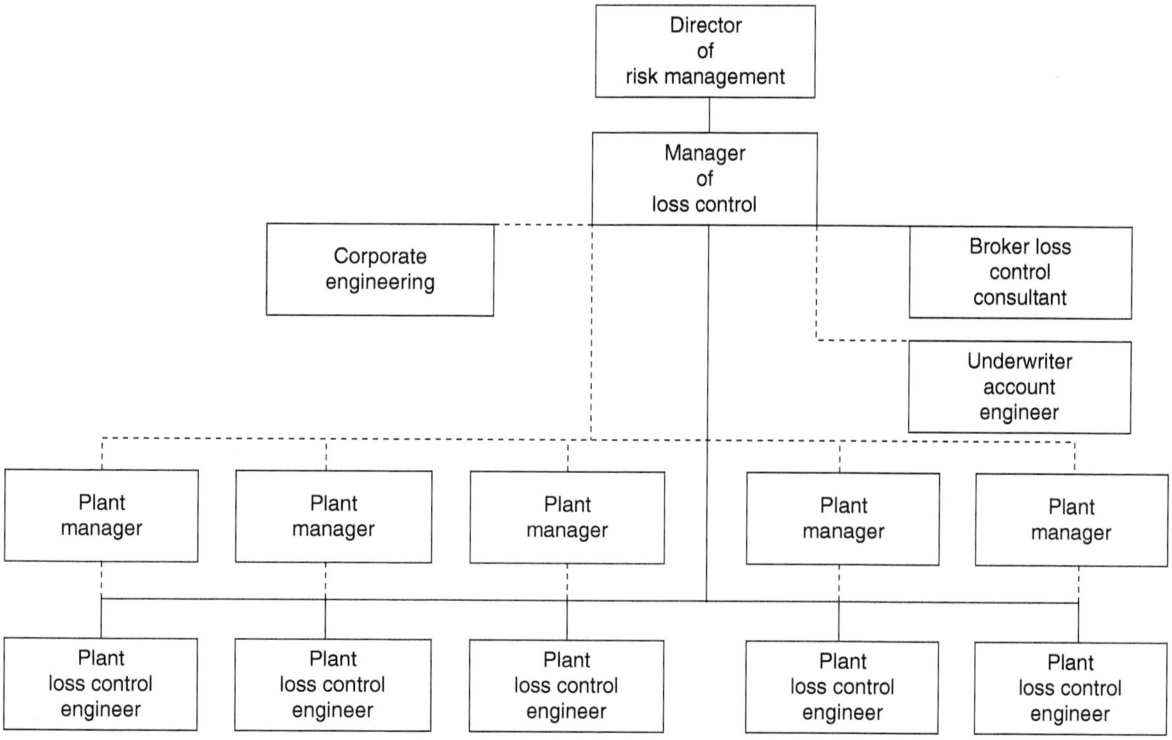


**Figure 17.1** A suggested loss control organization for a large multidivision corporation with a corporate loss prevention staff.

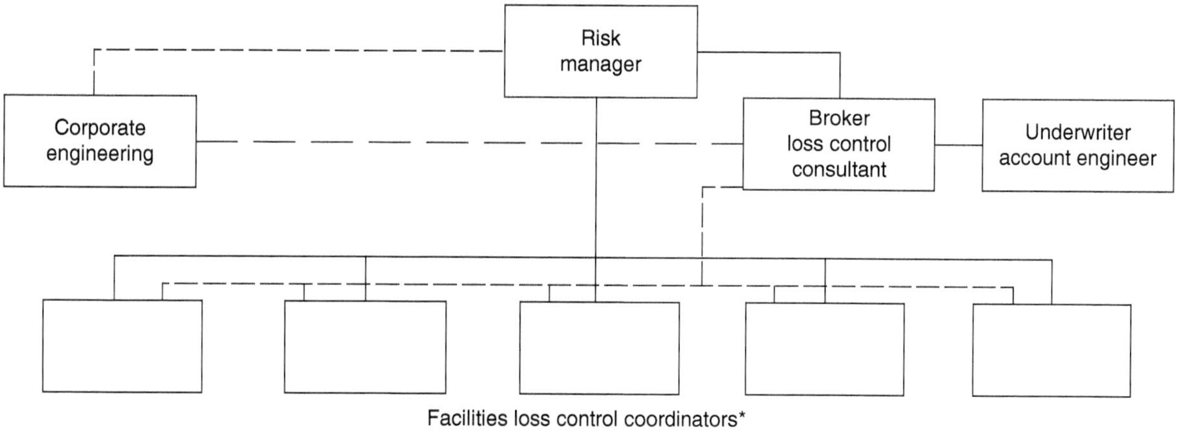


**Figure 17.2** A suggested loss control organization for a large corporation depending principally upon others for loss control efforts. The coordinators (*) will normally have principal responsibilities other than loss prevention. Ideally, coordinators will be members of facilities managers' staffs.

control manager reports to the facility manager and is responsible for employee safety and health management, fire protection management, property insurance management, and security management.

Basically, as a minimum, there should be some in-house loss control organization coupled with the assistance of outside sources, such as consulting fire protection and safety engineers and industrial hygienists, insurance brokerage personnel, and insurance company inspectors and engineers.

### 17.4.3 Identification

Once a loss prevention organization has been developed, or even if such an organization is not yet in effect, an audit of the facilities should be conducted. This audit is suggested as a first step in identifying hazards so that they can be evaluated and so that priorities can be established for the concentration of efforts in the loss control program.

Quite often, the corporate staff will require assistance in the evaluation of their facilities from outside sources. This assistance can be from the insurance personnel, either in a detailed review of property loss prevention reports received from insurance sources with some special visits to particular sites or from technical services personnel of insurance brokers. Brokers will supply this service using either their reports or the services of their control personnel accompanying corporate personnel in the physical inspection of various sites. An audit form should be developed to be used in the loss control program. A suggested form is shown in Figure 17.3. In addition to identifying the problems at the facility, the audit form can be of considerable value when kept on file at a corporate level for answering questions that frequently develop concerning construction, utilities, various occupancy details, etc.

Identification of hazards and their probability is facilitated by the following:

*Experience with similar units.* Some chemical processes have an inherent hazard such as high pressure or temperature, an exothermic, or "runaway," type of reaction, involve the formation of a toxic material, or use hazardous reactants such as corrosive acids.

*Plant accident reports.* Most companies maintain plant accident reports, which may provide a good reference for anticipating hazards and estimating probability. Many companies are computerizing these accident reports to facilitate access and compile probability indices.

*Step-by-step review of process.* A detailed review of each step of the process should be made by following through process flow and piping and instrument drawings. This review can be more effective if made with the process engineer familiar with the process and the operator familiar with operations.

*Seminars and meetings.* Seminars and meetings periodically conducted by the trade organizations and professional societies such as the American Petroleum Institute, the National Petroleum Refiners Association, the American Society of Mechanical Engineers, American Institute of Chemical Engineers, and the Society of Fire Protection Engineers (SFPE) and American Society of Safety Engineers (ASSE). These organizations provide access to a cross section of experiences and practices throughout similar industries, and each has a journal such as *Professional Safety* from ASSE and *Journal of Fire Protection Engineering* from SFPE.

*Published papers.* Papers published in journals often identify problems at the inception stage. This permits correction of problems without having to repeat the "learning experience."

*Insurers.* The insurer of a particular plant usually has compiled a history of losses with documentation of event causing these losses of all the sites insured by the company. Thus the insurer's experience may point out similar problems that occurred in other plants.

### 17.4.4 Evaluation

Once the information has been collected through the use of the audit program identifying potential fire hazard areas, it is important to develop and improve the recommendations list based upon the following priorities in descending order of importance:

1. *Life safety exposures.* The potential for fire casualties in industrial types of occupancies is directly related to the hazard of the operations or processes. Most of the multiple fatalities are a result of (i) flash fires in highly combustible materials or (ii) explosions of dusts, flammable liquids and vapors, and gases.
2. *Continuity of operations.* This includes hazards in areas that will have a significant effect upon the continuity of operations in the event of their loss, thus affecting possible corporate income and loss of corporate resources. Management is extremely conscious of financial return on investment, and those locations that affect production, such as computer facilities controlling production processes, should be highlighted.
3. *Other hazard areas.* This includes problems of protection from hazards not necessarily of a business continuity nature such as woodworking or machine shops.

Frequently, large firms will need numerous and similar facilities such as distribution locations, theaters, warehouses, etc. Where these situations are encountered, it may be advisable to develop corporate loss control guidelines outlining structural, building services, and protection requirements as related to property loss. The desirability of incorporating guidelines of this nature may be highlighted by an examination of past loss experience, which may show a need for improvement.

While the prime responsibility for the development of the suggested guidelines will rest with the loss control personnel, the project should be coordinated with both operating and facilities personnel. Input and approval should be obtained from all those involved, particularly those having architectural or engineering responsibilities so that the guidelines will carry the necessary weight and will command the respect and attention of both corporate employees and outside firms who may have some interface with the guidelines.

XYZ Corporation
Facilities Audit Form*

Location: Date:

Construction

Walls: Roof:

Floors: Partitions:

Unusual features (combustible interior finishing, insulation, etc.):

Boilers

Description and rating:

Fuel:

Combustion controls (including details on interlocks, purge cycle, flame safeguard, etc.):

Controls testing program:

Electrical

Power supply (including capacity, no. of feeders, etc.):

Transformers: No. Capacity

Major motors: HP Spares?

Hazardous electrical equipment (Class I or II):

Emergency generators:

Data processing: Description

Functions

Location and envelope construction

Tape storage

Air conditioning

Protection

Detectors

Protection:

Computer area

Underfloor

Tape library

Hazardous operations (flammable liquids, dust, etc.):

Plant protection

Water supplies (include water test data):

Underground mains and valving:

Automatic sprinklers: Full Partial Type

Design

Alarms

Special protective systems ($Co_2$, halon, etc.):

Security

Watchman service: Alarms:

Environs

Flood: Seismic zone:

Human-element programs (give date of programs)

Fire brigade:

Self–inspections:

Emergency planning:

Cutting and welding (hot work):

*Attach plan of facility.

**Figure 17.3** A form suggested for use in an audit of corporate facilities.

The components of protection guidelines should include:

1. *Construction details.* The construction details include those having a bearing on the structure's ability to withstand a fire or to provide fuel in a fire. This may involve the specific design of a desired fire resistance rating that may be important guidelines for locations; for example, specifications for insulated metal-deck roofs so that the roof might be classified as noncombustible rather than combustible and specification of mechanical fastening of the rigid insulation of a metal-deck roof at the roof perimeter to avoid windstorm losses.
2. *Safe installation of building services.* This will include the location and enclosure of switch gear; the location, types, and protection of transformers; the designation of wiring means; the designation of combustion controls for heating and processing equipment and the enclosure of this equipment; the proper installation of flues; the proper installation of air-conditioning systems for sensitive areas such as those in housing computers.
3. *Required fire protection features.* This could take the form of specifying the hydraulic design requirements for a hydraulically calculated sprinkler system and engineering requirements for similar installations; for theaters, the protection of storage areas and projection rooms.

When formulating loss control recommendations and guidelines, an important question to ask is "What loss expectancy can be tolerated?" This is particularly significant when the trend is toward higher insurance deductibles and self-retention of losses. When considering loss expectancy, particularly for lesser value but numerous locations such as supermarkets or theaters, the desirability of the formulation of loss control guidelines takes on significant importance.

### 17.4.5 Control

***17.4.5.1 Human Elements*** Despite excellent engineering and a good loss control program at the corporate level, problems at facilities and poor loss experience can still exist if the facility manager or first-line supervisor is not involved in the overall effort. The human elements of a loss control program refer to the continuing actions of personnel at the local level in the loss prevention and control effort in the formulation and updating of emergency planning procedures; in the constant readiness of the emergency organization (fire brigade); in the conduct of self-inspection programs; in the exercise of an established impairment notification system or fire protection system shutdowns; in the use of a cutting, welding, and a hot-work permit system; in the limiting of smoking to designated areas; and in good housekeeping procedures. The implementation of human-element programs does not involve extensive expenditures. Such programs provide for greater safety and have the further benefit of making local employees more aware of their environment, coupled with a feeling of being a part of the action.

***17.4.5.2 Hazard Elimination*** After a hazard has been identified, the potential results should be assessed to determine which risks are acceptable and which are not. Risks involving human life and health are not acceptable. Some risks involving property only may be acceptable to a limited degree. Providing no fire protection for a plant and letting it burn down on the premise that it is fully insured and can be rebuilt is *not* an acceptable risk. This is true even when economics are against fire protection.

Some methods used to eliminate or control hazards are as follows:

1. *Changing the process conditions where possible.* For instance, reducing the pressure or temperature, substituting a less hazardous material for one or more of the reactants or using dilution ventilation or local exhaust ventilation may be possible.
2. *Changing the design of process equipment.* Using pumps with double seals or can-type pumps for hazardous materials; using storage tanks with no bottom piping connections for refrigerated hydrocarbon storage, such as LNG; eliminating piping sliding and bellows-type expansion joints and proprietary coupling shaving soft elastomeric ring gaskets; using self-reinforcing forged nozzles rather than nozzle reinforcing pads for high-pressure vessels and exchangers; and upgrading equipment materials, such as using higher alloys for high-temperature equipment and equipment exposed to highly corrosive liquids and gases.
3. *Installing remote operated safety shutoff valves or interlocks that automatically operate such valves when an emergency occurs.* A few examples are remote operated valves on bottom outlets of vessels and storage tanks containing large volumes of flammable or toxic liquids; check valve in large-diameter piping to prevent backflow if a break occurs (only when flow is always in the same direction); excess flow valves that automatically close at high flow velocity resulting from a pipe rupture; automatic valves interlocked with a pressure sensor to shut in a

line when the pressure drops because of a piping failure; and air-operated or spring-loaded valves equipped with a fusible plug or link to close when exposed to fire.

4. *Installing fixed automatic firefighting equipment to control or extinguish a fire and minimize losses.* Some protective systems are automatic fixed halon extinguishing systems in computer rooms or inside gas turbine generator housings, sprinkler systems in buildings, high-density fixed water sprays designed for fire intensity control, and low-density water sprays to cool shells of pressure vessels when exposed to fire to prevent rupture due to overheating the metal.

## 17.5 FIRE PROTECTION

The overall plant fire protection system should be designed to meet the requirements to contain and extinguish the most serious fires that could occur. Where there is adequate manpower and a well-trained plant fire department is maintained, firefighting systems may be manual. Where the manpower is limited and high hazards exist, automatic and remote systems are prudent. The basic elements of firefighting systems are as follows:

### 17.5.1 Fire Water Supply

The first step is to provide a reliable fire water supply. For small low-hazard chemical plants, the municipal water supply may be adequate. The installation of a fire water storage tank is usually advisable. The quantity of water stored should be determined by the design flow rate to the fire main and the length of time expected to fight a major fire. This time varies from four to eight hours, depending upon the size of the plant and the exiting hazards. About 3000 gal min$^{-1}$ (680 m$^3$ h$^{-1}$) normally would be adequate for a small chemical plant, and a minimum of 4000–5000 gal min$^{-1}$ (909–1135 m$^3$ h$^{-1}$) would be required for a petrochemical plant or refinery. The design flow should be based on the assumption that only one process unit at a time will be on fire. Although there is a possibility of a fire spreading from one unit to another, the probability is low if units are adequately spaced. It is also unlikely that adequate manpower and firefighting equipment would be available to fight two process unit fires of major magnitude. In determining the flow rate, the total of fixed monitors, water sprays, and hose streams should be determined for each process unit. However, this flow should be based on the equipment expected to be required for any single type fire and not the total firefighting equipment installed onsite.

### 17.5.2 Fire Pumps

Fire water pumps should be selected on the basis of reliability and meet the requirements of NFPA 20, *Standard for the Installation of Centrifugal Fire Pumps* (NFPA 20 1993). Electric motor-driven pumps require little maintenance and are easily started but should not be relied upon where power failures can occur. Diesel-driven pumps are very reliable, but cost more than electric driven and require more maintenance. Steam turbine-driven pumps should be used only where there is a reliable source of steam, especially during a fire when steam requirements for spare pumps may be a high.

It is good practice to install more than one fire pump as backup. Often one electric pump and one or more diesel pumps are installed. Typical pump sizes vary between 1500 and 3000 gal min$^{-1}$ (340–680 m$^3$ h$^{-1}$).

When a fire water system is required to be pressured constantly for immediate use, then a jockey pump (which maintains the pressure in the system) should be installed. When the fire main pressure cannot be maintained by the jockey pump, then the main fire pumps should start automatically and sequentially upon demand. A constantly running jockey pump is preferred to an intermediate running pump with an air pressure tank. Jockey pumps should be electric with a minimum capacity of 100–200 gal min$^{-1}$ (23–45 m$^3$ h$^{-1}$) to avoid frequent start-up of fire pumps. Fire pumps should be installed in a fire safe area at least 200 ft from any fire hazard.

### 17.5.3 Fire Mains and Hydrants

The fire main system is usually laid out in a grid pattern. With this design, water supply is from two or more directions, which permits smaller diameter pipe. Reliability is also improved by installing diversion valves, which can be used to isolate a main break at any location and continue fighting a fire.

Steel pipe is the preferred material for resistance to mechanical damage. Pipe sections should be welded together and flanges used at valves, hydrants, and other fittings. Friction-type flexible couplings with elastomeric ring gaskets should be avoided where possible. Underground mains should have exterior surfaces coated and wrapped for corrosion protection. In addition, cathodic protection is usually provided. When saltwater is used, the pipe interior should be cement or epoxy lined.

Plastic pipe should be used with caution. Use a good quality plastic pipe and make sure by careful inspection that all joints are sound and do not leak. Fire mains around process units preferably should be underground to protect them against fires, explosions, and mechanical damage. In less hazardous off-site areas, fire mains are often above grade. Mains generally are 8-in. minimum size except that in low-hazard off-site areas 6-in. mains may be used when the required flow rate is well known.

Fire hydrants usually have two $2\frac{1}{2}$-in. hose connections and a $4\frac{1}{2}$ or 5-in. fire truck pumper connection. They should be located on fire mains looped around process units and spaced 150–200 ft (45–60 m) apart. They should also be located on mains looped around tank storage areas on about 200–300 ft (60–90 m) spacing. Hydrants around buildings and other off-site areas should be located as the hazard demands. They should be located near plant roads to be accessible to fire trucks. Hose threads should be compatible with municipal fire department and mutual aid services. When this is not feasible, hose coupling adapters should be available. Wet barrel hydrants, without underground shutoff valves, should have a valve on each hose connections.

### 17.5.4 Fire Monitors

Fixed water monitor nozzles (water nozzle with separate shutoff valve) should be installed around process units to protect specific pieces of equipment such as heaters, exchanger banks, vertical and horizontal pressure vessels, and compressors. This can best be accomplished by installing permanent monitor nozzles around such equipment at grade level, on special trestles, or on the roofs of buildings. They are also used in off-site areas to protect marine tankers and equipment on marine wharfs, at truck terminals to protect the loading facilities and trucks, in tank cars at tank car loading racks, in tank storage farms to protect LPG tanks, at gasoline treating plants, around TEL (tetraethyl lead) facilities and cooling towers.

Around process units it is preferable to locate monitors outside the battery limits, where firefighters are in a less exposed location. Where the process area is congested, it may be necessary to install some monitors inside the unit and also in elevated locations. Elevated monitors usually are operable from grades so that firefighters can readily escape.

The commonly used monitor size is 500 gal min$^{-1}$ (113 m$^3$ h$^{-1}$). This size is adequate for cooling most equipment and does not drain the water supply from other firefighting equipment. Larger monitors, 1000 gal min$^{-1}$ (227 m$^3$ h$^{-1}$), are sometimes used to protect such high hazards as marine wharfs and tankers. For flammable liquid and gas storage areas, foam monitor nozzles are used.

Two-wheeled portable foam/water monitors are useful for towing to a fire as needed. The 500 and 100 gal min$^{-1}$ sizes are common and are usually kept in the plant firehouse when not needed. The monitor generally is supplied with the foam/water solution properly proportioned by a fire truck.

### 17.5.5 Fixed Water Sprays

Fixed water sprays have the following uses:

- Cooling metal to prevent distortion, rupture, or buckling when fire exposed.
- Controlling fire intensity.
- Dispersing combustible and toxic vapor clouds.
- Maintaining the integrity of electrical cable exposed to fire.
- Cooling Class A materials below their ignition temperature to extinguish a fire.

Water sprays should be designed in accordance with NFPA 15, *Standard for Water Spray Fixed Systems for the Fire Protection* (NFPA 15 1990).

Water sprays for cooling metal are applied to

- Uninsulated pressure vessels containing flammable and toxic liquids.
- Uninsulated pressure storage vessels containing light ends such as propane and butane.
- Compressors.

Water sprays to control fire intensity are applied to

- Pumps.
- Hot piping manifold, particularly when the liquid is above autoignition temperature.

Vapor cloud dispersal applications are

- Toxic liquids or gases that could escape from a vessel, pump, compressor, manifold, or other equipment.
- Flammable liquids under very high pressure, which could leak out of flanges and pump seals to form a large vapor or mist cloud.

Electrical cable and transformer protection are used to

- Limit the loss of grouped electrical cable runs in tunnels or other areas difficult to reach for firefighting.
- Maintain the operability of grouped electrical cables during a fire.
- Maintain integrity of transformers.

Fire extinguishment applications are

- Extinguish fires on conveyor belts carrying combustible materials such as coal.
- Protect long conveyor belts from being destroyed by fire caused by hot roller bearings.
- Extinguish fires in wooden cooling towers.

Activation of water sprays may be manual or automatic. Manual activation is usually by a quarter-turn ball valve located in a fire safe area. Automatic activation may be by heat or smoke detectors.

Advantages of manual operation:

- Less expensive than automatic.
- More reliable.
- Less maintenance required.
- Permits operator to exercise judgment and prevents false alarms.

Advantages of automatic operation:

- Limit losses by quick activation.
- Frees operator to shut down the unit.
- Protects areas normally unattended or during off-hours.

When considering water spray protection, keep in mind that sufficient water must be available so that the fire water supply will not be depleted in an emergency. Also, drainage of spray water must be considered; otherwise a hydrocarbon fire could be spread throughout the area.

### 17.5.6 Live Hose Reels

Live hose reels hold 50–100 ft. (15–30 m) of noncollapsible hard rubber hose, usually $1\frac{1}{2}$ in. diameter. The water line is pressured up to the hose inlet, and to change the hose all that is required is to open the valve at the hose inlet. With hard rubber hose, not all of the hose needs to be removed from the reel in order to use it. Live hose reels have a flow rate of only about 100 gal min$^{-1}$ (23 m$^3$ h$^{-1}$) and are not suitable for fighting large fires. Their primary purpose is in fighting small incipient fires to provide holding action until the fire department arrives.

Some applications for live hose reels are as follows:

- At process units, where they are usually installed under the unit pipe rack for easy access and can be deployed on either side of the rack.
- On marine wharfs, to protect loading areas and piping manifolds.
- Inside buildings, such as warehouses and storehouse, where immediate fire hose response is desired.
- At truck loading racks, where manpower is limited and quick response is desired.
- At areas where small gas containers, such as propane, are filled.

In cold climates winterizing is required, usually by electric heat tracing the aboveground portion of the water supply piping.

### 17.5.7 Portable Fire Extinguishers

Portable fire extinguishers for industrial use range from the small 5-lb (3.2-kg) handheld unit to the 350-lb (160-kg) wheeled dry chemical unit. NFPA 10, *Standard for Portable Fire Extinguishers* (NFPA 10 1994), covers selection, installation, inspections, maintenance, and testing. The frequently used types of extinguishing materials are as follows:

- Potassium and sodium bicarbonate dry chemical, used for Class B fires such as most flammable liquids and gases.
- Multipurpose dry chemical, used for Class A, B, and C fires.
- Halon 121 (BCF), used for electrical equipment.
- $CO_2$, used for electrical equipment.
- Pressurized water, used for Class A fires, usually in office areas containing paper and other combustible materials.
- Aqueous film forming foam (AFFF), used where securement of small flammable liquid fire is desired.

Underwriters Laboratories (UL) tests and classifies fire extinguishers regarding the types of fires and fire area that can be extinguished. A UL classification, such as 120-BC, indicates that a type B or C fire, 120 ft in area, can be put out by a trained operator using the extinguisher. For a more complete explanation of the UL rating, see NFPA 10, Paragraph A-1-42 (NFPA 10 1994).

Some plant applications for extinguishers are as follows:

- *Office buildings*. Five-pound (2.3 kg) multipurpose dry chemical extinguishers located a maximum of about 75 ft (23 m) from any hazard. Pressurized water extinguishers, $2\frac{1}{2}$ gal (101), also may be used but are not as effective as dry chemical.
- *Warehouses and maintenance buildings*. Thirty-pound (9-kg) multipurpose dry chemical extinguishers. Where Class B materials are stored and handled, use 30-lb potassium bicarbonate. Locate extinguishers not more than 50 ft (15 m) from any hazard.
- *Laboratories*. At least 9-lb BCF and one 30-lb potassium dry chemical extinguisher or more, depending upon the size of the laboratory.
- *Process areas*. Locate 30-lb potassium bicarbonate extinguishers so that at least one extinguisher is not more than 50 ft (15 m) from all hazards. Also provide one or two 350-lb potassium bicarbonate wheeled extinguishers in each process unit. Extinguishers mounted on pipe rack columns are accessible and easily located. Also locate extinguishers at each fired heater, compressor, in pump areas and at transformers. At least one 22-lb BCF extinguisher for electrical fires should be located in control rooms and outside substations.
- *Off-site areas*. One 30-lb multipurpose extinguisher at top deck stairway landing of cooling towers. Locate one or more 30-lb potassium bicarbonate extinguishers at boilers and at off-site pump areas.
- *Marine terminals*. One or two 30-lb potassium bicarbonate extinguishers at each cargo loading area, plus one 30-lb unit in the control room and near piping manifolds. Also locate one 350-lb potassium bicarbonate wheeled extinguisher near cargo holes and loading arms. Large capacity 2000-lb (909 kg) skid-mounted dry chemical units are frequently used on wharfs, particularly where access for mobile firefighting equipment is poor.

The use of dry chemical extinguishers in areas where computers and other electrical equipment is located should be avoided because the cleanup required may be more expensive than fire damage.

Extinguishers must be properly mounted and the location clearly marked for quick identification. They should be periodically inspected and tagged. After using, they should be immediately recharged and replaced.

### 17.5.8 Fire Hose

Soft fire hose is used for hand hose line, to supply water from hydrants or afire truck to portable firefighting equipment such as wheeled monitors, and to supply water from hydrants to fire trucks. The basic hose sizes are $1\frac{1}{2}$, $2\frac{1}{2}$, and 3 in. Three-inch hose is used primarily to supply water from hydrants to fire trucks and is equipped with $2\frac{1}{2}$ in. couplings. Fire hoses should meet the requirements of NFPA 1961, *Standard on Fire Hose* (NFPA 1961 1992). Hose couplings should be in accordance with NFPA 1963, *Standard for Fire House Connections* (NFPA 1963 1963).

Single jacket hose consisting of all synthetic fiber with rubber tube liner is commonly used. Some hose is impregnated with polyvinyl chloride, Hypalon, or other plastic coating to reduce wear.

Hose is often stored in hose houses or cars in the plant area where it is expected to be used. The disadvantage of this practice is that a considerable inventory of hose may be stored throughout the plant and often is not maintained, due to lack of inspection. In large plants with a well-trained fire department and one or more fire trucks, all hose is carried on fire trucks. This is the preferred practice, because the hose is tested and maintained in good condition and is available wherever it is needed.

### 17.5.9 Fire Foam

Fire foams are used to extinguish and secure hydrocarbon pool fires such as spills and atmospheric storage tank fires. Types of foam concentrates are as follows:

- *Regular protein foam.* Used for spill fires and storage tank fires for top application only. Protein foam is less used since the development of synthetic foams with superior properties.
- *Fluoroprotein foam.* This foam serves the same uses as protein foam but is generally more effective and can be used for subsurface application for cone roof storage tank fires.
- *AFFF.* This foam is a combination of fluorocarbon surfactants and synthetic foaming agents that quickly spread across the surface of a hydrocarbon pool fire. It has quicker control and extinguishment than fluoroprotein foam but does not have as good sealing properties against hot metal, such as the shell of a storage tank on fire.
- *Alcohol-type foam.* Used for fires involving alcohol and other polar solvent liquids (such as acetone, methylethyl ketone, enamel, and lacquer thinners) that break down other foams. It can also be used against hydrocarbon fires.
- *High-expansion foam.* This foam is used at expansion rates of 800 to 1. It is intended primarily for use on Class A materials in confined areas such as warehouses and other buildings. Because of its low density, it can be blown away easily when applied outdoors.

Foam concentrate is supplied in two concentrations, 3 and 6%. The quality and cost of both foams is equivalent, but 3% foam has the advantage of making double the quantity of foam when stored in equal-size containers. This becomes important on mobile firefighting equipment where allowable weight is limited. NFPA 11, *Standard for Low Expansion Foam and Combined Agent Systems* (NFPA 11 1994), covers the characteristics of foam producing agents and the requirements for design, installation, and operation of foam extinguishing systems.

Foam solution can be proportioned and generated by the following:

- *Fixed proportioning systems.* One such fixed system, balanced pressure proportioning, consists of a foam concentrate storage tank, a foam concentrate pump, and a control valve to proportion the foam. Another fixed system, pressure proportioning, consists of a foam concentrate storage tank with an internal membrane or bladder and a venturi for proportioning foam. Water pressure on the bladder is used to force foam out of the liquid storage tank.
- *Mobile proportioning systems.* These are installed on fire trucks and have the advantage of being able to proportion foam solution wherever there is a water supply such as from hydrants or a pond. Balanced pressure proportioning is usually used on fire trucks.
- *Fixed and portable line proportioners.* This is an inexpensive proportioning method, designed for a predetermined water discharge at a predetermined water pressure. Water flowing through a venturi creates a vacuum that picks up foam concentrate through a connection on the side of the venturi. In a fixed installation, a foam concentrate tank supplies one or more venturis. In the portable system, a tube connect to the venturi is inserted into a drum of foam concentrate. The portable pickup tube is useful for converting a hose water line to a foam line by using a foam-making nozzle on the end of the hose. This arrangement is useful for small hydrocarbon spill fires.

Some applications for foam are as follows:

- *Cone roof and covered floating roof storage tanks.* Foam chambers installed at the top of the shell expand foam solution to extinguish tank fires. The system must be designed to cover the entire liquid surface at the prescribed rate. A fixed proportioning system or a fire truck may be used to supply foam solution to the proportioners.

- *Subsurface foam.* A high-back-pressure foam maker may be used to generate foam and inject it into the bottom of a cone roof storage tank. Fluoroprotein foam is preferred for this application. Subsurface application has the advantage of being simpler to install, because a product line to the tank may be used without having to weld a new connection onto the tank. Foam may be proportioned by a fixed system or a fire truck.
- *Catenary and over-the-top foam systems.* These systems are used to extinguish rim fires in open-top floating roof tanks. In the catenary system, foam is deposited at the floating roof seal by rigid pipe and flexible metal hose to adjust for movement of the floating roof. The over-the-top system deposits foam at the top of the shell and on the inside, and the foam runs down the shell and around the roof seal. Foam may be supplied by fixed system or a fire truck.
- *Foam/water sprinkler systems.* These systems are used mainly beneath marine wharfs and at truck loading racks. They are usually supplied by a fixed foam proportioning system.
- *Fire trucks.* Trucks are used to proportion foam and apply it to fires in process units or in storage tanks, marine terminals, and other plant off-site areas. Foam is generated and applied by hose lines with air aspirating foam nozzles or from turret monitors mounted on the truck. In this capacity, the foam fire truck becomes a very effective fire-fighting apparatus.

### 17.5.10 Halon and Carbon Dioxide Extinguishing Systems

Halon and carbon dioxide total flooding systems are used where valuable equipment is located, in unattended indoor locations, and where a clean agent is required for extinguishment.

The two basic extinguishing agents used are Halon 1301 and carbon dioxide. Halon 1211 is also used sometimes, but it has a greater toxicity and should be avoided wherever personnel could be present. Halon 1301 is nontoxic and is preferred to $CO_2$ but is considerably more expensive.

NFPA 12A, *Standard for Halon 1301 Fire Extinguishing Systems* (NFPA 12A 1992), contains minimum requirements for halogenated agent extinguishing systems.

NFPA 12, *Standard on Carbon Dioxide Extinguishing Systems* (NFPA 12 1993), contains requirements for $CO_2$ extinguishing systems.

Halon and carbon dioxide flooding systems are used for:

- Computer rooms, where a clean agent is required.
- Gas turbine enclosures, which would not be safe to enter during a fire.
- Storage vaults.

### 17.5.11 Fire Detectors

Fire detectors are used to give warning that a fire has occurred in unattended places and also to activate automatic firefighting equipment. They are usually located in remote high-risk areas or in high-investment areas. Typical locations are computer rooms, grouped shipping pumps, tank truck loading terminals at automatic water spray installations, warehouses and other storage areas, and gas compressors of combustion. Types of detectors and characteristics are as follows:

*Heat detectors*

- *Fusible link or plug.* Very reliable and needs to no power source but is slow to respond.
- *Fixed temperature.* Reliable and low cost but slow and affected by wind.
- *Rate of rise.* Will detect slow temperature rises and rapid rise faster than fixed temperature detectors; affected by wind and should be confined to indoor use.

*Smoke detectors*

- *Ionization.* Early warning, will detect smoldering fires, but subject to false alarms where internal combustion equipment is used or personnel are smoking; limited to indoor use.
- *Photoelectric.* Early warning, will detect smoldering fires, subject to false alarms from dusts; limited to indoors.
- *Ultraviolet (UV).* Fast response, high sensitivity, self-testing, but response retarded by thick smoke and false alarms from welding, lighting, etc.

- *Infrared (IR)*. Fast response, but very prone to false alarms from hot surfaces such as furnace firebrick and hot engine manifolds.
- *Combined UV/IR*. High-speed response and sensitivity, low false alarms, but thick smoke obscures range, is expensive.

### 17.5.12 Plant Fire Alarms

A plant fire alarm system is necessary to notify all personnel that a fire has occurred and to report the location. Methods used to report a fire are as follows:

- *Telephone*. The preferred system utilizes a dedicated number to report a fire. The call is usually received in the control building and sometimes also the gatehouse. The coded number dialed also activates audible fire horns or sirens in the plant.
- *Radio*. Walkie-talkie radios may be used to report a fire to the control room and also activate the plant fire alarm.
- *Manual fire alarm stations*. Pull box or pushbutton alarms located throughout the plant can be used to report a fire and sound the audible alarm. In large plants the location of the activated alarm station often appears on a dedicated computer CRT.

### 17.5.13 Fire Trucks

Medium-sized and large plants often have one or more fire trucks and a well-trained firefighting crew. The designated firefighters usually perform other duties in the plant and respond to the firehouse when the plant fire alarm is sounded. The design of plant fire trucks varies with the types of fires expected. Some types of trucks used are as follows:

- *Foam truck*. Used primarily to fight hydrocarbon spill fires and atmospheric storage tank fires, it carries a large volume, up to 1000 gal (3800 l), of foam concentrate. Water is supplied to the truck from hydrants, and the pressure is boosted by a fire water pump (usually 1000 gal min$^{-1}$) driven by the truck engine. Foam solution is proportioned by a foam concentrate pump and proportioning system.
- *Dry chemical truck*. Used primarily for rapid fire knockdown and extinguishment of process unit fires. A large capacity of potassium bicarbonate dry chemical is carried on the truck, up to 2000 lb (900 kg). The truck also usually contains a fire water pump to supply cooling water streams.
- *Combination foam/dry chemical truck*. Combines the advantages of both the foam and the dry chemical trucks. It contains 200 lb of dry chemical and 100 gal of foam concentrate and is effective for fighting both process unit and tank fires.
- *Triple agent truck*. Contains a large quantity of dry chemical, foam concentrate, and premixed AFFF solution. This truck is very effective in fighting various types of fires, but because of weight limitations, some compromises have to be made. The dry chemical capacity is 1800–2000 lb, the foam concentrate tank holds about 500 gal (1900 l), and the premix tank holds about 200 gal (760 l) of AFFF/water solution, which is pressurized for immediate use by nitrogen cylinders when activated. This truck can quickly extinguish a spill fire followed by securing with premixed AFFF. It can also be used to fight a tank fire when an outside backup source of foam concentrate is provided by a foam trailer or nurse truck.

### 17.5.14 Fireproofing

Fireproofing consists of a passive insulating coating over steel support elements and pressure-containing components to protect them from high-temperature exposure by retarding heat transfer to the protected member.

The objectives of fireproofing are as follows:

- To prevent the collapse of structures and equipment and limit the spread of fire.
- To prevent the release of flammable or toxic liquids from failed equipment.
- To secure escape routes for personnel.

Fireproofing is more critical in fire exposed areas where flammable liquids are processed, stored, or shipped and where a fire could occur from a leak or spill. In determining the need for fireproofing, it is necessary to consider the characteristics of the materials being handled, the severity of operating conditions, the quantity of fuel contained in equipment and piping, and the replacement value of plant facilities and business interruption.

Some types of fireproofing materials available are as follows:

- *Dense concrete.* This consists of Portland cement, sand, and aggregate having a dried density of 140–150 lb ft$^{-3}$ (2240–2400 kg m$^{-3}$). Concrete may be formed, troweled, or gun applied. Concrete requires no unusual skill to apply and has high resistance to impact and other abuse. Its high density makes it an inefficient insulator and requires more structural support than lighter-weight materials.
- *Lightweight concrete.* Also called cementitious materials, they consist of a lightweight aggregate such as perlite or vermiculite and a cement binder. The density dry is from 25 to 80 lb ft$^{-3}$ (400–1280 kg m$^{-3}$). Since they are lightweight, they are very good insulators. However, the low density also results in reduced strength, less resistance to impact than dense concrete, and high porosity, which leads to moisture penetration. These materials usually require a weather protective coat to prevent corrosion of substrate and spalling in freezing climates. In selecting the density of the mix, the properties of fire protection versus strength must be considered.
- *Intumescent and subliming mastics.* These mastics provide heat barriers through one or more of the following mechanisms:
  - Intumescence: Materials expand to several times their volume when exposed to heat and form a protective insulating char at the barrier facing the fire.
  - Subliming: Materials that absorb large amounts of heat while transferring directly from a solid to a gaseous state.
  - Mastics are sprayed on the substrate to form a thin coat, usually $\frac{3}{16}-\frac{1}{2}$ in. (5–13 mm). The main advantages are light weight, thin coats, and speed of application. Since proper bonding to the substrate and careful control of coat thickness are very important, only vendor-approved and experienced applicators should be used. Some mastics contain a flammable solvent and should not be applied near such ignition sources as fired heaters and boilers during operation.
- *Preformed inorganic panels.* Precast or compressed fire-resistant panels composed of a lightweight aggregate and a cement binder or a compressed inorganic insulating material such as calcium silicate. Panels are attached to the substrate by mechanical fasteners designed to withstand fire exposure without appreciable loss of strength. When used outdoors, and external weather coating system may be required. Also, all joints should be caulked with a mastic. These materials have the advantage of requiring no time-consuming curing or drying cycle. However, they are labor intensive when applied to existing units having instruments or other equipment supported on structural steel columns.

The thickness of fireproofing required for protection of a substrate is expressed in hours of protection, that is, the time until the steel reaches a temperature of 1000 °F (538 °C). Failure occurs when the average temperature reaches 1000 °F (538 °C). This temperature is perceived as the critical temperature for structural steel. Until recently, the fire test used to determine the hours of protection was ASTM E119 (ASTM 1983). Several years ago tests conducted by some oil companies determined that the E119 time–temperature curve rate of rise was too slow to predict accurately the hours of protection in a high-temperature-rise hydrocarbon fire (Warren and Corona 1975). UL has recently developed a high-rise fire test, UL 1709, which gives a more realistic evaluation of fireproofing materials in a hydrocarbon type fire (UL 1984).

In determining fireproofing requirements, the following factors should be considered:

- Volume of flammable or combustible liquid likely to be involved.
- Characteristics of material spilled.
- Ability of drainage system to remove spill.
- Congestion of equipment.
- Severity of operating conditions.
- Importance of the facility.
- Ability to isolate the leak with safety shutoff valves.

Some applications of fireproofing in fire exposed areas are as follows:

- *Multilevel equipment structures*. Structures supporting equipment containing hazardous materials should be fireproofed, usually up to the equipment support level. Structures supporting nonhazardous equipment are usually fireproofed from grade to a height of 30–40 ft (9–12 m).
- *Pipe racks*. Process unit pipe racks generally should have columns and at least the first level of beams supporting pipe fireproofed. When a quantity of large-diameter pipe (10 in. and larger) is carried on a second or third level, then consideration should be given to extending fireproofing to upper levels.
- *Equipment supports*. Pressure vessel skirts and legs, air cooler legs, legs of fired heaters, and high saddled supports over 3 ft high (1 m) are often fireproofed.
- *Grouped instrument cable*. Preferably, this should run underground. When located on the pipe rack, fireproofing should be considered.
- *Emergency valves*. The preferred design is to fail in the safe position. When this cannot be done, the valve operator and power supply should be fireproofed. Steel valve bodies do not require fireproofing. In designing fireproofing for electrical cable, it should be kept in mind that the insulation deteriorates at about 320 °F (160 °C).

### 17.5.15 Fire Prevention Plan

With any discussion of a loss control program, it is important to ensure that any local, state, or federal regulations in the areas of fire prevention and protection are followed. OSHA has established a number of general industry regulations, including fire protection requirements covering means of egress, fire brigades, fixed and portables fire suppression equipment, fire detection systems, fire prevention plans, and emergency action plans (EAP).

This section discusses a model plan to prepare for emergencies and fire prevention and outlines the specific OSHA criteria to follow (including references to OSHA regulations). This plan was adapted from the model plan outlined in the North Carolina Department of Labor's *A Guide to Occupational Fire Prevention and Protection* (Smith 1989). By auditing the work area, by training the employees, by acquiring and maintaining the necessary equipment, and by assigning responsibilities and preparing for an emergency, human life and facility resources will be preserved. At the beginning, this model plan requires management decisions as to whether employees will be employed to fight fires. Management's selection of a course of action regarding employees and fire protection depends on the requirements and the needs of each individual facility. This decision, usually made by top management, requires careful consideration. The most important factors in providing adequate safety in a fire are the availability of proper exit facilities to ensure ready access to safe areas and the proper education of employees as to the actions to be taken in a fire.

There are two basic options available to management: Option A where employees will fight fires and Option B where employees will not fight fires.

***17.5.15.1 Option A: Employees Will Fight Fires*** The selection of Option A entails two additional decisions. First, who will fight fires: (1) all employees, (2) designated or selected employees, (3) fire brigade/emergency organization, or (4) any combination of these? Second, what types of fires will be fought: (1) incipient stage fires only, or (2) interior structural fires?

Once these decisions have been made, management should follow one of the following plans:

*Plan 1*. All employees. Provide education in fire extinguisher use and hazards involved with incipient stage firefighting upon initial employment and annually thereafter as required by OSHA in 29 CFR 1910.157{g} (OSHA 1996). Should an employer except all employees to fight an incipient stage fire in their immediate work areas, as a designated group (such as a fire squad), those employees should receive hands-on training with the appropriate firefighting equipment.

*Plan 2*. Designated or selected employees. Provide education in fire extinguisher use and hazards involved with incipient stage firefighting upon initial employment and annually thereafter. Provide an EAP that designates specific employees to use firefighting equipment (29 CFR 1910.38(a)) (OSHA 1996). Provide annual training with firefighting equipment.

*Plan 3*. Fire brigades (emergency organization, incipient-stage fires only). Prepare a fire brigade organizational statement that establishes the existence of a fire brigade (organizational structure, training, number of members, functions) as required by OSHA 29 CFR 1910.156 (OSHA 1996). Provide training for duties designated in organizational statement. Provide hands-on training annually in the use of extinguishers, $1\frac{1}{2}$ in. hose lines (such as Class II standpipe system), and small hose lines ($\frac{5}{8}$ to $1\frac{1}{2}$ in.). Train and educate members in special hazards and provide standard operational procedures. Provide a higher level of training and education for leaders and instructors.

[Or] Fire brigade (emergency organization, interior structural fires). Prepare a fire brigade organizational statement. Ensure physical capability of members. Provide training for duties assigned in fire brigade organizational statement. Provide educational or training sessions at least quarterly and handson training with appropriate firefighting equipment at least annually. Train members in special hazards and provide standard operational procedures. Provide higher level of training and education for leaders and instructors. Provide required protective clothing and breathing apparatus.

**17.5.15.2 Option B: Employees Will Not Fight Fires** The selection of Option B entails one additional decision – whether to have portable fire extinguishers and hoses in the facility. Note that specific regulation may require that extinguishers be provided.

One of the following plans should be followed:

*Plan 1.* Provide fire extinguishers and hoses. Provide fire extinguishers and hoses required by regulation or insurance carrier. Provide EAP and fire prevention plan as required by OSHA 29 CFR 1910.38{a} (OSHA 1996). Maintain and test this equipment. Provide for critical operations shutdown and evacuation training.

*Plan 2.* Fire extinguishers and hoses not provided. Provide EAP and fire prevention plan. Provide for critical operations shutdown and evacuation training.

**17.5.15.3 Elements of Fire Prevention Plan** With the elements of fire prevention plans seen earlier, depending on what option is chosen, a written fire prevention plan may be required. Obviously, if a good fire loss control program is established, the required elements of the fire prevention plan will be included automatically. If there is no fire loss control program, then at a minimum a fire prevention plan should be developed. Facility managers and/or loss control managers should be responsible for developing this plan and keeping it current. This section discusses the necessary elements of the plan.

*Element 1.* Names of persons responsible for control of fire protection and maintenance equipment and ignition sources. Persons who are responsible for the control and maintenance of equipment related to fire control or for the control of particular hazards should be clearly identified. An example form for personnel assignments is shown in Figure 17.4.

*Element 2.* Control of major workplace fire hazards. Major fire hazards peculiar to the facility and a plan for control of such hazards should be identified. The plan should include proper handling and storage procedures, potential ignition sources and their control procedures, and the type of fire protection equipment available. An example of a type of form that could be used is shown in Figure 17.5.

Fire Protection and Prevention Assignments

| Name | Work Location | Job Title | Assignment |
|---|---|---|---|
| ________ | ________ | ________ | Responsible for maintenance of equipment and systems installed to prevent or control ignition of fires. |
| ________ | ________ | ________ | Responsible for control of fuel source hazards. |
| ________ | ________ | ________ | Responsible for regular and proper maintenance of equipment and systems installed on heat-producing equipment to prevent fires. |

Emergency Plan and Fire Protection Plan Coordinator:

Name: ____________ Date: ____________

**Figure 17.4** Example of a form for personnel assignments. *Source:* Courtesy of North Carolina Department of Labor.

*Element 3.* Housekeeping procedures established. Proper housekeeping is an element in ensuring effective fire prevention.

*Element 4.* Fire prevention for heat-producing equipment. Heat-producing equipment, such as heaters, furnaces, and temperature controllers for such equipment, need maintenance to ensure proper operation. An example form for keeping track of heat-producing equipment and controls is shown in Figure 17.6.

*Element 5.* Review plan with employees. Once established, the plan should be reviewed by all employees covered by the plan to ensure that they are aware to the types of fire hazards of the materials and processes to which they might be exposed.

### 17.5.16 Elements of Emergency Action Plan

An EAP should be an integral part of the life safety element in a fire loss control program. Many emergencies at facilities – such as fire, explosion, bomb threats, chemical releases, and natural disasters – require that employees and people evacuate a building. History has shown that an EAP and adequate employee and occupant familiar with a building can prevent disasters. A written EAP is recognized as the best way to plan evacuation and, in most occupational situations, is required by OSHA in 29 CFR 1910.38, *Employee emergency plans and fire prevention plans* (OSHA 1996). Facility managers and/or loss control managers are responsible for developing an EAP and keeping it up to date. This section discusses the necessary elements and development of an EAP.

#### *17.5.16.1 Responsibilities*

*17.5.16.1.1 Loss Control Manager* The loss control manager (or if there is no loss control manager, the facility manager) should be responsible for (i) overseeing the development, implementation, and maintenance of the overall facility EAP; (ii) designating and training evacuation wardens (or fire wardens, floor safety officers, or similar designation); (iii) reviewing the EAP with employees and building occupants, arranging to train new personnel, and notifying employees of plan changes; and (iv) relaying applicable information to public fire department personnel, emergency organization personnel, employees, and evacuation wardens in event of an emergency.

*17.5.16.1.2 Evacuation Wardens* The evacuation wardens are normally appointed by the loss control manger to ensure evacuation, together with other emergency duties. The evacuation wardens may also be members of the

***Control of Major Workplace Fire Hazards***

The following major fire hazards are present in the work area, and proper control procedures are described.

| | Major Fire Hazard | Location | Company Controls |
|---|---|---|---|
| 1. | | | |
| 2. | | | |

**Figure 17.5** Example of a form for listing fire hazards and control procedures. *Source:* Courtesy of North Carolina Department of Labor.

***Fire Prevention for Heat-Producing Equipment***

| Heat-Producing Equipment and Controls | Routine Maintenance | Frequency of Maintenance | Assigned Responsibility |
|---|---|---|---|
| | | | |

**Figure 17.6** Example of a form for listing heat-producing equipment and controls and maintenance procedures. *Source:* Courtesy of North Carolina Department of Labor.

emergency organization. They should be trained in the complete facility layout and the various evacuation routes. Responsibilities include (i) checking for complete evacuation of their designated area and notifying the public fire department and emergency organization of missing persons or location of the fire and of the location of any trapped persons; (ii) performing emergency operations as needed and detailed in the EAP, such as closing windows and doors or turning off certain electrical appliances; (iii) reporting any malfunctioning alarms; and (iv) assisting any disabled persons with emergency evacuation procedures.

*17.5.16.1.3 Elements of Emergency Action Plan* The EAPs are specific to each building and should be coordinated among the various departments by the loss control manager. The following elements should be included in an EAP:

1. Preferred means of reporting fires and other emergencies such as chemical spills or personnel injury,
2. Emergency evacuation procedures and emergency evacuation route assignments.
3. Procedures to be followed by employees who remain to perform critical plant operations before they evacuate.
4. Procedures to account for all building occupants after emergency evacuation has been completed.
5. Rescue and medical duties for those employees who are to perform them.
6. Names or regular job titles of persons or departments who can be contacted for further information or explanation of duties under the plan.

*Element 1.* The EAP should direct employees to activate the nearest alarm box or other alerting mechanism in the event of a fire. It should state that it may be necessary to activate additional boxes or shout the alarm if people are still in the building and the alarm has stopped sounding or if the alarm does not sound. Any unusual alarm notification procedures should be explained. Examples might include areas that have no audible alarm systems, noisy areas, areas with deaf employees, and places of assembly where a public address (PA) system should be used. The plan should instruct any person discovering a fire to notify the public fire department and the emergency organization stating the location of the fire. The plan should direct employees and occupants to call security and the emergency organization to report all other emergencies. It should direct people to give their name, location of the building, and nature of emergency and should advise people to mention specific features of the building, such as toxic substances, critical operations, disabled persons, etc.

*Element 2.* All employees and occupants should know where the primary and alternate exits are located and be familiar with the various evacuation routes available. Primary and alternate evacuation routes and exit locations should be described in the EAP. If a floor plan is posted, it should display the various evacuation routes and exits. The EAP should direct people *not* to used elevators as an evacuation route in the event of a fire. In buildings where departments share mutual areas and/or evacuation routes, the department EAPs should be coordinated. It may be necessary to combine the department EAPs into an overall building EAP. Personnel who are regularly transient because of the nature of their jobs (e.g. facilities maintenance personnel) should be trained in general evacuation procedures.

*Element 3.* Critical operation shutdown is dependent on the building use. It may be as complicated as dealing with pressure vessels or fuel supply or as simple as turning off bunsen burners, hot plates, or electrical equipment. Persons assigned to these duties should be either facility management and/or evacuation wardens or those who cannot leave immediately because of danger to others if the area is abandoned, such a chemists who are conducting potential runaway reactions. These responsible people should be identified in the EAP. Ventilation system shutdown in some facilities will be controlled by automatic protection systems or at a central control station. If facility management or evacuation wardens are required to perform necessary shutdown operations, this should be identified in the EAP.

*Element 4.* The EAP should direct groups working together in the same area to congregate at a prearranged location identified in the EAP. It is a good idea to have an alternate location for inclement weather, if an outside location, or an alternate safe refuge area. A department organization list should be developed to provide a roster of personnel to ensure that everyone has evacuated.

*Element 5.* Those personnel who will perform rescue and medical duties should be identified in the EAP.

*Element 6.* An EAP organization list should contain the names of employees, managers, or other personnel and their job titles, positions, and relative EAP collateral duties. The EAP should include appropriate numbers to call for emergency: fire, police, medical rescue, chemical, biological or radioactive release, water service, fire protection systems, utilities, etc.

Once established, the plan should be reviewed by each employees and occupant covered by the plan to ensure that they know what it expected of them in all emergency possibilities that have been planned, to ensure their safety.

A well-designed EAP will anticipate possible impediments to quick and complete evacuation and will provide solutions and options for these problems. Management should make every effort to develop an EAP that includes necessary contingency planning.

### 17.5.17 Evacuation Drills

Just providing exits and an EAP is not enough to ensure life safety during a fire. Evacuation drills are essential so that employees and occupants of the building will know how to make an efficient and orderly evacuation.

The frequency of drills should be determined by the amount and type of hazardous operations present in the facility and by the complexity of shutdown or evacuation procedures. All elements of the EAP should be practiced during the drill.

After each drill, the loss control manager, facility managers, and evacuation wardens should discuss and evaluate the drill to fine tune the EAP and solve any problems that may have occurred.

### 17.5.18 Life Safety Principles

Generally, life safety requires the following principles which are also covered in the *Life Safety Codes* (NFPA 101 2018):

- Provide sufficient number of unobstructed exits.
- Protect exits from growth.
- Provide alternate exists.
- Subdivide areas and construct to provide safe areas of refuge in occupancies where evacuation is last resort.
- Protect vertical openings to limit spread of fire.
- Ensure early warning of fire.
- Provide adequate lighting.
- Ensure exists and routes of escape are clearly marked.
- Provide emergency evacuation procedures.
- Ensure construction is adequate to provide structural integrity during a fire while people are evacuating.

The *Life Safety Code* deals with occupancy groups according to their life safety hazard. These groups are (i) assembly, (ii) educational, (iii) healthcare and penal, (iv) residential, (v) mercantile, (vi) business, (vii) industrial, (viii) storage, and (ix) unusual structures. Each group has life safety requirements specific to that occupancy.

When discussing the hazards of contents, the three classifications are low, ordinary, or high. *Low hazard* covers contents with low combustibility such that no self-propagating fire can occur, with the only probable danger being from panic, fumes, smoke, or fire from an external source. *Ordinary hazard* covers contents that are liable to burn with moderate rapidity or develop a considerable volume of smoke, but with no poisonous fumes or threat of explosion. *High hazard* covers contents that are liable to burn with extreme rapidity with poisonous fumes and the threat of explosion.

***17.5.18.1 Exiting*** Reference should be made to the *Life Safety Code* (NFPA 101 2018) for detailed requirements on exiting and life safety features. Also state and local building codes should be queried for any requirements peculiar to the local area. Basically the requirements in the *Life Safety Code* for means of egress in general industrial occupancies include many of the features required in any structure. The travel distance to an exit is 100 ft except with a complete automatic sprinkler system the distance is increased to 150 ft. In most large industrial complexes, this travel distance is hard to maintain due to the vast open areas in some plants. In some cases exit tunnels, overhead passageways or travel through fire walls with horizontal exits may be necessary. In most cases, though, it is not practical or economical to build exit tunnels or overhead passageways. In these cases travel distances of 400 ft may be permitted if approved by the local fire officials and meet the following:

- Limit to one-story buildings.
- Limit interior finish to Class A or B (flame spread <75 and smoke developed <450) in accordance with NFPA Std. 255, *Method of Test of Surface Burning Characteristics of Building Materials*.
- Provide emergency lighting.

- Provide automatic sprinkler or other automatic fire extinguishing system that is supervised for malfunctions, closed valves, and water flow.
- Provide smoke and heat venting designed so employees will not be overcome by smoke and heat within 6 ft of the floor before reaching exits.

For high-hazard industrial facilities, the travel distance to an exit is limited to 75 ft, with no common path of travel. All high-hazard facilities are required to have at least two separate exits from each high-hazard area. In general industry facilities a 50-ft common path of travel is allowed to the separate exits.

## 17.6 EMERGENCY PLANNING

While the need for emergency planning may vary somewhat with different facilities, there should be general agreement on the importance of providing at least some level of planning for emergencies that may be encountered at a particular facility. Previously, the value of an audit of the hazards or problems that could possibly be encountered by local facility managers was discussed. This is the first step in emergency planning: the recognition of the nature of possible emergencies. The second step is determining the impact of these emergencies. Finally, it is important to evaluate the effect that these possible emergencies may have on the interruption or continuity of operations.

Often, reviews of large loss situations indicate confusion and lack of effective actions by local facility personnel at the time of a fire, explosion, or other emergency, which may have added considerably to the adverse impact of disasters. These catastrophes may include, but will not necessarily be limited to, the following:

| Fire | Earthquake |
|---|---|
| Explosion | Civil disturbances |
| Water damage | Bomb threats |
| Storms/high wind | Hazardous material releases |

Once the type of emergencies have been recognized, emergency planning programs should be established to provide a way to address these problem areas properly by emergency response and thus eliminate or lessen any potential loss. The initial actions will involve the emergency organization and outside response. The second action will be the plans that are put into effect after the emergency to lessen its impact by providing ways to continue to provide safety for the employees and occupants and to continue production, maintain the integrity of the facility, etc., by instituting contingency plans.

The third part of emergency planning is to ensure that there is an emergency organization available to implement the plans that have been formulated. The formation of emergency plans must be specific to the individual facility.

While the emphasis for the formation for these plans should be at a local level, corporate management should establish policies requiring the formation of plans and guidelines containing a general outline and areas that should be included in the planning. It may be possible for corporate staff members to provide a general emergency plan if there is a great repetitiveness in locations, which could conceivably be encountered in distribution facilities, a theater chain, or similar types of facilities. The need for emergency planning will, to a great extent, depend upon whether large loss factors exist and upon an analysis of what type of response might be anticipated at the time of a specific emergency.

The emergency plan should be coordinated with any emergency response procedures and hazardous chemical facility profiles that may be required by community emergency planning commissions under the Environmental Protection Agency (EPA) community right to know regulations (EPA 1996). A well-thought-out emergency plan will help plant management to meet the emergency response procedures necessary under the regulations and to ensure that the plant personnel will be protected during a community-wide hazardous chemical disaster.

### 17.6.1 Elements of Emergency Planning

Elements of emergency planning will include:

- Assessment of risk.
- Facility organization for dealing with emergencies.
- Assessment of resources.
- Training.

- Emergency headquarters.
- Security.
- Public relations.

***17.6.1.1 Risk Assessment*** This assessment must enumerate all the potential problems that may be encountered at a facility, including those that could occur in the surrounding community and affect the plan. The magnitude of the potential problem must be established as well as the probability of the event occurring, enabling management to prioritize its effort in plan development.

***17.6.1.2 Facility Organization*** A management team must be established to cope with emergencies. Ideally, it should include staff members from production, utilities, fire and safety, security, medical, and maintenance. In small facilities some organization members may have multiple roles. In any event, roles and authority must be clearly defined. In addition, the organization listing must include contact phone numbers when staff members are not on site. More importantly, each emergency organization member must have a staff alternate in the event that the primary team member cannot be contacted or is otherwise unaccessible.

***17.6.1.3 Assessment of Resources*** A factual current knowledge of plan emergency response equipment is essential to any emergency preparedness program. Not only must the capability and utility be known, but plant personnel must also know how to use it. In addition, to the extent possible, plant personnel must access the capability, availability, and utility of the emergency organization that may be called upon to assist in an emergency. Following this, a meeting involving all potential respondents to an emergency should be held. The objectives of the meeting are to familiarize them with the facility, resources at the facility, and resources in the community and to define the role of various groups.

***17.6.1.4 Training*** Plant personnel must be proficient in several areas, including knowledge of the emergency alerting system, familiarity with respect to the emergency organization, shutdown procedures, firefighting, first aid, etc. Each of these must be outlined in concise clear terms. Plant personnel must be thoroughly trained in the use of emergency equipment and the carrying out of the emergency response procedures. Community response personnel should also be informed of these plant operations that may have an impact on their role in the response effort.

***17.6.1.5 Headquarters*** In large facilities there should be at least two widely separated areas that would serve as group headquarters for dealing with emergencies. The locations should be equipped with all the necessary equipment to enable the staff to be in contact with plant personnel as well as mutual aid organizations. These access lines should be dedicated so that essential communication will be readily available to the response group.

Once the above elements are in place, it is essential that joint exercises be held to evaluate the effectiveness of the plan and to correct any shortcomings that may be present.

***17.6.1.6 Security Considerations*** Provisions should be made to have a security component to the emergency response team. Typically, security deals with such hazards as sabotage, terrorism, and so on; however, more typically the primary function in most emergencies is to limit access to those people and that equipment that will assist in coping with and resolving the emergency. To accomplish this requires:

- Identifying key plant and community personnel who can be activated instantaneously.
- Establishing a group having security responsibility or other plant personnel who assume security roles in an emergency.

***17.6.1.7 Public Relations*** An important element of any emergency plan is appointing an individual who is responsible for informing both plant and community personnel about aspects of the emergency. A person must be designated to deal with the public/media. To accomplish this effectively, he or she must have access to the highest levels of management dealing with the emergency.

## 17.7 AUDITS

Periodic audits are recommended to assure that engineering controls are being maintained and also to determine if any upgrading of safety controls is advisable. Preferably an operator familiar with unit operation, and sometimes a process design engineer, should accompany the safety or loss prevention engineer during the audit. The safety audit should include:

- Piping and instrument drawings to verify that safety shutdown systems have been installed and maintained in accordance with the latest drawings. Process unit plot plans and equipment lists to identify equipment.
- Checklist of emergency safety shutdown systems so that an inspection of these facilities can determine if they are maintained in good operating condition.
- Review of plant safety procedures to assure that they are kept current with any changes in plant operations. Visual inspection and testing of plant fire protection systems, such as fixed water sprays, water monitors, fire pumps, fire mains, etc.
- Review of plant accident and fire reports to determine what types of accidents occur most frequently.

### 17.7.1 Self-Inspection Program

Another important part of a fire loss control program is the facility self-inspection. This is recognized by insurance underwriters as being extremely valuable and can help satisfy the OSHA fire protection requirements. The purpose of the self-inspection is basically to ensure that the facility fire protection systems are operable and therefore capable of performing their function and to detect hazard areas that may create fires, such as poor housekeeping, the unsafe handling of flammable liquids, poorly maintained electrical equipment, and others.

The first item to consider is the assignment of the inspection task. The loss control manager should establish inspection schedules, the types of inspection to be conducted, and the routing of inspection reports. It is also the manager's responsibility to see that these inspections are conducted properly. In a large facility, it is important that the individual delegated with this responsibility have a good mechanical knowledge and, specifically, a knowledge of the facility fire protection equipment. The most logical individual will be maintenance oriented and, in addition, may be involved in the emergency organization operations. This individual should be reliable, possess the necessary knowledge and have the confidence of the loss control manager and the upper management.

Next, the inspection frequency should be determined. Normally, where sprinkler protection is involved, the recommended frequency is weekly, with particular emphasis on ensuring that water is available and sprinkler control valves are open. Where the facility is of small or moderate size and there are not fixed protection systems, monthly inspections are usually sufficient, since unusual conditions should be readily apparent to facility personnel.

Facility management should make sufficient time available to the inspector to complete the task. This inspector must have confidence that management will correct the deficiencies that he or she finds and give them proper attention on a priority basis.

Fire insurance underwriters usually furnish forms that are generally all inclusive and can be used in the performance of self-inspections. On the other hand, the loss control managers at many facilities have found it advisable to adapt a form that is specific to their facility. As an example, a self-inspection form from one of the major insurance underwriters is shown in Figure 17.7. This type of form is used extensively throughout industry. An example of an inspection form that is specific to a facility is shown in Figure 17.8.

While this section does not cover all possible elements that may be included in an inspection program, some of the items highlighted here are sprinkler system control valves, extinguishers, fire pumps, electrical deficiencies, and the handling of flammable liquids.

Every control valve should be recorded, showing whether each valve is open or closed and sealed or unsealed and including notes about conditions to be remedied. When the seal on a post indicator valve (PIV) is found to be broken or missing, it is advisable to make a drain test at the nearest sprinkler riser to ensure that the valve is open. Even though a target or supervisory switch indicates that the valve is open, PIVs should be tested by turning the valve wide open or until the "spring" is felt, thus ensuring that the valve is in an open position. The outside screw and yoke (OS&Y) valve does not require more than a visual inspection that the valve is open; however, if a seal is broken, the reason for the broken seal should be investigated. With dry pipe valves, checks should be made to ensure that the proper air pressure is provided in the system and, where "quick opening" devices (accelerators and exhausters) are provided, that the equipment is operable. An examination of the sprinkler piping and heads will indicate whether any heads are damaged, corroded, or loaded with deposits, dust, or lint and whether or not sprinklers are obstructed.

Self-inspection programs should include drain tests of sprinkler systems when weather conditions permit. The inspector should know the normal pressure and record it on the inspection form. This will indicate a problem such as a partially shut valve or other obstructions when the pressure is considerably lower than normal.

Industrial Risk Insurers

OVERVIEW FORMS PACKET
(See Section 12 in the OVERVIEW Manual)

Overview
A Total Management Program for Loss Prevention and Control.

# FIRE PROTECTION EQUIPMENT INSPECTION REPORT

Facility: ______________________ Inspector: ______________________

Location: ______________________ Date: ______________________

## The Following Items Should Be Checked At Least Weekly.

**Any "No" response should be explained.**

### WATER SUPPLY, SECTIONAL, AND SPRINKLER SYSTEM CONTROL VALVES

| Valve ID | Open | Shut | Sealed | Valve ID | Open | Shut | Sealed | Valve ID | Open | Shut | Sealed | Valve ID | Open | Shut | Sealed |
|---|---|---|---|---|---|---|---|---|---|---|---|---|---|---|---|
| | | | | | | | | | | | | | | | |
| | | | | | | | | | | | | | | | |
| | | | | | | | | | | | | | | | |

### PUBLIC WATER

Public water supply in service? ☐ Yes ☐ No ______________________ Pressure: ________ psi

Fire department connection accessible, caps in place, couplings free to rotate? ☐ Yes ☐ No ______________________

### FIRE PUMPS

| Pump ID | Type | Set For Auto.? | | Operated Today? | | Checklist Completed? | | Comments |
|---|---|---|---|---|---|---|---|---|
| | | Yes | No | Yes | No | Yes | No | |
| | | | | | | | | |
| | | | | | | | | |

### WATER SUPPLY TANKS

| Tank ID | Tank Full? | | Heater Working? | | Water Temp. | Comments |
|---|---|---|---|---|---|---|
| | Yes | No | Yes | No | | |
| | | | | | | |
| | | | | | | |

### AUTOMOTIVE FIRE APPARATUS

Each fully in service? ☐ Yes ☐ No ______________________

Checklist completed? ☐ Yes ☐ No ______________________

### SPECIAL EXTINGUISHING SYSTEMS

| System ID | Type | In Service? | | Date Last Serviced | Date Last Tested | Comments |
|---|---|---|---|---|---|---|
| | | Yes | No | | | |
| | | | | | | |
| | | | | | | |
| | | | | | | |

## The Following Items Should Be Inspected At Least Monthly.

**Any "No" response should be explained.**

### WET PIPE, DRY PIPE, DELUGE, AND PRE-ACTION SPRINKLER SYSTEMS

| System ID | Alarm Tested? | | Water Pressure | | | Heat Adequate? | | Air/Supv. Press. | Comments |
|---|---|---|---|---|---|---|---|---|---|
| | Yes | No | Static | Flow | Differ-ential | Yes | No | | |
| | | | | | | | | | |
| | | | | | | | | | |
| | | | | | | | | | |

 (over) 

**Figure 17.7** Example of a self-inspection form. *Source:* Reproduced with permission of Industrial Risk Insurers.

**FIRE EXTINGUISHERS, INSIDE HOSE CONNECTIONS, AND STANDPIPES**

Each unit in service? ☐ Yes ☐ No ____________________

Checklist completed? ☐ Yes ☐ No ____________________

**HYDRANTS, HOSE HOUSES, AND MONITOR NOZZLES**

| Monitor Nozzle/ Hydrant ID | Accessible? | | Drained? | | Equipment | | | | Comments |
|---|---|---|---|---|---|---|---|---|---|
| | | | | | Adequate? | | Cond. OK? | | |
| | Yes | No | Yes | No | Yes | No | Yes | No | |
| | | | | | | | | | |
| | | | | | | | | | |
| | | | | | | | | | |

**FIRE DOORS**

Fire doors and shutters in good condition? ☐ Yes ☐ No ____________________

Automatic closing devices operable? ☐ Yes ☐ No ____________________

**SMOKE AND HEAT, AND EXPLOSION-RELIEF VENTS**

Vents operable? ☐ Yes ☐ No ____________________

Areas around vents unobstructed? ☐ Yes ☐ No ____________________

**PROTECTIVE SIGNALING SYSTEMS**

All systems been tested satisfactorily? ☐ Yes ☐ No ____________________

**<u>OTHER PROTECTION DEFICIENCIES</u> FOUND DURING THE COURSE OF EACH INSPECTION SHOULD BE REPORTED BELOW:**

| | Yes | No | If "Yes," note location. |
|---|---|---|---|
| Stock within 36 inches of sprinkler heads? . . . . . . . . . . | ☐ | ☐ | |
| Sprinkler heads or piping bent? . . . . . . . . . . . . | ☐ | ☐ | |
| Sprinkler heads painted? . . . . . . . . . . . . . . | ☐ | ☐ | |
| Sprinkler heads or piping corroded? . . . . . . . . | ☐ | ☐ | |
| Sprinkler heads loaded with debris? . . . . . . . . | ☐ | ☐ | |
| Items hanging from, or supported by sprinkler heads? . . . . . | ☐ | ☐ | |
| Sprinkler heads obstructed by partitions? . . . . . . . . . . . | ☐ | ☐ | |
| Signs of internal sprinkler piping obstruction? . . . . . . . . | ☐ | ☐ | |
| Fire doors blocked by materials? . . . . . . . . . . . . . | ☐ | ☐ | |

**ADDITIONAL COMMENTS AND RECOMMENDATIONS**

____________________

____________________

____________________

Report reviewed by: ____________________ (signed) Position: ____________________

Has prompt action been initiated? ☐ Yes ☐ No ____________________

**FILE FOR REVIEW BY IRI REPRESENTATIVE**

**Figure 17.7** (*Continued*)

DISTRIBUTION: ORIGINAL TO: FACILITIES SERVICES
COPY TO: INSURANCE DEPARTMENT

If condition is satisfactory, enter a check ✓. If unsatisfactory enter O and explain on back of this form.

| | | FLOOR B | 1 | 2 | 3 | 4 | 5 | 6 | |
|---|---|---|---|---|---|---|---|---|---|
| Housekeeping | General orderliness & cleanliness | | | | | | | | |
| Portable Fire Extinguishers | Water units tagged within 12 mos. | | | | | | | | |
| | $Co_2$ & dry chem. units tagged within 6 mos. | | | | | | | | |
| | Gage pressure in operating range | | | | | | | | |
| | Properly mounted & sealed | | | | | | | | |
| | Accessible | | | | | | | | |
| Exits | All aistes & doors clear | | | | | | | | |
| | Doors open freely | | | | | | | | |
| | No storage in stairways | | | | | | | | |
| | No stairway doors blocked open | | | | | | | | |
| | All exit signs & stairway lights lit | | | | | | | | |
| Fire Hose | In rack | | | | | | | | |
| | Connected | | | | | | | | |
| | In good condition | | | | | | | | |
| Fire Doors | None blocked open | | | | | | | | |
| | No damaged doors | | | | | | | | |
| | All operate freely | | | | | | | | |
| | Fusible links missing or painted | | | | | | | | |
| Electrical | Junction & panel boxes closed | | | | | | | | |
| | No temporary wiring | | | | | | | | |
| | No frayed or unsafe wiring | | | | | | | | |

| Sprinkler Control Valves | Open | Locked |
|---|---|---|
| 2½" O.S. & Y.'s in 1st floor rear passageway | | |

Prepared by: ______________________ Name Title

Mgmt. review by: ______________________ Name Title

Deficiencies Corrected ☐ Yes ☐ No

*IF "NO". EXPLAIN ON BACK OF SHEET

**Figure 17.8** A self-inspection form specific to a facility.

Where water supply depends upon fire pumps, the pumps should be visually checked to determine that power is supplied to the pump where electrical drives are provided, that steam is available where steam is the driving force, and that fuel is available and that batteries are in operation with internal combustion engine drives.

Other parts of the overall facility protection that are also involved in the inspection program include water tanks (properly heated during extremely cold weather), hydrants (not obstructed), hydrant wrenches (available), and hose houses (readily accessible). Accessibility also applies to extinguishers and hose stations within the facility. A check of extinguishers will also indicate whether pressure is satisfactory and whether extinguishers are properly located and serviced.

Part of the inspection program should be to check that proper containers are being used for the handling of flammable liquids, that electrical grounds and bonding are in place, and that bulk quantities of flammable liquids are stored in the flammable-liquids vault.

Unsafe electrical items include overloading and improper overcurrent protection of circuits, poorly maintained electrical cords that are frayed or cracked, combustible storage located within switch-gear rooms, etc.

Other areas involved in the self-inspection program are safe welding and cutting practices, housekeeping practices, fire door, storage practices, control of smoking, heating, water supplies, special hazards, and special types of protection.

## 17.8 PLANT SITE SELECTION

Fire protection input plant design should be provided as early as possible and preferably at preengineering conferences. Although fire protection generally represents a small portion of the overall project cost, this expense can be substantial. Early involvement of the fire protection engineer can possibly avoid the necessity of costly change orders or addenda. While the fire protection engineer is, or should be, cost conscious, factors in addition to cost–benefit ratios and cost-effectiveness must be included in the decision-making process to determine whether or not to provide certain levels of protection or structural features. One major consideration is the uniqueness of a facility or operation and the importance of maintaining continuity of operation of a profit center. It is suggested that this type of consideration is often of greater importance than a cost–benefit ratio or immediate cost effectiveness.

### 17.8.1 Systems Approach

Fire safety can be incorporated into facility design in several different ways. One way is to require that the building be designed strictly to specification codes, such as building codes or military and government specifications. These are normally very stringent and have very little flexibility. They often do not take into account overall facility designs and often need to be interpreted because of confusion in certain specifications and lack of coverage of certain design features.

Another way to incorporate fire safety is to use performance codes. These codes try to overcome the inflexibility of specification codes by defining the expected fire safety performance of a separate component of the design and allowing alternative ways to meet the design. This approach has problems when one looks at the building as a whole design, since some components have a better chance of successful performance than do others.

The final approach is to look at fire safety as an integrated subsystem of the building, along with the functional, structural, electrical, or mechanical subsystems. This method relies on the use of the best engineering methodology rather than on strict compliance with codes. In other words, this approach provides an "equivalent" alternative to code requirements that provide equivalent fire safety. In the long run, this is the best way to incorporate fire safety into the overall design.

To effectively incorporate the facility's fire protection considerations into the design, the fire safety objectives must be identified. One descriptive tool to help designers identify the fire safety objectives in a systems approach is the systems tree of fire protection. This concept was first developed by the NFPA in its "fire safety concepts tree" (NFPA 550 1995) and by the General Services Administration in its systems tree (Nelson 1972). A trimmed-down version developed by the Architecture Life Safety Group, University of California at Berkeley, for the US Fire Administration, Department of Commerce, is shown in Figure 17.9. Reference should also be made to the *Guide to the Fire Safety Concepts Tree,* NFPA 550, for a more detailed fire safety concepts tree and explanation and examples on how to use it.

By moving through the various elements of the tree, the facility can be analyzed or designed for fire safety. As an example, the "self-termination" objective (Figure 17.9) can be easily controlled by the designers. Geometry, fuel, and ventilation can be manipulated through the design process. The height and volume of the room; the amount, volume, and distribution of furnishings, the flame spread factor, and the type of wall finishes will have effects on a fire.

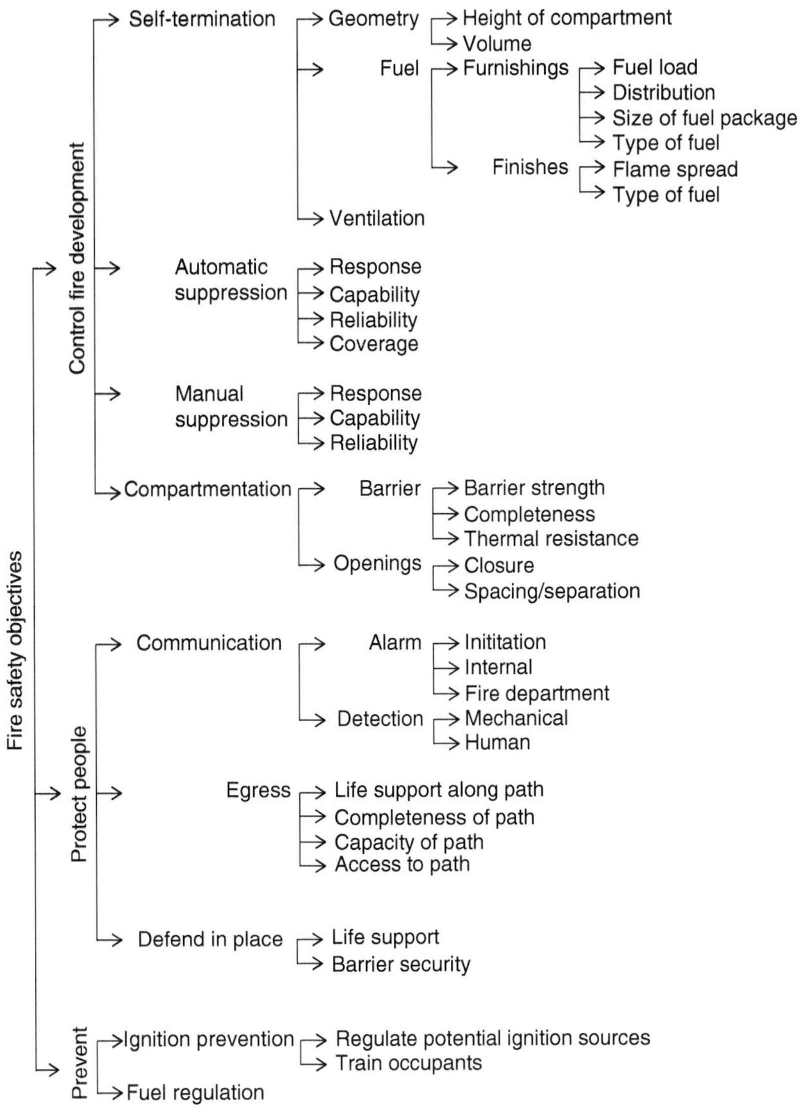


**Figure 17.9** Fire protection tree. *Source:* Courtesy of the Department of Commerce.

Design decisions have an effect on the facility's fire protection, which creates a compelling argument for the early introduction of fire protection engineering into the design process. Conversely, the "Prevent" section has an operational component that stresses workplace procedures such as storage, maintenance, etc.

### 17.8.2 Site Selection

Site features should be considered in the overall fire safety design, as well as in the other design subsystems. The site characteristics can play a major roll in the type of fire defenses used for the facility. The major site features follow.

***17.8.2.1 Water Supplies*** From the viewpoint of a fire protection engineer, one of the main considerations in plant site selection is the availability and strength of the water supply. Although water supply information is only one of many inputs that must go into site selection, this input can be of utmost importance because costly expenditures may be required if an adequate municipal supply is unavailable.

***17.8.2.2 Traffic and Transportation*** Time plays a major role in the response of the public fire department to the facility. Traffic and the arrangement of the streets in the community have an effect on the response time. Long fire department response times may signal a need for a fully equipped internal fire brigade.

***17.8.2.3 Public Fire Protection*** The type of protection, whether it is paid, part-paid, or voluntary; the location of the fire stations and the anticipated response time; and the nature and condition of the firefighting equipment and the alarm system need to be considered. Access to the site is very important. Built-up congested areas may include other properties, limiting access to the site. In addition, the topography of the site may limit fire department access.

***17.8.2.4 Exposures to the Site*** The facility could be exposed to conflagration hazards from a large concentration of combustibles within close proximity, such as hazardous chemical risks, oil refineries, large quantities of piled lumber, etc. During a fire, exposure can occur from horizontal radiation and from flames coming from the roof or top of a lower burning building. A number of factors – such as the intensity of the exposing fire, the duration, the total heat produced, the features of exposed and exposing buildings, the air temperature and humidity, the wind velocity and direction, and fire department access – significantly influence the danger of an exposing fire. A number of detailed guides have been established to quantify the exposure hazard. One of these is the *Recommended Practice for Protection of Buildings from Exterior Fire Exposures* (NFPA 80A 1993).

***17.8.2.5 Exposures to the Community*** As stated earlier, the EPA has established regulations concerning the community's right to know of the processing, storing, manufacturing, and use of hazardous materials by industry. In addition, the community may also have noise control ordinances, explosives ordinances, and other types of environmental regulations. The site should be planned to take into account the total effect on the surrounding community.

***17.8.2.6 Flood Plain*** Fire hazards are increased during flooding, and fire department access is severely restricted. Careful attention should be directed to whether the site is in a known flood plain. For instance, being in a 100-year flood zone may seem fairly secure, but, in fact, works out to be a 63% chance of being flooded within the 100-year period.

Information on flood-plain areas can be obtained from the US Geological Survey flood maps, US Army Corps of Engineers reports, Tennessee Valley Authority reports, and Federal Insurance and Mitigation Administration flood-hazard boundary maps and flood insurance studies. Tied in with the general subject is the nature of surface drainage in the area, which could possibly create flashflooding conditions.

***17.8.2.7 Earthquake Zone*** A large amount of the total damage caused by earthquakes may be due to the fires and explosions that follow. The site should be assessed for its location in a possible earthquake zone. A number of design modifications would be necessary if the site is located in such a zone. In a recent earthquake in the San Francisco Bay area (October 1989) that measured 7.0 on the Richter scale, the buildings designed to be earthquake resistant had little or no damage, particularly the high-rise buildings, while the conventionally designed older buildings had considerable damage.

Fault locations can be identified from fault and seismic risk maps available from the US Geological Survey, *Uniform Building Code,* National Science Foundation, and the American National Standards Institute.

### 17.8.3 Plant Layout

Fire protection generally requires compromises. Compromise, which may seem inappropriate, is based upon the necessity of being able to operate a plant efficiently incorporating engineering design features that, at the same time, will make the facility as fire safe as practical. For example, fire protection engineers often must be content with having a hazardous operation located in the center of a plant building, where explosion-relief venting can be obtained only through the roof, rather than at the periphery of a building, where explosion relief can be more readily obtained through side walls and the process can be more readily enclosed. Other factors such as process flow may dictate the exact location of the hazardous operation.

While the interior layout of a single-building plant can present some problems, the layout is relatively simple and clear-cut from a fire safety standpoint. Layouts become somewhat more complicated, again from a fire safety standpoint, when one is involved with complex, multibuilding plants involving hazardous processing, such as petrochemical facilities. Here, consideration must be given to the concentration of these hazardous processes; the accessibility; the location and protection of the vital utilities and control rooms; and the location of employee amenity areas, such as offices, locker rooms, lunch rooms, etc.

Important considerations involved in plant layout are as follows:

1. The separation of hazardous operations or processing by distance or building separation. If explosion hazards are involved, it must be determined that separation by distance is adequate, or as a less desirable alternative,

protective measures should be considered, including the construction of substantial explosion-resistant barricades. The adequacy of the distance must be determined by considering the nature of the hazard involved from a fire and explosion standpoint (possibly a vapor cloud analysis) and also by considering the exposure to loss of unusual value concentrations and specific problems involving the possibility of disruption to the continuity of operations, that is, the criticality of a part or unit of the production chain. The NFPA Handbook (NFPA 2008) is a good source for additional information on this topic.

2. Accessibility for manual fire suppression. This will determine the adequacy of roadways or other means of access by fire equipment to production units. The inability of fire equipment to reach certain locations and provide effective firefighting will often dictate the need for further automatic protection, such as automatic water spray systems and monitor nozzles either remotely or manually controlled.
3. The separation of important and vital plant utilities such as boiler rooms, electrical substations, and control rooms, from exposure to fire or explosion. In multibuilding operations, it is generally considered that the safest location for such facilities will be on the periphery of the plant site.
4. Location of employee amenity areas, as mentioned earlier, such as locker rooms, cafeteria, office areas, etc. From a standpoint of life safety, these areas should be located on the periphery of the facility.
5. The extent and concentration of valuable areas. While it is generally desirable to group more hazardous operations together, consideration should be given the concentration of values and separation to avoid the possibility of unacceptable losses. In single-building locations, this will usually involve the separation by fire walls of high-valued storage areas from processing areas.
6. Topography. This is an extremely important consideration where plant operations involve the use of extensive volumes of flammable liquids. In these instances, consideration must be given to locating vital production units and utilities so that they will not be exposed by possible flows from the release of flammable liquids under emergency conditions. Where is it impractical to locate these units at elevations not subject to flammable liquid flows, the use of diversionary curbs or walls to direct flammable liquid flows to safe impounding areas or confinement of flammable liquids by diking is advisable.

Large plants, and indeed every facility, require careful consideration of the plant layout, taking into account the elements discussed above. At the same time, one must recognize the problems brought about by modern technology, the increasing tendency to a greater concentration of values and hazards, the need for a high degree of employee training in handling complex processing, and the very important need, particularly in complicated, high-valued chemical plants, for preventative maintenance programs of the highest quality.

## 17.9 PROCESS EQUIPMENT AND FACILITIES DESIGN

### 17.9.1 Designing Safety into Process Equipment Operation

The safe design of most process equipment is usually assured by recognized codes such as the ASME *Boiler and Pressure Vessel Code*, ASTM and ANSI standards, and Tubular Exchanger Manufacturers Association (TEMA). The finished product is hydrostatically tested, inspected by a registered inspector, and stamped. However, after the equipment is installed, it is the responsibility of the user to ensure safe operation. This section discusses design features to protect equipment from poor operation, faulty fabrication, inadequate maintenance, and circumstances beyond the control of the user.

### 17.9.2 Fired Heaters

Fired heaters are either (i) direct (flue gas, kilns) or (ii) indirect (coils and jackets) and are used in chemical plant processing equipment.

Most heater firebox explosions occur during lightoff, usually due to the failure of the operator to follow the recommended lightoff procedure. After an unscheduled shutdown, the temptation is great to take a "shortcut" to make up for lost production. Therefore, it is mandatory that the company establish a clear operating procedure and see to its enforcement. A "prover"-type system to assure that all pilot and main burner fuel valves are completely closed may be installed. This involves a 1-in. valved bypass around the pilot control valve. When the bypass valve is opened, the pilot control valve cannot be opened unless the pressure in the bypass reaches operating pressure. After about 30 seconds, a timer automatically closes the fuel bypass valve. Such systems can improve safety on lightoff but are also no substitute for a good procedure enforced.

Other safety features normally recommended are as follows:

*Low-fuel-pressure shutdown.* A valve on both the pilot and main burner fuel lines can be interlocked to shut down sequentially on main burner fuel low pressure and low pilot fuel. These valves, commonly called "chopper valves," require resetting in the field so that an operator must go to the site and determine what caused the valve to trip.

*Coil backflow prevention.* The most effective way to extinguish a heater fire resulting from a process coil rupture is to shut off flow in the coil. This can be done by the use of a check valve on the coil outlet and a remote operated valve on the inlet to the coil. In fouling service, where deposits such as coke are laid down inside the coil, it is necessary to install a remote operated valve on the coil outlet, because a check valve may fail to close. Ability to close valves remotely, that is, from the control room or in the field (at least 75 ft, or 23 m, away from the heater) may be provided.

*Snuffing steam.* A provision for injecting steam for purging flammable vapors in the firebox prior to lightoff should be present. This steam can be used also for snuffing out a fire resulting from a ruptured process tube. Valves for injecting steam may be manual and are usually located a minimum of 50 ft (15 m) from the heater to permit access during a fire. On large heaters it may be advisable to install remotely operated valves for quick steam injection.

*Remote damper control.* Remote control of the heater stack damper should be considered, at least for large-duty heaters (over 20 million BTU per hour) and especially for heaters sharing a common stack. Besides helping to control the fire and limit damage inside ducting, improved operating economy will result.

*Stack thermocouple.* A thermocouple with high-temperature alarm may be used to detect a fire resulting from a ruptured coil. A thermocouple becomes more important with a windowless control room and little or no manpower in the field, because it provides quick fire detection.

### 17.9.3 Pressure Vessels

Some pressure vessels contain large quantities of flammable or toxic liquids that would create a serious hazard if released to the atmosphere. Although pressure vessels are inherently safe, precautions must be taken to prevent leaks and spills. Some protective systems are as follows:

*Shutoff valves.* A shutoff valve is usually provided on all bottom connections so that the vessel contents can be controlled when piping has holed through due to corrosion or damaged by an accident. A shutoff valve should seriously be considered when the vessel contents of flammable or combustible liquid exceeds 50 barrels (8 m$^3$). The valve is usually operated manually, but in the case of large-diameter piping a remote operated valve should be considered.

*Depressuring valves.* A depressuring valve is often used to reduce pressure inside an uninsulated vessel when it is exposed to fire. The vessel relief valve, set to open at the vessel maximum allowable working pressure, will not protect the vessel during a fire of long duration. If the vessel metal becomes overheated (above 1000 °F, 540 °C), the shell may rupture at the normal operating pressure. The valve is usually remotely operated and discharges to the flare header (where gases are allowed to burn off). When the valve is located close to the vessel, the valve operator may require fireproofing to assure that it functions properly when exposed to fire. Guidelines for designing depressuring systems are provided in API Recommended Practice 521 (API RP 521 2012).

Unless the normal operating pressure of the vessel is over 100 psig (690 kPa), it becomes difficult to reduce pressure effectively before the metal temperature rises to a high level; otherwise, the valve and depressuring line would have to be of very large diameter.

*Pressure relieving systems.* To prevent pressure vessels, exchangers, pumps, and other pressure-containing equipment from being overpressured and damaged, relief valves are installed. These valves are generally spring-loaded devices that to open when the pressure of the system reaches the maximum allowable working pressure. Relief valves should be located as specified in the ASME *Boiler and Pressure Vessel Code* (ASME 1996), API *Recommend Practice* 521 (API RP 521 2012), and API *Recommend Practice* 520 (API RP 520 2014).

The relieving load for each valve should be calculated based on upset conditions such as electrical or steam power failure, cooling water failure, blocked outlet, or fire. The valve is sized for a single maximum load. The total release from valves in the process unit is collected in the unit flare header, which then goes off-site to the flare knockout drum and is burned at the flare.

It is desirable to release only clean, nontoxic flammable vapors into the atmosphere. This can be done safely only if the vapors are

- Lighter than air.
- Heavier than air but with a molecular weight not exceeding 80 and with a discharge velocity of 500 ft s$^{-1}$ (150 m s$^{-1}$) to disperse vapors.

*Water draws.* Water accumulates in the bottom of some pressure vessels and periodically must be drawn off to an open sewer. This operation may be performed by manually operating a valve or by a control valve set to open automatically on high water level and close on low level. Both methods can be hazardous. The manual system requires that an operator remain at the valve whenever water is being drawn, to prevent the escape of flammable liquids or gases when the water level falls too low. If the operator is not alert, a combustible vapor cloud can form, when the vessel contains a light-vapor-pressure material, and expose both the operator and equipment. This hazard can be reduced by keeping the water draw line small, preferably 1 in., installing a low-level alarm to warn the operator, and keeping the water draw valve as far as practical from the point of emergence (usually an open drain hub going to the sewer). When the product is a high-vapor-pressure material that auto-refrigerates (freezes when released), there should be two water draw valves at least 2 ft apart. The downstream valve is opened first, and then the upstream valve is opened. If the downstream valve freezes in the open position, the upstream valve can be used to shut off flow.

While the precautions mentioned above enhance safety during manual water drawing, the responsibility for safety lies with management. The following items may be utilized to assure the required protection:

- Training personnel.
- Warning signs.
- Written procedures.
- Fresh-air breathing equipment.
- Color coding piping to identify hazardous materials.
- Plugging or capping drains.
- Toxic or combustible gas analyzers with visual and audible alarms.

When water is drawn automatically from vessels to an open sewer, the control valve may fail in the open position, permitting combustible vapors to escape. When automatic water draws are used, the vessel usually should have a low water level gauge to alarm both in the field and in the control room.

Whenever a vessel contains a toxic liquid material, the water draw should discharge to a closed system, but never to the flare header, which is generally designed for vapor releases only. Even with other less hazardous materials, it is preferable to discharge to a closed system when feasible.

### 17.9.4 Compressors

Compressors generally require protection from fire exposure from more hazardous equipment. However, some inherent hazards should be guarded against. The most serious is a seal leak that could cause the formation of a vapor cloud, resulting in a fire or deflagration. The severity of the event can be minimized by installing a remote shutoff valve on the compressor inlet an a check valve on the outlet to stop the flow of process gas.

If possible, compressors should not be installed inside closed buildings. When a shelter is advisable, the sides should be open to facilitate ventilation. Sometimes it becomes necessary to install compressors indoors in very cold climates or to install enclosures around them for noise suppression. In these cases, the building or enclosure should be ventilated, sometimes using fans, to prevent the accumulation of combustible vapors. The installation of combustible gas analyzers should be considered, to detect any accumulation in the early stages and permit shutting down the compressor. An inerting and fire extinguishing system using halon or $CO_2$ may be advisable inside noise suppression enclosures.

### 17.9.5 Pumps

Most pump fires result from seal leaks, which usually are a result of lack of proper maintenance. An experienced operator or maintenance person can detect a bad seal or bearing and shut down the pump before a failure occurs.

Pumps should be located outdoors where possible, rather than indoors, where combustible vapors can accumulate.

All pumps handling flammable liquids should have steel casings. Cast iron can crack readily when exposed to fire, releasing additional fuel.

### 17.9.6 Air Coolers

Fin Fan Coolers are prone to damage from any nearby fire, because the fan will induce the hot combustion gases, exposing the tubes. A clearly identified fan shutoff switch located at grade, about 50 ft (15 m) from the cooler, will permit shutting down in an emergency.

### 17.9.7 Exchangers

Shell and tube heat exchangers sometimes require cleaning, while the process unit remains on stream. If this is anticipated, provisions should be made to isolate the exchanger using blinds (barriers) to protect personnel. A valve alone should never be relied on for tight shutoff.

### 17.9.8 Piping

Piping should be adequately supported, guided, and anchored. The practice of hanging pipe from rods should be avoided, since such supports are vulnerable to failure in a fire. The use of sliding and bellows expansion joints and proprietary couplings using rubber rings should also be avoided, because they are prone to failure. Preferably, expansion loops should be used. Small size piping can be damaged easily by vibration and impact and should be adequately supported.

### 17.9.9 Emergency Shutdown Valves

Emergency shutdown (ESD) valves may be manual, solenoid, pneumatic, or motor operated. Solenoid and pneumatic-operated valves should be designed to fail in the safe position. The safe position is not always obvious. Should a valve for depressuring a pressure vessel fail in the open or closed position? The safety engineer prefers the valve to fail *open* in case it becomes inoperable in a fire, but the operator wants the valve to fail *closed,* to prevent all depressuring valves from opening upon loss of instrument air. Both conditions create a hazard. This situation may be resolved by installing backup air cylinders or installing depressuring valves in non-fire-exposed areas.

Where these valves cannot be made to fail safe, they should be fireproofed to assure operation during a fire. Fireproofing should include electrical power and valve operator, but not the steel valve body. When stainless steel instrument air tubing is used, it usually requires no fireproofing. Electric-motor-operated valves fail in the last position and must be fireproofed to assure operation.

ESDs are used to isolate systems where a hazard exists, such as fired heaters, compressors, vessels containing large quantities of flammable materials or toxic materials, or long pipelines where a break could occur.

### 17.9.10 Battery Limit Valves

Battery limit valves are used to isolate a process unit from the main pipe rack and other process units to permit safe shutdown. In some applications they are used to shut in the entire plant. They can also be used in an emergency to shut off feed to a process unit during a fire. The valves usually are located on top of the unit pipe rack before joining the main pipe rack. Since they are used infrequently, they are generally manually operated and should be located in a fire-safe area to be tenable in an emergency.

### 17.9.11 Drainage

Grading within the process unit is important, not only for the removal of stormwater but also for the removal of flammable and combustible liquid spills. The preferred design of the concrete slab is to have the high point of paving beneath the centerline of the process unit pipe rack with catch basins located on both sides of (but not under the rack). This design will cause any piping spill to run from beneath the rack and also prevent a spill from some other source from running beneath the rack. No catch basin should be located beneath equipment, especially fired heaters. Catch basins should be fire stopped by a turned down elbow or over–under weir, both having a 6-in. liquid seal. These liquid seals will prevent a fire from going through the entire sewer and also stop an explosion at the first seal. The seal or fire stop is usually located at the inlet to the catch basin, with a straight-through outlet so that flammable and combustible liquids will not be tapped in the catch basin.

Stormwater and oily water may be combined in a single sewer system, but it is becoming common to have separate storm and oily water sewers. With secondary and tertiary water treatment systems being installed, economics dictate segregating the two types of water to reduce the quantity of water being treated.

The stormwater system should be designed to handle both the maximum rate of rainfall expected and the fire water rate anticipated, but not simultaneously. The fire water drainage requirement should take into account the fixed monitor, fixed water sprays, and hose lines that are likely to be utilized. This does not include the total fire water that could be used by activating all fire water systems simultaneously, but the equipment expected to be required to fight a fire. A total of 4000 gal $min^{-1}$ (908 $m^3$ $h^{-1}$) is often used to fight a fire in one process unit. Catch basins are often designed to handle 500 gal $min^{-1}$ and the maximum drainage area for one catch basin usually does not exceed 3000 $ft^2$ (279 $m^2$).

### 17.9.12 Electrical Classification of Areas

The classification of hazardous locations for electrical equipment is defined in NFPA 30, *Flammable and Combustible Liquids Code* (NFPA 30 1993) and NFPA 70, *National Electric Code*, Article 500 (NFPA 70 1996), and API 500A, *Classification of Locations for Electrical Installations in Petroleum Refineries* (API 500A 1982). Once an area has been classified, the *National Electric Code* specifies the type of electrical equipment that can be installed safely.

Most areas inside plants handling flammable liquids will be classified as Division II. Division I areas are those where heavier-than-air gases are present in below-grade areas such as sumps, open trenches, and sewers; enclosed areas containing equipment handling flammable such as compressor enclosures and pump houses; and immediately around vents. With lighter-than-air gases, Division I areas might be found in the roof portion of an inadequately ventilated compressor house.

Electrical switches and other sparking devices located in Division I and II areas should be explosions proof or intrinsically safe. Motors located in Division II areas should be nonsparking type rated for Division II service; motors in Division I areas should be explosion proof.

"Explosion proof" does not mean that an explosion cannot occur. If flammable vapors penetrate into the electrical housing, an explosion will occur, but it will be contained within the housing and not ignite vapors outside. However, the electrical equipment inside the housing may be destroyed.

### 17.9.13 Control Buildings

Since control buildings contain numerous electrical sparking devices, it is important to exclude flammable gases. This is accomplished by keeping the inside of the building at a slightly higher pressure (usually $\frac{1}{4}-\frac{1}{2}$ in. of water). The pressurization air must be taken from a gas-free area, and the intake is usually at least 25 ft (7.6 m) above grade. It is preferable to use two air blowers with an interlock so that if the operating blower fails the other blower will operate. An alarm should warn the operator when a blower has failed so that it may be repaired. A vestible at each doorway is preferred to maintain pressure when a door is opened.

Pressurization is preferable to installing explosion-proof electrical equipment in the control room. Not only is explosion-proof equipment expensive, but the explosion proof rating of such equipment is also difficult to maintain over a period of time.

Many central control buildings constructed today are designed to be blast resistant. The objective of blast-resistant construction is to protect personnel in the building from an explosion in a process unit and also to permit an orderly shutdown of all other units. Such buildings are designed for the usual maximum anticipated wind and snow loads and also a short-time peak blast overpressure. Determining the design peak overpressure may be difficult. The pressure anticipated will depend upon the distance to the control building from the process units and a realistic determination of what flammable material and how much should be involved. A hazards analysis of all involved units may be advisable in order to make a realistic determination of the possible blast severity. When the critical material processed has a tendency to form a high-velocity (above the speed of sound) detonation and the control building is close to the source (50–100 ft). Then a blast-resistant design may not be practical. The only solution in this is to locate the control building further from the blast source (200 ft or more).

Blast-resistant construction usually requires the elimination of windows. Although some operators object to running a plant they cannot see, it has been proven that an operator can "see" almost anything almost anything that goes wrong in the process unit from instrument data. With the use of computerized data processing, instantaneous information on what is happening in the unit has been greatly expanded.

## 17.10 TRAINING

Continuous training of plant personnel is very important to assure that proper actions are take in any emergency. A good training program will teach personnel to take the correct action instinctively in any emergency, because all emergencies that can arise have been anticipated and carefully thought out prior to their occurrence. The following types of preplanning and training are necessary:

- Training of process unit operators regarding corrective measures to take in all unit upset conditions and in ESD of the unit of plant. The unit operation manual should contain detailed operational procedures for all emergency conditions.

- Training of the plant firefighting crew in the operation of all plant firefighting equipment. A fire training ground should be established where various types of fires can be started and extinguished by the plant firefighting crew. It is usually beneficial to invite the local municipal fire department to train with you in order to familiarize them with the types of hazards in the plant and the use of plant firefighting equipment. Other plant personnel should be trained in the use of first-aid firefighting equipment, such as fire extinguishers.
- Training of the plant security force regarding what action to take in emergencies, such as making the accident area accessible to municipal and mutual aid fire companies, and keeping the area free of people who have no useful task to perform in the emergency.
- Training personnel in first aid so that they can administer the proper minor treatment until trained medical help arrives. This training serves as a useful purpose not only in the plant but also in the home and elsewhere outside the plant.
- Instructing the medical department (nurse, doctor and ambulance crew) in actions to be taken in emergencies and accidents. They should be familiar with the effects of all toxic and hazardous materials used in the plant and the recommended treatment.
- The affected plant personnel should be trained in the proper use of all plant safety equipment such as air masks, safety showers, and fire blankets.

## BIBLIOGRAPHY

Anyakora, S.N., Engel, G.F.M., and Lees, F.P. (1981). Some data on the reliability of instruments in the chemical plant environment. *Chemical Engineer* (November).

API 500A (1982). *Classification of Locations for Electrical Installations in Petroleum Refineries.* Washington, DC: American Petroleum Institute.

API RP 2003 (2003). *Protection against Ignitions Arising out of Static, Lightning, and Stray Currents.* Washington, DC: American Petroleum Institute.

API RP 520 (2014). *Recommended Practice for the Design and Installation of Pressure-Relieving Systems in Refineries.* Washington, DC: American Petroleum Institute.

API RP 521 (2012). *Guide for Pressure Relieving and Depressuring Systems.* Washington, DC: American Petroleum Institute.

ASME (1996). Section VIII, Pressure Vessels, Division 1 and 2. In: *Boiler and Pressure Vessel Code.* New York City, NY: American Society of Mechanical Engineers.

ASTM (1983). *E119, Fire Tests or Building Construction and Materials.* West Conshohocken, PA: American Society of Testing and Materials.

Bochnak, P.M. (1976). Developments in fire detection. *Proceedings from the Mutual Engineers' Conference, American Mutual Insurance Alliance*, Atlanta, GA (5–7 April 1976).

Bochnak, P.M. (1977). Smoke and fire detectors for wood heating systems. *Proceedings from the Wood Heating Seminar I, Wood Energy Institute*, Cambridge, MA (20–21 April 1977).

Bochnak, P.M. (1980). State of the art: fire alarm technology related to protecting life in work places. *Proceedings from the 52nd Fire Department Instructors Conference, the International Society of Fire Service Instructors*, Memphis, TN.

Bochnak, P.M., NIOSH Member and Project Officer (1982). *Classification of Gases, Liquids, and Volatile Solids Relative to Explosion-Proof Electrical Equipment.* NMAB 353-5. Committee on Evaluation of Industrial Hazards. Performed by National Academy of Sciences under sponsorship of NIOSH (August).

Bochnak, P.M., NIOSH Member and Project Officer (1982). *Rationale for Classification of Combustible Gases, Vapors, and Dusts with Reference to the National Electrical Code.* NMAB 353-6. Committee on Evaluation of Industrial Hazards Performed by National Academy of Sciences under sponsorship of NIOSH (July).

Bochnak, P.M. (1991). *Fire Loss Control: A Management Guide.* New York: Marcel Dekker, Inc.

Bochnak, P.M. (1986). *NIOSH ALERT: Request for Assistance in Prevention Fatalities Due to Fires and Explosion Oxygen-Limiting Silos.* DHHS (NIOSH) Publication No. 86-118. Cincinnati, OH: National Institute for Occupational Safety and Health.

Bochnak, P.M. and Moll, M.B. (1984). *Relationships Between Worker Casualties, Other Fire Loss Indicators, and Fire Protection Strategies (internal report).* Morgantown, WV: U.S. Department of Health and Human Services, National Institute for Occupational Safety and Health.

Bochnak, P.M. and Pettit, T.A. (1983). *Occupational Safety in Grain Elevators and Feed Mills.* DHHS (NIOSH) No. 83-126. Cincinnati, OH: National Institute for Occupational Safety and Health.

Bochnak, P.M. and Pettit, T.A. (1983). *Comprehensive Safety Recommendations for Land-Based Oil and Gas Well Drilling.* DHHS (NIOSH) No. 83-127. Cincinnati, OH: National Institute for Occupational Safety and Health.

Bochnak, P.M., Pizatella, T.J., Lark, J.J., and NIOSH (1981). *LP-Gas Emergencies: A Review and Appraisal of Selected Ignited Leaks during Liquid Transfer Operations*. Morgantown, WV: National Institute for Occupational Safety and Health.

Browning, R. (1969). Reactive probabilities of loss incidents. *Chemical Engineering* (15 December): 134.

Cote, A.E. (ed.) (1991). *Fire Protection Handbook*, 17e. Quincy, MA: National Fire Protection Association.

Cote, R. (ed.) (1994). *Life Safety Code Handbook*, 6e. Quincy, MA: National Fire Protection Association.

Cote, A.E. (ed.) (1997). *Fire Protection Handbook*, 18e. Quincy, MA: National Fire Protection Association.

Earthquake Concerns (1988). *P8805*. Norwood, MA: Factory Mutual Engineering Corporation.

EPA (1996). *Hazardous Chemical Reporting: Community Right to Know*, 40 CFR 370. Washington, DC: Environmental Protection Agency.

Factory Mutual Engineering Corporation (1993). *The Handbook of Property Conservation*. Norwood, MA: Factory Mutual Engineering Corporation.

Factory Mutual Engineering Corporation (1996). *Approval Guide*. Norwood, MA: Factory Mutual Engineering Corporation.

Flood! And Whey You're Not as Safe as You Think (1980). *P8001*. Norwood, MA: Factory Mutual Engineering Corporation.

Lerup, L., Cronwrath, D., and Liu, J.K.C. (1977). *Learning from Fire: A Fire Protection Primer for Architects*. Washington, DC: U.S. Fire Administration.

McCoy, C.S. and Hanly, F.J. (1985). Fire-resistant lubes reduce danger of refinery explosions. *Oil and Gas Journal* **10**: 191–197.

National Disasters (1988). *P8812*. Norwood, MA: Factory Mutual Engineering Corporation.

Nelson, Harold E. (1972). *The Application of Systems Analysis to Building Fire Safety Design*. Accident and Fire Prevention Division, General Services Administration.

NFPA (1990). *Industrial Fire Hazards Handbook*, 3e. Quincy, MA: National Fire Protection Association.

NFPA (1994). *NFPA Inspection Manual*, 7e. Quincy, MA: National Fire Protection Association.

NFPA (1998). *National Fire Codes*. Quincy, MA: National Fire Protection Association.

NFPA (2008). *Fire Protection Handbook*, 20e. Quincy, MA: National Fire Protection Association.

NFPA (2018). *NFPA 101 – Life Safety Code Handbook*. Quincy, MA: NFPA.

NFPA 10 (1994). *Standard for Portable Fire Extinguishers*. Quincy, MA: National Fire Protection Association.

NFPA 11 (1994). *Standard for Low Expansion Foam and Combined Agent Systems*. Quincy, MA: National Fire Protection Association.

NFPA 12 (1993). *Standard on Carbon Dioxide Extinguishing Systems*. Quincy, MA: National Fire Protection Association.

NFPA 12A (1992). *Standard for Halon 1301 Fire Extinguishing Systems*. Quincy, MA: National Fire Protection Association.

NFPA 15 (1990). *Standard for Water Spray Fixed Systems for Fire Protection*. Quincy, MA: National Fire Protection Association.

NFPA 1961 (1992). *Standard on Fire Hose*. Quincy, MA: National Fire Protection Association.

NFPA 1963 (1963). *Standard for Fire Hose Connections*. Quincy, MA: National Fire Protection Association.

NFPA 20 (1993). *Standard for the Installation of Centrifugal Fire Pumps*. Quincy, MA: National Fire Protection Association.

NFPA 214 (1992). *Standard on Water-Cooling Towers*. Quincy, MA: National Fire Protection Association.

NFPA 30 (1993). *Flammable and Combustible Liquids Code*. Quincy, MA: National Fire Protection Association.

NFPA 321 (1991). *Standard on Basic Classification of Flammable and Combustible Liquids*. Quincy, MA: National Fire Protection Association.

NFPA 550 (1995). *Guide to Fire Safety Concepts Tree*. Quincy, MA: National Fire Protection Association.

NFPA 70 (1996). *National Electric Code*. Quincy, MA: National Fire Protection Association.

NFPA 80A (1993). *Recommended Practice for Protection of Buildings from Exterior Fire Exposures*. Quincy, MA: National Fire Protection Association.

NFPA 8502 (1995). *Prevention of Furnace Explosions/Implosions in Multple Burner Boilers*. Quincy, MA: National Fire Protection Association.

NFPA 86 (1995). *Ovens and Furnaces*. Quincy, MA: National Fire Protection Association.

OSHA (1996). *General Industry Standards – 29 CFR 1910*. Washington, DC: Occupational Safety and Health Administration.

SFPE (1995). *Handbook of Fire Protection Engineering*, 2e. Quincy, MA: National Fire Protection Association.

Smith, M.R. (ed.) (1989). *A Guide to Occupational Fire Prevention and Protection*, NC-OSHA Industry Guide, vol. **4**. Raleigh, NC: North Carolina Department of Labor.

Underwriters Laboratories (1984). *UL 1709, Structural Steel Protected for Resistance to Rapid Temperature Rise Fires*. Northbrook, IL: Underwriters Laboratories, Inc.

Underwriters Laboratories (1996a). *Fire Protection Equipment Directory*. Northbrook, IL: Underwriters Laboratories, Inc.

Underwriters Laboratories (1996b). *Publications Stock, Fire Resistance Directory*. Northbrook, IL: Underwriters Laboratories, Inc.

Warren, J.H. and Corona, A.A. (1975). This method test fire protective coatings. *Hydrocarbon Processing*.

CHAPTER 18

# PHILOSOPHY AND MANAGEMENT OF ENGINEERING CONTROL

PAMELA GREENLEY and WILLIAM A. BURGESS

*Massachusetts Institute of Technology, Bldg. 56-235, 77 Massachusetts Ave., Cambridge, MA, 02139-4307*

Adapted from "Philosophy and management of engineering control" by William A. Burgess, Chapter 5 in R.L. Harris, L.V. Cralley, and L.J. Cralley (eds.), *Patty's Industrial Hygiene and Toxicology*, 3e, Vol. IIIA. New York: Wiley, 1994, pp. 129–180.

## 18.1 INTRODUCTION

This chapter will cover the engineering methods widely used to minimize adverse effects on health, well-being, comfort, and performance at the workplace and in the community. Major attention will be given to controls for chemical airborne contaminants, although many of the control principles are also applicable to biological and physical hazards. Brief case studies will illustrate the applications of the techniques. The target for control may be a hand tool, a piece of equipment or device, an integrated manufacturing process such as an electroplating line, or a complete manufacturing facility in a dedicated building such as a foundry.

The critical zones of contaminant generation, dispersion, and exposure are shown in Figure 18.1. Ideally the goal is to design each element of the process to eliminate contaminant generation. If it is impossible to achieve this goal and a contaminant is generated, the second defense is to prevent its dispersal in the workplace. Finally, if we fail in that defense and the material released from the operation results in worker exposure, the backup control is collection of the air containing the contaminant by exhaust ventilation. When engineering controls are not feasible or as a supplement to them, personal protective equipment such as respiratory protection and clothing are used. See Chapters 16 and 17 for guidance on their selections and use. The contaminant is frequently removed from the exhaust airstream by air cleaning before returning the air to the workplace or the general environment. Until recently, little emphasis was given to the prevention of contaminant generation and release, and ventilation was relied on as a principal remedy. Since the late 1980s a cooperative effort between industry and federal and state regulatory agencies to prevent pollution has emerged. Although the main thrust of the pollution prevention approach is to reduce environmental releases, there are frequently benefits to the occupational environment as well. In this chapter we will confirm the importance of ventilation, but emphasis will be placed on the primary controls associated with process and material design to minimize generation and release of the contaminant.

The Occupational Safety and Health Administration (OSHA) considers engineering controls as any modification of plant, equipment, processes, or materials to reduce employees' exposure to toxic materials and harmful physical agents. The zones of generation and control methods for each are listed in Figure 18.2. Each of these approaches should be considered when designing exposure controls. These approaches provide a useful outline for discussion and act as checklists to ensure a comprehensive review. A comprehensive algorithm for control of air contaminants has been designed by Sherwood and Alsbury (1983). Brandt (1947) states: "Measures for preventing the inhalation of excessively contaminated air have been discussed by many authors, and there are as many classifications of these methods as there are papers on the subject. The principles expounded, however, are always essentially the same." Brandt also reminds the reader that "The control of an atmospheric health hazard is rarely accomplished by a single measure. It usually involves a combination of measures."

*Handbook of Occupational Safety and Health*, Third Edition. Edited by S. Z. Mansdorf.

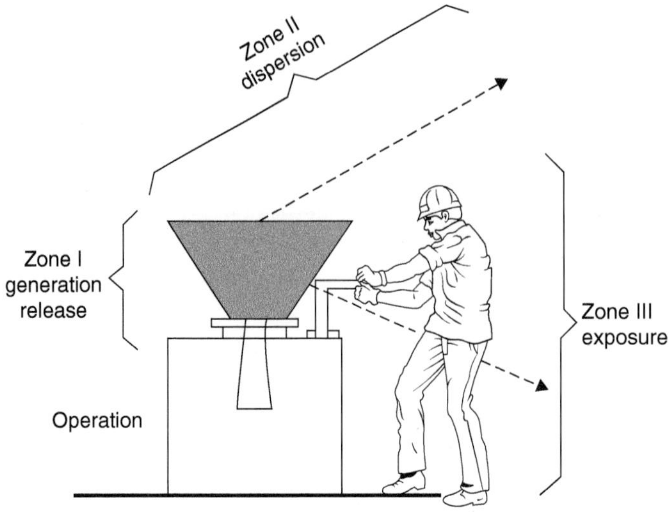


**Figure 18.1** Contaminant generation, release, and exposure zones in the workplace.

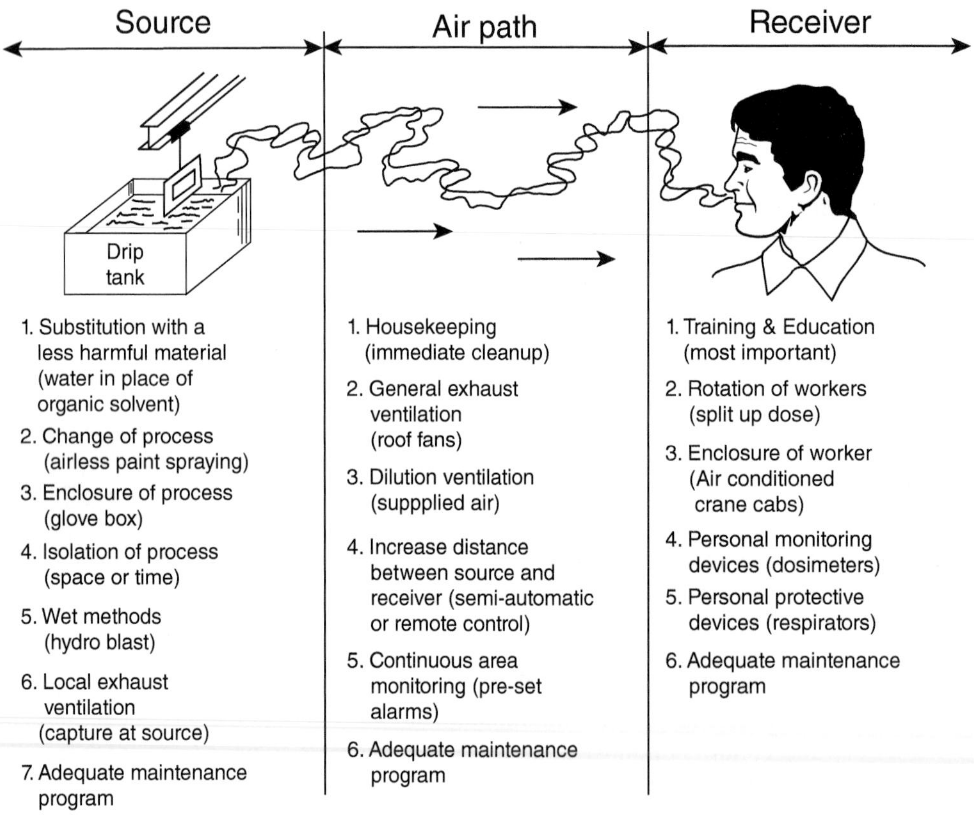


**Figure 18.2** Generalized diagram of methods of control. *Source:* From Olishifski (1988).

In this chapter engineering controls will include all techniques except personal protective devices and administrative controls and those changes made in the work schedule to reduce the time-weighted average exposure of the worker.

## 18.2 GENERAL GUIDELINES

The strategy to be used in solving a control technology problem varies depending on the setting. As described above, the control problem may involve a specific piece of equipment, a process, or an integrated manufacturing plant. In a common situation faced by many in the 1990s, a shop repairing electric motors must consider a change in solvents in a vapor-phase degreaser. The number of persons participating in the review is small, the decision will be made in a short time, and it will probably rely heavily on vendor information. The range of options presented in this chapter probably will not be thoroughly reviewed. An industrial hygienist will not be involved unless the service shop is a support facility operated by a major company or consultation is available from an insurance carrier, consulting industrial hygienist, or state OSHA consulting program.

A far more complex situation is the scale-up of a chemical process to manufacture a photographic dye. The research to develop the new molecule and synthesize a small quantity for initial tests is completed by a research chemist at a laboratory bench. At this stage little is known about the chemical or its toxicity. However, since gram quantities are handled in a well-controlled laboratory hood, the risk is negligible. The next step involves a scale-up of the operation in a pilot plant facility designed to manufacture kilogram quantities of the material for testing purposes. This testing will provide limited data on the physical properties of the new chemical, its product potential, and toxicity data. If the results are encouraging, production scale-up is considered. A premanufacturing notification (PMN) must be submitted to the Environmental Protection Agency (EPA) prior to production. This submission, which requires a description of air contaminant control technology and an estimation of worker exposure, involves a comprehensive review of the operation as noted in Table 18.1. It is decided to manufacture the chemical at an existing plant using conventional chemical processing equipment. If the industrial hygienist has been involved in the pilot plant operation, he or she is well placed to participate in a review of the control technology necessary for the full-scale plant. As noted in Table 18.1, the range of talent available to define the engineering controls is extensive, and the time frame is extended. Hopefully, step-by-step review to identify the appropriate engineering controls will minimize worker exposure. If exposure problems are not anticipated, controls must be retrofitted at significant cost, whereas early integration of controls is most easily accepted and economically implemented.

The most complex engineering control package is encountered in the design and construction of a major production facility. The project is complex due to the range of issues that are encountered; however, the design process is started with a clean design sheet, and state-of-the-art controls can be included in the design. In the example in Table 18.1, a large number of specialists are included in the review team. If the design is done in-house, the industrial hygienist should be a key participant. If the design work is contracted to a design and construction firm, plant personnel are one step removed from the design process. The company will have a project engineer to interface between the outside team and company engineering group. In this case clear communication must be established by the industrial hygienist to ensure involvement in the design review.

For the health and safety professional facing his or her first process hazard review, the job may seem awesome. An initial step is to identify your responsibility. Does it include merely health issues at the workplace, or are environmental and fire protection problems to be included in your review? Once your responsibility has been defined, the issues of concern for this particular plant should be identified. Mature industries such as metallurgical, glass, and ceramics have been studied, and extensive information is available on standard operations and sources of contamination. The EPA has published emission rates for many processes that are useful in process hazard reviews. Another source of information on control technology for mature industries is the series of over 50 technical reports published by the National Institute for Occupational Safety and Health (NIOSH) (see Appendix A).

If the company has similar operations elsewhere, obtain material and process data, industrial hygiene survey data, and air sampling information from that plant to assist in the review. A step-by-step review of the operations is then undertaken. This health and safety review should be as rigorous as the review of percent yield on the process and should produce a detailed process flow diagram showing the locations of contaminant loss in addition to the major physical stresses such as noise, heat, and ergonomics. It is important to also consider the ancillary systems including compressed air, refrigeration, cooling towers, and water treatment. Bulk handling of chemicals and granular minerals frequently presents a major industrial hygiene problem and should be given special attention. Finally, a detailed description of the workers' tasks including required emergency response actions is required, since work location and movement have a major effect on contaminant control. A comprehensive checklist for review of environmental and occupational health and safety issues in planning major plant construction is given by Whitehead (1987).

An element that is frequently overlooked in the design of engineering controls is the worker. A review of a specific operation conducted by several workers frequently will identify subtle differences in the way the task is performed. A worker inspired by a wish to reduce lifting, provide "cooldown" in the heat, or reduce dust exposure may modify the tasks to accomplish this end. Workers frequently have insight into details of the operation that the design engineer will overlook. The video real-time monitoring approach discussed in Section 18.4.1 is useful in defining these modifications in tasks.

It is difficult to recommend engineering controls for emerging technologies. Hopefully each industrial group will develop and publish its own engineering controls. Such is the case with advanced composite manufacturing, a technology that has seen expanding applications in recent years in the manufacture of aircraft. An association of companies involved in this work has published a volume describing the major hazards and the controls that should be applied (SACMA 1991).

**TABLE 18.1 Specialists Participating in Risk Assessment of New Chemical Manufacture**

| | Administrative | Legal | Sales | Marketing | Chemists | Engineers | Toxicologists | Occupational hygienists | Environmental scientists | Statisticians | Risk analyst | Transportation systems analysts | Economists | Information specialists |
|---|---|---|---|---|---|---|---|---|---|---|---|---|---|---|
| Part I: General information | | | | | | | | | | | | | | |
| A. Manufacturer identification | × | × | | | | | | | | | | | | |
| B. Chemical identity | | | | | × | | | | | | | | | × |
| C. Marketing data | | | × | × | | | | | | × | | | × | |
| D. *Federal Register* notice | × | × | | × | × | × | × | × | × | | | | | |
| E. Schematic flow diagram | | | | | × | × | | × | × | | | | | |
| Part II: Risk assessment data | | | | | | | | | | | | | | |
| A. Test data | | × | | | × | | × | | × | × | × | | | |
| B. Exposure from manufacture | | | | | | | | | | | | | | |
| 1. Worker exposure | | | | | × | × | × | × | | × | × | | | |
| 2. Environmental release | | | | | × | × | | | × | × | × | | | |
| 3. Disposal | | | | | × | × | × | × | × | | × | | | |
| 4. By-products, etc. | | | | | × | × | | | | | × | | | × |
| 5. Transportation | | × | | × | | × | | | | | × | × | | |
| C. Exposure from operations | | | | | | | | | | | | | | |
| 1. Worker exposure | | | | | × | × | × | × | | × | × | | | |
| 2. Environmental release | | | | | × | × | | | × | × | × | | | |
| 3. Disposal | | | | | × | × | × | × | × | | × | | | |
| D. Exposure from consumer use | × | × | × | × | | × | | | × | | × | | | |
| Part III: Risk analysis and optional data | | | | | | | | | | | | | | |
| A. Risk analysis and optional data | × | × | | | × | | × | | × | | × | | | |
| B. Structure–activity relationships | | | | | | | × | | | | × | | | |
| C. Industrial hygiene | × | × | | | × | × | × | × | | | × | | | |
| D. Engineered safeguards | | | | | × | × | × | × | × | | × | | | |
| E. Industrial process and use restriction data | | | | | × | × | | | | | | | | |
| F. Process chemistry | | | | | × | × | | | | | | | | |
| G. Nonrisk factors: economic and noneconomic benefits | | | | | × | × | | | | | | | | |
| | × | × | × | × | × | × | | | | | × | | × | |

*Source:* Arthur D. Little, Inc.

Frequently the best control cannot be achieved directly on the design board. In the 1970s a number of cases of asbestosis in shipyard workers fabricating asbestos insulating pads for steam propulsion plants prompted the plant to install an integrated workstation incorporating a number of control features. Initial designs of the downdraft work table were fabricated and evaluated by workers, resulting in a number of recommendations that were incorporated in the final design, improving its acceptance and performance (Figure 18.3).

## 18.3 CONTROL STRATEGY

The general strategy as noted earlier is to eliminate the generation source. If that is not possible, then prevent escape of the contaminant, and if it does escape, collect and remove the contaminant before the worker is exposed. This strategy will be reviewed by considering action on toxic materials, equipment, processes, job tasks, plant layout, and exhaust ventilation.

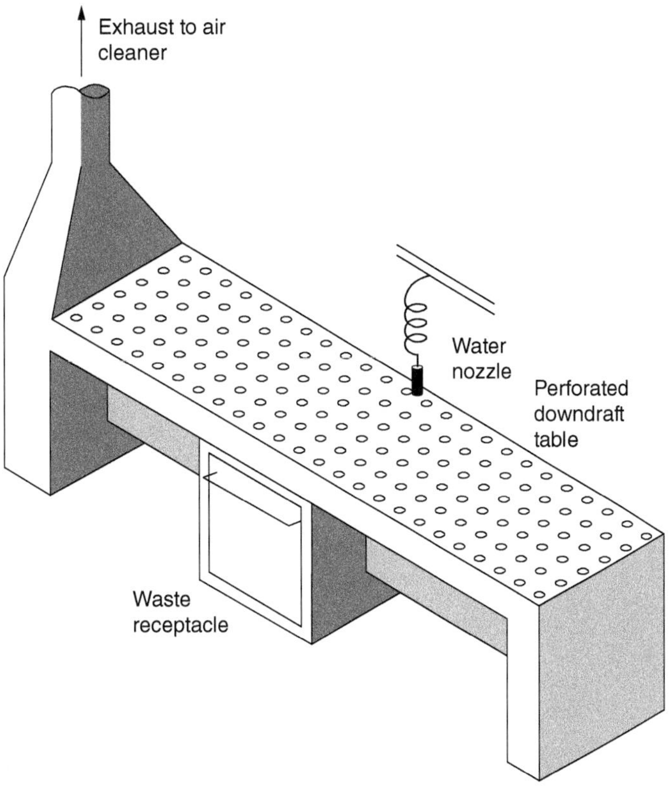


**Figure 18.3** Integrated controls on an asbestos insulation fabrication workbench.

### 18.3.1 Toxic Materials

The use of toxic materials ranges in complexity from simple wet degreasing to the synthesis of new chemicals. See Chapter 3 for a review of the hazards associated with various industries. In degreasing the exposure is associated with the use of the cleaning agent or handling its waste. In synthesis of a new chemical, exposure may occur to raw materials, intermediates, by-products, and the new chemical and waste streams. In minimizing potential worker exposure to these materials, it is necessary to consider the following options.

***18.3.1.1 Eliminate the Toxic Material*** This option refers to elimination without replacement. Frequently toxic chemicals are used in a process to improve yield or in a product to improve function or appearance. In review, it may be possible to eliminate the chemical. The pressure to do so may come from potential health effects, the cost of workplace, and environmental controls to introduce the toxic chemical into the plant, or it may arise from pressure from the marketplace. An example of the latter resulted from legislation in Switzerland requiring precautionary labeling of batteries containing mercury. The conventional Leclanche dry cell battery utilizes electrodes of manganese dioxide and zinc with an electrolyte of ammonium chloride. A characteristic of the cell is zinc corrosion with subsequent power loss. A mercury compound is added to inhibit this corrosion and resulting performance degradation. At least one manufacturer has determined that the performance enhancement from mercury is not necessary and the mercury compound has been removed from the battery formulation (Ahearn et al. 1991). This change has eliminated exposure to mercury for several hundred workers in this company. The Montreal Protocol calling for the elimination of ozone-depleting compounds has resulted in the use of water and detergents as cleaning agents in many areas. Other possible substitutes for ozone-depleting compounds are discussed in Section 18.3.2.4.

### 18.3.2 Replace with Alternate Material

This control approach holds the greatest potential for significant reduction in worker exposure to toxic chemicals. Historically, the approach has had success in the United States. One of the most effective steps was the cooperative action taken by the US Public Health Service, the State of Connecticut Department of Public Health, industry, and labor unions in 1941 to prevent mercurialism by replacing mercuric nitrate with nonmercury compounds in the carroting process in felt hat manufacture.

EPA's ability to ban the use of toxic materials, in this case asbestos, was severely curtailed in the case *Corrosion Proof Fittings v. EPA*. The court ruled that EPA must fully evaluate the relative safety of known alternatives to asbestos and choose the least burdensome regulatory measure for each application (Rossi and Geiser 1994). Another environmental concern is the exposure of children to lead from lead-based paint and potable water supplies containing lead–tin–copper sweat joints. In 1986 the amendments to the Safe Drinking Water Act effectively banned the use of lead-based solder. The acceptable alternative solders are alloys of tin in the range of 90–95%, and the balance is copper, silver, zinc, or antimony (Kireta 1988). The introduction of the alternative solder will minimize exposure of manufacturing, construction, and maintenance personnel to lead. The banning of mercury in heat manufacture demonstrates an action taken specifically for occupational health reasons; in the second and third examples, the action was prompted by environmental concerns but will have a positive impact on the health of all workers.

A total of 10 materials have been banned for specific operations in the United Kingdom (HSE 1988). An important element of the UK regulation is the prohibition of silica sand for abrasive blasting. Although there have been attempts to restrict the use of sand for this purpose in the United States, as the most frequently used blasting material, it continues to present a major risk to workers.

Greater emphasis will be given to replacement control technology as we enter the next century. As discussed below, the impetus will continue to result not from the workplace health issue but from the environmental health and ecological impact. Health and safety characteristics including toxicity, smog contribution, ozone depletion, and fire potential must be evaluated. In addition to the review of the performance of the replacement material, the changes in process and product design to satisfactorily change over to the new material must be considered in the final decision.

***18.3.2.1 National Toxic Use Reduction Programs*** Until the 1980s the EPA emphasis was on pollution control and dealt principally with "end-of-pipe" abatement technology. In 1984 the Hazardous and Solid Waste Amendments to the Resource Conservation and Recovery Act of 1976 redirected environmental quality efforts from the conventional emphasis on waste treatment and disposal to waste minimization. This approach to waste management is characterized by source reduction, recycling, and reuse. In 1986 the Office of Technology Assessment published a report that described the concept of prevention versus control (OTA 1986). Since these environmental initiatives have great impact on the replacement of toxic chemicals in industry and, therefore, worker exposure, it is important that the health and safety professional understand the environmental management nomenclature (Table 18.2). The Pollution Prevention Act of 1990 consolidated this approach by specifying the following waste management hierarchy as national policy: first source reduction as the most desirable approach, then recycling and reuse, then treatment, and finally disposal as the least desirable approach (U.S. Congress 1990).

To initiate this strategy, in 1989 the EPA established the Waste Reduction Innovative Technology Evaluation (WRITE) Program. This program funded programs in six states and one county over a three-year period to (i) provide engineering and economically feasible solutions to industry on specific pollution prevention problems, (ii) provide performance and cost information on these techniques, and (iii) promote early introduction of these pollution prevention programs into commerce and industry. A review of completed and ongoing WRITE programs that demonstrates the importance of replacement technology is shown in Table 18.3 (Harten and Licis 1991). Although initiated for environmental quality reasons, each of these changes will have a major impact on worker health.

**TABLE 18.2 Environmental Control Approaches**

| Form of control | Emphasis | End rule |
|---|---|---|
| Pollution control | "End-of-pipe" control | Does not eliminate pollution, but merely transfers contaminant from one medium to another. Incurs an environmental risk from transporting toxic chemicals |
| Pollution prevention | Reduction in the production of contaminants | Direct waste reduction; convers all pollutants |
| Toxic use reduction | Reduction in pollutants generated in manufacturing by changes in plant operation procedures, production processes, materials, or end products<br>Focus on a target list of chemicals | Priority setting permits control of high potential risk chemicals |

**TABLE 18.3 Examples of Material Replacement from EPA WRITE Program**

| Operation | Original material | Replacement |
|---|---|---|
| Anodizing aluminum parts | Chromic acid anodizing, hexavalent chromium exposure | Sulfuric acid anodizing eliminates exposure to chromium |
| Removal of paint on defective parts | Methylene chloride-based paint stripper | Abrasive blasting with tic beads, no vapor exposure |
| Flexographic printing | Application of alcohol-based ink labels | Water-based inks eliminate vapor exposure |
| Cleaning and deburring small metal parts | Vapor degreasing and alkaline tumbling | Automatic aqueous rotation washer. Eliminates vapor and caustic mist exposure |
| Cleaning flexographic plates | Solvent cleaning with Stoddard solvent, acetone, toluene, and alcohol | Cleaning with terpenes or aqueous cleaners |

*Source:* Adapted from Harten and Licis (1991).

In 1991, the EPA's Office of Pollution Prevention and Toxics initiated the Design for the Environment Program. This is a cooperative effort between EPA and industry where EPA uses its technical resources to help an industry segment eliminate the use of a toxic material. Many industries have recognized that to produce a product that can compete effectively on the global market, the design effort must go beyond minimizing the use of toxic materials. Areas also included in the design for the environment approach include minimizing the use of natural resources and producing a finished product that is recyclable. On a conceptual level, design for the environment evolved out of the field of industrial ecology.

Industrial ecology, introduced to the general public in 1989 by Robert Frosch and Nicholas Gallopoulos (1989), promotes the concept that the traditional linear manufacturing model of raw materials in, produce and waste materials out, be modified to mimic the biological ecosystems. Natural resource consumptions and waste generations and minimized, and wastes that are generated are used as raw materials for other processes. Even the final product, when its useful life has ended, is reused. An example of this is BMW's roadster produced in 1988, which can be quickly disassembled and whose coded body plastic panels can be easily recycled.

Examples of the EPA–industry cooperative efforts include the printed wiring board and screen printing industries. Pollution prevention efforts with the printed wiring board focused on six areas: reducing drag-out between the process baths, reducing chemical use, reducing copper buildup on plating racks, reducing chemical losses from evaporation by using polypropylene balls, using chemical substitutes that produce less sludge, and recovering metals from the wastewater treatment process for recycling. Pollution prevention efforts must be examined closely to ensure both environmental releases and occupational exposures are controlled. The use of polypropylene balls to reduce evaporation achieves both goals. Recycling metals from wastewater, if handling of a dry filter cake is required, may not.

Plant-specific case studies are available on the EPA's Design for the Environment World Wide Web page (http://es.epa.gov/index.html). The case studies typically involve smaller plants using low capital investment solutions to produce a quick rate of return on investment.

Denmark, one of the many European countries embarking on a national toxic use reduction program, based their work on the simple step model shown in Table 18.4 (G. Goldschmidt, personal communication). The simplicity of this approach suggests that replacement technology can be done on an ad hoc basis; however, Goldschmidt warns that chemical replacement for occupational health reasons is a complex task and requires a systematic review of both the existing conditions and those prevailing after the replacement action. A review of 162 individual replacement actions by the Danish Occupational Health Services confirms that replacement is successful if done in a rational manner following the step model shown in Table 18.4. Among the impressive examples of this program is the replacement with water-based systems of the solvent-based systems in all indoor paints and most outdoor paints.

**TABLE 18.4 Step Model for the Substitution Process**

| Step | |
|---|---|
| 1 | Problem identification |
| 2 | Identification and development of a range of alternatives |
| 3 | Identification of consequences of the alternatives |
| 4 | Comparison of the alternatives |
| 5 | Decision |
| 6 | Implementation |
| 7 | Evaluation of the result |

*Source:* From Goldschmidt (personal communication).

The principal advances in worker protection in the next decade will probably result from replacement technology accomplished as a result of such worldwide environmental health initiatives.

***18.3.2.2 State Toxic Use Reduction Programs*** The recent emphasis on pollution prevention and specifically on toxic use reduction in manufacturing is reflected in recent legislative action in 17 states and by the federal government (Rossi et al. 1991). The Massachusetts Toxics Use Reduction Act of 1989 (Massachusetts General Laws 1989) is described as a seminal program in that it does focus on toxic use reduction and excludes off-site recycling, off-site nonproduction unit recycling, transfer, or treatment. Rossi identifies the five toxic use reduction techniques acceptable under the Massachusetts Act as:

- In-process recycling or reuse. Recycling, reuse, or extended use of toxics by using equipment or methods that become an integral part of the production unit of concern, including, but not limited to, filtration and other closed-loop methods.
- Improved operations and maintenance. Modification or addition to existing equipment or methods including, but not limited to, such techniques as improved housekeeping practices, system adjustments, product and process inspections, or production unit control equipment or methods.
- Changes in the production process. Includes both the modernization of production equipment by replacement or upgrade based on the same technology and the introduction of new production equipment of an entirely new design.
- Input design. Replacement of a chemical used in production with a chemical of a lower toxicity.
- Product reformulation. Redesign of the product to produce a product that is nontoxic or less toxic on use, release, or disposal.

Rossi proposes two divergent routes the manufacturing group may take. The route for ease of implementation includes first the recycling and improved operations and maintenance, then production process changes, and finally replacement of toxic chemicals and product reformulation. However, for effective toxic use reduction, the first step is to eliminate or replace the toxic chemical in the process or reformulate the product, then initiate production process changes, and finally recycle and apply operations and maintenance control techniques.

An institute funded by industry assessment has been established at the University of Massachusetts at Lowell with the function to provide a teaching and research facility to encourage aggressive action in toxic use reduction. Ellenbecker (1996) reviews the experiences of the Toxics Use Reduction Institute (TURI) to date by presenting three case studies involving process changes. The first case study involves investigating the use of multiprocess wet cleaning to replace the dry cleaning process. Perchloroethylene, the primary solvent used by the dry cleaning industry, creates both occupational health and environmental release problems for the many small dry cleaning establishments. Wet cleaning includes a variety of techniques including steaming, immersions and gentle handwashing in soapy water, spot cleaning, and tumble drying. The more labor-intensive wet cleaning was found to be cost competitive with the capital-intensive dry cleaning.

In the second case study, blanket washes containing small percentages of volatile organic compounds (VOCs) were found to remove ink as effectively as high VOC products. Blanket washing is an open type of process where even with local exhaust ventilation, printers are exposed to high concentrations of the wash. In the final case study, the replacement of chlorinated solvents (methylene chloride, 1,1,1-trichloroethane [TCA], and trichloroethylene) used in vapor degreasers to clean oils from metal parts with an immersion tank using an aqueous cleaner was evaluated. The operating costs of the immersion tank were one-sixth the costs of the vapor-phase degreaser. Unfortunately the aqueous cleaners recommended by the immersion tank manufacturer did not work as well in cleaning all the metal parts. Research was conducted in conjunction with TURI's Surface Cleaning Laboratory to find the most effective aqueous cleaner.

In all cases where water was incorporated into the process, occupational chemical exposure potential would be reduced. In the cases where water becomes contaminated, costs of treatment must be considered in the economic evaluation of the process.

***18.3.2.3 Industry Programs*** A number of company initiatives have been described that respond to specific legislation or broad environmental concerns. One company program has multiple goals to reduce the use of toxic chemicals, minimize toxic emissions to all environmental media, encourage recycling of chemicals, and reduce all waste (Ahearn et al. 1991). In the design of this program, the company followed many of the recommendations of the Office of Technology Assessment report on the reduction of hazardous waste (OTA 1986). In this company program-specific goals for toxic use and waste reduction are set for all production managers, and their annual performance reviews include this issue.

**TABLE 18.5 Categories for a Corporate Toxic Use Reduction Program**

| |
|---|
| Categories I and II: Use to be reduced |
| Human or animal carcinogens, teratogens, or reproductive agents |
| Highly toxic chemicals |
| Chemicals with human chronic toxicity |
| Chemicals for which there are adverse environmental impacts |
| Category III: Transportation off-site to be reduced via source reduction or recovery and reuse on-site |
| Suspected animal carcinogens |
| Moderately toxic materials |
| Chemicals that cause severe irritation of eyes and respiratory tract at low concentrations |
| Chemicals with limited chronic toxicity |
| Corrosive chemicals |
| Chemicals for which there are environmental considerations |
| Category IV: Disposal volume to be reduced via waste reduction or reuse following on-site recycling |
| All other chemicals not included in category V |
| Category V: Any material such as paper, metal parts, and so on that is identified as rubbish, rubble, or trash |

*Source:* From Karger and Burgess (personal communication).

Each chemical used in the company is assigned to one of five environmental risk categories (ERC). The materials in ERC I pose the greatest risk to the environment, and ERC II, ERC III, and ERC IV represent progressively less risk. A fifth category includes plastic, steel, paper, and other waste identified as rubbish, rubble, and trash. A critical issue in the design of the program was the decision logic for assigning categories. A number of options were considered including ranking chemicals based on workplace exposure guidelines such as threshold limit values (TLVs) or permissible exposure limits (PELs), label signal words (ANSI 1988), regulatory lists published by OSHA and EPA, and the mathematical weighting of a range of risk factors including toxicity and chemical and physical properties (E. Karger and W.A. Burgess, personal communication).

The company chose to assign the ERCs based on a "wise person" approach based on a battery of risk factors reflecting the materials' total occupational and environmental impact from release as an isolated incident such as a spill or the chronic release of small amounts of the chemical to air, water, and soil. The majority of risk factors are toxicity based and chosen to identify significant adverse health and environmental impact. Included in the group of risk factors are acute and chronic toxicity based on animal studies or structural analysis, human case history and epidemiological studies, carcinogenicity, mutagenicity, teratogenicity, and reproductivity effects. A limited number of physical and chemical properties associated with the "releasability" of the chemicals were also included in a second group of risk factors. The third group of risk factors reflected broad environmental impacts. The specific criteria for assignment to the five categories are shown in Table 18.5.

The company has a major chemical production facility involved in the manufacture of photographic dyes, which require frequent changes in processes. In the design of new processes and products, the evaluation of toxic use and waste minimization technology is given equal importance to considerations of cost and performance of the new process or product. The engineering group developing the new processes rigorously reviews all materials and process alternatives to ensure that the target reductions are met. In this company the "end-of-pipe" abatement strategy is given secondary attention relative to optimization of process and product design for toxic use reduction.

An important part of the program is the goal of a 10% reduction in the use of toxic materials per unit of production each year for the first five years following the general strategy of elimination of use, replacement with a material of lower risk, reduction in quantity used, and a hierarchical program of waste handling. The goal is to eliminate, replace, or reduce the use of category I and II materials and assure the maximum reuse of category III and IV materials.

In describing the success of the program, the authors (Ahearn et al. 1991) present an example of a process in which a change resulted in elimination of a carcinogenic material. The original process required an aqueous oxidation step using a hexavalent chromium compound; the new process is based on a catalyzed air oxidation step, which does not require the chromium compound (Figure 18.4). As a result of this change, worker exposure to hexavalent chromium was eliminated, and this process change saved the company over one million dollar. It is the conclusion of this company that the process that uses the least quantity of toxic chemicals is usually the most economical.

As industries seek new ways to compete on the global market, many are implementing international management consensus standards. The International Organization for Standardization recently published ISO 14000, the Environmental Management Systems Standard (International Standards Organization 14000 1996). To become registered as an ISO 14000 company, a corporation is visited by an outside auditor, who, through an interview process, determines whether the company has the required management system in place to make the appropriate decisions

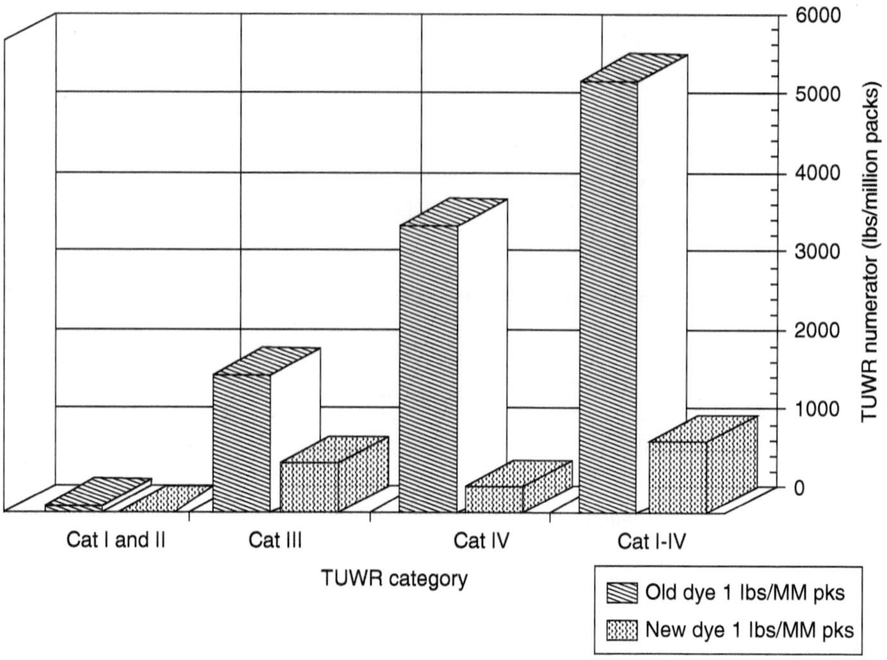


**Figure 18.4** Impact of replacement technology on a chemical process resulting in the elimination of hexavalent chromium from the workplace and its impact on toxic use and waste reduction categories (TUWR). *Source:* Courtesy of Polaroid Corp.

regarding their impact on the environment. An analogous health and safety standard is being considered by ANSI International Advisory Council/Occupational Health and Safety Advisory Group and is expected within the next few years. Implementation and evaluation of engineering controls will certainly be part of this standard.

***18.3.2.4 Outcome of Replacement Technology*** Although the replacement of toxic chemicals may provide impressive gains, the replacement of a target material must be done with great caution. If the replacement technique is to be used successfully, it must be carefully reviewed to define its impact on worker health and environmental impact, product yield, and product quality. It is important that the impact be evaluated in both pilot plant operation and field trials before making the change.

Industrial hygienists practicing from the 1950s to the 1990s devoted more attention to solvent replacement for metal cleaning than any other control problem. It is useful to track the history of such cleaning agents during those years since it shows the complexity of the replacement technology approach and indicates pitfalls that may arise in using this technique. Wolf et al. (1991) traced the major decisions on general degreasing solvents during this period as follows.

In the 1960s and 1970s in particular, chlorinated solvents were widely substituted for flammable solvents because they better protected workers from flammability. As smog regulations were promulgated in the late 1960s and early 1970s, users moved from the photochemically reactive flammable solvents and trichloroethylene to the other "exempt" chlorinated solvents. In the 1980s there was increased scrutiny of the chlorinated solvents. Trichloroethylene, perchloroethylene, and methylene chloride were considered undesirable because of their suspect carcinogenicity, and trichloroethylene and freon 113 (trichlorotrifluoroethane) were being examined for their contribution to stratospheric ozone depletion.

Each change starting in the 1950s, progressing from flammable solvents to chlorinated solvents to nonphotochemical reactive materials to chlorofluorocarbon (CFC) solvents, introduced a new set of problems. The six major classes of solvents in use in 1992 that can be considered for application as general solvents and the critical characteristics that determine their acceptability are shown in Table 18.6 (Wolf et al. 1991). The alternatives to the common chlorinated solvents are not encouraging. The flammable solvents require rigorous in-plant fire protection controls. Both flammable and combustible solvents including terpenes are smog producing and therefore are subject to tight federal and local air district regulations. The chlorinated solvents will continue to see pressure for replacement due to the array of human toxicity and environmental effects. The increased use of CFCs in the 1970s and 1980s to the present level of 180 000 metric tons per year is based on performance and reduced chronic toxicity; however, the impact of these chemicals on ozone depletion has resulted in the ban on manufacturing of CFC-113 and TCA worldwide in 1996. The hydrochlorofluorocarbons are considered interim candidates for

**TABLE 18.6 Characteristics of Generic Solvent Categories**

| Generic solvent category | ODP[a] | Photochemical reactivity | GWP[b] | Flash point[c] | Tested for chronic toxicity |
|---|---|---|---|---|---|
| Flammable solvents | — | Yes | — | F | |
| Isopropyl alcohol | | | | | Rule issue |
| Mineral spirits | | | | | No |
| Combustible solvent[d] | — | Yes | — | C | |
| Terpenes | | | | | Limited[e] |
| DBE | | | | | No |
| NMP | | | | | Rule issue |
| Alkyl acetates | | | | | No |
| Chlorinated solvents | | | | | |
| TCE | — | Yes | — | — | Yes |
| PERC | — | Yes[f] | — | — | Yes |
| METH | — | No | — | — | Yes |
| TCA | 0.1 | No | 0.02 | — | Yes |
| Chlorofluorocarbons | | | | | |
| (CFCs) | | No | | — | |
| CFC-11 | 1.0 | | 1.0 | — | Yes |
| CFC-113 | 0.8 | | 1.4 | — | Yes |
| Hydrochlorofluorocarbons | | | | | |
| (HCFCs) | | No | | | |
| HCFC-123 | 0.02 | | 0.02 | — | In testing |
| HCFC-141b | 0.1–0.18 | | 0.09 | — | In testing |
| HCFC-225 | NA | | NA | — | In testing |
| Hydrofluorocarbons | | | | | |
| (HFCs) and fluorocarbons (FCs) | | | | | |
| Pentafluoropropanol | — | NA[g] | NA | NA | No |

*Source:* From Wolf et al. (1991). Reprinted by permission of *Journal of the Air and Waste Management Association.*

[a] OPD: The oxygen depletion potential is the potential for ozone depletion of 1 kg of a chemical relative to the potential of 1 kg of CFC-11, which has a defined ozone depletion potential of 1.0.

[b] GWP: The global warming potential of a chemical is the potential of 1 kg of the chemical causing global warming relative to the potential of 1 kg of CFC-11 to cause global warming. CFC-111 a GWP of 1.0.

[c] F, flammable; C, combustible.

[d] DBE, dibasic esters; NMP, *N*-methyl-2-pyrrolidone.

[e] One of the terpenes, D-limonene, has been tested.

[f] Although PERC is not photochemically reactive, it is not exempt under the Clean Air Act.

[g] NA, not available.

industrial degreasing but are considered for banning in 2020 and 2040. The hydrofluorocarbons and the other candidate solvents not containing chlorine do not present an ozone depletion risk but may present unacceptable toxicity concerns.

A vigorous research effort is being pursued by dozens of companies to develop alternatives to the cleaning agents listed in Table 18.6. Presented in Table 18.7 are the advantages and disadvantages of three new cleaning techniques compared with CFC-113 and TCA from the perspective of a company performing precision cleaning for the military.

A flow diagram of an aqueous and semiaqueous cleaning process is shown in Figure 18.5. When aqueous cleaners are not sufficient, a hydrocarbon such as a terpene is used to provide cleaning. This is followed by an emulsion rinse stage to limit the amount of hydrocarbon that mixes with the water rinse. These are the systems where regulatory pressure in the environmental area may decrease occupational health and safety. Flammability of the hydrocarbons and limited toxicology data are the two main areas of concern for these compounds. It is doubtful that a single agent will be found that will have the broad application of the chlorinated solvents; however, progress is being made in non-ozone-depleting cleaning techniques.

Burgess (1996) identified powder coating as an example of a new technology that significantly improves worker health and safety by reducing solvent exposure. Powder coatings are composed of a finely pulverized powder of thermoplastic or thermosetting resins with little or no solvent. Powder coatings produce a durable, high-quality finish and can be applied to a variety of materials. The powder guns have a high transfer efficiency of 50%. As with any replacement technology, new process hazards, in this case inhalation exposure to the powder, must be understood and controlled. The advantages and limitations of powder coatings and other alternatives to solvent-based coatings are presented in Table 18.8.

**TABLE 18.7 Advantages and Disadvantages of Cleaning Methods Compared with CFC-113 and 1,1,1-Trichloroethane**

| Cleaning method | Advantages | Disadvantages |
|---|---|---|
| Aqueous cleaning (water with an acid, alkaline, or neutral pH soap) | Good for bulk processing | Specialized equipment required |
| | Limited health hazards | |
| | Water component of cleaner recycled | Cleaning cycle time increased over CFCs |
| | | Rinse cycle required |
| | | Drying time required |
| Semiaqueous and terpenes | Excellent solvents that clean a wide range of materials | Compatibility issues |
| | | Terpenes have low flash point |
| | Low VOC emissions | Residues |
| | Low toxicity | Odors |
| | | Some terpenes contain D-limonene, an irritant |
| Supercritical $CO_2$ cleaning | Good for bulk processing | High-pressure operation |
| | Removes trapped fluids from parts | High pressure can damage delicate parts |
| | Fluid and contaminants can be recovered from SCF/$CO_2$ separator | High up-front capital expenditures for the extractor |
| | $CO_2$ inexpensive | |

*Source:* Adapted from a presentation by Dr. Bill Agopovich, Draper Labs, Semiconductor Safety Association Meeting, 28 October 1993, Cambridge, MA.

Employees affected by process changes may not always agree with industrial hygienists and process engineers that changes have improved their work environment. An interesting example of this occurred in the semiconductor industry in the substituting of a less hazardous, although more odorous, photoresist solvent. Many photoresists used in the 1980s contained ethylene glycol monomethyl ether acetate (EGMEA). OSHA reduced the PEL for EGMEA from 50 to 5 ppm based on reproductive health effects. Photoresist manufacturers began marketing a photoresist product containing propylene glycol monomethyl ether acetate (PGMEA), which had been tested and found not to produce the reproductive health effects. Digital Equipment Corporation was eager to try the new product since the original studies investigating concerns over increased spontaneous abortion rates were conducted at their semiconductor fabrication plants. Process engineers and industrial hygienists were surprised to see OSHA-reportable illnesses increase by 30% when the PGMEA-based photoresist was introduced into production (R. Nill, personal communication). The increased headache and related illnesses may have resulted from the lower odor threshold of PGMEA compared with EGMEA. The employees' perception was that health risks were increased by the substitution rather than decreased (Hallock et al. 1993). Engineering controls satisfactory for transfer of EGMEA were not satisfactory for PGMEA. Canister filling had to be conducted outside the fab area, and process delivery lines had to be upgraded. In this case, if production employees had been involved in the acceptance testing of the new photoresist, the odor issue may have come up and illnesses avoided due to early implementation of engineering controls.

### 18.3.3 Dust Control

A number of techniques have been proposed to minimize worker exposure to pneumoconiosis-producing and toxic dusts based on changes in physical form or state of the material. The common techniques to reduce dustiness in this manner are discussed below.

***18.3.3.1 Moisture Content of Material*** The relationship between moisture content of granular materials and worker exposure to dust is well known. The use of water as a dust suppressant has been of great interest to the mining, mineral processing, and foundry industries. In the 1960s research by the British Cast Iron Research Association demonstrated that silica in air concentrations in foundries was maintained below existing exposure standards if the moisture content of foundry sand was kept above 30% (BCIRA 1977). A sand handling technique that involves direct blending of new moist sand directly from the mixer with shakeout sand to add moisture to cool the shakeout sand and reduce dustiness is described by Schumacher (1978). The author also demonstrated a correlation between results of a simple laboratory dustiness test and in-plant dust exposure. Goodfellow and Smith (1989) identify the importance of a dustiness index: "Dustiness testing can be a useful testing tool in establishing the type and efficiency of dust control

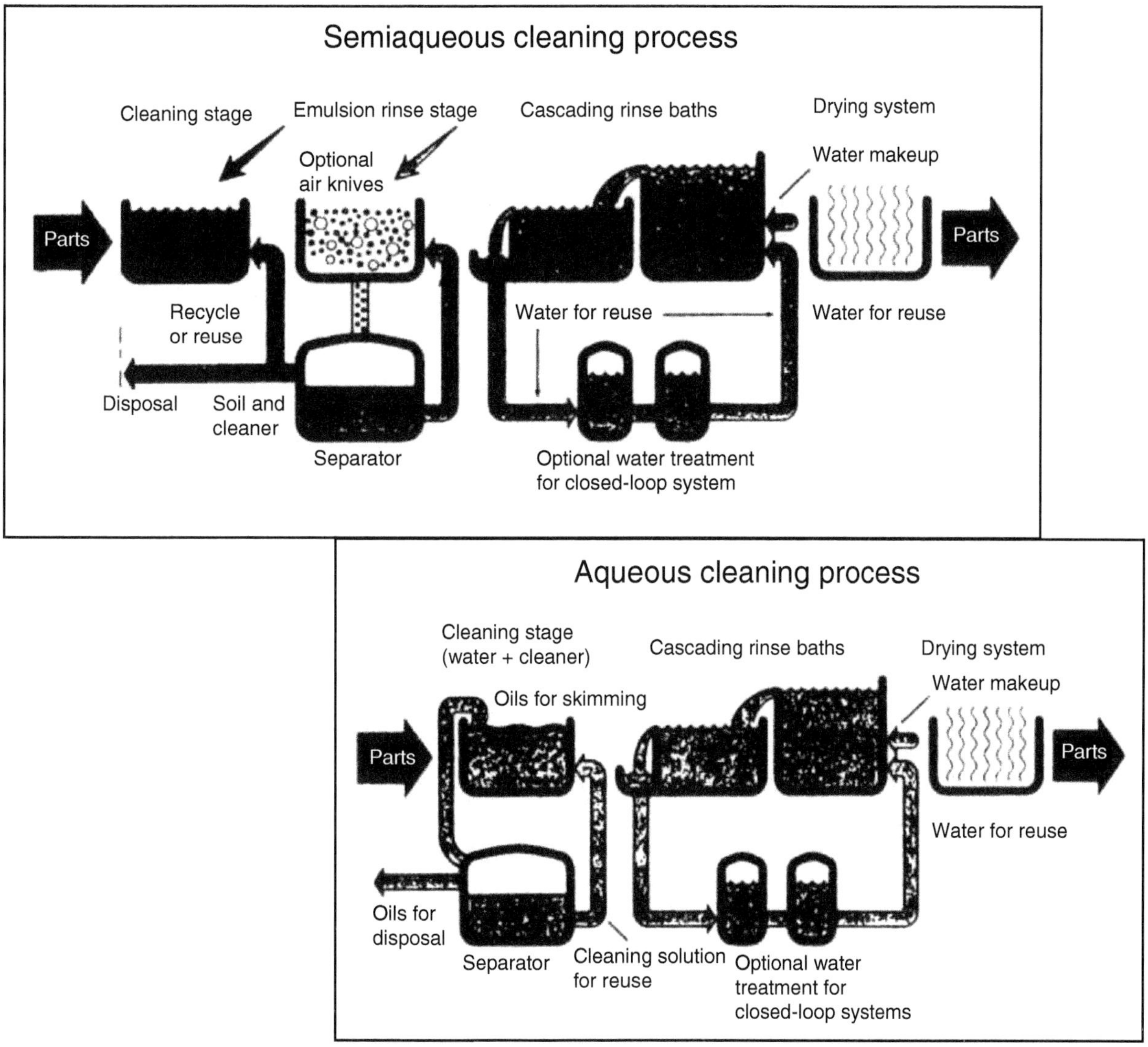


**Figure 18.5** Semiaqueous and aqueous cleaning process. *Source:* From Sprow (1993).

required for different materials and materials handling systems. Further field data and verification are required to develop this procedure into a useful tool for practitioners."

The Bureau of Mines and the National Industrial Sand Association have investigated various moisture application techniques for dust control in mineral processing plants (Volkwein et al. 1989). The various methods investigated included the use of foams, steam, and water sprays. The type of moisture added and the percentage in dust reduction are presented in Table 18.9. Foam was found to be about 20% more effective than water alone or water with a surfactant. Steam experiments showed dust reductions of 65%. While both foam and steam are generally more effective, they are more costly to produce than water sprays. The authors also recommended processing partially wet product rather than complete drying after crushing, which is now the common practice.

The British Occupational Hygiene Society (BOHS 1985) has also explored various dustiness estimation methods that may be useful in design of control systems. In chemical processing a solid product is frequently isolated in cake form that is wet with solvent or water. Frequently this material must be dried to ensure product quality and shelf life and to permit packing and shipping. However, if the product is to be used "in-house," one should determine if the drying operation can be omitted permitting direct handling of the wet cake, thereby saving money and reducing dustiness.

***18.3.3.2 Particle Size*** In general, the more extensive the grinding or comminution of a granular material, the greater the dust hazard the material will present during transport, handling, and processing. When possible, purchase the most coarse form of the chemical that is suitable for the process. There are production implications that may override considerations of worker exposure. As an example, if the material must be placed in solution, the large particle size will slow this process.

**TABLE 18.8 Summary of Alternatives to Solvent-Based Coatings**

| Alternative | Applications | Toxic use reduction benefits | Operational advantages | Operational disadvantages | Cost | Product quality | Limitations |
|---|---|---|---|---|---|---|---|
| Water-based coatings | Metal, wood, plastics, concrete, paper, leather | Reduced VOC emissions; reduced fire and explosion hazards; reduced hazardous waste; solvent not required for clean-up | Most formulations can be applied with conventional nonelectrostatic spray equipment and techniques; overspray easily recovered and reused; equipment may be cleaned with water; decreased drying time with drying oven; low odor levels | Require careful temperature and humidity control; require careful surface preparation; may require longer drying time; corrosion inhibitor may be needed on metal substrate; bacterial sensitivity reduces shelf life; may become unstable if frozen; emulsion coatings susceptible to foaming | Higher costs per gallon; special equipment and techniques needed for electrostatic application; may require special pumps and piping; may require drying oven | Reduced gloss; may cause grain raising in wood; some resins may cause water spotting; impact resistance may be reduced; some forms may have reduced corrosion resistance | Reductions in VOCs may be offset by use of solvents in surface preparation; additives to control water spotting may present worker safety hazards |
| High solids coatings | Metal, wood, plastics | Reduced VOC emissions; reduced fire and explosion hazards; reduced hazardous waste | Can increase paint transfer efficiency; lower-viscosity coatings compatible with conventional equipment | Narrow "time–temperature–cure" window; require careful surface preparation; generally require high cure temperatures; generally shorter pot life; may require worker retraining | Lower-viscosity coatings applied with conventional equipment; higher-viscosity coatings may require special equipment; reduced paint waste and supply needs; reduced energy use | Similar to that with solvent-based coatings | Reductions in VOCs may be offset by use of solvents in surface preparation; solvents still needed for clean-up |
| Powder coatings | Mostly metals, but also wood, plastics, glass, and ceramics | VOC emissions and exposure eliminated or significantly reduced in application; no solvent required for clean-up; reduced fire hazard; reduced hazardous waste | High transfer efficiency; minimal solid waste; no dripping or running during application; thick coatings can be applied in one operation; overspray easily retrieved and recycled; no overspray with fluidized bed application; no mixing or stirring; requires little operator expertise | Color changes and matches can be difficult; potential for explosion must be minimized; some difficulty in applying thin coatings; requires handling of heated parts | Higher equipment and materials costs offset by savings in labor, maintenance, energy, waste, and pollution control | Durable, high-quality finish with good corrosion resistance | May present skin contact or dust inhalation hazards; good ventilation and protective equipment required; potential for explosion must be minimized; resins may still produce low VOC emissions |

| | | | | | | | |
|---|---|---|---|---|---|---|---|
| Radiation-cured coatings | Plastics, wood, paper, metal | VOC emissions and fire and explosion hazards eliminated or greatly reduced; reduced hazardous waste | Rapid curing; high transfer efficiency; low heat requirement for drying, useful on heat-sensitive substrates; consistent performance; low maintenance; unreacted overspray can be collected for reuse | Requires new equipment and operating procedures; curing of pigmented coatings may be difficult; may be difficult to strip | High capital investment costs – considerably higher for EB systems; lower energy requirement; lower materials use; less waste | Similar to that with solvent-based coatings | Solvent still needed for clean-up; acrylate materials in most coatings present worker safety concerns and require protective equipment |
| Supercritical fluid spray application | Metal, plastics, wood | Reduced VOC emissions; reduced fire hazard; reduced hazardous waste | Easily retrofitted into existing facilities; higher viscosity allows thicker coatings without runs and sags | High-pressure gas and operating temperature require care in operation; lower fluid delivery rates than airless or spray guns | Replacement of fluid handling equipment; potentially reduced operating costs | Thicker coatings may be applied without runs and sags | Require care in working with high pressure and high temperature; still in testing phase for some industries |
| Surface-coating-free materials | Metals and plastic; other substrate materials under development | Elimination of VOC emissions, fire and explosion hazards, and hazardous material use | Stripping and repainting not required throughout service life; elimination of coating operation | | Initial increased cost may be offset by reduced operating costs | Surface finish appearance limited | Substrate may contain other materials of concern |

*Source:* From The Massachusetts Toxic Use Reduction Institute (1994).

**TABLE 18.9 Type of Moisture Added and Resulting Dust Reductions at One of the Study Sites**

| Test condition | Volume of liquid (ml) | Dust reduction (%) |
|---|---|---|
| Foam | 1420 | 91[a] |
| | 1300 | 73[a] |
| | 764 | 68 |
| Water | 757 | 46 |
| | 1324 | 58 |
| Water with 1.5% surfactant | 1324 | 54 |
| Water with 2.5% surfactant | 1324 | 54 |

*Source:* From Volkwein et al. (1989).
[a] Average reduction.

**TABLE 18.10 Dust-Controlled Forms of Rubber Chemicals: Comparative Performance[a]**

| | | Wax or otherwise bound | | | | Polymer bound | |
|---|---|---|---|---|---|---|---|
| Property | Untreated powder | Soft paste | Putty | Prills | Pellets/granules | Slab | Pellets or granular |
| Active content | 5 | 3 | 4 | 5 | 4 | 4 | 4 |
| Convenience of handling | 3 | 1 | 2 | 4 | 5 | 2 | 5 |
| Freedom from dust | 1 | 5 | 5 | 4 | 4 | 5 | 5 |
| General cleanliness and safety | 1 | 1 | 2 | 3 | 3 | 5 | 5 |
| Suitability for automatic weighing | 3 | 1 | 1 | 4 | 4 | 1 | 5 |
| Wastage | 3 | 2 | 4 | 4 | 4 | 5 | 5 |
| Ease of disposal of containers | 3 | 1 | 3 | 3 | 4 | 4 | 4 |
| Identification | — | — | — | — | | — | — |
| Mill mixing behavior | 3 | 1 | 3 | 3 | 3 | 5 | 4 |
| Internal mixing behavior | 5 | 2 | 3 | 5 | 5 | 5 | 5 |
| Dispersion in rubber | 5 | 5 | 4 | 3 | 4 | 4 | 4 |
| Total | 32 | 22 | 31 | 38 | 40 | 40 | 46 |

*Source:* From Hammond (1980). Reprinted by permission of the *British Occupational Hygiene Society*.
[a] 5 = excellent; 4 = good; 3 = average; 2 = below average; 1 = poor.

***18.3.3.3 Dust-Controlled Forms*** In the past two decades, the rubber, pharmaceutical, pigments, and dyestuff industries have given attention to the dustiness of the raw materials they produce and use. Dustiness testing has been extended, and significant product design changes have been adopted to minimize dustiness, worker exposure, and product acceptance. In tire manufacture at least a dozen chemicals in granular form are added in small quantities to the batch mix. In an effort to minimize worker exposure to dust from these chemicals, the British Rubber Manufacturers' Association has sponsored design of low dusting forms of these common chemicals. The properties of seven dust-controlled forms of rubber chemicals are reviewed by Hammond (1980) in Table 18.10. The author notes that the disadvantages of the most effective approach, coating the chemical with a polymer, include the variability in active chemical content based on bulk weight, the reduced chemical content, and the unsuitability of the polymer in the formulation.

***18.3.3.4 Slurry Form*** This application has limited application, but in those cases where it can be used, it does have great impact on dust concentrations. In the tire industry, "masterbatch" rubber, rubber processed with all chemicals except the vulcanizing agent, is processed from the Banbury to the drop mill, where it is "sheeted off" for storage. Until 1960 dry talc or limestone was dusted on the slabs of masterbatch material to keep it from sticking. This operation resulted in poor housekeeping and a significant exposure to talc. The present technique, adopted in the 1960s, involves dipping the stock in a slurry of talc in water before racking for storage. This simple change resulted in a significant reduction in worker exposure to talc.

### 18.3.4 Impurities in Production Chemicals

In low concentrations impurities or unreacted chemicals in raw or final product may represent a potential exposure that warrants attention.

***18.3.4.1 Residual Monomer in Polymer*** In polymer manufacture there is frequently unreacted monomer in the final product. Residual monomer had not been given much attention until the early 1970s when angiosarcoma, a rare liver cancer, noted in workers manufacturing the vinyl chloride polymer was attributed to the monomer exposure. Investigation showed that the polymer used in subsequent fabricating operations had unreacted monomer present in concentrations as high as 0.4% (Braun and Druckman 1976). This level of contamination prompted concern about the monomer exposure of workers handling the bulk polymer in plastic fabrication operations such as injection molding. As a result of this concern, the vinyl chloride manufacturers modified the manufacturing process to reduce the monomer concentration to less than 1 ppm (Berens 1981), thereby eliminating significant worker exposure. Residual monomer is frequently present in concentrations up to 1% in many of the common polymers and may warrant attention. If significant air concentrations are noted when handling the polymer, engineering control is first based on the removal or reduction of the monomer content in the polymer with other controls considered later if this is not adequate.

***18.3.4.2 Solvent Impurities*** Impurities may pose an unrecognized risk, especially in solvents of high volatility. In the manufacturing of automobile tires, the various rubber components are "laid up" on a tire building machine. To effect good bonding between the components, the rubber is made tacky by applying a small amount of solvent to the surface with a pad. For several decades this solvent was benzene. The worker exposure, probably in the range of 1–10 ppm, may be responsible for the excess leukemia seen in older tire builders. Starting in the 1950s the industry started to replace benzene with white gasoline. In studies completed in the 1970s by the Harvard School of Public Health Joint Rubber Studies Group, the residual benzene content in white gasoline was 4–7%, and air sampling indicated that one-third of the air samples on tire builders exceeded 1 ppm (R.L. Treitman, personal communication). The PEL for benzene at that time was 10 ppm, although it was anticipated that it would be dropped to 1 ppm (this change did occur in 1987). Technical-grade chemicals commonly have significant impurities. The level of contamination should be identified, and if sufficiently high, worker exposure should be evaluated. At that time the necessity for reduction in the impurity level can be determined.

## 18.4 EQUIPMENT AND PROCESSES

In Section 18.3.1 a variety of engineering control options are focused on the choice of materials to minimize the generation and release of airborne contaminants. An equally important step is a review of the various alternatives in the choice of equipment and processes.

In the discussion on dusty materials in Section 18.3.3.3, techniques to determine the relative index of dustiness are mentioned to assist in the choice of material form and the dust suppression treatment. We do not have such an index for equipment and processes; however, there are a number of operational insights that should be considered in choice of facility. Wolfson (1993) has emphasized the importance of this step in stating that the removal of the dispersal device should be a first step in the engineering control of air contamination.

### 18.4.1 Diagnostic Air Sampling

Contaminant control cannot be achieved until the significant operational elements of the process that generate and release the contaminant are identified. Occasionally this can be done simply by a critical review of the operations, but usually diagnostic air sampling is necessary. In Chapter 7, the traditional exposure monitoring for compliance purposes is discussed. The value of short-interval, task-oriented air sampling using conventional integrated sampling with subsequent analysis has been clearly stated by Caplan (1985a). In describing this approach, illustrated in Table 18.11 Caplan states:

> For the job analyzed in Table 5.11 presumably a single sample would have shown a concentration of 0.21 mg m$^{-3}$. The task-oriented sampling, however, reveals several interesting things. Column 5 shows that tasks B and C are the major contributors to the day's exposures and that a significant reduction in the concentration at either of those tasks would be adequate to bring the 8 hour exposure well below the TLV. This is true even though task C in itself is below the TLV. In addition, it shows that task F, well above the TLV concentration, is of such short duration that significant improvement in that part of the exposure would not have a large effect on the 8-hour exposure.

The difficulty in this approach is that frequently it is not possible to measure the air concentrations during brief individual tasks and activities due to the low air sampling rate and the limited sensitivity of the analytical procedures. To reveal important generation points in the job, it is necessary to resolve the air concentration profile in a time frame of seconds.

**TABLE 18.11 Task-Oriented Air Sampling**

| Task | GA or BZ[a] | Minutes per day | Concentration ($mg\,m^{-3}$) | Minutes per day × concentration |
|---|---|---|---|---|
| Charge pot | BA | 40 | 0.12 | 4.8 |
| Unload pot | BZ | 80 | 0.50 | 40.0 |
| General survey | GA | 250 | 0.16 | 40.0 |
| Lab – sample trips | GA | 20 | 0.05 | 1.0 |
| Change room | GA | 30 | 0.08 | 2.4 |
| Pump room – repack | BA | 30 | 0.32 | 9.6 |
| Lunch room | GA | *30* | 0.07 | *2.1* |
| | | 480 | | 99.9 |

LV = 0.2 $mg\,m^{-3}$; Wt. avg. = 99.9/480 = 0.21 $mg\,m^{-3}$

*Source:* From Caplan (1985a). Reprinted by permission of John Wiley and Sons from *Patty's Industrial Hygiene and Toxicology*, Cralley and Cralley, Eds.

[a] GA = general air; BZ = breathing zone.

The advent of real-time direct reading air sampling instruments for particles, gases, and vapors permits the investigator to identify these critical contaminant elements in process events and work practices. In the 1980s this air sampling technique saw expanded application with coincidental videotaping of the worker during a work cycle. In its most sophisticated form, the time-coupled, real-time contaminant concentration at the workers' breathing zone is superimposed on the video display, permitting the viewer to analyze the data display to identify the specific time and location of release. Control technology is then applied to those tasks or incidents.

The video display may also be used to identify work practices that may either positively or adversely affect worker exposure. In a talc bagging operation reviewed in Vermont in the 1970s, one individual consistently had the lowest dust exposure, although visual inspection did not reveal any differences in equipment or work practice. If the real-time technique were available, the worker's "secret" could have been identified and applied to the other workers. If a "correct way of doing the job" can be identified, the video concentration format is an excellent educational tool for workers.

A series of studies by NIOSH investigators describe the application of this technique to a range of in-plant tasks (Gressel et al. 1987; O'Brien et al. 1989). The air sampling instrument is a real-time monitor with a response time much shorter than the period of the shortest worker activity or movement. The instrument is equipped with a data logger with a clock "locked-in" or synchronized with the video. The data logger is downloaded to a computer, and the data file is analyzed, permitting a graphical overlay of air contamination data on the videotape. In Figure 18.6 the video display from a study of a bag dumping operation is re-created in a line drawing for clarity to show the time-coupled concentration on the video screen display (Cooper and Gressel 1992). Graphical representation of the air sampling data for three jobs with the concentration from the video overlay is shown in Figure 18.7 (Gressel et al. 1988).

The application of this technique to control technology was shown by Gressel and Fischbach (1989) in a study of a chemical weighing and transfer station. The information obtained permitted the investigators to redesign the workstation controls based on a perimeter exhaust hood and an air shower to eliminate eddies induced by the worker's body. Effective control of the dust exposure was obtained with one-third the airflow of the original system. In addition to improved worker protection, the cost savings of the new system resulted in a payback period of 4.5 years.

A variation of this technique first developed by Sweden's National Institute of Occupational Health and adapted for use in this country by NIOSH allows real-time mixing of the concentration and video portions by the use of radio telemetry (Kovein and Hentz 1992). Rather than recording the concentration data on a data logger, the direct reading instrument signal is converted to a radio frequency and broadcast with a telemetry transmitter to a receiver for immediate processing by a microcomputer (Figure 18.8). Researchers and employees can get immediate feedback on the effect of engineering controls and their work practices on exposure levels. This allowed for immediate investigation into the cause of exposure and allowed NIOSH investigators to work with employees to reduce exposure levels.

In addition to using air sampling to determine where controls are needed, air sampling can also be used to initiate controls. This approach is common in the semiconductor industry where continuous monitoring of toxic gases is required (Uniform Fire Code 1988). If the gas is detected above some predetermined level in a gas storage cabinet, emergency shutoff valves for the toxic gases are activated. Through the use of sensitive, specific air monitoring devices, the leak is controlled before any adverse occupational or environmental releases occur.

These diagnostic tools will see expanded application to workstation design in the next decade since it does permit the engineer to identify the specific tasks, equipment function, or worker movement that contribute to worker exposure.

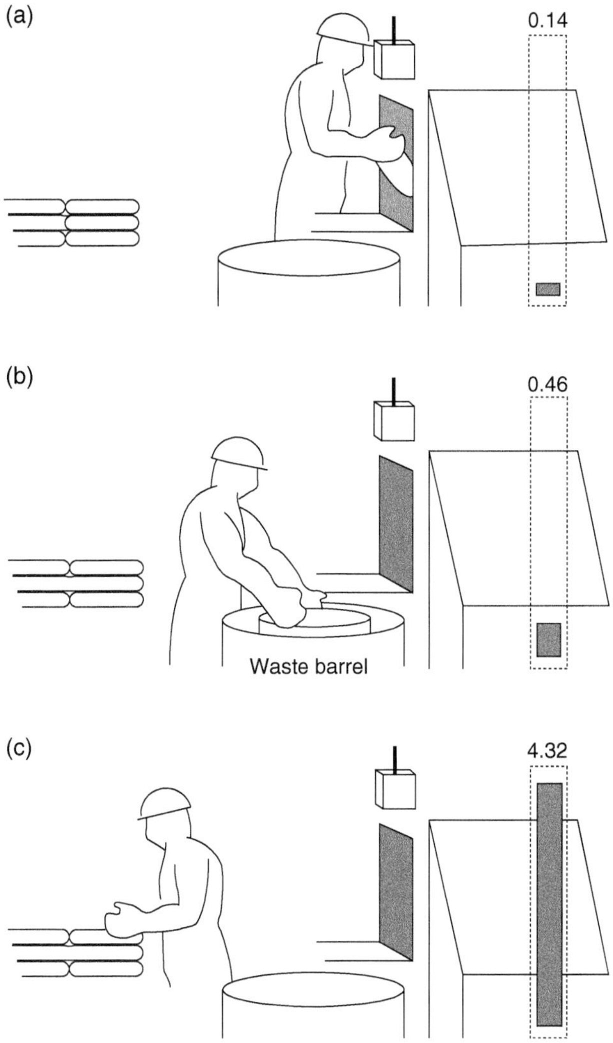


**Figure 18.6** Relative dust exposure during bag dumping as determined by video air sampling technique. (a) Operator slits bag and dumps the granular material into a hopper equipped with an exhaust hood with minor dust exposure. (b) Operator drops the bag into the waste barrel; there is a significant increase in dust concentration. (c) As the operator pushes the bag into the waste barrel, a cloud of dust is released, and the relative dust concentration increases to 4.32. *Source:* Reprinted with permission from Cooper and Gressel (1992).

## 18.4.2 Equipment

***18.4.2.1 Ancillary Equipment*** The stepwise review of the dispersal potential noted above should include all equipment, not only the major machinery. The importance of this approach is highlighted when one looks at the history of the simple *air nozzle* in industry. For most of the 1900s, the widespread application of the air nozzle operating at line pressure (100 psi) was used to remove chips and cutting oil from machined parts and to dry parts after cleaning with solvent. In the 1970s OSHA required that the operating pressure for these nozzles for cleaning purposes be dropped from 100 to 30 psi. This change had many positive effects, including a reduction in noise, eye injuries, solvent mist, and vapor exposures.

Another common piece of equipment, albeit much more complex, is the *centrifugal pump*. The difficulty of equipment selection is typified by the experience of a chemical engineering team designing a modem chemical plant or refinery having hundreds of centrifugal pumps. A major issue in such plants is the impact of fugitive losses from rotating machinery such as pumps on the workplace and general environment (BOHS 1984; NIOSH 1991). The total loss percentage from rotating machinery varies from 8 to 24%, with overall uncontrolled emissions ranging from 31 to 3231 tons per year for a model chemical plant or a large refinery (Lipton and Lynch 1987). The authors describe two possible control programs for pump fugitive losses – either a monitoring and maintenance program or engineering control by installation of dual seal pumps, seal-less pumps, and a closed exhaust hood with air cleaning.

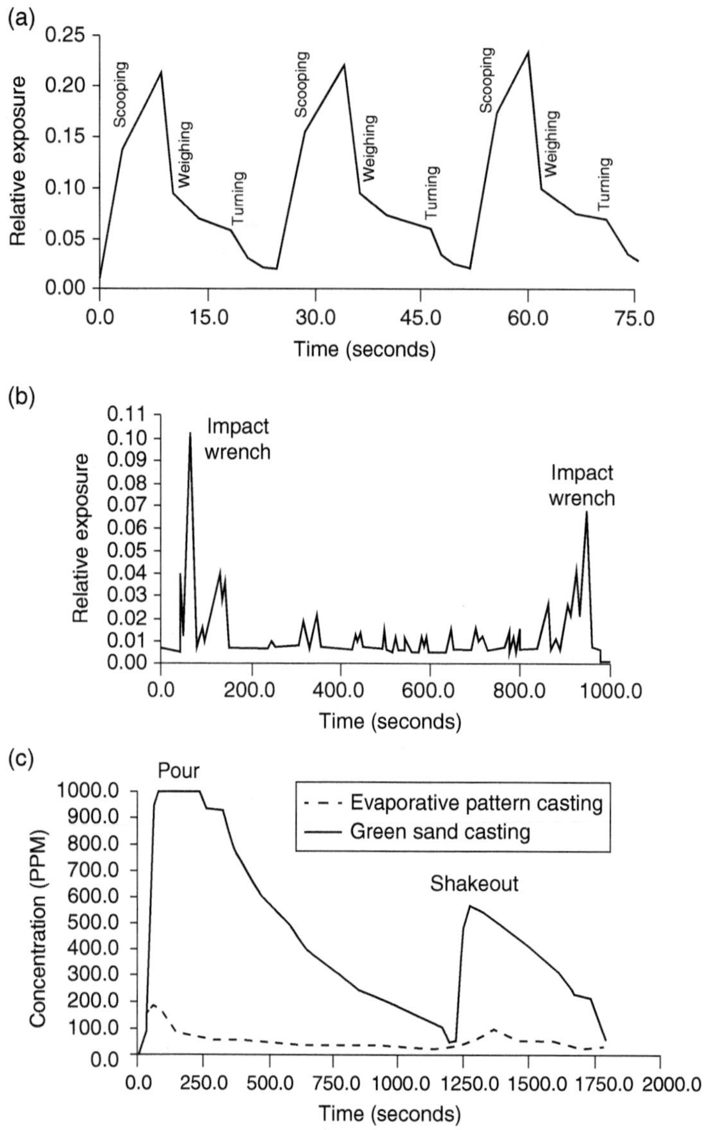


**Figure 18.7** Peak dust exposures identified by video real-time monitoring. (a) Relative exposure versus time for three bags during manual weigh-out of powders. (b) Relative dust exposure during automotive brake servicing. (c) Carbon monoxide emissions from evaporative pattern casting and green sand processes. *Source:* From Gressel et al. (1988). Reproduced by permission of *Applied Industrial Hygiene.*

The phaseout of TCA and trichlorotrifluoroethane solvents in vapor-phase degreasers in the 1990s has accelerated the search for alternative solvents, as described in Section 18.3.2.4. Initial steps to reduce emissions from this common equipment in the 1980s included the use of covers, increased freeboard height, refrigerated condenser fluids, and lower hoist speeds. Although these changes do reduce losses and therefore worker exposure, the survival of this type of equipment rests with the availability of a closed system unit as described by Mertens (1991). Such *vapor-tight degreasers* are available in Europe and are under development in the United States. In the past, closed system technology has been associated with major processing facilities. In the future individual job shop equipment will utilize this approach to meet critical workplace and environmental constraints.

***18.4.2.2 Major Processing Equipment*** The complexity of the control technology problem becomes apparent when one moves from choices of ancillary equipment to decisions on major pieces of processing equipment. As an example, a small chemical processing plant for organic synthesis is a multifloor plant with a series of major operations staffed by six chemical technicians. The condensed version we will review (Figure 18.9) permits discussion of equipment alternatives to minimize the generation and release of air contaminants. The three-step process includes, first, a reaction step to form a solid in suspension, conducted in a pressure vessel. In this case there is no alternative equipment.

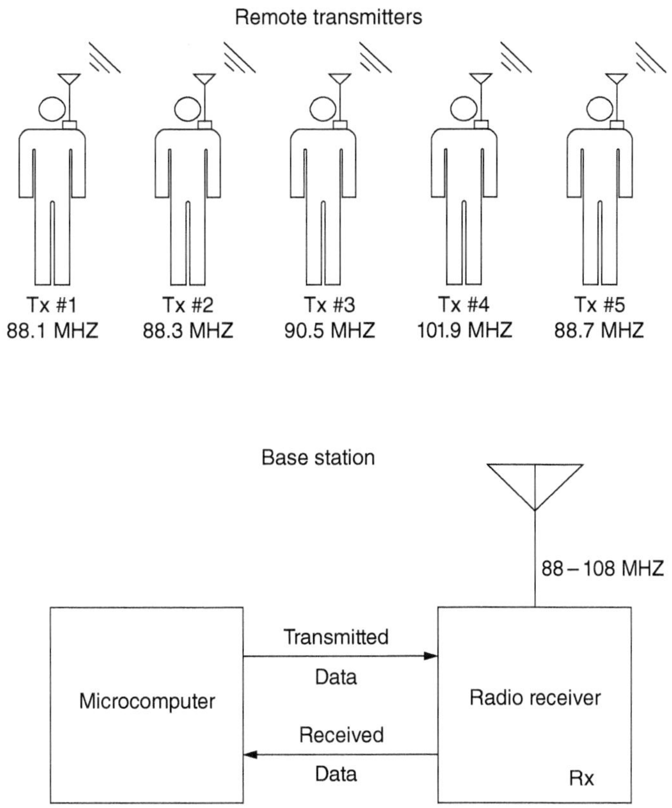


**Figure 18.8** Radio telemetry block diagram. *Source:* From Kovein and Hentz (1992).

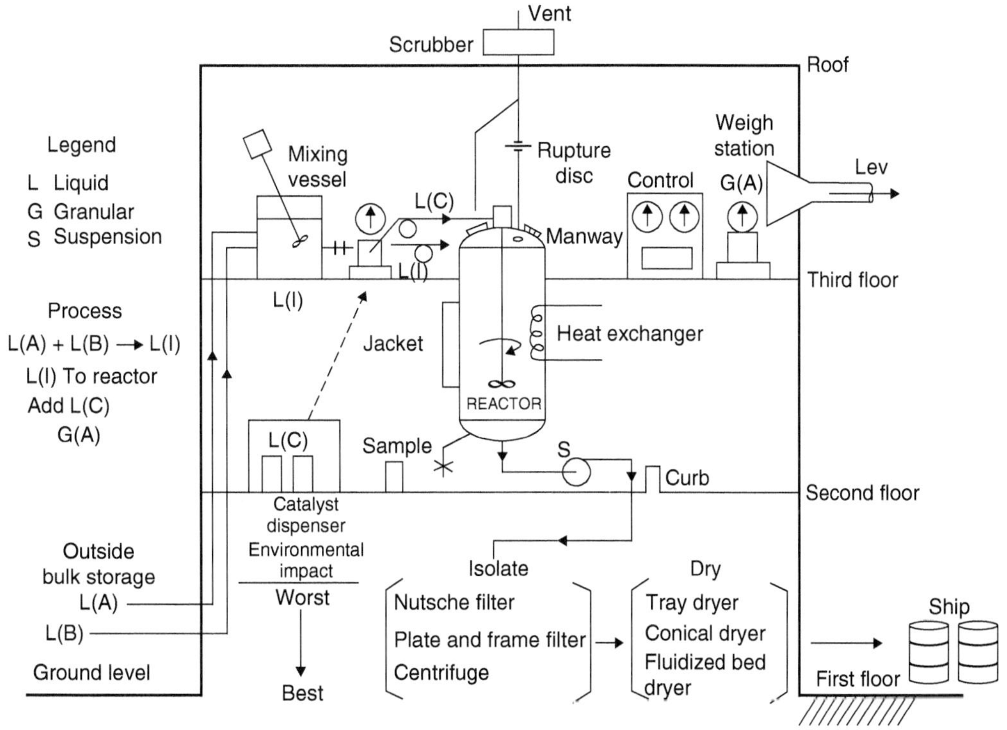


**Figure 18.9** Chemical processing facility. Liquid reagents A and B are transferred to a mixing vessel. After mixing, the resulting liquid C is transferred to the reactor by vacuum. Granular chemical X is weighed under local exhaust ventilation (LEV) and dumped into the reactor through a change port also equipped with LEV. A catalyst D is transferred to the reactor by a positive displacement pump. After the reaction is complete, the new chemical is isolated by one of three techniques – Nutsche filter, plate and frame filter, or centrifuge. The resulting wet cake is then dried by one of three methods – tray drier, rotary drier, or fluidized bed drier. The isolation and drying techniques are ranked by environmental impact with the worst process noted first.

The reaction must be conducted in a closed pressure vessel equipped with the safety controls shown in Figure 18.9 to minimize the possibility of a chemical accident. The second step, the isolation of the solid reaction product, can be done by at least three different techniques. The environmental contaminant potential of these processes varies considerably. However, the choice cannot be made on this basis alone, since these techniques do have specific production capabilities that may be the major determinant in the choice. Finally, the reaction product is dried and drummed for sale or use in a subsequent operation. Again there are several ways to accomplish the drying. The ranking of these isolation and drying operations based on their environmental impact is shown in Figure 18.9.

The responsibility for the choice of specific equipment in the processing plant will rest with the project design group and will be based on production capacity, ancillary services required, maintenance history, cost, and delivery, in addition to health and safety considerations. It is not expected that the health and safety professional will have complete operational and application data on all chemical unit operations, but their environmental impacts must be understood and made known to the engineering group.

It would be extremely helpful to the health and safety professional if the individual equipment were assigned an index of contaminant generation/dispersal similar to the index of dustiness or fugitive losses from a pump. This approach is not currently realistic; however, a better characterization of the performance of such equipment will emerge in the next decade.

### 18.4.3 Processes

A review of the production engineering literature shows a range of processes that can be considered for a manufacturing facility. The choice is usually made based on production output and cost data. Industrial processes designed to accomplish a given task have one other parameter, the ability to contaminate the workplace; this characteristic of the process should be given equal weight to production criteria (Peterson 1985).

The development of technology in a given area such as welding is pushed by production needs, not by health concerns. Since 1900 the technology has become more complex and frequently more environmentally challenging. Infrequently, studies are conducted to influence the process design to minimize air contamination. This was the case with Gray and Hewitt (1982), who defined the welding operating parameters that influence fume generation and proposed operational configurations that minimize fume formation rate and the chemical composition of the fume, thereby permitting improved efficiency of controls.

An exercise that demonstrates the value of a ranking approach for process environmental impacts is shown in the manufacture of an electronic cabinet (Figure 18.10). As in the case of the chemical process shown above, the fabrication steps are simplified for ease of discussion. The four enclosure components are fabricated from lightweight aluminum stock using sheet metal shear, brake, and punch. These operations may present a noise hazard but do not generate air contaminants and will not be included in our review. For a discussion of controls used for noise, see Chapter 21. The enclosure can be assembled and painted by a number of techniques; the processes are ranked from worst to best in terms of adverse effect from air contamination.

Another general approach that applies to many processes is the use of containment. This control approach was considered by an engineering control panel sponsored by the EPA, and the following containment ranking was proposed (EPA 1986).

Standard control measures may often be categorized for each process: a sealed and isolated system where airborne contaminant levels are very low (ppb range for vapors); a substantially closed system where airborne contaminant levels are still relatively low (fraction of ppm or low ppm range for vapors); a semiclosed system, which is typical of nondedicated equipment used in job shop chemical processing facilities; and an open system.

An example of a sealed and isolated system is the modern refinery, where such an approach is required for both production reasons and health and safety. A substantially closed system is typified by a semiconductor facility, and a semiclosed system by the chemical processing plant in Figure 18.9. Most common facilities, that is, foundries, electroplating, painting, welding, and machining, are examples of open systems.

The contaminant released as a mist from electrolytic plating operations can be trapped by a layer of plastic chips or a persistent foam blanket on the surface of the bath. Also surfactants may be added to the electrolyte to reduce mist escape. The plate and frame filter in the chemical processing industry is difficult to handle with local exhaust ventilation. Frequently a solvent wash of the cake with a highly toxic material results in high exposure when the filter is broken and the solvent-wet cake is removed. In some cases after the initial wash, the cake can be washed with isopropanol to strip out the toxic solvent. It is then washed with water so that when the filter is broken and the product is removed, exposure is nil.

A series of process changes developed in the WRITE Program is shown in Table 18.12.

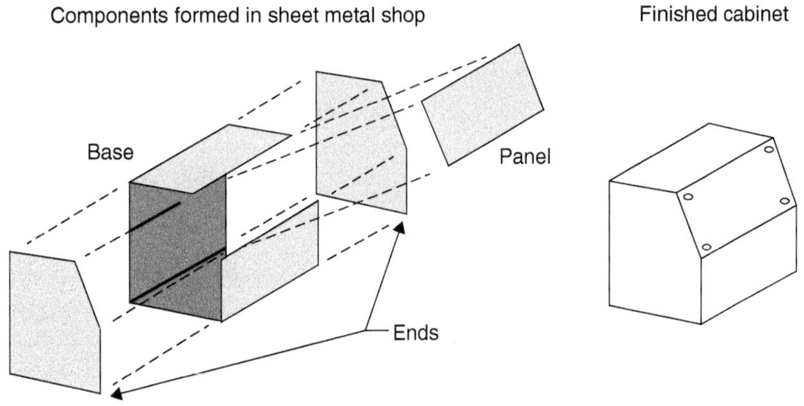


Finishing options

| Metal cleaning | Welding | Painting |
|---|---|---|
| Wet degreasing with chlorinated HC | Torch welding | Air atomization |
| Vapor with chlorinated HC | Inert gas shielded arc welding | Airless spraying |
| Wet degreasing with terpenes | Submerged arc | Hot paint |
| Water with detergent | Resistance | Electrostatic |
| | | Powder |
| | | Dipping |

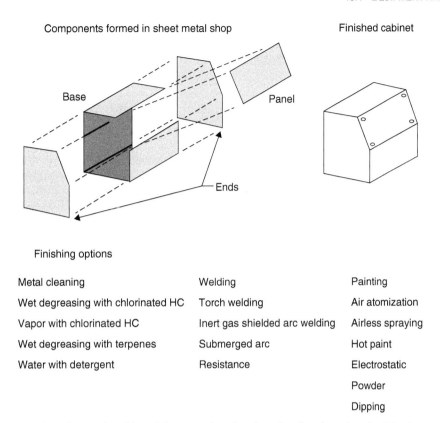

**Figure 18.10** Manufacture of an electronic cabinet. The operational options for cleaning, abrasive blasting, and painting are ranked in terms of environmental acceptance with the worst process noted first.

**TABLE 18.12 Examples of Process Change Technology from EPA WRITE Program**

| Operation | Process change |
|---|---|
| Hand mixing of paint | Proportional mixer blends paint at the gun, thereby eliminating handling paint and solvent at a mix operation |
| Conventional manual air spray painting | Computer-controlled robotic painting with an electrostatic spray to reduce worker exposure to paint mist and solvent vapors |
| Performance testing of electronic arts with CFC-based cooling system | Installed compressed air cooling system |

*Source:* Adapted from Harten and Licis (1991).

### 18.4.4 Work Task Modification, Automation, and Robotics

It is well known that simple changes in work tasks may have significant impact on job outcome. In the early 1900s, workplace time and motion studies were used to improve productivity. Later a job placement technique devised by Hanman (1968) based on a detailed analysis of the time and effort of the job tasks was effective in reducing on-the-job injuries. More recently, the analytical tools of the ergonomic specialist permits identifying difficult tasks contributing to occupational injuries and illness.

As indicated in Section 18.4.1, the health and safety professional has techniques to investigate the source of contaminants and the generation mechanism and to make a semiquantitative assessment of the generation rate. This approach has tremendous value in analyzing not only the critical generation points on the machine but in viewing the impact of specific worker actions and movement on air concentrations. Preliminary studies reveal that minor changes in work position and movement may offer significant reductions in exposure.

Occasionally the specific modification in work practice is dictated by knowledge of the mechanisms of generation and release of the contaminant. Such is the case with the flow of granular material at material transfer points.

The generation mechanism is the airflow induced by the falling granular material (Anderson 1964). The induced airflow can be minimized by restricting the open area of the upstream face, reducing the free-fall distance, and reducing the material flow rate as defined by Anderson's equation.

Automation is a general technique that separates the worker from the individual process. In the 1960s and the 1970s, this was usually done by simple electromechanical equipment design. A good example of this procedure is the manufacture of asphalt roof shingles. In the 1970s competition resulted in the automation of all parts of the process from the dipping of the stock to the bundling of the package.

Another example of a simple automation process that reduces worker exposure is tire curing. In the early plants the worker lifted the tire out of a curing press at the end of the curing cycle and placed another tire in the mold for curing. During this period the worker was directly exposed to the emissions released from the press. By 1970 most plants had automated this process. The worker now places the uncured tire on a holding rack in front of the curing press line. When the curing cycle is completed, the tire is ejected from the press to a belt conveyor, and the next tire to be cured is transferred from the rack to the press without exposure of the worker.

The advent of robotic techniques has permitted almost all industrial procedures to be candidates for automation. The movement of the worker who buffs rubber boots can now be captured by a robotic system, as shown in Figure 18.11. The ultimate application of robotic techniques is to spray painting. In this case the robotic system can reproduce the movement of a skilled painter. Although ventilation is still required on this job, the worker overseeing the operation is separated from the point of release of the paint mist and solvent.

Robotics may have been a solution to a difficult problem in the 1950s. In a large generator shop, the generator coils were preformed of copper bar stock. The coils were then wrapped with insulation tape and painted with an asphalt compound. Protective clothing notwithstanding, the workers had skin exposure to the asphalt, which required aggressive cleaning at the end of the shift. Dermatitis and photosensitivity were frequent occurrences. At that time coil winding experts devoted time and money to developing machine wrapping concepts without success. This type of problem can be solved by robotics in the 1990s.

### 18.4.5 Facility Layout

As stated earlier, the most efficient way to do a job is probably the one that impacts the least on the workplace environment. Certainly this is true insofar as overall plant layout is concerned (Caplan 1985b). In the fabricating shop, the desired flow of materials is from the incoming truck dock or railroad siding to sheet metal fabrication and then to assembly, finishing, inspection, and finally shipping. In a facility manufacturing pharmaceuticals, the input chemicals are transported to bulk storage and then to chemical processing, packaging, inspection, and shipping. In all manufacturing processes from handling metal to fine chemicals, it is important to minimize the distance the material is moved and the number of times it is picked up and transferred. If this rule is followed, worker exposure to air contaminants will be minimized.

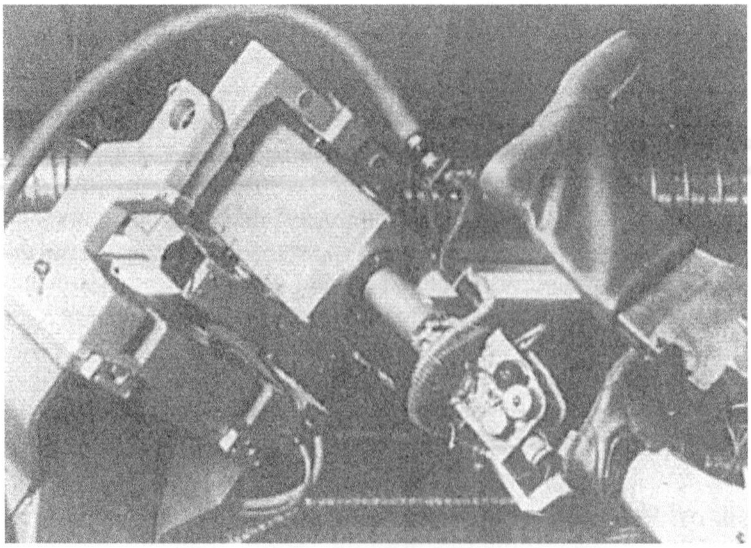

**Figure 18.11** Robotic buffing of rubber boots. *Source:* Photograph reproduced by permission of Matti Koivumaki.

***18.4.5.1 Material Transport*** This issue is given major attention in industry where large quantities of raw material, intermediates, and final products are handled. Examples are injection molding of children's toys, manufacture of automobile tires, and paint manufacture. In some cases material handling alone defines the plant layout. Frequently bulk storage is in large silos located outside the plant with delivery to the workstations by mechanical or pneumatic conveyors. This is true in the manufacture of plastic tape, where tons of PVC granules are used each day (Figure 18.12). This plant also requires large quantities of solvent delivered from bulk storage by piping to mix tanks and then directly to the coating heads. The intent of this system is to have all material handling to the individual coaters done in closed systems.

In organic synthesis it is common practice to have outside bulk storage of at least a dozen solvents. The solvents are transferred to an inside reaction vessel by piping to a reactor manifold with necessary valving and meters. Granular materials used in small quantities are stored in adjoining storage areas and delivered to the reactor and charged by manual dumping, with a dumping fixture, or occasionally with a transfer lock. When a large number of drums must be dumped, a bulk handling system is used with direct delivery to the reactor vessel.

***18.4.5.2 General Considerations*** The location of the process within the facility may influence the worker's exposure. This is true of operations such as foundries, where jobs may be easily classed as clean or dirty and exposures vary greatly. The conventional iron foundry provides a good example of the importance of plant layout. It is common practice to define the optimal plant layout as one with airflow from the cleanest to the dirtiest operations, with the exhaust focal point establishing this gradient (Figure 18.13a). The layout of a similar foundry that does not follow this guideline is shown in Figure 18.13. In the latter case the relatively clean molding line is positioned adjacent to the shakeout, an area where the control of airborne foundry dust and thermal degradation products is difficult. This poor layout results in silica and other contaminants released by shakeout moving into the molding area and exposing this work population. If the desired concentration gradient cannot be achieved by ventilation or distance, then structural walls or plastic barriers are a possibility. Although such barriers define the space, compartmentalizing complicates the design and application of local exhaust ventilation, replacement air, HVAC, and other important services.

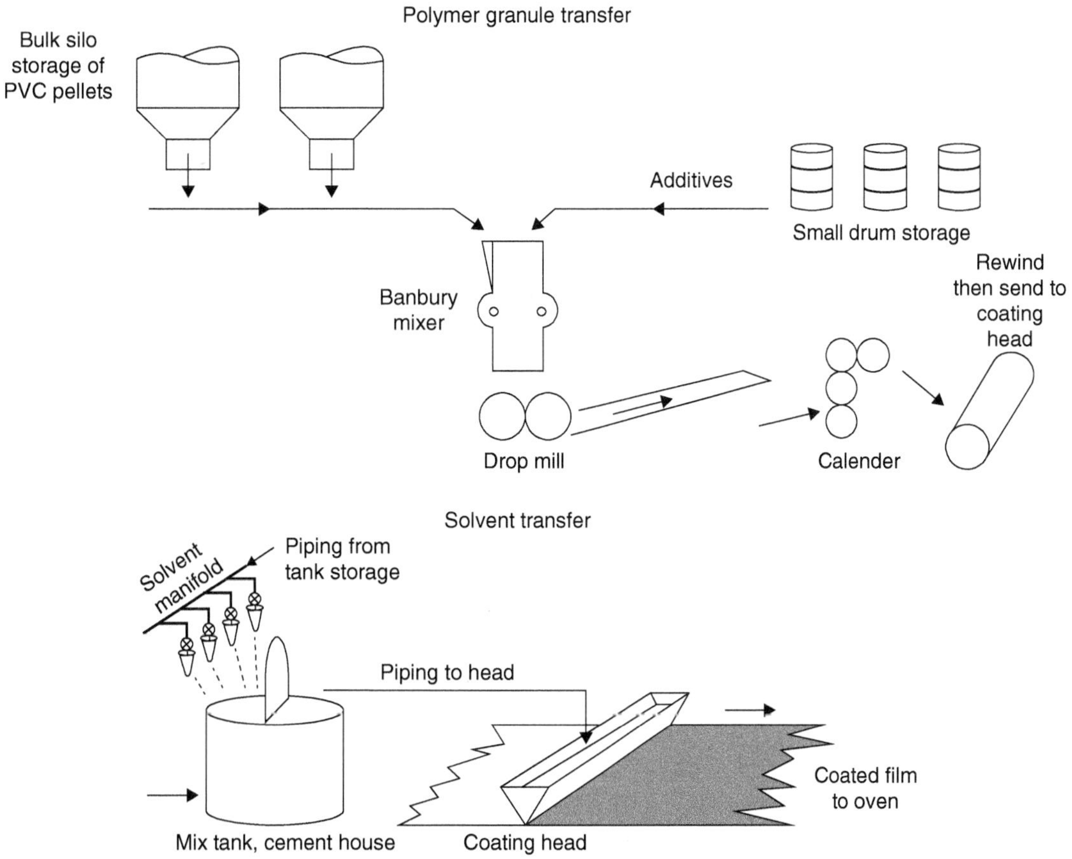


**Figure 18.12** Manufacture of electrical tape showing the closed bulk transport of granular plastic and the piping of solvents to the mix tank.

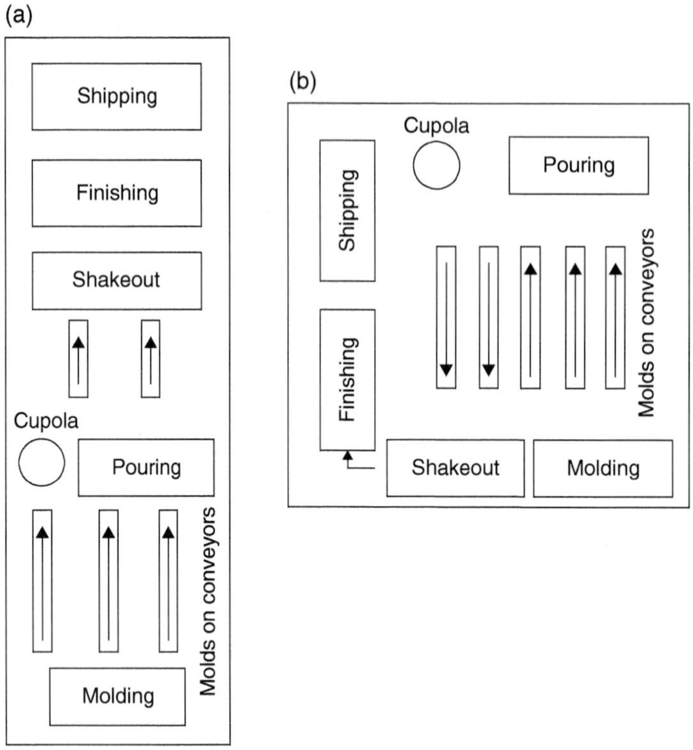


**Figure 18.13** Layout of two gray iron foundries. Foundry in (a) is designed for straight process flow and gradual transition from activities that represent limited contamination to those of greater contamination. In foundry (b) the process flow reverses and the dirty shakeout ends up adjacent to the molding line.

***18.4.5.3 Workstations*** A foundry in Finland faced problems not only of airborne contaminants but also of heat, noise, and housekeeping in casting cleaning. The problems were resolved by redesigning the open work space to provide individual workstations designed with integrated services positioned for ease of work (Figure 18.14). This layout resulted in improved material flow, better housekeeping, reduced air contamination, better ergonomics, and improved productivity. The success of such an installation was due, in part, to close employee–employer involvement in the design.

***18.4.5.4 Service and Maintenance*** Frequently, close attention is given production area layout as noted in the above example, but rarely is attention given the working environment of the maintenance group. An example of an area where maintenance was not considered was tire curing presses in the plants of the 1970s. The presses were arranged in double rows back to back with limited space between rows for steam, water, and electrical services. While this layout and density represented optimal use of space, repairs of steam and water piping were done in an extremely tight space, and controls for the worker were difficult to set up.

The best example of a planned layout that provides proper work space for the trades is noted in the semiconductor industry (Figure 18.15). In this layout all transfer pumps, distribution lines, and vacuum pumps are located in a service alley separated from the fabrication bay. This design reflects principal concern, not for the comfort of the maintenance worker, but rather to protect the fabrication area against contamination. Notwithstanding, it does result in adequate space for the trades to carry out their work in a safe manner.

An extension of this concern is the installation of a small field service bench with necessary tools and equipment in plant production areas where maintenance is frequently done. This arrangement permits the maintenance person to do many repairs at the site and not transport equipment back to a main facility with the potential for chemical spills.

***18.4.5.5 Segregation of Operation*** It is frequently necessary to segregate or remove an operation from the main production area as an engineering control measure. In the semiconductor and fiber optics industry each time a new facility is designed, the engineers must choose whether to store small amounts of highly toxic gases close to the

(a)

(b)

**Figure 18.14** Well-integrated workstation in a foundry. (a) Overall layout of foundry finishing area. (b) Individual workstation with exhaust hood. *Source:* Photograph reproduced by permission of Matti Koivumaki.

production tool or store large quantities at a segregated position some distance from the plant to be distributed by double-walled piping to the production tools. This latter segregation technique permits the plant to reduce the number of persons at risk while providing extensive controls at the worksite. This option must be weighed against storing small quantities of gas directly at the tool, thereby minimizing failures in the transfer systems but requiring frequent change of gas cylinders with the entailed risks.

***18.4.5.6 Isolation of the Worker*** The practice of isolating the worker as a control measure is placed under this section since it should be considered in conjunction with layout of the major equipment. This widely used technique is mandated by OSHA coke oven regulations, which requires enclosures with clean, conditioned air. The rail cars that travel above the coke ovens have controlled environment cab enclosures; workers can also retreat to enclosures designed to remove them from the hostile coke oven environment during available rest periods.

In other industries this approach has been chosen to eliminate worker exposure to air contaminants while providing a comfortable working environment. These applications include enclosed booths on a variety of construction equipment, front-end loaders in smelters, crane cabs in metallurgical industries, operator cabs in steel rolling mills, and pouring stations in foundries. Unfortunately off-the-shelf control booths are not available for the range of applications seen in industry, nor have engineering guidelines been published.

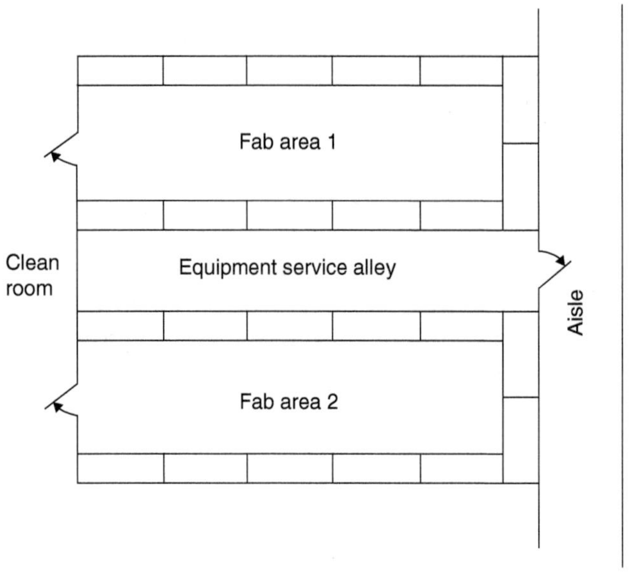


**Figure 18.15** Layout of semiconductor facility showing service aisle isolated from, but adjoining, the fabrication bays.

**TABLE 18.13 Ventilation Control Hierarchy**

| Type of ventilation[a] | Hood type | Example |
|---|---|---|
| LEV | Total enclosure | Glove box |
| LEV | Partial enclosure | Laboratory hood |
| LEV | Low-volume, high-velocity tool integrated | Portable grinder |
| LEV | Exterior hood | Welding hood |
| GEV | Mechanical exhaust | Roof ventilators |
| GEV | Natural | Wind induced |

*Source:* Adapted from BOHS (1987).
[a] LEV, local exhaust ventilation; GEV, general exhaust (dilution) ventilation.

### 18.4.6 Ventilation

In the introduction to this chapter, we stated that ventilation is the third step in the control hierarchy: (i) do not generate the contaminant; (ii) if you do generate a contaminant, do not allow it to be released; and (iii) if it is released, collect the contaminant before it reaches the worker. The design goal of industrial ventilation is to protect the worker from airborne contamination in the workplace. To the newcomer this may suggest installing a system that will reduce exposure below the permissible exposure guidelines or an appropriate action level. This is not the case. The professional will design the system to meet the goal of "as low as reasonably practical" (BOHS 1987).

Within this control approach the effectiveness of the major ventilation techniques is shown in Table 18.13 (BOHS 1987). There is no agreement on the position of low-volume, high-velocity systems since its effectiveness varies greatly depending on the degree of integration with the tool. When designing a hood for a particular operation, Burton (1991) recommends starting with a complete enclosure concept and then removing only those portions of the enclosure necessary to provide access to the equipment. This will ensure the most effective containment with the lowest airflow. When it is not possible to enclose the equipment or operation, an exterior hood must be used. These are hoods that direct the contaminant away from the breathing zone of the operator once the contaminant is released. Enclosing hoods are preferred over capture hoods by designers because they provide a better containment with less airflow and are less affected by crossdrafts. Users may prefer capture hoods because they tend to be less restrictive. For a capture hood, the amount of air required for effective capture is a function of the square of the distance between the release point and the hood. Thus the distance between the hood and the release point must be determined during design and the hood used accordingly once installed. Some hood designs can take advantage of the momentum of the contaminant in ensuring its capture. These are called *receiving hoods*, and the best example is a canopy hood, which receives rising hot air and gases.

**TABLE 18.14 Application of Local Exhaust and General Exhaust Ventilation**

| Local exhaust ventilation | General exhaust ventilation |
|---|---|
| Contaminant is toxic | Contaminant has low order of toxicity |
| Workstation is close to contaminant release point | Contaminants are gases and vapors not particles |
| Contaminant generation varies over shift | Uniform contaminant release rate |
| Contaminant generation rate is high with few sources | Multiple generation sources, widely spaced |
| Contaminant source is fixed | Generation sites not close to breathing zone Plant located in moderate climate |

*Source:* Adapted from Soule (1991).

**TABLE 18.15 Information Needs for Problem Characterization**

Emission source behavior
- Location of all emission sources or potential emission sources
- Which emission sources actually contribute to exposure?
- What is the relative contribution of each source to exposure?
- Characterization of each contributor: chemical composition, temperature, rate of emission, direction of emission, initial emission velocity, continuous or intermittent, time intervals of emission

Air behavior
- Air temperature
- Air movement
- Mixing potential
- Supply and return flow conditions
- Air changes per hour
- Effects of wind speed and direction
- Effects of weather and season

Worker behavior
- Worker interaction with emission source
- Worker location
- Work practice
- Worker education, training, cooperation

*Source:* From Burton (1991). Reprinted with permission of D. Jeff Burton.

Soule (1991) reviews the application of the two major ventilation control approaches, dilution and local exhaust ventilation (Table 18.14). In most industries dilution is not the primary ventilation control approach for toxic materials. It is accepted that local exhaust ventilation will not provide total capture of contaminant and dilution ventilation is frequently applied to collect losses from such systems. In addition, it is used for multiple, dispersed, low-toxicity releases.

***18.4.6.1 Design Phase*** The specific design methods for both general exhaust ventilation and local exhaust ventilation are presented by Soule (1991) and in other volumes dedicated to ventilation control (ACGIH 1996; Burgess et al. 1989; Burton 1991). An important predesign phase identified by Burton (1991) as problem characterization is frequently given little attention (Table 18.15). This is a topic that can be best addressed by an industrial hygienist, who can provide data on emissions, air patterns, and worker movement and actions.

If the process is new, a videotape of a similar operation with the same unit operations may be available. The best of all worlds would be the availability of a video concentration tape as discussed in Section 18.4.1. If the facility is a duplicate of one in the company, the industrial hygienist should visit the operation with the ventilation designer. Frequently the designer is an outside contractor. In this case it is important that the problem characterization approach be followed and an information package be provided to the designer.

The precautions that should be reflected in the design have been discussed in detail elsewhere, but should include worker interface, access for maintenance, and routine testing. Computer-aided manufacturing and design (CAM/CAD) technology now permits precise placement of equipment and ductwork so that ad hoc placement by the installer should be a thing of the past. As discussed earlier, it may be worthwhile to "mock up" a specific design solution prior to final design and construction. This is especially true when a large number of identical workstations are to be installed, as was the case in a shipyard asbestos insulation workroom, as noted in Figure 18.3.

General cautions are appropriate on the use of available design data for control of industrial operations. The ACGIH Industrial Ventilation Manual provides the most comprehensive selection of design plates for general industry (ACGIH 1996). Each of these plates provides four specific design elements: hood geometry, airflow rate, minimum duct velocity, and entry loss. The missing element is the performance of the hood in terms of percent containment. Roach (1981) has recommended such an index as a minimum performance specification for local exhaust hoods.

As noted by Burgess (1993), there has been little consolidation and publication of successful designs by industrial groups. With the exception of the Steel Mill Ventilation volume published by the American Iron and Steel Institute in the 1960s (AISI 1965) and the Foundry Environment Control manual in 1972 (AFS 1972), there is no evidence that companies in the United States wish to share ventilation control technology. One exception to this is the semiconductor industry, which mostly through information exchange promoted by the Semiconductor Safety Association has published some information on ventilation design for their industry. Table 18.16 from the *Semiconductor Industrial Hygiene Handbook* lists the types of local exhaust ventilation systems commonly used in the semiconductor industry. Some of the control equipment is very specific to the industry, such as wet stations and diffusion furnaces. The original source of exhaust recommendations is frequently the equipment manufacturer, who may or may not have fully investigated optimal enclosure design and exhaust rates. In the case of wet stations, the concern on the part of manufacturers and process engineers is to provide chemical exposure (personnel protection) and particle contamination control (product protection) with the same device. Users of the equipment must ensure that the manufacturer's concern for product protection does not outweigh concerns for personnel protection. Verification of recommended exhaust rates by the customer or a third party (prepurchase containment testing) is frequently required.

Major sources of information on the performance of ventilation systems are the technical reports on engineering control technology published by NIOSH (Appendix A). Many of these reports couple ventilation assessment with measurement of worker exposure.

This discussion indicates that ventilation control designs on standard operations in the mature industries have been published, although performance has usually not been reported. It is important to evaluate performance by diagnostic air sampling, both to ensure the worker is protected and to prevent overdesign. The latter was shown to be the case in the design for control of a push–pull system for open surface tanks (Sciola 1993). A mock-up of one tank demonstrated that satisfactory control could be achieved at minimal airflow. Operating at the reduced airflow rate saved \$100 000 in installation costs and \$263 000 in annual operating costs.

A substantial portion of the operating costs for a local exhaust system are the costs to heat or cool the makeup air required by the exhaust system. These heating and cooling costs may range from \$3–5 per cfm per year. One approach that has been used to reduce these costs is to recycle the air back into the workplace once the contaminant has been removed from the exhaust stream. This technique is most successful when a relatively nontoxic material, such as wood dust, that can be reliably and efficiently cleaned from the airstream is present. Recirculation is less viable when multiple contaminants and/or high-toxicity contaminants are present. The American National Standard for the Recirculation of Air from Industrial Process Exhaust Systems (ANSI 1996) must be followed when considering recirculation of air from a local exhaust system. This standard requires that a thorough hazard evaluation be conducted before designing a local exhaust system that may include recirculation. It also requires that a continuous monitoring device be used to ensure the exhaust air contains less than 10% of the acceptable level of a highly toxic substance before it is recirculated back into the building. The system must be designed to divert the air to the outside or initiate backup air cleaning if the primary air cleaner fails.

The types of air cleaning devices available depend on the physical state of the contaminant present in the local exhaust system. The most common types of air cleaning devices are for dusts, and there is a wide variety of filtration media available. For heavy loading of dusts, a baghouse that may contain upward of a 100 large vacuum cleaner-type bags may be used. This may be followed by high-efficiency particulate arrestor (HEPA) filtration, which provides a high removal efficiency (99.97% at 0.3 μm) for a light dust loading. For vapors and gases, carbon adsorption, wet scrubbers, or incineration are possible choices. The air cleaning device usually represents a significant pressure drop in the local exhaust system. To add air cleaning to an existing system that does not already have it, upgrading the fan and motor is usually required. As a minimum, upgrading the fan and motor will be required if air cleaning is added onto a system.

### 18.4.7 Acceptance and Start-Up Testing

It is important that new ventilation systems be inspected and tested on completion of the installation to ensure the system conforms to the design specifications and that worker exposure be kept as low as reasonably possible. Unfortunately it is the author's experience that only 10% of the new systems in general industry undergo such scrutiny. The situation is baffling given the importance of contaminant control, the general practice of industry to test other services before acceptance, and the cost of ventilation systems (\$10–30 per cfm to install and \$3–5 per cfm per year to operate).

**TABLE 18.16 Types of LEV Systems Commonly Used in the Semiconductor Industry**

*Laboratory-type hoods*. Used in many locations for storage/use of chemicals, parts storage, or in QA/QC/reliability labs for analytical procedures; ensure materials of construction are compatible with starter and intermediate chemicals and sash adjustments are marked for minimum face velocities of 80–120 lf $min^{-1}$ as per ASHRAE/ANSI Z9.5-1992, Lab Ventilation; crossdrafts from general HVAC and especially PCS/VLS can disperse air contaminants, as well as foot traffic in front of hoods; pay attention to compatibility of materials stored in the hood

*Wet sinks (primarily for etching or cleaning of wafers/boats)*. Plenum exhaust with shroud that fits over the deck surface to allow better capture at the etch/clean baths and to direct overhead vertical flow from laminar flow hood (making a push–pull system), with face velocities at shroud of 100–125 lf $min^{-1}$, in addition to compatibility of materials in the sink; pay attention to their compatibility with drains (e.g. cyanide solutions should not be kept in sinks with acid aspirators). Another approach used within the industry is to provide open wet sinks without shrouds with an exhaust volume of 125–150% of the laminar flow supply

*Gas cabinets*. Use 150 cfm per bottle in cabinet design; UFC Article 51.107(c)3 requires gas cabinet be ventilated with 200 FPM average face velocity at the access port, with minimum velocity of 150 FPM at any point; ensure vent lines are routed to gas conditioning system or silane burnoff system; make sure exhaust from one cabinet is not intermingled with other cabinet exhausts unless chemically compatible

*Gas jungles or gas control valves*. Must be exhausted; some companies have set internal standards, for example, a minimum of 5 air changes per minute (ACM) and 125 ft $min^{-1}$ of face velocity if accessible, or 5 ACM if not accessible

*Open surface tanks (plating or degreasing)*. Commonly used in "back-end" processing for adding precious metal layers; depending on the plating tank constituents may require use of push–pull systems; use of cyanides require stringent precautions with the use of acids (possible HCN formation); general HVAC system can cause crossdrafts that will dilute air contaminants into the general plating room air

*Diffusion furnaces*. Local exhaust provided at "source end" jungle, and at "load end" of tube, using collector-type end caps to route tube exhaust products to vestibule exhaust opening; ensure VLF and HLF systems do not cause eddy currents and transient leads/odors

*Chemical storage cabinets*. Minimum of 50 cfm per standard sized cabinet; ensure compatibility of materials stored within cabinet; use rated flammable storage vessels if flammables are transferred from original container

*Equipment cleaning hoods*. Used for cleaning equipment parts that are contaminated with arsenic, antimony, or spinner/coater residues, pumps, etc.; typically want a minimum face velocity of 125 lf $min^{-1}$ at hood face; housekeeping in the hood is very important; also important to ensure the hoods are big enough to handle the largest parts to be cleaned

*Pump and equipment exhaust lines*. Chemical compatibility with effluents, ability to handle pyrophoric materials, and potential for duct fires; blockage of ducting is possible from reaction products (such as CVD products from $Si_3N_4$); monitor exhaust duct/line pressure closely

*Portable chemical hoods*. Use with nonflammable materials or as temporary exhausts for equipment or materials that are cleaned sporadically; they require preventative maintenance more than in-place chemical hoods and are more limited in the chemicals that can be effectively used in them

*Glove boxes*. Used for the containment of highly toxic source materials in vials or jars (antimony, arsenic) or flammable metals (phosphorus); need LEV to provide negative pressure at load door and HEPA filtration on LEV duct to capture fugitive particulate

*Burn-in testers*. Temperature testing for ICs; need LEV for heat and smoke detectors for ICs or boards that may start to decompose from dropping on heating elements or over-temperature shutoff malfunctioning on tester

*Drying ovens including Blue M®-type ovens*. Used primarily to drive off moisture after cleaning of equipment parts; other volatiles can come off IC package materials and cause odors if not purged properly prior to door opening; sometimes due to temperature control problems, local exhaust is provided via an exhaust duct damper opened immediately prior to door opening; sometimes exhausted for heat only

*Welding/brazing/cutting hoods*. LEV control velocities vary depending on the composition of the materials being worked on; LEV exhaust duct must be situated to avoid re-entrainment into general HVAC system

*Abrasive blasting hoods*. Commonly used for cleaning contaminated parts from metal deposition areas (copper, platinum, gold, titanium/tungsten, aluminum, etc.); segregate "bead blasters" that are used for cleaning ion implanter parts (arsenic contaminated or phosphorus contaminated – which is a fire hazard); and other metallization equipment; ensure fine particle leakage is contained within the bead blaster by closing leakage points; establish controls/procedures for changing the contaminated bead blasting media (usually glass bead or silicon carbide)

*HEPA portable and house vacuum systems*. Portable HEPA vacuums are used extensively for cleaning the interiors of ion implanters or collector deposition in diffusion furnaces, cleaning up residues in parts cleaning hoods or around bead blasters; care must be taken in using the standard HEPA vacuum without impregnated activated charcoal (designed for vacuuming up mercury) as arsenic vapors or arsine gas may be generated during implanter and molecular beam epitaxy – the impregnated activated charcoal was very effective in stripping out contaminants from the vacuum exhaust stream; house vacuum systems are very susceptible to contamination with hazardous materials in the system; they should not be used to clean surface contaminated with hazardous materials; however, because their use is difficult to control, they must be treated as being contaminated (i.e. full protective equipment and procedures)

*Source:* From Williams et al. (1995).

**18.4.7.1 *Construction Details*** The ventilation system design, completed by the plant facilities and engineering group or a consulting firm, is based on certain assumptions and specifications on hardware details such as elbows, entries, expansions, contractions, and so forth. The hood construction has been detailed by the draftsperson. If the hardware elements provided for installation are supplied by a standard manufacturer, the losses will be approximately those used by the designer. If the system is large, it may be worthwhile obtaining samples of the components to test for losses. The hoods warrant special attention. The entry losses on standard hood designs as noted in the ACGIH ventilation manual are quite reliable. However, if the hood design is not typical, it may be difficult to estimate entry loss. In such a case the best solution is to have a single hood fabricated and tested by a laboratory to define the entry loss. This is especially important when a large number of hoods are to be installed. With a sample hood the coefficient of entry can be defined for use in routine hood static suction measurements to calculate airflow. This approach will be discussed in the next section.

An initial physical inspection of the completed system should be conducted to determine if the system is built according to the design. The size and type of fittings should be checked, and the hood construction reviewed. The fan type and size should be as specified, and the direction of rotation, speed, and current drain should be noted. A detailed inspection sheet should be completed. This type of inspection may seem redundant, but in the author's experience it is worthwhile. In one case a miter elbow was added in a large submain to save space; the single elbow presented high losses that caused the system to malfunction.

The air cleaning component should be inspected to determine if it is installed according to specifications. If ancillary equipment such as pressure sensing or velocity measuring equipment is included in the systems, it should also be inspected and calibrated.

All conventional ductwork and air cleaning equipment should be mounted on the suction side of the fan. If this is not done, toxic contaminants may leak to the occupied space. In one such case an installation handling a volatile and odorous organic chemical with the duct under positive pressure contaminated the workplace. Due to the complexity of the system, the cost of rectifying this problem was $50000. This problem should have been picked up in the design review. If a duct run must be under positive pressure, special design features must be utilized by the engineer.

**18.4.7.2 *Airflow*** The general format for testing a new local exhaust ventilation system balanced without blast gates is shown in Figure 18.16. The airflow rate through each branch servicing a hood should be evaluated by a pitot-static traverse (ACGIH 1996). The pitot-static tube is a primary standard, and, if used with an inclined manometer, it does not require calibration. Care is required in the choice of traverse location, and the device is limited at low velocities. In the past decade electronic manometers with microprocessor-based instruments are available for direct reading of pressure and velocity. When used in conjunction with a small portable recorder, the pitot traverse can be conducted by one person with ease.

One of two critical design velocities will be specified depending on the hood type. A partial enclosure such as a paint spray hood has a design velocity or control velocity at the plane of the hood face. Exterior hoods such as a simple welding hood will have a capture velocity specified at a certain working distance. The control and capture velocities

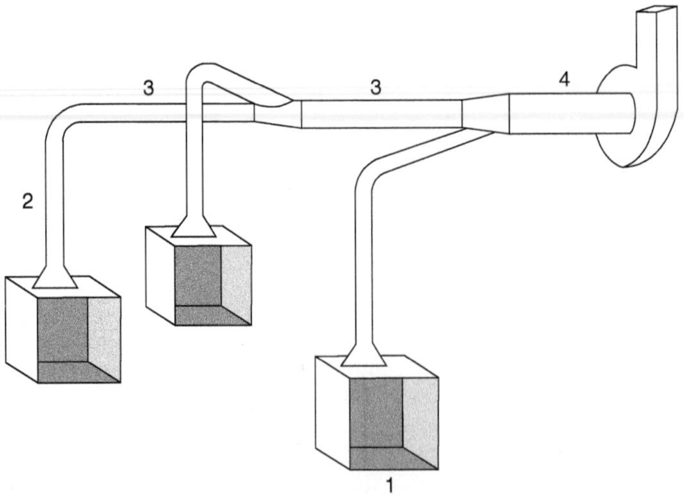


**Figure 18.16** Locations for ventilation measurements: (1) At the face of booth-type hoods, (2) just downstream of a hood at a branch location, (3) in the main to define hood exhaust rate by difference, and (4) in the main to define total system airflow.

may be calculated from the airflow rate and the hood dimensions. Frequently the actual velocities must be measured to satisfy plant or regulatory requirements. Face velocities at partial enclosures are usually evaluated by one of three direct reading anemometers: rotating vane, swinging vane, or heated element. These instruments have a wide range of applications, but care should be taken to observe their limitations (Burgess et al. 1989). These simple direct reading instruments can be used for the required inspections.

The measurement of hood static suction is also of value in periodic inspections and should be measured at the acceptance tests. The measurement, made two to four diameters downstream of the hood, provides a simple method of calculating airflow (Burgess et al. 1989).

The data collected in the acceptance tests are important baseline information for subsequent periodic inspections.

**18.4.7.3 *Performance*** The term *performance*, as used in this discussion, describes the ability of the local exhaust ventilation to minimize the release of air contaminants from equipment serviced by the system. This measurement can be done by qualitative, semiquantitative, and quantitative methods.

**18.4.7.4 *Qualitative Methods*** Smoke sticks, smoke candles, and theater fog generating devices allow one to visualize airflow patterns around a hood. The use of smoke tracers released at the generation point, if done properly, provides an excellent qualitative method of exploring the performance of the hood. The control boundary for capture hoods can be established with this technique. In addition, it can be an excellent teaching tool for the worker who can view the impact of his own body and actions and that of external disturbances such as drafts from windows, doors, and traffic on the ventilation. Corrosive smoke and theater fog cannot be used in semiconductor facilities; dry ice, liquid nitrogen, and water vapor wands have been evaluated for this purpose. In Europe, another qualitative technique, light scatter with backlighting, is widely used to identify the source and generation mode of particle contamination. Exquisite photographs of dust release based on this technique have been used as a design input for ceramic industry dust control.

**18.4.7.5 *Semiquantitative*** The local exhaust ventilation system is first assessed to determine if it meets good practice in terms of general design and airflow rate. If that is acceptable, the system is checked qualitatively by smoke and then semiquantitatively using a tracer (Figure 18.17). A series of release grids are designed that model the actual release area. An oil mist generator operating with corn oil is used to generate a submicrometer-sized aerosol. The generator is positioned to release the mist deep in the hood so that all mist is collected. The probe of a forward light scatter photometer samples the duct stream laden with the mist, and the reading is defined as 100% containment. The grid is then positioned at the actual work release point and a new reading is taken in the duct. If all the mist is collected, the reading will be 100%; if only one-half the oil mist is collected, the reading of containment is 50%.

This system and others now available using tracers such as sulfur hexafluoride are semiquantitative. The most common of these is the ASHRAE 110-1995 used to test performance in laboratory fume hoods (ASHRAE 1995). This test procedure can be modified to test other types of enclosing hoods. These tests do not attempt to closely model the specific chemical or its generation rate. However, the approach provides insight into performance of specific systems and permits "tune-up" of operating parameters such as airflow rate, hood geometry, baffles, and so forth before the quantitative studies are conducted.

**18.4.7.6 *Quantitative*** The containment efficiency of a local exhaust ventilation system is usually impossible to establish in a plant using the actual chemical. To conduct such a test, the system must either be modified so that all the

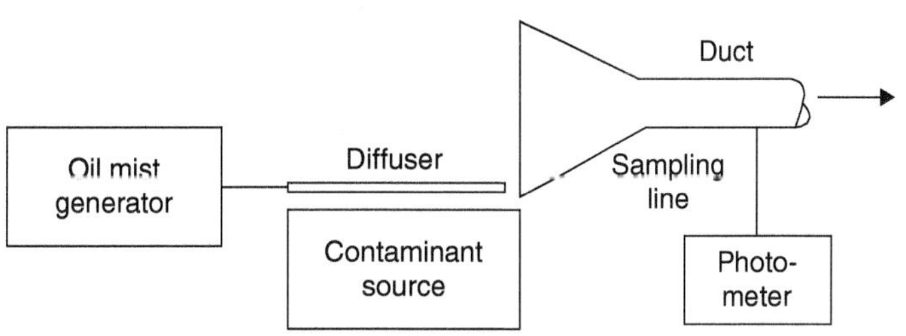


**Figure 18.17** Containment fraction for a hood can be determined in the field using a tracer introduced first inside the hood to establish the 100% benchmark and then at the normal contaminant release point. *Source:* From Ellenbecker et al. (1983). Reproduced with permission of the *Journal of the American Industrial Hygiene Association*.

contaminant enters the hood to establish 100% containment, or the fraction lost from the hood must be identified and that fraction added to the material collected by the hood. The fraction lost to the room usually cannot be measured.

In most cases, however, quantitative measurement of performance is made by air sampling on the worker. In this case the demonstrated performance is not specifically of hood containment but rather the ability of the local exhaust ventilation system to achieve control at the breathing zone of the worker. The most realistic test involves the operator using the actual chemical. Occasionally, as in the case of a PMN chemical, it may be useful to demonstrate performance of the controls using a substitute material. In one such test a simulant chemical was used for the actual PMN chemical. The simulant was a nontoxic blue dye that was inexpensive, reasonably dusty, and easy to sample and analyze. A later test with the actual chemical confirmed the information obtained with the tracer.

The Semiconductor Equipment and Materials International published the Test Method for Enclosures Using Sulfur Hexafluoride Tracer Gas and Gas Chromatography (SEMI F15-93 1993). The test method requires that tracer gas be released at a flow rate that would be expected during an accidental toxic gas release. This is based on the trip point of the emergency shutoff (excess flow) valve or the flow-restricting orifice located just downstream of the main cylinder valve.

While tracer gas is released, air samples are collected outside of the enclosure in polypropylene syringes and analyzed with a gas chromatograph. An equivalent release concentration for the process gas is calculated, and the enclosure is considered acceptable if the TLV or PEL of the process gas is not exceeded.

### 18.4.8 Periodic Testing and Maintenance

The local exhaust ventilation system must be inspected periodically to ensure that the system meets performance and design standards and complies with various regulations. The inspection frequency varies and may range from a monthly to a semiannual interval. The system is inspected visually. The motor-fan components are inspected to determine the condition of the fan blades, direction of rotation, belt tension, guarding, and lubrication. The hoods and ductwork are inspected for corrosion damage, plugging, leaks, and any local modification by the plant. The airflow rate at each station is determined, preferably by the hood static suction method since it is accurate and fast. On semienclosing hoods such as paint spray booths, the face velocity must be evaluated. Certain regulations may require measurement of capture velocity, that is, the velocity at the contaminant release point. This measurement is difficult to conduct accurately. The measurement of airflow rate to achieve that velocity is the more accurate measurement.

The records management system for the inspection program is of great importance especially for facilities with hundreds of hoods. The need for such a system is described by Stott and Platts (1986).

The recording and manipulation of the necessary data is time consuming, tedious, and prone to inaccuracies. Most is recorded in notebooks, which must be transcribed onto official forms, which means that comparisons with previous measurements are often difficult. So at this stage, although a wealth of information is available regarding the extraction systems, the opportunity to analyze the data, initiate maintenance work, or diagnose faults is wasted.

These authors developed a computer-based system to satisfy these needs. The person conducting the field test uses a handheld data terminal that has been loaded with a program on the ventilation system under test. Each ventilation station is named, the position to be tested is identified, and a space is provided for entry of the test results. In the plant the tester calls up the station and conducts the velocity or pressure test measurement. The data are inserted into the portable data logger. On return to the office, the data logger is downloaded to a personal computer. The data are scanned, and those stations where a preset minimum value is not achieved are identified by the computer for follow-up. Additional measurements are requested, and based on these data, a diagnosis of the ventilation system problem is identified by the computer for referral to plant engineering.

The importance of well-qualified personnel to design, install, operate, and inspect local exhaust ventilation systems has been discussed in detail (BOHS 1987; Burton 1991). It is especially important to develop an in-house training program for personnel who conduct the periodic inspections (Burgess 1993; Johnson et al. 1993).

The maintenance problems encountered in system operation and hopefully identified by the acceptance and periodic inspection schedule have been discussed by Burton (1991). The cost of conducting an inspection program has not been published, but in evaluating the cost/benefits of such a program, one should recall the installation cost of the system and its operating cost in addition to the important role it plays in worker health. In many locations the required inspection frequency has prompted the installation of direct reading monitoring equipment, such as pressure sensors at a hood static suction location, a pitot-static tube in the hood branch, a heated element sensor, or a swinging vane anemometer at the hood face. In a nuclear energy facility, an in-line orifice meter is installed to measure flow through a glove box. This mass flow measuring technique has limited application due to energy loss, erosion of the orifice, and contamination.

In discussing the performance of local exhaust ventilation, DallaValle (1952) stated:

> Whereas hoods are installed to eliminate a health hazard, tests should be conducted to establish their effectiveness. The ultimate criterion is not the provision of a 'strong' suction but the handling of an air volume which reduces the concentration of the contaminants in question below the MAC (maximum acceptable concentration) level.

## 18.5 SUMMARY

In this chapter the advances in control technology in the past decade are reviewed. The ability to identify critical exposure conditions has been enhanced by the coupling of sophisticated direct reading air sampling instruments with videotaping of the job tasks. This technique permits the investigator to characterize the nature and origin of emissions in great detail with the subsequent application of effective controls. A second major shift during the decade is the emphasis on the reduction in the use of toxic materials with replacement by materials "kinder" to the worker and the environment. Equipment advances during this period include robotic techniques that permit the worker to be separated from the point of greatest exposure. Finally the application of ventilation control is enhanced by improved knowledge of airflow patterns into hoods, new techniques for evaluating hood containment, and management tools for ventilation systems.

# APPENDIX A

## NIOSH CONTROL TECHNOLOGY REPORTS*

RN: 00192317
TI: Engineering Health Hazard Control Technology for Coal Gasification and Liquefaction Process. Final Report
AU: Anonymous
SO: Division of Physical Sciences and Engineering, NIOSH, U.S. Department of Health and Human Services, Cincinnati, Ohio, Contract No. 210-78-0084, 105 pages, 78 references
PY: 1983

RN: 00092812
TI: Control Technology Assessment: The Secondary Nonferrous Smelting Industry
AU: Burton-DJ; Coleman-RT; Coltharp-WM; Hoover-JR; Vandervort-R
SO: Division of Physical Science and Engineering, NIOSH, Cincinnati, Ohio, NIOSH Contract No. 210-77-0008, 393 pages
PY: 1979

RN: 00092805
TI: An Evaluation of Occupational Health Hazard Control Technology for the Foundry Industry
AU: Scholz-RC
SO: Division of Physical Sciences and Engineering, NIOSH, Cincinnati, Ohio, DHEW (NIOSH)
Publication No. 79-114, Contract No. 210-77-0009, 436 pages, 56 references
PY: 1978

RN: 00145199
TI: Engineering and Other Health Hazard Controls in Oral Contraceptive Tablet Making Operations
AU: Anastas-MY
SO: Division of Physical Sciences and Engineering, NIOSH, U.S. Department of Health and Human Services, Cincinnati, Ohio, 94 pages

RN: 00080675
TI: Development of an Engineering Control Research and Development Plan for Carcinogenic Materials
AU: Hickey-JLS; JJ-Kearney
SO: Applied Ecology Department, Research Triangle Institute, Research Triangle Park, North Carolina NIOSH Contract No. 210-76-0147, 168 pages, 19 references
PY: 1977

RN: 00074602
TI: Engineering Control of Welding Fumes
AU: Astleford-W
SO: Division of Laboratories and Criteria Development, NIOSH, Cincinnati, Ohio, DHEW (NIOSH) Publication No. 75-115, Contract No. 099-72-0076, 122 pages, 13 references
PY: 1974

RN: 00074249
TI: Engineering Control Research and Development Plan for Carcinogenic Materials
AU: Hickey-J; Kearney-JJ
SO: Division of Physical Sciences and Engineering, NIOSH, Cincinnati, Ohio, Contract No. 210-76-0147, 167 pages, 19 references
PY: 1977

---

* *Source:* Courtesy of National Institute for Occupational Safety and Health.

RN: 00052419
TI: Engineering Control Research Recommendations
AU: Hagoapian-JH; Bastress-EK
SO: Division of Physical Sciences and Engineering, NIOSH, Cincinnati, Ohio, Contract No. 099-74-0033, 210 pages, 181 references
PY: 1976

RN: 00182292
TI: Control Technology Assessment of Enzyme Fermentation Processes
AU: Martinez-KF; Sheehy-JW; Jones-JH
SO: Division of Physical Sciences and Engineering, NIOSH, U.S. Department of Health and Human Services, Cincinnati, Ohio, DHHS (NIOSH) Publication No. 88-114, 81 pages, 34 references
PY: 1988

RN: 00177551
TI: Minimizing Worker Exposure during Solid Sampling: A Strategy for Effective Control Technology
AU: Wang-CCK
SO: Division of Physical Sciences and Engineering, NIOSH, U.S. Department of Health and Human Services, 28 pages
PY: 1983

RN: 00177548
TI: A 3-E Quantitative Decision Model of Toxic Substance Control through Control Technology Use in the Industrial Environment
AU: Wang-CCK
SO: NIOSH, U.S. Department of Health and Human Services, Cincinnati, Ohio, 53 pages, 4 references
PY: 1982

RN: 00168700
TI: The Illuminating Engineering Research Institute and Illumination Levels Currently Being Recommended in the United States
AU: Crouch-CL
SO: The Occupational Safety and Health Effects Associated with Reduced Levels of Illumination, Proceedings of a Symposium, July 11–12, 1974, Cincinnati, Ohio, NIOSH, Division of Laboratory and Criteria Development, HEW Publication No. (NIOSH). 75-142, pages 17–27
PY: 1975

RN: 00133474
TI: Control Technology Assessment of the Pesticides Manufacturing and Formulating Industry
AU: Fowler-DP
SO: Division of Physical Sciences and Engineering, NIOSH, Cincinnati, Ohio, Contract No. 210-77-0093, 667 pages
PY: 1980

RN: 00148166
TI: NIOSH Technical Report: Control Technology Assessment: Metal Plating and Cleaning Operations
AU: Sheehy-JW; Mortimer-VD; Jones-JH; Spottswood-SE
SO: NIOSH, U.S. Department of Health and Human Services, Cincinnati, Ohio, Publication No. 85-102, 115 pages, 71 references
PY: 1984

RN: 00144935
TI: A Study of Coal Liquefaction and Gasification Plants: An Industrial Hygiene Assessment, a Control Technology Assessment, and the Development of Sampling and Analytical Techniques, Volume II
AU: Cubit-DA; Tanita-RK
SO: Division of Respiratory Disease Studies, NIOSH, U.S. Department of Health and Human Services, Morgantown, West Virginia, Contract No. 210-78-0101, 179 pages, 15 references
PY: 1983

RN: 00144934
TI: A Study of Coal Liquefaction and Gasification Plants: An Industrial Hygiene Assessment, a Control Technology Assessment, and the Development of Sampling and Analytical Techniques: Volume I
AU: Cubit-DA; Tanita-RK
SO: Division of Respiratory Disease Studies, NIOSH, U.S. Department of Health and Human Services, Morgantown, West Virginia, Contract No. 210-78-0101, 192 pages, 22 references
PY: 1983

RN: 00136362
TI: Control Technology Assessment of Selected Petroleum Refinery Operations
AU: Emmel-TE; Lee-BB; Simonson-AV
SO: Division of Physical Sciences and Engineering, NIOSH, U.S. Department of Health and Human Services, Cincinnati, Ohio, NTIS PB83-257-436, Contract No. 210-81-7102, 122 pages
PY: 1983

RN: 00136196
TI: Proceedings of the Second Engineering Control Technology Workshop, June 1981
AU: Konzen-RB
SO: Division of Training and Manpower Development, NIOSH, U.S. Department of Health and Human Services, Cincinnati, Ohio, NTIS PB-83-112-755, NIOSH Report No. 80-3794, 138 pages
PY: 1982

RN: 00133178
TI: Principles of Occupational Safety and Health Engineering, Instructor's Guide
AU: Zimmerman-NJ
SO: Division of Training and Manpower Development, NIOSH, U.S. Department of Health and Human Services, Cincinnati, Ohio, P.O. No. 81-3030, 262 pages, 17 references
PY: 1983

RN: 00135180
TI: Health Hazard Control Technology Assessment of the Silica Flour Milling Industry
AU: Caplan-PE; Reed-LD; Amendola-AA; Cooper-TC
SO: Division of Physical Sciences and Engineering, NIOSH, U.S. Department of Health and Human Services, Cincinnati, Ohio, 60 pages, 18 references
PY: 1982

RN: 00135171
TI: Engineering Noise Control Technology Demonstration for the Furniture Manufacturing Industry
AU: Hart-FD; Stewart-JS
SO: NIOSH, U.S. Department of Health, Education, and Welfare, Grant No. 1-ROH-OH-00953, 123 pages, 8 references
PY: 1982

RN: 00132081
TI: Control Technology Assessment for Chemical Processes Unit Operations
AU: Van-Wagenen-H
SO: NIOSH, Cincinnati, Ohio, Contract No. 210-80-0071, NTIS PB83-187-492, 19 pages
PY: 1983

RN: 00134239
TI: Engineering Control of Occupational Health Hazards in the Foundry Industry. Instructor's Guide
AU: Scholz-RC
SO: NIOSH, Cincinnati, Ohio, NTIS PB82-231-234, 156 pages, 49 references
PY: 1981

RN: 00133884
TI: Demonstrations of Control Technology for Secondary Lead Reprocessing
AU: Burton-DJ; Simonson-AV; Emmel-BB; Hunt-DB

SO: NIOSH, U.S. Department of Health and Human Services, Rockville, Maryland, Contract No. 210-81-7106, 291 pages
PY: 1983

RN: 00132005
TI: Pilot Control Technology Assessment of Chemical Reprocessing and Reclaiming Facilities
AU: Crandell-MS
SO: Engineering Control Technology Branch, NIOSH, Cincinnati, Ohio, NTIS PB83-197-806, 11 pages
PY: 1982

RN: 00130228
TI: Occupational Health Control Technology for the Primary Aluminum Industry
AU: Sheehy-JW
SO: Public Health Service, NIOSH, U.S. Department of Health and Human Services, Cincinnati, Ohio, DHHS Publication No. 83-115, 59 pages, 6 references
PY: 1983

RN: 00130213
TI: Control Technology Assessment in the Pulp and Paper Industry
AU: Schoultz-K; Matthews-R; Yee-J; Haner-H; Overbaugh-J; Turner-S; Kearney-J
SO: NIOSH, Public Health Service, U.S. Department of Health and Human Services, Cincinnati, Ohio, Contract No. 210-79-0008, 974 pages
PY: 1983

RN: 00106373
TI: Mechanical Power Press Safety Engineering Guide, Wilco, Inc., Stillwater, Minnesota
AU: Anonymous
SO: NIOSH, Division of Laboratories and Criteria Development, U.S. Department of H.E.W., Cincinnati, Ohio, 207 pages
PY: 1976

RN: 00123450
TI: Assessment of Engineering Control Monitoring Equipment. Volume II
AU: Anonymous
SO: Enviro Control, Inc., Rockville, Md., NIOSH, Cincinnati, Ohio, 393 pages, 23 references
PY: 1981

RN: 00123377
TI: Control Technology Assessment of Selected Process in the Textile Finishing Industry
AU: Collins-LH
SO: Bendix Launch Support Division, Cocoa Beach, Florida, NIOSH, Cincinnati, Ohio, 227 pages, 44 references
PY: 1978

RN: 00119991
TI: Phase I Report on Control Technology Assessment of Ore Beneficiation
AU: Todd-WF
SO: NIOSH, U.S. Department of Health and Human Services, Cincinnati, Ohio, 82 pages, 80 references
PY: 1980

RN: 00117613
TI: Symposium Proceedings, Control Technology in the Plastics and Resins Industry
AU: Anonymous
SO: NIOSH, U.S. Department of Health and Human Services, Cincinnati, Ohio, 333 pages, 23 references
PY: 1981

RN: 00117359
TI: Proceedings of the Symposium on Occupational Health Hazard Control Technology in the Foundry and Secondary Non-Ferrous Smelting Industries
AU: Anonymous

SO: NIOSH, U.S. Department of Health and Human Services, 401 pages, 45 references
PY: 1981

RN: 00116023
TI: Control Technology Summary Report on the Primary Nonferrous Metals Industry, Vol. IV, appendix C
AU: Hoover-JR
SO: NIOSH, Center for Disease Control, Public Health Service, U.S. Department of Health, Education, and Welfare, 120 pages
PY: 1978

RN: 00116022
TI: Control Technology Summary Report on the Primary Nonferrous Metals Industry, Vol. 5, appendix D
AU: Hoover-JR
SO: NIOSH, Center for Disease Control, Public Health Service, U.S. Department of Health, Education, and Welfare, 180 pages
PY: 1978

RN: 00115792
TI: Control Technology for Primary Aluminum Processing
AU: Sheehy-JW
SO: Department of Health and Human Services, Public Health Service, Center for Disease Control, NIOSH, Division of Physical Sciences and Engineering, Cincinnati, Ohio, pages 1–22
PY: 1980

RN: 00115530
TI: An Evaluation of Engineering Control Technology for Spray Painting
AU: O'Brien-DM; Hurley-DE
SO: NIOSH, Center for Disease Control, Public Health Service, U.S. Department of Health and Human Services, 117 pages, 52 references
PY: 1981

RN: 00114224
TI: Control Technology Summary Report on the Primary Nonferrous Metals Industry, Vol. II, appendix A
AU: Coleman-RT; Hoover-JR
SO: Division of Physical Science and Engineering, NIOSH, 291 pages, 10 references
PY: 1978

RN: 00114223
TI: Control Technology Summary Report on the Primary Nonferrous Metals Industry, Vol. III, appendix B
AU: Coleman-RT
SO: Division of Physical Science and Engineering, NIOSH, Cincinnati, Ohio, 102 pages
PY: 1978

RN: 00113373
TI: Control Technology Summary Report on the Primary Nonferrous Smelting Industry, Volume 1: Executive Summary
AU: Coleman-RT; Hoover-JR
SO: Division of Physical Science and Engineering, NIOSH, Cincinnati, Ohio, Contract No. 210-77-0008, Radian Corporation, Austin, Texas, 36 pages, 8 references
PY: 1978

RN: 00112367
TI: Control Technology Assessment of Raw Cotton Processing Operations
AU: Anonymous
SO: Envirocontrol Inc., Rockville, Md., NIOSH, Cincinnati, Ohio, Contract No. 210-78-0001, 351 pages
PY: 1980

RN: 00112366
TI: Engineering Control Technology Assessment for the Plastics and Resins Industry
AU: Anonymous
SO: Division of Physical Sciences, NIOSH, U.S. Department of Health, Education and Welfare, Contract No. 210-76-0122, Cincinnati, Ohio, 234 pages, 60 references
PY: 1977

RN: 00102819
TI: Proceedings of the Symposium on Occupational Health Hazard Control Technology in the Foundry and Secondary Non-Ferrous Smelting Industries
AU: Scholz-RC; Leazer-LD
SO: Department of Health, Education, and Welfare, Public Health Service Center for Disease Control, NIOSH, Cincinnati, Ohio, 447 pages, 10 references
PY: 1980

RN: 00094260
TI: Assessment of Selected Control Technology Techniques for Welding Fumes
AU: Van-Wagenen-HD
SO: Division of Physical Sciences and Engineering, NIOSH, Cincinnati, Ohio, NIOSH Publication No. 79-125, 29 pages, 17 references
PY: 1979

RN: 00094228
TI: Control Technology for Worker Exposure to Coke Oven Emissions
AU: Sheehy-JW
SO: Division of Physical Sciences and Engineering, NIOSH, Cincinnati, Ohio, NIOSH Publication No. 80-114, 29 pages, 26 references
PY: 1980

RN: 00092888
TI: Proceedings of NIOSH/University Occupational Health Engineering Control Technology Workshop
AU: Talty-JT
SO: Proceedings of the Workshop on Occupational Health Engineering Control Technology, May 16–17, 1979, Division of Physical Sciences and Engineering, NIOSH, Cincinnati, Ohio, 149 pages
PY: 1979

RN: 00092810
TI: Control Technology Summary Report on the Primary Nonferrous Metals Industry, Volume V, appendix D: Review of the Testimony Presented at the 1977 OSHA Public Hearing on Sulfur Dioxide
AU: Hoover-JR
SO: Division of Physical Science and Engineering, NIOSH, Cincinnati, Ohio, NIOSH Contract No. 210-77-0008, 183 pages
PY: 1978

RN: 00092809
TI: Control Technology Summary Report on the Primary Nonferrous Metals Industry, Volume IV, appendix C: Review of the Testimony Presented at the 1977 OSHA Public Hearings on Inorganic Lead
AU: Hoover-JR
SO: Division of Physical Science and Engineering, NIOSH, Cincinnati, Ohio, NIOSH Contract No. 210-77-0008, 123 pages
PY: 1978

RN: 00092808
TI: Control Technology Summary Report on the Primary Nonferrous Metals Industry, Volume III, appendix B: Review of the Testimony Presented at the 1975/6 OSHA Public Hearing on Inorganic Arsenic
AU: Coleman-RT
SO: Division of Physical Science and Engineering, NIOSH, Cincinnati, Ohio, NIOSH Contract No. 210-77-0008, 105 pages
PY: 1978

RN: 00092807
TI: Control Technology Summary Report on the Primary Nonferrous Metals Industry, Volume II, appendix A
AU: Coleman-RT; Hoover-JR
SO: Division of Physical Science and Engineering, NIOSH, Cincinnati, Ohio, NIOSH Contract No. 210-77-0008, 298 pages, 9 references
PY: 1978

The following reports are not yet listed in NIOSHTIC:

Control Technology for Ethylene Oxide Sterilization in Hospitals, V. D. Mortimer and S. L. Kercher, Division of Physical Science and Engineering, NIOSH, Cincinnati, Ohio, Pub No. 89-120, 167 pages, 82 references, 1989

Control of Asbestos Exposure During Brake Drum Service, J. W. Sheey, T. C. Cooper, D. M. O'Brien, J. D. McGlothlin, and P. A. Froelich, Division of Physical Science and Engineering, NIOSH, Cincinnati, Ohio, Pub No. 89-121, 69 pages, 48 references, 1989

## BIBLIOGRAPHY


ACGIH (1996). Committee on industrial ventilation, industrial ventilation. In: *A Manual of Recommended Practice*, 25e. Cincinnati, OH: American Conference of Governmental Industrial Hygienists.

AFS (1972). *Foundry Environmental Control*. Des Plaines, IL: American Foundrymen's Society.

Ahearn, J., Fatkin, H., and Schwalm, W. (1991). Case study: Polaroid Corporation's systematic approach to waste minimization. *Pollution Prevent. Rev.* **1** (13): 257–271.

AISI (1965). *Steel Mill Ventilation*. New York: American Iron and Steel Institute.

Anderson, D.M. (1964). Dust control by air induction technique. *Ind. Med. Surg.* **34**: 168.

ANSI (1988). *Precautionary Labeling of Hazardous Industrial Chemicals*. ANSI Z129.1-1988. New York: American National Standards Institute.

ANSI (1996). *American National Standard for the Recirculation of Air from Industrial Process Exhaust Systems*. ANSI/AIHA Z9.7-1996. New York: American National Standards, Institute.

ASHRAE (1995). *ANSI/ASHRAE 110-1995, Performance Testing of Laboratory Hoods*. Atlanta, GA: American Society of Heating Refrigeration and Air Conditioning Engineers.

BCIRA (1977). *Proceedings of the Working Environment in Iron Foundries*. Birmingham: British Cast Iron Research Association.

Berens, A.R. (1981). Vinyl chloride monomer in PVC: from problem to probe. *Pure Appl. Chem.* **53**: 365–375.

BOHS (British Occupational Hygiene Society) (1984). *Fugitive Emissions of Vapors from Process Equipment*, Technical Guide No., vol. **3**. Northwood, Middlesex: Science Reviews Ltd.

BOHS (British Occupational Hygiene Society) (1985). *Dustiness Estimation Methods for Dry Materials, Their Uses and Standardization*, Technical Guide No., vol. **4**. Northwood, Middlesex: Science Reviews Ltd.

BOHS (British Occupational Hygiene Society) (1987). *Controlling Airborne Contaminants in the Workplace*, BOHS Technical Guide No., vol. **7**. Northwood, Middlesex: Science Reviews Ltd.

Brandt, A.D. (1947). *Industrial Health Engineering*. New York: Wiley.

Braun, P. and Druckman, E. (eds.) (1976). Public health rounds at the Harvard School of Public Health. Vinyl chloride: can the worker be protected. *N. Engl. J. Med.* **294**: 653–657.

Burgess, W.A. (1993). The international ventilation symposia and the practitioner in industry. *Ventilation '91, Proceedings of the Third International Symposium on Ventilation for Contaminant Control*, ACGIH, Cincinnati, OH (16–20 September 1991).

Burgess, W.A. (1996). *Cummings Award Lecture at American Industrial Hygiene Conference and Exhibition*, Washington, DC.

Burgess, W.A., Ellenbecker, M.J., and Treitman, R.D. (1989). *Ventilation for Control of the Work Environment*. New York: Wiley.

Burton, D.J. (1991). *Industrial Ventilation Work Book*. Salt Lake City, UT: IVE, Inc.

Caplan, K. (1985a). Philosophy and management of engineering controls. In: *Patty's Industrial Hygiene and Toxicology* (ed. L.J. Cralley and L.V. Cralley). New York: Wiley.

Caplan, K.C. (1985b). Building types. In: *Industrial Hygiene Aspects of Plant Operations*, vol. **3** (ed. L.V. Cralley, L.J. Cralley, and K.C. Caplan). New York: Macmillan.

Cooper, T. and Gressel, M. (1992). Real-time evaluation at a bag emptying operation: a case study. *Appl. Ind. Hyg.* **7** (4): 227–230.

DallaValle, J.M. (1952). *Exhaust Hoods*. New York: Industrial Press.

Ellenbecker, M.J. (1996). Engineering controls as an intervention to reduce worker exposure. *Am. J. Ind. Med.* **29**: 303–370.

Ellenbecker, M.J., Gempel, R.J., and Burgess, W.A. (1983). Capture efficiency of local exhaust ventilation systems. *Am. Ind. Hyg. Assoc. J.* **44**: 752–755.

EPA (1986). Workshop: predicting workplace exposure to new chemicals. *Appl. Ind. Hyg.* **1** (3): R-11–R-13.

First, M.W. (1983). Engineering control of occupational health hazards. *Am. Ind. Hyg. Assoc. J.* **44** (9): 621–626.

Frosch, R.A. and Gallopoulos, N.E. (1989). Strategies for manufacturing. *Scientific American* (September).

Gideon, J., Kennedy, E., O'Brien, D., and Talty, J. (1979). *Controlling Occupational Exposures: Principles and Practices*. Cincinnati, OH: National Institute for Occupational Safety and Health.

Goodfellow, H.D. and Smith, J.W. (1989). Dustiness testing: a new design approach for dust control. In: *Ventilation '88, Proceedings of the Second International Symposium on Ventilation for Contaminant Control* (ed. J.H. Vincent). Oxford: Pergamon Press.

Gray, C.N. and Hewitt, P.J. (1982). Control of particulate emissions from electric-arc welding by product modification. *Ann. Occup. Hyg.* **25** (4): 431–438.

Gressel, M. and Fischbach, T. (1989). Workstation design improvements for the reduction of dust exposures during weighing of chemical powders. *Appl. Ind. Hyg.* **4** (9): 227–233.

Gressel, M., Heitbrink, W.A., McGlothlin, J., and Fischbach, T. (1987). Real-time, integrated, and ergonomic analysis during manual materials handling. *Appl. Ind. Hyg.* **2** (3): 108–113.

Gressel, M., Heitbrink, W., McGlothlin, J., and Fischbach, T. (1988). Advantages of real time data acquisition for exposure assessment. *Appl. Ind. Hyg.* **3** (11): 316–320.

Hallock, M.F., Hammond, S.K., Kenyon, E., Smith, T.J., and Smith, E.R. (1993). Assessment of task and peak exposures to solvents in the microelectronics fabrication industry. *Appl. Occup. Environ. Health* **8** (11): 945–954.

Hammond, C.M. (1980). Dust control concepts in chemical handling and weighing. *Ann. Occup. Hyg.* **23** (1): 95–109.

Hanman, B. (1968). *Physical Abilities to Fit the Job*. Wakefield, MA: American Mutual Liability Insurance Company.

Harten, T. and Licis, I. (1991). Waste reduction technology evaluations of the U.S. EPA WRITE program. *J. Air Waste Manag. Assoc.* **41** (8): 1122–1129.

HSE (1988). Health and Safety Executive, *Control of Substances Hazardous to Health Regulations 1988, Approved Codes of Practice*. London: HMSO.

International Standards Organization 14,000 (1996). *International Organization for Standardization*. Geneva, Switzerland: ISO Central Secretariat.

Johnson, G.Q., Ostendorf, R., Claugherty, D., and Combs, C. (1993). Improving dust control system reliability. *Ventilation '91, Proceedings of the Third International Symposium on Ventilation for Contaminant Control*, ACGIH, Cincinnati, OH (16–20 September 1991).

Kireta, A.G. (1988). Lead solder update. *Heat/Pip./Air Cond.*, 119–125.

Kovein, R.J. and Hentz, P.A. (1992). Real-time personal monitoring in the workplace using radio telemetry. *Appl. Occup. Environ. Health* **7** (3): 168–173.

Lipton, S. and Lynch, J. (1987). *Health Hazard Control in the Chemical Process Industry*. New York: Wiley.

Massachusetts General Laws (1989). Massachusetts Toxics Use Reduction Act. Chapter 211 (24 July).

Mertens, J.A. (1991). CFCs: in search of a clean solution. *Environ. Prot.*, 25–29.

NIOSH (1991). *Technical Report, Control of Emissions from Seals and Fittings in Chemical Process Industries*, DHHS (NIOSH) Publication No. 81-118. Cincinnati, OH: National Institute for Occupational Safety and Health.

O'Brien, D., Fischbach, T., Cooper, T. et al. (1989). Acquisition and spreadsheet analysis of real time dust exposure data: a case study. *Appl. Ind. Hyg.* **4** (9): 238–243.

Olishifski, J.B. (1988). Methods of control. In: *Fundamentals of Industrial Hygiene*, 3e (ed. B. Plog). Chicago, IL: National Safety Council.

OTA (Office of Technology Assessment) (1986). *Serious Reduction of Hazardous Waste for Pollution and Industrial Efficiency*. Washington, DC: U.S. Government Printing Office.

Peterson, J. (1985). Selection and arrangement of process equipment. In: *Industrial Hygiene Aspects of Plant Operations*, vol. **3** (ed. L.V. Cralley, L.J. Cralley, and K.C. Caplan). New York: Macmillan.

Roach, S.A. (1981). On the role of turbulent diffusion in ventilation. *Ann. Occup. Hyg.* **24** (1): 105–133.

Rossi, M. and Geiser, K. (1994). A proposal for managing chemical restrictions at the state level. *Pollut. Prev. Rev.* (Spring).

Rossi, M., Geiser, K., and Ellenbecker, M. (1991). Techniques in toxic use reduction: from concept to action. *New Solut.* **2** (2): 25–32.

SACMA (1991). *Safe Handling of Advanced Composite Materials*. Arlington, VA: Suppliers of Advanced Composite Materials Association.

Schumacher, J.S. (1978). A new dust Control system for foundries. *Am. Ind. Hyg. Assoc. J.* **39** (1): 73–78.

Sciola, V. (1993). The practical application of reduced flow push pull plating tank exhaust systems. *Ventilation '91, Proceedings of the Third International Symposium for Contaminant Control*, Cincinnati, OH (16–20 September 1991).

SEMI F15-93 (1993). *A Test Method for Enclosures Using Sulfur Hexafluoride Tracer Gas and Gas Chromatography*. Semiconductor Equipment and Materials International Organization.

Sherwood, R.J. and Alsbury, R.J. (1983). Occupational hygiene, systematic approach and strategy of exposure control. In: *Encyclopaedia of Occupational Health and Safety* (ed. L. Parmeggiani). Geneva: International Labor Organization.

Soule, R.D. (1991). Industrial hygiene engineering controls. In: *Patty's Industrial Hygiene and Toxicology* (ed. G. Clayton and F.E. Clayton), 1, Part B. New York: Wiley.

Sprow, E. (1993). How to be solvent free in 1993. *Manuf. Eng.* **110** (2 February).

Stott, M.D. and Platts, P.J. (1986). The Ventdata ventilation plant monitoring and maintenance system. In: *Ventilation '85, Proceedings of the First International Symposium for Contaminant Control* (ed. H.D. Goodfellow). Amsterdam: Elsevier.

The Massachusetts Toxic Use Reduction Institute (1994). *Fact Sheet 5, Alternatives to Solvent-Based Coatings, 5*. Lowell, MA: University of Massachusetts.

Uniform Fire Code (1988). Article 51, Semiconductor Fabrication Facilities Using Hazardous Production Materials.

U.S. Congress (1990). Pollution Prevention Act of 1990, Congressional Record, Section 6606 of the Budget Reconciliation Act, p. 12517 (26 October).

Volkwein, J.C., Cecala, A.B., and Thimons, E.D. (1989). Moisture application for dust control. *Appl. Ind. Hyg.* **4** (8): 198–200.

Whitehead, L.W. (1987). Planning considerations for industrial plants emphasizing occupational and environmental health and safety issues. *Appl. Ind. Hyg.* **2** (2): 79–86.

Williams, M., Baldwin, D., and Manz, P. (1995). *Semiconductor Industrial Hygiene Handbook*. Park Ridge, NJ: Noyes Publications.

Wolf, K., Yazdani, A., and Yates, P. (1991). Chlorinated solvents: will the alternatives be safer? *J. Air Waste Manage. Assoc.* **41** (8): 1055–1061.

Wolfson, H. (1993). Is the Process Fit to Have Ventilation Applied? *Proceedings of the Third International Symposium on Ventilation for Contaminant Control*, ACGIH, Cincinnati, OH (16–20 September 1991).

CHAPTER 19

# ENVIRONMENTAL HEALTH AND SAFETY (EHS) AUDITING

ANDREW MCINTYRE

*2150 North First Street, Suite 450, San Jose, CA, 95131*

HARMONY SCOFIELD

*BSI EHS Services and Solutions, 1400 NW Compton Dr., Suite 203, Hillsboro, OR, 97005*

and

STEVEN TRAMMELL

*BSI EHS Services and Solutions, 110 Wild Basin Rd., Suite 270, Austin, TX, 78746*

## 19.1 INTRODUCTION

Environmental health and safety (EHS) auditing has become an important and regular business practice for companies of all sizes. These audits determine the level of compliance to regulatory requirements, standards, goals, and other business benchmarks including codes of conduct. A well-conducted EHS audit will result in a prioritized list of nonconformances and opportunities for improvement, which can be used as a roadmap for improving safety and business practices. It conceptually serves many of the same purposes as internal financial auditing.

This chapter will provide a short background and history of EHS auditing, particularly as it is used currently in the United States, and then lead into a practical walkthrough of the auditing process, from pre-audit activities and fieldwork through recordkeeping and observations and then through evaluating and reporting on findings and recommendations. The final step is a corrective action plan that includes conformance priorities as well as best management practices and opportunities for improvement.

### 19.1.1 A Brief History

Auditing of environmental and occupational health and safety compliance with regulatory requirements and company standards has become a routine practice for the majority of large companies, especially multinationals. A few progressive organizations have performed routine inspections or audits since the early 1950s; however, auditing did not become a regular practice in business until the early 1980s, in the wake of large industrial accidents like the Union Carbide Bhopal incident of December 1984 and increased environmental scrutiny following SEC actions against U.S. Steel, Allied Chemical, and Occidental Petroleum. Corporate audits were required for these companies because the SEC was concerned that large organizations were understating their EHS liability in the wake of several significant environmental incidents including Love Canal. Following the Bhopal tragedy, Union Carbide (now Dow) developed a comprehensive assurance program to help reduce the risk of another incident. This was followed by on-site inspections (audits) at their worldwide facilities, which were repeated annually and became a model for regular global site auditing.

Many other companies followed suit as the overall culture of compliance and responsibility meant they were under increased scrutiny by federal and state regulatory bodies as well as stakeholders. This was strengthened in 2002 with the passage of the Sarbanes–Oxley Act, which, while primarily finance based, nonetheless alerted companies to the

*Handbook of Occupational Safety and Health,* Third Edition. Edited by S. Z. Mansdorf.

need to improve governance at a corporate level for all areas of compliance, including EHS. Benchmarking surveys now indicate that approximately 80% of major multinational corporations practice some form of routine auditing for EHS concerns (Cahill and Kane 2011).

### 19.1.2 Value to the Organization

As stakeholders become more aware of the costs (both financial and reputational) of EHS risks, companies are increasingly pressured to disclose audit findings, nonconformances, and other risk-based metrics to stakeholders as well as to the general public.

Like financial auditing, EHS auditing provides both a level of assurance for corporate officers and a benchmark for the management performance of operating companies. It can also provide the basis for risk analysis of operations and asset allocation for correction of deficiencies. It is considered a management tool essential to the proper governance of modern-day operations by many business and EHS management experts. Auditing can provide an effective means to monitor the performance and management of EHS at local operations as well as establish a path to continuous improvement. Audits are also an effective means to determine and evaluate risk profiles across the enterprise.

## 19.2 TYPES OF AUDITS

While compliance audits are the most widely practiced and common form of EHS auditing, there are a large number of variations. These are listed below (not in order of prevalence):

- Corporate compliance audits.
- OHSMS audits (management systems audits).
- Special emphasis program audits.
- Enterprise risk assessments (audits).
- EHS sustainability reporting audits (assurance audits).
- Due diligence audits (not ASTM phase 1–3).
- Third-party manufacturing audits.
- Supplier audits.

A brief discussion of each type of these EHS audits is as follows.

### 19.2.1 Corporate Compliance Auditing

Auditing of EHS programs can be done at several levels and in several forms. Self-evaluations of compliance with regulatory requirements and corporate directives by the individual operations are an essential element of any audit program. Self-auditing can be accomplished most effectively with the establishment of specific performance requirements for the operations. The internal self-audits are typically followed by auditing against these established criteria using a team of personnel outside of the local operation being audited. This may be on a divisional level or a corporate level or both. Many companies use a combination of plant-level, divisional, and corporate personnel to form the corporate auditing team. This provides the added benefit of cross-fertilization, peer involvement, and training. Depending on the level of comfort desired by corporate officers and the board of directors, auditing may also be conducted by private firms outside of the corporate structure. An optimal approach is to have internal, divisional, corporate, and outside auditing conducted such that all operations are audited at least annually at some level. Multinational companies that want to assure compliance with local regulations typically use local experts (internal or external) on the audit team and may also subscribe to one of the regulatory updating services such as Enhesa (provides key international regulatory updates).

The sophistication of corporate auditing programs for EHS varies considerably. The audit findings can be scored. Many organizations use the scores to determine a portion of the performance bonus paid to local plant managers and divisional heads. Most organizations that use a scoring system publish the scores among their operations so that every local manager knows their performance against their peers. This also provides a strong incentive for competition and improvement. Many organizations also risk rank recommendations (e.g. the risk from a missing sign versus the risk from an improperly designed pressure relief system). This allows for prioritization of corrective actions.

A number of US-based companies use attorney–client privilege for auditing to shield self-incrimination although many environmental regulations require self-reporting of noncompliance and the protection can be very limited.

It is also important to note that most auditing programs include an evaluation of the management of EHS. This segment of the audit is intended to promote the self-governance of EHS issues and concerns leading to both continuous improvement and best practices. This segment typically includes elements such as vision, staffing, employee participation, communications, accountability, management systems, and other such management elements.

The essential elements of a successful corporate auditing program include:

- Establishing the objectives of the auditing program.
- Establishing performance objectives and specific criteria.
- Skilled auditors.
- Allowing for a reasonable level of effort by the auditing team.
- Documentation of deficiencies found during the audit.
- Root cause analysis of deficiencies.
- Clear, concise, and prompt reporting of audit results.
- Follow-up on the resolution of deficiencies found.

### 19.2.2 Management Systems Audits

There are national and international management standards and guidelines for safety and health and environment. Safety and health has been separated from environmental as different but similar management standards. Routine auditing of the program by those adopting EHS standards or guidance is a requirement. This is in addition to the accreditation and reaccreditation management systems audits that are required. The health and safety management standard most used internationally is Occupational Health and Safety Management Standard (OHSMS) (also known as OHSMS 18000 first published in 1999) and in the United States, ANSI Z10. The International Organization for Standardization (ISO) has just recently finalized their new OHSMS standard ISO 45001 (March, 2018). Environmental management systems have had an ISO standard since 1996 known as the ISO 14000 series. The management systems approach includes the auditing function under the "check" concept of the overall management systems scheme of "plan–do–check–act." For example, the ISO 45001 standard implies the need for an annual assessment of compliance with the management systems, internal standards, and regulatory requirements. The ISO 14000 standard is written in a similar fashion.

### 19.2.3 Special Emphasis Audits

It is common to implement a special emphasis audit when there is an incident, new regulation, regulation requiring narrow scope audits or area of concern. For example, a new requirement from the EPA or OSHA could trigger an assessment of the organization's situation by conducting inspections or audits of a substance or process or other areas of concern. This audit scheme is narrow in scope but typically done organization-wide. Process safety management audits fall within this definition. Reassessment of the process safety management programs is most commonly done by outside firms every three to five years.

### 19.2.4 Enterprise Risk Assessments

Enterprise risk assessments are a new trend that has evolved from recent regulatory requirements in Europe. Because the enterprise risk assessment is broad in scope, encompassing business risk of all types, the focus is not on regulatory compliance, though this can be included. These assessments are not audits in the conventional sense, and the objective is a relatively wide range review and prioritization of exposures across a business' portfolio of operations, taking into account traditional internal and external environmental, safety, and health aspects in additional to financial, governance, reputational, and supply chain risks.

### 19.2.5 EHS Sustainability Reporting Audits

Third-party validation of the data and statements in sustainability reports is a best practice – driven by nongovernmental organizations (NGOs) – and is done in part or whole by most of the organizations recognized as sustainability leaders. In integrated reporting, this is typically required by the financial auditors providing assurance for annual reports. All of these validations require some level of auditing of a statistical sampling of sites and data and may actually include all data for some areas of interest (such as greenhouse gas emissions).

### 19.2.6 Due Diligence Audits

Most organizations will perform a safety, health, and environmental audit for sites being acquired to determine their potential liability and resources needed to bring the organization up to corporate standards. This is in addition to the typical lender required "phase 1" (the first of three stages) audits for signs of historical environmental pollution. This is a direct result of "Superfund" legislation in the United States and is widely practiced worldwide.

### 19.2.7 Third-Party Manufacturing Audits

There are a number of companies and sectors that rely on third-party manufacturing for all or a portion of the production of their products. The garment and electronics sectors are especially well known for outsourcing essentially all of their manufacturing. Corporate social responsibility (CSR) principles and a large number of CSR NGO's suggest that the companies contracting their manufacturing have a special responsibility for safety, occupational health, and environment as well as other social issues. Horrific incidents (e.g. Rana Plaza in Bangladesh with 1132 garment workers killed) have further driven large companies to implement CSR audit protocols for their third-party manufacturers. Like regular auditing, these audits are typically done annually by internal and external experts.

### 19.2.8 Supplier Audits

Pressure from CSR NGO's has spurred the expansion of the auditing noted above to top tier suppliers. Additionally, many of the rating agencies and reporting organizations such as the Carbon Disclosure Project request information from major suppliers on EHS performance.

## 19.3 PRE-AUDIT ACTIVITIES

Many of the important activities associated with an audit should occur in the days or weeks preceding the actual on-site visit. Careful pre-audit preparation is one of the fundamental keys to providing a good audit and will greatly improve efficiencies of the auditor and audit team. A clearly defined objective and scope statement needs to be developed, to which all parties to the audit agree. The objective statement should include determination of the audit purpose (e.g. mandated by regulation, internal management systems, obtaining or retaining certifications) and the communication expectations (e.g. formal written report, oral report, and/or presentation). Once objectives are set, the audit scope can be defined. During scope discussions with the auditee, agreements and determination of functional areas to be audited, audit timeline (e.g. look-back vs. snapshot reviews), and information needed can be accomplished in addition to estimation of audit resources and audit length. The audit scope should be appropriately sized to comprehend geography (multiple sites), operational complexity, and organizational size to ensure that appropriate resources from both the auditor and site perspectives can be effectively managed. It may be appropriate to divide the audit into multiple events over several individual site visits to ensure audit objectives and quality can be maintained.

Initial arrangements for the audit, such as timing and resource requirements, need to be discussed as soon as scope is defined. From the objectives, the auditor can suggest required expertise needed for the audit, both external and internal. Early identification of resources is important to ensure that site supplier personnel or specialized external expertise can be arranged for to be present during the specific audit dates.

Pre-audit documentation reviews are important and help familiarize the auditor with the specific activities, operations, and organizations that will be the subjects of the audit. This information will help the auditor gain a sense of the hazards associated with the operations and provide clues as to what specific regulatory requirements may apply. Becoming somewhat familiar with the processes and operations will allow the auditor to prepare knowledgeable questions prior to and during the audit, which helps to establish initial credibility with the site teams. Documentation that may provide early insight and help in preparation can include:

- Description of the processes, operations, and activities within the audit scope.
- Drawing or pictures of the processes or operations.
- Process safety information.
- List of operational permits.
- Operational procedures and maintenance plans.
- Personnel training requirements.

- Results of previous audits or inspections, findings, and status of applicable corrective actions.
- Incident reports.

It is important that the auditor ask only for information that he/she intends to review in advance. Otherwise, the information should be requested to be available at the start of the audit. A pre-audit checklist can be a useful tool for both the auditor and the site to use for efficient preparation for the audit. Preparatory information requests as described above can be included in such a checklist. An example of a pre-audit checklist is provided as Attachment 19.1.

The audit team leader should determine the level of confidentiality required for the audit. The majority of audits will be company confidential; however some legal or compliance audits, or audits performed in the aftermath of an accident, may be classified as attorney–client privileged. This may influence the documentation requirements during and after the audit and limit distribution of audit reports and communication of results. All working papers, electronic communication, draft, and final reports should be marked appropriately per the confidentiality category.

### 19.3.1 Audit Protocols/Checklists

Pre-prepared audit protocols or checklists to be used during an audit can be extremely helpful tools for organizing the structure of the review. Such tools provide reminders to the auditors to help ensure that all key items within the scope of the audit are addressed. For larger audits, checklists can be divided between the audit team based on auditor specialty or to evenly divide the audit workload.

Checklists should not be used as a substitution for clear understanding, planning, and preparation for the audit. They should be customized as much as possible to address areas of audit as described in the scope. A generic checklist with large sections of "not applicable" items uses up valuable auditor time and may lead to a premature conclusion that the operations are of low risk.

More discussion on checklists is provided in the Working Papers section of this document, and sample sections from two difference audit checklists are provided as Attachments 19.2 and 19.3.

### 19.3.2 Team Member Selection, Staffing, and Characteristics

Audits that meet goals and objectives are highly dependent upon resources applied, including inputs from all team members. Proper team selection is critical for an effective audit activity and should include all functions associated with activities contained in the audit scope. Audit personnel are ideally a balance of internal and external sources, which provides for familiarity with internal processes along with inputs derived from external experiences.

The makeup of audit teams will vary depending on the type of audit and the activities being audited. Audit team members should be able to provide knowledge specific to the processes reviewed and their responsibilities associated with those processes. For audits focused on specialized manufacturing activities, engineering, operations, and maintenance personnel associated with those activities are critical to the team. For audits with heavy emphasis on management systems and regulatory compliance or permitting, personnel assigned to those functions should be included. Audits focused on EHS aspects should involve EHS specialists and management, while those on quality aspects should involve quality specialists and management.

All audit team members should have familiarity with the audit process and protocols and, if necessary, have previous training to build auditing skills. The team members must know the laws, regulations, and permits applicable to the site, along with site and company policies. It is good practice for the team leader to brief all members of the audit team on the audit process; however, in-depth audit training should be a separate event from the audit itself.

An audit team leader is generally a technically trained person who has had experience in EHS management and extensive experience in at least one functional area of the audit. It is important to choose the audit team leader wisely. This person needs to have the support and respect of management, the audit team, and site personnel. The leader must be credible to site management and have a willingness and ability to help the facility understand why the facility is being audited, what the audit is all about, and what the audit team is doing. Beyond these skills, the audit team leader should possess the characteristics below that represent the ideal audit team member.

Audit team members should have a strong combination of the following characteristics:

- Training and experience
  - Technical training and experience in some aspects of the audit.
  - General experience in auditing.
  - Independent audit team training or "on-the-job" audit experience.

- Objective and unbiased approach
- Balance approach to findings
    - Sensitive to issues and concerns about identifying potential violations of law and regulations or findings about inherent potential hazards that have not yet been controlled.

Key auditor qualities include:

- Observant and alert.
- Curious and inquisitive.
- Honest and diplomatic.
- Persistent, yet patient.
- A professional attitude and appearance.
- Confident and enthusiastic.
- Disciplined with strong planning and goal-setting skills.
- Focused, able to hone in on important issues and get questions answered.
- A good listener and communicator, with the ability to conduct conversations at all levels.
- Friendly, good natured, and empathetic, with sensitivity of the feelings of the auditee.
- An ability to set the auditee at ease and give them the sense that the process was pleasant and beneficial to the organization.

### 19.3.3 Scheduling the Visit

Once the objectives and scope of the audit are established, audit length and resource requirements can be accurately defined. A face-to-face or telephone meeting should be scheduled with the site contact to confirm audit dates and times and to discuss a preliminary daily agenda. Clear and specific expectations should be set regarding resource requirements needed from the site, which include personnel representing specific functional areas and external suppliers and such logistics requirements as an adequately sized conference room (fitted with a computer, projector, whiteboards, etc.), arrangements for any meals required, and understanding timing of scheduled employee shifts and breaks. The initial meeting should be followed by a memorandum confirming all agreed-upon items.

Approximately two weeks prior to the site audit, a follow-up telephone call and discussion should be arranged with the site contact to finalize the agenda, confirm resource availability, review location security requirements (i.e. badging, identification requirements, camera/laptop restrictions), understand personal protective equipment requirements, and make any last-minute logistics adjustments.

## 19.4 ON-SITE ACTIVITIES

### 19.4.1 Opening Meeting

An opening meeting should be scheduled at the beginning of the on-site audit. Attendees at the meeting should include site upper management, management for each functional or process area to be audited, and the audit team. Representatives from Human Resources may also attend, especially if confidential employee interviews are part of the audit scope. During this meeting, the audit team leader will facilitate introductions, discuss roles of the audit team, and describe the overall audit schedule. On-site activities should be discussed, which might include an orientation tour, documentation reviews, employee interviews, and schedules for daily debriefings and the closing meeting. If specialists are needed for only specific portions of the audit, appropriate times for those resources should be determined. A typical agenda for an opening meeting might be:

- Welcome and introductions.
- Review of the audit scope and objectives.
- Overview of the audit process.
- Review of the audit agenda.
- Communication of imminent hazard findings.

- Communications plan (daily out-briefings, confirmation of site points of contact).
- Closing meeting date/time.

### 19.4.2 Negative Impressions During On-Site Activities

Because this may be the first time the audit team has met the organization's attendees in person, it is important to give a good first impression and to follow that impression with a professional and favorable approach to the working team for the duration of the audit. Some action and attitudes to avoid during on-site activities include the following:

- Arriving late to the audit.
- Being critical of the efforts made by the site under review.
- Having a negative or pessimistic attitude.
- Preaching theories about how the employees of the site ought to handle things the way you do it.
- Getting too personal in discussions, including pointing fingers or other confrontational gestures or body language.
- Being sarcastic.
- When working as a team, allowing differences of opinion between audit team members to become obvious to auditees.
- Not being respectful.
- Arguing about findings.
- Using derogatory language.
- Giving the impression that your task is top secret and that the auditee must not be aware of your observations.

Employees and other on-site personnel may already be nervous about meeting with external auditors. To set the attendees at ease and establish rapport, auditors should avoid giving an impression that the audit intention is to police the site to find problems, but instead use a more value-added and personable approach.

Here are some examples of each approach.

| The "police officer" approach | The "value-added" approach |
|---|---|
| We're there to find problems | We are there to help the site improve |
| They will hide things from us; we must be suspicious | They will be open and candid; still, we should be thorough |
| We'll hit the ground running on Monday; real-time planning is the way to go | We must be sensitive to the site's scheduling needs; we should set an agenda prior to the audit |
| The findings are mine; the site doesn't have to agree | We can negotiate findings, but the final call is ours |
| Findings can be based on my opinion | Findings must be defensible and based on specific regulatory and company policies |
| We are secretive with the findings until the closing conference | New findings are very unlikely; where it happens we will alert the site prior to the report |
| It's ok if there are new findings in the report that were not discussed on-site | |
| Overall, we were relentless in identifying, justifying, and reporting the deficiencies | Overall, we made a substantial difference in improving the safety and environmental performance at the site |

### 19.4.3 Site Orientation and Comprehensive Walkthrough

Adequate time should be set aside on the first of the audit for the auditors to tour the site to become familiar with the operations. Auditors should refrain from detailed auditing during the tour, but instead focus on understanding the operations, listening to information provided by the process experts, asking generalized questions, and recording brief notes to jog memory for follow-up review and questions later on in the audit. For large or complex sites, it is useful to have a site layout map available to review as you work your way through the site tour. From this familiarization tour, a prioritized list of areas can be generated for a more comprehensive walkthrough.

### 19.4.4 Documentation Review

Review of documentation is a key component to an audit, as this provides the paper trail verification for many elements of both regulatory and internal management systems compliance. As previously mentioned, a list of documents to be reviewed should be included in a request during the pre-audit preparation stage. Although the audit team may not be fully aware of all relevant documentation in advance, typical documents and records to be reviewed as part of any audit may include:

- EHS policies, management systems documents, and aspect/impact analysis reports.
- Process hazard analysis reports.
- Employee training programs and training records.
- Equipment operating instructions.
- Equipment preventive maintenance plans and records.
- Regulatory permits.
- Accident/incident records, near-miss reports, and root cause analysis studies.

An issue found in many audits is that the documentation lags behind the actual work practice. Sampling of the documentation, especially areas where key risk controls are involved, should be compared against work practice to validate that they are current.

Preventive maintenance (PM) plans should be accompanied by applicable records documenting performance of the maintenance. Lack of such records could mean a program deficiency and a potential serious finding.

### 19.4.5 Interviews

Some audits may include interviews of site personnel to gauge insight into actual operational/maintenance practices, understanding of EHS program requirements, and efficacy of training. Interviews should be done with small groups or preferably one-on-one between the auditor and employee. Unless required by site management, supervision or other management should not be present during the interview. Specific interview results from individuals should be kept confidential, and results only reported in a manner that protects confidentiality of the interviewee. Results may also be consolidated or aggregated with other interviewee responses to ensure identification of individuals is not possible.

Using an "open" question technique is typically the best way to elicit informative responses from interviewees. These are questions such as "Tell me about your job responsibilities" or "What are the main safety risks associated with the equipment?" This is in contrast to the "closed" questioning technique, which asks questions with determinative (yes or no) response expectations. These are questions such as "Do you think this equipment is safe?" or "Does this process use hazardous chemicals?" However, at times, closed questions may be required, such as when you need to see an employee's operating license or to get them to show a specific feature of interest in the process or equipment.

Auditors should avoid giving a perception that an interview is an interrogation session.

### 19.4.6 Working Papers

Working papers created during the audit will contain all information that the auditor believes may be necessary to support the audit findings. These documents may include auditor checklist and field notes during site visits, notes from personnel interviews, and program compliance evidence such as copies of permits, training records, EHS programs, PM schedules, etc. If allowed by site policy, photographs are also helpful in recording key information.

A precise format for working papers is not as important as completeness and legibility. All of these documents should be easy to read and understood by the auditor and anyone else reviewing the papers. Another person looking at the working papers should be able to tell, with little difficulty, the scope of the work performed, sources of information, findings and observations, what was not reviewed, and conclusions reached. All notes prepared by the auditor should be complete, simple, and factual. Do not add speculative opinion into these written records, as they will be taken as fact if a subsequent review of the audit occurs.

Observations should be documented while performing the fieldwork. Do not wait until the end of the audit – or even end of the day – to document findings or recommendations. If checklists are used, an effective method is to document notes directly on the checklist at the appropriate question. However, if separate field notes are kept, reference to the checklist question associated with the observation should be made. Follow-up actions should be highlighted within the notes or kept in a separate log so closure of these items is not forgotten.

### 19.4.7 Finding Evaluation and Classification

An audit finding is a documented conclusion of noncompliance with the audit evaluation criteria. Although audits are generally a snapshot in time of compliance, the scope of some audits may also include a look back to determine how long the compliance program has been in effect (e.g. a review of "wet ink" programs). Audit findings are based on objective and verifiable evidence documented by the auditor during the audit activity.

Classification of audit finding can be done by a variety of methods. Some sites desire a graduated qualitative or semiquantitative ranking scheme, such as "high, medium, low" risks, or risk ranked 1–5. Other methods of classification can be simply "compliant" or "noncompliant." Graduated ranking schemes have an advantage in that varying levels of control types can be applied based on seriousness, along with completion time frames. It may also be useful to separate out findings associated with physical risk from those associated with management systems. This will help the auditors and the site derive effective corrective action types for these different types of findings. An example of a findings and corrective action matrix for a graduated qualitative ranking system is provided below.

| Finding classification (5 is highest risk) | Minimum control by type | Completion timeline |
|---|---|---|
| 5 | Engineering controls or process substitution | Within 30 days |
| 4 | Engineering controls or safety devices | Within 60 days |
| 3 | Safety devices | Within 90 days |
| 2 | Safety devices or administrative controls (with management approval) | Within 120 days |
| 1 | Administrative controls | Within 120 days |

Often an additional ranking layer is used for regulatory compliance issues, which include very short completion timelines regardless of the type of control to be applied (e.g. management systems or physical safeguard).

### 19.4.8 Closing Meeting Discussion

Appropriate time should be set aside for preparation for the closing meeting. This is important, as this meeting will likely be the last opportunity for face-to-face discussion with much of the site leadership and will leave the last impression of the audit team and the process. The audit leader should prepare and lead the meeting, and all members of the audit team should attend if possible. The closing audit meeting should be relatively short, focused on the key audit results, and include a high-level summary of audit activities. Details of each finding are not necessary at the closing meeting; however repeated findings or trends should be mentioned, even if they classified as lower risks. Regulatory noncompliance and high-level risks should be the focus, along with specific discussion on potential or already agreed to corrective actions.

In addition to the audit findings, the audit team leader should also mention identified areas of good practice and other positive aspects discovered during the audit.

Time allowance for management to ask questions and engage the audit leader and team in discussions should be provided. The audit team leader should be prepared to directly answer challenges to findings with discussion and perhaps display of evidence gathered during the audit. Detailed discussions or challenges to findings categorization or the findings themselves should be suggested as a separate meeting outside of the closing meeting.

The meeting should end with an explanation of the overall reporting process and timing, including the site's role in this process. Description of how the report is to be prepared, what it will include, schedule, and the reports' review cycle should be given. A written summary list of findings and recommendations should be left with the site and marked as DRAFT for the site to review. This provides guidance to the site to begin working on high priority actions as appropriate. The site team should be encouraged to contact the audit team leader if questions arise post-audit.

## 19.5 REPORTING

The final report will document all findings; however, they should also be reported to the site audit leader as they are found. This is especially important for serious findings where there may be imminent danger. It also allows the site an opportunity to clarify or bring in additional information that may not have been readily available during the initial visit or discussions. It is important that the auditors do not discuss findings with anyone outside of the audit team or designated site contacts.

Quality assessment and control is a critical part of the audit process. A technical reviewer with subject matter expertise should review the report narrative and all findings and recommendations for technical accuracy and clarity. Industry- or equipment-specific technical language should be reviewed with special care and defined when necessary (especially when using acronyms). If possible, a technical editor with an understanding of the report's audience should review and edit for clarity and consistency. The report should be carefully spell-checked, formatted cleanly and consistently throughout, and presented in a format that is easily readable and printable.

A draft of the final audit report may be requested by the site for review and discussion prior to publication of the final version. Site personnel may dispute findings and/or their classification, and although it is important for the auditor to carefully consider the site position, a legitimate finding that can be justified with evidence should not be changed at this point. The site may also have completed some of the corrective actions from findings during the audit. These completed items should not be removed from the final report; however, record of the completion should be noted.

The final report format should include the following minimum information and sections:

- Purpose, date, scope of audit, and description of areas audited.
- An executive summary.
- Identity of the auditors and auditees, including management contacted.
- Description of the audit protocols or checklists used.
- Any disclaimers, including deviations from the scope.
- Tabular report of findings, which can be divided into functional sections, such as EHS, management systems, etc. Each of these sections should contain:
  - A summary of the requirements
  - Observations made by the auditor
  - Recommendations

An example follows related to a hazardous energy control finding within a manufacturing location subject to California OSHA (Cal/OSHA) requirements.

| Health and safety compliance issues | Regulatory citation | Requirements/findings/recommendations | Priority |
|---|---|---|---|
| *Hazardous energy control:* Lockout/tagout (LOTO) equipment-specific procedures annual review | 8 CCR §3314 (g), (h)<br>8 CCR §3314(g)(1) | *Requirements*<br>An energy control procedure must be developed and utilized when employees are engaged in the cleaning, repairing, servicing, or adjusting of prime movers, machinery, and equipment.<br>The procedure must clearly and specifically outline the scope, purpose, authorization, rules and techniques to be utilized for the control of hazardous energy, and the means to enforce compliance, including but not limited to the following:<br>• A statement of the intended use of the procedure.<br>• The procedural steps for shutting down, isolating, blocking, and securing machines or equipment to control hazardous energy.<br>• The procedural steps for the placement, removal, and transfer of lockout devices or tagout devices and the responsibility for them.<br>• The requirements for testing a machine or equipment to determine and verify the effectiveness of lockout devices, tagout devices, and other energy control devices.<br>The hazardous energy control procedures must also be documented in writing and inspected and certified annually. Training must also be conducted for those authorized to perform lockout under COMPANY's program, and contractors must be provided with the written lockout procedures during servicing and maintenance of the equipment. The employer's hazardous energy control procedure shall include separate procedural steps for the safe lockout/tagout of each machine or piece of equipment affected by the hazardous energy control procedure. | *High* |

**(Continued)**

| Health and safety compliance issues | Regulatory citation | Requirements/findings/recommendations | Priority |
|---|---|---|---|
| | | Exception to subsection (g)(2)(A): The procedural steps for the safe lockout/tagout of prime movers, machinery, or equipment may be used for a group or type of machinery or equipment, when either of the following two conditions exist:<br>• Condition 1:<br>  • The operational controls named in the procedural steps are configured in a similar manner.<br>  • The locations of disconnect points (energy isolating devices) are identified.<br>  • The sequence of steps to safely lockout or tagout the machinery or equipment is similar.<br>• Condition 2: The machinery or equipment has a single energy supply that is readily identified and isolated and has no stored or residual hazardous energy.<br>*Findings*<br>A written hazardous energy control program is in place; equipment-specific procedures have been developed; however the annual review of equipment-specific procedures is not formalized, nor is it documented.<br>In addition, equipment-specific procedures do not specify a "zero-energy state" verification step using appropriate testing equipment for tasks conducted by facility maintenance staff. Currently, the procedures require only a step to attempt to turn on the equipment. This is acceptable for production-type activities where shutdown is required; however activities do not require direct physical contact with hazardous energy parts and/or components such as cleaning a mixer. For maintenance staff, "zero-energy state" verification is imperative given the nature of maintenance type activities.<br>*Recommendations*<br>• Add "zero-energy state verification using testing equipment" in equipment-specific procedures for tasks involving maintenance staff accessing, maintaining and/or repairing equipment.<br>• Develop a written process to ensure all procedures are inspected and certified annually by an authorized employee. | |

Appropriate phrasing of the report narrative will convey conclusive and unambiguous information and instruction based on the outcome of the audit. Because audits are based on observation and are therefore limited to what could be observed and reviewed during the audit time frame, it is not generally possible to make absolute statements about the site or processes. As a best practice, use phrasing that appropriately echoes the experience during the audit.

Examples of report phrasing to avoid, and some better phrasing options, follow.

| Phrasing to avoid | Appropriate phrasing |
|---|---|
| The site does not have... | We were unable to confirm that...<br>We were unable to determine that...<br>We were unable to verify that...<br>Plant personnel were unable to locate copies of... |
| I found such and such to be true...<br>The site is in compliance... | We understand that...<br>We were told that...<br>It appears that....<br>Based on our review, we observed the plant to be in compliance with...<br>Based on our review it appears that...<br>On the basis of x samples, we found that... |
| A few SDSs were missing from the R&D laboratory...<br>(Minimize use of terms such as a few, many, several, some, etc.) | SDSs for four out of the five chemicals we randomly selected from the R&D laboratory could not be found in the MSDS binder for the area... |

The report narrative should avoid indirect or passive expressions (e.g. *The site was reviewed.* vs. *XXX reviewed the site*). Use short, familiar words in digestible sentences (15–18 words) and spell out any acronyms. Be complete and accurate, taking into consideration the needs of the reader/client, and make sure there are no inconsistencies within the report.

## 19.6 USING TECHNOLOGY TO MANAGE AND TRACK AUDIT DATA, CORRECTIVE ACTIONS, AND FOLLOW-UP

There are many software options for managing and tracking risk data and corrective actions. Using technology to assist in tracking and reporting on audit data can be a key time- and resource-saving strategy for many larger organizations, especially as the scope of audits scales up to include multiple facilities or sites. A tracking database on a computer tablet can be used throughout the audit process to record findings while on the site. Software solutions can also provide direct reporting of audit findings and recommendations, reducing reporting turnaround time.

A comprehensive software solution that compiles audit findings, assigns corrective actions, and provides reminders for repeating requirements such as permit applications can increase efficiency while reducing the time needed to effectively track and maintain an effective audit program. A successful software solution will also provide up-to-the-minute dashboards that show performance against goals, overdue actions, or rising risks in charts or other visual formats. These dashboard elements can be effective tools in management presentations to highlight the results of audits and actions.

## 19.7 CORRECTIVE ACTION FOLLOW-UP

Actions that are closed prior to the end of the audit should be noted as such in the final report; however, they should not be omitted from the report. The site will generally be responsible for action item closure; however, the audit team leader may continue to be engaged in a follow-up role to ensure the corrective actions are implemented.

Attachment 19.1: Pre-Audit: Sample Health and Safety Checklist
Attachment 19.2: Hazardous Energy Control: Sample California-Based Checklist
Attachment 19.3: Stormwater: Sample California-Based Checklist

**Attachment 19.1**

### HEALTH AND SAFETY AUDIT PROGRAM PRE-AUDIT SAMPLE QUESTIONNAIRE

#### Privileged and Confidential (If Audit Is Being Conducted via Internal or External Counsel)

The pre-audit provides background information from facilities about their health and safety management activities. This background information will assist our auditors in planning and conducting the facility audit; thus, please complete this questionnaire accurately and on a timely basis. Please return the completed questionnaire to the person designated below at least *five business* days prior to the scheduled audit, if possible.

Company name ______________________________
Person completing this questionnaire ______________________________
Street address ______________________________
City/county ______________________________
State/province ______________________________
Zip code/postal code ______________________________
Phone/fax number ______________________________
Date questionnaire completed ______________________________
SIC code for facility ______________________________

## General

What types of operations are or have been conducted at this location (e.g. extruding, refining, chemical manufacturing, assembly of parts, etc.)?

What are the primary products and services supplied by this location?

How many employees work at the site (full/part time, salaried/hourly)?

How many shifts does the plant operate? What are the hours of operation?

How large is the facility or site (e.g. number of buildings, square footage, acres, etc.)?

## Facility health and safety management

| Position | Name | Telephone number |
|---|---|---|
| Facility manager | ______ | ______ |
| Health and safety Manager/coordinator | ______ | ______ |

In the table below, indicate who is responsible for the development and implementation of programs for compliance with applicable state and/or government and company/business requirements for each of the following functional areas. Indicate NA for areas that you believe are not applicable.

| Functional area | Program development | Program implementation |
|---|---|---|
| Hazardous energy control programs (i.e. lockout/tagout) | ______ | ______ |
| Electrical safety | ______ | ______ |
| Injury and illness prevention | ______ | ______ |
| Confined space entry | ______ | ______ |
| Hot work | ______ | ______ |
| Laboratory safety | ______ | ______ |
| First aid or medical services | ______ | ______ |

| Functional area | Program development | Program implementation |
|---|---|---|
| Respiratory protection | ________ | ________ |
| Hearing conservation | ________ | ________ |
| Hazard communication | ________ | ________ |
| Blood-borne pathogens | ________ | ________ |
| Biosafety | ________ | ________ |
| Fire protection equipment | ________ | ________ |
| Fire brigade | ________ | ________ |
| Emergency response | ________ | ________ |
| Radiation safety | ________ | ________ |
| Incident investigation | ________ | ________ |
| Accident recordkeeping/OSHA 300 Log | ________ | ________ |
| Ergonomics | ________ | ________ |

## Employee safety

| | Y | N | N/A |
|---|---|---|---|
| 1. Are there written programs for: | | | |
| Injury and illness prevention (California only)? | ☐ | ☐ | ☐ |
| Hazardous energy control (i.e. lockout/tagout)? | ☐ | ☐ | ☐ |
| Confined space entry? | ☐ | ☐ | ☐ |
| Hot work? | ☐ | ☐ | ☐ |
| Contractor safety? | ☐ | ☐ | ☐ |
| Electrical safety-related work practices? | ☐ | ☐ | ☐ |
| Ergonomics? | ☐ | ☐ | ☐ |
| 2. Does the facility have programs (e.g. audits, inspections, etc.) to monitor? | | | |
| Safe work practices? | ☐ | ☐ | ☐ |
| Contractor safety? | ☐ | ☐ | ☐ |
| 3. Does the facility use or have any of the following: | | | |
| Powered industrial trucks (e.g. forklifts)? | ☐ | ☐ | ☐ |
| Cranes? | ☐ | ☐ | ☐ |
| Hoisting equipment and slings? | ☐ | ☐ | ☐ |
| Aboveground walking/working surfaces? | ☐ | ☐ | ☐ |
| Portable ladders? | ☐ | ☐ | ☐ |
| Fixed ladders? | ☐ | ☐ | ☐ |
| Scaffolding? | ☐ | ☐ | ☐ |
| Emergency eyewashes? | ☐ | ☐ | ☐ |
| Safety showers? | ☐ | ☐ | ☐ |
| 4. Does the facility have equipment on-site requiring guarding? | ☐ | ☐ | ☐ |

If yes, describe the types of equipment (e.g. grinder saws, drill presses, robotic equipment, lathes, etc.) and where a majority of the equipment located.

| |
|---|
| |

| | Y | N | N/A |
|---|---|---|---|
| 5. Does the facility have routine inspection/preventive maintenance programs for any of the above equipment? | ☐ | ☐ | ☐ |

If yes, specify the frequency for each piece of equipment.

| |
|---|
| |

| | Y | N | N/A |
|---|---|---|---|
| 6. Does the facility maintain live electrical equipment (e.g. transformers)? If yes, specify the highest voltage with which the facility employees have to work. | ☐ | ☐ | ☐ |
| 7. Has the facility conducted job safety analyses? | ☐ | ☐ | ☐ |
| 8. Does the facility contain asbestos-containing building materials (ACBM) in any capacity? | ☐ | ☐ | ☐ |
| 9. Does the facility have the following type of laboratory on-site: | | | |
| Use of clinical/biological samples? | ☐ | ☐ | ☐ |
| Quality control? | ☐ | ☐ | ☐ |
| Research and development? | ☐ | ☐ | ☐ |
| 10. Does the facility have medical services on-site? **If yes**, specify the type of coverage (e.g. full time, all shifts, part time, first shifts only, etc.). **If not**, how far away are the nearest medical services? | ☐ | ☐ | ☐ |

### Employee safety

| | Y | N | N/A |
|---|---|---|---|
| 1. Does the facility employ contract workers? | ☐ | ☐ | ☐ |
| If yes, are they covered under the facility's health and safety programs? | ☐ | ☐ | ☐ |

### Ergonomics

| | Y | N | N/A |
|---|---|---|---|
| 1. Has the company had any repetitive strain injuries or other injuries due to poor ergonomics? If so, explain job task and injury (review OSHA logs or accident investigations). | ☐ | ☐ | ☐ |
| 2. Is there a formal written program? | ☐ | ☐ | ☐ |
| 3. Are training records available? | ☐ | ☐ | ☐ |
| 4. Have ergonomic risk assessments been completed and are they available? | ☐ | ☐ | ☐ |
| 5. Have ergonomic risk assessments been completed and are they available? | ☐ | ☐ | ☐ |

### Industrial hygiene

| | Y | N | N/A |
|---|---|---|---|
| 1. Are there formal written programs for: | | | |
| Respiratory protection? | ☐ | ☐ | ☐ |
| Hearing conservation? | ☐ | ☐ | ☐ |
| Hazard communication? | ☐ | ☐ | ☐ |
| Laboratory safety? | ☐ | ☐ | ☐ |
| Personal protective equipment? | ☐ | ☐ | ☐ |
| 2. Has a site-wide sound level survey been conducted? | ☐ | ☐ | ☐ |

| | Y | N | N/A |
|---|---|---|---|
| 3. Has the facility conducted a workplace hazard assessment for the need for personal protective equipment? | ☐ | ☐ | ☐ |
| 4. Does the facility provide any of the following personal protective equipment for employee use: | | | |
| Hearing protection? | ☐ | ☐ | ☐ |
| If yes, where? | | | |
| Respirators? | ☐ | ☐ | ☐ |
| If yes, what types of respirators are used (e.g. two-strap dust mask, half-mask [negative pressure], fullface [negative pressure], PAPRs, supplied air, SCBA, etc.), and where? | | | |
| Protective eyewear (e.g. safety glasses, goggles, face shields, etc.)? | ☐ | ☐ | ☐ |
| If yes, where? | | | |
| Safety shoes? | ☐ | ☐ | ☐ |
| If yes, where? | | | |
| Head protection? | ☐ | ☐ | ☐ |
| If yes, where? | | | |
| Hand protection? | ☐ | ☐ | ☐ |
| If yes, where? | | | |
| Fall protection? | ☐ | ☐ | ☐ |
| If yes, where? | | | |
| Back braces/belts? | ☐ | ☐ | ☐ |
| If yes, where? | | | |
| 5. Does the facility provide employees with medical evaluation/surveillance for the following: | | | |
| Employees working in high noise areas? | ☐ | ☐ | ☐ |
| Respirator wearers? | ☐ | ☐ | ☐ |
| Industrial powered truck operators (e.g. forklift drivers, DOT operators)? | ☐ | ☐ | ☐ |
| ERT/spill response team members? | ☐ | ☐ | ☐ |
| Fire brigade members? | ☐ | ☐ | ☐ |

| | Y | N | N/A |
|---|---|---|---|
| 6. Employees working with regulated substances (e.g. benzene, lead, formaldehyde, methylene chloride, etc.)? | ☐ | ☐ | ☐ |
| If yes, please specify. | | | |
| Other? | ☐ | ☐ | ☐ |
| 7. Has the facility conducted a hazard assessment to determine priorities for exposure monitoring? | ☐ | ☐ | ☐ |
| 8. Does or has the facility conducted any exposure monitoring? | ☐ | ☐ | ☐ |
| If yes, for what contaminants? | | | |
| 9. Does the facility have an exposure monitoring schedule/plan? | ☐ | ☐ | ☐ |
| 10. Does the facility have any of the following sources? | | | |
| Nonionizing radiation (ultraviolet, radio frequency, magnetic fields)? | ☐ | ☐ | ☐ |
| If yes, specify type and location. | | | |
| Ionizing radiation (machines, sealed sources)? | ☐ | ☐ | ☐ |
| If yes, specify type and location. | | | |
| Lasers? | ☐ | ☐ | ☐ |
| If yes, specify classification and location. | | | |
| 11. Are ventilation systems (fume hoods, local exhaust trunks) used to control chemical exposures? | ☐ | ☐ | ☐ |
| 12. Are other engineering controls used for control of hazards (e.g. noise enclosures, controlled atmosphere control rooms)? | ☐ | ☐ | ☐ |
| If yes, please specify. | | | |
| 13. Are any administrative controls used for hazard control (e.g. limiting exposure time). | ☐ | ☐ | ☐ |
| If yes, please specify. | | | |

## Loss prevention and emergency response

| | Y | N | N/A |
|---|---|---|---|
| 1. Does the facility have the following? | | | |
| Fire prevention plan? | ☐ | ☐ | ☐ |
| Emergency action plan? | ☐ | ☐ | ☐ |
| Emergency response plan? | ☐ | ☐ | ☐ |
| If yes, list the type of response (s). | | | |

| |
|---|
| |

| | Y | N | N/A |
|---|---|---|---|
| 2. Does the facility train designated employees to use portable fire extinguisher and/or hoses? | ☐ | ☐ | ☐ |
| 3. Does the facility have the following fire protection equipment: | | | |
| Automatic sprinkler system? | ☐ | ☐ | ☐ |
| If yes, specify whether they are inspected by vendors or in-house personnel. | | | |

| |
|---|
| |

| | Y | N | N/A |
|---|---|---|---|
| Fixed extinguisher system? | ☐ | ☐ | ☐ |
| If yes, what type (e.g. CO, halon, etc.) and indicate whether they are inspected by vendor or in-house personnel. | | | |

| |
|---|
| |

| | Y | N | N/A |
|---|---|---|---|
| Fire detection system? | ☐ | ☐ | ☐ |
| If yes, specify whether they are inspected by vendors or in-house personnel. | | | |

| |
|---|
| |

| | Y | N | N/A |
|---|---|---|---|
| Fire alarm system? | ☐ | ☐ | ☐ |
| If yes, specify what type (e.g. PA system, sounding alarm, etc.) and indicate whether they are inspected by vendor or in-house personnel. | | | |

| |
|---|
| |

| | Y | N | N/A |
|---|---|---|---|
| 4. Does the facility have the following: | | | |
| Fire brigade? | ☐ | ☐ | ☐ |
| Emergency response team? | ☐ | ☐ | ☐ |

## Loss prevention and emergency response

| | Y | N | N/A |
|---|---|---|---|
| 1. Does the facility store flammable liquids in: | | | |
| Bulk storage tanks? | ☐ | ☐ | ☐ |
| If yes, specify the type of equipment (e.g. flame arresters, internal floating roofs, foam dispensers, etc.) the tanks are supplied with. | | | |

| |
|---|
| |

| | Y | N | N/A |
|---|---|---|---|
| Inside storage areas/rooms? | ☐ | ☐ | ☐ |
| Outside storage buildings? | ☐ | ☐ | ☐ |
| Warehouse? | ☐ | ☐ | ☐ |
| 2. Does the facility dispense flammable liquids? | ☐ | ☐ | ☐ |
| 3. Does the facility conduct spray painting operations? | ☐ | ☐ | ☐ |

## GENERAL HEALTH AND SAFETY MANAGEMENT

| | Y | N | N/A |
|---|---|---|---|
| 1. Does the facility have a safety manual and/or safety guidelines? | ☐ | ☐ | ☐ |
| 2. Are new and modified processes reviewed for occupational safety and health issues? | ☐ | ☐ | ☐ |
| 3. Does the facility track health and safety national regulations? | ☐ | ☐ | ☐ |

If yes, how?

**Attachment 19.2**

## HAZARDOUS ENERGY CONTROL AUDIT: CALIFORNIA-BASED CHECKLIST

**Facility:** ____________ **Inspected by:** ____________ **Date:** ____________

| EHS Requirements | | | | | |
|---|---|---|---|---|---|
| **General requirements** | **Citation** | **True** | **False** | **N/A** | **Web link** |
| 1. Facility has a written hazardous energy control program for multiple forms of energy, including electrical energy. | | | | | |
| 2. Written procedures should include the following information: | | | | | |
| (a) Statement of management support | | | | | |
| (b) Scope and application | | | | | |
| (c) Regulatory applicability | | | | | |
| (d) Requirement for equipment-specific hazardous energy control (LOTO) procedures | | | | | |
| (e) Definitions, including "affected" persons, "authorized" persons, "qualified" electrical worker, key regulatory definitions | | | | | |
| (f) List of responsibilities for employer, employee, and contractor | | | | | |
| (g) List of lockout devices | | | | | |
| (h) Documentation of lockouts that last longer than one work shift | | | | | |
| (i) Written procedure for abandoned locks | | | | | |
| (j) Group lockout procedures | | | | | |
| (k) Adjustments or tests requiring temporary removal of a locked-out condition | | | | | |
| (l) Normal production operations (e.g. minor tool changes or adjustments) | | | | | |
| (m) Cord and plug situations and details describing how the plug is to be controlled | | | | | |
| (n) Means to enforce compliance | | | | | |
| (o) Disciplinary measures in the event of noncompliance with procedures | | | | | |
| (p) Actions to take if deficiencies are discovered | | | | | |
| (q) List of audit criteria | | | | | |
| (r) Recordkeeping | | | | | |

| | | | | | |
|---|---|---|---|---|---|
| (s) Documentation of training and training programs | | | | | |
| (t) If program includes electrical energy, qualified person is defined | | | | | |
| (u) If the program includes electrical energy, "hot work" or energized electrical work is discussed. | | | | | |
| 3. There is a management of change program in place. | | | | | |
| **Comments:** | | | | | |
| **Equipment: specific procedures** | **Citation** | **True** | **False** | **N/A** | **Web link** |
| 1. Written lockout procedures are prepared specifically for each piece of equipment. | | | | | |
| 2. Procedures must be in writing and include the following: | | | | | |
| (a) Specific energy isolation points | | | | | |
| (b) Specific energies involved | | | | | |
| (c) Specific details on | | | | | |
| (d) Shutting down | | | | | |
| (e) Isolating | | | | | |
| (f) Verifying isolation | | | | | |
| (g) Restoring the systems or equipment | | | | | |
| (h) Detail on how flowable energies to be isolated | | | | | |
| 3. A formal hazard analysis is used to develop the lockout procedures for complex equipment. | | | | | |
| 4. The procedures are present where the workers are working. | | | | | |
| **Comments:** | | | | | |
| **Use of locks** | **Citation** | **True** | **False** | **N/A** | **Weblink** |
| 1. Each worker who enters the hazard zone has their own lock and tag on each energy isolation device. | | | | | |
| 2. Locks and tags must be: | | | | | |
| (a) Standardized throughout the system | | | | | |
| (b) Uniquely identified | | | | | |
| (c) Not used for any other purpose | | | | | |
| (d) Substantial and durable | | | | | |
| 3. Each worker has the **only** key to his/her lock. | | | | | |
| 4. Locks and tags are unique and standardized in regard to appearance and color. | | | | | |
| 5. Group lockout devices are available. | | | | | |
| 6. If tags are used instead of locks, additional measures are taken to ensure worker safety (e.g. additional training, additional isolation steps equivalent to the protection level of a lock). | | | | | |
| **Comments:** | | | | | |
| **Training** | **Citation** | **True** | **False** | **N/A** | **Web link** |
| 1. Hazard energy control training should include all of the following elements required in the standard: | | | | | |
| (a) Description and requirements of the energy control program | | | | | |
| (b) Elements of the energy control procedures relevant to the employee's duties | | | | | |

| | | | | | |
|---|---|---|---|---|---|
| (c) The pertinent requirements of the standard | | | | | |
| (d) The training includes instruction of authorized, affected, and other people on recognition and means of control of hazardous energies | | | | | |
| (e) The application and release of any controls | | | | | |
| (f) Contractors | | | | | |
| (g) Shift changes | | | | | |
| (h) Other pertinent operational details | | | | | |
| 2. "Qualified" electrical worker (QEW) per electrical safety standards training provided to all workers in the hazard zone through a separate training class. | | | | | |
| 3. Personnel who change job assignments to other equipment with different hazards are being retrained. | | | | | |
| 4. Personnel are being retrained if the procedures or hazardous energies change on the equipment they are authorized to perform lockout on. | | | | | |
| 5. Personnel are being retrained when the annual inspection reveals that they are not adequately qualified. | | | | | |
| 6. Retraining is provided when necessary. | | | | | |
| 7. Additional training is provided for affected persons when a tagout (instead of a lockout) is used. | | | | | |
| **Comments:** | | | | | |
| **Program audit** | **Citation** | **True** | **False** | **N/A** | **Web link** |
| 1. The lockout/tagout program is audited at least annually by an authorized and qualified person not participating in the lockout process. | | | | | |
| 2. The following should be identified: | | | | | |
| (a) Responsibilities | | | | | |
| (b) Lockout devices used | | | | | |
| (c) Equipment-specific written procedures are available | | | | | |
| (d) Equipment-specific written procedures are effective | | | | | |
| (e) Equipment-specific written procedures are followed | | | | | |
| (f) Training is adequate and documented | | | | | |
| 3. The auditors use a standardized checklist of program requirements during the audit. | | | | | |
| 4. The auditor questions the affected persons when tagout is used. | | | | | |
| 5. Action items resulting from the annual audits are promptly addressed and corrected. | | | | | |
| 6. The workers are following the necessary lockout/tagout procedures. | | | | | |
| 7. Written audit records are collected and retained. | | | | | |
| **Comments:** | | | | | |
| **Contractors** | **Citation** | **True** | **False** | **N/A** | **Weblink** |
| 1. Contractors who perform lockout on-site are included in the hazardous energy control program. | | | | | |
| 2. The required coordination between the facility and the contractor is being accomplished: | | | | | |
| (a) The company ensures the adequacy of the contractors' hazardous energy control program through auditing their work practices. | | | | | |

| | | | | | |
|---|---|---|---|---|---|
| (b) Adequate procedures are available and used by contractors for each piece of equipment and situation. | | | | | |
| (c) Employees are notified about the affected area of the contractors' lockout procedures and requirements. | | | | | |
| **Comments:** | | | | | |

**Attachment 19.3**

## STORMWATER AUDIT: CALIFORNIA-BASED CHECKLIST

**Facility:** ____________________ **Inspected by:** ____________________ **Date:** ______________

| EHS Requirements | | | | | |
|---|---|---|---|---|---|
| **Stormwater general issues** | **Citation** | **True** | **False** | **N/A** | **Web link** |
| 1. All stormwater flow directions and sampling points are located on a map from Stormwater Pollution Prevention Plan (SWPPP). | | | | | |
| 2. If found to be in violation of receiving water limitations, Regional Water Board has asked for report of additional best measure of practices (BMPs). | | | | | |
| 3. There are non-stormwater discharge sources (i.e. hydrant flushing, potable water sources, drinking fountain water, atmospheric condensate, landscape watering, springs, etc.). | | | | | |
| 4. SWPP includes BMPs to prevent or reduce the contact of non-stormwater discharges with equipment and materials and minimize flow or volume of non-stormwater discharge. | | | | | |
| 5. Visual observations have been conducted quarterly. | | | | | |
| 6. Non-stormwater discharge included in annual report. | | | | | |
| **SWPPP general permit requirements** | **Citation** | **True** | **False** | **N/A** | **Web link** |
| 1. Permit contains a compliance activity sheet. | | | | | |
| 2. Permit identifies specific individuals as members of the stormwater team. | | | | | |
| 3. Site map size is 8.5″ × 11″ or bigger. | | | | | |
| 4. Site map contains the following information: | | | | | |
| (a) Facility boundaries, drainage areas, flow directions, and other features | | | | | |
| (b) Location of stormwater collection, conveyance systems, and associated points of discharge | | | | | |
| (c) An outline of impervious areas | | | | | |
| (d) Location of materials directly exposed to precipitation and areas of previous significant spills | | | | | |
| 5. There is a list of significant materials handled and stored at the site. | | | | | |
| 6. Description of potential pollution sources: | | | | | |
| (a) Description of each industrial process utilized on the site | | | | | |
| (b) Material handling and storage areas on the site | | | | | |
| (c) Dust and particulate generating activities on the site | | | | | |

| | | | | | |
|---|---|---|---|---|---|
| (d) Any significant spills or leaks on-site | | | | | |
| (e) Non-stormwater discharges | | | | | |
| (f) Presence of soil erosion being caused by on-site activities | | | | | |
| 7. An assessment of potential pollution sources has been conducted. | | | | | |
| **Comprehensive site compliance evaluation for SWPPP requirements** | **Citation** | **True** | **False** | **N/A** | **Web link** |
| 1. The visual inspection records, sampling, and analysis results have been reviewed. | | | | | |
| 2. All potential pollution sources have been visually inspected. | | | | | |
| 3. All BMPs have been reviewed and inspected. | | | | | |
| 4. If an evaluation report has been conducted, it should contain the following information: | | | | | |
| (a) ID of personnel performing the evaluation | | | | | |
| (b) Date of evaluation | | | | | |
| (c) Necessary SWPPP revisions | | | | | |
| (d) Schedule for implementing revisions | | | | | |
| (e) Any incidents of noncompliance or corrective actions taken | | | | | |
| (f) Certification that the facility is in compliance with the requirements of the general permit | | | | | |
| (g) Evaluations that have been submitted as part of the annual report and retained for the last five years | | | | | |
| 5. The SWPPP has been updated within 90 days after the operator determined that the SWPPP was in violation of the requirement of the general permit. | | | | | |
| **Stormwater best management practices (BMPs)** | **Citation** | **True** | **False** | **N/A** | **Web link** |
| 1. The following must be considered in the discussion for the potential effectiveness of each BMP: | | | | | |
| (A) Nonstructural BMPs: | | | | | |
| (a) Good housekeeping | | | | | |
| (b) Practice manual (PM) | | | | | |
| (c) Spill response | | | | | |
| (d) Material handling and storage | | | | | |
| (e) Employee training | | | | | |
| (f) Waste handling and recycling | | | | | |
| (g) Recordkeeping and internal reporting | | | | | |
| (h) Erosion control and site stabilization | | | | | |
| (i) Inspections | | | | | |
| (j) QA | | | | | |
| (B) Structural BMPs: | | | | | |
| (a) Overhead coverage | | | | | |
| (b) Retention ponds | | | | | |
| (c) Control devices | | | | | |
| (d) Secondary containment structures | | | | | |
| (e) Treatment | | | | | |

| Monitoring program and reporting requirements | Citation | True | False | N/A | Web link |
|---|---|---|---|---|---|
| **Non-stormwater discharge:** | | | | | |
| 1. Quarterly visual inspections of all unauthorized non-stormwater discharge drainage areas have been conducted. | | | | | |
| 2. Quarterly visual inspections of all authorized non-stormwater discharge drainage areas have been conducted. | | | | | |
| 3. All visual observations/inspections documented any discoloration, stains, odors, floating material, and any identified sources. | | | | | |
| 4. Records include date, time, and observations of inspections. | | | | | |
| **Stormwater discharge:** | | | | | |
| 5. Visual observations of stormwater discharges have been taken from one storm event per month during the wet season (1 October to 30 May). | | | | | |
| 6. Visual observations been taken during the first hour of discharge and at all discharge locations. | | | | | |
| 7. In the event of a missed first storm event, two storm events should have been sampled with an explanation in the annual report. | | | | | |
| 8. All visual observations/inspections documented any discoloration, stains, odors, floating material, and any identified sources. | | | | | |
| 9. Records include date, time, and observations of inspections. | | | | | |
| **Sampling and analysis:** | | | | | |
| 10. Stormwater samples have been collected during the first hour of discharge from the first storm event of the wet season and at least one other storm event. | | | | | |
| 11. Samples have been analyzed for: | | | | | |
| (a) TCC | | | | | |
| (b) pH | | | | | |
| (c) TOC (oil and grease can be substituted for TOC) | | | | | |
| (d) Specific conductance | | | | | |
| 12. Facility is subject to Federal Stormwater Effluent Limitations Guidelines. | | | | | |
| 13. Samples are preserved and analyzed. | | | | | |
| 14. Sampling and analysis conditions are met for one of the following: | | | | | |
| (a) No Exposure Certification (NEC) | | | | | |
| (b) Regional Water Board Certification Programs | | | | | |
| (c) Local Agency Certification | | | | | |
| 15. Records have been retained of all stormwater monitoring information and copies for at least five years. | | | | | |
| 16. These records include the following: | | | | | |
| (a) Date, place, time of inspection, sampling, visual observations, and measurements | | | | | |
| (b) The individual(s) conducting the work | | | | | |
| (c) Flow measurements (if required) | | | | | |
| (d) Date and time of analysis | | | | | |
| (e) Individual(s) who performed the analysis | | | | | |

| | | | | | |
|---|---|---|---|---|---|
| (f) Analytical results, diction limits, and lab methods | | | | | |
| (g) QA/QC documents | | | | | |
| (h) Non-stormwater discharge inspection and visual observations and stormwater discharge visual observation records | | | | | |
| (i) Visual observation and sample collection exception reports | | | | | |
| (j) Calibration and maintenance records | | | | | |
| (k) Sampling and analysis exemption and reduction certifications and supporting documentation | | | | | |
| (l) Records of corrective actions and follow-up activities resulting from visual observations | | | | | |
| 17. Annual reports have been submitted by 1 July of each year. | | | | | |
| 18. Annual reports must include the following information: | | | | | |
| (a) Summary of visual observations and sampling results | | | | | |
| (b) An evaluation of the visual observations and sampling/analysis results | | | | | |
| (c) Laboratory reports | | | | | |
| (d) Annual comprehensive site compliance evaluation | | | | | |
| (e) An explanation of why the facility did not implement any actions required by the permit | | | | | |
| **Discharges of stormwater** | **Citation** | **True** | **False** | **N/A** | **Web link** |
| 1. All indirect or direct stormwater discharges from industrial facilities going to waters of the United States are covered by an individual waste discharge/National Pollutant Discharge Elimination System (NPDES) permit or by an NOI included as part of a general stormwater release permit. | | | | | |
| 2. If any of the discharge stormwater goes to municipal sanitary sewer systems or combined sewer systems, it complies with the stormwater discharge requirements issued by a Regional Water Quality Control Board (RWQCB). | | | | | |
| 3. All stormwater discharge points are identified and unauthorized non-stormwater discharges to stormwater systems have been eliminated. | | | | | |
| 4. The facilities under the general permit meet certain general requirements. | | | | | |
| 5. The facilities under the general permit developed a SWPPP. | | | | | |
| 6. The facilities under the general permit implemented a monitoring program. | | | | | |
| 7. The facilities are subject to Emergency Planning and Community Right-To-Know Act (EPCRA) section 313 reporting Form R and section 313 water priority chemicals. (Note: If yes, facilities are subject to additional requirements). | | | | | |
| 8. Even if not subjected to stormwater permitting requirements, facility that handles and stores significant materials should enact certain controls. | | | | | |

## REFERENCE

Cahill, L.B. and Kane, R.W. (2011). *Environmental Health and Safety Audits*. Lanham, MD: Government Institutes.

PART III

# MANAGEMENT APPROACHES

CHAPTER 20

# ADDRESSING LEGAL REQUIREMENTS AND OTHER COMPLIANCE OBLIGATIONS

THEA DUNMIRE

*ENLAR Compliance Services, 3665 E Bay Dr. #204C, Largo, FL, 33771*

One of the important considerations for organizations in establishing their safety and health programs is ensuring that they are in compliance with applicable legal requirements. An organization's compliance obligations can arise as a result of statutory law, as set out in laws and regulations, or common law. Of increasing importance are the legal obligations that can arise when organizations make a commitment to conform to the requirements set out in national or international consensus standards.

Most countries have a mix of both statutory and common laws. This chapter is focused primarily on understanding compliance obligations in the United States. Although the laws and legal framework may be similar in other countries, the specific laws and the resulting compliance obligations may be very different.

Given the fact-specific nature of many laws, the evolution of the law over time, and the differences in legal requirements from jurisdiction to jurisdiction, no attempt is made in this chapter to define specific legal obligations. Instead, the goals of this chapter are to provide an overview of the US legal and regulatory framework, to discuss some of the legal issues associated with occupational health and safety data management, and to highlight some specific circumstances when safety and health professionals may want to consider seeking legal advice.

## 20.1 UNDERSTANDING THE LAW

Most safety and health professionals are familiar with the Occupational Safety and Health Administration (OSHA) regulations. Many are also familiar with regulations promulgated by federal agencies such as the Mine Safety Administration, the Environmental Protection Agency, the Department of Transportation, and various state and local agencies. These regulatory programs are not, however, the only legal issues that may impact safety and health professionals. In addition to federal regulations, there are also the laws passed by Congress and state legislatures and an entire body of law that is simply referred to as "common law." There are also requirements set out in consensus standards that may be incorporated by reference into either regulations or into legally enforceable documents such as contracts.

### 20.1.1 Statutory Law

Statutory law is law that is written down, or codified. In the United States, federal law is set out in the US Code. State law is similarly set out in various state statute books. Both federal and state statutes are available for review on various governmental websites.[1]

[1] For a listing of websites with listings of federal and state environmental agencies and links to environmental laws and regulations, go to http://www.environmentallawnet.com/lawsregs.html.

*Handbook of Occupational Safety and Health*, Third Edition. Edited by S. Z. Mansdorf.

Although statutory laws are written down, interpreting their meaning is not simply a matter of reading the words. The meaning of a particular law can be significantly altered because of applicability requirements, use of defined terms, and subsequent court decisions.

A state statute or regulation may be unenforceable because it is determined to be preempted by a federal law. For example, the Occupational Safety and Health Act (OSH Act) preempts state occupational safety and health legislation. Without an approved state plan, a state is precluded from enacting or enforcing state laws, regulations, or standards relating to issues covered by the OSH Act. This preclusion, however, does not extend to areas that are not covered by an OSHA standard, for example, boilers and elevators.

An example of the preemptive effect of a federal law is provided by the case of *Gade v. National Solid Wastes Management Association.*[2] In 1988, Illinois enacted the Hazardous Waste Crane and Hoisting Equipment Operators Licensing Act. Under the act, persons working at hazardous waste cleanup sites were required to obtain a state license. In order to get the license, persons had to pass a written examination prescribed by the Illinois Environmental Protection Act. The act authorized employers to be fined if they permitted employees to work without having the required state licenses. The Supreme Court invalidated the state law by finding that this state regulation of workers at hazardous waste sites had been preempted by the Occupational Safety and Health Act and the regulations OSHA had promulgated to protect workers at hazardous waste sites.

### 20.1.2 Common Law

The common law is based on a set of legal precedents determined by analyzing how and why judges have decided individual cases. As such, the common law is not something one can simply look up in a book. Instead, common law evolves over time and may vary from state to state depending on the decisions made by individual judges in various courts.

Because case decisions are based on specific disputes between individual parties, they are case or fact specific. Therefore, similar disputes can have very different outcomes either because of differences in the underlying facts or because a court's view of what the law requires is different. These differences in the interpretation of the law are, in some instances, resolved by the highest state court or the US Supreme Court.

An example of such a circumstance involves the admissibility of scientific testimony in a federal court. Following the enactment of the Federal Rules of Evidence in 1975, several federal district courts came to different conclusions about when scientific theories were sufficiently credible to be admissible as evidence or, in the alternative, were inadmissible as "junk science." Different courts used different criteria until the Supreme Court resolved the issue in 1991 in a case called *Daubert v. Merrell Dow Pharmaceuticals.*[3]

### 20.1.3 Consensus Standards

Requirements set out in consensus standards are not legal requirements, but they can become legally enforceable. This can happen either because a standard is incorporated into another document that is legally enforceable, such as a regulation or a contract, or because the consensus standard is found to establish a "standard of care" for how organizations, or safety and health professionals, are expected to behave. Determination of the appropriate standard of care is one of the important components of a negligence case.

An example of an instance where consensus standards have been incorporated into an OSHA regulation is the references in the process safety regulations[4] to "Recognized and Generally Accepted Good Engineering Practices" (RAGAGEP). The PSM regulations require an organization to document that certain equipment complies with RAGAGEP and that inspections and tests of process equipment subject to mechanical integrity requirements are performed in accordance with RAGAGEP.

## 20.2 EVOLUTION OF THE LAW

Under both statutory and common laws, legal requirements can change over time. This change in compliance obligations may be the result of one or more of the following factors: society's views about certain issues may have changed; there may be changes in technology; new information may be available about an issue; or the courts may interpret a law differently.

---

[2] 60 U.S.L.W. 4587.

[3] 113 S Ct 2786 (1993).

[4] 20 CFR 1910.119.

An example of an issue that has been treated differently at different times is the protection of wetlands. Prior to the 1960s, wetlands were believed to be useless, nonproductive swamps that bred snakes and mosquitoes. Public policy and the laws encouraged projects to drain and fill wetlands to convert them to "better" uses. This policy was reversed in the 1970s, and laws were enacted to protect wetlands and to prevent them from being filled in.

## 20.3 FEDERAL STATUTORY LAW IMPORTANT TO OCCUPATIONAL HEALTH AND SAFETY PROGRAMS

There are a number of different federal laws in the United States that are important to safety and health professionals. Several of these are outlined below. There are also a number of other federal laws and regulations that address health and safety issues. It is important for organizations to identify all of the laws and regulations that are applicable to their activities, products, and services.

### 20.3.1 Occupational Safety and Health Act

The Occupational Safety and Health Act sets out the primary framework for regulating occupational health and safety in the United States. It is administered by the OSHA. OSHA has the authority to promulgate regulation, as well as inspection and enforcement authorities. It can impose fines on organizations who do not comply with its regulations.

The intent of the Occupational Safety and Health Act is "to assure so far as possible every working man and woman in the Nation safe and healthful working conditions." Under the OSH Act, employers are required to furnish places of employment that are free from recognized hazards that are causing or are likely to cause death or serious physical harm to their employees. This requirement is sometimes referred to as the General Duty Clause.

### 20.3.2 Mine Safety and Health Act

The Federal Mine Safety and Health Act of 1977 consolidated all of the existing federal health and safety mining regulations and transferred enforcement authorities to the Mine Safety Administration (MSA). The mission of MSA is to promote improved safety and health conditions in the nation's mines. Like OSHA, MSA has the authority to promulgate regulations. It also has inspection and enforcement authorities and can impose fines on organizations that do not comply with its regulations.

### 20.3.3 National Labor Relations Act

The National Labor Relations Act (NLRA) was passed in 1935 to protect the rights of employees and employers. It is administered by the National Labor Relations Board. This law gives employees the right to act together, with or without a union, to try to improve their working conditions.

For example, in 2008 a group of employees posted a YouTube video outlining their concerns about being required to handle contaminated soil at a construction site. Their employer fired them. Following an investigation, the NLRB regional director determined that the YouTube video was protected because employees were voicing concerns about safety in the workplace and the video accurately described their concerns about working conditions. As the hearing on the case opened, the case settled and workers received full back pay.[5]

### 20.3.4 Toxic Substances Control Act

The Toxic Substance Control Act (TSCA) provides the regulatory framework for controlling exposure and use of chemical substances. This act requires chemicals be evaluated before use to determine whether they pose risks to human health or the environment.

Under TSCA, the manufacture, use, import, or disposal of chemicals may be banned, controlled, or restricted. Substances that are subject to regulation under TSCA include polychlorinated biphenyls (PCBs), asbestos, and radon- and lead-based paint.

---

[5] NLRB Case 19-CA-31580. Referenced at https://www.nlrb.gov/rights-we-protect/protected-concerted-activity.

### 20.3.5 Safe Drinking Water Act

Under the Safe Drinking Water Act, EPA is authorized to set two types of drinking water standards. Primary standards apply to substances that may have an adverse effect on health. Secondary standards provide guidelines on substances or conditions that affect color, taste, smell, and other physical characteristics of drinking water. These standards can apply in a workplace if an employee provides drinking water that is not from a public drinking water supply.

### 20.3.6 Hazardous Materials Transportation Act

The Hazardous Materials Transportation Act requires the secretary of the Department of Transportation (DOT) to regulate hazardous materials transportation in intrastate, interstate, and international commerce. Hazardous materials that are shipped must be packaged, labeled, marked, and placarded in accordance with DOT regulations. There are specific training requirements for employees who prepare hazardous materials for off-site shipment.

### 20.3.7 Americans with Disabilities Act

The Americans with Disabilities Act of 1990 (ADA) was enacted to establish a national program to eliminate discrimination against individuals with disabilities.

This law is administered by the Equal Employment Opportunity Commission (EEOC) that has developed regulatory standards dealing with equal employment opportunities for disabled individuals. No employee may discriminate on the basis of disability with regard to any terms, conditions, or privileges of employment.

An individual is qualified if he or she can perform the essential functions of a job. In addition, employers must make reasonable accommodation for the known physical or mental limitations of an otherwise qualified individual with a disability. As a result, it is important for employers to define the essential functions of each job position. For many jobs, written job descriptions need to address safety and health issues such as range of motion abilities, the requirement that certain personal protective equipment be worn or that periodic medical testing is required (e.g. audiometric testing).

## 20.4 STATE STATUTORY LAW IMPORTANT TO OCCUPATIONAL HEALTH AND SAFETY PROGRAMS

States promulgate their own laws and regulations, and they administer a variety of federal regulatory programs.

### 20.4.1 OSHA-Approved State Plans

Under section 18(a) of the act, if the Secretary of Labor determines that a state has standards comparable to OSHA's and has an enforcement plan meeting the criteria of 18(c), jurisdiction over occupational health and safety is given to the state.[6] States with an approved state plan can have regulations that are more stringent than those contained in the OSHA regulations. For example, both California and Washington adopted state ergonomic standards even though the Congress prohibited OSHA from issuing its proposed ergonomic rule.

### 20.4.2 Other State Health and Safety Laws and Regulations

States can also develop and enforce public safety laws and regulations that may impact occupational health and safety programs. An example would be state and local fire codes.

Another example is the California Safe Drinking Water and Toxic Enforcement Act of 1986, better known by its original name of Proposition 65. Proposition 65 requires the state to publish a list of chemicals known to cause cancer or birth defects or other reproductive harm. Businesses are required to provide a "clear and reasonable" warning before knowingly and intentionally exposing anyone to a listed chemical. One common way that warning is done is by posting signs in the workplace.

### 20.4.3 Workers' Compensation Laws

All states have workers' compensation laws that have, to some extent, replaced common law claims on the part of injured workers. Under workers' compensation, employees are compensated for injuries or illnesses arising out of or in the course of their employment without regard to the fault of the worker or the employer. Except in narrow

---

[6] Information on OSHA-approved state plans can be found at https://www.osha.gov/dcsp/osp/index.html.

circumstances, a worker's sole remedy against his or her employer is workers' compensation benefits. The award to an injured employee is limited to lost wages, medical expenses, and set compensation levels based on the state's workers' compensation schedule.

Intentional misconduct on the part of the employee or the employer can, however, eliminate the exclusive nature of workers' compensation. Other situations where workers' compensation may not apply include product liability actions, fraudulent concealment, and negligence conduct by third parties (e.g. contractors).

There are several cases where employees have alleged that their employers engaged in fraudulent concealment by failing to inform them about the dangers of chemicals they were exposed to in their work. An example is *Palestini v. General Dynamics Corporation*[7] where the court found that a worker's allegations were sufficient to allow him to proceed in a suit against General Dynamics. Factors that the court cited in reaching its decision were that Mr. Palestini reported his health impacts to his employer and that General Dynamics knew that these health impacts were well known to be aggravated by continued exposure to the chemicals he was working with.

## 20.5 OVERVIEW OF COMMON LAW CLAIMS

Court cases may be brought under a variety of legal theories. Some are based on the rights and obligations created as the result of federal or state statutes. Others are brought based on common law claims. Two categories of common law claims are tort claims and contract claims.

### 20.5.1 Tort Claims

A tort is "a private or civil wrong or injury, other than a breach of contract, for which the court will provide a remedy in the form of an action for damages." The tort claims that most commonly arise in the safety and health field are negligence, misrepresentation, and strict products liability.

The one element common to all tort cases is that the plaintiff has suffered some damage or injury. Therefore, one of the most effective ways of avoiding tort liability is to prevent injury to persons or damage to property. Safety and health professionals can help to prevent tort claims being brought against their companies by developing programs to identify and then eliminate safety and health hazards that cause injuries and illnesses.

***20.5.1.1 Negligence*** The most frequently alleged tort claim is negligence. In order to establish negligence, the plaintiff must prove that the defendant owed him or her a duty, the defendant breached that duty, and the breach proximately caused injury or damage to the plaintiff. Whether a duty exists on the part of the defendant is dependent on both the relationship between the parties and the existence of laws or regulations imposing some sort of legal obligation. The types of relationships that may create legal duties include landlord–tenant, consultant–client, and employer–employee.

An example of a negligence action brought because of a landlord–tenant relationship is *Wright v. McDonald's Corporation*.[8] In this case, Mr. Wright alleged he suffered permanent neurological injury caused by carbon monoxide exposure resulting from a defectively installed water heater in the McDonald's restaurant he operated under a franchise and lease agreement. In a decision on a summary judgment motion, the court stated that a jury could find McDonald's liable if they concluded that McDonald's "designed and constructed a ventilation system with a latent dangerous defect of which it should have been aware and which Mr. Wright could not uncover by reasonable inspection."

Under certain circumstances, OSHA or environmental regulations may be used to establish the existence of the legal duty giving rise to a negligence claim. An example of such a circumstance is the case of *Arnett v. Environmental Science & Engineering, Inc.*[9] In this case, Mr. Arnett alleged he was injured by the fumes generated as the result of his job duties involving the pouring of chemicals into 55-gallon drums during an asbestos abatement project. The suit was filed against Environmental Science and Engineering (ESE) and an individual, Charles Jenkins, in their capacities as the asbestos project managers for the project. ESE was not Mr. Arnett's employer. In the case, ESE argued that it did not have any legal duty to protect Mr. Arnett from exposure to fumes because that was his employer's responsibility. The court disagreed. In reaching its decision, the court referred to the provisions of the Illinois Asbestos Abatement Act, which imposed certain duties and obligations on the project manager. Based on those provisions, the court stated

[7] Accessible at http://caselaw.findlaw.com/ca-court-of-appeal/1075982.html.

[8] 1993 U.S. Dist. LEXIS 2635.

[9] 275 Ill. App. 3d 938; 657 N.E.2d 668, 1995.

that the project manager was not free to ignore the conditions existing at the abatement worksite and, further, that it was the duty of the project manager to require the contractor to take appropriate safety precautions.

***20.5.1.2 Misrepresentation*** Another tort claim a plaintiff might allege is misrepresentation – either fraudulent or negligent. The elements of fraudulent misrepresentation are as follows: a misrepresentation of the fact that the defendant knew was false, an intent on the part of the defendant that the plaintiff rely on the misrepresentation for some action, and damage or injury to the plaintiff caused by that person's justifiable reliance on the misrepresentation. The kind of acts that could be considered fraudulent misrepresentations includes removing manufacturers' warning labels on toxic substance containers, knowingly providing inadequate safety equipment, and misrepresenting the danger or extent of the toxicity of materials and the need for proper safety equipment.

An example of a case alleging fraudulent misrepresentation is *Altair Strickland Inc. v. Chevron U.S.A. Products Co.*[10] On 6 March 1995, Altair Strickland Inc. (ASI) began work as the prime contractor for maintenance and construction work to be done during a refinery shutdown at Chevron's El Paso, Texas, refinery. Almost immediately ASI's workers started complaining of rashes and respiratory problems. On 17 March, ASI refused to continue work, claiming that its workers were being exposed to high levels of sulfur dioxide. On 18 March, Chevron locked ASI out of the refinery and replaced it as the contractor for the project. ASI sued Chevron, charging it has committed fraud "by intentionally misrepresenting that work conditions in and around the crude towers at the refinery were safe, when they were not." The jury awarded ASI $5 million in compensatory damages and $38.5 million in punitive damages.

A defendant would engage in negligent misrepresentation, as set out in the Restatement of Torts (a summary of the common law often referred to by the courts), if "in the course of his or her business, profession, or employment, he or she supplied false information for the guidance of others in their business transaction and failed to exercise reasonable care or competence in obtaining or communicating the information."

In both fraudulent and negligent misrepresentation, the information transmitted and relied on was false. The difference lies in whether the defendant knew the information being transmitted was false (fraudulent misrepresentation) or the defendant was careless in obtaining and communicating the information (negligent misrepresentation). A number of cases have been brought against consultants alleging negligent misrepresentation. These cases have involved misrepresentation of laboratory data,[11] misrepresentations concerning the condition of property in environmental assessment reports, and misrepresentation of the regulatory compliance status of facilities in audit reports.

***20.5.1.3 Strict Products Liability*** Strict liability is a legal concept often applied by the courts in product liability cases. In these cases, the seller of "any product in a defective condition unreasonably dangerous to the user" is liable for any damages caused by the defective product. The difference between negligence and strict liability is that strict liability is liability without fault. Lawsuits alleging strict liability focus on the product that caused the injury rather than the actions of the defendants.

A product may be held to be defective because it does not carry an adequate warning to prevent it from being inherently dangerous. "Failure to warn" is one of the primary claims made in the lawsuits brought alleging asbestos-related injuries of workers. One of the earliest cases holding that manufacturers had a duty to warn workers of the dangers associated with the use of their products was *Borel v. Fibreboard Paper Products Corp.*[12] In this case, the court stated that a manufacturer "must keep abreast of scientific knowledge, discoveries and advances and is presumed to know what is imparted thereby. A product must not be made available to the public without disclosure of those dangers that the application of reasonable foresight would reveal."

### 20.5.2 Contract Claims

A contract is an agreement between two or more parties to engage in certain acts in the future. Except in certain circumstances, this agreement does not have to be in writing to be valid. It can consist of the statements made during a series of telephone calls. It is also possible for a "contract" to be the agreement reached as the result of a series of separate documents (e.g. a proposal, response, purchase order, and invoice for a particular project). It can

---

[10] http://www.morelaw.com/verdicts/case.asp?n=b152339%20&s=TX&d=1109 bl52339 (Dist. Ct. Jefferson Co. Texas). Information based on case summary reported in *National Law Journal,* Monday, 27 January 1993, A13.

[11] One case alleging the misrepresentation of laboratory test data is *Metal Finishing Technologies Inc. V. Fuss and O'Neill, Inc.* (1991 WL 172833 (Conn. Super 1991)). In this case, Fuss and O'Neill, an environmental consulting firm, was hired to help the plaintiff comply with an environmental consent decree and federal and state environmental laws. In a decision on a summary judgment motion, the court stated that the plaintiff could bring a claim of negligent misrepresentation based upon the allegations that Fuss and O'Neill knew or should have known that sludge test results provided by its subcontractor were false, the plaintiff would rely on the results, and the plaintiff did, in fact, reasonably rely on the representations.

[12] 493 F.2d 1076 (5th Cir 1973).

even be memorialized in a series of e-mails. Common contract claims include breach of contract, breach of warranty, or indemnification claims.

***20.5.2.1 Breach of Contract*** A breach of contract is a failure, without legal excuse, to comply with a provision, or obligation, of the contract. Typical contract obligations include promises to perform certain tasks within certain time periods, obligations to meet certain conditions (e.g. licensing requirements), and promises that the work will meet certain specifications. Other contract provisions include warranties, indemnification, and specification of dispute resolution mechanisms such as arbitration. Contract problems often arise because of vague contract language or because the contract is incomplete.

***20.5.2.2 Breach of Warranty*** A warranty is a contractual promise that a product or service will attain a particular level of quality. If it does not, then there is a potential "breach of warranty" claim. An example of a warranty would be:

> Consultant warrants to Company that all services supplied by Consultant in the performance of this Agreement shall be supplied by personnel who are licensed or certified, as required by applicable law, and experienced and skilled in their respective professions.

If personnel used on the job are not licensed and/or certified, the company would have a breach of warranty claim against the consultant.

***20.5.2.3 Indemnification Claims*** An indemnification provision in a contract is an agreement on the part of one party that it will compensate the other party for any loss or damage arising from some specified act or acts. A typical indemnification provision reads:

> [Party A] will indemnify and hold harmless [Party B] from and against any claims, damages, losses and expenses arising out of or resulting from the performance of the Work as defined in the agreement.

Such an indemnification provision might be limited by adding certain conditions to the indemnification obligation, such as requiring that the party providing the indemnification be negligent before compensation would be required.

Indemnification provisions are similar to insurance policies in that they shift the risk of loss to other parties. Like insurance policies, indemnification provisions will not provide compensation if the party providing the indemnification has no assets to cover the potential loss.

## 20.6 CRIMINAL LIABILITY

Both federal and state prosecutors may seek criminal indictments when they believe there has been sufficiently egregious misconduct or violations of safety, health, or environmental laws. Almost all federal environmental statutes, as well as the Occupational Safety and Health Act, have criminal penalty provisions. In addition, Title 18 of the US Code sets out criminal offenses that are generally applicable to a wide variety of activities (e.g. making false statements to government officials). In addition, criminal charges may be brought under general criminal statutes such as those prohibiting involuntary manslaughter or criminally negligent homicide.

In particular, safety and health professionals should be aware that many state and federal statutes contain provisions imposing substantial penalties for making false statements and for destroying, altering, or concealing records.

For example, in 2016, two state officials in Michigan had criminal charges filed against them for "willfully and knowingly misleading" the federal Environmental Protection Agency and the local health department about the dangers associated with drinking water in Flint, Michigan.[13]

## 20.7 HOW AN OSHA ENFORCEMENT ACTION PROCEEDS

Typically, a regulatory enforcement action starts with an inspection. If an OSHA inspection results in findings of noncompliance, the agency will issue a citation and notification of penalty. Employees are given a set period of time in which to contest the citation and/or penalty.

---

[13] For information on the criminal charges brought related to contamination of drinking water in Flint, Michigan, go to http://www.nytimes.com/2016/04/21/us/first-criminal-charges-are-filed-in-flint-water-crisis.html?_r=0.

There are two general types of OSHA inspections, programmed and unprogrammed. Programmed inspections of worksites are scheduled based upon predetermined selection criteria. Unprogrammed inspections are conducted in response to alleged hazardous working conditions that have been identified at a specific worksite. This type of inspection results from reports of imminent dangers, fatalities, employee complaints, and referrals from other agencies.

If any violations are found during an inspection, OSHA issues a Citation and Notification of Penalty describing the violations and proposed penalties. The citation will set out a proposed time period within which to correct any violations.

Upon receiving a Citation and Notification of Penalty, the employer must post the citation (or a copy of it) at or near the place where each violation occurred to make employees aware of the hazards to which they may be exposed.

An employer who has received a citation may either (i) agree to the citation, correct the condition by the date set in the citation, and pay the penalty or (ii) formally contest the citation. Employees or their authorized representatives may also contest any or all of the abatement dates set for violations if they believe them to be unreasonable. A formal contest to a citation must clearly state what is being contested. If the written Notice of Intent to Contest is properly filed within the required time period, the case is forwarded to the Occupational Safety and Health Review Commission for resolution.

## 20.8 HOW A LEGAL CASE PROCEEDS

Just as there are differences between regulatory requirements and common law obligations, there are also significant differences between a regulatory enforcement action and a court case.

The conduct of a court case is governed by the procedural rules of the court in which it is filed. In the federal courts, the Federal Rules of Civil Procedure govern all civil cases. These rules govern how documents are prepared, how many pages a document filed with the court may have, and the time periods the parties are required to meet in answering pleadings and motions. Once a case is filed, the court has substantial power over the parties. The court can order that certain documents be released, that parties appear in front of the court at specific times, and that certain case preparation deadlines be met. If the court's orders are not complied with, the court can impose penalties on the parties.

All legal cases start out in a trial court; this could be a municipal court, state court, or federal district court. If one of the parties believes that the judge has committed some error in the case, the case may be appealed to a higher court. In the federal court system, cases are appealed to the court of appeals and, finally, the US Supreme Court. Appellate courts do not, however, retry cases. If they decide some mistake has been made, they send the case back to the trial court for further proceedings.

### 20.8.1 Initial Pleadings

In most cases, the first formal notice of a lawsuit is the receipt of a complaint. The purpose of the complaint is to allege some legal cause of action and put the opposing party on notice concerning the allegations the plaintiff believes supports his or her claim for damages. Unless some sort of motion to dismiss the complaint is filed, the defendant files a response called an answer. In the answer, the defendant typically responds to each of the allegations in the complaint by admitting it, denying it, or stating that there is insufficient information to either admit or deny the allegation. The answer will often also contain what are called affirmative defenses and counterclaims. Counterclaims are claims for damages that the defendant makes against the plaintiff. It is not uncommon for other parties to be brought into the action at this point by the defendant, the plaintiff, or both. Typically there will also be a variety of motions filed with the court that will need to be resolved before the case can proceed.

### 20.8.2 Discovery

Once the initial pleadings are filed and the preliminary motions are resolved, the parties enter into an exchange of information concerning the issues in the case. This is typically called discovery. In the course of discovery, each party tries to obtain detailed information concerning the other party's claims. The Federal Rules of Civil Procedure set forth a number of different discovery mechanisms, including depositions, interrogatories, and requests for production of documents.

One of the primary discovery tools is depositions. Lawyers for the parties can subpoena individuals associated with the case to give oral testimony under oath. The witness is sworn to tell the truth, and a transcript of the testimony is prepared. This testimony is used by the parties to help them prepare their cases. The deposition transcript can be used during the trial to impeach a witness if his or her testimony conflicts with the statements made during the deposition.

### 20.8.3 Trial and Decision

Once discovery is completed and all the pretrial motions are resolved, the case is set for trial. Depending on the case, the case may either be decided by a jury or by the judge.

## 20.9 RECORDKEEPING LEGAL ISSUES

Developing and maintaining an occupational health and safety program typically involves a substantial amount of document creation, data collection, and record management. For example, safety and health professionals are required to develop various written policies and procedures, maintain information on toxic materials (e.g. safety data sheets), maintain training records, generate exposure data, and keep records concerning employee injuries. As a result, many safety and health professionals soon find they are dealing with vast amounts of data. This "ocean of data" necessitates the development of a data management program. There are several legal considerations that need to be kept in mind in developing such a program.

### 20.9.1 Good Data Management Helps Prevent Legal Actions

A good data management program can help avoid legal problems. First, by having all legally mandated records maintained so they are easily retrieved, a legal action can be avoided because records are available when they were needed (e.g. when requested by a government inspector or when an emergency happens). Second, it is much easier to identify gaps in safety and health program documentation when documents are well organized. Once such gaps are identified, they can be corrected.

### 20.9.2 Dealing with Sensitive, Privileged, and Confidential Documents

A safety and health data management program needs to adequately protect data. This includes privileged documents (those subject to attorney–client or attorney–work product privileges), documents containing confidential business information (e.g. trade secrets), and documents containing personal information (e.g. employee medical information).

Written procedures should be established to prevent the dissemination of sensitive or confidential documents. In particular, confidential documents should never be filed with documents that are required to be provided to governmental agencies upon their request (e.g. permits, OSHA logs, emergency procedures).

### 20.9.3 Attorney–Client and Attorney–Work Product Privileges

Certain documents are protected from disclosure in a legal proceeding. These are typically referred to as privileged documents. There are two well-recognized legal privileges: attorney–client correspondence and attorney–work product.

The purpose of the attorney–client correspondence privilege is to promote the freedom of discussion between a lawyer and client without the fear that disclosure can later be compelled from the lawyer without the client's consent. In order to establish the privilege, the correspondence must be prepared by the attorney, acting in a legal role, at the request of the client. The correspondence must include a legal analysis (not just a discussion of a factual investigation) and should include a specific reference to the likelihood of litigation. It should be noted that the privilege does not apply to all documents addressed or sent to a lawyer nor can it be used to protect documents prepared prior to the lawyer being consulted.

When an attorney directs specific information gathering "in anticipation of litigation," the documents produced during that information gathering process may be protected as attorney–work product. Again, the important point in establishing this privilege is that the investigation is directed by an attorney. The privilege cannot be applied to documents prepared without an attorney's involvement nor does it apply to documents prepared in the regular course of business.

This privilege may protect the reports of experts (for example, an accident investigation report); however, this protection may not be absolute. For example, disclosure may be compelled if the report contains factual information that cannot be obtained in any other way by the opposing party. Therefore, experts' reports should typically be divided into two parts – data compilation and opinions. The data compilation may be required to be disclosed, whereas the opinion of the expert will typically remain privileged.

It should be noted that simply marking a document as "confidential" does not protect it from disclosure. Specific steps must be taken in order to establish these privileges. In addition, if privileged documents are released to other parties, their privileged status will be lost.

### 20.9.4 Self-Evaluative Privilege and Self-Audit Privilege

Several states have passed the socalled privilege laws. In addition, the US EPA has issued its own policy concerning the handling of environmental audit results.

It should be noted, however, that although many of these laws and the EPA policy are referred to privilege laws, they often do not have the same effect as the legal privilege laws discussed above. Rather than protecting disclosure of information, they typically create a limited immunity if proper disclosure is made. Although there may be occasions where taking the steps to qualify for a limited immunity would be appropriate, these laws do not offer the same protection against disclosure as the traditional attorney–client and attorney–work product privileges.

### 20.9.5 Confidential Business Information

Companies may have certain information that is considered "trade secret." A trade secret is defined as "any confidential formula, pattern, process, device, information or compilation that is used in one's business and which gives the business an opportunity to obtain an advantage over competitors who do not know or use it." There are specific legal protections available for trade secrets under both common law and certain safety, health, and environmental regulations. In a court action, a trade secret may be protected under a protective order.

### 20.9.6 Employee Privacy Issues

In the performance of their job functions, safety and health professionals may have access to private information concerning employees (e.g. medical test results). There are state and federal laws relevant to the protection of employee's personal privacy. In general, no personal information about employees should be disclosed to others without an evaluation of the privacy issues raised.

### 20.9.7 Record Retention Policies

An important component of a records management program is a records retention policy. A records retention program improves an organization's ability to handle valuable information. By getting rid of unnecessary data, valuable information is more easily retrieved and less likely to get lost, misfiled, or accidentally destroyed. Without such a program, it can be extremely difficult to locate critical information when it is needed or establish that, if information is unavailable, it has been properly destroyed. A written record retention program also helps to ensure that records that are required to be kept for legally prescribed periods of time are not improperly destroyed.

## 20.10 WHEN TO CONSIDER SEEKING LEGAL ADVICE

Safety and health professionals should consider seeking legal advice in the following circumstances: when they are uncertain of their legal obligations, when entering into contracts that raise significant legal issues, when determining legal compliance status, and when served with a complaint or other legal summons.

### 20.10.1 To Determine Legal Obligations

As discussed above, it is sometimes difficult to determine an organization's legal obligations. Yet "ignorance of the law" is not considered a defense in a legal action. Since both common law and some occupational health and safety and environmental statutes can impose individual criminal liability, it may be prudent to seek legal advice to help clarify what you are legally required to do.

### 20.10.2 When Negotiating Contracts

There are certain legal protections, such as the impact of noncompete limitations, which may be impacted by the terms of contracts you execute. In addition, certain contractual provisions, such as limitations of liability and indemnifications, may greatly increase your potential legal liability. If the terms of a contract are unclear to you, consider consulting with an attorney before you execute the contract.

### 20.10.3 When Determining Compliance Status

Although there are several benefits associated with evaluating whether a company is complying with safety, health, and environmental requirements, the documentation created as a result of such an evaluation, or audit can end up being used as evidence used against a company in a subsequent legal proceeding. Consulting with an attorney before conducting compliance audits can help in structuring the audit to minimize the potential for increasing the company's legal liability.

### 20.10.4 When Served with Legal Documents

As discussed above, there are documents that will be sent to, or served on, parties when a legal action is initiated. Typically, these must be responded to within certain time periods, or there can be serious legal consequences. It is always advisable to consult with an attorney if you are served with such a document.

CHAPTER 21

# OCCUPATIONAL SAFETY AND HEALTH MANAGEMENT

FRED A. MANUELE

*Hazards Limited, 200 W Campbell #603, Arlington Heights, IL, 60005*

## 21.1 SAFETY AND HEALTH RESOURCES

Only a few of the organizations that provide resources available to safety and health professionals are mentioned in the following list. It should be recognized that the explosive growth of the Internet has made access to a broad range of up-to-date information faster and better. Exchanges of information through informal networking have also been good resources for safety and health professionals:

- American Society of Safety Engineers (ASSE)
- American Industrial Hygiene Association (AIHA)
- National Safety Council (NSC)
- American National Standards Institute (ANSI)
- Human Factors and Ergonomics Society (HFES)
- National Fire Protection Association (NFPA)
- Occupational Safety and Health Administration (OSHA)
- National Institute for Occupational Safety and Health (NIOSH)

## 21.2 OCCUPATIONAL SAFETY AND HEALTH HISTORY

### 21.2.1 The Past

In 1760 BCE, King Hammurabi, who belonged to the first dynasty of Babylonia, set in motion the collection of laws and edicts that came to be called the Hammurabi Code. It was believed that the code had been received by the king from the sun god, Shamash. The code included the procedures to be applied regarding property rights, personal rights, and debts. It provided for, among many other things, damages to be paid caused by neglect in various trades. For example:

- If a builder has built a house for a man, and has not made his work sound, and the house he built has fallen, and caused the death of its owner, that builder shall be put to death (Rule 229).
- If a boatman has built a boat for a man, and has not made his work sound, and in that same year that boat is sent on a voyage and suffers damage, the boatman shall rebuild that boat, and, at his own expense, shall make it strong, or shall givc a strong boat to the owner (Rule 235).

Hippocrates, the celebrated Greek physician who was practicing medicine circa 400 BCE, has been called the father of medicine. H is credited with helping check a great plague around Athens, as well as prescribing treatment for injuries to the head caused by accidents.

*Handbook of Occupational Safety and Health*, Third Edition. Edited by S. Z. Mansdorf.

During the early and up to the late Middle Ages, a variety of occupational hazards were identified including the effects of lead and mercury exposure, burning of fires in confined spaces, and the need for personal protective equipment. However, there were no organized or established safety standards or requirements during that time.

Workers were generally independent craftsman or part of a family-run shop or farm and were solely responsible for their own safety, health, and well-being. In the early part of the eighteenth century and on the cusp of the Industrial Revolution, Bernardino Ramazzini wrote his classic "Discourse on the Diseases of Workers." Ramazzini is considered to be the father of occupational medicine. He identified the causes of occupational diseases exhibited by chemists working in laboratories. His great admiration for the chemists, however, led him to believe it would be an insult to their profession if he proposed any safety interventions. He also described the pain exhibited in the hands of scribes, foretelling our modern-day interventions regarding repetitive strain injuries. He suggested that this question be asked as an addition to the standard patient history questionnaire: "What is your occupation?"

In the late 1700s, the factory system exposed workers to new and unknown hazards. Textiles led the way with power looms, the cotton gin, and the spinning jenny, along with the associated risks of machinery, noise, and dust. Deaths and injuries were accepted as part of the industrial landscape. The management of safety and health, then, was an unthought of concern or need. Labor was plentiful, and the workers were glad just to have a paying job.

Through the early part of the 1800s, as the Industrial Revolution swept through the United States, the labor force consisted mostly of untrained immigrant labor and children. Common laws of the day favored the owners, and there was virtually no compensation for occupational illnesses or injuries. Nor were there any agreed-upon standards for workplace safety.

Early in the twentieth century, governmental interest in occupational health and safety increased considerably. Workers' compensation requirements were established federally and throughout the states. Concurrently, interested groups began the development of basic safety and health standards. Some organizations initiated what we know today as safety and health management systems.

Impetus was given to the further development of safety and health standards by the infamous Triangle Shirtwaist Factory Fire in 1911, which resulted in the deaths of 146 garment workers. Formation of the NSC during this time was an indication of the greater interest in occupational safety and health.

Up until 1931, most of the safety and health intervention efforts were directed toward improving factory conditions. Then H. W. Heinrich published a book entitled *Industrial Accident Prevention*. He is sometimes called the father of modern safety because he proposed the first organized set of safety principles.

Heinrich's accomplishments deserve the recognition he has achieved as a pioneer in the field of accident prevention. Publication of the four editions of his book spanned nearly 30 years. From the 1930s to today, Heinrich has had more influence than any other individual on the work of occupational safety and health practitioners. For many years, Heinrich was considered the authority. His premises were broadly publicized and that resulted in greater attention being given to occupational safety and health.

But it should also be recognized that some of his premises led to misdirection of safety and health efforts. Two of those premises are:

1. Unsafe acts of workers are the principle causes of occupational accidents.
2. Reducing accident frequency will achieve an equivalent reduction in injury severity.

These premises were the foundation upon which many safety practitioners built the safety management systems they proposed. They believed that efforts to control the unsafe acts of employees should be the focus of safety and health management. Also, they acted on the premise that reducing the frequency of occurrence of less than serious injuries would also achieve an equivalent reduction in serious injuries and fatalities. This author's research with respect to these two premises is discussed in the following section.

### 21.2.2 The Present

In 1970, legislation was enacted for the US Occupational Safety and Health Act (OSHA) to take effect in 1971. Some would argue that one effect of the OSHA was the diversion of attention by management from the prevention of injuries to being compliant with the law. In many organizations, that occurred and the focus became being in compliance with OSHA standards and regulations.

However well intentioned, the initial safety regulations adopted by OSHA came from publications developed by standards-producing or approving entities such as the ANSI and the NFPA. In many cases, those standards were intended to be used as guidelines. Responsible application of safety guidelines was replaced with strict "how do we comply" attitudes to some extent.

Many of those standards, adopted by OSHA in the early 1970s, are out of date. Unfortunately, the standards making or revising system that OSHA must apply may require as much as 10 years to update or issue new standards. As an example, on several occasions, OSHA declared its intent to develop a standard for a safety and health management system. As of June 2015, OSHA has given up on developing a standard: instead, the Safety and Health Program Management Guidelines issued in 1989 are to be updated.

OSHA may apply sanctions in the form of fines should an employer's operations not be in compliance with its requirements. Initial OSHA standards and guidelines tended to be specification oriented and provided great detail on what needed to be done. But an OSHA standard issued in 1992 is a management system standard and allows for reasoned judgments and responsible application of its requirements. That example is OSHA's "Process Safety Management of Highly Hazardous Chemicals; Explosives and Blasting Agents," the designation for which is 29 CFR Part 1910.

While most major companies have incorporated the OSHA requirements into their health and safety management systems, progressive companies have moved beyond just compliance. They recognize that simply being in compliance with safety regulations is not sufficient and that workplace conditions, which are the primary focus of the regulations, are only one aspect of a well-managed safety system.

At the ASSE 2015 Professional Development Conference, Dr. David Michaels, who was the chief executive at OSHA, said precisely the same thing – merely being in compliance with OSHA standards and guidelines does not constitute an effective safety and health management system.

**21.2.2.1 Accomplishments** What benefits have been achieved since the OSHA became effective in 1971? Some say that compliance with OSHA had little to do with the benefits achieved. Those same people will acknowledge that the OSHA, along with its research partner, the NIOSH, created an extended interest and a new era in safety and health management. Consider the following statistics.

Data in the following table was taken from "Accident Facts," an NSC publication, for the years 1971 through 1995 and the "Census of Fatal Occupational Injuries" issued by the Bureau of Labor Statistics (BLS) for the years 1996 through 2014. The fatality rate is the number of fatalities per 100000 full-time equivalent workers.

**From 1971 through 2014,**

- *The number of full-time equivalent workers increased 75.3%.*
- *The number of fatalities was reduced 65.8%.*
- The fatality rate dropped 80.5%.

Taking into consideration that the final statistics for 2014 may show an increase of concern for the number of fatalities, it can still be concluded that the reduction in occupational fatalities from 1971 through 2014 is remarkable. Similarly, data produced by the National Council on Compensation Insurance indicates that worker injury claims have been reduced substantially.

That data also indicates that a larger part of the reduction has been for less serious injuries. It is important to note that the reduction in serious injuries has not been equivalent to the reduction in less serious injuries.

It cannot be said that these reductions resulted entirely from the passage of the OSHA. But taking a broad view of all types and sizes of organization, workplaces would more than likely be more hazardous if the legislation that created OSHA had not been enacted.

**21.2.2.2 A Systems Approach** As the practice of safety and health matured, it became apparent that if inquiry was sufficiently thorough, it would be found that the causal factors for accidents and illnesses were mainly systemic rather than principally being unsafe actions or inactions of employees.

This author undertook extensive research with respect to the two Heinrich premises previously cited. Heinrich professed that among the direct and proximate causes of industrial accidents, 88% are unsafe acts of persons, 10% are unsafe mechanical or physical conditions, and 2% are unpreventable.

Of all Heinrich's concepts, his thoughts on accident causation as expressed in the 88 : 10 : 2 ratios have had the greatest impact on the practice of safety and have resulted in the most misdirection. Why is that so? Because when basing safety efforts on the premise that unsafe acts of persons cause most accidents, the preventive efforts are directed at modifying worker behavior rather than on improving the operating system in which work is performed.

It is easier for supervisors and managers to be satisfied with taking superficial preventive action, such as retraining a worker, reposting the standard operating procedure, or reinstructing the work group, than it is to try to correct system problems.

A directly opposite view to that taken by Heinrich was expressed by W. Edwards Deming in his book *Out of the Crisis*. Deming is known throughout the world for the principles he developed to achieve superior quality. This author states that those principles are comparable to the concepts that must be applied to achieve superior results in safety. Deming wrote:

> The supposition is prevalent throughout the world that there would be no problems in production or service if only our production workers would do their jobs in the way that we taught. Pleasant dreams. The workers are handicapped by the system, and the system belongs to the management. (p. 134)

Safety and health practitioners should be aware of the importance of taking a systems approach to reduce risks. To achieve acceptable risk levels, attention must be given to the hazards and risks in the system, and the system is under the direction of management.

***21.2.2.3 Behavior-Based Safety*** In the late 1970s, behavior-based safety was introduced to the practice of safety and became popular among safety practitioners who focused principally on the unsafe acts of workers and modifying their behavior. Behavior-based safety activities became hugely popular.

Thomas Krause popularized this approach in his book *The Behavior-Based Safety Process*. Based on the work of Harvard psychologist B. F. Skinner, the methodology includes identifying critical behaviors, observing actual behaviors, and providing feedback that leads to changed and improved behaviors.

In his book *The Psychology of Safety*, E. Scott Geller, also a leader in behavior-based safety, recognized – to his credit – that the culture of an organization significantly influences the behavior of workers.

Initially, many behavior-based safety proponents made overly strong statements indicating that doing what they proposed – observing at-risk behavior and focusing on changing worker behavior – should be the core of a safety management system. While some safety practitioners still promote behavior-based safety, its prominence has diminished considerably.

In the August 2001 issue of *Industrial Safety & Hygiene News* (ISHN), an editorial appeared on behavior-based safety, written by Dave Johnson, the editor and publisher. Its title is "Beyond behavior – New labels for old medicine." Excerpts from that editorial follow:

> Bottom line pressures, management turnover and good old human nature just about guarantee a steady stream of safety issues to deal with. So the hunt never ends for the next, new thing to cure, or at least arrest, those chronic challenges. Behavior-based safety has been the medicine of choice in recent years. But interest shows signs of waning, and the field's experts now prescribe a cocktail mix of behavioral, systems and motivational remedies.
>
> Two days in February 1998 might have been the field's high-water mark. That's when the American Society of Safety Engineer's behavioral safety symposium in Orlando smashed all expectations in attracting more than 900 safety pros. The field's luminaries – go-getter personalities with engaging speaking styles and books and videos to their names – shared the stage for the first time, and the star power was palpable.
>
> But at several behavior-based safety sessions held this past June during ASSE'S annual conference, a head count of attendees revealed a different story. Topics and speakers that packed rooms five years ago were met with plenty of empty seats.
>
> Some of the well-known speakers shifted or expanded their focus at the ASSE meeting. Dr. Scott Geller, for example, discussed qualities for effective safety leadership. Other talks connected behavior to culture change and motivation. A recent ad for the pioneering safety consulting firm Behavioral Science Technology talks about "performance improvement", "organizational citizenship culture" and "safety systems," but nowhere in the copy do you find the word "behavior."

In the 27 May 2002 issue of ISHN's *E-News*, Mr. Johnson asks, "What's next?," and says the question "came up lately because signs point to the decline of behavior-based safety, which has been hailed as the latest, greatest elixir for easing safety headaches."

***21.2.2.4 Transitions in the Practice of Safety*** Several developments have occurred in recent years that illustrate how the practice of safety is evolving. Emphasis is moving toward recognition of the:

- Impact of original design and redesign decisions on occupational safety, health, and environmental management–thus the emergence of prevention through design.
- Risk assessment being a core component of a safety and health management system.
- Content and implications of a national occupational health and safety management system standard and of a pending international standard.

- Safety professionals as culture change agents.
- Transition in progress with respect to reducing human error – recognition that to reduce human error, inquiry must be made into the design of the workplace and work methods.

***21.2.2.5 Prevention Through Design*** In the early 1990s, the NSC created the Institute for Safety Through Design to encourage management to give greater attention to the design and redesign processes. Accomplishments of the Institute were notable. This activity arose from studies of incident investigation reports made by safety professionals who discovered that, in many instances, the incident descriptions implied that there were design shortcomings and that they were not being addressed. In accord with its sunset provision, the Institute was dissolved in 1995.

Subsequently, the NIOSH undertook a major national initiative with respect to prevention through design. Accomplishments at NIOSH are many. One noteworthy development is approval of a standard by the ANSI on prevention through design. The standard's designation is ANSI/ASSE Z590.3-2011. Its title is "Prevention through Design Guidelines for Addressing Occupational Hazards and Risks in Design and Redesign Processes."

Many universities giving safety science degrees have incorporated safety through design concepts into their course work. As an example, a very important activity is in progress in the engineering school at Arizona State University. Its vision is:

> To establish a global center for safety at ASU that supports world-class education, research, and outreach programs focused on both safety through design and integration of safety into the core business; it will focus on other important safety initiatives as defined.

Involvement in prevention through design activities by safety practitioners is increasing. Standards personnel at the ASSE say that sales of the prevention through design standard are noteworthy. Hazard identification and analysis and risk assessment are the core of prevention through design. In recognition of the need for its members to have knowledge of risk assessment techniques, ASSE has moved forward on risk assessment as shown in the next section.

***21.2.2.6 Risk Assessments*** European communities have been leaders in promoting the significance of risk assessments. In August 2008, the European Union launched a two-year health and safety campaign focusing on risk assessment. They said in an undated bulletin that is no longer available on the Internet:

> Risk assessment is the cornerstone of the European approach to prevent occupational accidents and ill health. If the risk assessment process – the start of the health and safety management approach – is not done well or not at all, the appropriate preventive measures are unlikely to be identified or put in place.

A briefer version of the above citation is available at https://osha.europa.eu/en/healthy-workplaces-campaigns/previous-healthy-workplaces-campaigns (accessed 17 August 2015).

This action by the European Union is highly significant. That risk assessment should be the cornerstone of an operations risk management system is foundational. William Johnson expressed a companion view in *MORT Safety Assurance Systems* – published in 1980. He wrote:

> Hazard identification is the most important safety process in that, if it fails, all other processes are likely to be ineffective. (p. 245)

Twenty years ago, the term risk assessment did not appear often in safety and health literature. Acceptance of the significance of risk assessment as an important element for safety and health management effectiveness is now prominent. In recognition of the knowledge need for its members, the ASSE has established the Risk Assessment Institute. And the Institute has developed an ASSE Risk Assessment Certificate Program and produced several educational pieces on the subject.

***21.2.2.7 A National US Health and Safety Management System Standard and a Pending International Standard*** While OSHA has not been successful in establishing a standard for occupational safety and health management systems, the AIHA has done so. AIHA was the secretariat for ANSI/AIHA Z10-2005. Its title is "American National Standard: Occupational Health and Safety Management Systems." A second edition was issued in 2012. The ASSE is now the secretariat. This standard focuses extensively on management responsibility for health and safety activities and employee involvement.

In recognition of the worldwide developments with respect to the significance of risk assessments, the 2012 version contains a "shall" requirement for risk assessments to be made. Comments are made on this standard later in this chapter. Sales of this standard have been noteworthy. That is an indication of recognition by safety and health professionals of its importance and the sound base on which it is built.

Work is in progress on the development of an international standard for occupational health and safety management systems under the auspices of the International Organization for Standardization (ISO). It is highly probable that this activity will be successful. The Table of Contents for this work is shown later in this chapter.

***21.2.2.8 Safety Professionals as Culture Change Agents*** Arguments for recognition of the premise that the overarching role of a safety professional is that of a culture change agent were convincingly stated by Dr. Martha Bidez, the professor who created the master's degree for Advanced Safety Engineering and Management at the University of Alabama at Birmingham.

A case can be logically made that all existing hazards and the risks that derive from them arose out of deficiencies in decision-making – deficiencies in management systems or processes. Every proposal made by safety professionals for hazard amelioration and risk reduction must also address those deficiencies.

Amelioration of hazards and reduction of risks without also modifying the decision-making that allowed their existence are inadequate as risk management measures. If revisions are not made in the *system of expected performance* – the culture – similar hazards and risks will be more than likely created. Thus, the primary role for a safety professional is that of a culture change agent.

What does the term "overarching" mean? A composite definition, as found in dictionaries, is "encompassing everything," "embracing all else," or "including or influencing every part of something."

Definitions of a change agent are numerous. The following definition is a composite that fits well with the safety professional's position:

> A change agent is a person who serves as a catalyst to bring about organizational change. Change agents assess the present, are controllably dissatisfied with it, contemplate a future that should be, and take action to achieve the culture changes necessary to achieve the desired future.

This premise – that the overarching role of a safety professional is that of a culture change agent – applies universally to all who give advice on improving health and safety management systems. There are no exceptions.

***21.2.2.9 Transitions with Respect to Human Error Prevention*** On 4 and 5 November 2010 in San Antonio, ASSE sponsored a symposium titled "Rethink Safety: A New View of Human Error and Workplace Safety." It was not surprising that speakers commented on cognitive theory, the properties of human cognition, variable errors and constant errors, imperfect rationality and mental behavioral aspects of error, etc.

But the suggestions made by several speakers on the sources of human error and the corrective actions to be taken when human errors occur were surprising. Several speakers said, in summary, that:

- The best solution when human errors occur is to examine the design of the workplace and the work methods.
- Managers may wish to address human error by "getting into the heads" of their employees with training being the default corrective action – which will not be effective if error potential is designed into the work
- It is management's responsibility to anticipate errors and to have work systems and work methods designed so as to reduce error potential.

An article written by this author titled "Reviewing Heinrich: Dislodging Two Myths from the Practice of Safety" was published in the October 2011 issue of *Professional Safety*. One of the letters to the editor about that article was written by E. Scott Geller. It was published in the December 2011 issue of *Professional Safety*. Scott Geller has been among the most prominent leaders in behavior-based safety. He said in his letter that "Behavior is an outcome and not a causal factor." He also said that:

> As the one who coined the term "behavior-based safety" in 1979 some *Professional Safety* readers may be surprised to learn that I agree completely with Manuele's analysis, conclusions, and recommendations.
>
> My Partners at Safety Performance Solutions (SPS) have taught BBS principles and procedures for 16 years, and they have never claimed behavior to be the cause of an injury. *We follow Dr. Deming's sage advice ("Don't blame people for problems caused by the system")* [Emphasis added] and assert behavior is an outcome of a number of cultural factors, including the work climate, the relevant equipment, the work process, and the management system.

Geller's comments, together with those made at the symposium on human error, prompted a further review because they depart significantly from having the nucleus of safety activities be directed to changing worker behavior. It was found that the writings of Sidney Dekker have particularly influenced how people think about reducing human error.

Dekker wrote *The Field Guide to Understanding Human Error*, a 2006 publication. Dekker's doctorate is in cognitive systems engineering – acquired at Ohio State University. Only a few excerpts from Dekker's book follow:

- Human error is not a cause of failure. Human error is the effect, or symptom, of deeper trouble. Human error is systematically connected to features of people's tools, tasks, and operating systems (p. 15).
- Sources of error are structural, not personal. If you want to understand human error, you have to dig into the system in which people work. You have to stop looking for people's shortcomings (p. 17).
- The Systemic Accident Model....focuses on the whole [system], not [just] the parts. It does not help you much to just focus on human errors, for example, or an equipment failure, without taking into account the socio-technical system that helped shape the conditions for people's performance and the design, testing and fielding of that equipment (p. 90).

Dekker quotes from James Reason's book *Human Error*: "Rather than being the main instigator of an accident, operators tend to be the inheritors of system defects created by poor design, incorrect installation, faulty maintenance and bad management decisions. Their part is usually that of adding the final garnish to a lethal brew whose ingredients have already been long in the cooking" (p. 88 in Dekker: p. 173 in Reason's *Human Error*).

Reason's premise is foundational and difficult to counter. It puts the focus on the system created by management and states that the focus should not be on the worker.

This transition in the human error field from attempting to change worker behavior to improving the design of the system in which people work also supports the proposal that prevention through design be a specifically defined element in an operational risk management system.

### 21.2.3 The Future

Although the future of occupational safety and health management is difficult to predict, some insights have emerged. Governmental activity will probably continue at its present level. As was previously stated, David Michaels was a principal speaker at the ASSE 2015 Professional Development Conference. His presentation on what OSHA was doing and contemplating was somewhat different from what was heard in the past.

Daniels recognized that many organizations, necessarily, have gone far beyond merely complying with OSHA requirements and have sophisticated safety and health management systems in place. And they have achieved very good results.

Early in his term as head of OSHA, there was a concerted effort to diminish emphasis on the Voluntary Protection Programs (VPP). This program outlines performance-based criteria for an effective safety and health management system. To achieve the VPP designation, an organization has to demonstrate that it meets the established criteria and have good results as measured by OSHA statistics. Having recognized that merely complying with OSHA standards and guidelines is inadequate, additional encouragement may be given by OSHA to management to review and improve their safety and health management systems and perhaps being in accord with the requirements of the VPP program.

Daniels spoke of how difficult it is to develop new standards and commented on the antiquity of some OSHA standards. Safety professionals should not expect that many OSHA standards will be updated in the near future.

Michaels recognized the importance of risk assessments within a safety management system and commented on their value. Also, he recognized the greater emphasis now being given to applying a systems approach to prevent occupational injuries and illnesses. Safety professionals can expect that focusing on the design of the workplace and the work methods will become more prominent.

One development in recent years whereby safety professionals at many locations have both occupational safety and health and environmental management responsibilities will continue to expand. These combinations are economics driven. In a brief survey made by this author of 6 Fortune 500 companies, it was found that at headquarter levels, at regional levels, and at very large locations, there still are separate specialists for occupational safety and environmental management. At other locations, the range for having one person responsible for giving advice on both activities was from a low of 50% to a high of 90+% with most responses being closer to 90%.

In recognition of this trend, Indiana University of Pennsylvania (IUP) now gives a bachelor's degree that includes occupational safety, health, and environment in its title. At the University of Central Missouri (UCM), major revisions have been made in course content for occupational safety, industrial hygiene, and environmental management, and the degree given will state the choices made.

Since there is a greater recognition of safety through design and risk assessment as valued elements within a safety management system, and since many safety professionals also have environmental responsibilities, it can be expected that the technical aspects of safety and health management will be more prominent. Additionally, as more safety professionals recognize that the proper approach to reduce human error is to reduce the risks that came to be because of inadequacies in the design and redesign processes, they will also be more involved technically.

## 21.3 SAFETY MANAGEMENT AND THE ORGANIZATION

### 21.3.1 What Types of Skills Will Employers Expect Safety and Health Practitioners to Have?

That will depend on many factors and the following questions have a relationship to the subject. What types of hazards and risks are present in the organization? What are the regulatory requirements imposed by governmental agencies? What type of technology drives the organization? Does the job require the management of other safety and health professionals? To what extent is the safety and health practitioner expected to influence operating managers? Does the position require technical skills for input into equipment and facility design? Is interpretation of regulatory issues an important part of the job?

As this chapter indicates, the scope and function of the safety and health practitioner's position is in transition. Besides having knowledge of a wide range of hazards, controls, and risk assessment methods, safety and health professionals may of necessity have to become cognizant with respect to physical, chemical, biological, and behavioral sciences, mathematics, business, training, and educational techniques, engineering concepts, and particular kinds of operations.

Unfortunately, some university safety degree programs have prepared students principally in the personnel aspects of safety – such as management direction, leadership, supervision, communication, and training. And their graduates are having difficulty in adjusting to what management now expects.

Many universities have recognized the evolution in progress and have adjusted course work to achieve a balance between technical and managerial subjects. While safety and health practitioners must have skills as enablers and be proficient in change management, they must also have broad technical knowledge with respect to the hazards and risks inherent in their places of employment.

### 21.3.2 Where in the Organization?

In some organizations, the chief executive officers make it known that they are also the chief safety officers. In that case, it is recognized that safety starts at the top. Other large organizations have vice presidents for environment, health, and safety at a staff level with professional health and safety personnel reporting within local operations. In many small or medium-sized enterprises, the safety professional wears a variety of hats and may have responsibilities for occupational safety and health and environmental issues and perhaps such as security, human resources, or facilities management.

In most entities, responsibility for health, safety, and environmental issues is integrated into strategic business units. A centralized matrix of functional experts may be on call when special hazards and risk situations arise. Establishing a strong centralized safety, health, and environmental function that has safety practitioners at locations reporting directly to the centralized group obviously does not make sense for an organization that works with decentralized business units.

In any case, wherever the environmental, health, and safety entity fits within an organization, its activities must be fully aligned with its business goals. Providing reasoned and professionally articulated advice up and down the organization on risk reduction establishes that the advice givers are providing value to the organization.

### 21.3.3 Effects of Organizational Culture

Several authors whose articles are published in the health and safety literature have postulated that the most important aspect of a successful safety and health management system is the culture of the organization. This author is one of those proponents.

An overabundance of definitions of safety culture can be found in the literature. In many of those definitions, terms such as the following appear:

- Shared beliefs, attitudes, values, and norms of behavior.
- Shared assumptions.

- Individual and group attitudes that are shared about safety.
- Entrenched attitudes and opinions that a group of people share.

An organization's safety culture – a subset of its overall culture – derives from the decisions made by the governing body and at the senior management level and occasionally at a location management level. Outcomes of those decisions may produce a good or a not-good safety culture.

Note the frequent use of the term "shared" in the citations above. In an organization where employee perceptions are that safety is not near the top among management's priorities, there may be few "shared" values. Employee perceptions may be positive or negative concerning the safety culture established by management. Those employee perceptions are, in effect, their reality. Realistic or unrealistic, perceptions employees have are their truths.

Management owns the culture that it translates into a *system of expected performance*. Strong emphasis is given to the phrase *a system of expected performance* because it defines what the staff believes management actually wants done. Although organizations may issue safety policies and procedure manuals, the staff's perception of what is expected of them and the performance for which they will be measured – its *system of expected performance* – may differ from what is written.

Only top management can provide the leadership, direction, and involvement needed to establish, implement, and maintain an effective occupational health and safety management system with positive shared values. Major improvements in safety – toward being free-er from unacceptable risks – will be achieved only if a culture change takes place, or only if major changes occur in the *system of expected performance*.

Dan Peterson spoke eloquently about culture in the third edition of *Human Error Reduction and Safety Management*. He wrote:

> The culture of an organization sets the tone for everything in safety. In a positive safety culture, it says that everything you do about safety is important. In a participative culture, the organization is saying to the worker – we want and need your help. Some cultures urge creativity and innovation: some destroy it. (p. 66)

Safety practitioners in locations where the safety culture has always been negative or has drifted into a negative state are usually faced with situations in which resources are limited. This author does not say that achieving significant improvement in safety and health management systems in such situations will be easy. But, ethically, the attempt must be made.

## 21.4 SAFETY AND HEALTH MANAGEMENT SYSTEM ELEMENTS

### 21.4.1 Introduction

Having the elements in a health and safety management fully integrated within an organization's culture should be the goal of all safety and health practitioners. Opinions differ on what elements should be included in a health and safety management system, as will be shown by the following examples. It will be noted that there are many similarities in these examples. These guidelines and standards were selected because they represent the variations in current practice. They are provided as references from which safety practitioners can choose what they believe is appropriate for the organizations to which they give counsel.

ISO standards for product quality management and environmental management are built on the plan–do–check–act (PDCA) structure popularized by W. Edwards Deming. So also are Z10, the US occupational health and safety management system standard, and the international standard for which work is in progress.

### 21.4.2 OSHA's Safety and Health Program Management Guidelines

As was stated previously, David Michaels said that it is the intent to have the 1989 version of OSHA's Safety and Health Program Management Guidelines updated and reissued.

Excerpts are provided here in some length because they will be the basis for an updated version. These excerpts are taken verbatim from the guidelines. Although the reissued guidelines will not have the effect of a standard, they could be referenced under OSHA's General Duty Clause as representing good practice.

***Excerpts from OSHA's Safety and Health Program Management Guidelines***

**Issued at 54 FR 3904, 26 January 1989**

**Scope and Application**

This guideline applies to all places of employment covered by OSHA, except those covered by the construction safety standard – 29 CFR 1926.

**Introduction**

Effective management of worker safety and health protection is a decisive factor in reducing the extent and severity of work-related injuries and illnesses. Effective management addresses all hazards whether or not they are regulated by government standards.

OSHA urges all employers to establish and maintain programs which meet these guidelines in a manner which addresses the specific operations and conditions of their worksites.

**The Guidelines: General**

Employers are advised and encouraged to institute and maintain in their establishments a program which provides systematic policies, procedures, and practices that are adequate to recognize and protect their employees from occupational safety and health hazards.

An effective program includes provisions for the systematic identification, evaluation, and prevention or control of general workplace hazards, specific job hazards, and potential hazards which may arise from foreseeable conditions.

Although compliance with the law, including specific OSHA standards, is an important objective, an effective program looks beyond specific requirements of law to address all hazards. It will seek to prevent injuries and illnesses, whether or not compliance is at issue.

The extent to which the program is described in writing is less important than how effective it is in practice.

As the size of a worksite or the complexity of a hazardous operation increases, however, the need for written guidance increases to ensure clear communication of policies and priorities and consistent and fair application of rules.

Next in the Guidelines comes item (b), Major Elements, of which there are four. In a following item (c), Recommended Actions are set forth for each of the Major Elements.

In these excerpts, the applicable Recommended Actions follow the Major Elements, directly.

**Major Element 1**

*Management commitment and employee involvement* are complementary. Management commitment provides the motivating force and the resources for organizing and controlling activities within an organization. In an effective program, management regards worker safety and health as a fundamental value of the organization and applies its commitment to safety and health protection with as much vigor as to other organizational purposes.

Employee involvement provides the means through which workers develop and/or express their own commitment to safety and health protection for themselves and for their fellow workers.

**Recommended Actions**

- State clearly a worksite policy on safe and healthful work and working conditions, so that all personnel with responsibility at the site and personnel at other locations with responsibility for the site:
- Understand the priority of safety and health protection in relation to other organizational values.
- Establish and communicate a clear goal for the safety and health program and objectives for meeting that goal.
- Provide visible top management involvement.
- Provide for and encourage employee involvement.
- Assign and communicate responsibility so that [all] know what performance is expected of them.
- Provide adequate authority and resources.
- Hold managers, supervisors, and employees accountable.
- Review program operations at least annually to evaluate their success so that deficiencies can be identified [and] objectives can be revised.

**Major Element 2**

*Worksite analysis involves* a variety of worksite examinations, to identify not only existing hazards but also conditions and operations in which changes might occur to create hazards.

Effective management actively analyzes the work and worksite, to anticipate and prevent harmful occurrences.

**Recommended Actions**

- Conduct comprehensive baseline worksite surveys for safety and health and periodic comprehensive update surveys.
- Analyze planned and new facilities, processes, materials, and equipment.
- Perform routine job hazard analyses.
- Provide for regular site safety and health inspections.
- Provide a reliable system for employees to notify management [of perceived hazards], without fear of reprisal.
- Provide for investigation of accidents and "near miss" incidents.
- Analyze injury and illness trends.

**Major Element 3**

*Hazard prevention and Control* are triggered by a determination that a hazard or potential hazard exists. Where feasible, hazards are prevented by effective design of the job site or job.

Where it is not feasible to eliminate them, they are controlled to prevent unsafe and unhealthful exposure.

**Recommended Actions**

- So that all current and potential hazards, however detected, are corrected or controlled in a timely manner, establish procedures for that purpose using the following measures:
  1. Engineering techniques where feasible and appropriate.
  2. Procedures for safe work which are understood and followed, as a result of training, positive reinforcement, and, if necessary, enforcement through a clearly communicated disciplinary system.
  3. Provision of personal protective equipment.
  4. Administrative controls, such as reducing the duration of exposure.
- Provide for facility and equipment maintenance, so that hazardous breakdown is prevented.
- Plan and prepare for emergencies, and conduct training and drills.
- Establish a medical program.

**Major Element 4**

*Safety and health training* requires addressing the safety and health responsibilities of all personnel.

**Recommended Actions**

- Ensure that all employees understand the hazards to which they may be exposed and how to prevent harm to themselves and others from exposure to these hazards.
- So that supervisors will carry out their safety and health responsibilities effectively, ensure that they understand those responsibilities and the reasons for them, including:
  1. Analyzing the work to identify hazards.
  2. Maintaining physical protections in their work areas.
  3. Reinforcing employee training on hazards [and] on needed protective measures, through continual performance feedback and, if necessary, through enforcement of safe work practices.
  4. Ensure that managers understand their safety and health responsibilities as described under Management Commitment and Employee Involvement, so that the managers will effectively carry out those responsibilities.

### 21.4.3 OSHA's Standard for Process Safety Management of Highly Hazardous Chemicals; Explosives and Blasting Agents

This standard was promulgated as 29 CFR 1910 in 1982. It was issued with the cooperation of many effected companies because of the adverse accident experience in chemicals at that time. Overall, OSHA incident statistics for chemical operations since then have been quite good. This is a performance standard. Although the requirements imply that there is to be management direction and involvement, no such statement is made as is the case in other standards. An outline of the subjects addressed follows.

This standard outlines the requirements for preventing or minimizing the consequences of catastrophic releases of toxic, reactive, flammable, or explosive chemicals. Elements of the standard are as follows:

(a) Application
(b) Definitions
(c) Employee participation
(d) Process safety information
(e) Process hazard analysis
(f) Operating procedures
(g) Training
(h) Contractors
(i) Pre-start safety review
(j) Mechanical integrity
(k) Hot work permit
(l) Management of change
(m) Incident investigation
(n) Emergency planning and response
(o) Compliance audits
(p) Trade secrets

### 21.4.4 ANSI/AIHA Z10-2012: A Standard for Occupational Health and Safety Management Systems

This standard is approved by the ANSI and, as such, is the American standard for occupational health and safety management systems. It was first issued in 2005. The 2012 issue is a major extension of the earlier version. This author recommends the standard as a reference to all safety and health professionals. Data on the Table of Contents page follow:

**Table of Contents**

Foreword
1.0 Scope, Purpose, and Application
- 1.1 Scope
- 1.2 Purpose
- 1.3 Application

2.0 Definitions
3.0 Management Leadership and Employee Participation
- 3.1 Management Leadership
  - 3.1.1 Occupational Health and Safety Management System
  - 3.1.2 Policy
  - 3.1.3 Responsibility and Authority
- 3.2 Employee Participation

4.0 Planning
- 4.1 Review Process
- 4.2 Assessment and Prioritization
- 4.3 Objectives
- 4.4 Implementation Plans and Allocation of Resources

5.0 Implementation and Operation
- 5.1 OHSMS Operational Elements
  - 5.1.1 Risk Assessment
  - 5.1.2 Hierarchy of Controls
  - 5.1.3 Design Review and Management of Change
  - 5.1.4 Procurement
  - 5.1.5 Contractors
  - 5.1.6 Emergency Preparedness

- 5.2 Education, Training, Awareness, and Competence
- 5.3 Communication
- 5.4 Document and Record Control Process

6.0 Evaluation and Corrective Action
- 6.1 Monitoring, Measurement, and Assessment
- 6.2 Incident Investigation
- 6.3 Audits
- 6.4 Corrective and Preventive Actions
- 6.5 Feedback to the Planning Process

7.0 Management Review
- 7.1 Management Review Process
- 7.2 Management Review Outcomes and Follow Up

On the page following the Table of Contents, titles are listed of fourteen appendices pertaining to subjects in the standard. A bibliography and a listing of references follow the appendices.

### 21.4.5 ISO/PC 283 Is the Designation for a Working Group for an ISO Standard, Which Is to Have the Designation ISO/CD 45001.2. Its Tentative Title Is "Occupational Health and Safety Management Systems – Requirements with Guidance for Use"

As of June 2015, approval was given to a draft of the standard, and that moved it up a level in the ISO approval procedures. If all progresses as expected, this international standard could be in place as early as 2017. Only the Table of Contents of the current version is duplicated here:

**Table of Contents: As of June 2015**

Foreword and Introduction

1. Scope
2. Normative references
3. Terms and definitions
4. Context of the organization
   - 4.1 Understanding the organization and its context
   - 4.2 Understanding the needs and expectations of workers and other interested parties
   - 4.3 Determining the scope of the OH&S management system
   - 4.4 OH&S management system
5. Leadership, worker participation and consultation
   - 5.1 Leadership and commitment
   - 5.2 Policy
   - 5.3 Organizational roles, responsibilities, accountabilities and authorities
   - 5.4 Participation, consultation and representation
6. Planning
   - 6.1 Actions to address risks and opportunities
   - 6.2 OH&S objectives and planning to achieve them
7. Support
   - 7.1 Resources
   - 7.2 Competence
   - 7.3 Awareness
   - 7.4 Information and communication
   - 7.5 Documented information
8. Operation
   - 8.1 Operation planning and control
   - 8.2 Management of change
   - 8.3 Outsourcing
   - 8.4 Procurement
   - 8.5 Contractors
   - 8.6 Emergency preparedness and response

9. Performance evaluation
   9.1 Monitoring, measurement, analysis and evaluation
   9.2 Internal audit
   9.3 Management review
10. Improvement
    10.1 Incident, nonconformity and corrective action
    10.2 Continual improvement
        10.2.1 Continual improvement objectives
        10.2.2 Continual improvement process

Annex A contains 36 dissertations pertaining to the subjects listed in the Table of Contents, followed by Bibliography.

### 21.4.6 ANSI/ASSE Z590.3: Prevention Through Design Guidelines for Addressing Occupational Hazards and Risks in Design and Redesign Processes

A major step forward with respect to the practice of safety was achieved when, on 1 September 2011, the ANSI approved a prevention through design standard. Safety practitioners should be aware of the content of this standard and the professional opportunities it presents. This is the definition of prevention through design as in the standard:

> Addressing occupational safety and health needs in the design and redesign process to prevent or minimize the work-related hazards and risks, associated with the construction, manufacture, use, maintenance, retrofitting, and disposal of facilities, processes, materials, and equipment. (p. 13)

This standard is built on the premise that hazards and risks are most effectively and economically avoided, eliminated, or controlled in the design and redesign processes. Thus, processes for hazard identification and analysis and risk assessment are prominent within the standard. The outcome of those processes is to achieve acceptable risk levels and to avoid bringing hazards and their accompanying risks into the workplace.

In the standard, the definition of acceptable risk incorporates the concept of ALARP as follows:

> *Acceptable risk* is that risk for which the probability of an incident or exposure occurring and the severity of harm or damage that may result are as low as reasonably practicable (ALARP) in the setting being considered.
>
> *As Low As Reasonably Practicable (ALARP)* is that level of risk which can be further lowered only by an increase in resource expenditure that is disproportionate in relation to the resulting decrease in risk.

Applications of the ALARP concept are commonly found in several countries. Figure 21.1 indicates that, ideally, hazards and their accompanying risks are to be considered and properly dealt with in the conceptual and the initial

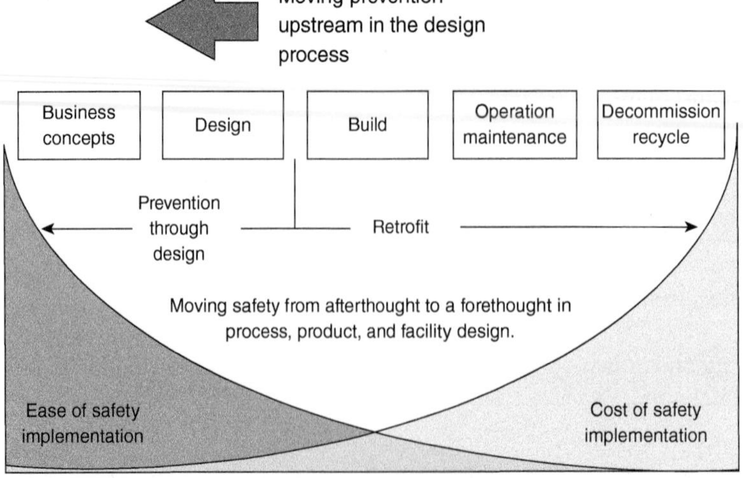


**Figure 21.1** Depiction – prevention through design. *Source:* Reproduced with Permission of Wayne C. Christensen, CSP, PE Christensen Consulting for Safety Excellence, Ltd.

design steps of the design process. It is impractical to assume that perfection can be achieved at the drawing-board stage. Thus, hazards and risks are also to be identified and resolved as development of a facility moves into the build, operation, and maintenance steps. Unfortunately, that requires retrofitting, which can be very expensive.

***21.4.6.1 History*** In 2008, the NIOSH announced that one of its major initiatives was to "Develop and approve a broad, generic voluntary consensus standard on Prevention through Design that is aligned with international design activities and practice." The ASSE agreed to sponsor development of the standard.

Activities at NIOSH are limited to occupational safety and health. Thus the focus of Z590.3 is for the occupational setting. But, by intent, the terminology was kept broad enough so that the guidelines could be applied for all hazard-based needs – environmental controls, property damage that could result in business interruption, product safety, etc.

**Table of Contents**

1. Scope, Purpose, and Application
   - 1.1 Scope
   - 1.2 Purpose
   - 1.3 Application
2. Referenced and Related Standards
   - 2.1 Referenced Standards and Guidelines
   - 2.2 Referenced American National Standards and Technical Reports
3. Definitions
4. Roles and Responsibility
5. Relationships with Suppliers
6. Design Safety Reviews
7. The Hazard Analysis and Risk Assessment Process
   - 7.1 Management Direction
   - 7.2 Select a Risk Assessment Matrix
   - 7.3 Establish the Analysis Parameters
   - 7.4 Anticipate/Identify the Hazards
   - 7.5 Consider the Failure Modes
   - 7.6 Assess the Severity of Consequences
   - 7.7 Determine Occurrence Probability
   - 7.8 Define the Initial Risk
   - 7.9 Select and Implement Risk Reduction and Control Methods
   - 7.10 Assess the Residual Risk
   - 7.11 Risk Acceptance Decision Making
   - 7.12 Document the Results
   - 7.13 Follow Up on Actions Taken
8. Hazard Analysis and Risk Assessment Techniques
9. Hierarchy of Controls

There are nine addenda and a bibliography. The core of prevention through design is hazard identification and analysis and risk assessment. The goals of applying prevention through design concepts are to:

- Achieve acceptable risk levels.
- Prevent or reduce occupationally related injuries, illnesses, and fatalities.
- Reduce the cost of retrofitting necessary to mitigate hazards and risks that were not sufficiently addressed in the design or redesign processes.

***21.4.6.2 Summary*** As the data in Table 21.1 shows, the fatality experience for the years 2009 through 2014 has plateaued. Statistically, both the number of fatalities and the fatality rate have remained within a narrow range. Data issued by the National Council on Compensation Insurance indicates that claims for occupational injuries and injuries have been substantially reduced, although the reduction in less than serious injuries is greater than the reduction for

**TABLE 21.1 All Fatalities: All Occupations for 1971 Through 2014**

| Year | No. of fatalities | Fatality rate | Number of workers millions |
|---|---|---|---|
| 1971 | 13700 | 17 | 81 |
| 1981 | 12500 | 13 | 96 |
| 1991 | 9800 | 8 | 123 |
| 2001 | 5900 | 4.3 | 137 |
| 2009 | 4551 | 3.5 | 130 |
| 2010 | 4690 | 3.6 | 130 |
| 2011 | 4693 | 3.5 | 134 |
| 2012 | 4628 | 3.4 | 136 |
| 2013 | 4585 | 3.3 | 139 |
| 2014 | 4679 | 3.3 | 142[a] |

[a] Statistics are preliminary for 2014. As is said in the BLS bulletin, a final report for 2014 will be issued in late spring of 2016. For the last five years, they say, net increases in the preliminary count for fatalities have averaged 173 cases.

serious injuries. It is suggested that safety and health professionals make analyses to identify opportunities for improvement in the organizations to which they give counsel. They could ask:

1. What safety management systems have been effective in achieving the results obtained and whether even they can be improved?
2. What safety management systems are notably deficient or have not been instituted so that advice can be given to management to additionally reduce the potential for the occurrence of injuries and illnesses?

It should not be assumed that one approach can be applied in all organizations. For a gap analysis, it is suggested that a comparison be made relating an existing safety and health management system to the provisions in ANSI/AIHA Z10-2012, "Occupational Health and Safety Management Systems," extending the gap analysis to include preventive maintenance and prevention through design.

As attempts are made to reduce serious injury and fatality potential, the enormity of the culture changes needed must be recognized as well as how deeply some deterring premises are embedded in many organizations.

## FURTHER READING


*Accident Facts (Now Injury Facts)* (1995). Itasca, IL: National Safety Council.

ANSI/AIHA Z10-2012 (2012). *American National Standard: Occupational Health and Safety Management Systems*. Fairfax, VA: American Industrial Hygiene Association ASSE in now the secretariat at http://www.aiha.org/marketplace.htm.

ANSI/ASSE Z590.3 (2011). *Prevention through Design: Guidelines for Addressing Occupational Hazards and Risks in Design and Redesign Processes*. Park Ridge, IL: ASSE.

ANSI/ASSE Z590.3-2011 (2011). *American National Standard: Prevention through Design: Guidelines for Addressing Occupational Hazards and Risks in Design and Redesign Processes*. Park Ridge, IL: ASSE.

ASSE (2010). *Rethink Safety: A New View of Human Error*. Park Ridge, IL: ASSE.

ASSE (2015). *Risk Assessment Institute and the Risk Assessment Certificate Program*. Park Ridge, IL: ASSE.

Bureau of Labor Statistics. *National Census of Fatal Occupational Injuries*. Washington, DC: US Department of Labor.

Code of Hammurabi. http://www.commonlaw.com/Hammurabi.html (accessed 30 June 2015).

Dekker, S. (2006). *The Field Guide to Understanding Human Error*. Burlington, VT: Ashgate Publishing Company.

Deming, W.E. (1986). *Out of the Crisis*. Cambridge, MA: Massachusetts Institute of Technology.

European Agency for Safety and Health at Work. Risk assessment: Europe. https://osha.europa.eu/en/healthy-workplaces-campaigns/previous-healthy-workplaces-campaigns (accessed 18 August 2015).

Franco, G. and Franco, F. (2001). Bernardino Ramazzini: the father of occupational medicine. **91** (9): 1382.

Geller, E.S. (1996). *The Psychology of Safety*. Radnor, PA: Chilton Book Company.

Geller, E.S. (2011). Letter to the editor. *Professional Safety* (December 2011).

Heinrich, H.W. (1931). *Industrial Accident Prevention*, 1e. New York: McGraw Hill.

Hippocrates. Hippocrates biography. http://www.notablebiographies.com/He-Ho/Hippocrates.html (accessed 18 August 2015).

ISO/PC 283 (2017). ISO/PC 283: Designation for a working Group for an ISO standard which is to have the designation ISO/CD 45001.2. Its tentative title is Occupational health and safety management systems – Requirements with guidance for use.

Johnson, W.G. (1980). *MORT Safety Assurance Systems*. New York: Marcel Dekker, Inc.

Johnson, D. (2001). Beyond behavior: new labels for old medicine. *Industrial Safety and Hygiene News* (August 2001).

Johnson, D. (2002). Comments on behavioral safety. *ISHN's E-News* (27 May 2002).

Krause, T.R. (1997). *The Behavior-Based Safety Process*, 2e. New York: Wiley.

Manuele, F.A. (2011). *Reviewing Heinrich: Dislodging Two Myths from the Practice of Safety*. Des Plaines, IL: Professional Safety.

Manuele, F.A. (2013). *On the Practice of Safety*, 4e. Hoboken, NJ: Wiley.

Manuele, F.A. (2014). *Advanced Safety Management: Focusing on Z10 and Serious Injury Prevention*, 2e. Hoboken, NJ: Wiley.

National Council on Compensation Insurance (2015). *NCCI's Annual Issues Symposium: Analysis of Workers Compensation Results*. Boca Raton, FL: National Council on Compensation Insurance https://www.ncci.com/Documents/AIS-2015-SOL-Presentation.pdf (accessed 25 August 2015).

OSHA (1970). *Occupational Safety and Health Act of 1970*. Washington, DC: OSHA.

OSHA (1982). *OSHA's Standard for Process Safety Management of Highly Hazardous Chemicals; Explosives and Blasting Agents, 29 CFR 1910*. Washington, DC: OSHA.

OSHA (1989). *OSHA's Safety and Health Program Management Guidelines, Issued at 54 FR 3904*. Washington, DC: OSHA.

OSHA (2003). OSHA's voluntary protection program (VPP). Section C. in CSP 03-01-002–TED 8.4 voluntary protection programs (VPP): policies and procedures. http://www.osha.gov/pls/oshaweb/owadisp.show_document?p_table=DIRECTIVES&p_id=2976 (accessed 18 August 15).

Petersen, D. (1996). *Human Error Reduction and Safety Management*, 3e. Hoboken, NJ: Wiley.

Triangle shirtwaist fire. http://www.history.com/topics/triangle-shirtwaist-fire (accessed 18 August 2015).

CHAPTER 22

# EFFECTIVE SAFETY AND HEALTH MANAGEMENT SYSTEMS: MANAGEMENT ROLES AND RESPONSIBILITIES

FRED A. MANUELE

*Hazards Limited, 200 W Campbell #603, Arlington Heights, IL, 60005*

## 22.1 INTRODUCTION: SCOPE OF THIS CHAPTER

Having management leadership, commitment, involvement, and accountability is an absolute requirement – a sine qua non – to achieve superior results with respect to preventing occupational injuries and illnesses. This chapter will emphasize that fact through comments made on the safety and health management systems in place in organizations that have achieved superior results.

American National Standards Institute (ANSI)/American Industrial Hygiene Association (AIHA)/American Society of Safety Engineers (ASSE) Z10-2012 (Z10), the American National Standard for Occupational Health and Safety Management Systems, is used as a base for discussion. Z10 provides senior management with a well-conceived concept and action outline for most of the elements needed in a safety and health management system.

Comments will also be made on preventive maintenance and the additional attention now given to addressing hazards and their accompanying risks in the design and redesign processes.

### 22.1.1 Significance of Management Leadership, Commitment, Involvement, and Accountability and the Culture Management Creates

Many large organizations have achieved and maintained superior safety results as evidenced by their excellent-in-class Occupational Safety and Health Administration (OSHA) incident rates and by phenomenally low workers' compensation costs. This chapter provides a composite review of the safety management systems in place in those organizations for the benefit of safety and health professionals who might ask – How did they achieve their superior records?

One could argue that superior OSHA incident rates are not absolutely indicative as performance measures, and they are not. Some of those companies with superior OSHA statistical records are occasionally faced with the dilemma of having serious injuries and fatalities.

Without question, there are inconsistencies, even within companies, in classifying and recording OSHA statistics. Still, if the inconsistencies in the reporting system are constant, the data produced can serve as useful performance and trend indicators if the statistical sample is large enough.

But while the data may be a predictor of the experience that will develop in the future, it has shortcomings that must be recognized. It does not provide hazard-specific information to which attention could be given. Nor does the data provide usable information on the occurrence potential for low probability/serious consequence events, which, by description, happen infrequently.

But OSHA incident rates continue to be used as comparative performance measures for locations within companies and within trade organizations of which companies are members. This author puts exceptionally low OSHA incident rates to the following test: Do the organization's workers' compensation costs match the excellent-in-class OSHA rates?

*Handbook of Occupational Safety and Health*, Third Edition. Edited by S. Z. Mansdorf.

As management provides the leadership and makes decisions directing the organization, the outcomes of those decisions establish its safety culture – a subset of its overall culture. Safety is culture driven, and management establishes the culture. Safety is defined as freedom from unacceptable risks. In many definitions of safety culture, terms such as the following appear:

- Shared beliefs, attitudes, and values and norms of behavior.
- Shared assumptions.
- Individual and group attitudes that are shared about safety.
- Entrenched attitudes and opinions that a group of people share.

Note the frequent use of the term "shared," which may be inappropriate in some instances. An organization's safety culture derives from the decisions made at an entity's governing level, at the senior management level, and occasionally at a location management level. An organization's culture with respect to occupational safety, environmental safety, the safety of the public, and product safety is determined by the outcome of decisions made by management as indicated by the risk levels in the technical and social aspects at a facility.

The culture created by management is the dominant factor with respect to the risk levels attained – acceptable or unacceptable. Management owns the culture that is translated into a *system of expected performance*. Over the long term, the injury and illness experience attained are a direct reflection of an organization's safety culture.

Strong emphasis is given to the phrase *a system of expected performance* because it defines what the staff believes, in reality, what management wants done. Although organizations may issue safety policies, manuals, and standard operating procedures, the staff's perception of what is expected of them and the performance for which they will be measured – its *system of expected performance* – may differ from what is written.

***22.1.1.1 The Role of Safety and Health Professionals with Respect to the Safety Culture*** What is the safety and health professional's role with respect to the safety culture? In an organization where safety is a core value and management at all levels walks the talk and demonstrates by what it does that it expects the safety culture to be superior, the role of the safety and health professional is easier in the role of a culture change agent as he or she gives advice that supports and maintains the culture.

In a large majority of organizations, an advanced safety culture does not exist. Then, the role of the safety and health professional as a culture change agent has greater significance and requires more diligence as attempts are made to influence management to move toward achieving a superior culture. In that endeavor, the possibility of being successful is enhanced if the safety professional attains the status of an integral member of the business team. That will result from giving well-supported, substantial, and convincing risk reduction advice that serves the business interests.

Admittedly, convincing management that safety should be one of the organization's core values may not be easily achieved. Safety and health professionals should understand that as steps forward are taken by management to improve on management system deficiencies, the result in each instance is a culture change. And the requirements to achieve a permanent culture change should be intertwined into each proposal made to improve on a management deficiency.

***22.1.1.2 Absolutes for Management to Attain Superior Results*** During a review of statements made in annual reports on safety, health, and environmental controls issued by five companies that consistently achieve outstanding results, a pattern became evident that defines the absolutes necessary to attain such results. Elements in that pattern are as follows:

- Safety considerations are incorporated within the company's culture, within its expressed vision and core values – within its *system of expected performance.*
- The governing group and senior management lead the safety initiative and make clear by what they do that safety is a fundamental within the organization's culture.
- There is a passion for and a sense of urgency for superior safety results.
- Safety considerations permeate all business decision-making, from the concept stage for the design of facilities and equipment through to their disposal.
- An effective performance measurement system is in place.
- All levels of personnel are held accountable for results.

Whatever the size of an organization – 10 employees or 100000 – the foregoing principles apply. Where there is a passion for superior results, management insists that it be informed about significant hazard and risk problems and that it be resolved. This insistence by senior management to be informed of the reality of management system deficiencies is vital to achieving stellar results.

Unfortunately, what Whittingham wrote in *The Blame Machine: Why Human Error Causes Accidents* defines actuality, sometimes. He wrote:

> Organizations, and sometimes whole industries, become unwilling to look closely at the system faults which caused the error. Instead the attention is focused on the individual who made the error and blame is brought into the equation. (Preface)

It is easier to define a culture that includes safety as a core value than it is to locate and describe a situation in which the culture drifts negatively and deteriorates over time, the effect the deterioration has on increasing risk, and the position in which such deterioration places a safety professional. But such a situation is depicted in the following excerpts taken from an internally produced report by BP Products North America pertaining to a fire and explosion that occurred on 23 March 2005 at an owned and operated refinery in Texas City, Texas, United States. As a result of that incident, 15 people were killed and over 170 were harmed.

It is important to note that these excerpts represent a self-evaluation. They are taken from an Internet entry titled "BP Issues Final Report on Fatal Explosion, Announces $1 Billion Investment at Texas City" (accessed on 18 August 2015):

- Over the years, the working environment had eroded to one characterized by resistance to change and lacking of trust, motivation, and a sense of purpose.
- Coupled with unclear expectations around supervisory and management behaviors, this meant that rules were not consistently followed, rigor was lacking, and individuals felt disempowered from suggesting or initiating improvements.
- Process safety, operations performance, and systematic risk reduction priorities had not been set and consistently reinforced by management.
- Many changes in a complex organization had led to the lack of clear accountabilities and poor communication, which together resulted in confusion in the workforce over roles and responsibilities.
- A poor level of hazard awareness and understanding of process safety on the site resulted in people accepting levels of risk that are considerably higher than comparable installations.
- Given the poor vertical communication and performance management process, there was neither adequate early warning system of problems, nor any independent means of understanding the deteriorating standards in the plant.

This BP report says, better than an outsider could say, how management decision-making can result in a culture that is conducive to catastrophe. Consider the difficulty that such a progressing deterioration in a safety culture presents to safety and health professionals. While they are obligated to bring to management's attention that the risk of serious injury and fatality is scaling upward, it is understandable that frustration would arise if the message conveyed was persistently ignored.

Readers will not find a statement in this chapter indicating that a deficient safety culture can be easily changed. Yet, the endeavor is worth undertaking and attaining positive results, perhaps in small steps, can be rewarding.

## 22.2 OVERVIEW OF ANSI/AIHA/ASSE Z10-2012: A STANDARD FOR OCCUPATIONAL HEALTH AND SAFETY MANAGEMENT SYSTEMS

On 25 July 2005, the ANSI approved a new standard titled *Occupational Health and Safety Management Systems*. Its designation was ANSI/AIHA Z10-2005. That was a major development.

For the first time in the United States, a national consensus standard for a safety and health management system applicable to organizations of all sizes and types was issued.

ANSI requires that an approved standard be reviewed every five years for revision or updating or just to be reaffirmed. Thus, the AIHA, the Secretariat, formed the necessary committee to review Z10's 2005 version. The outcome of its work is a revised and extended standard approved on 27 June 2012. Shortly after approval of the 2012 version, the Secretariat was transferred by AIHA to the ASSE.

Tim Fisher, Director for Standards at ASSE, has given permission for this author to cite from the Table of Contents in the standard. He asked that the document be referenced as ANSI/ASSE Z10-2012. He also provided the following link to a technical brief on the standard: http://www.asse.org/assets/1/7/Z10_Tech_Brief_2012_Revised.pdf.

This standard provides management with most of the elements required for a well-conceived concept and action outline for a safety and health management system and will serve partially as a base for the following comments. Also, some of the observations made in the following text relate to this author's experience with organizations that have achieved superior results with respect to occupational safety and health management.

In crafting the current version of Z10, the intent was to present management system requirements that, when effectively implemented, would achieve significant safety and health benefits but also to impact favorably on productivity, financial performance, quality, and other business goals. The standard is built on the well-known plan–do–check–act (PDCA) process for continual improvement.

### 22.2.1 A Major Theme

Throughout all of the sections of Z10, starting with section 3.0, which addresses management leadership and employee participation, through section 7.0, which pertains to the management review provisions, the following theme is prominent.

Processes for continual improvement are to be in place and implemented to assure that:

- Hazards are identified and evaluated.
- Risks are assessed and prioritized.
- Management system deficiencies and opportunities for improvement are identified and addressed.
- Risk elimination, reduction, or control measures are taken to assure that acceptable risk levels are attained.

### 22.2.2 Compatibility and Harmonization

One of the goals of the drafters of Z10 was to assure that it could be easily integrated into the management systems an organization has in place. That goal was met. As to structure, the standard is compatible and harmonized with quality and environmental management system standards at the time of the approval of Z10 – the ISO 9000 and ISO 14000 series, respectively.

### 22.2.3 Review of the Sections in Z10

For simplicity, reference is made in the following comments to the numerical designations of sections as they appear in the standard.

#### *Section 1.0 The Scope, Purpose, and Application of Z10*

**1.1 Scope** defines the *minimum requirements* (emphasis added) for occupational health and safety management systems (OHSMS). The intent is to provide a systems approach for continual improvement in safety and health management and to avoid specifications. To their credit, the writers of the standard recognized the uniqueness of the cultures and organizational structures of individual organizations and the need for each entity to decide upon operational specifications.

**1.2 Purpose** says that the primary purpose of this standard is to provide a management tool to reduce the risk of occupational injuries, illnesses, and fatalities.

**1.3 Application** states that this standard is applicable to organizations of all sizes and types. There are no limitations or exclusions in Z10 by industry or business type or number of employees. It is made clear that the structure of the standard is to allow integration with quality and environmental management systems. Doing so is a good idea.

#### *Section 2.0 Definitions*

As is typical in ANSI standards, definitions are given of certain of the terms used in the standard. Safety professionals should become familiar with them.

#### *Section 3.0 Management Leadership and Employee Participation*

Section 3.0 is the standard's most important section. Safety professionals will surely agree with the premise that top management leadership and effective employee participation are crucial for the success of an OHSMS.

An organization's safety culture derives from decisions made at the governing body and top management levels. And continual improvement cannot be achieved without effective top management direction.

In organizations with superior records, senior management provides the leadership necessary to achieve the results expected. Management leadership, commitment, direction, and visible involvement are the sine qua non, the prime requirement for effectiveness in safety. If superior results are desired, there must be a long-term commitment to long-term goals. That's an absolute.

What management does, rather than what management says, defines the actuality of commitment or noncommitment to safety. What management does permeates the thousands of decisions made that create the work environment, set design specifications for facilities and equipment, establish fire protection standards, respond to environmental needs, etc. What senior management does is interpreted by the organization as the role model to be followed. Supervisors in top performing companies convey the element of trust between management and operations employees. Participation by supervisors directly reflects the beliefs their superiors have of what the organization's culture expects and what they understand to be the actual performance measures.

Supervisors will do what they perceive to be important to their superiors. If their superiors convey, by what they do, that safety is important, supervisors will so respond. If supervisors are held accountable for the prevention and control of hazards and to achieve acceptable risk levels, success will result.

Supervisors must have a sound support structure to be successful. That support structure begins with the location manager and the staff immediately subordinate to the manager. It includes depth of training, a good communication system on hazards, up and down, and the resources of qualified safety professionals as consultants.

Safety professionals in superior performing companies say that management has recognized that effective employee involvement builds confidence and trust, develops more enthusiastic and productive employees, and supports the position that all employees are working together to achieve understood objectives.

Thus, employees are given the necessary training and opportunity, the result being that they make substantial contributions in hazard identification, propose solutions to problems, and participate in applying those solutions. Safety initiatives obviously are more effective if employees have "bought into them."

As an example, practitioners in ergonomics tell countless stories of work practice innovations originating from first-line employees. Many of those innovations are easy to apply, inexpensive, and effective and often result in greater productivity. This asset – employee involvement – is well utilized in the top performers to achieve effective risk management.

This section requires that management issue a written policy for occupational safety and health. In smaller organizations, the policy may be tacitly expressed. In either case, the policy should:

- Clearly state management's position on safety, health, and the environment and indicate that avoiding injury and illness to employees and to the public from operations or from products sold and damage to the environment is an organizational value.
- Be expressed by the senior executive or manager.
- Be appropriate to the nature of the organization's operations and their scope.
- Be current.
- State a commitment to comply with all applicable legislation and standards.
- Affirm that issued safety, health, and environmental policies are to be followed.
- Make clear that employees are to actively participate in all elements of the safety and health management system.
- Pledge to a continual improvement process to further reduce risks.

### *Section 4.0 Planning*

Planning is the first step in the PDCA process. As would be expected, this section sets forth the planning process management to implement the standard and to establish plans for improvement. The planning process goal is to identify and prioritize the issues within a safety management system that need improvement.

Those issues are defined in the standard as hazards, risks, management system deficiencies, and opportunities for improvement. Throughout the standard, the emphasis is on having systems and processes in place to identify hazards and assess their accompanying risk and to identify the management deficiencies related to them.

In the continual improvement process, information defining opportunities for further improvement in the safety and health management system, and thereby risk reduction, is to be fed back into the planning process for additional consideration.

**4.1** Requires that a comprehensive review be made under the direction of management to identify the differences between existing operational safety management systems and the requirements of the standard. This is a very important activity. If done well, the base is established for the activities to be undertaken to improve safety and health management systems and to set priorities.

**4.2 Assessment and Prioritization** – This requires that employers have processes in place to identify and analyze hazards, assess the risks deriving from those hazards, and establish priorities for amelioration that, when acted upon, will attain acceptable risk levels. Thus, employers are to have processes in place to identify and analyze hazards, assess the risks deriving from those hazards, and establish priorities for amelioration that, when acted upon, will attain acceptable risk levels.

**4.3 Objectives** – Management is to have processes in place to establish actionable objectives based on previously required studies and set priorities for improvements in systems for identified issues, as previously defined, so that the greatest improvements in risk reduction can be achieved.

**4.4 Implementation Plans and Allocation of Resources** – This follows logically in accord with a sound problem-solving procedure. After hazards, risks, and shortcomings in safety management systems have been identified and objectives have been outlined, a documented plan must be established and implemented to achieve the objectives.

One element in this section pertains to providing adequate resources. It is an absolute that to have a safe workplace, the necessary resources must be provided so that:

- There is adequate staffing, both as to the number of employees and the appropriateness of their skills.
- Facilities and equipment are designed and maintained to achieve acceptable risk levels.

### *Section 5.0 Implementation and Operation*

This section defines the systems that management must have in place to achieve effective risk reduction and control. It is said that this section "provides the backbone of an occupational health and safety management system and the means to pursue the objectives from the planning system." That statement established the significance of this section.

**5.1 OHSMS Operational Elements** – These operational elements are to be integrated into the management system. A new and important addition to Z10 was made in this section. It follows immediately.

### *5.1.1 Risk Assessment*

"The organization shall establish and implement a risk assessment process(es) appropriate to the nature of hazards and level of risk." Superior performers have risk assessment and risk management provisions within their OHSMS. They are aware that not all risks are equal and that prioritization is necessary for expenditure of funds and the timing of expenditures.

Generally, the Europeans are in the lead in promoting risk assessments as a core value in the prevention of injuries and illnesses. In August 2008, the European Union launched a two-year health and safety campaign focusing on risk assessment. They said the following in an undated bulletin that was previously available on the Internet (a briefer version of this citation, accessed on 19 August 2015, is available at https://osha.europa.eu/en/healthy-workplaces-campaigns/previous-healthy-workplaces-campaigns):

> Risk assessment is the cornerstone of the European approach to prevent occupational accidents and ill health. If the risk assessment process – the start of the health and safety management approach – is not done well or not at all, the appropriate preventive measures are unlikely to be identified or put in place.

That statement is foundational and difficult to counter. William Johnson expressed a companion view in *MORT Safety Assurance Systems* – published in 1980. He wrote:

> Hazard identification is the most important safety process in that, if it fails, all other processes are likely to be ineffective. (p. 245)

Hazards include all aspects of technology and activity that produce risk. Hazards are the generic base of, as well as the justification for the existence of, the practice of safety. If there are no hazards, or if there is no potential for harm, safety and health professionals need not exist. The entirety of purpose of those responsible for safety, regardless of their titles, is to manage with respect to hazards so that the risks deriving from the hazards are acceptable. Thus, a sound case can be made that risk assessment should be the core of an occupational safety and health management system.

This author's experience has been that if employees at all levels had more knowledge and awareness of hazards and risks, fewer serious injuries and fatalities would occur. Getting the required knowledge embedded into the minds of all employees requires a major, long-term endeavor, and a culture change must be achieved. Crafting specifically directed training and communication programs will be necessary to achieve the awareness and knowledge required.

Appendix F provides a six-page overview of risk assessment and includes data on many risk assessment techniques. Having knowledge and capability with respect to the following methods will satisfy most, but perhaps not all, of the needs of safety and health professionals as they give counsel on risk assessment:

- Preliminary hazard analysis
- What-if/check analysis
- Failure modes and effects analysis

A risk assessment matrix can serve as a valuable instrument when working with decision-makers on setting risk levels and prioritizing ameliorating actions. Variations in published risk assessment matrices are substantial. A safety professional should develop a matrix that is suitable to the organization to which counsel is given.

### *5.1.2 Hierarchy of Controls*

Provisions for the use of a specifically defined hierarchy of controls are outlined. A hierarchy is a system of persons or things ranked one above the other. The hierarchy of controls in Z10 is to provide a systematic way of thinking, considering steps in a ranked and sequential order, to choose the most effective means of eliminating or reducing hazards and the risks that derive from them. Acknowledging that premise – that risk reduction measures should be considered and taken in a prescribed order – represents an important step in the evolution of the practice of safety.

Z10 requires that risk reduction measures be taken in a prescribed order as in the following hierarchy:

A. Elimination.
B. Substitution of less hazardous materials, processes, operations, or equipment.
C. Engineering controls.
D. Warnings.
E. Administrative controls.
F. Personal protective equipment.

### *5.1.3 Design Review and Management of Change (MOC)*

Management is required to have processes in place to avoid or control hazards when design or redesign activities are undertaken and when changes are made.

#### *The Design Process*

For quite some time, this author and others have professed that the most effective and economical way to achieve acceptable risk levels is to have the hazards from which the risks derive addressed in the design process. That's what this standard requires. This is an exceptionally important element in this standard. Impact of its application can be immense.

Since prevention-through-design concepts are more frequently applied, emphasis is given to them in Section 22.3.2.

#### *Management of Change (MOC)*

Processes are to be established to identify and take appropriate steps to avoid or otherwise control hazards and reduce the potential risks associated with them when changes are made to operations, products, services, or suppliers. The management of change (MOC) process is to assure that:

- Hazards are identified and analyzed and risks are assessed.
- Appropriate avoidance, elimination, or control decisions are made so that acceptable risk levels are achieved and maintained throughout the change process.
- New hazards are not knowingly brought into the workplace by the change.
- The change does not impact negatively on previously resolved hazards.
- The change does not make the potential for harm of an existing hazard more severe.

This process is applied when modifications are made with respect to technology, equipment, facilities, work practices and procedures, design specifications, raw materials, organizational or staffing situations, and standards or regulations. In the MOC process, as applicable, consideration would be given to:

- Safety of employees making the changes.
- Safety of employees in adjacent areas.

- Safety of employees who will be engaged in operations after changes are made.
- Environmental aspects.
- Safety of the public.
- Product safety and product quality.
- Fire protection so as to avoid property damage and business interruption.

Reviews made by this author of over 1800 incident investigation reports, mostly for serious injuries and fatalities, support the need for and the benefit of having MOC systems in place. Those reviews showed that a significantly large share of incidents resulting in serious injury or fatality occurs:

- When unusual and nonroutine work is being performed.
- In nonproduction activities.
- In at-plant modification or construction operations (replacing a motor weighing 800 lb to be installed on a platform 15 ft above the floor).
- During shutdowns for repair and maintenance and start-ups.
- Where sources of high energy are present (electrical, steam, pneumatic, chemical).
- Where upsets occur: situations going from normal to abnormal.

Having an effective MOC system in place would have reduced the probability of serious injuries and fatalities occurring in the operational categories shown above. Superior performing companies are leaders with respect to MOC systems and have well-established MOC procedures embedded in their cultures. Safety professionals are an integral part of those procedures and are recognized as valuable resources in hazard recognition and risk reduction for the MOC process.

But getting effective MOC procedures in place is not easy to do. It should be understood that adopting such a process requires a culture change (Manuele, 2013, p. 166, 2014, p. 62).

It is suggested that safety professionals study thoroughly the MOC requirements of Z10 to determine how they might assist in achieving the culture change necessary for their implementation. Applying change management methods will be necessary. Note that sections 5.1.3.1 and 5.1.3.2 are extensions of *5.1.3 Design Review and Management of Change*.

### 5.1.3.1 Applicable Life Cycle Phases

This provision says that the entirety of the life cycle of the subject being considered in a design or redesign process must be considered.

### 5.1.3.2 Process Verification

"The organization shall have processes in place to verify that changes in facilities, personnel, documentation, and operations are evaluated and managed to ensure safety and health risks arising from these changes are controlled." This section speaks well for itself.

### 5.1.4 Procurement

An organization is to establish processes to avoid bringing hazards into a workplace with respect to purchased products and for contracted arrangements. Although the requirements for procurement are plainly stated and easily understood, they are brief in relation to the enormity of what will be required to implement them in many organizations. An interpretation of the Z10 requirements could be – safety practitioners, you are assigned the responsibility to convince management and purchasing agents that, in the long term, it can be very expensive to buy cheap.

But having safety specifications included in purchasing procedures is common in superior performing organizations. Safety professionals who are involved say that they are proud of the influence they have had on the purchasing provisions in effect in their companies. Working with design and engineering personnel and with senior executives, they have achieved a culture change that results in fewer hazards and their accompanying risks being brought into their operations.

Including safety-related specifications in procurement documents also assists in attaining corporate financial goals both because risks of injury are reduced and the cost of expensive retrofitting is avoided. With respect to achieving recognition and a sense of accomplishment, safety professionals would serve themselves and their organizations well if they found a way to get such provisions installed.

### 5.1.5 Contractors

This section requires that top management establish processes to avoid injury and illness to the organization's employees from activities of contractors and to the contractor's employees from the organization's operations. Many entities have such procedures in place. Literature on this subject is extensive. One of the provisions indicates that the process is to include "contractor health and safety performance criteria." That implies, among other things, vetting the contractor with respect to its previous safety performance before awarding a contract.

### 5.1.6 Emergency Preparedness

To meet the requirements of this provision, top management is to have processes in place "to identify, prevent, prepare for, and/or respond to emergencies." Also, periodic drills are to be conducted to test the emergency plans, and they are to be updated periodically. For this subject, literature giving guidance to management is also plentiful. Managements of companies that are dedicated to protecting their employees and their communities provide the resources necessary to establish and maintain sound emergency and disaster planning. But, with sympathy, it needs to be said that it's very difficult to maintain activities that are seldom used.

Expectations of emergency and disaster plans cannot be fulfilled without regularly testing their ability to deliver. Establishing communications with the community resources is necessary, without which the actions expected when an emergency occurs will not take place. Training and practice requirements are considerable.

### 5.2 Education, Training, Awareness, and Competence

An organization is required to establish the skill levels necessary for competence, to be sure that workers are familiar with the OHSMS requirements, and to remove structural or organizational barriers to participation in education. Also, it is specifically stated that training to be given in languages that trainees understand, that training is continuously given as needed, and that trainers are competent.

This section has six alpha-designated provisions. In three of them, the words "competence" and "competent" appear. Thus, competence is emphasized. Employees and contractors are to be competent to fulfill their responsibilities. Trainers are to be competent to train.

These provisions are applicable to contractors also. It is interesting that in the examples of the training that should be given, both safety by design and procurement are mentioned.

In companies with superior safety records, training is serious business. Unfortunately, safety training is often much talked and written about but poorly done. Senior management in the model companies is well trained. It all starts there. All levels of management become aware of the inherent risks in their businesses and acquire knowledge of risk management needs. They cannot be role models and provide the necessary leadership if they are not schooled in how the risk management job is to be done.

Training takes place in many ways – in formal classroom settings or perhaps on the job by demonstration and observation. In the superior performers, training is a never-ending process. Safety training must be well planned, continuous, and measured for results.

Too much emphasis cannot be given to the importance of the supervisor in employee training or to the priority given to training in those companies where successes in risk management are noteworthy. Supervisors, and experienced employees serving as lead persons, are the role models that new employees will follow.

But consider this situation as representative of a reality that is too prevalent. Early during a safety audit, an industrial relations director proudly reviewed with the auditor a marvelous indoctrination and safety training program for new employees. During the audit, an interview was arranged with an employee who had been in the shop for about three months. The intent was to determine what he thought of the indoctrination and safety training program.

His response was: "What indoctrination and safety training program?" This employee had bid up to his third job, had never gone through the indoctrination and safety training program, said that he never saw his supervisor, and did not know how to get anyone to pay attention to gearbox covers that had been removed and not replaced. Situations of that sort define the place safety has in an organization's culture.

Training needs are always in transition, and recent developments require different emphases. Safety professionals in superior performers have spoken of these situations:

- New technology is continuously developed that may not have been evaluated for safety. Thus, safety professionals are more often engaged in pre-operational hazard and job hazard analyses, and the additional training those analyses indicate is necessary.
- It is more common for employees with seniority to be assigned to new jobs without adequate training, and that requires particular attention by the safety staff.

- Demographics and the greater differences in spoken and written languages in the changing workforce stretch training capacity to its limits.

### 5.3 Communication

An organization is to institute processes to inform all levels of employment on the progress achieved on its implementation plan and that incidents, hazards, and risks are to be reported promptly and to encourage employees to communicate to their supervisors on hazards and risks. Particularly, contractors and "relevant external interested parties" are to be informed of changes made that affect them. Emphasis is given to eliminating any barriers that may interfere with any of these requirements.

In all organizations where safety expectations are high, rather exotic technical information systems exist to serve as resources on hazards and their control. Personnel at all levels come to rely on those resources. Management promotes a continuing and open discussion of hazards, incidents, and concerns about risks. Progress relative to established goals is published, discussed, and routinely communicated to employees. Two-way communications, upward and downward, exist throughout the organization. Thus, the knowledge and experience of employees is brought to bear to improve safety.

### 5.4 Document and Record Control Process

Documentation requirements for certain systems are specified in several places in Z10. As a performance standard would say, the document and record control processes are to fit the requirements of the safety and health management system in place. Documentation procedures established should be commensurate with the size of an organization as represented by the number of employees, the complexity of operations, and its inherent hazards and risks. The standard requires that documents be updated as needed, legible, adequately protected against damage or loss, and retained as necessary.

### Section 6.0 Evaluation and Corrective Action

This section outlines the process requirements to evaluate the performance of the safety management system, to take corrective action when shortcomings are found, and to provide feedback to the planning and management review processes. Communications on lessons learned are to be fed back into the planning process. The expectation is that new objectives and action plans will be written in relation to what has been experienced.

### 6.1 Monitoring, Measurement, and Assessment

Requirements in this section are those applicable in a sound business process. Processes are to be established to measure the effectiveness of all systems in place to identify issues, defined as hazards, risks, deficiencies in management systems, and opportunities for improvement. Shortcomings in those systems are to be referred to those responsible for planning so that improvements can be made.

### 6.2 Incident Investigation

The standard requires that processes be established so that incidents are reported, investigated, and analyzed in a timely order. The purpose is to identify shortcomings in management systems so that corrective action can be taken. This is easy to say but somewhat difficult to accomplish.

Mention was previously made of this author's review of over 1800 incident investigation reports. Most of those reports came from Fortune 500 companies. One of the findings was that the gap between issued procedures on incident investigation and what actually takes place can be enormous. For causal factor determination, on a scale of 10 with 10 being best, an average score of 5.7 would be the best that could be given, and that could be a bit of a stretch.

An inquiry was made using the Five Why analysis system to determine why there was such a huge gap. As the Five Why analysis proceeded, it became apparent that incident investigation procedures do not recognize the deterrent human factors that almost always exist:

- **When supervisors are required to complete incident investigation reports, they are asked to write performance reviews on themselves and on the people to whom they report – all the way up to the board of directors.**
- **So also are the managers in levels above the supervisor who participate in incident investigations in a position of having to evaluate their own performance and the results of decisions made at levels above theirs.**

It is understandable that supervisors will not comment on their shortcomings when they prepare incident investigation reports. The probability is close to zero that a supervisor will write:

> This incident occurred in my area of supervision and I take full responsibility for it. I overlooked.... I should have done....My boss did not forward the work order for repairs I sent him three months ago....

Logically, self-preservation dominates. This also applies to all management levels above the line supervisor. And upper-level managers will be reluctant to write about the management system deficiencies for which the people to whom they report may be responsible.

With respect to operators and incident causation, James Reason wrote this in *Human Error*:

> Rather than being the main instigator of an accident, operators tend to be the inheritors of system defects created by poor design, incorrect installation, faulty maintenance and bad management decisions. Their part is usually that of adding the final garnish to a lethal brew whose ingredients have already been long in the cooking. (p. 173)

Supervisors are one step above line employees. They also work in a "lethal brew whose ingredients have already been long in the cooking." They have little or no input to the original design of operations and work systems and may face difficulty in having major changes made in them. Nor do most supervisors have sufficient knowledge of hazards and risks to qualify them to offer recommendations to improve the operating systems.

Safety professionals should realize that constraints similar to those applicable to a supervisor also apply, in varying degrees, to all personnel within an organization who lead or are members of investigation teams.

Importantly, safety professionals should be aware of the importance of the guidance they give when incidents are investigated.

In some organizations, the procedure is to have a team investigate incidents that result in serious injury or illness if had the potential to do so. That's good practice. In the reviews made of incident investigation reports, those completed by teams were superior. To the extent feasible, investigation team leaders should have very good managerial and technical skills and not be associated with the area in which the incident occurred.

Chapter 7 in *Guidelines for Investigating Chemical Process Incidents* is titled "Building and Leading an Investigation Team." While the word "chemical" is in the book's title, the text is largely generic. This is the opening paragraph in chapter 7:

> A thorough and accurate incident investigation depends upon the capabilities of the assigned team. Each member's technical skills, expertise, and communication skills are valuable considerations when building an investigation team. This chapter describes ways to select skilled personnel to participate on incident investigation teams and recommends methods to develop their capabilities and manage the teams' resources. (p. 97)

This book is recommended as a good resource on incident investigation. Competence, objectivity, capability, and training are emphasized. The author's paper "Incident Investigation: Our Methods Are Flawed" is listed as a reference in this chapter.

### 6.3 *Audits*

Periodically, an organization shall have audits made. Audits are to measure the organization's effectiveness in implementing the elements of its OHSMS. Thus, audits are to determine whether the management systems in place do or do not effectively identify hazards and control risks.

Although many safety professionals are familiar with safety audit processes, it is suggested that they review what the Z10 standard requires and determine whether it will be to their benefit to revise their audit systems. It is made clear that audits are to be "system" oriented rather than "compliance" oriented.

Also, and importantly, while the standard says that those who make audits should not be connected with the operations being audited, it also says that the requirement does not imply that audits must be conducted by persons "external to the organization."

### 6.4 *Corrective and Preventive Actions*

Organizations are to identify hazards and risks and shortcomings in management systems and take corrective action as necessary to achieve acceptable risk levels. This relates particularly to section 4.0 – which says that the goal is to identify and prioritize the issues within a safety and health management system that need improvement. Those issues are defined as hazards, risks, management system deficiencies, and opportunities for improvement.

This section also includes a quality control requirement whereby the effectiveness of the risk management measures is to be determined. Also, it is emphasized that appropriate and immediate action is to be taken for hazard/risk situations that have the potential for serious injury or illness.

### 6.5 *Feedback to the Planning Process*

This is a communication provision pertaining to all shortcomings observed in the safety management system. Its purpose is to provide a base for revision in the planning process. The standard says that the communication process established shall ensure a proper flow of the information developed in the monitoring and measurement systems, audits, and incident investigations and in the corrective actions taken to those involved in the planning process to achieve continual improvement endeavors.

#### *Section 7.0 Management Review*

Processes are to be established so that OHSMS performance is reviewed periodically and action is taken by management on deficiencies cited in the review process. It was said earlier in this chapter that management leadership and employee participation is the most important section in Z10. This section on management review is a close second. Making periodic reviews of management systems' effectiveness is an important part of the plan–do–check–act process.

#### *7.1 Management Review Process*

These are some of the subjects to be reviewed, at least annually: progress in the reduction of risk; effectiveness of processes to identify, assess, and prioritize risk and system deficiencies; effectiveness in addressing underlying causes of risks and system deficiencies; the extent to which objectives have been met; and performance of the OHSMS in relation to expectations.

#### *7.2 Management Review Outcomes and Follow-Up*

This section requires that management determine whether changes need to be made in "the organization's policy, priorities, objectives, resources, or other OHSMS elements." Senior management is expected to provide the direction needed to implement the changes that should be made in safety and health systems and processes. This provision gives the needed importance to the management review process. Action items are to be recorded, communicated to those affected, and followed through to a proper conclusion.

### 22.2.4 On the Importance of the Standard's Advisory Content and Appendices

Z10 provides a large amount of exceptionally valuable explanatory and supportive data in the advisory column – the E column – and in the Appendices.

Alphanumerical pages 1 through 29 pertain to the requirements of the standard and the material in the advisory column. Pages 30 through 88 are devoted to the Appendices. That's about a 65% increase in the space devoted to Appendices over the 2005 version.

A safety professional must have a copy of the standard to appreciate the value of the guidance material provided.

This author has had to respond to many questions on how Z10 compares to other standards. A studied response is provided in Appendix N (Informative). It is titled "Management System Standard Comparison (Introduction)."

### 22.2.5 Observation

This revision in Z10 is important work. Prudent safety professionals will study the requirements of the standard to determine whether additional skills and capabilities are needed and move forward to acquire those skills. Having done so, they will be better equipped to give guidance to management to improve systems in place and to put in place safety management elements that may not exist in the organizations to which they give counsel.

## 22.3 ADDITIONAL ELEMENTS FOR AN OCCUPATIONAL SAFETY AND HEALTH MANAGEMENT SYSTEM

Two elements that are not included in the current version of Z10 that will be more than likely to be added to the next version are preventive maintenance and prevention through design. Having such elements in place will serve well to reduce risks of injury and illness. Comments are made here on those elements.

### 22.3.1 Preventive Maintenance: System Integrity

Maintenance of an operating system at a high level obviously impacts greatly the mechanical integrity of operations and of a safety management system. Maintenance done well or not done well sends messages to the staff informing them of the reality of an organization's intent with respect to controlling hazards and maintaining acceptable risk levels.

Visit a location where the culture demands good safety practice, and immediately, from the appearance of the exterior premises, you will get a "feel" for the quality of maintenance. That isn't necessarily an absolute indicator, but the opposite is almost always true. If the exterior of the premises is shabby, safety maintenance will likely be inadequate within the facility. In the best operations, cleanliness is truly a virtue, maintenance schedules are adhered to, and personnel are encouraged to report on and seek to eliminate hazards.

Consider this situation for an opposite and real picture. A safety professional is making an audit of a safety management system. The maintenance superintendent displays an elaborate computer-based maintenance program, of which he is very proud. During the plant tour, many hazardous conditions are observed. A supervisor is asked why work orders aren't being sent to the maintenance department to have those conditions corrected. And the response is: "We don't do that anymore. Safety work orders are the last priority for the maintenance department." Later it is determined that a great number of safety-related work orders are over six months old. But the maintenance program, on paper, was supposed to prevent that sort of thing from happening.

A negative message is delivered in a situation of that sort. If the staff is to believe that their safety is taken seriously by management, a safe environment must be maintained as a demonstration of its commitment to do so. Superior performers do that.

### 22.3.2 Management Responsibility for Safety in the Design and Redesign Processes

Management responsibility for safety begins in the design decision-making for new or altered facilities and operating systems. An important statement relative to that responsibility is made by Joe Stephenson in his book *System Safety 2000: A Practical Guide for Planning, Managing, and Conducting System Safety Programs*. In an update of that book, titled *System Safety for the 21st Century*, Richard A. Stephans repeated Stephenson's statement. They wrote:

> The safety of an operation is determined long before the people, procedures, and plant and hardware come together at the work site to perform a given task. (Stephenson p. 10; Stephans p. 13)

Their statement can easily be substantiated. Start from the beginning. Consider, first, the necessity for the development of a new operation to make a site survey for ecological considerations, and move into the construction and fitting out of the facility. Thousands of safety-related decisions are made in the design process, and their outcome determines the level of inherent risk. Usually, those decisions meet (or exceed) applicable safety-related codes and standards with respect to such as (to name but a few) the contour of exterior grounds; sidewalks and parking lots; building foundations; facility layout and configuration; floor materials; roof supports; process selection and design; determination of the work methods; aisle spacing; traffic flow; hardware; equipment; tooling; materials to be used; energy choices and controls; lighting, heating, and ventilation; fire protection; occupational health exposures; and environmental concerns.

Designers and engineers make design decisions that establish what they believe to be acceptable risk levels and their achievements are noteworthy. But analyses of employee injuries and illnesses indicate that a more effective job could be done in the design processes to identify and analyze potential hazards and their accompanying risks. For example, a study made by this author of incident investigation reports indicated that there were implications of workplace and work method design inadequacies in over 35% of the incidents. That study is supported by an analysis made in Australia.

In *Guidance on the Principles of Safe Design for Work*, comments are made on the "contribution that the design of machinery and equipment has on the incidence of fatalities and injuries in Australia." They say:

> Of the 210 identified workplace fatalities, 77 (37%) definitely or probably had design-related issues involved. Design contributes to at least 30% of work-related serious non-fatal injuries. (p. 6)

The practice of safety is hazards based. Thus, Johnson wrote appropriately that hazard analysis is the most important safety process (p. 245). Since all risks in an operational setting derive from hazards and since the intent of an operational risk management system is to achieve acceptable risk levels, it follows that risk assessment should be the cornerstone of an operational risk management system. The core of prevention through design is hazard identification and analysis and risk assessment.

In 2008, the National Institute for Occupational Safety and Health (NIOSH) announced that one of its major initiatives was to "Develop and approve a broad, generic voluntary consensus standard on Prevention through Design that is aligned with international design activities and practice."

The ASSE agreed to sponsor the standard development effort. A major step forward for the practice of safety was taken when, on 1 September 2011, the ANSI approved ANSI/ASSE Z590.3-2011, *Prevention through Design Guidelines for Addressing Occupational Hazards and Risks in Design and Redesign Processes*. Safety practitioners should be

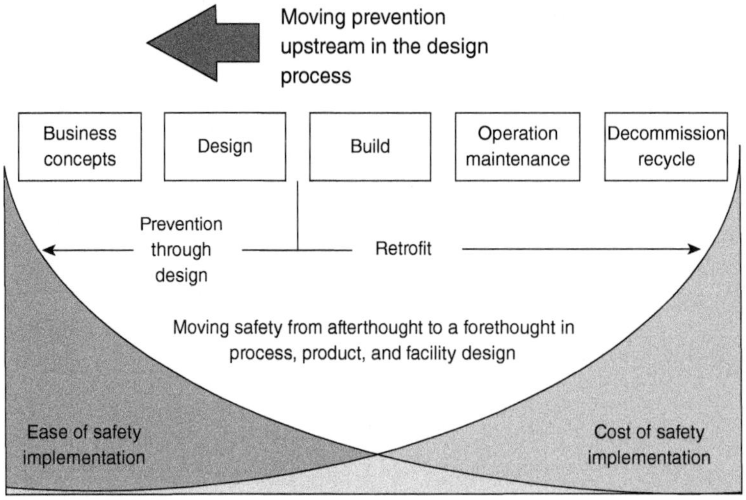


**Figure 22.1** Duplicated with Permission from Christensen Consulting for Safety Excellence.

aware of the standard's content and the opportunities it presents for them. This is the definition of prevention through design as in the standard:

> Addressing occupational safety and health needs in the design and redesign process to prevent or minimize the work-related hazards and risks, associated with the construction, manufacture, use, maintenance, retrofitting, and disposal of facilities, processes, materials, and equipment.

Activities at NIOSH are limited to occupational safety and health. Thus the focus of Z590.3 is on the occupational setting. But the terminology in the standard was kept broad so that the guidelines could be applicable to all hazard-based needs – product safety, environmental controls, property damage that could result in business interruption, etc. Also, this standard pertains to each of the four major stages of an occupational risk management system:

1. Pre-operational stage – In the initial planning, design, specification, prototyping, and construction processes, where the opportunities are greatest and the costs are lowest for hazard and risk avoidance, elimination, reduction, or control.
2. Operational stage – Where hazards and risks are identified and evaluated and mitigation actions are taken through redesign initiatives or changes in work methods before incidents or exposures occur.
3. Post-incident stage – Where investigations are made of incidents and exposures to determine the causal factors that will lead to appropriate interventions and acceptable risk levels.
4. Post-operational stage – When demolition, decommissioning, or reusing/rebuilding operations are undertaken.

Figure 22.1 depicts the theoretical ideal. Prevention through design is moved upstream in the design process rather than being an afterthought. The intent is to have all hazards and risks identified dealt with in the conceptual and design steps. But that requires unattainable perfection from the people involved. Hazards and risks will also be identified in the build and operation and maintenance steps for which redesign is necessary in a retrofitting process.

An example is given here for the wording of an element on prevention through design that could be included in a safety and health management system. It is for the benefit of imaginative safety professionals who recognize the need and want to promote having such an element installed.

Top management shall provide the leadership to institute a policy on prevention through design and to achieve effective application of it for the original design and redesign processes through which:

1. Hazards are anticipated, identified, and evaluated for avoidance, elimination, substitution, or control.
2. Risks deriving from identified hazards are assessed and prioritized.
3. Risks are reduced to an acceptable level through the application of a hierarchy of controls.
4. The knowledge, skills, experience, insight, and creativity of employees close to the hazards and risks are utilized in the risk assessment process.

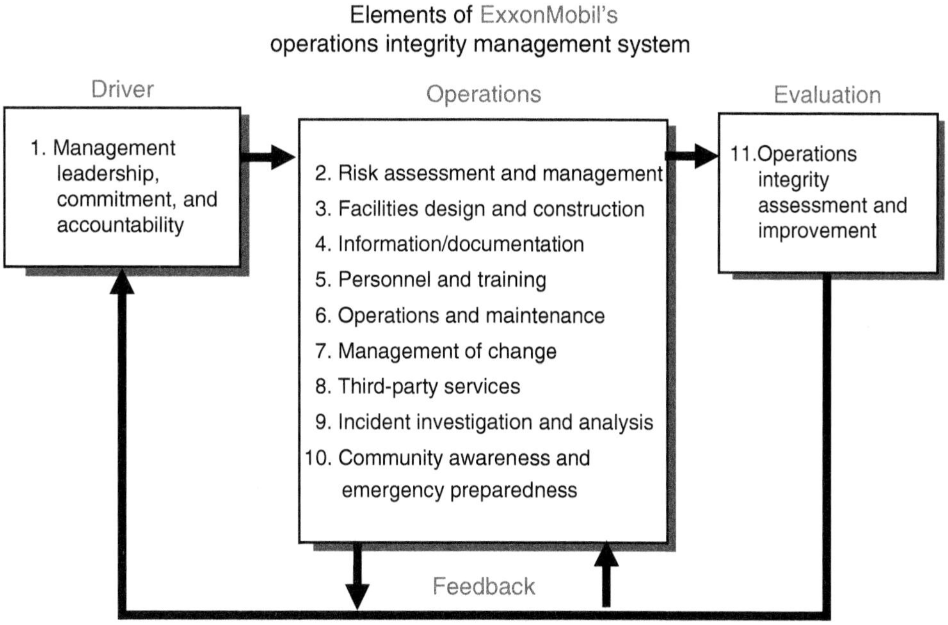


**Figure 22.2** Duplicated with Permission from ExxonMobil.

5. Original design and/or redesign process effectiveness is monitored through feedback between employees and management to provide for continual improvement.
6. Appropriate recordkeeping systems are developed and used to document design reviews and to track feedback and safety and health reports over the life cycle.

Enter "ExxonMobil Operations Integrity Management System" into a search engine and a brochure with that title will appear. On page 5, there is a depiction of the OIMS system, a variation of which follows in Figure 22.2.

For a safety and health management system to achieve its goals, management leadership, commitment, and accountability are absolutely necessary and should be the lead element as is shown. But note the first subject in the listing of operations.

Risk assessment and management is given the place that this author believes it deserves. A repetition follows from Chapter 1, which quotes from a European Community document on risk assessment. It would be difficult to argue against its message:

> Risk assessment is the cornerstone of the European approach to prevent occupational accidents and ill health. If the risk assessment process – the start of the health and safety management approach – is not done well or not at all, the appropriate preventive measures are unlikely to be identified or put in place.

Facilities design and construction follows immediately. That can be interpreted as prevention through design. It is obvious that the decision-makers at ExxonMobil have recognized the value of addressing hazards and risks in the design process. That is a good thing to do.

## 22.4 ACCIDENT COSTS

Presentations to management on the costs of worker injuries and illnesses can be attention-getting and convincing, provided the data is plausible and can be supported with suitable references. For many years, safety practitioners have used the ratio of indirect to direct costs of accidents to inform managements on total accident costs and to achieve improvements in safety management systems.

The most commonly used ratio is 4 : 1, but the literature contains a wide range of ratios and proposed cost categories and methods used for determining ratios. But no ratios published prior to 1995 can be considered valid because they could not have considered the fact that increase in direct costs in the past 20 years – indemnity payments made to workers and medical costs – has greatly exceeded the increase in indirect costs. To introduce this discussion, comments are made on the ratios published by only two authors. They were chosen because of their prominence.

### 22.4.1 On Heinrich's Indirect and Direct Cost Ratios

H. W. Heinrich, an influential leader in the practice of safety, is the author of *Industrial Accident Prevention: A Scientific Approach*. In that text, he said that the indirect to direct cost ratio for injuries and illnesses is 4 : 1. His is the most often ratio referenced in safety-related literature. Heinrich's first edition was published in 1931. His 4 : 1 ratio derived from a study made in 1926. Although the study was made about 90 years ago, his ratio has had staying power.

His ratio is repeated in the three editions of his book published after the first edition in 1941, 1950, and 1959. Unfortunately, there is no research data available that supports Heinrich's ratio (first edition, p. 17; second, third, and fourth editions, p. 50).

### 22.4.2 Frank E. Bird Jr. On Accident Costs

In his 1974 book titled *Management Guide to Loss Control*, Frank E. Bird Jr. presented what he called the iceberg theory of incident costs. Bird gave his ratios of insured and uninsured costs in an exhibit having the appearance of an iceberg. His exhibit is captioned "The Real Costs Of Accidents Can Be Measured And Controlled." In a 1985 book titled *Practical Loss Control Leadership, Revised Edition*, Bird and George L. Germain presented the same ratios.

Bird's insured costs – medical and compensation costs – are the same as the direct costs in some other presentations. The 1985 version of the iceberg is the base for the adaptation shown in Figure 22.3.

As a minimum, the Bird data implies that employers absorb uninsured costs at a ratio of \$6 (5 + 1) to \$1 of insured costs. At a maximum, the ratio is \$53 (50 + 3) to \$1. This author has not located research and hard data that supports such ratios. Depictions of Bird's "iceberg" have appeared in many texts and articles, and his cost ratios are frequently repeated.

### 22.4.3 A Construction Industry Study

Under contract to the Business Roundtable, the Department of Civil Engineering at Stanford University conducted research to provide guidance on reducing accident frequency and severity in the construction industry and the attendant indirect and direct injury and illness costs.

As a result, Technical Report No. 260 was published in August 1981, the title of which is "Improving Construction Safety Performance: The User's Role." In 1982, the Business Roundtable issued a condensed report titled "Improving Construction Safety Performance."

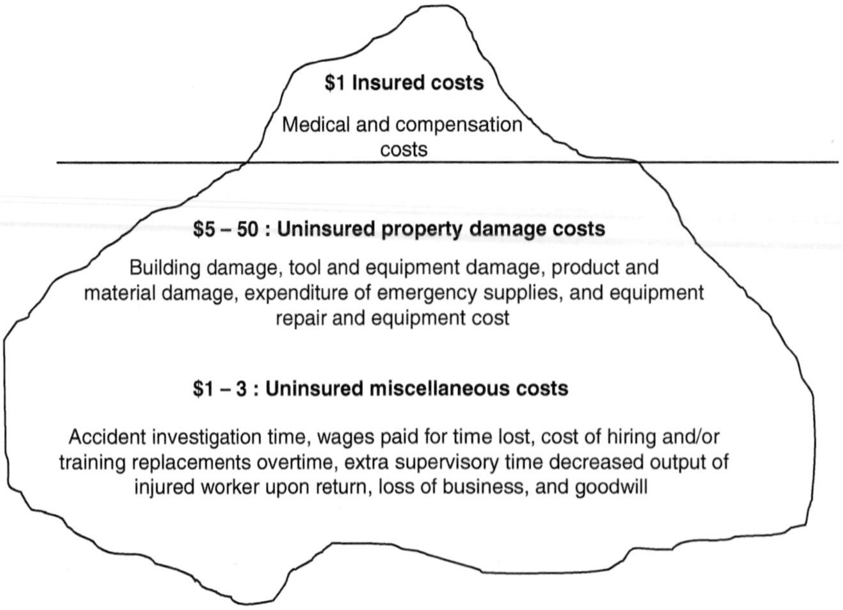


**Figure 22.3** Variation of Bird's cost of accidents data – the iceberg.

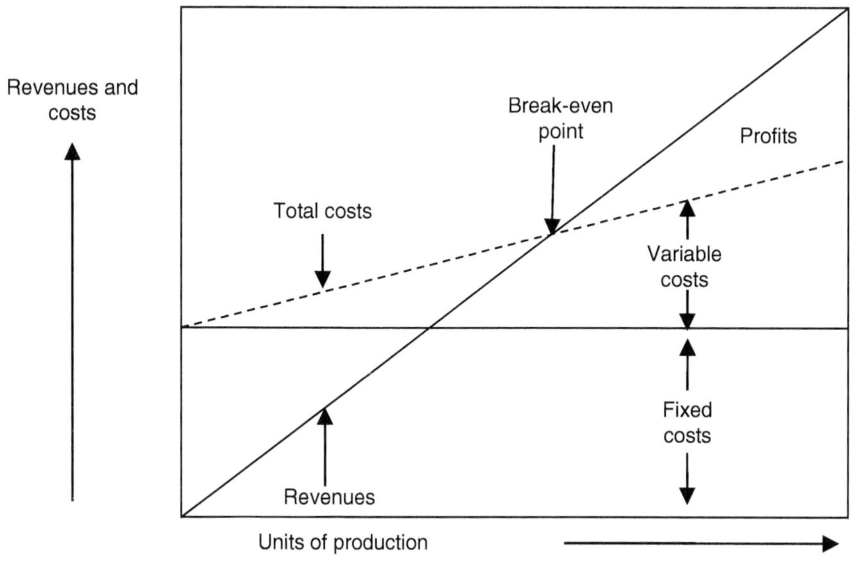


**Figure 22.4** Break-even chart.

The research done at Stanford University is important work because it establishes significant data points on indirect and direct injury and illness costs. Keep in mind that the money numbers in the report pertain to 1980/1981 – the time the research was conducted. Interesting discussion could take place about the details of the method, the definitions of terms, the content of the data collection forms, and the data collection system. Nevertheless, the work is significant because:

1. A study was actually made.
2. The report confirms the lack of data to support other published ratios.
3. Research determined that as injury severity increased, the ratio of indirect to direct costs decreased.
4. Computations indicate that, for that study, the indirect to direct cost ratio is 1.6 : 1.

That a study was actually made, following accepted research practices, is significant. And the report is available for review. It is noteworthy that the report confirms the lack of data to support other published ratios. It says: "For this aspect of the research [Hidden Cost of Accidents], much effort was expended trying to determine the source of the ratios that have appeared in various reports. No hard data was uncovered to substantiate any of the widely publicized ratios that had come to the attention of the researchers" (p. 14).

Table 22.1 is taken from Technical Report No. 260. Data in Table 22.1 supports the premise that as injury severity increased, the indirect cost ratio decreased. That premise will withstand a logic test for various ranges of injury costs.

Dividing the total of hidden costs for all claims by the total cost for benefits paid resulted in a 1.6 : 1 ratio of indirect to direct costs.

**TABLE 22.1 Analysis of Accident Costs by Size of Accident Stanford Research**

| Range of benefits paid | Number of cases | Average benefits paid | Average hidden cost | Average ratio hidden cost to benefits paid |
|---|---|---|---|---|
| *No lost time* | | | | |
| $0–199 | 13 | $125 | $530 | 4.2 |
| 200–399 | 7 | 250 | 1275 | 5.1 |
| 400 plus | 4 | 940 | 4740 | 9.2 |
| *Lost time* | | | | |
| $0–2999 | 9 | 869 | 3600 | 4.1 |
| 3000–4999 | 8 | 3694 | 6100 | 1.6 |
| 5000–9999 | 4 | 6602 | 7900 | 1.2 |
| 10000 plus | 4 | 17137 | 19640 | 1.1 |

### 22.4.4 Updating the Stanford Indirect and Direct Cost Ratios

Significant changes have occurred in the relationship between indirect and direct costs since 1980. An updating exercise was undertaken to estimate what the 1.6 : 1 average indirect to direct cost multiplier shown in the Stanford study might be as of 2015. If statisticians delved more precisely into the trending of indirect and direct costs, their computations could achieve a higher accuracy level.

Indirect costs derive primarily from the costs of time spent by persons other than the injured person, plus the cost of property damage. Assume that indirect costs increased at the same rate as inflation. The Cumulative Inflation Calculator on the Internet at http://inflationdata.com/inflation/Inflation_Calculators/Cumulative_Inflation_Calculator.aspx was used to determine the aggregate inflation increase from 1 January 1995 to 31 December 2013. It was 55.06%.

Kathy Antonello is the chief actuary at the National Council on Compensation Insurance (NCCI). Her report titled "NCCI's 2015 Annual Issues Symposium: Analysis of Workers Compensation Results" can be found at https://www.ncci.com/Documents/AIS-2015-SOL-Presentation.pdf.

From 1995 to 2013, the report shows that the cumulative increase in the "WC Average Medical Cost per Lost-Time Claim" was 226.76% (p. 48). For the same period, the cumulative increase in "WC Average Indemnity Cost per Lost-Time Claim" was 135.2% (p. 45). During that time span, the average:

- Medical cost per lost-time claim increased from $9100 to 28300.
- Indemnity cost per lost-time claim increased from $9800 to 22600.

In 1995, the average medical cost per lost-time claim was 48.15% of the total claim cost. In 2013, they were 55.6%.

Numbers in the NCCI report relate directly to direct costs of injuries. They dwarf the 55.06% increase in inflation – which relates largely to indirect costs. Computations were made to relate the 1.6 : 1 ratio shown in the Stanford report to the present. Taking into consideration the substantial increases in direct costs, the computations showed that the current the ratio of indirect to direct costs could be in the neighborhood of 0.8 : 1. That could be rounded to a 1 : 1 ratio.

### 22.4.5 On Computing the Additional Sales Necessary to Cover Injury Costs

In his *Management Guide to Loss Control*, Frank E. Bird Jr. proposed that, in reports to management, data should be included to show the additional sales required to pay for accident costs. Many organizations use adaptations of the Bird statistics showing profit levels, accident costs, and the additional sales necessary to cover the accident costs. As will be noted in the following discussion, the premise that "additional sales" are necessary to cover accident costs cannot be supported.

A composite of several variations of Bird's statistics is shown in Table 22.2. If an Internet search is made of accident costs, several similar examples will be found.

To determine the "additional sales" necessary, the accident costs are divided by the selected profit expected expressed as a decimal. In most similar exhibits, accident costs are a combination of direct and indirect costs.

**TABLE 22.2 Additional Sales Required to Pay for Accidents**

| | Profit expected | | | | |
|---|---|---|---|---|---|
| Accident costs ($) | 1% | 2% | 3% | 4% | 5% |
| 1000 | 100000 | 50000 | 33000 | 25000 | 20000 |
| 5000 | 500000 | 250000 | 167000 | 125000 | 100000 |
| 10000 | 1000000 | 500000 | 333000 | 250000 | 200000 |
| 25000 | 2500000 | 1250000 | 833000 | 625000 | 500000 |
| 50000 | 5000000 | 2500000 | 1667000 | 1250000 | 1000000 |
| 100000 | 10000000 | 5000000 | 3333333 | 2500000 | 2000000 |
| 150000 | 15000000 | 7500000 | 5000000 | 3750000 | 3000000 |
| 200000 | 20000000 | 10000000 | 6666000 | 5000000 | 4000000 |

Discussions were held on the statistics shown in Table 22.2 with an operations executive whose degree is in finance. They were highly instructive, particularly with respect to unit pricing methods and break-even charts. He said that making computations to determine the "additional sales" necessary to cover indirect and direct costs will not withstand a logic test. Points taken from the discussion, briefly summarized, are as follows:

- As a company budgets for its operations, both the indirect and direct costs of injuries (unidentified for indirect costs and sometimes identified for direct costs) are included in estimates of operating costs, and the unit prices for products or services are set to recover those costs and the profit margin expected.
- Thus, from the first dollar of sales onward, parts of the indirect and direct costs are recovered.
- When revenues equal total operating costs, as shown in Figure 22.3, a break-even chart, all indirect and direct costs incurred up to that time are recovered.
- As sales increase and operations become profitable, revenues obtained continue to encompass all additional indirect and direct costs.
- No additional sales are needed to cover indirect and direct costs.

Assume the following as an example:

- Sales are budgeted for a year – $10 000 000.
- A 4% profit – $400 000 – is expected.
- Operating costs are $9 600 000.
- Total direct and indirect costs are $300 000.
- Using the formula applied when this system is used, the "additional sales" necessary to cover total injury costs would be determined by dividing $300 000 by 0.04, the result being $7 500 000.

Senior management would unlikely accept that $7 500 000 of "additional sales" is required to cover accident costs. Since all of the indirect and direct injury costs would be contained in the budgeted operating costs, no "additional sales" are necessary.

It was suggested by the previously mentioned executive that relating total accident costs to dollars of profit could be conceptually supported and such a comparison might have significance. For example, a report prepared by a safety practitioner for management indicating that injury costs are equal to 100% of profits could get attention.

## 22.5 CONCLUSION

The literature on direct and indirect costs does not present a uniformly accepted computation method. Differences in the various systems are substantial. More importantly, no published ratios are currently valid because the increase in direct costs – indemnity and medical costs – has exceeded the increase in indirect costs substantially in the past 15 years.

Computations to update the Stanford study indicate that the ratio of indirect to direct accident costs could currently be about 0.8 : 1. That ratio is given as an approximation. Safety professionals who use a 1 : 1 ratio can be reasonably comfortable. This author recommends against using ratios for which there is no supporting data (4 : 1, 6 : 1, or 10 : 1, or higher).

In his book *Techniques of Safety Management*, Dan Petersen eloquently expressed concern over the use of an indirect to direct cost ratio for which supporting data is questionable:

> Although hidden costs are very real, they are very difficult to demonstrate. To say arbitrarily to management that they amount to four times the insurable costs is asking for trouble. If management asks for proof, you can only say, "Heinrich said so." Management wants facts – not fantasy. Without proof, hidden costs become fantasy. (p. 132)

If safety practitioners are to use an indirect to direct cost ratio, they need "proof" to support the ratios they use. At the present time, the computations shown in this chapter may be the best resource to provide such "proof." Also, from the preceding discussion, it should be obvious that computing the "additional sales" needed to cover total injury and illness costs produces invalid data.

## FURTHER READING


ANSI/AIHA/ASSE Z10-2012 (2012). *An American National Standard for Occupational Health and Safety Management Systems*. Park Ridge, IL: American Society of Safety Engineers. https://www.asse.org/shoponline/products/Z10_2005.php. Tech Brief: http://www.asse.org/assets/1/7/Z10_Tech_Brief_2012_Revised.pdf (accessed 25 August 2015).

ANSI/AIHA/ASSE Z590.3-2011 (2011). *Prevention Through Design: Guidelines for Addressing Occupational Hazards and Risks in Design and Redesign Processes*. Des Plaines, IL: American Society of Safety Engineers.

ANSI/ISO/ASQ Q9001-2000 (2000). *American National Standard: Quality Management Systems—Requirements*. Milwaukee, WI: American Society For Quality.

Antonello, K. (2015). *NCCI's Annual Issues Symposium: Analysis of Workers Compensation Results*. Boca Raton, FL: National Council on Compensation Insurance. https://www.ncci.com/Documents/AIS-2015-SOL-Presentation.pdf (accessed 25 August 2015).

Australian Safety and Compensation Council (2006). *Guidance on the Principles of Safe Design for Work*. Canberra, Australia: Australian Safety and Compensation Council.

Bird, F.E. Jr. (1974). *Management Guide to Loss Control*. Loganville, GA: Institute Press.

Bird, F.E. Jr. and Germain, G.L. (1985). *Practical Loss Control Leadership*, Revised Edition. Loganville, GA: Det Norske Veritas.

BP. BP Issues Final Report on Fatal Explosion, Announces $1 billion Investment at Texas City. http://www.bp.com/en/global/corporate/press/press-releases/bp-issues-final-report-on-fatal-explosion-announces-1-billion-investment-at-texas-city.html (accessed 16 July 2015).

Business Round Table (1981). *Improving Construction Safety Performance: The User's Role*, Technical Report No, vol. **260**. Palo Alto, CA: Department of Civil Engineering, Stanford University. (This report resides in a storage facility at the Department of Civil Engineering at Stanford University. It was made available through an interlibrary loan service.).

Center for Chemical Process Safety (2003). *Guidelines for Investigating Chemical Process Incidents*, 2e. New York: Center for Chemical Process Safety of the American Institute of Chemical Engineers.

EU-OSHA (2008). Risk Assessment. European Community. https://osha.europa.eu/en/healthy-workplaces-campaigns/previous-healthy-workplaces-campaigns (accessed 22 August 2015).

ExxonMobil (2009). Operations Integrity Management System. http://corporate.exxonmobil.com/~/media/global/Brochures/2009/OIMS_Framework_Brochure (accessed 26 August 2015).

Heinrich, H.W. (1931). *Industrial Accident Prevention: A Scientific Approach*. New York: McGraw-Hill Book Company, 1941, 1950, 1959.

Inflationdata.com Cumulative Inflation Calculator (2018). http://inflationdata.com/inflation/Inflation_Calculators/Cumulative_Inflation_Calculator.aspx.

ISO 14001:2004 (2004). *Environmental Management Systems: Requirements with Guidance for Use*. Geneva, Switzerland: International Organization for Standardization.

James, R. (1990). *Human Error*. New York: Cambridge University Press.

Johnson, W. (1980). *MORT Safety Assurance Systems*. Itasca, IL: National Safety Council (Also published by Marcel Dekker, NY).

Manuele, F.A. (2013). *On the Practice of Safety*, 4e. Hoboken, NJ: Wiley.

Manuele, F.A. (2014). *Incident Investigation: Our Methods Are Flawed*. Park Ridge, IL: American Society of Safety Engineers. *Professional Safety*.

Manuele, F.A. (2014). *Advanced Safety Management: Focusing on Z10 and Serious Injury Prevention*, 2e. Hoboken, NJ: Wiley.

Petersen, D. (2003). *Techniques of Safety Management*, 4e. Park Ridge, IL: American Society of Safety Engineers.

Stephans, R.A. (2004). *System Safety in the 21st Century*. Hoboken, NJ: Wiley.

Stephenson, J. (1991). *System Safety 2000*. New York: Van Nostrand Reinhold.

The Business Roundtable (1982). *Improving Construction Safety Performance*. Report **A-3**. New York: The Business Roundtable. Enter the title into a search engine on the Internet to access the paper (accessed 25 August 2015).

United States Department of Labor (1970). *Occupational Safety and Health Act of 1970*, Public Law **91-596**. Washington, DC: Department of Labor. https://www.osha.gov/pls/oshaweb/owasrch.search_form?p_doc_type=oshact (accessed 15 August 2015).

Whittingham, R.B. (2004). *The Blame Machine: Why Human Error Causes Accidents*. Burlington, MA: Elsevier Butterworth-Heinemann.

CHAPTER 23

# SAFETY AND HEALTH MANAGEMENT OF INTERNATIONAL OPERATIONS

S. Z. MANSDORF

*7184 Via Palomar, Boca Raton, FL, 33433*

This chapter is a revision of Mansdorf, S.Z., "Management of international EH&S programs," Chapter 45, in *Patty's Industrial Hygiene*, Vol. 4 (ed. V. Rose and B. Cohrssen). Hoboken: Wiley Interscience, 2011.

The responsibility for providing safety and health management of international operations whether stationed in the United States or elsewhere can be challenging but also very broadening. Challenges include operating in a different organizational structure, different management philosophies and approaches, different cultural and societal characteristics, language barriers, inadequate infrastructure in some countries, and finding local technical resources. Conflicts in approaches to management can be especially challenging where there are differences in compliance cultures.

Recognizing the limitations and benefits of organizational structures is the first step in deciding how to best manage a program. Companies that are centrally controlled lend themselves to a more centralized environment, health, and safety (EH&S) structure, while those that are more division, product, or geography based are more likely to have resources at the division or zone level with a smaller corporate structure. Establishing the best balance of centralized versus local or regional support is important in far-flung international operations.

Good governance in EH&S is viewed by most NGOs as requiring a management systems approach with ISO 14000 and BSI 18000 or equivalent accreditation as the minimum. Best practice for EH&S is to also have company standards for key EH&S aspects (e.g. occupational hygiene, waste management, fire protection, etc.) that are strictly applied worldwide regardless of any lack of local legal requirements or enforcement.

Cultural, societal, and language differences can present a range of challenges from the very minor to major obstacles. Some cultures favor strict adherence to rules, while others view rules as guidance as one example of culture differences. The ES&H rules in countries and regions themselves vary widely in terms of content, coverage, and enforcement. Physical stature and weight can be quite different in different parts of the world. Western societies tend to have much heavier and taller workers than Asian societies as one example. This can have a significant effect on workplace design and ergonomic considerations such that machine design may need to be altered rather than standardized. Cultural and societal differences also promote differences in business practices. This can result in more risk taking and leaner operations versus less risk taking and more staffing for the same or similar operations. Finally, language differences require a considerable amount of energy at making sure that the translation of the message is properly understood in all parts of the world.

Effective management of EH&S in the less developed world requires consideration of technical support issues such as a lack of highly qualified local technical consultants, accredited laboratories, and infrastructure (e.g. clean water, waste collection, wastewater treatment, etc.). This is also complicated by differences in work schedules, sometimes drastic differences in time zones (e.g. Asia and the United States), and lack of local human resources, requiring significant investments in talent development.

Identification, collection, and publishing key performance indicators (KPIs) for ES&H are a hallmark of all performance management programs. This follows the tenant of Peter Drucker's "What Gets Measured, Gets Done."

*Handbook of Occupational Safety and Health*, Third Edition. Edited by S. Z. Mansdorf.

Typically, this means 20–30 safety and industrial hygiene indicators and about 40–50 environmental indicators collected and consolidated monthly. This not only permits trend analysis and comparison to goals but also promotes a healthy competition among the reporting sites.

The nature of designing and implementing global programs leads to many new challenges but with an equally significant richness of diversity to allow one to reap the benefit of different ways of thinking and promoting the best of the best regardless of origin.

## 23.1 INTRODUCTION

The global nature of business is well recognized and commonplace. The most recognized brands can be seen in almost every country across the globe with business-to-business suppliers equally well distributed. It is common to have manufacturing in the country where the product is sold due to tariffs and trade restrictions even when the brand is "foreign."

The Fortune 500 list in 2016 showed that the majority of global companies are based in the United States (134), followed by China (103), Japan (52), France (29), Germany (28), and the United Kingdom (26) as the six leaders (Wikipedia 2016). All of the countries decreased in the number of companies represented except China from 2015. It is likely that a safety and health professional may find employment in one of these global giants or other companies that operate on multinational scale. This could be for a US-based multinational or a foreign-based multinational with operations in the United States. In 2010, multinational companies operating in the United States employed over 28 million Americans (Slaughter 2013). It is also quite common for companies to send promising professionals to overseas headquarters or on assignment in other countries to broaden their perspectives.

Working for one of these multinational companies requires a diverse range of technical and management skills. This may include the ability to work in not only safety and industrial hygiene but environment as well. This is most commonly the case in the less developed countries. Therefore, we would use the term ES&H with an emphasis on safety and health in the rest of this chapter.

There can be many challenges in the management of safety and health across borders. There are significant differences in regulations, regulatory compliance, infrastructure, cultures (company and personal), and social practices as well as the challenge of language barriers. Nevertheless, with proper planning, execution, and key performance metrics approaches, a rewarding and successful program can be developed.

## 23.2 COMPANY STRUCTURE VARIATIONS IN GLOBAL OPERATIONS

Corporate governance structures vary widely for US- and non-US-based companies. Companies may be organized with centralized control of staff and support functions (meaning large numbers of support personnel and functions) or a more decentralized corporate role where many of these functions are handled at the division level. In some cases, the corporate role may be as a holding company for many diverse subsidiary companies. This is important to the person responsible for safety and health since it will determine their organization of managerial and technical support.

Companies are typically structured geographically or by brand or product. Chemical companies sometimes organize by functional categories or end use such as surfactants. Some companies have strong corporate oversight on all operations (common in the United States), while other companies operate their businesses quite separately with the corporate office more of a "holding" company model, which is most common outside of the United States. These differences in corporate structure will largely determine the basis for organization of EH&S on a structural basis.

### 23.2.1 Geographical or Product Organization or Both

Geographical organization typically results in a division or zone that is devoted to the United States and sometimes North America or North and South America. Other common zones or divisions include Europe or quite commonly Europe, Middle East, and Africa (EMA), Asia, South America or Latin America, and Oceania Pacific (India, Australia, New Zealand, etc.). If companies are organized by product, brand, or chemical category, then they will have these zones or geographic divisions within the product categories. These organizational structures become important when the management of programs, corporate standards, and technical support are considered. All organizational structures require resources at the zone or division level. The difficultly is that normally the geographic and product organizations both favor the developed world (North America and Europe), since they are the most developed markets; hence sales, staffing, and resources are normally largest in these areas. Generally, staffing tends to be much less, especially EH&S staffing, in the less developed markets where sometimes the safety and health needs may be greater. The issue then becomes whether to staff centrally (corporate staff) to support the distant zones or to rely more on divisional or zone staffing with a smaller corporate staff. Minimized corporate staffing is the current trend influenced in part by the expense and difficulties in providing support centrally. This trend is further complicated by the lack of qualified personnel in the less developed world. In the past, the trend has been to use expatriate in the zones and divisions outside of the host country in this role. The most current thinking on "human capital" and corporate social responsibility (CSR) favors local hires over the difficulties and expense of expatriates. The problem this presents is finding sufficiently competent personnel. There is a further complication in retaining people in whom significant training and other resources have been invested due to increasing demand from other companies' venture in less developed countries and search for technical talent. This is especially a problem in Asia where it is common for technically competent employees to receive multiple offers of employment weekly.

Geographical divisions (e.g. Asia-Pacific) offer the most opportunity for ease of staffing since these centers tend to be largest, while product divisions are most challenging since the staffing in the central locations tends to be smallest (depending on the size of the product business).

Another model for staffing is the "shared services" model. In EH&S, this model is problematic because of the funding of these operations on a payment for services or charge-back basis. Aside from the governance and accounting issues, this type of arrangement typically leads to complaints about the lower costs available to the business from local consultants that do not have the policy overview and the expertise of long-term internal employees.

### 23.2.2 Centralized Versus Decentralized Control

Most companies will have an organizational culture, which either provides central direction or acts more like a holding company. There are many examples of the two extremes with most companies something of a hybrid of the two approaches. All companies exercise some centralized control over their subsidiaries (joint ventures being another approach). Some of the largest European companies have very large divisions or businesses that are essentially run independently. The United States has many examples of more centralized companies. The importance of this organization culture is the staffing of a centralized EH&S structure and the implications of a centralized set of operating rules for EH&S. Those that have fairly independent divisions have most of their staffing and operational requirements at the division level, while those that are more centrally controlled generally have more resources at the corporate level. Many companies operate on a matrix approach. This typically results in the EH&S competence leaders being located in the different divisions rather than in a central location. As noted previously, this structure can suffer from a lack of balance commonly found between the divisions with the largest ones having the most resources. There is a natural tendency for the most support to be in the division where the EH&S person is located. Ideally, a strong central core of experts is desirable with adequate staffing at the division or regional levels. Since most organizations will have been long established, the ability to change the support structure for EH&S may be limited, but over time a more balanced approach may be possible.

Whatever the organization structure, there is a need for a management systems "backbone" for the overall EH&S policy and programs.

## 23.3 MANAGEMENT STANDARDS AND SAFETY AND HEALTH PROGRAMS

Management standards can serve as an internationally recognized starting point for environmental, safety, and industrial hygiene programs. The best known of the standard setting organizations is the International Organization for Standardization (ISO).

ISO was founded in 1947 to help promote international trade through the internationalization of standards, specifications, and guidance. ISO is made up of more than 157 country members (with their own individual standard organizations). They have developed and published several "management system" standards as part of this effort. This includes ISO 14001 and others under development such as a standard for CSR. The ISO 14000 series management standard emerged as a consequence of the GATT trade negotiations and the RIO Summit on the Environment held in 1992. The first version of the Environmental Management Standard (14001) was adopted in 1996 with the latest version published in 2015 (Slaughter 2013). Some companies have fully integrated their safety and health aspects within this standard. Registration of companies to this standard or its equivalent is widely regarded by most outside stakeholders as evidence of commitment to responsible environmental management. Some countries provide further incentives for those holding this certification relative to permits and inspections.

Health (industrial hygiene) and safety as a management standard is just now being finalized under the ISO banner (ISO 45001 is in the final stages of adoption). Earlier, there were many earlier attempts, but because of the implications of labor regulation and other issues, they failed (Mansdorf 2007). As a consequence, the British Standards Institute (BSI) published their guide to Occupational Health and Safety Management System in 1996 and their standard OHSAS 18001 in 1999. This was followed by the International Labor Organization (ILO), which also published a voluntary guide for Occupational Safety and Health Management Systems (ILO 2001). The BSI standard has since been amended with the latest version published in 2007 (BSI 2007). Initially the American Industrial Hygiene Association under the banner of the American National Standards Institute (ANSI) published their ANSI/AIHA American National Standard for Occupational Health and Safety Management Systems in 2005 followed by the American Society of Safety Engineers taking over the Secretariat and revising the standard in 2017 (ANSI Z10-2012-R2017) (ASSE 2017).

The BSI standard has been aggressively marketed and adopted as a global rather than national standard of practice. It does not differ significantly from the US National Standard Z10 but has gained international acceptance by most nongovernmental organizations (NGOs) and CSR analysts. This has resulted in most large multinationals adopting the quality (ISO 9000 series), environment (ISO 14000 series), and occupational safety and health (OHSAS 18000 series) standards as evidence of good governance and programs in these areas. The ISO 14000 and OHSAS 18000 (or Z10) series do represent a good basis for an internationally accepted management system for EH&S. These management systems are best implemented in an integrated approach (either aligned or integrated) using similarities (all are essentially based on a plan–do–check–act and continuous improvement approach) for structuring the systems. Some companies have mapped their management approaches against the ISO and BSI standards to receive "equivalency" certifications to be able to state their programs comply with these management standards. Other approaches include individual site accreditations versus a company accreditation. It is still too early to judge what the impact of a new ISO management system standard for safety and health will mean (the proposed ISO 45001). However, it is more prescriptive than the OHSAS and Z10 versions and includes much more on risk assessment. There is also the possibility that the burden of the implementation of these standards will ultimately lead to their downfall. There are a number of major companies that have already decided to use these management systems approaches as guidance without formal certification or verification by outside registry bodies.

Table 23.1 shows a comparison of the key elements of ISO 14001 and OHSAS 18001. There are differences in some of the individual subsections, but the overall approach is similar.

### 23.3.1 Other Government-Sponsored Management Systems Approaches

The Occupational Safety and Health Administration Voluntary Protection Program (OSHA VPP) has a similar management structure to OHSAS and is the "system" most recognized by US-based stakeholders as showing commitment to occupational health and safety. However, it is not well recognized in most of the rest of the world. Other management systems exist such as the DNV ISRS system and Responsible Care Management System (well recognized in the chemical industry), but they also lack recognition by the lay public and key stakeholders such as CSR investor groups as a benchmark. Likewise, the various voluntary programs under the US Environmental Protection Agency are also not well recognized internationally.

The overall effectiveness of the OHSAS and environmental management standards is beyond the scope of this chapter. Nevertheless, they are widely recognized as the baseline management systems approach on a global basis, and

**TABLE 23.1 Summary of the Key Elements of ISO 14001 and OHSAS 18001**

| Section (clause) | ISO 14000 | OHSAS 18000 |
|---|---|---|
| 1 | Scope | Scope |
| 2 | References | References |
| 3 | Definitions | Terms and definitions |
| | Environmental management system – Requirements | OH&S management system requirements |
| 4 | 4.1 General requirements | 4.1 General requirements |
| | 4.2 Environmental policy | 4.2 OH&S policy |
| | 4.3 Planning | 4.3 Planning |
| | 4.4 Implementation and operation | 4.4 Implementation and operation |
| | 4.5 Checking and corrective action | 4.5 Checking and corrective action |
| | 4.6 Management review | 4.6 Management review |

as a consequence many major companies have either adopted them or had their own systems determined to be of equivalence. The biggest advantage for the person responsible for multinational operations is a consistent overall approach to safety and health.

Once the overarching management approach is developed, the next issues are the program elements and rules within the management framework. A key question confronting someone with a global responsibility for EH&S is what standards to apply. This may seem simple but is actually a very complex and a challenging problem. Ethically, is it reasonable to have workers in less developed countries with no standards or weak standards at more risk than those in developed countries? Simply applying US-based regulations will not work either given the extent of them.

## 23.4 LOCAL VERSUS GLOBAL STANDARDS

The United States is one of the most highly regulated countries in the world. The number of pages in the Federal Register was estimated to be around 180000 at the end of 2015 by the Regulatory Studies Center for the George Washington University (https://regulatorystudies.columbian.gwu.edu/reg-stats). While the exact number of pages of the Federal Register devoted to Occupational Safety and Health General Industry rules is difficult to determine, it is likely about a thousand pages. Beyond the written requirements (federal, state, and local), the United States has an enforcement mentality that is not shared by much of the rest of the world. This is also generally true for most of the English-speaking developed countries (e.g. Canada, United Kingdom, and Australia). Europe has an equal body of regulations, but the enforcement approach is quite different. Eastern Europe has very limited enforcement, while the "Latin" countries have more of a negotiation approach. Many countries lack the written rules and qualified inspectors necessary for an effective enforcement scheme. Because of these differences, it is quite common for companies that are US based to establish standards and requirements that reflect the US situation. Nevertheless, the issue becomes whether to "enforce" company standards uniformly on a global basis or to rely solely on local standards or both. On an ethical basis, workers should be equally protected no matter where they work. Said another way, entry into a confined space should be protected in the same way regardless of the geographic location. While this sounds logical and simple, it is quite difficult to actually implement. It requires the development of internal standards that can take considerable time and effort to publish in local languages and to find the training resources to carry out this approach. Many of those with global EH&S programs have established a core set of requirements in the most critical risk areas. Although it takes a considerable amount of effort and time, this can be done without regard to the specifics or lack of local regulatory requirements. Those companies with absolute standards (versus guidelines) require local legal compliance or adherence to company standards based on whichever provides the most protection. Local legal standards where conflicts exist would obviously always be followed. However, this is rarely a problem as most very large company standards are far more stringent than local rules in the less developed countries. Most international companies develop between 20 and 100 internal standards. Commonly, they include:

- Biological safety.
- Contractor safety.
- Confined space entry.
- Electrical safety.
- Fire protection.
- Hazardous chemicals.
- Hazard communications.
- Hazardous energy control (lockout/tag out).
- Hazardous waste management.
- Hazardous work permits and procedures.
- Incident and injury reporting.
- Machine guarding.
- Occupational hygiene.
- Process safety.
- Personal protective equipment.
- Powered industrial trucks.
- Reproductive hazards.

- Spill prevention and emergency response.
- Transportation of dangerous goods.
- Waste management.
- Air emission permits.
- Wastewater discharge permits.

The above list is not exhaustive but rather illustrative of global standards that are applied by some multinational companies.

## 23.5 CULTURAL, SOCIETAL, AND LANGUAGE DIFFERENCES

There are obvious and considerable differences in individual countries and regions based on culture and societal aspects of work. One example is the difference in average height and weight between Americans and Asians, which has obvious ergonomic implications. Another example is the number of hours of work per year (Asia having the most hours, especially in Korea, and Europe, especially Central Europe and the Nordic countries, having the least), which has chemical exposure implications. Sometimes these differences can be subtle such as the value of environmental measures (such as the handling of waste) or risk taking. It is not possible in this chapter to detail all of these differences, but some examples can be given to highlight the effects on ES&H programs.

A study was done in 2008 by Swiss Re on differences in the petrochemical industry based on their insured inspections worldwide (Straub and Zirngast 2008). They made some general conclusions that are quite interesting and pertinent to EH&S. They found that the developed English-speaking countries (United States, Canada, United Kingdom, and Australia) had much leaner staffing than the less developed countries and a much more aggressive push to productivity. They also found that maintenance was more minimized and driven by automated systems (breakdown versus preventive) rather than time based. Likewise, they found the highest use of contractors versus employees for maintenance and other work in the English-speaking countries. Countries in South America, Africa, Middle East, and less developed Asian countries (not Japan) were characterized as having a "conservative and careful" approach to operations, the lowest use of contractors, the least number of changes during operations, and a time-based maintenance system. They found that attitudes toward safety were company specific and focused on personal safety in the less developed countries versus more compliance-driven approach in the English-speaking countries. While these comments are quite broad and generalized, their loss ratios (costs of losses versus premiums) suggest that the less developed countries have a lower risk profile than the more regulated and advanced English-speaking countries. Europe, developed Asia (e.g. Japan and South Korea), the Gulf States, and Eastern Europe fall somewhere in between these differences. These conclusions are quite opposite of what one might suspect.

Personal experience has demonstrated that cultural differences result in much higher injury accident rates in Europe, especially Central and Southern Europe than in most of Asia, especially Japan and China. The reasons for this are not universal and probably multifactorial. However, it would seem that the attitude of workers that feel accidents are a personal responsibility and reflect a personal fault in Asia is quite different than the more European feeling that personal safety is more of a company responsibility and not the fault of the worker. Historically, it is common for workers in Europe to receive a typical two-week "injury recuperation" period from their personal doctor for any injury including very minor ones, while Asian workers tend to stay at work or return as quickly as possible. Many international companies have implemented "light-duty" policies that have helped to drive down these differences. Lost-time statistics alone will not demonstrate these differences. Use of a metric for light duty (restricted workday cases) and something similar to the OSHA recordable statistic can help to pinpoint these trends. Many large multinational companies headquartered outside of the United States use this approach although the equivalence is not required in most countries for legal reporting requirements. One additional factor to be noted is that it is common in the United States to track worker compensation costs as a metric. However, most countries with socialized medicine systems do not share this data.

There are many more apparent and subtle differences in behavior driven by societal and culture differences that can be characterized by region. These differences can also be intellectual. The Europeans are much more focused to "risk assessments" and control banding approaches than the Americans, as one example. Practically, this means much more effort at doing individual risk assessments of work situations or jobs than air sampling or documentation in Europe. Their approach is more of a "modeling" one rather documentation such as individual air monitoring. The European regulation titled Registration, Evaluation, Authorization, and Restriction of Chemicals (REACH) is a perfect example. The supplier is responsible for determining "safe" exposures and control measures for exposures to the downstream users for characteristic uses. Many of the recommendations will be based on a complex risk modeling software rather than field measurements. While the overall concept is laudable, it ignores the individual differences in the uses of chemicals that are sure to be found in individual companies and among individual workers.

Effective communication in multiple languages and dialects can be a very significant obstacle to effective management of a global program. It is quite common in the large multinational companies to have expatriates in their remote locations in a few key management positions. It is also common for these expatriates to have knowledge of the local language (obviously depending on the length of the assignment). This can facilitate the translation of some documents and provide aid whenever there are audits or other visits by someone from outside the region. However, programs are best implemented and monitored in the local language. English tends to be the universal business language, but as a second or third language for some, there can easily be misinterpretations especially at the supervisor or worker level. Additionally, there may be some situations where there is no direct translation of a term or intent from the headquarters language to the local language. There are also many locations where the majority of workers are foreign nationals (also called guest workers) working in that country (e.g. Saudi Arabia, Kuwait, and other Gulf States). This situation exists in much of the world today including the United States where there are large numbers of native Spanish speakers in some areas. In this case and where there are multiple languages or dialects at the same site, photographs, graphs, and pictograms should be used whenever possible. Additionally, this may complicate hazard communications where the Globally Harmonized System requires labeling and safety data sheets to be in the local language (should be the local language of the workers and users and not necessarily the country of location).

Translation is widely available even in the most remote locations (although the actual translation may be performed elsewhere) for documents. Nevertheless, it is very difficult for the originator of the policy or document to know for sure that the intent is the same in the translated document. For visits by teams for audits and other such measures, use of a local consultant that is fluent in native language of the visitor and the local workers can be helpful.

Ideally, the best global program will take advantage of the cultural, societal, and intellectual differences where they offer a benefit.

## 23.6 REGULATIONS

Regulations for EH&S and their enforcement vary considerably from country to country as might be expected. Anglo-Saxon regulations tend to be similar, but not the same. The example is the United States and Canada. The EH&S laws generally cover the same topics but have differing requirements and are organized much differently (the provinces and territories have more of a role in Canada).

European regulations are common where they are part of the European Union (EU) regulatory scheme; however, individual country differences can be significant. The EU regulations are not inclusive in the EH&S realm but cover some of the most important aspects. This general scheme has been further complicated by the Brexit (the United Kingdom leaving the EU and hence their requirements for EH&S). The rest of the world can be characterized by regional similarities, but with country differences. Enforcement can vary widely depending on the regulatory culture and level of sophistication of the inspectors. In some locations, there is strict enforcement of regulations for multinationals while essentially ignoring local business (sometimes viewed as being disadvantaged). Additionally, there are some countries where enforcement is seen as a way to exert authority over the multinational for something they want that is unrelated to the actual violation. Finally, it is common in most of the world to exceed the local legal requirements (especially environmental) with the full knowledge of the local authorities who consider the violation insignificant. In these cases, they will ignore the violation but will not provide any written verification of this derogation. What is universal is the application of regulations after any incident of significance that is public. This is true in the developed English-speaking world but even more so in the less developed countries. There are many examples of criminal charges being filed after an incident even though the authorities were well aware of the violations. It is also evident that multinationals are held to a higher standard than the local companies in most economically underdeveloped countries. ISO 14000 and OHSAS 18000 both require knowledge of legal requirements and programs for compliance (as will the new ISO 45000). Ideally, each operation and location would be required to stay current with all local requirements. This can be difficult in some locations and typically requires the use of local attorneys with dual reporting responsibilities (local and central EH&S structures). Many organizations use an "updating" service at the regional, division, or corporate level. There are several services available that can provide the regulations, changes proposed, and interpretation on what is required for most of the major countries of the world. Enhesa provides a quarterly updating summary service without cost (www.enhesa.com). Country-specific alerts and monthly reports are available on a fee-for-service basis from them and other providers. As might be imagined, the volume of new regulations and changes is so large that it is more a matter of alerting the affected locations than one of analyzing each regulation or change centrally.

## 23.7 TECHNICAL SUPPORT SERVICES AND INFRASTRUCTURE

The professional support systems that take for granted in the modern industrial world to facilitate our personal and professional lives will not be present in much of the rest of the world. This includes evaluation of soil or water contamination, high-quality technical support for issues related to fire and explosion risks, exposure monitoring sampling and analysis, and environmental fate and biological monitoring, to name a few. The obvious solution is to simply ship the sample or import the technical expertise. However, there are a very large number of obstacles to overcome such as customs duties, export restrictions, right-to-work permits, time-to-analysis constraints, and a large number of other issues. Mutual aid pacts and professional courtesies with companies that are already established in the area can be very beneficial. Fire protection is a simple example. The BRIC countries (Brazil, India, China) all lack local fire pumps, standby diesel engines, and piping that meet typical US standards. They all have local fire protection engineering standards, but they are not at present as stringent or specific a level as the National Fire Protection Association (NFPA) standards followed by most large companies in the United States or FM Global here and in other countries. If NFPA or FM Global fire protection standards are adopted worldwide, this can result in significant additional costs to import pumps and engines, for example, as import duties must be paid.

Infrastructure problems can have a big effect in the management of company programs. For example, there may be no wastewater treatment capability and no receiving stream where you could build a wastewater treatment station. In this situation, wastewater may need to be trucked to the nearest adequate facility for treatment (underground sewer piping is usually not present and is usually not a viable solution due to local housing). Recycling and the prohibition of land disposal are a common goal for many progressive companies. However, local infrastructure constraints may severely restrict any opportunity for this (e.g. only landfill and no recycling capacity). The best solution for this situation is for companies to build third-party start-up companies to carry this forward or to act together in establishing the needed infrastructure. This may include the need to prevent the direct recycling of some waste streams and solid wastes. For example, it is common in less developed countries for the local populations to use discarded 55-gallon drums and other chemical containers. This could obviously result in community issues, and hence the waste containers must either be carefully decontaminated or tightly controlled to their point of destruction.

## 23.8 TIME ZONES AND WORK SCHEDULES

An obvious but sometimes overlooked issue is the time differences between the "home office" and the local sites. Most dramatic is the time difference between Asia and the United States. It can vary from 12 to 14 hours. This greatly complicates communications by phone and video conference, especially when it involves Asians, North Americans, and Europeans at the same time. Communications with Europe from the United States are much less difficult, but still problematic. The difference can range from five hours (United Kingdom) to more than eight hours (Russia). Typically, at least one of the parties involved will be inconvenienced. Work schedules and holidays are another issue. One dramatic example is the summer holidays in Europe. From about the middle of July until the middle of August, most management in Europe is gone. Late morning start times and late evening departures in Europe versus the reverse habits in much of the United States are another example of differences. Effective planning is the best solution to these two difficulties. For emergencies, it is common to communicate via mobile, e-mail devices, or smartphone by e-mail or to have a list of mobile phone contact numbers established in advance.

## 23.9 HUMAN RESOURCES

Industrial hygiene is a well-developed professional career as is safety and environmental science and engineering in the United States. This is also generally true for the developed English-speaking countries. The international arena is quite different. The most abundant personnel available are those with an engineering background but typically with little experience in safety, industrial hygiene, or environment.

Specialized environmental areas (modeling, wastewater engineering, emissions control systems, etc.) are more difficult to staff. Of all the fields, industrial hygiene is one of the most difficult to fill in the less developed countries.

Industrial hygiene (occupational hygiene) is not well recognized in most countries outside of the English-speaking ones even though there may be some level of professional education available. This is especially true where there are social programs governing medical care and where there are legal requirements to have company medical doctors or nurses on-site based on the number of persons employed. In most of these countries, the physician becomes the key person responsible for occupational health, and the role of the hygienist tends to be diminished since most of them place their emphasis on treatment rather than prevention. This is true in much of the Spanish-speaking world, for example. When the profession is not well recognized by the government and other professionals, it tends to be driven by multinationals using staffing models that are used in countries where the industrial hygiene profession is more prominent. The problem is the staffing of a reasonable technical skill level of capability at a local, country, or regional level when there are no "native" resources from which to choose. There are several options – none of which are very good. The first option is to fill the position with an expatriate. This is most effective when one can find a native that is interested in returning to their country or region. It is least effective when the person has limitations of one type or another (spouse, children that will need special schooling, medical conditions, age that will limit the assignment, etc.). In some areas of the world, technically competent persons returning to their home area can command salaries that are equivalent or higher than in First World countries even without the other associated expatriate costs. The second option is to provide training to a technically oriented person. This can be a good option but is very long term and requires interim support while the candidate gains skills and knowledge.

Some companies even sponsor development programs for longer-term employees to attend colleges and universities in their native country or other countries to develop the specific expertise needed. This usually means a multi-year commitment and significant financial resources. The third option is to provide support through consultants (preferably local). This can be a good interim solution but is not at all viable for the best of programs in the long term, since there is no inherent skill and knowledge retained in the organization. It also risks the danger of "not knowing what we don't know."

The situation with safety and environmental science or environmental engineering professionals is a bit better (they are easier to find in both the developed and less developed world). Nevertheless, they are also quite hard to find in some locations. The solutions are the same as those for the hygienists.

## 23.10 ESTABLISHING KEY PERFORMANCE INDICATORS

KPIs are critical to success. All modern businesses rely on performance metrics. The old saying "What Gets Measured, Gets Done" (Peter Drucker) is absolutely true. Sustainability and CSR reporting requires public reporting of these KPIs. There are several guides for which KPIs should be collected and publically reported in the EH&S arena. These are helpful even if the organization decides to only use the metrics internally. One of the most recognized guides is the Global Reporting Initiative (GRI), which is currently in version 4 (https://www.globalreporting.org/Pages/default.aspx). The GRI covers the range of broad sustainability issues. For a narrower focus, the Center for Safety and Health Sustainability (CSHS) has recently published a best practice guide to safety and health in sustainability reporting (Mansdorf 2016). Whether the performance indicators are published externally or just used internally (or a combination of the two), they should be established and collected monthly or more frequently so that trends can be quickly analyzed and corrective actions established as needed. It is recommended and common for companies to collect 20–30 safety and industrial hygiene indicators and 40–50 environmental indicators. Some companies collect many more KPIs. These can be both retrospective and prospective. Common categories of indicators are as follows.

Safety and industrial hygiene

- Injury and illness rates.
- Persons at risk (related to exposure levels such as persons exposed above the OEL).
- Training hours.
- Observations (for behavior-based systems).
- Number of unresolved action items in safety and industrial hygiene.
- Fines and actions.

Environment

- Energy use.
- Water use.
- Wastewater (e.g. chemical oxygen demand, biological oxygen demand, dissolved solids, etc.).
- Waste (all aspects including recycling).
- Disposal methods by weight or volume.
- Air emission (e.g. SO2, NO*x*, CO2, etc.).

- Spills.
- Community complaints.
- Fines and actions.
- Reduction in use of specific chemicals.

A word of caution about these indicators is appropriate. It is very common for one to spend an inordinate amount of time assuring the correct understanding and interpretation of exactly what each indicator is based upon. It is critical that the definitions be simple and comprehensive. Even with that said, there will be issues. Waste provides a good example. What is a "waste"? Does it include returnable packaging? Does it include construction waste such as dirt that actually goes back to the ground? Does it include pallets that are reused? A simple definition might be anything that is not in or part of the finished product or intermediate product that leaves the site of manufacture. Even with this simple definition, there will still be questions such as the ones mentioned above. If there is any doubt about the difficulty this can present, one only has to look at the definition of a reportable accident under OSHA, which is the size of the average textbook.

All of these indicators can be indexed. Common indexes (or ratios) are hours worked, number of products produced, tons of product produced, sales, and others. This allows for reporting eco-efficiency measures (such as grams of waste per product produced).

It is common for these indicators to be consolidated using a wide variety of computerized databases. Most are Internet- or intranet-based systems. The "rolled-up" data permit monthly reporting by sector, country, region, division, and company. Wide distribution across the company to all those involved including the senior management enhances goal monitoring as well as friendly competition across the sites.

Effective management of international EH&S programs can be a challenging but immensely rewarding experience. Differences provide great opportunity to learn from others and to take advantages of the best practices whatever their origin.

## REFERENCES

ASSE (2017). *American National Standard for Occupational Health and Safety Management Systems*. ANSI/ASSE Z10-2012 (R2017). Park Ridge: ASSE. http://www.asse.org/ansi/asse-z10-2012-r2017-occupational-health-and-safety-management-systems (accessed 6 January 2016).

BSI (2007). *Occupational Health and Safety Management Systems: Requirements*. BS OHSAS 18001:2007. London: British Standards Institute. www.bsi-global.com (accessed 3 January 2016).

ILO (2001). *Guidelines on Occupational Safety and Health Management Systems: ILO-OSH 2001*. Geneva: International Labour Office. http://www.ilo.org/public/english/protection/safework/managmnt/guide.htm (accessed 3 February 2015).

Mansdorf, Z. (2007). The ISO Man Cometh, Almost! *The Synergist* (May 2007): 70.

Mansdorf, S.Z. (2016). *Best Practice Guide for Occupational Health and Safety in Sustainability Reports*. Park Ridge: Center for Safety and Health Sustainability, ASSE. www.centershs.org (accessed 3 February 2016).

Slaughter, M.J (2013). American companies and global supply networks. *Business Roundtable* (January 2013).

Straub, U. and Zirngast, E. (2008). Chemical plant accident rates in Eastern Europe and America. *Swiss Re Presentation for the European Conference Board Safety and Health Council*, Brussels (9 October 2008).

Wikipedia. Fortune Global 500. https://en.wikipedia.org/wiki/Fortune_Global_500 (accessed 3 March 2016).

CHAPTER 24

# THE SYSTEMS APPROACH TO MANAGING OCCUPATIONAL HEALTH AND SAFETY

VICTOR M. TOY

*Insyst OH&S, 7 West 41st Ave., #508, San Mateo, CA, 94403*

The popularity of Occupational Health and Safety Management Systems (OHSMS) has grown since the concept was first introduced in the mid-1990s. While a single global standard has been long in coming, a recent survey conducted by the OHSAS (Occupational Health and Safety Assessment Series) project group shows these types of standards are being used by organizations in 127 countries. Countries have now worked on the development of a single global OHSMS standard through the International Standards Organization process (ISO). ISO is the organization responsible for the quality (ISO 9001) and environmental (ISO14001) standards that have successfully increased focus in these areas.

The reasons for the growth of management systems in general and their application to OHS may well be tied to the interconnectivity in the global economy. Organizations, particularly large multidisciplinary corporations, have also sought the benefits of improved efficiencies and effectiveness with this approach for maintaining safe workplaces. More to the point is the realization of the need to manage risks and their impact on the safety of workers and organizational goals.

Management systems seek to systematize how objectives are met. In the past, the management of health and safety has mostly been hazard centric or issue specific. A look at regulatory requirements reveals management by issues, such as noise, electrical safety (lock out/tag out), machine guarding, and chemical exposures. Much of this is due to the writing of regulatory standards as a result of data showing large numbers of injuries, illnesses, and fatalities. Attempts to provide more comprehensive approaches for managing health and safety risks have been growing with the concept of health and safety program standards and, more specifically, management system standards.

Management systems are tools that center on an organization's approach to managing risks. It begins with leadership engagement and direction via a policy on health and safety and the need for continual improvement and compliance to laws and regulations. The job of safety in the workplace is multifaceted and interconnected. It involves not just one individual or group but a recognition that everyone in the organization has a role to play in the overall success of its health and safety performance. The system approach is holistic. It ties plans for managing risks with implementation, execution, and assessment of the organization's activities with a management review and the continual consideration of new opportunities to improve health and safety performance. This performance is not just a consideration of past injuries and illnesses but current and emerging risks to the organization and its objectives. It also considers the interest of external as well as internal parties to help provide for a sustainable organization.

The popularity of OHSMS parallels the consideration for a single global OHSMS standard and successful implementation by numerous organizations. It shifts the management of health and safety into alignment with the way successful organizations are managed.

## 24.1 INTRODUCTION

The International Organization for Standardization (ISO) said it best when it described management systems as a process for systemizing how things are done. In other words, a management system is a tool that integrates people, process, and other things (e.g. policies, business strategies, and materials) in order to achieve the goals of the organization.

*Handbook of Occupational Safety and Health*, Third Edition. Edited by S. Z. Mansdorf.

ISO made an important point stating that all organizations have some sort of a management system. Otherwise it could be argued that the organization could not survive as it would not meet its objectives.

> *ISO definition common to all of its Management system standard: set of interrelated or interacting elements of an organization to establish policies and objectives and processes to achieve those objectives.* (International Standards Organization 2014)

What is different in today's world is whether this systemization addresses and enhances health and safety performance. This is where the emergence of management system standards such as ISO 9001 for quality and ISO 14001 for environmental has created a model for today's occupational health and safety management systems (OHSMS) such as the American National Standards Institute (ANSI) Z10 (Occupational Health and Safety Management Systems) and the British Standards Institute's (BSI) 18001 (Occupational Health and Safety Management Systems) standards. The intent of this chapter is to provide an overview of the general concepts and elements of an OHSMS as opposed to specific requirements that may vary from one OHSMS standard to another.

In the past, the management of health and safety has been done more in silos as hazards that have led to injury and illnesses were identified and programs developed specifically to manage those risks. In many ways, these methods for managing risks coincided with a focus on lagging indicators where an incident occurred necessitating controls for a particular risk. The concern with this approach is that organizations were often "putting out fires" rather than looking at potential risks that could cause an event. In many ways, the silo mentality is how our regulations are set and why compliance to legal requirements alone is often insufficient.

There have been many cases where this hazard-centric approach has failed to prevent significant incidents, resulting in fatalities even with good historical performance based on lagging injury and illness statistics (Moure-Eraso 2015).

A review of global injury and illness statistics continue to show disconcerting figures related to the safety of workers. Figures from a 2017 study revealed 2.78 fatalities a year due to work related activities representing an increase of 26% over a similar study conducted in 2005 statistics released at the 17th World Congress on Safety and Health at Work showed that 2.2 million workers were fatally injured as the result of a work-related accident, which was an increase of 10% over the three previous years. These figures equate to around 6000 cases per day. A published report in 2013 shows the figures are still hovering around 2.3 million fatalities a year (Takala et al. 2014). It is this type of data that has driven the need for a single global OHSMS standard through the ISO standards setting process. One of the principles behind an OHSMS is that it is the organization's, not a department's (e.g. health and safety), safety management system. Secondly, the focus is more on identifying and managing risks and continual improvements rather than simply assessing and addressing incidents from injury and illness statistics.

## 24.2 HISTORY AND EVOLUTION

A 2011 review by the Occupational Health and Safety Assessment Series (OHSAS) project group shows OHSMS standards are being used in 127 countries (International Standards Organization 45001 2018) with a steady increase over the past 10 years. There have also been estimates suggesting approximately 100000 organizations have also chosen certification to an OHSMS standard such as OHSAS 18001, ANSI Z10, or the Canadian Z1000 standard. While it is good to see this trend, it pales in comparison with quality and environmental standards that have 1.2 million and 300000 certificants, respectively. It is likely that the actual number of users of these standards is much higher as organizations can adopt and implement management systems without seeking certification. What this pattern shows is that there is room for growth along with a demand for an OHSMS approach to managing health and safety, a fact supported by a 2013 vote by ISO member countries to develop a single global OHSMS standard.

The development of a single global standard for health and safety management has taken a rather long journey since the first proposal was raised in 1996. Interests in such a standard has been growing beginning with the release of BSI 8800 in 1996, OHSAS 18001 in 1999, and the International Labour Organization's (ILO) 2001 Guidelines for Occupational Health and Safety Management Systems adopted by many countries. The US ANSI Z10 standard was released in 2005.

The Occupational Safety and Health Administration (OSHA) began their first effort in 1996 for a safety and health program standard containing many of the same core elements seen in current OHSMS standards. This proposal included (i) management leadership and employee participation, (ii) hazard assessment, (iii) hazard prevention and control, (iv) training, and (v) evaluation of program effectiveness. The proposed standard did not result in a regulatory requirement but emerged as a guideline most recently published in 2016 (Occupational Safety and Health Administration 2016). OSHA's Voluntary Protection Program, which began in 1982, also uses many of the same concepts.

Perhaps the challenge with OSHA's attempts is that most management system standards are written at a high level, so the organization can apply this tool in a way that best fits its organizational culture. Many of the prescriptive requirements that you might find in a regulation are left to the organization to determine so long as their management system conforms to the system level requirements. This allows the organization to design its process while still meeting the intent of the requirement and in a way that shows improvements in its occupational health and safety (OHS) performance.

## 24.3 APPLICATION

For the most part, a management system is adopted by an organization. Certain countries require them, in varying forms, in their regulations with several states in the United States being an example in the form of an injury illness prevention program. Whether required by regulation or legislation, the focus is always on an organization's desire to protect and improve the health and safety of its workforce and the value that it has in helping the organization meet its business objectives. In doing so, the importance of an OHSMS is not so much in compliance with legal requirements, but more so in its desire to manage its health and safety risks. The standards for OHSMS are essentially written for organizations of all types and sizes who want to improve their health and safety performance. However, this presents a significant challenge for standard writers who need to balance the basic needs of small businesses with the larger enterprises who have deeper access to health and safety resources. This is one of the reasons why OHSMS standards are written at a high level so it can apply to the small 10-person operation as well as the large organizations with tens of thousands if not hundreds of thousands of employees.

While an OHSMS clearly applies to the organization, it is the extent of that coverage that is sometimes called into question. Most organizations would agree the OHSMS applies to its employees, both management and nonmanagement. However, the extent of coverage when an employee leaves the workplace controlled by the organization is sometimes unclear. For example, how much of the management system applies when an employee is traveling or enters another workplace controlled by another company or even in an employee's home. In some cases, country laws provide some level of clarity, while in others it is left to the organization to determine. Often, the decision centers on the concept of whether the organization has control. However, even that at times is unclear since an organization may choose not to do business with another organization that puts their employees at risk. There is also the question of whether contractors, visitors, or supply chains are covered. Generally, if not specified by the OHSMS standard, it is left to the organization to determine within the scope of their management system. Part of the growth in management system certifications has been the requirement by organizations with certification that their suppliers have it as well.

## 24.4 SYSTEMS THINKING

The ISO concept of management systems is as a set of interconnected processes. This differs from the concept of a program that tends to have a more linear connotation. Rather, a process is one that has inputs and outputs and a feedback loop that cycles back as input into one or more interconnected processes. This is analogous to a team concept in sports where each member plays a role, but each role is complementary and interactive as the team drives toward a successful outcome. Dr. Russell Ackoff, one of the leading thinkers and theorists in management systems, describes a system as a whole that consists of parts, each of which can affect its behavior or properties of the other (Ackoff 1994). Each part of the system is then dependent on the other and is therefore interdependent. As this applies to the workplace, the outputs of the organization are dependent on the safety of the workers. It is with this concept that a management system that seeks to protect the safety of the worker is interdependent with the rest of the organization including its resources (including skills and knowledge), operational processes, and strategic initiatives. A program typically manages one aspect.

The basic elements of most management system includes:

1. Policy
2. Leadership
3. Employee engagement
4. Planning
5. Operations and controls
6. Monitoring and assessment
7. Management review

In fact, most systems are fairly similar with differences attributable to local cultures, experiences, and needs identified by the developers. This similarity, no matter if the discipline is health and safety, environmental, quality of any of the other 50, or so ISO management standards, has given rise to a harmonization effort to ease the development and revision of management systems standards. More so, it is intended to make it easier for organizations to adopt such standards. The common framework is known as Annex SL and is a part of an ISO directive for its management system standards (International Standards Organization 2014). It can account for 50% or more of the text of a management system standard such as that found in ISO 45001 OHSMS. The hope is that, with the use of common language, organizations will recognize the value of a using the same integrated approach for managing OHS as they do other business processes such as quality.

As tools for managing risks and opportunities, management system standards generally do not state specific criteria beyond the framework of a management system. The aspects on how these requirements are met are left to the organization that provides desired flexibility since no two organizations, or their cultures, are alike.

## 24.5 CHOICE OF A MANAGEMENT SYSTEM FOR OCCUPATIONAL HEALTH AND SAFETY

The primary objective behind choosing a management system is to improve the management of health and safety whether that is in reference to OHS performance or the system itself. There are a number of choices currently available, and the selection may depend on country practices, the global nature of the organization, and even the structure or size of the organization. Smaller organizations may choose systems that have closer ties to regulatory authorities such as the injury illness prevention or voluntary protection programs in the United States. Larger organization may select systems that are more advanced in managing risks, correcting system issues, or leveraging opportunities to improve the performance of the system itself to help meet business objectives.

Prior to the adoption of ISO 45001, OHSAS 18001, widely used in Europe and Asia, served as a de facto international standard and as a template for a number of Country OHSMS standards. However, other standards such as the American ANSI Z10 have found use even outside its borders. In some cases, the selection of which standard to adopt is influenced by an organization's clients who may request a certificate or registration to one of these standards. When selecting or designing a management system, particular attention should be paid to other management systems currently in use in their organizations. This could simplify implementation while allowing for greater integration with other discipline-specific management systems. This again is where the ISO directive for common language (International Standards Organization 2014) for management systems should help in facilitating a multidisciplinary management system approach within an organization.

### 24.5.1 Scope of the Occupational Health and Safety Management System

When implementing a management system for the first time, it is helpful to conduct a gap analysis against the requirements of the chosen standard. Generally, the first step is to determine the scope of the system. This is not to be confused with the scope of an OHSMS standard. Rather, this includes organizational boundaries as well as what is covered topically by the system itself. Staring with the subject matter, it is rather straightforward that the OHSMS covers OHS. Many standards specifically mention other related aspects such as employee wellness that can be included at the organization's choosing. As another example, some organizations cover nonoccupational return to work programs as an element similar to case management for occupationally related injuries and illnesses. Product and property safety is usually not covered by OHSMS standards except where these issues impact the worker. OHSMS standards typically do not cover the environment, which is the domain of the ISO 14001 family of standards on the environment. However, due to a common plan–do–check–act (PDCA) framework, some organizations have decided to coordinate and align their approach to implementing their environmental and occupational health and safety management systems. This is being made easier with ISO's common management system framework (International Standards Organization 2014).

The organization also decides the organizational boundaries covered by the management system. The intent is to help the organization decide where to begin. While small business may elect to cover their entire organization, larger entities may start with one region or division. A successfully implemented OHSMS may then be more easily implemented elsewhere in the organization where the model and culture behind the adopted OHSMS has been initiated. When a portion of the organization has been chosen, all of the requirements of the system described below and in OHSMS standards must be implemented. This is critical as it takes all of the elements working together as a whole to make for an effective system.

The decision on the scope of the OHSMS must consider the internal and external issues that affect the performance of the system. Internal issues include business strategies, people, structure, culture, hazards, policies and processes, resources, and organizational requirements. External issues include legal/regulatory, technological, economic, supply chain, social, and community issues. The interests of the parties impacted by the OHSMS relative to these issues should also be addressed relative to these issues. For example, a fire department needs to understand the hazards it may need to address in an emergency. Issues of culture and language are important when designing training and communications. All of these considerations affect the management of the system that cannot be implemented in isolation.

### 24.5.2 The Elements

OHSMS standards outline the required elements for a management system. Most, if not all, are based on the PDCA cycle popularized by Dr. Edwards Deming who is widely considered the father of modern quality control. Figure 24.1 illustrates the PDCA cycle with the general management system elements covered in this chapter.

Quality is defined as meeting a client's expectations, so it can be argued that the safety of workers is also an aspect of quality and the relationship to the PDCA cycle applies. What this cycle stresses is the concept of continual improvement as the cyclical nature of the model promotes learnings that feeds back into the planning phase and extends knowledge and actions further within the system. In Plans, objectives are established with goals or targets identified that are then carried out on the Do phase where plans are implemented. Plans may be operational as well as strategic particularly when generating improvements. These activities are then assessed and evaluated in the Check phase to determine whether plans have been implemented as designed and the outputs or results were generated as expected or targets met. Act is where the results of the assessment lead to additional actions such as a management review to provide input back into the planning phase in order to manage and improve OHS and management system performance against objectives.

### 24.5.3 Planning

***24.5.3.1 The Art of Planning*** The Oxford dictionary refers to an accident as an incident that happens unexpectedly and unintentionally. In the case of a managing health and safety, the adverse outcome is one that can result in injury, illness, or death. With that in mind, the concept of planning is core and central to the management system in order to manage risks in addition to identifying opportunities for continual improvement. The process for planning traditionally begins with the identification of hazards, risks, and opportunities considering the internal and external context of the organization and the potential impact(s) for meeting the intended outcomes of the management system. These risks and opportunities are assessed and controlled, or new objectives identified, and plans made to support risk

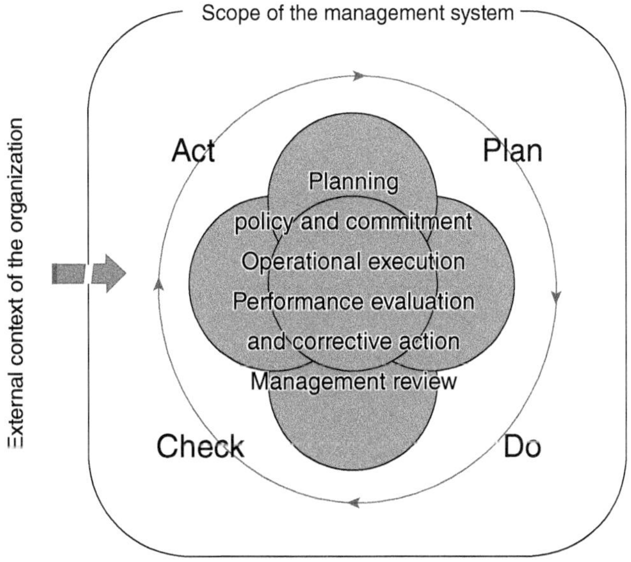


**Figure 24.1** PDCA cycle.

reduction and continual improvements through the creation of objectives. Risks include identifying and meeting legal requirements and other requirements to which the organization subscribes in addition to managing changes within the organization.

Planning is essential because it requires the identification of risks as well as opportunities within the context of the organization that potentially impacts OHS. This requires an understanding not only of your internal activities but also of the community, the local infrastructure (such as access to the fire department and clinics), culture, and language. These are all important considerations for understanding your risks that is central to planning.

The organization identifies hazards as well as system issues in order to manage risks and opportunities to improve overall OHS performance. At the same time, improvement objectives are set while other remaining or residual risks are managed to ensure they continue to be controlled.

The art of planning ensures that the right inputs and feedback from the other parts of the management system are fed into the process so that planning decisions are made to optimize OHS and OHSMS performance. This includes the active involvement from top management and the engagement of employees to maintain current controls and to implement new plans for reducing risks within the OHS management system. To complete the loop, the outputs from planning are linked with other parts of the management system (such as operations, evaluation, and management review) to ensure successful implementation and beneficial outcomes that enhance overall OHS performance.

## 24.6 THE ROLE OF THE TEAM

To be effective, planning cannot be done in vacuum. In other words, the process needs to be open. Traditional health and safety management may have placed the responsibilities for planning squarely on the shoulders of its health and safety staff who in turn may have leveraged traditional metrics such as injury/illness statistics and regulations to help with its decision-making. However, such methods often miss emerging risks. For every accident, there are likely many more near misses that by chance did not result in an occupational injury or illness. This way of thinking often reveals that attention paid to actual incidents often misses potential risks that could be identified and addressed by others with more advanced management system approaches.

As with any process, a critical aspect for planning are the inputs. Often times, an organization will adopt a team approach consisting of line workers, management, support staffs (such as engineering), and business leaders to gain proper perspectives of potential risks that can be better managed. Workers who are often the closest to the hazard provide firsthand knowledge of potential risks of their work activities. Management provides insight into business strategies that potentially change risk profiles. The process typically includes consideration of past incidents, including near misses, but also includes a discussion of changes in the organization that include new knowledge of hazards, materials, processes, and business initiatives as well as personnel changes.

## 24.7 INTEGRATING THE EFFORTS FOR COMPLIANCE WITH LEGAL AND OTHER REQUIREMENTS

All management systems standards have a commitment to compliance that is a fundamental input into plans that must be addressed. An organization not only needs access to all relevant laws and regulations but also other standards and practices to which it subscribes. This includes the OHSMS standard, if one is chosen, in addition to technical standards such as exposure limits and fire life safety codes. Often, the OHSMS is thought of as an umbrella standard that covers all other requirements, both legal and self-subscribed, and the collective way in which it is managed. As such, these requirements need to be identified as compliance is a fundamental commitment within an OHSMS.

There are different perspectives toward simply a “commitment” to compliance as required in most management system standards rather than an absolute requirement to comply with the law (and other requirements to which an organization subscribes). This seemingly contradictory requirement can best be explained as the standard itself is a tool to improve performance as opposed to a regulation. Undoubtedly, there are organizations that are not yet in full compliance with local regulations that may be attributed to several reasons such as absence of knowledge, resources, and worker engagement. For this reason, the OSHMS provides the framework a way not just to ensure there are plans for meeting legal and other requirements but also to provide a constant focus to continue to meet those requirements. It should be noted that management systems are not advocating noncompliance but helping organizations to establish a commitment to meet and continue to meet legal and other requirements to which the organization subscribes.

## 24.8 OHS ISSUES TO CONSIDER

The question is often how or where to begin the planning process or what the issues are related to OHS that should be considered. What comes to mind are the hazards, risks, legal (and other) requirements, and incidences of injuries and illnesses. Just as important are the issues related to system performance. However, it may be easiest to begin with a list of the hazards and safety programs as the OHS aspects to be considered before moving on to systems issues. System issues include not only hazards and risks but also deficiencies and opportunities related to the function of the system. An example are the resources or communication channels required to manage hazards and risks. There is no set way for how these issues are identified. The intent behind planning is that issues are identified so that an assessment of the related risks can be conducted with a decision for what actions should be planned that presents the most significant opportunities for improvement.

As a system matures, other inputs become increasingly more important as the organization becomes more strategic rather than operational. Rather than planning for objectives due to a negative event, considerations are made to anticipate changes along with other opportunities within the system to improve overall performance. This may include changes in the organization's structure, its strategies for new business opportunities, or simply improving existing processes such as plan reviews and procurement of goods and services. The simple approach to managing safety is to look at the hazards, including related incidents, and design plans to manage those risks in order to lower injury and illness rates. The systems approach extends beyond that to include anticipating the introduction of new hazards and risks so they can be managed accordingly. For example, an organization may introduce a new chemical but may not have the adequate process in place to ensure the right procedures, engineering controls, and information and training are in place at the time of introduction. The system ensures, at the concept stage, that the requester of the chemical, the buyer, the process engineer who designs controls and the business strategists are working together along with the health and safety professional to ensure safe handling before changes are made or that the risks are otherwise properly addressed. System considerations include all of the elements discussed in this chapter.

## 24.9 DETERMINING OBJECTIVES AND PLANS

Once OHS and management system issues are identified, an assessment of the risks is conducted. Some standards refer to this as a risk assessment. This is done so that plans and resources can be allocated to manage those risks. In some cases, the hazards relative to the risks are managed adequately. In this case, plans are made to maintain those controls while new opportunities are identified. The principle of continual improvement in a management system requires the identification of opportunities in order to improve the performance of the OHSMS in managing its risks.

The methodology for assessing risks (or a risk assessment) is generally left to the organization to allow flexibility according to need. A chemical manufacturer's process may differ greatly, perhaps being more complex and quantitative in its risk assessment methods, than with that of a small radiator repair shop or a call center. Oftentimes, a simple risk assessment matrix is used.

The assessment is used to determine objectives and plans for managing current risks and improvements. Objectives are typically statements of what results are expected. An example of an OHS objective could be to increase the level of "awareness of risks and improvements in safe work practices related to workstation ergonomics." To be effective, the objective should be measurable either qualitatively or quantitatively. Sometimes a target is included that might say

**TABLE 24.1 Overview of General Planning Steps**

1. Gather input, including hazards and other risks to the performance of the management system such as business directions, current and new operational (process) issues, regulations and standards, health and safety data, and worker input
2. Assemble the planning team assuring the right decision-makers are included in the process who impact or are impacted by the OHS objectives including provisions of resources
3. Identify the methods for assessing hazards and risks including determining current controls and the level of adequacy for managing risks
4. Identify opportunities for improving OHS and OHSMS performance that includes reducing risks from hazards in addition to how well the system is working overall to manage risks
5. Determine objectives and goals/targets, and establish documented plans with management support
6. Communicate plans, and ensure coordination and cooperation with affected parties to ensure successful outcomes and effective implementation of the plans

"by 50%." Stating such an objective would require a way to measure against the target that could be a survey or a decrease in incidents.

Plans, which should be documented, include steps to be taken, resources, dates, and responsible parties. It is not unusual to seek approval (and is often advisable) by top management in the organization to demonstrate and secure support.

## 24.10 POLICY AND COMMITMENT BY THE ORGANIZATION

The management system is guided by an OHS policy that provides direction for the OHSMS from top management. The policy requires a framework for setting OHS objectives, a commitment to satisfying legal and other requirements to which an organization subscribes, and prevention of injury and illnesses and continual improvements.

The importance of the policy is quite simple. It is the highest-level statement by the organization by which the management system defines its intended outcomes. The design and outputs of the system must be aligned with the policy. The policy itself is communicated to everyone, internal or external, who has an interest in the organization's OHSMS, so that roles and responsibilities against these commitments are well understood.

The policy also sets the tone for the culture of the organization relative to health and safety. It must be reviewed and renewed periodically. It also provides a good calibration point as OHS objectives are determined to ensure the intended outcome of the management system is met for the prevention of occupationally related injury and illnesses and the provision of a safe workplace.

The connection of top management to the policy is a statement that health and safety is a priority in meeting the overall goals and objectives of the organization. It is typical to review the performance of the management system against the policy and the resulting OHS objectives when determining the adequacy, suitability, and effectiveness of the management system.

OSH policies typically include commitments (American National Standards Institute 2017; International Labour Organization 2001; Occupational Health and Safety Assessment Series Project Group 2001) to:

- The prevention of injury and illnesses and the provision of a safe workplace.
- Resources that support the OHSMS.
- OSH roles, participation, and responsibilities of employees, including management.
- Meeting legal and other requirements determined by the organization.
- Setting objectives and continual improvement of the OSH management system.

### 24.10.1 Leadership and Worker Engagement

The OHSMS ensures that leadership provides the direction for OHS performance linked to the success of its business. While it may start with the OHS policy, there is recognition that everyone under the control of the organization has an active role to play in the management of risks particularly those closest to the hazards. These roles and responsibilities are determined and defined so that the entire organization works together in meeting OSH objectives.

Management system standards typically refer to the responsibilities of "top" management in defining leadership requirements. Top management is normally defined as a person or group of people who directs and controls an organization at the highest level (International Standards Organization 2014). This is the level of management that has overall responsibility and the authority to provide direction and resources for the management of health and safety. Often, the size and structure (i.e. subdivisions) of the organization play a role in the determination of these individuals. Consideration should be given to an individual who not only carries the responsibility and accountability for worker health and safety but also those who have intimate knowledge of business operations and strategies to promote the inclusion and integration of health and safety initiatives. It is critical for top management to be actively engaged in the OHSMS and not simply to delegate responsibility.

The management system is all about preventing worker injury, illnesses, and fatalities by providing safe work environments. Each and every person under the control of the organization has a role and responsibility for safety. This includes management, nonmanagement, and others who are directly connected with the organization's activities. The organization, depending on the standard or OHSMS practice it adopts, determines whether it covers workers in general beyond employees who have a role in activities controlled by the organization. This may include contractors, visitors, employee representatives, or employees working at another organization's workplace. Employee or worker participation is key to a high-performing system. It has been said that a sustainable management system is one where dependencies do not rest solely on the shoulders of one employee or, for that matter, one group such as the health and safety department. It is important to consider the required level of participation where the impact of the related activities affects the outcome of the management system. For example, a worker handling hazardous energy may have great insight on how the activity can be conducted safely since they have direct and intimate knowledge of the potential risks. The level of worker participation within the management system may include the development and management of plans, implementation of operations, and review of performance.

Participation is not effective without provision of resources that is tied to management's commitment in its health and safety policy. Resources are not just physical or financial. It includes human and operational resources such as access to knowledge, information, and expertise within the organization. This is closely tied to defining roles, responsibilities, and authorities and requires OHS competencies with associated job functions. Good leadership seeks to integrate the OHSMS into the business rather than a standalone initiative that works off to the side. By integrating the OHSMS with the organization's activities, the responsibilities for providing a safe workplace becomes everyone's responsibilities and an integral part of their work and responsibilities.

### 24.10.2 Operational Execution

Undoubtedly, planning is core to an effective system. However operational execution is the required element to ensure these plans are carried out. These plans require the implementation of methods typically using an accepted hierarchy of controls. However, plans as described earlier are not just about managing OHS risks, but also the risks and opportunities of and to the management system. This is often referred to as OHSMS risk vs. OHS risks. The operations and implementation element of the OHSMS are where plans are carried out whether they are directly related to reducing risks of a particular hazard or improving the management system to take advantage of an opportunity. This latter concept of management system risks might include initiatives to promote deeper worker involvement, adopting new risk assessment methodologies, or better means for education and awareness using new technologies for communications.

## 24.11 ROLES AND RESPONSIBILITIES

A management system should be less dependent on any specific single or group of individuals that, if missing, would cause the system to fail. In other words, the design of the system should be such that everyone in the organization has a role to play and it is the overall shared responsibilities that makes the system work effectively. For this to occur, the roles and responsibilities from top management to other workers including contractors must be determined and communicated. Processes are also in place to ensure these responsibilities are not only defined, but accountability is built into the activities that impact worker health and safety.

Participation includes an understanding of a person's role and the potential impact it has on preventing injuries and illnesses. Barriers to participation should be identified and addressed. Examples include language, knowledge, time, and resources. Participation should be considered for the relevant parts of the system to promote the integration of the process to produce the best possible outcome. For example, workers who are closest to the hazards or those involved with managing activities with potential risks should play a role in identifying controls and evaluating its performance. This in turn should feed back into the planning process and be tracked as to the effectiveness of the implemented controls, the identification of residual risks, and the unintended impacts resulting from the change. Engineers should make sure that the design of suggested controls is optimized and feasible.

## 24.12 MAINTAINING WHILE IMPROVING

Implementation is the "do" phase of the PDCA cycle. Some standards infer that new improvement initiatives are done within the planning phase where the improvement plan(s) are created that defines the who, what, where, when, and how it is to be completed. Still other standards suggest that improvement plans are written in planning and implemented under the operations and implementation (or "do"). In reality, it does not matter so long as the plans, whether operational, tactical, or strategic, are implemented somewhere within the management system usually in planning, operations, or both. Much of this will be dictated by the organization's culture for ensuring these activities are tracked and completed.

Operations and implementation is one of the more traditional elements in managing health and safety practices typically by programs (Toy 2012), whereas more recent management system requirements refer to management by process. A program may outline the steps used to manage a particular issue such as electrical safety. In some cases, specific regulations form the basis for what controls must be in place to achieve an acceptable level of risk. Often, this is referred to as a program. A program defines specific measures to control the risk in addition to outlining roles and responsibilities. Programs are necessary since the management of certain hazards can be quite different such as the need for work organization to manage musculoskeletal injuries versus a lockout/tagout procedure for electrical safety or exhaust ventilation to control chemical exposures. A process, on the other hand, looks at the hazards and considers all the inputs in relation to the outputs and whether the intended outcomes have been met.

Processes consider the interactions and interrelationships of these activities. Programs are still necessary, but processes looks at how the programs work together to meet OHS objectives and the intended outcome of the OHSMS. Some of the process requirements in OHSMS standards include those for managing changes within the organization, its contractors, outsourcing, and emergency preparedness.

## 24.13 RISK ASSESSMENT AND THE HIERARCHY OF CONTROLS

In the planning phase of the management system, hazards are assessed and actions taken to manage risks. Some refer to this as a risk assessment, while others reserve this term for a deeper quantitative assessment. In reality, there are different levels of risk assessments. Some standards avoid the confusion by referring to an assessment or risk in planning activities that leverage the output of detailed risk assessments conducted as a part of normal operations. The organization decides what methods are used and where this activity is conducted, whether in planning, operations, or both.

Often a simple matrix is used with frequency on one axis and severity on the other. If the risk is deemed unacceptable by the organization, additional controls are added to achieve an acceptable level of risk. This follows the use of the hierarchy of controls where elimination or substitution at the top of the pyramid is preferred over the use of personal protective equipment, which is a lower order control. An example of following the hierarchy is the use of a respirator during an unplanned maintenance activity where exhaust ventilation is not available.

## 24.14 CHANGE MANAGEMENT

Keeping an eye on changes internal or external to the organization helps to ensure potential impacts on the performance of the management system are monitored and addressed appropriately. These changes could be either negative or positive such as the introduction of new control technology or improvements in the organizational structure. More often than not, the changes are assessed for potentially negative impacts such as with new acquisitions or tools and materials that introduces new hazards or risks. For example, it may be that adding additional machines could increase the ambient noise level, the maintenance required, or the electrical load. Within the OHSMS is a need to establish a process to detect changes initiating a review with possible actions. The review should be completed, and controls implemented before the changes, including those to related processes, are made.

Less apparent are the unplanned changes that are sometimes considered as a part of normal operations. An example of such a change is one connected with facilities or a process equipment where there is a sudden loss of electricity or

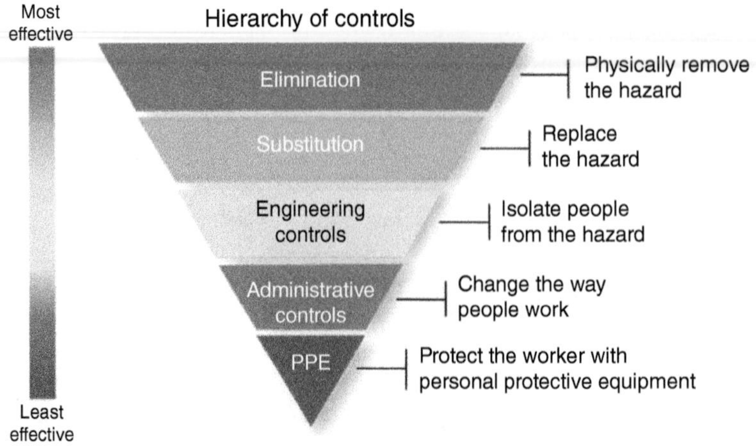


**Figure 24.2** NIOSH hierarchy of control. *Source:* National Institute of Occupational Safety and Health (2015).

a breakdown in the tool itself. These types of "what-if" scenarios should be included in the change management or operational processes to ensure hazards and related risks outside of normal operations are managed.

## 24.15 EMERGENCY PREPAREDNESS

While considering changes that can impact an organization's activities and workforce, the possibility of an emergency event must be included. These types of requirements are typically found in all OHSMS standards. Examples include fires, chemical spills, and explosions as well as natural events such as an earthquake or a hurricane. While it may not be possible to control natural disasters, it is possible to develop a response plan in anticipation of actions required to mitigate the risks.

These risks are considered in the planning stage to determine what actions are required to protect a worker's health and safety during an emergency. It includes communication throughout the organization so that workers know what to do when an event occurs. Responders are trained and provided with personal protective equipment, response equipment, and supplies. Procedures are developed, tested, and updated as necessary. Plans are also coordinated with external response agencies and the community as appropriate.

## 24.16 KNOWLEDGE MANAGEMENT

Knowledge is a key aspect and a fundamental requirement for all management systems. This includes the knowledge required to run the management, system in addition to the awareness and consequences of the potential hazards and risks, and the competencies required to work safely and to provide for a safe workplace.

The organization determines the knowledge based on the hazards, risks, and compliance obligations of its activities. The provision of information and training is assessed as to the level of competencies attained to support management system objectives and repeated as required. This includes updating training and awareness as changes occur within the system such as with the introduction of new materials and equipment. It also requires an assessment of the effectiveness of the transfer of knowledge that includes addressing barriers due to language.

Knowledge is also required to implement the management system. Individuals with these responsibilities must be trained on the system's requirements relative to their roles. Top management is informed and educated on the importance of the management system and their leadership responsibilities, workers on their contributions to the system, contractors to their impact on potential risks, and others, such as procurement, on job-specific considerations. Much of the information transfer begins with awareness of the system, its requirements, and subsequent hazards and risks.

There must also be a process for communicating information to internal and, as appropriate, external interests. This covers the four W's of who, what, when, and why for what information is covered. Perhaps a little more obvious is the internal communications, the changes in risks, and the actions required of an organization's workers. However, external information is also required to those with a need to know. This includes the emergency response agencies, building departments, and OHS authorities and may even include the community to satisfy an organization's social responsibility initiatives or legal responsibilities.

## 24.17 MANAGING EXTERNAL INFLUENCES

As noted earlier, the OHSMS is an open system meaning that external influences that are not always under the direct control of the organization can affect OHS performance. Aside from the local natural impacts, such as weather conditions and earthquakes, there are external parties that impact the OHS objectives. These include contractors, suppliers, and outsourced activities critical to the organization's business objectives. The challenge with these types of arrangements is the nature of thc control that is often indirect. As a result, the organization must often coordinate its system with those of another organization.

## 24.18 PROCUREMENT

New equipment, materials, and services potentially introduce new hazards and risks. As a result, an assessment is required with resulting health and safety criteria to ensure these risks are mitigated. The same hierarchy of controls is used in the assessment to determine if the hazard can be eliminated or engineering controls added to the design of equipment and processes to ensure new risks are well managed. The process may include acceptance testing and sign off to ensure the controls are in place and working before the activity is placed into regular service.

## 24.19 CONTRACTORS AND VENDORS

OHSMS standards, as well as many legal and regulatory requirements, include provisions to ensure contractors are well aware of the hazards to which they may be exposed from organizations' activities. In addition, the contractor, in conducting their activities, may introduce risks to the organization's employees. These requirements include a process to coordinate safety practices to ensure the organization's objectives are met while working with the contractor's employer's own health and safety requirements. As a minimum, the organization provides information on the hazards and controls in the workplace to the contractor while the contractor discloses of the hazards present in its activities. The organization's requirements, such as a fire watch for hot work, are made known to the contractor. Other workers who are potentially exposed to the hazards posed by the contractor's activities are informed of the relevant hazards and associated controls to prevent a health and safety incident.

OHSMS Standard may also require control over outsourced activities where another organization performs part of an organization's function or process. This is not the same as a supply chain management, but the same principles may apply to those who manage, use, or benefit from global supply chains. A broader view links the management of supplier health and safety to the organization's sustainability initiatives.

## 24.20 DOCUMENTATION

Documentation requirements, sometimes called documented information (International Standards Organization 2014), to imply records and documents, exist for two reasons. One is to ensure process continuity. The other is to provide proof that an action has been completed. In assuring process continuity, a system seeks to ensure efficiencies, so it makes sense that procedures and the related processes are managed consistently and without loss of information over how things are to be done. It is not always that a document is required if the knowledge in the organization is well known. For example, in a small business of 15 people, it may be that everyone is aware and can recite the steps needed in case the building needs to be evacuated. However, in a large building with hundreds of employees, written procedures are often necessary in order to ensure the organization can evacuate its people, critical supplies are maintained, information such as assembly points are disseminated, and procedures are tested. The level of documentation will always depend on the needs of the organization.

The second type of documents is records. Records show that the management system is operating as designed and provides proof that activities were completed. It also provides data that can be analyzed for possible improvements to the system. An injury/illness log is one form of a record that provides this type of information. Organizations who desire to show conformity to an OHSMS standard will need both types of records: first, that all the required elements are in place and second, that the system is working as intended. Table 24.2 provides a list of documents typically retained within an OHSMS

Just as important as the document itself are the requirements to preserve them for use when needed. This includes dates, revisions, approvers, access and availability, protection against damage, defined retention periods, and control of obsolete documents.

**TABLE 24.2 Typical OHSMS Documentation**

- Scope of the OHSMS
- Policy
- Plans and objectives
- Roles/responsibilities
- Legal and other requirements
- Records of process completion
- Evidence of competence/training
- Monitoring, measurement, analysis and evaluation results
- Internal audit program implementation and results
- Corrective action results
- Nonconformities and actions taken

Oftentimes, documentation is seen as creating unnecessary bureaucracy with little to no benefit. While some of this is necessary and required of the organization meeting the requirements of an OHSMS standard, it is important to keep in mind the intent of records and documents in contributing to the performance of the OHSMS. As such, it may be necessary for the organization to define what additional documentation beyond the standard is required.

## 24.21 PERFORMANCE EVALUATION

An OHSMS is only as good as the value it delivers in meeting its objectives. This can only be determined by evaluating its performance in meeting policy commitments. A system requires a process or procedures to assure that it is meeting or exceeding expectations. This includes not just the traditional health and safety inspections but other methods of monitoring and measurements to check if the controls are working effectively to manage risks and that objectives, including those related to continual improvements, are met. An organization must also assess the conformance of an organization's OHSMS to any management system standard it adopts. Hazards and risks requiring monitoring and measurement include workplace inspections, equipment performance, exposure assessments, legal and other requirements to which the organization subscribes, health and exposure assessments, and injury and illness events. The system should leverage recognized monitoring methods and ensure the use of calibrated equipment.

### 24.21.1 Assessment of Controls

In addition to improvements, objectives can be set to maintain an existing level of performance. Hazards that are identified in planning are assessed for risks, and the appropriate level of controls identified and maintained. These controls are monitored to ensure compliance with requirements and to check for any degradation in performance. Operational objectives in this case may be to simply maintain those controls. That is possible only if the control itself is evaluated for the expected level of performance based on the criteria established.

Plans related to improvement objectives are also monitored to ensure completion and attainment of results. It is the outputs that are evaluated in order to determine if plans were completed, but more importantly, if the expected outcomes were attained. The review provides input shared with top management and feedback into planning to determine if other changes are warranted if the results were not achieved or if other opportunities have been identified.

### 24.21.2 Management System Audit

If an OHSMS standard is adopted, an assessment is conducted to evaluate whether the requirements in that standard have been met. Often called a conformity assessment, this review can be completed internally, or if seeking registration or certification, an external provider. Organizations often choose third-party certification to demonstrate conformance to these systems as a social responsibility and a business practice to demonstrate their commitment to worker health and safety as a responsible supplier.

Whether the audits and inspections are done internally or externally, it is important and many times a requirement to use an impartial auditor who is not connected to the process being evaluated.

### 24.21.3 Metrics

Monitoring and measurement can include both qualitative and quantitative metrics. Either way, criteria are required to assess performance. A good example of a qualitative metric is hazard communication and the level of awareness being low, medium, or high. Some criteria may be lagging metrics that are indicators of past performance, whereas a system strives for an approach using more leading indicators of risks particularly those that have not, but could have, resulted in an injury or illness. If a control is defined, then performance criteria are identified so that it can be evaluated for effectiveness. Criteria are also identified for processes such as required actions or activities.

## 24.22 CORRECTIVE AND PREVENTATIVE ACTION

Once a nonconformance in the system has been detected, a corrective action plan is created. A conformance is defined as the fulfillment of a requirement. A nonconformance may include a breakdown of a requirement that resulted in an injury or illness or a requirement of an OHSMS standard that has not been met. A corrective action, versus a correction, also requires determining and correcting the source of the defect commonly referred to as a root cause. A required follow-up to incidents include not only accidents but also near misses along with a determination of underlying causes to fix not just the hazard but other potential risks that share or could share the same fate.

There must be a process for corrective action that includes a scheme for incident investigation to prevent further occurrences or reoccurrence. The time frame for implementing the corrective action depends on the severity of the incident. Those occurrences that expose workers to significant risk and harm are corrected immediately or other intermediate actions taken to control the risks until a permanent fix is implemented. Other findings, such as inadequate work on improvement objectives to meet intended outcomes, may wait until the next planning cycle to be assessed with other risks to the management system.

There are always lessons learned with any corrective action. A key aspect of the process is to communicate these actions and the issues behind them to prevent the same or similar incidents. Workers are informed to avoid similar injuries or illnesses. Engineers are told so they can design and include additional controls. Leaders are advised to assure proper attention is paid toward prevention. All of this feeds back into the relevant parts of the management system including planning, operations, and knowledge management to continually improve the performance of the system. Outputs on corrective and preventative action are also tied to development of new objectives as determined in the planning phase.

## 24.23 MANAGEMENT REVIEW

The leadership, or top management, sets the tone for the organization on the importance of the system in preventing injuries and illnesses. So it must be that in the system, leadership has the ultimate responsibility for determining whether the OHSMS is suitable, adequate, and effective. Suitability has to do with fit and how well the system works to meet the overall objectives of the organization. Determination of adequacy addresses whether the management system is meeting requirements. Effectiveness describes the extent to which plans are met and as intended in preventing injuries and illnesses.

In order for top management to make this determination, they must be provided with information on the performance of the OHSMS and the issues that impact the system that require changes to or within the system. When considering performance, the relevant data from the evaluation of performance is presented. This may be given at a high level, with greater details provided for the more significant issues especially those causing or that could have caused worker injury, illness, or fatality. The presentation of the data is to help drive consideration for improvements to the OHSMS. Discussion topics can include incidents supported by injury/illness data, results of inspections and audits relative to performance criteria, worker input and concerns such as those determined by surveys and questionnaires, and changes in business direction and strategies. Information on current, new, or emerging risks that need to be addressed are discussed that require senior management attention. Risks include where the organization stands relative to compliance with legal and other requirements as well as performance in meeting OHS objectives.

Just as important as incidents is a report on how well the system is working as whole. This may be the results of the audit of the management system. This should not be viewed as defects so much as opportunities within the system that can improve OHS performance. It may be improvements in communicating hazards and methods of control or moving OHS reviews and assessments into the concept stage for any changes. It can also include benchmarking efforts with other high-performing organizations. Adopting an OHSMS standard may itself open up new opportunities. The review can be informal or based on a structured approach such as a score card (American National Standards Institute 2017). The idea is to engage top management and secure their direction on improvements needed to the system that is then fed back into planning for creating new improvement objectives. While many of the OHSMS standards may leave the frequency to the organization, some have defined it as an annual review (American National Standards Institute 2017) that can be tied to the strategic and financial planning cycles. More frequent review may be required where there are numerous issues of noncompliance and nonconformance.

## 24.24 THE FUTURE OF MANAGEMENT SYSTEMS

It has been an evolution of nearly three decades to reach an agreement for developing a single global OHSMS standard (International Standards Organization 45001 2018). The reasons for such a move may be less well understood than simply the market forces and the global economy and sustainability initiatives driving demand for improvements to worker and workplace health and safety. The evidence of the growing popularity of management systems to meet business objectives is clear. Adding to the need for countries working together to develop a single OHSMS standard is the growth of management systems in general and the challenges faced by organizations tasked with implementing various system frameworks within the same management structure. There are currently more than 50 ISO management system standards. ISO's directive (International Standards Organization 2014) to use a common framework and language in its management system standards will serve to ease the burden on organizations. The benefits to worker and workplace health and safety may be found in the integration of health and safety into management and business objectives. Regardless of whether an ISO standard is adopted as the single global OHSMS, it is clear that this is the approach the progressive organizations are taking to manage and make improvements in its health and safety affairs.

## REFERENCES

Ackoff, R. (1994). Beyond Continual Improvement. https://www.youtube.com/watch?v=OqEeIG8aPPkl (accessed 1 December 2018)

American National Standards Institute (2017). *ANSI/ASSP Z10 Occupational Health and Safety Management Systems*. Park Ridge, IL: American Society of Safety Professionals.

International Labour Organization (2001). *Guidelines on Occupational Safety and Health Management Systems (ILO-OSH 2001)*. Geneva: ILO.

International Standards Organization (2014). Annex SL, Proposals for Management Systems, ISO/IEC Directives, Part 1 Consolidated ISO Supplement, Procedures Specific to ISO 5th Edition

International Standards Organization 45001 (2018). *Occupational Health and Safety Management Systems – Requirements* (March 2018).

Moure-Eraso, R. (2015). *Oversight of the Implementation of the President's Executive Order on Improving Chemical Facility Safety and Security, Written testimony to the Joint Committee: Senate Committee on Environment and Public Works and the Senate Committee on Health, Education, Labor, and Pensions, December 11, 2014.* Washington, DC: U.S. Government Publishing Office.

National Institute of Occupational Safety and Health (2015). Hierarchy of Controls. http://www.cdc.gov/niosh/topics/hierarchy (accessed 1 December 2018).

Occupational Health and Safety Assessment Series Project Group (2001). *OHSAS 18001 Occupational Health and Safety Management Systems: Requirements*. London, UK: British Standards Institute.

Occupational Safety and Health Administration (2016). *Recommended Practices for Safety and Health Programs*. https://www.osha.gov/shpguidelines (accessed 1 December 2018).

Takala, J., Hämäläinen, P., Saarela, K.L. et al. (2014). Global estimates of the burden of injury and illness at work in 2012. *Journal of Occupational and Environmental Hygiene* **11** (5): 326–337.

Toy, V.M. (2012). The industrial hygiene program. In: *Fundamentals of Industrial Hygiene*, 6e (ed. B. Ploy and P. Quinlan), 791–803. Itasca, IL: National Safety Council.

# INDEX




*Handbook of Occupational Safety and Health*, Third Edition. Edited by S. Z. Mansdorf.